Anonymous

Manual of American Waterworks

Volume 3

Anonymous

Manual of American Waterworks
Volume 3

ISBN/EAN: 9783337142834

Printed in Europe, USA, Canada, Australia, Japan

Cover: Foto ©berggeist007 / pixelio.de

More available books at **www.hansebooks.com**

MANUAL OF AMERICAN WATER-WORKS.

DENNIS LONG & COMPANY,

LOUISVILLE, KY.,

MANUFACTURE EXCLUSIVELY

CAST IRON
GAS AND WATER
PIPE

———AND———

SPECIAL CASTINGS

OF ALL SIZES.

INDEX TO ADVERTISEMENTS.

AND

BUSINESS DIRECTORY.

ACCUMULATORS.
Builders' Iron Foundry, Providence, R. I.. xi
AIR COMPRESSORS.
Gordon Steam Pump Co., Hamilton, O..............................Inside front cover
AQUEDUCT LININGS.
Builders' Iron Foundry, Providence, R. I.. xi
ARTESIAN WELL MACHINERY.
Pierce Artesian & Oil Well & Supply Co., 80 Beaver St., New York City......xxiii
ARTESIAN WELL PUMPS.
Barr Pumping Engine Co., Germantown Junction, Philadelphia, Pa xxiv
Gordon Steam Pump Co., Hamilton, O.......................Inside front cover page
AXLES (Steel, for Mine Cars).
Jackson & Woodin Mfg. Co., Berwick, Pa.. xv
BANKERS.
Coffin & Stanton, New York and London.. xi
BOILERS.
Hawkins, R. F., Springfield, Mass.. xxv
Heine Safety Boiler Co., 706 Bank of Commerce Bldg., St. Louis, Mo............xxiv
Pancoast & Rogers, 28 Platt St., New York.. iv
Pond Engineering Co., 707 Market St., St. Louis, Mo.................................xxv
Porter Mfg. Co., Ltd., Syracuse, N. Y... x
Sharon Boiler Works, Sharon, Pa... xxv
Tippett & Wood, Phillipsburg, N. J.. xxvi
Valk & Murdoch Iron Works, 14 to 20 Hasell St., Charleston, S. C............... xi
BOILER TUBES.
Pancoast & Rogers, 28 Platt St., New York.. iv
Ripley & Bronson, 800 N. 2nd St., St. Louis, Mo... x
BRASS AND IRON GOODS (for Steam, Water and Gas).
Michigan Brass and Iron Works, Detroit, Mich.. viii
CARS (and Wheels for Freight Cars).
Jackson & Woodin Mfg. Co., Berwick, Pa.. xv
CASTINGS (Air Furnace and Gun Iron).
Builders' Iron Foundry, Providence, R. I... xi
CASTINGS (Brass).
Stebbins Mfg. Co., E. Brightwood, Mass... xii
CASTINGS (Iron).
Builders' Iron Foundry, Providence, R. I... xi
Jackson & Woodin Mfg. Co., Berwick, Pa... xv
Hawkins, R. F., Springfield, Mass.. xxv
Mellert Foundry & Machine Co., Reading, Pa.. xv
Millar & Son, Charles, Utica, N. Y... xv
South Pittsburg Pipe Works, South Pittsburg, Tenn................................xxiv
Valk & Murdoch Iron Works, 14 to 20 Hasell St., Charleston, S. C............... xi
CASTINGS (Loam).
South Pittsburg Pipe Works, South Pittsburg, Tenn................................xxiv
Wood & Co., R. D., 400 Chestnut St., Philadelphia, Pa................................. i
CEMENT.
The Lawrence Cement Co., 67 William St., New York............................... xiii
COCKS (Corporation, Round Way Stop, Self-Closing and Service).
Hays Mfg. Co., Erie, Pa.. vi
Mueller Mfg. Co., H., 220 E. Main St., Decatur, Ill....................................... v
Ripley & Bronson, 800 N. 2nd St., St. Louis, Mo... x
Stebbins Mfg. Co., E. Brightwood, Mass... xii
Union Water Meter Co., Worcester, Mass.. xix

INDEX TO ADVERTISEMENTS.

COCKS (Stop and Waste, Check and Waste).
Hays Mfg. Co., Erie, Pa vii
Mueller Mfg. Co., 220 E. Main St., Decatur, Ill. vi
Stebbins Mfg. Co., E. Brightwood, Mass. xii
Union Water Meter Co., Worcester, Mass. xix

CONTRACTORS (Water, Gas, Sewers, Pavements).
Carson Trench Machine Co., 16 Dorrance St., Charleston District, Boston, Mass. v
Harlow & Co., Jas. H., 108 4th Ave., Pittsburg, Pa. xxv
Holmes, A. L., Grand Rapids, Mich. xxiii
Millar & Son, Charles, Utica, N. Y. xv
Moffett, Hodgkins & Clarke Co., Syracuse, N. Y. xvii
Pierce, Charles D., 127 Pearl St., New York xxiii
Woltmann, Keith & Co., 11 Wall St., New York xxii

COPPER AND BRASS GOODS.
Hays Mfg. Co., Erie, Pa vi
Millar & Son, Charles, Utica, N. Y. xv
Mueller Mfg. Co., H., 220 E. Main St., Decatur, Ill. vi
Stebbins Mfg. Co., E. Brightwood, Mass. xii

CRANES (to be Attached to Hydrants to Conduct Water to Sprinkling Carts).
Ludlow Valve Co., Troy, N. Y. ii

CURB BOXES.
Millar & Son, Charles, Utica, N. Y. xv

DESIGNERS OF WATER-WORKS MACHINERY AND SUPPLIES.
Charles Carr, 7 Exchange Place, Boston, Mass. ix
Fanning, J. T., Minneapolis, Minn. xxiii
Hill, John W., 21 and 22 Glenn Building, Cincinnati, O. xxiii
Pearsons, C. W., 1611 Baltimore Ave., Kansas City, Mo.

DIGESTORS.
Tippett & Wood, Phillipsburg, N. J. xxvi

ELECTRIC LIGHTING AND STREET RAILWAYS.
Moffett, Hodgkins & Clarke Co., Syracuse, N. Y. xvi

ELEVATORS (Hydraulic).
Tuerk Hydraulic Power Co., 237 Broadway, N. Y., and Chicago xxv

ENGINEERS (Consulting Mechanical).
Charles Carr, 7 Exchange Place, Boston, Mass. ix
Fanning, J. T., Minneapolis, Minn. xxiii
Hill, John W., 21 and 22 Glenn Building, Cincinnati, O. xxiii
Pearsons, C. W., 1611 Baltimore Ave., Kansas City, Mo. xxiii

ENGINEERS (Civil, Hydraulic, Sanitary, and Contracting).
Babcock, Stephen E., Little Falls, N. Y. xxii
Croes, J. J. R., 13 William St., New York xxii
Fanning, J. T., Minneapolis, Minn. xxiii
Harlow & Co., Jas. H., 108 4th Ave., Pittsburgh, Pa. xxv
Harman, Jacob A., Room 100, Y. M. C. A. Building, Peoria, Ill xxii
Hering, Rudolph, 277 Pearl St., New York xxii
Hill, John W., 21 and 22 Glenn Building, Cincinnati, O. xxiii
Holmes, A. L., Grand Rapids, Mich. xxiii
MacDougall, Alan, 32 Adelaide St., Toronto xxii
Moffett, Hodgkins & Clarke Co., Syracuse, N. Y. xvii
Pearsons, C. W., 1611 Baltimore Ave., Kansas City, Mo. xxiii
Pierce, Charles D., 127 Pearl St., New York xxiii
Pond Engineering Co., 707 Market St., St. Louis, Mo. xxv
Woltmann, Keith & Co., 11 Wall St., New York xxii
Wynn, George W., Cedar Rapids, Iowa xxii

ENGINES.
Porter Mfg. Co., Ltd., Syracuse, N. Y. x
Valk & Murdoch Iron Works, 14 to 20 Hasell St., Charleston, S. C xi

EXTENSION SERVICE BOXES.
Hays Mfg. Co., Erie, Pa. vii
Millar & Son, Charles, Utica, N. Y. xv
Ripley & Bronson, 800 N. 2nd St., St. Louis, Mo. x

FILTERS.
National Water Purifying Co., 145 Broadway, N. Y. xiii

FIRE HOSE (Cotton, Linen and Rubber).
Boston Belting Co., 256 Devonshire St., Boston, Mass. xii
Millar & Son, Charles, Utica, N. Y. xv
Ripley & Bronson, 800 N. 2nd St., St. Louis, Mo. x

FITTINGS (Malleable and Cast Iron).
Hays Mfg. Co. Erie, Pa. vii

MANUAL OF AMERICAN WATER-WORKS.

FLEXIBLE JOINT.
Holmes, A. L., Grand Rapids. Mich.. xxiii

FORGINGS.
Jackson & Woodin Mfg. Co., Berwick, Pa.. xv

FOUNDERS (Brass).
Stebbins Mfg. Co., E. Brightwood, Mass xii

FOUNTAINS (Lawn, Drinking).
Ripley & Bronson, 800 N. 2nd St., St. Louis, Mo. x

FURNACES.
Builders' Iron Foundry, Providence, R. I.. xi

GAS HOLDERS.
Wood & Co., R. D., 400 Chestnut St., Philadelphia, Pa...................... i

GATES (Sluice).
Bourbon Copper & Brass Works, 198-204 E. Front St., Cincinnati, O......... v
Ludlow Valve Mfg. Co., Troy, N. Y... ii

GENERAL FOUNDERS AND MACHINISTS.
South Pittsburg Pipe Works, South Pittsburg, Tenn............................ xxiv

GLOBE SPECIAL CASTINGS FOR WATER-WORKS.
Builders' Iron Foundry, Providence, R. I.. xi
Drummond, M. J., 192 Broadway, New York................................... xiv

GROUND KEY WORK.
Millar & Son, Charles, Utica, N. Y... xv
Stebbins Mfg. Co., E. Brightwood, Mass....................................... xii

HOSE (Rubber).
Boston Belting Co., 256 Devonshire St., Boston, Mass........................ xii
Ripley & Bronson, 800 N. 2nd St., St. Louis, Mo.............................. x

HYDRAULIC CRANES.
Ludlow Valve Mfg. Co., Troy, N. Y... ii
Tuerk Hydraulic Power Co., New York and Chicago........................ xxv
Wood & Co., R. D., 400 Chestnut St., Philadelphia, Pa...................... i

HYDRANTS (Fire).
Bourbon Copper & Brass Works, 198-204 E. Front St., Cincinnati, O......... v
Chapman Valve Mfg. Co., Indian Orchard, Mass............Opposite preface
Charles Carr, 7 Exchange Place, Boston, Mass................................. ix
Drummond, M. J., 192 Broadway, New York................................... xiv
Hays Mfg. Co., Erie, Pa... vii
Holyoke Hydrant & Iron Works, Holyoke, Mass........... x and opposite contents
Michigan Brass & Iron Works, Detroit, Mich.................................. viii
Ludlow Valve Mfg. Co., Troy, N. Y.. ii
Mellert Foundry & Machine Co., Reading, Pa................................. xv
Millar & Son, Charles, Utica, N. Y.. xv
Pancoast & Rogers, 23 Platt St., New York.................................... iv
Ripley & Bronson, 800 N. 2nd St., St. Louis, Mo.............................. x
Wood & Co., R. D., 400 Chestnut St., Philadelphia, Pa...................... i

INVESTMENT SECURITIES.
Coffin & Stanton, New York & London.. xi

IRON BRIDGES.
Hawkins, R. F., Springfield, Mass... xxv

IRON WORK (Structural and Ornamental).
Builders' Iron Foundry, Providence, R. I.. xi
Tippett & Wood, Phillipsburg, N. J... xxvi

LAMP POSTS.
Drummond, M. J., 192 Broadway, New York................................... xiv
Mellert Foundry & Machine Co., Reading, Pa................................. xv

LAYING SUBMERGED PIPES (for Water, Gas and Sewerage).
Holmes, A. L., Grand Rapids, Mich.. xxiii
Thacher & Breymann, 306 Water St., Toledo, O............................... xx

LEAD AND TIN (Pig).
Millar & Son, Charles, Utica, N. Y.. xv
Ripley & Bronson, 800 N. 2d St., St. Louis, Mo............................... x

MACHINERY.
Mellert Foundry & Machine Co., Reading, Pa................................. xv
Valk & Murdoch Iron Works, 14 to 20 Hasell St., Charleston, S. C......... xi

INDEX TO ADVERTISEMENTS.

METERS (Water).

Builders' Iron Foundry, Providence, R. I. xi
Ripley & Bronson, 800 N. 2nd St., St. Louis, Mo. x
Thomson Meter Co., Temple Court Building, N. Y. xviii
Union Water Meter Co., Worcester, Mass. xix

MOTORS (Water).

Roberts & Co., Geo. J., Dayton, O. xxv
Tuerk Hydraulic Power Co., New York and Chicago xxv

ORDNANCE.

Builders' Iron Foundry, Providence, R. I. x

PACKING (Round and Square Piston Rod).

Boston Belting Co., 256 Devonshire St., Boston, Mass. xii

PACKING (Steam and Water).

Boston Belting Co., 256 Devonshire St., Boston, Mass. xii
Millar & Son, Charles, Utica, N. Y. xv
Ripley & Bronson, 800 N. 2d St., St. Louis, Mo. x

PIPE (Artesian Well).

Pancoast & Rogers, 28 Platt St., New York lv

PIPE AND SPECIALS (Cast Iron).

Donaldson Iron Co., Emaus, Pa. xv
Builders' Iron Foundry, Providence, R. I. xi
Drummond, M. J., 192 Broadway, New York xiv
Jackson & Woodin Mfg. Co., Berwick, Pa. xv
Long & Co., Dennis, Louisville, Ky. .... Opposite inside front cover page
Mellert Foundry & Machine Co., Reading, Pa. xv
Millar & Son, Charles, Utica, N. Y. xv
Pancoast & Rogers, 28 Platt St., New York lv
Radford Pipe & Foundry Co., Cincinnati, O. xv
Ripley & Bronson, 800 N. 2nd St., St. Louis, Mo. x
South Pittsburg Pipe Works, South Pittsburg, Tenn. xxiv
Warren Foundry & Machine Co., 160 Broadway, New York xiv
Wood & Co., R. D., 400 Chestnut St., Philadelphia, Pa. i

PIPE (Culvert).

Blackmer & Post, St. Louis, Mo. xvi

PIPE (Flume).

Valk & Murdoch Iron Works, 14 to 20 Hasell St., Charleston, S. C. xi

PIPE (Lead and Tin Lined).

Millar & Son, Charles, Utica, N. Y. xv

PIPE (Vitrified Sewer).

Blackmer & Post, St. Louis, Mo. xvi
Millar & Son, Charles, Utica, N. Y. xv

PIPE (Wrought Iron).

Millar & Son, Charles, Utica, N. Y. xv
Pancoast & Rogers, 28 Platt St., New York lv
Ripley & Bronson, 800 N. 2nd St., St. Louis, Mo. x

PLUGS (Brass).

Mueller Mfg. Co., H., 220 E. Main St., Decatur, Ill. vi

PRESSURE REGULATORS (Water).

Mueller Mfg. Co., H., 220 E. Main St., Decatur, Ill. vi
Union Water Meter Co., Worcester, Mass. xix

PULVERIZERS.

Builders' Iron Foundry, Providence, R. I. xi

PUMPS AND PUMPING MACHINERY AND FITTINGS.

Barr Pumping Engine Co., Philadelphia, Pa. xxiv
Builders' Iron Foundry, Providence, R. I. xi
Deane Steam Pump Works, Holyoke, Mass. .... Opposite title page
Gordon Steam Pump Co., Hamilton, O. .... Inside front cover
Holly Mfg. Co., Lockport, N. Y. .... Inside back cover
Millar & Son, Charles, Utica, N. Y. xv
Pond Engineering Co., 707 Market St., St. Louis, Mo. xxv
Valk & Murdoch Iron Works, 14 to 20 Hasell St., Charleston, S. C. xi
Roberts & Co., Geo. J., Dayton, O. xxv

MANUAL OF AMERICAN WATER-WORKS.

ROUGH STOPS AND SUPPLIES (Plumbers).
Stebbins Mfg Co., E. Brightwood, Mass. xii

RUBBER BELTING AND SUCTION HOSE.
Boston Belting Co., 256 Devonshire St., Boston, Mass. xii

SEWER EXCAVATING MACHINES.
Carson Trench Machine Co., 16 Dorrance St., Charlestown District, Boston, Mass. v

SHAPED IRON.
Builders' Iron Foundry, Providence, R. I. xi

SHEET IRON WORK.
Valk & Murdoch Iron Works, 14 to 20 Hasell St., Charleston, S. C. xi

SOLDERING NIPPLES (Brass).
Mueller Mfg. Co., H., 220 E. Main St., Decatur, Ill. vi

SPECIAL MACHINERY AND SUPPLIES (Builders' Iron).
Builders' Iron Foundry, Providence, R. I. xi

STACKS.
Sharon Boiler Works, Sharon, Pa xxv
Tippett & Wood, Phillipsburg, N. J xxvi
Valk & Murdoch Iron Works, 14 to 20 Hasell St., Charleston, S. C xi

STAND PIPES.
Hawkins, R. F., Springfield, Mass. xxv
Pancoast & Rogers, 28 Platt St., New York iv
Porter Mfg. Co., Ltd, Syracuse, N. Y. x
Sharon Boiler Works, Sharon, Pa xxv
Tippett & Wood, Phillipsburg, N. J. xxvi

STONE CRUSHERS.
Porter Mfg. Co., Ltd, Syracuse, N. Y. x

STREET WASHER.
Hays Mfg. Co., Erie, Pa vii
Ripley & Bronson, 800 N. 2nd St., St. Louis, Mo. x

SUBMARINE ENGINEERS AND DIVERS.
Thacher & Breymann, 306 Water St., Toledo, O. xx

SUPPLIES (Contractors).
Millar & Son, Charles, Utica, N. Y. xv
Ripley & Bronson, 800 N. 2nd St., St. Louis, Mo. x

SUPPLIES (Steam Engineers).
Valk & Murdoch Iron Works, 14 to 20 Hasell St., Charleston, S. C xi

TANKS.
Hawkins, R. F., Springfield, Mass. xxv
Pancoast & Rogers, 28 Platt St., New York iv
Sharon Boiler Works, Sharon, Pa xxv
Tippett & Wood, Phillipsburg, N. J. xxvi
Valk & Murdoch Iron Works, 14 to 20 Hasell St., Charleston, S. C xi

TAPPING MACHINES (Water and Dry Pipe).
Builders' Iron Foundry, Providence, R. I. xi
Charles Carr, 7 Exchange Place, Boston, Mass. ix
Millar & Son, Charles, Utica, N. Y. xv
Mueller Mfg. Co., H., 220 E. Main St., Decatur, Ill. vi
Ripley & Bronson, 800 N. 2nd St., St. Louis, Mo. x

TURBINE WHEELS.
Mellert Foundry & Machine Co., Reading, Pa xv
Wood & Co., R. D., 400 Chestnut St., Philadelphia, Pa i

VALVES.
Bourbon Copper & Brass Works, 198-204 E. Front St., Cincinnati, O. v
Chapman Valve Mfg. Co., Indian Orchard, Mass. .... Opposite preface
Charles Carr, 7 Exchange Place, Boston, Mass. ix
Drummond, M. J., 192 Broadway, New York xiv
Hays Mfg. Co., Erie, Pa. vii
Ludlow Valve Mfg. Co., Troy, N. Y. ii
Mellert Foundry & Machine Co., Reading, Pa xv

INDEX TO ADVERTISEMENTS.

Michigan Brass & Iron Works, Detroit, Mich. viii
Millar & Son, Charles, Utica, N. Y. xv
Pancoast & Rogers, 28 Platt St., New York. iv
Peet Valve Co., 163 Albany St., Boston, Mass. xxi
Rensselaer Mfg. Co., Troy, N. Y. iii
Ripley & Bronson, 800 N. 2nd St., St. Louis, Mo. x
Wood & Co., R. D., 400 Chestnut St., Philadelphia, Pa. i

VALVES (Indicator).
Chapman Valve Mfg. Co., Indian Orchard, Mass. .... Opposite preface

VALVES (Radiator).
Chapman Valve Mfg. Co., Indian Orchard, Mass. .... Opposite preface

VALVES (Rubber).
Boston Belting Co., 256 Devonshire St., Boston, Mass. xli

VALVES (Stand Pipes).
Chapman Valve Mfg. Co., Indian Orchard, Mass. .... Opposite preface

VALVES (Steam Yacht).
Chapman Valve Mfg. Co., Indian Orchard, Mass. .... Opposite preface

VALVES (to Stand from 125 to 200 Lbs. Live Steam Pressure).
Chapman Valve Mfg. Co., Indian Orchard, Mass. .... Opposite preface

WATER CONNECTIONS (Brass).
Mueller Mfg. Co., H., 220 E. Main St., Decatur, Ill. vi

WATER FILTERS.
National Water Purifying Co., 145 Broadway, New York. xiii

WATER GATES.
Chapman Valve Mfg. Co., Indian Orchard, Mass. .... Opposite preface
Peet Valve Co., 163 Albany St., Boston, Mass. xxi
Rensselaer Manufacturing Co., Troy, N. Y. iii

WATER METERS.
Builders' Iron Foundry, Providence, R. I. xi
Ripley & Bronson, 800 N. 2nd St., St. Louis, Mo. x
Thomson Meter Co., Temple Court Building, New York. xviii
Union Water Meter Co., Worcester, Mass xix

WATER MOTORS.
Roberts & Co., Geo. J., Dayton O. xxv
Pelton Water-Wheel Co., San Francisco, Cal., and New York City. xx

WATER PURIFIERS.
National Water Purifying Co., 145 Broadway, N. Y. xiii

WATER TOWERS.
Porter Mfg. Co., Ltd., Syracuse, N. Y. x
Tippett & Wood, Phillipsburg, N. J. xxvi

WATER WHEELS.
The Pelton Water-Wheel Co., San Francisco, Cal., and New York City. xx

WATER-WORKS BUILDERS.
Harlow & Co., Jas. H., 108 4th Ave., Pittsburg, Pa. xxv
Moffett, Hodgkins & Clarke Co., Syracuse, N. Y. xvii
Pearsons, C. W., 1611 Baltimore Ave., Kansas City, Mo. xxiii
Woltmann, Keith & Co., 11 Wall Street, New York. xxii

WATER-WORKS' SUPPLIES.
Harlow & Co., Jas. H., 108 Fourth Avenue, Pittsburg, Pa. xxv
Peet Valve Co., 163 Albany St., Boston, Mass. xxi

WATER, STEAM AND GAS WORKS' SUPPLIES.
Bourbon Copper & Brass Works, 198-204 E. Front St., Cincinnati, O. v
Harlow & Co., Jas. H., 108 4th Ave., Pittsburg, Pa. xxv
Hays Mfg. Co., Erie, Pa. vii
Millar & Son, Charles, Utica, N. Y. xv
Peet Valve Co., 163 Albany St., Boston, Mass. xxi
Ripley & Bronson, 800 N. 2nd St., St. Louis, Mo. x
Stebbins Mfg. Co., E. Brightwood, Mass xii

WELL CASINGS AND TUBING.
Millar & Son, Charles, Utica, N. Y. xv
Ripley & Bronson, 800 N. 2nd St., St. Louis, Mo. x

WELL MAKING MACHINERY.
Pierce Artesian & Oil Well Supply Co., 80 Beaver St., New York City. xxiii

MANUAL OF AMERICAN WATER-WORKS.

THE DEANE
(OF HOLYOKE)

Steam Pumps,

Water-Works Engines

The DEANE Water-Works Engines Are Pumping 475,000,000 Galls. Daily.

THE DEANE STEAM PUMP CO.,
HOLYOKE, MASS.

OFFICES AND WAREROOMS:

New York.   Boston.   Chicago.   St. Louis.   Philadelphia.   Denver.

Write for Catalogue.

The Manual

OF

American Water-Works

COMPILED FROM SPECIAL RETURNS.

Containing the History, Distribution, Consumption, Revenue and Expenses, Cost, Debt and Sinking Fund, etc., etc., of the Water-Works of the United States and Canada,

WITH SUMMARIES
For Each State and Group of States,

AND CLASSIFICATION.
By Size, of Towns Having Works.

M. N. BAKER, Ph. B., Editor.

1891.

ENGINEERING NEWS PUBLISHING CO.
TRIBUNE BUILDING, NEW YORK.
1892.

COPYRIGHT 1892, BY THE ENGINEERING NEWS PUBLISHING CO.

THE "HOLYOKE" FIRE HYDRANT,

MANUFACTURED BY

HOLYOKE HYDRANT AND IRON WORKS, Holyoke, Mass.

Unexcelled for SIMPLICITY, DURABILITY and LOW COST OF MAINTENANCE.

All sizes made, from 4-in. with single outlet to 10-in. with six outlets, and independent gate for each outlet if desired.

GATE OPEN.

GATE CLOSED.

Points of Excellence.

The gate opens down, leaving an unobstructed water way.

The gate works easily, and does not produce water hammer in closing.

All wearing surfaces are of composition.

Stones, sand or gravel in the hydrant cannot interfere with operating the gate.

It cannot freeze, as it drips below the main.

The drip valve is positive in action, and cannot become clogged.

The gate remains closed in case of accident to post whereby it is broken, and a new one can be substituted without shutting the water from main.

It is carefully designed, and well constructed of the best material, so that cost of maintenance is practically nothing.

CONTENTS.

	PAGE
Advertisements	Inside front cover and 1.
Abbreviations (*Following Preface*).	
Introduction	10.

Water-works completed, in process of construction, projected, or reported projects since become indefinite in:

United States.

		Massachusetts	20	Tennessee	182
		Michigan	213	Texas	302
Alabama	175	Minnesota	267	Utah	350
Arizona	357	Mississippi	180	Vermont	15
Arkansas	307	Missouri	300	Virginia	165
California	312	Montana	297	Washington	350
Colorado	320	Nebraska	283	West Virginia	10
Connecticut	51	Nevada	368	Wisconsin	
Delaware	151	New Hampshire	8	Wyoming	206
District of Columbia	135	New Jersey	99		
Georgia	168	New Mexico	328	**Dominion of Canada.**	
Florida	173	New York	61		
Idaho	300	North Carolina	164	British Columbia	378
Illinois	230	North Dakota	295	Manitoba	377
Indiana	205	Ohio	191	New Brunswick	362
Iowa	255	Oklahoma Territory	320	Newfoundland	385
Kansas	273	Oregon	337	Northwest Territory	378
Kentucky	186	Pennsylvania	100	Nova Scotia	363
Louisiana	181	Rhode Island	48	Ontario	369
Maine	1	South Carolina	166	Prince Edward Island	345
Maryland	152	South Dakota	291	Quebec	366

Group 1—New England States	1	Group 7—Southwestern States	300
2—Middle "	61	8—Pacific "	310
3—South Atlantic "	155	9—Canadian Maritime Provinces	362
4—South Central " (Gulf & Mississippi Valley)	175	10—Canadian Interior and Western Provinces	366
5—Middle States	191		
6—Northwestern States	255		

Classification, by population, of towns having works 379

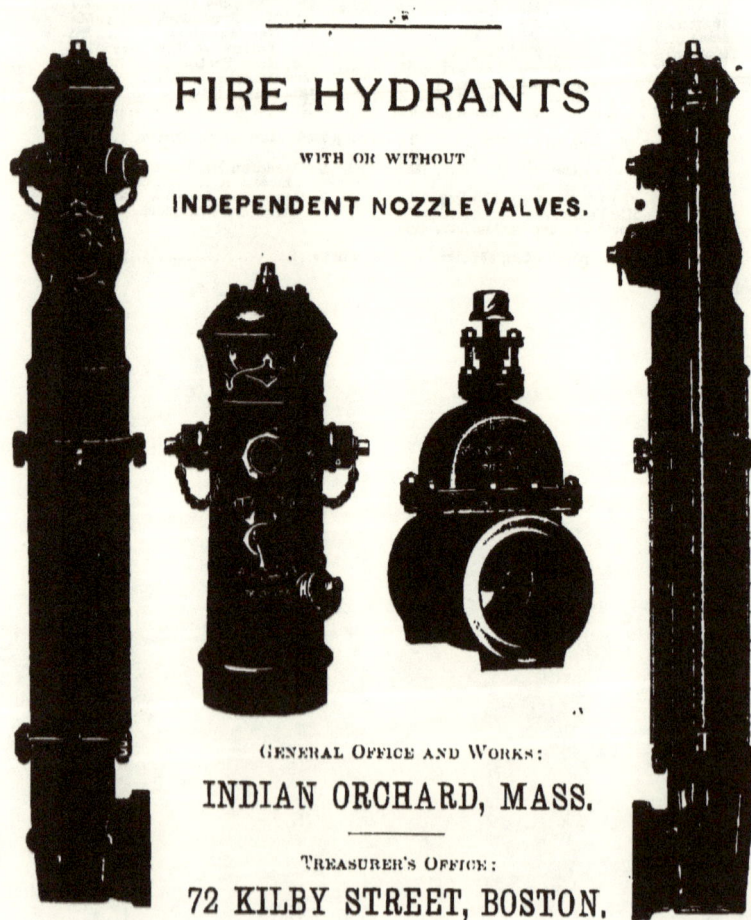

PREFACE.

This volume contains descriptions of all water-works known to be in operation or under construction July 1, 1891. In addition brief descriptions are given of such projects as bid fair to develop into works.

As the descriptive matter relating to old works has been printed in one or both of the previous editions of the MANUAL the greater part of it is omitted from this volume, which, so far as the main descriptive matter goes, should be used in connection with the issue for 1889-90. All changes and improvements to existing water-works made since the last issue, and all new works, are included in the present volume, the descriptions of new works being as complete as the information at hand would allow.

The figures under "Distribution" and "Financial" have been corrected to include the fiscal year 1890. All other information has been corrected and supplemented to as late a date as the compilation and press-work would permit, extending, in general, from July to November, 1891. The fiscal year 1890 includes all years closing between July 1, 1890, and June 30, 1891. The date upon which the several fiscal years close is stated whenever known, and figures are always for the fiscal year 1890 unless otherwise indicated. When so late figures could not be obtained the nearest available ones were generally substituted and their date stated.

In place of the matter mentioned above as omitted much statistical matter for previous years is given, thus presenting a means of comparing the yearly growth of the more important works. These figures are taken from a large and recently made collection of annual reports and from previous issues of the MANUAL and from the "Statistical Tables of American Water-Works," which was published by ENGINEERING NEWS for several years prior to the first issue of the MANUAL, and of which the latter work is an outgrowth. The reduction in the size of the present volume is largely owing to more compact typography.

Hearty thanks are extended to all who have furnished information for this and previous issues.

For the sake of convenience and brevity, the abbreviations on the following page are used in the MANUAL.

M. N. BAKER, *Editor.*

TRIBUNE BUILDING, New York, December, 1891.

ABBREVIATIONS.

Asst. Engr. Assistant Engineer.
Av. Average.
Av. dy. Average daily.
Cap. Capacity.
cem. Cement.
c. i. cast-iron.
Com. Committee.
Comr. Commissioner.
Co. Company.
Const. Engr. Constructing Engineer.
Consult. Engr. Consulting Engineer.
Contr. Contractor.
Co. County.
Dy. cap. Daily capacity.
Des. Engr. Designing Engineer.
diam. diameter.
Engr. Engr.
est. estimated.
exp. expenses.
ft. feet.
galls. gallons.
galv. i. galvanized iron.
Gen. Mgr. General Manager.

HP. Horse-Power.
Imp. Imperial.
in. inch.
ins. inches.
Incorp. Incorporated.
int. interest.
kal. i. kalamein iron.
lbs. pounds.
Mgr. Manager.
p. page.
pp. pages.
Pop. Population.
Prest. President.
Pumpg. Engr. Pumping Engineer.
Secy. Secretary.
sq. ft. square feet.
Subcontr. Sub-contractor.
Supt. Superintendent.
Treas. Treasurer.
V.-Prest. Vice-President.
vol. volume.
w. i. wrought iron.

The Manual

OF

American Water-Works.

1891.

INTRODUCTION.

The shortage of water supply in several cities and towns during the past few months, especially in the Northeast, and notably in this city, and the comparatively recent failure or suspension of several prominent firms of water-works promoters and dealers in water-works securities bring this class of municipal works before the public mind with more than usual prominence, and render especially timely the following study of the water-works of the United States and Canada, which is based upon the facts and figures presented in the body of this volume. The facts presented are all the more significant, because the figures given are, for the most part, for the year 1890, and are used in connection with the census for the same year. In addition, there is a growing interest in the study of municipal problems and the regulation and management of city works. This is especially true as regards the comparative merits of the public or private ownership of such natural monopolies as water, electric lighting and rapid transit systems.

For no class of city works has so large and so orderly a body of information been gathered as for that under consideration; as regards public *versus* private ownership no other class of works is so well suited for study, since the ownership of water-works in this country is very nearly equally divided, so far as mere numbers goes, between the two classes.

As a matter of general interest it may be stated that prior to 1878 but little had been published regarding the water-works of this country, and that little, naturally, referred to a few of the larger works only. In 1878 there began to appear in ENGINEERING NEWS a series of descriptions of all the water-works of the country, so far as it was possible to locate the works and obtain the necessary information. These descriptions were prepared by J. J. R. CROES, C. E., and were continued for some years. The statistical matter was finally published in book form in 1883, revised, enlarged and republished in 1885 and again in 1887, as the "Statistical Tables of American Water-Works." In 1888 the descriptive and historical matter was combined, with necessary condensations, with the statistical information, the whole was corrected from official reports and supplemented from every available source and in place of the "Tables" there was published the MANUAL OF AMERICAN WATER-WORKS. A second issue of the MANUAL, further enlarged and with figures a year later, appeared in 1890. The third issue is presented herewith, and contains figures for works existing in 1890, brought down to the close of their latest completed fiscal year, whenever possible, the figures in some cases being for May, or even June, 1891. The volume also contains descriptions of all works put under construction prior to July 1, 1891, and of all promising projects to the same date for New England, and to a later date for the other groups of states, corrections and additions being made for old, new and projected works to the latest possible moment, in the western states, in a few instances, as late as November, 1891. Most of the descriptive matter which appeared in the last MANUAL has been omitted from the present to give room for statistics showing the growth of many individual works, and also to reduce the size of the volume, which was becoming very bulky.

If the previous records have been in any particular incomplete, as any one who has collected similar information knows must be the case. it is not owing to any lack of persistency on our part, as many as a dozen inquries having been sent to some parties, sometimes with, and sometimes without, response.

While the many who have promptly replied to each inquiry deserve, and have, hearty thanks, it is true that a large number utterly fail to realize two facts: First, that, for their own interests, the collection and publication of such a diversity of information as is given in the MANUAL saves them, not all, but a very large number of inquiries from this and from other countries; second, that the interests of others and the relation which each individual, and water-works, bears to all others, and the public character of the information sought demands a reasonable response to such solicitations as are made for a book like the MANUAL.

The classification of the water-works in the various issues of the "Statistical Tables" and the MANUAL has differed somewhat, although the main difference will be found in the forthcoming work, where, for the first time, what is considered as a satisfactory classification has been adopted. In the earlier records some plants which were built and used for fire protection only were included, and others were recorded and counted which were for the exclusive supply of such large public institutions as reform schools and insane asylums, and again for the exclusive supply of military posts. In the present record all the public institutions and the posts have been dropped entirely, and the fire protection systems have been, except for some new ones, listed merely. Further than this, a distinction has been made between plants supplying water both for consumption and fire protection, and those supplying for consumption only. The former are termed Water-Works, the latter Water-Supplies.

The various states and provinces are arranged in the same groups as in the last MANUAL, viz.:

GROUP 1.—*New England States*. Maine, New Hampshire, Vermont, Massachusetts, Rhode Island, Connecticut.

GROUP 2.—*Middle States* New York, New Jersey, Pennsylvania, Delaware, Maryland, District of Columbia.

INTRODUCTION. (iii)

GROUP 3.—*South Atlantic States.* Virginia, West Virginia, North Carolina, South Carolina, Georgia, Florida.

GROUP 4.—*South Central States.* (Gulf and Mississippi Valley.) Alabama, Mississippi, Louisiana, Tennessee, Kentucky.

GROUP 5.—*North Central States.* Ohio, Indiana, Michigan, Illinois, Wisconsin.

GROUP 6.—*Northwestern States.* Iowa, Minnesota, Kansas, Nebraska, South Dakota, North Dakota, Wyoming, Montana.

GROUP 7.—*Southwestern States.* Missouri, Arkansas, Texas, Oklahoma, Colorado, New Mexico.

GROUP 8.—*Pacific States.* Washington, Oregon, California, Arizona, Nevada, Utah, Idaho.

To this, which includes the whole United States, there is added:

GROUP 9.—*Canadian Maritime Provinces.* New Brunswick, Nova Scotia, Prince Edward Island, Newfoundland. (Newfoundland is not practically a part of the Dominion of Canada, but geographically it belongs with Canada.)

GROUP 10.—*Canadian Interior and Western Provinces.* Quebec, Ontario, Manitoba, Northwest Territory, British Columbia.

We now proceed to consider the water-works of the United States and Canada under the following head:

PRESENT STATUS.

The total number of water-works in operation or under construction in the United States on July 1, 1891, was 2,037. Deducting from this total the number of duplicate and triplicate works, 29 in all, we have 2,008 cities and towns supplied from the 2,037 works within their limits. In addition the total number of works reported supply 179 towns outside of the municipalities in which the works are located, making a total of 2,187 cities, towns and villages having water-works facilities. There are doubtless more outside towns supplied than appear in the record, but most of them are small and many are mere hamlets included within larger bodies. The important point is the number of works. It should be noted that, of the 17 outside towns supplied in New Jersey, 11 are supplied from one company, the Hackensack Water Co., of Hoboken.

In Canada there are 95 works, no duplicates, and 7 outside towns supplied, making a total of 102 towns with the advantages afforded by water-works. For the United States and Canada together there are 2,132 works, 29 duplicate and triplicate works and 186 outside supplies, making 2,289 towns supplied in both countries. In addition there are 4 works listed for Newfoundland and 131 water-supply systems in the United States and 3 in Canada.

The distribution of these works and towns by states and groups of states is shown in detail by Table 1-S. It will be seen that Pennsylvania has the largest number of works, 216, with 245 towns supplied, and New York the next largest number, 199 works, with 218 towns supplied. Massachusetts has 128, Michigan 113, California, 103 and Illinois 102 works, making six states with over 100 works in each. Louisiana and Arizona each have but 4 works. Delaware 6, South Carolina and Mississippi, North Dakota and Utah but 7 each. In Canada, Ontario has 48 of the 95 works, one more than half, and Quebec 24, making 72, or about 75 per cent. of the total, in these two provinces, and leaving Nova Scotia but

(iv) MANUAL OF AMERICAN WATER-WORKS.

9, British Columbia 8, New Brunswick 6, and one each to three other great areas.

TABLE 148.

UNITED STATES AND CANADIAN WATER-WORKS. — SUMMARY OF NUMBER OF WORKS, OUTSIDE AND TOTAL TOWNS SUPPLIED, AND NUMBER OF WATER SUPPLY SYSTEMS BY STATES AND GROUPS OF STATES.

	No. of w'ks.	Outside towns supplied.	Deduct dupli-cate works.	Total towns sup-plied.	Wtr. sup-plies.		No. of w'ks.	Outside towns supplied.	Deduct dupli-cate works.	Total towns sup-plied.	Wtr. sup-plies.
Me...	37	20	1	36	3	Mo....	36	1	0	37	1
N.H..	36	5	0	41	6	Ark...	12	0	0	12	1
Vt....	27	0	0	27	8	Tex..	60	0	1	81	3
Mass.	128	12	1	139	14	Col...	57	9	2		4
R.I...	14	17	0	31	2	N.M.	9	0	0	9	0
Conn..	48	7	1	54	2						
Tot. N.E..	290	61	3	348	35	Tot. S.W.	174	10	3	181	10
N.Y..	199	22	3	218	16	Wash.	38	1	2	37	5
N.J..	58	17	1	74	3	Ore...	26	0	2	24	6
Pa...	216	33	4	245	25	Cal...	103	10	5	108	9
Del...	6	0	0	6	0	Ariz..	4	0	0	4	0
Md...	18	0	0	18	0	Nev...	9	1	0	10	1
D.C...	1	1	0	2	0	Utah.	7	0	0	7	0
						Ida...	10	0	0	10	1
Tot. Mid...	498	73	8	563	44						
Va....	41	2	0	43	4	Tot. Pac...	197	12	9	200	22
W.Va.	10	1	0	11	1						
N.C...	15	0	0	15	0						
S.C...	7	0	0	7	0	Tot. U.S.	2,037	179	29	2,187	131
Ga....	21	0	0	21	3						
Fla....	12	0	0	12	1	N.B...	5	3	0	8	0
						N.S...	9	0	0	9	0
Tot. S.A..	106	3	0	109	9	P.E.I.	1	0	0	1	0
Ala...	20	1	1	20	0						
Miss..	7	0	0	7	4	Tot. East-					
La....	4	0	0	4	0	ern Prov.	15	3	0	18	2
Tenn..	16	2	1	17	1						
Ky....	23	1	0	24	1	Que...	21	4	0	24	0
						Ont...	44	0	0	44	1
Tot. S.C...	70	4	2	72	6	Man..	5	0	0	5	0
O......	86	3	0	89	0	N.W.T	1	0	0	1	0
Ind....	50	0	0	50	0	B.C...	6	0	0	6	0
Mich..	113	1	2	112	0						
Ill.....	102	3	2	103	0	Tot. Int. &					
Wis...	50	3	0	53	0	W. Prov.	80	4	0	84	1
Tot. N.C...	401	10	4	407	0						
Ia.....	77	1	0	78	0						
Minn.	34	0	0	34	2	Tot. Can'a.	95	7	0	102	3
Kan...	79	2	0	81	0						
Neb..	62	1	0	63	0						
S.Dak	22	0	0	22	1	Tot. U.S.					
N.Dak	7	0	0	7	0	and Can..	2132	186	29	2289	134
Wyo..	9	2	0	11	0						
Mont..	11	0	0	11	2	New-					
						f'land.	4	0	0	4	....
Tot. N.W.	301	6	0	307	5						

In number of works the several United States groups rank as follows:

	Works.	Towns supplied.		Works.	Towns supplied.
Middle................	498	563	Pacific............	197	200
North Central........	401	407	Southwest'n......	174	181
Northwestern........	301	307	So. Atlantic......	106	109
New England.........	290	348	South Central....	70	72
Total, U.S............				2,037	2,187

The Middle group, it will be seen, has almost 25 per cent. of the total number of works in the United States.

INTRODUCTION. (v)

It should be noticed that of the 179 outside towns supplied, 61 are in the New England and 73 are in the Middle group, making 134, or about 78 per cent., in these two groups, while the South Atlantic group has but three outside supplies and none of the remaining four more than 12.

The duplicate and triplicate plants, of which there are 29 in 25 cities, are located as follows, the cities containing them being given instead of the group, and the total number of works in each city being given:

Springvale, Me	2	Grand Haven, Mich	2
Great Barrington, Mass	2	Grand Rapids, Mich	2
Ansonia, Conn	2	Chicago, Ill	2
Brooklyn, N. Y	2	Waco, Tex	3
Jamaica, N. Y	2	Denver, Colo	2
Jordan, N. Y	2	Pueblo, Colo	2
Atlantic City, N. J	2	Seattle, Wash	2
Pittsburg, Pa	2	Spokane, Wash	2
Scrant't, Pa	2	Portland, Ore	3
Troy, Pa	2	Los Angeles, Cal	3
Tunkhannock, Pa	2	Monrovia, Cal	2
Mobile, Ala	2	Pasadena, Cal	3
Chattanooga, Tenn	2		

Most of the duplicate works are owned by companies, and many are in small places. Several of the plants supply sections not reached by others, and with the exception of two or three very notable instances there is little or no competition between these works. In city lighting, for example, competing plants have played a very disturbing part, although consolidation has followed in the most notable instances. Water-works are probably as complete natural monopolies as are in existence, and competition among them is rare and almost impossible. The subject will be discussed more fully further on.

The last column of Table 1-S relates to water-supplies, that is, to systems furnishing water for consumption only, and that mostly for domestic use, and never for fire protection. Nearly all of these are in small places. Of the 131 water-supply systems in the United States 35 are in the New England and 44 are in the Middle group, while 22 more, making 101 out of 131, are in the Pacific group. These are nearly all gravity systems, of small cost and slight importance. Some are owned by a few families who divide the light yearly expenses equally.

Table 2-S shows by states and provinces, groups of the same, and for both the United States and Canada, the population on works, according to the 1890 census, the number of works, miles of mains, number of taps, meters and hydrants and cost of works. As the cost was not reported for all works, the number of works reporting is first given, then the corresponding cost, then the estimated total cost for all works.

TABLE 2—S.

UNITED STATES AND CANADIAN WATER-WORKS.—SUMMARY OF POPULATION ON WORKS, NUMBER OF WORKS, MILES OF MAINS, NUMBER OF TAPS, METERS AND HYDRANTS AND COST BY STATES AND GROUPS OF STATES.

State.	Population on works, 1890.	No. of works.	Miles of m'ns.	Taps.	Meters.	Hydrants.	No. reporting.	Reported cost.	Est. total cost.
Me	254,334	37	566	15,127	330	2,114	26	$5,879,888	$8,130,000
N. H	174,810	36	334	17,056	1,955	1,943	33	3,850,704	4,975,000
Vt	90,002	27	322	9,553	613	1,031	26	1,660,302	1,715,000
Mass	1,884,715	128	3,621	249,499	31,808	25,706	121	64,713,574	67,515,000
R. I	272,485	14	522	27,055	15,036	3,534	13	4,905,167	4,965,000
Conn	492,115	48	847	43,361	1,102	5,041	43	10,211,216	10,640,000
N. E. Group	3,169,461	290	6,212	361,651	50,874	39,369	262	91,229,851	97,940,000

TABLE 2–S.—Continued.

State.	Population on works, 1890.	No. of works.	Miles of m'ns.	Taps.	Meters.	Hydrants.	No. reporting.	Reported cost.	Est. total cost.
N. Y........	4,041,487	199	4,001	328,246	37,869	34,281	168	89,037,776	97,075,000
N. J........	959,540	58	1,358	111,883	8,543	9,519	51	20,637,118	21,635,000
Pa........	2,664,124	216	3,395	355,309	2,172	20,597	180	48,495,531	53,835,000
Del........	73,602	6	100	13,489	28	792	6	1,091,366	*1,091,336
Md........	500,311	10	550	70,551	1,030	2,427	10	12,078,238	12,540,000
D. C........	230,392	1	215	33,270	87	1,080	1	8,500,000	*8,500,000
Mid. Group	8,469,485	498	9,628	921,748	49,782	68,696	422	180,640,050	195,076,366
Va........	291,836	41	421	27,382	294	2,065	35	4,278,074	5,280,000
W. Va.....	79,105	10	123	10,980	6	777	10	1,338,322	*1,338,322
N. C........	98,444	15	128	2,944	495	978	13	1,302,689	1,300,000
S. C.........	91,827	7	88	1,582	33	862	4	196,685	1,150,000
Ga........	243,578	21	255	15,269	3,806	2,107	16	2,756,439	3,015,000
Fla........	57,797	12	96	4,055	454	824	11	981,942	1,110,000
S.Atl. Group	863,587	106	1,111	62,212	5,088	7,613	80	10,914,151	13,283,322
Ala........	141,749	20	304	11,158	1,304	1,754	15	4,591,988	5,390,000
Miss.....	51,102	7	65	1,550	20	526	6	524,000	725,000
La........	267,617	4	98	5,165	30	1,418	3	2,770,783	2,845,000
Tenn........	238,124	16	275	17,511	351	1,591	12	5,692,852	5,900,000
Ky........	337,379	23	415	28,029	1,702	2,624	20	9,442,365	10,440,000
S. C'nt.Gro'p	1,035,971	70	1,157	63,413	3,425	7,913	56	23,021,988	25,300,000
O........	1,412,020	86	2,439	127,880	5,143	12,745	79	29,614,549	30,915,000
Ind........	537,680	50	556	17,860	927	6,133	42	8,921,087	9,640,000
Mich........	891,501	113	1,371	83,345	2,157	9,089	111	13,269,135	13,610,000
Ill........	1,704,877	102	2,099	199,237	6,121	18,696	100	24,887,191	25,285,000
Wis........	541,051	50	711	46,212	6,835	6,520	42	8,677,835	9,975,000
N. C. Group	5,085,129	401	7,176	474,234	21,183	54,083	374	85,309,797	89,425,000
Ia........	438,962	77	510	21,086	2,439	4,373	74	7,226,267	7,375,000
Minn........	452,239	34	548	26,759	1,291	5,358	32	7,577,284	7,635,000
Kan........	316,094	79	491	17,910	999	4,395	66	4,313,368	5,615,000
Neb........	318,577	62	441	14,266	1,485	3,446	58	6,019,616	6,370,000
S. Dak........	41,448	22	92	2,716	78	579	22	961,019	*961,019
N. Dak........	18,990	7	32	1,401	204	221	7	394,283	*394,283
Wyo........	29,671	9	103	1,598	195	268	5	314,299	405,000
Mont........	44,804	11	126	2,740	149	548	9	846,906	1,305,000
N. W. Group	1,690,805	301	2,343	88,476	6,840	19,188	273	27,683,042	30,150,302
Mo........	829,406	36	837	61,046	5,921	7,329	30	21,769,666	22,445,000
Ark........	72,838	12	106	4,399	132	643	11	1,395,000	1,470,000
Tex........	363,162	60	631	33,818	1,753	3,734	57	5,433,379	6,535,000
Colo........	218,352	57	713	24,657	285	3,600	49	7,535,525	8,935,000
N. M........	19,783	9	60	1,436	69	186	9	633,048	*633,048
S. W. Group	1,502,571	174	2,347	125,356	8,163	15,392	156	36,766,618	40,018,048
Wash........	151,044	38	263	7,199	332	953	33	2,593,472	2,745,000
Ore........	83,114	26	223	11,631	83	651	24	1,882,676	1,990,000
Cal........	647,815	103	1,652	84,051	16,641	4,810	86	36,229,554	38,230,000
Ariz........	71,936	4	50	1,088	55	342	4	900,000	*900,000
Nev........	18,160	9	104	5,520	13	254	8	5,697,790	6,045,000
Utah........	71,922	7	116	4,684	0	373	5	729,389	1,210,000
Idaho........	14,030	10	41	1,670	0	231	10	377,105	*377,105
Pac. Group...	998,051	197	2,449	115,823	17,124	7,614	170	48,409,986	51,487,105
Total U. S...	22,814,061	2,037	32,423	2,212,913	162,479	219,868	1,802	504,035,492	542,770,143
N. B........	66,997	5	105	6,955	186	606	4	1,148,369	1,525,000
N. S........	61,443	9	105	17,065	41	592	9	1,157,000	*1,157,000
P. E. I........	11,374	1	17	1,347	1	87	1	195,521	*195,521
East'n Group	139,814	15	227	25,367	228	1,285	14	2,500,890	2,877,521
P. Q........	383,541	24	399	68,148	1,031	2,380	24	11,318,869	*11,318,869
Ont........	479,769	48	672	86,179	1,660	5,617	43	9,882,290	10,330,000
Man........	25,642	1	16			102	1	130,000	*130,000
N. W. T........	3,876	1	5		0	30	0		75,000
B. C........	41,762	6	106	3,504	271	313	5	1,373,042	1,400,000
W'st'n Group	934,593	80	1,198	157,831	2,971	8,442	73	22,704,201	23,253,869
Total Canada	1,074,407	95	1,425	183,198	3,199	9,727	87	25,505,091	26,131,390
T't'l U. S. and Canada...	23,888,468	2,132	33,848	2,306,111	165,676	229,595	1,889	529,540,583	568,901,533
Newf'dl'd...(est.)	52,300	4	30	1,998	0	219	2	446,000	500,000

* Actual cost.

INTRODUCTION. (vii)

Taking up first the populations, we see that of a total of 62,622,250 people in the United States, 22,814,001, or about 36 per cent., live in towns having public water-works. The total population of towns having 8,000 or more inhabitants is 18,235,670, or 29 per cent. of the total population. There are but a few towns of 8,000 or more without works, and some of these are townships rather than towns, and contain a large rural population. It should be stated that the New England populations contain a large rural element, the village, in many cases, not being separable from the rural population. Many of the villages with works have a population of from 1,000 down to 300, or even less, the number of works in small places having rapidly increased in the past few years, as will be shown in some detail later. Water-works, once a luxury for our smaller towns, are now considered almost a necessity in small villages and hamlets, both for fire protection and sanitary uses. This necessity will become more rather than less prevalent in the future, although for a time there is a check in water-works construction, owing to the stringency of the money market, and the consequent difficulty of floating stock and bonds. Of the total population on works 8,469,486, or about 37 per cent., are in the Middle group. The North Central group has over 5,000,000 people on works, while the South Atlantic has but 863,587. New York State, alone, has over 4,000,000 people potentially supplied, and Arizona has but 11,936.

The Middle group, likewise, leads in all the particulars included in the table, with the single exception of meters, it having but 49,782 meters against 50,874 in New England, or 1,092 less.

Of the 2,037 works in the United States and Canada, the cost of 1,802 is known with a fair degree of accuracy, the total, as shown in the table, being $504,035,492. In some cases the figures for cost are for a year previous to 1890, it seeming more accurate to include these figures instead of estimates, and it being known that in some cases the costs reported are excessive. Where no cost has been available, the same has been estimated for each state by multiplying the total number of miles of mains included in such works by the average cost per mile of main of all works in the state, as worked out in detail in the introduction to the MANUAL for 1888. These estimates raise the estimated total cost to about $543,000,000 for all works in the United States, which may be taken as fairly correct, the figures probably being too low rather than too high. Of Canadian works, 87 of the 95 have cost $25,505,091, and the estimates raise the total by only $625,000.

The reports included in the MANUAL were quite generally complete so far as miles of mains, number of taps, meters and hydrants are concerned. In making up the totals given in table 2—S., where no figures for 1890 were available, those for an earlier year were included, as in the case of cost of works. There were but very few omissions which could not be supplied in the above named way, and both these and the increase since the earlier figures were given are thought to be offset by the tendency of some correspondents to exaggerate when reporting these items

Those interested in any particular state or group of states will find the particulars for the same given in the table. In concluding this subject the following comparison of the mileage of water-works mains and railway lines will be found of interest, the figures for railways being taken from the last issue of "Poor's Manual of Railroads":

Group.	Water-works, miles mains.	Rail-ways, miles.		Water-works, miles mains.	Rail-ways, miles.
N. England	6,212	6,841	Pacific	2,449	12,020
Middle	9,628	20,115			
S. Atlantic	1,111	17,308	T'tal U. S.	32,423	166,817
S. Central	1,157	13,388	Canada	1,425	*21,624
N. Central	7,176	36,945			
Northwestern	2,343	36,194	Total U. S. and Canada	33,848	*188,441
Southwestern	2,347	24,006			

* Including Newfoundland.

It will be seen from the above that in the New England group alone does the mileage of water mains approach that of railway lines, the figures being 6,212, against 6,841. For the whole United States there are 32,423 miles of water mains, against 166,817 miles of railways, the former being nearly one-fifth of the latter. There is no comparison, however, between the two except as a matter of curiosity. "Poor's Manual" gives the total investment in railways for 1890 as $10,122,635,900, this sum being the total of the stock and bonds of the various lines at par value, and the total cost as about $8,970,000,000, this total being based on the "'Cost of Road and Equipment' per mile of completed road," and therefore being somewhat excessive. This cost must include some watered stock, as it is only about $1,200,000,000, or about 12 per cent. less than the total investment given above, while the estimated cost of water-works in the United States, as given in Table 2-S, $542,770,143, is for the most part actual cost of construction. But taking the cost of railways as given, the cost of water-works is about 6 per cent. as much, a surprising percentage when we consider the magnitude of our railway interests and the prodigal waste in connection with their construction.

GROWTH BY NUMBER OF WORKS AND POPULATION SUPPLIED.

At the close of the last century there were in the United States but 16 cities or towns having water-works, while in Canada there were none, the first works in Canada having been built at Montreal in 1801 by a company. The Montreal works were bought by the city in 1845. The 16 works built in the United States prior to the close of 1801, together with the date built, builders and present owners—that is, whether company, individual or city—are as follows:

	Date.	Original ownership.	Present ownership.
Boston, Mass.	1652	Co.	City.
Bethlehem, Pa.	1761	Co.	City.
Providence, R. I.	1772	Co.	City.
Morristown, N. J.	1779	Co.	Co.
Geneva, N. Y.	1787	Co.	Co.
Peabody, Mass.	1790	Co.	City.
Plymouth, Mass.	1796	Co.	City.
Worcester, Mass.	1796	Indv.†	City.
Hartford, Conn.	1797*	Co.	City.
Portsmouth, N. H.	1798	Co.	Co.
Salem, Mass.	1799	Co.	City.
New York, N. Y.	1799‡	Co.	City.
Lynchburg, Va.	1799	Co.	City.
Newark, N. J.	1800	Co.	City.
Albany, N. Y.	Prior to 1800	Co.	City.
Winchester, Va.	Prior to 1800	City.	City.

* Probable date.
† Individual.
‡ Works started by city in 1774, but stopped by Revolutionary War.

INTRODUCTION. (ix)

As will be seen from the foregoing Boston was the first city supplied, by over 100 years, and Bethlehem, Pa., next. The Boston supply was from springs, by gravity to a box, or tank, while the Bethlehem supply was pumped from a creek to a tank. The Boston works were designed for domestic use and fire protection, but it cannot be stated at this time whether water was distributed to consumers, or was conveyed from the tank in buckets or barrels. At Bethlehem water was distributed at least as early as 1769, as at that date some of the pitch pine distributing logs were renewed, and in 1786 some were replaced with lead pipe. The works in this city were built in 1799 by the Manhattan Co., which was organized to do a banking business and supply water. The city subscribed for 2,000 shares of the company's stock. The charter of the company provided that it must keep a tank filled with water in order to hold its charter, and although the company long since stopped supplying water it still keeps a tank filled, on or near Centre St., the same company now conducting the Manhattan Bank.

Starting with 16 works at the close of 1800 there were in the United States 50 years later but 67 in addition, or 83 in all, while Canada then had but 4 works. The growth of water-works has, therefore, been almost wholly in the last 40 years, and largely in the last 10 or 10½ years, there having been but 598 works in this country at the close of 1880, while at the close of 1890 there were 1,878, and on July 1, 1891, 2,037 works had been completed or put under construction, including 14 of unknown date.

The details of this wonderful growth are shown by Table 3–8, the first, or upper, part of which gives the number of works built in each state, the United States and Canada for each half decade from 1801 to 1870, and for each year and half decade from 1871 to date. The lower part of the table shows the number of works at the end of each decade.

TABLE 3–8.

UNITED STATES AND CANADIAN WATER-WORKS—SUMMARY OF DATES OF CONSTRUCTION BY GROUPS OF STATES FOR EACH HALF DECADE UNTIL 1870, AND FOR EACH YEAR FROM 1871 TO AND INCLUDING THE PARTIAL YEAR 1891.

	N. E.	Mid.	S. Atl.	S. C.	N. C.	N. W.	S. W.	Pac.	Total U.S.	Can.	U.S.& Can. Tot'l
Before 1801...	8	6	2	..	..	..	..	..	16	..	16
1801–5........	3	4	..	..	..	..	..	..	7	1	8
1806–10.......	..	3	..	..	..	..	..	..	3	..	3
1811–15.......	..	..	..	..	..	..	..	..	..	..	..
1816–20.......	1	2	..	..	1	..	..	..	4	..	4
1821–5........	..	1	..	1	..	..	..	..	2	..	2
1826–30.......	1	5	3	..	1	..	1	1	12	..	12
1831–5........	1	3	2	3	1	..	..	..	10	..	10
1836–40.......	1	4	1	2	2	..	..	..	10	1	11
1841–5........	1	2	..	1	2	..	..	..	6	1	7
1846–50.......	3	9	..	..	1	..	..	..	13	1	14
1851–5........	2	14	1	..	2	..	..	4	23	2	25
1856–60.......	6	13	2	1	3	..	..	5	30	2	32
1861–5........	5	9	..	1	1	..	..	10	26	..	26
1866–70.......	23	30	1	3	16	3	..	8	81	..	81
1871..........	2	14	1	..	6	2	1	5	31	..	31
1872..........	6	12	1	2	8	..	1	3	33	1	34
1873..........	15	5	..	2	8	..	2	4	39	3	42
1874..........	8	11	1	..	9	2	.	5	36	2	38
1875..........	6	14	2	..	9	1	3	5	40	4	44
Total, '71–5..	37	59	5	4	40	5	7	22	179	10	189
1876..........	9	11	1	1	9	6	1	7	45	3	48
1877..........	1	5	1	..	5	2	..	3	17	1	18
1878..........	7	11	3	2	5	4	3	4	38	2	40
1879..........	10	9	..	1	10	1	6	1	39	2	41
1880..........	8	10	1	..	5	3	5	4	37	4	41
Total, '76–80.	35	46	6	5	34	16	15	19	176	12	188

MANUAL OF AMERICAN WATER-WORKS.

TABLE 3–S.—*Continued.*

	N. E.	Mid.	S. Atl.	S. C.	N. C.	N. W.	S. W.	Pac.	Total U. S.	Can.	U.S. & Can. Tot'l
1881	5	12	..	1	6	5	4	3	36	2	38
1882	10	17	1	1	21	13	9	7	82	4	86
1883	8	18	1	2	17	10	14	4	85	5	90
1884	21	21	2	..	17	20	17	7	105	4	109
1885	17	28	5	2	25	15	12	3	107	1	108
Total, '81–5	61	96	15	6	86	71	56	24	415	16	431
1886	15	31	7	4	35	24	10	7	133	5	138
1887	29	36	7	4	34	38	19	14	181	6	187
1888	13	33	11	8	47	47	18	21	198	9	207
1889	23	40	16	12	37	41	17	14	200	7	207
1890	14	25	12	1	33	32	11	25	153	6	159
Total, '86–90	94	165	53	29	186	182	75	81	865	33	898
Unknown	..	4	..	1	..	.	3	6	14	2	16
*1891	11	23	15	13	25	24	17	17	145	14	159
Total	290	498	106	70	401	301	174	197	2,037	95	2,132

NUMBER OF WORKS AT THE END OF EACH HALF DECADE.

	N. E.	Mid.	S. Atl.	S. C.	N. C.	N. W.	S. W.	Pac.	Total U. S.	Can.	U.S. & Can. Tot'l
1800	5	6	2	..	..	..	..	..	16	..	16
1805	11	10	2	..	..	..	..	..	23	1	24
1810	11	13	2	..	..	..	..	..	26	1	27
1815	11	13	2	..	..	..	..	..	26	1	27
1820	12	15	2	..	..	..	..	..	30	1	31
1825	12	16	2	1	1	..	..	..	32	1	33
1830	13	21	5	1	2	..	1	1	44	1	45
1835	14	24	7	4	3	..	1	1	54	1	55
1840	15	28	8	6	5	..	1	1	64	2	66
1845	16	30	8	7	7	..	1	1	70	3	73
1850	19	39	8	7	8	..	1	1	83	4	87
1855	21	53	9	7	10	..	1	5	106	6	112
1860	27	66	11	8	13	..	1	10	136	8	144
1865	32	75	11	9	14	..	1	20	162	8	170
1870	52	105	12	12	30	3	1	28	243	8	251
1875	89	164	18	16	70	8	3	50	422	18	440
1880	124	210	23	21	104	24	23	69	598	30	628
1885	185	306	38	27	190	95	79	93	1,013	46	1,059
1890	279	471	91	56	376	277	154	174	1,878	79	1,957
Unknown	..	(4)	..	(1)	..	(3)	(6)	(14)	(2)	(16)
*1891	290	475	106	70	401	301	174	197	2,037	95	2,132

* Completed or under construction prior to July 1.

This table needs little comment. It will be seen that at the close of 1800 there were works in only the New England, Middle and South Atlantic groups of states, 16 in all, being those already named above. Prior to 1820 these three groups were the only ones having works, while not until the decade 1866–70 does the Northwest appear in the table, and then with but three works at the close of 1870. In the Southwest one town, St. Louis, Mo., was supplied between 1825 and 1830, but not another until after 1870. Since the latter date both of these groups have built works at a rapid rate, as will be seen from the following comparison, to which has been added corresponding figures for the United States:

	Northwest.		Southwest.*		United States.*	
	No. w'ks.	Inc., No. p. c.	No. w'ks.	Inc., No. p. c.	No. w'ks.	Inc., No. p. c.
Close of '70	3	....	1	....	243	....
" '75	8	5 60	8	7 875	422	179 74
" '80	24	16 200	23	15 158	598	176 42
" '85	95	71 338	79	56 244	1,013	415 69
" '90	277	182 192	154	75 95	1,878	865 85
July 1, '91†	301	24 48	171	17 11	2,023	145 7

* Three of unknown date not included in Southwest and 14 not included in United States.
† Completed or under construction.

Looking at the column, just above, giving the total number of works for the United States, we see that 415 works were added during the five years 1881–5, and 865 during the five years 1886–90. This increase was continual year by year, until 1890 (see table 3–S), when there was a falling off of nearly 25 per cent., from 200 to 153, in the total for the year. For the period covered in 1891 145 works have been completed or put under construction, so that the total for the present year will be above that for 1890, but probably considerably below that for 1889.

It is significant that in the decade 1811 to 1815, which includes the war of 1812, not a works was built in either the United States or Canada. Again, in 1861-5, the civil war period, but 26 works were built, against 80 in the preceding and 81 in the succeeding half decade. In 1863 not a single system was built. Although the five years 1866-70 saw 81 works built the boom of the reconstruction period soon died out, and from 1871 to 1881, inclusive, 11 years, an average of only 36 works per year was built, the number reaching 45 in 1876, but sinking to 17 in the following year. As has already been stated, construction has increased since the close of 1881, with the exception of the present year and last year, when another period of depression has retarded water-works development. Of course the towns still unsupplied are limited, if not so much in numbers most certainly in size, as is shown further on.

While the increase in number of works built in the decade 1881-90 shows the great activity in water-works construction during that period and indicates the number of municipal units that have been supplied since the close of 1881, these figures are not as significant as those given in Table 4-S. This table shows the population in cities and towns having works at the close of 1880 and of 1890, respectively. the actual and per cent of increase during the period, and in addition indicates what part of this increase is attributable to the growth of the works supplied before the close of 1890 and what part to the population on works added from 1881 to 1890, inclusive, the percentage being given here, also, in each case. For convenience the number of works at the two dates, the increase and increase per cent. are given in the last columns of the table. All the particulars are shown by states and groups of states and for the whole country.

TABLE 4-S.

UNITED STATES AND CANADIAN WATER WORKS.—GROWTH OF WORKS FROM 1880 TO 1890 IN POPULATION SUPPLIED AND IN NUMBERS, BY STATES AND GROUPS OF STATES.

	Population supplied.				Increase in pop'n on works built.					No. of wo'k's at close of		Increase.	Inc. p. ct.
	1880.	1890.	Incr.	In.p.c	Before end of 1880.	In. p. c.	During 1881-90.	In. p. c.	1880.	1890.		
Me...	128,359	250,143	121,784	95	16,826	13	104,958	82	9	35	26	289
N. H..	99,320	167,652	68,332	69	26,797	27	41,535	42	11	33	22	200
Vt....	45,755	88,240	42,485	93	7,703	17	34,782	76	13	26	13	100
Mass..	1,129,554	1,881,307	751,753	67	352,269	32	399,484	35	58	126	68	117
R. I..	156,680	249,656	92,976	59	58,385	37	34,591	22	6	13	7	117
Conn.	291,785	492,115	200,330	68	109,435	37	90,895	31	27	46	19	74
N. E..	1,851,453	3,129,113	1,277,660	69	571,415	31	706,245	38	124	279	155	12
N. Y..	2,669,935	4,038,074	1,368,139	51	910,523	34	457,616	17	74	192	118	159
N. J..	524,724	942,235	417,511	79	198,634	38	218,877	41	22	54	32	145
Pa....	1,720,460	2,641,352	920,892	54	578,594	34	342,298	20	107	202	95	89
Del...	46,178	73,602	27,424	59	19,263	42	8,161	17	2	6	4	200
Md...	351,671	498,023	146,352	42	102,369	29	43,983	13	4	16	12	300
D. C..	159,871	230,392	70,521	44	70,521	44	0	0	1	1	0	0
Mid...	5,472,839	8,423,678	2,950,839	52	1,879,904	33	1,070,935	19	210	471	261	123
Va...	166,247	278,763	112,516	68	44,800	27	67,716	41	11	35	24	218
W. Va	37,072	76,535	39,463	106	14,676	39	24,787	67	2	9	7	350
N. C..	4,825	94,253	89,428	1,853	2,168	45	87,260	1,888	2	14	12	600
S. C..	49,984	81,831	31,847	64	4,968	10	26,879	54	1	5	4	400
Ga...	107,502	228,598	121,096	113	65,320	61	55,776	52	6	16	10	166
Fla ..	7,650	57,797	50,147	655	9,551	125	40,596	530	1	12	11	1,100
S. Atl.	373,280	817,777	444,497	120	141,483	39	303,014	81	23	91	68	296

MANUAL OF AMERICAN WATER-WORKS.

TABLE 4-S.—Continued.

	Population supplied.				Increase in pop'n on works built.				No. of work's at close of		Increase.	Inc. p. ct.
	1880.	1890.	Incr.	In.p.c.	Before end of 1880.	In. p. c.	During 1881-90.	In. p. c.	1880.	1890.		
Ala..	53,918	140,996	87,078	162	33,214	62	53,864	100	4	18	14	350
Miss..	3,995	47,816	43,821	1,097	564	13	43,257	1,084	1	6	5	500
La...	216,090	264,496	48,406	23	25,949	12	22,457	11	1	3	2	200
Tenn...	98,159	220,508	131,349	134	86,377	88	44,972	46	6	12	6	100
Ky....	214,225	315,695	101,470	47	65,128	30	36,342	17	9	17	8	80
So.C't.	586,387	998,511	412,124	70	211,232	36	200,892	34	21	56	35	167
O......	742,789	1,381,079	638,290	86	289,495	39	348,795	47	31	77	36	113
Ind...	263,357	523,200	259,843	99	100,999	38	158,844	61	20	47	27	135
Mich..	280,122	878,247	589,125	204	186,797	65	402,328	139	23	108	85	369
Ill.....	692,750	1,694,673	1,001,923	145	658,289	95	343,634	50	26	.97	81	312
Wis...	138,908	532,704	393,796	283	100,313	72	293,483	211	4	47	43	1,075
No.C..	2,126,926	5,009,903	2,882,977	136	1,335,893	63	1,547,084	73	104	376	272	262
Ia......	156,099	409,453	253,354	163	72,766	47	180,588	116	16	64	48	300
Minn..	107,623	443,086	335,463	312	219,730	204	115,724	108	4	30	26	650
Kans..	18,440	315,189	296,749	1,609	33,996	184	262,753	1,425	2	78	76	3,800
Neb...	......	344,735	344,735	†	......	...	344,735	†	0	58	58	†
S.Dak.	3,777	40,347	36,570	938	*1,411	*37	37,981	1,005	1	21	20	2,000
N.Dak	......	18,990	18,990	†	......	...	18,990	†	0	7	7	†
Wyo ..	2,696	20,671	26,975	1,001	3,602	137	23,283	864	1	9	8	800
Mont..	......	41,954	41,954	†	......	...	41,954	†	0	10	10	†
N.W...	288,635	1,643,425	1,354,790	470	328,782	114	1,926,008	356	24	277	253	1,054
Mo ..	466,407	817,317	350,910	75	207,271	41	143,639	31	6	31	25	417
Ark...	13,138	70,383	57,245	436	12,736	97	44,509	339	1	11	10	1,000
Tex...	65,729	351,784	286,055	435	66,989	102	219,066	333	5	52	47	940
Colo...	74,623	218,023	143,403	192	78,909	106	64,494	86	11	51	40	364
N.M...	......	19,763	19,783	†	......	...	19,783	†	0	9	9	†
S.W...	619,891	1,477,290	857,306	139	365,905	50	491,491	80	23	154	131	567
Wash.	6,512	142,402	135,890	2,087	6,440	99	120,450	1,988	3	29	26	867
Ore...	22,317	70,775	48,428	218	24,038	107	24,390	109	4	22	18	450
Cal...	410,658	642,535	231,877	56	189,711	46	42,166	10	51	64	43	84
Ariz...	......	11,936	11,936	†	......	...	11,936	†	0	4	4	†
Nev...	24,594	18,160	*6,434	*27	*7,584	*31	1,150	4	8	9	1	13
Utah..	25,706	71,922	46,216	179	26,552	103	19,664	76	3	7	4	133
Idaho.	......	13,181	13,181	†	......	...	13,181	†	0	9	9	†
Pac...	489,817	970,911	481,094	98	239,157	40	241,937	49	69	174	105	153
Total U.S.	11,809,231	22,470,608	10,661,377	90	5,073,771	43	5,587,606	47	598	1,878	1,280	211
N.B..	48,360	66,997	18,637	38	4,019	8	14,618	30	2	5	3	150
N.S..	39,561	61,443	21,882	55	4,097	10	17,785	45	2	8	6	300
P.E.I.	......	11,374	11,374	†	......	...	11,374	†	0	1	1	†
P.Q..	256,441	383,544	127,103	50	68,892	27	58,211	23	12	21	9	75
Ont...	236,997	438,187	201,190	85	92,898	39	108,292	46	13	38	25	192
Man..	......	25,642	25,642	†	......	...	25,642	†	0	1	1	†
N.W.T...	......	3,876	3,876	†	......	...	3,876	†	1	1	1	†
B.C..	5,925	35,121	29,196	493	10,916	184	18,280	309	1	4	3	300
Total Can.	587,284	1,026,184	438,900	75	180,822	31	258,078	44	30	79	49	163
Total U.S. & C.	12,396,515	23,496,792	11,100,277	90	5,254,593	43	5,845,684	47	628	1,957	1,329	212

*Decrease. † Infinity.

As will be seen by referring to Table 4-S, the total population on works in the United States at the close of 1880 was 11,809,231. At the close of 1890 the population on works had increased to 22,470,608, a gain of 10,661,377, or 90 per cent. Of this gain 5,073,771, or 43 per cent. of the total gain, was due to the natural increase of the population in the towns supplied at the close of 1880, and 5,587,606, or 47 per cent. of the total increase, was due to the building of works in towns not previously supplied. Considering these figures in connection with the increase in number of works, we see that while the latter increased from 598 to 1,878, a gain of 1,280, or 211 per cent., the population added by this large increase in numbers was but 47 per cent. of that previously supplied, the increase in numbers and in population added thus being to each other as 4½ to 1. Of course this is accounted for by the fact that all of the large cities were supplied prior to 1880.

By groups the greatest increase in population supplied is in the Middle group, where 2,950,839 was added. By percentages the greatest increase is in the Northwestern group, 470 per cent., the population supplied increasing from 288,635 at the close of 1880 to 1,643,425 at the close of 1890, a gain of 1,354,790. This group also has the greatest percentage of increase of population due to growth of towns supplied prior to the close of 1880 and works built since that time, the increase of 470 being divided into 114 per cent. due to the former and 356 to the latter cause. By states and provinces the increase cannot well be compared, as five states and three provinces had no works in 1880, and several others had but one, so that in some cases the increase is designated by infinity and in others by thousands, as South Dakota, which had but one works in 1880 and 21 in 1890, an increase of 20 works or 2,000 per cent.

For further details of growth in the past decade, either in numbers of works or population supplied, reference may be made to the table, where the figures are given for each state, and for the larger divisions and the whole country.

The practical value of this table, or at least a part of it, will be evident to all engineers engaged in providing for the future water supply of cities, as the figures given show for each state the percentage of increase in the total population of all towns and cities having works. Of course these figures have only a general bearing upon any particular city, but as they are the averages for places big and small, under varying conditions, they may serve as a rather broad basis of comparison in connection with the known growth and conditions of the particular locality in question.

Unfortunately the figures showing the cost of the works at the close of 1880 and other figures indicating their extent are not at hand for comparison with the same figures for 1890. It may, however, be of some interest to know that in the past three years the cost of all the works in the United States has increased from about $430,000,000 to some $545,000,000, or 26 per cent. In the same period the mileage increased from about 24,000 to some 32,500 miles, or 35 per cent. These figures are only approximate, as the figures for three years back, taken from the first issue of the MANUAL, contained many estimates, while the present cost contains estimates for some 200, or about one-tenth of all the works, these generally being in the smaller places.

The total increase in population supplied shows with what rapidity the demand for water-works material has increased in the 10 years under consideration. While the like population will, in all probability, not be added in the next 10 years, it should be remembered that in most, if not all, of the cities and towns included, not all the population is yet supplied with water, while in many the population taking water is less than half of the total. In addition, many systems are entirely inadequate to supply the population already dependent upon them, while for a vast number of works there must be provided not only more water and increased facilities for handling, storing and distributing it, but also a better source of supply, which in many instances will involve a large outlay of money for material and labor, and require marked engineering skill.

The coming field for engineers and all others interested in water-works construction will be in connection with the improvement, extension and general enlargement of existing works, especially with the improvement of the quality of water supplied. Filtration, aeration and sedimentation present a large field, and one which, except in the first particular, has been scientifically worked only here and there in this country. The use of meters and other means of reducing the excessive consumption of water in most of our cities awaits more extended attention than it has yet received. While this is more especially within the province of the water-works manager, there is no reason why it should be left so entirely to him or be entirely ignored, as it has been in the past. Now and then an engineer, when called upon to remedy an alleged inadequate water supply, suggests that the abolition of waste is all that is necessary, but generally no mention is made of this phase of the question, or, if made, only in an incidental way.

In addition to the above, there will in the future be an increased demand for experienced engineers to fill the offices of superintendents of water-works. Late years have seen some advance along this line.

SIZE OF TOWNS SUPPLIED.—PROJECTED WORKS.

References to Tables 5 and 6-S, and especially the latter, will show that new water-works construction will in the future be largely confined to small towns. Table 5-S shows by groups of states the size of towns supplied with water at the close of 1880, and also those supplied or about to be supplied July 1, 1891. Table 6-S shows the prospects for construction in the near future, the figures being for states and groups of states. In Table 5-S duplicate works are deducted, and the figures given are the number of cities having works, but do not include the number of outside towns supplied.

TABLE 5—S.

UNITED STATES AND CANADIAN WATER-WORKS.—SUMMARY BY POPULATIONS OF TOWNS HAVING WORKS COMPLETED AT THE CLOSE OF 1880 AND HAVING WORKS COMPLETED OR UNDER CONSTRUCTION JULY 1, 1891, BY GROUPS OF STATES.

	Above 20,000.		10,001 to 20,000.		5,001 to 10,000.		3,001 to 5,000.		1,001 to 3,000.		1 to 1,000.		Unknown.		Total.	
	'80.	'90.	'80.	'90.	'80.	'90.	'80.	'90.	'80.	'90.	'80.	'90.	'80.	'90.	'80.	'90.
New England..	19	30	28	37	24	60	30	65	18	59	3	7	1	29	123	287
Middle..........	31	45	20	44	46	75	35	79	54	166	14	35	6	46	206	490
S. Atlantic.....	8	9	3	13	5	18	3	21	3	32	1	4	..	8	23	105
S. Central......	7	13	3	7	5	12	5	16	1	15	..	3	..	3	21	69
N. Central......	17	32	19	53	20	93	20	64	23	116	4	27	..	12	103	397
Northwestern..	5	13	5	17	5	30	3	45	4	128	2	61	..	7	24	301
Southwestern..	5	12	6	10	3	27	4	22	4	63	1	19	..	18	23	171
Pacific..........	4	10	5	7	2	7	11	27	32	70	10	43	3	24	67	188
United States.	96	164	89	188	110	322	111	339	139	649	35	199	10	147	590	2,008
Canada..........	8	10	1	10	10	21	6	19	4	23	1	2	..	10	30	95

For convenience the total number of works existing in towns of various populations in the entire United States, with the corresponding percentages, are arranged in a different form below, as follows:

Number of works in towns with population:	Number.		Per cent.	
	1880.	1891.	1880.	1891.
Above 20,000	96	164	16	8
10,001 to 20,000	89	188	15	9
5,001 to 21,000	110	322	18	16
3,001 to 5,000	111	339	19	19
1,001 to 3,000	139	649	24	32
1 to 1,000	35	199	6	9
Unknown	10	147	2	7
Total	590	2,008	100	100

INTRODUCTION. (xv)

From the above it will be seen that in both 1880 and 1891, towns of from 1,000 to 3,000 had the greatest number of works, the totals being 139 and 649, or 24 and 32 per cent., respectively, of the totals for each period. The number of towns of this size supplied in 1891 was greater than the total of all sizes supplied in 1880, the actual numbers being 649 and 590. Down to towns above 5,000 the percentages decrease from 1880 to 1891. In towns from 3,001 to 5,000 the percentages are the same in both periods, while they increase, as we have seen, in the next class and yet again in the next. Of the large number of 1891 works for which the 1890 census was not given, nearly all are known to be in small towns, most of which are below 3,000. A classification of all towns above the several sizes gives the following:

Number of works in *all* towns with population:	Number. 1880.	Number. 1891.	Per cent. 1880.	Per cent. 1891.
Above 20,000	96	164	16	8
" 10,000	185	352	31	17
" 5,000	295	674	49	33
" 3,000	406	1,013	68	52
" 1,000	545	1,662	92	84
1,000 or below	580	1,861	98	93
Unknown	590	2,008	100	100

Those wishing to study works in towns of any special size will find a list beginning on page 379 of all towns having works, arranged in accordance with the results of the 1890 census. Public and private ownership is designated by symbols. It is thought that this last will be of great practical value to many, as a person studying the needs of a particular town can select from the list as many towns of about the same size as he chooses, and separate them into pumping or gravity supplies, public or private ownership, for comparison.

Table 6-S., as has been stated, relates to projected works, of which there are 348 in the United States and 15 in Canada, or a total of 363, against 265 in the second and 159 in the first issue of the MANUAL OF AMERICAN WATER-WORKS. This increase in projected works from year to year is largely due to better facilities for getting track of the projects, and partly to a less rigid classification in the present record, but few works having been classed as "indefinite" this year, it seeming better to present briefly the facts regarding each project, unless the scheme had entirely fallen through.

TABLE 6-S.

UNITED STATES AND CANADIAN WATER-WORKS.—SUMMARY BY STATES AND GROUPS OF STATES OF WORKS PROJECTED WITH FAIR PROSPECTS OF CONSTRUCTION.

Population, 1890.	Number of Projects. Total.	Pub.	Co.	Unk.	Population, 1890.	Number of Projects. Total.	Pub.	Co.	Unk.
Me........ 10,213	6	..	6	..	Ia........ 21,485	16	11	4	1
N. H...... 19,150	8	2	6	1	Minn...... 12,291	10	10	..	..
Vt........ 6,930	4	1	3	..	Kan.......	..	..	..	..
Mass...... 19,197	7	3	3	1	Neb....... 11,611	11	7	1	3
R. I...... 1,700	2	..	1	1	So. Dak... 2,663	3	1	1	1
Conn...... 14,031	4	..	2	2	No. Dak...	..	..	..	..
					Wyo....... 1,069	2	1	1	..
					Mont...... 2,453	3	..	2	1
Total N. E.. 17,230	31	5	21	5	Total N. W. 51,572	45	30	9	6
N. Y...... 51,620	34	12	18	4					
N. J...... 6,473	7	..	5	2	Mo........ 14,302	6	3	3	..
Pa........ 46,590	31	2	26	3	Ark....... 12,991	5	1	2	2
Del....... 1,226	1	1	..	..	Tex....... 35,223	26	7	16	3
Md........	..	..	..	..	Ok........ 6,939	2	1	1	..
D. C......	..	..	..	..	Colo...... 4,914	10	6	2	2
					N. M...... 1,208	1	..	..	1
Total Mid...105,909	73	15	49	9	Total S. W.. 75,577	50	18	24	8

TABLE 6–S.—*Continued.*

Population.	1890.	Number of Projects Total.	Pub.	Co.	Unk.	Population.	1890.	Number of Projects Total.	Pub.	Co.	Unk.
Va.	8,235	6	..	6	..	Wash.	7,940	8	4	4	..
W. Va.	9,578	6	2	3	1	Ore.	2,874	6	1	5	..
N. C.	16,792	6	1	3	2	Cal.	2,632	3	..	3	..
S. C.	14,269	5	2	3	..	Ariz.	1,773	1	..	1	..
Ga.	25,109	16	9	7	..	Nev.	....	..	..	..	..
Fla.	27,300	7	2	5	..	U.	9,248	3	2	..	1
						Ida.	....	..	..	..	..
Total S. Atl.	101,373	46	16	27	3						
Ala.	19,025	16	7	7	2						
Miss.	8,349	4	4	..	..	Total Pac.	24,467	21	7	13	1
La.	6,631	2	1	..	1						
Tenn.	15,986	11	1	9	1						
Ky.	18,259	9	3	6	..	Total U. S.	595,331	348	132	175	41
Total S. Cent	69,153	42	16	22	4						
O.	35,322	12	7	3	2	Canada.	18,342	15	13	2	..
Ind.	21,372	8	4	3	1						
Mich.	18,266	11	8	1	2						
Ill.	10,822	1	3	1	..						
Wis.	10,267	5	3	2	..	Total U. S.					
Total N. Cent	96,049	40	25	10	5	and Canada	613,673	363	145	177	41

The total population in the towns reporting projects is 613,673 for the United States and Canada, against 646,910 in the last Manual, but the latter figures were all estimated and were known to be too high.

The Middle group has the most projects, 73, and the Pacific the least, 21. While the ownerships, as projected, are necessarily very indefinite, they are included in the table for each state, as of some interest. The totals for the United States were 132 public and 175 private, with 41 undecided or unknown. In Canada 13 public and 2 private works are projected.

The populations for projected works are estimated in some cases, the 1890 census populations not being available for some of the smaller towns, or these latter not being separately returned. The classification of projected works by size of towns for each group and for the United States and Canada is as follows:

	Above 5,000.	3,001 to 5,000.	1,001 to 3,000.	125 to 1,000.	Unknown or included.*	Total
New England	0	10	15	5	1	31
Middle	1	3	38	28	3	73
South Atlantic	2	9	18	15	2	46
South Central	0	6	22	11	3	42
North Central	1	13	22	4	..	40
Northwestern	0	1	25	19	..	45
Southwestern	1	7	21	18	3	50
Pacific	1	0	6	13	1	21
United States	6	49	167	113	13	348
Canada	1	1	5	..	8	15

* "Included" designates towns already supplied, the populations of which are included under completed works.

From the above it will be seen that 113 projects are in towns of less than 1,000 and 167 in towns of 1,000 to 3,000, or 280 out of 348 in towns below 3,000. In towns with a population of between 3,000 and 5,000 there are but 49 projects for the whole United States and one in Canada. The number of projects in places with a population of above 5,000 is but six in the United States and one in Canada, the former projects being in the following cities, which severally have the populations given: Key West, Fla., 18,000 (estimated); Steelton, Pa., 9,250; New Berne, N. C., 7,843; Noblesville, Ind., 5,274; Helena, Ark., 5,189; Provo City, Utah, 5,159. The Canadian project is at Dartmouth, N. S., the population of which is 6,249.

INTRODUCTION.

The small size of the towns where the projected works are located substantiates what has already been said regarding the field for future water-works construction. Unfortunately the results of the last census are not wholly available; but if they were, a list of towns above 5,000 still without works would be seen to be a small one, while many of the towns of from 1,000 to 5,000, and several smaller ones, would be shown to be covered by the projects included above.

MANAGEMENT OF PUBLIC WATER-WORKS AND TENURE OF OFFICE OF GOVERNING BODIES.

Inquiries regarding the board or other body in whom the management of public water-works is vested elicited responses from about one-fourth of the works in the United States, or 225 out of 879. Of this 225, 23 works are managed directly by the general municipal government, 12 by a board of public works and 190 by a separate body, having a variety of names all of which may be designated for present purposes by the title of water board. The practice in the several groups of states, the total number of public water-works, the number of works reporting and the manner of selecting water boards is shown in detail as follows:

	No. of works.			Board of Public Wks.	Management of public water-works vested in: Water board.			
	Total.	Reporting.	Council.		Appointed by Mayor.	Elected by Council.	Elected by People.	Total.
N. E......	134	41	..	..	1	6	34	41
Mid......	166	47	4	1	11	18	13	42
S. Atl....	42	14	3	1	6	4	..	10
S. Cent....	10	6	..	..	2	4	..	6
N. Cent....	255	74	8	10	15	13	28	56
N. W......	177	32	6	..	15	11	..	26
S. W......	58	8	2	..	4	2	..	6
Pac........	37	3	..	..	2	..	1	3
U. S......	879	225	23	12	56	58	76	190

In general it may be said that the municipal government has direct charge of works only in the smaller places. Boards of public works seldom have control, and of the 12 instances 10 are in the North Central States. Of these 12 it may be stated that the mayor appoints five, the council elects four and the people three.

In New York, Chicago, Philadelphia and Brooklyn, and it is believed in some other of the largest cities of the country, the management of the water-works instead of being vested in a board is intrusted to one man who is either directly responsible to the mayor or to a commissioner of public works, so responsible. This is the ideal method, as under it the people may look to one man, the mayor, for a good administration of all city affairs and this official will be likely to appoint better men than would be elected by the people or by the city council, and may often reappoint incumbents from previous administrations. Like all other methods this one is, of course, open to abuses which need not be discussed here.

It has been shown above that of the 190 water boards reported 56 are appointed directly by the mayor, while 58 are elected by the council, the election in many cases, doubtless, being approved by the mayor. Boards elected by the councils are directly responsible to the general city government and are generally to be preferred to those elected by the people, on much the same grounds as those given for appointments by the mayor. Thus 30 per cent. of the boards reported are raised to office in the most desirable way, another 30 per cent. in the next desirable, and the minority or 40 per cent., in the least desirable way. This classification, of course, does not take into account local reasons favorable to one or another method, but is based upon the general principle that one official, or as few as possible, should be directly responsible to the people, and all others to the one or to the few comprising the council.

It should be borne in mind, however, that nearly half of the boards elected by the people are in New England, where the good old-fashioned town meetings still prevail, where affairs are largely managed, as one might say, by committees of the whole, and where the direct responsibility to the people of all intrusted with public business naturally follows.

The North Central States, the local government of which, in some particulars, is similar to that of New England, have 28 out of the 76 boards elected by the people, the Middle States have 13, and for all the remainder of the country but one board is reported, and that is in the Pacific States. While information regarding the management of all public works might alter the above ratios it is thought that the reports received are a fair index of the general practice of the whole country.

Next in importance to forms of management and methods of appointment is tenure of office. This is shown below for 188 public works: the total number of public works, number reporting, and number of managing bodies serving for various terms being given in detail. The length of the terms are from one to six years, for life, or perpetual (see note), and the number under each, by groups and for the country, is as follows:

	No. Public Works		One	Two	Three	Four	Five	Six	Life	Perpetual*
	Total	Reporti'g								
N. E........	134	38	4	..	8	..	..	1	..	25
Mid.........	166	41	16	1	7	1	3	2	1	10
S. Atl.......	42	10	3	4	3	..	..	..	..	..
S. Cent.....	10	6	1	4	1	..	..	..	..	..
N. Cent.....	255	65	23	4	13	..	2	..	..	23
N. W........	177	21	15	5	..	1	..	..	..	..
S. W........	58	4	1	2	..	1	..	..	..	..
Pac..........	37	3	2	1	..	..	..	..	..	..
U. S........	879	188	65	21	32	3	5	3	1	58

* That is, where the managing body is renewed after the same general plan as the U. S. Senate, it never consisting entirely of new members.

From the above it will be seen that for the whole United States 65 out of 188 managing bodies, or over one-third, hold office for one year only, 21 for two and 32 for three years. Eleven bodies hold office from four to six years, and one holds office for life. The number of managing bodies holding office for only one year is surprisingly small, but over against these 65 may be set almost as large a number, 58, designated as perpetual, which are never wholly composed of new members, a fixed number, generally a minority, giving place, unless re-elected, to new members at recurring periods. The three to six year bodies stand a fair chance of having, through re-election, at least one old member continually among them, so that the showing under tenure of office is, after all, a good one, since the three to six year bodies together number 43, and these, with the bodies chosen for life and the perpetual bodies, number 102, or about 55 per cent. of the total number reporting. In addition it must be borne in mind that many of the works are under the same superintendent year after year, and that upon him the successful operation of the works often almost wholly depends. In many instances the superintendent is the only man who really knows anything about the works, and woe unto that city who loses such a superintendent.

INTRODUCTION. (xix)

CONSUMPTION OF WATER AND USE OF METERS.

In no department of water-works construction and management, it seems safe to say, is there a greater chance for improvement and economy than in connection with an increased use of meters. In submitting reports and estimates for new works engineers seldom, if ever, include the item of meters, although in the long run it would undoubtedly prove wise if this were done and meters considered as a part of the works. It is true that meters, like service connections, are put in after the main works are completed, and, also like the services, are perhaps more often than otherwise paid for directly by the consumer, or else are rented to him. But meters play so important a part in reducing or keeping down the consumption of water, and consequently the necessary size and corresponding cost of water-works, that they should be introduced at the start, considered a part of the system, and their cost included in the general construction account.

Works already in existence would do well when adding meters to charge their cost to the general construction account. In the case of both old and new works the opposition on the part of consumers would undoubtedly be greatly lessened if the course just named were adopted and the heavy charge to the consumer for the introduction of a meter, or the irksome annual rental often imposed, were thus done away with or avoided from the start. An additional consideration of great importance is that if the meters were considered the absolute property of the water department or company, it would naturally follow that they would be repaired without direct charge to the consumer, and thus unpleasant words would be avoided when the meters were inspected and repairs found necessary, and opposition to their use would be still further lessened. Perhaps of still greater importance would be the city's or company's unrestricted and unquestioned privilege to select just such a meter as it saw fit for any particular service connection, instead of, as is often the case, allowing the consumer to select such a meter as he sees fit, in which case cheapness rather than efficiency is the basis of the choice.

That there is a wide field open for the introduction of meters is shown by the facts regarding their use or non-use presented in the body of the MANUAL and set forth below, especially in Tables 7 and 8-S. The first named table gives a summary by states and groups of states regarding the use of meters and their ratio to the total number of taps in use. The second table is designed to aid in a study of the effects of meters upon the consumption of water.

Before considering the tables in detail it may be said that the statement regarding the field for the introduction of meters is borne out by the following facts regarding the United States as a whole : Of 2,037 waterworks only 98 have 20 per cent. or over of their taps metered and these 98 works have about 61 per cent. of all the meters reported, or 109,474 out of a total of 163,178. It might be thought that these works taken together include a majority of all the taps in the United States, but this is not the case, the 98 works reporting but 281,967 out of a total of 2,212,913 taps, or only 13 per cent., against the 61 per cent. of meters on the same 98 works. The very small number of meters in use by some of the largest cities is shown by the first part of Table 8-S, from which it will be seen that but eight of the 50 largest cities of the United States have 20 per cent. or over of their taps metered.

TABLE 7—S.

UNITED STATES AND CANADIAN WATER-WORKS. USE OF METERS BY STATES AND GROUPS OF STATES.

	Number of		P. ct. taps metr'd.	Number of works				P. ct. of wks. rep't'ng metrs.
	Meters.	Taps.		Using metrs.	Not using metrs.	Not report-ing.	Total.	
Maine	362	15,127	2.4	15	14	8	37	40.5
New Hampshire	1,956	17,056	11.5	15	7	14	36	41.7
Vermont	615	9,553	6.4	7	14	6	27	25.9
Massachusetts	31,859	249,499	12.8	92	28	8	128	71.9
Rhode Island	15,036	27,055	55.6	10	1	3	14	71.4
Connecticut	1,110	43,361	2.6	27	14	7	48	56.3
New England	50,938	361,651	14.1	166	78	46	290	57.2
New York	38,469	328,246	11.7	93	72	34	199	47.7
New Jersey	8,546	111,883	7.6	37	14	7	58	63.8
Pennsylvania	2,194	355,309	.6	74	102	40	216	34.3
Delaware	28	13,489	.2	1	4	1	6	16.6
Maryland	1,093	79,551	1.4	12	3	3	18	66.6
District of Columbia	87	33,270	.3	1	0	0	1	100
Middle	50,417	921,748	5.5	218	195	85	498	43.8
Virginia	294	27,362	1.1	12	14	15	41	29.3
West Virginia	6	10,080	.½	3	6	1	10	33.3
North Carolina	495	2,944	16.5	8	3	4	15	53.3
South Carolina	33	1,582	2.1	4	1	2	7	57.1
Georgia	3,806	15,269	24.9	9	5	7	21	42.9
Florida	454	4,055	11.2	6	2	4	12	50
South Atlantic	5,088	62,212	8.2	42	31	33	106	39.6
Alabama	1,304	11,158	11.7	9	7	4	20	45
Mississippi	29	1,550	1.8	3	1	3	7	42.9
Louisiana	39	5,165	.8	3	0	1	4	75
Tennessee	351	17,511	.2	8	2	6	16	50
Kentucky	1,702	28,029	6.1	15	3	5	23	65.2
South Central	3,425	63,413	5.4	38	13	19	70	54.3
Ohio	5,143	127,880	4.2	51	19	16	86	59.3
Indiana	927	17,860	5.2	28	13	9	50	56
Michigan	2,157	83,345	2.6	28	24	61	113	24.7
Illinois	6,121	198,937	3.1	45	13	44	102	44.1
Wisconsin	6,835	46,212	14.8	25	7	18	50	50
North Central	21,183	474,234	4.5	177	76	148	401	44.1
Iowa	2,439	21,086	11.1	32	20	25	77	41.6
Minnesota	1,291	26,759	4.1	13	9	12	34	38.2
Kansas	999	17,910	5.6	33	22	24	79	41.8
Nebraska	1,485	14,266	10.4	18	20	24	62	29.3
South Dakota	78	2,716	2.8	3	12	7	22	13.6
North Dakota	204	1,401	14.5	4	3	0	7	57.1
Wyoming	195	1,598	12.2	3	2	4	9	33.3
Montana	149	2,740	5.5	5	2	4	11	45.4
Northwestern	6,840	88,476	7.7	111	90	100	301	36.8
Missouri	5,924	61,046	9	23	2	11	36	63.9
Arkansas	132	4,399	3	7	5	0	12	58.3
Texas	1,753	33,818	5.2	28	16	16	60	46.7
Colorado	285	24,657	1.1	14	23	20	57	24.6
New Mexico	69	1,436	4.8	3	0	6	9	33.3
Southwestern	8,163	125,356	6.5	75	46	53	174	43.1
Washington	332	7,199	4.6	6	2	30	38	15.6
Oregon	83	11,631	.7	5	11	10	26	19.2
California	16,641	84,081	19.8	36	25	42	103	35
Arizona	55	1,038	5.3	2	0	2	4	50
Nevada	13	5,520	.2	1	3	5	9	11.1
Utah	0	4,684	..	0	4	3	7	0
Idaho	0	1,670	..	0	4	6	10	0
Pacific	17,124	115,823	14.8	50	49	98	197	25.4
Total United States	163,178	2,212,513	7.4	877	578	582	2,037	43.1
Total Canada	3,198	183,849	1.7	32	17	46	95	33.7

INTRODUCTION.　　　　　　　(xxi)

The general practice in the several States is shown by Table 7-S, referred to above. This table gives for each State the number of meters and taps and per cent. of taps metered; also the number of works reporting meters in use, number reporting no meters and number not reporting; also the total number of works in each State and the percentage of these which report meters.

For the whole country it will be seen that 877 works report meters in use (one or more), or 43.1 per cent. of all the works; 578 works report no meters, and 582 make no statement. While the number of works reporting meters in use is fairly large compared with the total, it is evident from the percentage of taps metered, 7.4 for the United States, and the fact that 61 per cent. of all meters are on the works having 20 per cent. or more of their taps metered, that many of the works reporting meters have but few in use ; in fact, there are at least 100 with only one, and several others with but two or three meters.

By groups of States the Pacific has the largest percentage of its taps metered, 14.8, and New England next, its percentage being nearly the same, 14.1. The percentages in the other groups range from 8.2 in the South Atlantic to 4.5 per cent. in the North Central, the figures for the whole United States being, as stated, 7.4 per cent. Canada has but 1.7 per cent. of its taps metered, only one works, Cote St. Antoine, P. Q., reporting over 20 per cent. of its taps metered.

By states Rhode Island has by far the largest percentage of taps metered, 55.6 per cent. But a large part of all the taps in the state, 78 per cent., are on three works, the figures for these works and the state being as follows :

	Taps.	Meters.	Per cent. of taps metered.
Providence	14,896	9,286	62.4
Pawtucket	5,322	3,539	66.5
Woonsocket	1,117	924	82.7
Total	21,335	13,749	64.4
Whole state	27,055	15,036	55.6

Pennsylvania has but 0.6 per cent. of its taps metered, 74 works, or 34 per cent. of the total, reporting but 2,194 meters, or an average of less than 30 each, and no works having as many as 20 per cent. metered. Some of the smaller states report a less percentage of meters and Utah and Idaho report none, but from the great state of Pennsyvlvania with its 216 works better things might well have been expected.

Philadelphia, with its 170,911 taps, nearly half of the total for the state, has but 522 meters, 0.3 per cent. of its taps. Chicago, also, reports but a few meters, 3,924 for at least 170,000 taps, the exact number not being known. New York has 20.2 per cent. of its taps metered and an average daily consumption per capita of but 79 galls., while Chicago has a per capita consumption of 138 and Philadelphia of 132 galls. True, New York would have used more water in the first half of 1890 had it been available to consumers, but during the latter half of the year, or a great part of it, an ample supply was furnished.

Chicago and Philadelphia have the largest water-works pumping plants in the country, if not in the world, the combined daily capacity of all the pumps at Philadelphia at the close of 1890 having been 185,290,000 and at Chicago 218,000,000 galls., the latter figures including pumps with a capacity of 20,000,000 galls. for which water was not always available. In addition, Chicago had 42,000,000 galls. pumping capacity not yet accepted, and both cities let contracts in 1891 for more pumps, those at Chicago to be primarily for use at the Columbian Exposition. In 1890 the total average daily consumption at Chicago was 152,000,000 and at Philadelphia 138,000,000 galls.

It may well be asked why do not these cities, and many others with a showing proportionately bad or worse, reduce their consumption by the use of meters and thus lessen their outlay for pumping machinery, new and larger mains, and other appurtenances and thus lessen, in turn, the fixed and current expenses of the works?

While the field for the introduction of meters is very large, yet advance has been made in the last three years, as is shown in some detail below. This advance is due to the growing enlightenment of the people on the subject, which has been furthered by the experience of cities where meters have been largely used with excellent results and through the educational work done by the American and New England Water-Works Associations at their conventions and in their publications, and, it may be added, by the technical journals. The advance is also very largely due to improvements in meters and possibly to a reduction in their price, together with the business enterprise of the meter manufacturers.

The figures relating to taps and meters given three years ago in the first issue of the MANUAL may be compared with those in the present volume. The number of meters reported in the 1888 and the 1891 MANUALS, the increase and increase per cent., and the per cent. of taps metered, as reported in the two volumes for the several groups of states and for the whole country, is as follows:

	Taps.				Meters.			Per cent. of taps metered.	
	1888.	1891.	Increase.	Increase per cent.	1888.	1891.	Inc.	Inc. per cent.	1888. 1891.
N. E....	315,404	361,651	46,247	14.6	37,913	50,938	13,025	34.3	12. 14.1
Mid....	890,021	921,748	31,727	3.6	34,346	50,417	16,071	46.7	3.9 5.5
S. Atl..	48,334	62,212	12,878	26.6	2,889	5,088	2,199	76.1	6. 8.2
S. Cent.	50,302	63,413	13,111	26.1	2,365	3,425	1,060	44.8	4.7 5.4
N. Cent.	352,463	474,234	121,771	34.6	12,085	21,183	9,098	75.3	3.4 4.5
N. W...	55,338	88,476	32,038	59.3	3,447	6,840	3,407	98.8	6.2 7.7
S. W...	89,756	125,356	45,600	50.8	4,425	8,163	3,738	84.5	4.1 6.5
Pac.....	87,926	115,823	27,897	31.7	9,045	17,124	7,179	72.2	11.3 14.8
U. S.....	1,889,744	2,212,013	323,109	17.1	107,415	163,178	55,763	51.9	5.7 7.4
Canada..	172,947	183,849	10,902		2,077	3,198	1,121	54.	1.2 1.7

The last two columns are of special interest, showing that in each group there has been in the past three years an increase in the percentage of taps metered, the Pacific group having increased from 11.3 to 14.8 per cent., and the United States from 5.7 to 7.4 per cent., a gain of 1.7 per cent., or about one-third in the three years.

In numbers the meters have increased 51.9 per cent. in the United States, or from 107,415 to 163,178, against an increase of but 17.1 per cent. in number of taps, or in other words the number of meters has increased at a rate three times as fast as the taps, which agrees with figures given at the close of the last paragraph. In the Northwestern group the number of meters has nearly doubled in the three years, the percentage of increase being 98.8, but the original number was only 3,447. Other details may be seen by referring to the table.

Passing from the use of meters alone to their effect upon consumption, reference may be made to Table 8-S for facts relating to the 50 largest cities of the United States and to all cities of smaller size having more than 50 per cent. of their taps metered. This table shows the population

INTRODUCTION. (xxiii)

TABLE 8–8.

CONSUMPTION OF WATER AND USE OF METERS IN THE FIFTY LARGEST CITIES OF THE UNITED STATES AND IN ALL OTHER CITIES, TOWNS OR VILLAGES HAVING FIFTY PER CENT. OR MORE OF THEIR TAPS METERED.

	Population. 1890.*	Owner-ship.	Source.†	Mode.†	Number taps.	Number meters.	P. c. taps metered.	Pop'n per tap.	Daily consumption.		
									Total.	Per inhabitant.	Per tap.
1. New York	1,515,301	Pub.	S.	G.	108,581	22,072	20.2	13.9	121,000,000	79	1,111
2. Chicago¹	1,099,850 (1,085,000)	Pub.	L.	P.	170,911	3,924	0.3	6.1	152,372,288	140	806
3. Philadelphia²	1,046,964 (1,040,000)	Pub.	S.	P.	89,493	2,253	2.5	8.7	137,736,703	132	616
4. Brooklyn³	776,638	Co.		P.	3,732			7.9	55,000,000	72	
	29,505	Pub.	S., U.	P.							
5. St. Louis⁴	451,770	Pub.	S.	P.	38,163	3,115	8.2	11.8	32,479,000	72	851
6. Boston⁵	448,477 (527,606)	Pub.	L., S.	G., P.	80,328	4,018	5	6.6	42,173,100	90	525
7. Baltimore	434,439	Pub.	S.	G.	74,728	913	0.1	5.8	40,974,299	94	548
8. San Francisco⁶	298,997	Co.	S.	P.	30,200	12,505	41.4	9.8	18,350,000	61	608
9. Cincinnati⁷	296,908 (302,581)	Pub.	S.	P.	35,439	1,451	4.1	8.5	33,907,007	112	959
10. Cleveland⁸	261,353 (270,055)	Pub.	L.	P.	30,688	1,794	5.8	8.7	27,757,158	103	936
11. Buffalo	255,664	Pub.	L.	P.	40,331	94	0.2	6.3	47,517,137	186	1,178
12. New Orleans⁹	242,039	Pub.	S.	P.	4,450	20	0.4	54	8,976,715	37	2,017
13. Pittsburg¹⁰	238,617 {	Pub.	S.	P.	25,000	37	0.2		36,400,000		
		Co.	S.	P.	7,651	"Very few."					
14. Washington¹¹	230,392	Pub.	S.	G., P.	35,404	98	0.3	8.2	11,530,000	114	1,467
15. Detroit	205,876	Pub.	S.	P.	40,351	886	2.1	5.1	36,588,629	158	1,033
16. Milwaukee	204,468	Pub.	L.	P.	18,422	5,876	31.9	11.1	33,906,067	161	823
17. Newark¹²	181,830 (185,317)	Pub.	S.	P.	21,632	520	2.4	8.6	22,290,783	110	1,215
18. Minneapolis	164,738	Pub.	S.	P.	9,900	623	6.3	16.3	14,070,703	76	654
									12,416,117	75	1,243

* Populations are according to the 1890 census, except in two instances near the foot of the second part of the table. They are for the whole city, regardless of the proportion of the population supplied. When outside populations are supplied the total populations of all cities and towns supplied is given in parentheses at the right.

† Sources of supply are divided into three classes, lakes, streams and underground, denoted by L., S. and U., respectively. Lakes are designed to include supplies from all bodies of water not artificial; streams, all supplies from springs, running and surface water, and from artificial ponds; and underground, supplies from wells or from filter galleries. Modes of supply are divided simply into gravity and pumping.

¹ Chicago. Figures are for main city works. Estimated populations supplied by small public plant built by former village of Pullman, Washington Heights and of former village of Pullman, supplied by a company, are excluded. Number of taps is unknown, but it must be at least 170,000.

² Philadelphia. Estimated populations of Lolmesburg and Tacony, supplied by companies, are excluded.

³ Brooklyn. Long Island Water Supply Co. supplies 26th Ward. Figures for city are first given, then those for company. Total population, 806,343.

⁴ St. Louis. Figures are for year closing Apr. 9, 1890, as these are nearer census consumption. Consumption for succeeding year was: Total average daily, 36,001,000; per capita, 80; per tap, 871.

⁵ Boston. Supplies Somerville, with population of 40,152; Chelsea, 27,909; Everett, 11,068. About 65% of total supply is by gravity.

⁶ San Francisco. Figures are for June 30, 1890. Special report made Dec. 16, 1891, gives following figures: Taps, about 33,000; meters, 11,812, or 42.1% of taps; total average daily consumption for calendar year 1891, with estimates for last 16 days of December, 19,372,000 galls.

⁷ Cincinnati. Supplies Avondale and Clifton, with populations of 4,173 and 1,270, latter estimated.

⁸ Cleveland. Supplies Brooklyn and West Cleveland, with populations of 4,565 and 4,117.

⁹ New Orleans. In addition to taps given there were, April, '91, 5,680 not in use.

¹⁰ Pittsburg. Monongahela Water Co. supplies "South Side" and outside towns with estimated aggregate population of 75,000.

¹¹ Washington. Figures are for July 1, 1891, and population is for the whole District of Columbia, the government of which and of Washington is now coextensive.

¹² Newark. Supplies Belleville, through meter, population of which is 3,487.

(xxiv) MANUAL OF AMERICAN WATER-WORKS.

TABLE 8—S—CONTINUED.

	Population, 1890.*	Owner-ship.	Source.†	Mode.†	Number taps.	Number meters.	p. c. taps metered.	Pop'n per tap.	Daily consumption.		
									Total.	Per inhabitant.	Per tap.
19. Jersey City[12]	163,003 (197,438)	Pub.	S.	P.	20,456	240	1.2	11.9	19,300,000	97	929
20. Louisville[14]	161,129	Co.	S.	P.	13,512	792	5.9	74	11,871,683	74	2,261
21. Omaha[15]	140,452 (148,478)	Co.	S.	P.	6,188	1,200	19.4	28.1	14,000,000	91	431
22. Rochester[16]	133,896	Pub.	L., S.	G., P.	24,698	2,844	11.4	5.4	8,000,000	60	765
23. St. Paul	133,156	Pub.	L.	G., P.	10,498	440	4.2	12.7	8,000,000	60	763
24. Kansas City[17]	132,716 (171,032)	Co.	S.	P.	11,488	1,971	17.6	15.3	12,000,000	71	1,071
25. Providence[18]	132,146 (140,900)	Pub.	S.	G., P.	14,988	9,298	62.4	9.4	6,745,682	48	453
26. Denver	106,713 {	Co.	P., G.		10,792	85	0.8		15,000,000		1,381
27. Indianapolis	106,436	Co.	S.	P.	2,988	225	7.6	35.6	7,500,000	71	2,517
28. Allegheny	105,287	Pub.	S.	P.	15,375	0		7	7,500,000		1,066
29. Albany[19]	91,923	Pub.	S.	G., P.	7,619	60	0.4	6.2	25,000,000	288	
30. Columbus	88,150	Pub.	U.	P.	4,100	491	6.4	21.5	Unknown.		
31. Syracuse[20]	88,143	Pub.	S.	G., P.	9,450	600	6.4	21.5	6,000,000	68	1,464
32. Worcester	84,055	Pub.	S.	G.	8,451	4,871	14.6	8.9	4,971,840	59	588
33. Toledo	81,434	Pub.	S.	P.	4,371	411	9.4	18.6	5,482,708	72	1,255
34. Richmond	81,388	Pub.	S.	P.	10,383	143	1.4	7.9	13,497,102	167	1,310
35. New Haven	81,298	Co.	S.	"No record."		69				135	
36. Paterson	78,347	Co.	S.	P.	6,948	2		11.8	11,000,000	128	1,568
37. Lowell	77,696	Pub.	S.	P.	8,471	1,935	22.9	9.2	3,127,139	40	1,005
38. Nashville[21]	76,168	Pub.	S.	P.	5,096	40	0.8	14.9	11,153,585	146	2,188
39. Scranton	75,215 {	Co.	G.	G.	1,761	0			2,500,000		1,417
40. Fall River	74,398	Pub.	S.	G.	4,060	3,717	74.6	14.9	2,138,162	29	425
41. Cambridge	70,028	Pub.	S.	G., P.	10,554	254	2.4	6.6	4,459,130	64	425
42. Atlanta	65,533	Pub.	S.	P.	3,373	2,994	88.9	20.0	2,559,664	36	721
43. Memphis	64,495	Co.	U.	P.	5,400	290	3.7	11.9	8,000,000	123	1,462
44. Wilmington	61,431	Pub.	S.	P.	12,528	28	0.2	5	6,841,912	113	557
45. Dayton	61,220	Pub.	U.	P.	3,044	117	3.8	20.1	2,942,936	47	935
46. Troy	60,956	Co.	S.	P.	5,786	150	3.9	10.5	7,666,468	125	1,325
47. Grand Rapids[22]	60,278 {	Pub.	S.	P.	1,700	459	15		4,392,193		1,150
48. Reading	58,661	Pub.	S.	P.	3,819	6	12.0	5.8	5,000,000	75	1,005
					10,000		0.1				439

* See note p. xxiii.
† See note p. xxiii.
[12] Jersey City. Figures are for 1889. Supplies Bayonne, population of 19,033, Harrison, 8,338 and Kearney, 7,061. All towns supplied by meter measure, and Bayonne has large percentage of taps metered, all of which reduces consumption for Jersey City. Report *did not state* whether total average daily consumption includes supply to above places, but it is assumed that it did.
[14] Louisville. City owns practically all of company's stock.
[15] Omaha. Supplies South Omaha; population, 8,025.

[16] Rochester. Domestic supply by gravity from lakes; remainder by direct pumping from river.
[17] Kansas City. Supplies Kansas City, Kan., with population of 38,316.
[18] Providence. Supplies population in adjacent towns, estimated at 10,000, which estimate is low rather than high.
[19] Albany. Taps and meters are for December, 1891.
[20] Syracuse. Figures are for December, 1891, and are approximate.
[21] Nashville. Meters, Dec. 29, 1891, had increased to 172.
[22] Grand Rapids. Company figure, first line, are for 1889.

INTRODUCTION. (xxv)

TABLE 8-S—CONTINUED.

CITIES	Population, 1880.*	Owner-ship.	Source.†	Mode.‡	Number taps.	Number meters.	P. c. taps metered.	Pop'n per tap.	Daily consumption. Total.	Daily consumption. Per inhabitant.	Daily consumption. Per tap.
49. Camden.[23]	58,313	Pub.	S.	P.	12,336	10	62.4	12.7	7,960,000	131	782
50. Trenton.[24]	57,458	Pub.	S.	P.	9,590	46		6.	3,569,150	62	376
	NOT INCLUDED ABOVE HAVING FIFTY PER CENT. OR MORE OF THEIR TAPS METERED.										
Hoboken, N. J.,[25]	48,546 (87,163)	Co.	S.	P.	7,249	4,635	62.4	12.7	5,627,000	63	782
Des Moines, Ia.	50,093	Co.	S.	P.	2,500	1,500	60	20	2,753,000	55	1,100
Pawtucket, R. I.[26]	23,667 (50,000)	Co.	S.	P.	5,332	3,539	66.5	9.4	3,215,555	64	604
Utica, N. Y.	44,007	Pub.	S.	G.	3,000	2,500	83.3	14.6			
Yonkers, N. Y.	32,033	Pub.	S.	G.	3,685	2,212	82.4	10.	2,174,386	68	811
Brockton, Mass.	27,294	Pub.	S.	G.	2,700	1,675	60.2	10.1			
Newton,[28]	24,370	Pub.	S.	G.	4,410	2,995	67.4	5.5			224
Joliet, Ill.[27]	23,264	Pub.	S., U.	P.	365	365	100		965,596	40	
Lexington, Ky.[1]	21,567	Co.	S.	P.	866	536	69.2	26.7	968,000	75	905
Woonsocket, R. I.[28]	20,830	Pub.	S.	P.	1,117	621	87.7	16	326,455	37	292
Bayonne, N. J.[]	19,033	Co.	Jersey City.	P.	870			18.7		16	
Columbus, Ga.[29]	17,303	Co.	S.	G.	818	722	88.3	21.1	631,286	40	772
San Diego, Cal.[30]	16,159	Co.	S.	P.	2,394	1,232	52.3	6.8			
Alameda, Cal.[31]	11,165	Co.	L.	U.	1,600	1,600	100				
Jacksonville, Ill.	10,740	Co.	S.	P.	550	430	52.7	19.5	500,000	47	909
El Paso, Tex.	10,338	Co.	S., U.	P.	990	500	56.3	12.9	640,000	58	750
Flushing, N. Y.	9,140	Pub.	L.	P.	918	450	54.5	9.9	317,739	35	346
Laconia & Lake Village, N. H.[31]	8,436	Co.	U.	P.	1,297	675	52	6.5	512,000	61	395
Paris, Tex.	8,254	Pub.	U.	P.	196	121	61.7	42.1	150,000	18	765
New Rochelle, N. Y.	8,217	Co.	S.	P.	608	560	92.1	13.5	130,000	16	213
Winston, N. C.	8,018	Co.	S.	P.	155	100	64.5	51.7	130,000	19	838
Westerly, R. I.	6,813	Co.	U.	P.	380	225	59.2	17.9	203,500	30	534
Gainesville, Tex.	6,594	Co.	S.	P.	380	202	54.8	17.9			
Pasadena, Cal.[32]	4,882	Co.	S.	P.	300	300	100		130,000		500
Tarrytown, N. Y.	3,562	Pub.	S.	G.	300	300	100	11.8			
Princeton, N. J.	3,472	Co.	S.	P.	292	265	92	11.9	65,900	19	226
Salem, N. C.	2,711	Co.	S.	P.	124	105	84.7	22.4			
Clyde, N. Y.[33]	2,688	Co.	S., U.	P.	70	38	54.3	38.4	40,600	15	572
Irvington, N. Y.[33]	2,299 est.	Pub.	S., U.	P.	100	100	100	25	15,000	7	
Bridgeport, Ala.	1,000 est.	Co.	S.	G.	40	25	62.5	4.8	75,000	75	250
Wakefield, R. I.	1,000 est.	Co.	S.	P.	209	143	68.4	6.8	40,000	54	264
Union, Md.	743	Co.	S.	P.	110	115		2.3			
Green River, Wyo.	723	Co.	S.	P.	310	190	61.3		500,000	692	1,613

* See note p. xxiii.   † See note p. xxiii.   ‡ See note p. xxiii.
[23] Camden. Taps are for 1888.
[24] Trenton. Taps approximate. Meters for 1887.
[25] Hoboken. Supplies 11 outside towns with combined population of 43,615.
[26] Pawtucket. Supplies several outside towns with population estimated as 23,467.
[27] Joliet. Taps and meters for 1886.

[28] Bayonne, N. J., and Union, Md. Taps and meters are probably unmetered and metered taps, respectively, since reports give more meters than taps.
[29] San Diego. Works are now leased and operated by the city.
[30] Alameda. Supplies Fitchburg, of unknown population.
[31] Laconia and Lake Village. Population estimated.
[32] Pasadena. North Pasadena Land & Water Co.
[33] Irvington. Taps and meters for 1887.

of the several cities; the population on works (not the actual population supplied, but the population of the city in question and other towns supplied by it); the ownership of the works, whether by the public or by a company; the source and mode of supply; number of taps and meters and per cent. of taps metered, together with the population per tap; and under daily consumption the total amount, amount per inhabitant and per tap.

Attention is called to the fact that the populations and average daily consumptions correspond as nearly as possible in point of time, the 1890 census having been taken in June and the consumption being the average for the whole year. Some of the fiscal years do not correspond exactly with the calander years, but with two or three exceptions, indicated in the foot notes, there is not much divergence in this particular.

Unfortunately the population actually supplied by each works cannot be given, as only the total population is known. To supply this lack so far as possible the population per tap has been included for the several works. While this is far from being an absolute guide as to the relative proportion of population supplied, it helps materially in understanding many of the figures for consumption per inhabitant. For instance, New Orleans has the very low daily consumption of 37 galls. and but 0.4 per cent. of its taps metered. But it also has a population of 54 per tap, four times as great as New York with its dense population.

Some of the figures under average daily consumption are evidently estimated and that not very closely, as they are given in millions of gallons. All figures, however, are included as reported. The percentage of taps metered, of course, is more satisfactory than some of the other figures, as the exact number of service connections can always be known and, it is believed, has generally been given.

The above general remarks, together with the foot notes to the table, will enable the reader to form an opinion regarding the value of all figures given. The figures are the best that can be obtained, and are more complete, uniform and more nearly up to date, it is believed, than any ever before published. They give certain facts regarding the consumption of water and use of meters in the 50 largest cities of the United States and in others both large and small which, although they must be taken with qualifications, are for comparative purposes of great general value and are unsurpassed as relating to the several cities in question.

Aside from its value as a convenient reference table the figures above are of chief use in connection with a study of the effect of meters upon the consumption of water. It is true that they do not show what has been done in any of the cities in question by means of inspection of fixtures and other measures to prevent waste, but the use of meters alone has a sufficient effect upon consumption to warrant a study from which all other waste-preventing factors are eliminated. To facilitate this study the 50 largest cities, those in the first part of the table, which are there arranged by size, are given below, first in order of consumption, greatest to least, and then in order of percentage of taps metered. The latter arrangement is placed beside the former and the order reversed from least to greatest, in order that it may be seen how nearly consumptions and meter percentages correspond. The percentage of taps metered is given in connection with the first and the consumption with the last arrangement, and with each the population per tap is included, as well as the

INTRODUCTION. (xxvii)

rank of the city in size and also in the classification under immediate consideration. The figures are as follows:

Works arranged in order of

Consumption, greatest to least. City.						Taps metered, least to greatest. City.					
Rank in:						Rank in:					
Size.	Consumption.		Consumption per inhabitant.	Per cent. taps metered.	Population per tap.	Size.	Taps metered.		Per cent. taps metered.	Consumption per inhabitant.	Population per tap.
28	1	Alleghany	238	0	7	28	1	Alleghany	0	238	7
11	2	Buffalo	186	0.2	6.3	40		Camden	small	131	....
34	3	Richmond	167	1.4	7.9	36	2 to 5	Paterson	small	128	11.8
15	4	Detroit	161	2.1	5.1	50		Trenton	small	62	6
14	5	Washington	158	0.3	6.5	13		Pittsburg (Co)	small	153	8.2
13	6	Pittsburg (Co.)	153	small	8.2	48	6	Reading	0.1	75	5.8
38	7	Nashville	146	0.8	14.9	7	6	Baltimore	0.1	94	5 8
13	8	Pittsburg(Pub)	144	0.2	....	13	7	Pittsburg(Pub)	0.2	144	....
2	9	Chicago	140	....	....	11	7	Buffalo	0.2	186	6.3
35	10	New Haven	135	....	....	44	7	Wilmington	0.2	113	5.
3	11	Philadelphia	132	0.3	6.1	3	8	Philadelphia	0.3	132	6.1
49	12	Camden	131	small	....	14	8	Washington	0.3	158	6.5
36	13	Paterson	128	small	11.8	12	9	New Orleans	0.4	37	54.
46	14	Troy	125	3.9	10.5	29	9	Albany	0.4	unk'n	6.2
43	15	Memphis	124	3.7	11.9	38	10	Nashville	0.8	146	14.9
44	16	Wilmington	113	0.2	5	26	11	*Denver	0.8	....	....
9	17	Cincinnati	112	4.1	8.5	19	12	Jersey City	1.2	97	....
16	18	Milwaukee	110	31.9	11.1	34	13	Richmond	1.4	167	7.9
10	19	Cleveland	103	5.8	8.7	15	14	Detroit	2.1	161	8.7
19	20	Jersey City	97	1.2	....	41	15	Cambridge	2.4	64	6.6
7	21	Baltimore	94	0.1	5.8	17	15	Newark	2.4	76	8.6
21	21	Omaha	94	19.4	24	4	16	Brooklyn	2.5	72	8.7
6	22	Boston	80	5	6.6	43	17	Memphis	3.7	124	11.9
1	23	New York	79	20.2	13.9	45	18	Dayton	3.8	47	20.1
30	24	Columbus	78	6.4	11.5	46	19	Troy	3.9	125	11.5
17	25	Newark	76	2.4	8.6	9	20	Cincinnati	4.1	112	8.6
48	26	Reading	75	0.1	5.8	23	21	St. Paul	4.2	60	12.7
18	26	Minneapolis	75	6.3	10.5	6	22	Boston	5.	80	6.6
20	27	Louisville	74	5.9	11.9	10	23	Cleveland	5.8	103	8.7
33	28	Toledo	72	9.4	18.6	20	24	Louisville	5.9	74	11.9
4	28	Brooklyn	72	2.5	8.7	18	25	Minneapolis	6.3	75	16.5
5	28	St. Louis	72	8.2	11.8	30	26	Columbus	6.4	78	11.5
27	29	Indianapolis	71	7.6	35 6	27	27	Indianapolis	7.6	71	35.6
24	29	Kansas City	71	17.6	15.3	5	28	St. Louis	8.2	72	11.8
31	30	Syracuse	68	14.6	21.5	33	29	Toledo	9.4	72	18.6
37	31	Lowell	66	22.9	9.2	22	30	Rochester	11.4	66	5.4
22	31	Rochester	66	11.4	5.4	47	31	Gr. R'pids(Pub)	12	....	....
41	32	Cambridge	64	2.4	6.6	31	32	Syracuse	14.6	68	21.5
50	33	Trenton	62	small	6	47	33	Gr'd R'pids(Co)	15	....	....
8	34	San Francisco	61	41.4	9.9	24	34	Kansas City	17.6	71	15.3
23	35	St. Paul	60	4.2	12.7	21	35	Omaha	19.4	94	24
32	36	Worcester	59	89.4	8.9	1	36	New York	20.2	79	13.9
25	37	Providence	48	62.4	9.4	37	37	Lowell	22.9	66	9.2
45	38	Dayton	47	3 8	20.1	16	38	Milwaukee	31.9	110	11.1
12	39	New Orleans	37	0.4	54	8	39	San Francisco	41.4	61	9.9
42	40	Atlanta	36	89.6	20	25	40	Providence	62.4	48	9.4
40	41	Fall River	29	74.6	14.9	40	41	Fall River	74.6	29	14.9
47	..	Gra'd Rapids(Co)..	15	....	....	32	42	Worcester	89.4	59	8.9
47	..	G'd Rapids (Pub)..	12	....	....	42	43	Atlanta	89.6	36	20
39	..	Scranton (2 Co.'s)..	....	....	....	35	..	New Haven	....	135	....
29	..	Albany	....	0 4	6.2	39	..	Scranton	....	....	....
26	..	Denver (2 Co.'s)	....	....	....	2	..	Chicago	....	140	....

*Denver City Water Co.

It will be seen from the foregoing that Allegheny has the highest consumption and that it has no meters. It will also be seen that all the places with a high consumption have but a few meters, none of the 17 highest on the list having more than 4.1 per cent. of their taps metered, and Milwaukee being the only city with a consumption of more than 100 galls., which has over 6 per cent. of its taps metered, this percentage at Milwaukee being notably large, 31.9. Glancing down the list it will be seen that as the consumption decreases the percentage of taps metered shows a general increase until with the two lowest consumptions, 36 and 29 at Atlanta and Fall River, respectively, the percentages of taps metered are 89.6 and 74.6. There are some instances of a low consumption in connection with a very few meters, the most notable one being New Orleans, with a consumption of only 37 galls. and with but 0.4 per cent. of its taps metered. This low consumption, however, is apparent rather than real, the population per tap, as noted above, being 54 against 13.9 in New York, which shows that a very large part of the population is not supplied, a statement which is further proven by the fact that there are more taps in the city not in use than in use.

The classification by percentage of taps metered makes clear, from another point of view, the facts already considered, but does not need further comment than has already been made.

It should be stated before leaving this part of table 8-S that some of the works in question have a rigid system of inspection of fixtures by which the consumption is reduced, as for instance Boston.

The second part of table 8-S shows the same facts as are given in the first part for such of the remaining works of the United States as have over 50 per cent. of their taps metered, arranged in order of size. There are 33 such works, and in addition, four of the 50 largest cities come in this class, as follows: Atlanta, 89.6 per cent.; Worcester, 89.4; Fall River, 76.6, and Providence, 62.4 per cent.

No attempt is made to classify works with more than 50 per cent. of taps metered except by size, owing to the wide differences between the several towns and the incompleteness of some of the returns upon which the table is based. It may be said, however, that all the consumptions are low, with the exception of Green River, where, it is believed, a large number of collieries are supplied, beside several railway shops and buildings. None of the other consumptions go above 75 and 12 out of 23 are at or below 30, with 6 below 20. But these figures are often much too low, owing to the large population not supplied, as shown by the high figures in the population per tap column. In addition some of the total average daily consumptions reported are probably not even estimates but mere guesses.

OWNERSHIP.

Those who attended the earlier meetings of the American or New England Water-Works Associations will remember that a favorite subject for discussion was the relative advantages of public and private ownership of water-works. In later years the subject has not been so much discussed by these bodies, possibly because it has been seen that neither side has been making converts, actual connection with public or private works being more forcible than arguments.

Only a few years ago even the total number of water-works was not known, and the ownership still less completely. In collecting information for this issue of the MANUAL, a special effort, which proved successful, was made to close the comparatively few gaps still remaining in the list of ownerships. In addition great pains were taken to ascertain how, in the case of private works, the interests of the public had been guarded in the granting of franchises.

The final results of the above investigations are summarized and commented upon below, but with idea of presenting facts rather than of pleading for either private or public works. It is believed that the facts given have an important bearing upon the whole question of natural monopolies, of course from the historical side.

TABLE 9-S.

UNITED STATES AND CANADIAN WATER-WORKS.—SUMMARY BY STATES AND GROUPS OF STATES OF THE NUMBER OF PUBLIC AND PRIVATE WORKS, TOTAL NUMBER OF WORKS, AND PER CENT. OF PUBLIC AND PRIVATE TO THE WHOLE NUMBER.

	No. of works.			Per cent. of total.			No. of works.			Per cent. of total.	
	Public.	Private.	Total.	Public.	Private.		Public.	Private.	Total.	Public.	Private.
Me........	2	35	37	5.4	94.6	Ia...........	51	26	77	66.2	33.8
N. H.......	11	25	36	30.6	69.4	Minn.......	21	13	34	61.8	38.2
Vt.........	19	8	27	70.4	29.6	Kan.........	32½	46½	79	41.1	58.9
Mass.......	*82½	*45½	128	64.5	35.5	Neb.........	48	14	62	77.4	22.6
R. I........	5	9	14	35.7	64.3	S. Dak......	17	5	22	77.3	22.7
Conn.......	14	34	48	29.2	70.8	N. Dak......	5	2	7	71.4	28.6
						Wyo........	3	6	9	33.3	66.7
New Eng..	133½	156½	290	46	54	Mont.......	0	11	11	0	100
N. Y.......	81	118	199	41.2	58.8	N. W......	177½	123½	301	59	41
N. J........	22	36	58	38	62						
Pa.........	51½	164½	216	23.8	76.2	Mo.........	4	32	36	11.1	88.9
Del........	5	1	6	83	17	Ark.........	1	11	12	8.3	91.7
Md.........	4	14	18	22.2	77.8	Tex.........	20	40	60	33.4	66.6
D. C........	1	0	1	100	0	Colo........	33	24	57	57.9	42.1
						N. Mex.....	1	8	9	11.1	88.9
Mid........	161½	333½	498	33.2	66.8						
						S. W........	59	115	174	33.9	66.1
Va.........	20	21	41	48.8	51.2						
W. Va......	5	5	10	50	50	Wash........	10	28	38	26.3	73.7
N. C.......	3	12	15	20	80	Ore.........	11	15	26	42.3	57.7
S. C........	1	6	7	14.3	85.7	Cal.........	10½	92½	103	10	90
Ga..	11	10	21	52.4	47.6	Ariz........	1	3	4	25	75
Fla........	2	10	12	16.6	83.4	Nev.........	0	9	9	0	100
						Utah........	3	4	7	42.9	57.1
S. Atl......	42	64	106	39.6	60.4	Ida.........	1	9	10	10	90
Ala.........	1	19	20	5	95	Pac.........	36½	160½	197	18.6	81.4
Miss.......	1	6	7	14.3	85.7						
La.........	1	3	4	25	75	Total U. S..	878	1,159	2,037	43.1	56.9
Tenn.......	2	14	16	12.5	87.5						
Ky.........	5	18	23	21.7	78.3	N. B........	3	2	5	60	40
						N. S........	9	0	9	100	0
S. Cent....	10	60	70	14.3	85.7	P. E. I......	1	0	1	100	0
						P. Q........	13	11	24	54.2	45.8
O..........	58	28	86	67.4	32.6	Ont.........	31	17	48	64.6	35.4
Ind........	27	23	50	54	46	Man........	0	1	1	0	100
Mich.......	85	28	113	75.2	24.8	N. W. T.....	0	1	1	100	100
Ill.........	69	33	102	67.7	32.3	B. C........	2	4	6	33.3	66.7
Wis........	16	34	50	32	68						
						Tot'l Canada	59	36	95	62.1	37.9
N. Cent....	255	146	401	63.6	36.4						

* Fractional ownership denotes works owned jointly by city and company, of which there are but few instances.

The present ownership is shown in detail for both the United States and Canada in Table 9-S, where the number of public, private and all works, together with the percentages of public and private are given for each state and group of states.

By percentages the number of public and private works in each group is as follows, the South Central group leading in private and the North Central in public works:

	Private.	Public.		Private.	Public.
South Central...........	85.7	14.3	South Atlantic.........	60.4	39.6
Pacific.............	81.4	18.6	New England.........	54	46
Middle.............	66.8	33.2	Northwestern.........	41	59
Southwestern........	66.1	33.9	North Central........	36.4	63.6

MANUAL OF AMERICAN WATER-WORKS.

By states the percentages of private and public works are as follows:

Private.	Public.	Private.	Public.	Private.	Public.	Private.	Public.
Mont..100	0	Fla.... 83.4	16.6	Tex.... 66.6	33.4	Minn... 38.2	61.8
Nev...100	0	N.C... 80	20	R.I.... 64.3	35.7	Mass... 35.5	64.5
Ala.... 95	5	Ky..... 78.3	21.7	N.J.... 62	38	Ia...... 33.8	62.2
Me ... 94.6	5.4	Md..... 77.8	22.2	Kan.... 58.9	41.1	O...... 32.6	67.4
Ark. . 91.7	8.3	Pa..... 76.2	23.8	N.Y .. 58.8	41.2	Ill..... 32.3	67.7
Cal.... 90	10	La..... 75	25	Ore.... 57.7	42.3	Vt..... 29.6	70.4
Idaho.. 90	10	Ariz.. 75	25	Utah.. 57.1	42.9	N. Dak 28.6	71.4
Mo ... 88.9	11.1	Wash.. 73.7	26.3	Va 51.2	48.8	Mich.. 24.8	75.2
N. M.. 88.9	11.1	Conn... 70.8	29.2	W. Va. 50	50	S. Dak. 22.7	77.3
Tenn.. 87.5	12.5	N. H .. 69.4	30.6	Ga..... 47.6	52.4	Neb.... 22.6	77.4
S. C.... 85.7	14.3	Wis... 68	32	Ind.... 46	54	Del.... 17	83
Miss... 85.7	14.3	Wyo... 66.7	33.3	Colo... 42.1	57.9	D. C... 0	100

In general it may be said that those sections of the country that are new or slow in their development, or have been greatly retarded at any time, have the largest percentage of private works, as note Montana and Nevada at head of the list with no public works, and all the Southern states but Georgia with 50 per cent. or more of private works, while in some the percentage is quite high. Maine, Arkansas, Missouri and California might not be considered as new states. But Maine is peculiarly favorable to corporations. Arkansas does not allow cities to incur a bonded indebtedness. Corporations obtained immense power in Caliornia while the state was new. Missouri is backward, at least in developing cities. Maryland may with fairness be classed with the Southern states. Pennsylvania has many low-cost works in small places where companies have taken the initiative, and in late years a few capitalists have secured several charters covering several adjacent towns with a promise of future growth.

It is interesting to note in passing that Virginia, with 51.2 per cent. of private works, has four of the nine public gas plants in the United States, and that West Virginia, with public and private works half and half has one public gas plant. The other public gas plants are two in Ohio, one each in Kentucky and Pennsylvania, the latter at Philadelphia.

While in numbers the private plants are ahead the showing is sharply reversed when populations are considered, 66.2 per cent., or 15,018,552 out of a total of 22,678,354, being supplied by public works. These figures differ from those previously given, since estimated populations are included for about 150 small places. The distribution of population on public and private works and the percentage of each to the total in the several groups is as follows:

Group.	Public.	Private.	Total.	Per cent. Public.	Private.
New England	2,205,357	1,050,936	3,256,303	67.7	32.3
Middle	6,547,342	1,754,982	8,302,324	78.8	21.2
South Atlantic	487,534	391,414	878,948	55.5	44.5
South Central	183,625	856,651	1,040,276	17.7	82.3
North Central	3,896,632	1,098,556	4,995,188	78	22
Northwestern	772,926	880,053	1,652,979	46.8	53.2
Southwestern	673,019	837,956	1,510,975	44.5	54.5
Pacific	252,117	789,154	1,041,271	24.2	75.8
United States	15,018,552	7,659,802	22,678,354	66.2	33.8
Canada	866,460	197,003	1,063,463	81.5	18.5

The several groups rank as follows in percentage of public and private works:

	Percentage of Public.	Private.		Percentage of Public.	Private.
Middle	78.8	21.2	Northwestern	46.8	53.2
North Central	78	22	Southwestern	44.5	54.5
New England	67.7	32.3	Pacific	24.2	75.5
South Atlantic	55.5	44.5	South Central	17.7	82.3

INTRODUCTION. (xxxi)

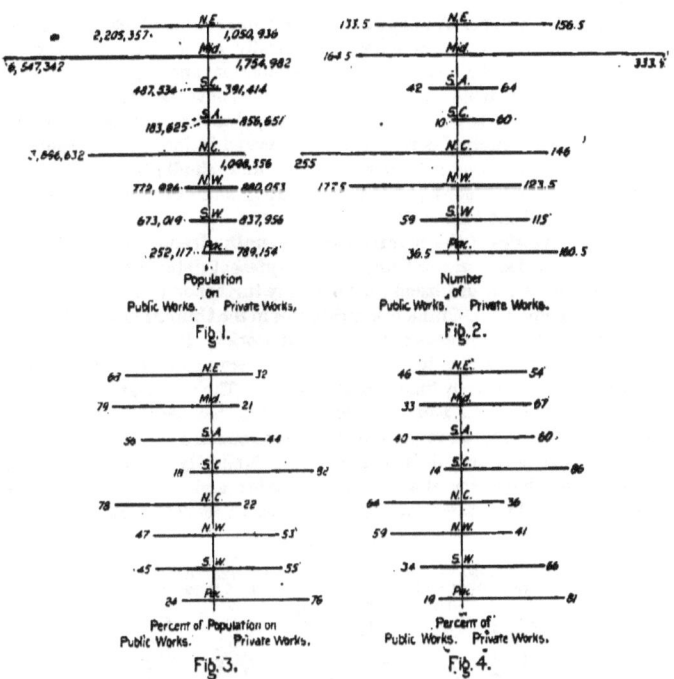

United States Water-Works.—Diagrams Showing by Groups of States 1890 Population on Public and Private Works in Operation or Under Construction July 1, 1891; Number of Public and Private Works; also Corresponding Percentages.

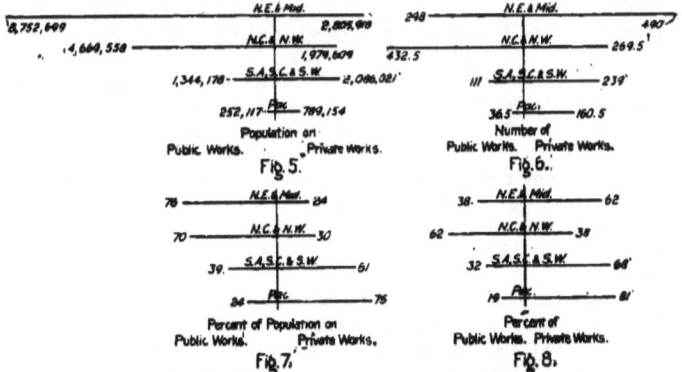

United States Water-Works.—Diagrams Showing by Four Great Groups of States 1890 Population on Public and Private Works in Operation or Under Construction July 1, 1891; Number of Public and Private Works; also Corresponding Percentages.

The relative number of public and private works and of populations on each, by groups of states, is shown graphically by the diagrams Figs. 1 to 8 (see p. xxxi). Fig. 1 shows the 1890 population, and Fig. 2 the number of public and private works. Figs. 3 and 4 show the percentages corresponding to the above. In Figs. 5 to 8 populations, number of works and the corresponding percentages are massed in four great sections of the country, including : (1) New England and Middle States; (2) North Central and Northwestern; (3) South Atlantic, South Central and Southwestern, and (4) the Pacific States.

The largest cities of the country have generally deemed it best to own their water-works, sooner or later, the exact present status in this respect being given below and also discussed more fully further on in connection with changes in ownership. Of the 10 largest cities in the United States all but San Francisco and New Orleans now own their works. New Orleans bought the private works there in 1868 and sold them again in 1878, the sale probably being caused by financial depression. The works at San Francisco have cost too much and the company has too secure control over available water supplies to permit of the duplication of works by the city, while perhaps purchase of the existing plant is equally out of the question.

Table 10-S shows for this country by states and groups of states the number of works in towns of the size named at its head. In towns with

TABLE 10-8.

UNITED STATES AND CANADIAN WATER-WORKS.—OWNERSHIP CLASSIFIED BY SIZES OF TOWNS FOR EACH GROUP OF STATES, THE UNITED STATES AND CANADA.

Group	Above 100,000		50,001 to 100,000		20,001 to 50,000		5,001 to 20,000		Below 5,000		Unknown		All works		
	Pub.	Priv.	Pub.	Priv.	Pub.	Priv.	Pub.	Priv.	Pub.	Priv.	Pub.	Priv.	Pub.	Priv.	Total
New England...	1	..	7	1	16	5	48	50	57½	75½	4	25	133½	156½	290
Middle.........	11	2	6	4	12	12	49	73	80½	202½	7	39	164½	333½	498
S. Atlantic.....	..	..	2	1	5	2	13	18	21	36	1	7	42	64	106
S. Central......	..	2	1	1	2	8	4	15	3	31	..	3	10	60	70
N. Central.....	6	2	5	1	13	8	67	80	156	51	8	4	255	146	401
Northwestern..	2	1	1	1	1	7	12	35	154½	79½	6	1	177½	123½	301
Southwestern..	1	3	..	1	3	6	4	34	46	58	4	14	58	116	174
Pacific..........	..	1	..	3	4	6	3	12	21½	119½	5	19	36½	160½	197
United States	21	11	22	13	56	54	200	317	543	653	35	112	877	1,160	2,037
Canada........	2	..	1	..	6	1	18	13	27	17	6	4	59	36	95

over 100,000 inhabitants there are 21 public, against 11 private works. From 50,000 to 100,000, the relation is nearly the same, 22 to 13. In the next class, from 20,000 to 50,000, the public works are nearly overtaken by the private, there being 56 of the former to 54 of the latter. In towns of from 20,000 down private works have the strongest footing. The works in the "unknown" column are in towns not returned separately by the last census, so far as it is yet available, but it is known that they are mostly in places below 5,000. In Canada private works prevail in no class of towns.

The smaller number of private works in the larger towns is due to the feeling that the works can be best managed for all interested by the public, and to the additional fact that the money to build or buy works can be secured. The smaller places often lack power or credit sufficient to build works and are supplied by private capital. Very often extreme conservatism in the incurring of debt prevents the town from building works, or the privilege is granted to a company when the town is small without thought except for the immediate future.

The relative growth of public and private works during the nine decades of this century is shown by Table 11-S., and graphically by Fig. 9. The table gives for each half decade the number of works built, public, private and total; the total number of works of each class which had been built to the close of the several periods; changes in ownership; and finally the actual status at the end of each decade. In connection with the above, proper percentages are given. The diagram shows the number of public and private works built to the close of each half-decade, this being indicated by the broken line, and also the actual number of works of each class at the several dates. In addition the heavy line in the diagram shows by ownership the present population of all towns having works at the close of each decade. In this part of the diagram the works in operation at the end of each period are represented by their present importance in point of population, it being assumed that the several towns have had the same population as now since their works were built. In other words, the element of growth of towns has been eliminated, and the number of works of each class has been weighted by the present population of the towns in which they are located.

Starting with but one public against 15 private works (see p. (viii.) for names and dates built) at the beginning of this century, the numbers have increased, until in July, 1891, there were completed or under construction 878 public and 1,159 private plants. The corresponding percentages are 6.3 against 93.7, and 43.1 against 56.9. The relative changes in number of works has been occasioned both by the increase of public works and by the changes in ownership. Of the latter there have been 83 private plants which have become public, against only 17 changes from public to private. There were no changes from public to private ownership until between 1876 and 1880, but the opposite changes began as early as 1810, when the works at Wilmington, Del., built in 1804 by the Wilmington Spring Water Co., were bought by the borough for $10,000. The number of changes in each half-decade is shown in Table 11-S. The places in which changes have occurred, dates of changes, dates built, and 1890 populations, are as follows:

CHANGES FROM PRIVATE TO PUBLIC OWNERSHIP.

Works at:	Date of change.	Date built.	Population, 1890.	Works at:	Date of change.	Date built.	Population, 1890.
Boston, Mass.	1848	1652	448,477	Baltimore, Md.	1854	1807	434,439
Hartford, Conn.	1854	1797	53,230	Trenton, N. J.	1855	1803	57,458
Plymouth, Mass.	1855	1796	7,314	Cohoes, N. Y.	1857	1848	22,509
Danbury, Conn.	1861	1833	19,473	Newark, N. J.	1860	1800	181,830
Worcester, Mass.	1864	1796	84,655	Reading, Pa.	1865	1821	58,661
Salem, Mass.	1865	1799	30,801	Allentown, Pa.	1865	1827	25,228
Cambridge, Mass.	1865	1837	70,028	Buffalo, N. Y.	1868	1852	255,664
New London, Ct.	1872	1802	13,757	Oxford, Pa.	1870	1868	1,711
Peabody, Mass.	1873	1799	10,158	Camden, N. J.	1870	1853	58,313
Springfield, Mass	1875	1843	44,179	Bethlehem, Pa.	1871	1761	6,762
Providence, R. I.	1870-6	1722	132,146	New Brunswick, N. J.	1873	1868	18,603
Spencer, Mass.	1884	1883	8,747	Burlington, N. J.	1877	1804	7,264
Woonsocket, R. I.	1885	1884	20,830	Bradford, Pa.	1877	1830	10,511
Dover, N. H.	1888	1826	12,790	Edenburgh, Pa.	1882	1830	1,202
Windsor, Vt	1888	1850	1,846	Slatington, Pa.	1883	1861	2,716
Swanton, Vt	1889	1887	3,231	Schenectady, N.Y	1885	1371	19,902
Uxbridge, Mass.	1890	1880	3,408	Gouverneur, N.Y.	1890	1867	3,458
Rochester, N. H.	1891	1885	7,396	Richfield Sp'ngs, N. Y.	1879		1,621
Braintree, Mass.	1891	1887	4,848				
Milford, N. H.	1891	1889	3,014				
Franklin, N. H.	1891	1891	•4,085				
Wilmingt'n, Del.	1810	1804	61,431	Lynchburg, Va.	1828	1799	19,709
New York, N. Y.	1843	1774	1,515,301	Augusta, Ga.	1861	1828	33,390
Frederick, Md.	1844		8,193				
Kennet Sq're, Pa.	1850	1842	1,326	Huntsville, Ala.	1858	1830	7,995
Albany, N. Y.	1851	17—	94,923	*New Orleans, La.	1868	1833	242,039

"In 1878 city sold works to Company.

Changes from Private to Public Ownership—Continued.

Works at:	Date of change	Date built.	Population, 1890.	Works at:	Date of change	Date built.	Population 1890.
Detroit, Mich.	1836	1827	205,876	Abilene, Tex.	1887	1886	3,194
Cincinnati, O.	1839	1820	296,908	Central Cy., Colo.	1889	1873	2,480
Chicago, Ill.	1854	1842	1,099,850	Galveston, Tex.	1889	1884	29,084
Jackson, Mich.	1871	1870	20,798	Berthoud, Colo.	1889	1886	228
Moline, Ill.	1886	1883	12,000	Lamar, Colo.	1890	1887	566
Joliet, Ill.	1888	1881	23,264	Georgetown, Colo.	1891	1874	1,927
Bessemer, Mich.	1890	1888	2,566	E. Dallas, Tex.	1891	1886	38,067
Lewiston, Ill.	1890	1888	2,166	Vernon, Tex.	1891	1890	2,857
Hayward, Wis.	1890	1888	1,349	Cañon Cy., Colo.		1880	2,825
Galesburgh, Il.	1891	1883	15,264	Ft. Worth, Tex.		1883	23,076
Huntington, Ind.	1891	1891	7,328				
Kenton, O.		1882	5,557	Portland, Ore.	1887	1853	46,385
				Seattle, Wash.	1889	1884	42,837
St. Paul, Minn., since	1882	1870	133,156	The Dalles, Ore.	1890	1862	est., 3,000
Sauk Center, Minn.	1888	1884	1,695	Oceanside, Cal.	1890	1886	est., 1,200
Bunker Hill, Kans.	1889	1876	157	Aberdeen, Wash.	1891		1,038
Springfield, Kans.	1889	1889	347	Portland, Ore*.	1891	1887	
Fargo, N. D.	1890	1887	5,664	Santa Cruz, Cal.	1891	1876	5,596
				Gilroy, Cal.		1872	1,691

CHANGES FROM PUBLIC TO PRIVATE OWNERSHIP.

Works at:	Date of change	Date built.	Population, 1890.	Works at:	Date of change	Date built.	Population 1890.
Hoboken, N. J.	1882	1855	43,648	Peoria, Ill.	1889	1868	41,024
Chester, Pa.	1888	1868	20,226				
Green Island, N.Y.	1888	1878	1,463				
Beaver, Pa.	1891	1886	1,552	McPherson, Kan.	1890	1883	3,172
				Beloit, Kan.		1887	2,455
Manchester, Va.	1889	1840	9,244				
Orangeb'gh, S.C.	1889	1887	2,966				
				Sedalia, Mo.	1887	1871	4,068
†New Orleans, La.	1878	1833	242,039	Cleburne, Tex.	1891	1884	3,278
Lexington, Ky.	1884	1832	21,567				
Columbia, Tenn.	1884	1840	5,370				
Frankfort, Ky.	1886	1841	7,892	Ogden, Utah.	1890	1882	14,889
Jacksonville, Ala.	1889	1883	1,237				

* East Portland, now part of Portland, first supplied in 1887 by a company. In 1891 another company built works, which were bought by town.

† Built in 1833 by company; bought by city in 1868; sold to company in 1878.

For convenience the number of changes in each group and the present populations involved in these changes may be brought together as follows:

	Private to Public.		Public to Private.	
	Number.	Population.	Number.	Population.
New England	21	984,513	....	......
Middle	23	2,849,029	4	69,880
South Atlantic	2	53,009	2	12,210
South Central	2	250,034	5	278,105
North Central	12	1,692,926	1	41,024
Northwestern	5	141,019	2	5,627
Southwestern	10	104,304	2	7,346
Pacific	8	102,350	1	14,889
Total, U. S.	83	6,177,184	17	429,090

The Middle States lead both in number of changes from private to public and in the present population involved. The New England States come next in point of numbers, and the North Central in population. Boston is included in the New England, New York in the Middle, New Orleans in the South Central, and Chicago in the North Central changes. Other large cities, named further on, were involved in these changes, so that the present population of changes to public ownership is 6,177,184, against 429,090 on the other side. In changes from public to private the South Central group leads, New Orleans also being included on this side, its works having been built by a company in 1833, bought by the city in 1868 and sold to a company in 1878.

As will be seen by referring to Table 11-S, the greatest number of works to change from private to public in any one period was 23 in the last half-decade. The same may be said for the opposite changes, 11 works having changed from public to private in 1886-90.

INTRODUCTION. (xxxv)

TABLE II–8.

UNITED STATES WATER-WORKS.

SUMMARY BY HALF-DECADES OF PUBLIC, PRIVATE AND ALL WORKS BUILT; OF PUBLIC AND PRIVATE WORKS BUILT TO CLOSE OF EACH PERIOD; CHANGES FROM PRIVATE TO PUBLIC AND FROM PUBLIC TO PRIVATE; AND ACTUAL NUMBER OF WORKS AT END OF EACH HALF-DECADE, PUBLIC, PRIVATE AND TOTAL.

	Works built.										Changes in ownership.				Actual number of works at end of each half-decade.				
	In each half-decade.					To close of each half-decade.					Priv. to pub.	Pub. to priv.	P. c. of total.					Per c't of total.	
	Pub.	Priv.	Total.	P. c. pub.	P. c. priv.	Pub.	Priv.	P. c. pub.	P. c. priv.				Pub.	Priv.	Pub.	Priv.	Total.	Pub.	Priv.
Before 1801	1	15	16	6.3	93.7	1	15	6.3	93.7		…	…	…	…	1	15	16	6.3	93.7
1801–5	1	6	7	14.3	85.7	2	21	8.7	91.3		…	…	…	…	2	21	23	8.7	91.3
1806–10	2	1	3	66.7	33.3	4	22	15.4	84.6		1	…	…	…	5	21	26	19.2	80.8
1811–15	…	4	4	…	100	4	26	15.4	84.6		…	…	…	…	5	25	30	16.6	83.4
1816–20	…	4	4	…	100	4	28	12.5	87.5		…	…	…	…	5	27	32	15.6	83.4
1821–25	3	9	12	25	75	7	37	15.9	84.1		…	…	…	…	9	32	41	20.5	79.5
1826–30	6	4	10	60	40	13	41	24.1	75.9		2	…	2.3	…	15	39	54	27.8	72.2
1831–35	6	4	10	60	40	19	45	29.7	70.3		1	…	…	…	23	41	64	35.9	64.1
1836–40	2	6	8	33.3	66.6	21	49	30	70		2	…	3.1	…	27	43	70	38.6	61.4
1841–45	4	9	13	30	70	25	58	30.1	69.9		3	…	2.9	…	33	50	83	39.7	60.3
1846–50	10	13	23	43.3	56.5	33	71	31.7	67		2	…	2.4	…	48	58	106	45.3	54.7
1851–55	7	23	30	23.3	76.7	42	116	30.9	69.1		5	…	4.7	…	57	79	136	41.9	58.1
1856–60	4	22	26	15.4	84.6	46	154	28.4	71.6		2	…	1.5	…	68	94	162	42	58
1861–65	43	38	81	53.1	46.9	89	229	36.5	63.4		5	…	4.3	…	116	127	243	47.7	52.3
1866–70	104	75	179	58.1	41.9	193	310	43.1	56.8		7	…	2.1	…	227	195	422	53.8	46.2
1871–75	65	111	176	36.9	63.1	258	342	43.2	56.8		2	1	1.7	0.2	293	305	598	49	51
1876–80	150	265	415	36.1	63.9	408	605	40.2	59.8		7	3	0.7	0.3	447	566	1,013	44.1	55.9
1881–85	317	518	835	40.1	59.9	755	1,123	40.2	59.8		23	11	1.2	0.6	806	1,072	1,878	42.9	57.1
1886–90	0	14	14	…	100	812	1,225	39.9	60.1		5	2	0.6	0.1	878	1,159	2,037	43.1	56.9
Unknown	57	88	145	39.3	60.7	…	…	…	…		12	17	4.1	0.9	…	…	…	…	…
July 1, '91																			
Total changes	…	…	…	…	…	…	…	…	…		83								

Coming now to the actual status at the end of each decade we find public works relatively increasing from 1801-10, and gaining one by the change in ownership at Wilmington, Del., already noted. In 1811-15 all works were stationary and in 1816-25 no more public works were added, causing a fall in the percentage of public works. In 1830 public works had regained their relative importance and they continued to gain until 1855, when they had 45.3 per cent. of the total, or 48 against 58 private plants.

The years 1856-65, preceding and including the civil war saw but 11 public works built against 55 private, so that deducting these and the 9 works which changed from private to public ownership we have a net gain for private works of 35 which increased their percentage to 58. In the war period many of the cities North and South were contributing money directly and all indirectly to carry on the war, and consequently there were but few communities that had the means, and perhaps fewer still the heart, to engage in local improvements. In 1863, as in the whole of the 1812 war period, not a works was built.

In the ten years after the war public works gained rapidly in relative numbers, increasing from 42 to 53.8 per cent. of the total, standing above private works for the first and only period, the numbers being 227 public and 195 private in 1875. In each of the last two half decades public works have shown a relative decrease, standing at 42.9 per cent. of the total in 1890.

From the foregoing review it seems that public works gain on private with the general prosperity of the country. Public works are built almost wholly from the proceeds of bond issues, and these are generally sold at a lower rate of interest than company's bonds, and in some instances cannot by law be sold below par, and still less can they be sold at a low figure, because of popular feeling regarding the city's credit. Many private water-works put a fair proportion of cash from capital stock into their plants, and are probably not so averse to selling bonds below par, indeed, can sometimes hold bonds until the market is better. Besides, hard times, real or threatened, deter the mass of the citizens from incurring further municipal indebtedness when the money market is but little affected, while capitalists making a business of building works are not always so much influenced by financial depressions.

In the last 10 years, also, promoters and parent companies have been active in constructing private works, one parent company controlling nearly or quite 40 local companies, another some 25 to 30, and several others from two or three to ten or thereabouts.

But just now the tide is setting strongly for municipal ownership, although the table does not show it, and with the building of new and the purchase of old works the next few years will probably see a considerable increase in both the relative number and importance of public works, unless indeed the late financial depression sets the tide more strongly the other way, which, considering its strength, is hardly probable.

In Canada up to 1845 there were but two works, both of which were public. In 1870, and again in 1875, there were 4 public and 4 private works. In 1880 public works had risen to 72.2 per cent., but they have since fallen and now stand at about 62 per cent. of the total. Altogether there have been 9 changes in ownership, all from private to public, as follows:

	Date of change.	Date built.	Population, 1891.
Montreal, P. Q.	1845	1801	216,650
St. John, N. B.	1855	1836	39,179
Halifax, N. S.	1861	1848	38,556
Toronto, Ont.	1873	1841	144,028
Kingston, Ont.	1887	1851	24,260
Valleyfield, P. Q.	1887	1886	5,516
Niagara Falls, Ont.	1889	1856	3,349
Owen Sound, Ont.	1890	1880	7,497
St. Henri, P. Q.	1891	1880	13,415
Total population			492,459

The total number of works in Canada is 95, so that nearly 10 per cent. of all works have changed from private to public ownership, and it may be added that some 45 per cent. of the present population on works is supplied by the plants included above.

The growth of public and private works and changes of ownership in the United States is best shown by the diagram, Fig. 9 (p. xxxviii) especially the relative importance of the two classes at any time, this being indicated by the heavy lines in the diagram. As has already been explained the broken lines in the diagram represent public and private works as they would have been without change in ownership, while the full light lines show the actual status of each class, with changes. The changes began in 1830, but cannot be shown on the diagram until 1840, owing to the small number of works at this time. The population diagram shows the changes from 1830 to 1860 in a very forcible way. It was in these 30 years that the largest cities (see the list below) in the country went over from private to public ownership, causing, together with a more rapid increase in population supplied by public works, private populations to fall from 4 millions in 1830 to 2.4 millions in 1860, and public populations to rise from 1.9 to 8.5 millions. It took until 1880, or 50 years, for private works to regain, present populations considered, it must be remembered, the standing they had in 1830, while in 1880 public works had risen to 12.9 and in 1890 to 14.8 against 7.5, a little over half, on private works, the figures for 1890 being the actual populations in that year. Of the 50 largest cities of the United States 17 have changed from private to public ownership, as follows, the places being arranged in order of dates of changes:

	Date of change	Present pop'lat'n		Date of change	Present pop'lat'n
Wilmington, Del.	1810	61,431	Newark, N. J.	1860	181,830
Detroit, Mich.	1836	205,876	Worcester, Mass.	1864	84,655
Cincinnati, O.	1839	296,908	Cambridge, Mass.	1865	70,028
New York, N. Y.	1843	1,515,301	Reading, Pa.	1865	58,661
Boston, Mass.	1848	448,477	Buffalo, N. Y.	1868	255,664
Albany, N. Y.	1851	94,923	New Orleans, La.	1868	242,039
Chicago, Ill.	1854	1,099,850	Camden, N. J.	1870	58,313
Baltimore, Md.	1854	434,439	St. Paul, Minn.	since 1882	133,156
Trenton, N. J.	1855	57,458			

It is interesting to note in connection with the above that only 12 of the 50 largest cities of the United States are now supplied wholly by companies, the cities being as follows, in order of size:

San Francisco, Cal.	298,907	Indianapolis, Ind	105,436
New Orleans, La.	242,039	Syracuse, N. Y.	88,143
Louisville, Ky. (city owns practically all the stock.)	161,129	New Haven, Conn.	81,298
		Paterson, N. J.	78,347
Omaha, Neb.	140,452	Scranton, Pa.	75,215
Kansas City, Mo.	132,716	Memphis, Tenn.	64,495
Denver, Colo.	106,713		

Thus 29 of our 50 largest cities were once supplied by private companies and now all but 12 have their works under public ownership.

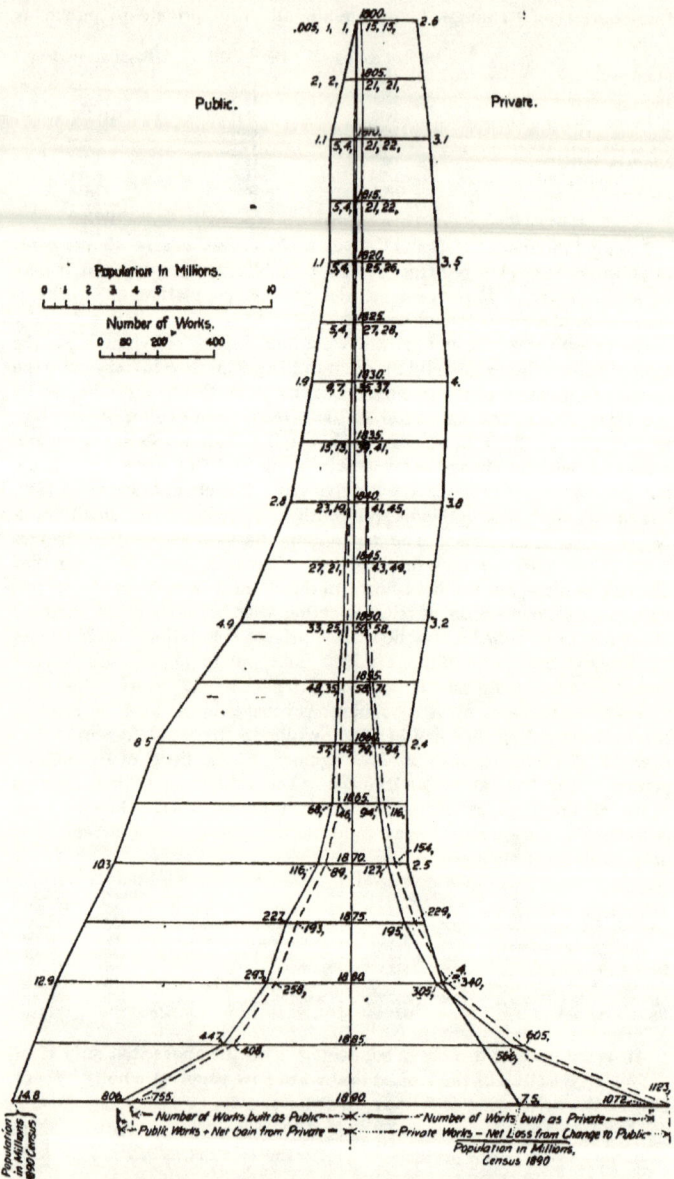

Fig. 9.—Growth of Public and Private Water-Works in the United States.—Diagrams Showing the Number of Works of Each Class Built to the Close of Each Half-Decade and the Actual Standing After Changes in Ownership; also the Relative Importance of Works of Each Class at the End of Each Decade, According to Present, or 1890, Populations.

INTRODUCTION. (xxxix)

In addition to the above, Brooklyn, Pittsburg and Grand Rapids have a part of their population supplied by companies, but the part so supplied is less than 4 per cent. of the total population in Brooklyn, probably less than 25 per cent. in Pittsburg, and less than half at Grand Rapids.

A list of 29 duplicate and triplicate plants in 25 cities was given on p. (V). and further comment upon competing plants was promised.

These duplicate works are so circumstanced that there is, with two or three exceptions, little or no real competition between them, or the works in themselves are very small. In Brooklyn the Long Island Water Supply Co. supplies an annexed part of the city, the 26th ward. These works were built before annexation—it has been hinted to sell to the city, as those familiar with what the New York papers termed the " Brooklyn water-works scandal" will remember. At Jamaica the two companies are practically under the same management and have been from the first, the second company supplying water to the first. At Pittsburg the Monongahela Water Co. supplies the citizens on one side of the river and an outside population, the whole aggregating about 75,000, while the city supplies the greater part of its inhabitants. At Mobile the first company built works in 1840, and a larger plant was built by a company in 1888. These are evidently competing, as the second company went to the courts to prove its exclusive right to supply the city, but failed in the attempt. At Chattanooga the second company supplies a different section of the city, or at least is on amicable terms with the first, since it buys water from it and repumps it. At Grand Haven and Grand Rapids, each, there is a public and private plant, and in the latter city the two are in lively competition. At Chicago the private plant is in the former village of Pullman, now a part of Chicago, and was built to supply various industries and the dwellings of the Pullman Palace Car Co. At Denver there is bitter rivalry between the two companies, the second being the outgrowth, according to reports, of a quarrel among the stockholders of the first.

At Seattle the present city plant was built by a company and recently bought by the city. The other, and private plant, is small, and may yet be bought by the city. In Spokane the private plant supplies a small suburb and in Portland the companies operate in districts annexed within a very short time, while one of the companies had a public works for a rival before annexation. At Los Angeles the third company is small and does not compete with the other two, while it seems that there may be some rivalry between the first two.

From the above review it will be seen that among American water companies competing plants play scarcely any figure, the works in most cases being small, or occupying independent territory, or both. There has been some competition and consolidation, but a water supply system, of any size, is one of the most complete of natural monopolies and almost defies competition.

FRANCHISES OF WATER-WORKS COMPANIES.

Over half of the water-works of the United States, about 57 per cent., are owned by companies. These companies are, in reality, public servants clothed with more or less authority and performing public functions with the privileges of condemning private property, using public streets and making profit from consumers. As the public surrenders these privileges to companies and has a right to expect and demand good service in return, it may well be asked in what way and to what extent do our cities and towns when granting franchises provide for good service, fair charges, resumption of the franchises and the ultimate ownership by the municipality when desired.

The question has almost as many answers as there are water companies, for while in some states nearly all the powers granted are conferred upon the company directly by the state when it is chartered, in others each municipality fixes most of the privileges which the company may enjoy and the duties which it must perform when the franchise is granted, and all have more or less opportunity to secure proper agreements from the company, either in granting the franchise or in making the contract for hydrant rental. All franchises should contain provisions for regulation of rates and for the purchase of the works by the town at a fair price. The price should never include the value of the franchise itself, this always remaining the property of the city, leased or granted to the company for a time, but reverting to it at no distant date, especially when for any cause purchase by the city will not be easy, in order that a new franchise may be granted, if desired, on conditions suitable to the time and changed circumstances.

Provisions from the franchises of over one-third of the water companies in the United States are abstracted below from the information collected for the body of the MANUAL. The facts presented are of unusual interest owing to their bearing upon important and practical subjects and to the large number of companies concerned, no extended information of this character ever before having been collected and published. The points covered are length of franchises, regulation of rates and purchase of works.

Table 12-S shows, by groups of states, the length of the franchises of 391 out of 1,159 companies. Two companies have 5-year, 15 have 10-year and 9 have 15-year franchises. The largest number in any one class is 132 in the 20-year. The 20, 21, 25 and 30-year classes together have 267 out of the 391 companies reporting. Of 50-year franchises there are 46: of 99-year, 11, and of perpetual, 39, or 10 per cent. of the number reported. Many of the long-term franchises are, doubtless, really state charters, in addition to which there is some form of contract between the city and company, but each company was included as reported. Over half of the New England companies included and about one-third of those in the Middle States report perpetual franchises, while of all the 50-year franchises reported over half are in the Pacific States. In all but the New England, Middle and Pacific States 20 to 30-year franchises are in the majority.

TABLE 12-S.
LENGTH OF FRANCHISE IN YEARS OF 391 WATER COMPANIES.

	Years.										Perpetual.	Total reporting.	Total number of companies.
	5	10	15	20	21	25	30	40	50	99			
N. E.	..	..	1	3	..	2	2	1	1	..	13	23	156½
Mid	1	9	..	4	..	3	3	..	2	6	13	41	333½
S. Atl	..	..	..	3	..	7	6	1	2	3	..	22	64
S. Cent	..	..	..	9	..	5	8	..	..	..	5	27	60
N. Cent	1	4	5	35	1	12	21	..	4	..	2	85	146
N. W	..	..	1	36	9	13	12	..	1	2	4	78	123½
S. W	..	1	..	36	..	10	5	..	12	..	..	64	115
Pac	..	1	2	6	..	14	2	..	24	..	2	51	160½
Total U. S.	2	15	9	132	10	66	59	2	46	11	39	391	1,159

INTRODUCTION. (xli)

TABLE 13-S.

PROVISIONS OF WATER-WORKS FRANCHISES FOR THE REGULATION OF RATES AND PURCHASE OF WORKS BY CITY.

	No. of companies.		Rates.				Purchase.							
				Regulation.			Time.					Price.		
	Total.	Reporting.	None.	Subj't to approv'l of cit'y.	Max'm rates fixed.	Rates fixed.	No report'g.	No provis'n.	Any time.	Fixed int'ls.	Not given.	No report.	Cost.	Nam'd.	Fixed by ar'tion.
Me......	35	20	20	..	..	..	18	14	2	2	..	3	1	..	2
N. H.....	25	6	6	..	..	..	6	6	..	..	..	..	..	..	..
Vt.......	8	3	3	..	..	..	3	2	..	1	..	..	..	..	..
Mass.....	45½	21	19	..	..	2	20	5	8	3	4	9	..	2	7
R. I......	9	6	6	..	..	..	6	3	..	3	..	1	..	..	1
Conn.....	34	14	14	..	..	..	14	12	2	..	..	2	..	..	2
N. E.	156½	70	68	..	..	2	67	42	12	9	4	15	1	2	12
N. Y.....	118	42	32	1	6	3	42	26	4	9	3	5	1	..	4
N. J.....	36	13	12	..	1	..	12	11	..	..	1	2	1	..	1
Pa.......	164½	56	50	2	2	2	56	50	2	4	..	5	1	1	3
Del......	1	..	..	..	..	..	..	..	..	..	..	..	..	..	..
Md......	14	8	7	1	..	..	8	6	1	1	..	..	..	..	..
Mid....	333½	119	101	4	9	5	118	93	7	14	4	12	3	1	8
Va.......	21	3	2	..	1	..	3	2	..	1	..	..	..	..	..
W. Va...	5	3	1	..	1	1	3	1	..	2	..	1	..	..	1
N C.....	12	4	2	..	..	2	4	1	..	3	..	..	..	..	..
S. C.....	6	3	1	..	1	1	3	..	..	3	..	..	..	..	..
Ga.......	10	3	1	..	2	..	3	1	..	2	..	1	..	..	1
Fla......	10	7	3	..	4	..	7	2	1	4	..	4	..	..	4
S. Atl.	64	23	10	..	9	4	23	7	1	15	..	6	..	..	6
Ala.....	19	3	..	..	1	2	3	..	..	3	..	1	..	..	1
Miss....	6	4	1	..	3	..	3	1	..	2	..	..	..	..	..
La......	3	2	..	..	2	..	2	..	..	2	..	2	1	..	1
Tenn....	14	9	2	..	4	3	6	1	1	4	..	1	..	..	1
Ky......	18	5	4	..	1	..	5	4	..	1	..	1	..	..	1
S. Cent.	60	23	7	..	11	5	19	6	1	12	..	5	1	..	4
O.......	28	10	3	..	4	3	8	1	..	6	1	4	..	..	4
Ind.....	23	11	2	..	7	2	11	1	1	9	..	7	..	1	6
Mich....	28	13	7	1	3	2	..	..	..	..	..	..	..	..	..
Ill......	33	10	3	1	3	3	12	3	7	..	2	2	..	..	2
Wis.....	34	13	6	..	3	4	13	2	..	11	..	5	..	..	5
N.Cent.	146	57	21	2	20	14	44	7	8	26	3	18	..	1	17
Ia.......	26	13	3	..	4	6	12	4	1	6	1	4	1	..	3
Minn....	13	..	..	..	..	..	1	..	..	1	..	..	..	..	..
Kans....	46½	11	2	..	2	7	24	1	17	6	..	4	..	..	4
Neb.....	14	1	..	..	1	..	2	..	..	2	..	2	..	..	2
S. Dak...	5	1	..	..	1	..	2	..	..	2	..	2	..	..	2
N. Dak..	2	..	..	..	..	..	..	..	..	..	..	..	..	..	..
Wyo.....	6	..	..	..	..	..	2	..	..	2	..	1	..	..	1
Mont....	11	4	2	1	1	..	4	2	..	2	1	1	..	..	1
N. W...	123½	30	7	1	9	13	47	7	18	20	2	14	1	..	13
Mo......	32	17	1	4	4	8	18	..	1	17	..	4	..	..	4
Ark.....	11	2	..	2	..	..	2	..	..	2	..	1	..	..	1
Tex.....	40	10	4	..	1	5	10	1	1	4	1	3	..	1	3
Colo....	24	5	4	..	..	1	5	3	..	1	1	1	..	1	..
N. M....	8	4	3	..	1	..	3	3	..	..	..	..	..	..	..
S. W....	115	38	12	6	6	14	38	10	2	24	2	9	..	1	8
Wash....	28	8	3	..	2	3	7	5	..	..	2	..	..	..	..
Ore.....	15	6	5	..	..	1	4	4	..	..	..	1	..	..	1
Cal.....	92½	..	..	..	..	..	35	33	1	2	..	..	..	..	..
Ariz ...	3	3	1	..	1	1	3	1	..	1	1	1	..	..	1
Nev.....	9	3	3	..	..	..	3	1	1	..	..	..	..	..	..
Utah....	4	1	1	..	..	..	2	1	..	..	..	..	..	..	..
Idaho ..	9	5	4	..	1	..	5	5	..	..	..	..	..	..	..
	160½	26	17	..	4	5	60	52	2	3	2	2	..	..	2
U. S....	1,159	386	243	13	68	62	416	224	51	123	18	81	6	5	70

* Constitution of California provides that rates shall be fixed annually by County Board of Supervisors.

Where the right to purchase at reasonable intervals is reserved and the franchise regulates, or unquestionably provides that the city may regulate, the rates a long-term franchise is not so very objectionable. Otherwise the franchise should be short, although not so short as to hurt the sale of the company's securities.

Table 18-S shows by states and groups of states details regarding the regulation of rates and provisions for the purchase of works. As to regulation of rates, reports have been received regarding 386 franchises, and of these 243, or about 63 per cent., contain no regulation. Of the remaining 143 it is stated that in 13 cases the rates are subject to the approval or regulation of the city, 68 have maximum rates fixed, and in 62 cases it is merely stated that the rates are fixed. The 93 companies in California, none of which are included in Table 13-S, have their rates fixed yearly by County Boards of Supervisors in accordance with the state constitution.

In one or more states aside from California, it is believed, local authorities have power to establish rates. New England shows but two companies, both in Massachusetts, whose franchises fix their rates in any way, against 68 with no regulations; but it may be that the general laws in these states make easy the securing of fair rates through the courts or otherwise.

Passing to provisions for purchase we find that of 416 franchises reported on 224 contain no provisions for purchase, 51 provide for purchase at any time, 123 name intervals and for 18 others the time is not stated. The New England and Pacific States have the largest number of franchises without provision for purchase, there being 42 out of 67 in New England and 52 out of 60 in the Pacific States. In Kansas over a third of all the companies, 17 out of 46, have franchises providing for purchase at any time, and only one company reports no provision. Of the 192 companies reporting some provision for purchase but 81 state how the price is to be determined. Of these 81 six works may be bought at cost, five for a price named in the franchise and 70, or nearly all, by arbitration, which is probably the most desirable method when proper provision is made for the selection of the arbitrators.

On the whole, the above figures show, so far as mere abstract figures can, that the majority of our cities and towns have been so anxious to secure water, or what is worse, so careless and ignorant, that they have not properly protected their interests in granting franchises, having failed to retain for themselves a reasonable control over water rates, and the privilege of purchasing the works at a fair price whenever, within proper limits, they desire to do so. In addition, some franchises have been given away for all time and others for more than a generation. In the last particular, however, the showing is not so bad, for, as will be seen by referring again to Table 12-S, only 98 out of 391 franchises are for more than 30 years, and 158 are for 20 years or less, there being 132 franchises running just 20 years each. Twenty years seems plenty long enough for an ordinary franchise, and in many cases 15 might be better, and, generally speaking, the privilege of purchasing should not come less often than once in five years, or every five years after the first 10 years have elapsed.

Taking into consideration the looseness and faultiness of many franchises both the companies and towns in which the latter operate are to be congratulated on having no more difficulties than they do, although injustice often enough results to both sides, more often, however, to the public, since a water-works company has a practically complete natural monopoly. The lesson to be derived from the foregoing is that in granting franchises to water companies more care should be taken to preserve and secure to the public their natural rights in accordance with past experience and present enlightenment.

NOTICE.

Water-Works Officials

will greatly oblige us by mailing to us their latest **Annual Reports** and **Schedules of Water Rates,** and by giving notice of any errors or omissions they may find in the description of their works.

Hydraulic Engineers and Contractors

will likewise favor us by noting errors or omissions, and by informing us of new water-works with which they are connected, or of which they have knowledge.

The Manual

OF

American Water-Works.

1891.

GROUP I.--New England States.--Water-Works Completed, in Process of Construction or Projected, in Maine, New Hampshire, Vermont, Massachusetts, Rhode Island and Connecticut.

MAINE.

Water-Works Completed or in Process of Construction.

1. **AUBURN,** *Androscoggin Co.* (Pop., 9,555—11,250.) Built in '69 by Auburn Aqueduct Co. Supply.—Lake Auburn, by gravity, with pumping for fire protection. Distribution.—*Mains,* 25 miles. *Taps,* 1,100. *Meters,* 12. *Hydrants,* 60. *Consumption,* 500,000 galls. Financial.—*Cost,* $250,000. *Debt,* $40,000 at 4%. *Expenses:* operating, $4,000; int., $1,600. *Revenue:* consumers, $18,500; city, $2,300; total, $20,800. Management.—Prest. and Supt., F. M. Jordan. Secy. and Treas., M. J. Jordan.

NOTE.—June 20, '91, commission had been appointed to confer with company concerning purchase price of plant, city having decided to secure more adequate water supply, either by buying company's works or by building new system. The 1891 session of State Legislature passed an act enabling city to construct water-works provided that it first offered to buy company's system.

2. **AUGUSTA,** *Kennebec Co.* (Pop., 8,665—10,527.) Built in '73 by Co., extended by Augusta Water Co. Supply.—Kennebec river, pumping to reservoir; filtered. Fiscal year closes May 31. In '90 a 20-in. conduit, 750 ft. long, was laid through a canal, and in '91 to be extended 500 ft. farther, from canal head gates to middle of river, in order to obtain purer water supply. A 3,000,000-gall. Holly pump and 300 H. P. Holyoke water-wheel were also added. Distribution.—*Mains,* 22 miles. *Taps,* 1,000. *Meters,* 5. *Hydrants,* 82. *Consumption,* 1,000,000 galls. Financial.—Withheld. Management.—Prest., Geo. P. Westcott, Portland. Secy., J. H. Manley. Supt., W. H. Williams.

3. **BANGOR,** *Penobscot Co.* (Pop., 16,856—19,103.) Built in '75-6 by city. Supply.—Penobscot River, pumping direct; filtered. Fiscal year closes March 13.

Year.	Mains.	Taps.	Meters.	Hydrants.	Consumption.	Bonded debt.	Int. ch'gs.	Exp.	Rev.
'77-8	22.1	1,096	0	152	unk'n	$50,000	......	......	$11,836
'78-9	22.4	1,223	0	154	713,819	500,000	$30,015	$14,612	13,217
'79-80	23.1	1,348	0	155	656,011	500,000	......	......	13,600
'80-1	23.1	1,458	0	155	1,015,484	500,000	29,515	7,754	18,023
'81-2	23.6	1,577	0	155	945,235	500,000	30,360	9,581	17,900
'82-3	23.7	1,693	0	155	1,141,745	500,000	30,015	8,017	18,050
'83-4	24.7	1,849	0	158	1,365,025	500,000	29,910	13,502	18,378
'84-5	25.3	1,960	0	160	1,545,319	500,000	29,805	9,937	19,470
'85-6	25.7	2,096	0	161	1,760,112	500,000	30,015	12,362	27,150
'86-7	26.9	2,263	0	161	1,767,290	500,000	29,925	......	24,219
'87-8	28.5	2,452	0	167	1,874,132	500,000	30,000	......	27,888
'88-9	29.6	2,610	0	172	1,843,000	500,000	30,000	6,500	29,860
'89-90	31.9	2,795	0	174	2,500,000	500,000	30,000	......	30,513

Financial.—See above. Management.—Clerk, G. W. Snow. Supt., Frank E. Sparks.

FOR FURTHER DESCRIPTIVE MATTER see Manual for 1889-90. CHANGES since 1890 are here given. POPULATIONS are for 1880, and 1890 respectively.

4. BAR HARBOR, *Hancock Co.* (Pop., 624.) Built in '74 by Bar Harbor Water Co. **Supply.**—Eagle Lake, by gravity. **Distribution.**—(In '87) *Mains,* 15 miles. *Taps,* 400. *Hydrants,* 20. **Financial.**—(In '87) *Cost,* $150,000. *Cap. Stock,* $50,000. *No debt.* **Management.**—Prest., S. H. Rodick. Mgr., F. Rodick.

NOTE.—No report has been received since '87. In '90 Eden Water Co., John T. Higgins, Clerk, was incorporated to supply Bar Harbor, but June 4, '91, project was reported as indennite.

5. BATH, *Sagadahoc Co.* (Pop., 7,874—8,723.) Built in '86-7 by National Water-Works Syndicate, Boston, Mass., under 30-years franchise; now controlled by Maine Water Co.; city neither controls rates nor has right to purchase works. Fiscal year closes Dec. 31. **Supply.**—Thompson's Brook, pumping to stand-pipe. **Distribution.**—*Mains,* 18 miles. *Taps,* 1,100. *Meters,* 8. *Hydrants,* 123. *Consumption,* 380,000 galls. **Financial.**—*Cap. stock:* authorized, $500,000, all paid up. *Bonded Debt,* $300,000 at 5%. **Management.**—Prest., Arthur Sewall. Secy. and Treas., Josiah S. Maxcy, Gardiner. Supt., J. W. Wakefield.

6. BELFAST, *Waldo Co.* (Pop., 5,308—5,294.) Built in '87 by Belfast Water Co., under perpetual franchise. **Supply.**—Little River, pumping to stand-pipe. Fiscal year clo.es Dec. 31. **Distribution.**—*Mains,* 8 miles. *Taps,* 285. *Meters,* 5. *Hydrants,* 50. *Consumption,* 72,800 galls. **Financial.**—*Cost,* $171,330. *Cap. Stock:* common, authorized, $80,000, all paid up; preferred, authorized, $20,000, all paid up, 5% div'd guaranteed. *Bonded Debt,* $70,000 at 5%, due in 1907. *Expenses:* operating, $2,679; int., $3,417; taxes, $544; total, $6,640. *Revenue:* consumers, $5,549; city, $2,087, of which $1,800 is for fire protection; total, $6,836. **Management.**—Prest., Chas. F. Parks, Boston, Mass. Clerk, Chas. Baker. Treas., Elbert Wheeler, 24 Washington St., Boston, Mass. Supt., D. N. Bird.

7. BETHEL, *Oxford Co.* (Pop., 2,077—2,209.) **History.**—Built in '90 by Bethel Water Co.; put in operation Sept. 23. Des. Engr., E. C. Jordan, Portland. Const. Engr., J. J. Moore, Hingham, Mass. **Supply.**—Impounding reservoir on mountain stream, by gravity. **Reservoir.**—Formed by dam of stone masonry. **Distribution.**—*Mains,* 8-in. c. i., 8 miles. *Services,* ¾-in. galv. i. Pipes and specials from R. D. Wood & Co., Philadelphia, Pa. *Taps,* 75. *Hydrants,* Mathews, 25. *Valves.* Eddy. *Pressure,* 100 to 240 lbs. **Financial.**—*Cost,* $42,000. *Cap. stock,* authorized, $60,000. *Bonded debt,* $35,000 at 5%. *Hydrant rental,* $32. **Management.**—Prest., Enoch Foster. Secy. and Treas., A. E. Herrick.

8. BIDDEFORD and SACO, *York Co.* (Pop.: Biddeford, 12,651—14,443; Saco, 6,389—6,057.) Built in '84-85 by Biddeford & Saco Water Co.; city neither controls rates nor has right to purchase works. **Supply.**—Saco River, pumping to reservoir; filtered. Fiscal year closes Dec. 31. During '90 a 4,500,000-gall. Worthington comp., dup. pump was added.

Year.	Mains.	Taps.	Moters.	Hydrants.	Consumption.	Bonded debt.	Int. charges.	Rev.*
'86	10	400	0	102	unk'n	$300,000	$15,000	$5,100
'87	23	1,300	6	109	778,074	300,000	15,000	5,450
'88	24	1,365	8	111	1,100,000	3,0,000	15,000	5,700
'89	29	1,549	10	118	1,382,491	300,000	15,000	6,844
'90	30.3	1,708	10	126	1,820,921	291,000	14,550	7,308

* From city.

Financial.—*Cost,* $596,774. *Cap. Stock:* Authorized, $200,000, all paid up. *Bonded Debt,* $291,000 at $5%. *Revenue;* City, $7,308. **Management.**—Prest., D. W. Clark. Treas., Geo. P. Westcott. Both of Portland. Supt., James Burnie.

Booth Bay Harbor, *Lincoln Co.,* (Pop., 1,000-1,600.) Jesse W. Starr, Jr., Philadelphia, Pa., contracted to build works as described in 1889-'90 Manual, but failed to fulfill his contract and project has been abandoned.

9. BRADLEY, *Penobscot Co.* (Pop., 829—823.) Supply being introduced in '01 by Public Works Co., Brewer.

9. BREWER, *Penobscot Co.* (Pop., 3,170—4,193; pop. supplied, 10,995—14,603).
PUBLIC WORKS CO.

In '91 Public Works Co. acquired plants of Brewer Water Co. and Penobscot Water & Power Co., see below. Co. also owns electric street railways and electric light and power systems of Bangor and Brewer. **Management.**—Prest., F. M. Laughton, Bangor. Secy., F. H. Clergue, Bangor. Treas., N. H. Wardwell.

BREWER PLANT.

Built in '83 by Brewer Water Co.; now owned by Public Works Co., see above. Franchise does not regulate rates, but provides that town may buy works at any time by paying actual cost. **Supply.**—Penobscot Water & Power Co., described below under Oldtown and Veazie plants. Until '90 supply was from Bangor waterworks. Distribution.—*Mains*, 12 miles. *Taps*, 59.5. *Meters*, 1. *Hydrants*, 49. **Financial.**—*Revenue*, consumers, $4,500; city, $1,500; total, $6,000. **Management.** —See Public Works Co., above.

OLDTOWN AND VEAZIE PLANTS.

Built in '90 by Penobscot Water & Power Co., 20-year franchise; now controlled by Public Works Co., see above. Furnishes water to Brewer plant and supplies Oldtown, Veazie and Upper Stillwater, and by Oct. 1, '91, expects to supply Orono, Milford and Bradley from Oldtown plant, Upper Stillwater also being supplied from this plant. Veazie plant put in operation Oct. 1, '90, and Oldtown, Nov. 1, '90. **Supply.**—Penobscot River, filtered, pumping direct; stand-pipe was to be added in '91. **Filters.**—"Home-made structures" were to be added in '91. **Pumping Machinery**—Dy. cap., 6,000,000 galls. *Veazie Plant*, 2,000,000 Holly rotary; 2,000,000 Worthington, added in '91; water power, but steam was to be added in '91. *Oldtown Plant*, 2,000,000 Worthington, driven by 150 HP. Holmes water wheel, all in Bangor electric light and power station. **Distribution.**—*Mains*, 12 to 4-in., 27 miles. *Services*, galv. i. *Hydrants*, Mathews, 116. *Valves*, Eddy. *Pressure*, ordinary, 100 lbs.; fire, 116 lbs. **Financial.**—Withheld. **Management.**—See Public Works Co., above.

10. **BRUNSWICK,** *Cumberland Co.* (Pop., 5,384-6,012.) Built in '86-7 by Pejepscot Water Co., now controlled by Maine Water Co.; franchise does not regulate rates, but provides for purchase of works by city in '91-2. **Supply.**—Androscoggin River and springs, collected in reservoir, pumping to stand-pipe; filtered. Fiscal year closes Dec. 31. **Distribution.**—*Mains*, 11 miles. *Taps*, 419. *Hydrants*, 61. *Consumption*, 226,358 galls. **Financial.**—*Cap. Stock*: authorized, $100,000. *Bonded debt*, 100,000 at 5%. **Management.**—Prest., S. J. Young. Clerk, Geo. F. West. Treas., Geo. P. Westcott. Supt., J. R. Simpson. Gen. Mgr., A. S. Hall.

11. **BUCKSPORT,** *Hancock Co.* (Pop., 3,947—2,921.) Built in '89 by Bucksport Water Co.; franchise provides neither for control of rates nor purchase of works by town. **Supply.**—Great pond, by gravity. Works were put in operation Dec. 1, '89. **Distribution.**—*Mains*, 3 miles. *Hydrants*, 10. **Financial.**—*Cost*, $20,000. *Cap. Stock*: authorized, $40,000; paid-up, $20,000. *No debt*. *Operating expense*, $300. *Revenue*: consumers, $1,200; city, none; hydrant rental, free. **Management.**—Prest. and Treas., Parker Spofford. Secy., O. P. Cunningham. Supt., Wm. H. Lee.

12. **CALAIS,** *Washington Co.* (Pop., 6,173—7,290.) Built in '86 by Calais Water Co., now controlled by Maine Water Co.; city neither controls rates nor has right to purchase works. **Supply.**—St. Croix River, by gravity; filtered. Fiscal year closes Dec. 31. **Distribution.**—*Mains*, 15 miles. *Taps*, 380. *Meters*, 0. *Hydrants*, 65. *Consumption*, 140,000 galls. **Financial.**—*Capital Stock*: authorized. $200,000. *Bonded Debt*: $120,000 at 5%. *Revenue*: city, $3,250. **Management.**—Prest., Weston Lewis. Secy. and Treas., Josiah S. Maxcy. Both of Gardiner. Supt., W. E. McAllister.

30. **CAMDEN,** *Knox Co.* (Pop., 4,386—4,621.) Supply from Rockland.

13. **CARIBOU,** *Aroostook Co.* (Pop., 2,756—4,087.) Built in '89 by the Caribou Water Co., under 20 years' franchise; town neither controls rates nor has right to buy works. **Supply.**—Aroostook river, pumping to tank and direct. Fiscal year closes Dec. 31. **Distribution.**—*Mains*, 4 miles. *Taps*, 86. *Meters*, 0. *Hydrants*, 25. *Consumption*, 80,000 galls. **Financial.**—*Cost*, $100,000. *Cap. Stock*: authorized, $60,000; all paid up. *Bonded Debt*, $80,000 at 5%. *Revenue*: city, $2,000. **Management.**—Prest., Weston Lewis, Gardiner. Treas., J. S. Maxcy, Gardiner. Secy., H. M. Heath, Augusta. Supt., A. B. Fisher.

27. **CUMBERLAND MILLS,** *Cumberland Co.* (Pop. of town, 1,619—1,487.) Supply from Portland.

27. **DEERING,** *Cumberland Co.* (Pop., 1,619—1,487.) Supply from Portland.

14. **DIAMOND ISLAND,** *Cumberland Co.* (Pop., est., 400.) Built in '87 by Diamond Water Co., under absolute and perpetual charter. **Supply.**—Springs, pumping to stand-pipe. **Distribution**—*Mains*, 4 miles. *Consumption*, 5,000 galls. **Financial.**—*Cost*, $10,000. *Cap. Stock*, $5,000. *Bonded Debt*, $5,000 at 6%. **Management.**—Prest., F. M. Lawrence, Sec. and Treas., Seth L. Larabee, Portland.

For FURTHER DESCRIPTIVE MATTER see Manual for 1889-90. CHANGES since 1890 are here given. POPULATIONS are for 1880 and 1890, respectively.

15. DOVER and FOXCROFT, *Piscataquis Co.* (Pop.: Dover, 1,647—1,942; Foxcroft town, 1,263—1,726.) Built in '87 by Dover and Foxcroft Water Co. under 40 years' franchise, now controlled by Maine Water Co.; city neither controls rates nor has right to purchase works. Supply.—Piscataquis River, pumping to reservoir Fiscal year closes Dec. 1. Distribution.—*Mains,* 7 miles. *Taps,* 240. *Meters,* 0. *Hydrants,* 30. *Consumption,* 47,261 galls. Financial.—*Cap. Stock:* authorized, $100,000. *Bonded Debt,* $90,000 at 5%. Management.—Prest., Weston Lewis. Treas., Josiah S. Maxcy. Both of Gardiner. Secy. and Supt., Jno. P. Crocker, Dover.

16. EASTPORT, *Washington Co.* (Pop., 4.006—4,908.) Built in '89 by Eastport Water Co.; franchise does not regulate rates, but provides that city may purchase works at any time within 20 years from date of construction. Prest., N. B. Nutt. Address C. A. Lockwood, 52 Broadway, N. Y. City.
NOTE.—Co. ignored all inquiries sent in '90 and '91.

17. ELLSWORTH, *Hancock Co.* (Pop., 5,052—4,804.) Built in '89-'90 by Ellsworth Water Co. Supply.—Branch Pond stream, pumping to stand-pipe; filtered through sand and gravel. Fiscal year closes Dec. 31. Distribution.—*Mains,* 7 miles. *Taps,* 54. *Meters,* 0. *Hydrants,* 53. Financial.—*Cap. Stock:* authorized, $100,000; all paid up. *Bonded Debt,* $75,000 at 5%. *Revenue:* city, $1,800 for 50 hydrants, $40 for each additional. Management.—Prest., Nath. Cleaves. Secy., Geo. P. Westcott. Both of Portland. Supt., L. H. Cushman.

36. FAIRFIELD, *Somerset Co.* (Pop., 3,044—3,510.) Supply from Waterville.

18. FORT FAIRFIELD, *Aroostook Co.* (Pop., est., 2,807—3,526.) Built in '89 by Frontier Water Co.; franchise neither regulates rates nor provides for purchase of works by city. Supply.—Springs, by gravity. Fiscal year closes Sept. 10. Distribution.—*Mains,* 4 miles. *Taps,* 66. *Meters,* 2. *Hydrants,* 17. Financial.—*Cost,* $30,000. *Cap. Stock:* authorized, $30,000. *Bonded Debt,* $30,000 at 5%. *Expense:* operating, nominal; int., $1,500; taxes, none. *Revenue:* consumers, $827; city, $850; total, $1,677. Management.—Prest., Nicholas Fessenden. Secy. and Supt., Edward L. Houghton. Treas., J. F. Hocker.

15. FOXCROFT, *Piscataquis Co.* (Pop., 1,263—1,726.) Supply from Dover.

19. FREEPORT, *Cumberland Co.* (Pop., 2,579—2,482.) History.—Construction began about July 1, '91, by Freeport Water Co., under franchise granted May 23; to be in operation by Dec., '91. Town has right to buy at intervals of five years at appraised value. Supply.—Frost Gully Brook, filtered, pumping from reservoir to stand-pipe. Dam.—Stone, 60 ft. long, 5 ft. high. Filter.—Gravel. Pumping Machinery.—Two pumps. Stand-Pipe.—Capacity, 130,000 galls., 35x60 ft. *Mains,* 10 to 4-in. c. i., not less than 2.7 miles. *Hydrants,* 18. Financial.—*Cap. Stock,* $50,000. Hydrant rental, total, $1,000. Management.—Prest., E. B. Mallett, Jr. Treas., W. H. Soule.

20. FRYEBURG, *Oxford Co.* (Pop. 1,632—1,418.) Built in '82 by Fryeburg Water Co. Supply.—Mountain stream, by gravity from reservoir. Fiscal year closes May 1. Distribution.—*Mains,* 6½ miles. *Taps,* 130. *Meters,* 0. *Hydrants,* 20. *Consumption,* 17,000 galls. Financial.—*Cost,* $17,000. *Cap. Stock:* paid up, $17,000; paid 10% div'd in 1890. *Expenses:* operating, $60; no int. nor taxes. *Revenue:* consumers, $130; city, $100; total, $230. Management.—Prest. and Secy., T. C. Shirley. Treas., D. L. Lamson. Supt., A. R. Jenness.

21. GARDINER and RANDOLPH, *Kennebec Co.* (Pop.: Gardiner, 4,439—5,491; Randolph, 1,281). Built in '85 by Gardiner Water Co.; city neither controls rates nor has right to purchase works. Supply.—Cobbossee Contec River, pumping from impounding to distributing reservoir. Fiscal year closes Dec. 31.

Year.	Mains.	Taps.	Meters.	Hydrants.	Consumption.	Bonded debt.	Int. ch'gs.	Rev.*
'86	10	350	0	17	60,000	$100,000	$5 000	$2,500
'88	14	726	4	64	200,000	120,000	6,000	3,200
'90	14	870	0	64	208,493	120,000	6,000	3,200

*From city.

Financial.—*Cap. Stock:* authorized, $250,000; paid no div'd in 1890. *Bonded Debt:* $120,000 at 5%. Management.—Prest., Weston Lewis. Secy. and Treas., Josiah S. Maxcy. Supt., Gustavus Moore.

37. GOULDSBORO, *Hancock Co.* See Winter Harbor.

37. GRINDSTONE NECK, *Hancock Co.* See Winter Harbor.

MAINE.

22. HALLOWELL, *Kennebec Co.* (Pop., 3,154—3,181.) Built in '76 by Hallowell Water Co. **Supply.**—Springs, collected in reservoir, thence by gravity to village; filtered. Fiscal year closes July 1. **Distribution.**—*Mains*, 5½ miles. *Taps*, 160. *Meters*, 0. *Hydrants*, 1. **Financial.**—*Cost*, $21,000. *Cap. Stock:* authorized, $4,000. *Debt:* bonded, none; floating, $3,400 at 4%. *Expenses:* int.. $114; taxes, $65. *Revenue:* consumers, $1,400. **Management.**—Prest., Jas. F. Bodwell. Secy. and Supt., Orlando Currier.

23. HOULTON, *Aroostook Co.* (Pop., 3,228—4,015.) Built in '85 by Houlton Water Co., under perpetual charter; city neither controls rates nor has right to purchase works. **Supply.**—Surface water, pumping to stand-pipe; filtered through gravel crib in stream at point of supply. Fiscal year closes March 31.

Year.	Mains.	Taps.	Meters.	Hydrants.	Consumption.	Cost.	Exp.	Rev.
'86	5	100	0	32	18,630	$50,000	$2,500	$2,400
'89-90	5	182	2	33	81,910	55,000	2,245	4,656
'90-1	51	190	5	33	92,425	55,000	2,441	4,800

Financial.—*Cap Stock*: authorized, $50,000; all paid up; paid 5% dividend 1890. *No Debt*. *Expenses*: operating, $2,200; taxes, $45; total, $2,245. *Revenue:* consumers, $3,300; city, $1,500; total, $4,800. During '90 hydrant rental was increased to $1,800, under 10 years' contract. **Management.**—Prest., Walter Mansur. Treas. and Supt., J. F. Holland.

Kennebunk, *York Co.* (Pop., 2,852—3,172.) Built by town for fire protection only. See 1889-90 Manual, p. 8.

1. LAMOINE BEACH, *Hancock Co.* (Pop., not given.) Summer resort. Built in '86-7 by East Lamoine Water Co. **Supply.**—Blunt's Pond, by gravity. Pond is about 1¼ miles from first tap and 164 ft. above tide. About ¾ miles street mains supply water to 2 hotels, 8 houses and steamers at ferry wharf during summer. *Cost*, $3,000. *Cap. Stock:* authorized, $10,000; paid-up, $1,700. *No Revenue*, water being supplied to stockholders only. Prest., W. S. Hodgkins, E. Lamoine. Secy., C. F. Walker. Supt., Geo. F. Jordan.

24. LEWISTON, *Androscoggin Co.* (Pop., 19,083—21,701.) Built in '78-9 by city Supply.—Androscoggin River, pumping to reservoir. Fiscal year closes March 1.

Year.	Mains.	Taps.	Meters.	Hydrants.	Consumption.	Cost.	Exp.	Rev.
'79-80	22	1,788	..	141	1,000,000	$498,062	$4,008	$14,626
'81-2	23	1,800		149	1,016,260	516,513	4,717	24,386
'85-6	24.5	1,946	75	149	1,640,000	540,616	4,855	30,581
'88-9	28	2,064	70	155	2,252,148	563,464	7,315	34,236
'89-90	28.2	2,131	67	156	2,462,231	570,175	9,451	35,542
'90-1	28.4	2,176	63	156	2,511,535	573,281	8,211	38,075

Financial.—See above. Int. charges in '90-1, $24,870. **Management.**—Six comrs., two elected every year. Chn., J. H. Day. Clerk, Walter A. Goss. Supt., Isaac C. Downs.

27. LITTLE FALLS, *Cumberland Co.* (Pop., not given.) Supply from Portland.

9. MILFORD, *Penobscot Co.* (Pop., 731-835.) Supply being introduced in '91 by Public Works Co. Brewer.

North Anson, *Somerset Co.* (Pop. of Anson town, 1,444—1,555.) Built in '83 by town and Potter & Watson, jointly, for fire protection. See 1889-90 Manual, p. 9.

25. NORWAY and SOUTH PARIS, *Oxford Co.* (Pop : Norway, 1,467—928; South Paris, 815—1,164.) Built in '86 by National Water-Works Syndicate, Boston, Mass. **Supply.**—Lake and river, pumping to reservoir. Fiscal year closes Dec. 31. **Distribution.**—*Mains*, 9 miles. *Taps*, 450. *Meters*, 2. *Hydrants*, 50. **Financial.**—*Cost*, $62,000. *Cap. Stock*, authorized, $60,000. *Bonded Debt*, $50,000 at 5%, due in 1906. *Expenses:* operating, $1,300; int., $2,500. *Revenue:* consumers, $3,300; city, $2,500; total, $5,800. **Management.**—Prest., W. H. Whitcomb. Secy., H. D. Smith. Supt., W. W. Whitcomb.

26. OLD ORCHARD, *York Co.* (Pop., 877.) Built in '87 by Old Orchard Water Co. under 20-years' franchise. **Supply.**—Spring flowing into reservoir; by gravity, with direct pumping for fire protection. Fiscal year closes Dec. 31. **Distribution.**—*Mains*, 7¼ miles. *Taps*, 144. *Meters*, 1. *Hydrants*, 25. **Financial.**—*Cap. Stock*: authorized, $75,000; paid up, $61,000. *Bonded Debt*, $64,000 at 6%. **Management.**—Prest., A. B. Turner, Boston, Mass. Secy., Clarence Hale, Portland. Treas., J. W. Brown, Boston. Supt., J. W. Duff.

FOR FURTHER DESCRIPTIVE MATTER see Manual for 1889-90. CHANGES since 1890 are here given. POPULATIONS are for 1880 and 1890, respectively.

9. **OLD TOWN,** *Penobscot Co.* (Pop., 3,295—5,312.) Supply from Public Works Co., Brewer.

9. **ORONO,** *Penobscot Co.* (Pop , 2,245—2,790.) July 21, '91, Public Works Co., Brewer, reported supply was being introduced.

27. **PORTLAND,** *Cumberland Co.* (Pop., 33,810—36,425.) Built in '67-9 by Portland Water Co.; franchise provides for purchase of works by city at any time for price mutually agreed upon, or, failing to agree, by the court; no provision is made for regulation of rates, except for large consumers. Supply.—Lake Sebago, by gravity to city. Fiscal year closes Dec. 31. In '90 high-service supply was adopted, and 20,000,000-gall. reservoir completed.

Year.	Mains.	Taps.	Meters.	Hydrants.	Bonded debt.	Int. ch'ges.
'80	65	4,414	...	202	......	......
'82	68.7	4,983	45	262	$1,000,000	$60,000
'84	76.4	5,283	55	248	1,000,000	60,000
'86	90	5,700	134	276	1,300,000	78,000
'88	98	6,700	199	293	1,500,000	90,000
'89	105.5	....	200	299	1,500,000	75,692
'90	107.5	....	237	307	1,500,000	73,734

Financial.—*Cost,* $2,848,503. *Cap. Stock:* authorized, $100,000; all paid up; paid 5% dividend in 1890. Bonded debt, $1,500,000: $200,000 at 6%, due in '91: $200,000 at 6%, due in '99; $1,025,000 at 4%, due in 1927; $75,000 at 5%. *Sinking Fund,* $20,000. *Expenses:* operating, $23,131; int., $73,734; taxes, $10,619; total, $107,488. *Revenue:* consumers, $143,102; city, $4,000; total, $147,102. Water for fire protection is furnished free, in return for privilege of using streets for laying and repairing mains. **Management.**—Prest., D. W. Clark. Secy., E. R. Payson. Treas. and Supt., Geo. P. Westcott.

28. **PRESQUE ISLE,** *Aroostook Co.* (Pop., 2,446—3,046.) Built in '87 by National Water-Works Syndicate, Boston, Mass. Supply.—Kennedy brook, by gravity, with pumping in case of emergency. Fiscal year closes Dec. 31. Distribution.—*Mains,* 4.3 miles. *Taps,* 84. *Meters,* 0. *Hydrants,* 10. *Consumption,* unk'n. Financial.—*Cost,* $30,000. *Cap. Stock,* authorized, $30,000; all paid up. *Bonded Debt,* $30,000 at 5%. *Expenses,* operating, $200; int., $1,500; no taxes; total, $1,700. *Revenue:* consumers, $1,100; city, $600; total, $1,700. **Management.**—Prest. and Supt., G. H Truman. Secy., Sidney Graves. Treas., D. A. Boodey.

21. **RANDOLPH,** *Kennebec Co.* (Pop., in '90, 1,281.) Supply from Gardiner.

29. **RICHMOND,** *Sagadahoc Co.* (Pop., 2,658—3,082.) Built in '86 by Co ; city neither controls rates nor has right to purchase works. Supply.—Kennebec River, pumping to reservoir. Fiscal year closes Jan. 1. Distribution.—*Mains,* 6 miles. *Taps,* 240. *Meters,* 0. *Hydrants,* 41. *Consumption,* 47,225 galls. Financial.—*Cap. Stock,* authorized, $100,000; paid no div'd in 1890. *Bonded Debt,* $60,000 at 5%. **Management.**—Prest., Weston Lewis. Treas., Josiah S. Maxcy, both of Gardiner. Secy. and Supt., Jno. L. Pushard.

30. **ROCKLAND, CAMDEN AND THOMASTON,** *Knox Co.* (Pop., Rockland, 7,559—8,174; Camden, 4,386—4,621; Thomaston, 3,017—3,000.) Built in '51 by Rockland Water Co ; franchise provides neither for control of rates nor purchase of works by city. Supply.—Lake, by gravity to reservoir. Distribution.—*Mains.* 80 miles. *Hydrants,* 112. Financial.—Withheld, except hydrant rental, $54. **Management.**—Prest., A. G. Crockett. Secy. and Treas., W, S. White. Clerk and Supt., J. W. Crocker.

27. **SACARAPPA,** *Cumberland Co.* (Pop., not given.) Supply from Portland.

8. **SACO,** *York Co.* (Pop., 6,389—6,075.) Supply from Biddeford.

31. **SANFORD,** *York Co.* (Pop., 4,201.) Built in '87 by Sanford Light & Water Co. Supply.—Springs, by gravity; and pumping to tank and direct for fire protection. Distribution.—*Mains,* 2 miles. *Taps,* 300. *Meters,* 0. *Hydrants,* 16. Financial.—*Cost,* $12,000. *Cap. Stock,* $10,000. *Debt,* $3,000 at 6%. *Expenses,* operating; $100, int., $180. *Revenue:* consumers, $2,400; city, none. **Management.**—Prest. and Supt., E. M. Goodall. Secy., E. E. Hussey. Treas., Geo. H. Howell.

32. **SKOWHEGAN,** *Somerset Co.* (Pop., 3,860-5,068.) First supplied in '50. See note. Built in '87-8 by Skowhegan Water Co.; city neither controls rates nor has right to purchase works. Supply.—Springs, pumping to stand-pipe. Fiscal year closes July 1. Distribution.—*Mains,* 10 miles. *Meters,* 3. *Hydrants,* 75. *Consumption,* 125,000 galls. Financial.—*Cost,* $116,000. *Total Debt,* $86,500; floating, $8,500 at 6%; bonded, $78,000 at 5%. *Expenses:* operating, $1,656; int., $1,250; taxes, $56; total, $5,962. *Revenue:* consumers, $2,100; city, $1,734; other purposes, $200; total, $4,034. **Management.**—Prest., R. B. Shepherd. Secy. and Treas., C. M. Brainard.

NOTE.—For description of Aqueduct Co.'s line, see MANUAL for '89-90, p. 13.

MAINE.

33. SORRENTO, *Hancock Co.* (Pop., not given.) Built in '87 by Frenchman's Bay & Mt. Desert Land & Water Co.; franchise provides neither for control of rates nor purchase of works by city. **Supply.**—Long Pond, by gravity, to and from reservoir. **Distribution.**—*Mains*, 6 miles. *Meters*, 0. *Hydrants*, 10. **Financial.**—*Cost*, $20,000. *Expenses*: operating, $1,000; **Management.**—Prest., Chas. H. Lewis; Treas., J. Wesley Kimball, both of Boston. Sec., Chas. P. Simpson.

1. SOUTH NORRIDGEWOCK, *Somerset Co.* (Pop., not given.) Built in '88 by Henry K. Sawyer. **Supply.**—Spring, pumping to tank. In '88 there was one mile of pipe and 50 consumers. No later statistics have been received. Address, Henry K. Sawyer.

25. SOUTH PARIS, *Oxford Co.* (Pop., not given,) Supply from Norway.

31, 35. SPRINGVALE, *York Co.* (Pop., 1,116.)

SPRINGVALE AQUEDUCT CO.

Built in '76; franchise provides neither for control of rates nor purchase of works by city. **Supply.**—Pond, by gravity. **Distribution.**—*Mains*, 4 miles. *Taps*, 150. *Meters*, 0. *Hydrants*, 8. **Financial.**—*Cost*, $11,000. **Management.**—Prest., C. H. Frost. Secy. and Supt., J. B. Ricker.

BUTLER SPRING WATER CO.

Built in '89. **Supply**—Springs, pumping to tank. Fiscal year closes Dec. 31 **Distribution.**—*Mains*, 3½ miles. *Taps*, 150. *Meters*, 0. *Hydrants*, 2. *Consumption*, unknown. **Financial.**—*Cost*, $9,000. *Cap. stock:* Authorized, $9,000; all paid up; no dividend declared, net revenue being used for extensions and improvements. *No debt*, operating expenses or int. *Taxes*, $6. *Revenue:* Consumers, $900; city, none. **Management.**—Prest., J. A. Butler. Secy. and Supt., F. A. Clark.

30. THOMASTON, *Knox Co.* (Pop., 3,017–3,009.) Supply from Rockland.

9. UPPER STILLWATER, *Penobscot Co.* (Pop., not given.) Supply introduced Nov. 1, '90. See Public Works Co., Brewer.

9. VEAZIE, *Penobscot Co.* (Pop., 622–650.) Supply from Public Works Co., Brewer.

36. WATERVILLE and FAIRFIELD, *Kennebec Co.* (Pop.: Waterville, 4,672–7,104; Fairfield, est., 5,000.) Built in '87 by Waterville Water Co., under 20 years franchise; city neither controls rates nor has right to purchase works. **Supply.**—Messalonskee River, pumping to reservoir. Fiscal year closes Jan. 1. **Distribution.**—*Mains*, 20 miles. *Taps*, 780. *Meters*, 6. *Hydrants*, 80. *Consumption*, 300,000 galls. **Financial.**—*Cap. Stock*, authorized, $200,000; paid no dividend in 1890. *Bonded debt*, $200,000 at 5%. **Management.**—Prest., Arthur Sewall, Bath. Secy. and Supt., J. F. Nash. Treas., Josiah S Maxcy Gardiner.

27. WESTBROOK, *Cumberland Co.* (Pop., 3,981–6,632.) Supply from Portland.

37. WINTER HARBOR AND GRINDSTONE NECK, *Hancock Co.* (Pop. of town, 1,825–1,709.) **History.**—Built in '91 by Grindstone Neck Water Co., put in operation June 15. Engr. and Contr., I. S. Cassin, Philadelphia, Pa. **Supply.**—Brick-Harbor Pond, pumping to stand-pipe. **Pumping Machinery.**—Dy. cap., 350,000 galls.; Holly comp. dup. **Stand-Pipe.**—Cap., 28,000 galls.; 8 × 75 ft. **Distribution.**—*Mains*, 8 and 6-in. c. i. *Hydrants*, Cassin. *Valves*, Cassin. **Financial.**—*Cost*, $25,000. *Cap. Stock*, $25,000. No debt. **Management.**—Prest., Jno. B. Lunig. Clerk, Chas. C. Hutchings. Treas., C. B. Taylor, 903 Walnut St., Philadelphia, Pa.

Winthrop, *Kennebec Co.* (Pop., 2,146–2,111.) Built in '83 by village and private subscription, for fire protection only. See 1889-90 Manual, p. 15.

11. YORK BEACH, *York Co.* (Pop. of town, 2,463-2,444.) Built in '83 by York Beach Water Co. **Supply.**—Artesian wells, pumping direct and to tank. **Distribution.**—*Mains*, 3 miles. *Taps*, 60 *Hydrants*, 0. **Financial.**—*Cost*, $8,000. *No Debt.* **Management.**—Supt., H. E. Evans.

Water-Works Projected with Fair Prospects of Construction.

KENNEBUNK, *York Co.* (Pop., 2,852–3,172.) Mousam Water Co. was organized in '91 under charter granted by Legislature. Prest., Sydney T. Fuller. Clerk, Walter L. Dane. Treas., Frank M. Ross.

FOR FURTHER DESCRIPTIVE MATTER see Manual for 1889-90. CHANGES since 1890 are here given. POPULATIONS are for 1886 and 1890, respectively.

MADISON, *Somerset Co.* (Pop., 1,315–1,815.) Village meeting was to be held July 25 to see if $1,200 yearly would be raised for hydrants.

PHILLIPS, *Franklin Co.* (Pop., 1,437–1,394.) In '91 charter was granted, and Phillips Water & Electric Co. organized to construct water and electric-lighting plant. Address Mason Parker or J. W. Brackett.

PITTSFIELD, *Somerset Co.* (Pop., 1,909–2,503.) Pittsfield Water Co. organized July, '91, with W. R. Hunnewell, Prest., J. F. Connor, Secy. and Treas. Town may build, otherwise Co. proposes to do so. Town already has small fire protection works.

SEAL HARBOR, *Hancock Co.* (Pop., not given.) Seal Harbor and Shore Front Water Co. secured charter from State in '90. Supply to be taken from Jordan's Lake. June 4, '91, it was reported that work would not be commenced at present. Prest., J. C. Gardner, Albany, N. Y. Engr., D. E. Kimball. See Manual for '88, p. 12; also Manual for '89–90, p, 15.

SULLIVAN HARBOR AND WEST SULLIVAN, *Hancock Co.* (Pop. of town, 1,023–1,379.) **Present Supply.**—Wells, pumping from large spring to tanks on high ground. **Proposed Works.**—See Manual for '89–90, p. 15. June 4, '91, Sullivan Harbor Land Co., 95 Milk St., Boston, Mass., reported that construction of water-works was being considered by committee.

Water-Works Projects Reported in '90 and '91, but now Indefinite.

Cape Elizabeth, Cumberland Co.

NEW HAMPSHIRE.
Water-Works Completed or in Process of Construction.

1. **BARTLETT,** *Carroll Co.* (Pop., 1,041–1,247.) Built in '86-7, by Bartlett Water Co. **Supply.**—Mountain brook, by gravity. **Distribution.**—*Mains,* 1.8 miles. *Taps,* 70. *Hydrants,* 17. **Financial.**—*Cost,* $4,000. *Cap. Stock,* $3,000. *Expenses,* $100. **Management.**—Prest., N. P. Watson. Secy. and Treas., C. H. George. Supt., Frank George.

2. **BETHLEHEM,** *Grafton Co.* (Pop., 1,400–1,267.) Built in '79 by Crystal Springs Water Co. **Supply.**—Springs, by gravity, from receiving and distributing reservoirs. **Distribution.**—*Mains,* 4.7 miles. *Consumers,* 55. *Hydrants,* 9. **Financial.**—*Cost,* $21,000. *No Debt. Operating Expenses,* $400. *Revenue:* total, $1,400. **Management.**—Supt., J. S. Blandin.

3. **BRISTOL,** *Grafton Co.* (Pop., 1,352–1,524.) Built in '86 by Bristol Aqueduct Co.; franchise provides neither for control of rates nor purchase of works by city. **Supply.**—Newfound lake, by gravity. Fiscal year closes June 1.

Year.	Mains.	Taps.	Hydrants.	Cost.	Bonded Debt.	Exp.	Rev.
'87-8	5	69	32	$22,000	none	$400	$1,900
'88-9	5	103	32	22,230	"	200	2,023
'90-1	5.5	135	33	23,370	"	460	2,287

Financial.—*Cap. Stock:* authorized, $22,000; all paid up; paid 6% div'd in 1890. *No Debt. Expenses:* operating, $159; no int.; taxes, $301; total, $460. *Revenue:* consumers, $1,357; city, $930; total, $2,287. **Management.**—Prest., Benj. F. Perkins. Treas., Geo. M. Davis. Secy. and Supt., Jno. H. Brown.

4. **CANAAN,** *Grafton Co.* (Pop., 1,762–1,417.) **History.**—Built in '90 by Crystal Lake Water Co.; put in operation Jan. 1, '91. Engrs. and Contrs., Bartlett, Gay & Young, Manchester. **Supply.**—Crystal Lake, by gravity. **Distribution.**—*Mains,* 12 to 6-in, c. i., 1.5 miles. *Hydrants,* Ludlow, 9. *Valves,* Ludlow. **Financial.**—*Cost,* $13,000. *Cap. Stock:* authorized, $20,000; paid up, $13,000. *Hydrant Rental,* $25. **Management.**—Prest., G. W. Murray. Secy., C. O. Barney. Treas., A. E. Barney.

5. **CLAREMONT,** *Sullivan Co.* (Pop., 4,704–5,595.) Built in '87-8 by Claremont Water-Works Co. **Supply.**—Two brooks, by gravity. Fiscal year closes Dec. 31. **Distribution.**—*Mains,* 10 miles. *Taps,* 322. *Meters,* 32. *Hydrants,* 44. **Financial.**—*Cap. Stock:* authorized, $150,000; paid-up, $55,000; paid no div'd in 1890. *Bonded Debt,* $75,000 at 5%, due '92–'97. *Expenses:* operating, $560; int., $3,750; taxes,

$495; total. $4,805. *Revenue:* consumers, $2,556; city, $1,435; other purposes, $75; total, $4,066. **Management.**—Prest., Chas. H. Bartlett. Secy., Edw. H. Carpenter. Treas., Jas. A. Weston. All of Manchester. Supt., Jas. L. Rice.

6. **COLEBROOK,** *Coos Co.* (Pop., 1,580–1,736.) **History.**—Built in '79–'82 by G. W. Annis. **Supply.**—Springs, by gravity to reservoirs. **Reservoirs.**—Combined cap., 60,000 galls.; one 42,000-gall. and one 18,000-gall. **Distribution.**—*Mains*, 5 to 3-in. w. i., 3 miles. *Services*, galv. i. *Taps*, 70. *Hydrants*, Chapman, 9. *Valves*, Chapman. *Pressure:* ordinary, 20 lbs.; fire, 50 lbs. **Financial.**—Cost, $10,000. *Expenses*, $150. *Revenue:* consumers, $700; city, $100; total, $800. **Management.**—Owner, G. W. Annis.

7. **CONCORD, WEST CONCORD AND PENACOOK,** *Merrimack Co.* (Pop., Concord, 13,483–17,004; West Concord and Penacook, not given.) Built in '72-3 by City of Concord and extended to West Concord and Penacook in '87. **Supply.**—Penacook Lake, by gravity. Fiscal year closes Dec. 31. In '91, plan was submitted by Percy M. Blake, Engr., Hyde Park, Mass., for improving and increasing water supply. In July, '91, it was unofficially reported that a 2,000,000-gall. high-service pumping station would be built, a 20-in. force and 14 and 12-in. supply main laid.

Year.	Mains.	Taps.	Meters.	Hydr'ts.	Cost.	B'd'd debt.	Int Ch'gs.	Exp.	Rev.
'74	23.25			92	$351,293	$350,000	$21,000		$13,035
'76	24.75	1,261		93		350 000	20,898	$2,476	14,501
'78	25.5	1,575		94		350,000	21,000	2,317	17,370
'80	26.23	1,691		95		350,000	20,922	2,474	22,215
'82	29.84	1,774		100		350,000	21,000	3,267	21,243
'84	30.7	1,877		104		393,000	22,720	2,614	22,916
'86	32.35	1,920		113	371,335	393,000	22,720	3,142	24,863
'88	42.33	2,297	30	170	487,798	435,000	24,400	3,172	32,441
'89	42.94	2,309	30	174	492,265	435,000	24,455	3,315	34,238
'90	43.89	2,427	30	183	499,967	435,000	24,400	3,390	35,727

Financial.—*Bonded Debt*, $135,000; $350,000 at 6%, due '92-5; $85,000 at 4%, due '95-1912. *No Sinking Fund. Expenses:* total, $27,790. **Management.**—Seven comrs., appointed every three years. Supt., V. C. Hastings, elected by mayor and aldermen.

8. **CONWAY,** *Carroll Co.* (Pop., 2,094–2,331.) **History.**—Built in '90 by Conway Aqueduct Co; put in operation Nov. 1. Engr. and Contr., F. A. Snow, Providence, R. I. **Supply.**—Impounding reservoir on Cold Brook, by gravity. Impounding **Reservoir.**—Cap., 101,000 galls.; 90 × 40 ft., with av. depth of 4 ft.; paved with stone. *Dam*, of earth, 70 ft. long, 8 ft. high, 6 ft. wide on top, sloping 1½ to 1. Spillway is in center of dam and carries entire stream that feeds reservoir. **Distribution.**—*Mains*, c. i., 3.5 miles. *Services*, galv. i. Pipes and specials from R. D. Wood & Co., Phillipsburgh, N. J. *Taps*, 60. *Hydrants*, Mathews, 21. *Valves*, Eddy. *Pressure*, 43 lbs. **Financial.**—*Cost*, $13,000. *Cap. Stock:* authorized, $15,000; paid-up, $8,500. *Bonded Debt*, $4,500 at 6%. *Expenses:* operating, $150; int., $270; total, $420. *Revenue:* consumers, $600; city, $400; total, $1,000. *Hydrant rental*, $20 **Management.**—Prest., Wm. S. Abbott. Sec. and Treas., Jno. C. L. Wood. Supt., B. F. Clark.

9. **DERRY,** *Rockingham Co.* (Pop., 2,140–2,604.) **History.**—Built in '90, by Derry Water-Works Co., under perpetual franchise. Put in operation Nov. 22. Engrs. and Contrs., Wheeler & Parks, Boston, Mass. **Supply.**—Driven wells, pumping to stand-pipe. **Pumping Machinery.**—Dy. cap., 1,500,000 galls.; two 750,000 Knowles dup., one compound and one high-pressure. *Boilers:* from Cunningham Iron Works, Charlestown, Mass. **Stand-Pipe.**—Cap., 185,000 galls.; 30 × 35 ft.; from E. Hodge & Co., E. Boston, Mass. **Distribution.**—*Mains*, 10 to 4-in. c. i., 4.8 miles. Pipe of M. J. Drummond, N. Y. City. *Taps*, 93. *Meters*, Hersey, 2. *Hydrants*, Coffin, 40. *Valves*, Coffin. *Pressure*, 40 to 80 lbs. **Financial.**—*Cost*, $85,000. *Cap. Stock:* authorized, $50,000; all paid up. *Bonded Debt:* authorized, $50,000; issued, $35,000 at 5%. *Expenses*, not determined. *Revenue:* consumers, $1,354; city, $2,000; total, $3,354. **Management.**—Prest., Wm. S. Pillsbury. Clerk, Fred. R. Felch. Treas., F. J. Shepard, Supt., F. C Haake.

10. **DOVER,** *Stafford Co.* (Pop., 11,687–12,790.) Bought by city in '88. First supplied in '26. When city bought works three co's supplied town. **Supply.**—Springs, gathered in two wells, reinforced by Willard pond, pumping to reservoir. Fiscal year closes Dec. 31

FOR FURTHER DESCRIPTIVE MATTER see Manual for 1889-90. CHANGES since 1890 are here given. POPULATIONS are for 1880 and 1890, respectively.

10   MANUAL OF AMERICAN WATER-WORKS.

Year.	Mains.	Taps.	Meters.	Hy-drants.	Con-sumption.	Cost.	Bonded debt	Int. Ch'gs.	Exp.	Rev.
'89	21.3	1,378	66	146	382,679	$352,004	$555,000	$11,695	$9,567	$21,774
'90	21.8	1,461	122	149	436,347	356,846	355,000	15,100	8,020	29,420

Financial.—$67,500 was paid to Cocheco Aqueduct Co., $20,000 to Dover Aqueduct Co., and $13,000 to Dover Landing Aqueduct Co. for their works. *Bonded Debt*, $355,000 at 4½%. *Expenses and Revenue*, see above. Net revenue, after all expenses are paid, is applied to debt. **Management.**—Three comrs. Clerk, Chas. A. Fairbanks. Supt., W. D. Taylor, appointed by comrs.

I. **DREWSVILLE**, *Cheshire Co.* (Pop., est., 100.) First supplied in '04, rebuilt in '76 by Drewsville Water Co. **Supply.**—Spring, by gravity. **Distribution.**—*Mains*, logs, 300 rods. *Consumers*, 20. *Hydrants*, none. **Financial.**—*Cost*, $1,000. *Op. exp.*, $10. *No rev.*, expenses being divided among shareholders. **Management.**—Prest., B. Lowell. Secy., J. Fisher.

33. **EAST ROCHESTER**, *Strafford Co.* In town of Rochester, which see.

11. **EXETER**, *Rockingham Co.* (Pop. 3,569—4,284.) Built in '86-7 by Exeter Water Co. **Supply.**—Well and stream, pumping to stand-pipe; National filter. Fiscal year closes Dec. 31. **Distribution.**—*Mains*, 10.5 miles. *Taps*, 391. *Meters*, 8. *Hydrants*, 51. *Consumption*, 88,611 galls. **Financial.**—*Cost*, $148,041. *Cap. Stock*, $90,000. *Total Debt*, $56,000; floating, $7,000, at 6%; bonded, $49,000, at 5%. One $1,000 bond canceled in '90. *Expenses:* operating, $2,115; int., $2,982; taxes, $325; total, $5,622. *Revenue:* consumers, $4,920; city, $2,040; meters, $342; total, $7,302. **Management.**—Prest., E. G. Eastman. Secy., A. O. Fuller. Treas., Elbert Wheeler, 24 Washington St., Boston, Mass. Supt., Chas. H. Johnson.

12. **FARMINGTON**, *Strafford Co.* (Pop., 3,044—3,064.) Built in '71 by city. **Supply.**—Springs, pumping to reservoir. Fiscal year closes March 31. **Distribution.**—*Mains*, 3 miles. *Taps*, 160. *Meters*, 31. *Hydrants*, 28. **Financial.**—*Expenses*, $900. *Revenue*, $1,100. **Management.**—Three Comrs., Wm. F. Thayer, W. H. Plummer, F. E. Farwell.

13. **FRANKLIN and FRANKLIN FALLS**, *Merrimack Co.* (Pop., 3,265—4,085.) **History.**—Town works to be put in operation Oct. 1, '91. First supplied by Cold Brook Spring Co., date built not given. Town bought works for $25,000, taking possession July 1, '91. Engr., F. L. Fuller, Boston, Mass. **Supply.**—Springs, pumping to reservoir. **Pumping Machinery.**—Dy. cap., 1,000,000 galls.; water power pump, with 11½-in. plunger, 10-in. stroke, made by Holyoke Machine Co.; also a Deane steam pump. **Reservoir.**—Cap., 484,000 galls.; covered; 69 ft. in diam. at bottom, 71 at top, water 17 ft. deep. Masonry wall, covered with brick arches. Arches sprung from two concentric rows of piers, about 12 ft. on centers, and central portion covered with a "dump" or dome about 24 ft. in diam.; arches have 1½ ft., dome 3¼ ft. rise; arches have 8-in. ring. Above subject to correction, as description was given before completion. See Journal N. E. W. Wks. Assn., June, 1891. **Distribution.**—*Mains*, 12 to 4-in. c. i. Pipe from McNeal Pipe and Foundry Co., specials from Builder's Iron Foundry, Providence, R. I. *Hydrants*, 55. *Valves*, Ludlow. **Financial.**—Cost, $80,000 of which $25,000 was paid for Cold Brook Spring system and water supply. *Bonded Debt*, $80,000 at 4%. **Management.**—Prest., Warren P. Daniell, Secy., Frank W. Parsons.

13. **FRANKLIN FALLS**, *Merrimack Co.* See Franklin.

14. **GOFFSTOWN**, *Hillsborough Co.* (Pop., 1,699-1,981.) **History.**—Built in '91 by village. Contrs., Bartlett, Gay & Young, Manchester. **Supply.**—Impounding reservoir on mountain stream, by gravity. **Distribution.**—*Mains*, 12 to 4-in. c. i. Pipe from Donaldson Iron Co., Emaus, Pa. *Hydrants*, Ludlow, 35. **Financial.**—*Cost*, about $40,000. **Management.**—Board of four. Chn., G. W. Colby.

NOTE.—Description is subject to correction, as it was furnished immediately after letting of contracts, before construction had begun.

15. **GORHAM**, *Coos Co.* (Pop., 1,383—1,710.) Built in '73, by Alpine Aqueduct Co.; city neither controls rates nor has right to purchase works. **Supply.**—Springs, by gravity, to and from reservoir. Fiscal year closes Nov. 30. **Distribution.**—*Mains*, 3¼ miles. *Taps*, 250. *Meters*, 0. *Hydrants*, 11. **Financial.**—*Cost*, $22,033. *Cap. Stock*: authorized, $29,000; div'd, $1,305. *No Debt*. *Expenses:* Operating, $150; taxes, $214. *Revenue:* consumers, $2,244; city, $165; total, $2,409. **Management.**—Acting Prest., E. L. Knight. Secy. and Treas., Thos. Fiske.

16. **GREAT FALLS**, *Strafford Co.* (Pop., 5,660.) Built in '64, by Great Falls M'fg. Co. In '89, Great Falls Water Co. was organized, management being same as

before. **Supply.**—River, pumping to reservoir, stand-pipe and direct. During '90, stand-pipe was erected, 20 × 70 ft., increasing fire pressure to 85 lbs. **Distribution.** —*Mains,* 4 miles. *Taps,* 350. *Hydrants,* 5). *Consumption,* 390,000 galls **Financial.** —*Cost,* $125,000 *No Debt.* **Management.**—Treas., J. Howard Nicholls, Boston, Mass.

NOTE.—In '87 Somersworth and Rollinsford Water Co. was incorporated to supply Gt. Falls and Salmon Falls, villages located in towns of Somersworth and Rollinsford, respectively, but no further steps had been taken June 18, '91. In '91 town of Somersworth was authorized by State Legislature to construct system of water-works, and contemplated so doing.

Greenville, *Hillsborough Co.* (Pop., 1,072-1,255.) Built in '90 by town and Columbia Mfg. Co. for fire protection only. Supply, Sowhegan River, pumping to tank. Cost, $3,900. Chn., Wm. C. Greene.

17. **GROVETON,** *Coos Co.* (Pop., not given.) **History.**—Built in '90 by Groveton Water Co.; put in operation Nov. 15. **Supply.**—Springs, by gravity. **Distribution.**—*Mains,* 3-in., 3 miles. *Taps,* 100. *Consumption,* 10,000 galls. *Pressure:* ordinary, 90 lbs.; fire, 98 lbs. **Financial.**—*Cost,* $6,000. *Cap Stock:* authorized, $6,000; all paid up. *Bonded Debt,* $6,000. **Management.**—Prest., C. C. O'Brien. Secy. and Treas., G. M. Soule. Gen. Mgr., C. T. McNally.

II. **HANOVER,** *Grafton Co.* (Pop., 2,147—1,817.) Built in '20 by Hanover Aqueduct Co.; franchise provides neither for control of rates nor purchase of works by city. **Supply.**—Springs, collected in wells, by gravity. Fiscal year closes first Tuesday in June. **Distribution.**—*Mains,* 2 miles. *Taps,* 150. *Meters,* 0. *Hydrants,* 0. *Consumption,* 6,000 galls. **Financial.**—*Cost,* $12,000. *Cap. Stock,* authorized, $10,000; all paid up; paid 6% div'd in 1890. *No debt. Expenses:* $300. *Revenue:* $1,200. **Management.**—Prest. and Supt., Jno. K. Lord. Secy., Edw. R. Ruggles.

18. **HILLSBOROUGH BRIDGE,** *Hillsborough Co.* (Pop., 1,646—2,120.) Built in '87 by Hillsborough Water Co. **Supply.**—Loon Pond, by gravity. Fiscal year closes May 1. **Distribution.**—*Mains,* 5 miles. *Taps,* 180. **Financial.**—*Cost,* $50,000. *Cap. Stock:* authorized, $25,000. *Bonded Debt,* $25,000 at 5%. Other information withheld. **Management.**—Prest., Jno C. Campbell. Secy. and Supt., G. W. Holman. Treas., C. L. Goodhue.

Hinsdale, *Cheshire Co.* (Pop., 1,803—2,258.) Built in '84 by village, for fire protection only. See 1889-90 Manual, p. 20.

19. **HUDSON,** *Hillsborough Co.* (Pop., 1,045—1,092.) **History.**—Built in '91 by Hudson Water-Works Co.; to be put in operation Oct. 1. **Supply.**—Wells, near pond, pumping to stand-pipe. Pumping Machinery.—Dy. cap., 75,000 galls.; Deane. Stand-Pipe.—Cap., 50,000 galls.; 12 × 60 ft. **Distribution.**—*Mains,* 6 and 4-in., c i., 1.5 miles. *Families Supplied,* about 100. *Hydrants,* 10. **Financial**— *Cost,* $20,000. **Management.**—Prest., G. O. Sanders. Treas., L. P. Sanders.

20. **KEENE,** *Cheshire Co.* (Pop., 6,784—7,446.) Built in '69 by town and enlarged in '89. **Supply**—Two lakes, by gravity. Fiscal year closes Dec. 1.

Year.	Mains.	Taps.	Meters.	Hy-drants.	Con-sumption.	Cost.	Bonded debt.	Int. ch'gs.	Exp.	Rev.
'80	25	724	0	121	Unkn.	$162,700	$150,400	.....	$995	$9,376
'83	26	1,000	0	125	"	166,482	150,400	.....	1,261	10,134
'86	26	1,101	0	131	"	168.632	......	$5,908	1,140	11,485
'87	27	1,187	0	138	"	211,096	131,000	6,765	1,292	12,917
'88	28	1,226	0	144	"	223,501	133,590	6,735	1,600	13,200
'89	28.5	1,261	0	148	"	232,230	113,800	5,553	2,123	13,841
'90	29.1	1,295	9	155	1,300,000	246,379	101,800	4,785	3,051	14,028

Financial.—*Bonded Debt,* $101,800; $29,800 at 6%; $35,000 at 5%; $37,000 at 3½%. *Sinking Fund,* $39,453, deposited in bank at 5%. *Expenses and Revenue,* see above. **Management**—Three comrs., one appointed every year by city council. Chn., E. S. Foster. Supt., P. F. Babbidge.

21 **LACONIA AND LAKE VILLAGE,** *Belknap Co.* (Pop., Laconia, 3,790— 6,143; Lake Village, est., 3,000.) **History.**—Built in '85 by Laconia & Lake Village Water-Works Co.; city neither controls rates nor has right to purchase works. **Supply.**—Lake Winnipiseogie, pumping to reservoir. Fiscal year closes May 31.

FOR FURTHER DESCRIPTIVE MATTER see Manual for 1889-90. CHANGES since 1890 are here give. POPULATIONS are for 1880 and 1890, respectively.

During '90-'91, force-main was laid from pump to reservoir and 2,000,000-gall. Davidson pump added.

Year.	Mains.	Taps.	Meters.	Hydrants.	Consumption.	Cost.	Bonded debt.	Int. ch'g's.	Exp.	Rev.
'86-7	12.14	451	4	76	....	$34,919	$25,000	$1,062	......	$6,521
'87-8	13.13	502	9	78	$333,091	88,789	25,000	1,291	$3,375	8,565
'88-9	13.8	653	390	79	360,982	102,269	40,000	1,359	3,320	9,300
'89-90	14.32	772	432	80	304,241	105,972	45,000	2,125	3,681	10,456
'90-91	16	918	500	84	317,739	124,155	50,000	2,454	4,027	12,939

Financial.—*Cap. St ck:* authorized, $60,000; all paid up; 5% div'd in 1891. May 31, '91, stockholders voted to increase cap. stock to $75,000. *Total Debt:* $57,204; floating, $7,204; bonded, $50,000, at 5% int. *Expenses:* total, $7,081. *Revenue:* consumers, $7,528; city, $2,922; total, $10,450. **Management.**—Prest., Jno. C. Moulton. Treas., Edmund Little. Secy. and Supt., Frank P. Webster.

21. **LAKE VILLAGE,** *Belknap Co.* (Pop., not given.) Supply from Laconia.

III. **LANCASTER,** *Coos Co.* (Pop., 2,721—3,373.)

LANCASTER AQUEDUCT CO.

Rebuilt in '72. **Supply.**—Springs by gravity. **Distribution.**—*Mains*, 6 miles. *Taps*, 100. *No hydrants.* **Financial.**—Withheld. **Management.**—Treas., C. E. Allen.

VILLAGE PLANT.

Built for fire protection. Address W. Marshall.

NOTE.—June 24, '91, it was reported that a company had been chartered to furnish water for fire protection and domestic use. Address J. G. Williams or H. O. Kent.

22. **LEBANON,** *Grafton Co.* (Pop , 3,354—3,763.) Built in '87 by village. **Supply.**—Mascoma River, pumping to reservoir. Fiscal year closes April 1.

Year.	Mains.	Taps.	Meters.	Hydrants.	Consumption.	Cost.	Bonded Debt.	Int. Charges.	Exp.	Rev.
'87-8	5.5	80	0	52	229,000	$45,185	$40,000	$1,800	......	......
'90-1	6.9	287	1	56	300,000	57,863	40,000	1,800	$826	$2,529

Financial.—*Total Debt*, $42,775; floating, $2,775 at 6%; bonded, $40,000 at 4½%, due in 1907. *Expenses:* total, $2,792. *Revenue:* consumers, $2,529; city, $1,680; total, $4,209. **Management.**—Three comrs., one elected every year. Chn., C. E. Lewis. Supt., H. P. Goodrich, appointed by comrs.

23. **LISBON,** *Grafton Co.* (Pop., 1,897—2,060.) Built in '83 by Lisbon Water Works Co. **Supply.**—Pearl Lake, by gravity. Fiscal year closes Dec 31. **Distribution.**—*Mains*, 4 miles. *Taps*, 88, *Hydrants*, 32. **Financial.**—*Cost*, $40,000. *Cap. Stock:* authorized, $20,000; all paid up. *Bonded Debt:* $22,000 at 5%, due in 1908. *Expenses:* operating, $209; int., $1,000; no taxes; total, $1,209. *Revenue:* consumers, $892; city, $775; total, $1,667. **Management.**—Prest., Chas. H. Bartlett. Secy., Edwin H. Carpenter. Treas., Jas. A. Weston. All of Manchester. Supt., L. C. Payne.

24. **LITTLETON,** *Grafton Co.* (Pop. of town, 2,936—3,365.) Built in '79 by Littleton Water & Electric Light Co. Original works built by Apthorp Reservoir Co., in 1880, now controlled by Debenture Guarantee & Assurance Co., Chicago, Ill. **Supply.**—Source not given, by gravity and pumping to reservoir. Fiscal year closes Dec. 31. **Distribution.**—*Mains*, not given. *Taps*, 371. *Hydrants*, 40. *Consumption*, 192,000 galls. **Financial.**—*Cost*, $100,000. *Cap. Stock:* Common; authorized, $75,000, all paid-up. *Bonded Debt*, $75,000, at 6%, due in 1907. *Expenses:* Op. and taxes, $1,500; int., $4,500; total, $6,000. *Revenue:* Consumers. $5,000; city, $3,000; total, $8,000. **Management.**—Prest. and Supt., B H. Corning: Secy. and Treas., Chas. B. Ludwig, 72 Broadway, N. Y. City.

26. **MERIDEN,** *Sullivan Co.* (Pop., not given.) **History.**—Built in '90 by Meriden Water Co. Des. Engr., Downs & Son, Lebanon. Const Engr., E. K. Muller. **Supply.**—Springs, by gravity to reservoir. Springs are 200 ft. above highest point in village. Reservoir.—*Cap.*, 256,000 galls.; lined with granite and cem. **Distribution.**—*Mains*, w. i., 2.5 miles. *Services*, ¾-in., w. i. *Taps*, 46. *Hydrants*, 5. *Consumption*, 1,200 galls. **Financial.**—*Cost*, $3,750. *Cap. Stock*, $4,000. *No Debt.* **Management.**—Prest., A. B. Chellis. Secy., J. S. Wood. Treas., Jno. T. Duncan. Supt., E. K. Miller.

25. **MANCHESTER,** *Hillsborough Co.* (Pop., 32,630—44,126). Built in '72-4 by city. **Supply.**—Lake Massabesic, pumping to reservoir. Fiscal year closes Dec. 31. In '91, high service system was contemplated. M. M. Tidd, Engr., Boston, Mass., was appointed to draw up plans.

NEW HAMPSHIRE.

Year.	Mains.	Taps.	Meters.	Hydrants.	Consumption.	Cost.	Bonded debt.	Int. ch'gs.	Exp.	Rev.*
'74	23.12	625	....	215	1,305,056	$614,010	$......	$40,679	$......	$27,119
'76	30.6	1,239	166	289	1,216,380	708,792	......	......	7,038	38,879
'78	32.9	1,571	226	303	1,289,837	731,808	......	40,679	7,740	48,874
'80	33.4	1,807	280	307	1,180,930	745,717	......	40,679	10,282	57,655
'81	36.8	1,904	328	328	1,180,520	773,841	......	40,679	9,434	60,216
'82	38.5	2,127	371	339	1,213,531	788,438	750,000	38,000	11,506	67,630
'83	39.83	2,294	404	350	1,211,278	799,107	750,000	38,000	9,939	73,458
'84	43.55	2,476	416	371	1,281,265	824,989	......	......	10,878	75,580
'85	45.87	2,630	486	389	1,452,664	851,323	600,000	36,000	15,169	80,404
'86	47.94	2,809	613	404	1,581,483	882,472	600,000	36,000	17,714	75,130
'87	49.41	2,944	730	418	1,635,082	901,274	600,000	30,000	20,542	80,518
'88	51.56	3,087	842	426	1,822,726	924,007	600,000	32,000	14,283	85,397
'89	53.5	3,220	951	441	2,010,023	954,240	600,000	34,186	17,006	86,700
'90	54.9	3,392	1,135	461	1,932,073	965,554	600,000	40,679	22,089	90,232

* Including allowance for fire protection, which was $60 per hydrant per annum until close of '84; then $50 until April 1, '89; then $40.

Financial.—*Bonded Debt*, $600,000; $400,000 at 6%; $200,000 at 4%. *No Sinking Fund.* Expenses and Revenue, see above. **Management.**—Seven comrs., elected by aldermen for 6 years. Prest., Wm. P. Fiske. Secy., Jas. A. Weston. Registrar, Arthur E. Stearns. Supt., Chas. K. Walker.

27. MILFORD, *Hillsborough Co.* (Pop., 2,398—3,014.) Now owned by town. Built in '89 by Milford Water-Works Co.; bought by town in '91. **Supply.**—Driven wells, pumping to reservoir. **Distribution.**—*Mains*, 6 miles. *Taps*, 204. *Meters* 7. *Hydrants*, 42. Financial.—*Cost*, $65,000. *Bonded Debt*, $65,000 at 4%, redeemable at any time after 1911. *Sinking Fund*, $1,200 annually. **Management.**—Three Comrs.; B. R. Crane, F. W. Sawyer, F. W. Sargent.

28. NASHUA, *Hillsborough Co.* (Pop., 13,397—19,311.) Built in '55, by Pennichuck Water Co.; franchise provides neither for control of rates nor purchase of works by city. **Supply.**—Pennichuck brook; pumping direct and to reservoir. Fiscal year closes Oct. 1. Reported early in '91 that a 3,000,000-gall. Gaskill pump and 43-in. Itis bon turbine had been bought.

Year.	Mains.	Taps.	Meters.	Hydrants.	Consumption.	Cost.
'80	23	1,600	1	82	1,000,000	$225,000
'88	35	2,500	50	90	2,000,000	400,000
'90	37	2,987	76	103	2,067,819	413,684

Financial.—*Cap. Stock:* authorized, $250,000, all paid up; paid 8% dividend in 1890. *No Debt. Expenses:* operating, withheld; no int.; taxes, $4,300. *Revenue:* consumers, withheld; city, $5,124; other purposes, $2,624. **Management.**—Prest., Edw. Spaulding. Secy., F. A. Andrews. Treas., H. M. Hobson, Supt., H. G. Holden.

IV., V. **NEWPORT**, *Sullivan Co.* (Pop., 2,612—2,623).

S. H. EDES' AQUEDUCT LINE.

Laid in '60. Supply, springs, by gravity. About 1 mile of 3 and 2-in. i and cem. pipe supplying 20 families. *Cost*, $4,000. *Revenue*, $300.

M. S. JACKSON'S AQUEDUCT LINE.

Laid in '84. Supply, springs, by gravity. About 1¼ miles of pipe supplying 14 families. *Revenue*, $130.

29. NORTH CONWAY, *Carroll Co.* (Pop. of Conway town, 2,094—2,331.) Built in '83 by North Conway Water Co.; franchise provides neither for control of rates nor purchase of works by town. Supply, Artists brook, by gravity. Extensions were to be made during '91, in July. Pipes were being laid to conduct water from Kearsarge Mountain. **Distribution.**—*Mains*, ¾ mile. *Taps*, 98. *Meters*, 0. *Hydrants*, 7. Financial.—*Cost*, $10,000. *Cap. Stock*, $20,000. *No Debt. Revenue:* consumers, withheld; city, free. **Management.**—Prest., W. M. Pitnam. Secy. and Supt., Lycurgus Pitnam.

31. NORTHFIELD, *Merrimack Co.* (Pop., 918—1,115.) Supply from Tilton.

7. **PENACOOK**, *Merrimack Co.* (Pop., not given.) Supply from Concord.

Petersborough, *Hillsborough Co* (Pop., 2,206—2,507.) Built by town for fire protection only. See 1889-90 Manual, p. 23.

30. PITTSFIELD, *Merrimack Co.* (Pop., 1,974—2,605.) Built in '84 by Pittsfield Aqueduct Co. **Supply.**—Berry Pond, by gravity to reservoir, thence to village. **Distribution.**—*Mains*, 7 miles. *Taps*, 326. *Meters*, 1. *Hydrants*, 27. *Consumption*, 200,000 galls. Financial.—*Cost*, $33,000. *Cap. Stock*, $33,000. *No debt.*

FOR FURTHER DESCRIPTIVE MATTER see Manual for 1889-90. CHANGES since 1'90 are here given. POPULATIONS are for 1880 and 1890, respectively.

Expenses: $500. *Revenue:* consumers, $3,500; city, $945; total, $4,445. **Management.**—Prest., R. L. French. Treas., Jno. A. Goss. Supt., S. J. Winslow.

31. **PLYMOUTH**, *Grafton Co.* (Pop., 1,719—1,852.) Built in '81 by Plymouth Aqueduct & Water Co.; franchise provides neither for control of rates nor purchase of works by city. **Supply.**—Springs, by gravity, water being collected in two small reservoirs. Fiscal year closes Dec. 31. **Distribution.**—*Mains*, 4 miles. *Taps*, 160. *Hydrants*, 27. *Consumption*, 60,000 galls. **Financial.**—*Cost*, $40,000. *Cap. Stock:* authorized, $20,000; all paid up; paid 4% dividend in 1890. *Bonded Debt*, $20,000 at 5%, due in 1910. *Sinking Fund*, 4% of income. *Expenses:* operating, $200; int., $1,000; no taxes; total, $1,200. *Revenue:* consumers, $1,750; city, $675; total, $2,425. **Management.**—Prest., Alvin Burleigh. Secy., Geo. W. Adams. Treas., Jno. Mason. Supt., W. F. Langdon.

32. **PORTSMOUTH**, *Rockingham Co.* (Pop., 9,690—9,827.) Built in 1798 by Portsmouth Aqueduct Co. **Supply.**—Twenty springs, by gravity, and pumping to reservoir for high service. Fourteen fire reservoirs about town. In '90 city council voted to buy works for $150,000; great dissatisfaction prevailed among tax-payers on ground that price is excessive, and Jan. 12, '91, it was reported that petition would be sent to State Legislature that Supreme Court fix just price for works. June 2, '91, question had not been settled. No statistics have been received from Co. **Management.**—Prest, J. J. Pickering. Secy. and Supt., Jno. J. Ayers.

33. **ROCHESTER**, *Strafford Co.* (Pop., 5,784—7,396.) Built in '85 by Rochester Aqueduct & Water Co.; bought by town in '91 for $118,000. **Supply.**—Brook, by gravity, through canal to reservoir, thence to town. Fiscal year closes Dec. 31. In July, '91, reported that mains would be extended to East Rochester and 35 × 40 ft. stand-pipe added. **Distribution.**—*Mains*, 8 miles. *Taps*, 297. *Meters*, 1. *Hydrants*, 55. *Consumption*, unknown. **Financial.**—*Cost*, $100,000. *Cap. stock*, $100,000; all paid up; paid 5% div'd in 1890. *No debt. Expenses:* Operating, $875; no int.; taxes, $1,450; total, $2,325. *Revenue:* Consumers, $4,135; city, $2,650; total, $6,965. **Management.**—Elected every May. Prest., Frank Jones. Secy., T. D. Wentworth. Supt. and Treas., Albert Wallace.

NOTE.—Above statistics were furnished by Co. before system was bought by town.

35. **SOUTH WOLFBOROUGH**, *Carroll Co.* Supply from Wolfborough.

34. **TILTON AND NORTHFIELD**, *Belknap and Merrimack Cos.* (Pop., Tilton, town, 1,282—1,521; Northfield, town, 918—1,115.) Built in '88 by Tilton & Northfield Aqueduct Co. **Supply.**—Knowles Lake and Forest Pond, by gravity. Fiscal year closes Feb. 2. **Distribution.**—*Mains*, 6 miles. *Taps*, 180. *Meters*, 0. *Hydrants*, 24. **Financial.**—*Cost*, $48.000 *Cap. Stock:* authorized, $100,000; paid up, $39,000; stock is all owned by Chas. E. Tilton, Pillsbury Bros. and Peabody Estate. *Bonded Debt*, $9,000 at 6%; all held by stockholders. *Expenses:* operating, $500; int., $540; total, $1,040. *Revenue:* consumers, $2,699; city, $700; total, $3,398. **Management.** Prest., Alpha J. Pillsbury. Secy., W. B. Fellows. Treas., Jno. J. Pillsbury. Supt., Mark Keiser.

VI **WEST CANAAN**, *Grafton Co.* (Pop., not given.) Built in '86 by West Canaan Aqueduct Co., for domestic use only. **Supply.**—Spring, by gravity. **Distribution.**—*Mains*, cem., ¾ mile. Families supplied, 5. *Hydrants*, 0. **Financial.**—*Cost*, $500, secured by cash subscriptions for shares. *Expenses*, about $10. No rev. **Management.**—Clerk, M. F. Colby.

35. **WOLFBOROUGH**, *Carroll Co.* (Pop., town, 2,222—3,020.) Built in '89 by town. **Supply.**—Beach Pond by gravity. Fiscal year closes March 1. Extended in '90 to South Wolfborough, two miles. **Distribution.**—*Mains*, 11.6 miles. *Taps*, 56. *Hydrants*, 51. **Financial.**—*Cost*, $54,855. *Bonded Debt*, $47,000. *Int.*, $1,671. *Rev.*, $1,643. **Management.**—Three Comrs., John L. Peavy, H. W. Furber I. B. Manning.

36. **WOODSVILLE**, *Grafton Co.* (Pop., not given.) Built in '86 by Woodsville Aqueduct Co. **Supply.**—Ammanoosuc River, pumping direct. Fiscal year closes Jan. 1. **Distribution.**—*Mains*, about 4 miles. *Taps*, 177. *Hydrants*, 20. **Financial.**—*Cost* (in '89), $20,602. *Cap. Stock*, authorized, $50,000; paid up, $29,600; paid 4% dividend in 1890. *Floating Debt*, $10,000 at 5%. *No bonds. Int.*, $1,187. *Revenue:* consumers, $2,483; city, $400; total, $2,883. **Management.**—Secy. and Treas., Chester Abbott. Supt., Ezra B. Mann.

NEW HAMPSHIRE.

Water-Works Projected with Fair Prospects of Construction.

ANTRIM, *Hillsborough Co.* (Pop., 1,172—1,248.) At annual meeting in March, '90, fire precinct appointed committee " to investigate matter of supplying village with water." Committee reported adversely to construction of works by village, and was ordered to continue investigations with view of organizing company. Address committee: Eben Bass, G. D. Dresser, C. B. Gardner.

BERLIN FALLS, *Coos Co.* (Pop. of town, 1,144—3,729.) Project described in 1889-'90 Manual, p. 26, was abandoned. At town meeting in March, '91, proposition for town to build water-works and sewerage system was defeated and selectmen were ordered to appoint committee to secure bids from outside parties. No action had been taken June 4. Address Chn. of Com., Moses Hodgdon.

BOSCAWEN AND PENACOOK, *Merrimack Co.* (Boscawen; pop., 1,381—1,487; Penacook; pop., not given.) See 1889-'90 Manual, p. 26. System was still projected June 4, '91, by Boscawen & Penacook Water-Works Co., but nothing was to be done before '92. Address Isaac K. Gage. July 10, '91, it was unofficially reported that following committee on new works had been appointed: C. H. Amsden, E. S. Harris, E. E. Graves.

LANCASTER, *Coos Co.* (Pop., 2,721—3,373.) Lancaster Water Co. has charter. In June, '91, elected G. R. Eaton temporary Secy. and appointed investigating committee. Co. talked of buying plant of Aqueduct Co. See p. —. June 16, '91, nothing definite had been done.

NORTHWOOD NARROWS, *Rockingham Co.* (Pop. of town, 1,345—1,478.) Effort was being made to construct works in '91, but work had not begun June 16. See 1890-91 Manual, p. 26.

SALMON FALLS, *Strafford Co.* (Pop. of town, 1,713—2,003.) In '91 State Legislature authorized town of Rollinsford, in which village of Salmon Falls is located, to construct system of water-works. See note under Gt. Falls.

THE WEIRS, *Belknap Co.* (Pop., not given.) Charter obtained in '89 by Weirs Water-Works Co., and survey made by local engineer. Nothing further had been done June 5, '91.

WHITEFIELD, *Coos Co.* (Pop., 1,828—2,041.) In '91 bill was before State Legislature, calling for charter to put in water-works. July, '91, it was unofficially reported that fire wardens had been authorized to make 15 yrs.' contract for hydrant rental with Whitefield Aqueduct Co., or any other co., to pay $2,000 annually. Address C. J. Parcher, G. H. Morrison or B. C. Garland.

VERMONT.

Water-Works Completed or In Process of Construction.

1. **BAKERSFIELD,** *Franklin Co.* (Pop., 1,248—1,162.) Built in '40; ½ mile of 1-in. lead pipe supplies about 40 families, who jointly own system. *Cost,* $2,000. Supt., R. H. Barlow.

1. **BARRE,** *Washington Co.* (Pop., 2,060—6,812.) Built in '88 by Barre Water Co.; franchise does not regulate rates, but provides for purchase of works by town at intervals of 5 yrs. **Supply.**—Stevens branch of Winoski River, by gravity. Fiscal year closes April 1. **Distribution.**—*Mains,* 12 miles. *Taps,* 250. *Meters,* 0. *Hydrants,* 61. **Financial.**—*Cost,* $75,000. *Cap. Stock,* $50,000. *Bonded Debt,* $25,000 at 4%, due in 1909. *Revenue:* city, $1,875. **Management.**—Prest., J. Henry Jackson. Secy., Edw. W. Bisbee. Supt., W. D. Phelps. NOTE.—In '91 it was reported that Reynolds & Thompson were laying 2 miles of pipe from spring to supply certain houses.

2. **BARTON,** *Orleans Co.* (Pop., 742—778.) Built in '89 by village. **Supply.**—Mountain brook by gravity; filtered at entrance of main, through 18 inches of charcoal. Fiscal year closes Dec. 31. **Reservoir.**—Cap., 1,000,000 galls. **Distribution.**—*Mains,* 12 to 4-in. c. i., 3.5 miles. *Services,* w. i. *Hydrants,* Ludlow, 22. **Financial.**—*Cost,* $17,000. *Bonded Debt,* $16,000 at 4½%, due in '99. *Expenses,* operating, $10; int., $720; total, $730. *Revenue:* consumers, $750. **Management.**—Three Coms., one elected every year. Chn., Geo. H. Blake. Secy., E. F. Dutton.

FOR FURTHER DESCRIPTIVE MATTER see Manual for 1889-90. CHANGES since 1890 are here given. POPULATIONS are for 1880 and 1890, respectively.

3. BARTON LANDING, *Orleans Co.* (Pop. 378—432.) Built '84 by town. Supply.—Springs, by gravity, to reservoir. Reservoir.—Cap., 30,000 galls.; 23 ft. in diam. × 10½ ft. deep. Distribution.—*Mains,* 2 and 1¾-in., 2 miles. *Taps,* 80. *Hydrants,* 2. *Valves,* Globe. *Consumption,* 1,800 galls. *Pressure,* 40 lbs. Financial. —*Cost,* $3,100. *Debt,* $2,000, *Expenses,* $95. *Revenue,* $440. Management.—Prest., E. L. Chandler. Treas., J. D. R. Collins. Secy. and Supt., C. E. Joslyn.

4. BELLOWS FALLS, *Windham Co.* (Pop., 2,229 3,092.) Built in '48 by village. Supply.—Lake Minard, by gravity, Fiscal year closes first Monday in April. Distribution.—*Mains,* 7 miles. *Taps* (as reported), 1,200. *Meters,* 7. *Hydrants,* 47. Financial.—*Cost,* $75,000. *Debt,* $18,000 at 5%. *Expenses,* $800. Management.—Secy., F. B. Bolles. Supt., F. A. George.

5. BENNINGTON, *Bennington Co.* (Pop., 6,333—6,391.) Built in '85-7 by Bennington Water Co. Supply.—Springs, by gravity. Distribution.—(In '88). *Mains*- 7 miles. *Taps,* 250. *Meters,* 3. *Hydrants,* 33. Financial.—(In '88), *Cost,* $60,000. *Cap. Stock,* $30,000. *Bonded Debt,* $48,000 at 5%. *Revenue:* ci y, $1,100. Management.—Prest., H. W. Putnam. Secy., J. T. Shurtleff. Treas., C. H. Putnam. NOTE.—No report could be obtained for '89 or '90.

6. BRANDON, *Rutland Co.* (Pop., 3,280—3,310.) Built in '79 by fire district. Supply.—Hitchcock's pond, by gravity. Fiscal year closes Dec. 31. Distribution. —*Mains,* 10 miles. *Taps,* 300. *Meters,* 0. *Hydrants,* 38. *Consumption,* unknown. Financial.—*Cost,* $40,000. *Bonded Debt:* $35,000 at 4%, due in annual payments of $1,000 for 9 yrs. and balance in annual payments of $1,500. Management.—Com. of three, elected every year by popular vote. Report by Chas. W. Briggs.

7. BRATTLLEBORO, *Windham Co.* (Pop., 4,471—5,466.) Built in '84 by Geo. E Crowell. Supply.—Springs, by gravity. Distribution.—*Mains,* 10 miles. *Taps,* 250. *Meters,* 0. *Hydrants,* 5. *Consumption,* 50,000 galls. Financial.—*Cost,* $100,000. *No Debt, Revenue,* $3,250. Management.—Owner, Geo. E. Crowell.

II. BRISTOL, *Addison Co.* (Pop., 1,579—1,828.) Small system for domestic use only, built in '87-'90 by N. H. Munsill, Malone, N. Y. Supply.—Springs, by gravity to reservoir. *Mains,* 4 miles. *Takers,* 100. *Cost,* $5,000. *Revenue,* $500.

8. BURLINGTON, *Chittenden Co.* (Pop., 11,365—14,590.) Built in '67 by city. Supply.—Lake Champlain, pumping to reservoirs and to tank, for high service. Fiscal year closes Dec. 31.

Year.	Mains.	Taps.	Meters.	Hydrants.	Consumption.	Cost.	Bonded debt.	Int. ch'g's.	Exp.	Rev.
'80	..	1,200	..	138	600,000	......	......	......	$6,443	$23,923
'82	20	1,785	90	118	593,881	......	......	......	7,627	22,407
'83	20	1,843	134	161	617 462	$208,092	$214,900	......	9,307	23,733
'85	28.7	2,115	239	162	621,812	......	244,900	$14,694	14,238	23 484
'87	29.58	2 338	374	162	616.743	324,351	231,500	11.490	13,200	30,661
'88	30.1	2,433	445	153	696,877	349,335	241,600	11,191	12,836	33,100
'89	30.3	2,511	525	163	705,637	351,152	215,000	11,192	......	32,629
'90	30.41	2,540	584	164	756,101	363 919	215,000	11,192	10,406	33,290

Financial. — *Bonded Debt,* $215,000, at 4, 5, and 6%. *Sinking Fund,* $111,600. *Expenses,* total, $21,598. *Revenue:* Domestic, consumers, $29,996; manufactures, $2,- 763; city fire protection, $1,200; other purposes, $400; total, $33.290. Revenue from works did not pay running expenses till 1879; since that date, the excess of receipts over expenditures has been $26,000. Management.—Comrs. Chn., F. H. Parker. Supt., F. H. Crandall.

9. FAIR HAVEN, *Rutland Co.* (Pop., 2.211—2,791.) Built in '80 by village. Supply. —Inman pond, by gravity. Fiscal year closes April 1. Distribution.—*Mains,* 4.3 miles. *Taps,* 450. *Meters,* 0. *Hydrants,* 29. *Consumption,* unk'n. Financial.— *Cost,* $36,000. *Bonded Debt,* $35,000 at 4%. *Sinking Fund,* $5,700 at 4%. *Expenses:* operating, $264; int., $1,400; total, $1,634. *Revenue:* consumers, $3,449. Management.—Three Comrs.; Will V. Roberts. O. A. Bryant. Frank E. Allen.

III. HARDWICK, *Caledonia Co.* (Pop., 1,484—1,547.) First supplied by Hardwick Aqueduct Co., water being for domestic use only. In '91, village bought this system for $1,500. June 19, village voted to build an entirely new plant for fire protection and domestic supply. Est. cost, $10,000. Address M. V. B. Hathaway, Village Clerk.

IV. HYDE PARK, *Lamoille Co.* (Pop., 1,715—1,633.) Built in '80 by Hyde Park Water Co., now owned by C. S. Page. Supply.—Springs, by gravity and pumping. Distribution.—*Mains,* 2 miles. *Taps,* 45. *No Meters. No Hydrants.* Financial.—

VERMONT.

Cost, $3,500. *No Debt. Expenses*, $150. *Revenue*, $450. **Management.**—Owner, C. S. Page.

10. ISLAND POND, *Essex Co.* (Pop., est., 1,200.) Built in '79 by Island Pond Aqueduct Co. **Supply.**—Springs, by gravity. Fiscal year closes April 15. **Distribution.**—*Mains*, 3 miles. *Taps*, 100. *Meters*, 0. *Hydrants*, 5. *Consumption*, 30,000 galls. **Financial.**—*Cost*, $12,000. *No Debt. Expenses*, $200. *Revenue*, $1,250. **Management.**—Owner, Geo. W. Noyes.

11. MIDDLEBURY, *Addison Co.* (Pop., 1,834–1,762.) **History.**—Built in '91 by village; to be in operation in fall. Engr., Wm. H. Lang. Contrs., Lang, Goodhue & Co., Burlington. **Supply.**—Otter Creek, pumping direct; filtered through gravel. **Pumping Machinery.**—Lang water-power pump. **Distribution.**—*Mains*, 1., 7.5 miles. **Financial.**—*Cost*, $36,000. *Bonded Debt*, $30,000 at 4%. **Management.**—Chn., C. M. Wilds.

V., 12. MONTPELIER, *Washington Co.* (Pop., 1,847–3,617.)

THE HUBBARD WORKS.

Built in '20 by Roger Hubbard; town neither controls rates nor has right to purchase works. **Supply.**—Springs connected with small reservoir. Fiscal year closes Nov. 1. **Distribution.**—*Mains*, 8.5 miles. *Taps*, 190. *Hydrants*, 0. **Financial.**—*Cost*, unk'n. *No Expenses. No Revenue.* **Management.**—Owner, Jno. E. Hubbard.

VILLAGE WORKS.

Built in '84. **Supply.**—Berlin pond, by gravity; filtered. Fiscal year closes Oct. 1.

Year.	Mains.	Taps.	Meters.	Hydrants.	Cost.	B'd'd debt.	Int. ch'g's.	Exp.	Rev.*
'84	9.67	96	0	51	$58,892	$50,000	.....	.....	.....
'85	10.45	238	0	53	66,483	50,000	$2,250	$3,535	$1,654
'86	11.75	404	0	54	72,074	50,000	2,250	1,405	3,798
'87	12.43	465	4	54	74,727	50,000	2,250	2,653	4,549
'88	15.14	503	4	54	101,343	50,000	2,250	2,708	8,599
'89	16.41	491	5	55	114,015	50,000	2,250	2,031	9,524
'90	17.19	531	5	58	119,983	50,000	2,250	2,050	9,782

*Including allowance for public uses of water.

Financial.—*Bonded Debt*, $50,000 at 4½%. *Sinking Fund*, $6,380, loaned to the corporation at 6%. *Expenses;* total, $2,050. *Revenue:* consumers, $7,709; city, $2,080; total, $9,789. **Management.**—Six trustees, elected every year by popular vote. Supt., Joel Foster, appointed by trustees.

13. MORRISVILLE, *Lamoille Co.* (Pop., est., 1,500.) Built in '85 by Morrisville Aqueduct Co., now owned by W. W. Warren. **Supply.**—Springs, by gravity. Fiscal year closes Dec. 31. **Distribution.**—*Mains*, 8 miles. *Taps*, 300. *Meters*, 8. *Hydrants*, 10. **Financial.**—*Cost*, $25,000. *No Debt.* **Management.**—Owner and Supt., W. W. Warren, Waterbury.

14. NEWPORT, *Orleans Co.* (Pop., 920–1,730.) Built in '85 by city. **Supply.**—Lake Memphremagog, pumping direct. Fiscal year closes Dec. 31. **Distribution.**—*Mains*, 2 miles. *Taps*, 80. *Meters*, 0. *Hydrants*, 17. **Financial.**—*Cost*, $8,000. *No Debt. Expenses*, $500. *Revenue*, $1,000. **Management.**—Operated by Prouty & Miller.

VI. NORTHFIELD, *Washington Co.* (Pop., 1,813–1,222.) Built in '54 by several co.'s; now owned by Northfield Aqueduct Co. **Supply.**—Springs, by gravity. **Distribution.**—*Mains*, 5 miles. *Taps*, 90. *Hydrants*, 0. **Financial.**—*Cost*, $10,000. *No Debt. Expenses*, $100. **Management.**—Prest., Ed. S. Steffins. Secy., Ira Beard. Supt., Wm. M. Davis.

15. POULTNEY, *Rutland Co.* (Pop., 2,717–3,031.) Built in '89 by fire district. **Supply.**—Crystal lake, by gravity. Fiscal year closes first Monday in Jan. **Distribution.**—*Mains*, 4½ miles. *Taps*, 60. *Hydrants*, 40. **Financial.**—*Cost*, $30,000. *Bonded Debt*, $30,000 at 4%. *Expenses:* operating, $50; int., $1,200; total, $1,250. *Revenue:* consumers, $787. Net revenue is applied to int. on bonds and balance raised by taxation. **Management.**—Committee of three; P. B. Clark, H. Herrick, M. Costello.

FOR FURTHER DESCRIPTIVE MATTER see Manual for 1889-90. CHANGES since 1890 are here given. POPULATIONS are for 1880 and 1890 respectively.

16. **READSBOROUGH**, *Bennington Co.* (Pop., 743—910.) **History.**—Built in '89 by fire district. Des. Engr., A. W. Locke. Const. Engr. and Contr., A. W. Hunter. Both of North Adams, Mass. Fiscal year closes Dec. 31. **Supply.** Brook, by gravity to reservoir; strained through wire net. **Reservoir.**—Cap., 150,000 galls.; of stone and earth. **Distribution.**—Mains, 6-in., 1., 2.5 miles. *Takers*, 50. *Hydrants*, 10. *Pressure*, 80 lbs. **Financial.**—*Cost*, $10,000. *Bonded Debt*, $10,000 at 5%. *Expenses*, nominal. *Revenue:* Consumers, $300. **Management.**—Prudential Com.: W. D. Howe, F. M. Sprague, M. J. Ross.

17. **RICHFORD**, *Franklin Co.* (Pop., 789—1,116.) Built in '89 by village. **Supply.**—Springs, by gravity. **Distribution.**—*Mains*, 4.5 miles. *Taps*, 125. *Hydrants*, 29. **Financial.**—*Cost*, $16,609. **Management.**—Comr., J. C. Baker.

18. **RUTLAND**, *Rutland Co.* (Pop., 7,502—8,239.) Built in '56 by village. **Supply.**—Mendon River to and from settling and distributing reservoirs. Fiscal year closes Feb. 28. During summer of '91 new settling reservoir was in course of construction; cap., 6,000,000 galls.; 400 ft. long; av. width, 210 ft.; greatest depth, 14½ ft. Connected with Mendon River by open canal 8 ft. wide, 380 ft. long. Contract price, $17,988. Des. Engrs., Chappell & Burke. Contrs., Flood & Sherrell, Sandy Hill, N. Y. Contracts for furnishing and laying 14,700 ft. of 6-in. c. i. supply main were let to Thos. Regan, Philadelphia, and B. S. Chew, Gloucester, for $37,897. It is expected that improvements will be completed by winter.

Year.	Mains.	Taps.	Hydrants.	Cost.	B'd'd Debt.	Int. Ch'g's.	Exp.	Rev.
'86-7	15	1,050	56	.....	$72,500	.....	$3,000	$14,849
'88-9	16.9	1,350	60	.....	67,500	$3,350	3,000	17,000
'89-90	17.7	1,449	65	$131,108	67,500	3,350	3,321	18,964
'90-1	18.4	1,500	65	140,086	62,500	3,350	4,106	20,170

Bonded Debt, $62,000 at 5½%, due '93-99. *Sinking Fund*, $21,837, invested in town or village orders at 5½%. *Expenses:* total, $7,456. *Revenue*, $20,170. Bonds are redeemed from net income. **Management.**—Comrs., J. M. Davis, N. L. Davis, W. C. Landon.

19. **SAINT ALBANS**, *Franklin Co.* (Pop., 7,193—7,771.) Built in '73 by village. **Supply.**—Surface water, by gravity. Fiscal year closes March 1.

Year.	Mains.	Taps.	Meters.	Hydrants.	Cost.	Bonded debt.	Int. charges.	Exp.	Rev.
'80-1	13	590	0	54	$165,000	$153,000	.....	$1,501	.....
'84-5	13.5	650	0	54	175,000	150,000	$7,500	1,500	$7,500
'86-7	14	710	0	54	175,000	150,000	7,500	1,506	7,224
'88-9	14	900	0	65	175,000	150,000	7,500	1,640	8,942
'90-1	14	965	0	70	175.000	95,800	3,832	1,963	9,490

Financial.—*Bonded Debt*, $95,800 at 4%. *Sinking Fund*, $687, on deposit. *Expenses:* total, $5,798. *Revenue:* consumers, $9,490. **Management.**—Committee: Chn., G. C. Green. Secy., B. D. Hopkins. Supt., Marshall Mason.

VII., 20. **SAINT JOHNSBURY**, *Caledonia Co.* (Pop., 3,360—3,857.)

ST. JOHNSBURY AQUEDUCT CO.

Built in '60. **Supply.**—Springs, by gravity. July, '91, it was reported that Co. would lay additional pipe at expense of $25,000. First cost reported as $50,000. No further information could be obtained. Address E. & F. Fairbanks & Co.

VILLAGE WORKS.

Built in '76. **Supply.**—Passumpsic River, pumping direct; filtered. Fiscal year closes Dec. 31. **Distribution.**—*Mains*, no record, about 8 miles. *Taps*, unk'n. *Hydrants*, 78. **Financial.**—*Cost*, $90,000. *Bonded Debt*, $50,000 at 5½%. *Expenses:* operating, $1,152; int., $2,750; total, $3,902. *Revenue*, consumers, $3,664. **Management.**—Two comrs. Chn., M. J. Coldbrook, Supt., O. W. Orcutt.

VIII. **SPRINGFIELD**, *Windsor Co.* (Pop., 1,586—1,512.) Built in '84 by Springfield Village Water Supply Co.; town neither controls rates nor has right to purchase works. **Supply.**—Four springs, by gravity. Fiscal year closes Nov. 1. **Distribution.**—*Mains*, ½ mile. *Taps*, 24. *Hydrants*, 0. **Financial.**—*Cost*, $2,000. *No Debt*. *Expenses*, $15. *Revenue*, $250. **Management.**—Owner, Albert Brown.

21. **SWANTON**, *Franklin Co.* (Pop., 3,079-3,231.) Built in '87 by Swanton Water-Works Co.; now owned by village. **Supply.**—Missisquoi, pumping direct. Fiscal year closes March 31. **Distribution.**—*Mains*, 6 miles. *Taps*, 134. *Hydrants*, 34. **Financial.**—*Cost*, $20,000. *Bonded Debt*, $14,000 at 4%. *Expenses:* operating, 300; int., $550; total, $860. *Revenue:* consumers, $1,160; city, $300; total, $1,460. **Management.**—Prest., E. W. Jewett. Secy., H. A. Burt. Treas., A. J. Ferris. Supt., H. D. Thomas.

VERMONT.

22. VERGENNES, *Addision Co.* (Pop., 1,782—1,773.) Built in '69 by city. **Supply.**—Otter Creek, pumping direct; filtered through sand and gravel. Fiscal year closes March 5. **Distribution.**—*Mains,* 6 miles. *Taps,* 300. *Hydrants,* 35. **Financial.**—*Cost,* $75,000. *Bonded Debt,* $52,000; $26,000 at 6%; $26,000 at 4%. *Expenses:* operating, $749; Int., $2,600; total, $3,349. *Revenue:* consumers, $3,000. **Management.**—Three Comrs., S. D. Miner, H. Stevens, C. A. Hoffnagle.

23. WATERBURY, *Washington Co.* (Pop., 756-955.) Built in '79 by Warren & Somerville. Supply, springs and gravity. **Distribution.**—*Mains,* 7 miles. *Taps,* 300. *Hydrants,* 35. **Financial.**—Withheld. **Management.**—Supt., C. C. Warren.

24. WEST RANDOLPH, *Orange Co.* (Pop., 1,069—1,573.) Built in '78 by village. Supply, springs and brook, by gravity. Fiscal year closes May 15. **Distribution.**—*Mains,* 6 miles. *Taps,* 250. *Hydrants,* 37. **Financial.**—*Cost,* $23,000. *Bonded debt,* $22,000 at 4%. *Expenses:* Operating, $517; int., $880; total, $1,397. *Revenue:* consumers, $1,846. All expenses paid from revenue. **Management.**—Five Comrs., one elected each year. Chn., Wm. H. Du Bois; Supt., Lemuel Richmond, appointed by Comrs.

25. WINDSOR, *Windsor Co.* (Pop., 2,175—1,846.) Built in '88 by village; first supplied in '50 by Windsor Aqueduct Co. **Supply.**—Dudley brook, by gravity from impounding reservoir. Fiscal year closes March 4.

Year.	Mains.	Taps.	Hyd'ts.	Cost.	B'd'd debt.	Int. ch'ges.	Exp.	Rev.
'89-90	4.6	136	40	$31,605	$30,000	$1,250		
'90-91	4.6	146	40	34,605	30,000	1,250	$145	$1,989

Financial.—*Total Debt:* $32,885; floating, $2,885 at 5%; bonded, $25,000 at 4%; $5,000 at 5%. *Expenses*: total, $1,395. *Revenue:* consumers, $1,989. All expenses paid from revenue. **Management.**—Five Comrs., elected every three years. Chn., S. N. Stone; Supt., C. D. Penniman, appointed by Comrs. NOTE.—For cross-section of dam see Manual for '89-90, p. 33.

26. WINOOSKI, *Chittenden Co.* (Pop., 2,833—3,659.) Built in '73 by Winooski Aqueduct Co. **Supply.**—Springs, by gravity. Fiscal year closes Dec. 31. **Distribution.**—*Mains,* 6.5 miles. *Taps,* 417. *Meters,* 6. *Hydrants,* 29. **Financial.**—*Cost,* $35,000. *Cap. Stock:* Authorized, $15,000; all paid up. *No debt.* Taxes, $124. *Revenue:* consumers, $700; city, $870; other purposes, $2,670; total, $4,240. **Management.**—Prest., Jos. Sawyer, 68 Chauncy St., Boston, Mass. Secy., F. C. Kennedy. Engr., G. D. Nash.

27. WOODSTOCK, *Windsor Co.* (Pop., 1,266—1,218.) Built in '87 by Woodstock Aqueduct Co.; franchise provides neither for control of rates nor purchase of works by town. **Supply.**— Streams, by gravity.

Year.	Mains.	Taps.	Meters.	Hydrants.	Consumption.	Cost.	Exp.	Rev.
'88	7	100	0	37	unk'n.	$36,711	$573	$1,584
'89	7	122	0	37	"	38,000	500	2,024
'90	7.5	166	0	38	"	40,000	488	2,206

Financial.—*Cap. Stock:* authorized, $40,000; paid up, $36,000; paid no div'd in 1890. *No Debt. Expenses:* operating, $409; taxes, $79; total, $488. *Revenue:* consumers, $1,366; city, $840; total, $2,206. **Management.**—Prest., F. W. Billings. Secy., H. C. Philips. Mgr., F. W. Wilder.

Water-Works Projected with Fair Prospects of Construction.

BRADFORD, *Orange Co.* (Pop., 619—610.) July 18, '91, Village Clerk reported that village proposed to build water-works and that committee had been appointed. Supply, Brook, by gravity to reservoir. Est. cost, $20,000. Engr., Joel Foster, Montpelier. Address Col. J. S. Stearns, or A. F. Colburn, Village Clerk.

PITTSFORD, *Rutland Co.* (Pop., 1,982—1,775.) In '90, Co. obtained charter to build works but nothing further had been done June 6, '91. See 1889-90 Manual, p. 34.

ROCHESTER, *Windsor Co.* (Pop., 1,362—1.257.) Dec. 13, '90. Committee on Water-Works were securing options on water rights, but nothing definite had been decided upon June 6, '91. Address Q. M. Ford.

TOWNSEND, *Windham Co.* (Pop., 1,099—365.) During '91, committee was appointed which submitted plan for system of water-works at estimated cost of $2,500, to be built by a company. June 5, sufficient stock had not been raised and it was doubtful if it could be. Address H. B. Kenyon.

FOR FURTHER DESCRIPTIVE MATTER see Manual for 1889-90. CHANGES since 1890 are here given. POPULATIONS are for 1880 and 1890 respectively.

WEST RUTLAND, *Rutland Co.* (Pop. in '90, 3,680.) Co. will probably be formed with cap. stock of $25,000. Address Jno. O. Rourke, J. D. Hourahan, J. M. Davis.

MASSACHUSETTS.
Water-Works Completed or in Process of Construction.

1. ABINGTON, *Plymouth Co.* (Pop., 3,697–4,260.) Built in '86 by town. Pumping machinery, 6 miles of mains and stand-pipe, owned jointly with Rockland. **Supply.**—Big Sandy pond, pumping to stand-pipe. Fiscal year closes Dec. 21.

Yr.	Mains.	Taps.	Meters.	Hydr'ts.	Cost.	B'd d'd'bt.	Int.chs.	Exp.	Rev.
'87	13.36	390	25	97	$132,592	$147,000	.....	.....	.....
'88	17.6	502	44	122	155,478	147,000	$5,880	$1,412	$6,958
'89	17.9	580	73	124	158,089	146,000	5,840	2,785	5,762
'90	18.3	625	102	127	160,534	146,000	5,840	3,493	6,385

Financial.—*Bonded Debt,* $146,000 at 4%, due in '91–1918. *Expenses:* total, $9,133. *Revenue:* consumers, $6,385; city, $2,480; total, $8,865. **Management.**—Three comrs. Secy., Geo. A. Beal. Supt., Aug. H. Wright. Chn., E. P. Reed.

2. ADAMS, *Berkshire Co.* (Pop., 5,591–9,213.) Built in '73 by village. **Supply.**—Bassett and Dry brooks, by gravity from impounding reservoir; also emergency supply from wells at Renfrew Mills, pumping to reservoirs. Fiscal year closes March 12. **Distribution.**—*Mains,* about 11 miles. *Hydrants,* 170. **Financial.**—*Revenue,* $12,161. **Management.**—Prudential Committee: Clerk, T. L. A. Hall. Chf. Engr., R. N. Richmond.

I. AGAWAM, *Hampden Co.* (Pop., 2,216–2,352.) Built in '77 by C. L. Goodhue, Springfield. **Supply.**—Springs, by gravity. **Distribution.**—*Mains,* 5 miles. *Taps,* 74. *Meters,* 0. *Hydrants,* 0. **Financial.**—*Cost,* $6,000. *No Debt.* **Management.**—Owner, C. L. Goodhue, Springfield.

3. AMESBURY, *Essex Co.* (Pop., 3,355–9,798.) Built in '84 by Powow Hill Water Co.; franchise provides that works may be purchased by city at any time; city has no control of rates. Supply, driven wells, pumping to reservoir. Fiscal year closes Jan. 1. **Distribution.**—*Mains,* 9 miles. *Taps,* 800. *Meters,* 70. *Hydrants,* 120. **Financial.**—*Cost,* $96,623. *Cap. Stock,* authorized, $40,000; all paid up; no div'd, net earnings used for extensions. *Total Debt,* $47,600; floating, $7,600 at 5½%; bonded, $40,000 at 5%. *Expenses:* operating, $3,500; int., $2,000; taxes, $641; total, $7,899. *Revenue:* consumers, $10,153; city, $3,500; total, $13,653. **Management.**—Prest., R. F. Briggs. Clerk, O. S. Baley. Treas., E. R. Sibley.

4. AMHERST, *Hampshire Co.* (Pop., 4,298–4,512.) Built in '79 by Amherst Water Co.; franchise provides neither for control of rates nor purchase of works by city. Supply, Amethyst brook and springs. Fiscal year closes May 6. **Distribution.**—*Mains,* 12 miles. *Taps,* 350. *Hydrants,* 54. **Financial.**—*Cost,* $82,338. *Cap. Stock:* authorized, $40,000; all paid up. *Total Debt,* $40,126; floating, $126; bonded, $40,000 at 5%. *Revenue:* city, $1,000. **Management.**—Prest., W. A. Dickinson. Treas., E. D. Bangs. Supt., W. W. Hunt.

5. ANDOVER, *Essex Co.* (Pop., 5,169–6,142.) Built in '89–90 by town; fire protection system built in '86. **Supply.**—Haggett's pond, pumping to reservoir. Fiscal year closes Feb. 3.

Year.	Mains.	Taps.	Meters.	Hydrants.	Consumption.	Cost.	B'd'd debt.	Int. ch'g's.	Exp.	Rev.
'89–90	20	44	..	..	.....	.....	$150,000	$1,009	$273	.....
'90–91	20.5	329	80	161	135,932	$159,602	155,000	5,252	2,835	$3,106

Financial.—*Bonded Debt,* $155,000 at 4%. *Sinking Fund,* $1,062, deposited in savings bank. *Expenses;* total, $8,137. *Revenue:* consumers, $3,106; city, $3,000; total, $6,106. **Management**—Water comrs. Chn., Jno. H. Flint. Secy., F. G. Haynes.

6. ARLINGTON, *Middlesex Co.* (Pop., 4,100–5,629.) Built in '73 by town. **Supply.**—Brooks, by gravity from storage reservoir; filtered. Fiscal year closes Dec. 31. Additional filter gallery built in '89–90; 12 × 12 × 100 ft.; of stone, against dam of reservoir, puddled on three sides, filtering through sand and gravel.

Year.	Mains.	Taps.	Meters.	Hydrants.	Cost.	B'd d'd't.	Int.ch'g's.	Exp.	Rev.
'73	12.17	245	0	86	$154,722	$155,000	.....	.....	$7,034
'75	12.9	259	0	91	295,026	155,000	$9,750	$1,067	6,600
'80	13.5	356	0	98	305,177	300,000	17,200	1,102	8,330
'82	14	430	0	98	305,950	300,000	17,200	1,182	9,092
'84	14	492	0	98	308,808	300,000	17,200	3,084	8,632
'86	14.3	567	0	99	311,773	300,000	17,200	2,557	9,972
'88	16	699	1	106	312,931	300,000	17,200	3,223	11,118
'89	16	720	1	108	313,305	291,000	16,720	2,867	11,671
'90	16	760	2	109	313,712	291,000	16,720	2,583	12,276

MASSACHUSETTS. 21

Financial.—*Total Debt*, $314,000; floating, $23,000 at 5 and 7%; bonded, $217,000 at 6%, due in '92; $74,000 at 5%, due in '98; $200,000 at 6% to be refunded in '92. *Sinking Fund*, $20,164, invested at 4%. *Expenses:* total, $19,303. *Revenue:* consumers, $12,276; city $1,200; total, $14,276. Income applied to interest and floating debt; deficiencies met by taxation. **Management.**—Three comrs., one elected every year. Chn., Alfred D. Hoitt. Secy., B, Delmont Lock. Supt., Thos. Rodner.

7. ASHBURNHAM, *Worcester Co.* (Pop., 1,666—2,074.) Built in '70 by Ashburnham Reservoir Co. **Supply.**—Springs, by gravity from impounding reservoir. **Distribution.**—*Mains*, 3 miles; cem. pipe has been replaced with c. i. *Taps*, 30. *Meters*, 0. *Hydrants*, 17. *Consumption.* unk'n. **Financial.**—*Cap. stock*, $3,200. *Debt*, $1,189. **Management.**—Prest., Ivers W. Adams. Secy., Walter R. Adams. Supt., Chas. A. Robbins.

8. ATHOL, *Worcester Co.* (Pop., 4,307—6,319). Built in '76 by Athol Water Co.; town does not control rates, but may purchase works after '93; price determined by arbitration. **Supply.**—Brooks, by gravity from impounding reservoirs to distributing reservoir; filtered. Fiscal year closes Dec. 31. An 8-ft. filter having cap. of 250 galls. per minute was built in '90 by Albany Steam Trap Co.

Year.	Mains.	Taps.	Meters.	Hyd'ts.	Cons'pt'n.	Cost.	B'd d't.	Int.ch'gs.	Exp.	Rev.
'80	8	250		50	unk n.	$89,000			$500	....
'86	12	400	5	55	"	140,000	$40,000	$2,400	....	....
'88	12	482	15	56	"	140,000	40,000	2,400	1,200	$7,674
'89	14.5	540	15	56	"	150,000	40,000	2,400	1,500	10,500
'90	14.5	516	15	56	"	150,000	40,000	2,400	1,500	10,700

Financial—*Cap. Stock*, authorized, $80,000; all paid up. *Bonded Debt*, $40,000, bonded at 6%, due in 15 yrs. *Expenses:* Operating, $1,200; int., $2,400; taxes, $300; total, $3,900. *Revenue:* Consumers, $7,900; city, $2,800; total, $10,700. **Management.**—Prest., S. L. Wiley. Secy. and Supt., A. MacDonald.

9. ATTLEBOROUGH, *Bristol Co.* (Pop., 11,111—7,577). Built in '73 by town. **Supply.**—Well and Ten Mile River, pumping to tank, with direct pumping from river for fire purposes. Fiscal year closes April 30. During '90, iron standpipe was erected. Cap., 660,000 galls. Engr., M. M. Tidd, Boston. Contrs., Porter Mfg Co., Syracuse, N. Y. Total cost, $17,867.

Year.	Mains.	Taps.	Meters.	Hydrants.	Cost.	B'd'd Debt.	Int. Ch'g's.	Exp.	Rev.
'80-1	6	500	125	70	$80,000	$61,000		....	....
'86-7	10	755	254	99	88,742	88,000	$6,160	$4,040	$5,450
'88-9	16	800	328	111	113,578	72,000	4,280	3,600	7,500
'90-1	17	793	391	121	140,910	90,000	....	....	10,355

Financial.—*Total Debt*, $95,000; floating, $5,000; bonded, $90,000. *Sinking Fund*, $4,123. *Revenue:* consumers, $10,355; city, none. **Management.**—Three comrs ; Wm. M. Stone, G. A. Deane, L. Z. Carpenter. Supt., H. A. Bodman.

10. AVON, *Norfolk Co.* (Pop., 1,384.) Built in '89 by town. **Supply.**—Surface wells, fed by spring, pumping direct and to stand-pipe. Fiscal year closes Dec. 31. **Distribution.**—*Mains*, 4 miles. *Taps*, 127. *Meters*, 2. *Hydrants*, 41. **Financial.**—*Cost*, $48,000. *Bonded Debt*, $50,000 at 4%. *Expenses:* operating, $786. *Revenue:* consumers, $748; city, $750; total, $1,498. **Management.**—Three comrs. Chn., C. H. Felker. Treas., Nathan Tucker.

11. AYER, *Middlesex Co.* (Pop., 1,881—2,148.) Built in '88 by town. **Supply.**—Collecting well, pumping to reservoir. **Distribution.**—(In '88.) *Mains*. 5.4 miles. *Taps*, 197. *Meters*, 5. *Hydrants*, 43. *Consumption*, 40,000 galls. **Financial.**—(In '89.) *Cost*, $63,000 *Bonded Debt*, $63,000 at 4%. *Expenses:* operating, $700; int., $2,520; total, $3,220. *Revenue*, $3,950. **Management.**—Three comrs. Chn., L. J. Spaulding. Secy., C. C. Bennett. Supt., Arthur Fenner. NOTE.—No report for '89 or '90 could be obtained.

12. BELMONT, *Middlesex Co.* (Pop., 1,615—2,098.) Built in '87 by town. **Supply.**—Watertown Water Supply Co. Fiscal year closes Dec. 31.

Year.	Mains.	Taps.	Meters.	Hydrants.	Cost.	B'd'd Debt.	Int. Ch'g's.
'88	7	108	..	54	$11,121	$11,250	$1,650
'90	9.5	188	13	69	56,814	45,250	1,710

Financial.—*Bonded Debt*, $45,250 at 4%, due '91-1909, in annual payments to be provided for by taxation. Contract with Watertown Water Supply Co. provides that company shall make all collections, and, for the use of pipes, return 50% of same, including hydrant rental, to town. Hydrant [rental, $17.50. **Management.**—Comr., W. L. Chenery.

FOR FURTHER DESCRIPTIVE MATTER see Manual for 1889-90. CHANGES since 1890 are here given. POPULATIONS are for 1880 and 1890 respectively.

22   MANUAL OF AMERICAN WATER-WORKS.

13. BEVERLY, *Essex Co.* (Pop., 8,456—10,821.) Distribution system built in '67; supply taken from Salem works till '87, when town built reservoir and pumping station. Supply.—Wenham Lake, pumping direct and to reservoir. Fiscal year closes Jan. 31.

Year.	Mains.	Taps.	Meters.	Hydrants.	Exp.	Rev.
'86	47	1,895	18	130	$12,655	$19,019
'88	50	2,150	50	162	17,000	22,350
'90	55.5	2,240	20	170	......	26,000

Financial.—No statistics given except those above. Management.—Board of three, one elected every year. Chn., Peter E. Clark. Secy., Geo. Swan. Supt., Chas. Picket, appointed by board.

14. BOSTON, *Suffolk Co.* (Pop., 362,839—448,477. Pop. supplied, 419,840—521,479.) First supplied in 1652 by "Water-Works Co." In 1796 the "Aqueduct Corporation" was incorp. and soon after began to supply water. See Manual for '89-90, p. 39. Cochituate supply introduced by city in '48, and supplemented by Sudbury supply in '78. In '74 annexation of Charlestown gave city Mystic works, built in '64; these works also supply Chelsea, Everett and Somerville. Fiscal year closes Dec. 31.

COCHITUATE AND SUDBURY DEPARTMENT.

Supply.—Lake Cochituate, by gravity, with pumping for high service.

Lake Cochituate has an available storage capacity of 1,508,000,000 galls. In '90-1 a concrete and stone dam, backed with earth, was being built at outlet to replace dams described in '89-90 Manual, p. 39. Dam, 200 ft. long, 16.64 ft. high; on fine sand. Concrete 9 ft. thick at bottom, has granite steps from apron to top, at spillway, down stream side. Sheet-piling, 8-in., driven on inner side of dam some distance below bottom; sheeting covered some distance above top end and down side with clay. Spillway, at middle of dam, 63 ft. long, and discharging weir, 15 ft. wide, controlled by gate, for ordinary flow. Dam designed to waste 6 ins. of water in 24 hours over whole drainage area of 18 sq. miles. Chief Engr., Wm. Jackson, City Engr. Des. and Const. Engr., Desmond Fitzgerald. Contr., Thos. A. Rowe. Cost, about $30,000.

Sudbury River Storage Basins, Nos. 1 to 4, have available capacities of 288,400,-000, 479,700,000, 1,074,200,000 and 1,404,800,000 galls., or total of 3,247,100,000 galls., which, with Lake Cochituate, gives available storage capacity, Cochituate and Sudbury department, of 4,755,100,000 galls.

Dam No. 4, built in '82-6 at cost of $521,998. Engr., A. Fteley. Built largely by days' work. Earth, with concrete heart-wall. Dam, 1,857 ft. long, 51 ft. high above meadow, 83 ft. above bed-rock, including depth of heart-wall; 20 ft. wide on top, 244 at bottom of embankment. Inner slope, 1.65 to 1 from bottom to berm, 1½ to 1 remaining distance; cobbles or broken stone from near bottom of slope, increasing in thickness to berm; from berm to near top paved, on cobbles or broken stone. Berm, 6 ft. wide about 14 ft. below top, and about 7 ft. below water line, which is 7 ft. below top. Outer slope, 2½ to 1, turfed. Heart-wall, 8 ft. wide at bottom, stepped to 3 ft. at top; backed with clay puddle from natural surface up. Spillway, at end of dam, free, lined with masonry, 30 ft. long, 7 ft. deep; stepped away from dam. Two 48-in. discharge pipes. Forms reservoir with area of 162 acres. Drainage area of basin, 6.06 sq. miles. Gate-house of masonry, in dam; two influent, 32½ and 57 ft. deep below gate-house floor, and one effluent chamber, also 57 ft. deep. For further details see Jour, N. E. W. Wks. Assn., Dec., '87, paper by W. F. Learned, Watertown.

Construction of Basin No. 5 was begun in '90, and will be completed about '94. Will be formed by earth dam about 60 ft. high, with heart-wall carried about 100 ft. below top of dam. Will have available capacity of 1,300,000,000. Est. cost, $400,000. Chief Engr., Wm. Jackson, City Engr. Des. and Const. Engr., Desmond Fitzgerald.

Yr.	M'ins.	Taps.	M't'rs.	Hy-dr'ts.	Con-sumpt'n.	Cost.	Bonded debt.	Int. ch'rges.	Exp.	Rev.
'50	......	13,463	....	....	....	$3,998,052	....	....	....	$161,053
'60	129.7	23,245	100	1,415	....	5,661,072	....	....	....	334,515
'70	170	30,471	1,076	1,984	....	9,341,909	....	....	....	708,784
'80	384.5	46,315	1,219	4,204	20,312,900	16,750,518	....	....	....	1,039,897
'84	388.5	50,632	4,439	4,573	25,090,500	17,775,956	$13,045,473	$668,657	$336,578	1,203,193
'85	400	51,810	3,114	4,681	25,607,200	18,567,279	13,217,474	676,142	321,137	1,239,758
'86	414	53,400	4,417	4,806	26,627,900	18,973,616	13,890,274	680,994	336,507	1,206,065
'87	436.5	55,235	4,059	4,990	29,852,100	19,527,483	13,217,474	721,585	339,693	1,244,192
'88	456.68	56,947	3,133	5,008	33,310,700	20,049,615	14,941,274	735,791	383,638	1,317,385
'89	479.72	58,810	3,456	5,225	32,070,000	20,432,974	15,476,274	755,268	345,987	1,357,738
'90	498.73	60,718	3,627	5,459	33,871,700	20,994,561	16,246,274	765,079	381,147	1,382,423

MASSACHUSETTS. 23

Financial. — *Bonded Debt*, $16,246,274 at 3½ to 6%. *Sinking Fund*, $5,854,530. *Expenses:* total, $1,146,226. *Revenue:* consumers, $1,382,423.

THE MYSTIC DEPARTMENT.

Built in '64. Supply, Mystic Lake, by gravity to pumping station, thence pumping to reservoir.

Yr.	Mains.	Taps.	Meters.	Hy-drts.	Consumption.	Cost.	Bonded debt.	Int. ch'g's.	Exp.	Rev.
'81	117.7	.....	420	757	7,194,700	$1,634,109	$1,127,000	$67,620	$78,824	$254,359
'83	.....	14,453	501	770	7,093,500	1,641,762	840,000	.....	116,572	259,791
'84	129.2	14,939	567	794	6,209,700	1,548,452	839,000	48,960	128,126	262,244
'85	131	15,929	567	781	6,737,350	1,656,805	839,000	48,775	122,858	276,658
'86	133.2	16,110	469	818	7,309,800	1,857,459	839,000	45,324	134,439	219,610
'87	136.1	16,809	420	935	7,629,000	1,650,639	839,000	44,928	153,345	203,019
'88	142.2	17,607	387	965	8,258,400	1,690,757	839,000	41,993	102,086	293,056
'89	147.7	18,527	386	908	7,830,560	1,695,281	839,000	41,643	125,660	306,637
'90	152.3	19,520	391	1,073	8,301,400	1,708,782	739,000	42,358	144,184	317,197

Financial.—*Bonded Debt*, $739,000. *Sinking Fund*, $719,723. *Expenses:* total, $186,542. *Revenue:* consumers, $317,197. **Management.**—Board of three. City Engr., Wm. Jackson. Registrar, Wm. F. Davis. Asst. Engr. in charge construction, Eastern and Mystic Divisions, Dexter Bracket. Supt. of Western Division, Cochituate Dept., Desmond Fitzgerald; of Eastern Division, W. J. Welch; of Mystic Dept., E. S. Sullivan.

15. **BRADFORD**, *Essex Co.* (Pop., 2,643—3,720.) **History.**—Built in '89—90 by Bradford Water Co. Contrs., Goodhue & Birnie, Springfield. **Supply.**—Wells, pumping to reservoir and direct. **Pumping Machinery.**—Dy. cap., 1,000,000 galls.; Worthington comp. cond. **Reservoir.**—Cap., 1,000,000 galls.; of earth, 70 ft. sq., 11 ft. above ground line, 16 ft. water; bank 10 ft. wide on top, with 3-ft. puddle wall in center; inner slope, 2 to 1, battered; outer slope, 1 to 1. **Distribution.**—*Hydrants*, Chapman. *Valves*, Chapman. **Financial.**—Withheld. **Management.**—Prest., Chas. L. Goodhue, Springfield. Secy. and Treas., Thos. N. Birnie, Springfield. Supt., Dennis Gilderson.

16. **BRAINTREE**, *Norfolk Co.* (Pop., 3,855—4,848.) Built in '57 by Braintree Water Supply Co.; purchased by town in '91. In '87 town voted to buy works, but, owing to disagreement as to extent of Co.'s contracts, rescinded the decision and began construction of independent system. Supreme Court decided that town must purchase Co.'s system, and appointed comrs. to determine price. Town, after having suspended operations on its own system, and thereby become involved in lawsuit with contrs., took possession of Co.'s system June 1, '91, paying $159,610. Town had paid contrs., John Cavanagh & Son, Boston, $30,000 before work on its system was stopped, and finally compromised for $17,000 additional on account of breaking contract. Before this work was stopped a pumping station had been built, two miles of 12-in. mains laid, and other material purchased. **Supply.**—Filter gallery, pumping to stand-pipe. **Distribution.**—*Mains*, 1¼ miles. *Taps*, 490. *Hydrants*, 100. **Financial.**—*Bonded Debt*, $200,000 at 4%. *Annual Sinking Fund*, $4,000 at 4%. **Management.**—Three comrs.: J. T. Stevens, T. H. Dearing, J. V. Scollard.

17. **BRIDGEWATER and EAST BRIDGEWATER**, *Plymouth Co.* (Pop.: Bridgewater,|3,620—4,249; E. Bridgewater, 2,710—2,911.) Owned by Mass. Loan & Trust Co., Boston. Built in '87 by B. C. Mudge & Co. Contractors failed before works were completed. Bonds are held by one person, and company is a nominal affair kept up for benefit of bondholder. **Supply.**—Two wells, pumping to stand-pipe. Fiscal year closes Jan. 1. **Distribution.**—*Mains*. 23¼ miles. *Taps*, 361. *Hydrants*, 60; 10 added during '91. *Consumption*, 85,213 galls. **Financial.**—*Cost*, $206,551. *Cap. Stock:* authorized, $100,000; all paid up; no divi'd. *Expenses:* operating, $2,420; int., $5,273; taxes, $290; total, $7,983. *Revenue:* consumers, $5,700; city, $1,800; total, $7,500. **Management.**—Prest., B. W. Harris. Secy., R. O. Harris. Treas., I. N. Nutter. Supt., H. H. Thorndike.

NOTE.—In June, '91, it was unofficially reported that system had been bought by towns of Bridgewater and East Bridgewater. Chn. of comrs., F. D. King; Secy., F. F. Murdock.

18. **BROCKTON**, *Plymouth Co.* (Pop., 13,608—27,294.) Built in '80-1 by town. **Supply.**—Salisbury Brook, by gravity from impounding reservoir. Fiscal year

---

FOR FURTHER DESCRIPTIVE MATTER see Manual for 1859-90. CHANGES since 1890 are here given. POPULATIONS are for 1880 and 1890 respectively.

24 MANUAL OF AMERICAN WATER-WORKS.

closes Dec. 1. High service system was completed in '91, consisting of 3,000,000-gall. Worthington comp. cond. pump and 1,300,000-gall. stand-pipe. Stand-pipe is 59x62 ft., of riveted steel plates, from Cunningham Iron Works, East Boston, Mass. Reported in '91 that National filter was to be added.

Year	Mains	Taps	M't'rs	Hydrants	Cost	Bonded debt	Int. charges	Exp	Rev
'80	9	121	....	....	$166,386	.....	.....	.....	$1,610
'81	11.75	005	96	127	203,915	$200,000	.....	$1,932	3,935
'82	15.7	665	180	167	235,364	200,000	.....	.....	.....
'83	18.2	....	389	183	263,700	225,000	.....	1,526	9,150
'84	22.82	....	633	216	309,065	245,000	.....	2,068	16,707
'85	25.6	1,087	875	241	335,053	295,000	.....	3,178	18,775
'86	30.1	1,668	1,030	279	404,782	310,000	$13,950	2,751	22,604
'87	32.42	2,100	1,180	295	429,983	332,000	14,940	2,578	25,423
'58	34.15	2,450	1,283	306	452,803	405,000	17,075	4,376	27,511
'89	35.66	2,619	1,438	318	465,740	415,000	.....	3,282	28,636
'90	38.1	2,700	1,626	336	508,082	488,000	20,365	3,586	30,365

Financial.—*Bonded Debt*, $488,000 at 5, 4 and 3½%. *Sinking Fund*, $103,000. *Expenses:* total, $23,960. *Revenue:* consumers, $30,365; city, $3,000; total, $33,365. Management.—Chn., E. H. Reynolds. Secy., F. B. Gardner. Supt., W. F. Cleaveland.

19. **BROOKFIELD,** *Worcester Co.* (Pop., 2,820–3,352.) Built in '88-9 by town. Supply.—Springs, by gravity. Distribution.—*Mains*, 3 miles. *Taps*, 31. *Hydrants*, 12. Financial.—*Cost*, $18,000. *Debt*, $16,000 at 4%. *Expenses:* operating, nominal; int., $640. *Revenue:* consumers, $500. Management.—Board of three. Chn., C. L. Vizard. Clerk, E. D. Isendok.

20. **BROOKLINE,** *Norfolk Co.* (Pop., 8,057–12,103) Built in '75 by town. Supply.—Charles River, pumping to reservoir and tank; filtered. Fiscal year closes Jan. 31. During '90 supply was increased by driving additional wells. Work was to be continued during summer of '91.

Year	Mains	Taps	Meters	Hyd'ts	Consump.	Cost	B'd'd debt	Exp	Rev
'76	22.5	587	...	...	336,452	$490,901	$501,000	$5,985	$6,601
'77	23.5	767	...	...	352,041	501,172	591,000	6,428	8,777
'78	24.9	838	...	...	462,868	510,487	591,000	7,212	10,816
'79	25.5	910	...	...	490,349	591,190	591,000	6,727	11,884
'80	26.3	974	...	...	570,738	598,961	591,007	7,579	13,106
'81	28.5	1,035	...	...	477,113	621,531	591,000	8,000	14,199
'82	29.5	1,082	...	...	493,000	627,996	591,000	8,999	15,384
'83	30	1,134	63	169	549,898	636,197	591,000	8,744	16,885
'84	30.3	1,191	81	173	575,741	641,888	591,000	9,487	17,536
'85	35.1	1,278	95	189	665,295	731,175	591,000	11,255	27,100
'86	37.2	1,336	121	204	703,700	753,382	591,000	14,500	25,004
'87	38.4	1,501	156	212	768,900	768,323	591,000	13,333	23,256
'88	44	1,631	414	258	837,627	802,703	591,000	13,292	21,537
'89	44.6	1,754	475	263	779,000	818,752	651,000	15,225	26,383
'90	47.6	1,908	694	263	884,000	879,960	651,000	12,250	31,592

Financial.—*Bonded Debt*, $691,000. *Expenses:* operating, $12,250; int., withheld. *Revenue:* consumers, $31,592; city, $13,487; total, $45,079. Management.—Board of three members, one elected each year. Chn., William K. Melcher. Secy., F. H. Hunnewell. Supt., F. F. Forbes, appointed by the board.

21. **CAMBRIDGE,** *Middlesex Co.* (Pop., 52,669–70,028.) Owned by city since '65. Built in '37 by Cambridgeport Aqueduct Cc. Supply.—Stony brook, impounded, pumping to reservoir; filtered. Fiscal year closes Nov. 30.

Year	Mains	Taps	Meters	Hydrants	Consumption	Cost	B'd'd debt	Int. ch'g's	Exp	Rev
'65	26.4	2,191	8	120	974,032	$291,480	.....	.....	$7,196	$32,367
'70	60	4,191	68	243	1,739,869	885,007	.....	$49,704	19,014	92,607
'75	90	6,570	93	496	2,718,484	1,552,567	$1,372,467	85,404	28,652	138,880
'76	90	6,820	105	567	2,438,564	1,656,710	1,470,610	83,730	28,587	179,167
'77	....	6,956	118	575	2,631,732	1,672,592	1,486,492	90,000	22,614	154,844
'78	....	7,058	126	581	2,257,190	1,692,172	1,507,772	90,980	25,229	157,444
'79	....	7,183	136	579	2,432,386	1,696,597	1,510,197	91,585	25,993	164,681
'80	....	7,322	156	579	2,423,220	1,721,831	1,535,431	91,585	25,865	173,326
'81	....	7,523	157	582	2,472,108	1,728,655	1,478,500	91,585	23,834	170,063
'82	....	7,724	157	596	2,474,616	1,745,614	1,478,500	88,585	2,870	177,431
'83	....	8,004	173	599	2,703,261	1,751,345	1,459,500	88,585	32,976	179,362
'84	....	8,338	184	601	2,633,885	1,782,865	1,459,500	84,451	25,765	161,526
'85	98.5	8,731	195	609	3,082,884	2,087,378	1,747,500	92,725	20,915	185,544
'86	100.8	9,091	213	612	3,385,093	2,294,043	1,865,500	99,015	29,853	190,404
'87	102.5	9,370	219	613	3,861,400	2,436,496	1,825,500	97,355	37,611	204,719
'88	104.7	9,730	221	617	4,137,947	2,539,780	1,723,500	99,765	41,951	221,125
'89	107	10,158	231	648	4,182,025	2,610,333	1,631,500	84,735	42,603	211,156
'90	109.2	10,554	254	663	4,489,180	2,867,735	1,771,500	82,195	45,663	231,117

Financial.—*Sinking Fund*, $765,756. *Expenses:* total, $127,858. *Revenue:* con-

MASSACHUSETTS. 25

sumers. $231,117. **Management.**—Prest., C. W. Kingsley. Registrar, W. H. Harding. Supt., Hiram Nevons.

22. CANTON, *Norfolk Co.* (Pop., 4,516—4,538.) Built in '88-9 by town. Supply.—Filter well, pumping to stand-pipe. Fiscal year closes Feb. 28.

Year.	Mains.	Taps.	Meters.	Hy-drants.	Con-sumption.	Cost.	Bonded debt.	Int. charges.	Exp.	Rev.
'89-90	15.32	160	0	138	52,945	$126,643	$120,000	$2,400	$1,650	$1,608
'90-91	15.32	347	15	150	79,107	136,000	118,000	2,360	1,800	3,629

Financial,—*Bonded Debt,* $118,000, due in annual payments of $4,000 at 4%. *Expenses:* total, $3,100. *Revenues:* consumers, $3,629. **Management,—**Three comrs., one elected every year. Chn., Jno. Everett. Secy., Robt. Bird. Supt., Chas. Galligan, appointed by comrs.

14. CHELSEA, *Suffolk Co.* (Pop., 22,782—27,909.) Statistics included under Boston.

23. CHESHIRE, *Berkshire Co.* (Pop., 1,537—1,308.) Built in '75 by Cheshire Water Co.; franchise provides neither for control of rates nor purchase of works by town. Supply.—Brooks, fed by mountain spring, by gravity from impounding reservoir; filtered. Fiscal year closes second Tuesday in May. Distribution.—*Mains,* 4 miles. *Taps,* 125. *Hydrants,* 17. Financial,—*Cost,* $17,000. *Cap. Stock:* authorized, $15,000, all paid up; paid 6% dividend in '90. *No Debt. Expenses:* operating, $100; no int.; taxes, $223; total, $323. *Revenue:* consumers, $700. Management.—Prest., J. D. Northup. Secy., G. Z. Dean. Treas., J. G. Northup. Supt., H. A. Northup.

24. CHICOPEE, *Hampden Co.* (Pop., 11,286—14,050.) Built in '87 by Chicopee Water Co. Supply.—Brook and 6 large wells, pumping to tank. Fiscal year closes Nov. 30. Distribution.—*Mains,* 12 miles. *Taps,* 875. *Meters,* 6. *Hydrants,* 38. Financial.—*Cost,* $86,252. *Cap. Stock:* authorized, $75,000, all paid up; paid 6% div'd in 1890. *Total Debt,* $18,600; floating, $600; bonded, $18,000. *Expenses:* operating, $2,500; int., $890; taxes, $1,125; total, $4,515. *Revenue:* consumers, $8,300; city, $1,500; total, $9,300. Management.—Prest., Emerson Gaylord. Treas. and Supt., E. R. Stickney.

25. CHICOPEE FALLS, *Hampden Co.* (Pop., 4,000.) Built in '81-3 by village and Chicopee Manufacturing Co. Supply.—Chicopee River and springs, pumping to reservoir. Fiscal year closes April 30. During 1891 Chicopee and Chicopee Falls were consolidated, but the two systems of water-works remain separate. A 4,000,-000-gall. pump was added in 1891.

Year.	Mains.	Taps.	Hydrants.	Cost.	Rev.
'84	4	200	40	$20,000	$803
'89-90	4.24	209	50	22,919	3,321
'90-1	4.78	210	50	25,611	3,686

Financial.—*Total Debt,* $12,350, floating, at 4%. *Expenses,* $126. *Revenue:* consumers, $3,321; city, $365; total, $3,686. Co. has one-half revenue from consumers and is paid $365 for hydrant rental. Management.—Prudential Com Prest., N. R. Wood. Secy., Henry J. Boyd.

26. CLINTON, *Worcester Co.* (Pop., 8,029—10,424.) Built in '82 by town. Supply.—Springs and surface water, by gravity from storage and impounding reservoir. Fiscal year closes Jan. 31.

Year.	Mains.	Taps.	Meters.	Hy-drants.	Cost.	Bonded debt.	Int. charges.	Exp.	Rev.
'82	11.26	....		71	$194,663	$200,000	$8,000	....	
'83	16.61	451	15	99	238,319	213,000	8,520	....	$16,958
'84	22.32	672	25	180	265,361	240,000	8,960	$2,164	17,673
'85	23.11	754	..	184	274,850	248,000	9,500	2,909	19,576
'86	24.05	854	..	188	280,926	250,000	9,640	3,340	20,060
'87	25.2	951	..	189	287,710	253,000	10,310	3,618	20,109
'88	25.73	1,023	37	190	304,715	270,000	6,760	5,376	20,923
'89	25.9	1,082	47	190	311,490	271,000	10,680	....	
'90	26.5	1,148	50	192	316,699	275,000	11,160	3,782	22,369

Financial.—*Bonded Debt,* $275,000 at 4%. In '90 legislature authorized issue of additional bonds to amount of $50,000. *Sinking Fund,* $28,209; $19,000 invested in town bonds. *Expenses:* total, $14,942. *Revenue:* consumers, $22,369. Management.—Three comrs.: J. E. Howe, Jno. W. Corcoran, H. A. Thissell. Supt., H. H. Lowe.

27. COCHITUATE, *Middlesex Co.* (Pop., 1,161—not given.) Built in 78 by town. Supply.—Snake River, by gravity from impounding reservoir, with direct pumping for fire protection. Fiscal year closes March 1. Distribution.—*Mains,*

FOR FURTHER DESCRIPTIVE MATTER see Manual for 1889-90. CHANGES since 1890 are here given. POPULATIONS are for 1880 and 1890 respectively.

6 miles. *Taps*, 290. *Meters*, 0. *Hydrants*, 33. **Financial.**—*Cost*, $30,000. *Bonded Debt*, $30,000; $25,000 at 5%; $5,000 at 4%. *Sinking Fund*, $8,000 at 4%. *Expenses:* operating, $250; int., $1,450; total, $1,700. *Revenue:* consumers, $2,600; city, $584; total, $3,184. **Management.**—Secy., W. H. Bent. Supt., Ralph Bent. Chn., A. H. Bryant.

28. **COHASSET**, *Norfolk Co.* (Pop., 2,182--2,448.) Built in '86 by Cohasset Water Co. **Supply.**—65 driven wells, pumping to reservoir. Fiscal year closes July 15. **Distribution.**—*Mains*, 7 miles. *Taps*, 194. *Meters*, 5. *Hydrants*, 35. **Financial.**—*Cost*, $76,167. *Cap. Stock*, authorized, $50,000; paid up, $30,100. *Total Debt*, $47,091; floating, $17,091; bonded, $30,000. *Revenue*: city, $1,750. **Management.**—Prest., C. A. Welsh, Boston. Treas., Waldo Higginson. Secy., Caleb Lathrop. Supt., D. N. Tower.

29. **CONCORD**, *Middlesex Co.* (Pop., 3,922--4,427.) Built in '74 by town. **Supply.**—Sandy pond, by gravity. Fiscal year closes March 1. **Distribution.**—*Mains*, about 23.3. *Taps*, about 600. *Meters*, 5. *Hydrants*, 112. **Financial.**—*Cost*, $150,000. *Bonded Debt*, $111,000 at 4%. *Sinking Fund*, $22,090. *Expenses*, operating, $1,250; int., $4,440; total, $5,690. *Revenue*, consumers, $12,500. Revenue pays all expenses. **Management.**—Five comrs., elected for three years. Chn., Sam'l Hoar. Secy., Jno. C. Friend. Supt., Jno. O. Haskell, appointed by comrs.

30 **COTTAGE CITY**, *Dukes Co.* (Pop., 672—1,080.) **History.**—Built in '90 by Cottage City Water Co., under perpetual franchise which fixes rates and provides for purchase of works by town at any time, price determined by arbitration or by three appraisers appointed by court. Put in operation June 15. Engrs. and contrs., Wheeler & Parks, Boston. Fiscal year closes April 30. Fire-protection system was built by fire district in '84. **Supply.**—Flowing springs, pumping to stand-pipe. Springs have dy. cap. of 1,500,000 galls. **Pumping Machinery.**—Dy. cap., 2,000,000 galls.; one 1,200,000 Deane comp. dup. and one 800,000 Knowles high-pressure. *Boilers:* from E. Hodge & Co., E. Boston. **Stand-Pipe.**—Cap., 165,000 galls.; 20 × 70 ft.; from E. Hodge & Co., E. Boston. **Distribution.**—*Mains*, 12 to 4-in., c. i., 8.8 miles. *Services*, enamel. Pipe from McNeal Pipe & Foundry Co.; specials from Builders Iron Foundry, Providence, R. I. *Taps*, 195. *Meters*, 0. *Hydrants*, Coffin, 40. *Valves*, Coffin. **Financial.**—*Cost*, $130,000. *Cap. Stock*: authorized, $80,000; all paid up. *Bonds:* authorized, $75,000; issued, $50,000 at 5%, due in 1910. *Expenses:* operating, $2,317; int., $1,250; taxes, $61; total, $3,648. *Revenue:* consumers, $2,054; city, $1,467; other purposes, $619; total, $4,140. **Management.**—Prest., L. H. Fuller, Putnam, Conn. Clerk, J. F. J. Mulhall; Treas., Wm. Wheeler; both of Boston. Supt., H. J. Greene.

31. **DALTON**, *Berkshire Co.* (Pop., 2,052—2,885.) Built in '84 by fire district. **Supply.**—Mountain brook, by gravity from reservoir. Fiscal year closes April 18. **Distribution.**—*Mains*, about 12 miles. *Taps*, 498. *Hydrants*, 69. **Financial.**—*Cost*, $57,000. *Total Debt*, $54,500; floating, $4,500; bonded, $50,000 at 4%, due in 1914. *Expenses;* operating, $623; int., $2,225; total, $2,853. *Revenue:* total, $4,544. **Management.**—Three comrs. Secy., W. B. Clark. Treas., Jno. D. Carson. Supt., Geo. Pike.

32. **DANVERS**, *Essex Co.* (Pop., 6,598—7,454.) Built in '76 by town. Supplies villages of Tapleyville, Danvers Centre and Danversport, in town of Danvers and Middleton, on pipe line. **Supply.**—Middleton and Swan ponds, pumping to reservoir. Fiscal year closes Jan 31.

Yr.	Mains.	Taps.	Meters.	Hy-drants.	Con-sumption.	Cost.	Bonded debt.	Int. ch'g's.	Exp.	Rev.*
'82	29.2	881	20	196	400,000	$222,479	$205,000	......	$6,562	$11,205
'84	31.7	1,008	20	200	463,500	236,170	220,000	$10,600	5,993	18,853
'86	33.1	1,148	20	201	430,930	226,960	220,000	10,600	5,487	19,929
'88	37.5	1,262	30	207	450,000	256,257	220,000	10,600	6,011	22,499
'90	38.9	1,332	35	216	500,000	259,491	220,000	11,512	5,931	22,419

*Including allowance for hydrant rental.

Financial.—*Bonded Debt*, $220,000 at 5%. *Sinking Fund*, $42,970. *Expenses:* operating, $5,931; int., $11,512; sinking fund, $5,156; total, $22,599. *Revenue:* consumers, $18,019; town, $4,400; total, $22,419. **Management.**—Chn., R. R. Lewis. Secy., C. F. Putnam. Registrar, Henry Newhall.

32. **DANVERS CENTRE**, *Essex Co.* (Pop. not given.) Supply from Danvers.

.32. **DANVERSPORT**, *Essex Co.* (Pop. not given.) Supply from Danvers.

MASSACHUSETTS.

33. DEDHAM, *Norfolk Co.* (Pop., 6,233—7,123.) Built in '81 by Dedham Water Co. Supply.—Well, pumping to stand-pipe. Fiscal year closes Jan. 13.

Year.	Mains.	Taps.	Hydrants.	Consumption.	Cost.	B'd'd debt.	Int. ch'g's.	Exp.	Rev.
'82	10	247	80	80,189	$130.511	.....	.....	$3,472	$9,886
'85	13.5	536	....	129,474	142,263	$35,000	$1,404	4,875	12,461
'88	17	726	127	180,063	163,028	65,000	2,550	7,243	14,423
'89	20	766	128	203,005	186,008	90,000	3,875	5,631	16,190
'90	20.3	814	129	224,541	198,414	95,000	4,625	6,026	16,812

Financial.—*Cap. Stock:* authorized, $100,000; all paid up; paid 4% dividend in 1890. *Bonded Debt:* $95,000 at 5%. *Expenses:* operating, $6,026; interest, $4,625; taxes, $1,412; total, $12,063. *Revenue:* total, $16,810. Management.—Prest., Winslow Warren. Treas., Jno. R. Bullard. Supt., Wm. F. Hill.

17. EAST BRIDGEWATER, *Plymouth Co.* (Pop., 2,710—2,911.) Supply from Bridgewater.

34. EASTHAMPTON, *Hampshire Co.* (Pop., 4.206—4,395.) Built in '73 by town. Supply.—Williston pond and springs, pumping direct from former and gravity from latter. Authority to issue $50,000 of bonds for additional supply was granted in '91. Fiscal year closes Jan. 31. Distribution.—*Mains,* about 5 miles. *Hydrants,* 75. Financial.—*Cost,* $25,000. *Bonded Debt,* $15,000 at 4%. *Expenses:* operating, $1,210; int , $600; total, $1,810. *Revenue,* consumers, $2,051. Management.—Committee: Jno. Mayher, L. S. Clark. A. B. Lyman, Geo. L. Manchester, A. J. Fargo.

14. EVERETT, *Middlesex Co.* (Pop., 4,159—11,068.) Supply from Mystic System, Boston.

35. FALL RIVER, *Bristol Co.* (Pop., 48,961—74,398.) Built in '71-4, by city. Supply.—Watuppa Lake, pumping to stand-pipe and direct. Fiscal year closes Dec. 31.

Year.	Mains.	Taps.	Meters.	Hydrants.	Consumption.	Cost.	Bonded debt.	Int. ch'g's.	Exp.	Rev.
'74	....	672	53	....	507,108	$828,190	....	$57,695	$23,262	$9,337
'75	....	1,147	193	....	810,980	1,241,885	....	67,660	29,422	21,439
'76	....	1,660	585	472	1,057,704	1,320.539	....	89,630	31,924	29,003
'77	48.43	2,060	881	502	1,173,601	1,381,700	....	95,910	17,802	36,814
'78	51.66	2,324	1,165	534	1,204,207	1,415,972	....	96,220	18,025	41,980
'79	51.66	2,497	1,372	537	1,263,925	1,432,906	....	102,275	15,408	44,691
'80	52.05	2,685	1,583	537	1,353,641	1,451,364	....	100,620	15,526	49,706
'81	....	2,906	1,780	....	1,488,247	....	....	....	22,536	56,587
'82	54.2	3,120	1,966	555	1,830,801	....	....	....	22,105	61,301
'83	55.5	3,370	2,187	575	1,640,481	1,510,000	....	....	38,282	66,561
'84	56.61	3,611	2,421	586	1,425 861	1,555,819	1,700,000	95,650	22,876	66,768
'85	57.84	3,818	2,569	599	1,488,137	1,585,640	1,700,000	97,920	29,821	68,475
'86	58.79	3,986	2,725	606	1,903,482	1,719,546	1,700,300	97,030	26,522	75,759
'87	59.73	4,197	2,941	613	1,590,960	1,754,442	1,700,000	96,410	25,045	83,425
'88	61.37	4,412	3,138	625	1,768,524	1,814,004	1,700,000	96,970	26,009	85,108
'89	62.6	4,698	3,128	634	1,877,937	1,861,439	1,700,000	96,540	23,927	91,909
'90	63.6	4,980	3,717	642	2,136,182	1,901,406	1,700,000	96,640	37,207	100,249

Financial.—*Bonded Debt,* $1,700,000; $1,300,000 at 6%, due '92-1900, 1904-6; $300,000 at 5%, due 1908-9; $100,000 at 4%, due in 1900. *Sinking Fund,* $200,495 *Expenses:* total, $133,847. *Revenue:* Consumers, $100,349; city, $30,000; total, $130,349. Revenue from private consumers and annual appropriation from city pay operating expenses and interest. Extensions are paid for by special appropriations. Management.—Three trustees, one elected every year. Registrar, Wm. W. Robertson. Supt., Patrick Kieran.

36. FITCHBURG, *Worcester Co.* (Pop., 12,429—22,037.) Built in '71-2, by city. Supply.—Scott and Falulah brooks, by gravity from storage reservoirs. Fiscal year closes Dec. 1.

Year.	Mains.	Taps.	Meters.	Hydrants.	Cost.	B'd'd debt.	Int. ch'g's.	Exp.	Rev.
'72	....	....	....	....	$173,000	....	....	$1,474	$2,108
'81	26.	1,255	93	203	439,500	$400,000	$24,000	4,813	16,267
'83	29.3	1,515	197	212	510,646	400,000	....	6,001	20,529
'85	32.2	1,825	360	235	531,569	400,000	....	17,397	25,600
'86	33.52	1,999	360	245	531,569	485,000	16.612	11,011	24,999
'87	34.7	2,203	600	256	621,701	500,000	27,930	17,574	30,203
'88	38.11	2,456	657	268	643,332	500,000	27,930	22,342	33,761
'89	41.06	2,624	793	281	644,941	500,000	27,930	25,577	40,526
'90	43.66	2,905	952	297	658,949	525,000	27,930	29,657	41,587

Financial.—*Bonded Debt,* $525,000; av. int., 3½%. *Sinking Fund,* $215,257. *Expenses:* total, $57,527. *Revenue:* consumers, $41,587; city, $16,000; total, $57,587. Management.—Comrs. Prest , S. D. Sheldon. Secy. and Supt., F. C. Lovell.

For Further Descriptive Matter see Manual for 1889 90. Changes since 1890 are here given. Populations are for 1880 and 1890 respectively.

36. **FLORENCE**, *Hampshire Co.* (Pop., not given.) Supply from Northampton.

37. **FRANKLIN**, *Norfolk Co.* (Pop., 4,051—4,831.) Built in '84 by Franklin Water Co., under 25 yrs. franchise, which does not regulate rates but provides for purchase of works by town at price determined by the courts. Supply.—Two wells, pumping to tank. Fiscal year closes Nov. 30. Distribution.—*Mains*, 9 miles. *Taps*, 300. *Meters*, 6. *Hydrants*, 67. *Consumption*, 78,000 galls. Financial.—*Cost*, $131,592. *Cap. Stock:* authorized, $75,000; all paid up. *Bonded Debt*, $50,000 at 5%. *Revenue:* city, $3,600. Management.—Prest., G. Ray. Secy., Geo. W. Wiggin. Supt., Wm. E. Nason. Treas., G. N. Weaver.

38. **GARDNER**, *Worcester Co.* (Pop., 4,998—8,424.) Built in '82 by Gardner Water Co.; franchise provides that works may be purchased by city in 1902; city has no control of rates. Fiscal year closes May 1. Supply.—Crystal Lake, pumping to reservoir. Distribution.—*Mains*, 15½ miles. *Taps*, 800. *Meters*, 32. *Hydrants*, 81. Financial.—*Cost*, $151,144. *Cap. Stock:* authorized, $100,000. *Total Debt*, $76,100; floating, $26,100; bonded, $50,000 at 6%. Management.—Prest. and Treas., C. H. Green, Northfield. Secy., N. W. Howe. Supt., H. W. Conant.

39. **GLOUCESTER**, *Essex Co.* (Pop., 19,329—24,661.) Built in '85 by Gloucester Water Supply Co.; franchise does not control rates, but provides for purchase of works by town. Supply.—Dyke's pond, pumping to reservoir. Fiscal year closes April 3. Distribution.—*Mains*, 28 miles. *Taps*, 925. *Meters*, 30. *Hydrants*, 175. *Consumption*, 300,000 galls. Financial.—*Cost:* For '89, $525,732; for '90, $560,310. *Cap. Stock:* Authorized, $250,000, all paid up. *Bonded Debt*, $250,000 at 5%. *Expenses:* Operating, withheld; int., $12,500. *Revenue:* Consumers, withheld; city, $8,500. Management.—Prest., J. O. Proctor. Treas., G. Norman Weaver, Newport, R. I. Secy., D. W. Low. Supt., Jno. Moran.

NOTE.—In '91 city was considering question of purchasing works.

40. **GRAFTON and NORTH GRAFTON**, *Worcester Co.* (Pop., 4,030—5,002.) Built in '86-7 by National Water-Works Syndicate, Boston; franchise does not regulate rates, but provides for purchase of works by town; price, in '91, $86,000, $1,500 added annually to price thereafter. Supply.—Surface and artesian wells, pumping to stand-pipe. Fiscal year closes June 11. Distribution.—*Mains*, 10.1 miles. *Taps*, 260. *Hydrants*, 97. *Consumption*, 100,000 galls. Financial.—*Assets*, $81,314. *Cap. Stock:* authorized, $30,000; all paid up. *Total Debt*, $81,314; floating, $1,314; bonded, $80,000; $65,000 at 5%, $15,000 at 6%. *Expenses:* operating, $2,500; int., $4,150; taxes, $100; total, $6,750. *Revenue:* city, $2,500. Management.—Prest., F. S. Parnell. Secy., C. F. Ferris. Treas. and Supt., Solon F. Smith.

41, 42. **GREAT BARRINGTON**, *Berkshire Co.* (Pop., 4,653—4,612.)

GREAT BARRINGTON WATER CO.

Built in '67. Supply.—Mountain stream, by gravity from reservoir, and by pumping. Distribution.—*Mains* (in '87), 3 miles. *Taps* (in '87), 315. *Hydrants*, 29. Financial.—*Cap. Stock*, $28,000. No debt. Management.—Treas., F. T. Whiting.

BERKSHIRE HEIGHTS WATER CO.

Built in '86-9. Supply.—Green River, pumping to reservoir. Distribution.—*Mains*, 1 mile. *Taps*, 15. *Hydrants*, 1. Financial.—*Cost*, $40,319. *Cap. Stock:* authorized, $50,000; paid-up, $20,000. *Floating Debt*, $13,006 at 4%. *Expenses:* operating, withheld; int., $520; taxes, $58. *Revenue:* "no collections." Management.—Prest. and Supt., E. D. Brainard. Secy., M. E. Tobey.

NOTE.—In 1890 town offered to buy plants of both co.'s, but no price being agreed upon, comrs. were appointed by Supreme Court; final hearing was to be July 1, '91.

43. **GREENFIELD**, *Franklin Co.* (Pop., 3,903—5,252.) Built in '70 by fire district. Supply.—Brook, by gravity from impounding reservoir. Fiscal year closes May 1. Distribution.—*Mains*, about 22 miles. *Taps*, 1,000. *Meters* (in '88), 6. *Hydrants*, 71. Financial.—*Cost*, $161,227. *Expenses:* Operating, $989; int., not given. *Revenue:* Consumers, $12,986. Management.—Three comrs., one elected every year. Chn., Chas. J. Day. Clerk and Supt., Jas. Porter. Appointed by comrs.

44. **HAVERHILL**, *Essex Co.* (Pop., 18,472—27,412.) Built in 1801 by Haverhill Aqueduct Co.; franchise does not fix rates, but provides for purchase of works by town, price determined by arbitration or by commissioners appointed by Supreme Court. Supply.—Springs and ponds, by gravity and pumping to stand-pipe. Fiscal year closes April 30. Distribution.—*Mains* (in '87), 40 miles. *Taps*, 3,250. *Meters*, 8.

Hydrants, 200, owned by city. Financial.—Withheld. Management.—Prest., L. C. Wadleigh. Secy. and Supt., C. W. Morse. Treas., J. H. Castleton.

NOTE.—In July, '91, city government passed an order for purchase of works, and authorized the mayor to petition for appraisers.

45. HINGHAM and HULL, *Plymouth Co.* (Pop.: Hingham, 4,485—4,564; Hull, 383—289.) Built in '79-80 by Hingham Water Co.; town does not control rates, but has right to buy works. Supply.—Ponds, by gravity, and pumping to stand-pipe and receiving basins. Fiscal year closes June 30. Distribution.—*Mains*, 41.3 miles. *Taps*, 1,255. *Meters*, 2. *Hydrants*, 150. Financial.—*Cost*, $265 610 *Cap. Stock:* authorized, $120,000; all paid-up; paid 6% div'd in 1890. *Total Debt*, $127,000; floating $7,000; bonded, $120,000 at 6%. Management.—Prest., E. L. Ripley. Socy., Starkes Whitin. Supt., C. M. S. Seymour.

46. HINDSDALE, *Berkshire Co.* (Pop., 1,595—1,739.) Built in '89 by fire district. Supply.—Springs and brooks, by gravity from reservoir. Fiscal year closes April 1. Distribution.—*Mains*, 4 miles. *Taps*, 160. *Meters*, 0. *Hydrants*, 43. Financial.—*Cost*, $36,750. *Bonded Debt*, $35,000 at 4%. Management.—Three comrs., one elected every year. Chn., G. T. Plunkett.

47 HOLBROOK, *Norfolk Co.* (Pop., 2,130—2,474.) Built in '87-8 by towns of Randolph and Holbrook jointly, each town owning its own distributing system. Supply.—Lake, pumping to a stand-pipe. Fiscal year closes Feb. 28. Feb. 28, '91, new boiler had been erected and town was receiving proposals for new pump.

Year.	Mains.	Taps.	Meters.	Hyd'ts.	Cost.	B'd'd debt.	Int. ch'g's.	Exp.	Rev.
'88-9	9	270	0	64	$92,717	$100,000	$2,000	$2,496	$3,143
'89-90	9.2	315	0	65	97,352	100,000	4,000	2,403	3,260
'90-1	9.3	345	0	67	102,593	100,000	4,000	3,322	3,735

Financial.—Cost of joint works, of which one-half is paid by each town, $133,615; additional cost to Holbrook, $33,128; total cost to Holbrook, $102.593. *Bonded Debt*, $100,000 at 4%. *Sinking Fund*, $7,501. *Expenses:* total, $7,322. *Revenue:* consumers, $3,735; city, $2,500; total, $6,235. Management.—Three comrs.: W. F. Gleason, F. G. Morse, Geo. W. Wilde. Supt., E. J. Chadbourne.

48. HOLLISTON, *Middlesex Co.* (Pop., 3,098—2,619.) History.—Built in '91 by Holliston Water Co.; to be put in operation during summer of '91. Supply.—Wells and springs, pumping to stand-pipe. Pumping Machinery.—Dy. cap., 1,000,000 galls.; two Deane. Stand-Pipe.—Is 30 x 60 ft., from Cunningham Iron Works, Boston. Distribution.—Mains, 12 and 10-in., i. Pipes and specials from R. D. Wood & Co., Philadelphia, Pa. *Service*, lead. Financial.—Withheld. Management.—Prest. and Supt., John D. Stupper. Treas., Z. Talbott.

49. HOLYOKE, *Hampden Co.* (Pop., 21,915—35,637.) Built in '73 by city. Supply.—Wright's and Ashley's ponds, by gravity. Fiscal year closes Dec. 31. Whiting Street dam and reservoir built '89-90; cap., est., 600,000,000 galls.; floods 112 acres. Dam of masonry, 1,773 ft. long, 21 ft. high above natural surface; foundation carried 5 ft. below natural surface. Engr., E. A. Ellsworth, Holyoke. Contrs. for masonry, Delaney Bros., Holyoke. *Cost*, $80,000; $60,000 for masonry, $20,000 for other work. For cross-section and further description see ENGINEERING NEWS, Vol. XXIV., p. 204.

Year.	Mains.	Taps.	Meters.	Hy-drants.	Con-sumption.	Cost.	Bonded debt.	Int. ch'gs.	Exp.	Rev.
'80	23.1	...	...	150	...	$291,438	$250,000	$15,000	$3,942	$32,293
'85	36.4	2,019	126	248	2,000,000	432,664	250,000	15,000	11,213	57,335
'86	39.9	2,038	140	261	2,100,000	462,928	250,000	15,000	12,500	58,300
'87	42.4	2,173	149	274	2,246,622	480,373	250,000	15,000	8,713	57,634
'88	45	2,285	143	284	2,380,000	513,663	250,000	15,000	10,340	58,955
'89	46.5	2,454	145	301	2,482,393	576,739	250,000	15,000	9,209	61,739
'90	49.1	2,583	149	315	2,548,045	619,365	250,000	15,000	9,910	63,121

Financial.—*Bonded Debt*, $250,000 at 6%, due in 1900. *Sinking Fund*, $112,030, invested in first mortgage on real estate at 5%. *Expenses:* total, $24,910. *Revenue:* consumers, $63,121; city, $3,298; total, $66,419. Revenue covers all expenses, including redemption of bonds. Management.—Three comrs., appointed every three years by aldermen. Chn., Maurice Lynch. Secy., Moses Newton. Supt., J. D. Hardy, appointed by comrs.

73. HOPEDALE, *Worcester Co.* (Pop., 2,499—2,623.) Supply from Milford.

50. HOPKINTON, *Middlesex Co.* (Pop., 4,607—4,068.) Built in '84 by town. Supply.—Four artesian wells, pumping to tank. Fiscal year closes Feb. 1. In '89

FOR FURTHER DESCRIPTIVE MATTER see Manual for 1889-90. CHANGES since 1890 are here given. POPULATIONS are for 1880 and 1890 respectively.

30   MANUAL OF AMERICAN WATER-WORKS.

supply was increased by additional well, at cost of $1,000. Distribution.—*Mains,* about 5 miles. *Taps,* 379. *Meters,* 9. *Hydrants,* 34. Financial.—*Cost,* $36,000. *Floating Debt,* $34,041. *Sinking Fund,* $4,753. *Expenses:* operating, $1,297; int., $1,283; total, $2,580. *Revenue:* consumers, $2,392; city, $900; total, $3,392. Management.—Comrs., J. A. Woodbury, Bernard Mahon, Henry McFarland.

51. **HOUSATONIC,** *Berkshire Co.* (Pop., not given.) Built in '88 by Housatonic Water Co. Supply.—Long Lake, by gravity. Fiscal year closes June 30. Distribution.—*Mains,* 6.5 miles. *Taps,* 110. *Meters,* 3. *Hydrants,* 16. *Consumption,* 100,000 galls. Financial.—*Cost* (in '88), $60,500. *Cap. Stock:* authorized, $40,000; all paid up. *Total Debt,* $40,917; floating, $947; bonded, $40,000 at 6%, due in 1908. Management.—Prest., M. J. Drummond, 192 Broadway, N. Y. City. Secy., O. C. Houghtaling. Supt., T. Z. Potter.

52. **HUDSON,** *Middlesex Co.* (Pop., 3,739—4,670.) Built in '84 by town. Supply.—Gates Pond, by gravity. Fiscal year closes Dec. 31.

Yr.	Mains.	Taps.	Metrs.	Hyd'ts.	Cost.	Debt.	Int. ch'g's.	Exp.	Rev.
'84	8.2	104	0	60	$38,178	$75,000	$1,280	$69	none
'85	8.35	241	0	82	71,441	77,727	4,171	522	$2,047
'86	10.12	323	1	79	83,290	79,290	4,360	722	4,121
'87	10.93	380	1	85	87,383	81,383	3,220	600	4,896
'88	11	422	1	85	88,928	80,927	3,113	563	5,628
'89	12.81	482	1	99	98,726	86 726	3,823	588	6,102
'90	13.61	548	1	103	103,221	91,221	4,057	679	7,214

Financial.—*Total Debt,* $91,221; floating, $23,221; bonded, $68,000 at 4%, due in annual payments of $2,000 for 5 yrs., $2,500 for 10 yrs. and $3,000 for 10 yrs. thereafter. *No Sinking Fund. Expenses:* total, $4,057. *Revenue:* consumers, $7,214; city, $1,000; total, $8,214. Management.—Three comrs., one elected every year. Chn., Edw. M. Stone. Secy., Herman C. Tower. Supt., D. W. Stratton.

45. **HULL,** *Plymouth Co.* (Pop., 383—989.) Supply from Hingham.

53. **HYDE PARK,** *Norfolk Co.* (Pop., 8,088—10,193.) Built in '85 by Hyde Park Water Co.; franchise does not fix rates but provides for purchase of works by city. Supply.—Driven wells, pumping to reservoir. Fiscal year closes March 31. During '90 pumping capacity was increased by a 1,000,000 Blake pump; a 500 000-gall. steel tank was also erected.

Year.	Mains.	Taps.	Meters.	Hydrants.	Cost.	B'd'd debt.
'86-7	18	400	5	104	$150,000	none
'87-8	22	1,000	30	104	......	none
'88-9	23	1,150	40	104	......	......
'89-90	23 9	1,254	66	104	......	$75,000
'90-1	24.9	1,369	66	104	189,568	100,000

Financial.—*Cap. Stock:* authorized, $100,000, all paid up; paid 6% divd. in 1891. *Total Debt,* $101,472; floating, $1,472; bonded, $100,000 at 5%. Management.—Prest., Robert Bleakie. Clerk, B. F. Allen. Treas., Chas. F. Allen. Supt., Albert S. Adams.

54. **KINGSTON,** *Plymouth Co.* (Pop., 1,524—1,659.) Built in '86 by town; first supplied in '04 by Co. Supply.—Well and river, pumping direct, surplus water passing to reservoir. Fiscal year closes Dec. 30.

Year.	Mains.	Taps.	Meters.	Hyd'ts.	Cost.	B'd'd d't.	Int. ch'g's.	Exp.	Rev.
'86	7.75	81	0	50	$42,964	$40,000	$1,600	$650	$407
'87	8	137	0	56	43,953	40,000	1,600	1,546	608*
'88	8	202	0	58	46,183	40,000	1,600	1,729	1,704
'89	8	226	0	59	52,269	38,500	1,440	1,095	2,264
'90	8.5	236	0	60	53,460	38,000	1,520	819	2,423

*For six months.

Financial.—*Bonded Debt,* $38,000 at 4%. *Expenses:* total, $2,339. *Revenue:* consumers, $2,423; city, none. Management.—Three comrs., one elected every year. Chn. and Supt., Green Evans.

55. **LANCASTER,** *Worcester Co.* (Pop., 2,003—2,201.) Built in '85 by Lancaster Water Co., under 15-years' franchise. Supply.—Clinton Water-Works. Fiscal year closes March 1. Distribution.—*Mains,* 8 miles. *Taps,* 100. *Meters,* 10. *Hydrants,* 25. Financial.—*Cost,* $40,000. *Cap. Stock,* $20,000. *Bonded Debt,* $20,000 at 5%. *Expenses:* operating, withheld; int., $1,000. *Revenue:* consumers, withheld; city, $1,250. Management.—Prest., E. V. R. Thayer. Treas., J. T. Langford. Boston. Supt., J. F. Philbin, of Clinton Water-Works.

56. **LAWRENCE,** *Essex Co.* (Pop., 39,151—44,654.) Built in '73-5 by city. Sup-

MASSACHUSETTS.

ply.—Merrimack River, pumping to reservoir and direct. Fiscal year closes Dec. 31.

Year.	Mains.	Taps.	Meters.	Hydrants.	Consumption.	Cost.	B'd'd debt.	Int. ch'g's.	Exp.	Rev.
'75	...	...	...	...	...	$1,303,735	$1,300,000	...	...	$2,095
'76	33.2	1,973	...	414	...	1,461,786	1,300,000	$50,267	$10,884	15,205
'77	36.6	2,413	89	442	1,554,058	1,543,728	1,300,000	78,000	20,673	44,626
'78	39.6	2,754	138	455	1,250,290	1,613,998	1,300,000	78,000	13,728	43,558
'79	40.6	3,011	200	461	1,456,029	1,659,138	1,300,000	78,000	13,342	57,473
'80	42.17	3,224	272	464	1,866,440	1,706,072	1,300,000	78,000	22,196	62,670
'81	43.87	3,459	332	466	1,837,553	1,744,777	1,300,000	78,000	18,808	68,332
'82	45.3	3,653	421	470	1,950,658	1,792,017	1,300,000	78,000	19,702	71,281
'83	48.3	3,896	555	479	2,162,919	1,833,843	1,300,000	78,000	18,715	71,711
'84	49.71	4,059	609	481	2,152,567	1,861,908	1,300,000	78,000	16,822	77,121
'85	51.8	4,215	680	490	2,206,265	1,894,654	1,300,000	78,000	16,173	72,814
'86	51.8	4,356	792	494	2,572,666	1,924,015	1,300,000	78,000	17,609	76,418
'87	53.64	4,456	918	496	2,520,115	1,953,840	1,300,000	78,000	23,242	79,563
'88	55.1	4,586	1,050	500	2,667,259	1,983,325	1,300,000	78,000	18,410	71,908
'89	56.15	4,713	1,213	502	2,633,010	2,013,900	1,300,000	78,000	17,867	71,158
'90	57.3	4,861	1,402	506	2,770,502	2,037,855	1,100,000	72,000	18,097	73,980

Financial.—*Bonded Debt*, $1,100,000 at 6%; sinking fund, $237,408. *Expenses:* total, $90,097. *Revenue:* consumers, $73,989. **Management.**—Prest., Jesse Moulton. Registrar, H. F. Whittier. Supt., A. H. Salisbury.

57. LEE, *Berkshire Co.* (Pop., 3,939—3,785.) Built in '81 by Berkshire Water Co. **Supply.**—Mountain stream, by gravity, from impounding reservoir. Fiscal year closes April 1. **Distribution.**—*Mains*, 28 miles. *Taps*, 200. *Meters*, 0. *Hydrants*, 21. *Consumption*, unk'n. **Financial.**—*Cost*, $40,319. *Cap. Stock:* authorized, $20,000; all paid up; paid 6% div'd in 1891. *Total Debt*, $21,500; floating, $1,500; bonded, $20,000 at 5%. *Expenses:* operating, $100; int., $1,000; taxes, $375; total, $1,475. *Revenue:* consumers, $3,200; city, $400; total, $3,600. **Management.**—Prest., C. L. Goodhue. Treas., Thos. N. Birnie. Secy., R. D. Anable. Supt., J. C. Chaffee.

58. LEEDS, *Hampshire Co.* (Pop., not given.) Supply from Northampton.

Leicester, *Worcester Co.* (Pop., 2,770—3,120.) Construction started in '89 by water supply district. Works not completed time of last report, April 13 '91. Proposed supply, wells, pumping to stand-pipe. **Financial.**—*Cost*, April 13, '91, $10,230. *Debt*, $10,000 at 4%. **Management.**—Three comrs.: H. O. Smith, David Bemis, C. W. Warren.

58. LENOX, *Berkshire Co.* (Pop., 2,(43—2,889.) Built in '75 by Lenox Water Co. **Supply.**—Springs and river, by gravity from impounding reservoirs, and by pumping to reservoirs in case of need. Fiscal year closes July 1. **Distribution.**—*Mains*, 12.3 miles. *Taps*, 18.3. *Meters*, 2. *Hydrants*, 8. **Financial.**—*Cost*, $43,142. *Cap. Stock:* authorized, $30,000; paid up, $22,000. *Total Debt*, $26,013; floating, $43; bonded, $26,000. In '90 legislature authorized issue of additional bonds. **Management.**—Prest., Thos. Port. Secy., Edw. McDonald. Treas., Wm. D. Curtis. Supt., I. I. Newton.

59. LEOMINSTER, *Worcester Co.* (Pop., 5,772—7,269.) Built in '73 by town. **Supply.**—Morse and Haynes brooks, by gravity, from impounding and distributing reservoirs; filtered. Fiscal year closes March 1. About July 1, '91, contract for new reservoir was let to A. Lancier, South Framingham.

Year.	Mains.	Taps.	Meters.	Hydrants.	Cost.	B'd'd debt.	Int. ch'g.s.	Exp.	Rev.*
'80	18	510	0	109	$170,000	$150,000	$7,500	$1,683	......
'84-5	19.4	731	0	117	175,512	150,000	7,500	2,352	$17,096
'88-9	22.8	997	0	145	189,298	150,000	8,679	2,000	20,744
'89-90	24.9	1,077	0	155	195,069	150,000	7,400	3,432	23,522
'90-1	25.9	1,137	0	160	199,051	150,000	7,700	2,973	21,516

* Including allowance for hydrant rental.

Financial.—*Bonded Debt*, $150,000 at 5%. *No Sinking Fund. Expenses,* total, $10,673. *Revenue:* consumers, $15,009; city, $6,507. **Management.**—Water Board of three, one elected each year. Chn., Cephas Derby. Clerk, Geo. Hall. Treas. and Supt., J. G. Tenney, elected by board.

60. LEXINGTON, *Middlesex Co.* (Pop., 2,460—3,197.) Built in '84 by Lexington Water Co.; franchise does not regulate rates, but provides for purchase of works, price determined by arbitration or by the courts. **Supply.**—Ground water, pumping to tank. Fiscal year closes May 11. **Distribution.**— *Mains*, 6 miles. *Taps*, 327. *Meters*, 1. *Hydrants*, 50. **Financial.**—*Cost*, $99,780. *Cap. Stock:* Authorized, $60,000; all paid up. *Bonded debt*, $37,500; $17,500 at 5%, $20,000 at 5%. *Sinking Fund,*

FOR FURTHER DESCRIPTIVE MATTER see Manual for 1889-90. CHANGES since 1890 are here given. POPULATIONS are for 1880 and 1890 respectively.

$4,133. *Revenue:* City, $1,400. **Management.**—Prest., W. F. Draper, Hopedale. Secy., C. E. Tingley ; Treas. and Supt., E. L. Wires ; both of 105 Summer street, Boston.

61. **LINCOLN**, *Middlesex Co.* (Pop., 907—987.) Built in '75 by town. Supply.—Sandy pond, pumping to reservoir. Fiscal year closes July 1. Distribution.—(In '89.) *Mains*, 3.5 miles. *Taps*, 35. *Hydrants*, 11. Financial.—(In '89.) *Cost*, $31,-000. *Bonded Debt*, $30,000 at 6%. *Expenses:* operating, $651; int., $1,800; total, $2,451. *Revenue :* consumers, $819. **Management.**—Supt., Geo. L. Chapin.

62. **LOWELL**, *Middlesex Co.* (Pop., 59,475 -77,696.) Built in '72 by city. Supply.—Merrimack River, pumping to reservoir; filtered. In '91 a 10,000,000-gall. Worthington pump was added and a new 30-in. conduit, 4,335 ft. long, was laid from Beaver Brook to pumping station, affording additional supply of 4,000,000 galls. A Jewell filtering plant was also proposed, and, as an alternative, there was talk, in August, of obtaining a better supply from driven wells.

Year	Mains	Taps	Meters	Hydrants	Consumption	Cost	Bonded debt	Int. ch'g's	Exp.	Rev.
'73	...	...	...	...	511,452	$1,629,635	...	...	...	$39,169
'78	57.5	4,714	377	...	1,890,181	2,682,403	...	...	$19,900	98,534
'80	...	...	...	...	2,250,715	2,986,932	...	...	...	117,107
'84	76.96	6,659	1,348	752	3,157,936	3,746,753	...	...	...	144,614
'85	79.26	6,913	1,421	760	3,560,904	3,921,296	$1,890,000	$111,850	47,157	151,903
'86	81.29	7,127	1,461	773	3,949,651	4,089,239	1,837,000	115,535	35,253	102,289
'87	83	7,440	1,530	779	4,319,164	4,270,303	1,834,000	110,790	33,527	175,058
'88	85.7	8,115	1,630	823	4,978,000	4,453,583	1,831,000	113,400	44,431	173,548
'89	89.16	8,471	1,757	844	4,633,165	4,647,119	1,828,000	112,981	65,335	174,741
'90	92.2	8,471	1,945	854	5,127,199	4,841,227	1,125,000	115,013	45,524	192,711

Financial.—*Bonded Debt*, $1,125,000: $100,000 at 6.6%, $150,000 at 6¼, $200,000 at 6, $675,000 at 4%. *Sinking Fund*, $53,874. *Expenses*, total, $160,537. *Revenue*, consumers, $192,711. **Management.**—Committee of five, one elected every year. Prest., M. F. Brennan. Secy., Chas. L. Knapp. Supt., S. P. Griffin.

105. **LUDLOW**, *Hampden Co.* (Pop., 1,526—1,939.) Supply from Springfield See '89-90 Manual, p. 65. Ludlow Mfg. Co. first supplied in '75.

63. **LYNN**, *Essex Co.* (Pop., 38,274—55,727.) Built in '70-72 by city. Supply.—Brooks, by gravity, to pump wells, thence by pumping to reservoir. Fiscal year closes Dec. 31.

Year	Mains	Taps	Meters	Hydrants	Consumption	Cost	Bonded debt	Int. ch'g's	Exp.	Rev.
'72	41.8	2,191	...	327	...	$639,938	...	...	...	$17,568
'75	48.6	3,218	102	361	1,291,741	930.552	$921,500	51,790	11,807	42,553
'80	55.9	4,488	121	411	1,238,290	981,510	921,500	52,230	10,278	69,635
'81	60.2	4,948	137	453	1,238,289	1,007,655	...	53,190	13,590	75,968
'82	65.1	5,439	156	498	1,510,764	1,041,220	...	54,066	12,577	76,420
'83	69.8	5,989	178	527	1,557,974	1,139,728	...	55,437	21,456	90,891
'84	73.1	6,505	168	543	1,739,083	1,230,685	...	60,326	27,843	105,833
'85	75.2	6,919	193	557	1,920,519	1,292,510	...	64,088	19,329	108,584
'86	78.7	7,374	220	572	2,116,199	1,342,144	...	64,807	19,299	110,089
'87	81.1	7,800	236	590	2,379,031	1,373,472	1,083,395	65,981	21,952	116,375
'88	82.2	8,233	254	602	2,474,556	1,479,204	1,422,500	67,309	36,121	123,507
'89	87.2	8,606	324	620	2,450,413	1,659,283	1,612,500	73,613	21,249	129,480
'90	91.9	9,490	340	659	2,656,890	1,833,348	1,677,500	80,184	22,182	134,788

Financial.—*Bonded Debt*, $1,677,500 at 3½ to 6%; 1890 legislature authorized city to issue additional bonds to amount of $150,000. *Sinking Fund*, $311,123. *Expenses:* operating, $22,182; int., $80,184; rebate to Saugus, $1,078; total, $103,444. *Revenue:* consumers, $134,788; city, $20,000; total, $154,788. **Management.**—Comrs. Prest., David Knox. Registrar, W. O. Mudge. Supt., John C. Haskell.

64. **MALDEN**, *Middlesex Co.* (Pop., 10,127—13,805.) Built in '69 by city. Supply.—Spot pond, by gravity, and additional pumping through stand-pipe; also driven wells, pumping to tank. Fiscal year closes Dec. 31.

Year	Mains	Taps	Meters	Hydrants	Cost	B'd'd debt	Int. ch'g's	Exp.	Rev.
'70	29.7	...	...	137	$155,161	...	$4,800	$470	$187
'75	33.6	1,718	7	150	340,183	$300,000	18,458	14,872	33,239
'80	35.93	2,372	19	170	353,694	350,000	21,000	5,951	32,277
'81	36.7	2,468	26	193	362,102	350,000	21,000	12,077	34,301
'82	36.8	2,487	31	193	375,124	350,000	21,000	8,498	37,588
'83	39.9	2,632	36	193	388,780	350,000	21,000	11,505	38,231
'84	41.25	2,700	43	199	400,316	350,000	21,000	7,616	38,993
'85	44.32	2,903	42	200	435,599	390,000	21,438	6,199	41,245
'86	47.85	3,279	47	201	464,743	400,000	22,675	8,561	46,308
'87	50.8	3,580	51	230	492,871	440,000	23,375	12,200	51,822
'88	53.75	3,777	53	233	510,020	440,000	24,575	14,192	56,133
'89	61.78	3,988	55	243	667,723	440,000	27,375	21,678	62,532
'90	64.9	4,360	54	252	710,493	440,000	25,675	27,577	69,823

MASSACHUSETTS. 33

Financial.—Bonded Debt, $440,000; 1890 legislature authorized increase of bonds to $600,000. *Sinking Fund*, $172,931. *Expenses:* total, $53,552. *Revenue:* consumers, $69,823. **Management.**—Three comrs.: N. A. Fisk, E. A. Stevens, Jr., A. R. Turner, Jr. Supt., Solon M. Allis.

65. MANCHESTER, *Essex Co.* (Pop., 1,640—1,789.) **History.**—Built in '91 by town; to be in operation Nov. 1. Des. Engr., Percy M. Blake. **Supply.**—Saw Mill brook, pumping to stand-pipe. **Stand-Pipe.**—Cap., 517,000 galls.; 35 × 75 ft.; from Cunningham Iron Works, Boston. **Distribution.**—*Mains*, 12-in. c. i., 12 miles. Pipes and specials from Camden (N. J.) Iron Works; trenching and pipe-laying done by J. P. Langford, Boston. *Hydrants*, Ludlow, 80. *Valves*, Ludlow. **Financial.**—*Cost*, $125,000. *Bonded Debt*, $125,000 at 4%. **Management.**—Address Sam'l Knight.

66. MANSFIELD, *Bristol Co.* (Pop., 2,762—3,432.) Built in '88 by Mansfield Water Supply District. **Supply.**—Cate springs, pumping to stand-pipe and direct. Fiscal year closes Feb 1.

Year.	Mains.	Taps.	Meters.	Hy-drants.	Consumption.	Cost.	Bonded debt.	Int. charges.	Exp.	Rev.
'88-9	6.5	128	22	46	130,000	$67,459	$69,000	$2,443	.....	.....
'90-1	7.4	243	67	50	176,563	75,060	74,000	2,060	$1,775	$4,450

Financial.—*Total Debt*, $74,853; floating, $853; bonded, $74,000 at 4%, due in 1918. *Sinking Fund*, $3,287, invested in real estate mortgages at 6%. *Expenses:* total, $4,735. *Revenue:* consumers, $4,450; city, $1,500; total, $5,950. **Mangement.**—Three comrs.: D. S. Spaulding, A. B. Day, Wm. B. Rogerson.

67. MARBLEHEAD, *Essex Co.* (Pop., 7,467—8,202.) Distribution put in in '87 by town; in '89 independent supply and pumping station secured; see Manual for '89-90, p. 67. Fire protection system built in '78. **Supply.**—Driven wells, pumping to stand-pipe. Fiscal year closes May 7. **Distribution.**—*Mains*, 15. *Taps*, 300. *Hydrants*, 159. *Consumption*, 80,000 galls. **Financial.**—*Cost*, $150,000. *Bonded Debt*, $132,000 at 4%. *Expenses and Revenue*, withheld. **Management.**—Three comrs. Chn., Wm. J. Goldthwait. Supt., Thos. Main.

68. MARLBOROUGH, *Middlesex Co.* (Pop., 10,127—13,805.) Built in '83 by town. **Supply.**—Lake Williams, pumping to reservoir and direct Fiscal year closes Dec. 31.

Year.	Mains.	Taps.	Meters.	Hy-drants.	Consumption.	Cost.	B'd'd debt.	Int. ch'g's.	Exp.	Rev.
'84	10.5	362	32	136	137,485	$165,174	$200,000	$8,000	.....	.....
'88	24.8	1,476	116	245	292,412	252,764	250,000	10,000	$1,694	$17,105
'89	25.9	1,627	143	253	316,706	259,214	.....	.....	6,710	19,612
'90	27	1,794	198	262	347,885	269,604	270,000	9,820	5,445	23,063

Financial.—*Bonded Debt*, $270,000 at 4%. *Sinking Fund*, $14,658, invested in 4% bonds. *Expenses:* total, $15,265. *Revenue:* consumers, $23,663; city, none. **Management.**—Prest., Jno. A. McConnell. Secy., Wm. McNully. Supt., Geo. A. Stacy.

69. MAYNARD, *Middlesex Co.* (Pop., 2,291—2,700.) Built in '89 by town. **Supply.**—Receiving well, pumping to reservoir. Fiscal year closes March 1. **Distribution.**—*Mains*, 5 miles. *Faucets*, 328. *Meters*, 7. *Hydrants*, 60. **Financial.**—*Cost*, $103,178. *Bonded Debt*, $105,000 at 4%; 1890 legislature authorized an additional loan of $125,000. *Expense:* operating, $458; int., $4,200; total, $4,658. *Revenue:* consumers, $1,019. **Management.**—Comrs. Chn., Thos. Naylor. Secy., O. S. Fowler.

70. MEDFORD, *Middlesex Co.* (Pop., 7,573—11,079.) Built in '70 by city. **Supply.**—Spot pond, by gravity and by pumping to reservoir when water in pond is low; filtered through screens. Fiscal year closes Feb. 1.

Year.	Mains.	Taps.	M'trs.	Hy-drants.	Cost.	Bonded debt.	Int. charges.	Exp.	Rev.
'71	23.5	619	..	110	$206,692	$190,000	$11,341	$844	$7,641
'75	30.8	1,114	..	144	297,984	295,000	17,354	9,181	12,131
'80	32.5	1,328	..	153	311,067	300,000	18,000	4,139	13,287
'81	32.9	1,375	..	153	313,839	300,000	16,800	3,410	11,952
'82	33.1	1,371	9	155	316,761	300,000	19,170	3,888	14,470
'83	34.1	1,441	8	158	324,044	300,000	18,030	5,531	16,580
'84	34.9	1,538	9	166	328,087	300,000	18,030	4,144	16,720
'85	35.6	1,621	7	169	332,265	300,000	17,790	5,305	17,457
'86	36.5	1,761	8	173	335,768	300,000	17,970	4,699	18,352
'87	37.5	1,875	9	181	340,592	300,000	17,970	6,771	20,956
'88	38.5	1,932	11	191	345,375	300,000	18,120	6,418	22,020
'89	40.3	2,109	10	200	353,306	300,000	17,850	7,087	25,080
'90	41.7	2,233	10	224	360,921	200,000	14,710	6,968	25,861

FOR FURTHER DESCRIPTIVE MATTER see Manual for 1889-90. CHANGES since 1890 are here given. POPULATIONS are for 1880 and 1890, respectively.

Financial.—*Bonded Debt*, $200,000 at 6%. *Sinking Fund*, $16,765. *Expenses:* total, $21,678. *Revenue:* consumers, $15,861. **Management.**—Three comrs., one elected every year. Chn., Dan'l A. Gleason. Clerk, E. W. Hayes. Supt., R. M. Gow, appointed by comrs.

71. **MELROSE**, *Middlesex Co.* (Pop., 4,560—3,519.) Built in '70 by city. **Supply.**—Spot pond, by gravity and pumping to reservoir. Fiscal year closes Dec. 31.

Year.	Mains.	Taps.	Meters.	Hydrants.	Consumption.	Cost.	Bonded debt.	Int. ch'gs.	Exp.	Rev.
'80	16.3	835	0	85	unk'n.	$175,829	$150,000	$9,000	$2,036	$9,882
'85	19.51	1,213	0	109	"	194,728	150,000	9,000	5,315	15,193
'86	23.76	1,324	0	115	"	240,939	174,000	10,400	8,321	17,301
'87	25.4	1,426	2	123	"	252,942	174,000	10,720	9,023	19,170
'88	26.8	1,547	4	132	"	262,698	217,000	11,280	9,660	21,688
'89	27.85	1,632	0	142	"	268,786	217,000	11,680	6,812	22,776
'90	28.45	1,989	0	146	"	279,244	167,000	9,893	9,868	24,764

Financial.—*Bonded Debt*, $167,000 at 5%; additional loan of $25,000 was authorized in '91. *Sinking Fund*, $18,992. *Expenses:* total, $19,761. *Revenue*, consumers, $24,764; city, none. **Management.**—Comrs. Registrar, E. H. Goss. Supt., Jas. W. Riley.

Methuen, *Essex Co.* (Pop., 4,392—4,814.) Built in '76 by village for fire protection. See Projected Works.

72. **MIDDLEBOROUGH**, *Plymouth Co.* (Pop., 5,237—6,065.) Built in '85 by fire district. **Supply.**—Well, pumping to stand-pipe and direct. Fiscal year closes Dec. 31.

Year.	Mains.	Taps.	Meters.	Hydrants.	Consumption.	Cost.	Bonded debt.	Int. ch'g's.	Exp.	Rev.
'85	8	119	24	64		$71,024	$70,000		$655	$803
'86	8	207	25	64	....	77,075	70,000	$2,800	......	......
'87	9.6	375	78	78	72,707	79,330	75,000	3,000	2,209	4,121
'88	9.9	412	85	79	89,885	81,584	74,500	3,000	3,039	5,300
'89	10.5	467	104	81	93,912	84,008	74,500	2,930	2,483	5,618
'90	11.3	519	125	86	128,775	87,894	75,000	2,930	2,505	6,775

Financial.—*Bonded Debt*, $75,000 at 5%, due '90—1917. *Sinking Fund*, $1,749. *Expenses*; total, $5,185. *Revenue:* consumers, $6,775; city, $2,000; total, $8,775. **Management.**—Three comrs., one elected each year. Chn., Eugene P. Le Baron Secy. and Supt., Jos. E. Beals.

32. **MIDDLETOWN**, *Essex Co.* (Pop., 1,000—934.) Supply from Danvers.

73. **MILFORD and HOPEDALE**, *Worcester Co.* (Pop., Milford, 9,310—8,780; Hopedale, 399—1,176.) Built in '81 by Milford Water Co. **Supply.**—Charles River for fire purposes and wells for domestic use, pumping direct. Fiscal year closes Feb. 10. In '91 Co. renewed its contract for hydrant rental with city, and decided to add a 2,000,000-gall. Worthington pump and two 100 H.P. boilers. Cost of improvements, $13,000. **Distribution.**—(In '88.) *Mains*, 19 miles. *Taps*, 851. *Meters*, 67. *Hydrants*, 93. *Consumption*, 325,000 galls. **Financial.**—*Cost*, $257,789. *Cap. Stock:* authorized, $150,000; all paid up. *Bonded Debt*, $130,000. **Management.**—Prest., Chas. W. Shippee. Supt., Henry W. Rogers.

NOTE.—No statistics have been received from Co. since '88. Financial figures from State Corporation Report.

74. **MILTON**, *Norfolk Co.* (Pop., 3,206—4,278.) **History.**—Built in '90 by Milton Water Co., which has since absorbed Brush Hill Water Co., of which it was an extension. Engr., Percy M. Blake, Hyde Park, Mass. **Supply.**—Hyde Park Water Co.'s system; Co. is looking for independent supply. **Distribution.**—*Mains*, 12 to 6-in., c.i., 18 miles. *Hydrants*, 150. *Pressure*, 20 to 100 lbs. **Financial.**—*Cost*, $160,000. *Cap. Stock:* authorized, $150,000; paid-up, $64,000. *Bonded Debt:* authorized, $150,000; issued, $60,000. *Hydrant rental*, $40. **Management.**—Prest., E. P. Whitney, Blue Hill. Secy., C. S. Rackemann, Readville. Treas., Henry Wainwright, 40 State St., Boston.

75. **NAHANT**, *Essex Co.* (Pop., 808—880.) Built in '85 by town. Supply from Marblehead Water Co.'s works. Fiscal year closes Feb. 28.

Year.	Mains.	Taps.	Meters.	Hydrants.	Cost.	Expenses.	Revenue.
'86-7	7	62	2	50	$53,000	$3,000	$1,500
'89-90	8	191	7	50	55,000	4,820	3,730
'90-1	8	206	8	51	55,000	4,213	3,904

Financial.—*No Debt*. *Expenses:* total, $4,213; $275 per month is paid Marblehead Water Co. for water. *Revenue:* consumers, $3,904; city, $800; total, $4,704. New construction is met by direct taxation. All other expenses are paid from revenue.

MASSACHUSETTS.

Management.—Three selectmen, elected every year. Chn., J. C. Wilson. Supt., R. L. Cochran.

76. **NANTUCKET,** *Nantucket Co.*; franchise provides neither for control of rates nor purchase of works by town. (Pop., 3,727–3,268). Built in '78 by Wannacomet Water Co. **Supply.**—Wannacomet Pond, pumping to tank and direct. Fiscal year closes June 30.

Year.	Mains.	Taps.	Meters.	Hyd'ts.	Cons'n.	B'd'd debt.	Int. ch'g's.	Exp.	Rev.
'80–1	5	85	0	2	....	$20,000	$1,200	....	....
'85–6	6.6	368	0	31	34,000	20,000	1,200	$2,500	$3,800
'86–7	6.6	416	1	31	43,000	20,000	1,200	3,100	4,400
'87–8	7.3	447	0	32	41,000	20,000	1,200	3,800	5,400
'88–9	7.5	479	0	34	53,000	20,000	1,200	3,300	6,300
'89 90	7.6	514	0	36	59,000	20,000	1,200	2,500	6,500
'90–1	8.3	565	0	48	57,000	20,000	1,200	3,260	7,540

Financial.—*Cost*, $61,068. *Cap. Stock*, authorized, $30,000; all paid up. *Total Debt*, $32,935; floating, $12,935 at 5%; bonded, $20,000 at 6%, due in 1900. *Sinking Fund*, $2,000 at 0%. *Expenses:* operating, $3,070; int., $1,582; taxes. $190; total, $5,142. *Revenue*, consumers. $6,557; city, $992; total, $7,549. **Management.**—Prest., R. Gardner Chase. Secy. and Supt., Wm. F. Codd.

77. **NATICK,** *Middlesex Co.* (Pop., 8,479–9,118.) Built in '74 by city. **Supply.**—Dug pond, pumping direct and to reservoir. Fiscal year closes Feb. 28.

Year.	Mains.	Taps.	Meters.	Hydrants.	Consumption.	Cost.	B'd'd debt.	Int. ch g's.	Exp.	Rev.
'80–1	22	837	19	122	213,000	$200,000	$171,000	....	$3,925	$11,299
'83–4	22.3	927	26	126	277,469	201,225	165,000	....	4,375	14,068
'87–8	25	1,230	41	112	265,000	220,000	118,400	$4,736	4,380	15,520
'88–9	26.1	1,347	45	145	251,488	233,034	115,700	4,028	4,675	17,562
'89–90	26.7	1,349	46	146	253,655	286,002	117,700	4,708	4,150	18,246
'0–1	28.2	1,358	47	148	263,830	347,398	101,200	4,018	4,629	18,130

Financial.—See above. **Management.**—Three comrs., Francis Bigelow, Royal Ellis, Chas B. Felch. Supt., J. W. Morse.

78. **NEEDHAM,** *Norfolk Co.* (Pop., 5,252–3,035.) History.—Built in '90 by town. Engr., Louis E. Hawes, 75 State St., Boston. **Supply.**—Well, fed by springs, pumping to stand-pipe. Well is 22 ft. in diam., 24 ft. deep; built of rubble lined with brick and cement. **Pumping Machinery.**—Dy. cap., 750,000 galls.; two Blake comp., dup., direct-acting. *Boilers:* two horizontal tubular. **Stand-Pipe.**—Contrs., F. Hodge & Co., E. Boston. **Distribution.**—*Mains.* 12 to 4-in, c. i., 7.7 miles. *Services:* cem., i. and lead. Pipes and specials from Gloucester Iron Wks.; trenching and pipe-laying done by Chas. H. Eglee, Flushing, N. Y. *Hydrants*, Ludlow, 59. *Valves*, Ludlow. **Financial.**—*Cost*, $63,752. *Bonded Debt*, $75,000 at 3½%. **Management.**—Three comns.: Jas. Mackintosh, John Mosely, John M. Hodge.

79. **NEW BEDFORD,** *Bristol Co.* (Pop., 26,845–40,733.) Built in '68–9 by city. **Supply.**—Acushnet River and Little Quittacus Pond, by gravity from storage reservoir to city, thence by pumping to distributing reservoir and stand-pipe. Fiscal year closes Dec. 1.

Year.	Mains.	Taps.	Meters.	Hydrants.	Consumption.	Cost.	Bonded debt.	Int. ch'gs.	Exp.	Rev.
'70	17.4	553	....	127	329,376	$641,774	$600,000	......	......	$7,974
'75	35	2,323	9	247	1,136,821	927,809	700,000	$43,000	$16,944	16,977
'80	42.27	3,708	22	295	2,011,200	1,013,325	700,000	45,388	14,484	17,644
'81	43.3	3,935	34	304	2,213,012	1,030,539	700,000	45,388	18,857	18,024
'82	44.2	4,203	41	313	2,326,352	1,059,440	700,000	45,388	23,694	30,741
'83	46.4	4,465	49	332	2,326,191	1,169,940	790,000	45,388	21,879	35,037
'84	48.6	4,691	60	352	2,371,080	1,170,062	780,000	58,300	20,665	36,674
'85	50.3	4,967	67	367	2,876,167	1,217,592	750,000	51,600	24,199	40,510
'86	52.1	5,225	82	388	2,976,807	1,259,402	720,000	49,700	36,315	43,918
'87	54.37	5,475	102	410	3,047,404	1,298,461	790,003	17,800	26,208	45,880
'88	59.2	5,785	108	445	3,360,323	1,355,321	780,000	46,3 0	31,249	49,584
'89	61.1	6,104	120	460	3,590,379	1,396,663	750,000	46,400	27,092	56,019
'90	62.5	6,394	143	477	4,036,200	1,436,910	720,000	43,500	30,701	62,482

Financial.—*Bonded Debt*, $720,000: $130,000 at 7%, $280,000 at 6%, $100,000 at 5%, $120,000 at 4%, $100,000, trust fund at 6%. *Sinking Fund*, $1,911. *Expenses:* total, $74,201. *Revenue:* consumers, $62,482; city, $12,000; total $74,782. **Management.**—Water Board of five members; the mayor, president, and three members, one elected each year. Prest., Walter Clifford. Clerk and Supt., R. C. P. Coggeshall, elected by Water Board.

80. **NEWBURYPORT,** *Essex Co.* (Pop., 13,538–13,917.) Built in '81 by New-

For Further Descriptive Matter see Manual for 1889-90. Changes since 1890 are here given. Populations are for 1880 and 1890, respectively.

buryport Water Co. Supply, wells, pumping to tank. Fiscal year closes Jan. 14. Distribution —*Mains*, 25 miles. *Taps*, 1,450. *Meters*, 20. *Hydrants*, 168. *Consumption*, 400,000 galls. Financial.—*Cost*, $351,617. *Cap. stock:* Authorized, $300,000; all paid up. *Debt*, $9,536. Management.—Prest., H. B. Little. Secy. and Supt., J. E. McCusker. Treas., Geo. H. Norman, Newport, R. I. NOTE.—The city talks of buying the works.

81. **NEWTON,** *Middlesex Co.* (Pop., 16,995—24,379.) Built in '76 by city. Supply, ground water, intercepted on its way to Charles River, by conduits, pumping to reservoir from filter basin. Fiscal year closes Dec. 31. *Underground Covered Reservoir* completed in '91; cap., 2,500.000 galls., being first section of proposed 10,000,000-gall. reservoir; 126x173 ft . x15.75 ft. deep; 325 ft. above tide; for high pressure. Has rubble masonry walls and brick piers spanned one way with brick arches to support flooring, which is said to be of brick, arched, with 4-in. ring and 4 ins. of concrete above. Walls, 7 ft. 2 ins. to 7 ft. 10 ins. thick for 4 ft., then diminish to 5 ft. 2 ins. at top. Piers 10 ft. apart, 20 ins. sq., on masonry foundations 3 ft. sq. Arches have 10-in. rise. Circular gate-chamber, 30 ft. diam., at northwest corner, with steel receiving tank 7½x12 ft.; 24-in. influent and effluent pipe, latter extending to center of reservoir and terminating in ¼-bend. Cost, about $18,000.

[Above subject to correction, as promised information was not received in time for insertion, and description was condensed from Boston *Herald*, July 18, '91. If description is received in time it will be given in ENG. NEWS, vol. xxv.—ED.]

Wa'er tower 42x140 ft. said to be projected. Engr., A. F. Noyes, City Engr. Consult. Engr., A. Fteley.

Reported that old 48-in. filtering conduit has been rebuilt at cost of $23,500, and new conduit or extension constructed for $52,900. Also that 5,000,000-gall. Blake pump and two 7x30-ft. Belpaire boilers have been added.

Year.	Mains.	Taps.	Meters.	Hydrants.	Consumption.	Cost.	B'd'd debt.	Int. ch'g's.	Exp.	Rev.
'78	53.3	1,685	268	269	328,312	$774.052	....	$14,800	$9,778	$17,598
'80	59.7	2,145	458	314	452,032	841,906	....	47,750	9,224	24,816
'81	61.7	2,412	515	334	450,987	865,223	....	48,990	12,682	26,953
'82	65	2,581	664	366	594,930	911,190	....	49,630	13,812	32,703
'83	67.6	2,710	761	398	624,381	942,992	$676,013	50,900	16,266	36,477
'84	69.6	2,919	868	438	532,804	966,583	....	52,500	13,352	37,611
'85	72.9	3,134	998	451	614,968	999,316	....	53,100	12,873	39,563
'86	77.1	3,432	1,501	471	675,298	1,043,489	997,000	51,380	12,987	43,184
'87	83.2	3,767	2.370	540	703,702	1,119,544	1,115,000	55,740	13,952	42,948
'88	86.8	4,050	2,506	588	703,491	1,168,951	1,165,000	59,600	13,938	48,692
'89	90.2	4,203	2,708	606	853,435	1,218,648	1,217,000	61,340	15,421	51,297
'90	93.6	4,440	2,995	612	985,396	1,440,532	1,400,000	66,840	14,791	59,011

Financial.—*Bonded Debt*, $1,400,000; $600,000 at 6% due in 1905; $250,000 at 5%, due in 1906; $550,000 at 4%, due in 1910–18. *Sinking Fund*, $322,520, invested in 4 and 5% water bonds. *Expenses:* total, $81,634. *Revenue:* consumers, $59,011; city, $16,315; total, $75,327. Management.—Water board of five members, three appointed by aldermen for indefinite term; one alderman and one councilman appointed by mayor annually. Prest., F. A. Dewson. Registrar, J. C. Whitney. Supt., H. W. Hyde, Jr., appointed by aldermen, "for life, good behavior, or change in ordinance."

82. **NORTH ADAMS,** *Berkshire Co.* (Pop., 10,191—16,074.) Built in '65 by fire district. Supply.—Brook and two artesian wells, by gravity from brook and by pumping from wells. Fiscal year closes May 1. Distribution.—*Mains*, about 14 miles. *Taps*, 1,000. *Meters*, 26. *Hydrants*, 125. *Consumption*, 1,500,000 galls. Financial.—*Total Debt:* $419,700: floating, $86,200; bonded, $128,000 at 3½%, $235,500 at 3¼%. *Expenses:* operating, $47,577; int., $15,585; total, $63,162. *Revenue:* consumers, $27,384. Management.—Chn., V. A. Whitaker. Secy., Jas. W. Hardenbergh. Supt., Wm. F. Hodge.

83. **NORTHAMPTON,** *Hampshire Co.* (Pop., 12,172—14,990.) Built in '71 by city. Supplies villages of Florence and Leeds on pipe line. Supply.—Roberts-meadow brook, by gravity from impounding reservoir. Fiscal year closes Dec. 1.

Year.	Mains.	Taps.	Meters.	Hydrants.	Cost.	B'd'd debt.	Int. charges.	Exp.	Rev.
'72	18	200	....	....	107	$193,674	$200,000	$12,000	$524	$5,125
'75	20.3	....	....	....	117	205,196	200,000	12,000	2,933	11,878
'80	22.2	975	....	....	127	210,757	200,000	12,000	1,904	13,081
'81	21.3	1,174	....	....	139	215,817	200,000	11,563	1,568	13,702
'82	25.1	1,259	3	....	144	220,314	200,000	....	1,746	14,846
'83	25.7	1,429	9	....	146	238,526	200,000	....	2,747	16,412
'84	27	1,491	10	....	150	241,263	200,000	....	2,084	18,238
'85	28.9	1,574	10	....	154	245,604	200,000	11,000	1,910	19,046
'86	31.1	1,961	10	....	160	248,802	200,000	....	1,910	21,041

MASSACHUSETTS. 37

Year.	Mains.	Taps.	Meters.	Hyd'ts.	Cost.	B'd'd d't.	Int ch'g's.	Exp.	Rev.
'87	32.5	2,032	9	172	255,907	170,000	..	2,592	21,643
'88	35	2,100	10	184	267,565	165,000	8,625	2,218	22,365
'89	36.6	2,176	11	193	273,032	165,000	8,466	2,696	23,531
'90	38.5	2,257	11	204	279,132	165,000	8,183	2,547	21,416

Financial.—See above. **Management.**—Three comrs., one elected every year. Prest., D. W. Bond. Clk., H. A. Kimball. Supt., J. M. Clark.

84. **NORTH ATTLEBOROUGH,** *Bristol Co.* (Pop., 6,727.) Built in '84 by fire district. Supply, well, pumping to stand-pipe. Fiscal year closes March 31.

Year.	Mains.	Taps.	Meters.	Hy-drants.	Con-sumption.	Cost.	B'd'd debt.	Int. ch'g's.	Exp.	Rev.
'84-5	7.3	160	55	63	.....	$109,172	$100,000	$2,000	$2,009	$ 951
'86-7	9.8	304	107	88	75,500	132,869	125,000	5,000	3,500	3,710
'88-9	10.7	449	171	95	119,990	137,324	125,000	5,000	3,642	5,134
'89-'90	11.3	496	200	101	123,555	140,968	125,000	5,000	3,455	5,295
'90-1	11.6	524	261	102	166,575	143,989	125,000	5,000	4,373	6,371

Financial.—Bonded Debt, $125,000 at 4%. Sinking Fund, $27,016, invested in notes and mortgages. *Expenses:* total, $9,373. *Revenue:* consumers, $6,371; city, $4,175; total, $10,546. **Management.**—Comrs.: O. B. Bestor, A. H. Bliss, W. P. Whittemore. Supt., W. P. Whittemore.

85. **NORTHBOROUGH,** *Worcester Co.* (Pop., 1,676–1,952.) Built in '82 by town. **Supply.**—West branch of Cold Harbor brook, by gravity from impounding reservoir. Fiscal year closes Feb. 1. During 1889 a new reservoir was added.

Year.	Mains.	Taps.	Hydrants.	Cost.	Bonded debt.	Int. charges.	Exp.	Rev.
'82-3	6.6	42	40	$54,150	$50,000	$2,000		$706
'84-5	7	147	40	64,372	60,000	2,400	$351	2,969
'86-7	8.8	170	51	67,915	61,500	2,460	384	3,419
'88-9	9	218	51	68,142	61,500	2,460	355	3,440
'89-90	9	230	51	69,516	61,500	2,460	361	2,797
'90-1	9.5	244	51	70,006	61,500	2,460	195	3,544

Financial.—Bonded Debt, $61,500 at 4%, due in 1912. Sinking Fund, $18,560. *Expenses:* total, $2,655. *Revenue:* consumers, $3,544; city, none. **Management.**—Three comrs., one elected every year. Prest., G. P. Heath. Secy., Sumner Small. Supt., R. R. Yates.

86 **NORTH EASTON,** *Bristol Co.* (Pop., not given.) Built in '87-8 by fire district. **Supply.**—Collecting well, pumping to stand-pipe and direct. **Distribution.**—(In '88) *Mains,* 6.5 miles. *Taps,* 240. *Meters,* 9. *Hydrants,* 60. *Consumption,* 87,000 galls. **Financial.**—(In '88) Cost, $78,000. Bonded Debt, $75,000 at 6%. *Expenses:* Operating, $1,300; int., $3,000; total, $4,300. *Revenue:* Consumers, $1,260. **Management.**—Three comrs. Clerk, Geo. K. Davis.

NOTE.—Early in '91 Clerk wrote that it was impossible to give desirable report for 1890, owing to improvements and changes that would not be completed till close of '91.

II. **NORTHFIELD,** *Franklin Co.* (Pop., 1,603–1,869.) Built in '85 by Northfield Aqueduct Co. **Supply.**—Brook, by gravity. **Distribution.**—*Mains,* 1¼ miles. *Taps,* 12. *Hydrants,* 0. **Financial.**—Withheld. **Management.**—Address Co.

40. **NORTH GRAFTON,** *Worcester Co.* (Pop. not given). Supply from Grafton.

87. **NORWOOD,** *Norfolk Co.* (Pop., 2,345–3,733.) Built in '85 6 by town. **Supply.**—Pond, pumping to reservoir. Fiscal year closes Jan. 31.

Year.	Mains.	Taps.	Meters.	Hy-drants.	Consump-tion.	Cost.	B'd'd debt.	Int. ch'g's.	Exp.	Rev.
'87	12	333	70	84	200,000	$106,975	$100,000	$4,000	$2,500	$4,500
'88	12.5	371	85	86	150,117	108,980	100,000	4,000	2,915	5,040
'89	14	430	96	98	160,485	117,409	100,000	4,000	2,011	6,777
'90	16	478	114	103	168,842	123,957	90,000	3,600	2,213	7,051

Financial.—Bonded Debt, $90,000 at 4%. Sinking Fund, $3,408. *Expenses:* total, $5,813. *Revenue:* consumers, $7,051. **Management.**—Chn., E. J. Shattuck. Supt., Geo. A. Bucknam.

III. **ORANGE,** *Franklin Co.* (Pop., 3,169–4,568.) Built in '75 by New Home Sewing Machine Co. **Supply.**—Spring, by gravity from reservoirs, with pumping from Miller's River, in dry weather. **Distribution.**—*Mains,* 2 miles. *Consumers,*

FOR FURTHER DESCRIPTIVE MATTER see Manual for 1889-90. CHANGES since 1890 are here given. POPULATIONS are for 1880 and 1890, respectively.

100. *Hydrants*, 0. **Financial.**—*Expenses*, nominal. *Revenue*, $1,200. **Management.**—Address New Home Sewing Machine Co.

NOTE.—More extensive system is projected, but cannot be built till authorized by Legislature. Supply, pond, pumping to reservoir. Ownership not determined. Address E. A. Goddard, Chn., or W. M. Wright, Secy., of Water Board.

88. **PALMER**, *Hampden Co.* (Pop., 5,504—6,520.) Built in '86 by Palmer Water Co. **Supply.**—Springs, by gravity from storage to distributing reservoir, thence by gravity to village. Fiscal year closes Jan. 1. **Distribution.**—*Mains*, 6 miles. *Taps*, 190. *Meters*, 5. *Hydrants*, 47. **Financial.**—*Cost*, $42,642. *Cap. Stock:* authorized, $20,000; all paid up. *Total Debt*, $22,642; floating, $2,642; bonded, $20,000 at 5%. *Expenses:* operating, $1,200; int., $1,000; taxes, $217; total, $2,417. *Revenue:* consumers, $3,560; city, $910; total, $4,500. **Management.**—Prest., C. S. Goodhue, Springfield. Treas. and Supt., J. H. Gamwell. Secy., F. L. Gamwell, Worcester.

89. **PEABODY**, *Essex Co.* (Pop., 9,022—10,158.) Built in 1790 by Salem & Danvers Aqueduct Co., now owned by town. **Supply.**—Spring and pond, by gravity and pumping to tank. Fiscal year closes Jan. 15.

Year.	Mains.	Taps.	Hydrents.	Consumption.	Cost.	B'd'd debt.	Int. ch'g's.	Exp.	Rev.
'82	14.6	1,293	130	186,359	$233,580	$215,000	......	$3,010	$17,504
'86	19.5	1,560	163	617,114	287,135	223,000	$8,920	6,002	24,606
'87	25.1	1,440	171	617,114	299,640	223,000	8,920	6,602	23,815
'88	25.7	1,510	176	688,346	292,552	223,000	11,467	5,235	23,055
'89	26	1,560	178	753,320	293,807	223,000	10,932	5,173	22,910
'90	27	1,617	188	826,940	298,594	223,000	12,552	5,536	22,684

Financial.—*Total Debt*, $223,000: notes, $138,000; bonds, $85,000 at 4%. *Sinking Fund*, $158,000. *Expenses:* total, $17,788. *Revenue:* consumers, $22,684. **Management.**—Comrs.: D. S. Littlefield, N. M. Quint, J. E. T. Bartlett Engr., J. R. Petley.

90. **PITTSFIELD**, *Berkshire Co.* (Pop., 13,364—17,251.) Built in '55 by fire district. **Supply.**—Ashley Lake and brooks, by gravity to two distributing reservoirs. Fiscal year closes Jan. 1.

Year.	Mains.	Taps.	Meters.	Hydrants.	Cost.	Exp.	Rev.
'80	20	1,600		75	$193,835	$1,800	$16,668
'86	36	2,100	8	83	228,000	2,413	23,047
'88	37.2	2,250	8	85	242,504	2,924	24,038
'89	42	2,290	8	96	281,364	5,099	25,331
'90	42.9	2,340	8	101	284,615	5,500	*22,653

* For 9 months.

Financial.—*Bonded Debt*, $195,000 at 5%. *Expenses:* operating, $5,590; int., $9,750; total, $15,340. *Revenue:* consumers, $22,653, for nine months. **Management.**—Three comrs., one elected every year. Supt., Jno. M. Hatch.

91. **PLYMOUTH**, *Plymouth Co.* (Pop., 7,093—7,314.) Built in '55 by town. Old works built in 1796 by Joshua Thomas. **Supply.**—Natural ponds, by gravity and pumping direct. Fiscal year closes Dec. 31.

Year.	Mains.	Taps.	Hydrants.	Cost.	B'd'd debt.	Int. ch'g's.	Exp	Rev.
'80	18	1,083	62	$195,551	$90,000	$5,310	$3,905	$11,242
'82	20.3	1,123	63	198,992	90,000	5,040	4,904	12,136
'84	21.2	1,212	70	202,866	90,000	5,280	4,705	13,842
'86	23.5	1,234	75	217,257	102,000	4,570	5,937	14,331
'87	27.3	1,317	83	231,114	141,000	4,000	6,169	14,883
'88	28.2	1,364	87	234,822	139,700	6,130	5,532	16,097
'89	28.7	1,389	91	236,702	138,400	6,318	4,198	16,045
'90	29	1,415	93	238,219	172,300	5,820	4,832	16,129

Financial.—*Bonded Debt*, $122,300; $20,000 at 6%, due in '94; $102,300 at 4%, due in equal annual payments for 20 yrs. *Sinking Fund*, $1,651. *Expenses:* total, $10,516. *Revenue:* consumers, $16,129. Bonds are redeemed from net income. **Management.**—Five comrs., elected every three years. Registrar, Jno. H. Harlow, Supt., R. W. Bagnell, appointed by comrs.

92. **QUINCY**, *Norfolk Co.* (Pop., 10,570—16,723.) Built in '84 by Quincy Water Co.; franchise does not regulate rates, but provides for purchase of works at any time, price determined by comrs. appointed by Supreme Court. **Supply.**—Storage reservoir on Town brook pumping to stand-pipe. Fiscal year closes Aug. 31. Filter bed described in 1889-90 Manual has been abandoned. New filter contemplated.

Year.	Mains.	Taps.	Meters.	Hydrants.
'84	22	200	0	85
'86	23.5	700	25	90
'88	30	1,000	45	98
'90	35	1,198	53	103

Financial.—*Cost*, $541,888. *Cap. Stock:* authorized, $250,000; all paid-up. *Total Debt*, $292,135; floating, $42,135; bonded $250,000 at 6%. *Revenue:* city, $3,605. Man-

MASSACHUSETTS.

agement.—Prest., J. A. Gordon. Secy., W. L. Faxon. Supt., F. E. Hall. NOTE.—In '91 town was authorized to buy the works.

93. **RANDOLPH**, *Norfolk Co.* (Pop., 4,027—3,946.) Built in '87-8 by towns of Randolph and Holbrook. **Supply.**—Great pond, pumping to stand-pipe. In '91 bids were asked for a 1,500,000-gall. pump; reported in July that contract had been let. Fiscal year closes Feb. 28.

Year.	Mains.	Taps.	Meters.	Hydrants.	Cost.	B'd'd debt.	Int. ch'g's.	Exp.	Rev.
'87	6.4	172	0	90	$95,693	$100,000	$4,000	$1,495	
'88	8.7	305	0	104	109,359	104,000	4,100	3,273	$3,283
'89	10.3	389	0	116	120,843	110,000	4,400	3,444	3,794
'90	11	446	0	119	125,825	115,000	4,600	6,120	4,525

Financial.—*Bonded Debt*, $115,000 at 4%. *Sinking Fund*, $14,524, invested in Ry. bonds at 4½ and 4%. *Expenses*: total, $10,729. *Revenue*: consumers, $4,525. **Management.**—Three comrs.: J. W. Belcher, P. B. Hand, Chas. A. Wales. Supt., E. J. Chadbourne.

94. **READING**, *Middlesex Co.* (Pop., 3,181—4,088.) **History.**—Built in '90 by town. Engr., M. M. Tidd, Boston. Fiscal year closes March 31. **Supply.**—Ipswich River, pumping to stand-pipe: filtered. **Pumping Machinery.**—Dy. cap., 1,200,000 galls.; two 600,000 Blake. **Stand-Pipe.**—Cap., 229,800 galls.; 30 x 100 ft.; from E. Hodge & Co., E. Boston. **Distribution.**—*Mains*, 12 to 6-in. c. i., 14 miles. *Services*, cem. lined i. Pipes and specials from Gloucester Iron Works. *Taps*, 60. *Hydrants*, Chapman. 80. *Valves*, Chapman. **Financial.**—*Cost*, $123,009. *Bonded Debt*, $150,000 at 4%. **Management.**—Three comrs.: L. M. Bancroft, Geo. E Abbott, E. C. Nichols. Supt., L. M. Bancroft.

95. **REVERE and WINTHROP**, *Suffolk Co.* (Pop.: Revere, 2,263—5,668; Winthrop, 1,043—2,726.) Built in '84 by Revere Water Co. under 20-years franchise, which provides neither for control of rates nor purchase of works by the city. **Supply.**—Artesian wells, pumping to reservoir. Fiscal year closes Dec. 31. **Distribution.**—*Mains*, 33 miles. *Taps*, 1,707, *Meters*, 1. *Hydrants*, 123. *Consumption*, 428,448 galls. **Financial.**—*Cost*, $231,283. *Cap. Stock:* authorized, $125,000, all paid up. *Total Debt*, $200,287; floating, $287; bonded, $200,000. *Revenue:* city, $5,856 **Management.**—Prest., Andrew Burnham. Treas., Thos. N. Birnie, Springfield. Secy. and Supt., A. S. Burnham.

Richmond Furnace, *Berkshire Co.* (Pop., 796—1,124.) Built in '82 by Richmond Iron Works for fire protection.

IV. **RIVERSIDE**, *Franklin Co.* (Pop., not given.) Owned by Riverside Water Co.; bought from individuals Sept., '89. **Supply.**—Springs by gravity. There are 4,000 ft. of pipe and 40 taps. *No hydrants.* Cost, about $4,000. Rev., about $500. **Management.**—Prest., T. M. Stoughton. Secy., R. Stoughton. Supt., M. A. Ward.

96. **ROCKLAND**, *Plymouth Co.* (Pop., 4,553—5,213.) Built in '86 by town. Pumping machinery, 6 miles mains and stand-pipe owned jointly with Abington. **Supply.**—See Abington. Fiscal year closes Dec. 31.

Year.	Mains.	Taps.	Meters.	Hydrants.	Consumption.*	Cost.	Exp.	Rev.
'84	14	269	....	90	....	$120,510	$319	....
'87	15	546	50	99	163,791	147,837	2,963	$6,800
'88	16	652	61	103	248,143	154,300	2,903	5,785
'89	17.5	728	....	109	337,695	161,602	4,106	6,557
'90	18	801	120	112	311,827	166,087	3,631	7,102

* Including Abington.

Financial.—*Total Debt*, withheld. *Revenue:* Consumers, $7,102; city, $2,150; total, $9,352. Bonds are redeemed by direct taxation. Revenue pays operating expenses, and balance is applied to new construction. **Management.**—Three comrs., one elected every year. Chn., Geo. W. Kelley, Secy. and Supt., W. R. Groce.

97. **SALEM**, *Essex Co.* (Pop., 27,563—30,801.) Built in '65-9 by city. First supplied in 1799 by Salem Aqueduct Co. **Supply.**—Wenham Lake, pumping direct and to reservoir. Fiscal year closes Dec. 1. In '90-1 a 30-in. main, nearly 6 miles long, was laid from Dodge and Salem Sts., N. Beverly, to connect with a 20-in. main in

---

FOR FURTHER DESCRIPTIVE MATTER see Manual for 1889-90. CHANGES since 1890 are here given. POPULATIONS are for 1880 and 1890, respectively.

Salem. Eight-in. connections were made with both the Beverly and Danvers works, and 5 hydrants set in Danvers and 2 in Peabody. Cost, $187,809.

Year.	Mains.	Taps.	Meters.	Hydrants.	Consumption.	Cost.	Exp.	Rev.
'70	34.1	1,520	....	230	....	$5,067,746	$3,623	$8,709
'76	42.2	4,157	78	340	1,962,433	1,370,944	16,483	38,734*
'80	43.4	4,856	127	344	2,006,774	1,391,492	22,519	41,143*
'82	44.3	4,557	134	351	2,300,000	1,406,648	27,126	53,512*
'84	44.9	4,914	136	357	2,209,128	1,417,731	24,124	56,196*
'86	50	5,195	140	374	2,579,675	1,435,047	22,975	66,080*
'87	50	5,305	130	377	2,255,242	1,444,378	25,195	80,656*
'88	50	5,412	137	378	2,111,272	1,450,072	26,270	52,614
'89	....	5,500	127	383	2,761,494	1,458,429	28,473	51,079
'90	....	5,583	129	390	2,135,600	1,641,186	26,785	53,244

* Including receipts from Beverly.

Financial.—See above. **Management.**—Wenham Water Board. Prest., A. H. Smith. Clerk, Nath M. Brown. Supt., Henry W. Rogers.

98. **SAUGUS**, *Essex Co.* (Pop., 2,625–3,673.) Built in '78 by city. **Supply.**—Lynn water-works.

Year.	Mains.	Taps.	Meters.	Hyd'ts.	Cost.	B'd'd d't	Int.ch'gs.	Exp.	Rev.
'87	9	90	0	80	$45,800	$40,800	$1,632	None.	$362*
'89	9	248	..	83	45,800	40,800	1,632	"	854*
'90	9	292	0	84	45,800	40,800	1,632	"	1,078*

* Rebate from Lynn.

Financial.—See above. Lynn collects water rates from Saugus and gives rebate of 37½%, after deducting hydrant rental of $28 per hydrant. **Management.**—Comr., Wilbur Newhall, E. Saugus.

99. **SHARON**, *Norfolk Co.*; franchise fixes rates and provides for purchase of works by town at any time that two-thirds of voters at town-meeting vote to that effect, price determined by arbitration or by comrs. appointed by Supreme Court. (Pop., 1,492–1,643.) Built in '85 by Sharon Water Co. **Supply.**—Well, near Beaver Brook, pumping to tank. Fiscal year closes July 1.

Year.	M'ns.	Taps.	Meters.	Hy-dr'nts.	Con-s'mpt'n.	Cost.	Bonded debt.	Int. charges.	Exp.	Rev.
'86	2.5	87	8	17	20,000	$40,000	$20,000	$1,200	....	$1,000*
'88	6.3	125	8	21	50,090	52,000	20,000	1,200	$3,000	3,000
'90	6.3	150	8	26	30,137	52,008	20,000	1,200	1,589	3,510

*From city.

Financial.—*Cap. Stock:* authorized, $20,000; all paid up. *Total Debt,* $33,797; floating, $13,797; bonded, $20,000 at 6%. *Expenses:* operating, $1,503; int, $1,450; taxes, $86; total, $3,039. *Revenue:* consumers, $2,310; city, $1,200; total, $3,510. **Management.**—Prest., E. W. Bond. Treas., E. R. Stickney, both of Springfield. Supt., W. B. Wickes.

V. **SHELBURNE FALLS**, *Franklin Co.* (Pop., 1,621–1,553.) Built in '84 by Covell Aqueduct Co. **Supply.**—Mountain springs, by gravity. Fiscal year closes May 1. **Distribution.**—*Mains,* 23 miles. *Taps,* 70. *Meters,* 0. *Hydrants,* 0. *Consumption,* 9,090 galls. **Financial.**—*Cost,* $5,000. *No Debt. Expenses:* operating, $20; taxes, $28; total, $48. *Revenue:* consumers, $423; city, $28; total, $451. **Management.**—Prest., G. G. Merrill. Secy., Herbert Newall. Supt., L. T. Covell.

100. **SOMERVILLE**, *Middlesex Co.* (Pop., 24,933–40,152.) Partially supplied as early as '58 from Cambridge; present supply introduced in '68 by city. **Supply.**—Mystic system of Boston water-works. Fiscal year closes Dec. 31. High service system was introduced in '89-90, consisting of a 2,000,000-gall. Worthington dup. pump; horizontal boiler; an iron tank, 30 × 100 ft., and 14-in. force main.

Year.	Mains.	Taps.	Meters.	Hy-drants.	Cost.	Bonded debt.	Interest charges.	Exp.	Rev.
'80	45.5	1,227	31	283	$342,608	$335,000	......	$14,000	$14,678
'88	54	5,410	..	375	361,947	......	......	15,570	47,431
'89	56.8	5,885	86	401	443,065	334,500	$19,845	16,000	51,171
'90	61.3	6,417	89	446	494,868	380,000	21,830	17,098	55,880

Bonded Debt, $380,000; $242,000 at 4%, $127,500 at 5%, $10,500 at 5½%. *Expenses:* total, $39,528. *Revenue:* consumers, $55,880. Rates are collected by Boston and a rebate of 50% given to Somerville. **Management.**—Board of three, one appointed every year by mayor. City Engr., Horace L. Eaton.

NOTE.—In '91 Council requested comr. on water to consider expediency of city owning its water supply by purchase of Mystic system or otherwise. March 23 matter was in hands of committee.

MASSACHUSETTS. 41

101. SOUTHBRIDGE, *Worcester Co.* (Pop., 6,464—7,655.) Built in '80 by South Bridge Water Supply Co. **Supply.**—Springs and surface water, by gravity. Fiscal year closes May 26. **Distribution.**—*Mains*, 5 miles. *Taps*, 300. *Hydrants*, 35. **Financial.**—*Cost*, $34,675. *Cap. Stock*: authorized, $18,000, all paid up; paid 6% div'd. In 1891. *Total Debt*, $9,000. *Expenses:* Operating, $159; int., $477; taxes, $277; to tal, $913. *Revenue:* consumers, $2,633; city, $1,150; total, $3,783. **Management.**—Prest., F. L. Chapin. Secy. and Treas., F. W. Eaton.

102. SOUTH FRAMINGHAM, *Middlesex Co.* (Pop., 6,235—9,239.) Built in '85 by Framingham Water Co.; franchise provides that town may buy works at any time, on terms to be mutually agreed upon, or by three comrs., appointed by court, upon application of one or both parties. **Supply.**—Pond and Sudbury River, pumping direct; filtered. Fiscal year closes Dec. 31.

Year.	Mains.	Taps.	Meters.	Hydrants.	Consumption.	Cost.	Bonded debt.	Int. charges.	Exp.	Rev.
'87	11	425	54	72	260,000	$225,000	$125,000	$7,500	$6,000	$13,520
'89	12	600	80	75	170,000	253,000	125,000	7,800	6,000	13,525
'90	15	720	110	83	204,000	266,0 0	125,000	7,500	7,086	14,445

Financial.—*Assets*, $257,258. *Cap. Stock:* authorized, $125,000; all paid up; paid no div'd in 1890. *Total Debt*, $132,258; floating, $7,258 at 5%; bonded, $100,000 at 6%, due 1905; $25,000 at 6%, due in 1908. *Expenses:* operating, $3,500; int., $8,100; taxes, $586; total, $14,586. *Revenue:* consumers, $11,000; city, $3,445; total, $14,445. **Management.**—Prest., W. M. Ranney. Secy., A Phillips. Treas., G. E. Cutler. Supt., A E. Martin.

103. SOUTH HADLEY FALLS, *Hampshire Co.* (Pop., 3,538—4,261.) Built in '72 by village. **Supply.**—Brook, by gravity from storage reservoir. Reported, summer of '91, that construction of reservoir on Leaping Well Brook had been voted. Fiscal year closes March 1.

Year.	Mains.	Taps.	Meters.	Hydrants.
'89-90	8	400	1	56
'90-1	8.5	425	1	60

Financial.—*Revenue*, $4,100. **Management.**—Six comrs., elected every three years. Chn., H. P. Street. Secy., F. M. Smith Supt., M. L. Barnes.

104. SPENCER, *Worcester Co.* (Pop., 7,466—8,747.) Now owned by town; built in '83 by Spencer Water Co. **Supply.**—Shaw's pond, by gravity. Fiscal year closes March 1. **Distribution.**—*Mains*, 13.3 miles. *Taps*, 950. *Meters*, 0. *Hydrants*, 87. *Consumption*, 165,000 galls. **Financial.**—*Cost*, $243,000. *Bonded Debt:* $240,000 at 4%, due in 1914. *Sinking Fund*, $35,166, invested in mortgages and bonds at 5 and 5½%. *Expenses:* operating, $722; int., $9,000; total, $10,322. *Revenue:* consumers, $17,202. **Management.**—Comrs.: A. H. Ames, A. G. Pease, Dexter Bullard. Supt., Dexter Bullard.

105. SPRINGFIELD, *Hampden Co.* (Pop., 33,340—44,179.) Present works built in '73-5 by city; first supply introduced in '43 by Chas. Stearns. **Supply.**—Impounding reservoir, by gravity. Water furnished to Ludlow, which see. Fiscal year closes Dec. 31.

In '90-91 extensive improvements were made at and about Ludlow reservoir and supply canals entering it, and additional supply was secured. Dam was built across east branch of lower or south end of Ludlow reservoir, cutting off 12 acres and forming 45,000,000-gall. basin. This is to shut out swampy water from large reservoir above and to receive water from canals. Previously water from Higher Brook was brought by canal into west branch of south end of reservoir. Dam was built across canal near its old entrance to reservoir, and water conveyed to new basin by cast-iron pipe. Water from Broad and Axe Factory Brooks was formerly conveyed by canal from Belchertown reservoir, on Broad Brook, to east side Ludlow reservoir, at north end of Cherry Valley dam; short canal connects Axe Factory Brook with Broad Brook Canal close by reservoir. Broad Brook Canal was extended to the north through Belchertown reservoir and swamp to Jabish Brook, east of Broad Brook, where a stone diverting dam, 200 ft. long and 16 ft. high, was built. Water from Broad Brook Canal is now carried across Cherry Valley dam through 54-in. iron pipe 1,400 ft. long, and then canal continued to discharge into new basin just below cut-off dam. Total length of canal about 8 miles, of which 11,930 ft. was old canal. Latter has been enlarged to size of new, 8 ft. wide on bottom, 22 at

FOR FURTHER DESCRIPTIVE MATTER see Manual for 1889-90. CHANGES since 1890 are here given. POPULATIONS are for 1880 and 1890, respectively.

top, 4.66 ft. deep, slopes being 1½ to 1. Engrs., E. C. and E. E. Davis, Northampton; with P. M. Blake, Hyde Park, A. Fteley and Clemens Herschel, New York, as Consult. Engrs. Contrs.: stone dam on Jabish Brook, and canal to connect above with old Broad Brook Canal, W. P. Latham, Northampton; 54-in. iron pipe, R. F. Hawkins, Springfield; new canal around reservoir, G. M. Atkins & Co., Palmer; cut-off dam across branch of Ludlow reservoir, and part of Higher Brook Canal work, H. E. Stanton, Huntington; remaining work, including new waste-way at Cherry Valley dam, done by city. Most of work completed by Dec. 31, '90. Appropriation for above improvements to Dec. 10, '90, $161,000, which it was thought would complete them.

Year.	Mains.	Taps.	Meters.	Hyd'ts.	Cost.	B'd'd d't.	Int.ch'g's.	Exp.	Rev.
'75	55.2	.....	...	310	$1,146,279	$1,200,000	$82,000	$16,611	$46,109
'80	65	3,290	44	381	1,216,847	1,200,000	82,000	10,369	65,702
'81	66.1	3,530	77	397	1,227,025	1,200,000	82,000	11,064	71,981
'82	67.8	3,737	101	424	1,258,752	1,500,000	82,000	19,436	77,407
'83	70.2	3,793	145	402	1,276,915	1,200,000	82,000	21,863	89,033
'84	73.2	4,013	218	434	1,294,260	1,200,000	82,000	16,116	87,445
'85	75	4,237	285	457	1,306,187	1,200,000	82,000	22,136	93,422
'86	78.7	4,519	349	490	1,320,935	1,200,000	82,000	21,817	100,538
'87	81.5	4,819	472	528	1,335,535	1,200,000	82,000	25,148	108,140
'88	85	5,147	556	557	1,349,574	1,200,000	82,000	28,339	115,146
'89	91.1	5,497	751	601	1,378,185	1,200,000	82,000	18,462	122,512
'90	92.3	5,770	928	685	1,382,763	1,325,000	82,000	34,800	149,913

Financial.—*Bonded Debt*, $1,325,000: $1,000,000 at 7%, $200,000 at 6%, $125,000 at 3½%. *No Sinking Fund. Expenses*: total, $116,800. *Revenue*: consumers, $149,913 for water; $19,281 for services and repairs; city, fire protection, $12,160; other purposes, $5,030; total, $186,384. **Management.**—Com'rs. Chn., H. L. Sanderson. Supt., J. C. Hancock.

106. **STOCKBRIDGE,** *Berkshire Co.* (Pop., 2,357–2,132.) Built in '82 by Stockbridge Water Co.; franchise provides that city may purchase works at any time on payment of total cost of its works, including 7% int. on each expenditure from its date to date of purchase; city has no control of rates. **Supply.**—Springs, surface well and artesian well, by gravity and pumping to reservoir; filtered. Fiscal year closes last Saturday in June. **Distribution.**—*Mains*, 3 miles. *Taps*, 100. *Meters*, 0. *Hydrants*, 6. **Financial.**—*Cost*, $28,000. *Cap. Stock:* authorized $40,000; paid up, $7,000; paid no div'd in 1890. *Total Debt*, $8,900; floating, $500 at 6%; bonded, $8,400 at 5% due in 1905. *Expenses:* operating, $655; int., $458; no taxes; total, $1,113. *Revenue:* consumers, $1,486; city, none. **Management.**—Prest., D. R. Williams. Secy. and Treas., D. A. Kimball.

NOTE.—Town has authority to build works. In fall of '90 committee reported in favor of supply from Lake Averick; est. cost, $125,000.

113. **STONEHAM,** *Middlesex Co.* (Pop., 4,890–6,155.) Supply from Wakefield.

107. **STOUGHTON,** *Norfolk Co.* (Pop., 4,875–4,852.) Built in '86 by Jno. G. Phinney, now known as Stoughton Water Co. **Supply.**—Well, pumping to tank. **Distribution.**—*Mains*, 5 miles. *Taps*, 200. *Meters*, 0. *Hydrants*, 20. *Consumption*, 25,000 galls. **Financial.**—*Cost*, $90,000. *Cap. Stock:* $50,000. *Bonded Debt*, $50,000 at 6%. **Management.**—Prest., Chas. W. Lunn. Secy. and Supt., Frank F. Phinney. Treas., W. L. Faxan.

VI. **SUNDERLAND,** *Franklin Co.* (Pop., 755–663.) Built in '83 by Sunderland Water Co.; franchise provides neither for control of rates nor purchase of works by city. **Supply.**—Springs, by gravity to small reservoir. Fiscal year closes Oct. 1. **Distribution.**—*Mains*, 2 miles. *Taps*, 49. *Hydrants*, 0. **Financial.**—*Cost*, $2,200. *Cap. Stock:* authorized, $3,000; paid up, $2,200; $194 divided in 1890. *No Debt. Expenses:* operating, $12; taxes, $15; total, $27. *Revenue:* consumers, $224; other purposes, $3; total, $227. **Management.**—Prest., Joel Burt. Secy. and Treas., N. Austin Smith.

108. **SWAMPSCOTT,** *Essex Co.* (Pop., 2,500–3,198.)

TOWN PLANT.

Built in '83–'4. **Supply.**—Marblehead Water Co.'s plant. See Manual for '89-90, p. 85. **Distribution.**—*Mains*, about 5 miles. *Taps*, 425. *Meters*, 1. *Hydrants*, 65. No further information.

MARBLEHEAD WATER CO.

Built in '85; now supplies water to Swampscott and Nahant, each town owning its distribution. Town does not control rates, but has right to buy works. **Supply.**—

MASSACHUSETTS. 43

Surface and tubular wells, pumping to stand-pipe. Fiscal year closes March 31. Added 19 2½-in. wells In '90. A 12-in. main leads to Swampscott and another to Nahant. Financial.—*Cost*, $210 213. *Cap. stock*, $100,000. *Total Debt*, $122,701. *Expenses:* operating, $4,000. Int. on debt, 5%. Taxes, $300. Net profits, $10,302. Other information withheld. Management.—Prest., Thos. Appleton. Treas., E. W. Bond. Secy., E. N. Sterling. Supt., Kendall Pollard.

32. **TAPLEYVILLE,** *Essex Co.* In town of Danvers, which see.

109. **TAUNTON,** *Bristol Co.* (Pop., 21.213—25,448.) Built in '76 by city. Supply.—Ground water and seven driven wells, with Taunton River for emergency, pumping direct. Fiscal year closes Nov. 30. A 4,000,000-gall. Gaskill pump was added in '90.

Year.	Mains.	Taps.	Me'ers.	Hy-drants.	Consumption.	Cost.	Bonded debt.	Int. ch'g's.	Exp.	Rev.
'77	23.9	744	61	210	366,200	$253,933	$235,000	$12,000	$8,866	$15,706
'78	27.5	1,162	114	238	342,498	275,303	300,000	14,000	10,000	12,558
'79	27.6	1,355	182	239	393,850	276 876	300,000	15,000	7,643	16,167
'80	30.27	1,608	258	284	510,261	338,333	307,000	15,000	9,202	18,411
'81	35.7	1,813	332	320	566,512	365,221	364,000	17,166	10,090	21,865
'82	41.6	2,062	401	307	582,111	422,225	380,000	17,989	10,953	26,064
'83	47	2,289	464	403	611.816	485,592*	320,000	20,275	16,536	29,992
'84	47.2	2,422	527	414	509,208	472.313	440,000	22,284	14,618	31,376
'85	48.5	2,523	597	422	675,870	479,024	451,700	22,875	14,938	31,076
'86	52.9	2,727	680	457	773,531	504.839	478,700	23,115	13,729	33,349
'87	54.6	2,892	782	470	724,982	551.158	523,700	24,223	15 964	36,763
'88	55.86	3,027	858	487	751.887	558,229	533,700	26,123	14,168	37.814
'89	59.29	3,143	938	514	750,551	571,995	548,700	26,423	12 618	40,730
'90	61.35	3,253	990	536	796,716	602,607	593,700	27,108	13,061	40,658

*Includes cost of laying services on private property. Financial.—*Bonded Debt*, $593,700: $250,000 at 6%, due '96; $30,000 at 4½%, due '96 ; $313,700 at 4%, of which $153,000 is due in '96, $12,000 in 1901, $1,700 in 1905, $27,000 in 1906, $45,000 in 1917, $5,000 in 1897, $5,000 in 1908, $15,000 in 1909, $45,000 in 1910. *Sinking Fund*, $286,454, invested chiefly in municipal and railway bonds. *Expenses:* total, $40,619. *Revenue:* consumers, $40,658; city, $6,307; total, $46,965. There is an annual bonded appropriation for new construction, int. of which is paid from revenue. Management.—Three comrs., one appointed every year by City Council. Chn., P. J. Perrin. Secy. and Supt., Geo. F. Chace.

110 **TURNERS FALLS,** *Franklin Co.* (Pop., not given.) Built in '86 7 by fire district; first supplied in '73 by Co. Supply.—Lake Pleasant, pumping to reservoir. Fiscal year closes May 19. Pumping station was burned early in '91.

Yr.	Mains.	Taps.	Meters.	Hy-drants.	Consump.	Cost.	Bonded debt.	Int. charges.	Exp.	Rev.
'87-8	10	196	1	50	225,000	$100.211	$100,000	$4,000	$1.600	$4,890
'88-9	10	247	2	50	246 526	104,596	100,000	4,000	8,264	6,708
'89-90	10	289	2	50	247,207	105,756	100,000	4,000	3,728	7,606
'90-1	10	307	2	50	202.854	107,732	100,000	4,000	2,771	7,815

Financial.—See above. Management.—Comrs., L. J. March, H. D. Bardwell, E. J. Cassidy.

111. **UXBRIDGE,** *Worcester Co.* (Pop., 3,111-3,408.) Now owned by town. Built in '80 by Uxbridge Water Co.; bought by town April 11, '90. Supply.—Springs, by gravity through covered trenches to reservoir. Fiscal year closes Feb. 2. Distribution.—*Mains*, 3 miles. *Taps*, 119. *Meters*, 32. *Hydrants*, 24. *Consumption*, 25,000 galls. Financial.—*Cost*, $27,500. *Bonded Debt*, $27,500 at 5%, due in 1910. *Sinking Fund*, $945, deposited in bank at 4%. *Expenses:* operating, $945. *Revenue:* consumers, $1.536; city, $720; total, $2,256. Management.—Comrs.; G. M. Aldrich, W. P. Scott, Waldo Rawson.

112. **VINEYARD HAVEN,** *Dukes Co.* (Pop., not given.) Built in '87 by Vineyard Haven Water Co. Supply.—Tashmoo springs, pumping to stand-pipe. Fiscal year closes Jan. 1. Distribution.—(In '88.) *Mains*, 3 miles. *Meters*, 3, *Hydrants*, 11. Financial.—*Cost*, $59,228. *Cap. Stock:* authorized, $30,000; all paid up. *Total Debt*, $32,368; floating, $2,901; bonded. $29,967. Management.—Prest., Chas. C. Jackson. Treas., Francis Peabody. Supt., A. F. Cummings.

NOTE.—No statistics have been received from company since '88. Financial data taken from State Corporation Report.

113. **WAKEFIELD and STONEHAM,** *Middlesex Co.* (Pop., Wakefield, 5,547—6,982; S oneham, 4,890 - 6,155.) Built in '83 by Wakefield Water Co. Supply.—

FOR FURTHER DESCRIPTIVE MATTER see Manual for 1889-90. CHANGES since 1890 are here given. POPULATIONS are for 1880 and 1890, respectively.

44   MANUAL OF AMERICAN WATER-WORKS.

Crystal Lake, pumping to stand-pipe. Fiscal year closes Dec. 31. **Distribution.**—*Mains,* 50 miles. *Taps,* 1,800. *Meters,* 0. *Hydrants,* 152. **Financial.**—*Cost,* $503,-008. *Cap. Stock:* authorized, $75,000; all paid up. *Total Debt,* $598,897; floating, $39,516; notes, $57,380; bonded, $427,000. **Management.**—Prest., D. H. Darling. Secy., C. O. Burbank. Supt., F. W. Harrington. NOTE.—In '91 Wakefield talked of buying works.

114. **WALTHAM,** *Middlesex Co.* (Pop., 11,712—18,707.) Built in '73 by town. **Supply.**—Open filter basin near Charles River, pumping to reservoir. Fiscal year closes Dec. 31. Reported in '91 that authority had been secured to increase amount of water drawn from river from 1,000 000 to 3,000,000 galls daily.

Year.	Mains.	Taps.	Meters.	Hydrants.	Consum't'n.	Cost.	Bonded debt.	Int. charges.	Exp.	Rev.
'75	....	644	..	...	481,428	......	......	$19,050	$6,680	$10,732*
'80	....	1,135	..	...	410,017	......	......	20,119	4,984	25,133*
'85	....	1,802	..	...	529,661	......	......	15,064	8,800	34,401*
'86	32.7	1,924	20	178	677,637	$376,000	$361,000	21,659	9,046	32,838
'87	34.3	2,081	23	188	544,138	388,000	373,000	18,060	11,148	34,433
'88	36.4	2,248	27	197	534,063	412,000	397,000	18,800	18,593	30,622
'89	37 7	2,404	34	203	569,342	420,859	387,0 0	19,680	12,472	43,692
'90	38.9	2,489	40	211	625,779	432,000	387,000	18,747	10,029	45,810

*Including revenue from city.

Financial.—*Bonded Debt,* $387,000; $25,000 at 6.6%, due in '93; $262,000 at 4%. *Sinking Fund,* $43,341. *Expenses;* total, $28,777. *Revenue:* consumers, $45 810; city, 5,037; total, $50,817. **Management.**—Three comrs., one appointed every year by City Council. Chn., Jno. Harris. Secy., Laroy Brown. Supt., Geo. E. Winslow, appointed by board.

115. **WARE,** *Hampshire Co.* (Pop., 4,817—7,329.) Built in '86-7 by town. **Supply.**—Collecting well, pumping to reservoir. Fiscal year closes Jan. 31.

Year.	Mains.	Taps.	Meters.	Hydrants.	Consumption.	Cost.	Bonded debt.	Int. ch'g's.	Exp.	Rev.*
'86	6.5	77	9	49	40,000	$75,000	$75,000	$2,625	....	.....
'87	7.4	308	41	52	76,709	81,048	76,500	2 753	$4,471	$3,029
'89	7.75	350	87	56	133,359	34,395	73,800	2,780	5,282	5,100
'90	8.25	410	198	59	166,758	87,654	71,000	2,584	5,479	7,182

*Including allowance for hydrant rental.

Financial.—*Bonded Debt:* $71,000; $52,000 at 3½, $2,000 due each year; $19,100 at 3%, $800 due each year. *Expenses:* total, $8,063. *Revenue:* consumers, $5,803; city, $1,329; total, $7,132. Expenses are paid by revenue and surplus, if any, is used for construction, by vote of town; bonds are paid by taxation. **Management.**—Three comrs., elected every three years. Chn., Jno. Kennedy. Supt., Fred. A. Volk, appointed by comrs.

VII., VIII., IX. **WARREN,** *Worcester Co.* (Pop., 3,889—4,681.)

AQUEDUCT COS.

Warren Aqueduct Co. built in '37; *Aqueduct Co.* in '72; *J. W. Hasting* in '75. Each has supply from springs by gravity, and each supplies about 20 families.

FIRE PROTECTION PLANTS.

Town built about '86. See Manual for '89-90, p. 88. About '88 W. H. Fairbanks and others put in pipe to protect a few buildings.

116. **WATERTOWN,** *Middlesex Co.* (Pop., 5,426—7,073.) Built in '84 by Watertown Water Supply Co.; franchise does not regulate rates, but provides for purchase of works by town at any time between five and fifteen years after construction of plant. **Supply.**—Filter galleries near Charles River, pumping direct and to tank. Fiscal year closes Jan. 20. Supply was increased in '90 by constructing another filter gallery, or a well, 20 ft. in diam. and 24 ft. deep, at distance of 400 ft from old gallery, with which it is connected by conduit. An additional 1,500,000-gall. pump was contracted for in '91.

Year.	Mains.	Taps.	Meters.	Hydrants.	Consumption.
'86	18	600	.....	164	150,000
'88	23	750	20	179	275,000
'90	26	900	50	204	400,000

Financial.—*Cost,* $312,500. *Cap. Stock:* authorized, $150,000; all paid up; paid 6% div'd in 1891. *Bonded Debt,* $135,000 at 5%, due in 1900. *Expenses:* operating, $5,200; int., $7,500; taxes, $1,400; total, $14,100. *Revenue:* consumers, $17,000; city, $8,000; total, $25,000. **Management.**—Prest., A. O. Davidson. Secy., S. S. Gleason. Treas., Jno. H. Conant. Supt., Jno. H. Perkins.

MASSACHUSETTS. 45

117. WEBSTER, *Worcester Co.* (Pop., 5,696—7,031.) Domestic supply introduced in '82 from works built in '68 by H. N. Slater Co. for fire protection. **Supply.**—Lake Chaubunagungamug, pumping direct and to reservoir. Fiscal year closes Sept. 30.

Year.	Mains.	Taps.	Meters.	Hydrants.	Consumption.	Cost.	Rev.
'80	3	126	2	35	35,000	$30,000	$1,250
'88	6	270	6	..	200,000	35,000	3,860
'89	6	242	9	36	200,000	40,000	3,400
'90	7	273	10	40	300,000	40,000	4,518

Financial.—*No Stock or Bonds. Expenses,* cannot be separated from mill expenses. *Revenue:* consumers, $3,700; city, $758; other purposes, $30; total, $4,518. **Management.**—Owner, W. M. Slater. Supt., Thos. K. Bates. NOTE. Works are projected by town.

118. WELLESLEY, *Norfolk Co.* (Pop., 3,600.) Built in '84 by town. **Supply.**—Filter gallery and well, pumping direct and to reservoir. Fiscal year closes Dec. 31.

Year.	Mains.	Taps.	Meters.	Hydrants.	Consumption.	Cost.	Bonded debt.	Int. ch'g's.	Exp.	Rev.
'85	13.2	....	....	67	62,939	$140,047	$125,000	$4,266	$1,890	$1,664
'86	17	....	....	77	89,414	158,016	146,000	7,617	3,796	3,809
'87	17.1	332	48	103	101,172	164,363	151,000	5,840	2,660	5,651
'88	19.6	412	66	126	132,233	175,484	161,000	6,410	2,971	5,225
'89	20.6	465	83	134	194,647	185,320	171,000	6,540	3,935	6,160
'90	22.1	503	93	145	255,418	195,448	183,000	6,600	4,939	8,083

Financial.—See above. **Management.**—Three comrs. Chn., A. R. Clapp. Supt., W. H. Vaughn.

119. WESTBOROUGH, *Worcester Co.* (Pop., 5,214—5,195.) Built in '78 by town. **Supply.**—Sandra Pond, by gravity. Fiscal year closes Feb. 1.

Year.	Mains.	Taps.	Meters.	Hyd'ts.	Cost.	B'd'd d't.	Int. ch'g's.	Exp.	Rev.
'80	6.5	210	..	56	$53,000	......	......	$400	.....
'82	6.5	256	40	57	53,000	$53,000	$2,650	1,786	$3,119
'84-5	9	400	5	70	80,000	80,000	.....	550	5,500
'86-7	9.5	435	3	71	80,000	79,000	3,710	.....	6,354
'88-9	10.5	551	30	80	109,000	109,000	4,910	1,058	6,593
'89-90	11.89	600	36	80	195,000	108,000	4,910	2,223	7,419

Financial.—*Bonded Debt*, $180,000: $54,000 at 5%, due in 1908; $35,000 at 4%, due in 1908; $19,000 at 4%, due in 1916; *Sinking Fund*, $18,029. *Expenses*: total, $7,133. *Revenue*: consumers, $7,419. All expenses, including interest and sinking fund are paid from revenue; new construction is paid from taxation or proceeds of bonds. **Management.**—Three comrs., one elected every year. Chn., M. G. McCarty. Secy. and Supt., E. T. Gilmore, appointed by comrs.

X., XI., XII. WEST BROOKFIELD, *Worcester Co.* (Pop., 1,917—1,592.)

AQUEDUCT COS.

West Brookfield Aqueduct Co., built in '38; *Quaboag Aqueduct Co.*, in '51; *Cement Aqueduct Co.*, in '68-9. Each has supply from wells, by gravity, and each supplies about 40 families.

TOWN PLANT.

Built in '84 for fire protection. See '89-90 Manual, p. 90.

120. WESTFIELD, *Hampden Co.* (Pop., 7,587-9,805.) Built in '74 by town. **Supply.**—Moose Meadow Brook, by gravity from storage to distributing reservoir. Fiscal year closes Feb. 1. Distribution—*Mains*, about 14 miles. *Hydrants*. 136. **Financial.**—*Cost* (in '87), $361,000. *Expenses*, $2,441. *Revenue*, $17,790. **Management.**—Clerk, C. N. Oakes. Supt., C. H. Diehl.

XIII. WESTPORT HARBOR (*P. O. Adamsville, R. I.*), *Bristol Co.* (Pop., 2,804—2,599.) **History.**—Under construction summer of '91 by Westport Harbor Aqueduct Co., to supply summer cottagers. **Supply.**—Well, by gravity to reservoir. **Reservoir.**—Cap., 15,000 galls. **Financial.**—*Cap. Stock*, $3,000. **Management.** Prest., E. S. Brown. Clerk, Edward B. Jennings. Treas., A. Manchester.

121. WEST SPRINGFIELD, *Hampden Co.* (Pop., 4,149—5,077.) Built in '95 by West Springfield Aqueduct Co.; franchise provides neither for control of rates nor purchase of works by town. **Supply.**—Darby Brook, by gravity from impounding reservoir. Distribution.—*Mains*, 9 miles. *Taps*, 399. *Meters*, 0. *Hydrants*, 20. **Financial.**—*Cost*, $75,509. *Cap. Stock:* authorized, $50,000; all paid up.

FOR FURTHER DESCRIPTIVE MATTER see Manual for 1889-90. CHANGES since 1890 are here given. POPULATIONS are for 1880 and 1890, respectively.

46 • MANUAL OF AMERICAN WATER-WORKS.

Bonded Debt, $25,000. **Management.**—Treas., Thos. Birnie, Springfield. Supt., M. L. Tourtelotte.

XIV. **WEST STOCKBRIDGE,** *Berkshire Co.* (Pop. of town, 1,923—1,492. Built in '73 by East Mountain Water Co.; town neither controls rates nor has right to purchase works. **Supply.**—Spring, by gravity. Fiscal year closes July 1. **Distribution.**—About 2,000 ft. of pipe, supplying 30 families. **Finanial.**—*Cost,* $4,060. *Cap. Stock,* $4,000; div'd., 5%. *No Debt. Expenses:* operating, nominal; taxes, $35. *Revenue:* consumers, $217. **Management.**—Prest., W. W. Kniffin, Secy., C. W. Kniffin. Treas. and Supt., J. S. Moore.

122. **WEYMOUTH,** *Norfolk Co.* (Pop., 10,570—10,866.) Built in '85 by town. **Supply.**—Weymouth Great Pond, by gravity, and for South Weymouth, pumping to stand-pipe. Fiscal year closes Dec. 31.

Year.	Mains.	Taps.	Meters.	Hydrants.	Cost.	Bonded debt.	Int. charges.	Exp.	Rev.
'85	34.4	473	0	280	$264,681	$275,000	$9,700	....	....
'86	40.6	1,253	0	299	292,798	294,000	11,800	$5,113	$10,568
'87	43.7	1,310	0	303	323,907	290,000	11,610	14,160	16,338
'88	45.7	1,450	2	308	361,554	344,000	13,760	4,127	16,852
'89	48.5	1,609	2	322	373,383	371,000	13,840	6,790	19,167
'90	50.4	1,754	2	333	421,618	400,000	14,840	7,304	21,331

Financial.—*Bonded Debt,* $400,000, at 4%. *Sinking Fund,* $27,603, invested in bonds. *Expenses:* total, $22,144. *Revenue:* consumers, $21,331; city, $7,319; total, $28,650. **Management.**—Five comrs. Chn., Aug. J. Richards. Secy., Henry A. Nash. Treas., Jno. H. Stetson. Supt., J. C. Howe.

123. **WHITINSVILLE,** *Worcester Co.* (Pop., est., 2,500.) Built in '89 by Whitin Machine Co. **Supply.**—Springs, by gravity from reservoir. Fiscal year closes Dec. 31. **Distribution.**—*Mains,* 7 miles. *Taps,* 400. *Meters,* 0. *Hydrants,* 45. **Financial.**—Withheld. **Management.**—Whitin Machine Works.

124. **WHITMAN,** *Plymouth Co.* (Pop., 3,024—4,441.) Built in '83-4 by town. **Supply.**—Hobart pond, pumping to stand-pipe; filtered. Fiscal year closes Dec. 31.

Year.	Mains.	Taps.	Met'rs.	Hydrants.	Consumption.	Cost.	Bonded debt.	Int. ch'g's.	Exp.	Rev.
'85	7.8	205	87	72	Unk'n	$36,884	$55,000	....	....	$3,163
'86	8.5	276	125	83	"	63,841	60,000	$2,300	$2,033	3,584
'87	9.2	318	120	85	"	73,830	68,000	2,560	2,984	3,757
'88	10.5	373	126	92	93,329	82,384	73,000	2,800	3,219	5,306
'89	10.5	445	141	92	100,963	84,055	75,000	3,000	3,026	4,601
'90	11.6	528	168	105	87,141	91,744	81,000	3,100	2,897	5,145

Financial.—*Bonded Debt,* $81,000 at 4%. *Sinking Fund,* $7,738, invested in 4% bonds. *Expenses:* total, $5,997. *Revenue:* consumers, $5,145; city, $2,050; total, $7,195. **Management.**—Comrs. elected every three years. Chn., Albert Davis. Secy. and Supt., J. C. Gilbert, appointed by comrs.

125. **WILLIAMSTOWN,** *Berkshire Co.* (Pop., 3,394—4,221.) Built in '59 by Williamstown Aqueduct Co., now owned by Williamstown Water Co. **Supply.**—Spring, by gravity from reservoir. Fiscal year closes April 1. **Distribution.**—*Mains,* 10 miles. *Taps,* 200. *Hydrants,* 14. **Financial.**—Withheld. **Management.**—Prest., A. C. Houghton. Secy., W. B. Clark. Supt., G. H. Sanford.

126. **WINCHESTER,** *Middlesex Co.* (Pop., 3,802—4,861.) Built in '73 by town. **Supply.**—Surface water, by gravity from impounding reservoir; also high service for about 45 takers, pumping with wind-mill to reservoir. Fiscal year closes March 1.

South reservoir dam completed in '90 or '91, having been begun in '82 and abandoned in '83 after expenditure of $75,000. Cap., not given. Water area, 160 acres; max. depth of water, 10 ft.; dam of earth, with heart-wall; 900 ft. long, 45 ft. high, 12 ft. wide at top, 176 ft. at 40 ft. below top. Heart-wall 66 ft. high, masonry, carried to rock. Inner slope paved with 15-in. split rubble stone. Driveways begin at middle of foot of inner slope and extend each way, ascending to form a berm. A 24-in. discharge and 30-in. waste-pipe. Total cost about $135,000. Engr. for completing, P. M. Blake, Hyde Park, with E. Worthington, Jr., in charge. Contrs. for completing, Rowe & Hall, Boston.

Year.	Mains.	Taps.	Meters.	Hyd'ts.	B'd'd d't.	Int. ch'g's.	Exp.	Rev.
'80	16	630	....	73	....	....	$1,600	......
'84	17	650	0	75	$150,000	....	2,000	$10,819
'88	22	847	1	83	221,000	$11,460	2,000	14,006
'89	23	867	3	85	221,000	11,460	3,000	15,062
'90-'91	23	897	3	85	281,000	12,103	3,544	16,061

Financial.—See above. **Management.**—Three comrs. Supt., W. D. Dotten.

95. **WINTHROP,** *Suffolk Co.* (Pop., 1,043—2,726.) Supply from Revere.

MASSACHUSETTS.

127. WOBURN, *Middlesex Co.* (Pop., 10,931—13,499.) Built in '73 by town. Supply.—Ground water from filter gallery near Horn pond, pumping direct and to reservoir. Fiscal year closes Dec. 31. During '90 additional force-main was laid to reservoir.

Year.	Mains.	Taps.	Meters.	Hydrants.	Consump'tion.	Cost.	Bonded debt.	Int. ch'g's.	Exp.	Rev.
'73	21.5	493	..	180	906,841	$374,974	$100,000	$11,424	$1,977	$4,600
'75	24	835	..	200	730,530	....	....	24,500	7,166	11,917
'76	29	992	..	232	668,837	....	....	24,500	8,604	14.687
'77	29.5	1,081	27	234	513,335	....	422,500	25,109	6,849	16,508
'78	33	1,208	..	250	651,034	....	....	25,125	7,393	17,925
'79	33	1,290	..	252	733,546	....	451,300	25,798	7,527	17,610
'80	36	1,361	..	260	738,433	....	466,300	26,441	10,337	19,101
'81	36.5	1,413	16	271	772,578	522,551	466,300	26,741	8,291	20,624
'82	37	1,487	33	278	824,319	526,846	466,300	27,041	9,878	23,512
'83	37.1	1,550	35	280	600,704	527,465	....	27,041	9,276	24,515
'84	37.3	1,627	33	283	623,489	527,904	....	27,011	7,692	23,982
'85	38.1	1,730	34	288	688,295	532,568	....	27,041	8,974	23,017
'86	30.4	1,844	36	294	919,369	533,642	....	27,397	10,293	25,333
'87	40.2	1,966	40	298	870,977	539,504	....	27,074	13,327	26,752
'88	42.2	2,081	37	305	755,072	547,207	400,000	26,983	10,608	27,473
'89	43.5	2,154	33	307	674,758	550,853	400,000	26,036	11,135	28,647
'90	45.5	2,203	35	308	775,933	561,220	400,000	25,307	11,504	28,450

Financial.—*Total Debt,* $415,000; floating, $16,000 at 3½ and 4%; bonded, $400,000 at 6%, due '92-'94. *Sinking Fund,* $95,192 at 5⅛%. *Expenses:* total, $34,811. *Revenue:* consumers, $28,450; city, $8,305; total, $36,755. **Management.**—Three comrs., one appointed by city council every year. Chn., E. B. Parkhurst. Clerk and Registrar, A. P. Barrett. Supt., P. F. Cully, appointed by comrs.

128. WORCESTER, *Worcester Co.* (Pop., 58,291—84,655.) First supplied in 1796 by Dan'l Gording; present works built in '64 by city. Supply.—Impounding reservoirs in Leicester and Holden, by gravity. Fiscal year closes Nov. 30.

Year.	Mains.	Taps.	Meters.	Hyd'ts.	Cons'ption.	Cost.	Int. ch'g's.	Exp.	Rev.
'78	76.4	4,971	3,481	....	2,500,000	$1,237,966	....	$13,765	$73,573
'79	77.5	4,971	3,481	....	2,500,000	....	....	....	73,149
'80	80	5,200	....	633	....	....	....	....	....
'81	....	5,531	....	....	....	....	....	....	....
'82	85.8	5,908	4,709	680	3,250,000	....	....	23,071	92,007
'83	93.9	6,367	1,246	744	unk'n	....	....	21,594	100,636
'86	97.9	7,191	6,005	784	....	1,720,930	....	55,134	110,265
'87	110	8,110	6,854	914	4,000,000	1,776,178	....	94,146	131,610
'88	115	8,529	7,354	964	4,635,000	1,996,892	....	98,210	133,341
'89	114.1	9,001	7,923	935	4,753,680	2,279,523	$34,632	55,630	133,135
'90	117.9	9,450	8,451	1,009	4,971,340	2,304,389	41,988	59,876	140,294

Management.—Comr., Jno. G. Brady. City Engr., C. A. Allen. Registrar, Geo. E. Batchelder.

Water-Works Projected with Fair Prospects of Construction.

FAIRHAVEN, *Bristol Co.* (Pop., 2,875–2,919.) Fairhaven Water Co. was chartered in April, '88. In '91 mains were laid one block and one hydrant set, in order to hold charter, which otherwise would be annulled. Nothing further had been done July 16, '91. Address Jos. K. Nye, Prest.

FALMOUTH, *Barnstable Co.* (Pop, 2,422–2,567.) System to be built in '92-4 by trustees of "Falmouth Highlands Trust," town reserving right to purchase works at any time at net cost. Supply, Long Pond, pumping to reservoir. Cost $10,000. Address Geo. W. Parke, Trustee.

FOXBOROUGH, *Norfolk Co.* (Pop., 2,950–2,933.) In July, '91, Water Supply District voted to construct system of water-works. Supply, well, pumping to stand-pipe. Chn. comrs., E. P. Carpenter. Clerk, Geo. F. Williams.

HANOVER, *Plymouth Co.* (Pop., 1,897–2,033.) Hanover Water Co. was incorporated June 11, '91, cap. stock not to exceed $60,000. Address J. Dwelley, Lot Phillips, or Harvey H. Pratt.

HIGHLANDVILLE, *Norfolk Co.* (Pop., not given.) Address Jno. Hodge.

METHUEN, *Essex Co.* (Pop., 4,392–4,814.) In '91 town was authorized by legislature to construct water-works, but enabling act was not accepted by town. July 14 Town Clerk reported that matter would be brought up for reconsideration.

For Further Descriptive Matter see Manual for 1889-90. Changes since 1890 are here given. Populations are for 1880 and 1890, respectively.

NORTH BROOKFIELD, *Worcester Co.* (Pop., 4,459—3,871.) Town intends to build water-works during '91, but nothing had been done July 9. Address Theo. C. Bates, 29 Harvard St., Worcester.

SALISBURY BEACH, *Essex Co.* (Pop. of town, 4,079—1,316.) In '90 Aqueduct Co. extended pipe line from large well in Salisbury to tank at Salisbury Beach, to supply cottagers. Address E. B. Shaw, Newburyport.

Water-Works Projects Reported in '90 and '91, but Now Indefinite.
Hamilton, Essex Co. *Ipswich,* Essex Co. *Medway,* Norfolk Co. *Merrimac,* Essex Co. *Millbury,* Worcester Co. *Provincetown,* Barnstable Co. *Southborough,* Worcester Co. *Wenham,* Essex Co. *Westminster,* Worcester Co.

RHODE ISLAND.
Water-Works Completed or in Process of Construction.

12. **ARTIC,** *Kent Co.* (Pop., not given.) Supply from Warwick.

8. **ASHTON,** *Providence Co.* (Pop., not given.) Supply from Pawtucket.

1. **BARRINGTON,** *Bristol Co.* (Pop., 1,359—1,461.) Built in '87 by Barrington Water Co. Supply.—Bristol & Warren Water-Works Co.'s system. Distribution.—(In '88.) *Mains,* 3 miles. *Taps,* 30. *Meters,* 30. Financial.—(In '88.) *Cost* $15,000. *No Debt.* Management.—Prest., H. J. Steers. Treas., C. H. Merriman. NOTE.—Co. refuses to give statistics for '89 and '90.

8. **BERKELEY,** *Providence Co.* (Pop., not given). Supply from Pawtucket.

1. **BLOCK ISLAND,** *Newport Co.* (Pop., est., 1,300). Built in '88 by Block Island Water Co. Supply.—Pond, pumping to stand-pipe. Fiscal year closes July 1. Distribution.—*Mains,* 1½ miles. *Taps,* 35. *Meters,* 0. *Hydrants,* 10. Financial.—*Cost,* $12,000. *No Debt. Expenses,* $200. *Revenue,* $850. Management.—Prest., E. S. Payne. Treas., F. A. Rose.

2. **BRISTOL and WARREN,** *Bristol Co.* (Pop.; Bristol, 6,028—5,478; Warren, 4,007—4,489). Built in '82 by Bristol & Warren Water-Works Co.; franchise provides that works may be purchased by city after fifteen years; city has no control of rates. Supply.—Kickenmit River, pumping to tank; filtered through sand and gravel. Fiscal year closes Oct. 31. Distribution.—*Mains,* 10 miles. *Taps,* 625. *Meters,* 15. *Hydrants,* 130. *Consumption,* 300,000 galls. Financial.—*Cost,* $250,000. *Cap. Stock:* authorized, $250,000; all paid up; paid 6% div'd in 1890. *Expenses and Revenue,* withheld. Management.—Prest., I. F. Williams. Secy., Sidney Dean. Treas., G. Norman Weaver, Newport. Supt., B. B. Martin.

3. **CENTRAL FALLS,** *Providence Co.* (Pop., 13,765.) Built in '79 by fire district. Supply.—Pawtucket Water-works. Fiscal year closes Nov. 30. Distribution.—*Mains,* 14 miles. *Taps,* 988. *Meters,* 801. *Hydrants,* 81. Financial.—*Revenue:* consumers, $11,522; city, $1,800; total, $13,322. Works are being operated by Central Falls fire district. Pawtucket collects rates and gives Central Falls a rebate of 30%, allowing $20 each a year for hydrants and drinking fountains. Contract extends for 10 yrs. from 1890. Management.—Prest., Jethro Baker. Treas., C. P. Moies, Pawtucket. Supt., Jno. Booth.

4. **COWESETT,** *Kent Co.* (Pop., not given.) Supply from East Greenwich.

10. **CRANSTON,** *Providence Co.* (Pop., 5,940—8,099.) Supply from Providence.

11. **DROWNVILLE,** *Bristol Co.* (Pop., est., 500.) Built in '87 by Drownville Water Co.; town does not control rates. Supply.—Well and pond, pumping to tank. In '90 an 18-ft. Eclipse windmill, 4x8 double-acting pump and a Davidson steam pump having 3½x8 steam and 5x8 water cylinders; were added. Latter pump is used only in emergencies. Distribution.—*Mains,* 2.5 miles. *Taps* (in '88), 33. *Meters* (in '88), 15. *Hydrants,* 0. *Consumption,* 8,000 galls. Financial.—*Cost,* $9,605. *Cap. Stock:* authorized, $20,000; paid-up, $7,600; paid no div'd in 1890. *Floating Debt,* $1,800. *Expenses:* operating, $150; int., $100; taxes, exempt for 20 yrs.; total, $250. *Revenue:* consumers, $450; city, none. Management.—Prest., David H. Waldron. Secy. and Supt., Chas. F. Anthony. Treas., Geo. I. Baker.

RHODE ISLAND. 49

4. EAST GREENWICH, *Kent Co.* (Pop., 2,877—3,107.) Built in '86 by East Greenwich Water Supply Co.; franchise provides neither for control of rates nor purchase of works by city. **Supply.**—Hunt's River, pumping to stand-pipe. Fiscal year closes Nov. 30. During '90 main was extended to Cowesett, 1.2 miles, to supply village; water for domestic use only. Co. contemplates extending mains to supply village of Apponaug.

Year.	Mains.	Taps.	Meters.	Hy-drants.	Con-sumption.	Cost.	Bonded debt.	Int. ch'g's.	Exp.	Rev.
'87	9	180	....	52	100,000	$100,000	$73,000	$3,650	$3,000	$4,500
'88	9	230	15	53	160,000	$100,000	$73,000	$3,650	$5,700	$3,150
'90	11	351	60	54	200,000	$115,000	$88,000	$4,550	$3,000	$3,025

Financial.—*Cap. Stock:* authorized, $75,000; all paid-up. *Bonded Debt,* $88,000; $73,000 at 5%, due in 1906; $15,000 at 6%, due in 1910. *Expenses:* total, $7,550. *R-venue:* consumers, $6,500; city, $1,525; total, $8,025. **Management.**—Prest., Jos. H. Eldridge. Secy., Fred. S. Dewes. Supt., Jno. L. Congdon.

5. EAST PROVIDENCE, *Providence Co.* (Pop., 5,056—8,422.) Built in '80 by town. Operated by Pawtucket Water-Works, which pays rebate of 25% on all rates collected. Other data included under Pawtucket. In '91 there was talk of obtaining an independent supply.

9. FISKEVILLE, *Providence Co.* (Pop., not given.) Supply from Phenix.

9. HOPE, *Providence Co.* (Pop., not given.) Supply from Phenix.

6. JAMESTOWN, *Newport Co.* (Pop., 459—707.) **History.**—Built in '90 by Jamestown Light & Water Co., under 30-yrs. franchise. Engr., Geo. H. Norman, Newport. **Supply.**—Source not given, pumping to stand-pipe. **Pumping Machinery.**—Not given. **Stand-Pipe.**—Cap., 80,000 galls. **Distribution.**—*Mains,* 12-in. cem., 13,000 ft. *Meters,* Thompson. *Hydrants,* 20. *Valves,* Chapman. *Pressure,* 60 lbs. **Financial.**—Hydrant rental, $30. **Management.**—Prest., Geo. E. Can. Supt., Norman Weaver, Newport.

10. JOHNSTON, *Providence Co.* (Pop., not given.) Supply from Providence.

8. LONSDALE, *Providence Co.* (Pop., not given.) Supply from Pawtucket.

14. MANVILLE, *Providence Co.* (Pop., not given.) Supply from Woonsocket.

11. NARRAGANSETT PIER, *Washington Co.* (Pop., not given.) Supply from Wakefield.

I. NAYATT POINT, *Bristol Co.* (Pop., not given.) Supply from Barrington.

7. NEWPORT, *Newport Co.* (Pop., 15,693—19,467.) Built in '76-7 by Geo. H. Norman; now owned by Newport Water-Works Co.; franchise provides neither for control of rates nor purchase of works by city. **Supply.**—Easton's and Paradise ponds, pumping to reservoir; filtered. New pump and two new filters were put in operation in '91. **Distribution.**—*Mains,* 60 miles. *Taps,* 3,000. *Meters,* 12. *Hydrants,* 219. *Consumption,* 1,500,000 galls. **Financial.**—*Cost,* $800,000. *Cap. Stock,* $750,000. *No Debt. Expenses,* $20,000. *Revenue:* consumers, $70,000; city, $10,500. **Management.**—Prest., Geo. H. Norman. Secy., R. S. Franklin. Treas. and Supt., Wm. S. Slocum.

10. OLNEYVILLE, *Providence Co.* (Pop., not given.) Supply from Providence.

8. PAWTUCKET *Providence Co.* (Pop., 19,030—26,333.) Built in '77-9 by city. Ashton, Berkeley, Central Falls, East Providence, Lonsdale and Valley Falls are supplied. See Central Falls and East Providence. **Supply.**—Abbot's run, pumping to reservoir through distribution mains; filtered. Fiscal year closes Nov. 30.

Year.	Mains.	Taps.	Meters.	Hyd'ts.	Con-sumpt'n.	Cost.	B'd'd d'bt.	Int. chg's.	Exp.	Rev.*
'79	43	1,400	588	...	802,467	....	....	....	....	$23,396
'81	75	2,679	1,422	561	1,068,877	$774,993	....	$25,000	$12,356	47,043
'82	80	3,077	1,648	578	1,513,628	786,079	....	36,644	13,318	56,789
'83	84	3,456	1,919	536	1,868,603	898,137	....	43,325	18,828	64,101
'84	89.5	3,810	2,159	629	2,332,843	....	$1,000,000	44,480	18,239	70,224
'85	92	4,165	2,400	722	2,308,221	1,200,184	1,100,000	40,527	17,119	60,482
'86	95	4,452	2,596	753	2,256,281	1,288,713	1,100,000	54,321	16,912	79,013
'87	98	4,795	2,852	778	3,176,193	1,336,658	1,100,000	56,195	23,141	90,392
'88	104	5,084	3,098	795	4,258,187	1,477,705	1,100,000	56,200	26,051	93,919
'89	106	5,061	3,274	807	3,554,033	1,549,797	1,500,000	62,522	34,166	104,375
'90	108	5,322	3,539	821	3,216,553	1,580,823	1,500,000	65,510	26,183	104,555

* Including allowance for hydrant rental.

For Further Descriptive Matter see Manual for 1889-90. Changes since 1890 are here given. Populations are for 1880 and 1890, respectively.

Financial.—*Total Debt*, $1,541,663; floating, $41,663 at 5%; bonded, $500,000 at 5%; $1,000,000 at 4%. *Sinking Fund*, $290,000. *Expenses:* total, $91,693. *Revenue:* consumers, $90,543; other towns, $14,012; total, $104,555. Op. exp. and int. paid from revenue; new construction, from bonds. **Management.**—Three comrs. appointed every three years by city council. Chn., L. B. Darling. Secy. and Supt., E. Darling, appointed by comrs.

8. **PAWTUXET,** *Providence Co* (Pop., not given.) Supply from Pawtucket.

11. **PEACEDALE,** *Washington Co.* (Pop., not given.) Supply from Wakefield.

9. **PHENIX,** *Kent Co.* (Pop., 1,038— .) Built in '85-7 by Pawtuxet Valley Water Co.; franchise does not regulate rates, but provides that city may purchase works in 1911. **Supply.**—Lake brook, by gravity from reservoir. Fiscal year closes July 1. **Distribution.**—*Mains*, 12 miles. *Taps*, 132. *Meters*, 31. *Hydrants*, 118. **Financial.**— *Cost*, $100,000. *Cap. Stock:* authorized, $100,000; paid-up, $38,000. *Total Debt*, $66,000; floating, $16,000; bonded, $50,000. *Revenue:* city, $3,540. **Management.**—Prest., R. Howland. Secy., V. A. Bailey. Supt., A. F. Hill.

10. **PROVIDENCE,** *Providence Co.* (Pop., 101,857—132,146.) Built in 70-6 by city; first supply introduced in 1772 by Co. **Supply.**—Pawtuxet River, pumping to reservoirs. Fiscal year closes Dec. 31. Fruit Hill reservoir showed leaks after completion. In '90 J. B. Francis, B. S. Church and J. Herbert Shedd reported as probable cause of leakage seamy character of ledge beneath parts of embankment and porous nature of puddle used for lining. In '90-91 repairs were being made at est. cost of $75,000.

Year.	Mains.	Taps.	Meters.	Hydrants.	Consumption.	Cost.	Exp.	Rev.*
'72	13,7					$1,494,313	.....	$41,003
'80	155.2	9,091	4,036	1,137	3,110,279	5,084,182	.....	247,705
'81	160.1	10,151	4,424	1,161	3,716,937		.....	260,531
'82	169.8	10,357	5,279	1,186	3,665,427	.....	$45,180	269,319
'83	175.4	10,921	5,721	1,203	4,143,798	4,928,682	.....	283,633
'84	184.5	11,464	6,157	1,216	4,083,373		61,953	302,368
'85	190	11,989	6,648	1,234	4,730,556	.....	.....	313,561
'86	194.9	12,553	7,135	1,248	4,822,125	.....	69,828	323,085
'87	201.1	13,128	7,623	1,278	4,939,982	6,234,672	.....	339,514
'88	210.4	13,643	8,094	1,294	5,518,691		57,757	346,732
'89	218	14,281	8,627	1,316	5,780,961	6,976,798	.....	377,498
'90	226.1	14,800	9,286	1,363	6,743,092	7,156,333	76,885	403,165

*Including allowance for fire protec'ion.

Financial.—*Bonded Debt*, $5,983,000 : $4,000,000 at 5 and 6%, due in 1900 ; $1,500,000 at 5%, due in '96 ; $483,000 at 3½%, due in 1916. *Sinking Fund*, $551,583. *Expenses:* operating, $76,885 ; int., $316,263 ; total, $393,148. *Revenue:* consumers, $316,778 ; city, $56,387 ; total, $403,165. **Management.**—Comr. of Public Works, Robt. E. Smith. Secy., Walter F. Slade. City Engr., J. Herbert Shedd.

9. **RIVER POINT,** *Kent Co.* (Pop., not given.) Supply from Phenix.

8. **VALLEY FALL,** *Providence Co.* (Pop., not given.) Supply from Pawtucket.

11. **WAKEFIELD,** *Washington Co.* (Pop., not given.) Built in '88-9 by Wakefield City Water Co ; franchise provides neither for control of rates nor purchase of works by city. **Supply.**—Rocky brook, pumping from impounding reservoir to stand-pipe; National filter. Fiscal year closes July 1. **Distribution.**—*Mains*, 13.5 miles. *Taps*, 200. *Meters*, 143. *Consumption*, 75,000 galls. **Financial.**--*Cost*, $165,000. *Cap. Stock*, $150,000. *Bonded Debt*, $150,000 at 6%. *Revenue*, $2,352 for portion of one year. **Management.**—Prest., Richard Pancoast. Secy., A. B. Richards. Both of 28 Platt St., N. Y. City. Mgr., Willard Kent. Supt., E. A. Waterhouse.

2. **WARREN,** *Bristol Co.* (Pop., 4,007—4,489.) Supply from Bristol.

12. **WARWICK and COVENTRY,** *Kent Co.* (Pop., Warwick, 12,164—17,761; Coventry, 4,519—5,068.) **History.**—Built in '90-1 by Warwick & Coventry Water Co. to supply Warwick, Coventry and other towns; put in operation in July, '91. Engr., H. B. Barton, Centreville. **Supply.**—Caw's pond, situated in W. Greenwich, by gravity to tank. *Tank*.—Is 40 x 50 ft.; from Porter Mfg. Co., Syracuse, N. Y. **Distribution.**—*Mains*, c. i. and fire clay, 20 miles. *Services*, w. i. *Hydrants*, Holyoke, 155. *Valves*, Peet. *Pressure*, 80 lbs. **Financial.**—*Cap. Stock*, $100,000. *Bonded Debt*, $250,000. *Hydrant rental*, $30. **Management.**—Prest., Jno. J. Arnold, Phenix. Secy., Herbert B. Barton. Treas., Enos Lapham. Both of Centreville.

12. **WASHINGTON,** *Kent Co.* (Pop., not given.) Supply from Warwick.

RHODE ISLAND. 51

13. WESTERLY, *Washington Co.* (Pop., 6,104—6,813.) **History.**—First supply introduced in '76 by company; present works built in '86 by Westerly Water-Works Co.; franchise does not regulate rates, but provides for purchase of works by town in 1911, price determined by board of three, chosen by town and co. **Supply.**—Shunoc Brook, pumping to stand-pipe; filtered. Fiscal year closes Dec. 31. **Pumping Machinery**--Cap., 1,500,000 galls. **Stand-Pipe.**—Cap., 372,000 galls. **Distribution.**—*Mains,* 14 miles. *Taps,* 380. *Meters,* 225. *Hydrants,* 23. *Consumption,* 203,000 galls. **Financial.**—*Cap. Stock,* authorized, $100,000; all paid up. *Bonded Debt,* $100,000 at 5%. *Expenses:* operating, $2,844; int., $5,000; taxes, exempt till 1906; total, $7,844. **Management.**—Prest., Jonathan Chace. Secy., Chas. Perry. Supt., Everett Barnes.

NOTE.—Nov. 3, '90, Council voted to buy works, provided satisfactory price could be agreed upon. About June, '91, Council notified Co. that it wished to buy works.

14. WOONSOCKET, *Providence Co.* (Pop., 16,050—20,830.) Now owned by city; built in '84 by Co.; bought by city Apr. 1, '85, for $289,613. **Supply.**—Crook Falls brook, pumping to tank. Fiscal year closes Dec. 31. Additional pumps were to be added and other improvements made in '91.

Year.	Mains.	Taps.	Meters.	Hy'd'ts.	Consumption.	Cost.	Bonded debt.	Int. charges.	Exp.	Rev.
'85	18.3	425	364	270	147,029	$314,493	$300,000	$12,000	$6,934	$7,249
'86	19	595	536	284	243,565	735,925	300,000	12,125	7,316	10,940
'87	21.8	761	674	334	269,618	356,387	300,000	12,795	9,035	13,945
'88	23.5	861	761	349	269,528	368,810	300,000	8,889	4,889	11,453
'89	25.6	983	827	377	302,310	390,364	300,000	12,328	6,850	12,073
'90	27.7	1,117	924	403	326,495	4,6,011	300,000	14,095	8,331	18,992

Financial.—*Bonded Debt,* $300,000 at 4%. *Expenses:* total, $22,421. *Revenue:* consumers, $18,992; city, $14,768; total, $33,760. **Management.**—Three comrs. Chn., Jno. W. Ellis. Clerk, Chas. N. Brown. Supt., Byron I. Cook.

Water-Works Projected with Fair Prospect of Construction.

APPONAUG, *Kent Co.* (Pop., not given.) East Greenwich Water Supply Co. will probably extend its mains here. Address Jno. L. Congden, Supt.

RIVERSIDE, *Providence Co.* (Pop., not given.) In Dec., '90, question of introducing water supply at est. cost of $100,000 was referred to committee. No definite plan had been formed July 17, '91. Address L. S. Winchester.

CONNECTICUT.

Water-Works Completed or in Process of Construction.

1, 2. ANSONIA, *New Haven Co.* (Pop., 3,855—10,342.)

ANSONIA WATER CO.

Built in '69; franchise provides neither for control of rates nor purchase of works by town. Supply, surface water, by gravity. Fiscal year closes March 1. **Distribution.**—*Mains,* 8 miles. *Taps,* 500. *Meters,* 5. *Hydrants,* 88. *Consumption,* unk'n. **Financial**—*Cost,* $136,000. *Cap. Stock,* $30,750. *Total Debt,* $91,240: floating, $32,700; bonded, $58,500. *Expenses:* operating, $1,450; int., $5,111; taxes, $399; total, $6,910. *Revenue:* consumers, $4,161; city, $1,585; other purposes, $1,200; total, $10,946. **Management.**—Prest., Thos. Wallace. Secy. and Supt., Dana Bartholomew.

FOUNTAIN WATER CO.

Built in '72. Supply, springs, by gravity from reservoir. Fiscal year closes Oct. 1. **Distribution.**—*Mains,* 14 miles. *Taps,* 583. *Hydrants,* 34. **Financial.**—*Cap. Stock:* authorized, $20,000; all paid up; paid 6% div'd in 1890. *Floating Debt,* $5,000 at 5%. *No Bonds. Expenses:* operating, $1,290; int., $250; taxes, $133; total, $1,583. *Revenue:* consumers, withheld; city, $680. **Management.**—Prest., David Torrance, Birmingham. Secy. and Supt., J. A. Fiske.

3. BETHEL, *Fairfield Co.* (Pop., 1,767—2,335.) Built in '78 by Borough. Sup-

---

FOR FURTHER DESCRIPTIVE MATTER see Manual for 1889-90. CHANGES since 1890 are here given. POPULATIONS are for 1880 and 1890, respectively.

ply.—Surface water, by gravity from impounding reservoir on small stream. Fiscal year closes June 1.

In '91 filter bed with area of about 13,000 sq. ft. was built. Dam was built about 150 ft. below old impounding dam to form basin for downward filtration through bed. Gate chamber was placed in new dam, an old supply main, and gates arranged to turn water back through another pipe to bed. Drain tile carries filtered water from bottom of basin to well at corner, from which it passes through another pipe back to chamber, thence to village. Engrs. and Contrs., W. B. Rider & Son, South Norwalk. Cost, about $6,500. For full descriptions and illustrations see ENG. NEWS, vol. xxv., p. 213.

Distribution.—*Mains*, 11.3 miles. *Taps*, 460. *Meters*, 0. *Hydrants*, 60. *Consumption*, 150,000 galls. Financial.—*Cost*, $37,464. *Bonded Debt*, $25,000 at 6%. *Expenses:* operating, $437; int., $1,500; total, $1,957. *Revenue:* consumers, $1,392. Revenue covers all expenses and leaves surplus in treasury. Management.—Three Comrs., one elected each year. Chn., E. C. Oakley. Secy., A. W. Twiss. Supt., Chas. H. Hart, appointed by Comrs.

4. **BIRMINGHAM,** *New Haven Co.* (Pop., 3,026—4,413.) Built in '60 by Birmingham Water Co. Supply.—Springs and surface water, by gravity from reservoirs. Distribution.—*Mains*, 8 miles. *Taps*, 400. *Hydrants*, 52. Financial.—Withheld. Management.—Prest., Wm. H. Wooster. Secy., David Torrance. Treas. and Agt., Chas. H. Nettleton.

5. **BRIDGEPORT,** *Fairfield Co.* (Pop., 27,643—48,866.) First supplied in '48 by Rev. Elijah Waterman; in '53 Nathaniel Green built works, which in '57 passed to Bridgeport Hydraulic Co.; in '83 the Citizens' Water Co. started works, which were soon consolidated with and in '88 connected with Bridgeport Hydraulic Co.'s works. Supply.—Streams, by gravity and pumping. Fiscal year closes Dec. 1. In '91 new 300,000,000-gall. storage reservoir was being built. Unofficially reported that dam will be of earth, 400 ft. long, 30 to 35 ft. high, 150 ft. wide at base, 15 at top. Heart-wall of concrete, 6 ft. thick; total height, 35 ft. Two subsidiary earth dams about 100 ft. long. Contr., B. D. Pierce, Jr. Distribution.—*Mains*, about 90 miles. *Taps* (in '83), 3,810. *Hydrants*, 336. Financial.—Withheld. Management.—Prest., Chas. Sherwood. Treas., T. B. De Forest. Secy., H. Thorpe. Supt., Geo. Richardson.

6. **BRISTOL,** *Hartford Co.* (Pop., 5,347—7,382.) Built in '85 by Bristol Water Co.; town neither controls rates nor has right to purchase works. Supply.—Poland Brook, by gravity from impounding reservoir. Fiscal year closes March 31. In '89 second storage reservoir built on Poland Brook; cap., 98,000,000 galls.; area, 19 acres. Dam of earth, puddled, with heart-wall; 650 ft. long, 31.3 ft. high, 27 ft. water; 14 ft. wide on top, 115 ft. at bottom. Inner slope, 2 to 1, paved; outer, 1½ to 1, not paved. Heart wall, stone in cement, 9 ft. thick at bottom, 2 at top; carried to ledge. Spillway, 20 ft long, 4 ft. deep, lined with "river stone;" free; one 12-in. discharge pipe. Engr., H. W. Ayres, Hartford. No contr. Cost, $18,500.

Year.	Mains.	Taps.	Meters.	Hydr'ts.	Cost.	B'd'd debt.	Int ch'gs.	Exp.	Rev.
'86-7	11.5	260	..	67	$100,000	$40,000	$2,000	....	....
'87-8	13.9	250	2	67	100,000	40,000	2,000	....	$1,200*
'88-9	14.9	360	3	60	110,000	40,000	2,000	$1,000	1,200*
'89-90	15.4	453	3	60	120,000	40,000	2,000	1,211	8,920
'90-1	16	530	8	61	126,275	40,000	2,000	....	1,200*

* From city.

Financial.—*Cap. Stock* authorized, $60,000. all paid up; paid 5% div'd in '91. *Total Debt*, $65,000; floating, $25,000 at 6%; bonded, $40,000 at 5%. Management.—Prest., J. H. Sessions. Secy. and Treas., C. S. Treadway. Supt., J. H. Keirn.

7. **CANAAN,** *Litchfield Co.* (Pop., 1,157—970.) Built in '79 by Canaan Water Co.; franchise provides neither for control of rates nor purchase of works by city. Supply.—Springs, by gravity from impounding reservoir. Fiscal year closes Oct. 1. Distribution —*Mains*, 4 miles. *Taps* (in '88), 93. *Meters*, 0. *Hydrants*, 6. *Consumption*, unknown. Financial.—*Cost*, $10,000. *Cap. Stock:* authorized, $5,000; all paid up; div'd in 1893, $400. *No debt.* Taxes, $53. *Revenue:* consumers, $987; city, none; other purposes, $125; total, $1,112. Management.—Prest., Luman Foot. Secy., S. A. Eddy. Treas., J. W. Peet. Supt., S. A. Bennett.

8. **DANBURY,** *Fairfield Co.* (Pop., 11,666—19,473.) First supplied in '33 by Danbury Water Co.; built in '61 by borough. Supply.—Surface water, by gravity from impounding reservoir. Fiscal year closes April 1.

CONNECTICUT.

Year.	Mains.	Taps.	Meters.	Hydrants.	Cost.	Bonded debt.	Exp.	Rev.
'82-3	25	1,500	0	110	$108,405	....	$1,192	$13,977
'86-7	30	2,500	0	150	240,000	$140,000	2,801	19,500
'88-9	35	2,600	0	230	300,090	232,000	3,031	26,000
'89-90	39	3,896*	0	251	319,203	....	3,040	27,159
'90-1	41.1	......	0	273	363,943	....	2,379	29,142

*Consumers.

Financial.—See above. **Management**—Supt., Chas. B. Mason, appointed by City Council.

9. DANIELSONVILLE, *Windham Co.* (Pop., 3,118—not given.) Built in '86 by Crystal Water Co., under perpetual charter. **Supply.**—Higgins' Brook, by gravity from impounding reservoir. Fiscal year closes Dec. 31. In June, '89, fire reservoir was built, 133 ft. in diam. at top, 10 ft. deep, affording fire pressure of 75 lbs. **Distribution.**—*Mains,* 9.5 miles. *Taps,* 175. *Meters,* 2. *Hydrants,* 51. **Financial**—*Cost,* $131,082. *Cap. Stock:* common, authorized, $120,000; paid up, $60,000; preferred, $20,000, all paid up; 5% dividend guaranteed. *Bonded Debt,* $50,000 at 5%, due in 1914; $10,000 in treasury to be expended in extensions. *Expenses:* operating, $323; int., $2,540; taxes, $376; total, $3,242. *Revenue:* fixtures, $2,281; meters, $144; city, $1,530; total. $3,955. **Management.**—Prest., T. E. Hopkins. Secy., M. P. Dowe. Treas., Elbert Wheeler, 24 Washington St., Boston, Mass.

4. DERBY, *New Haven Co.* (Pop., not given.) Supply from Birmingham.

1. DURHAM, *Middlesex Co.* (Pop., 990—856.) Built in 1798; rebuilt in '32 by Durham Aqueduct Co. **Supply.**—Cold spring, by gravity. Fiscal ear closes Feb. 1. **Distribution.**—*Mains,* 1.8 miles. *Taps,* 41. *Hydrants,* 0. **Financial.**—*Expenses,* nominal. *No revenue.* **Management.**—Prest., F. Hubbard.

31. EAST NORWALK, *Fairfield Co.* (Pop. not given.) Supply from South Norwalk.

10. FARMINGTON, *Hartford Co.* (Pop., 3,017—3,179). **History.**—Built in '81 by Farmington Water Co. Engr., A. R. Wadsworth. **Supply.**—Impounding reservoir, by gravity. **Reservoir.**—Cap., 35,000,000 galls. **Distribution.**—*Mains,* cem., c. i., 2 miles. *Services,* galv. i. *Taps,* 263. *Hydrants,* 1. *Pressure,* 50 lbs. **Financial.**—*Cost,* $16,000. *Expenses,* $10. *Revenue,* $319. **Management.**—Owner, A. R. Wadsworth.

11. GREENWICH, *Fairfield Co.* (Pop., 7,892—10,131.) Built in '80 by Greenwich Water Co. **Supply.**—Putnam Lake, by gravity. **Distribution.**—(In '88.) *Mains,* 25 miles. *Taps,* 95. *Meters,* 6. *Hydrants,* 40. *Consumption,* 500,000 galls. **Financial.**—Withheld. **Management.**—Prest., A. Foster Higgins. Secy., Geo. G. Mc. Nall. Treas., Wm. Rockfeller, 56 Broadway, N. Y. City. NOTE.—No statistics were received for '89 or '90.

37. GROTON, *New London Co.* (Pop., 5,128—5,539.) Present supply from Stonington. Groton Water Co. was organized June 5, '91, to construct independent system. Prest., E. L. Baker. Secy. and Treas., Thos. F. Morgan.

12. HARTFORD, *Hartford Co.* (Pop., 42,551—53,230.) Built in '54 by city. First supplied, probably, in 1797 by Hartford Aqueduct Co. **Supply.**—Brooks by gravity from storage reservoir; also emergency supply from Connecticut River, pumping direct. Fiscal year closes March 1.

Year.	Mains.	Taps	M't'rs.	Hyd'ts.	Cost	B'd'd debt.	Int. ch'g's.	Exp.	Rev.
'56	14.8	650	...	101	$336,918	$152,000	$13,165	$6,578	$7,039*
'60-1	27.6	1,935	...	132	389,631	375,000	22,695	8,471	24,727
'65-6	32.7	2,511	...	162	456,186	375,000	22,698	26,223	54,611*
'70-1	47.6	3,550	...	209	997,300	750,000	44,920	21,591	73,779
'75-6	67.5	4,624	...	300	1,360,919	1,052,000	63,030	42,033	100,551
'80-1	71.6	4,962	99	360	1,495,519	1,027,000	67,245	36,297	108,927
'81-2	72.7	5,057	148	372	1,512,362	1,002,000	58,745	40,116	114,457
'82-3	73	.....	...	381	1,526,537	937,000	57,461	36,970	132,882*
'83-4	73.3	5,168	169	395	1,504,756	962,000	55,912	29,988	138,903*
'84-5	75.5	5,382	260	429	1,603,428	937,000	54,843	31,913	123,371
'85-6	77.1	5,525	280	454	1,623,485	912,000	54,089	30,587	130,138
'86-7	78.7	5,676	295	461	1,679,764	887,000	52,630	31,298	132,098
'87-8	80	5,836	316	476	1,697,339	857,000	51,433	28,416	136,946
'88-9	82	5,984	345	488	1,743,163	827,000	49,323	28,910	140,156
'89-90	83	6,216	378	511	1,772,129	797,000	48,412	31,032	142,702
'90-1	86.9	6,446	401	540	1,815,508	762,000	44,589	32,116	146,695

* Including allowance for hydrant rental and street sprinkling.

FOR FURTHER DESCRIPTIVE MATTER see Manual for 1889 90. CHANGES since 1890 are here given. POPULATIONS are for 1880 and 1890 respectively.

Financial.- *Total Debt*, $752,000 ; floating, $150,000; bonded, $602,000, due '91—1906. *No Sinking Fund. Expenses:* total, $76,705. *Revenue:* consumers, $146,605; city, fire protection, $10,200; street sprinkling, $10,200; total, $159,616. **Management.—** Six comrs. Prest. and Supt., Ezra Clark. Secy., J. Seymour Chase.

13. **KENT**, *Litchfield Co.* (Pop., 1,542—1,063.) Built in '82 by Kent Water Co. **Supply.—** Springs and surface water, by gravity from reservoir. **Distribution.—** (In '88.) *Mains*, 2.3 miles. *Taps*, 40. *Hydrants*, 10. Financial (in '88).—*Cost*, $15,500. *No Debt. Expenses* $55, *Revenue*, $500. **Management.**—Prest., Luther Eaton. Secy. and Supt., C. A. Eaton. Treas., C. H. Gaylord. NOTE.—No statistics could be obtained for '89 or '90.

14. **LITCHFIELD** *Litchfield Co.* (Pop., 452—1,058.) Built in '89-90 by Co. **Supply**—Griswold Brook, by gravity from impounding to distributing reservoir. April, '90, contract was let Cunningham Iron Works, Boston, for wrought iron stand-pipe to equalize pressure, 30 × 35 ft.; plates ½, ⅜, $\frac{7}{16}$ and ¼ ins. thick. **Distribution.—** *Mains*, 6 miles. *Hydrants*, 38. **Financial.—***Cost*, $61,000. *Cap. Stock*, $40,000. *Bonded Debt*, $20,000 at 5%. *Expenses* and *Revenue*, not determined. **Management.**—Prest., G. M. Woodruff. Secy. and Treas., C. B. Bishop. Supt., E. B. Allen. Report by T. H. McKenzie, Chf. Engr.

15. **MANCHESTER**, *Hartford Co.* (Pop., 6,462—8,022.) Built in '89 by Manchester Water Co. under perpetual franchise. **Supply.**—White's Brook, by gravity from impounding reservoir. Fiscal year closes March 7. **Distribution.—***Mains*, 7.5 miles. *Meters*, 5. *Hydrants*, 50. **Financial.**-*Cost*, $50,000. *Cap. Stock*, $20,000. *Bonded Debt*, $28,000 at 5%. **Management.**—Pres., Wm. Foulds. Secy. and Supt., W. H. Childs.

16. **MERIDEN**, *New Haven Co.* (Pop., 15,540—21,652.) Built in '69 by city. **Supply.**—Artesian wells and surface water, by gravity from impounding reservoir. Fiscal year closes Nov. 30. In '91 contract was let for new dam and storage reservoir. As originally planned dam was to be 33 ft. high and impound 186,000,000 galls. Additional drainage area of 2 sq. miles is secured. A 5,000,000-gall. pump, or pumps, will force water to distributing reservoir 280 ft. above town. As planned 20,500 ft. of 20-in. pipe would lead from storage reservoir to city and 1,700-ft. branch, 20-in., would extend to reservoir. This may have been changed. Reported that H. E. Stanton, Huntington, Mass., had received contract for dam, for $16,500; C. H. Eglee, Flushing, N. Y., for laying main, for $24,300.

Year.	Mains.	Taps.	Meters.	Hy'dts.	Cost.	B'd'd debt.	Int. ch'g's.	Exp.	Rev.*
'80	22.	1,500		180				$6,000	$32,094
'82	25	2,200	65	180	$350,000	$325,000	$22,750	5,000	38,500
'84	29.4	2,576	74	193	363,161	325,000	22,750	3,628	43,600
'86	30.4	2,576	74	203	374,028	325,000	22,750	3,628	47,957
'88	33	2,700	85	216	380,000	200,000	7,000	8,000	52,092
'90	33	2,950	89	232	400,000	200,000	7,000	10,525	49,420

* Including allowance for public use of water.

Financial.—*Bonded Debt*, $200,000, at 3½%. *Sinking Fund*, $12,672, loaned to city at 4%. *Expenses:* total, $17,525. *Revenue:* consumers, $16,000; city, fire protection, $3,420; other purposes, $200; total, $19,420. **Management.**—Prest., F. P. Evarts. Supt., H. L. Schleiter.

17. **MIDDLETOWN**, *Middlesex Co.* (Pop., 6,826—9,013.) Built in '66 by city. **Supply.**—Surface water, by gravity from impounding reservoir. Fiscal year closes Dec. 31.

Year.	Mains.	Taps.	Meters.	Hydrants.	Consumption.	Cost.	Bonded debt.	Int. ch'g's.	Exp.	Rev.*
'74	15.5	825		70				$10,620	$3,592	$14,673
'83	17.8	1,050	7	85		800,000	$227,408	$136,500	8,190	4,886	17,826
'85	18	1,150	7	95		750,000	237,996	139,000	7,170	7,947	19,123
'87	18	1,198	8	100		750,000	214,642	139,000	6,617	5,336	18,855
'89	18.8	1,266	10	103		850,000	251,585	139,000	6,647	7,314	22,938
'90	19	1,330	10	107		900,000	253,533	139,000	6,647	5,540	23,966

*Including revenue for public use of water.

Financial.—*Bonded Debt*, $139,000; $64,000 at 6%, due in '96; $20,000 at 4%, due in 1908; $55,000 at 3.65%, due in 1915. *Sinking Fund*, $25,111. *Expenses:* total, $12,187. *Revenue:* consumers, $20,704; city, $3,135; total, $23,966. **Management.**—Four comrs.: Jas. Lawton, C. W. Harris, H. L. Mansfield, W. F. Burrows. Supt., Jno. C. Broatch.

37. **MYSTIC**, *New London Co.* (Pop., not given.) Supply from Stonington.

18. **NAUGATUCK**, *New Haven Co.* (Pop., 4,274—6,218.) Built in '88-'89 by Naugatuck Water Co.; franchise provides neither for control of rates nor purchase

of works by city. **Supply.**—Straitsville brook, by gravity. New storage reservoir was constructed in '91. Dam, of masonry, 300 ft. long, 20 ft. high. Contr., O. M. Weand, Reading, Pa. Fiscal year closes Feb. 2. **Distribution.**—*Mains*, 12. *Taps*, (in '89), 175. *Meters*, 0. *Hydrants*, 63. **Financial.**—*Cost*, $85,000. *Cap. Stock.* authorized, $45,000; paid up, $41,425. *Bonded Debt*, $39,500 at 6%.. *Expenses:* operating, $3,200.; int., $2,370: taxes, $155; total, $5,725. *Revenue:* Total, $5,000. **Management.** —Prest., F. B. Tuttle. Secy. and Supt., E. C. Barnum.

19. **NEW BRITAIN,** *Hartford Co.* (Pop., 13,979—19,007.) Built in '57 by city Supply.—Brook, by gravity from impounding to distributing reservoir. Fiscal year closes March 31. Summer of '91 P. M. Blake, Hyde Park, Mass, was making plans for new dam at Shuttle Meadow Lake to raise water level and increase storage capacity. Also proposed to lay 20 or 24-in. supply main.

Year.	Mains.	Taps.	Meters.	Hyd'ts.	B'd'd d't.	Int. ch'g's.	Exp.	Rev.
'85	25	1,350	53	153	$190,000	$11,400	$3,018	$16,712
'86-7	25.3	3,500	48	177	171,000	10,470	3,862	27,561
'87-8	27	....	54	190	171,000	10,470	4,000	28,548
'88-9	29	3,700	54	202	171,000	10,470	5,000	31,026
'89-90	32		60	215	171,000	10,470	3,989	31,398
'90-1	33.5	....		229	171,000	10,470	6,524	32,653

Financial.—*Bonded Debt*, $171 000: $75,000 at 7%; $96,000 at 4%. *Sinking Fund*, $72,-362. *Expenses:* total, $16,994. *Revenue:* consumers, $32,653; city, $5,375; total, $38,028. **Management.**—Comrs. appointed by City Council. Chn., J. W. Ringrose. Supt., D. A. Harris.

20. **NEW HAVEN,** *New Haven Co.* (Pop., 62,883—86,045.) Built in '60-2 by New Haven Water Co. Supply.—River and lakes, by gravity and pumping to reservoir. Fiscal year closes Dec. 31. During '90-91, 160,000,000-gall. storage reservoir was built in the town of Woodbridge for new gravity supply, and 4 miles of 27-in. pipe laid as conduit. Dam is 500 ft. long, 40 ft. high, of earth and stone. Engr., H. B. Gorham. Reported that another dam and storage reservoir is to be built. **Distribution.**—*Mains*, 134 miles. *Taps*, unk'n. *Meters*, 60. *Hydr'nts*, 750. *Consumption*, 11,500,000 galls. **Financial.**—Withheld. **Management.**—Prest. and Treas., H. S. Danvers. Secy., E. I. Foote. Supt., S. E. Grannis.

II. **NEWINGTON,** *Hartford Co.* (Pop., town, 931—953.) Built in '66 by Newington Water Co. Supply.—Stream or springs on Cedar Mountain, by gravity. Fiscal year closes Sept. 1. **Distribution.**—*Mains*, 2.5 miles. *Taps*, 33. *Hydrants*, 0. **Financial.**—*Cap. Stock:* $3,000. Eleven stockholders use water and share in operating expenses. There is a prospect of the works being enlarged. **Management.**—Prest, Erastus Kilbourne. Secy., Jno. S. Kimball.

21. **NEW LONDON,** *New London Co.* (Pop, 10,537—13,757.) Built in '72 by city, first supplied in '02 by Aqueduct Co. Supply.—Lake Konowoc, by gravity, with pumping to tank for high service. Fiscal year closes Sept. 1. New 20-in. c. i. supply main, 21,185 ft. long, was laid in '89-90; pipe from Warren Foundry and Machine Co., Phillipsburg, N. J.; laid by C. H. Eglee, Flushing, N. Y. High service also added '89-90, consisting of pump driven by motor utilizing pressure from passage of water to low service, a force main and tank. Motor is on 20-in. supply main, 500 ft. from tank; works against pressure of 15 to 20 lbs. and causes loss of 2 or 3 lbs pressure to low service. Built by W. H. Lang, Goodhue & Co., Burlington, Vt., after designs by W H. Lang. Described in Report Water Comrs. for '90 and details of tests promised in '91 report. Force main, 10-in. to and from tank. Tank has capacity of 90,100 galls.; w. l., 25 × 26 ft., on tower 68½ ft. high. Tower consists of w. l. lattice posts on masonry foundations; contrs., Berlin Iron Bridge Co., E. Berlin.

Year.	Mains.	Taps.	Meters.	Hydrants.	Consumption.	Cost.	Bonded debt.	Int. ch'g's.	Exp.	Rev.
'75	20.3	819	25	99	......	$277,335	$250,000	$17,500	$3,651	$10,773
'80	21	1,256	23	109	......	292,206	250,000	17,500	4,119	16,310
'81	21.9	1,298	22	116	......	299,477	250,000	17,500	3,517	16,931
'82	22.5	1,371	24	118	......	310,174	250,000	17,500	3,875	16,959
'83	23.1	1,421	23	120	......	312,562	250,000	17,500	3,707	18,651
'84	23.5	1,501	26	123	700,000	317,513	250,000	17,500	3,652	19,030
'85	24	1,573	24	127	933,000	3 2,367	250 000	17,500	3,758	20,096
'86	24.9	1,659	36	130	822,000	327,846	250,000	17,500	4 528	21,584
'87	25.8	1,755	40	136	1,100,000	337,308	250.000	17,500	4,187	23,189
'88	26.2	1,833	37	137	1,342,000	341,817	250,000	17,500	7,174	23,579
'89	27	1,934	39	139	1 378,000	344,099	250,000	17,530	4,972	24,597
'90	33.6	2,078	43	148	1,500,000	469,415	365,000	22,140	5,339	26,694

For Further Descriptive Matter see Manual for 1889-90. Changes since 1890 are here given. Populations are for 1880 and 1890, respectively.

MANUAL OF AMERICAN WATER-WORKS.

Financial.—*Bonded Debt*, $306,000: $250,000 at 7%, due in 1900; $116,000 at 4%, due in 1917. *No Sinking Fund. Expenses:* total. $18,908. *Revenue*, consumers, $26,768. All expenses are met by revenue from consumers. **Management.**—Four comrs., one appointed by council, three elected by popular vote for three years. Prest., B. A. Armstrong. Secy., P. C. Dunford, Supt., W. H. Richards, appointed by common council.

22. **NEW MILFORD,** *Litchfield Co.* (Pop., 3,907-3,917.) Built in '74 by New Milford Water Co. under perpetual charter; town neither controls rates nor has right to purchase works. **Supply.**—Surface water, from mountain stream, by gravity from impounding reservoir. Fiscal year closes Aug. 1. **Distribution.**—*Mains,* 5.9 miles. *T'ps,* 206. *Meters,* 0. *Consumption,* unknown. **Financial.**—*Cost,* $48,398. *Cap. Stock:* authorized, $50,000; paid up, $25,000: paid 6% div'd, in 1893. *Floating Debt,* $3,000 at 5%. *Expenses,* operating, $842; int., $163; taxes, $185; total, $1,270. *Revenue:* consumers, $3,838; city, $667; total, $1,505. **Management.** —Prest., Chas. H. Booth. Secy., and Treas., Chas. H. Noble. Supt., H. O. Warner.

23. **NORWALK,** *Fairfield Co.* (Pop., 13,956—17,717.) Built in '72 by Borough. **Supply.**—Silver Mine stream, by gravity from impounding to distributing reservoir. During '90, supply was increased and new supply-main laid. Fiscal year closes May 1. **Distribution.**—*Mains,* 25 miles. *Taps,* 1,300. *Meters,* 1. *Hydrants,* 125. *Consumption,* 1,200,000 galls. **Financial.**—*Cost,* $280,000. *Total Debt,* $250,000; floating, $50,000 at 4%; bonded, $200,000 at 4%. *Expenses:* operating, $2,172; int., $10,000; total, $12,172. *Revenue:* consumers $13,356; city, none. **Management.**—Three comrs. Secy., A. A. Camp. Supt., F. H. Beers. Report by C. N. Wood.

24. **NORWICH,** *New London Co.* (Pop., 15,112—16,156.) Built in '66—8 by city. **Supply.**—Brook and springs, by gravity from impounding reservoir. Fiscal year closes March 31.

Year.	Mains.	Taps.	Meters.	Hyd'ts.	Cost.	B'd'd debt.	Int. ch'g's.	Exp.	Rev.
'82-3	33.5	1,969	153	257	$411,870	......	......	$5,327	$27,999
'85-6	34.8	2,129	..	273	421,134	$300,000	$16,500	12,694	42,058
'88-9	35.3	....	110	274	500,884	300,000	16,500	12,145	38,137
'90-1	37.3	2,184	77	299	513,416	300,000	16,500	19,627	41,119

Financial.—*Bonded Debt,* $300,000; $150,000 at 6%, due in '93; $150,000 at 5%, due in 1908-10. *Expenses,* (See above; $5,238. operating exp., proper balance repairs and int.) total, $36,127. *Revenue:* consumers, $41,119; city none. **Management.**—Chn., R. A. France. Secy., A. C. Hatch. Supt., E. T. Gardner.

PAWCATUCK, *New London Co.* (Pop., not given.) Supply from Westerly, R. I.

25. **PLAINVILLE,** *Hartford Co.* (Pop., 1,930—1,993.) Built in '84 by Plainville Water Co. **Supply.**—Surface water, by gravity from impounding reservoir. Fiscal year closes July 16. **Distribution.**—*Mains,* 6.3 miles. *Taps,* 162. *Meters,* 1. *Hydrants,* 24. *Consumption,* unk'n. **Financial.**—*Cost,* $30,330. *Cap. Stock.* authorized, $30,000; all paid up; paid 6% div'd in 1890. *No Debt. Expenses,* $315. *Revenue:* consumers, $1,927; city, $665; total, $2,592. **Management.**—Prest., Wm. L. Cowles. Secy., E. C. Chapman. Supt., H. W. Higgins.

32. **PLANTSVILLE,** *Hartford Co.* (Pop., not given.) Supply from Southington.

26. **PORTLAND,** *Middlesex Co.* (Pop., 4,157-4,678.) Built in '89 by Portland Water Co. under perpetual charter; town has option of buying works at any time, price determined by arbitration or by commissioners appointed by court. **Supply.**—Brook, by gravity from impounding reservoir. Fiscal year closes Dec. 31. **Distribution.**—*Mains,* 4.5 miles. *Taps,* 187. *Meters,* 18. *Hydrants,* 70. **Financial.**—*Cost,* $172,099. *Cap. Stock:* Authorized, $100,000; all paid-up. *Bonded Debt,* $70,000, at 5%, due in 1909. *Expenses:* Op., $937; int., $3,640; total, $4,577. Co. is exempt from taxation for 10 yrs. *Revenue:* Consumers, $2,369; city, $2,100; total, $4,469. **Management.**—Prest., Oliver Gildersleeve. Secretary, Harry Gildersleeve, Jr. Treas., Chas. F. Parks. Hoston, Mass. Supt., J. A. Butler.

27. **PUTNAM,** *Windham Co.* (Pop., 5,827—6,512.) Built in '85-6 by Putnam Water Co.; franchise does not regulate rates, but provides that town may purchase works at any time, price determined by arbitration. **Supply.**—Little river, pumping to stand-pipe. Fiscal year closes Dec. 31.

CONNECTICUT. 57

Year.	Mains.	Taps.	Meters.	Hydrants.	Consumption.	Cost.	Bonded debt.	Int. Ch'g's.	Exp.	Rev.
'86	11	190	7	81	100,000	$145,000	$50,000	$2,500	....	$5,700
'88	12.2	400	10	61	180,000	150,000	50,000	2,650	$1,850	10,000
'90	12.9	417	13	68	230,000	159,685	48,000	2,331	2,386	11,505

Financial.—*Cap. Stock:* Authorized, $200,000; paid-up, $110,000; paid 3½% dividend in 1890. *Bonded debt*, $48,000, at $5%, due in 1906. *Expenses:* Operating, $2,386; int., $2,331; taxes, $405; total, $5,122. *Revenue:* Fixtures $6,316; meters, $3,272; hydrants, $1,817; total, $11,505. **Management.**—Prest., L. H. Fuller, Secy., Geo. E. Shaw. Treas., Elbert Wheeler, 24 Washington St., Boston, Mass. Supt., C. D. Sharpe.

28. ROCKVILLE, *Tolland Co.* (Pop., 5,902—7,772.) Built in '47 by Rockville Aqueduct Co : town neither controls rates nor has right to purchase works. **Supply.**—Shenipsit lake, by gravity. Fiscal year closes Dec. 1. **Distribution.**—*Mains,* 8 miles. *Taps,* 600. *Meters,* 0. *Hydrants,* 60. **Financial.**—*Cost,* $65,000. *Cap. Stock,* paid-up, $22,050. No Debt. **Management.**—Pres., George Maxwell. Secy. and Treas., J. C. Hammond, Jr. NOTE.—In '91 there was talk of city works.

29. SHARON, *Litchfield Co.* (Pop., 2,580—2,149.) Built in '88 by Sharon Water Co.; franchise provides neither for control of rates nor purchase of works by city. **Supply.**—Mountain brook, by gravity from impounding and distributing reservoir; filtered. Fiscal year closes Oct. 1. **Distribution.**—*Mains,* 1.8 miles. *Taps,* 50. *Meters,* 0. *Hydrants,* 14. **Financial.**—*Cost,* $14,500. *Debt,* $2,500 at 6%. *Expenses:* operating, $75; int., $150; total, $225. *Revenue,* $475. **Management**—Prest., Milo B. Richardson, Lime Rock, Secy., Willard Baker. Supt., I. N. Bartram.

30. SHELTON, *Fairfield Co.* (Pop., 1,362.) Built in '79-81 by Shelton Water Co.; franchise provides neither for control of rates nor purchase of works by town. **Supply.**—Curtis brook, by gravity from impounding and distributing reservoirs. Fiscal year closes May 16. **Distribution.**—*Mains,* 5.5 miles. *Taps,* 200. *Meters,* 1. *Hydrants,* 34. *Consumption,* unk'n. **Financial.**—*Cost,* $13,470. *Cap. Stock:* authorized, $25,000; all paid-up; paid 6% div'd. in 1890. *Floating Debt,* $12,600 at 6%. **Management.**—Prest., D. N. Plumb. Secy., D. S. Brinsmade.

31. SIMSBURY, *Hartford Co.* (Pop., 1,830—1,874.) Built in '73 by Simsbury Water Co. **Supply.**—Streams, by gravity from reservoir. **Distribution.**—(In '88.) *Mains,* 2 miles. *Taps,* 50. *Hydrants,* 10. **Financial.**—*Cost,* $30,000. *Cap Stock,* $30,000. **Management.**—Address W. H. Whitehead.

32. SOUTHINGTON, *Hartford Co.* (Pop., 5,411—5,501.) Built in '83-4 by Southington Water Co.; supplies Plantsville; franchise provides neither for control of rates nor purchase of works by town, but town took 150 of 600 shares of company's stock. **Supply.**—Brook, by gravity from impounding and distributing reservoirs; passed through two copper wire screens, first of one-ninth and second of one-eighth-inch meshes. Fiscal year closes Oct. 1.

Year.	Mains.	Taps	Meters.	Hyd't's.	Cost.	B'd'd debt.	Int. ch'g's.	Rev.
'85	11.5	290	0	69	$88,753	$27,500	$1,650	$1,380
'87	12.3	350	0	70	93,000	32,200	1,610	6,400
'89	14	495	0	74	94,199	30,500	1,625	6,728
'90	14	500	0	74	91,200	32,500	1,625	7,000

Financial.—*Cap. Stock:* authorized, $60,000; all paid up; paid 5% dividend in 1890. *Debt,* $32,500 at 5%, payable on demand; loan is from Southington Savings Bank. *Taxes,* $347. *Revenue:* (in '89) consumers, $5,405; city, $1,324; total, $6,729. **Management.**—Prest., J. B. Savage. Secy., Treas. and Supt., T. H. McKenzie.

NOTE.—See paper entitled "Water-works of Southington, Conn.," by T. H. McKenzie, Engr., in Trans. Am. Soc. C. E., vol. XV., p. 285; also company's report for 1885.

33. SOUTH MANCHESTER, *Hartford Co.* (Pop., not given.) Built in '91 by Cheney Bros. **Supply.**—Impounding reservoir on Porter Brook, covering area of 9 acres, with average depth of 9 ft. *Conduit,* 12-in. 1. No further information could be obtained.

34. SOUTH NORWALK, *Fairfield Co.* (Pop., 3,726—not given.) Built in '75 by town. East Norwalk supplied in '89 or '90 by laying main across harbor; cost about $12,000. **Supply.**—Surface water, by gravity from three impounding reservoirs on stream. Third reservoir was built in '91; Cap., 700,000,000 galls.; has gravel like

FOR FURTHER DESCRIPTIVE MATTER see Manual for 1889-90. CHANGES since 1890 are here given. POPULATIONS are for 1880 and 1890 respectively.

built across it 800 ft. long, through which water is filtered. A 16-in. main was laid from reservoir to city. Engr., W. B. Rider. **Distribution** —*Mains*, 14 miles. *Taps*, 451. *Meters*, 21. *Hydrants*, 100. *Consumption*, 1,225,000 galls. **Financial.**—*Cost*, $124,902. *Bonded Debt*, $88,000; $70,000 at 6%; $18,000 at 4%. *Expenses:* operating, $1,209; int., $4,920; total, $6,120. *Revenue:* consumers, $11,000; city, $2,000; total, $13,-000. **Management.**—Three comrs.: C. S. Trowbridge, W. B. Reed, Sam'l Raymond. Engr., J. B. Rider.

35. **STAFFORD SPRINGS,** *Tolland Co.* (Pop., 2,081—2,353.) Built in '87 by Stafford Springs Water Co.; franchise provides neither for control of rates nor purchase of works by town. **Supply.**—Roaring brook, by gravity from storage and distributing reservoirs. Fiscal year closes Dec. 1. Capacity of reservoir is 25,000,000 galls., instead of 5,000,000 galls. as stated in 1889-90 Manual. **Distribution** —*Mains*, 10.5 miles. *Taps* (in '88), 150. *Hydrants*, 30. *Consumption*, 500,000 galls. **Financial.**—*Cap. Stock*, $75,000. *Total Debt*, $36,200: floating, $1,200 at 6%; bonded, $35,000 at 5%, due 1906. *Expenses:* operating, $279; int., $1,750; taxes, $218; total, $2,265. *Revenue:* consumers, $2,285. City: fire protection, $900; total, $3,185. **Management.**—Prest., Alfred Birnie. Secy., A. S. Hicks. Treas., T. N. Birnie. Supt., Anthony Adams.

36. **STAMFORD,** *Fairfield Co.* (Pop., 11,297—15,700.) Built in '60 by Stamford Water Co. **Supply.**—Trinity Lake, by gravity to and from Simsbury distributing reservoir; also pumping. Fiscal year closes Oct. 31. A 2,000,000-gall. Knowles dup. pump was added in '90.

Year	Mains	Taps	Meters	Hydrants	Cost	B'd'd debt	Int. ch'g's	Exp	Rev
'86	18	835	2	80	$205,685	None	None	$4,473	$20,050
'88	34	1,173	4	86	353,243	$75,000	$3,500		27,578
'90	44	1,404	12	87	385,709	70,000	3,500	7,130	2,000*

*From city.

Financial.—Paid 5% div'd in 1890. *Bonded Debt*, $70,000 at 5%, due 1917. *Expenses:* operating, $5,244; int., $3,500; taxes, $1,886; total, $10,630. *Revenue:* city, $2,000. **Management.**—Prest., Jas. B. Williams. Secy., Walton Ferguson. Treas., Geo. H. Hoyt. Supt., Geo. E. Whitney.

37. **STONINGTON,** *New London Co.* (Pop., 7,335—7,184.) Built in '88 by Mystic Valley Water Co. Supplies Groton and Mystic. **Supply.**—Mistuxel Brook, pumping from storage to distributing reservoirs. **Distribution**—(In '89.) *Mains*, 13 miles. *Taps*, 175. *Hydrants*, 102. *Consumption*, 200,000 galls. **Financial.**—(In '89) *Cost*, $240,000. *Bonded Debt*, $100,000 at 5%. *Expenses:* operating, $1,200; int., $5,000; total, $6,200. *Revenue:* $7,380. **Management** (in '89)—Prest., T. E. Packer; Secy., Geo. E. Grinnell, both of Mystic. Treas., D. B. Spalding, Stonington. Supt., A. G. Watrous, Mystic Bridge. NOTE.—No report for '90 could be obtained.

38. **TERRYVILLE,** *Litchfield Co.* (Pop., not given.) **History.**—Built in '90-91 by Terryville Water Co. Des. Engr., T. H. McKenzie, Southington. Const. Engr., Edw. H. Phipps. **Supply.**—Springs, by gravity to reservoir. **Reservoir.**—Cap., 900,000 galls. **Distribution.**—*Mains*, cem., 3 miles. *Hydrants*, Ludlow, 7. *Valves*, Ludlow. **Financial and Management.**—Not given.

39. **THOMASTON,** *Litchfield Co.* (Pop., 3,225-3,278.) Built in '80 by Thomaston Water Co.; town neither controls rates nor has right to purchase works **Supply.**—Springs, by gravity from impounding reservoir. Fiscal year closes Dec. 31. **Distribution.**—*Mains*, 6 miles. *Taps*, 205. *Meters*, 1. *Hydrants*, 36. *Consumption*, unk'n. **Financial.**—*Cost*, $40,500. *Cap. Stock*, authorized, $40,000; paid up, $36,000; paid 6% div'd in 1889 and 7% in 1890. *No Debt*. *Expenses:* operating, $600; taxes, $150; total, $750. *Revenue:* consumers, $2,884; city, $600; total, $3,484. **Management.**—Prest., Aaron Thomas. Secy., W. F. Woodruff. Treas., A. J. Hine.

40. **THOMPSONVILLE,** *Hartford Co.* (Pop., 3,791.) Built in '85 by Thompsonville Water Co. Supplies Warehouse Point. **Supply.**—Springs, pumping to tanks. Fiscal year closes Dec. 31. **Distribution.**—*Mains*, 18 miles. *Taps*, 500. *Hydrants*, 55. *Consumption*, 185,000 galls. **Financial.**—*Cost*, $80,000. *Cap. Stock*, $50,000. *No Debt* **Management.**—Prest., Chas. H. Briscoe; Secy., F. N. Birnie; both of Springfield. Supt., H. R. Cooper.

41. **TORRINGTON,** *Litchfield Co.* (Pop., 3,327—6,048.) Built in '78 by Torrington Water Co.; town neither controls rates nor has right to purchase works. **Supply.**—Surface water, by gravity from impounding reservoirs on stream; filtered through sand and gravel. Fiscal year closes March 1. April 7, 1991, an additional 120,000,000-

CONNECTICUT.

gall. storage reservoir was in course of construction. Dam 9 ft. high is erected at mouth of pond and water diverted from natural channel to open trench 600 ft. long, from which it is conveyed in 12 and 10-in. c. i. pipe to stream on which old reservoirs are located. Engr., W. B. Rider, So. Norwalk. Distribution.—*Mains* 9.5 miles. *Taps*, 535. *Meters*, 0. *Hydrants*, 34. Financial,—*Cost*, $58,000. *Cap. Stock:* authorized, $40,000; all paid up; paid 10% div'd in 1890. In '91 cap. stock was increased to $60,000. *Taxes*, $311. *Revenue:* consumers, $7,000; city, $1,160; total, $8,160. **Management.**—Prest., Isaac W. Brooks. Secy. and Treas., L. McNeil. Supt., O. R. Tyler.

42. **WALLINGFORD,** *New Haven Co.* (Pop., 3,017—4,230.) Built in '82 by borough. **Supply.**—Lake Pistapaugh, by gravity. Fiscal year closes Nov. 1. Summer of '91 contracts were let for new supply main, to be [as planned] 16, 14 and 12 ins in diam.; from gate-house to Chestnut Hill, about 1½ miles, cement-lined pipe from Phipps Cement Co., New Haven, is used at cost of $17,934; remaining distance, c. i., from Warren Foundry & Machine Co., Phillipsburg, N. J., at cost of about $16,512; prices for material only.

Year.	Mains.	Taps.	Hydrants.	Cost.	Bonded debt.	Int. ch'g's.	Exp.	Rev.
'83	12	326	60	$96,329	$85,000	$3,000	$899	$795
'85	16	508	69	104,139	105,000	4,200	971	5,371
'87	16.3	598	69	106,575	105,000	4,200	894	6,095
'89	17.7	719	70	109,901	105,000	4,177	1,058	7,227
'90	18.1	682	70	111,659	105,000	4,220	1,591	7,204

Financial.—*Total Debt*, $106,092 : *floating*, $1,092; *bonded*. $105,000 at 4%, due in 1912. *Expenses:* total, $5,721; *Revenue:* consumers, $7,204; city, $545; total, $7,749. **Management.**—Three comrs., one elected each year. Chn., M. O. Callaghan. Secy. and Supt., W. J. Morse, elected by comrs.

40. **WAREHOUSE POINT,** *Hartford Co.* (Pop., not given.) Supply from Thompsonville.

43. **WATERBURY,** *New Haven Co.* (Pop., 17,806—28,646.) Built in '63 by city. **Supply.**—Springs and surface water, by gravity from impounding and distributing reservoirs; also emergency supply by pumping from Mad River. Fiscal year closes Dec. 31.

Year.	Mains.	Taps.	Meters.	Hyd'ts.	Cost.	B'd'd debt.	Int ch'g's.	Exp.	Rev.
'68	15.8	...	..	96	$168,700	$175,000	$12,250	......	$1,346
'75	23.2	1,250	.	148	237,824	190,000	13,300	$4,917	14,590
'80	29	1,863	..	166	327,278	255,000	13,300	2,901	21,538
'81	30	2,038	65	173	342,507	255,000	15,675	4,293	25,321
'82	31.1	2,217	84	180	361,885	265,000	15,675	3,797	29,785
'83	32.5	2,375	96	193	381,735	265,000	16,675	6,208	31,405
'84	33.3	2,493	104	204	401,236	265,000	16,675	9,257	33,273
'85	34	2,630	112	219	416,236	265,000	16,675	9,349	36,152
'86	35	2,751	123	224	432,486	265,000	16,675	9,038	41,341
'87	35.2	2,910	128	236	473,886	265,000	16,675	10,009	44,878
'88	36	3,056	134	238	493,011	265,000	16,675	13,335	49,309
'89	36.5	3,194	151	241	505,011	265,000	16,675	12,141	53,431
'90	37.5	3,339	162	248	520,011	265,000	16,675	11,719	58,793

Financial.—*Bonded Debt*, $265,000; $150,000 at 7%, due in '97; $40,000 at 7%, due in '90; $75,000 at 4½%; due, $5,000 each year after '91. *Sinking Fund*, $75,626. *Expenses:* total, $28,815. *Revenue:* consumers, $58,791; city, $4,160; total, $62,951. **Management.**—Two comrs., one elected every year. Chn., N. J. Welton. Secy. and Supt., F. B. Merriman.

44. **WEST HAVEN,** *New Haven Co.* (Pop., 1,975—not given.) Built in '83 by West Haven Water Co. **Supply.**—Cove River, pumping to reservoir. Distribution.—*Mains*, 12.5 miles. *Taps*, 412. *Meters*, 7. *Hydrants*, 18. Financial.—*Cost*, $90,000. *Cap. Stock*, $100,000. *No Debt*. *Revenue:* city, $600. **Management.**—Prest., D. Goffe Phipps, New Haven. Supt., E. H. Phipps.

45. **WILLIMANTIC,** *Windham Co.* (Pop., 6,508—8,648.) Built in '85-6 by city. **Supply.**—River Natchaug, pumping to reservoir. Fiscal year closes Nov. 1.

Y'r.	Mains.	Taps.	M'tr's.	Hy-drants.	Consump-tion.	Cost.	B'd'd debt.	Int. ch'g's.	Exp.	Rev.
'87	14	686	35	134	350,000	$232,097	$200,000	$8,000	$10,500	$9,830
'88	14	791	37	141	193,238	239,459	......	. .	4,336	10,060
'89	15.5	972	39	141	......	......	......	......	......	10,223
'90	17	1,018	64	143	383,393	......	200,000	8,888	3,792	11,198

For Further Descriptive Matter see Manual for 1890-91. Changes since 1890 are here given. Populations are for 1880 and 1890, respectively.

Financial.—See above. **Management.**—Three comrs., one elected each year. Chn., S. C. Smith. Secy., H. McCullock. Supt., H. S. Moulton.

46. WINDSOR, *Hartford Co.* (Pop., 3,056—2,954.) Built in '69-70 by Windsor Water Co.; city neither controls rates nor has right to purchase works **Supply.**—Springs, by gravity. Distribution.—*Mains*, 2 miles. *Taps* (in '88), 20 *Meters*, 0. *Hydrants,* 7. *Consumption,* 20,000 galls. **Financial.**—*Cost*, $15,000. *Cap. Stock:* authorized, $40,000. *No Debt. Revenue,* $1,200. **Management.**—Owner, H. Sydney Hayden.

47. WINDSOR LOCKS, *Hartford Co.* (Pop. of town, 2,332—2,758). Built in '89 by Windsor Locks Water Co. System was built under a state charter; town neither controls rates nor has right to buy works. **Supply.**—Springs, pumping to stand-pipe. Fiscal year closes Nov. 1. In '90 a contract for fire protection was made with town, and to meet this increased demand for water Co. built new brick pumping station in which a 1,000,000-gall. comp. dup. pump and a 54-in. boiler were placed and provision made for another pump and boiler; fire pressure, 80 lbs. Engrs., Dunham & Paine, N. Y. City. Contrs., S. P. Townsend & Son. Distribution —*Mains,* 7.8 miles. *Taps,* 125. *Meters,* 2. *Hydrants.* 37. **Financial.**—*Cost,* $66,000. *Cap. Stock:* authorized, $80,000. *Bonded Debt,* $50,000 at 5%. *Expenses.* not determined. *Revenue:* city, $1,075. **Management.**—Prest., Ed. D. Cogden, Secy., J. W. Johnson. Supt., S. P. Townsend, Hartford.

48. WINSTED, *Litchfield Co.* (Pop., 4,195—4,846). Built in '62 by borough. **Supply.**—Lake, by gravity; passes through wire sieves. Fiscal year closes May 1. Distribution.—*Mains,* 10 miles. *Taps,* 725. *Hydrants,* 90. Fin n ial.—*Cost* (in '88), $128,656. *Bonded Debt,* $25,000 at 4½%; $15,000 due in '91, when $10,000 will be paid; $10,000 due in '91. *Sinking Fund,* $2,111. *Expenses:* operating, $4,678: int., $1,150; total, $3,823. *Revenue:* consumers, $6,628. **Management.**—Comrs.; Michael M. Ryan, T. C. Richards, Otis Tupper.

Water-Works Projected with Fair Prospects of Construction.

EAST HARTFORD, *Hartford Co.* (Pop., '3,500—4,455.) East Hartford Water Co. contracted with fire district. July, '91, to have works completed by Sept. 1, '92. Co. must have $50,000 cap. stock; must not issue bonds to more than 75% of stock or pay more than 5% int. on bonds. Fire district may buy works on 30 days' notice, paying cost and 6% int., less profits. Prest., E. J. McKnight. Treas., L. S. Forbes P. S. Bryant, Hartford, and L. S. Forbes are interested.

NEW CANAAN, *Fairfield Co.* (Pop., 2,673—2.701.) System was to be built during the summer of '91. Ownership not determined. Supply, Five Mile River. Chn. comrs , L. M. Monroe.

NEW HARTFORD, *Litchfield Co.* (Pop., 3,302—3,160.) Co. to build system. Capital not subscribed. Preliminary surveys made by T. H. McKenzie, Southington. Address Wm. S. Seymour.

WESTPORT, *Fairfield Co.* (Pop., 3,477—3,715.) Preliminary surveys made in '91 for construction of works. Est. cost, $20,000. Address M. W. Wilson or John S. Jones.

Water-Works Projected in '90 a d '91, but now indefinite.

Unionville, Hartford Co.

GROUP 2.--Middle States.--Water-Works Completed, In Process of Construction or Projected, in New York, New Jersey, Pennsylvania, Delaware, Maryland and District of Columbia.

NEW YORK.
Water-Works Completed or In Process of Construction.

1. **ADAMS,** *Jefferson Co* (Pop., 1,250—1,360.) Built in '86 by Adams Water Works Co.; controlled by Moffett, Hodgkins & Clarke, Syracuse. Supply.—Springs, pumping to stand-pipe, with direct pumping in case of fire. Distribution—(In '88.) *Mains*, 3½ miles. *Taps*, 79. *Meters*. 3. *Hydrants*, 25. *Consumpt on*, 31,249 galls. Financial.—(In '88.) *Cost*, $30,000. *Debt*, $25,000, at 6%. *Expenses:* operating, $1,000. *Revenue:* consumers, $2,100; city, $1,182; total, $3,282. Management.—Address Moffett, Hodgkins & Clark, Syracuse.

2. **ADDISON,** *Steuben Co.* (Pop., 1,596—2,166.) Built in '89 by Addison Water Co.; franchise does not regulate rates, but gives town option of buying works every five years. Supply.—Brook, pumping to tank. Distribution.—*Mains*, 4.7 miles. *Hydrants*, 60. Financial.—*Cost*, $45,000. *Cap. Stock:* authoriz d, $10,000; all pa'd-up. *Bonded Debt*, $35,000 at 6%. *Hydrant rental*, $25. Management.—Prest., E. M. Welles. Secy. and Treas., Geo. C. Howard.

3. **ALBANY,** *Albany Co.* (Pop., 90,575—94,923.) First supplied from small gravity works built by co. last century; bought by city in '51 and extended. Supply.—Patroon and Sand creeks, by gravity, and Hudson River by pumping to reservoir and repumping to high service reservoir. Financial year closes Nov. 1, but statistics for distribution are for Dec. 31.

In '85 Special Water Commission was appointed to obtain a new supply. Attempt was made to secure 15,000,000 galls. of water daily from driven wells, and in Sept., '89, 390 2-in. wells had been driven. Attempt was a failure, and effort is being made to recover money paid contractors. In connection with driven well supply four 5,000,000-gall Allis engines were bought. For further description of wells see Manual for '89-'90, p. 127. Under Commission three reservoirs have been finished and were accepted by city in March, '90. Reservoirs store water from Patroon and Sand creeks for gravity supply to city east of Pearl St. Act appointing Special Commission authorized $1,200,000 bonds, of which $600,000 had been issued Nov. 16, '91, and $532,618 expended.

Special Commission has recommended new gravity supply from stream called Normanskill, and during '91 stream was being gaged. Plan includes a dam in town of Guilderland, above French's Mills, forming a storage reservoir with an available est. storage cap. of 1,955,000,000 galls., area of about 600 acres and drainage area of about 114 sq. miles. At site of proposed dam stream flows through channel in rock 250 ft. wide and about 50 ft. deep. A 4-ft. c. i main would deliver water to Bleecker Reservoir, connecting at a point ¾ mile west of Allen St. with present brick conduit from Rensselaer Lake. I. M. de Varona, Brooklyn, is Supt. and Engr. for Special Water Commission, but had nothing to do with driven well experiment Surveys for Normanskill supply were made by J. Herbert Shedd, Providence, R. I., for or in conjunction with R. D. Wood & Co., Philadelphia, who have offered to introduce supply in accordance with above plan for $735,000, not including land damages and water rights, or for $980,000 including these. By agreement between Commission and other parties interested Engr. Varona appointed John Bogart and Rudolph Hering, N. Y. City, Consult. Engrs., to report on project, and they have reported favorably.

Year.	Mains.	Taps.	Meters.	Hyd'ts.	B'd'd d't.*	Int. ch'g's.	Exp.	Rev.
'80	77.3	....	....	493	....	......	....	....
'81	76.5	unkn.	12	500	$1,110,000	......	$28,250	....
'83	80	11,882	37	508	1,099,000	......	30,586	$156,665
'86	81.7	13,652	47	525	1,089,000	......	......	162,600
'87	90.65	14,246	152	575	1,089,000	$65,230	92,836	173,510
'88	92.15	....	....	603	1,289,000	72,130	96,117	183,739
'89	96.85	....	....	628	1,269,000	95,780	98,346	301,774

*Does not include $600,000 of bonds issued for special Water Commission since '85; in '88-9 $30,000 of these bonds were redeemed, and in same year $20,000 of regular bonds were also redeemed.

FOR FURTHER DESCRIPTIVE MATTER see Manual for 1889-90. CHANGES since 1890 are here given. POPULATIONS are for 1880 and 1890, respectively.

62   MANUAL OF AMERICAN WATER-WORKS.

Financial.—*Cost* (in '82), $1,550,000. *Sinking Fund* (Nov. 1, '89), $82,190; invested in bonds and deposited in banks. *Expenses.*—Operating ('88–9), $98,346; int., $95,780, of which $72,230 was on regular bonds, $23,550 on special Commission bonds; total, $204,973. *Revenue* (in '88-9), consumers, $304,775; other purposes, $555; total, $305,330. **Management.**—Five comrs. Prest., R. L. Banks. Supt., Geo. W. Carpenter. Chf. Engr. of Steam Power, J. J. Greagon.

NOTE.—Financial figures as nearly correct as can be compiled from written and printed reports.

4. ALBION, *Orleans Co.* (Pop., 5,147–4,586.) Built in '88-9 by Albion Water-Works Co.; franchise fixes rates, and provides that town may purchase works in '99, price determined by appraisers, but not to be less than cost. **Supply.**—Wells, pumping to stand pipe. Fiscal year closes Dec. 31. **Distribution.**—*Mains,* 8.5 miles. *Taps,* 400. *Meters,* 190. *Hydrants,* 80. **Financial.**—Cap stock: authorized, $100,-000; all paid up. *Bonded Debt,* $100,000 at 6%. *Revenue:* city, $3,000. **Management.**—Prest., G. M. Wood. Secy., E. M. Bassett, Buffalo. Supt., C. R. Sawyer.

5. AMSTERDAM, *Montgomery Co.* (Pop., 9,466–17,366.) Built in '52 by city. Supply, surface water, by gravity. Fiscal year closes May 1. During '90 a conduit was constructed connecting Hans and Bunn creeks. Line consists of 12.8 miles of vitrified and iron pipe, 18 and 20 ins. in diam. A dam, 5 ft. high, was constructed on Hans Creek.

Year	Mains	Taps	Meters	Hyd'ts	Cost	B'd'd d't.	Int. ch'gs.	Exp.	Rev.
'83-4	12.7	616	24	118	$254,284		$10,600		$14,833
'85-6	13.8	1,091	33	123	293,102	$257,000	12,810	$4,382	17,325
'86-7	14.4	1,199	37	130	300,621	257,000	12,850	4,436	18,518
'87-8	15.11	1,314	43	140	304,853	257,000	12,850	4,318	20,237
'88-9	19.75	1,586	54	194	330,610	280,000	13,225	5,089	25,095
'89-90	20.5	1,754		206	369,736	330,000	13,600	5,237	27,416
'90-1	20.7	1,840		207	470,019	407,000	16,975	5,001	28,446

Financial.—*Bonded Debt,* $407,000; $257,000 at 5%, $150,000 at 3%. *Expenses:* total $21,976. *Revenue:* consumers, $28,446. **Management.**—Nine comrs., three elected every year. Prest., James R. Snell. Secy., Jno. McClumpha. Supt., A. H. DeGraff.

6. ANDES, *Delaware Co.* (Pop in '90, 416.) Built in '76 by Andes Water Co.; town neither controls rates nor has right to purchase works. **Supply.**—Springs, by gravity. Fiscal year closes Dec. 31. **Distribution.**—*Mains,* 1.5 miles. *Taps,* 60. *Meters,* 0. *Hydrants,* 16. **Financial.**—*Cost,* $7,000. *Cap. stock:* authorized, $6,000; all paid up; paid 8% in '90. *N. Debt. Expenses:* operating, $100; taxes, $40; total, $140. *Revenue,* $650. **Management.**—Prest., Jas. F. Scott. Secy. and Supt., David Ballantine.

7. ATTICA, *Wyoming Co.* (Pop., 496–1,994.) Built in '79-80 by Attica Water Supply Co.; franchise provides that rates must be approved by trustee; works may be bought according to general state law. **Supply.**—Spring, by gravity. Fiscal year closes Dec. 31. **Distribution.**—*Mains,* 7.5 miles. *Taps,* 150. *Meters,* 1. *Hydrants,* 32. *Consumption,* 250,000 galls. **Financial.**—*Cap. Stock,* $60,000. *Total Debt,* $45,-500; floating, $5,500 at 6%; bonded, $40,000 at 6%. *Revenue,* city, $1,120. **Management.**—Prest., C. E. Loomis. Secy., G. G. Loomis.

8. AUBURN, *Cayuga Co.* (Pop., 21,924–25,858.) Built in '65 by Auburn Water-Works Co. **Supply.**—Owasco Lake, pumping direct. For description of suction and siphon pipe, with illustrations, see ENG. NEWS, vol. XXIV., p. 383. Following correction is made to error in 1889-90 Manual: Pumping Machinery—Dy. cap., 27,000,000 galls.; one 8,000,000 Gaskill at the lake pump house, to use in case of accident to siphons; one 8,000,000 Gaskill, one 5,000,000 Holly quadruplex, and two 3,000,000 rotary. **Distribution**—*Mains,* 53 miles. *Hydrants,* 302. **Financial.**—Withheld. *Cost* (in '88.) $500,000. **Management.**—Prest., A. G. Beardsley. Secy. and Supt. N. B. Eldred. NOTE.—In '91 city talked of buying works.

54. ARVERNE-BY-THE-SEA, *Queens Co.* See Far Rockaway.

9. AVOCA, *Steuben Co.* (Pop., 547–953.) Built in '89 by village. **Supply.**—Springs and creek, by gravity. Fiscal year closes Feb. 1. **Distribution.**—*Mains,* 5.2 miles. *Taps,* 70. *Meters,* 0. *Hydrants,* 31. *Consumption,* unk'n. **Financial.**—*Cost,* $19,650. *Total Debt,* $18,730; floating, $730; bonded, $18,000 at 3½%; due in 1909. *Expenses:* operating, $75; int., $630; total, $705. *Revenue,* $450. **Management.**—Four comrs.,

NEW YORK.

appointed every two years from trustees. Chn., A. J. Arnold. Secy., T. C. Chase. No Supt.

10, 11. AVON, *Livingston Co.* (Pop., 1,617—1,653.)

STEPHEN HOSMER'S WORKS.

Built in '57. **Supply.**—Springs, by gravity. In '85 there were 7 miles of mains, 200 taps and 9 hydrants.

VILLAGE WORKS.

Built in '88. Supply.—Conesus Lake, by gravity to distributing reservoir. Distribution.—(In '88.) *Mains*, 12 miles. *Hydrants*, 32. Financial.—(In '88.) *Cost*, $63,-000; bonded debt, $50,000, at 3.7%. **Management.**—Secy., Orange Sackett. No Supt.

12 BAINBRIDGE, *Chenango Co.* (Pop., 788—1,049.) Built in '83 by J. M. Roberts; town neither controls rates nor has right to purchase works. Supply.—Artesian wells, springs and reservoirs, by gravity. Fiscal year closes Dec. 31. Distribution.—*Mains*, 6 miles. *Taps*, 60. *Meters*, 0. *Hydrants*, 20. Financial.—*Cost*, $20,000. *Bonded debt*, $10,000 at 6%, due in '91. *Expenses*: operating, $50; int., $600; total, $650. *Revenue*: consumers, $500; city, $510; total, $1,010. **Management.**—Owner and Mgr., J. M. Roberts.

13. BALDWINSVILLE, *Onondaga Co.* (Pop., 2,121—3,040.) Built in '89-90 by village. Supply.—Surface well, pumping to stand-pipe. Fiscal year closes Jan. 31. Distribution.—*Mains*, 9 miles. *Taps*, 206. *Meters*, 2. *Hydrants*, 69. *Consumption*, 80,000 galls. Financial.—*Cost*, $60,000. *Bonded Debt*, $60,000 at 3½%. *Expenses*: operating, $5,000; int., $1,890; total, $6,890. *Revenue*: consumers, $2,000; city, none. **Management.**—Seven comrs., elected every three years. Chn., Chas. M. Bliss. Secy., Chas. B. Baldwin. Supt., J. E. Connell.

14. BALLSTON SPA, *Saratoga Co.* (Pop., 3,011—3,527.) Built in '70 by village. Supply.—Springs, by gravity. Distribution.—*Mains*, 12 miles. *Taps*, 420. *Meters*, 1. *Hydrants*, 121. *Consumption*, 600,000 galls. Financial.—*Cost*, $100,000. *Bonded Debt*, $76,250, at 5 and 7%. *Expenses*: operating, $500. *Revenue*: consumers, $6,000. **Management.**—Prest., A. Comstock. Secy., C. H. Groos. Treas., Thos. Curley.

15. BATAVIA, *Genesee Co.* (Pop., 4,815—7,221.) Built in '65 by village. Supply.—Tonawanda Creek, pumping to tank and direct in case of fire. Fiscal year closes March 14. A Gaskill pump was purchased in '90. Distribution.—*Mains*, 7 miles. *Taps*, 250. *Hydrants*, 70. *Consumption*, 1,000,000 galls. Financial.—*Cost*, $125,000. *Bonded Debt*, $35,000 at 3½%. *Expenses*: operating, $3,500. *Revenue*: consumers, $2,500. **Management.**—Board. Prest., Jno. M. Leonard. Supt., Geo. Perrin.

NOTE.—In '91 the Pure Water Co. was organized to introduce a new supply, but to June nothing had been done, and it is not probable that the co. will ever build.

16. BATH, *Steuben Co.* (Pop., 3,183—3,261.) Built in '87 by Bath Water Works Co. Supply.—Wells and springs, pumping to stand-pipe. Fiscal year closes Dec. 31. Distribution.—*Mains*, 7 miles. *Taps*, 180. *Meters*, 51. *Hydrants*, 73. *Consumption*, 100,000 galls. Financial.—*Cost*, $100,000. *Cap. Stock*: authorized, $100,-000; all paid up; paid no div'd in 1890. *Bonded Debt*, $60,000 at 6%. *Revenue*, city, $2,482. **Management.**—Prest., C. M. Harrington. Secy., E. M. Bassett, both of Buffalo. Supt., H. W. Lattin.

74. BATH BEACH, *Kings Co.* See Gravesend.

75. BATH-ON-THE-HUDSON, *Rensselaer Co.* (Pop., 2,046—2,309.) See Greenbush.

90. BAYPORT, *Suffolk Co.* See Islip.

74. BAY RIDGE, *Kings Co.* See Gravesend.

90. BAYSHORE, *Suffolk Co.* See Islip.

FOR FURTHER DESCRIPTIVE MATTER see Manual for 1889-90. CHANGES since 1890 are here given. POPULATIONS are for 1880 and 1890, respectively.

1. BERLIN, *Rensselaer Co.* (Pop. 465—Est., 600.) Built in '88 by Berlin Water Supply Co. for domestic use. **Supply.**—Springs, by gravity. **Reservoir.**—Cap. 1,000,000 galls. **Distribution.**—*Mains*, 4 and 6-in c. i., 3 miles. *Taps*, 20. *Meters*, none. *Hydrants*, 1. *Consumption*, 75,000 galls. *Pressure*, 80 to 85 lbs. **Financial**—*Cost*, $3,000. *Capital Stock*: authorized, $5,000; paid up, $1,000. *Debt*, floating $2,000 at 10%. Bonds to the amount of $5,000 have been authorized but none issued. *Expenses:* operating, $100. **Management.**—Prest., W. F. Taylor. Secy. and Treas., H. C. Gifford. Report Jan. 17, '91.

17. BINGHAMTON, *Broome Co.* (Population 17,317—35,005.) Built in '67-3 by city. **Supply.**—Dug wells and Susquehanna River, pumping direct. Fiscal year closes Dec. 31. A 12,000,000 gall. Gaskill pump was added in '90.

Year.	Mains.	Taps.	Meters.	Hydrants.	Consumption.	Cost.	B'd'd debt.	Int. ch'g's.	Exp.	Rev.
'81	24	1,935	15	175	2,068,368	$200,000	$191,000	.....	$16,233	$29,280
'83	24.3	2,745	112	181	2,058,233	258,605	192,500	$13,380	12,172	34,468
'85	25.7	2,968	137	207	2,393,422	.....	179,500	12,565	12,227	41,047
'87	29	3,036	174	257	2,514,978	471,000	179,000	7,140	14,977	54,726
'89	34.9	4,081	257	318	2,996,569	550,000	177,000	7,105	10,593	66,166
'90	38.5	4,403	316	358	3,290,490	614,354	177,000	7,000	14,631	70,714

Financial.—Total Debt, $153,780: floating, $8,780; bonded, $22,000 at 7% due $1,500 each year until 19.1; $153,000 at 3½%, of which $91,000 are due in 1907, $62,000 in 1918. *No Sinking Fund.* *Revenue:* consumers, $70,714. *Expenses:* total, $21,631. **Management.**—Five Comrs.; one elected each year. Prest., Jno. Anderson. Secy. Horace E. Allen. Supt. Darwin Felter.

18. BROCKPORT and HOLLEY, *Monroe and Orange Cos.* Pop. (In '90.) Brockport, 3,742; Holley, 1,381. **History.**—Built in '90 by Brockport Water-Works Co. Engrs. and contrs. for trenching, pipe-laying and stand-pipe, Bassett Bros., Buffalo. **Supply.**—Springs, pumping to stand-pipe. **Pumping Machinery.** Dy. cap., 1,000,000 galls., 1,000,000 water power and 1,000,000 dup. steam pump, as reserve, both from Buffalo Steam Pump Co. *Boilers:* two 80 HP., Phœnix Boiler Works, Syracuse. **Stand-Pipe.**—Cap., 250,000 galls. **Distribution.**—*Mains*, 16 miles. *Taps* (June 24, '91), 290. *Meters* (June 24, '91), 90. *Hydrants*, 101. *Pressure*, ordinary, 70 lbs. **Financial.**—*Cost*, $175,000. *Capital Stock*, authorized, $35,000; all paid up. *Bonds:* $135,000 issued; $15,000 more authorized; int., 6%. *Revenue:* (June 24, '91) Consumers, $6,100; public uses of water, $3,775, $2,650 from Brockport, $1,125 from Holley. **Management.**—Prest., C. M. Harrington. Secy., E. M. Bassett. Both of Buffalo, Mass., C. R. Bassett, Brooklyn. Supt., A. T. Throop. NOTE.—For more detailed description see *Fire and Water*, Feb. 21, '91.

19, 20. BROOKLYN, *Kings Co.* (Pop., 566,663—306,343.) Built in '56-0 by city. **Supply.**—Surface water impounded in reservoirs, open wells and Andrews' system of driven wells, pumping to reservoir and re-pumping for high service. Fiscal year, for distribution closes in May; for finances, Dec. 31. Late in '89 contracts were let for the extension of supply to secure 25,000,000 galls. daily, and to include five additional ponds, a storage reservoir, conduit extension and a pumping station. Ponds are to east of old ones. Names and capacities, from east to west, are as follows: Massapequa, 31,000,000 galls.; Ridgewood, 23,500,000 galls.; Newbridge, 16,500,000 galls.; East Meadow, 23,500,000 galls.; Millburn, 19,500,000 galls.; Baldwin storage reservoir, 413,700,000 galls.; total storage cap., 527,700,000 galls. With exception of Millburn, ponds are all on conduit, which passes through embankment at pond outlets. From most easterly pond, Massapequa, brick conduit extends 37,437 ft. to Millburn pumping station. Millburn pond is connected with station by short main. Here three 10,000,000-gall. Davidson pumping engines will force water through 48-in. c. i. force mains, 7,938 ft. long, to Baldwin storage reservoir. From reservoir 36-in. c. i. main extends 7,618 ft. and connects with old conduit, and 48-in. c. i. main extends to connecting gate-house, then parallels old conduit for 12¼ miles to Ridgewood pumping station. Further details regarding the conduit are given in Tables 1 and 2, and more information, with numerous illustrations, will be found in ENG. NEWS, vol. XXV., p. 225, *et seq*. The above extensions have been made under the direction of Robt. Van Buren, Chf. Engr. Dept. City Works, with I. M. de Varona Assistant Engr. in charge and J. P. Adams at the head of the department. In addition to the above work a new pumping station and 20,000,000-gall. engine is being erected at Ridgewood and the Ridgewood reservoir is being enlarged.

NEW YORK.

TABLE I.—LOCATION, GRADE AND DIMENSIONS OF THE SEVERAL CONDUIT SECTIONS, BROOKLYN WATER-WORKS EXTENSION.

Section.	LOCATION.	Length in ft.	Fall per mile, ins.	Max. depth of cutting, ft.	Min. depth of cutting, ft.	Av. depth of cutting, ft.	Cu. yds. excavation, per lin. ft.	Depth of foundation below spring line, ft.	Vers. sin. of invert, ft.	Rad. of invert, ft.	Chord of invert, ft.	Radius of upper arch, ft.	Height of side walls, ft.	Clear inside height, ft.	Total area of section, sq. ft.	Diam. of equivalent circle, ft
1	Millburn to East Meadow...	3,991.0	0.326	23.9	5.1	17.7	9.86	5.5 to 11.0	1	11.39	9.33	4.667	1.25	6.917	52.131	1.116
2	East Meadow to New Bridge	5,375	0.593	17.0	3.3	11.05	6.43	4.5 to 10.8	1	9.18	8.33	4.167	1.25	6.417	43.237	7.400
3	New Bridge to Ridgewood...	3,163	0.593	17.7	5.7	12.16	6.04	5.4 to 9.2	1	8.50	8.00	4.000	1.25	6.250	40.557	7.200
4	Ridgewood to Massapequa...	13,900	0.593	21.4	5.7	15.06	6.94	5.4 to 10.2	1	7.22		3.337		5.917	35.223	6.700

TABLE II.—LINE OF MAXIMUM FLOW IN CONDUIT, THE RESULTING CAPACITY AND VELOCITY AT EACH SECTION, ELEVATION OF THE FLOW LINE, ETC., BROOKLYN WATER-WORKS EXTENSION.

	Section.	Names of streams flowing into each section.	Clear inside height, ft.	Depth of water in conduit, ft	Grade and hydraulic slope, in.	Area of flow, sq. ft.	Wetted perimeter, ft.	Hydraulic mean radius.	Velocity (ft.) per sec.	Cu. ft. per second.	U. S. gallons in 24 hours.	Elevation of flow line at inlet.	Clear height in conduit above flow line.
1	East Meadow...		6.917	5.25	0.528	43.8650	18.6337	2.3557	2.0809	91.2662	58,996,897	7.705	1.667
2	Newbridge...		6.417	5.25	0.528	36.5963	17.8481	2.1025	1.9946	78.9616	51,756,458	8.690	1.167
3	Ridgewood...		6.250	5.25	0.528	35.3106	17.6137	2.0056	1.9635	72.4740	46,841,178	9.866	1.030
4	Massapequa		5.917	5.25	0.528	33.3095	17.2158	1.9348	1.8867	62.8447	40,617,596	11.244	0.667

Capacity.

Yr.	M'ns.	Taps.	Mtrs.	Hy-dr'ts.	Con-sumpt'n.	Cost.	Bonded debt.	Int. ch'g's.	Exp.	Rev.
'58	120.1	......	....	800	....	....	$165,302	$44,681	$133,733	
'59	123.9	......	....	805	....	....	321,000	64,507	256,400	
'60	136.5	9,302	....	843	3,439,265	....	354,000	72,889	239,356	
'61	145.6	12,856	....	981	4,244,781	....	330,000	71,997	303,296	
'62	157.5	15,105	....	1,094	5,244,125	....	333,000	95,027	302,750	
'63	165.6	17,145	....	1,168	6,778,843	....	339,000	129,221	386,416	
'64	171.7	18,935	....	1,271	8,385,000	....	340,350	137,578	419,106	
'65	176.7	20,382	....	1,334	9,643,148	....	368,880	167,501	462,619	
'66	183.8	22,244	....	1,403	11,389,142	....	406,500	188,484	528,539	
'67	199.9	24,888	....	1,443	12,896,156	....	435,630	213,537	617,986	
'68	213.9	28,173	....	1,496	11,640,800	....	459,180	315,866	582,656	
'69	237.7	32,097	....	1,606	17,630,400	....	476,130	348,226	642,769	
'70	258.5	35,930	....	1,753	18,682,219	....	503,610	250,052	884,580	
'71	277.9	39,760	....	1,941	19,353,689	....	545,130	268,616	971,415	
'72	290.8	42,906	....	2,092	22,711,751	....	573,212	308,282	931,822	
'73	306.7	45,876	....	2,235	21,895,955	....	609,668	3 5,784	910,884	
'74	323.8	49,791	....	2,326	24,772,467	....	647,717	343,730	895,203	
'75	327.5	51,102	....	2,427	27,170,949	....				

FOR FURTHER DESCRIPTIVE MATTER see Manual for 1889-90. CHANGES since 1890 are here given. POPULATIONS are for 1880 and 1890, respectively.

MANUAL OF AMERICAN WATER-WORKS.

Yr.	M'ns.	Taps.	Mtrs.	Hy-dr'ts.	Con-sumpt'n.	Cost.	Bonded debt.	Int. ch'g's.	Exp.	Rev.
'76	332	53,083	....	2,531	28,104,514	....	....	662,587	308,924	887,333
'77	336.3	54,879	....	2,589	30,342,112	....	....	680,207	299,551	900,967
'78	343.4	56,685	756	,645	30,500,871	....	$11,265,000	687,380	303,658	995,205
'79	347.7	58,295	859	2,823	32,912,149	....	11,265,000	687,380	224,710	940,631
'80	351.7	59,880	1,085	2,874	30,744,591	$11,379,500	11,379,500	697,160	272,321	977,903
'81	354.9	60,558	1,208	2,916	32,731,499	11,530,500	9,830,500	547,695	314,431	924,297
'82	358.7	63,286	1,462	3,017	34,618,834	11,664,507	9,858,500	615,710	037,013	971,199
'83	365.3	64,160	1,566	3,105	36,131,403	11,990 592	10,452,000	628,252	423,759	1,028,070
'84	371.5	65,918	1,712	3,113	38,879,034	12,858,000	11,158,000	643,631	486,578	1,130,101
'85	379.9	67,341	1,904	3,158	43,411,270	....	11,645,500	667,538	535,504	1,173,817
'86	387.9	72,402	1,984	3,335	45,322,066	....	10,893,000	653,890	541,836	1,212,019
'87	397.3	75,592	2,035	3,486	15,256,212	....	11,073,500	635,238	567,174	1,246,147
'88	406.8	79,073	2,137	3,845	48,901,771	....	11,682,500	641,063	548,986	1,296,645
'89	416.4	85,961	2,263	3,948	52,191,128	15,483,340	11,632,500	672,970	638,938	1,335,760
'90	430.8	89,493	2,263	4,251	55,000,000	....	....	....	....	....

Management.—Dept. City Works. Comr., J. P. Adams. Secy., D. L. Northrup. Chf. Engr., Robert Van Buren.

LONG ISLAND WATER SUPPLY CO.

History.—Built in '81 to supply town of New Lots, which in '86 was annexed to Brooklyn as 26th ward; 50-years charter, 25-years hydrant contract. Act annexing New Lots to Brooklyn provided that latter might buy works, and that until city did so, or co.'s charter expired, city should not lay mains in this section. Above act also provided that in case co. and city could not agree on purchase price latter could take property by eminent domain within two years; also that in case works came into possession of city latter should assume bonded indebtedness of $500,000 at 6%; $250,000 secured by first mortgage, dated Dec., '81, and $250,000 secured by second mortgage, July, '85. Dec. 22, '90, Mayor, Comptroller and Auditor contracted with co. to buy works, assuming the $500,000 of 6% bonds mentioned above, and paying $750,000, or $300 per $100 share, in 10-20 years 3% bonds. Wm. Ziegler, a taxpayer, obtained injunction to prevent purchase. Matter was carried to Court of Appeals, where it was decided that the purchase was illegal, owing to the fact that the two years named in annexation act (see above) had expired.

Water Supply.—Wells, pumping to reservoir; well 25 × 30 ft., with 7 6-in. driven wells.

Pumping Machinery.—Cap., 3,000,000 galls.; two Davidson pumps and two boilers.

Reservoir.—Cap., 4,000,000 galls.

Distribution.—*Mains*, c. i., 33 miles; also 16½ miles w. i. mains owned by individuals, but used by Co. *Consumers:* Jan. 1, '89, 2,414; do., '90, 3,081; do., '91, 3,732. *Hydrants*, 419.

Financial.—*Capital Stock:* $250,000. *Bonded Debt*, $500,000; $250,000 due in '95, $250,000 due in 1901. *Expenses:* operating, $16,869; int., $30,000; taxes, $517; total $47,186. *Revenue:* consumers, $55,854; city, fire protection, $18,720; misc., $449; total, $65,023.

Management.—Agent, Cyrus E. Staples.

21. **BUFFALO**, *Erie Co.* (Pop., 155,134—255,664.) Built in '52 by Co., now owned by city. **Supply.**—Niagara river, pumping direct and to reservoir. Fiscal year closes Dec. 31. In '89 a 20,000,0000-gall. pump was added, and in Aug., '91, contract was let for additional 20,000,000-gall. Gaskill pump. '90-2 a new 135,000,000-gall. storage and distributing reservoir was built. Reservoir in excavation and embankment, with water 31 to 40 ft. deep. No heart-wall. For illustrated description see ENG. NEWS, vol. xxv., p. 26. Des. and Consult. Engr., W. J. McAlpine. Const. Engr., L. H. Knapp. Cost, about $500,000.

Year.	Mains.	Taps.	M't'rs.	Hy-drants.	Con-sumption.	Cost.	Bonded debt.	Int. charges.	Exp.	Rev.
'71	67.8	4,227	0	497	7,604,801	$1,693,821	....	....	$77,457	$124,865
'75	91.2	......	..	.	10,449,298	2,824,369	....	....	64,862	184,516
'80	102.1	9,137	57	1,118	16,369,802	3,169,180	....	....	60,984	216,214
'82	131.1	13,379	140	1,400	19,386,885	3,531,284	$2,950,000	....	79,106	362,762
'85	191.1	......	...	1,783	28,606,557	3,245,857	2,778,382	....	92,438	444,441
'86	213.2	22,735	91	1,963	33,068,172	4,588,351	2,778,382	....	95,031	477,210
'87	232.5	29,467	55	2,140	38,484,936	4,797,893	3,110,882	$166,742	127,065	476,412
'88	252.1	32,655	115	2,286	42,590,042	5,226,398	3,110,882	168,185	143,126	494,778
'89	277.3	36,161	143	2,408	45,071,911	5,822,422	3,435,882	172,692	145,823	543,952
'90	302.5	40,531	94	2,589	47,517,137	6,142,085	3,435,882	172,054	143,092	595,457

Financial.—*Bonded Debt*, $3,135,882; $1,429,382 at 7%; $100,000 at 6%; $250,000 at 5%; $404,000 at 4%; $1,252,560 at 3½% *No Sinking Fund. Expenses:* total, $318,146.

Revenue: consumers, $495,457; city, $100,000. All expenses are covered by revenue except redemption of bonds; bonds, when due, are refunded at a lower rate of interest. **Management.**—Three comrs., for six years, appointed by mayor and approved by council. Chn., Louis P. Reichert. Secy., James Ryan. Supt. and Engr., Louis H. Knapp, appointed by comrs.

II. **CADOSIA,** *Delaware Co.* (Pop., est., 50.) **History.**—Built in '90 by Cadosia Water Co.; 90-years franchise. Engr. and Contr., G. N. Cowan, Stamford. **Supply.**—Spring Brook, by gravity. **Reservoir,**—Cap., 70,000. **Distribution.**—*Mains,* 6 and 4-in. c. i., 1 mile. *Taps,* 6. No *Meters* or *Hydrants.* *Pressure,* 38 lbs. **Financial.**—*Cost,* $13 000. *Bonded Debt,* $5,000 at 6%. *Cap. Stock:* authorized, $5,000; all paid-up. *Operating expenses,* $50. *Revenue:* consumers, $930; city, none. **Management.**—Prest., G. N. Cowan. Secy. and Treas., Jessie B. Cowan. Both of Stamford. Supt., C. H. May.

Caldwell, *Warren Co.* Lake George P. O., which see.

22. **CAMBRIDGE,** *Washington Co.* (Pop., 1,483—1,598.) Built in '86 by Cambridge Water-Works Co.; under 20 years franchise; town neither controls rates nor has right to purchase works. **Supply.**—Springs, pumping direct from collecting reservoir. Fiscal year closes April 1. **Distribution.**—*Mains,* 6.5 miles. *Taps,* 90. *Meters,* 1. *Hydrants,* 33. *Consumption,* unk'n. **Financial**—*Cost,* $45,000. *Cap. Stock:* authorized, $40,000; all paid-up; paid no div'd in 1890. *Total Debt,* $28,900; floating, $1,500; bonded, $26,500 at 5%, due in 1906. *Expenses:* operating, $253; int., $1,430; taxes, $226; total, $1,909. *Revenue:* consumers, $1,600; city, $1,350; total, $2,950. **Management.**—Prest., Wm. McKie. Secy. and Treas., C. Hawley. Supt., Jno. Larmon.

23. **CAMDEN,** *Oneida Co.* (Pop., 1,598—1,902.) Built in '86 by village. **Supply.**—Small stream, by gravity. Fiscal year closes Dec. 10. **Distribution.**—*Mains,* 6 8 miles. *Taps,* 231. *Meters,* 0. *Hydrants,* 40. **Financial.**—*Cost,* $40,000 *Bonded Debt,* $40,000 at 3½%, due in annual payments of $2,000, after '96. *Sinking Fund,* $3,122, deposited in savings bank at 4%. *Expenses:* Operating, $122; int., $1,300; total, $1,422. *Revenue:* Consumers, $1,440, city, none. **Management.**—Chn., A. C. Woodruff. Secy., Fred S. Gamble. Supt., Thos. Park.

24. **CANAJOHARIE,** *Montgomery Co.* (Pop., 2,013—2,089.)

CANAJOHARIE WATER WORKS CO.

Built in '52. **Supply**—Springs, by gravity and pumping to tank. In '88 franchise and property were sold in foreclosure proceedings to R. Spraker and Wm. Hatter. (See below.)

COLD SPRING WATER CO.

Built in '81. **Supply.**—Springs, by gravity for domestic use, ponds, by gravity, for fire protection. In '89, franchise and property were sold to R. Spraker, Wm. Hatter, and others, who organized under name of Canajoharie Consolidated Water Co., described below.

CANJOHARIE CONSOLIDATED WATER CO.

Organized in '89; see above. **Supply.**—For domestic use, springs collected in catchment basin, by gravity and pumping to tank for high service; for fire protection, pond by gravity, with spring, pumping to tank for high service. Catchment Basin.—Cap., 100 000 galls ; 353 × 6 × 6 ft.; 89 ft. of length is tunnelled into mountain. **Pumping Machinery.**—There are two stations, in one of which is a pump made by Rumsey & Co., in other pump made by Knowlson & Kelley. **Reservoir.**—Cap., 260,500 galls ; 50 × 50 × 14 ft.; filled by overflow from tank. **Distribution.**—*Mains,* 3.5 miles. *Services,* lead, *Taps,* 121. *Meters,* 0. *Hydrants,* 17. *Consumption,* 577.592 galls. *Pressure,* 50 to 60 lbs. **Financial**—*Cap. Stock:* authorized, $25,000 ; all paid up ; paid 2% div'd in 1890. *No debt :* expenses : operating, $384 ; taxes, $25; total, $409. *Revenue:* consumers, $1,712 , town, $400 ; total, $2,112. **Management.**—Prest., Jas. Arkell. Vice Prest., Wm Hatter. Secy. and Supt., Randolph Spraker.

25. **CANANDAIGUA,** *Ontario Co.* (Pop., 5,726—5,858) Built in '84 by Co. **Supply.**—Lake Canandaigua, pumping to tank. **Distribution.**—*Mains,* 14 miles.

FOR FURTHER DESCRIPTIVE MATTER see Manual for 1889-90. CHANGES since 1890 are here given. POPULATIONS are for 1880 and 1890 respectively.

Taps, 425. *Meters*, 6. *Hydrants*, 100. *Consumption*, 600,000 galls. **Financial.**—*Cost*, $185,000. *Debt*, $161,000 at 6%. *Expenses:* operating, $5,100; int., $9,660; total, $14,720. *Revenue:* consumers, $17,500; city, $4.600; total, $22,100. **Management.**—Prest. and Treas., F. B. Merrill. Secy., H. H. Lane.

26. CANASTOTA, *Madison Co.* (Pop., 1,569—2,774.) Built in '85 by village. Supply.—Springs, by gravity. **Distribution.**—*Mains,* 10 miles. *Taps,* 250. *Meters* 0. *Hydrants,* 54. *Consumption,* unk'n. **Financial.**—*Cost,* $70,000. *Bonded Debt,* $70,000 at 4%. *Expenses:* operating, $500; int., $2,800; total, $3,300. *Revenue,* consumers, $1,800. **Management.**—Supt., Decatur Gee.

27. CANISTEO, *Steuben Co.* (Pop., 1,907--2,071.) Built in '87-8 by Canisteo Water-Works Co.; franchise provides that city may purchase works in '97 and every 10 years thereafter, price decided by arbitration; rates are fixed by franchise. Supply.—Springs, by gravity. Fiscal year closes Dec. 31. **Distribution.**—*Mains,* 6.3 miles. *Taps,* 137. *Meters,* 0. *Hydrants,* 51. *Consumption,* unknown. **Financial.**—*Cap. Stock* authorized, $25,000, all paid up; paid no div'd in 1890. *Bonded Debt,* $30,000 at 6%, due 1902-7. **Management.**—Prest. and Supt., O. O. Lane. Secy. and Treas., G. R. Harlow.

28. CANTON, *St. Lawrence Co.* (Pop., 2,049—2,580.) Built in '89 by village. Supply.—De Grasse River, pumping to stand-pipe. Fiscal year closes Nov. 20. **Distribution**—*Mains,* 5.6 miles. *Taps,* 123. *Hydrants,* 50. *Consumption,* 75,000 galls. **Financial.**—*Cost,* $40,000. *Bonded Debt,* $40,000 at 3½%. *Expenses:* operating, $1,200; int., $1,400; total, $2,600. *Revenue,* consumers, $1,475; city, none. **Management.**—Five comrs., appointed by trustees for one year. Chn., J. H. Rushton. Secy., E. L. Heaton. Supt., E. S. Lewis, appointed by comrs.

29. CASTILE, *Wyoming Co.* (Pop., 965—1,146.) Built in '88-9 by Castile Water-Works Co.; franchise does not regulate rates, but provides that village may purchase works at any time. Supplies Silver Springs. Supply –Springs, by gravity. Fiscal year closes Dec. 1. **Distribution.**—*Mains,* 7.5 miles. *Taps,* 80. *Meters,* 0. *Hydrants,* 30. *Consumption,* unk'n. **Financial.**—*Cost,* $36,000. *Capital Stock,* authorized, $15,000, all paid up; paid no dividend in 1890. *Bonded Debt,* $20,000, at 6%. *Sinking Fund,* $300. *Expenses:* operating, $300; int., $1,200; taxes, $100; total, $1,600. *Revenue:* consumers, $650; city, $1,200; other purposes, $100; total, $1,950. **Management.**—Prest., Miles Ayrault, Tonawanda. Secy. and Sup*., M. N. Cole. Treas., Geo. H. Bush.

30. CATSKILL, *Greene Co.* (Pop., 4,320—4,920.) Built in '84 by village. Supply.—Hudson river, pumping to reservoir. Fiscal year closes April 15. **Distribution.**—*Mains,* 10.3 miles. *Taps,* 500. *Meters,* 9. *Hydrants,* 102. *Consumption,* 536,000 galls. **Financial.**—*Bonded Debt,* $154,000 at 4%. *Revenue:* consumers, $5,349. **Management.**—Chn., Percy Golden. Secy., N. S. Shaler. Supt., Wm. Comfort.

31. CATTARAUGUS, *Cattaraugus Co.* (Pop., 706—878.) **History.**—Built in '90 by Cattaraugus Water Co.; put in operation Dec. 1. Engr., G. N. Cowan. Contrs., G. N. & W. H. Cowan, both of Stamford. **Supply.**—Spring Brook and springs, by gravity. **Reservoir.**—Cap., 1,000,000 galls. **Distribution.**—*Mains,* 8-in. c. i. down, 4½ miles. *Services,* lead and galv. i. *Taps.* 142. *Meters,* 1. *Hydrants,* Holyoke, 20. *Valves,* Kennedy. *Pressure,* 40 to 100 lbs. **Financial.**—*Cost,* $43,000. *Capital Stock,* $25,000, all paid up. Bonded debt, $18,000 at 6%. Hydrant rental, $22.50. **Management.**—Prest., G. N. Cowan. Secy., Jessie B. Cowan. Treas., W. H. Cowan. All of Stamford. Supt., S. Burger. Report, Nov. 20, '90. NOTE.—Second co. incorporated in '91. See Projected Works.

32. CAZENOVIA, *Madison Co.* (Pop., 1,918—1,987.) **History.**—Built in '90 by village; put in operation Oct. 1. Fiscal year closes May 1, '91. Engr., C. W. Knight, Rome. Contrs., Moore & Shaler, Newburg. Supply.–Springs, by gravity. **Reservoir.**—Cap., 2,000,000 galls. **Distribution.**—*Mains,* c. i., about 5 miles., from Charles Millar & Sons, Utica. *Taps.* 16. *Hydrants,* 35. *Pressure,* av., 60 lbs. **Financial.**—*Cost,* $33,000. *Bonded Debt,* $33,000 at 3½%. *Expenses:* operating, including cost of making house connections, $322, int., $938. *Revenue:* consumers, $71; making connections, $240; total, $322. **Management.**—Four comrs. Prest., E. C. Bass. Secy, B. H. Stanley. Clerk, J. W. Hall.

33. CHATEAUGAY, *Franklin Co.* (Pop., 680—1,172.) Built in '80 by Chateaugay Water-Works Co.; franchise provides neither for control of rates nor purchase

NEW YORK.

of works by town. Supply.—Spring, by gravity. Fiscal year closes April 1. Distribution.—*Mains*, 1.5 miles. *Taps*, 150. *Meters*, 6. *Hydrants*, 25. Financial.—*Cap. Stock*. $10,000. *No debt. Expenses:* operating, $100; taxes, $138; total, $238. *Revenue*, city, $500. Management.—Prest., H. D. Farnsworth. Secy., George O. Bentley. Treas., A. Johnston. Supt., Barney Haney.

31. CHATHAM, *Columbia Co.* (Pop., 1,765–1,912.) Built in '76 by Chatham Water Co.; city neither controls rates nor has right to purchase works. Supply.—Wm. D. Andrews & Bros.'s system of gang-wells, pumping to reservoir. Distribution.—*Mains*, 7 miles. *Taps*, 125. *Hydrants*, 36. *Consumption*, 281,000 galls. Financial.—*Cap. Stock*, $100,000. Management.—Prest., H. W. McClellan. Secy. and Supt., Philo B Blinn.

35. CLINTON, *Oneida Co.* (Pop., 1,236–1,269.) Built in '84 by village. Supply.—Miller brook, by gravity from impounding reservoir. Fiscal year closes May 1.

Year.	Mains.	Taps.	Hyd'ts.	Cost.	B'd'd d't.	Int. ch'g's.	Exp.	Rev.
'85-6	5	72	40	$37,257	$35,000	$1,750	$854	$1,479
'89-90	5	93	40	37,455	30,000	1,750	328	1,016
'90-1	5	103	40	37,455	27,000	1,500	154	1,107

Financial.—*Bonded Debt*, $27,000 at 5%. *Expenses:* total, $1,954, *Revenue:* consumers, $1,107. Management.—Prest., David Anderson. Secy. and Treas., T. T. Thompson.

36 CLYDE, *Wayne Co.* (Pop., 2.826–2,633.) Built in '88-9 by Clyde Water Supply Co. Supply.—(Originally) springs, pumping to stand-pipe; in '91 suction pipe laid to river for new or additional supply. Fiscal year closes Dec. 31. Distribution.—*Mains*, 5.5 miles. *Taps*. 70. *Meters*, 38 *Hydrants*, 40. *Consumption*. 40,000 galls. Financial.—*Cap. Stock:* authorized, $50,000; all paid up. *Bonded Debt*, $50,000 at 6%. *Revenue:* city, $1,600. Management.—Prest., E. M. Harrington. Secy., E. M. Bassett. Both of Buffalo. Supt., A. M. Van Buskirk.

37. COBLESKILL, *Schoharie Co.* (Pop., 1,222–1,822.) Built in '86-7 by village, and operated by Cobleskill Water Co. Supply.—Springs and surface water, by gravity from impounding reservoir; filtered. Fiscal year closes Aug. 1. Distribution.—*Mains*, 5 miles. *Hydrants*, 45. Financial.—*Cost*. $61,043. *Bonded debt*, $60,000. *Expenses:* Operating, $254; int., $2,031; total, $2,285. *Revenue*, $2,408. Management.—Seven Comrs. Prest., M. D. Borst. Secy., G. D. Harder. Treas., DeWitt C. Dow.

38. COHOES, *Albany Co.* (Pop., 19,416–22,509.) Owned by village. Built in '58 by Cohoes Mfg. Co., bought by city in '57 and extended. Supp'y.—Mohawk River, pumping to reservoir. Fiscal year closes March 1. Distribution.—*Mains*, 27 miles. *Taps*, 1,500. *Meters*, 70. *Hydrants*, 236. *Consumption*, 3,000,000 galls. Financial.—*Cost*, unk'n. *Bonded Debt*, $269,000; $130,000 at 7%; $139,000 at 4%. *No Sinking Fund. Expenses:* operating, $13,147; int., $12,385; total, $25,532. *Revenue:* consumers, $33,953; city, none. Management.—Six comrs, one appointed every year by Common Council. Prest., Wm. E. Thorn. Supt., A. T. Kniffin, elected by comrs.

39. COLLEGE POINT, *Queens Co.* (Pop., 4,182–6,127.) Built in '74-5 by village. Supply.—Springs, pumping direct from impounding reservoir.

Yr.	Mains.	Taps.	Meters.	Hydrants.	Consumption.	Cost.	Bonded debt.	Int. ch'g's.	Exp.	Rev.*
'80	13.8	410	...	54	364,562	......	......	$5,000	......	......
'87	13	600	100	60	475,000	$213,000	$213,000	$14,910	5,000	$12,000
'88	15	629	130	60	450,000	213,000	213,000	14,910	5,000	15,000

Financial.—Int., 7%. No allowance by village for fire protection. Management.—Supt., Chas. Fuchs.

74. CONEY ISLAND, *Kings Co.* (Pop., 1,184–3,313.) See Gravesend; also '89-90 Manual.

40. COOPERSTOWN, *Otsego Co.* ((Pop., 2,199–2,657.) Built in '35 by Cooperstown Aqueduct Co. In '76 small gravity works were built by Ed. Clark. Supply.—Otsego lake, pumping direct. Franchise does not regulate rates, but provides that town may purchase works. In July, '91, town voted adversely on proposed purchase of co.'s works. Distribution.—(In '88.) *Mains*, 5.5 miles. *Taps*, 280. *No meters.*

FOR FURTHER DESCRIPTIVE MATTER see Manual for 1889 90. CHANGES since 1890 are here given. POPULATIONS are for 1880 and 1890, respectively.

Hydrants, 33. *Consumption*, 250,000. **Financial.**—*Cost* (in '88), $17,000. *Debt* (in '88), $25,000 at 6%. *Revenue:* consumers (in '88), $4,000; city (in '90), $900. **Management.** —Prest., E. M. Harris. Chf. Engr., J. Warren Lamb.

41. CORNING, *Steuben Co.* (Pop., 4,802—8,550.) Built in '70 by town; leased for 30 yrs. from '78 by Heermans & Lawrence; rates fixed in lease. **Supply.**—Creek and springs, pumping direct, and to reservoir; filtered through sand and gravel. Fiscal year closes Feb. 1. **Distribution.**—*Mains*, 7 miles. *Taps*, 600. *Meters*, 75. *Hydrants*, 70. *Consumption*, 400,000. **Financial.**—*Cost*, $75,000 *No Debt*. *Hydrant rental*, free for use of plant. *Total Debt*, none. **Management**—Lessees, Heermans & Lawrence. Supt., H. C Heermans.

42. CORTLAND, *Cortland Co.* (Pop., 4,055—3,590.) Built in '84 by Cortland Water-Works Co ; controlled by Moffett, Hodgkins & Clarke, Syracuse; maximum rates fixed in franchise, which provides for purchase of works by city, price determined by an appraiser. **Supply.**—Springs, pumping to tank. Fiscal year closes Dec. 31. **Distribution.**—*Mains*, 16 miles. *Taps*, 800. *Meters*, 126. *Hydran's*, 125. *Consumption*, 375,000 galls. **Financial.**—*Cost*, $150,000, *Bonded Debt*, $130,000 at 5%, due in 1908. *Expenses:* Operating, $6,000; int., $6,500; total, $12,500. *Revenue:* Total, $12,500. **Management.**—Prest. and Supt., B. F. Taylor. Secy., H. C. Hodgkins, Syracuse.

43. CROTON, *Delaware Co.* (Pop., ?23—est. 400.) **History.**—Built in '85 by Croton Water-Works Co.; put in operation Oct. 1. No engr. **Supply.**—Springs, by gravity. *Reservoir*—Cap., about 30,000 galls. **Distribution.**—*Mains*, 4 in., 4 miles. *Taps*, 50. *Meters*, none. *Hydrants*, 5. *Pressure*, 72 lbs. **Financial.**—*Cos* , $6,000. *Debt*, none. *Capital Stock*, $6,000, all paid up. *Operating Expenses*, nominal. **Management.**—Prest., C. H. Treadwell. Secy., L. Ogden. Supt., L. Saunders.

44. CUBA, *Allegany Co.* (Pop., 1,251—1,386.) Built in '82 by village. **Supply.**—Springs, by gravity, through distribution to reservoir. Fiscal year closes Dec. 1. During '90 supply was increased by extending 2,200 ft. of 3-in. c. ', pipe to another spring. **Distribution.**—*Mains*, 8.9 miles. *Taps*, 128. *Meters*, 0. *Hydrants*, 39. *Consumption*, 100,000 galls. **Financial.**—*Cost*, $42,892. *Bonded Debt* $33,000 at 3½%. *Expenses*, not determined. *Revenue:* consumers, $2,025; city, none. Works had not been in operation a full year at time of report. **Management.**—Village Prest., I. N. Sheldon. Secy , W. J. Penney. Supt., S. C. Bradford, appointed by Village Trustees.

Dansville, *Livingston Co.* (Pop., 3,625—3,758.) Built in '74 by village for fire protection.

45. DELHI, *Delaware Co.* (Pop., 1,381—1,564.) Built in '72 by Delhi Water Co.; franchise provides neither for control of rates nor purchase of works by town. **Supply.**—Surface water, by gravity from impounding reservoir on brook. **Distribution.**—*Mains*, 3.5 miles. *Taps*, 200. *Meters*, 0. *Hydrants*, 42. *Consumption*, unk'n. **Financial.**—*Cost*, $26,000. *No Debt*. *Expenses*, $250. *Revenue:* consumers, $2,000; city, none. **Management.**—Prest., Geo. E. Marvin. Secy., D. G. Arbuckle. Supt., Jas. Middlemist.

46. DEPOSIT, *Broome Co.* (Pop., 1,419—1,530.) Built in '85 by Deposit Water Co.; 10-years franchise. **Supply.**—Butler brook, by gravity. Fiscal year closes Sept. 30. **Distribution.**—*Mains*, 6 miles. *Taps*, 100. *Meters*, 0. *Hydrants*, 34. *Consumption*, unk'n. **Financial.**—*Cost*, $30,000. *Cap. Stock:* authorized, $25,000; all paid up; paid 8% div'd in 1890. *No Debt*. *Expenses*: operating, $213; taxes, $226; total, $439 *Revenue:* consumers, $1,272; city, $600; other purposes, $350; total, $2,222. **Management.**—Prest. and Supt., Jno. B. Perry. Secy., Chas. K. Brown. Treas., Jas. G. Minor.

125. **DOBBS FERRY,** *Westchester Co.* (Pop., in '90, 2,083.) See North Tarrytown.

47. DUNKIRK, *Chautauqua Co.* (Pop., 7,248—9,416.) Built in '71 by village. **Supply.**—Lake Erie, pumping direct. Fiscal year closes March 1.

Year	Mains.	Taps.	Hy'dts.	Cons'pt'n.	Cost.	B'd'd d't.	Int. ch'g's.	Exp.	Rev.
'80	14	510	96	900,000	$110,000	$100,000	$7,000	$7,300	$5,600
'86	168	722	103	1,400,000	137,091	105,000	7,350	7,000	9,300
'88	20	955	112	1,500,000	150,000	101,000	7,050	8,200	8,860
'89	22	1,004	116	1,149,516	150,000	100,000	7,000	8,359	9,059
'90	23	1,085	119	1,805,311	152,849	100,000	7,000	10,896	10,092

Financial.—*Bonded Debt,* $100,000 at 7%. *No Sinking Fund. Expenses:* total, $17,896. *Revenue:* consumers, $10,092. No allowance is made for water for public uses, but when net revenue does not pay int. on bonds, balance is made up by taxation. **Management.**—Eight comrs., appointed for life except on resignation or removal. Chn., M. L. Hinman. Secy., John Madigan. Supt., Geo. M. Abell. appointed by comrs.

75. **EAST ALBANY,** *Rensselaer Co.* See Greenbush.

48. **EAST AURORA,** *Erie Co.* (Pop., 1,100—1,582.) **History.**—Built in '89-90 by town. Contr., J. C. Shuttleworth, Springville. **Supply.**—Springs, direct pumping. **Pumping Machinery.**—Dy. cap, 750,0·0 galls. **Distribution.**—*Mains,* 8 to 4 ins., about 7 miles. *Hydrants,* about 50. **Financial.**—Contract price, $34,996. **Management.**—Village Prest., F. R. Whaley.

49. **EAST RANDOLPH,** *Cattaraugus Co.* (Pop., 286—est., 1,000.) Built in '89 by village; was to be in operation Aug. 10. Des. Engr., E. A. Curtis, Fredonia. Const. Engr. S. J. Benedict. Contrs., Renn & Chapman, Castile. **Supply.**—Springs, by gravity to and from reservoir. *Reservo r.*—Cap., about 30,000 galls. *Distribution.*—*Mains,* c. i., 4 miles. *Hydrants,* Eddy, 16. *Pressure,* heavy. **Financial.**—*Cost,* $11,500. Bonded debt, $14,000 at 4%. **Management.**—Prest., M. V. Benson. Secy., A. H. Helms. Supt., F. J. Benedict.

50. **ELIZABETHTOWN,** *Essex Co.* (Pop., 445—573.) Built in '63 by Elizabethtown Water Co.; franchise provides neither for control of rates nor purchase of works by village. **Supply.**—Springs, by gravity. **Distribution.**—*Mains,* 4 miles. *Taps* 80. *Meters,* 0. *Hydrants,* 20. **Financial.**—*Cost,* $18,500. *Cap. Stock,* $8,000. *Bonded Debt,* $7,000 at 6%. *Expenses:* operating, $150; int., $420; total, $570. *Revenue,* city, $150. **Management.**—Prest., R. Hand. Secy. and Supt., R. C. Kellogg.

51. **ELLENVILLE,** *Ulster Co.* (Pop. 2,75)—2,881.) Built in '71 by city. **Supply.**—Surface water, by gravity from impounding reservoir. Fiscal year closes Feb. 12.

Year.	Mains.	Taps.	Meters.	Hyd'ts.	B'd'd d't.	Int. ch'g's.	Exp.	Rev.
'80	3.7	177	0	40	....	....	$584	$977
'84	4.8	187	0	..	$35 000	$2,450	675	1,182
'89	8	291	0	56	51.000	3,010	1.122	2,164
'90	8	317	0	57	51.000	3,010	871	2.190

Financial.—*First Cost,* $33,996. *Bonded Debt,* $51,000: $31,000 at 7%, due in '91; $5,000 at 7%, due in '92; $16,000 at 3½%, due at option, after '92. *Sinking Fund,* $27,679, invested in mortgage loans and town bonds. *Expenses:* total, $3,881. *Revenue:* consumers, $2,190; city, none. Int. on debt and yearly sinking fund of $1,300 is raised by taxation; remaining expenses paid from rev. **Management.**—Three comrs. appointed by Council for three years. Chn., B. B. Demarest. Secy., J. A. Tice. Supt. D. S. Williams, appointed by comrs.

52. **ELLICOTTSVILLE,** *Cattaraugus Co.* (Pop., 748—552.) **History.**—Built in '90 by Ellicottsville Water Co.; put in operation Dec. 1. Engr., Halsey F. Northrup. Contrs., Ayrault Bros. & Co., Tonawanda. **Supply.**—Spring, by gravity. *Reservoir.*—Cap., about 18.000 galls. **Distribution.**—*Mains,* wood, 5 miles. *Taps,* 60. *Hydrants.* 15. *Pressure,* 80 lbs. **Financial.**—*Cost,* $10,000. *Capital Stock,* authorized, $10,000, all paid up; hydrant rental, $20, 5-years' contract. **Management.**—Prest., C. Perry Vedder. Secy., E. S. King. Treas., C. A. Case. Report Dec. 17, '90.

53. **ELMIRA,** *Chemung Co.* (Pop., 20,541—29,708.) Built in '60-1 by Elmira Water Co. Franchise is perpetual, but provides that city can purchase works at any time; city has no control of rates. **Supply.**—Carr's Creek, filter galleries near Chemung River, and Chemung River, by gravity from storage reservoir on creek to distributing reservoir, and by pumping from filters and river to reservoir. Fiscal year closes Dec. 31. During '90-1 a 6,000,000-gall. Worthing on pump, 16-in. main from gravity supply and 20-in. force main were added.

Year.	Mains.	Taps.	Meters.	Hyd'ts.	Cons'pt'n.	Cost.	Bonded debt.	Int. ch'g's.	Exp.	Rev.
'80	27	772	..	137	....	$342,456	....	....	$9,000	$30,033
'86	34	1,124	227	104	1,500,000	393,167	$129,000	....	12,600	35,300
'89	34.7	1,637	307	231	2,250,000	610,186	315,000	$18,900	13,217	41,556
½'90	38,4	2,016	352	267	2,250,000	647,102	335,000	20,100	14,638	46,000

Financial.—*Cap. Stock:* authorized, $1,000,000; paid up, $200,000; paid 4% dividend in 1889, 3½% dividend in '90. An uncapitalized surplus of $150,000 is applied to construc-

FOR FURTHER DESCRIPTIVE MATTER see Manual for 1889-90. CHANGES since 1890 are here given. POPULATIONS are for 1880 and 1890, respectively.

tion. *Total Debt*, $353,000; floating, $18,000 at 6%; bonded, $335,000 at 6%, due in 1913. *Expenses:* operating, $12,159; interest, $20,615; taxes, $2,479; total, $35,053. *Revenue:* consumers, $35,000; city, $11,000; total, $46,000. **Management.**—Prest., Geo. M. Diven. Secy., Treas. and Supt., J. M. Diven.

54. **FAR ROCKAWAY,** *Queens Co.* (Pop., 2,244—2,288.) Built in '84 by private co. Supplies Rockaway Beach and Arverne-by-the-Sea. **Supply** —About 50 gang wells, pumping to stand-pipe. **Distribution.**—(In '87) *Mains*, 12 miles. *Hydrants*, 80. *Consumption*. 600,000 galls. *Hydrant rental*, $4,000. **Management.**—" Principal Owner." F. E. Dutois. Supt., T. D. Smith.

55. **FISHKILL-ON-THE-HUDSON and MATTEAWAN,** *Dutchess Co.* (Pop., Fishkill-on-the-Hudson, in '90, 3,617; Matteawan, in '90, 638.) Built in '86-7 by Fishkill & Matteawan Water Co. **Supply.**—Surface water, by gravity from impounding reservoir. During '90 a 6,000,000-gall. reservoir was built. **Distribution.**—*Mains*, 430. *Taps*, 430. *Meters*, 5. *Hydrants*, 61. *Consumption*. 700,000 galls. **Financial.**—Withheld. **Management.**—Prest., G. E. Taintor; Secy., G. A. Schreifer; Treas., G. D. L. Hullier, all 11 Wall St., N. Y. City. Supt., H. F. Brinkerhoff.

56. **FLATBUSH,** *Kings Co.* (Pop. of town, 7,634—est., 10,000.) Built in '82 by Flatbush Water-Works Co. **Supply.**—Wells, "Stephens' patent system," pumping to stand-pipe. Fiscal year closes May 31. **Distribution.**—*Mains*, 37.5 miles. *Taps*, 1,166. *Meters*, 15. *Hydrants*, 158. **Financial.**—Withheld. **Management.**—Prest., Jno. Lefferts. Secy., Jno. Z. Lott. Supt., Jeremiah Lott.

57. **FLUSHING,** *Queens Co.* (Pop., 6,683—8,436.) Built in '74 by village. **Supply.**—Seventeen bored wells, and large pond used only for fire purposes, pumping direct. Fiscal year closes Dec. 31.

Year.	Mains	Taps.	Meters.	Hy-drants.	Con-sumption.	Cost.	Bonded debt.	Int. ch'g's.	Exp.	Rev.
'86	16.3	862	4	98	750,000	$200,000	$175,000	$12,250	$1,100	......
'89	24.1	1,168	518	154	505,000	224,202	217,202	13,630	7,023	$16,651
'90	25.1	1,297	675	164	512,000	226,976	219,976	12,650	5,859	18,216

Financial.—*Total Debt*, $223,302: floating, $3,226; bonded, $150,000 at 7%; $13,000 at 5%; $26,976 at 6%. *Expenses:* total, $18,509. *Revenue:* consumers, $18,216; city, $3,025; total, $21,241. **Management.**—Com., appointed from village board, by its president. Chn., E. V. W. Rossiter. Secy., C. B. Smith. Supt., G. A. Roullier, appointed by board.

58. **FONDA,** *Montgomery Co.* (Pop., 944—1,190.) Built in '85 by village. **Supply.**—Springs, by gravity. Fiscal year closes March 1.

Year.	Mains.	Taps.	Meters.	Hy'd'ts.	Cost.	B'd'd debt.	Int. ch'g's	Exp.	Rev.
'86-7	3.5	57	0	32	$26,000	$26,000	$1,300	$450	$600
'90-1	3.5	123	0	31	26,000	18,000	900	250	1,200

Financial.—*Bonded Debt*, $18,000 at 5%, due in annual payments of $2,000. *Expenses:* total, $1,150. *Revenue:* consumers, $1,200; city, none. **Management.**—Chn., J. N. Cole. Secy., J. O. Schuyler. Supt., U. M. Ansman.

111. **FORT EDWARD,** *Washington Co.* (Pop., 2,988—est., 3,200.) Built in '55 by Geo. Taylor, for domestic supply only. **Supply.**—Springs, by gravity. *Cost* (in '88), $10,000. *No Debt.* Address Mrs. Geo. H. Taylor. Supt. A. W. Powell.

74. **FORT HAMILTON,** *Kings Co.* (Pop. in 1890, 2,617.) Supply from Gravesend.

59. **FORT PLAIN,** *Montgomery Co.* (Pop., 2,413—2,864.) Built in '85 by Fort Plain Water-Works Co.; town does not control rates, but has right to purchase works at end of every five years. **Supply.**—Springs, by gravity; filtered. Fiscal year closes March 26. **Distribution.**—*Mains*, 9 miles. *Taps*, 165. *Meters*, 1. *Hydrants*, 52. **Financial.**—*Cap. Stock* (in '88), $10,000. *Bonded Debt* $60,000 at 6%. *Revenue:* consumers, $1,800; city, $2,080; total, $3,880. **Management.**—Prest., J. W. Low, Kingston. Secy. and Treas, Samuel Denison. Supt., Chas. Heineberg.

60. **FRANKLIN,** *Delaware Co.* (Pop., 660—581.) **History.**—Built in '89-90 by village; put in operation in Feb. '90. Des. Engr., W. H. De Chonk, for Kendall Bros., Reading, Pa. **Supply**—Bored well, pumping to reservoir; well is 303 ft., deep. **Pumping Machinery.**—Dy. cap., 40,000 galls.; Knowles *Boiler*: 25 HP. from Shapley & Wells, Binghamton. **Reservoir.**—Cap., 725,000 galls.; 10 × 50 × 60 ft. **Distribution.**—*Mains*, 6 and 4-in. c. i., 2 miles. *Services*, galv. i. and w. i.; pipe, and specials from Mellert Foundry & Machine Co., Reading, Pa.; trenching

and pipe-laying done by J. Vaughn. *Taps,* 51. *Meters,* none. *Hydrants,* Galvin., 18. *Consumption,* 6,000 galls. *Pressure,* 90 lbs. **Financial.**—*Cost,* $20,175. *Debt,* about $20,000; $19,000 bonded at 3½%; balance at 5%. *Operating Expenses,* $100. *Revenue:* Consumers, $500; city, $950. **Management.**—Pre*t.,* S. W.Potter. Secy., Geo. F. Salland.

61. **FRANKLINVILLE,** *Cattaraugus Co.* (Pop., 672—1,021.) **History.**—Built in '90 by village; put in operation Jan. 1. '91. Engr., E. A. Curtis, Fredonia. Contrs., Michigan Pipe Co., Bay City, Mich. **Supply.**—Springs, by gravity to and from reservoir. **Reservoir.**—Can., about 750,000 galls.; earth and puddle embankment. 110 × 111 ft., × 11 ft. deep. **Distribution.**—*Mains,* 8 to 4-in. c. i., 10 miles; from R., D. Wood & Co.. Philadelphia. *Hydrants,* Mathews, 35. *Valves,* Eddy. *Pressure,* 80 lbs. **Financial.**—*Cost,* $33,000. *Bonded Debt,* authorized, $32,000 at 4%, all issued. **Management.**—Four comrs. Prest., A. O. Holmes. Secy., Alfred Spring. Treas. S. A. Spring. Report Jan. 1. '01.

62. **FREDONIA,** *Chautauqua Co.* (Pop., 2,602—3,399.) Built in '84 by village. **Supply.**—Creek, by gravity from impounding reservoir. **Distribution.**—(In '88) *Mains,* 9 miles. *Taps,* 119 *Meters,* 7. *Hydrants,* 50. **Financial.**—(In '88.) *Cost,* $90,000. *Bonded Debt,* $84,000. *Expenses:* operating, $245; int.. $3.360. *Revenue,* $4,000. **Management.**—Comrs. Prest., A. Colburn. NOTE.—Report for '90 stated that there were no records furnishing desired information.

63. **FULTON,** *Oswego Co.* (Pop., 3,911—4,214.) Built in '85 by company. Supplies Oswego Falls. **Supply.**—Springs, pumping to stand-pipe. Fiscal year closes June 15. **Distribution.**—*Mains,* 9. 5 miles. *Taps,* 398. *Meters,* 5. *Hydrants,* 84. **Financial.**—Withheld. **Management.**—Prest., H. E. Nichols. Secy., G. S. Riper. Treas., F. A. Emerick

64. **FULTONVILLE,** *Montgomery Co.* (Pop.. 881—1,122.) Built in '75 by J. H. Starin. **Supply.**—Springs, by gravity to small reservoir, thence by pumping to large reservoirs. Fiscal year closes Dec. 31. **Distribution.**—*Mains,* 2 miles. *Takers,* 90. *Hydrants,* 12. **Financial.**—*Cost,* $20,000. *Expenses,* $50. *Revenue,* $300. **Management.**—Owner, J. H. Starin.

65. **GARDEN CITY,** *Queens Co.* (Pop., 574—est., 1,000.) Built in '76 by A. T. Stewart's estate. **Supply.**—Well, pumping direct. **Distribution.**—*Mains,* 11 miles. *Taps,* 100. *Meters,* 0. *Hydrants,* 35. *Consumption,* 500,000 galls. **Financial.**—Withheld. **Management.**—Supt., A. H. Cunliffe.

66. **GENESEO,** *Livingston Co.* (Pop. 1,925—2,236) Built in '87 by village. **Supply.**—Conesus Lake, pumping to reservoir. Fiscal year closes Dec. 1.

Year.	Mains.	Taps.	Meters.	Hyd'ts.	Cons'n'n.	Cost.	B'd'd d't.	Int.ch'g's.	Exp.	Rev.
'88	10.5	180	1	60	150,000	$75,000	$70,000	$2,450	$2,000	$2,433
'89	10.5	200	0	64	80,000	75,000	70,000	2,450	3,202	2,920
'90	11	220	0	73	90,000	76,335	70,000	2,450	2,238	3,019

Financial.—*Bonded Debt,* $70,000 at 3½%, due in '97-1917. *No Sinking Fund, Expenses:* total, $4,088. *Revenue:* consumers, $3,019; city, $2,450; total, $5,469. Net revenue pays about one-fifth of interest on bonds, in addition to new construction; balance paid by taxation. **Management.**—Three comrs, one elected each year. Prest., W. A. Wadsworth. Secy., T. F. Olmstead. Supt., J. B. Harris, appointed by comrs.

67. **GENEVA,** *Ontario Co.* (Pop., 5,878—7,577.) Built in 1787 by Geneva Water-Works Co.; franchise provides that works may be bought by city, time and place to be determined by a commission; city does not control rates. **Supply.**—Springs, by gravity and pumping to reservoir. Fiscal year closes May 1. **Distribution.**—*Mains,* 14 miles. *Taps,* 350. *Meters,* 32. *Hydrants,* 108. *Consumption,* unk'n. **Financial.**—*Cost,* $50,000. *Cap. Stock,* authorized, $20,000; all paid up. *Bonded debt,* $130,000 at 6%. *Expenses:* operating, $2,585; int., $7,800; taxes, $171; total, $10,853. *Revenue:* consumers, $7,727; city, $4,860; total, $12,587. **Management.**—Prest., S. H. Hammond. Secy., E. Kingsland. Treas., A. L. Chew. Supt., L. Graves•

68. **GLEN'S FALLS,** *Warren Co.* (Pop., 4,500—9,509.) Built in '72 by village. **Supply.**—Springs, by gravity from impounding and distributing reservoirs. Fiscal year closes May 1. **Distribution.**—*Mains,* 24 miles. *Taps,* 1,400. *Meters,* 0. *Hydrants,* 155 *Consumption,* unk'n. **Financial.**—*Cost and Expenses,* unk'n. *Bonded*

FOR FURTHER DESCRIPTIVE MATTER see Manual for 1889-90 CHANGES since 1890 are here given. POPULATIONS are for 1880 and 1890, respectively.

Debt, $100,000 at 7%. *Revenue*, $10,000. **Management.**—Prest., S. D. Kendrick. Clerk, A. Armstrong, Jr,

69. **CLOVERSVILLE**, *Fulton Co*. (Pop., 7,133—13,864.) Built in '78 by village. **Supply.**—Springs and Rice Creek, by gravity from reservoir. Fiscal year closes Feb. 1. **Distribution.**—(In '87) *Mains*, 17 miles. *Taps*, 1,400. *Meters*, 22. *Hydrants*, 102. **Financial.**—*Cost*, $189,830. *Bonded Debt*, $142,500. *Expenses:* operating, $4,012; int., $7,333; total, $11,378. *Revenue*, $19,641. **Management.**—Five comrs., one elected every year. Prest., J. H. Richardson. Secy., Z. B. Whitney. Supt., G. Guiwodda.

70. **GOSHEN**, *Goshen Co*. (Pop., 2,557, est., 3,000.) Built in '71 by village. **Supply.**—Springs, by gravity from impounding reservoir. Fiscal year closes March 13. **Distribution.**—*Mains*, 5.1 miles. *Taps* (in '88), 269. *Meters*, 0 *Hydrants*, 61. **Financial.**—*Cost*, $75,000. *Bonded Debt*, $31,500 ; $6,000 at 7%; $20,500 at 5%, due in annual payments of $1,000; $5,000 at 4%. *Expenses:* operating. $30; int., $1,625; total, $1,655. *Revenue:* consumers, $1,900; city, none. **Management.**—Trustees, elected every two years. Water Comr., E. Dikeman. Report by Village Prest., H. W. Nanny.

71. **GOUVERNEUR**, *Saint Lawrence Co*. (Pop., 2,071—3,153.) **History.**—Present works owned by village. In '67 Gouverneur Water-Works Co. built works. In '90 village built works, putting them in operation Jan 1, '91 ; village bought 2.16 miles of co.'s 4-in. iron pipe and 12 hydrants, paying same price as paid contractor for new pipe, laid. About 2 miles of co.'s wood and cement pipe and 12 hydrants were abandoned. Engr., F. S. Pecke, Watertown. Contr., L. J. Richardson, Cortland. **Supply.**—Oswegatchie River, pumping to stand-pipe. There is a 12-in. intake pipe, 283 ft. long, and two river crossings, one 8-in., 317 ft. long, one 6-in, 235 ft. long, all of spiral weld steel; 6-in river crossing on bridge, 8-in. on river bottom. **Pumping Machinery.**—Dy. cap., about 1,150,000 galls. ; two comp. dup. non-cond. Doane, with 12-in. h. p., 18¼-in. l. p. cylinders, 11-in. plungers and 18-in. stroke. *Boilers:* two 5 × 14-ft.; built by Phœnix Foundry & Machine Co., Syracuse. **Force Main.**—Of 10-in. spiral weld steel pipe, 2,700 ft. long. *Stand-pipe.*—Cap., 210,000 galls ; 30 × 40 ft.; from Porter Mfg. Co., Syracuse. **Distribution.**—*Mains*, 7.36 miles, including hydrant branches; from Jackson & Woodin Mfg. Co., Berwick, Pa. *Hydrants:* Ludlow, 26. *Valves*, Galvin. **Financial.**—*Cost*, $60,000. *Bonded Debt*, $30,000 at 3½%. **Management.**—Comrs. Prest., L. J. Whitney. Secy. and Treas., H. Sudds. Supt. A. L. Althouse.

NOTE.—For further details see ENGINEERING NEWS, vol. XXV., p. 99; also, for illustrations, *Fire and Water*, June 6, '91.

72. **GOWANDA**, *Erie and Cattaraugus Co*. (Pop., 1,243—est., 2,000.) Built in '87 by Gowanda Water-Works Co.; city neither controls rates nor has right to purchase works. **Supply.**—Springs, by gravity. Fiscal year closes Oct. 1. During summer of '90 supply was increased by addition of several springs at cost of $1.500. **Distribution.**—*Mains*, 3.5 miles. *Taps*, 100. *Meter*, 1. *Hydrants*, 17. *Consumption*, 100.000 galls. **Financial.**—*Cap. Stock:* authorized, $20,000; all paid-up; paid 10% dividend in 1890. *Floating Debt*, $500 at 7%. *No Bonds. Expenses:* operating $125; int., $12; taxes, $15; total, $152. *Revenue:* consumers, $1,500; city, $680; total, $2,480. **Management.**—Prest., W. W. Welch. Secy. and Supt., Fred. J. Blackman. Treas., Wm. H. Bard.

73. **GRANVILLE**, *Washington Co*. (Pop., 1,071—est., 1,700.) Built in '81 by village. **Supply.**—Mettawee River, pumping from storage reservoir direct and to distributing reservoir. **Distribution.**—*Mains* (in '88), 4 miles. *Taps* (in '88), 100. *No Meters. Hydrants* (in '88), 22. **Financial.**—(In 88.) *Cost*, $22,000. *Bonded Debt*, $20,000 at 4%. *No Sinking Fund. Operating Expenses*, $10,000; int., $800. *Revenue:* consumers, $1,000. **Management.**—Comr., E. R. Norton, O. L. Goodrich, J. Marvin Gray. Supt., O. L. Goodrich.

74. **GRAVESEND**, *Kings Co*. (Pop. supplied varies from 10,000—100,000 Now owned by Kings County Water Supply; original works built in '80 by Coney Island Water Works Co.; came into hands of above Co. in '85 and extended into township of New Utrecht; also supplies Coney Island, Unionville, Bath Beach, Fort Hamiton, Bay Ridge and Sheepshead Bay. Franchise provides neither for control of rates nor purchase by town. **Supply.**—Open wells, "Stephens' Patent System," pumping to stand-pipe and direct. **Distribution.**—*Mains* (in '87), 37 miles. *Taps* (in '87), 803. *Hydrants*, 27. **Financial.**—(In '87.) *Cost*, $300,000. *Bonded Debt*, $300,-

NEW YORK.

000 at 6%, *Hydrant rental*, $40 each. **Management.**—Prest., Jno. L. McKane. Secy., J. L. Voorhees.

75. GREENBUSH, *Rensselaer Co.* (Pop., 3,295—7,301.) Built in '87 by Greenbush Water Works Co.; controlled by Moffett, Hodgkins & Clarke, Syracuse. Franchise provides neither for control of rates nor purchase of works by village. Greenbush, East Albany and Bath-on-Hudson are one municipality and are all supplied by these works. **Supply.**—Well and Hudson River, pumping to stand-pipe. **Distribution.**—(In '88.) *Mains*, 10 miles. *Taps*, 382. *Meters*, 9. *Hydrants* (in '91), 83. *Consumption*, 323,466 galls. **Financial.**—(In '88.) *Cost*, $200,000. *Bonded Debt*, $155,000 at 6%. *Operating Expenses*, $3,600. *Revenue:* consumers, $8,400; city; fire protection, $4,800; total, $13,200. **Management.**—Prest., J F. Moffett. Supt., Jno. Winn.

76. GREEN ISLAND, *Albany Co.* (Pop., 4,160—4,403.) Present works built by Green Island Water Co. in '88 under 10-years' franchise; controlled by Moffet, Hodgkins & Clarke, Syracuse, First supplied in '78 by extension of West Troy works. **Supply.**—Hudson River, pumping to stand-pipe and direct; filtered. **Distribution** —(In '88.) *Mains*, 3 miles. *Taps*, 210. *Meters*, 3. *Hydrants*, 53. **Financial.** -(In '88.) *Cost*, $125,000. *Debt*, $78,000 at 6%. *Expenses:* (operating), $5,000; interest, about $4,680. *Revenue:* consumers, $8,000; city, $3,000; total, $11,000. **Management.**--(In '88.) Prest., J. V. Clarke. Supt., C. C. Hill.

IV. GREENPORT, *Suffolk Co.* (Pop , 2370--est., 2.500.) Built in '89 by Greenport City Water Co. **Supply.**—Four driven wells, pumping to reservoir. Fiscal year closes Nov. 1. **Distribution**—*Mains*, 5 miles. *Taps*, 50. *Meters*, 0. *Hydrants*, 0. **Financial.**—*Cost*, $60,000. *Cap. Stock:* authorized, $40 000: all paid up. *Bonded Debt:* $40,000 at 6%, due in 1909. **Management.**—Prest., C. N. Hoagland, 410 Clinton Ave.; Secy . Chas O. Gates, 100 Greene Ave., Treas., Geo. P. Tangeman, 274 Berkeley Place. All of Brooklyn.

77. GREENWICH, *Washington Co.* (Pop. 1,231—1.663) Built in '87 by Co. **Supply.**—Springs by gravity from reservoir. **Distribution.**—*Mains*, 6 miles. *Taps*, 140. *Meters*, 0. *Hydrants*. 25. *Consumption*. unk'n. *Financial.*—*Cost*, $38,000. *Cap. Stock*, $30,000. *Floating Debt*, $3,000 . *No Bonds. Revenue:* city. $1,050. **Management.**—Prest., Henry Gray. Treas., Leroy Thompson. Secy. and Supt., R. J. Wait.

78. GROTON, *Tompkins Co.* (Pop., 913—1,280.) Built in '88 by village. **Supply.** —Spring, by gravity. Fiscal year closes May 1. **Distribution.**—*Mains*, 5.5 miles. *Taps*, 69. *Meters*. 0. *Hydrants*, 45. *Consumption*. unk'n. **Financial.**—*Cost*, $23,000. *Bonded Debt*, $23,000 at 3½%. *Sinking Fund*, $1,200 at 3½%. **Management.** —Five comrs.. elected every three years. Prest., Ben. Conger. Secy, and Supt., Wm. D. Baldwin.

79. HAMBURGH, *Erie Co.* (Pop., 758—1,331.) Built in '89 by Hamburgh Water & Electric Light Co. **Supply**—Surface well, pumping to stand-pipe. Fiscal year closes July 1. **Distribution.**—*Mains*, 4.5 miles. *Taps*, 125. *Meters*, 0. *Hydrants*, 28 *Consumption*, 190,000 galls. **Financial.**—*Cost*, $17,032, for well and pipes. *Cap. Stock:* authorized, $50,000; paid up, $25,000'. *Debt*, $21,050 at 6%. *Expenses:* operating, $575; int., $1,260; taxes, $50; total, $1,885. *Revenue:* consumers. $1,200; city, $1,060; total, $2,260. **Management.**—Prest., F. L. Bunting. Secy. and Treas., H. S. Spencer. Supt., Jacob Peffer.

80 HANCOCK, *Delaware Co.* (Pop. of township, 686 1,279.) Built in '88 by Hancock Water Co., under 99-years franchise: town neither controls rates nor has right to purchase works. **Supply.**—Springs, by gravity from reservoir. Fiscal year closes Aug. 20. **Distribution.**—*Mains*, 2 miles. *Taps*. 50. *Meters*, 0. *Hydrants*, 21. **Financial.**—*Cost*, $21,000 *Cap. Stock:* authorized, $12,000; all paid up; no div'd in '90. *Total Debt*, $8,700: floating, $700; bonded, $8,000 at 6%, due in 1904. *Exp uses:* operating, $50; int , $480; taxes, $43; total, $573. *Revenue:* consumers, $425; city, $375; other purposes, $55; total, $855. **Management.**—Prest., Thos. Kerry. Supt., S. Wheeler.

125. HASTINGS-UPON-HUDSON, *Westchester Co.* (Pop., 1,290—1.466.) Supply from No. Tarrytown.

81. HAVERSTRAW, *Rockland Co.* (Pop., 3,506—5,170.) Built in '86-7 by Haverstraw Water Co. Franchise does not fix rates, but provides for purchase of **works**

FOR FURTHER DESCRIPTIVE MATTER see Manual for 1889 90. CHANGES since 1890 are here given. POPULATIONS are for 1880 and 1890, respectively.

76   MANUAL OF AMERICAN WATER-WORKS.

by town in 1902, at cost of construction. **Supply.**—Dorman stream and Minisceongo Creek, by gravity from impounding reservoir. Fiscal year closes May 1. In '90 co. leased works of West Haverstraw Water Co., guaranteeing int. on their bonded debt of $67,000. **Distribution.**—*Mains*, 16 miles. *Taps*, 400. *Hydrants*. 65. *Consumption*, 900,000 galls **Financial.**—*Cost*, $206,000. *Cap. Stock:* authorized, $100,000; all paid up. *Bonded Debt*, $40,00 at 6%. *Expenses:* operating, $1,500; int. $2,400; taxes, $400; total, $4,300. *Revenue:* consumers, $10,000; city, $2,000; other purposes, $1,000; total, $13,000. **Management.**—Prest., Jno. Lockwood; Treas., Jno. C. Lockwood; both of 52 Broadway, N. Y. City. Supt., D. D. Williams.

82. **HEMPSTEAD,** *Queens Co.* (Pop., 2,521—4,831.) **History.**—Built in '90 by Hempstead Water Co.; put in operation Jan. 1, '91. Engrs. and Contrs., John Lockwood & Son, 52 Broadway, N. Y. City. **Supply.**—Wells, pumping to standpipe. **Pumping Machinery.**—Cap., 750,000 galls.; Gaskill. **Stand-Pipe.**—Cap. 258,000 galls.; from Cunningham Iron Works, Boston. **Distribution.**—*Mains* 10 to 4-in. c. i., 5 miles; from M. J. Drummond, N. Y. City. *Services*, galv. i.—**Financial.**—*Cost*, $100,000. *Bonded Debt*, authorized, $40,000 at 6%, all issued. *Cap Stock:* authorized, $60,000. *Revenue:* from city, $1,200. **Management.**—Prest., John Lockwood. Treas., J. C. Lockwood, all of 52 Broadway, N. Y. City. Supt., W. Christopher.

83. **HERKIMER,** *Herkimer Co.* (Pop., 2,359—est., 3,150.) Built in '88 by village. **Supply.**—Driven wells, pumping through distribution to tank. Fiscal year closes Feb. 28.

Year.	Mains.	Taps.	Meters.	Hyd'ts.	Cons'pt'n.	Cost.	B'd'dd't.	Int. ch'g's.	Exp.	Rev.
'88	7	100	0	52	—	80,000	$50,000	$50,000	$1,750	...
'89	8	230	0	52	60,000	57,000	57,000	1,995	$1,900	$1,461
'90	8	320	0	56	70,000	50,000	57,000	1,995	2,550	2,744

Financial.—*Total Debt*, $58,000; floating, $1,000; bonded, $57,000 at 3½%, due 1905-9. *Expenses:* total, $4,545. *Revenue:* consumers, $2,744; city, $2,000; total, $4,744. **Management.**—Four comrs., one appointed every year by mayor. Chn., H. M. Quackenbosh. Supt., C. Spinner.

V. **HILLBURN,** *Rockland Co.* (Pop., 350—est., 500.) Built in '73 by Ramapo Wheel & Foundry Co. **Supply.**—Stream, by gravity from reservoir. There are ¾ miles of pipe, supplying 75 families, with no fire hydrants. Pressure, 25 lbs. *Revenue:* nominal. Address W. W. Snow, Supt.

84. **HOBART,** *Delaware Co.* (Pop., 390—561.) Built in '87 by Hobart Water Co. **Supply.**—Brook, by gravity; filtered through charcoal. Fiscal year closes second Tuesday in September. **Distribution.**—*Mains*, 1.5 miles. *Taps*, 27. *Meters*, 0. *Hydrants*, 12. **Financial.**—*Cost*, $11,300. *Cap. Stock:* authorized, $12,000; paid up, $11,300; paid 4% dividend in 1890. *Floating Debt*, $1,600. *No Bonds. Expenses:* operating, $50; int., $90; taxes, $69; total, $209. *Revenue:* consumers, $431; city, $2 0; total, $631. **Management.**—Prest., Chas. A. Hanford. Secy., Jas. A. Scatt. Treas., J. S. McNaught. Supt., A. S. Carroll.

18. **HOLLEY,** *Orleans Co.* Brockport and Holley.

85. **HOMER,** *Cortland Co.* (Pop., 2,331—est., 3,000.) Built in '86 by co.; controlled by Moffett, Hodgkins & Clarke, Syracuse; franchise provides neither for control of rates nor purchase of works by town. **Supply.**—Springs, pumping to tank. **Distribution.**—(In '88.) *Mains*, 5 miles. *Taps*, 95. *Meters*, 6. *Hydrants* (in '91), 54. *Consumption*, 855,542. **Financial.**—(In '88.) *Cost*, $50,000, *Debt*, $45,000 at 6%. *Expenses:* operating, $1,000; int., about $2,700. *Revenue:* consumers, $2,400; city, $1,650. **Management.**—(In '88.) Prest., J. F. Moffett. Supt., L. J. Richardson.

VI. **HONEOYE FALLS,** *Monroe Co.* (Pop., 1,098—1,128.) Built in '88 by Honeoye Falls Water-Works Co.; town neither controls rates nor has right to purchase wo ks. **Supply.**—Wells, pumping to tank. Fiscal year closes April 1. **Distribution.**—*Mains*, 1 mile. *Taps*, 50. *Hydrants*, 0. *Consumption*, 1,000 galls. **Financial.**—*Cost*, $3,319. *Cap. Stock:* authorized, $2,500; all paid up. *Debt*, $ 00 at 6%. *Expenses:* operating, $14; int., $12; taxes, $13; total, $39. *Revenue:* consumers, $506; other purpo es, $18; total, $524. **Management.**—Prest., W. G. Storm. Secy. and Treas., A. M. Holden. Supt., Robt. Crennel.

86. **HOOSICK FALLS,** *Rensselaer Co.* (Pop., 4,530—7,014). Built in '86-7 by Hoosick Falls City Water Supply Co.; franchise provides neither for control of rates nor purchase of works by town. **Supply.**—Wells, pumping to tank. Fiscal year closes Oct. 31. **Distribution.**—*Mains*, 8 miles. *Taps*, 400. *Meters*, 2. *Hydrants*, 85. *Consumption*, 153,313 galls. **Financial.**—*Cap. Stock:* authorized,

NEW YORK.

$100,000; all paid up; paid no div'd in 1890. *Bonded Debt*, $100,000 at 5%, due in 1905. *Revenue:* city, $1,040. **Management.**—Prest., Jno. Lockwood. Secy. and Supt., E. R. Estabrook. Treas., G. Norman Weaver, Newport, R. I.

NOTE.—In '91 town talked of building an independent plant.

87. **HORNELLSVILLE,** *Steuben Co.* (Pop., 8,195—10,996). Built in '81 by Hornellsville Water Co.; controlled by American Water·Works and Guarantee Co., Ltd., Pittsburg, Pa.; city has no right to buy works, but maximum rates are fixed in franchise. **Supply.**—Mountain stream, by gravity from impounding reservoir. Fiscal year closes Dec. 31. **Distribution.**—*Mains*, 9 miles. *Taps*, 973. *Meters*, 0. *Hydrants*, 127. *Consumption*, unk'n. **Financial.**—Withheld. **Management.**—Prest., W. S. Kuhn, Pittsburg. Secy., J. H. Purdy.

88. **HUDSON,** *Columbia Co.* (Pop., 8,670—9,970.) Built in '71-5 by city. **Supply.** —Hudson River, pumping to filters; filtered water passing to storage and distributing reservoir. Fiscal year closes Sept. 30.

Year.	Mains.	Taps.	Meters.	Hyd'ts.	Cons'pt'n.	Cost.	B'd'd d't.	Exp.	Rev.
'80	12.6	858	0	181	827,475	$257,651	.....	$7,820	.....
'82	14	1,021	0	181	1,068,750	960,075	.....	13,641	.....
'86	13	1,211	0	183	1,253,707	262,334	$160,000	10,327	$1,518
'87	14.3	1 251	0	185	1,267,825	282,735	200,000	11,532	2,178
'88	15 9	1,303	9	186	1,483,389	301,851	180,500	11,736	2,148
'89	15.9	1,341	..	186	1,515,910	309,961	170,500	10,259	2,557
'90	15.9	1,368	..	186	1,742,195	309 961	160,500	13,292	3,128

Financial.—*Bonded Debt*, $160,500: $115,000 at 7%; $45,500 at 3½%. No allowance for public uses of water. **Management.**—Secy., P. Miller. Supt., Jno. S. Roy.

89. **IRVINGTON,** *Westchester Co.* (Pop., 1,904—2,299.) Built in '84 by village. **Supply.**—Stream and artesian well, pumping to reservoir.

Yr.	Mains.	Taps.	Meters.	Hy-drants.	Con-sumption.	Cost.	Bonded debt.	Int. ch'g's.	Exp.	Rev.
'86	4	47	47	23	5,000	$40,000	$43.000	$1,720	$ 750	$1,200
'87	5	100	100	25	15,000	45,000	40 000	1,600	2,500	.....

Financial.—Int., 4%. *Revenue:* city, $1,000. **Management.**—Unknown.

90. **ISLIP,** *Suffolk Co.* (Pop., town, in '80, 6,453.) Built in '89-90 by Great South Bay Water Co.; town neither controls rates nor has right to purchase works. Supplies villages of Bayport, Bay Shore, Islip, Oakdale and Sayville, in town of Islip. **Supply.**—Wells, pumping to stand-pipe. Fiscal year closes May 1. **Distribution.**—*Mains*, 24 miles. *Taps.* 100. *Hydrants*, 200. *Consumption.* 200,000 galls. **Financial.**—*Cost*, $300,000. *Cap. Stock:* authorized, $100,000, all paid up. *Bonded Debt*, $200,000 at 5%. *Expenses:* operating, $1,000; int., $10,000; taxes, $300; total, $11,300. *Revenue:* consumers, $4,000; city, $6,000; other purposes, $1,000; total, $11,000. **Management.**—Prest., Jno. Lockwood; Secy. and Supt., Jno. C. Lockwood; both of 52 Broadway, New York City.

91. **ITHACA,** *Tompkins Co.* (Pop., 9,105—11,079.) Built in '72 by Ithaca Water Works Co.; town does not control rates. **Supply.**—Buttermilk Falls Creek, by gravity from impounding to distributing reservoirs. A Deane high service pump was added in '91. Plans have been made for an increased supply. Fiscal year closes May 1. **Distribution.**—*Mains*, 18 miles. *Taps*, 475. *Meters*, 8. *Hydrants*, 70. *Consumption.* 200,000 galls. **Financial.**—*Cost*, $224,000. *Cap. Stock:* $100,000. *Bonded Debt*, $75,000 at 6%. **Management.**—Prest., Leonard Treman. Secy. and Supt., E. M. Treman. Treas., L. L. Treman.

92, 93. **JAMAICA,** *Queens Co.* (Pop., 3,922—5,361.)

JAMAICA WATER SUPPLY CO.

Built in '87. Franchise provides that rates must not be higher than in Brooklyn; it does not provide for purchase of works. **Supply.**—Wells, pumping to tank. Fiscal year closes May 1. **Distribution.**—*Mains*, 11 miles. *Taps*, 250. *Meters*, 3. *Hydrants*, 240. *Consumption*, 240,000 galls. **Financial.**—*Cost*, $181,000. *Cap. Stock:* authorized, $75,000; all paid up. *Bonded Debt*, $100,000 at 6%. *Expenses:* operating, $3,000, including cost of operating Jamaica Township Co.'s Plant; int., $6,000; taxes; $250; total, $9,250. *Revenue:* consumers, $4,500; city, $3 000; other purposes, $2,500, total, $10,000. **Management.**—Prest., Jno. Lockwood; Secy. and Treas., Jno. C. Lockwood; both of 52 Broadway, N. Y. City.

JAMAICA TOWNSHIP CO.

Built in '88-9; leased in '89 to Jamaica Water Supply Co., which guarantees interest on bonds. **Supply.**—Jamaica Water Supply Co. Fiscal year closes May 1.

FOR FURTHER DESCRIPTIVE MATTER see Manual for 1889-90. CHANGES since 1890 are here given. POPULATIONS are for 1880 and 1890, respectively.

Distribution.—*Mains*, 10 miles. *Taps*, 150. *Meters*, 1. *Hydrants*, 75. *Consumption*, 10',000 galls. Financial.—*Cost*, $60,000; *Cap. Stock:* authorized, $10,000; all paid up. *Bonded Debt*, $50,000 at 6%. *Expenses:* operating, none; int., $3,000 *Revenue:* consumers, $2,000; city, $1,500; other purposes, $500; total,$4,000. Management.—Same as above.

94. **JAMESTOWN,** *Chautauqua Co.* (Pop., 9,357—16,038.) Owned by American Water-Works & Guarantee Co., Ltd., Pittsburg, Pa. Built in '82 by Jamestown Water Supply Co.; maximum rates fixed in franchise; no provision made for purchase of works by city. Supply.—Seven 8-in. artesian wells, flowing into one brick and one iron tank, thence pumping direct. Only three of these wells are required for present consumption, Pumping Machinery.—*Old Pumping Station.*—Dy. cap., 3,000,000 galls.; one Holly high-pressure having four 19-in. steam and four 9-in. water cylinders of 26-in. stroke. *Boilers:* Two horizontal, tubular, 60 HP. *New Pumping Station.*—Dy. cap., 5,500,000 galls.; one 3,500,000 Deane comp., 20 × 36 × 16 × 24, and one 2,000,000 Volker & Felthousen comp., 18½ × 29 × 13¼ × 18. *Boilers:* Two 300-HP., horizontal, tubular.

Year	Mains	Taps	Meters	Hydrants
'82	13	175	3	100
'86	14	750	7	102
'90	26	1,436	12	134

Financial.—Withheld. Management.—Prest., W. S. Kuhn. Secy., J. H. Purdy. Treas. and Supt., W. A. Kent.

95. **JOHNSTOWN,** *Fulton Co.* (Pop., 5,013—7,768.) Built in '78 by village. Supply.—Springs and streams, by gravity from impounding and distributing reservoir. Fiscal year closes Jan. 31. Bids for construction of 2,000,000-gall. reservoir were received July 25, '91.

Year	Mains	Taps	Hydrants	Cost	B'd debt	Int. ch'g's	Exp.	Rev.
'79	4.6	109	53	$59,806	$60,500			
'84	8	501	60	67,000	60,0 0	$3,690	$1,723	$6,394
'86	9.9	658	65	121,500	60,500	3,630	2,051	10,458
'88	12.6	944	74	124,121	60,500	3,630	2,035	10,516
'89	14.3	1,087	86	128,909	60,500	3,630	2,316	12,152
'90	15 3	1,155	99	133,897	60,500	3,630	2,050	12,500

Financial.—See above. Management.—Four comrs. Prest., Oliver Getman. Secy., A. J. Wilde. Supt., J. J. Buchanan.

96, 97. **JORDAN,** *Onondaga Co.* (Pop., 1,344—1,271.)

S. L. ROCKWELL & CO.'S SYSTEM.

Built in '80. Supply.—Outlet of Skaneateles lake, pumping to reservoir. Distribution.—*Mains*, ¾ miles. *Taps*, 20. *Hydrants*, 12. Financial.—*Expenses*, $50. *Revenue*, $150. Management.—Owner, S. L. Rockwell & Co.

JORDAN WATER CO.'S WORKS.

Built in '89. About 3 miles mains and 20 hydrants. Supt., C. C. Cole.

98. **KEESEVILLE,** *Essex and Clinton Co's.* (Pop., 2,181—2,103.) Built in 83 by village. Supply.—Ausable River, pumping direct. Fiscal year closes April 30. Distribution.—*Mains*, 6 miles. *Taps*, 225. *Hydrants*, 50. *Consumption*, unk'n. Financial.—*Cost*, $38,000. *Bonded Debt*, $26,000 at 5%, due in annual payments of $1,000, which, with int., is met by taxation. *Expenses:* operating, $500; int., $1,300; total, $1,800. *Revenue*, $1,480. Management.—Comrs. Prest., W. Harper. Secy. H. M. Mould.

99. **KINGSTON,** *Ulster Co.* (Pop., 18,344—21,26L) Built in '83 by Kingston Water Co. Supply.—Sawkill Creek, by gravity from impounding reservoir; partially filtered. Fiscal year closes March 10.

Year	Mains	Taps	Meters	Hydrants	B'd'd debt	Int. ch'g's	Exp.	Rev.
'84-5	23.5	600	20	230	$300,000	$18,000		
'86-7	32.5	1,600	35	242	300,000	18,000	$3,197	$30,803
'87-8	33.5	1,760	35	255	300,000	18,000	4,580	27,000
'90-1	36.9	1,907	35	286	300,000	18,000	4,228	9,004*

*From city.

Financial.—*Cap. Stock:* authorized, $300,000; all paid up. *Bonded Debt*, $300,000 at 6%, due in 1903. *Expenses:* operating, $4,228; int., $18,000; taxes, $377; total, $23,-105. *Revenue:* consumers, withheld; city, $8,514; other purposes, $389. Management.—Prest., Jas. G. Lindslay. Secy., Chas. D. Bruyn. Supt., A. Hudler.

VII. **LAKE GEORGE,** *Warren Co.* (Pop. of Caldwell, in '80, 319; est., 450.) History.—Built in '89 by Lake George Water Co. Supply.—Mountain stream, by gravity from reservoir about two miles from and 500 ft. from village. *Takers*, about 150. Address co. or S. M. Bates.

NEW YORK. 79

100. LANSINGBURGH, *Rensselaer Co.* (Pop., 7,132—10,550.) Built in '81 by village. **Supply.**—Oil Mill Creek, by gravity from impounding and distributing reservoir. Fiscal year closes Feb. 10.

Year.	Mains.	Taps.	Meters.	Hydrants.	B'd'd debt	Int. ch'g's.	Exp.	Rev.
'86-7	13	315	2	119	$160,000	$6,400	$8,000	$4,058
'89-90	14	884	18	130	200,000	8,000	....	....
'90-1	14	956	20	130	200,000	8,000	12,500	20,876

Management.—Three comrs. Chn., Richard De Freest. Secy., C. S. Holmes. Treas., Wm. Lea.

101. LARCHMONT, *Westchester Co.* (Pop., est., 1,000.) Built in '89-90 by Larchmont Water Co. **Supply.**—Shelldrake Lake, by gravity. Fiscal year closes Aug. 10. **Distribution.**—*Mains,* 5 miles. *Taps.* 80. *Meters.* 70. *Hydrants,* 20. **Financial.**—Cost, $120,000; cap. stock, authorized, $33,000; all paid up; paid no dividend in '90. *Total debt,* $99,000. first mortgages. *Expenses:* Operating, $1.20; int., withheld; taxes, $152. *Revenue:* Not determined. Figures subject to correction, as works had not been in operation a year at time of report. **Management.**—Prest., G. R. Wright. Secy., C. E. Keene. Treas., Wm. Murray. Supt., W. H. Campbell.

Le Roy, *Genesee Co.* (Pop., est., 2,900.) Built in '75 by village for fire protection.

VIII. LIBERTY, *Sullivan Co.* (Pop., 683—734.) Built in '87 by W. Winner; now owned by J. C. and Geo. Young. **Supply.**—Spring, by gravity through logs. *Families supplied,* 20.

102. LITTLE FALLS, *Herkimer Co.* (Pop., 6,910—8,783.) Built in '85-8 by village. **Supply.**—Beaver Creek, by gravity from impounding and distributing reservoirs; filtered. Fiscal year closes Dec. 31.

Year.	Mains.	Taps.	Meters.	Hyd'ts.	Cost.	B'd'd d't.	Int. ch'g's.	Exp.	Rev.
'88	26.5	581	26	131	$300,000	$305,000	$9,644	$4,500	$20,090
'89	26.7	756	..	132	300,885	305,000	9,644	5,917	14,798
'90	26.7	868	..	132	310,245	305,000	9,644	5,602	12,968

Financial.—*Bonded Debt,* $305,000: $275,000 at 3½%; $30,000 at 3½%. *No Sinking Fund. Expenses:* total, $15,246. *Revenue:* consumers, $12.988; city, $6,500; total, $19,488. **Management.**—Three comrs. Chf. Engr., S. E. Babcock.

IX., 103. LITTLE VALLEY, *Cattaraugus Co.* (Pop., 566—698.)

UNION WATER CO.

Management.—Prest., D. F. Rundell. Secy., C. Z. Lincoln. Treas., S. L. Sweetland. Supt., S. S. Benedict.

VILLAGE WORKS.

Built in '88. **Supply.**—Springs, by gravity. **Distribution.**—(In '88.) *Mains,* 1.5 miles. *Taps,* 115. No Meters. *Hydrants,* 16. **Financial.**—(In '88) Cost, $12,500. *Bonded Debt,* $12,500 at 4%. *Operating Expenses,* nominal. *Revenue:* consumers, $390. **Management.**—Three comrs. Prest., W. H. Lee.

104. LOCKPORT, *Niagara, Co.* (Pop., 13,522—16,038.) Built in '63 by city. Supply, Erie Canal, pumping direct, water being drawn through canal by tunnel. Fiscal year closes Dec. 31. In '90, 5,000,000-gall. Gaskill water power pump was added. Early in '91 committee was appointed to report on improved supply.

Year.	Mains.	Taps.	Hydrants.	Cost.	B'd'd debt.	Int. ch'g's.	Exp.	Rev.
'80	7	430	110	$.....	$.....	$.....	$2,000	$.....
'84	7.5	510	140	......	none	none	7,000	4,500
'88	13.5	680	206	40,100	40,000	1,400	10,063	7,437
'89	14.5	1,000	226	41,951	40,000	1,400	8,404	8,568
'90	18.3	1,100	284	53.651	40,000	1,400	7,145	8,777

Financial.—*Total Debt.*—$46,000. bonded, at 3½%. *Bonded Debt,* $40,000 at 3½%. *No Sinking Fund. Expenses:* total, $9,545. *Revenue:* consumers, $8,777; city, $5,000; total, $13,777. Expenses and improvements are paid from revenue; int. on bonds and their redemption provided for by taxation. **Management.**—Comrs. appointed by mayor every three years. Chn., Louis Vieds. Secy., W. C. Olmstead. Supt., R. J. Sterrett, appointed by comrs.

105. LONG ISLAND CITY, *Queens Co.* (Pop., 17,129—30,506.) Built in '74 by city. Supply, dug and driven wells, pumping direct. **Distribution.**—(In '82) *Mains,* 16 miles. *Taps,* 1,050. *Meters,* 27. *Hydrants,* 250. *Consumption,* 1,500,000 galls. **Financial.**—*First Cost,* $350,000. *Revenue,* consumers (in '82), $37,000. **Management**—Water comrs.

FOR FURTHER DESCRIPTIVE MATTER see Manual for 1889-90. CHANGES since 1890 are here given. POPULATIONS are for 1880 and 1890, respectively.

80   MANUAL OF AMERICAN WATER-WORKS.

106. **LYONS,** *Wayne Co.* (Pop., 3,820—4,475.) Built in '87 by Lyons Water-Works Co.; first supplied in '77; franchise provides neither for control of rates nor purchase of works by city. Supply.—Wells, pumping to stand-pipe. Fiscal year closes May 31. Distribution.—*Mains,* 9 miles. *Taps,* 200. *Meters,* 5. *Hydrants,* 41. Financial.—*Cost,* $50,000. *No Debt. Operating Expenses* (in '88), $2,460. *Revenue:* consumers (in '88), $1,700; city, $1,600. Management.—Prest., C. J. Ryan. Secy. and Supt., E. A. Browere. Treas., Daniel Moran.

108. **MALONE,** *Franklin Co.* (Pop., 4,193—4,986.) Built in '57 by Malone Water-Works Co.; franchise provides neither for control of rates nor purchase of works by city. Supply.—Springs, by gravity to and from reservoir; fall from springs to reservoir, 101 ft.; from reservoir to city, 225 ft. Fiscal year closes May 1. Distribution. —*Mains,* 21 miles. *Hydrants,* 62. Financial.—*First Cost,* $40,000. *Cap. Stock* (in '88), $75,000. *Bonded Debt,* $65,000. *Revenue:* city, $2,480. Management.—Prest., C. W. Brewer. Supt., Geo. Sabin. Treas., F. D. Kilburn.

108. **MAMARONECK,** *Westchester Co* (Pop. of town, 1,863—est., 2,000).) Built in '85 by Mamaroneck Water Co.; franchise provides neither for control of rates nor purchase of works by town. Supply.—Mamaroneck River, pumping to reservoir. Fiscal year closes Dec. 31. During '91 a 650,000-gall. Worthington comp. pump and a 48-in. boiler were added.

Year.	Mains.	Taps.	Meters.	Hyd'ts.	Cons'pt'n.	B'd'd d't.	Int. ch'g's.	Exp.	Rev.
'86	6	36	10	10	10,000	$30,000	$1,800	$1,2 0	$1,200
'88	6	100	10	14	30,000	30,000	1,800	1,200	2,000
'90	7	160	6	12	50,000	31,000	1,860	1,422	2,500

Financial.—*First Cost,* $65,000. *Capital Stock*: authorized, $25,000; all paid up; paid no dividend in '89. *Bonded Debt,* $31,000 at 6%. *Expenses:* operating, about $1,200; int., $1,860 due, none paid; taxes, $222; total, $3,282. *Revenue:* consumers, $2,5 0; city, none. Management.—Prest., J. M. Constable. Secy., W. T. Cornell. Supt., Louis Ottenmann.

109. **MANLIUS,** *Onondaga Co.* (Pop., 834—942.) Built prior to '17 by private parties. Supply.—Springs, by gravity. One-half mile of pipe supplies 19 families, 4 hydrants and 13 fire-cisterns. Address G. W. H. Wood.

110. **MARGARETTVILLE,** *Delaware Co.* (Pop., 418—est., 700.) Works for general supply owned by co. Secy. and Supt., C. C. Kaufman. No further information.

55. **MATTEAWAN,** *Dutchess Co.* See Fishkill and Matteawan.

Medina, *Orleans Co.* (Pop., 3,632—4,492.) Built in '60 by village for fire protection only. See projected works.

111. **MIDDLETOWN,** *Orange Co.* (Pop., 8,494—11,977.) Built in '67 by village. Supply.—Springs and surface water, by gravity from impounding reservoir, supplemented by pumping from Shawangunk Creek to reservoir. Fiscal year closes March, 1.

In '90-91 new 350,000,000-gall. reservoir, with area of 110 acres, was built, 8,500 ft. southwest and 80 ft. above high water mark in old reservoir. Reservoirs connected by 12-in c. i. pipe, discharging into paved waterway about 400 ft. long, with fall of 35 ft. Reservoir formed by two dams of earth, with masonry heart-wall, each 38 ft. high; max. depth of water, 34 ft. Mapes dam 516 ft. long on crest, 22 ft. wide at top, 152 at bottom; inner slope, about 1.3 to 1, heavily rip-rapped, outer 2.3 to 1; heart-wall 8 ft. thick from bottom to natural surface, then tapers to 2 ft. at top. Corey dam 437 ft. long on crest, 12 ft. wide at top, 75 ft. wide at roadway, on outer side, 23 ft. from top; inner slope, about 1 to 1, heavily rip-rapped, outer about 2 to 1; heart wall 6 ft. wide at from bottom to level of roadway, then tapers to 2 ft. at top; masonry gate-house on inner slope; two 20-in. pipes extend through dam. Contr. for dams and new reservoir, Jehiel Vaughn; contract price, $57,371. Old reservoir has cap. of 250,000,000 galls., and area of 86 acres. Summer of '91, $40,000 additional bonds, 4%, were issued for new supply main, 20; 14 and 12-in. c. i., and for replacing cement pipe.

Year.	Mains.	Taps.	Meters.	Hyd'ts.	Cost.	B'd'd d't.	Int. ch'g's.	Exp.	Rev.
'80	8	400		88				$2,000	
'82		546	6		$170,000	$160,000	$9,600	1,506	$16,637
'84	14	500		103	180,721	154,000	9,240	1,669	12,470
'87	16.25	550	40	115	180,721	154,000			16,479
'88	17.1	920	18	123	196,761	136,000	4,940	11,531	17,872
'89	18.5	987	16	129	198,147	134,000	4,960	6,367	18,000
'90	20.2	1,158	26	136	234,988	209,000	6,173		18,490

NEW YORK. 81

Financial.—*Total Debt*, $209,000; floating, none; bonded, $34,000 at 4%, due in 1911; $100,000 at 3½%, due in 1918; $75,000 at 3½% issued in '90 or '91. *Sinking Fund:* $2,500 is raised by taxation each year for sinking fund; all other expenses are paid from revenue. *Expenses:* operating (in '89), $6,367; int. (in '89), $4,960; total (in '89), $11,327. *Revenue :* consumers, $19,490; city, none. **Management.**—Comrs., elected every five years. Prest., J. W. Canfield. Clerk, D. F. Seward. Supt., I. F. VanDuzer. Clerk and Supt. are appointed by comrs.

112. **MILLERTON**, *Dutchess Co.* (Pop., 600—638.) **History.**—Built in '91 by village; put in operation July 1. Engr., E. D. Smalley, Syracuse, N. Y. Contr., John Marsden, Utica. **Supply.**—Spring and creek, by gravity. Small pump may be added for emergency use. **Reservoir.**—Cap., 500,000 galls. **Distribution.**—*Mains*, 8 to 4-in. c. i., 2½ miles. *Hydrants:* Eddy, 22. *Valves*, Kennedy. **Financial.**—*Cost*, $15,000. *Bonded Debt:* authorized, $15,000 at 4%, all issued. **Management.**—Comrs. Prest., E. H. Thompson. Secy., P. N. Paine. Reports June 20 and July 5, '91.

113. **MOHAWK**, *Herkimer Co.* (Pop., 1,441—est., 2,000). Built in '90-91 by village; finished Feb. 1. Engineer, E. D. Smalley, Syracuse. **Supply.**—Well, 15 ft. in diam., 12 ft. deep, pumping to stand-pipe. **Pumping Machinery.**—Cap., 1,000,000 galls.; one Gordon high pressure. *Boilers:* from Phœnix Foundry, Syracuse. **Stand-Pipe.**—Cap., about 75,000 galls.; 16×50 ft.; from Phœnix Foundry, Syracuse. **Distribution.**—*Mains*, 8 to 4-in. c. i.; 6 miles; from Utica Pipe Foundry; laid by Chas. Brown, Mohawk. *Services*, w. i.—*Taps* (in June, '91), about 125. *Hydrants:* Galvin, 45. *Valves*, Galvin. *Pressure*, 80 lbs. **Financial.**—*Cost*, $33,000. *Debt:* $33,000, bonded at 4%. **Management.**—Not given. Report by E. D. Smalley, Syracuse, Aug. 25, '91.

114. **MORAVIA**, *Cayuga Co.* (Pop., 1,540—1,486.) Built in '84 by Moravia Water-Works Co.; franchise provides neither for control of rates nor purchase of works by town. **Supply.**—Springs, by gravity. Fiscal year closes Jan. 1. **Distribution.**—*Mains*, 4 miles. *Taps*, 90. *Hydrants*, 30. *Consumption*, 64,000 galls. **Financial.**—*Cost*, $36,000. *Cap. Stock:* authorized, $30,000; all paid up. *Bonded Debt*, $20,000 at 6%. *Expenses:* operating, $100; int., $1,200; taxes, $65; total, $1,365. *Revenue:* consumers, $1,200; city, $900; total, $2,100. **Management.**—Prest., S. H. Morgan. Secy., A. A. Morgan. Treas. and Supt., F. C. Reynolds.

115. **MOUNT MORRIS**, *Livingston Co.* (Pop., 1,899—2,236. Built in '79 by Mills Water-Works Co. under 30-yrs. franchise, which provides neither for control of rates nor purchase of works by city. **Supply.**—Nine springs, by gravity to and from reservoir. **Distribution.**—*Mains*, 5 miles. *Taps*, 584. *Meters*, 15. *Hydrants*, 28. *Consumption*, 70,000 galls. **Financial.**—*Cost*, $60,000. *Cap. Stock:* authorized, $80,000; all paid up. *No Debt*. *Revenue:* city, $1,120. **Management.**—Prest. and Supt., M. H. Mills.

116. **MOUNT VERNON**, *Westchester Co.* (Pop., 4,586-10,830.) Built in '86 by New York & Mt. Vernon Water Co., now controlled by Debenture Guarantee & Assurance Co., Chicago, Ill. **Supply.**—Hutchinson's Creek, pumping from impounding reservoir to stand-pipe. Fiscal year closes Dec. 31. **Distribution.**—*Mains*, 26 miles. *Taps*, 1,539. *Meters*, 30. *Hydrants*, 246. **Financial.**—*Cost*, $525,000. *Bonded Debt*, $500,000 at 6%. *Expenses :* operating and taxes, $5,300; int., $30,000; total, $35,300. *Revenue :* total, $46,000. **Management.**—Prest., F. Hopkinson Smith. Secy. and Treas., T. L. Walker.

117. **NEWARK**, *Wayne Co.* (Pop., 2,450—2,824.) Built in '78 by Newark Water-Works Co.; controlled by Moffett, Hodgkins & Clarke, Syracuse; rates are fixed in franchise, which provides for purchase of works by village at intervals of five years. **Distribution.**—(In '88.) *Mains*, 9 miles. *Meters*, 2. *Hydrants*, 94. *Consumption*, 100,000 galls. **Financial.**—(In '88.) *Cost*, $120,000. *Cap. Stock*, $100,000. *Debt*, $93,000 at 6%. *Expenses:* operating, $1,500. *Revenue:* consumers, $4,600; city, $2,500. **Management.**—(In '89.) Prest., J. F. Moffett. Secy., C. T. Moffett. Supt., F. W. Kellogg.

118. **NEW BERLIN**, *Chenango Co.* (Pop., 937—979.) Built in '85 by village. **Supply.**—Springs and brook, by gravity from impounding reservoir.

FOR FURTHER DESCRIPTIVE MATTER see Manual for 1889-90. CHANGES since 1890 are here given. POPULATIONS are for 1880 and 1890, respectively.

82   MANUAL OF AMERICAN WATER-WORKS.

Year.	Mains.	Taps.	Meters.	Hydrants.	Consumption.	Cost.	Bonded Debt.	Int. Charges.	Exp.	Rev.
'86	3	65	0	24		$36,000	$36,000	$2,160	..	$800
'87	4	125	0	27	25,000	36,000	36,000	1,440	$50	1,775

Financial,—Int., 4%. **Management.**—Water Board. Prest. and Supt. (in '88) G. G. Beers.

119. **NEW BRIGHTON and PORT RICHMOND,** *Richmond Co.* (Pop.: New Brighton, 12,679—16,423; Port Richmond, 3,561—6,290.) Built in '80-81 by Staten Island Water Supply Co.; franchise provides neither for control of rates nor purchase of works by city. **Supply.**—Forty driven wells, which syphon into large receiving wells, pumping direct and to reservoir. Fiscal year closes Dec. 31. **Distribution.**—*Mains,* 40 miles. *Taps,* 2,350. *Meters.* 26. *Hydrants,* 264. *Consumption,* 2,500,000 galls. **Financial.**—*Cost,* $355,000. *Cap. Stock* (in '88), $200,000. *Bonded debt,* $200,000 at 6%. *Expenses:* operating, withheld; int., $10,500; taxes, $2,800. *Revenue,* city, $7,200. **Management.**—Prest., H. Brightman, Secy., J. F. Francis. Treas., H. L. Horton. Supt., J. S. Warner.

120. **NEWBURGH,** *Orange Co.* (Pop., 18,049—23,087.) Built in '53 by town. **Supply**—Washington Lake and Silver Creek, by gravity and pumping to reservoir. Fiscal year closes March 10.

Year.	Mains.	Taps.	Meters.	Hydrants.	Cost.	B'd'd Debt.	Int.	Exp.	Rev.
'54-5	9.2				$107,376	$112,300	$9,451	$1,186	$8,369
'60-1	11.2				128,747	112,300	6,600	1,168	13,145
'65-6	14.6				142,115	129,300	5,844	1,854	13,775
'70-1	18	1,386		115	187,044	126,300	6,478	3,373	19,680
'75-6	19.2	1,654	2	177	383,327	263,000	18,381	8,678	29,882
'80-1	20.1	1,789		181	432,801	284,500	17,008	4,264	27,334
'81-2	20.1	1,833		160	435,548	279,500	17,133	8,886	27,420
'82-3	20.3	1,903	10	185	448,501	280,500	16,378	3,966	30,427
'83-4	22.6	2,016		181	464,745	285,500	16,059	5,975	30,883
'84-5	24	2,131	5	185	476,143	270,500	16,138	7,003	29,944
'85-6	25	2,262		194	491,570	209,500	15,898	9,172	30,882
'86-7	29.9	2,379		211	498,817	294,500	16,858	6,242	31,384
'87-8	33	2,621		227	513,560	292,500	15,708	4,571	32,885
'88-9	33	2,853	5	242	524,108	290,500	14,938	4,990	36,054
'89-90	33	2,892	4	261	537,141	288,000	14,858	5,425	36,536
'90-1	35.3	2,032		276	554,555	286,000	14,560	6,115	40,273

Financial.—*Bonded Debt,* $286,000; $75,000 at 7%, due in '91-2; $23,000 at 6%, due in '95-6; $77,000 at 5%, due in '97; $66,000 at 4%, due in 1900-'05; $45,000 at 3½ and 3%, due in 1894-1907. *Sinking Fund,* $2,000 to $5,000 annually, used at once in paying off bonds. *Expenses:* total, $20,675. *Revenue:* consumers, $40,273; city, none. **Management.**—Five comrs., one elected every year. Prest., Robt. Huddelson. Secy. and Supt., Wm. Chambers.

121. **NEW DORP,** *Richmond Co.* (Pop., est., 400.) **History.**—Built in '89 by South Shore Water Co., of Staten Island. Engr., W. T. Hiscox, 13 Gold St., N. Y. City. Contr., C. K. Moore, N. Y. City. **Supply.**—Artesian wells, pumping to stand-pipe. **Pumping Machinery.**—Cap., 250,000 galls.; Worthington. **Stand-Pipe.**—Cap., 100,000 galls.; from Logan Iron Works. **Distribution.**—*Mains,* about 3 miles; from M. J. Drummond, N. Y. City. **Financial.**—*Cap. Stock:* authorized, $20,000, all paid up. *Bonded Debt:* authorized, $20,000; issued, $12,000. **Management.**—Prest., Howard Willett. Secy., R. W. Underhill. Treas., J. F. Halsey. Supt., E. W. Drinker. Report by W. T. Hiscox, Director, Aug. 1, '91.

122. **NEW ROCHELLE,** *Westchester Co.* (Pop., township, 5,276—8,217.) Built in '85-7 by New Rochelle Water Co.; franchise both regulates rates and provides for purchase of works by city. **Supply.**—Hutchinson's River, by gravity from impounding reservoir; National filters added in '90. Fiscal year closes Dec. 31. **Distribution.**—*Mains,* 25 miles. *Taps,* 608. *Meters,* 560. *Hydrants,* 77. *Consumption,* 130,000 galls. **Financial.**—*Cost,* $300,000. *Cap. Stock* (in '87) $80,000. *Bonded Debt* (in '87) $210,000. *Expenses:* operating, $6,136; taxes, $1,945. *Revenue:* city, $2,310. **Management.**—Prest., Adrian Iselin. Secy., Henry Fallo. Treas., Adrian Iselin, Jr. Supt., F. H. Davis.

123. **NEW YORK,** *New York Co.* (Pop., 1,206,299—1,515,301.) First supplied in 1774; Manhattan Co. built works about 1799. Present works built in '43 by city. **Supply.**—Croton and Bronx rivers, from storage and impounding to receiving and distributing reservoirs, with repumping for high service.

July 15, '90, water was passed through new Croton aqueduct, and structure has since been in use with the exception of a few stoppages to complete sections of the aqueduct or make repairs.

NEW YORK.

June 24, '91, completed sections were turned over to Department Public Works, but there then remained a comparatively small amount of work to be completed on some sections. Aqueduct starts at Croton Lake and extends southerly to point 7,000 ft. north of Jerome Park with uniform grade of 0.7 ft. per mile, est. dy. cap. of 318,000,-000 galls. and general form of horseshoe, with curved invert, 13.53 ft. high and 13.6 ft. wide. At above point it is proposed, eventually, to take out water for distributing reservoir to supply annexed district, and aqueduct now has various grades and is reduced to dy. cap. of 250,000,000 galls. and to circular form with diam. of 12 ft. 2 ins., and 10.5 ft. beneath Harlem River, which it passes by an inverted siphon. At 135th st. aqueduct changes to 12 48-in. c. i. mains, 4 each to be connected with old aqueduct, city distribution and large reservoir in Central Park. Aqueduct in tunnel for 29.63 miles, open trench for 1.12 miles, making 30.75 miles of aqueduct proper, which, with pipes, 2.37 miles to Central Park Reservoir, makes total length of 31.12 miles. For further details see Manual for '89-90, references to articles given there, in '88 Manual, and ENG. NEWS, vol. xxiv., p. 455.

Oct. 8, '90, A. Fteley, Chf. Engr. Aqueduct Commission reported on the "Building of a High Masonry Dam on the Lower Part of the Croton River," at the site known as Cornell's. This dam was to be a substitute for the proposed Quaker Bridge Dam (see '89-'90 Manual). Aqueduct Commissioners adopted report, but no further action has been taken. See ENG. NEWS, vol. xxiv., pp. 438, 455 and 480, for further details, and for general history of the New York Water Supply. It is expected that reservoirs now under construction will be completed as follows : Sodom, or Reservoir I, in '91; Bog Pond, an adjunct of Reservoir I, in '92; New Reservoir A, on Muscoot River, in '93; Reservoir M, on Titicus River, in '94; Reservoir D, on west branch of Croton River, in '95. These reservoirs will add a total of 30,000,000,000 galls. to the present storage capacity, which is now about 10,000,000,000 galls.

In '90 contract was let for new 10,000,000-gall. Worthington pump and for boilers to be placed at 98th st. Station.

Further information regarding the New York water-works, especially the dams and reservoirs, may be had by referring to the following articles in ENG. NEWS :

"Reservoir M., on Titicus River," with inset, vol. xxiii., p. 271; "New York City's Water Supply," vol. xxiv., 50; "Progress at the Sodom Dam," with view of structure taken about Oct. 1,'90, vol. xxiv., p. 379; "The New Croton Dam," with inset, dam recommended to take place of proposed Quaker Bridge Dam, vol. xxiv., p. 438; "A History of the New York Water Supply, and Recent Explorations for New Reservoirs," with inset, by E. P. Roberts, Ass't Eng. Aqueduct Commission, vol. xxiv., pp. 455, 480.

Year.	Mains.	Taps.	M't'rs.	Hyd'ts.	Cons't'n.*	Cost.† $	Debt.† $	Exp. $	Rev.‖ $
'81	512	82,509	5,293	6,496	....	33,712,197	....	347,439	1,604,314
'82	531.6	85,000	6,817	6,944	95,000,000	....	....	....	1,747,947
'83	546.4	.....	9,012	7,152	....	....	....	....	2,039,078
'84	562.9	92,700	11,625	7,400	....	36,419,122‡	11,429,700	....	2,150,019
'85	579.9	.....	13,680	7,581	....	....	....	....	2,239,453
'86	604.4	96,581	14,582	7,833	107,000,000	....	....	....	2,509,218
'87	619.8	100,258	16,552	7,984	110,000,000	38,833,196	....	....	2,668,387
'88	638.2	102,872	18,211	8,191	112,570,000	30,231,662	13,752,500	....	2,587,063
'89	657.19	105,815	19,870	8,420	112,670,000	39,500,000§	13,952,700	530,000†	2,782,051
'90	667.71	108,881	22,072	8,576	121,000,000	39,900,000§	13,906,300	530,000†	2,926,311

* Consumption since '86 includes supply from Bronx and Bryam rivers, which in '90 was 16,200,000 galls. per day. In '90 there was repumped for high service 6,079,959 galls. per day at High Bridge and 11,873,634 galls. at 98th st. pumping station. During '90 an average of 28,508,100 galls. per day passed through meters.

† Exclusive new Croton Aqueduct and appurtenances. Cost of new aqueduct to July, '90, about time water was first passed through aqueduct, was $23,558,030, including land damages. Total debt, account new aqueduct and appurtenances, on Dec. 31, '90, $23,945,000.

‡ Nov., '84.
§ Approximately.
‖ Total revenue, 1840 to 1890, inclusive, $59,786,077.

Bonded Debt.—Rates of int. on total bonded debt, including new aqueduct bonds, on Dec. 31, '89, were as follows: $3,881,500 at 7%; $2,057,400 at 6%; $3,004,800 at 5%; $21,410,000 at 4%; $2,040,000 at 3½%; $21,004,000 at 3%; $1,100,000 at 2½%. Management.— Comr. Pub. Wks., T. F. Gilroy. Chf. Engr., G. F. Birdsall.

FOR FURTHER DESCRIPTIVE MATTER see Manual for 1889-90. CHANGES since 1890 are here given. POPULATIONS are for 1880 and 1890, respectively.

84   MANUAL OF AMERICAN WATER-WORKS.

124. **NIAGARA FALLS**, *Niagara Co.* (Pop., 3,320–5,502.) Built in '77 by Niagara Falls Water-Works Co.; in '90 or early in '91 Cataract Construction Co. bought at least 95% of Cos. stock and proposed to enlarge and improve works, including new supply in connection with Construction Co.'s tunnel. **Supply.**—Suspension Bridge water-works. Fiscal year closes Dec. 31. **Distribution.**—*Mains*, 7 miles. *Taps*, 471. *Meters*, 0. *Hydrants*, 56. *Consumption*, 500,000 galls. **Financial.**—*Cost*, $51,379. *Cap. Stock:* authorized, $50,000; paid-up, $48,875; paid 8% div'd. in 1890. *No Debt. Expenses:* operating, $4,762; taxes, $874; total, $5,636. *Revenue:* consumers, $6,704; city, $2,800; total, $9,504. **Management.**—Prest., C. B. Gaskill. Secy., A. J. Porter.

125. **NORTH TARRYTOWN**, *Westchester Co.* (Pop., 2,684–3,179.) Built in '88 by Pocantico Water-Works Co., under 30 years franchise. Supplies Dobbs' Ferry and Hastings-on-Hudson. **Supply.**—Pocantico River, pumping direct and from impounding to distributing reservoir. Fiscal year closes Nov. 1. **Distribution.**—*Mains*, 23 miles. *Taps*, 300. *Meters*, 0. *Hydrants*, 105. *Consumption*, 600,000 galls. **Financial.**—*Cap. Stock*, $500,000. *Bonded Debt*, $250,000. **Management.** —Prest., J. S. Ellis, N. Y. City. Secy., A. V. H. Ellis. Treas., T. J. Lawrence. Supt. A. L. Betts.

178. **NORTH TONAWANDA**, *Erie Co.* See Tonawanda.

126. **NORTHVILLE**, *Fulton Co.* (Pop., 763-792.) **History.**—Built in '91 by village; was to be put in operation Aug. 15. Engr., E. B. Baker, Gloversville. Contr., Dennis Sullivan, Flushing. **Supply.**—Springs, by gravity. **Reservoir.**— Cap., 3,000,000 galls. **Distribution.**—*Mains*, 10 to 4-in. c. i., about 5½ miles; from Chas. Millar & Son, Utica. *Hydrants*, Ludlow, 30. *Valves*, Rensselaer. *Pressure*, 225 lbs. **Financial.**—*Cost*, $30,000. *Debt*, $30,000 at 3½%. **Management.**—Comrs. Prest., Ray Hubbell. Secy., J. R. Willard. Clerk, J. R. Van Ness. Report, June 20, '91.

127. **NORWICH**, *Chenango Co.* (Pop., 4,553-5,212.) Built in '81 by Norwich Water-Works Co.; franchise neither controls rates nor provides for purchase of works by city. **Supply.**—Hopkins' brook, by gravity from impounding reservoir. Fiscal year closes Dec. 31. A new impounding reservoir was completed in Dec., '90; cap., 60,000,000 galls.; situated above the old reservoir on the same stream and designed to be used conjointly with it or separately. Engr., W. F. Franklin, Lansingburg. Contrs., Troy Public Works Co. Cost, $60,000. **Distribution.**—*Mains*, 9.5 miles. *Taps*, 364. *Meters*, 3. *Hydrants*, 29. *Consumption*, unk'n. **Financial.**— *Cost*, $130,090. *Cap. Stock:* authorized, $50,000, all paid up; paid no div'd in 1890, net earnings used for extensions. *Total Debt*, $55,000: floating, $5,000 at 6%; bonded, $50,000 at 5%. *Expenses:* operating $551; int., $353; taxes, $625; total, $1,529. *Revenue:* consumers, $7,081; city, $1,026; total, $9,107. **Management.**—Prest., Jno. Mitchell. Secy., T. D. Miller. Treas., C. B. Martin. Supt., Jno. Seward.

128. **NUNDA**, *Livingston Co.* (Pop., 1,037–1,010.) Built in '86-7 by Nunda Water Co. **Supply.**—Springs, by gravity and pumping to reservoir for reserve. Fiscal year closes Dec. 31. **Distribution.**—*Mains*, 7 miles. *Taps*, 35. *Hydrants*, 25. *Consumption*, 20,000 galls. **Financial.**—*Cost*, $25,000. *No Debt. Expenses:* operating, $300; taxes, $100; total, $400. *Revenue:* consumers, $300; city, $1,000; total, $1,300. **Management.**—Pres., Miles Ayrault. Sec. and Supt., H. Willard.

129. **NYACK**, *Rockland Co.* (Pop. 3,881-4,111). Built in '73 by Nyack Water-Works Co. **Supply.**—Springs and Hackensack River, by gravity and pumping to reservoir. Latest report for year ending Aug. 1, '91. A 750,000-gall. Worthington comp. dup. pumping engine has been added.

Year.	Mains.	Taps.	Meters.	Hyd'ts.	Cost.	B'd'd d't.	Int. ch'g's.	Exp.	Rev.
'87	11	315	1	27	$180,000	$50,000	$3,000	.....	$10,500
'88	13	425	1	33	180,000	50,000	3,000	.....	11,000
'90-'91	..	550	0	30	....	50,000	3,000	$5,600	12,275

Financial.—*Capital Stock:* common, authorized, $100,000; all paid up; paid no dividend in '90-1. No preferred stock. *Total Debt:* $62,000; floating, $12,000 at 6%; bonded, $50,000 at 6%, due in 1901. *No Sinking Fund. Expenses:* operating, $5,600; int., $3,720; taxes, $972; total, $10,292. *Revenue:* consumers, $10,500; city, fire protection, $1,775; total, $12,275. **Management.**—Pres. and Treas., A. M. Voorhis. Secy. and Supt., W L. Voorhis.

90. **OAKDALE**, *Suffolk Co.* See Islip.

130. **OGDENSBURGH,** *St. Lawrence Co.* (Pop., 10,341—11,662.) Built in '68 by city. Supply.—Oswegatchie River, pumping direct; water power evidently used. Fiscal year closes March 31.

Year.	Mains.	Taps.	Meters.	Hyd'ts.	Cons'pt'n.	B'd'd debt.	Int. ch'g's.	Exp.	Rev.
'80-1	14	1,030	0	82	....	$135,000	$5,950	$2,181	7,939
'81-2	14.5	1,055	0	82	770,246	135,000	5,950	2,784	8,180
'82-3	14.6	1,078	0	83	864,548	135,000	5,950	2,592	8,032
'83-4	14.6	1,130	0	83	947,943	135,000	5,950	1,977	8,445
'84-5	14.8	1,170	0	83	1,030,921	135,060	5,950	2,068	8,674
'85-6	15.1	1,226	0	85	1,214,034	135,000	5,950	2,159	8,891
'86-7	15.2	1,291	0	85	1,753,377	135,000	5,950	2,620	8,689
'87-8	15.6	1,355	0	88	....	135,000	5,950	3,526	9,058
'88-9	16.1	1,374	0	91	....	135,000	5,950	4,240	10,480
'89-90	17.6	1,475	0	97	....	132,000	5,897	3,522	11,370

Financial,—(In '89-90.) *Cost*, about $175,000. *Bonded debt:* $132,000; $32,000 at 7% were due Aug. 1, '90; $97,000 at 3½% due $3,000 yearly until 1803, then $1,000 yearly till 1818. *Sinking fund*, see above. *Expenses*, see above; total, $9,419. *Revenue*, see above; city makes annual appropriation from taxes ($1,945 in '89-90) which, with revenue, pays all expenses, interest and construction. Construction account from '80-1 to '89-90, inclusive, has been about $22,600. Management.—Five comrs. Supt., A. H. Lord.

131. **OLEAN,** *Cattaraugus Co.* (Pop., 3,036—7,358.) Built in '78 by village. Supply.—Well, pumping to reservoir. Fiscal year closes Feb. 1.

Year.	Mains.	Taps.	Meters.	Hydr'ts.	Cost.	B'd'd debt.	Int. Ch'gs.	Exp.	Rev.
'84-5	10	254	6	75	$83,000	$80,000	$3,600	$3,776	$3,438
'86-7	10	411	3	74	....	80,000	3,600	3,791	8,850
'88-9	11.3	714	4	83	85,254	77,500	3,875	5,887	8,625
'89-90	11.3	807	9	93	97,754	77,500	3,875	9,968	8,756
'90-1	11.3	912	11	93	....	82,500	3,875	10,163	9,968

Financial.—*Bonded Debt,* $82,500: $70,000 at 4½%, $12,500 at 4%. *Expenses*, total, $14,038. *Revenue:* consumers, $9,968 ; city, none. Management.—Secy. and Supt., Jno. F. Le Fevre.

132. **ONEIDA,** *Madison Co.* (Pop., 3,964—6,083.) Built in '83 by Warner Water-Works Co. Supply.—Large springs, by gravity from impounding reservoir. Distribution.—*Mains*, 12 miles. *Taps*, 300. *Meters*, 0. *Hydrants*, 73. Financial.—*Cost*, $142,000. No debt. Management.—Prest. and Supt., J. W. Warner. Secy., I. C. Chapin. Treas., D. Chapin. Note.—Above figures are only approximately correct.

133. **ONEONTA,** *Otsego Co.* (Pop., 3,002—6,272.) Built in '82 by Oneonta Water-Works Co. Supply.—Oneonta creek, by gravity from impounding reservoir. Distribution.—*Mains*, 17 miles. *Taps* (in '87), 425. *Meters*, 8. *Hydrants*, 74. *Consumption*, unk'n. Financial.—*Cost*, $200,000. *Cap. Stock*, $160,000. *Revenue*, city, $2,000. Management.—Prest., Munro Westcott. Secy., Jas. Stewart. Treas., E. A. Scramling. Supt., T. A. Norton.

X. **ORWELL,** *Oswego Co.* (Pop., est., 325.) A. E. Olmstead and Mrs. B. Thomas own a pipe-line supplying about 10 houses.

134. **OSWEGO,** *Oswego Co.* (Pop., 21,116—21,842.) Built in '68 by Oswego Water-Works Co. Supply.—Oswego River, pumping to reservoir. Fiscal year closes Feb. 1. During '90 pump-house was repaired, filters rebuilt, and entire system thoroughly overhauled.

Year.	Mains.	Taps.	Meters.	Hydr'ts.	Cons'n.	Cost.	B'd'd debt.	Exp.	Rev.
'85-6	30	850	12	182	2,500,000	$400,000	$230,000	$13,500	$31,000
'87-8	38	1,050	12	185	......	500,000	237,000	10,500	31,000
'88-9	38	1,265	40	193	......	521,030	237,000	16,500	33,250
'90-1	42.6	1,378	57	209	2,392,640	585,991	......	......	......

Financial.—*Cap. Stock* (in '88), $229,500. Management.—Prest., T. S. Mott. Vice-Prest. and Treas., J. T. Mott. Secy., H. H. Lyman. Supt., T. H. Blunett.

63. **OSWEGO FALLS,** *Oswego Co.* (Pop., in '90, 1,821.) Supply from Fulton.

135. **OTEGO,** *Otsego Co.* (Pop., 749—est., 860.) Built in '89 by Otego Water Co. Supply, Otsdawa creek and springs, by gravity. Distribution.—*Mains*, 3.3 miles. *Taps*, 47. *Meters*, 0. *Hydrants*, 13. Financial.—*Cost*, $17,000. *Cap. Stock*, authorized, $21,000. *No Debt. Expenses:* operating, nominal; taxes, $25. Management.—Prest., W. H. Barker. Secy. and Treas., Lottie A. Field. Supt., F. D. Shurnway.

For Further Descriptive Matter see Manual for 1889-90. Changes since 1890 are here given. Populations are for 1880 and 1890, respectively.

136. **OWEGO,** *Tioga Co.* (Pop., 5,525—est., 6,200.) Built in '79 by Owego Water Co.; 10-years franchise; in '91 works came into hands of R. D. Wood & Co. Supply, Barnes Creek by gravity from impounding to distributing reservoir; filtered. Fiscal year closes Dec. 31. **Distribution.**—*Mains,* about 12 miles. *Taps,* 465. *Meters,* 99. *Hydrants,* 50. *Consumption,* 300,000 galls. **Financial.** *Cost,* $135,000. *Cap. Stock,* authorized, $100,000. *Bonded Debt,* $80,000 at 6%, due in 10 yrs. *Expenses:* operating, $3,000; int., $4,800; taxes, $675; total, $8,475. *Revenue:* total, $8,500. **Management.**—Secy. and Treas., Walter Wood; Gen. Supt., A. W. MacCullum; both of 400 Chestnut St., Philadelphia, Pa. Supt., G. Y. Robertson.

XI. **OXFORD,** *Chenango Co.* (Pop., 1,209—1,477.) Built in '80 by individuals for domestic supply. Supply, spring, by gravity. From ½ to 1 mile of mains.

NOTE.—Holly pump in mill and short length of pipe affords limited fire protection independent of above.

137. **PALMYRA,** *Wayne Co.* (Pop., 2,308—2,131.) Built in '89-90 by Palmyra Water-Works Co.; city has option of buying works at end of five years. **Supply.**—Surface well, pumping to stand-pipe. Fiscal year closes Dec. 31. **Distribution.**—*Mains,* 5 miles. *Taps,* 60. *Meters,* 29. *Hydrants,* 40. **Financial.**—*Cap. Stock,* authorized, $50,000; all paid up. *Bonded Debt,* $50,000 at 6%, due in 1909; there are $5,000 of reserved bonds in the treasury. *Revenue:* city, $1,600. **Management.**—Prest., C. M. Harrington; Secy., E. M. Bassett; both of Buffalo Supt., R. A. Vanderboget.

138. **PATCHOGUE,** *Suffolk Co.* (Pop., 2.503—1,695.) Built in '87 by Suffolk County Water Co. **Supply.**—Six-in. pipe wells. Fiscal year closes May 1. **Distribution.**—*Mains,* 6 miles. *Taps,* 170. *Meter,* 1. *Hydrants,* 24. *Consumption,* 100,000 galls **Financial.**—*Cost,* $105,000. *Cap. Stock:* authorized, $60,000; all paid up. *Bonded Debt,* $40,000 at 6%. *Expenses:* operating, $1,800; int., $2,400; taxes, $150; total, $4,700. *Revenue:* consumers, $2,600; city, $690; other purposes, $2,050; total, $5,340. **Management.**—Prest., Jno. Lockwood; Secy., Jno. C. Lockwood; both of 52 Broadway, N. Y. City.

139. **PEEKSKILL,** *Westchester Co.* (Pop., 6,893—9,676.) Built in '75 by city. **Supply.**—Van Cortland Creek, pumping to reservoir. Fiscal year closes Feb. 20.

Year.	Mains.	Taps.	Meters.	Hydrants.	Cost.	B'd'd debt.	Int. ch'g's.	Exp.	Rev.
'82-3	0.9	428	19	78	$138,300			$2,678	$5,431
'87-8	9.8	612	58	86	150,000	$150,000	$10,500	3,655	12,774
'89-90	10.5	742	35	104	154,360	158.360	10,834	7,160	10,061
'90-1	12	778	21	103	159,313	159,373	10,883	3,961	10,883

Financial.—*Bonded debt,* $159,373; $150,000 at 7%, $9,373 at 4%. *Sinking fund,* $15,547, invested in Peekskill Savings Bank at 3½% and in 7% water bonds. *Expenses:* total, $14,550. See above. *Revenue:* consumers, $10,883; city, $6,253; total, $17,136. **Management.**—Comrs., elected every two years. Secy., Frederick Lent. Supt., O. J. Loder.

140. **PELHAM MANOR,** *Westchester Co.* (Pop. of town in '80, 2,540.) **History.**—Construction begun Mar. 1, '91, by Pelham Heights Co., in connection with sewers and street improvements; to be completed by Nov. 1, '92. Engrs., J. F. Fairchild and G. H. Eldridge. Contrs., Fogg & Scribner, Mt. Vernon. **Supply.**—New Rochelle Water Co.'s works. **Distribution.**—*Mains,* 8 to 4-in. c. i., about 6 miles; from R. D. Wood & Co., Philadelphia. *Hydrants,* 35. **Financial.**—*Cap. stock:* authorized, $350,000. **Management.**—Prest., Benj. Fairchild, Pelhamville. Secy. and Treas., Paul Gorham, 155 Broadway. Report by C. E. Fogg, July 31, '91.

Penn **Yan,** *Yates Co.* (Pop., 34,75—4,254.) Built in '84 for fire protection by village and Russell & Birkett.

141. **PHELPS,** *Ontario Co.* (Pop., 1,369—1,336.) Built in '89 by Phelps Water-Works Co. **Supply.**—Springs, by gravity. Fiscal year closes Dec. 31. **Distribution.**—*Mains,* 4 miles. *Taps,* 50. *Meters,* 14. *Consumption,* 20,000 galls. **Financial.**—*Cap. Stock:* authorized, $20,000; all paid up. Paid no div'd in 1890. *Bonded Debt,* $20,000, at 6%, due in 1909; $5,000 of bonds reserved in treasury. **Management.**—Prest., E. M. Harrington; Secy., E. M. Bassett; both of Buffalo. Supt., W. D. Norton.

142. **PHOENIX,** *Oswego Co.* (Pop., 1,312—1,466.) First supplied in '79; present works built in '88 by Phoenix Water Co. **Supply.**—Oswego River, wells and springs, pumping to stand-pipe. Fiscal year closes Dec. 31. **Distribution.**—*Mains,* 6 miles. *Taps,* 97. *Meters,* 0. *Hydrants,* 27. *Consumption,* 50,000 galls. **Financial.**—*Cost,* $82,000. *Cap. Stock:* authorized, $40,000; all paid-up. *Total Debt,* $41,375: floating, $1,375;

NEW YORK. 87

bonded, $10,000 at 6%. *Expenses:* operating, $541; int., $2,400; taxes, $55; total, $2,999. *Revenue:* consumers, $2,075; city, $1,025; total, $3,100. **Management.**—Prest., J. I. Van Doren. Secy. and Treas., Van R. Sweet.

143. PLATTSBURGH, *Clinton Co.* (Pop., $5,245—7,010.) Built in '70 by village. **Supply.**—Surface water, by gravity from impounding and distributing reservoirs. **Distribution.**—*Mains,* 15 miles. *Meters,* 0. *Hydrants,* 68. *Consumption,* 500,000 galls. **Financial.**—*Cost* (in '81), $180,000. *Expenses,* operating, $3,072. *Revenue,* consumers, $17,565. **Management.**—Comrs.: N. Baker, C. E. M. Everett, H. L. Inman. Supt., Andrew Williams.

144. PORT BYRON, *Cayuga Co.* (Pop., 1,146—1,105.) Built in '72 by village. **Supply.**—Well, pumping to reservoir. **Distribution.**—*Mains,* 3 miles. *Taps,* about 60. *Meters,* 0. *Hydrants,* 26. *Consumption,* 25,000 galls. **Financial.**—*Cost,* $14,000. *No Debt. Expenses,* $300. *Revenue,* $300. **Management.**—Village Board. Report by John L. Davis.

145. PORT CHESTER, *Westchester Co.* (Pop., 3,254—5,274.) Built in '84 by Port Chester Water-Works Co.; supplies Rye and Harrison. **Supply.**—Greenwich, Conn., Water Co.'s works; filtered. **Distribution.**—*Mains,* 20 miles. *Taps* (in '87), 500. *Meters,* 6. *Hydrants,* 60. *Consumption,* 600,000 galls. **Financial.**—Withheld. **Management.**—Prest., John A. Lansbury. Secy., John W. Diehl. Treas., Wm. P. Abendroch.

146. PORT JERVIS, *Orange Co.* (Pop., 8,678—9,327.) Built in '70 by Port Jervis Water-Works Co. **Supply.**—Surface water, by gravity from storage reservoir. Fiscal year closes March 31. There are two reservoirs. Jansen's Lake has storage cap. of 175,000,000 galls., area of 25 acres; max. depth, 24 ft.; about 300 ft. above village and about 145 ft. above and 3 miles from lower reservoir. Formed by earth dam 400 ft. long, 29 ft. high; inner slope 2½ to 1, paved with thin stone on edge; outer slope, 2 to 1, unpaved. Embankment of clay loam and gravel; top soil removed from site for depth of 2 ft. Sheet-piling of 3 in. chestnut plank, driven 4 ft. beneath bottom of dam. Spillway in end of dam, 5 ft. deep, lined with masonry. free; slope, 1 in 4. Dam does not leak. Built in '80 without contr. Engr., Chas. Rose. Cost, $20,000.

An 8-in. c. i. pipe, 1,200 ft. long, serves as siphon to carry water over divide to stream leading to lower reservoir. Short leg of siphon is 300 ft. long and its lower end is 26 ft. below summit; long leg has lower end 60 ft. below summit. Summit is but 2 ft. above normal water line, but siphon sometimes draws water down to within 2 ft. of inlet. Inlet in upper end of reservoir, and has ditch leading from lowest part of reservoir to inlet-well; at well ditch is 16 ft. deep-lined with masonry. Ludlow valve at each end of siphon, and 2½-in. w. i. pipe connects with siphon at well and rises 2 ft. above summit. There is a ¾-in. air-cock at summit. Siphon is charged by means of hand-pump in well-house, the charge-pipe having a 2½-in. valve. Three feet of cutting for about 200 ft. at summit, and lowering of pipe, would have made pipe self-charging, but cut would have been in hard rock.

Lower reservoir, called Reservoir Lake, has area of 17 acres; cap., not stated; formed by clay and loam embankment 375 ft. long, 26 ft. high, 21 ft. water. Inner slope, 2 to 1, paved; outer 3 to 1, unpaved. Chestnut sheeting spiked to timber frame, in center of embankment, and carried 2 ft. below. Spillway 8 ft. deep, 20 ft. wide, in end of dam; bottom planked, sides lined with masonry; controlled by gates; slope, 6 in 100. A 16 and a 12-in. pipe, with slope of 2 in 100, extend through dam. Leaks very little. Built in '70. Des. Engr., John B. Jervis. Const. Eng. and Contr., H. H. Farnum. Cost, $20,000. **Distribution.**—*Mains,* 17 miles. *Taps,* 900. *Meters,* 0. *Hydrants,* 166. *Consumption,* 1,500,000 galls. **Financial.**—*Cost,* $120,000. *Cap. Stock,* $145,000. *Bonded Debt,* $20,000 at 5%. *Expenses:* operating, $500; int., $1,000; taxes, $112; total, $1,612. *Revenue:* total, $20,200. **Management.**—Prest., P. E. Farnum. Secy., W. H. Crane. Treas., W. E. Scott. Supt., E. H. Ellis.

119. PORT RICHMOND, *Richmond Co.* See New Brighton.

147. POTSDAM, *St. Lawrence Co.* (Pop., 2,762—3,691.) Built in '71 by village. **Supply.**—Racket River, pumping direct. Fiscal year closes Dec. 31. During '90 an 870,000-gall. Buffalo pump and 50 HP. boiler were added.

For Further Descriptive Matter see Manual for 1889-90. Changes since 1890 are here given. Populations are for 1880 and 1890, respectively.

Year.	Mains.	Taps.	Meters.	Hyd't's.	Cons'n.	Cost.	B'd'd d't.	Int.	Exp.	Rev.
'82	4	245	0	..	......	$50,000	......	......	$391	$1,889
'89	5	350	0	36	532,935	50,000	$47,000	$3,290	850	4,000
'90	5	373	0	35	492,752	50,000	47,000	3,290	903	4,995

Financial.—Bonded Debt, $47,000 at 7%, due in '91. *No Sinking Fund. Expenses*, total, $4,283. *Revenue*, consumers, $4,225. **Management.**—Three comrs., appointed every year by council. Chn., Geo. Lewis. Secy., S. C. Crane. Supt., W. F. P. Sealy, appointed by comrs.

148. POUGHKEEPSIE, *Dutchess Co.* (Pop., 20,207—22,206.) Built in '70 by city. **Supply.**—Hudson River, pumping to settling basin, thence through filters to reservoir, thence pumping to distributing reservoir. Fiscal year closes Dec. 31.

Year.	Mains.	Taps.	Meters.	Hy-drants.	Con-sumption.	Cost.	Bonded debt.	Int. ch'g's.	Exp.	Rev.
'72	9.3	651	0	136	508,661	$437,014	......	......	$22,128	......
'78	16.7	1,100	..	....	1,610,897	....	....	....	23,650	....
'80	16.7	1,326	122	280	1,104,861	552,388	$550,000	$38,500	18,626	$19,117
'82	16.7	1,436	176	285	1,392,616	553,546	550,000	38,500	15,286	22,083
'84	17.5	1,584	276	298	1,358,163	560,801	550,000	38,500	16,437	24,450
'85	18	1,652	358	310	1,620,014	580,869	550,000	38,500	16,822	25,106
'86	18	1,699	433	315	1,738,284	581,481	550,000	38,500	18,972	24,343
'87	18.4	1,761	530	315	1,766,735	583,055	550,000	38,500	18,011	24,883
'88	18.6	1,852	652	319	1,619,358	584,641	550,000	38,500	18,875	25,937
'89	18.8	1,961	812	323	1,567,867	585,552	550,000	38,500	19,941	26,073
'90	19.3	2,058	964	335	1,680,362	588,846	550,000	38,500	20,021	29,364

Management.—Four Comrs. Prest., Edgar C. Adriance. Supt., Chas. C. Fowler.

149. PRATTSVILLE, *Greene Co.* (Pop., 398—384.) Built in '85 by Prattsville Water Co., under 10-years franchise. **Supply.**—Stream, by gravity. **Distribution.**—*Mains*, 2 miles. *Taps*, 18. *Hydrants*, 10. **Financial.**—*Cost*, $7,000. *Expenses*, $50. *Revenue*, $100. **Management.**—Prest. and Treas., Francis Leonard, Norwalk, Conn. Secy. and Supt., Martin Chamberlain.

150. PULASKI, *Oswego Co.* (Pop., 1,501—1,517.) Built in '87 by village. **Supply.**—Spring brook, pumping direct from impounding reservoir. Fiscal year closes May 1.

Yr.	Mains.	Taps.	Meters.	Hydr'ts.	Cons'n Cost.	B'd'd d't.	Int ch'gs.	Exp.	Rev.
'88-9	3	150	0	30	6,000 $30,000	$25,000	$1,000	$200	$900
'89-90	3.5	168	0	32	7,000 34,000	25,000	1,000	200	900
'90-1	3.5	172	0	32	8,000 31,000	25,000	1,000	200	1,160

Financial.—*Total Debt*, $30,000: floating, $5,000 at 5%; bonded, $25,000 at 4%. *No Sinking Fund. Expenses:* total, $1,200. *Revenue:* consumers, $1,160; city, none. **Management.**—Chn., W. B. Dixon. Secy., Lorenzo Ling. Pumping Engr., E. D. Fitch.

151. RANDOLPH, *Cattaraugus Co.* (Pop., 1,111—1,201.) Built in '85 by village. **Supply.**—Spring, by gravity. **Distribution.**—*Mains*, 6 miles. *Taps*, 180. *Hydrants*, 18. **Financial.**—*Cost*, $21,000. *Bonded Debt*, $20,000 at 4%. *Expenses:* operating, none; int., $800. *Revenue*, $1,200. **Management.**—Prest., Warren Dow. Secy., C. C. Sheldon. Treas., Wm. L. Rathbone.

152. RICHFIELD SPRINGS, *Otsego Co.* (Pop., 1,307—1,623.) Owned by village. Built in '79 by Richfield Springs Water-Works Co. **Supply.**—Creek, pumping to reservoir; filtered. Fiscal year closes May 1. **Distribution.**—*Mains*, 5 miles. *Taps*, 220. *Meters*, 0. *Hydrants*, 35. *Consumption*, 250,000 galls. **Financial**—*Cost* (in '88), $22,500. *Bonded Debt*, $98,500: $49,000 at 7%, $22,000 at 6%, $27,000 at 5%. *Sinking Fund*, $2,500, invested in village bonds. *Expenses*, operating, $800. *Revenue:* consumers, $4,000; city, none. **Management.**—Prest., L. S. Chase. Secy., H. C. Watson. Treas., N. Gilman. Supt., C. D. Rishey.

153. ROCHESTER, *Monroe Co.* (Pop., 89,366—133,596.) Built in '72-6 by city. **Supply.**—Hemlock and Canadice Lakes and Genesee River, by gravity to storage and distributing reservoirs: from Genesee River, pumping direct for independent supply in business parts of town for fire protection and hydraulic motors; Apr. 6, '91, there were 13.9 miles of pipe in this, or "Holly" system; av. dy. consumption, this system, for year, 1,800,000 galls. In Aug., '91, temporary supply from wells or from river, filtered, was under consideration. At same time surveys were being made for new gravity conduit from Hemlock Lake, present source. Fiscal year closes first Monday in April.

NEW YORK. 89

Yr.	Mains.	Taps.	M'trs.	Hyd'ts.	Cons'n.	Cost.	B'd'd d't.	Int.ch'g's.	Exp.	Rev.
'76*	73	2,700	....	651	2,227,953	$3,250,000	....	....	....	$13,040
'80	112.8	7,800	125	930	4,311,432	3,119,749	....	....	....	96,693
'81	121.5	....	....	*1,048	4,237,078	....	....	....	....	75.766
'82-3	130.1	11,225	650	1,130	4,525,926	3,313,781	$3,182,000	$222,740	$47,792	93,380
'84-5	154	13,700	807	1,260	6,077,300	3,664,749	3,182,000	222,710	30,000	....
'85-6	....	15,810	1,094	1,461	6,342,292	....	....	....	....	....
'86-7	175.5	17,454	1,000	1,567	6,735,923	3,741,123	3,182,000	222,740	30,000	145.503
'87-8	187.6	17,000	1,590	1,713	7,200,090	3,805,123	3,182,000	222,740	30,000	150,572
'88-9	204.4	21,358	1,700	1,963	10,750,000	3,980,123	3,182,000	222,710	30,000	163,177
'89-90	215.1	23,146	2,115	2,049	9,090,000	4,214,153	3,592,000	250,000	61,209	196,830
'90-1	221.8	24,838	2,841	2,131	8,800,000	4,320,304	3,592,000	260,030	87,843†	309,513

* Year '76 closed Dec. 31; all others close first Monday in April.
† Of this sum $23,585 was for "extraordinary expenses," not occurring annually.
‡ Includes revenue from consumers, only, except for '90-1, when revenue from frontage tax, sale of meters, etc., is included.

Financial.—Total Debt, $3,592,000 : *floating*, none ; *bonded*, $3,182,000 at 7%, due in 1903; $410,000 at 7%, due in 1905. *Sinking Fund*, none. *Expenses:* operating, $64,263; extraordinary, $23,585; int., $260,000; total, $347,848. *Revenue:* consumers and frontage tax, $283,955; sale of meters, making taps, etc., $24,558; city, fire protection. $100,000. Total cost of construction and interest not paid from receipts, $7,734,216. *Management.*—Executive Board of three members, one elected every year. Chn., T. W. Aldridge. Clerk, Thos. J. Neville. Chf. Engr. and Gen. Supt., Emil Kuichling, appointed by board.

Note.—For illustrated description of parts of works, see *Engineering Record* for Nov. 29, '90, *et seq.*

54. ROCKAWAY BEACH, *Queens Co.* (Pop. (in '9)), 1,502.) Supply from Far Rockaway.

154. ROME, *Oneida Co.* (Pop., 12,194—14,991.) Built in '73 by city. Supply.— Mohawk River, pumping to reservoir and direct in case of fire. Fiscal year closes Oct. 5.

Year.	M'ns.	Taps.	Meters.	Hydrants.	Con- sumpt'n.	Annual Cost.	B'd'd debt.	Int. ch'g's.	Exp.	Rev.
'80	16	600	....	125	900,000	....	....	....	$2,500	....
'82	16	652	0	130	1,000,000	....	....	....	3,000	$11,000
'84	18	800	0	118	1,000,000	....	$160,000	$11,200	2,500	12,000
'86	18	1,000	6	135	1,400,000	....	160,000	11,200	....	13,000
'87	19	998	0	131	1,760,000	2,252	160,000	11,270	9,455	14,265
'88	19	1,050	0	147	2,144,372	9,535	160,000	11,200	9,858	15,470
'89	20.5	1,104	0	152	2,005,000	12,875	160,000	11,200	4,870	17,006
'90	22	1,228	0	160	2,063,648	19,492	160,000	11,200	4,363	17,893

Financial.—First Cost, $160,000. *Bonded Debt*, $160,000, at 7%, due in '92, will be renewed when due. *Expenses*, total, $15,563. *Revenue:* consumers. $17,893; city, none. *Management.*—Four comrs., elected every year by mayor, who is Chn. Secy. and Supt., Theo. S. Comstock.

155. ROUND LAKE, *Saratoga Co.* (Pop., est., winter, 300; summer, 1,500.) Built in '87 by Round Lake Ass'n. Supply.—Springs, collected in reservoir, pumping to tanks. Fiscal year closes Nov. 1. During '90 a 4,000,000-gall. storage reservoir was built. Distribution.—*Mains*, 2 miles. *Taps*, 150. *Meters*, 0. *Hydrants*, 14. *Consumption*, 60,000 galls. Financial.—*Cost*, $15,000. *No Debt. Expenses*, $300. *Revenue*, $1,600. Management.—Executive Com. of Ass'n. Prest., Wm. Griffin. Secy., W. S. Kelley. Supt., Jno. S. Rogers.

156. ROUSES POINT, *Clinton Co.* (Pop., 1,184—1,856.) Built in '89 by village. Supply.—Lake Champlain, pumping to stand-pipe. Fiscal year closes April 30. Distribution.—*Mains*, 3.5 miles. *Taps*, 135. *Meters*, 8. *Hydrants*, 40. *Consumption*, 65,515 galls. Financial.—*Cost*, $26,444. *Bonded Debt*, $24,000 at 4%, due in 1909. *Expenses:* operating, $1,850; int., $960; total, $2,810. *Revenue:* consumers, $2,139; city, none. Management.—Chm., W. T. Crook. Secy., S. H. Newton. Supt., J. Pardy.

157. ROXBURY, *Delaware Co.* (Pop., 335—est., 500.) History.—Built in '90 by Roxbury Village Water Co.; put in operation Nov. 1; contr., B. B. Bouton, Roxbury. Supply.—Springs, by gravity. Reservoir.—Cap., 200,000; 50 × 50 ft., × 12 ft. deep; bottom 200 ft. above village. Distribution.—*Mains*, 8 to 4-in. c. i., about 3 miles; from R. D. Wood & Co., Philadelphia. *Services*, w. i. *Taps*, 25. *Hydrants*, Matthews, 21. *Valves*, Eddy. Financial.—*Cost*, $20,000. *Cap. Stock:* authorized,

For Further Descriptive Matter see Manual for 1889-90. Changes since 1890 are here given. Populations are for 1880 and 1890, respectively.

90   MANUAL OF AMERICAN WATER-WORKS.

$20,000; all paid-up. *No Debt. Operating Expenses,* $150. *Revenue:* consumers, $400; city, $420. total. $820. **Management.**—Prest., C. G. Kestor. Secy., J. R. Dart. Treas., B. Brewster. Supt., B. B. Bouton.

145. **RYE,** *Westchester Co.* See Port Chester.

158. **SAG HARBOR,** *Suffolk Co.* (Pop., 1,996—est., 3,000.) Built in '89, by Sag Harbor Water Co. Supply.—Wells, pumping to stand-pipe. Fiscal year closes Dec. 1. Distribution.—*Mains,* 5 miles. *Hydrants,* 30. Financial.—*Cost,* $60,000. *Cap. Stock, authorized,* $60,000; all paid up; paid no div'd in 1890. *Bonded Debt,* $40,000 at 6%. **Management.**—Prest., C. N. Hoagland, 410 Clinton Ave., Brooklyn. Secy., Wm. B. Hill, 52 Wall street, N. Y. City. Treas., Chas. O. Gates, 100 Greene Ave., Brooklyn.

XII. **SAINT JOHNSVILLE,** *Montgomery Co.* (Pop., 1,072—1,263.) Built in '79 by village. Supply.—Springs, by gravity. Distribution.—*Mains,* 3 miles. *Taps,* 175. *Hydrants,* 0. Financial.—*Cost,* $5,000. *Debt,* $500 at 6%. *Expenses,* $130. *Revenue,* $500. **Management.**—Village trustees. Chn., G. Hough. Supt., M. S. Schram.

159. **SALAMANCA,** *Cattaraugus Co.* (Pop., 2,531—3,692.) Built in '81 by Salamanca Water-Works Co, Supply.—Springs and surface water, by gravity and by pumping to reservoir in dry weather. Fiscal year closes Jan. 5. Distribution.—*Mains,* 5 miles. *Taps,* 245. *Meters,* 2. *Hydrants,* 43. Financial.—*Cost,* $35,000. *Cap. Stock,* $20,000; paid 10% div'd in 1890. *Bonded Debt,* $20,000 at 6%. *Expenses:* operating, $1,130; int., $1,200; taxes, $481; total, $2,811. *Revenue:* consumers, $2,475; city, $1,350; total, $3,825. **Management.**—Prest., Hudson Ansley. Secy., H. A. Wait. Treas., W. W. Wellman. Supt., A. H. Krieger.

160. **SANDY HILL,** *Washington Co.* (Pop., 2,487—2,895.) Built in '88 by Spring Brook Water Co. under unlimited franchise. Supply.—Brook, fed by springs, pumping to stand-pipe. Fiscal year closes June 14.

Year.	Mains.	Taps.	Meters.	Hyd'ts.	Cons'n.	Cost.	B'd'd d't.	Int.ch'g's.	Exp.	Rev.
'88	6.75	149	37	41	100,000	$80,000	$50,000	$2,600	$1,500	$5,66
'90	6.75	200	40	41	125,000	80,000	40,000	2,600	2,594	5,26

Financial.—*Cap Stock:* authorized, $30,000; all paid up; paid no dividend in 1890 *Total Debt:* $55,000; floating, $15,000 at 5%; bonded, $40,000 at 5%, due in 1909. *Expenses:* operating, $2,160; int., $2,658; taxes, $433; total, $5,251. *Revenue:* consumers, $2,700; city, $2,562; total, $5,263. **Management.**—Prest., Loren Allen. Vice-Prest., S. H. Kenyon. Treas., Chas. T. Beach. Secy., G. M. Ingallebee. Supt., J. Doubleday.

161. **SARATOGA SPRINGS,** *Saratoga Co.* (Pop., 8,421—11,975.) Built in '71 by village. Supply.—Loughberry Lake, pumping direct. Fiscal year closes May 1. Distribution.—*Mains,* 32 miles. *Taps,* 2,500. *Meters,* 12. *Hydrants,* 240. *Consumption,* 3,500,000 galls. Financial.—*Cost,* $330,000. *Bonded Debt,* $266,300 at 3½, 4 and 5%. *Expenses:* operating, $13,000; int., $12,510; total, $25,510. *Revenue:* consumers, $43,582. **Management.**—Chn., Jas. M. Ostrander. Secy., A. S. Browne. Supt., D. Franklin.

162. **SAUGERTIES,** *Ulster Co.* (Pop., 3,923—4,257.) Built in '87 by Saugerties Water Co. under 30 years franchise. Supply, Plotull Creek, by gravity. Distribution. (In '88.)—*Mains,* 13.5 miles. *Taps,* 250. *Meters,* 3. *Hydrants,* 58. Financial. (In '88.)—*Cap. Stock,* $100,000. *Bonded Debt,* $100,000 at 6%. Hydrant rental, $10 each. **Management.**—Prest., Matthew Trumpbower. Secy., E. B. Walker. Treas., C. D. Bruyn. Supt., F. M. Murphy.

90. **SAYVILLE,** *Suffolk Co.* See Islip.

163. **SCHENECTADY,** *Schenectady Co.* (Pop., 13,655—19,902.) Owned by city. Built in '71 by Schenectady Water Co.; bought by city in '85. Supply.—Mohawk River, pumping direct. Fiscal year closes Nov. 1. Intake and crib put in in '87 were abandoned in '90, having broken apart in several places; a new 24-in. intake was put in; also a 5,000,000-gall. Gaskill pump.

Year.	Mains.	Taps.	Hydr'ts.	Cons'n.	Cost.	B'd'd d't.	Int.chg's.	Exp.	Rev.
'84	13	750	138	1,000,000	$98,507	$90,000	....	$12,000	$27,350
'86	15.1	1,011	149		123,791	107,000	....	15,967	14,117
'87	16.1	1,200	154	1,486,715	130,558	107,000	$4,110	12,388	19,744
'88	16.5	1,250	155	2,120,000	131,766	107,000	4,110	12,243	20,651
'89	17.1	1,250	161	2,274,143	150,678	163,000	4,110	19,464	22,814
'90	20.5	1,458	188	2,246,907	200,687	163,000	5,400	19,627	30,034

Financial.—*First Cost,* $115,000. *Total Debt.* $163,000 : floating, none ; bonded, $90,000 at 4%, due in 1902-15; $73,000 at 3%, due in 1902-16. *Sinking Fund,* $3,279, invested in Union Trust Co., N. Y., at 3% int. *Expenses:* Operating, $10,629; int. and sinking fund, $6,000. *Revenue:* consumers, $13,146; city, fire protection, $19,514, raised by taxation on improved and unimproved property called "fire tax." **Management.**—Three comrs., appointed every year by common council. Chn., Edgar M. Jenkins. Secy., Edw. C. Angle. 277 State St. Supt., Geo. T. Ingersoll.

164. **SCHENEVUS**, *Otsego Co.* (Pop. in '90, 665.) Built in '88 by Schenevus Village Water-Works Co. Fiscal year closes Jan. 14. During '90 a reservoir was built; cap., 10,000,000 galls.; dam of earth, 25 ft. high, with rubble heart-wall; cost, over $5,000. A system was purchased which supplies 25 families with spring water only; two systems may be connected, whenever there is a lack of water in either. Supply.—Springs forming small stream, by gravity from reservoir, filtered through stone, gravel and sand. Distribution.—*Mains*, 4 miles. *Taps*, 50. *Meters*, 0. *Hydrants*, 13. *Consumption*, 180,000 galls. Financial.—*Cost*, $25,000. *Cap. Stock:* authorized, $7,500; all paid up. Profits to date have been used in completing works. *Bonded Debt*, $10,000 at 5%, due in 1901. *Expenses:* operating, $100; int., $780; taxes, $75; total, $955. *Revenue:* consumers, $375; city, $325; Ry., $800; total, $1,500. **Management.**—Prest., Lyron Cass. Secy., Lottie A. Field, Cooperstown. Supt., M. E. Baldwin.

165. **SCHOHARIE**, *Schoharie Co.* (Pop., 1,188—1,028.) **History.**—Built in '90 by Schoharie Water Co.; put in operation Dec. 1. Supply.—Springs, by gravity. Distribution.—*Mains*, 8 to 4-in. Wyckoff, 1 mile. *Hydrants*, Holly. 7. Financial.—*Cost*, $10,000. *Cap. Stock*, authorized, $10,000. **Management.**—Prest., D. S. Mayham. Secy., E. H. Heck. Treas. and Supt., C. C. Kromer. Report, Nov. 8, '90.

XIII. **SEA CLIFF**, *Queens Co.* (Pop., 554; est., 1,000.) Built in '72 by Sea Cliff Grove & Metropolitan Camp Ground Ass'n of N. Y. and Brooklyn. Supply.—Seven driven wells, pumping to tank. *Mains* (in '87), 5 miles. No *Meters* or *Hydrants*. Financial.—(In '87). *Cost*, $5,000. *Operating Expenses*, $1,200. **Management.**—(In '88.) Prest., J. T. Pirie. Secy., J. K. Myers.

XIV. **SENECA FALLS**, *Seneca Co.* (Pop., 5,880–6,166.) Built in '87 by Co.; franchise provides neither for control of rates nor purchase of works by town. Supply.—Cayuga Lake, pumping to stand-pipe. Distribution.—*Mains*, 12 miles. *Taps*, 400. *Meters*, 0. *Hydrants*, 0. Financial—Withheld. **Management.**—Prest., Cornelius Hood. Secy., Frank Westcott. Treas., G. Norman Weaver, Newport, R. I. Supt., A. C. Martin.

74. **SHEEPSHEAD BAY**, *Kings Co.* See Gravesend.

166. **SHERBURNE**, *Chenango Co.* (Pop., 944—960.) Built in '83 by village; supplies Sherburne Quarter. Supply.—Surface water, by gravity from impounding reservoir. Fiscal year closes May 1. New storage reservoir and filter bed were to be constructed during '91.

Year.	Mains.	Taps.	Meters.	Hydrants.	B'd'd Debt.	Int Ch'g's.	Exp.	Rev.
'86-7	4	100	0	18	$32,000	$1,280	$500	$600
'88-9	4.75	150	0	21	31,000	1,240	500	1,900
'89-90	5	175	0	23	31,000	1,240	150	1,950
'90-1	5	180	1	24	30,000	1,240	375	2,000

Financial.—*Bonded Debt*, $30,000 at 4%, due in annual payments of $1,000. *Expenses:* total $1,615. *Revenue*, consumers, $2,000. **Management.**—Three comrs., one appointed every year. Chn., D. B. Ames. Secy., C. H. Todd. Supt., W. E. Davis, appointed by comrs.

166. **SHERBURNE QUARTER**, *Chenango Co.* (Pop., 337—est., 500.) Supply from Sherburne.

167. **SIDNEY**, *Delaware Co.* (Pop., in '90, 1,358.) Built in '89 by Sidney Water-Works Co. Fiscal year closes Dec. 31. Supply.—Mountain stream, by gravity from impounding and distributing reservoir; filtered. Distribution.—*Mains*, 5 miles. *Taps*, 71. *Hydrants*, 27. *Consumption*, unk'n. Financial.—*Cost*, $33,000. *Cap. Stock:* authorized, $30,000; paid-up, $16,500. *Total Debt:* $15,500; floating, $500 at 6%; bonded, $15,000 at 4%, $5,000 due in '99, in 1904, and in 1909. *Expenses:* operating, withheld; int., $600; taxes, $150. *Revenue:* consumers, $1,300; city, $435; total, $1,735. **Management.**—Prest., H. W. Clark. Secy., Jno. A. Clark. Supt., W. E. Barker

29. **SILVER SPRINGS**, *Wyoming Co.* See Castile.

For Further Descriptive Matter see Manual for 1889-90. Changes since 1890 are here given. Populations are for 1880 and 1890, respectively.

168. **SING SING,** *Westchester Co.* (Pop., 6,578—9,352.) Built in '88-9 by village. **Supply.**—Indian brook, pumping from impounding to distributing and high pressure reservoirs. Fiscal year closes May 1. **Distribution.**—(Jan. 1, '89) *Mains,* 6 miles. *Hydrants,* 93. **Financial.**—(In '88.) *Cost,* about $150,000. *Bonded Debt,* $150,000 at 4%. **Management.**—Comrs.

169. **SKANEATELES,** *Onondaga Co.* (Pop., 1,669—1,559.) **History.**—Built in '90 by Skaneateles Water-Works Co.; put in operation Dec. 15. Des. Engr., C. W. Knight, Rome. Const. Engr., Thos. Regan. Contrs., New York Pipe Mfg. Co., New York City. **Supply.**—Skaneateles Lake, pumping to stand-pipe; not filtered. **Pumping Machinery.**—Dy. cap., 1,000,000 galls.; Barr; works against 200-ft. head. *Boilers,* Erie. **Stand-Pipe.**—Cap., 94,000 galls.; 20 × 40 ft. **Distribution.**—*Mains,* 10 to 4-in., about 6 miles. *Hydrants,* Ludlow, 40. *Pressure:* ordinary, 20 to 80 lbs.; fire, 80 to 100 lbs. **Financial.**—*Cost,* about $35,000. *Cap. Stock:* authorized $40,000; all paid up. *Bonded Debt:* authorized, $40,000 at 6%; all issued. Hydrant rental, $30 each; 5-years' contract. **Management.**—Prest., C. H. Jackson. Secy., J. P. McQuade. Treas., E. S. Perot. Vice-Prest. and Mgr., George Borrow. Report, Dec. 23, '90.

170. **SPRINGVILLE,** *Erie Co.* (Pop., 1,227—1,883.) Built in '88 by Springville Water-Works Co. **Supply.**—Springs, pumping direct from collecting reservoir. **Distribution.**—(In '88.) *Mains,* 2 miles. *Hydrants,* 19. **Financial.**—Not given. **Management.**—Address J. Shuttleworth or C. J. Shuttleworth.

171. **STAMFORD,** *Delaware Co.* (Pop., 764—819.) Built in '83 by Stamford Water-Works Co. **Supply.**—Headwaters of Delaware River, by gravity from impounding reservoir. Fiscal year closes Jan. 6. **Distribution.**—*Mains,* 3 miles. *Taps,* 78. *Meters,* 0. *Hydrants,* 16. *Consumption,* unk'n. **Financial.**—*Cost,* $24,500. *Cap. Stock:* authorized, $20,000; all paid up; paid no div'd in '90. *Floating Debt,* $3,000. *Expenses:* operating $59; int., $126; taxes, $110; total, $295. *Revenue:* consumers, $1,352; city, $320; total, $1,672. **Management.**—Prest., M. Fredenburgh. Secy., V. Z. Wyckoff. Treas., J. H. Merchant. Supt., J. Hamilton.

172. **STAPLETON and EDGEWATER,** *Richmond Co.* (Pop.: Stapleton, est., 2,500; Edgewater, 8,044—14,265.) Built in '84 by Crystal Springs Water Co. **Supply.**—Driven wells, pumping to reservoir and stand-pipe. **Pumping Machinery.**—Cap., 2,500,000 galls., two Deane and one 1,000,000 Worthington. **Reservoir.**—*Cap.,* 5,000,000 galls. **Distribution.**—*Mains,* "sheet iron spiral pipe," 40 miles. *Taps,* 1,500. *Meters,* Crown, 30. *Hydrants,* 290. **Financial.**—Withheld. **Management.**—Prest., Calvin Detrick. Secy., J. W. Cummings. Treas., E. H. York.

173. **SUSPENSION BRIDGE,** *Niagara Co.* (Pop., 2,476—4,405.) Built in '75 by village. **Supply.**—Niagara Falls, pumping direct. Fiscal year closes June 1. **Distribution.**—*Mains,* 8.5 miles. *Taps,* 722. *Meters,* 0. *Hydrants,* 46. *Consumption,* 4,500,000 galls. **Financial.**—*Cost,* $180,000. *Bonded Debt,* $122,000 at 4, 5, 6 and 7%. *No Sinking Fund. Expenses:* operating, $8,000; int., $7,000; total, $15,000. *Revenue:* consumers, $12,700; city, $2,300; total, $15,000. Works are said to be self-sustaining, bonds being redeemed from net income. **Management.**—Five comrs., elected every four years. Prest., Jacob Bingenheimer, Sr. Secy., Jacob H. Bingenheimer, Jr. Supt., H. A. Keller, appointed by comrs.

174. **SYRACUSE,** *Onondaga Co.* (Pop., 51,792—88,143.) Owned by Syracuse Water Co.; franchise does not regulate rates, but provides for purchase of works at any time. Built in '29 by three individuals. **Supply.**—Springs and surface water, by gravity and pumping to reservoir. **Distribution.**—*Mains,* 42.5 miles. *Taps,* 3,438. *Hydrants,* 381. **Financial.**—*Cost* (in '88), $860,000. *Cap. Stock* (in '88), $440,000. *Bonded Debt* (in '88), $200,600. *Revenue:* city, $26,000. **Management.**—Prest., Dwight H. Bruce. Secy. and Treas., Jno. G. Butler.

NOTE.—City has right to issue $3,000,000 of bonds, purchase co's works and introduce new supply from Skaneateles Lake, 17 miles distant. Attempt was made in courts to prevent such a course, but in '91 suits were decided in favor of city and commission began appraisal of co's works.

175. **TARRYTOWN,** *Westchester Co.* (Pop., 3,025—3,562.) Built in '87 by village. **Supply.**—Surface water, pumping to reservoir and tank, latter for high service. Fiscal year closes May 1. During '91 plant was to be thoroughly repaired and a pump added. **Distribution.**—*Mains,* 14 miles. *Taps,* 300. *Meters,* 300. *Hydrants,*

73. *Consumption*, 35,000 galls. **Financial.**—*Cost* (in '88), $75,000. *Total Debt*, $99,-000; floating, $14,000 at 4½%; bonded, $85,000 at 4%. *Expenses:* operating, $6,177. *Revenue:* consumers, $4,292; city, none. **Management.**—Prest., D. D. Tallman. Secy., F. R. Pierson. Supt., Alex. Meginley.

176. **THOUSAND ISLAND PARK,** *Jefferson Co.* (Pop., in summer, est., 3,000.) Built in '85 by Co. Engr. and Contr., D. A. Bailey. **Supply.**—St. Lawrence River, pumping to tank. **Pumping Machinery.**—Dy. cap., not given; from Empire Pump Co., Carthage. *Boiler*, from Watertown Steam Engine Co. *Tank.*— Cap., 12,800 galls. **Distribution.**—*Mains*, w. l., 1 mile. *Taps*, about 150. *Meters*, none. *Hydrants* and *Valves*, Ludlow. *Pressure:* ordinary, 60 lbs.; fire, 100 lbs. **Financial.**—*Cost*, about $5,000. **Management.**—Supt., D. A. B. Bailey, Findlay, O.

177. **TICONDEROGA,** *Essex Co.* (Pop., about 1,800—2,267.) Built in 75 by Ticonderoga Water-Works Co. **Supply.**—Lake George, by gravity. Fiscal year closes April 30. Filter was to be added in '90-'91. **Distribution.**—*Mains*, 3.5 miles. *Taps*, 200. *Meters*, 0. *Hydrants*, 10. **Financial.**—*Cost*, $20,000. *No Debt. Expenses*, $1,200. *Revenue:* consumers, $2,200; city, $100; total, $2,300. **Management.** —Prest., J. B. Ramsay. Secy. and Supt., C. E. Bennett.

178. **TONAWANDA and NORTH TONAWANDA** *Erie and Niagara Cos.* (Pop.: Tonawanda, 3,864—7,145; North Tonawanda, in '80, 1,492.) Built in '87 by Tonawanda City Water-Works Co.; town has right to purchase works according to laws of New York State. **Supply.**—Niagara River, pumping direct. Fiscal year closes Nov. 1. **Distribution.**—*Mains*, 20 miles. *Taps*, 608. *Meters*, 3. *Hydrants*, 204. *Consumption*, 1,500,000 galls. **Financial.**—*Cost*, $275,000. *Cap. Stock:* authorized, $50,000; all paid up; paid no dividend in 1890. *Bonded Debt*, $240,000 at 6%. *Total Debt*, $240,000, bonded; $120,000 at 6%, due in 1906; $120,000 at 6%, due in 1908. **Management.**—Prest., Benj. L. Rand. Secy. and Supt., H. M. Fales. Treas., G. P. Smith.

NOTE.—July 21, '91, H. S. Wende, E. H. Rogers, Sr., and others were appointed com. to negotiate with Co. for purchase of works.

179. **TROY,** *Rensselaer Co.* (Pop., 56,747—60,967) Built in '33 by city. **Supply.**— Five artificial storage reservoirs in valley of Piscawen River, by gravity; also Hudson River at Lansingburgh, pumping to Oakdale reservoir on above creek. Fiscal year closes Feb. 28.

Year.	Mains.	Taps.	Meters.	Hyd'ts.	Cons'p't'n.	Cost.	B'd d'd'bt.	Int.	Exp.	Rev.
'60	17.3	....	....	55	....	$226,154	....	......	*7,150	$22,603
'70	22.5	....	....	101	....	392,764	....	......	13,599	38,892
'80	41.1	3,500	5	456	3,453,296	926,103	$300,000	......	39,291	74,424
'84	43.7	4,500	25	501	....	1,109,672	392,500	......	50,093	81,702
'85	50.3	4,950	35	545	....	1,149,064	443,500	$17,740	61,063	82,034
'86	52.1	5,012	43	578	8,000,000	1,162,380	421,000	17,285	57,479	89,158
'87	53.2	5,250	46	600	8,000,000	1,171,692	416,000	16,860	50,871	89,420
'88	53.5	5,453	58	629	8,000,000	1,177,310	406,000	16,610	58,676	91,503
'89	55.7	5,654	223	661	7,839,835	1,196,267	393,500	15,848	54,675	85,511
'90	55.9	5,786	226	684	7,608,468	1,208,723	378,500	15,423	37,874	91,862

Financial.—Int., 5, 4½, 4, and 3½%. *Sinking fund*, $30,119. *Expenses*, total, $53,297. *Revenue:* consumers, $91,862; city, none. **Management.**—Four comrs. Prest., Samuel O. Gleason. Clerk, Geo. B. Fales. Supt., Edward Dolan.

180. **TUXEDO PARK,** *Orange Co.* (Pop., est., 1,000.) **History.**—Built in '85-6 by Tuxedo Park Association to supply its property. Des. Engr., E. W. Bowditch, Boston. Const. Engrs., D. W. Pratt and F. M. Hersey. Since some time in '88 J. S. Haring has been Engr. **Supply.**—Tuxedo Lake, pumping to tank. **Pumping Machinery.**—Dy. cap., 1,500,000 galls.; 1,000,000 and 500,000 Worthington. *Tank.*—Cap., 740,000 galls.; 50 × 50 ft., from Whittier Machine Co., Boston. **Distribution.**—*Mains*, 12- and 4-in. c. i., 10 miles; from Warren Foundry & Machine Co., Phillipsburg, N. J., and R. D. Wood & Co., Philadelphia; laid by C. H. Eglee, Flushing. *Services*, galv. i. *Taps*, 94. *No Meters*. *Hydrants*, Mathews, 30. *Valves*, Eddy, Kennedy, Coffin. *Consumption*, 200,000 galls. *Pressure*, max., 150 lbs. **Financial.**—*Withheld.* **Management.**—Prest., P. Lorillard. Secy., Wm. Kent. Treas., G. D. Finlay, 111 First St., Jersey City. Engr. and Supt., J. S. Haring. Report, Dec. 17, '90.

NOTE.—In '90 co. was organized by about same interests as above to supply Tuxedo Village and villages down Ramapo River.

FOR FURTHER DESCRIPTIVE MATTER see Manual for 1889-90. CHANGES since 1890 are here given. POPULATIONS are for 1880 and 1890, respectively.

94 MANUAL OF AMERICAN WATER-WORKS.

181. **UNADILLA**, *Otsego Co.* (Pop., 922—1,157.) Built in '86 by S. S. North. Supply.—Springs, by gravity from collecting reservoirs, and formerly by pumping to reservoir. Latter part of '90 new dam and reservoir was built on Martin's Brook. Reservoir formed by curved masonry dam 100 ft. long, on and abutting against solid rock. Dam has radius of 100 ft., is 15½ ft. thick at base and 3½ at top; stepped spillway in or near center, 33 ft. long, 14½ ft. high above foundation, with top 4 ft. below top of dam, or wing walls. Dam backed with 2-ft. puddled blue clay wall, against which there is a gravel embankment sloping 2 to 1, paved. Sides of reservoir protected by riprap wall carried 3 ft. above top of spillway. A valve chamber is located back of each wing for supply and waste pipe. Water enters main through oval filter so arranged that water can be taken at different elevations. Engr., G. A. Wright, Walton. Contr., Wm. Kelley, Herrick Center, Pa. In '91 settling basin was completed on same stream, above one just described. New reservoirs have capacities of 1,932,000 and 816,000 galls., and old reservoirs combined cap. of 1,746,000 galls. New reservoirs connect with street mains at east, old at west end of village; gates are arranged so village can be supplied from either or both reservoirs. Pressure increased to 87 lbs. Distribution.—(In '88.) *Mains*, 2 miles. *Taps*, 100. *Meter*, 1. *Hydrants*, 16. *Consumption*, 100,000 galls. Financial.—*Withheld*. Management.—Owner, S. S. North. Supt., D. H. Loomis.

74. **UNIONVILLE**, *Kings Co.* See Gravesend.

182. **UTICA**, *Oneida Co.* (Pop., 33,914—44,007.) Built in '49 by Utica Water-Works Co.; franchise provides neither for control of rates nor purchase of works by city. Supply.—Surface water, by gravity from impounding to distributing reservoir. Fiscal year closes in April. Distribution.—*Mains*, 61 miles. *Taps*, 3,000. *Meters*, 2,500. *Hydrants*, 362. Financial.—*Cost*, $1,000,000. *Cap. Stock*, $500,000. *Revenue:* city, $24,000. City pays 6% on all mains ordered by Council, and in addition $10,000 yearly for fire purposes, and one-half of all taxes above $10,000. Management—Prest., Thos. Hopper. Secy., J. K. Chamberlayne. Treas., P. V. Rogers. Supt., C. W. Pratt. NOTE.—In '91 com. was appointed to confer with co. regarding purchase of its works.

183. **WALTON**, *Delaware Co.* (Pop., 1,389—2,299.) Built in '70 by Walton Water Co. Supply.—Surface water, by gravity from impounding reservoir on stream; filtered. Fiscal year closes May 31. In '91 new reservoir was proposed. Distribution. —*Mains*, 6.5 miles. *Taps*, 185. *Hydrants*, 48. Financial.—*Cost*, $25,000. *Cap. Stock*, $25,000. *No debt. Expenses*, $80. *Revenue*, $2,500. Management.—Chn., Geo. O. Mead. Secy., Geo. W. Fitch. Supt., J. W. St. John.

XV. **WARRENSBURGH**, *Warren Co.* (Pop., 748—893.) Built in '84 by Warrensburgh Water Co. Supply.—Brook, fed by springs, by gravity. Fiscal year closes Sept. 1. Distribution.—*Mains*, 6 miles. *Taps*, 70. *Meters*, 0. *Hydrants*, 1. Financial.—*Cost*, $6,500. *Debt*, $4,000 at 6%. *Expenses:* operating, $75; int., $210; total, $315. *Revenue:* consumers, $700. Management.—Owners, Bates & Co. Supt., Ira Cole.

184. **WARSAW**, *Wyoming Co.* (Pop., 1,910—3,120.) Built in '70 by Warsaw Water-Works Co.; franchise provides neither for control of rates nor purchase of works by village. Supply.—Springs, by gravity from collecting reservoir. Distribution.—*Mains*, 10 miles. *Taps*, 400. *Meters*, 10. *Hydrants*, 60. Financial. —*Cost*, $10,000. *Cap. Stock*, $25,000. *Debt*, $4,000 at 6%. *Revenue:* consumers, $3,700; city, contract reported as made in '90 for rental of $1,000 per year. Management.—Prest., W. H. Bradley. Secy. and Supt., E. H. Morris.

185. **WARWICK**, *Orange Co.* (Pop., 1,043—1,537.) Built in '72 by village. Supply.—Surface water, by gravity from impounding reservoir on small stream. Fiscal year closes March 1. March 5, '91, $15,000 was voted for new storage reservoir; contract was let to O'Hiere & Fagen, to be completed in Nov.; price, about $10,000; Engrs., Caldwell & Garrison, Newburg. Distribution.—*Mains*, 4.5 miles. *Taps*, 166. *Meters*, 0. *Hydrants*, 38. Financial.—*First Cost*, $25,000. *Bonded Debt*, $4,000 at 5%, due in annual payments of $1,000. *Expenses:* operating, $795; int., $240; total, $1,035. *Revenue:* consumers, $1,800; city, none. Management.—Prest., Sylvester Case. Clerk, F. C. Carey. Supt., Wm. A. Hulse.

186. **WATERFORD**, *Saratoga Co.* (Pop., 1,822—est., 3,500.) Built in '86 by Waterford Water Co.; controlled by Moffet, Hodgkins & Clarke, Syracuse. Supply.—Hudson River, pumping to tank.

NEW YORK. 95

Year.	Mains.	Taps.	Meters.	Hydrants.	Consumption.	Cost.	Bonded debt.	Int. ch'g's.	Rev.
'86	9.3	150	15	60	120,000	$85,000	none.	none.	$6,000
'87	6.5	256	45	57	170,000	90,000	$72,000	$4,320	2,725*
'88	8	326	60	60	164,857	72,000	72,000	4,320	....

*From city.

Financial.—*Debt* (in '88), $72,000 at 6%. **Management.**—(In '88.) Prest., M. P. Powell. Supt., W. A. Dennis.

187. **WATERLOO,** *Seneca Co.* (Pop., 3,893—4,350.) Built in '86 by Waterloo Water Co. **Supply.**—Seneca River, pumping to stand-pipe and direct. Fiscal year closes Dec. 31.

Yr.	Mains.	Taps.	Meters.	Hydrants.	Consumption.	Cost.	Bonded debt.	Int. ch'g's.	Exp.	Rev.
'86	9	45	0	50	.....	$100,000	$60,000	$3,600	.....	.....
'88	9	210	2	50	103,030	100,000	60,000	3,600	$1,800	$7,500
'00	10	270	3	60	260,000	150,000	100,000	6,000	2,620	10,540

Financial.—*Cap. Stock:* authorized, $40,000, all paid up. *Bonded Debt,* $100,000 at 6%, due 1904-10. *Expenses:* operating, $2,100; int., $6,000; taxes, $520; total, $8,620. *Revenue:* consumers, $8,500; city, $2,040; total, $10,540. **Management.**—Prest., Francis Bacon. Secy. and Treas.. A. G. Mercer. Supt., Geo. Clark.

188. **WATERTOWN,** *Jefferson Co.* (Pop., 10,697—14,725.) Built in '53 by village. **Supply.**—Black River, pumping to reservoir. Fiscal year closes July 1. **Distribution.**—*Mains* (in '88), 25 miles. *Taps,* 2,500. *Meters,* 20. *Hydrants,* 205. **Financial.**—*Cost,* $300,000. *Bonded Debt:* $235,000: $50,000 at 7%; $45,000 at 5%; $85,000 at 4%; $15,000 at 3½%; $40,000 at 3%. *Expenses:* operating, $6,500; int., $11,378; total, $17,884. *Revenue:* consumers, $23,916; city, $4,800; total, $28,716. **Management.**—Comrs., elected every five years. Chn., J. C. Knowlton. Secy., N. O. Wardwell. Supt., A. Salisbury, appointed by comrs.

189. **WATERVILLE,** *Oneida Co.* (Pop., in '90, 2,024.) Built in '88 by village. **Supply.**—Surface water, by gravity from storage and distributing reservoirs; filtered. Fiscal year closes March 15. **Distribution.**—(In '88.) *Mains,* 7 miles. *Taps,* 130. *Meters,* 0. *Hydrants,* 50. **Financial.**—*Cost,* $65,000. *Total Debt,* $70,500: floating, $650, at 4%; bonded, $64,000, at 4%. *Revenue:* consumers, $2,100. **Management.**—Chn., E. W. Buell. Secy., H. N. Cander. NOTE.—For cross-section of distributing reservoir dam see Manual for '89-90, p. 189.

190. **WATKINS,** *Schuyler Co.* (Pop., 2,716—est., 3,000.) **History.**—Built in '90-91 by village; finished July 1. Engr., E. D. Smalley, Syracuse. Contr., Chas. Brown, Mohawk. **Supply.**—Springs, by gravity; emergency supply by pumping from Seneca Lake; pumping done temporarily by private party. Springs are 400 ft. above village. Water [evidently] collected in reservoir at springs, flows to upper, then overflows to lower reservoir; latter reservoirs 300 and 200 ft. above village, respectively. *Reservoirs.*—Cap., 30,000, 120,000, and 500,000 galls. **Distribution.**—*Mains,* 10 and 4-in., c. i., 6 miles. *Services,* galv. i. *Hydrants,* Galvin, 57. *Valves,* Galvin. **Financial.**—*Cost,* $52,000. *Bonds:* authorized, $75,000 at 4%; issued, $52,000. **Management.**—Five comrs. Prest., W. N. Love. Secy., C. M. Woodward. Supt., F. Davis. Report, June 20, '91.

191. **WAVERLY,** *Tioga Co.* (Pop., 2,767—4,123.) Built in '80 by Waverly Water Co. **Supply.**—Surface water and springs, by gravity from impounding reservoir. Fiscal year closes April 30. **Distribution.**—*Mains,* 10 miles. *Taps,* 400. *Meters,* 11. *Hydrants,* 25. **Financial.**—Withheld. *Cost* (in '86), $60,000. **Management.**—Prest. and Treas., J. T. Sawyer. Supt., A. J. Vanatta.

192. **WELLSVILLE,** *Allegany Co.* (Pop., 2,049—3,435.) Built in '83 by W. Kuhn, of American Water-Works & Gaurantee Co., Pittsburg, Pa., under 25 years' franchise, which fixes maximum rates and provides for purchase of works by city in '93, price determined by arbitration. **Supply.**—Storage reservoir, by gravity. Reservoir is among the hills, one mile from city, supplied by streams, covers 1½ acres and has cap. of 6,000,000 galls. Also emergency supply from Genesee River. **Pumping Machinery.**—Dy. cap., 1,500,000 galls.; Worthington comp., 14 × 20 ×-12½ × 10. *Boiler,* 60 HP. Pumps are but rarely used.

Year.	Mains.	Taps.	Meters.	Hyd'ts.	Cost.	B'd d d't.	Int. ch'g's.	Rev.
'84	3.5	180	0	31	$40,000	$30,000	$1,800	$3,000
'86	6	210	0	31	50,000	30,000	1,800	3,800
'90	8	304	0	47	.....	.....	.....	.....

FOR FURTHER DESCRIPTIVE MATTER see Manual for 1889-90. CHANGES since 1890 are here given. POPULATIONS are for 1880 and 1890, respectively.

Financial.—*Withheld.* **Management.**—Prest., Jas. Macken. Vice-Prest., J. H. Purdy. Secy., W. S. Kuhn, Pittsburg. Supt., Simons & Voorhees.

193. **WESTFIELD,** *Chautauqua Co.* (Pop., 1,924—1,983.) Built in '89-91 by village. **Supply.**—Chautauqua Creek, filtered, by gravity to reservoir; filtered again, thence by gravity. **Distribution.**—*Mains,* 13 miles. *Hydrants,* 70. **Financial.**—*Bonded Debt,* $60,000 at 3½%. **Management.**—Chn., G. W. Paterson. Secy., S. T. Nixon. Treas., R. G. Wright. No. Supt.

NOTE.—Data is incomplete, as works were not finished at time of last report, Jan. 3, '91.

194. **WEST TROY,** *Albany Co.* (Pop., 8,820—12,967.) Built in '73-6 by West Troy Water-Works Co.; controlled by Moffett, Hodgkins & Clarke, Syracuse. **Supply.**—Green Island Water-Works.

Year.	Mains.	Taps.	Meters.	Hyd'ts.	Cons'p'n.	Cost.	B'd'd d't.	Int.ch'g's.	Exp.	Rev.
'80	14	270	..	100	1,000,000					
'82	15	300	1	110	1,000,000	$250,000	$170,000	$11,760	$17,000	$18,000
'86	16	400	0	108	1,000,000	275,000	195,000	11,475	5,000	18,000
'87	11	547	0	87	1,000,000	300,000	200,000		5,000	21,000
'88	11	584	0	87	1,000,000	300,000	200,000		3,000	22,000

Financial.—*Cap. Stock* (in '88), $100,000. *Bonded Debt* (in '88) $200,000 at 6 and 7%. *Revenue:* consumers, $12,000; city, $10,000. **Management.**—(In '89.) Prest., H. C. Hodgkins.

195. **WHITEHALL,** *Washington Co.* (Pop., 4,270—4,434.) Built in '52, rebuilt in '84 by village. **Supply.**—Springs, by gravity, pumping to reservoir and direct. Fiscal year closes April 1.

Year.	Mains.	Taps.	M't'rs.	Hydrants.	Consumption.	Cost.	Bonded debt.	Int. charges.	Exp.	Rev.
'84-5	10	45	..	52		$88,000	$79,000	$3,160	$600	$600
'86-7	7.9	235	5	51	160,000	81,800	79,000	3,160	2,200	4,345
'88-9	8.2	340	17	53	200,000	83,765	79,000	3,160	4,300	6,900
'89 90	8.7	378	36	52	200,000	83,928	79,000	3,160	4,500	5,000

Financial.—*Bonded Debt,* $79,000: $25,000 at 4% due in '94; $25,000 at 4% due in 1904; $29,000 at 4% due in 1914. *Expenses,* total, $7,660. *Revenue:* consumers, $5,000; city, none. **Management.**—Chn., O. F. Davis. Secy. and Supt., C. W. Hotchkiss.

196. **WHITE PLAINS,** *Westchester Co.* (Pop., 2,381—4,042.) Built in '85 by Westchester County Water-Works Co., controlled by Moffett, Hodgkins & Clarke, Syracuse. **Supply.**—Springs, pumping to stand-pipe.

Year.	Mains.	Taps.	Meters.	Hydrants.	Consumption.	Cost.	Bonded debt.	Int. ch'g's.	Exp.	Rev.
'86	6.8	...	...	65			$85,000	$5,150		$2,500*
'87	7	128	...	71		$120,000	86,000	5,160		3,250*
'88	9	237	104	77	26,480	120,000	86,000	5,160	$2,000	3,800

*From city.

Financial.—Int., 6%. *Revenue:* consumers, $6,000; city, $2,800. **Management.**—Prest., J. V. Clarke. Supt., A. L. Fassett.

XVI. **WHITESBORO,** *Oneida Co.* (Pop., 1,371—1,663.) Built in '27 by Lyman L. Wight. **Supply.**—Springs, by gravity from collecting reservoir. Small system, built by owner for his own use and extended for the convenience of neighbors. There are 2½ miles of pipe supplying 64 families and 1 fire hydrant. *Cost,* $3,500. *Expenses,* $50. *Revenue,* $800.

197. **WHITESVILLE,** *Allegany Co.* (Pop., 297—est., 575.) Built in '88 by Whitesville Water Co. City neither controls rates nor has right to buy works. **Supply.**—Springs, by gravity to reservoir. **Distribution.**—*Mains,* 1.3 miles. *Taps,* 35. *Hydrants,* 10. **Financial.**—*Cost,* $3,500. *No Debt. Revenue:* consumers, $175; city, $100; total, $275. **Management.**—Prest., J. L. Crittenden. Secy., A. C. Howe. Treas., A. D. Howe. Supt., J. B. Wiley.

198. **WORCESTER,** *Otsego Co.* (Pop., 682—est., 1,000.) Built in '87 by Worcester Water Co. **Supply.**—Caryl Lake, by gravity. Fiscal year closes May 6. **Distribution.**—*Mains,* 6 miles. *Taps,* 37. *Meters,* 0. *Hydrants,* 17. **Financial.**—*Cost,* $35,000. *Total Debt,* $15,500: floating, $500 at 6%; bonded, $15,000 at 6%. *Expenses:* operating, $300; int., $900; total, $1,200. *Revenue:* consumers, $600; city, $600; total, $1,200. **Management.**—Prest. and Supt., Wm. Fern. Secy., S. W. Ferguson.

199. **YONKERS,** *Westchester Co.* (Pop., 18,892—32,033.) Built in '76 by city. **Supply.**—Sprain and Grassy Sprain brooks, pumping from impounding to distributing reservoir and direct. Fiscal year closes Nov. 30.

NEW YORK. 97

Yr.	M'ns.	Taps.	Mtrs.	Hy-dr'ts.	Con-sumpt'n.	Cost.	Bonded debt.	Int. ch'g's.	Exp.	Rev.
'76	20.6	146	....	216	....	......	$600,000	.....	$2,972	$1,372
'77	21.9	352	164	230	280,000	$352,196	630,000	$43,750	11,210	6,530
'78	22.3	471	238	236	570,000	662,869	630,000	43,750	9,261	7,942
'79	22.4	652	318	238	631,272	633,920	630,000	43,750	7,696	11,090
'80	23	782	402	246	860,000	690,300	630,000	43,750	9,463	14,061
'81	23.4	947	496	250	840,000	488,059	645,000	45,000	10,494	19,078
'82	24.5	1,133	605	263	1,036,000	702,193	650,000	45,500	12,100	25,087
'83	28.7	1,344	732	310	1,246,200	755,713	730,000	48,550	14,474	29,684
'84	31	1,534	914	331	1,257,900	787,310	730,000	48,627	17,491	34,927
'85	31.6	1,727	1,085	350	1,406,900	795,884	745,000	49,000	19,294	49,132
'86	32.5	1,921	1,296	359	1,628,000	817,718	770,000	49,850	19,195	55,744
'87	33.4	2,123	1,538	372	1,565,000	828,750	785,000	50,450	20,065	53,000
'88	34.8	2,304	1,776	387	2,015,775	831,423	815,000	51,650	26,113	68,973
'89	36.5	2,503	2,036	412	2,016,699	866,476	8300,00	52,250	27,050	61,316
'90	38.5	2,683	2,212	436	2,176,393	904,173	850,000	52,930	26,132	09,536

Financial.—Bonded debt, $850,000; $25,000 at 7%, due in 1903-15; $30,000 at 5%, due in 1914; $175,000 at 4%, due in 1914-17; $20,000 at 3½%, due in 1917. *Sinking Fund*, $80,548, invested in water bonds at 4%. *Expenses*, total, $79,082. *Revenue:* consumers, $69,- 536; city, $12,360; total, $31,836. Revenue is applied to operating expenses, interest, and sinking fund; new construction is paid by a new issue of bonds; 2% of bonds is paid each year from revenue and by taxation. **Management.**—Five comrs., one appointed every year by Common Council. Chn., R. Elchemeyer. Clerk and Supt., Jos. A. Lockwood.

Water-Works Projected with Fair Prospects of Construction.

AFTON, *Chenango Co.* (Pop., 734,653.) Trustees talking of work. Clerk, G. L. Church.

AMENIA, *Dutchess Co.* (Pop., 393—est., 600.) Amenia Water-Works Co. incorp. in '91. Proposed supply, springs and stream, by gravity from small impounding reservoir, 200 ft. above village, to 200,000-gall tank. Est. cost, $10,000. Prest., B. H. Fry. Secy. and Treas., Chas. Walsh. Engr., J. R. Thompson.

ARCADE, *Wyoming Co.* (Pop., 762—est., 1,100.) Works talked of, with supply from Skim Lake, four miles from village. About Aug. 1, '91, G. A. Barnes, Village Clerk, wrote: "We do not intend to put in water-works this year."

CARTHAGE, *Jefferson Co.* (Pop., 1,912—2,278.) Reported July, '91, that Moffett, Hodgkins & Clarke, Syracuse, N. Y., had just made proposition to build, but that there was a strong sentiment in favor of public ownership. Supply, Black River, pumping to reservoir. See Manual for '88, p. 147, and for '90, p. 195.

CATTARAUGUS, *Cattaraugus Co.* (Pop., 705—878.) **Present Supply.**— See p. 68. Proposed Works.—July 17, '91, Cold Spring Water Co., incorp., proposed to have works in operation Dec. 1, '91. Supply, springs, by gravity. Est. cost, $12,000 to $15,000. Authorized capital stock, $20,000. Prest. E L. Johnson. Secy., L. H. Northrup. Treas., T. J. Farrar.

CHAMPLAIN, *Clinton Co.* (Pop., 1,509—1,275.) Village works projected summer of '91 at est. cost of $15,000 to $18,000.

CHERRY CREEK, *Chautauqua Co.* (Pop., 448—676.) G. N. Cowan, Stamford, has franchise, but July 20, '91, wrote that the works would not be built until '92. Supply, springs, by gravity.

CHITTENANGO, *Madison Co.* (Pop., 954—792.) Small system for domestic supply projected by village in '91. Supply, springs, by gravity. Cost, $1,000. Address O. A. Hitchcock.

COXSACKIE, *Greene Co.* (Pop., 1,631—1,611.) Works have been projected for some time. Address B. F. Stephens, Flatbush, or R. H. Van Begen.

DRYDEN, *Tompkins Co.* (Pop., 779—est., 900.) July 29, '91, F. S. Jennings stated that an engineer was to make estimates for works, after which construction by town would be voted on. Supply, springs, by gravity to and from reservoir.

FAIRPORT, *Monroe Co.* (Pop., 1,920—2,552.) Summer of '91 committee was investigating works. Est. cost, $28,000.

HERMON, *Saint Lawrence Co* (Pop., 522—473.) Contract let in '91 to U. S. Wind Engine & Pump Co., Batavia, Ill.; construction to start July 20 and be finished Sept. 1, '91. Engrs., Hinds & Bond, Watertown, N. Y. Supply, springs,

FOR FURTHER DESCRIPTIVE MATTER see Manual for 1889-90. CHANGES since 1890 are here given. POPULATIONS are for 1880 and 1890, respectively.

pumping to stand-pipe. About 2 miles mains and 9 hydrants. Cost, $7,500; same amount in bonds issued. O. A. Northrup, Secy., E. A. Conant, Treas., Village Trustees. Report by P. L. Doyle, July 14, '91.

HIGHLANDS, *Orange Co.* (Pop., 1,976—2,237.) Citizens' Water Co. were enjoined, July 30, '91, from taking water from Buttermilk Falls Channel.

HOLLAND, *Erie Co.* (Pop., in '90, 582.) June 22, '91, Secy. Holland Water-Works Co. reported expected to build fall of '91. Unofficially reported latter part of July that stock had been subscribed and that contracts had been let to Ayrault Bros. & Co., Tonawanda, for $9,987. Supply, springs, by gravity. Prest., W. B. Jackson. Secy., C. A. Button, Treas., A. Cutter.

HUNTINGTON, *Suffolk Co.* (Pop., 2,952—3,028.) Exclusive 30-years franchise granted Huntington Water-Works Co. in April, '91, for works in Union School District; construction to be started within 12 and completed within 18 months. Supply, wells, or "basins," pumping to reservoir with cap. of at least 250,000 galls. About 4½ miles of mains and 35 hydrants. Prest., J. M. Brush. Secy., Joseph Irwin. Engr., Oscar Darling, Huntington.

ILION, *Herkimer Co.* (Pop., 3,711—4,057.) In '91 investigating committee was appointed to report on works, after which people were to vote on question of building. Est. cost, $70,000 to $80,000. Chn. com., Thos. Ringwood. Secy., S. N. Russell.

LANCASTER, *Erie Co.* (Pop., 1,602—1,692.) Village works projected summer of '91; investigations were being carried on July 20. Address D. L. Ransom, Prest. village.

LAURENS, *Otsego Co.* (Pop., 252—255.) July 20, '91, W. H. Widger wrote that village was to vote on paying $300 hydrant rental yearly; if carried, Laurens Water Co. propose to bring water from lake, by gravity, through 1½ miles pipe, including supply main. Est. cost, $8,000.

LIVINGSTON MANOR, *Sullivan Co.* (Pop., est., 750.) July, '91, reported that surveys for works were being made. Supply, springs, by gravity from 70 × 120 ft. reservoir, through 6-in. main.

LOWVILLE, *Lewis Co.* (Pop., in '90, 2,511.) Projected in '90 and '91. Reported July, '91, that E. C. Cooke, Engr. for Moffett, Hodgkins & Clarke, Watertown, N. Y., was looking over ground. Gravity supply talked of at est. cost of $65,000. Address S. B. Richards.

MAYVILLE, *Chautauqua Co.* (Pop., about 1,050—1,164.) Aug. 1, '91, village was to vote on building works. Supply, Chautauqua Lake, pumping to reservoir. Est. cost, $10,000. Chn. Water Com., A. A. Van Dusen. Secy., G. E. Leet.

MECHANICSVILLE, *Saratoga Co.* (Pop., 1,265—2,679.) In '91 works were projected by village. Town Clerk, N. W. Kelso.

MEDINA, *Orleans Co.* (Pop., 3,632—4,492.) Medina Water-Works Co. propose to build, but matter is indefinite. Reported that Bassett Bros., Buffalo, were granted franchise, and that in July, '91, they were about to transfer it to Brownell & Foster, 92 Broadway, N. Y. City. Supply, Oak Orchard Creek, pumping to standpipe. Cap. stock, $66,000. Address above parties, or A. J. Hill, Medina. NOTE.—Fire protection works built in '60; see Manual for '89-90, p. 160.

PORT JEFFERSON, *Suffolk Co.* (Pop., 1,724—2,026.) A supply from driven wells is talked of; nothing definite done July 18, '91. Address J. E. Overton.

SANDY CREEK, *Oswego Co.* (Pop., 951—est., 1,150.) Summer of '91 about $20,000 was voted for gravity works. Address D. E. Ainsworth.

SHERMAN, *Chautauqua Co.* (Pop., 731—785.) G. N. Cowan, Stamford, secured franchise in '90 or early in '91, but works will not be built before '92.

SINCLAIRVILLE, *Chautauqua Co.* (Pop., 540—510.) Sinclairville Water Co. proposes to build fall of '91. Supply, springs, by gravity. About 2½ miles mains proposed. Est. Cost, $36,000. Cap. Stock, $20,000 authorized; all paid up. Bonds: $16,000 authorized, but none issued. Engr. and Prest., G. N. Cowan; Secy. and Treas., Jessie B. Cowan, both of Stamford. Report, July 20, '91.

SOUTHAMPTON, *Suffolk Co.* (Pop., 949—est., 1,000.) Works were projected in '89, but not built. July, '91, reported that there was talk of forming a co.

NEW YORK.

UNION, *Broome Co.* (Pop., 737—821.) Unofficially reported as under construction latter part July, '91. Proposed supply, spring, one mile from village, with reservoir. Aug. 27, '91, Union Water Co. incorporated by S. H. Morgan, Cuba; F. C. Reynolds, Moravia, and others. Cap. stock, $25,000.

WALDEN, *Orange Co.* (Pop., 1,804—2,132.) Water Board organized May 12, '91, after which plans were made for supply by water power, pumping from river. June 8, '91, nothing further had been done. Village Clerk, E. H. Nichols.

WALTON, *Delaware Co.* (Pop., 1,389—2,290.) Aug. 8, '91, Walton Water Supply Co. was incorporated by E. R. Howland, C. E. Ogden and others; authorized cap. stock, $15,000. For description of existing works see MANUAL for '89-'90, p. 187.

WESTPORT, *Essex Co.* (Pop., 364—563.) Projected by Portland parties. During summer '91 flow of spring was being gaged. Address Alice Lee.

WHITESTONE, *Queens Co.* (Pop., 2,520—2,808.) Village proposed to build in '91, pumping from wells to stand-pipe, with pond as reserve. Est. cost, $50,000. Engr., J. A. Roullier, Flushing. In July, '91, injunction was granted forbidding comrs. to issue $45,000 of bonds, on the ground that this amount was above legal limit. Chn. Comrs., Edw. Bleecker.

·YORKSHIRE CENTRE, *Cattaraugus Co.* (Pop.. 430—511.) Yorkshire Water Co. proposed to begin construction Sept. 1, and complete Dec, '91. Supply, springs, by gravity, with 200,000-gall. reservoir. About 2½ miles of mains and 15 hydrants. Est. Cost, $30,000. Cap. Stock, $18,000, paid-up. Bonds: authorized, $12,000; none issued. Prest.. Engr. and Cont'r, G. N. Cowan, Secy., Jessie B. Cowan, both of Stamford. Reported July 20, '91.

Water-Works Projects Reported in '90 and '91, but Now Indefinite.

Babylon, Suffolk Co. *Cohocton,* Steuben Co. *Fayetteville,* Onondaga Co. *Hewlets,* Queens Co. *Montgomery,* Orange Co. *New Paltz,* Ulster Co. *Weedsport,* Cayuga Co.

NEW JERSEY.

Water-Works Completed or in Process of Construction.

25. **ARLINGTON,** *Hudson Co.* In Kearney township.

1. **ASBURY PARK,** *Monmouth Co.* (Pop., 1,640—est., 3,500.) Built in '86 by village. Supply.—Artesian wells, pumping to stand-pipe. Fiscal year closes March 31.

Year.	Mains.	Taps.	Meters.	Hyd'ts.	Cost.	B'd'd debt.	Int. ch'g's.	Exp.	Rev.
'87-8	10	460	60	60	......	$92,000	$1,600	......	......
'8.-9	13.7	547	80	60	$92,000	92,000	4,6 0	$6,000	$11,362
'89-90	13.7	689	84	60	108,000	92,000	4.600	5,500	12,000
90-1	13.7	776	86	60	117,729	92,000	4,600	5,920	10,867

Financial.—*Bonded Debt,* $92,000 at 5%, for 20 years. *Expenses:* total, $10,520. *Revenue:* consumers, $10,867; city, none. **Management.**—Three comrs. Supt. and Secy., G. H. Coffin.

2, 3. **ATLANTIC CITY,** *Atlantic Co.* (Pop., 5,477—13,055.)

ATLANTIC CITY WATER-WORKS CO.

Built in '83. Supply.—Wells and streams, pumping to stand-pipe. Fiscal year closes Sept. 30. In '90, supply increased by sinking 50 additional wells. Distribution.—*Mains,* 28 miles. *Taps,* 2,044. *Meters,* 235. *Hydrants,* 150. *Consumption,* 1,281,467 galls. Financial.—*Cost,* $817,553. Other information withheld, "on account of legal complications with city." **Management.**—Prest., Walter Wood, Philadelphia. Secy. and Supt., G. T. Prince.

CONSUMERS' WATER CO.

Built in '88-9. Supply.—Wells, pumping to stand-pipe.

Year.	Mains.	Taps.	Meters.	Hydrants.	Consumption.	Cost.	Debt.
'88	15	500	0	100	500,000	$175,000	None

FOR FURTHER DESCRIPTIVE MATTER see Manual for 1889-90. CHANGES since 1890 are here given. POPULATIONS are for 1880 and 1890, respectively.

100   MANUAL OF AMERICAN WATER-WORKS.

4. BAYONNE CITY, *Hudson Co.* (Pop., 9,372—19,033.) Built in '82 by city. Supply.—Jersey City Water-Works. Fiscal year closes Apr. 30.

Year.	Mains.	Taps.	Meters.	Hyd'ts.	Cost.	B'd'd d't.	Int. ch'g's.	Exp.	Rev.
'86-7	9	425	350	68	$120,000	$120,000	$6,000	$40,000	$37,000
'89-90	17	678	1,000	145	173,000	136,000	6,775	55,198	59,168
'90-1	18	870	1,100	157	190,000	136,000	6,800	55,339	66,053

Financial.—*Bonded Debt*, $136,500 at 5%. *Expenses*, total, $62,139. *Revenue:* consumers, $66,053; city, none. **Management.**—Committee appointed annually from Board of Council. Chn., J. A. Cadmus. Supt., M. O. Connor.

30. BELLEVILLE, *Essex Co.* (Pop., 3,004—3,487.) Supply from Newark.

5. BELVIDERE, *Warren Co.* (Pop., 1,773—1,768.) Built in '77-8 by Belvidere Water Co. **Supply.**—Delaware River, pumping to stand-pipe. **Distribution.**—*Mains*, 5 miles. *Taps*, 175. *Meters*, 3. *Hydrants*, 25. **Financial.**—*Cost*, $23,500. *Cap. Stock:* authorized, $23,000; all paid up; paid 5% div'd. in 1890. *No Debt.* *Expenses:* operating, $1,935; taxes, $247; total, $2,182. *Revenue:* consumers, $2,000; city, $600; total, $2,600. **Management.**—Prest., D. C. Blair. Secy. and Supt., A. McCammon.

6. BEVERLY, *Burlington Co.* (Pop., 1,759—1,957.) Built in '86-7 by Beverly Water Co. **Supply.**—Delaware River, pumping to stand-pipe. **Distribution.**—*Mains*, 6.5 miles. *Taps*, 310. *Meters*, 0. *Hydrants*, 44. **Financial.**—*Cap. Stock*, $50,000. **Management.**—Prest., W. Sexton. Secy., Sanford Murphy. Treas., Jas. P. Michelion. Supt., Jas. Finley.

7. BLAIRSTOWN, *Warren Co.* (Pop., 1,458—1,602.) Built in '89 by Jno. T. Blair. **Supply.**—Spring, pumping to stand-pipe. Fiscal year closes May 1. About one mile of pipe supplies 50 consumers and 10 fire hydrants. *Cost*, $50,000. *No Revenue.* Address Jno. T. Blair.

15. BLOOMFIELD, *Essex Co.* (Pop., 5,748—7,708.) Supply from East Orange.

8. BORDENTOWN, *Burlington Co.* (Pop., 4,258—4,232.) Built in '56 by Bordentown Reservoir & Water Co.; city neither controls rates nor has right to purchase works. **Supply.**—Delaware River, pumping to reservoir; filtered. **Distribution.**—*Mains*, 5 miles. *Meters*, 5. *Hydrants*, 42. **Financial.**—Withheld. **Management.**—Prest., Wm. Steele. Treas., J. B. Woodward. Supt., Jno. M. Steele.

9. BOUND BROOK, *Somerset Co.* (Pop., 931—1,462.) Built in '88-9 by Bound-Brook Water Co. **Supply.**—Middle Brook, by gravity. Fiscal year closes Oct. 1. **Distribution.**—*Mains*, 5.5 miles. *Taps*, 50. *Meters*, 0. *Hydrants*, 23. *Consumption*, 40,000 galls. **Financial.**—*Cost*, $38,000. *No Cap. Stock. No Bonds. Floating Debt*, $5,000 at 6%. *Expenses:* operating, $1,000; int., $300; total, $1,300. *Revenue:* consumers, $1,300; city, $675, total, $1,975. **Management.**—Prest., Jas. M. Thompson. Secy., W. B. Mason. Mgr., Sylvanus Ayres, Jr.

10. BRIDGETON, *Cumberland Co.* (Pop., 8,722—11,424.) Built in '77-8 by city. **Supply,** East Lake, springs and well, pumping to reservoir. Fiscal year closes Dec. 31. In Oct., '90, a wooden conduit or feeder to the well was built, increasing the supply two-thirds.

Year.	Mains.	Taps.	Hyd'ts.	Cons'pt'n.	Cost.	B'd'd d't.	Int. chgs.	Exp.	Rev.
'78	7.8	138	57	79,036	$71,033	$69,000	$3,491	$1,125	$1,616
'80	10.8	302	72	180,308	79,071	76,500	4,140	1,443	3,876
'82	12.7	594	86	141,621	86,259	76,500	1,602	1,765	5,716
'84	14.6	803	96	215,120	92,859	80,000	4,800	1,647	8,045
'85	15	809	98	216,000	97,391	80,000	4,800	1,801	8,837
'86	15	973	98	207,246	99,159	80,000	4,800	2,407	9,678
'87	15.8	1,070	103	247,008	101,625	77,000	4,800	2,226	11,687
'88	16.1	1,154	105	271,361	104,921	74,000	4,020	2,058	12,258
'89	16.8	1,250	110	280,165	107,250	71,000	4,440	2,695	12,903
'90	17.1	1,350	111	362,000	109,727	68,000	4,260	2,676	14,444

Financial.—Bonded debt, $68,000, payable in yearly installments of $3,000. *Expenses:* total, $6,936. *Revenue:* consumers, $14,444; city, none. **Management.**—Committee of three. Chr., Benj. F. Harding. Supt., Timothy Woodruff.

11. BURLINGTON, *Burlington Co.* (Pop., 6,090—7,264.) Now owned by city. Built in 1804 by an Aqueduct Co.; in '43 Thos. Dugdale laid pipe and in '48 bought Aqueduct Co.'s works; in '60 works passed into hands of Burlington Water-Works Co., and in '77 were bought by city. **Supply.**—Delaware River, pumping to tank and stand-pipe. Fiscal year closes March 1. **Distribution.**—*Mains*, 9.3 miles. *Taps*, 1,690. *Meters*, 2. *Hydrants*, 109. *Consumption*, 400,000 galls. **Financial.**—*Cost*, $107,000. *Bonded Debt*, $60,000 at 4%. *Expenses:* operating, $4,000; int., $2,400;

total, $6,400. *Revenue:* consumers, $10,000; city, none. Op. exp. and int. consume 40% of revenue, new construction 10%, balance goes for redemption of bonds. **Management.**—Five comrs. appointed every five years by common council. Prest., J. A. Vandynlt. Secy., Treas., and Supt., G. A. Allison.

12. CAMDEN, *Camden Co.* (Pop., 41,659—58,313.) Owned by city. Built in '53 by Co. **Supply.**—Delaware River, pumping to reservoir and stand-pipe. Fiscal year closes Dec. 31.

Yr.	Mains.	Taps.	Meters.	Hydrants.	Consump'n.	Cost.	Bonded debt.	Int. ch'g's.	Exp.	Rev.
'70	19.74	....	..	102	....	$365,000	......	......	.....	.....
'80	35.07	6,000	..	267	....	461,000	......	......	.....	$58,051
'81	36.5	7,070	..	285	2,111,000	464,000	$400,000	$28,000	$26,310	65,644
'83	38.1	7,633	..	295	3,087,419	464,000	400,000	28,000	20,709	67,697
'86	46.5	9,008	..	362	3,321,008	500,000	400,000	28,000	20,000	78,660
'88	55.8	12,336	18	404	5,567,436	1,100,000	500,000	32,000	20,000	112,302
'89	57	....	10	464	7,660,000	1,100,000	500,000	32,000	21,000	122,730
'90	61	....	10	504	6,990,000	1,100,000	500,000	32,000	17,000	130,367

Management—Com., chosen from City Council by its President, each year. Chf. Engr., W. B. Doyle, appointed by Council.

13. CAPE MAY CITY, *Cape May Co.* (Pop., 1,699—2,136.) Built in '74 by city. **Supply.**—Surface wells, by gravity to tanks, with pumping as auxiliary. **Distribution.**—*Mains*, 8 miles. *Taps*, 700. *Hydrants*, 80. *Consumption*, 350,000. galls. **Financial.**—*Cost*, $35,000. *Debt*, $2,800 at 5%. *Expenses*, $1,500. *Revenue:* consumers, $7,690; city, none. **Management.**—Committee of three. Supt., J. Ashton Williams.

14. DOVER, *Morris Co.* (Pop., 2,958—not given.) Built in '87 by Dover Water Co. **Supply.**—Springs by gravity from collecting reservoir. Fiscal year closes June 30. **Distribution.**—*Mains*, 6.4 miles. *Taps*, 134. *Meters*, 81. *Hydrants*, 50. *Consumption*, 25,000 galls. **Financial.**—*Expenses*, $1,200. *Revenue*, city, $2,000. **Management.**—Prest., Archer N. Martin. Secy., Chas. E. Kimball. Treas., Chas. C. Fomeroy. Supt., Philip J. H. Bassett.

15. EAST ORANGE AND BLOOMFIELD, *Essex Co.* (Pop., East Orange, 8,349—13,282; Bloomfield, 5,748—7,708.) Built in '82 by Orange Water Co., under perpetual charter. **Supply.**—Springs and wells, pumping direct. Fiscal year closes Dec. 31. In '89 three wells and a gallery 700 ft. long connecting them, were constructed. **Distribution.**—*Mains*, 54.1 miles. *Taps*, 2,300. *Meters*, 150. *Hydrants*, 375. *Consumption*, 1,500,000 galls. **Financial.**—*Cost*, $550,000. *Cap. Stock*, authorized, $600 000. *No Debt*. **Management.**—Prest., F. M. Shepard. Secy., F. M. Shepard, Jr. Treas., J. A. Minott. Supt., Geo. P. Olcutt.

16. ELIZABETH, *Union Co.* (Pop., 28,229—37,764.) Built in '54 by Elizabeth Water Co., under perpetual charter, which provides neither for control of rates nor purchase of works by city. Fiscal year closes in Nov. **Supply.**—Elizabeth River, pumping from impounding to distributing reservoir and direct; filtered through charcoal.

Year.	Mains.	Taps.	Meters.	Hyd'ts.	Cons'pt'n.	Cost.	B'd'd d't.	Int.ch'gs.	Exp.	Rev.
'80	40	2,000	....	200	....	....	....	....	$45,000	.....
'84	40	3,000	30	200	2,000,000	$1,000,000	$400,000	$23,000	40,000	$75,000
'88	46	3,250	30	220	2,500,000	1,000,000	500,000	23,000	25,000	85,500
'90	52	3,750	30	231	2,500,000	1,000,000	500,000	23,000	25,000	85,500

Financial.—*First Cost*, $250,000. *Cap. Stock:* authorized, $250,000. *Bonded Debt*, $500,000: $100,000 at 7%; $400,000 at 4%. *Expenses*, total, $48,000. *Revenue:* consumers, $80,000; city, $5,500; total, $85,500. **Management.**—Prest., Jno. Kean. Secy. and Treas., J. M. Ross. Supt., Wm. Whelan.

23. ENGLEWOOD, *Bergen Co.* (Pop., 4,076—4,785.) Supply from Hoboken.

17. FLEMINGTON, *Hunterdon Co.* (Pop., 1,751—est., 2,000.) Built in '64 by Flemington Water Co.; franchise provides neither for control of rates nor purchase of works by town. **Supply.**—Springs and river, by gravity and pumping to reservoir. **Distribution.**—*Mains*, 7.5 miles. *Taps*, 140. *Meters*, 0. *Hydrants*, 9. **Financial.**—*Cost*, $25,000. *Revenue*, city, $240. **Management.**—Secy. and Treas., C. C. Dunham. Supt., Jno. C. Hopewell.

18. FREEHOLD, *Monmouth Co.* (Pop., 2,432—2,932.) No sewers. Has electric lights. **History.**—Built in '90-1 by town; put in operation May 1. Des. Engr., A. Harvey Tyson, Reading, Pa. Const. Engr., J. J. Moore. **Supply.**—Flowing ar-

FOR FURTHER DESCRIPTIVE MATTER see Manual for 1889-90. CHANGES since 1890 are here given. POPULATIONS are for 1880 and 1890, respectively.

tesian wells, pumping to stand-pipe; 8 wells, 4½ ins. in diam., 50 ft. deep. Pumping **Machinery**. Cap., 750,000 galls. ; comp. non-cond. Deane *Stand-Pipe*—Cap. 235,000 galls.; 20 x 100 ft.; from Tippett & Wood, Phillipsburg, N. J. Distribution.—*Mains*, 10 to 4-in. c. i., 8½ miles; from Union Hydraulic Co., Philadelphia. *Taps*, (May 1, '91), 105. *No Meters. Hydrants*, Eddy, 53. *Pressure :* ordinary, 43 lbs.; fire, 90. Financial.—*Cost*, $46,000. *Debt*, $40,000 at 4%. Management.—Prest., A. C. Hartshorne. Secy., John Enright. Supt., Perrin Voorhees.

19. **GLOUCESTER CITY**, *Camden Co.* (Pop., 5,347–6,564.) Built in '83-4 by city. Supply.—Springs and Newtown creek, pumping from reservoir to stand-pipe. Fiscal year closes Jan. 31. At time of report, July 9, '91, were about to open proposals for gang-well system, from which supply is to be entirely derived. Distribution.—*Mains*, 4 miles. *Takers*, 994. *Meters*, 1. *Hydrants*, 80. *Consumption*, 500,000 galls. Financial.—*Cost*, $125,000. *Bonded Debt*, $78,000 : $20,000 at 6%, due 1914-28; $18,000 at 5%, due 1924-32; $40,000 at 4%, due in annual payments of $2,000 after '93. *Expenses :* operating, $1.881. *Revenue :* consumers, $5,203; city, none. Management.—Supt., Henry M. Hailey. Engr., Jno. Lane.

23. **GUTTENBERG**, *Hudson Co.* (Pop., 1,206–1,917.) Supply from Hoboken.

23. **HACKENSACK**, *Bergen Co.* (Pop., 4,248–6,004.) Supply from Hoboken.

20. **HACKETTSTOWN**, *Warren Co.* (Pop., 2,502–2,417.). Built in '60 by village. Supply.—Streams, by gravity. Distribution.—(In '88.) *Mains*, 10 miles. *Taps* (in '87), 350. *No Meters. Hydrants*, 40. Financial.—(In '87.) *Cost*, $54,000. *Bonded Debt*, $53,500 at 5, 6 and 7%. *Revenue*, consumers, $5,000.—Management.—Prest., Wm. Drake. Secy. and Treas., W. S. Pettenhouse.

21. **HADDONFIELD**, *Camden Co.* (Pop., 1,480–2,502.) Built in '86 by Haddonfield Water Supply Co. Supply.—Stream, pumping from impounding reservoir to stand-pipe. Distribution.—*Mains*, 4.5 miles. *Taps*, 190. *Meters*, 27. *Hydrants*, 21. *Consumption*, 25,000 galls. Financial.—*Cost*, $40,000. Management.—Prest., Wm. H. Snowden. Secy., J. S. Simms. Treas., C. H. Hillman. Supt., D. G. Barnard.

22. **HARRISON**, *Hudson Co.* (Pop., 6,898–8,338.) Built in '86 by village. Supply.—Jersey City water-works. Distribution.—*Mains*, 7 miles. *Taps*, 422. *Hydrants*, 70. Financial.—*Cost*, $45,000. *Expenses*, $8,965, unusually large on account of extension of mains. *Revenue*, $7,426. Management.—Town Clerk, Edward J. Grace.

23. **HOBOKEN**, *Hudson Co.* (Pop., 30,999–43,648.) Owned by Hackensack Water Co.; built in '82. First supplied in '55 from Jersey City. Co. supplies Hoboken, West Hoboken, towns of Union, Weehawken and Guttenburg; townships of Union and North Bergen, in Hudson Co.; also townships of Ridgefield, Lodi, New Barbadoes (including Hackensack), Englewood, Midland and Palisades. Supply.—Hackensack River, pumping to reservoirs and tanks, filtered. Fiscal year closes Nov. 1. Mar. 30, '91, contract was let for 10,000,000-gall. high duty Worthington engine.

Year.	Mains.	Taps.	Meters.	Hy-drants.	Consump-tion.	Cost.	Bonded debt.	Int. chg's.	Exp.	Rev.
'84	16.1	3,500	107	150	$3,500,000	..	..	..	..	..
'87	78,5	5,549	2,667	550	4,250,000	....	$675,000	$48,750	$54,000	$150,180
'88	81	6,250	3,500	575	4,500,000	$1,858,169	1,105,000	55,250	58,127	163,917
'90	113	7,249	4,635	651	5,527,000	....	1,100,000	55,000	....	......

Financial.—*Cap. Stock :* common: authorized, $1,500,000; paid up, $537,000; paid 6% dividend in 1890. Preferred: authorized, $500,000; paid up, $375,000; 6% dividend guaranteed. *Bonded Debt*, $1,100,000 at 5%. Management.—Prest., R. W. de Forest. Secy., Wm. Shippen. Treas., E. A. Stevens. Supt., C. B. Brush.

24. **JERSEY CITY**, *Hudson Co.* (Pop., 120,722–163,003.) Built in '52 by city. Supplies Bayonne, Kearney Township and Harrison. Bayonne will probably have new supply in '92. Supply.—Passaic River, pumping to reservoir.

Yr.	Mains.	Taps.	Meters.	Hyd'ts.	Cons'pt'n.	Cost.	B'd'd d't.	Exp.	Rev.
'82	144	...	281	1,415	16,212,098	$4,950,000	$4,838,000	$95,185	$134,525
'84	...	11,510	185	1,436	16,500,000	4,950,000	4,833,000	450,000	480,000
'87	160	16,936	217	1,551	16,500,000	4,950,000	4,833,000	539,170	514,428
'88	178	20,456	249	1,738	19,300,000	....	....	......	......

Management.—Chf. Engr., W. W. Ruggles.

25. **KEARNEY**, *Hudson Co.* (Pop., 777–7,064.) Built in '87 by township. Supply.—Jersey City Water-Works, by gravity from receiving reservoir at Belleville. Fiscal year closes Feb. 28. Distribution.—*Mains*, 13 miles. *Taps*, 230. *Meters*,

NEW JERSEY. 103

70. *Hydrants*, 91. Financial.—*Cost*, $31,400. *Bonded Debt*, $32,000. *Expenses:* operating, $4,959; city, $4,500; total, $9,459. *Revenue:* consumers, $6,946; city, none. Management.—Water Purveyor, Wm. Green.

26. **LAKEWOOD**, *Ocean Co.* (Pop., not given.) Built in '86 by Lakewood Water Co. Supply.—South branch of Metideconk River. Fiscal year closes Dec. 31. Distribution.—*Mains*, 3.1 miles. *Taps*, 60. *No Meters. Hydrants*, 30. Financial.—*Withheld*. Management.—Prest., Isaac A. Van Hise. Secy. and Supt., Saml. D. Davis.

27. **LAMBERTVILLE**, *Hunterdon Co.* (Pop., 4,183–4,142.) Built in '78 by Lambertville Water Co.; town has no control over rates. Supply.—Swans Creek, by gravity. Fiscal year closes March 1. Distribution.—*Mains*, 4 miles. *Taps*, 278. *Meters*, 2. *Hydrants*. 28. Financial.—*Withheld*. Management.—Prest., C. A. Skinner. Secy., W. N. Cooley. Supt., T. Conyell.

I. **LITTLE YORK**, *Hunterdon Co.* (Pop., not given.) Built in '83 by a Distilling Co. Distillery was abandoned in '68 and any one who wishes has free use of water. Supply.—Springs, by gravity. Distribution.—*Mains*, 500 ft. *Consumers*, 20. *Hydrants*, none. Management.—Address J. R. Fox.

23. **LODI**, *Bergen Co.* (Pop., 4,071–5,131.) Supply from Hoboken.

28. **LONG BRANCH**, *Monmouth Co.* (Pop., 3,833–7,231.) Built in '77 by Long Branch Water Supply Co. Works supplying Monmouth Beach were consolidated with these in '82, and those supplying Seabright in '89. Supply.—Brook and springs, pumping to stand-pipe; filtered. Fiscal year closes Dec. 31.

Year.	Mains.	Taps.	Hydrants.	Consumption.	Cost.	B'd'd. debt.	Int. ch'g's.
'84	8	250	50	625,000	$100,000		
'88	25	780	112	710,000	350,000	$100,000	$5,000
'90	25	853	114	750,000	400,000	100,000	5,000

Financial.—*Cap. Stock*, $250,000. *Bonded Debt*, $100,000 at 5%. *Revenue*, city, $3,800. Management.—Prest., Geo. T. Baker. Secy. and Supt., Whitney Conant.

29. **MADISON**, *Morris Co.* (Pop., est., 3,000.) No sewers; electric lights projected early in '91. History.—Built in '90-1 by borough; put in operation about May 1. Engr. S. B. Leach, Tarrytown. Contrs., Moffett, Hodgkins & Clarke, Syracuse, N. Y. Supply.—Open well, pumping to stand-pipe; well (as designed) 30 ft. in diam., 20 ft. deep, lined with 30-in. stone wall. Pumping Machinery—Cap., 1,000,000 galls.; Deane comp. dup. *Boilers*, from Phœnix Foundry, Syracuse, N. Y. Stand-Pipe.—Cap., about 270,000 galls.; 25 × 75 ft.; from Phœnix Foundry Co. Force main extends to top of stand-pipe. Distribution.—*Mains*, 10 to 4-in. c. i., 9½ miles; from Jackson & Woodin Mfg. Co., Berwick, Pa. *Hydrants*, Holyoke. *Valves*, Rensselaer. *Pressure*, 60 to 110 lbs. Financial.—*Cost*, about $62,000. *Debt*, $60,000. Management.—Not given.

30. **MERCHANTVILLE**, *Camden Co.* (Pop., 439–1,225.) Built in '87-8 by Merchantville Water-Works Co.; franchise neither fixes rates nor provides for purchase of works by city. Supply.—Stream, fed by springs, pumping to stand-pipe. Fiscal year closes March 18. In '89-90 an additional 750,000-gall. pump and boiler were put in, increasing dy. cap. of pumps to 1,500,000 galls.

Year.	Mains.	Taps.	Hyd'ts.	Consum'p'n.	Cost.	B'd'd debt.	Int. ch'g's.	Exp.	Rev.
'87 8	6.5	124	20		$32,000	$15,000	$750	$1,000	$1,800
'88-9	7	180	23	26,354	35,000	15,000	750	1,957	2,678
'89-90	8.4	225	28	33,918	38,755	15,000	750	1,063	3,477
'90-1	9.5	261	40	46,601	45,968	15,000	750	1,197	4,375

Financial.—*Cap. Stock*: authorized, $50,000; paid-up, $15,000; paid 6% div'd in 1890. Pref'd, authorized, $50,000. *Bonded Debt*, $15,000 at 5%, due in 1906. *Expenses:* operating, $1,442; int., $865; taxes, $54; total, $2,362. *Revenue:* consumers, $3,355; city, $1,020; total, $4,375. Management.—Prest., A. G. Cattell. Secy., E. S. Hall. Supt., E. S. Thorpe.

23. **MIDLAND**, *Bergen Co.* (Pop., 1,591–1,829.) Supply from Hoboken.

31. **MILLVILLE**, *Cumberland Co.* (Pop., 7,660–10,002.) Built in '78 by Millville Water Co. Supply.—Maurice River, pumping from impounding reservoir to stand-pipe. Fiscal year closes Feb. 23. Distribution.—*Mains*, 9.6 miles. *Taps*, 485. *Meters*, 7. *Hydrants*, 67. Financial.—Withheld. Management.—Prest.,

For Further Descriptive Matter see Manual for 1889-90. Changes since 1890 are here given. Populations are for 1880 and 1890, respectively.

Geo. Wood, Philadelphia, Pa. Sec., N. G. Livermore. Treas., N. D. Kemble. Supt., R. W. Meredith.

28. **MONMOUTH BEACH,** *Monmouth Co.* (Pop., not given.) Supply from Long Branch.

32. **MONTCLAIR,** *Essex Co.* (Pop., 5,147—8,656.) Built in '87 by Montclair Water Co. Supply.—Wells, pumping to tank. After '91, supply is to be taken from mains of East Jersey Water Co.; see Newark. Fiscal year closes Dec. 31. Distribution.—*Mains,* 27.9 miles. *Taps,* 717. *Meters,* 57. *Hydrants.* 250. *Consumption,* 189,480 galls. Financial.—*Cost* (in '87), $180,000. *Cap. Stock:* authorized. $1,000,- 000, *No debt. Expenses* (in '89), $7,853. *Revenue:* (in '89), $17,219; city (in '90), $6,500. Management.—Prest., W. G. Snow. Secy., A. P. Fisher. Treas., J. T. Baulett.

33. **MOORESTOWN,** *Burlington Co.* (Pop., not given.) Built in '87-8 by Moorestown Water Co.; franchise neither provides for control of rates or purchase of works by city. Supply.—Impounding reservoir, filled by five small springs, pumping to tank and direct; also, connection of pump with Pennsauken Creek for emergency supply. Fiscal year closes Sept. 1. Distribution.—*Mains,* 7 miles. *Taps,* 162. *Meters,* 4. *Hydrants,* 54. *Consumption,* 20,000 galls. Financial.—*Cost,* $44,552. *Cap. Stock:* authorized, $40,000; all paid-up; paid no dividend in 1890. *Floating Debt,* $2,000. *No Bonds. Expenses:* operating, $1,352; int., $142; taxes, $170; total, $1,664. *Revenue:* consumers, $1,835; city, $1,080; other purposes, $416; total, $3,331. Management.—Prest., N. N. Stokes. Secy., Sam'l K. Robbins. Supt., Jos. Holton.

34. **MORRISTOWN,** *Morris Co.* (Pop., 5,418—8,156.) Built in 1779 by the proprietors of the Morris Aqueduct ; town neither controls rates nor has right to purchase works. Supply.—Springs, by gravity from impounding to distributing reservoir; in time of drought water is pumped from two different springs. Fiscal year closes Dec. 1. The following corrections of data in 1889-90 Manual have been made : *Conduits,* two 6-in. and one 10-in. c. i, *Pumping Machinery,* one 1,000,000 and one 500,00) Worthington. Distribution.—*Mains,* 30 miles. *Taps,* 1,000. *Meters,* 400. *Hydrants,* 100. *Consumption,* 300,000 galls. Financial.—*Cost,* $250,000. *Cap. Stock:* authorized, $100,000; all paid-up. *Total Debt,* $140,000: floating, $40,000; bonded, $100,000, at 5 and 6%. *Expenses,* operating, $8,000. *Revenue:* consumers, $19,000; city, $2,500; total, $21,500. Management.—Prest., H. C. Pitney. Secy. and Treas., Edw. Pierson. Supt., Wm. A. Dunn.

35. **MOUNT HOLLY,** *Burlington Co.* (Pop., 4,621—not given.) Built in '46 by Mount Holly Water Co. Supply.—Rancocas Creek, pumping to reservoir and direct. Fiscal year closes first Saturday in October. Financial.—*First cost,* $74,000. Cap. stock, $75,000. *Floating debt,* $5,150 at 5%. *No Bonds.* Expenses: operating, $2,775; int., $119; total, $2,894. *Revenue:* consumers, $6,348; city, $340; total, $6,688. Management.—Prest., B. F. Shrew. Secy. and Supt., H. C. Risdon.

36. **NEWARK,** *Essex Co.* (Pop., 136,508—181,830.) Built in '67-'76 by city; first supplied in 1800 by Newark Aqueduct Co. Supply.—Passaic River, pumping to reservoir with repumping for high service. Fiscal year closes Nov. 30. A 5,000,000-gall. high service Worthington pumping engine was bought in '90.

Yr.	Mains.	Taps.	Me- ters.	Hy- drts.	Con- s'ption.	Cost.	Bonded debt.	Int. ch'g's.	Exp.	Rev.
'80	136	11,000		1,186	9,386,000	$2,671,580			$12,000	
'82	140.4	12,601	242	1,197	9,580,160	3,066,671	$3.240,000	$226,800	114,343	$223,368
'83	143.7	13,592		1,232	9,672,878	3,230,530	3,240,000	226,800	98,380	236,072
'86	152	15,008	364	1,273	11,389,544	....	3,485,000	...	91,806	271,454
'87	161	12,207	432	1,341	12,642,155	....	3,485.000	......	103,407	315,623
'88	166.8	19,854	501	1,381	13,531,356	2,532,000	3,500.000	......	114,760	341,505
'90	176.4	21,532	520	1,460	14,079,793	3,113,524	3,532,000	236,635	122,185	369,540

Financial,—*Bonded Debt,* $3,532,000: $3,040,000 at 7%, due '92-6; $100,000 at 6%, due in '97; $100,000 at 5%, due in 1910; $235,000 at 4½%; $57,000 at 4%. *No Sinking Fund. Expenses,* total, $359,170. *Revenue:* consumers, $369 540; city, none. Management.—Aqueduct Board of six, two comrs. elected annually. Chn., the mayor. Secy., W. E. Greathead. Supt., Geo. R. Gray.

NOTE.—Under contract with East Jersey Water Co. gravity supply of 25,000,000 galls. daily will be introduced in '92 and 25,000,000 additional after 11 years. Supply will be from headwaters of Pequannock River, impounded in Clinton and Oak Ridge reservoirs. From storage reservoirs water will pass in natural channels of streams to a 30,000,000-gall. intake reservoir, from which a 48-in. riveted steel conduit, about 21 miles long, will deliver water into present reservoir at Belleville; a 36-in. branch conduit will deliver water to stand-pipe at high service reservoir, 300 ft. above tide. During 11 years from '92 Co. has right to sell surplus above 25,000,000 galls. Newark pays $4,000,000 upon delivery of first 25,000,000 galls., and $2,000,000 for additional amount. For full description, with illustrations, see ENGINEERING NEWS, vol. xxvi., p. 96, *et seq.*; also see *Engineering Record*, Aug. 1, *et seq.*

23. **NEW BARBADOES,** *Bergen Co.* Coextensive with Hackensack, which see

37. **NEW BRUNSWICK,** *Middlesex Co.* (Pop., 17,166—18,603.) Owned by city since '73; built in '65-8 by Brunswick Water Co. Supply.—Lawrence Brook, pumping from impounding to distributing reservoir. Fiscal year closes Dec. 31.

Yr.	M'ns.	Taps.	Mtrs.	H'd'ts.	C'ns'pt'n.	Cost.	B'd'd d't.	Int. ch'g's.	Exp.	Rev.
'73	16.9	764	2	104	725.000	$344,931	$440,000	$30,800	$15,219	$15,096
'75	18.2	948	37	123	808,000	......	440,000	30,800	16,821	22,152
'80	19.8	1,213	....	....	959,934	......	440,000	30,800	15,002	31,801
'82	20.7	1,330	....	152	947,587	445,048	440,000	30,800	13,791	34,019
'83	21.1	1,396	77	152	999,194	468,523	440,000	30,800	11,234	34,210
'84	21.6	1,468	86	154	1,047,811	472,562	440,000	30,800	11,592	35,347
'85	22.3	1,547	86	158	1,059,101	478,767	440,000	30,800	11,381	36,059
'86	23.8	1,658	84	173	1,093,101	484,960	440,000	30,000	11,370	39,094
'87	25	1,802	80	176	1,083,006	487,494	440,000	30,000	11,363	41,028
'88	25.8	1,899	83	181	1,153,363	493,586	440,000	23,925	17,323	42,246
'89	27.1	1,985	92	186	1,251,065	496,734	440,000	23,450	12,127	42,633
'90	28.2	2,075	99	190	1,254,844	502,452	440,000	29,585	10,875	43,379

Financial.—*Total Debt*, $440,000: $121,500 at 6%; $318,500 at 7%. As fast as the 7% bonds fall due they are refunded at 6%. Expenses, total, $40,460. Revenue: consumers, $43,379; city, $9,575; total, $52,954. Management.—Six comrs., appointed every two years by common council. Prest., Jno. Runyon. Secy., G. T. Applegate. Supt., C. H. Cramer.

23. **NORTH BERGEN,** *Hudson Co.* (Pop., 4,268—5,715.) Supply from Hoboken.

38. **NUTLEY,** *Essex Co.* (Pop., est., 1.500.) Built in '89 by Jas. R. Hay and others, of 84 Broadway, New York City. Supply, pumped by 500,000-gall. Worthington pump. Mains, about 5 miles. Hydrants, 40. Other statistics withheld.

39. **OCEAN GROVE,** *Monmouth Co.* (Pop. 620—2,754.) Built in '84 by Ocean Grove Association. Supply.—Twenty-one artesian wells, pumping to tank. Fiscal year closes Sept. 30. During 1890 eight additional artesian wells were sunk, at a cost of $3,320.

Year.	Mains.	Taps.	Meters.	Hyd'ts.	Consumption.	Cost.	Expenses.	Revenue.
'87	12	501	3	27	376,000	....	......	......
'88	12.6	540	3	27	400,000	....	......	......
'89	13.5	669	..	41	137,000	$53,858	$1,391	$5,570
'90	13.5	864	..	41	186,300	58,713	1,731	8,159

Financial.—*Cap. Stock and Debt* not separate from other expenses of Ass'n. Other data, see above. Management.—Prest., E. H. Stoker, Secy., G. W. Evans. Treas., D. H. Brown. Supt., L. Rainear.

40. **ORANGE,** *Essex Co.* (Pop., 13,207—18,844.) Built in '83 by city. Supply.—West branch of Rahway River, by gravity from impounding reservoir. Fiscal year closes Feb. 28.

Year.	Mains.	Taps.	Meters.	Hydrants.	B'd'd Debt.	Int. Ch'g's.	Exp.	Rev.
'83-4	21	150	....	200	$400,000	$20,000	......	$3,282
'87-8	25	1,300	....	200	400,000	20,000	$4,761	16,783
'88-9	25.9	1 321	....	206	425,000	21,250	5,739	20,453
'89-90	26.7	1,563	270	213	420,000	21,000	5,115	27,410
'90-1	27.9	1,933	306	210	420,000	21,000	6,289	33,819

Financial.—*Bonded Debt*, $420,000 at 5%. *Sinking Fund*, $47,450. Expenses: operating and int., $27,289; construction, $4,378; total, $31,667. Revenue: consumers; $33,-819; city, none. Management.—Clerk, Horace Stetson. Supt., Thomas Dowd.

23. **PALISADES,** *Bergen Co.* (Pop., 2,302—2,590.) Supply from Hoboken.

49. **PALMYRA,** *Burlington Co.* (Pop., not given.) Supply from Riverton.

41. **PASSAIC,** *Passaic Co.* (Pop., 6,532—13,028.) Built in '72 by Aquacknononk Water Co. Supply.—Reservoir of Passaic Water Co., at Paterson, pumping to reservoir; formerly Vreeland Lake, by gravity, supplemented by Passaic River. Pumping machinery remains in case of need. Fire pressure is increased from 35 to 55 lbs. Fiscal year closes Dec. 31. Distribution.—*Mains*, 20 miles. *Taps*, 700.

FOR FURTHER DESCRIPTIVE MATTER see Manual for 1889-90. CHANGES since 1890 are here given. POPULATIONS are for 1880 and 1890, respectively.

Meters, 0. *Hydrants*, 162. *Consumption*, 400,000 galls. **Financial.**—*Revenue*, city, $7,290. **Management.**—Prest., J.A. Hobart. Secy., E. T. Bell. Treas., J. J. Brown. All of Paterson. Supt., Washington Paulison.

42. **PATERSON,** *Passaic Co.* (Pop., 51,030—78.347.) Built in '56 by Passaic Water Co.; franchise provides neither for control of rates nor purchase of works by city. **Supply.**—Passaic River, pumping to reservoirs. Fiscal year closes Jan. 1.

Year.	Mains.	Taps.	Meters.	Hydrants.	Consumption.
'80	40	2,635	..	480	....
'82	48	3,300	0	520	6,000,000
'84	50	3,900	0	582	7,000,000
'86	.	4,100	1	617	8,000,000
'88	60	5,900	1	767	9,000,000
'90	70	6,648	2	834	10,000,000

Financial.—*Cost* (in '84), $1,440,360. *Cap.* Stock (in '88), $425,000. *Bonded Debt* (i '88), $1,100,000. *Revenue*, city, $30,858. **Management.**—Prest., G. A. Hobart. Secy., J. C. Ryle. Treas., J. J. Brown. Supt., Wm. Ryle.

II. **PENNINGTON,** *Mercer Co.* (Pop. not given.) Built in '86-8 by Pennington Spring Water Co., under perpetual franchise. **Supply.**—Spring, by gravity. Fiscal year closes third Thursday in October. **Distribution.**—*Mains*, 3.5 miles. *Taps*, 60. *Meters*, 1. *Hydrants*, 1. *Consumption*, unknown. **Financial.**—*Cost*, $8,000. *No Stock. No Debt.* **Management.**—Prest., D. A. Clarkson. Secy., Jno. G. Muirhead. Supt., J. S. Bend.

43. **PERTH AMBOY,** *Middlesex Co.* (Pop., 4,808—9,512.) Built in '82 by Perth Amboy Water Co.; franchise does not regulate rates, but provides for purchase of works by city, price determined by arbitration. **Supply.**—Springs and surface water, pumping from impounding reservoir to stand-pipe **Distribution.**—*Mains*, 7.3 miles. *Taps*, 313. *Meters*, 68. *Hydrants*, 63. **Financial.**—*Revenue*, city, $3,150 **Management.**—Prest., Wm. Hall. Secy., C. C. Hommann. Treas., Wm. E. Stiger. Supt., Milton A. Brown.

NOTE.—June 20, '91, application was filed in court by Corporation Counsel, for a Commission of Condemnation to adjudicate upon value of plant, franchise, and appurtenances of Perth Amboy Water Co., city wishing to purchase system.

44. **PHILLIPSBURG,** *Warren Co.* (Pop., 7,181—8,644.) Built in '87 by People's Water Co.; town neither controls rates nor has right to purchase works. Lehigh Water Co., Easton, Pa., supplies part of town. **Supply.**—Well, pumping to reservoir. **Distribution.**—*Mains*, 11 miles. *Taps*, 423. *Meters*, 7. *Hydrants*, 65. **Financial.**—*Cost*, $100,000. **Management.**—Prest., S. Thomas. Secy., J. O. Carpenter. Treas., J. A. Bachman. Supt., G. G. Stryker.

45. **PLAINFIELD,** *Union Co.* (Pop., 8,125—11,267.) Sewers projected. No electric lights. **History.**—Built in '90-1 by Plainfield Water Supply Co., under legislative charter granted about '69; put in operation in August. Engr., C. B. Brush, N. Y. City. **Supply.**—Twenty driven wells, pumping to stand-pipe. **Pumping Machinery.**—Cap., 5,000,000 galls.; two Worthington. **Stand-Pipe.**—Cap., 515,000 galls.; 25 × 140 ft.; from Tippett & Wood, Phillipsburg. **Distribution.**—*Mains*, 16 to 6-in. c. i., probably about 25 miles will be put in; from 'Warren Foundry & Machine Co., Burlington, N. J.; laid by Dowd & Rice. *Hydrants*, Ludlow. **Financial.**—Withheld and not determined. **Management.**—Prest., Frank Bergen. Secy. and Treas., E. R. Pope. Report Aug. 31. NOTE.—In '91 Council granted franchise to J. M. Low, Kingston, N. Y.

46. **PRINCETON,** *Mercer Co.* (Pop., 3,209—3,422.) Built in '83 by Princeton Water Co. **Supply.**—Ground water, pumping to elevated tank. Fiscal year closes Oct. 1.

Year.	Mains.	Taps.	Meters.	Hyd'ts.	Cons'pt'n.	Cost.	B'd'd d't.	Int.ch'g's.	Exp.	Rev.
'84	5	130	20	36	80,000	$55,000	none	none	$2,873	$3,971
'86	4	175	160	31	70,000	70,000	none	none	3,400	5,320
'88	6	255	245	50	60,000	66,200	$20,000	1,000	3,500	6,700
'90	6.5	288	265	50	65,000	70,000	20,000	$1,000	4,350	6,940

Financial.—*Cap. Stock*, authorized, $50,000. *Bonded Debt*, $20,000, at 5%. *Sinking Fund*, $3,000, at 3%. *Expenses:* operating $3,500; int., $1,000; taxes, $850; total, $4,350. *Revenue:* consumers, $5,500; city, $1,200; other purposes, $240; total, $6,940. **Management.**—Prest., Chas. E. Green. Secy., Wm. Libbey, Jr. Supt., Jas. E. Bunce.

NOTE.—For illustrated description of 10-in. main, see *Engineering Record*, Sept. 13, 1890.

47. **RAHWAY,** *Union Co.* (Pop., 6,455—7,105.) Built in '71-2 by city. **Supply.** —Rahway River, pumping direct. Fiscal year closes Dec. 31.

NEW JERSEY. 107

Year.	M'ns.	Taps.	M't'rs.	Hydr'ts.	Cons'pt'n.	B'd'd d'bt.	Int. ch'g's.	Exp.	Rev.
'75	....	....	....	....	311,573	$185,000	$12,950	....	$2,463
'80	5	....	....	....	335,285	185,000	12,950	$1,500	6,236
'82	12.3	430	0	122	583,827	185,000	12,950	4,439	8,444
'84	12.3	460	0	122	678,073	185,000	12,950	4,506	10,811
'86	12.6	*608	0	123	726,829	185,000	12,950	5,650	10,482
'88	13	*534	0	125	800,000	185,000	12,950	6,772	10,883
'89	13	546	1	125	666,550	185,000	12,950	5,394	11,854
'90	13	596	1	125	690,000	185,000	12,950	6,517	11,483

* Discrepancy not explained in printed Annual Report.

Financial.—*Bonded Debt*, $185,000, at 7%, due in '91. *Expenses*, total, $19,467. *Revenue:* consumers, $11,483; city, none. **Management.**—Five comrs. Chn., E. S. Hyer. Secy. and Supt., H. B. Bunn, appointed by Comrs.

51. **RARITAN**, *Somerset Co.* (Pop., 2,046—2,556.) Supply from Somerville.

48. **RED BANK**, *Monmouth Co.* (Pop., 2,684—4,145.) Built in '85 by village. **Supply.**—Wells, pumping to reservoir and direct. During '90 supply was doubled by sinking two artesian wells, each 4 ins. in diam., 250 ft. deep, syphoned into brick well. *Distribution.*—*Mains*, 9 miles. *Meters*, 75. *Hydrants*, 94. **Financial.**—*Cost*, $85,000. *Bonded Debt*, $85,000: $60,000 at 5%; $25,000 at 4%. *Operating Expenses*, $2,500. **Management.**—Three comrs. Prest., Jno. H. Cook. Secy., Wm. T. Corlies. Treas., Jno. S. Applegate.

23. **RIDGEFIELD**, *Bergen Co.* (Pop., 3,952—5,477.) Supply from Hoboken.

49. **RIVERTON and PALMYRA**, *Somerset and Burlington Cos.* (Pop.: Riverton, 586—not given; Palmyra, 571—est., 1,000.) Built in '89 by Riverton & Palmyra Water Co.; franchise neither regulates rates nor provides for purchase of works by city. **Supply.**—Infiltration well near Delaware River, pumping to tank. Fiscal year closes Dec. 31. *Distribution.*—*Mains*, 9 miles. *Taps*, 165. *Meters*, 0. *Hydrants*, 53. **Financial.**—*Cost*, $50,000. *Cap Stock:* authorized, $50,000; all paid up. *No Debt. Expenses*, $1,600. *Revenue*, city, $530. **Management.**—Prest., E. H. Ogden. Secy. and Treas., Howard Parry.

50. **SALEM**, *Salem Co.* (Pop., 5,056—5,516.) Built in '82 by city. **Supply.**—Well and surface water, pumping direct from impounding reservoir on Laurel River. Fiscal year closes Dec. 31.

Year.	Mns.	Taps.	Me-ters.	Hy-dr'ts.	Con-sumption.	Cost.	Bonded debt.	Int. ch'g's.	Exp.	Rev.
'83	8.3	201	0	54	97,328	$77,650	$75,000	$3,750	$2,463	$1,845
'86	8.4	263	0	56	165,956	80,000	75,000	3,750	2,694	4,000
'90	9.8	418	0	66	328,088	90,000	67,000	3,350	3,805	4,700

Financial.—*Bonded Debt*. $67,000 at 5%. *Expenses*, total, $7,155. *Revenue*, consumers, $4,700. **Management.**—Chn., Jno. Watson. Supt., C. W. Casper.

23. **SEABRIGHT**, *Monmouth Co.* (Pop., est., 1,000 in winter and 3,000 in summer.) Built in '87 by Seabright Water Supply Co. In '89 Seabright Water Supply Co. sold their works to the Long Branch Water Co., which now operates system at Seabright.

III. **SHORT HILLS**, *Essex Co.* (Pop., est., 400.) Built in '80 by Stewart Hartshorn. **Supply.**—Springs, pumping to reservoir. Pump has cap. of 500,000 galls. *Distribution.*—*Mains*, 5 miles. *Taps*, 65. *Meters*, 10. *Hydrants*, 2. *Consumption*, 40,000 galls. **Financial.**—*Cost*, $72,000. *No Debt. Expenses*, $1,700. *Revenue*, $350. **Management.**—Owner, Stewart Hartshorn. Secy. and Treas., Geo. E. Crosscup.

51. **SOMERVILLE and RARITAN**, *Somerset Co.* (Pop., Somerville, 3,105—3,861; Raritan, 2,046—2,556.) Built in '82 by Somerville Water Co. **Supply.**—Raritan River, pumping to stand-pipe and direct; filtered. Fiscal year closes April 1. During '91, a 1,000,000-gall Smith & Vaile water power pump was added, and it was proposed to put in a 1,000,000-gall. steam pump; it was unofficially reported that stand-pipe 75 ft. high was being added.

Year.	Mains.	Taps.	Meters.	Hy'd'ts.	Cost.	Exp.	Rev.
'88-9	8	400	6	41	$64,613	$3,936	$3,182
'89-90	8	475	6	41	66,717	3,595	3,623
'90-91	8.5	500	5	44	69,922	3,457	9,118

Financial.—*Cap. Stock:* authorized, $50,000; all paid up; paid 6% div'd in 1891. In '91 cap. stock was increased to $75,000. *Floating Debt*, $3,205 at 5%. *No Bonds. Expenses:* operating, $2,817; int., $173; taxes, $640; total, $3,630. *Revenue:* consumers,

For Further Descriptive Matter see Manual for 1889-90. Changes since 1890 are here given. Populations are for 1880 and 1890, respectively.

$7,970; city, $1,148; total, $9,118. **Management.**—Prest., Adolph Mack. Secy., W. H. Taylor. Supt., J. Harper Smith

52. **SOUTH ORANGE,** *Essex Co.* (Pop., 2,178—3,106.) History.—Built in '90-1 by borough; put in operation Jan. 29. Engr., C. P. Bassett, Newark. Supply.—Mountain Water Co.'s plant, at Summit, through two sets of meters. Distribution.—*Mains,* 10 to 4-in. c. i. (Mar. 12, '91), 5.83 miles. *Services,* lead and galv. I. *Consumers* (June, '91), about 125. *Hydrants,* Mathews and Beaumont, 45. *Pressure,* av., about 100 lbs. Financial.—*Cost* (to Mar. 12, '91), $24,719, Debt (Mar. 12, '91), $25,000 at 5%. **Management.**—Comrs. Chn., H. H. Hart. Supt., J. B. Maxwell.

53. **SUMMIT**, *Union Co.* (Pop., 1,910—3,502.) Built in '89 by Summit Water Co. Supplies water through me'ers to South Orange plant. Supply.—Wells in Blue Brook Valley, pumping to stand-pipe. Fiscal year closes Dec. 31. Distribution.—*Mains,* 10. *Taps,* 201. *Meters.* 10. *Hydrants.* 51. *Consumption,* 75,000 galls. Financial.—*Cap. Stock:* authorized, $100,000; paid-up, $60,000. *Bonded Debt,* $63,000 at 5%. *Expenses:* int., $3,000; taxes, $350. *Revenue,* city, $2,550. **Management.**—Prest., Franklin Murphy. Secy., Eugene Vanderpool. Treas., Fred. Frelinghuysen. Supt., Fred'k Green. Const. Engr., C. P. Bassett, Newark.

54. **TRENTON**, *Mercer Co.* (Pop., 29,910—57,458.) Bought and rebuilt by city in '55. Built in 1803 by Co. Supply.—Delaware River, pumping to reservoir. Fiscal year closes Jan. 31. In '90, high service tank was constructed at cost of $21,749.

Yr.	Mains.	Taps.	Meters.	Hyd'ts.	Cons'pt'n.	Cost.	B'd'd d't.	Int.ch'g's.	Exp.	Rev.
'82	35	5,338	38	185	1,275,126	$327.000	$265,000	$15,900	$9,774	$47,575
'84	35	6,046	38	205	1,409,921	349,371	265,000	.....	9,238	59,035
'86	37.2	6,929	46	214	2,000,000	422,951	265,000	15,850	13,373	57,324
'87	40	8,102	46	242	2,256,000	445,946	265,000	15,850	14,917	74,883
'88	60	.....	.....	266	2,533,405	530,384	310,000	18,100	16,980	74,657
'89	66	8,692	.....	298	2,849,335	581,557	310,000	18,100	15,806	80,306
'90	71	.....	.....	339	3,569,150	650,134	210,500	18,100	18,817	91,563

Financial.—*Bonded* Debt, $210,500 at 5 and 6 per cent. *Sinking Fund,* $143,534. *Expenses,* total, $36,917. *Revenue:* consumers, $91,563; city none. **Management.**—Six comrs., two appointed every year for two years. Prest., Robert Aitken. Sec. and Treas., Chas. A. Reid. Supt., Jacob E. Carter, appointed by comrs.

23. **UNION,** *Bergen Co.* (Pop., 865—1,560.) Supply from Hoboken.

55. **VINELAND,** *Cumberland Co.* (Pop., 2,519—3,822.) Built in '86 by Chas J. Keighley. In '91 Mr. Keighley incorporated Vineland Water-Works Co. to protect his interests; franchise provides that city may purchase works by paying cost of construction and 6% int. from first date of construction; city has no control of rates. Supply—Driven wells, pumping to tank. Fiscal year closes Dec. 31. Distribution.—*Mains,* 12 miles. *Taps,* 83. *Meters,* 3. *Hydrants,* 30. *Consumption,* 125,000 galls. Financial.—*Capital stock,* authorized, $75,000. *No debt. Revenue,* $514. **Management.**—Owner, Chas. J. Keighley.

56. **WASHINGTON,** *Warren Co.* (Pop., 2,142—2,834.) Built in '82 by Washington Water Co. Supply.—Stream, by gravity. Distribution.—*Mains,* 5.8 miles. *Taps,* 150. *Hydrants,* 30. Financial.—*Cost,* $35,000, *Cap. stock,* $35,000. *No debt.* **Management.**—Prest., J. W. Fritts. Treas., P. H. Hann. Supt., J. E. Fulper.

57. **WENONAH,** *Gloucester Co.* (Pop., 166—est., 500.) Built in '85 by Wenonah Water Co.; city neither controls rates nor has right to purchase works. Supply.—Spring, pumping to tank and direct. Fiscal year closes Dec. 31. Distribution.—*Mains* (in '88), 2 miles. *Taps* (in '88), 75. *Meters,* 0. *Hydrants,* 12. Financial.—*Cap. Stock:* authorized, $25,000; paid-up, $12,500; paid 5% dividend in 1890. *No Debt.* **Management.**—Prest., Stephen Greene. Secy., E. L. Farr. Supt., Blair Smith.

23. **WEEHAWKEN,** *Hudson Co.* (Pop., 1,102—1,943.) Supply from Hoboken.

23. **WEST HOBOKEN,** *Hudson Co.* (Pop., 5,441—11,665.) Supply, from Hoboken.

58. **WOODBURY,** *Gloucester Co.* (Pop., 2,302—3,911.) Built in '86 by village. Supply.—Head of Mantua Creek, pumping from settling to distributing reservoir, with direct pumping in case of fire.

Year.	Mains.	Taps.	Meters.	Hyd'ts.	Cons'pt'n.	Cost.	B'd'd d't.	Int.chgs.	Exp.	Rev.
'88	11	353	2	59	100,090	$72,000	$72,000	$2,880	$1,400	$4,000
'89	11.85	450	2	69	146,000	72,060	72,000	2,880	2,153	4,535
'90	12.05	520	2	70	213,000	72,894	72,000	2,880	2,537	5,683

Financial.—See above. **Management.**—Water Committee. Chn., W. A. Glover. Secy., Edmund Du Bois. Supt., B. W. Cloud.

Water-Works Projected with Fair Prospects of Construction.

BRIGANTINE, *Atlantic Co.* (Pop., est., 150.) Brigantine Water Co. incorp. April 1, '91; nothing has been accomplished June 20, '91, save to organize with J. S. Dougherty, 107 Market St., President, and E. Clarence Miller, 435 Chestnut St., Secretary and Treasurer; both of Philadelphia. Proposed supply, artesian well, pumping to stand-pipe. Estimated cost, $15,000. Authorized cap. stock, $13,000.

COLLINGSWOOD, *Camden Co.* (Pop. in '90, 539.) Collingswood Water & Light Co. incorp. early in '91 by G. Frank Davis, G. W. Storrs and others. Proposed supply, artesian well, pumping to tank or stand-pipe. Estimated cost, $10,0 0.

CRANFORD, *Union Co.* (Pop. of township, 1,184—1,717.) Unofficially reported that company was organized about July 15, '91, under charter granted in '70. Previously works were projected by W. C. Foster, 52 Broadway, at est. cost of $50,000. Probable supply, wells, pumping to tank.

OCEAN CITY, *Cape May Co.* (Pop. in '90, 452; summer, est., 5,000.) Projected by co. in '91; June 25 matter was before Council. Proposed supply, artesian wells, pumping to stand-pipe. Est. cost, $50,000. Address J. S. Dougherty, 107 Market St., Philadelphia, or E. B. Lake, Ocean City.

RUTHERFORD, *Bergen Co.* (Pop., 2,299—2,293.) Projected in '91 by Hackensack Water Co., reorganized, of Hoboken, which proposes to extend its works. Attempt was made in July to prevent co. building works.

SEA ISLE CITY, *Cape May Co.* (Pop. in '90, 766.) Works talked of in '91.

WOODSTOWN, *Salem Co.* (Pop., 490—556.) Works have been projected since '90. July, '91, surveys had been made by S. S. Carson and it was proposed to advertise for bids to build works on franchise plan or with city ownership. Supply, springs, two miles from town, pumping to stand-pipe. Est. cost, $30,000. Address Mayor, C. H. Richman.

Water-Works Projected in '90 and '91, but Now Indefinite.
Newton, Sussex Co.

PENNSYLVANIA.

Water-Works Completed or in Process of Construction.

1. **ADRIAN MINES,** *De Lancey P. O., Jefferson Co.* (Pop., est., 1,200.) Built in '86 by Rochester & Pittsburg C. & I. Co. **Supply.**—Artesian well, pumping to tanks. Fiscal year closes Sept. 30. **Distribution.**—*Mains* (in '88), 3 miles. *Taps,* 17. *Hydrants,* 9. *Consumption,* 12,000 galls. **Financial.**—*Cost,* $5,000. *No Debt. Expenses,* $1,730. *No Revenue.* **Management.**—Supt., Jno. H. Bell.

2. **ALLEGHENY,** *Allegheny Co.* (Pop., 76,682—105,287.) Built in '47 by city. **Supply.**—Allegheny River, pumping to reservoir. Last fiscal year closed Feb. 28, '91.

Year.	Mains.	Taps.	Hydrants.	Consumption.	Cost.	B'd Debt.	Exp.	Rev.*
'82	60	12,492	860	10,000,000	$1,100,000	$593,000	$25,150	$168,000
'84	70	12,937	700	12,000,000	1,330,717	....	75,000	170,713
'88	76	14,000	1,100	21,000,000	1,930,000	1,084,000	112,000	230,000
'90-1	90	15,000	1,100	25,000,000	2,225,000	....	120,000	270,000

* Including allowance for fire protection.

Distribution.—No meters. **Financial.**—See above. **Management.**—Committee appointed every year by council. Chn., Moses Curteen. Secy., Wm. Haslitt. Supt., Thos. Brown.

3. **ALLENTOWN,** *Lehigh Co.* (Pop., 18,063—25,228.) Built in '27 by Northampton Water Co.; now owned by city. **Supply.**—Springs, pumping to reservoir and stand-pipe. Fiscal year closes first Monday in April.

For FURTHER DESCRIPTIVE MATTER see Manual for 1889-90. CHANGES since 1890 are here given. POPULATIONS are for 1880 and 1890, respectively.

Year.	Mains.	Taps.	Meters.	Hydrants.	Consumption.	Cost.	B'd'd debt.	Int. ch'g's.	Exp.	Rev.
'81-2	22	2,250	18	96	800,000	$195,650	$125,622	.....	$4,427	$34,544
'83-4	24	3,069	18	107	....	209,165	111,600	.....	4,097	41,444
'85-6	25.4	3,214	18	121	....	213,471	110,200	$5,430	10,292	44,337
'86-7	25.6	3,442	..	134	....	216,479	110,200	6,040	11,429	47,126
'87-8	27	3,777	1	147	2,000,000	258,582	110,200	5,026	12,513	53,462
'88-9	30	4,106	1	163	2,400,000	275,235	110,200	5,240	14,713	56,166
'90-1	32	4,700	0	191	2,706,409	359,669	110,200	5,430	29,000	42,592

Financial.—*Bonded Debt*, $110,200: $11,000 at 7%, due in '97; $69,200 at 5%, due in '99; $30,000 at 4%, due in 1905. **Management.**—Prest., W. J. Hartzell. Secy., L. E. Butz. Supt., Peter Brown.

4. **ALTOONA,** *Blair Co.* (Pop., 19,710—30,537.) Built in '71 by city. **Supply.**—surface water, by gravity from impounding and distributing reservoir. Fiscal year closes Dec. 31. May, '91, $220,000 was voted for new supply.

Year.	Mains.	Taps.	Meters.	Hyd'ts.	Cons'pt'n.	Cost.	B'd'd d't.	Int.ch'gs.	Exp.	Rev.
'80	16	1,400	0	98	500,000	$200,000	$200,000	.....	.....	.....
'86	26	2,500	0	98	1,500,000	272,000	.....	.....	$16,300	$30,280
'87	35.8	5,272	0	112	1,175,000	359,000	250,000	$10,000	16,000	43,614
'88	37.6	5,345	0	129	1,753,000	.....	.....	.....	16,000	46,920
'89	36.7	5,909	0	138	2,140,000	.....	259,000	10,360	37,000	.....
'90	39.2	6,300	25	153	2,140,000	.....	259,000	10,360	38,000	.....

Financial.—*First Cost*, $200,000. *Total Debt*, $324,000: floating, $65,000 at 4%; bonded, $259,000 at 4%. *No Sinking Fund. Expenses*, total, $39,360. *Revenue*, withheld; is sufficient to pay expenses. **Management.**—Comrs., appointed by city council every three years. Chn., T. H. Wiggins. Secy., A. H. Martin. Supt., S. A. Gailey.

5. **ANNVILLE,** *Lebanon Co.* (Pop., 1,431—1,283.) Built in '90 by Annville Water Co.; put in partial operation Jan. 1, '91. Engrs. and Contrs., Dean & Schm, Annville. **Supply.**—Springs, by gravity. **Reservoir.**—Cap., 200,000 galls. **Distribution.**—*Mains*, 8 to 4-in. c. i., about 2 miles. *Taps*, 12. *No Meters*. Hydrants were to be added in '91. *Pressure*, 52 lbs. **Financial.**—*Cost*, $17,000. *Cap. Stock:* authorized, $20,000; paid up, $5,000; all was to be paid up by Jan. 8, '91. **Management.**—Prest., H. H. Kreider. Secy., C. Dean. Treas., O. H. Henry. Report, Dec., '90.

I. **ANSONVILLE,** *Clearfield Co.* (Pop., 99—est., 500.) Built in '90 by Ansonville Water Co. **Supply.**—Springs, by gravity. **Reservoir.**—Cap., 20,000 galls. **Distribution.**—*Mains*, 3-in., ¾ miles. *Services*, galv. i. *Taps*, 45. *No meters. Pressure,* 20 lbs. **Financial**—*Cost*, $1,500. *Cap. Stock:* authorized, $2,000; paid up, $1,000. *Debt*, floating, $500 at 6%. **Management.**—Prest., S. J. Miller. Secy., J. C. Davison.

6. **APOLLO,** *Armstrong Co.* (Pop., 1,156—2,156.) Built in '88 by Apollo Water Co. **Supply.**—Kiskiminates River, pumping to tanks. **Distribution.**—(In '88.) *Hydrants*, 20. **Financial.**—*Cost*, $15,000. Hydrant rental, $25 each. **Management.**—Prest., Jno. Duff. Secy., O. Tinsman, both of Parker's Landing. Supt., A. Randolph.

7. **ARCHBALD,** *Lackawanna Co.* (Pop., 3,049—4,032.) Built in '75 by Archbald Water Co. **Supply.**—Laurel Run, by gravity from impounding reservoir. **Distribution.**—*Mains*, 4 miles. *Taps*, 225 *Hydrants*, 6. **Financial.**—*First cost* $14,000. *Cap. Stock:* authorized, $20,000, all paid up; paid 8% dividend in 1890. *Bonded Debt*, $8,000 at 5%, due in 20 years. *Expenses:* operating, $400; int., $400; total, $800. **Management.**—Prest., J. J. Williams. Secy., Wm. Law. Treas., Jno. Carroll. Supt., Justus Bishop.

8. **ASHLAND,** *Schuylkill Co.* (Pop., 6,052—7,346.) Built in '76 by borough. **Supply.**—Little Mahanoy Creek, by gravity. Fiscal year closes March 4. **Distribution.**—*Mains*, 9.8 miles. *Taps*, 831. *Hydrants*, 39. **Financial.**—*Cost*, $87,144. *Bonded Debt*, $48,100: $20,000 at 6%, due in '91; $28,000 at 4%, due in '06. *Expenses:* operating, $1,327; int., $2,408; total, $3,735. **Management.**—Committee of three, appointed by Common Council. Chn., Jas. R. Deekan. Secy., Frank Rentz. Supt., D. T. Evans. Appointed by council.

212. **ASHLEY,** *Luzerne Co.* (Pop., 2,799—3,192.) Supply from Crystal Springs Water Co., Wilkes-Barre.

166. **ATHENS,** *Bradford Co.* (Pop., 1,592—3,274.) Supply from Sayre.

PENNSYLVANIA. 111

9. AUDENRIED, *Carbon Co.* (Pop., est., in '90, 3,000.) Built in '59 by Lehigh & Wilkes-Barre Coal Co.; supplies collieries and village. Water now furnished by Honey Brook Water Co. **Supply.**—Source not given, by pumping. **Pumping Machinery.**—Cap., 375,000 galls.; Cameron. **Distribution.**—*Mains*, 6 to 4 in. c. i., 3 miles. *Hydrants*, about 10. *Consumption*, 300,000 galls. *Pressure*, 60 lbs. **Financial.**—*Withheld*. **Management.**—Supt., J. I. Hollenbeck.

10. AUSTIN, *Potter Co.* (Pop., not given—1,679.) Built in '90-1 by Lumber City Water Co.; put in operation Mar. 2. Des. Engrs., Ayrault Bros. & Co. Const. Engrs. and Contrs., Renn & Chapman, Ellicottville, N. Y. Fire protection works were built by the borough in '88-9. **Supply.**—Springs, by gravity for domestic use, and by pumping in extreme cases for fire. **Pumping Machinery.**—Dean. **Reservoir.**—Cap., about 75,000 galls. **Distribution.**—*Mains*, 6 to 4-in., 3½ miles. *Taps*, 100. *No Meters*. *Hydrants*, Chapman, Ludlow, 1L *Pressure*: ordinary, 50 lbs.; fire, 200 lbs. **Financial.**—*Cost*, $16,792. *Cap. Stock*: authorized, $15,000; all paid up. *Debt*, floating, $1,792 at 6%; bonds to the amount of $7,500 have been authorized. No hydrant rental time of report. **Management.**—Prest., G. D. Hellwig. Secy., Treas. and Supt., J. G. Corbett. Report, July 16, '91.

11. BANGOR, *Northampton Co.* (Pop., 1,328—2,509.) Built in '86 by Bangor Water Co. **Supply.**—Spring, by gravity. **Distribution.**—(In '88.) *Mains*, 1.25 miles. *Taps*, 100. *Hydrants*, 11. **Financial.**—(In '88.) *Bonded Debt*, $15,000 at 6%. Hydrant rental, $25 each. **Management.**—Prest., J. Marshall Young. Secy., P. J. Currell.

162. BEAVER, *Beaver Co.* (Pop., 1,178—1,552.) **History.**—Now owned by Co. Built in '86 by village. Sold to Valley Water Co., Rochester, in '91. See Manual for '89-'90, p. 220.

12. BEAVER FALLS, *Beaver Co.* (Pop., 5,104—9,735.) Owned by Union Water Co. Built in '75 by Beaver Falls Water Co. **Supply.**—Beaver River, filtered through bank to well, thence pumping direct and to reservoir.

May 15, '90, Beaver Falls Water Co. sold plant to H. W. Hartman, of Beaver Falls, who reorganized it under name of Union Water Co. In connection with Valley Falls Water Co., of Rochester, in which Mr. Hartman is interested, dam is being built on Beaver River and pumping works to be run by water power has been constructed. Contrs. for pumps, Holly Mfg. Co.; for valves, Rensselaer Mfg. Co., Troy; hydrants, Geneva; filters, Hyatt; trenching and pipe-laying, Chanley Bros. & Co., Beaver Falls.

Year.	Mains.	Taps.	Meters.	Hydrants.	Consumption.	Cost.	Exp.	Rev.
'81	6	....	0	32	30,000	$30,000	$1,000	......
'86	12.5	1,421	0	86	2,500,000	175,000	5,000	$2,150*
'88	15	....	1	101	3,000,000	180,000	6,500	......

*From city.

Management.—Prest., H. W. Hartman. Secy., Merritt Greene, Ellwood City. Treas., J. A. B. Melvin. Supt., J. W. Ramsey.

13. BEDFORD, *Bedford Co.* (Pop., 2,011—2,242.) Built in '72-76 by town. **Supply.**—Mountain springs, by gravity. **Distribution.**—(In 88.) *Mains*, 7 miles. *Taps*, 214. *Hydrants*, 16. **Financial.**—(In 88.) *Cost*, $34,000. *Debt*, $23,500, bonded at 6%. *Expenses*, operating, $500. *Revenue*, derived indirectly from taxation, $2,900. **Management.**—Mayor, R. C. McNamara. Comrs., S. F. Statler, Jos. Megill, Jas. Crouse. Jas. Corby.

14. BELLEFONTE, *Center Co.* (Pop., 3026—3,946.) Built in '07 by borough. **Supply.**—Springs, pumping to reservoir. **Distribution.**—(In 88.) *Mains*, 10 miles. *Hydrants*, 40. **Financial.**—(In 88.) *Cost*, about $120,000. *Debt*, $100,000; bonded at 5, 6 and 7%. *Operating Expenses*, about $4,000. *Revenue*: consumers, $8,000; city none. **Management.**—Prest., Isaac Mitchell. Sec., C. F. Cook. Supt., S. D. Ryan.

120. BELLEVUE (*P. O. Scranton*), *Lackawanna Co.* Supply from Minooka.

15. BERWICK, *Columbia Co.* (Pop., 2,095—2,701.) Built in '48 by Berwick Water Co.; city does not control rates; franchise does not provide for purchase of works. **Supply.**—Spring on river's bank, pumping to reservoir. Fiscal year closes Dec. 1. Supply has been changed from pumping to gravity from reservoir about three miles from town. New supply affords pressure of 72 lbs. There are now two 15,000,000-gall. reservoirs.

FOR FURTHER DESCRIPTIVE MATTER see Manual for 1889-90. CHANGES since 1890 are here given. POPULATIONS are for 1880 and 1890, respectively.

112   MANUAL OF AMERICAN WATER-WORKS.

Year.	Mains.	Taps.	Meters	Hyd't's.	Cost.	B'd'd Debt.	Int.Ch'g's.	Exp.	Rev.
'82	2	300	0	2	.....	none.	none.	.....	.....
'86	8	450	0	15	$22,500	$7,500	$300	$1,500	$3,950
'87	8	500	1	16	25,000	7,500	300	1,500	4,160
'88	8	570	1	16	26,000	6,500	260	1,500	4,890
'89	8	.....	1	21	40,000	36,000	2,100	.....	4,370
'90	11	590	1	21	60,000	35,000	2,100	.....	4,656

Financial.—*Capital Stock:* authorized, $25,000, all paid up; paid 6% dividend in '90. *Total Debt*, $35,000; bonded, at 6%. *Expenses:* (as reported Aug. 29, '91), operating, $1,450. *Revenue:* consumers, $4,446; city, fire protection, $210; total, $4,656. **Management.**—Prest., C. H. Zehnder. Secy., G. L. Reagan. Treas., F. H. Eaton. Supt., Geo. Depuy.

16. **BETHLEHEM,** *Northampton Co.* (Pop., 5,193–6,762.) Now owned by borough. Built 1754–61 by Hans Christopher Christiansen; bought by borough in '71. **Supply.**—Spring, pumping direct, with tanks as regulator of pressure. Fiscal year closes May 1. Tank added in '89 has cap. of 831,000 galls.; w. l., 51.8 ft. in diam., 52.8 ft. high.

Year.	Mains.	Taps.	Hydrants.	Cons'pt'n.	Cost.	Exp.	Rev.
'83-4	5	497	78	.....	$38,000	$3,215	$4,635
'87-8	8	772	72	550,000	75,000	4,800	8,400
'88-9	8	800	73	650,000	90,000	4,100	9,000

Financial.—*First cost*, $20,300. **Management.**—Comrs. Chn., L. F. Giering. Supt., Chas. Bodder. NOTE.—May, '91, Mr. Giering wrote: "No additions or alterations to report."

17. **BIRDSBOROUGH,** *Berks Co.* (Pop., 1,795–2,261.) Built in '84 by E. & J. Brooke Iron Co., which is principal consumer. **Supply.**—Indian Run, by gravity from impounding reservoir. **Distribution.**—*Mains* (in '88), 5.25 miles. *Taps* (in '86), 270. *Hydrants* (in '88), 36. **Financial.**—*Cost* (in '88), $55,000. *Bonded Debt*, none. **Management.**—(In '89.) Prest., Geo. Brooke. Treas., G. W. Harrison. Supt., Ward Harrison.

18. **BLAIRSVILLE,** *Indiana Co.* (Pop., 1,162–3,126.) Built in '72 by borough. **Supply.**—Conemaugh River, pumping to reservoir.

Year.	Mains.	Taps.	Meters.	Hyd'ts.	Cons'pt'n.	Cost.	B'd d't.	Int.ch'g's.	Exp.	Rev.
'81	2.8	270	0	18	125,000	$25,000	$19,350	$1,161	$1,000	$2,800
'88	3.5	340	0	33	160,000	29,000	15,000	600	1,200	4,800

Management.—Com.

19. **BLOOMSBURGH,** *Columbia Co.* (Pop., 3,702–4,635.) Built in '80 by Bloomsburgh Water Co.; city does not control rates; franchise does not provide for purchase of works by city. **Supply.**—Fishing Creek, pumping to reservoir and direct. Fiscal year closes Oct. 1. A 12-in. pumping main was laid to reservoir in '90, and arranged to connect with new reservoir, when it is necessary to construct one. A 12-in. main has also been laid connecting pumps with supply main, for direct pumping in case of fire.

Year.	Mains.	Taps.	Meters.	Hydrants.	Consumption.	Cost.	Exp.	Rev.
'80	4.5	94	..	40	.....	$36,000	.....	.....
'82	5.5	175	0	42	46,925	37,000	.....	.....
'84	6	300	5	42	160,000	40,000	$2,400	.....
'86	6	400	9	45	240,000	42,000	2,190	$5,839
'87	6	400	9	45	300,000	45,208	2,228	5,783
'88	6	500	20	45	350,000	50,000	2,200	6,500
'89	6	600	20	45	298,944	58,312	2,829	7,239
'90	7.5	470	23	48	268,978	63,276	2,291	7,730

Financial.—*Cap. Stock:* common, authorized, $90,000; paid up, $50,180; paid 6% dividend in '90. *No Debt*. *Expenses:* operating, $2,169; int., $427; taxes, $126; total, $2,722. *Revenue:* consumers, $6,770; city, fire protection, $960; total, $7,730. **Management.**—Prest., L. N. Myer. Secy. and Treas., F. B. Billmeyer. Supt., G. H. Welliver.

20. **BOYERTOWN,** *Bucks Co.* (Pop., 1,099—est., 2,000.) Built in —— by Boyertown Water Co. **Supply.**—Springs, pumping direct from reservoir, **Distribution.**—(In '87.) *Mains*, 3 miles. *Hydrants*, 6, or more. **Management.**—Prest., Jeremiah Sweinhart. Secy., F. H. Stauffer. Treas., E. E. Stauffer.

21. **BRADDOCK,** *Allegheny Co.* (Pop., 3,310–3,561.) Built in '84-5 by borough. **Supply.**—Monongahela River, pumping direct. Fiscal year closes Jan. 1. New reservoir was to build in '91 at cost of $50,000.

PENNSYLVANIA. 113

Year.	Mains.	Taps.	Mtrs.	H'd s.Cons'pt'n.	Cost.	L'd'd d't.Int. ch'g's.	Exp.	Rev.
'88	8	600	14	51 't 1,000 000	$34,000	$50,000 $2,000	$3,500	$9,344
'89	8	511	8	67 2,000,000	100,000	50,0 } 2,000	5,000	......

Financial.—*Total Debt:* (in '88) $55,000: floating, $15,000 at 4%; bonded, $50,000 at 4%. *Sinking Fund,* $2,500, not invested. *Expenses:* operating, $5,000. *Revenue:* city, fire protection, $2,900. All expenses, including redemption of bonds, are met from revenue. **Management.**—Com. of three. Supt., Wm. Harwood.

NOTE.—Pennsylvania Water Co., Wilkinsburg, supplies water in Braddock township.

22. BRADFORD, *McKean Co.* (Pop., 9,197—10,514.) Owned by city. Built in '77 by Bradford Water-Works Co.; bought by city in '81. **Supply.**—Marilla and Gilbert brooks, by gravity from impounding reservoirs. Fiscal year closes Dec. 31. Unofficially reported that new reservoir was to be built in '91

Year.	Mains.	Taps.	Meters.	Hyd'ts.	Cons'pt'n.	Cost.	B'd'd t.	Int.chgs.	Exp.	Rev.
'82	4.5	360	0	13	unkn'n	$25,000	none	none	$5,000	$7,500
'84	9	480	0	40	43,000	53,000	$33,000	$1,980	8,000	11,164
'85	12.3	649	..	49	....	......	......	......	4,059	9,336
'86	15.6	903	0	55	....	......	......	......	3,618	11,924
'87	15.8	994	0	57	1,120,000	100,000	60,000	3,600	4,108	13,643
'88	15.9	1,130	1	57	1,500,000	115,000	59,500	3,570	4,216	15,386

Financial.—Int., 6%. No allowance for hydrants. **Management.**—(In '89.) Comrs. Prest., A. Newall. Secy. and Supt., C. J. Lane.

23. BRIDGEPORT, *Montgomery Co.* (Pop., 1,802—2,651.) Built in '80-91 by Bridgeport Water Co.; put in operation Sept. 1. In '90 franchise passed into hands of P. D. Wanner, Reading. Engr., C. E. Albright. Contr., H. E. Ahrens, Reading. **Supply.**—Schuylkill River, pumping to reservoir and direct. *Pumping Machinery.*—Cap., 1,000,000 galls.; Dean. *Boilers* from G. C. Wilson & Co., Reading. *Reservoir,* cap., 800,000 galls. *Distribution.*—*Mains,* 12 to 4-in., c. i., about 7 miles; from Mellert Foundry & Machine Co., Reading. *Services,* w. i and lead. *Taps,* 200. *No Meters. Hydrants,* Mellert, 43. *Pressure,* 80 lbs. **Financial.**—*Cost,* $45,000. *Cap Stock,* authorized, $80,000. *Debt,* $40,000, bonded at 6%. *Expenses,* operating, $1,000. *Revenue:* consumers, $3,600; city, $700. **Management.**—Pres., F. H. Cheney. Secy., Henry Steck. Treas., G. S. Moyer. Supt., Samuel Coats. Report by P. D. Wanner, Reading, Sept. 21, '91.

24. BRISTOL, *Bucks Co.* (Pop., 5 273—6,553.) Built in '74-5 by Bristol Water Co. **Supply.**—Delaware River, pumping to stand-pipe. Fiscal year closes May 28.

Year.	Mains.	Taps.	Meters.	H'd'nts.	C'ns'ption.	Cost.	Exp.	Rev.
'82	3.5	330	0	0	160,233	......	$2,400	......
'84	3.5	374	1	0	....	$39,450	2,100	......
'86	4.5	400	1	0	450,000	40,000	......	......
'87	5	575	1	51	600,000	50,000	3,000	$9,600
'88	5.5	637	1	51	600,000	62,639	3,294	10,058

Financial.—*First Cost,* $39,450. *Capital Stock,* $50,000. *No Debt. Revenue:* consumers, $10,058; city, fire protection, $600; total, $10,658. **Management.**—Prest., A. Swain. Secy., G. M. Dorrance. Treas., L. A. Hoguet. Supt., A. K. Joyce.

NOTE.—In '90 it was stated that there had been no changes since report for '88.

25. BROOKVILLE, *Jefferson Co.* (Pop., 2,136—2,478.) Built in '83-5 by Brookville Water Co., town neither controls rates nor has right to purchase works, under terms of franchise. **Supply.**—North Fork Creek, pumping to tanks.

Year.	Mains.	Taps.	Meters.	H'd'nts.	C'ns'ption.	Cost.	Exp.	Rev.
'86	5.3	110	0	24	unknown	$25,000	$1,000	$2,482
'88	6	160	0	26	unknown	28,000	1,200	3,692

Financial.—See above. *No Debt.* **Management.**—Prest., Jackson Heber. Secy., B. M. Marlin. Supt., W. R. Ramsey.

26. BROWNSVILLE, *Fayette Co.* (Pop., 1,489—1,417.) Built in '90-1 by N. N. Madison. **Supply.**—Springs, by gravity. *Reservoir.*—Cap., about 600,000 galls. *Distribution.*—*Mains,* 6 in. w. i., 1¾ miles. *Taps,* 100. *Hydrants,* 13. *Pressure,* 118 lbs. **Financial.**—*Cost,* $8,000. **Management.**—Owner, N. N. Madison.

213. BRUSHTON, *Allegheny Co.* (Pop., in '80, 165.) Supply from Wilkinsburgh.

27. BUTLER, *Butler Co.* (Pop., 3,163—8,734.) Built in '78 by Butler Water Co. **Supply.**—Conoquenessing Creek, pumping to reservoir. Pumping cap., 3,000,000 galls.; Keystone and Holly, or Gaskill.

FOR FURTHER DESCRIPTIVE MATTER see Manual for 1889-90 CHANGES since 1890 are here given. POPULATIONS are for 1880 and 1890, respectively.

114 MANUAL OF AMERICAN WATER-WORKS.

Year.	Mains.	Taps.	Meters.	Hyd'ts.	Con'sp'tn.	Cost.	B'd'd Debt.	Int. Ch'gs'.	Rev.
'86	8	250	0	50	216,000	......	......	......	......
'87	14	350	0	..	....	....	....	....	....
'88	17	400	0	50	....	$119,000	$30,000	$1,803	$600*
'90	21	500	0	62	500,000	....	30,000	1,800	1,023*

*From city.

Financial.—*First Cost*, $49,000. *Cap. Stock*, $100,000. Int., 6%. **Management.**—Prest., W. G. Brandon. Secy., Treas., and Supt., W. R. Meredith, Kittaning.

28. **CANTON,** *Bradford Co.* (Pop., 1,194—1,393.) Built in '80 by Citizens' Water Co. **Supply.**—Mountain brook and lake, by gravity. **Distribution.**—(In '88.) *Mains,* 6 miles. *Taps,* about 260. *Meters,* none. *Hydrants,* 11. **Financial.**—(In '88.) *Revenue:* Consumers, $1,800; city, $200. *First cost,* $37,500. **Management.**—(In '89.) Prest., G. W. Maynard, Williamsport. Sec. and Treas., J. E. Cleveland.

29. **CARBONDALE,** *Lackawanna Co.* (Pop., 7,714—10,833.) Built in '68 by Crystal Springs Water Co. **Supply.**—Springs, by gravity from storage to distributing reservoirs. Fiscal year closes June 1.

Year.	Mains.	Taps.	Meters.	Hyd'ts.	Cost.	Exp.	Rev.
'80	2	274	..	......	......	$400	......
'82	2	250	0	30	$20,400	....	....
'84	6	400	0	28	......	1,000	$3,500
'86	2.5	550	0	28	24,500	1,500	6,650
'87	8	530	0	36	24,500	1,500	8,400
'88	8	550	0	36	......	......	......
'90	9	1,365	14	45	......	3,700	900*

*From city.

Financial.—*First cost,* $11,625. *Capital stock:* authorized, $25,000; paid up, $23,250; Paid 20% dividend in 1890. *Total debt,* $2,000, floating. *Expenses:* operating, $3,700. no int.; taxes withheld. *Revenue:* consumers, withheld; city, fire protection, $20 per hydrant. **Management.**—Prest., R. Manville. Secy., A. Pascoe. Supt., Wm. McMullen.

30. **CARLISLE,** *Cumberland Co.* (Pop.,6,209—7,620.) Built in '54 by Carlisle Gas & Water Co. City holds some of the stock, and appoints 3 of the 8 directors; no further regulation of rates by city and no provision for buying works. **Supply.**—Canodoquinet Creek, pumping to reservoir. Fiscal year closes May 1. **Distribution.**—*Mains* (in '88), 12 miles. *Taps,* 480. *Meters,* 3. *Hydrants* (in '91), 77. **Financial.**—(in '88.) *Revenue:* consumers, $7,200; city, 325. **Management.**—Prest., John Hays. Secy. and Treas., J. B. Landis. Supt., C. E. Ramsey.

31. **CATASAUQUA,** *Lehigh Co.* (Pop., 3,065—3,704.) Built in '75 by Crane Iron Co. **Supply.**—Lehigh River, pumping to reservoir and stand-pipe from feeder connecting river with Lehigh canal. Fiscal year closes Dec. 31.

Year.	Mains.	Taps.	Meters.	Hydrants.	Cost.	Exp.	Rev.
'80	2	256	..	......	......	$100	......
'86	3.5	350	0	42	50,000	....	3,104
'90	5	350	0	55	60,000	150	2,340

Financial.—*No Capital stock or Debt. Expenses:* operating, $150; no int.; taxes, withheld. *Revenue:* consumers, $1,890 ; city, fire protection, $450; total, $2,340. **Management.**—Supt., Leonard Peckitt.

II. **CENTRE HALL,** *Centre Co.* (Pop., 250—441.) Built by Centre Hall Water Co., date not given. **Supply.**—Springs, by gravity. **Reservoir.**—Cap. about 100,000 galls. **Distribution.**—*Mains,* 6 to 4-in. w. i., 1¼ miles. *Taps,* 80. *Pressure,* 100 lbs. **Financial.**—*Cost,* $5,000. *Cap. Stock,* $5,000. Debt, $1,700. Operating expenses, $40. **Management.**—Prest., Fred. Kurtz. Secy. and Treas., Wm. Wolf.

III. **CENTRALIA,** *Columbia Co.* (Pop., 1,886—2,761.) Built in '66-7 by Centralia Water Co. Franchise does not provide for purchase of works; city has no control of rates. **Supply.**—Springs on mountain, by gravity. **Distribution.**—(In '88.) *Mains,* 3 miles. *Taps,* about 250. *No meters or hydrants.* **Financial.**—(In '88. *Cost,* $10,000. **Management.**—Mgr., S. A. Miller. Secy., J. W. Fortner. Treas., A. K. Mensch.

32. **CHAMBERSBURGH,** *Franklin Co.* (Pop., 6,877—7,863.) Built in '76 by town. **Supply.**—Surface water from impounding reservoir on stream, pumping to reservoir and direct in case of fire. Fiscal year closes March 31. A new reservoir was to be built and pumping station moved in '90-1.

PENNSYLVANIA.

Year.	Mains.	Taps.	Mtrs.	Hyd'ts.	Cons'pt'n.	Cost.	B'd'd d't.	Int. chgs.	Exp.	Rev.
'81-2	7.5	602	0	66	365,145	$53,000	$17,000	$2,820	$1,700	$5,600
'85-6	8.2	736	0	64	431,089	53,000	45,000	2,700	2,300	7,000
'86-7	8.5	817	0	67	450,000	53,000	45,000	2,700	2,300	7,000
'87-8	8.5	862	9	74	509,582	76,410	45,000	2,990	3,465	8,400
'88-9	8.75	913	0	79	555,352	77,425	45,000	2,700	4,165	8,563

Financial.—*First Cost,* $55,000. *Total Debt:* $15,000, at 4½%; $1,000 worth of bonds and int. on total bonded debt is paid yearly by taxation. *Expenses:* operating, $4,165; int., $2,700; total, $6,865. *Revenue:* consumers, $3,503. **Management.**—Comr. appointed every year by council. Supt., Allan C. McGratt, appointed by council.

33. CHESTER, *Delaware Co.* (Pop., 14,597—20,226.) First supplied in '67. Now owned by New Chester Water Co.; works built in '67 by and to supply South Ward: extended in '68 to supply whole city. In '86 New Chester Water Co. built works, and in '88 bought city works and included in its plant. Supplies boroughs of Upland and So. Chester, having bought their plants. Fiscal year closes Dec. 31.

Year.	Mains.	Taps.	Mtrs.	H'd'ts.	Cons'pt'n.	Cost.	B'd'd d't.	Int. chgs.	Exp.	Rev.
'75	12	1,100	..	57	406,000	$79,000	......	......	$5,000	$28,280
'89	38	3,037	2	104	1,720,000	2,400,000	$800,000	$48,000	21,200	43,000
'90	33	5,000	30	100	3,000,000	2,490,000	800,000	45,000	25,000	60,000

Financial.—*Capital Stock,* authorized, $1,600,000. *Total Debt,* $800,000, bonded at 6%. *Expenses:* operating and taxes, $25,000; int., $48,000; total, $73,000. *Revenue,* total, $60,600. **Management.**—Prest., J. L. Forwood. Secy., W. H. Miller. Treas., Walter Wood, Philadelphia. Supt., W. B. Chadwick.

34. CLARION, *Clarion Co.* (Pop., 1,169—2,164.) Built in '75 by Clarion Water Co. Supply.—Clarion river, pumping to tanks. Fiscal year closes July 1. Iron tank 20 × 45, built in '90.

Year.	Mains.	Taps.	Meters.	H'd'ants.	C'ns'mption.	Cost.	Exp.	Rev.
'87	2.5	120	2	11	40,000	$15,000	$1,200	$3,132
'88	2.5	200	6	11	40,000	20,000	1,500	3,965
'89	2.5	200	7	11	40,000	20,000	1,200	3,965
'90	4	200	7	28	40,000	20,000	1,200	3,965

Financial.—*First Cost,* $14,000. *Capital Stock,* authorized, $12,000. *No Debt. Expenses,* see above. *Revenue:* consumers, $3,800; city, $165; total, $3,965. **Management.**—Prest., D. R. Crull. Secy., W. H. Ross. Supt., J. G. Messinger.

35. CLEARFIELD, *Clearfield Co.* (Pop., 1,809—2,248.) Built in '82-3 by Clearfield Water Co.; town has no control of rates. Supply.—Moose Creek, by gravity. Fiscal year closes Dec. 31.

Year.	Mains.	Taps.	Hydrants.	Cost.	Exp.	Rev.
'86	6	300	36	$10,000	$600	$4,000
'87	6	300	36	40,000	800	5,200
'88	6	400	42	40,000	600	5,200
'90	9	325	42	50,000	750	5,400

Financial.—*Cap. Stock:* authorized, $50,000; paid up, $50,000; paid 6% divd. in '90. *No Debt. Expenses,* operating, $750. *Revenue:* consumers, $4,700; city, fire protection, $700; total, $5,400. **Management.**—Prest., W. W. Betts. Secy. and Supt., H. F. Bigler.

36. COATESVILLE, *Chester Co.* (Pop., 2,766—3,680.) Built in '70-1 by borough. Supply.—Spring, by gravity from collecting to distributing reservoir. Fiscal year closes March 3.

Year.	Mains.	Taps.	Hyd'ts.	Cost.	B'd'd debt.	Int. chgs.	Exp.	Rev.
'82-3	4.7	411	52	$12,000	$40,000	$2,000	$475	$2,500
'87-8	4.7	867	53	60,000	82,000	3,280	250	3,802
'88-9	4.8	869	54	82,000	76,980	3,079	250	3,985
'89-90	4.8	426	60	82,000	76,980	3,079	711	4,799

Financial.—*Total Debt,* $76,980; bonded at 4%, due in 1911. *Sinking Fund,* none. *Expenses:* operating, $711; int., $3,079; total, $5,790. *Revenue,* consumers, $4,799. Bonds are redeemed from net income. **Management.**—Com. elected every year. Chn., Isaac Yearly. Secy., Isaac Speakman.

37. COLUMBIA, *Lancaster Co.* (Pop., 8,312—10,599.) Built in '23-6 by Columbia Water Co. Supply.—Susquehanna River and spring, by gravity, and pumping to reservoir. Fiscal year closes first Monday in Feb.

Year.	Mains.	Taps.	Mtrs.	Hyd'ts.	Cons'pt'n.	Cost.	B'd'd d't.	Int.chgs.	Exp.	Rev.
'86	12	1,600	2	62	400,000	......	$70,000	$3,500	$6,946	$15,000
'87	12	....	2	..	400,000	$150,000	70,000	2,500	5,000	....
'90-1	13	1,900	5	90	500,000	150,000	60,500	3,025	5,000	17,500

FOR FURTHER DESCRIPTIVE MATTER see Manual for 1889-90. CHANGES since 1890 are here given. POPULATIONS are for 1880 and 1890, respectively.

Financial.—*First Cost*, $15,500. *Cap. Stock*, $100,000, all paid up; paid 6% div'd in '90-1. *Total Debt*, $60,500, bonded at 5%. *Expenses:* operating, $5,000; int., $3,025; taxes, $500; total, $8,525. *Revenue:* consumers, $17,500; city, fire protection, none. **Management.**—Prest., J. A. Meyers. Secy. and Treas., J. C. Clark. Supt., W. B. Fasig.

CONEMAUGH, *Cambria Co.* Now part of Johnstown.

38. **CONNELLSVILLE,** *Fayette Co.* (Pop., 3,609—5,629.) Built in '83 by Connellsville Water Co.; controlled by American Water Works & Guarantee Co., Ltd., Pittsburg. Maximum rates fixed in franchise which provides for purchase of works by city in '98, price determined by arbitration. **Supply.**—Springs, by gravity from two impounding reservoirs. Fiscal year closes Dec. 31.

Year.	Mains.	Taps.	Meters.	Hydrants.	Cost.	B'd'd Debt.	Int. Ch'g's.	Rev.
'86	7.5	320	..	69	$60,000	$35,000	$2,100	$1,940*
'87	14.0	550	..	75	120,000	95,000	5,700	2,960*
'90	13	630	2	78	......	......	......	......

* From city.

Financial.—Withheld. **Management.**—Prest., A. E. Claney. Vice-Prest., J. H. Purdy. Secy., J. F. Cockburn. Supt., J. E. Stillwagon.

39. **CONSHOHOCKEN,** *Montgomery Co.* (Pop., 4,561—5,470.) Built in '73 by Conshohocken Gas & Water Co; franchise provides that city may purchase works in '93, paying 10% int. on cost of construction; city has no control of rates. **Supply.**—Schuylkill River, pumping to reservoir. Fiscal year closes Nov. 30.

Year.	Mains.	Taps.	Meters.	Hyd'ts.	Cons'pt'n.	Cost.	B'd'd d't.	Int.chgs.	Exp.	Rev.
'84	5	250	42	51	140,000	$75,000	$20,000	$1,200	....	$4,624
'86	5	300	20	46	125,000	75,000	20,000	1,200	$999	5,330
'87	6	370	20	51	175,000	80,000	20,000	1,200	1,698	6,373
'88	7	422	34	52	175,000	90,000	15,500	620	3,289	....
'90	4	450	48	51	161,230	90,158	15,500	620	1,489	7,031

Financial.—*Cap. Stock:* Common, authorized, $50,000; all paid up; paid 5½% div'd in 1890. Pref'd: authorized, $19,900; paid up, $19,900; 6% div'd guaranteed. *Total Debt*, $15,500, bonded, 5-20's issued in '88. *No Sinking Fund*. *Expenses:* operating, $1,489; int., $620; total, $2,109. **Management.**—Prest., Jaywood Lukens. Secy. and Treas., Alfred Craft. Supt., Wm. Ferrier.

40. **CONYNGHAM,** *Luzerne Co.* (Pop., 488—1,209.) Charter for present works secured Jan., '80. About 1830 water was introduced through logs by a mutual association. **Supply.**—Springs, by gravity. Reservoir.—Is 29 × 29 ft. × 8 ft. deep. **Distribution.**—*Mains*, iron, 4 to 2-in., about 1½ miles. *Taps*, about 50. *No Meters*, *Hydrants*, 5. *Pressure*, 150 to 300 ft. head. **Financial.**—*Cost*, $4,000. *Cap. Stock*, $4,000, all paid up. *Operating Expenses*, "trifling." *Revenue*, consumers, about $400. **Management.**—Prest., G. W. Drum. Secy. and Treas., S. Benner.

41. **CORRY,** *Erie Co.* (Pop., 5,277—5,677.) Built in '86 by Corry Water Co., under 10-years franchise, when city may purchase works. **Supply.**—Hare Creek, pumping from impounding to distributing reservoir. Fiscal year closes Dec. 31.

Year.	Mains.	Taps.	Mtrs.	Hd'ts.	Cons'pt'n.	Cost.	B'd'dd't.	Int.ch'g's.	Exp.	Rev.
'86	9	287	4	80	170,000	$150,000	$100,000	$6,000	....	$3,000*
'87	8.5	240	..	60	......	130,000	100,000	6,000	$3,500	7,800
'88	9.5	..	..	61	......	300,000	100,000	6,000	3,500	....
'89	9.5	285	2	62	unk'n	300,000	100,000	6,000	4,500	9,100

* From city.

Financial.—*Cap. Stock*, authorized, $200,000; all paid up. Paid no div'd in 1889. *Total Debt*, $112,000: floating, $12,000 at 6%; bonded, $100,000 at 6%. *Expenses:* operating, $4,500; int., $6,000; total, $10,500. *Revenue:* consumers, $6,000; city, fire protection, $3,100; total, $9,100. **Management.**—Address co. NOTE:—Report for '90 states: "No changes from last year."

42. **COUDERSPORT,** *Potter Co.* (Pop., 677—1,530.) Built about '82 by Citizens' Water Co. Contr., W. H. Simpson, Olean, N. Y. **Supply.**—Springs, by gravity. **Distribution.**—*Mains*, c. i., 4-in. *Hydrants*, 25. *Pressure*, 75 lbs. **Financial.**—*Cost*, about $12,000. *No Debt*. *Operating Expenses*, small. *Revenue:* consumers, about $1,200; city, $200; total, $1,400. **Management.**—Prest., A. G. Olmsted. Secy., Jas. Knox. Treas., R. L. Nichols. Report by W. K. Jones.

43. **CURWENSVILLE,** *Clearfield Co.* (Pop., 706—1,664.) Built in '90-1 by Raftman Water Co. Engr., J. H. Young; Contr., C. Jesse Young; both of 1,404 S. Penn Square, Philadelphia. **Supply.**—Springs, by gravity to and from reservoirs.

PENNSYLVANIA. 117

Distribution.—*Mains*, 8 to 6-in., c. i., 5½ miles. *Taps*, 75. *Hydrants*. Mathews, 32. *Valves*, Eddy. *Pressure:* ordinary, 140 lbs.; fire, 200 lbs. **Financial.**—*Cost*, $55,000. *Cap. Stock:* authorized, $35,000; all paid up. *Debt*, $30,000, bonded at 6%. *Hydrant Rental*, $700. **Management.**—Prest., A. V. Lamb. Secy., J. H. Young, as above. Treas., C. W. Boyce. Supt., H. B. Thompson. Latest report, June, '91.

44. DANVILLE, *Montour Co.* (Pop., 8,346–7,998.) Built in '73 by borough. Supply.—North branch of Susquehanna River, pumping direct from filter gallery near river. Fiscal year closes June 1. A 3,000,000-gall. Worthington pump was added in '90. Distribution.—*Mains*, 11 miles. *Taps*, 1,210. *No Meters. Hydrants*, 100. *Consumption*, 750,000. **Financial.**—Cost (in '88), $170,000. Bonded debt, $117,700: $700 at 4½%, $117,000 at 4%. *Expenses:* operating, $3,496; int., $4,708; total, $13,204. *Revenue:* consumers, $7,478; city, $7,900; total, $15,378. Other data, see above. There is a 12-mill water tax and a 3-mill sinking fund tax on property; these funds, combined with revenue, pay all expenses. **Management.**—Three comrs., appointed every three years by town council. Prest., W. J. Baldy. Secy., Benj. Harris. Supt., Henry Kearns.

45. DERRY STATION, *Westmoreland Co.* (Pop., 755-est., 2,000.) Built in '88-9 by Derry Water Co. Supply.—Two streams, by gravity from two reservoirs. Distribution.—*Mains*, 5½ miles. *Taps*, 160. *No Meters. Hydrants*, 28. *Consumption*, about 350,000. **Financial.**—*Hydrant rental*, $20 each. **Management.**—Prest., Clifford S. Sims. Treas., P. S. Janeway, Philadelphia. Supt., Murray Forbes.

DICKSON CITY, *Lackawanna Co.* (Pop., 5,151–8,351.) Supplied from one of the Scranton co.'s under name of Dickson Water Co. No further information.

IV. DOWNINGTON, *Chester Co.* (Pop., 1,480–1,920.) Built in '71-4 by Downington Gas & Water Co.; franchise neither regulates rates nor provides for purchase of works by city. Supply.—Springs, by gravity from reservoirs Fiscal year closes March 31. During '90 a 350,000-gall reservoir was built.

Year.	Mains.	Taps.	Meters.	Hyd'nts.	B'd'd debt.	Int. c'h's.	Exp.	Rev.
'86	6	310	0	0	$1,000	$280	$300	$2,200
'87	6	332	0	1	4,000	280	400	6,700
'88	6	332	0	1	4,000	280	400	7,000
'89	7	425	0	1	4,000	240	575	5,000
'90	7	435	0	1	4,000	240	1,150	5,000

Financial.—*First Cost*, $50,000. *Cap. Stock:* authorized, $50,000; all paid up. Paid no dividend in 1890. *Total Debt*, $4,000, floating at 6%. *Expenses:* operating, $900; int., $240; taxes, $250; total, $1,350. *Revenue*, consumers, $5,000. **Management.**—Prest., Frank Buck. Secy., L. Buck, 1,505 W. 5th St., Philadelphia. Supt., Jas. Hamilton.

46. DOYLESTOWN, *Bucks Co.* (Pop., 1,845–1,733.) Built in '09 by borough. Supply.—Springs, pumping to reservoir, and direct for fire purposes. Fiscal year closes April 1.

Year.	Mains.	Taps.	Hyd'ts.	Cons't'n.	Cost.	B'd'd debt.	Int. ch'gs.	Exp.	Rev.
'70	6	160	11	......	$11,000	......	......	......	$500
'80	7	325	26	......	16,000	......	......	......	3,000
'87	10	4°5	41	75,000	35,000	$30,000	$1,200	$1,300	4,500
'90	10	460	45	90,000	37,654	35,000	1,400	1,313	5,062

Financial.—*Total Debt*, $35,312; floating; $312 at 4%; bonded, $35,000 at 4%, $1,000 due April 1, yearly. No sinking fund. *Expenses:* operating, $1,343; interest, $1,400; total, $2,743. *Revenue*, consumers, $5,062. No allowance for public uses of water. Net revenue is used for maintenance of streets. **Management.**—Committee of three, one elected every year. Chn., S. Firman. Secy., S. J. Freed. Supt. and Engr., Wm. J. Wintyen, appointed by council.

V. DRIFTON, *Luzerne Co.* (Pop., 1,026–1,300.) Built in '65 by Coxe Bros. & Co. to supply private mining town. Supply.—By pumping. *Mains*, 3 miles. *Taps*, about 60.

47. DUBOIS, *Clearfield Co.* (Pop., 2,718–6,149.) Built in '89-90 by U. S. Water Supply Co. Supply.—Springs, by gravity to and from reservoir. Distribution.— (In '89.) *Mains*, 3½ miles. *Hydrants*, 53; two private. **Financial.**—*Hydrant rental*, $35. **Management.**—Prest., C. J. Shuttleworth. Supt., G. B. Campbell.

Duke Center, *McKean Co.* (Pop., est., 700.) Built for fire protection. See '89-90 Manual, p. 231.

For Further Descriptive Matter see Manual for 1889-90. Changes since 1890 are here given. Populations are for 1880 and 1890, respectively.

48. DUNMORE, *Lackawanna Co.* (Pop., 5,151—8,315.) Built in '69-70 by Dunmore Gas & Water Co.; city has no control of rates; no provision in franchise for purchase by city. **Supply.**—Surface water, by gravity. **Distribution.**—*Mains,* about 5 miles. *Taps,* 500. *Meters,* 2. *Hydrants,* 6; June 30, '91, contract for 10, to be put in at once, had been let. **Financial.**—*Capital Stock* (in '87), $150,000. No hydrant rental. **Management.**—Prest., J. B. Smith. Secy. and Treas., F. M. Byea. Clerk, C. P. Savage. Supt., W. J. Smith,

VI., VII. **EAST BRADY,** *Clarion Co.* (Pop., 1,242—1,228.)

J. M. CUNNINGHAM'S PLANT.

Built in '70. **Supply.**—Surface water, by gravity. *Mains* (in '87), 0.5 miles.

J. Y. FOSTER'S PLANT.

Built in '75. **Supply.**—Springs, pumping to tank. There are 1¼ miles of pipes, 350 consumers, no hydrants. *Cost,* $5,000. *Ann. op. exp.,* $250. *Ann. rev.,* $1,250.

107. EAST MAUCH CHUNK, *Carbon Co.* (Pop., 1,853—2,772.) Supply from Mauch Chunk.

VIII., 49. **EASTON,** *Northampton Co.* (Pop., 11,924—14,481.)

EASTON WATER CO.

Built in '18. **Supply.**—Springs, by gravity. **Distribution.**—*Mains,* 4 and 3-in, iron, about 1½ miles. *Taps,* about 200. *Meters,* 1. *Hydrants,* none. **Financial.**—*Cap. Stock,* authorized and paid up, $22,890. *No Debt. Operating Expenses,* $400. *Revenue:* consumers, $2,200; city, none. **Management.**—Prest., J. M. Hackett. Secy., Treas. and Supt., Edmund Lorey.

LEHIGH WATER CO.

Built in '54 by Lehigh Water Co. City can buy works at any time by paying cost of same and 6% interest from time of expenditure, less all dividends; three comrs., elected by people, have control over rates. **Supply.**—Lehigh and Delaware rivers, pumping through distributing mains to reservoir. Fiscal year closes Feb. 1. New 8,000,000-gall. reservoir built in '91.

Year.	Mains.	Taps.	Meters.	Hyd'ts.	Cons'n.	Cost.	B'd'd d't.	Int.ch'g's.	Exp.	Rev.
'80-1	20	1,200	20	50	1,000,000	$300,000	$40,000	$2,400	$10,000	$23,500
'84-5	20	1,300	15	70	1,000,000	300,000	26,000	2,160	10,000	26,000
'87-8	20	1,300	50	85	1,500,000	......	15,000	900	10,000	29,000
'89-90	21	1,200	46	96	......	......	......	......	......	......

Financial.—*Cap. Stock* (in '82), $20,000. Hydrant rental, none; Co. furnishes hydrants and keeps them in repair. **Management.**—(In '89,) Prest., J. S. Rodenbough. Supt., R. P. Roder.

50. EAST STROUDSBURGH, *Monroe Co.* (Pop., 1,102—1,819.) Built in '89 -90 by borough; put in operation Feb. 1. Engr., O. D. Shepherd, Scranton. Contr., Jos. Shiffer, Stroudsburgh. **Supply.**—Sambo Creek, by gravity. Reservoir.—Cap., 2,000,000 galls. **Distribution.**—*Mains,* 14 to 18-in. c. i., about 5.5 miles; from R. D. Wood & Co., Philadelphia. *Taps,* 90. *Hydrants,* Mathews, 30. *Consumption,* 100,000 galls. *Pressure,* 86 lbs. **Financial.**—*Cost,* $37,500. *Debt,* $37,500 at 5%. *Operating Expenses,* $100. *Revenue,* total, $2,000. **Management.**—Secy., A. R. Brittain. Treas., E. B. Morgan. Supt., Jas. Deemer. Report, Jan. 17, '91.

51. EBENSBURGH, *Cambria Co.* (Pop., 1,123—1,202.) Built in '74 by borough. **Supply.**—Artesian well, with auxiliary supply from creek; pumping to reservoir. March 16, '91, new supply was being introduced. New Worthington pump has been added.

Year.	Mains.	Taps.	Meters.	Hyd'ts.	Cons'p'n.	Cost.	B'd'd d't.	Int.ch'g's.	Exp.	Rev.
'82	1.5	40	0	16	unknown	$16,000	$8,200	$328	$100	......
'87	1.5	60	3	16	unknown	16,000	6,700	268	500	$1,300
'89	2	70	6	16	20,000	16,000	......	......	......	......
'90	3	86	6	16	27,000	16,000	4,800	192	577	1,647

Management.—Committee of two, appointed by borough council. Chn., J. T. Young. Secy., W. H. Connell. Supt., R. S. Thomas.

52. EDENBURGH (*Knox P. O.*), *Clarion Co.* Pop., 767—858.) Now owned by borough; built in '80 by Edenburgh Water Co.; "turned over" to borough in '82. **Supply.**—300-ft. well, pumping to tanks. **Distribution.**—*Mains,* 1 mile. *Taps,* 30. *No Meters. Hydrants,* 5. *Consumption,* 2,500 galls. *Pressure,* 65 to 75 lbs. **Financial.**—*First cost,* $2,000. *No debt. An op. exp.,* $200. **Management.**—Mgr., Robt. Woodling.

213. EDGEWOOD, *Allegheny Co.* Supply from Wilkinsburg.

PENNSYLVANIA. 119

Eldred, *McKean Co.* (Pop., 1,165—1,050.) Built for fire protection. See '89-90 Manual, p. 232.

53. ELIZABETHVILLE, *Dauphin Co.* (Pop., not given—676.) Built in '89 by Elizabethville Water Co.; put in operation Jan. 1, '90. Engr., S. W. Cooper. No contr. **Supply.**—Mountain springs, by gravity. **Reservoir.**—Is 60 ft. sq. by 20 ft. deep. **Distribution.**—*Mains,* about 2 miles. *Taps,* about 30. *Hydrants,* 15. *Pressure,* 65 lbs. **Financial.**—*Cost,* $6,000. *Cap. Stock:* authorized and paid up, $57,000. *No Debt. Operating Expenses,* about $50. *Revenue;* consumers, $250; city, $250. '*Hydrant Rental,* about one mill on assessed valuation. **Management.**—Prest., H. H. Weaver. Secy., C. B. Stroup. Treas., Cyrus Romberger. Report, Jan., '91.

54, IX. EMAUS, *Lehigh Co.* (Pop., 874—883.)

BOROUGH WORKS.

Built in '71 by borough. **Supply.**—Two springs on Lehigh Mountain, by gravity to and from reservoir. Fiscal year closes March 3.

Year.	Mains.	Taps.	Meters.	Hyd'ts.	Cost.	B'd'd d't.	Int. ch'g's.	Exp.	Rev.
'83	2	153	0	10	$15,000	$14,225	....	$21	$887
'87	..	200	0	..	17,500	11,000	$140	50	885
'90	3	200	0	12	18,500	11,500	465	50	801

Financial.—*Total Debt,* $11,500, bonded at 4%. **Management.**—Town council. Sec., H. W. Janett. Supt., Jonas Fritz, appointed by council.

MOUNTAIN WATER CO.

Built in '91; put in operation about July 1. Supplies annex to borough. **Supply.** —Mountain springs, by gravity. **Reservoir.**—Cap., about 140,000 galls.; 30 × 95 ft. × 10 ft. deep. **Distribution.**—*Mains,* c. l., 4,500 ft. *Taps,* 18. No *Meters* or *Hydrants.* *Pressure,* 60 lbs. **Financial.**—*Cost,* about $5,000. *Cap. Stock,* authorized, $5,000. **Management.**—Prest., F. T. Jobst. Secy., H. W. Jarrett. Treas., E. G. Jobst. Supt., H. G. Jobst. Report, June 20, '91.

55. EMLENTON, *Venango Co.* (Pop., 1,140—1,126.) Built in '78 by Emlenton Water Co.; city has no control over rates; no provision in franchise for purchase by city. **Supply.**—Springs and river, by gravity and pumping to reservoir. **Distribution.**—(In '88.) *Mains,* 4 miles. *Hydrants,* 17. **Financial.**—*Hydrant rental,* $17.50 each. **Management.**—(In '88.) Prest., C. C. Cooper. Secy. and Treas., J. J. Gosser. NOTE.—All inquiries ignored.

56. EMPORIUM, *Cameron Co.* (Pop., 1,156—2,147.) Built in '87-8 by Co. **Supply.**—Springs, by gravity. Fiscal year closes April 9. In '90, a Deane pump was added, to be used during the dry months of summer.

Year.	Mains.	Taps.	Meters.	Hydrants.	Cost.	Exp.	Rev.
'88	7	190	0	25	$33,000	$50	$2,600
'90	8	200	0	49	38,251	3,040*	2,502*

*For 9 months.

Financial.—*Cap. Stock,* authorized, $30,000. *Total Debt,* $600, floating, at 6%. *Expenses:* see above. *Revenue:* consumers, $1,682; city, $880; total, $2,562. **Management.**—Prest., J. F. Parsons. Secy., R. Seger. Supt., W. R. White.

57. EPHRATA, *Lancaster Co.* (Pop. of township, 284—606.) Built in '79 by Ephrata Water Co. **Supply.**—Springs, by gravity. **Distribution.**—*Mains* (in '88,) 4,000 ft. *Taps* (in '80), 40. *No Meters. Hydrants* (in '84), 6. **Financial.**—(In '88.) *Cost,* $4,000. *Revenue,* $265. **Management.**—Address Co.

58. ERIE, *Erie Co.* (Pop., 27,737—40,634.) Present works built in '67-9 by city. First supplied in '40. **Supply.**—Lake Erie, pumping through stand-pipe to reservoir. Fiscal year closes Dec. 31.

Year.	Mains.	Taps.	Mtrs.	Hyd'ts.	Cons'pt'n.	Cost.	B'd'd d't.	Int.ch'g's.	Exp.	Rev.
'70	..	....	..	....	675,750	....	....	....	..	$6,237
'75	..	....	..	....	1,451,890	....	....	....	..	29,630
'80	35	....	..	130	2,153,000	....	....	....	$19,397	37,385
'81	..	....	..	....	2,673,000	....	....	....	..	40,386
'82	40.8	3,174	20	148	2,274,405	$607,311	....	....	17,513	43,619
'83	44	3,387	41	196	2,252,177	819,677	$675,000	$47,250	17,308	48,369
'84	..	....	..	....	2,507,600	....	....	....	..	51,852
'85	52	4,000	52	270	2,879,574	891,548	$675,000	47,250	17,386	53,550
'86	53.3	4,513	64	230	3,063,243	921,394	675,000	47,250	17,241	58,725
'87	56.5	6,367	76	318	3,335,190	975,646	675,000	47,250	17,193	67,122
'88	60.45	5,376	80	333	3,675,912	1,006,743	633,000	44,310	17,366	73,074
'89	62.38	5,781	80	346	4,040,908	1,040,234	619,500	43,365	18,999	81,111
'90	68.2	6,434	75	372	4,546,919	1,088,539	599,500	....	19,232	87,280

FOR FURTHER DESCRIPTIVE MATTER see Manual for 1889-90. CHANGES since 1890 are here given. POPULATIONS are for 1880 and 1890, respectively.

Financial.—*First Cost*, $675,000. *Total Debt*, $619,500, bonded at 7%, due in '90, '92-'93. *Expenses:* operating, $18,999; int., $43.365; total, $62,364. *Revenue*, consumers, $81,111. No allowance for public uses of water. **Management.**—Three comrs., appointed by Court of Quarter Sessions of Erie Co. every three years. Prest., Geo. W. Starr. Secy. and Treas., B. F. Sloan. Supt., Wm. O'Lone, appointed by comrs.

59. **ETNA**, *Allegheny Co.* (Pop., 2,334—3,767.) Built in '88 by Etna Water Co. **Supply.**—Allegheny River pumping to tank and direct. In '91 mains of systems at Etna and Sharpsburgh were connected for use in case of emergency. **Distribution.**—*Mains*, 4 miles. *Taps*, 115. *No meters.* *Hydrants*, 20. *Consumption*, 72,000 galls. **Financial.**—*Cost*, $50,000. Hydrant rental, $5) each. **Management.**—Prest., G. O Chalfant. Secy. and Treas., C. C. Henderson. Supt., C. W. Drake.

125. **EVERETT**, *Westmoreland Co.* See Mt. Pleasant.

X. **FOREST CITY**, *Susquehanna Co.* (Pop., 990—2,319.) Built in '88 by Rock Cliff Water Co. **Supply.**—By pumping. **Management.**—Prest., W. A. May, Scranton. Secy., L. A. Patterson. Carbondale. Supt., L. H. May.

122. **FORTY FORT**, *Luzerne Co.* (Pop., 478-1,031.) Supply from Moosic.

60. **FOXBURGH**, *Clarion Co.* (Pop., 514-640.) Built in '77 by Foxburgh Water-Works Co. **Supply.**—Allegheny River, pumping to tanks. *Mains* (in '87), 2 miles. *Hydrants* (in '87), 17. **Management.**—Prest., J. M. Fox. Supt., Frank Dale.

XI. **FRACKVILLE**, *Schuylkill Co.* (Pop., 1,707—2,520.) Built in '82 by Mountain City Water Co. Owing to litigation regarding water rights (see Manual for 89 90, p. 235) works were not operated for a few years, but in '89 or '90 Supreme Court decided against co., and works were again put in operation under name of "Water Co., F. S. & John Haupt, Proprietors." Source (see above reference) not given. **Distribution.**—*Mains*, about 2 miles. *Taps*, about 50. No *Meters* or *Hydrants* in '88. **Financial.**—*Cost* (in '88), $20,000. No debt. **Management**—F. S. and John Haupt.

61. **FRANKLIN**, *Venango Co.* (Pop., 5,010—6,221.) Built in '65 by Venango Water Co.; franchise provides maximum rates and for purchase of works by city after 10 years, or since '75. **Supply.**—Springs, by gravity from collecting reservoir, French Creek, pumping direct from impounding reservoir. Fiscal year closes Jan. 1.

Year.	Mains.	Taps.	Meters.	Hydrants.	Cost.	B'd'd debt.	Int. c'g's.	Exp.	Rev.
'86	11.4	..	..	45	$100,000	$40,000	$1,600	$3,000	......
'87	20	1,200	0	50	150,000	46,000	2,300	3,500	$12,500
'58	21	1,250	0	52	166,000	46,000	2,300	2,050	14,500
'90	..	......	..	60	178,464	46,000	2,300	3,876	15,530

Financial.—*First cost*, $80,000. *Cap. Stock*, $130,000, all paid up; 6% dividend in '90. *Total Debt*, bonded, $40,000 at 5% due in 1900. *Sinking Fund*, none. *Expenses*, operating, $3,876; int., $2,300. *Revenue:* consumers, $15,980; city, $550 till new reservoir is finished, then $1,000 for as many as city wishes; total, $15,530. **Management.**—Prest., F. W. Mitchell. Secy., H. W. Bostwick. Treas., Thos. Alexander. Supt., R. J. Hanna.

62. **FREELAND**, *Luzerne Co.* (Pop., not given—2,398.) Built in '83 by Freeland Water Co. **Supply.**—Artesian wells, pumping to reservoir and direct. Fiscal year closes June 30.

Year.	Mains.	Taps.	Meters.	Hyd't's.	Cons'n.	Cost.	B'd'd d't.	Exp.	Rev.
'85	1.8	120	0	17	60,000	$15,000	$12,000	$1,100	$2,655
'87	3.5	392	0	18	200,000	21,000	......	2,500	4,210
'90	3.5	423	0	20	208,000	30,000	......	2,600	5,025

Financial.—*Cap. Stock*, authorized, $21,000, all paid up; paid 12% dividend in '90. *Total Debt*, floating, $500 at 6%. *Sinking Fund*, $1,370. *Expenses:* operating, $2,600; no int.; taxes, withheld. *Revenue:* consumers, $4.800; city, fire protection, $225; total, $5,025. **Management.**—Prest., Jas. Birkbeck, 68 Dana St., Wilkes-Barre. Treas., Thos. Birkbeck. Supt., Henry Fisher.

XII. **FREEPORT**, *Armstrong Co.* (Pop., 1,614-1,637.) Built in '83-4 by Freeport Water-Works Co.; town does not control rates; franchise does not give right to buy works. **Supply.**—Allegheny River, pumping to tanks; filtered. Fiscal year closes Nov. 1. **Distribution.**—*Mains*, 5 miles, *Taps* (in '88), 250, *Meters and Hydrants*, none. **Financial.**—(In '88.) *Cost*, $12,000. No debt. *Cap. Stock*, authorized, $20,000; all paid up. Other figures withheld. **Management.**—Prest., H. M. Breckenridge. Secy., Geo. M. Hill. Supt., David B. Golden.

PENNSYLVANIA.

63. GETTYSBURGH, *Adams Co.* (Pop., 2,814-3.221.) Built in '46 by Gettysburgh Water Co.; franchise provides neither for control of rates, or purchase of works by city. Supply.—Two driven wells, pumping direct and to reservoir. Distribution. —*Mains*, 2 miles. *Taps*, about 300. *No Meters. Hydrants*, 30. Financial.—*Cap. stock:* authorized, $25,000; paid-up, $22,280. *No bonded debt. Operating expenses*, about $1,100. *Revenue :* consumers, about $2,000 ; city. $100. Management.—Prest., C. M. McCurdy. Secy. and Treas., J. W. Kendlebart. Supt., Chas. Beach.

64. GILBERTON, *Schuylkill Co.* (Pop. of borough, 3,098—3,687.) Built in '79 by Anthracite Water Co., principally to supply railways, and incidentally to supply Gilberton; controlled by Philadelphia & Reading Coal & Iron Co. Supply.—Mud and Waste House Runs, by gravity from reservoirs. Distribution.—*Mains*, not given. *Taps*, 225. *No Meters. Hydrants*, 25. Financial.—*Cost*, about $150,000. *No Debt. Operating Expenses*, about $1,000. Management.—Prest.,R. C. Luther. Secy. and Treas., Frank Carter. Supt., G. S. Clemens. All of Pottsville.

XIII, **65. GIRARDVILLE**, *Schuylkill Co.* (Pop., 2,730—3,581.)

GIRARD WATER CO.

History.—Built in '79-80 by Girard Estate, held in trust by the city of Philadelphia. Supplies the public in West Mahanoy Township and contiguous territory, chiefly to collieries, and also the water-works of the borough of Girardville. In '83, for self protection, a charter was obtained, under the name of the Girard Water Co. The pipe-line is still the property of the estate and is leased by the co. at an annual rental of $4,000. **Supply.**—Impounding reservoirs on Lost Creek, by gravity. Fiscal year closes Dec. 31. **Distribution.**—*Mains*, 8 miles. *Taps*, 15. *Meters*, 14. *No Hydrants. Consumption.* 328,380 galls. **Financial** —*Cost*, $102,637, including $25,000 paid Girard Estate for land and improvements. *Cap. Stock:* authorized, $100,000, all paid up; paid 4% div'd. in '89. *No Debt.* Ann. op. exp., $4,806. Ann. rev.; consumers, $9,944. **Management.**—Secy., Geo. E. Kirkpatrick. Supt., C. E. Wagner. Gen. Mgr., H. S. Thompson, Pottsville.

NOTE.—Dams are described in 15th Annual Report Directors City Trusts (Philadelphia.)

BOROUGH WORKS.

History.—Built in '81. **Supply.**—Girard Water Co.'s Works, see above. **Distribution.**—(In '88.) *Mains*, 4 miles. *Taps*, about 300. *Hydrants*, 30. **Financial.**—(In '88.) *Cost*, $16,000. *Debt*, $16,000. Int. 4%. *Operating Expenses*, about $600. *Revenue*, consumers. $1,200. **Management.**—(In '89) Clerk, W. G. Parker. Supt., Patrick O'Brien.

81. GLENSIDE, *Montgomery Co.* Supply from Jenkintown.

XIV. **GORDON**, *Schuylkill Co.* (Pop., 752—1,194.) Built in '89 by Geo. W. Hadesty under name of Hadesty Water Co. No further information can be secured.

66. GREENSBURGH, *Westmoreland Co.* (Pop., 2,500—4,202.) Built in '88 by Westmoreland Water Co. of Unity Township; extended in '89 to supply Jeanette. Penn, Manor and Irwin; city does not control rates, but may buy works in 1908, under state law. **Supply.**—Streams on Chestnut Ridge, by gravity to storage and distributing reservoirs. Fiscal year closes first Monday in Nov.

Y'r.	M'ns.	Taps.	M't'rs.	H'd'ts.	Cons'p't'n.	Cost.	B'd'd d'bt.	Int. ch'gs.	Exp.	Rev.
'88	20	174	6	37	125,000	....	......	....	....	....
'89	36.2	335	6	39	500,000	$1,233,000	$111,000	$24,660	....	....
'90	38.4	711	10	40	850,000	1,350,000	450,000	24,660	$10,000*	$12,000*

* From April to Nov., '90.

Financial.—*Cap. Stock*, authorized, $900,000. *Total Debt*, $450,000, bonded; at 6%; due in 1908. *Expenses and Revenue*, see above. **Management.**—Prest., Geo. F. Huff. Secy. and Supt., Murray Forbes. Treas., H. W. Janeway, 242 So. 3d St., Philadelphia.

67. GREENVILLE, *Mercer Co.* (Pop., 3,007—3,674.) Built in '85-7 by Greenville Water Co.; city has no control of rates. **Supply.**—Springs, by gravity; also 500,000-gall. pumping station added in '90-1. **Distribution.**—(In '88.) *Mains*, 5 miles. *Taps*, 300. *No Meters. Hydrants*, 35. **Financial.**—(In '88.) *Cap. Stock*,

FOR FURTHER DESCRIPTIVE MATTER see Manual for 1889-90. CHANGES since 1890 are here given. POPULATIONS are for 1880 and 1890, respectively.

122   MANUAL OF AMERICAN WATER-WORKS.

$34,000. Bonded Debt, $42,000, at 6%. Operating Expenses (in '87), $600. Revenue: Consumers, about $2,300; city, $1,225. Management.—Prest., J. T. Blair. Secy., P. E. McCray. Supt., S. R. Cochran.

68. **HAMBURGH**, Berks Co. (Pop., 2,010—2,127.) Built in 89-90 by Windsor Water Co., under perpetual franchise. Engr., E. L. Neubling. Supply.—Furnace Creek, by gravity from impounding and distributing reservoirs. Reservoir.—Cap., 600,000 galls.; to be increased to 1,000,000 galls. Fiscal year closes Dec. 31. Distribution.—Mains. 6.25 miles. Taps, 200. No Meters. Hydrants, 25. Financial.—Cost, $45,000. Cap. Stock; authorized, $50,000; paid up, $20,000. Total Debt. $25,000, bonded at 5%, due, 1900-10. Expenses: operating, $500; int., $1,250; total, $1,750. Revenue: consumers, $1,500; fire protection, $500; total, $2,000. Management.—Prest., P. D. Wanner, Reading. Secy., R. B. Kinsey. Supt., M. M. Dreibelbis.

69. **HANOVER**, York Co. (Pop., 2,317—3,746.) Built in '74 by Hanover Water Co. Supply.—Mountain springs, by gravity from collecting reservoir. Fiscal year closes April 30.

Year.	Mains.	Taps.	Meters.	Hyd'ts.	C'ns'pt'n.	Cost.	Exp.	Rev.
'80	....	213	....	25	....	$37,071	$300	$2,400
'84	6	275	2	26	30.000	38,000	1,000	2,800
'86	6	360	3	26	40.000	38,325	820	3,570
'88	6	425	5	32	50,000	38,325	1,000	3,640
'90	6	430	9	36	45,000	38,325	2,056	3,895

Financial.—Capital Stock, authorized, $20,000, all paid up; paid 8% dividend in 1890. No Debt. Expenses: operating, $2,056, unusually large, account of repairs on reservoir; no int.; taxes, withheld. Revenue: consumers, $3,715; fire protection, $180; total, $3,895. Management.—Prest., S. Keefer. Secy., R. M. Wirt. Supt., D. D. Bixler.

70. **HARRISBURGH**, Dauphin Co. (Pop., 30,762—39,385.) Built in '40 by borough, rebuilt in '72-4 by city. Supply.—Susquehanna River, pumping through base of stand-pipe to reservoir. Fiscal year closes Dec. 31.

Year.	Mains.	Taps.	Meters.	Hydrants.	Cons'p'tn.	Exp.	Rev.
'80	19.91	5,000	....	250	....	$20,000	....
'81	21	8,000	0	250	4,7::0,000	16,723	$50,000
'83	23	6,000	0	400	4,134,415	19,629	58,593
'85	23	6,000	0	400	4,815,803	13,806	......
'87	28.4	7,926	75	404	6,400,000	24,000	70,000
'88	29.5	8,200	92	429	6,635,963	22,000	75,196
'89	32.2	8,616	236	406	5,557,338	21,464	73,966
'90	33.9	8,933	516	516	5,856,937	23.537	73,173

Financial.—First Cost, $600,000. Total Debt, about $800,000; floating. "very little;" bonded, $800,000 at 0%. Sinking Fund, "about to be started." Revenue: consumers, $73,173; city, fire protection, $14,385; other purposes, $2,023; total, $89,581. Management.—Three comrs., one elected every year by joint convention of council. Pres., Edmund Mather. Secy., C. D. Stucker.

71, XV. **HAZLETON**, Luzerne Co. (Pop., 6,935—11,872.)

HAZLETON WATER COMPANY.

Built in '60 by Hazleton Water Co.; town has no control of rates; franchise does not provide for purchase of works. Supply.—Springs, by gravity from collecting reservoir, with pumping as auxiliary. Fiscal year closes Nov. 30. New comp,. dup. pump has been added, having 14 × 20-in. and 26 × 20-in. steam cylinders and 10¼ × 20-in. plungers.

Year.	Mains.	Taps.	Meters.	Hydrants.	Cons'pt'n.	Cost.	Exp.	Rev.
'86	12	2,000	1	25	500,000	$200,000	$3,580	....
'87	12	2,000	6	26	500,000	200,000	......	$350*
'88	12	2,000	6	26	500,000	200,000	......	....
'89	20	2,500	1	26	1,200,000	250,000	......	....
'90	20	2,500	1	26	1,000,000	....	12,000	340*

*From city.

Financial.—Cap. Stock, authorized, $200,000, all paid up; paid 5% dividend in '89. No Debt. Expenses, about $12,000. Revenue: consumers, withheld; city, $340. Management.—Prest., E. G. Wilbur. Secy. and Treas., T. M. Santee. Supt., David Clark.

DIAMOND WATER CO.

Built in '87; put in operation Jan. 1, '88. Des. Engr., T. S. McNair, Prest. Const. Engr., E. H. Lowell. Contr., W. W. Taylor, Philadelphia. Supply.—Springs, pumping from impounding to distributing reservoir. Reservoirs.—Cap.: impounding, 3,000,000 galls.; distributing, 800,000 galls. Pumping Machinery.—Cap.,

PENNSYLVANIA. 123

750,000 galls.; dup. cond. Deane. **Distribution.**—*Mains*, 8 to 3 in. c. i.; length not given; from Warren Foundry & Machine Co., Phillipsburg, N. J. *Services*, galv. i. *Taps*, 400. *Hydrants*, Ludlow, 27. *Consumption*, 100,000 galls. *Pressure*, 70 lbs. **Financial.**—*Cost*, $40,000. *Debt*, $10,000, bonded at 4%. *Capital Stock*, $30,000, all paid-up. *Expenses*, operating, $1,500. *Revenue*: consumers, $3,003; city, $337; total, $4,000. **Management.**—Prest., T. S. McNair. Secy. and Supt., Matthew Long. Treas., N. C. Yost.

72. **HOLLIDAYSBURG**, *Blair Co.* (Pop., 3,150—2,975.) Built in '67 by borough. **Supply.**—Roaring Run, by gravity. Fiscal year closes Dec. 31.

Year.	Mains.	Taps.	Hydrants.	Consumption.	Cost.	Exp.	Rev.
'87	13	800	45	150,000	$73,000	$230	$7,000
'88	14	800	50	200,000	75,000	300	8,000
'89	14	825	48	150,000	75,000	1,000	4,516

Financial.—*First cost*, $47,600. *Total debt*, $15,400: floating, $5,900, at 4½%; bonded, $9,500, at 4½%. *No sinking fund*. Operating expenses and new construction are paid from revenue;' int. on bonds from taxation. **Management.**—Three Comrs. appointed by chief burgess every year. Prest., E. R. Baldridge. Secy., J. Horace Smith. Supt., C. A. McFarland. Appointed by Council.

73. **HOLMESBURGH**, *Philadelphia Co.* (Pop., included in Philadelphia.) Built in '87-'8 by Holmesburgh Water Co.; city has right to purchase works, but has no control of rates. **Supply.**—Sandy run, pumping from subsiding reservoir to stand-pipe. Fiscal year closes Sept. 15.

Year.	Mains.	Taps.	Meters.	Hydrants.	Consumption.	Cost.	Exp.	Rev.
'88	5.5	128	....	35	......	$14,000	$1,600	$1,200
'90	5.5	136	1	31	46,000	47,376	1,406	2,347

Financial.—*Cap. Stock:* authorized, $50,000; paid up, $46,000; paid no div'd in '90. *No Debt. Expenses:* operating, $1,266; no int.; taxes, $110; total, $1,406. *Revenue:* consumers, $2,347; city, none. **Management.**—Prest., J. H. Brown. Secy. and Treas., Wm. Sourby. Supt., I. S. Taylor.

74. **HOMESTEAD**, *Allegheny Co.* (Pop., 592—7,911.) Contracts let by borough about June 9, '91. Engr., Jacob Schenneller, Pittsburg. **Supply.**—Monogahela River, pumping to reservoir. **Pumping Machinery.**—Cap., 2,000,000 galls.; Wilson & Snyder. Reservoir.—Cap., 2,500,000 galls. **Distribution.**—*Mains*, 11 to 6-in. c. i. **Financial.**—*Cost*, est., $65,000. *Bonds*, authorized, $70,000. **Management.**—Five comrs. Chm., J. H. Hoover. Clerk, M. P. Schooley. Above subject to correction, as latest report was made just after contracts were let.

75. **HONESDALE**, *Wayne Co.* (Pop., 2,620—2,816.) Built in '68-75 by Honesdale Water Co. **Supply.**--Ponds, 5 miles distant, by gravity to reservoir. Fiscal year closes first Monday in May.

Year.	Mains.	Taps.	Mtrs.	Hyd'ts.	Cons'n.	Cost.	B'd'd. d't.	Int.	Exp.	Rev.
'82-3	8	350	0	8	120,000				$1,600	$3,000
'88-9	10	400	0	8	100,000	$28,000	$20,000	$1,400	......	......
'90-1	10	420	..	5	......	......	......	......	......	5,000

Management.—Pres., F. D. Thayer. Secy., H. C. Hand. Treas. and Supt., T. McNaer.

158. **HORATIO**, *Jefferson Co.* See Reynoldsville.

76. **HUGHESVILLE**, *Lycoming Co.* (Pop., 899—1,358.) Built in '91 by Hughesville Water Co.; was to be in operation Oct. 1. Electric lighting plant built in connection with water-works. Engr., C. E. Albright, 802 Girard Building, Philadelphia. Contr., C. T. Gorman, Philadelphia. **Supply.**—Wells, pumping to reservoir. **Pumping Machinery.**—Not determined. Reservoir.—Cap, 500,000 galls. **Distribution.**—*Mains*, c. i., 4 miles. *Hydrants*, Taylor. *Valves*, Eddy. *Pressure*, 75 lbs. **Financial.**—*Cost*, est., $26,000. Hydrant rental, $30 each. **Management.**—Pres., F. H. Cheyney, 216 S. Third St.; Secy., J. S. Kite, 519 N. 46th St.; Treas., G. S. Moyer; all of Philadelphia. Report, June 22, '91.

77. **HUMMELSTOWN**, *Dauphin Co.* (Pop., 1,043—1,186.) Built in '88 by Hummelstown Water Co.; run in connection with flouring mills. **Supply.**—Swatara Creek, pumping to stand-pipe; filtered through gravel. Fiscal year closes Dec. 31. **Distribution.**—*Mains*, 2½ miles. *Taps*, 104. *No Meters. Hydrants*, 22. *Consumption*, 52,400 galls. **Financial.**—*Cost*, $30,000, including mill, $50,000. *Cap. Stock:* Common; authorized, $45,000; paid-up, $22,500; paid no dividend in '90. Preferred, none. *Total Debt*, $28,480; floating, $5,990 at 6%; bonded $22,500 at 6%.

FOR FURTHER DESCRIPTIVE MATTER see Manual for 1889-90. CHANGES since 1890 are here given. POPULATIONS are for 1880 and 1890, respectively.

due in '98. *No Sinking Fund. Expenses:* operating, $50; int., $1,650; taxes, $385; total, $2,085 for water-works. (Add for mill, $1,500 op. exp. and $300 insurance.) *Revenue:* consumers, $1,075, including $375 from railway; city, fire protection, $630; total, $1,705; add $1,200 rent for mill. **Management.**—Prest., F. J. Shaffner. Secy., W. H. Siple. Treas. and Supt., G. H. Grove.

78. **HUNTINGDON,** *Huntingdon Co.* (Pop., 4,125—5,729.) Built '85-6 by Huntingdon Water Co. **Supply.**—Standing Stone Creek, pumping to reservoir and direct. In '90 or '91 new 3,900,000-gall. reservoir was built near old one; excavated in rock, with concrete bottom. Engr., l. W. Hoffman, Supt. of company. **Distribution.**—(In '88.) *Mains,* 12 miles. *Taps,* 337. *Meters,* 1. *Hydrants,* 42. *Consumption,* 200,000 galls. **Financial.**—(In '88.) *Cost,* $75,000. *Bonded Debt,* $60,000, at 4%. Hydrant rental, $1,600. **Management.**—(In '89.) Secy. and Treas., H. Hinckley, Williamsport. Supt., Irvin Hoffman.

79. **INDIANA AND WEST INDIANA,** *Indiana Co.* (Pop., Indiana, 1,907—1,953; West Indiana, 1,077—1,034.) Built in '87 by Clymer Water Co. Engr., G. C. Fink. No contr. **Supply.**—Driven wells, pumping to tank or direct. **Pumping Machinery.** – Cap., about 135,000 galls.; deep well pumps. *Mains,* iron, about 7 miles. *Services,* galv. i. *Taps,* 180. *Meters,* 3. *Hydrants,* Mathews, 68. *Valves,* Eddy. *Consumption,* 80,000 galls. *Pressure:* ordinary, 55 lbs.; fire, 125 lbs. **Financial.**—*Cost,* $40,000. *Cap. Stock,* authorized and paid up, $40,000. *Operating Expenses,* $3,000. *Revenue:* consumers, $3,300; city, $1,700; total, $5,000. *Hydrant Rental,* $25. **Management.**—Prest., G. W. Hood. Secy., F. Sampson. Treas., J. R. Dougherty, Jr. Supt., C. H. Vensel.

80. **IRWIN,** *Westmoreland Co.* (Pop., 1,444—2,428.) Built in '89 by Irwin Water Co. **Supply.**—Westmoreland Water Co.'s works, of Greensburgh, by gravity from one of the reservoirs. Fiscal year closes Nov. 1. A 1,000,000-gall. storage reservoir has been added, giving pressure of 110 lbs. **Distribution.**—*Mains,* 3.11 miles. *Taps,* 100. *No Meters. Hydrants,* 26. *Consumption,* 35,000 galls. **Financial.**—*Cost,* $10,000. *Bonded Debt,* $37,000 at 6%; due in 1909. Int. charges, $2,220. **Management.**—Prest., Jno. F. Wolf. Secy. and Supt., Murray Forbes, Greensburgh.

66. **JEANETTE,** *Westmoreland Co.* (Pop. in '90, 3,296.) Supply from Greensburgh.

81. **JENKINTOWN,** *Montgomery Co.* (Pop., 810—1,609.) Built in '89 by Jenkintown Water Co.; supplies Jenkintown, Wyncote, Noble, Glenside and Heaslck; city neither controls rates nor has right to buy works. Fiscal year closes second Friday in Dec.

Year.	Mains.	Taps.	Meters.	Hydrants.	Cost.	Exp.	Rev.
'89	8	40	0	26	$40,000	....	....
'90	8	125	4	26	.....	$3,046	$3,660

Financial.—*Cap. Stock,* authorized, $150,000, all paid up. **Management.**—Prest., M. L. Kohler. Secy., J. W. Ridpath. Treas., C. J. Wilson.

82. **JERMYN,** *Lackawanna Co.* (Pop., 1,541—2,150.) Built in '84 by Jermyn Water Co.; franchise provides neither for control of rates nor purchase of works by city. **Supply.**—Aylesworth Creek, by gravity.

Year.	Mains.	Taps.	Meters.	Hydrants.	Cost.	Exp.	Rev.
'86	5	....	0	10	$26,350	....	*$150
'88	7.5	250	0	11	34,000	$500	5,150
'90	9	400	1	15	.....	950	6,000

*From city.

Financial.—*Cap. Stock,* authorized, $30,000, all paid up; paid 11% dividend in '90. *No Debt.* For other data see above. **Management.** – Prest., J. J. Jermyn, Scranton. Secy. and Supt., W. S. Hutchings.

83. **JERSEY SHORE,** *Lycoming Co.* (Pop., 1,411—1,853.) Built in '84-5 by Jersey Shore Water Co. **Supply.**—Springs and surface water, pumping to reservoir. Fiscal year closes Oct. 1.

Year.	Mains.	Taps.	Mtrs.	Hd'ts.	Cons'pt'n.	Cost.	B'd'd d't.	Int. ch'g's.	Exp.	Rev.
'85-6	3	70	0	28	100,000	$40,000	$20,000	$1,000	$600	$3,100
'86-7	4	88	0	31	100,000	50,000	20,000	1,000	600	3,430
'87-8	4	110	0	31	500,000	50,000	20,000	1,000	600	3,390
'89-90	4	125	0	30	.....	50,500	20,000	1,000	800	3,590

Financial.—*Cap. Stock:* authorized, $75,000. *Bonded Debt,* $20,000 at 5%, due in '91. *Expenses:* operating, $800; int., $1,000; total, $1,800. *Revenue:* consumers, $2,500;

fire protection, $1,090; total, $3,590. **Management.**—Prest., E. I. Jump. Secy., H. C. Jump. Supt., E. D. Jump.

84. JOHNSTOWN, *Cambria Co.* (Pop., 8,380—21,805.) Built in '67 by Johnstown Water Co. Supplies water in 13 boroughs; franchise provides neither for control of rates nor purchase of works by city. In '90 Stony Creek Water Co., which built works in '88, sold property to Johnstown Water Co., and dissolved. **Supply** —Conemaugh and St. Clair rivers, Laurel Run and Mill creeks, by gravity. Fiscal year closes May 4. A 30-in. intake line, 5 mi les long, was to be completed in March '91, extending from Stony Creek, and affording additional supply of 48,000,000 galls. Year. Mains. Taps. Meters. Hyd'ts. Cost. B'd'd d't. Int. ch'g's. Exp. Rev.

'80-1	30	2,000	.....	54	.....	.....	.....	$3,000 .....
'84-5	35	2,600	0	95	$373,400	none	none	.....
'87-8	41	2,550	0	100	479,400	$85,000	$3,400	4,500 $60,000
'89-90	42	2,900	0	50	.....	80,000	3,200	.....

Financial.—*First Cost*, $375,000. *Capital Stock:* authorized, $600,000; paid up, $375,000; paid 8% dividend in 1889. *Total Debt*, $80,000; bonded at 4%, due in 1900. "Books having been lost in flood of May 31, '89, these figures are only approximately correct." *No Sinking Fund. Expenses and Revenue,* withheld. **Management.**—Prest., Jas. McMillen. Secy., A. H. Waters. Treas., J. D. Roberts. Supt., Jas. Williams.

85. KANE, *McKean Co.* (Pop., in '90, 2,944.) Built in '87-8 by Spring Water Co. Engr., G. H. Lyon. **Supply.**—Hubert Run and large springs connected by pipe, pumping to tanks. **Pumping Machinery.**—Cap., about 600,000 galls.; Worthington, driven by natural gas. **Tanks.**—Cap., about 50,000 galls. **Distribution.**—*Mains,* c. i., about 1½ miles. *Services,* lead and galv. i. *Taps* [probably faucets], 500. *Meters,* Crown, 14. *Hydrants,* 12. *Pressure* 40 to 50 lbs. **Financial.**—*Cost,* $22,500. *Cap. Stock:* authorized, $40,000; paid up, $20,750; paid 6% div'd in '90, and passed 6% to surplus. *No Debt. Operating expenses,* $1,500. *Revenue:* consumers, $2,500; city, about $300. Hydrant rental, $25. **Management.**—Prest., T. L. Kane. Secy., A. D. Clark. Treas., J. Davis. Supt., G. P. Weeks.

86. KENNETT SQUARE, *Chester Co.* (Pop., 1,021—1,326.) Owned by borough. Built in '42 by Co.; bought in '50 by borough. **Supply.**—Surface water and spring, pumping to reservoir and stand pipe. Fiscal year closes Feb. 25. Pumping machinery was to be changed in '91. **Distribution.**—*Mains,* 5 miles. *Taps* (in '80), 250. *Hydrants,* 19. **Financial.**—*First Cost,* $30,000; other data withheld. **Management.**—Three comrs. appointed by burghers every year. Chn., George W. Taft. Supt., E. P. Mercer.

87. KINGSTON, *Luzerne Co.* (Pop., 1,418—2,381.) **History.**—Now owned by Spring Brook Water Co.; supplied several years ago. (See Moosic.) **Supply.**—Spring Brook, by gravity. **Distribution.**—*Mains,* 10 to 4-in. *Pressure,* 120 lbs. **Financial.**—*Cap.'Stock:* authorized, $25,000; all paid up. *No Bonds.* **Management.**—Prest., L. D. Shoemaker. Secy., L. A. Waters. Supt., Abram Nesbitt.

88. KITTANING, *Armstrong Co.* (Pop., 2,264—3,095.) Built in '71 by Kittaning Water Co. **Supply.**—Alleghenv River, pumping to reservoir. **Distribution.**—*Mains,* 8 miles. *Taps,* 500. *No Meters. Hydrants,* 30. **Financial.**—(In '88.) *Cap. Stock,* $50,000. *Bonded Debt,* $20,000. *Hydrant Rental,* $21.66 each. **Management.**—Prest., G. C. Orr. Secy., Treas. and Supt., W. B. Meredith.

89. KNOX, *Clarion Co.* (Pop., 767—888.) Built in '82 by borough. **Supply.**—Wells, pumping to tanks. **Distribution.**—*Mains,* 1 mile. *Taps,* 36. *Hydrants,* 4. *Consumption,* 8,500 galls. **Financial.**—*Cost,* $2,000. *Operating Expenses,* $300. *Revenue,* consumers, $300. **Management.**—Secy., J. M. Brothers. Supt., Robt. Woodling.

90. KUTZTOWN, *Berks Co.* (Pop., 1,198—1,595.) Built in '90 by Kutztown Water Co.; put in operation about June 1. Eng'r., Emil L. Neubling. Cont'r., H. E. Ahrens, Reading. **Supply.**—Artesian well and springs, pumping to reservoir. **Pumping Machinery.**—Cap., 1,000,000 galls.; Deane comp. dup. **Reservoir.**—Cap., 600,000 galls.; mostly in excavation. **Distribution.**—*Mains,* 8 to 4-in. c. i., abou 5 miles; from Mellert Foundry & Machine Co., Reading, *Services,* galv. i. *Taps,* 100, *No Meters. Hydrants,* Mellert, 25. *Pressure,* 80 lbs. **Financial.**—*Cost,* about $35,000. *Cap. Stock:* authorized, $50,000; paid up, about $15,000. *Bonded Debt,*

FOR FURTHER DESCRIPTIVE MATTER see Manual for 1889-90. CHANGES since 1890 are here given. POPULATIONS are for 1880 and 1890, respectively.

authorized and issued, $25,000 at 5%. *Hydrant Rental*, $400; 2-years' contract. **Management.**—Prest., J. S. Trexler. Secy., Jas. Marx. Treas., J. D. Sharaden. Supt., H. H. Boyer. Report by P. D. Wanner, Reading, Jan., '91.

91. LANCASTER, *Lancaster Co.* (Pop., 25,769—32,011.) Built in '36 by borough. **Supply.**—Conestoga Creek, pumping to reservoir. Fiscal year closes May 31. City has secured authority to issue $150,000 in bonds to build new storage reservoir and make other improvements. Unofficially reported that in summer of '91 contracts for 15,000,000-gall. reservoir were let to Kitch & Gonder, Lancaster, for $80,000.

Year.	Mains.	Taps.	M't'r's.	Hd'ts.	Cons'ption.	Cost.	Exp.	Rev.
'80	33	3,500	..	350	2,903,706	......	$10,000	$38,327
'81	32	3,500	2	300	3,257,474	$300,000	9,934	41,247
'84	33	4,100	9	390	4,017,155	300,000	13,000	48,139
'85	35.2	....	..	403	3,785,923	......	......	49,760
'86	36.3	5,200	10	408	3,872,796	......	9,903	51,475
'87	32	5,200	10	390	4,000,000	......	11,500	50,000
'88	33.5	5,350	..	40	4,500,000	620,000	13,000	58,000
'90-'91	41.5	4,500	6	465	5,072,000	no data	22,356	70,715

Financial.—*Total Debt*, bonded, $376,958; $221,558 at 6%, of which $208,558 is in sinking fund and $13,000 due in 1920; $41,400 at 5%, in sinking fund; $614,000 at 4%, of which $108,600 is in sinking fund, $91,400 due 1891-1901, $100,000 due 1895-1905, $125,000 due 1893-1918, $189,000 due 1905-1920. *Sinking Fund*, $358,558; invested in bonds, as noted above. *Expenses*, operating, $22,356. *Revenue:* consumers, $61,000; city, $9,715; total, $70,715. **Management.**—Com. of 7. Chn., Robt. Clark, Mayor's Clerk. Secy., Jos. Halbach. Supt., E. F. Frailey.

92. LANSDOWNE, *Delaware Co.* (Pop., in '90, 875.) Now owned by Lansdowne Water Co. First works built in '87 by Casper Pennock. Dec. 8, '90, above company was organized and bought works. **Supply.**—Three artesian wells, pumping to tanks. **Pumping Machinery.**—Cap., about 40,000 galls.; Smith & Valle. **Tanks.**—Cap., 40,000 galls. **Distribution.**—*Mains.*, 4-in., about 1¾ miles. *Taps*, about 135. *Hydrants*, 5. *Consumption*, about 40,000 galls. **Financial.**—*Cost*, $20,000. *Cap. Stock*, authorized, $50,000; paid up, $23,000. *No Debt. Operating Expenses*, $900. *Revenue:* consumers, $2,200; city, none. **Management.**—Prest., Casper Pennock. Secy. and Treas., J. Van Zandt. Report, Jan. 1, '91.

93. LANGHORNE, *Bucks Co.* (Pop., 558—727.) Built in 86-7 by Langhorne Spring Water Co.; franchise neither controls rates nor gives city right to purchase works. **Supply.**—Springs, pumping from collecting reservoir to tank. Fiscal year closes Dec. 31. During '90 a 1,000,000-gall. Worthington comp. dup. pump was added; also a 75 HP. return tubular boiler from Hoff & Fountain, Philadelphia. A stone pump and boiler house was erected, in which both the new and old machinery was placed.

Year.	Mains.	Taps.	Meters.	Hyd'ts.	Cons'pt'n.	Cost.	Exp.	Rev.
'89	3.5	80	0	9	15,000	$30,000	$1,400	$1,408
'90	4.25	100	0	9	65,000	39,000	1,800	1,844

Financial.—*First Cost*, $30,000. *Cap. Stock:* authorized, $40,000; paid-up, $30,000; paid no div'd in 1890. *Total Debt*, $8,000, floating. *No Meters. Revenue:* consumers, $1,736; fire protection, $108; total, $1,844. **Management.**—Prest., Chas. Hill. Secy., J. S. Wright, Eden.

94. LANSDALE, *Montgomery Co.* (Pop., 798—1,858.) Built in '72 by Lansdale Water Co. **Supply.**—Artesian well, pumping to tank. Fiscal year closes Nov. 17. **Distribution.**—(Sept. 1, '91.) *Mains*, 3.8 miles. *Taps*, 217. *Hydrants*, 21. **Financial.**—*Cost*, $24,091. *Cap. Stock*, paid-up, $15,000; paid 5% div'd in '90. *Debt*, $6,700, bonded. *Sinking Fund*, $1,104. *Expenses:* operating, $993; int., $240. *Revenue*, city, none. **Management.**—Pres. and Supt., W. D. Heebner, Secy., J. C. Boorse, Kulpsville, Treas., A. G. Godshall.

95. LANSFORD, *Carbon Co.* (Pop., 2,206—4,004.) Built in '73 by Panther Creek Water Co.; city neither controls rates nor has right to buy works under franchise. **Supply.**—Streams on Broad Mountain, by gravity from impounding reservoir. Fiscal year closes Dec. 31. **Distribution.**—*Mains* (in '84), 8 miles. *Taps*, 175. *No meters. Hydrants*, 13. **Financial.**—(In '89) *Cost*, $52,000. *No Debt. Operating Expenses*, $500. *Revenue:* consumers, $3,000; city (in '90), $70. **Management.**—Prest., Geo. Ruddle. Sec. and Treas., J. E. Lauer. Supt., W. D. Zehner.

XVI. LA PLUME, *Lackawanna Co.* (Pop., 5,821—8,061.) Built in '90 by La Plume Water Co. **Supply.**—Spring, by gravity. **Distribution.**—*Mains*, 4-in. iron. **Financial.**—*Cap. Stock*, authorized, $3,000. **Management.**—Prest., J. B. Dickson, Scranton. Secy., A. C. Sisson. Treas., D. T. Bailey. Supt., D. W. Brown,

PENNSYLVANIA. 127

96. LATROBE, *Westmoreland Co.* (Pop., 1,815—3,598.) Built in '86 by Latrobe Water Co. **Supply.**—Loyalhanna river, filtered to a well, thence by pumping direct. Fiscal year closes Dec. 31. March 13, '91, company reported that it contemplated building a 4,000,000-gall. reservoir in the near future. Summer of '91 contract was let for 1,500,000 gall. Gordon pump.

Year.	Mains.	Taps.	M't'rs.	Hd'ts.	Cons'pt'n.	Cost.	B'd'd d't.	Int ch'g's.	Exp.	Rev.
'87	4	110	0	23	400,000	$100,000	$25,000	$1,500	$2,500	$5,120
'88	6	150	0	43	500,000	110,000	25,000	1,500	2,000	8,900
'90	9.5	400	0	43	600,000	......	25,000	1,500	4,349	12,365

Financial.—*Cap. Stock*, authorized, $150,000, all paid-up; paid no div'd in 1890. Net income used for extensions. *Total Debt,* $41,000; floating, $16,000 at 6%; bonded, $25,000 at 6%. *Expenses:* operating, $4,114; int., $2,254; taxes, $235; total, $6,603. *Revenue:* consumers, $10,245; fire protection, $1,920; other purposes, $200; total, $12,365. **Management.**—Prest., H. S. Donnelly. Secy., Treas. and Supt., W. A. Johnston.

97. LEBANON, *Lebanon Co.* (Pop., 8,778—14,664.) Built in '72 by borough. **Supply.**—Hammer Creek and feeders, by gravity from impounding reservoir. Fiscal year closes Jan. 1. Feb. '91, bids were wanted for building earth dam 280 ft. long, 42 ft. high, with inner slope 3 to 1, outer, 2 to 1 ; 16-in. drain, 12-in. effluent pipe; area of reservoir, 4¼ acres; cap., 35,000,000 galls.

Year.	Mains.	Taps.	Meters.	Hyd'ts.	Cost.	B'd'd Debt.	Int. Ch'g's.	Exp.	Rev.
'84	..	1,800	0	76	......	$200,000	$8,000	$1,000	$8,000
'87	22	1,762	21	98	250,000	100,000	......	1,200	13,000
'88	22	3,000	24	93	......	130,000	......	1,200	18,000
'89	22	3,000	25	..	......	195,000	......	2,800	.....
'90	22	3,092	26	95	250,000	200,000	8,200	2,800	.....

Financial.—*Total Debt,* $200,000, bonded, at 4%. Bonds are redeemed from net income. *Expenses,* see above. *Revenue,* withheld. No allowance for public uses of water. **Management.**—Three comrs., appointed by city council. Chn., W. D. Rauch. Secy., M. Strause. Supt., H. M. Allwin.

XVII. LEECHBURGH, *Armstrong Co.* (Pop., 1,123—1,921). Built in '89 by Leechburgh Water-Works Co.; put in operation in Oct. Eng'r., G. R. Stewart. **Supply.**—Kiskaminitas Creek, pumping to tanks. **Pumping Machinery.**—Dy. cap., 1,000,000 galls.; Smith & Vaile dup. **Tanks.**—Cap., about 60,000 galls. **Distribution.**—*Mains,* c. i., 5 miles. *Services,* galv. i. *Taps,* 150. *No Meters. Hydrants,* 1. *Consumption,* 80,000 galls. *Pressure,* 160 lbs. **Financial.**—*Cost,* $15,000. *No Debt. Operating Expenses,* $1,200. *Revenue:* consumers, $1,700; city, $300; total, $2,000. **Management.**—Prest., J. C. Kirkpatrick. Secy., John Bredin. Treas., C. Campbell. Supt., H. L. George.

98. LEHIGHTON, *Carbon Co.* (Pop., 1,937—2,959.) Built in '89 by Lehighton Water Co.; city does not control rates; franchise does not provide for purchase of works. **Supply.**—Mountain stream, by gravity to reservoir. Fiscal year closes July 8. It was attempted to obtain water supply from artesian well. Very little water was obtained after drilling 726 ft., and this source was abandoned. Present supply is a 2,000,000-gall. impounding reservoir on mountain stream; thence by gravity to distributing reservoir having cap. of 750,000 galls.

Year.	Mains.	Taps.	Meters.	Hd'ts.	Cost.	B'd d'bt.	Int. ch'g's.	Exp.	Rev.
'89-90	5.75	112	0	42	$29,000	$9,000	......	...	.....
'90-91	8.65	183	0	43	47,000	25,000	$1,250	$360	$1,656

Financial.—*Cap. Stock,* authorized, $20,000, all paid up. *Total Debt,* $27,000; floating, $2,000; bonded, $25,000 at 5%, due in 1900. *Expenses:* operating, $300; int., $1,250; taxes, $60; total, $1,910. *Revenue:* consumers, $606; fire protection, $1,050; total, $1,656. **Management.**—Prest., Jno. S. Lentz. Secy., Howard Seaboldt.

99. LEWISBURGH, *Union Co.* (Pop., 3,080—3,248.) Built in '83 by Lewisburgh Water Co.; franchise neither controls rates nor gives city right to purchase works. **Supply.**—West branch of Susquehanna River, pumping to stand-pipe. Fiscal year closes Dec. 31.

Year.	Mains.	Taps.	Meters.	Hd'ts.	C'n'pt'n.	Cost.	B'd'd d'bt.	Int. ch's.	Exp.	Rev.
'84	4	76	1	43	30,000	$34,947	$5,000	$250	......	.....
'86	4.7	187	2	45	108,292	37,577	5,000	250	$1,604	$3,284
'88	4.7	251	3	47	154,940	38,572	4,000	200	1,994	3,993
'90	5	270	4	47	160,233	39,204	3,000	200	2,221	3,439

For Further Descriptive Matter see Manual for 1889-90. Changes since 1890 are here given. Populations are for 1880 and 1890, respectively.

Financial.—*Cap. Stock:* authorized, $32,000; paid up, $31,240; paid 4% div'd in 1889. *Bonded Debt*, $4,000 at 5%. *Expenses:* operating, $1,875; int., $200; taxes, $92; total, $2,167. *Revenue:* consumers, $2,639; fire protection, $800; total, $3,439. **Management.**—Prest., Wm. C. Duncan. Treas., Alfred Hayes. Secy. and Supt., Cyrus Dreisbach.

100. **LEWISTOWN**, *Mifflin Co.* (Pop., 3,222–3,273.) Built in '35 by Lewistown Water Co. **Supply.**—Springs, by gravity, and two artesian wells, pumping direct. Fiscal year closes first Monday in Feb. *Distribution.—Mains* (in '88), 3.5 miles. *No Meters.* *Hydrants*, 25. **Financial.**—*Cost* (in '88), $30,000. *Debt* (in '88), none. *Operating Expenses* (in '88), $2,000. *Revenue:* consumers, about $3,500; city, none. **Management.**—Prest., R. H. Lee. Secy., T. F. McCoy. Report by C. A. Zerbe.

199. **LEWISVILLE**, *Potter Co.* P. O. Ulysses, which see.

Lititz, *Lancaster Co.* (Pop., 1,113–1,494.) Built in '87 for fire protection and street sprinkling by Lititz Water Co. Address Johnson Miller, Secy.

101. **LOCK HAVEN**, *Clinton Co.* (Pop., 5,845–7,358.) Built in '09 by town. **Supply.**—Two mountain streams, by gravity from impounding reservoirs on each. *Distribution.—Mains* (in '88), 15 miles. *Taps* (in '87), 750. *Hydrants* (in '87), 80. **Financial.**—*Cost* (in '87), $130,000. *Bonded Debt* (in '87), $114,000. *Operating Expenses* (in '87), $1,400. *Revenue* (in '88), $10,000. **Management.**—Chn., B. F. Geary. Supt., C. F. Keller.

102. **LOCUST GAP**, *Northumberland Co.* (Pop., est., 1,500.) In Mount Carmel Township. Built in '91 by Bear Gap Water Co.; was to be in operation Oct. 1 to supply collieries and town. Eng'r., W. H. Dechant, Reading. Labor done by co. **Supply.**—Roaring Creek, pumping to reservoir. Water will be pumped over hill 750 ft. high. **Pumping Machinery.**—Cap., 1,440,000 galls.; Worthington high duty. Boilers: Babcock & Wilcox. *Reservoir.*—Cap., 500,000 galls.; to be 1,500,000 galls. *Distribution.—Mains*, c. i., 10 miles; from Warren Foundry & Machine Co., Phillipsburg, N. J. *Pressure*, 150 lbs. **Financial.**—*Cost*, est., $200,000. *Capital Stock*, authorized and paid up, $100,000. *Bonded Debt*, authorized and issued, $100,000 at 5%. **Management.**—Prest., W. C. McConnell. Secy., G. O. Martz. Treas., C. D. McWilliams, Shamokin. Report Sept. 1, '91.

XVIII. **LOCUST MOUNTAIN WATER CO.'S PLANT**, *Columbia Co.* Built in '82-14 to supply Mt. Carmel and Centralia in dry season, Ry. Cos. and Collieries. **Supply.**—Roaring Creek, pumping to reservoir. *Distribution.*—(In '88.) *Mains*, 15 miles. *Taps and Meters*, 15. *No Hydrants. Consumption*, about 112,000 galls. **Management.**—(In '89.) Prest., J. T. Blakeslee. Supt., C. A. Blakeslee, Mauch Chunk.

103. **LYKENS**, *Dauphin Co.* (Pop., 2,154–2,150.) Built in '85 by Lykens Water Co,; also supplies Wiconisco. Engr., G. H. Pierson, N. Y. City. **Supply.**—Rattling Creek, by gravity. Fiscal year closes Dec. 31. *Distribution.—Mains*, 2.75 miles. *Taps*, 44, and 318 "family hydrants." *Hydrants*, 26. **Financial.**—*Cost*, $29,060. *No Debt. Operating Expenses*, $538. *Revenue*, $3,570. **Management.**—Prest., Isaac J. Wistar. Secy., Horace Whiteman. Supt., J. M. Williams.

104. **McKEESPORT**, *Allegheny Co.* (Pop., 8,212–20,741.) Built in '81-2 by borough. **Supply.**—Youghiogheny River, pumping to reservoir. Fiscal year closes Feb. 28. April 27, 1891, city contemplated adding two 3,000,000-gall. comp., cond. pumping engines, and two 1,500,000-gall. pumps and a stand-pipe for high service. At present high service consists of small pump, pumping direct into mains and affording no fire pressure.

Year	Mains	Taps	Mtrs.	Hd'ts.	Cons'p't'n	Cost	B'd'd't	Int.ch'g's	Exp.	Rev.
'82	9	206	10	60	80,000	$98,304	$75,000	$3,750	$2,600	$3,200
'84	9.7	660	70	70	450,000	100,000	75,000	3,750	4,500	10,000
'86	15.5	1,135	96	118	600,000	140,000	102,000	5,100	5,095	11,778
'87	17.5	1,535	120	130	1,100,000	160,000	102,000	5,100	4,500	13,778
'89	18.25	1,735	198	130	1,600,000	200,050	102,000	5,100	6,500	19,089
'90	25.1	2,313	263	180	2,864,763	262,221	102,000	5,100	15,469	25,395

Financial—*Total Debt*, $114,206; floating, $12,206 at 6%; bonded, $102,000 at 5%; $27,00 due in 1904; $75,000 due in 1911. *Sinking Fund*, $28,000, not invested. See above. No allowance for public used of water. All deficiencies paid from general fund. **Management.**—Com. of Council, Chn. R. J. Block. Clerk. G. B. Herwick. Supt., Jos. Ecoff.

PENNSYLVANIA.

105. MAHANOY CITY, *Schuylkill Co.* (Pop., 7,181—11,286.) Built in '68 by Mahanoy City Water Co.; franchise neither controls rates nor gives right to purchase works. **Supply.**—Springs and stream, by gravity from impounding reservoirs. Fiscal year '90-1 closed May 9.

Year.	Mains.	Taps.	Met's.	Hyd'ts.	C'n't'n.	Cost.	B'd'd d't.	Int. ch'g's.	Exp.	Rev.
'70	4.5	1,100	..	15	......	$89,000	......	....	......	$10,000
'80	3.5	1,350	..	20	......	100,000	......	....	$2,000	12,000
'86	...	1,350	0	23	unk'n.	100,000	$3,000	$180	......	......
'87	5	1,350	0	23	350,000	100,000	3,000	180	2,865	12,100
'88	7	1,500	0	26	400,000	120,000	15,300	790	3,250	14,100
'89	7	1,750	0	25	500,000	143,600	none	none	4,400	14,600
'90	6	1,500	1	29	400,000	......	none	none	3,000	......

Financial.—*First Cost*, $15,000. *Cap. Stock:* authorized, $100,000; paid up, $95,750. No debt. *Expenses:* operating, $3,000; taxes, $600; total, $3,600. *Revenue:* consumers in '89, $10,000; city, $100. **Management.**—Prest., E. S. Silliman. Secy. and Treas., H. M. Parmley. Supt., E. Silliman, Jr.

106. MANHEIM, *Lancaster Co.* (Pop., 1,666-2,070.) Built in '85 by Manheim Water Co.; franchise neither controls rates nor gives city right to buy works. **Supply.**—Lehman's Creek, pumping to reservoir. Fiscal year closes Jan. 1.

Year.	Mains.	Taps.	Meters.	Hyd'ts.	Consumption.	Cost.	Exp.	Rev.
'87	4.5	160	0	..	100,000	$27,000	$810	$2,125
'88	4.5	182	0	58	200,000	28,000	1,518	2,113
'90	...	197	..	80	......	......	913	2,280

Financial.—*Cap. Stock:* authorized, $28,000; paid up, $27,000; paid 3½% div'd in 1889 and 4% div'd in '90. *No Debt. Expenses:* operating, $832; no int.; taxes, $81; total, $913. *Revenue:* consumers, $1,680; fire protection, $600; total, $2,280. In '89 total expense was $1,035, and revenue $2,194. **Management.**—Prest., A. Kline. Secy., B. H. Hershy. Treas., Geo. H. Danver. Supt., J. B. Singer.

66. MANOR, *Westmoreland Co.* (Pop., in '90, 578.) Supply from Greensburgh.

107. MAUCH CHUNK AND EAST MAUCH CHUNK, *Carbon Co.* (Pop.: Mauch Chunk, 3,752—4,101; East Mauch Chunk, 1,853—2,772.) Built in '49 by Mauch Chunk Water Co., to supply Mauch Chunk; in '61 domestic supply was introduced in East Mauch Chunk; franchise neither controls rates nor gives city right to buy works. **Supply.**—Springs, by gravity from impounding reservoirs. Fiscal year closes March 31.

Year.	Mains.	Taps.	Meters.	Hydrants.	Cost.	Expenses.	Revenue.
'81	4	430	0	28	$25,000	$750	$280*
'87	4	500	4	25	37,500	1,000	5,000
'88	4	520	4	25	37,500	1,000	5,000
'89	4	550	4	25	37,500	1,200	5,500
'90	1	560	4	47	45,000	1,200	5,600

*From city.

Financial.—*Cap. Stock:* authorized, $50,000; paid up, $37,500; paid 10% div'd in 1890. *No Pref'd Stock or Debt. Expenses:* operating, $1,000; no int.; taxes, $200; total, $1,200. *Revenue:* consumers, $5,600. **Management.**—Prest., Robt. Klotz. Secy., S. S. Smith. Treas., C. O. S. Keer.

108. MEADVILLE, *Crawford Co.* (Pop., 8,860—9,520.) Built in '74-5 by Meadville Water Co. **Supply.**—French Creek, pumping to reservoir; filtered through gravel bed. Fiscal year closes Dec. 31.

Year.	Mains.	Taps.	Meters.	H'd'ts.	Cons'ption.	Cost.	Exp.	Rev.
'80	13.25	700	..	89	280,000		$3,000	.
'81	15.5	800	3	100	402,000	$145,000	4,000	$16,000
'84	16	900	5	102	600,000	150,000	4,000	24,000
'87	17.5	1,150	40	105	1,000,000	170,000	7,000	21,250
'88	18.84	1,142	36	105	734,520	174,359	5,634	15,320
'89	20	1,286	34	106	750,000	177,006	5,290	23,086
'90	20	1,350	40	106	728,219	177,520	6,643	21,172

Financial.—*First Cost*, $137,785. *Cap. Stock:* authorized, $150,000; all paid up; paid 7% div'd in '90. *Total Debt*, $5,500, floating, at 5%. *Expenses:* operating, $6,118; int., $165; taxes, $525; total, $6,808. *Revenue:* consumers, $18,872; city, fire protection, $5,300; total, $21,172. **Management.**—Prest., J. D. Gill. Secy., Edgar Hindekoper. Supt., G. T. Cullum.

109. MECHANICSBURGH, *Cumberland Co.* (Pop., 3,018—3,691.) Built in '57-8 by Mechanicsburgh Gas & Water Co. **Supply.**—Spring, by gravity from receiving and distributing reservoirs; filtered. Fiscal year closes April 1. In '89-9)

FOR FURTHER DESCRIPTIVE MATTER see Manual for 1889-90. CHANGES since 1890 are here given. POPULATIONS are for 1880 and 1890, respectively.

three miles of C-in. main was replaced by 10-in. pipe, and supply was increased by additional spring.

Year	Mains	Taps	Meters	Hyd'ts	Cons'pt'n	Cost	B'd'd d't	Int.ch'g's	Exp	Rev
'80	2	500	..	21	.....	.....	.....	.....	$100	....
'84	2	500	1	23	.....	.....	$15,000	$750	850	$4,050
'87	4	520	0	26	12,000	$50,000	18,000	900	800	4,000
'88	4	600	0	29	74,000	68,000	31,000	1,700	3,400	4,400
'90	10	550	0	38	.....	83,000	30,000	1,500	1,000	6,100

Financial.—*First Cost*, $25,000. *Bonded Debt*, $30,000 at 5%, due in '94. *Expenses:* operating, about $1,000; int., $1,500; total, $2,500. *Revenue:* consumers, $5,100; city, fire protection, about $1,000; total, $6,100. Management.—Secy. and Treas., J. B Bowman. Supt., Andrew Seifert.

110. **MEDIA,** *Delaware Co.* (Pop., 1,919—2,736.) Built in '71 by borough. Supply.—Ridley Creek, pumping to reservoir.

Yr	Mains	Taps	M'trs	Hyd'ts	Cons'n	Cost	B'd'd d't	Int.ch'g's	Exp	Rev
'82	6.5	385	0	25	125,000	.....	$65,000	$2,600	$1,200	$4,028
'87	6.6	400	0	75	150,000	$80,000	65,000	2,600	1,000	5,500

Financial.—Int., 4%. Management.—Council. Supt., Ralph Buckley.

111. **MERCER,** *Mercer Co.* (Pop., 2,344—2,136.) Built in '87 by Mercer Water Co. Supply.—Otter Creek, pumping to tank and direct in case of fire; filtered. Fiscal year closes April 1. New pump put in in '90.

Yr	Mains	Taps	M'trs	Hyd'ts	Cons'n	Cost	B'd'd d't	Int.ch'g's	Exp	Rev
'86	3	....	....	23	.....	$20,000	$12,000	$720	....	$770*
'87	3	180	0	24	250,000	20,000	12,000	720	$900	2,180
'88	3	250	0	24	300,000	22,000	12,000	720	1,000	2,840
'89	4,5	300	2	25	200,000	.....	12,000	720	1,000	3,005
'90	4 5	240	2	23	150,000	40,000	12,000	720	....	3,130

* From city.

Financial.—*Cap. Stock:* authorized, $24,000; no div'd in '90. *Total Debt*, $12,000, bonded at 6%, due in '96. *Expenses* (in '89), operating, $1,0 0; int., $720; total, $1,720. *Revenue:* consumers, $2,300; city, fire protection, $805; other purposes, $25; total, $3,130. No dividend has ever been declared, net revenue being applied to extensions. Management.—Secy., Henry Robinson. Supt , L. M. Hamerer.

XIX. **MESHOPPEN,** *Wyoming Co.* (Pop., 554—597.) Built in '73 by Meshoppen Borough Water Co. Supply.—Springs and brook, by gravity; filtered through gravel and charcoal, between stone walls, at dam. Fiscal year closes Oct. 31. Distribution.—*Mains*, 3.3 miles. *Taps*, 112. *No Meters. Hydrants*, 2. Financial.—*Cost* (in '88), $10,000. *Capital Stock:* authorized, $4,000, all paid up; paid 6% dividend in '90. *No Debt. Expenses:* operating, $225; no int.; taxes withheld. *Revenue:* consumers, $680; city, fire protection, none. Management.—Prest., E. H. Wells. Secy., N. A. Wells Treas. and Supt., A. H. Sterling.

112. **MEYERSDALE,** *Somerset Co.* (Pop., 1,423—1,847.) Built in '88 by Sand Spring Water Co. Supply.—Sand spring, by gravity. Fiscal year closes Dec. 1. Distribution.—*Mains*, 3 miles. *Taps*, 160; 138 in '88. *No Meters. Hydrants*, 22. Financial.—*Cost*, $17,450. *Capital Stock:* authorized, $16.000; paid up, $14,750; paid 6% dividend in 1890. *Total Debt*, $2,750; bonded at 5%, due in '93. *No Sinking Fund. Expenses:* operating, $250; int., $138; taxes, $44; total, $432. *Revenue:* consumers, $1,266; city, $600; other purposes, $143; total, $2,009. Management.—Prest., A. Chamberlin. Secy., S. B. Philson. Treas., W. B. Cook. Supt., C. W. Wixal.

113. **MIDDLETOWN,** *Dauphin Co.* (Pop., 3,351—5,080.) Built in '85 by Middletown Water Co.; franchise provides neither for control of rates nor purchase of works by city. Supply.—Swatara Creek, pumping to stand-pipe. '

Year	Mains	Taps	Meters	Hyd'ts	Cons'pt'n	Cost	B'd'd d't	Int. chgs	Exp	Rev
'87	4	190	2	35	100,000	$45,000	$25,000	$1,500	$2,500	$4,019
'88	5	257	2	37	125,000	57,000	25,000	1,500	500	5,197

Financial.—(In '88.) *Cap. Stock*, $50,000. *Revenue:* consumers, $3,600; city (in '90), $1,600. Management.—(In '89.) Prest., Jos. Campbell. Treas., C. W. Raymond.

114. **MILFORD,** *Pike Co.* (Pop., 983—793.) Built in '66 by Milford Water Co.; franchise neither controls rates nor gives city right to buy works. Supply.—Spring, by gravity. Fiscal year closes May 7.

Year	Mains	Taps	Meters	Hydrants	Cost	Exp	Rev
'87	3	125	0	..	$15,000	$300	$1,000
'88	3	90	0	21	16,500	75	800
'89	3	130	0	21	18,000	239	1,000
'90	3	133	0	21	18 000	260	1,000

Financial.—*Cap. Stock:* authorized, $18,000; paid-up, $7,900; paid 4% div'd in 1889, 16% in '90. Small div'd in '89 was due to litigation with borough. No debt. *Ex-*

PENNSYLVANIA. 131

penses: operating, $225; no int.; taxes, '$35; total. $260. *Revenue:* consumers, $1,000; city, none. **Management.**—Prest., John C. Wallace. Secy., Chas. P. Mott. Treas., J. H. Van Etten. Supt., Milton Armstrong.

115. **MILLERSBURG,** *Dauphin Co.* (Pop., 1,410—1,527.) Built in '91 by Millersburg Home Water Co.; put in operation Apr. 1. Contrs., Raymond & Campbell, Middletown. **Supply.**—Springs; mode not given; reservoir used. **Financial.**—*Cost,* $30,000. *Cap. Stock,* paid up, $20,000. *Bonded Debt,* $10,000. **Management.**—Prest., G. W. Gilbert. Secy., E. H. Leffler. Report Aug. 11.

116. **MILLERSTOWN** (*Barnhart's Mills P.O.*), *Butler Co.* (Pop., 1,108—1,162.) Built in '74 by borough and H. A. Leopold. **Supply.**—Artesian well, pumping to tanks. **Distribution.**—*Mains,* about 1 mile. *Taps,* 150. *No meters. Hydrants,* 11. *Consumption,* 40,000 galls. **Financial.**—(In '88.) *Debt,* $600 at 6%. *Operating Expenses,* $1,500. *Revenue:* consumers, $2,000; city, $600. **Management.**—H. A. Leopold, Mgr.

117. **MILLVILLE,** *Columbia Co.* (Pop., 375—est., 500.) Built in '90-1 by Bennett Water Co.; put in operation Jan. 26. Engrs. and Contrs., J. H. Harlow & Co., Pittsburg. **Supply.**—Allegheny River, filtered through bed in river, pumping to tank. **Pumping Machinery.**—Cap. not given, 18 × 10 × 10 ins.; Hall. **Tank.**—Cap., 280,000 galls., 40 × 30 ft.; from Riter & Conley, Pittsburg. **Distribution.**—*Mains,* 10 to 4-in. c. i., about 5 miles. *No Meters. Hydrants,* 30. *Pressure,* 100 lbs. **Financial.**—*Cost,* reported as $150,000. *Bonded Debt,* authorized and issued, $50,000 at 6%. Hydrant rental, total, $1,200. **Management.**—Prest., F. Klussman. Supt., Geo. Fitch. Report, Feb. 7, '91.

118. **MILTON,** *Northumberland Co.* (Pop., 2,102—5,317.) Built in '83 by Milton Water Co. **Supply.**—West branch of Susquehanna River, pumping to reservoir. Fiscal year closes Dec. 31.

Year.	Mains.	Taps.	Mtrs.	Hd'ts.	Cons'pt'n.	Cost.	B'd'dd't.	Int.ch'g's.	Exp.	Rev.
'86	7	305	0	40	300,000	$50,000	$6,000	$300	$1,500	$6,060
'87	7	350	0	41	300,000	50,000	6,000	300	2,850	6,080
'88	7	400	2	41	300,000	50,000	6,000	300	2,700	5,200
'89	7	480		41	300,000	53,220	6,000	300	2,984	5,447
'90	7	425	0	83	300,000	61,971	6,000	300	3,957	9,420

Financial.—*Cap. Stock:* authorized, $50,000; all paid up. Paid 6% div'd in 1890. *Total Debt,* $7,500; floating, $1,500 at 6%; bonded, $6,000 at 5%. *Expenses:* operating, $3,727; int., $300; taxes, $240; total, $4,167. *Revenue:* consumers, $6,600; city, fire protection, $820; total $9,420. **Management.**—Prest. and Supt., R. F. Wilson. Secy., H. R. Frick.

119. **MINERSVILLE,** *Schuylkill Co.* (Pop., 3,249—3,504.) Built in '57-61 by Minersville Water Co. **Supply.**—Mountain springs, by gravity. Fiscal year closes second Monday in January. One mile of 8-in. pipe taken up and replaced with 14-in pipe.

Year.	Mains.*	Taps.	Meters.	Hydrants.	Consumption.	Cost.	Exp.	Rev.
'88	9	350	2	30	120,000	$40,000	$1,500	$6,500
'89	9	...	2	30	115,000	50,000	1,500	7,000
'90	9	400	2	30	200,000	50,915	1,544	7,000

* In '80, 7.5 miles of mains; in '87, 9 miles.
Financial.—*First Cost,* $53,000. *Cap. Stock:* authorized, $150,000; paid up, $107,125; paid no div'd in 1890. *Floating Debt,* $6,900, at 6%, due in '91. *Expenses:* operating, $1,500; int., $414; taxes, $44; total, $1,958. *Revenue:* consumers, $7,000; city, none. **Management.**—Prest., R. F. Potter. Secy., Edw. R. Kear. Treas., J. S. Lawrence. Supt., Chas. R. Kear.

120. **MINOOKA,** *Lackawanna Co.* (Pop., not given.) Built in '88 under four charters, viz.: Minooka, Taylorville, Bellevue and Peoples (of Scranton) Water Cos. **Supply.**—Cold Spring reservoir, by gravity. **Distribution.**—(In. '89.) *Mains,* 29 miles. *Taps,* 500. *Meters* 2. **Financial.**—(In '89.) *Cost* (as given), $30,000. *No Debt. Revenue,* consumers, $3,687. Hydrant rental, $25 each. **Management.**—(In '90.) Prest., Lemuel Amerman; Secy and Treas., L. A. Waters; both of Scranton.

121. **MONONGAHELA CITY,** *Washington Co.* (Pop., 2,904—4,096.) Built in '88-9 by Monongahela City Water Co.; controlled by Kittaning Consolidated Water Co.; franchise provides neither for control of rates nor purchase of works by city. In '90 or '91 South View Water Co. was consolidated with above. **Supply.**—Monongahela River, pumping from filter well to reservoir. **Distribution.**—*Mains,*

FOR FURTHER DESCRIPTIVE MATTER see Manual for 1889-90. CHANGES since 1890 are here given. POPULATIONS are for 1880 and 1890, respectively.

9.5 miles. *Taps*, 149. *Hydrants*, 20. **Financial.**—*Cost* (to Jan. 1, '89), $63,000. *Cap. Stock*, authorized, $130,000. *Bonded Debt* (in '89), $30,000 at 6%. *Hydrant rental*, $500 for all. **Management.**—Prest., V. Neubert. Secy. and Supt., G. H. Fox, Kittanning. Treas., Wm. Pollock.

122. **MOOSIC**, *Lackawanna Co.* (Pop., 600—not given.) Built in '86-7 by Spring Brook Water Co. Charter covers Avoca, also called Pleasant Valley; under local name of Forty Fort Water Co. Forty Fort is supplied; said Co. incorp. in '91 with authorized cap. stock of $5,000. Spring Brook Co. leases Moosic Water Co.'s works, included below. **Supply.**—Spring Brook, pumping to reservoir. **Distribution.**—(In '88.) *Mains*, 12 miles. *Taps*, 375. *Meters*, 2. *Hydrants*, 8. *Consumption*. 500,000 galls. **Financial.**—(In '88.) *Cost*, $54,000. *Cap. Stock:* authorized, $50,000; paid up, $30,000. *Bonded Debt*, $15,000 at 6%. *Expenses:* operating, $1,700; int., $900. *Revenue*, consumers, $3,620. City, $280.

NOTE.—In '89, after above figures were reported, Co. was reorganized and system extended to supply several towns by laying 9 miles 2½ in., 6 of 2-in. and 4 of 10-in. pipe. See Kingston for further information regarding above Co.

123. **MOUNT CARMEL**, *Northumberland Co.* (Pop., 2,378—8,254.) Built in '85 by Mount Carmel Water Co.; franchise provides neither for control of rates nor purchase of works by city. **Supply.**—Springs, by gravity. Fiscal year closes Dec. 31. **Distribution.**—*Mains*, 5.5 miles. *Taps*, 350. *Meters*, 42. *Hydrants*, 36. **Financial.**—*Cost* (in '88), $85,000. *Cap. Stock*, $25,000. *Total Debt* (in '87), $5,625; floating, $625; bonded, $5,000 at 4%. *Operating Expenses*, $500. *Revenue:* consumers (in '87), $4,400; city, $300. *Hydrant Rental*, $10 each for 24, $5 each for additional. **Management.**—Prest., T. M. Righter. Secy., M. K. Watkins. Treas., L. W. Johnson. Supt., C. L. Johns.

124. **MOUNT JOY**, *Lancaster Co.* (Pop., 2,058—1,848.) Built in '74 by town. **Supply.**—Chiquesalunga Creek, pumping to reservoir, with direct pumping for fire protection. Fiscal year closes Dec. 31.

Year.	Mains.	Taps.	M't'rs.	Hd'ts.	Cost.	B'd'd. debt.	Int. ch'g's.	Exp.	Rev.
'90	3.3	159		25				$500	
'86	3.3	250	0	27	$48,000	$40,000	$1,600	500	$2,600
'88	3.35	270	0	28	50,000	40,000	1,560	600	2,650
'89	3.5	278	0	28	50,000	39,000	1,594	800	2,681
'90	3.5	284	0	28	50,000	38,000	1,520	1,385	2,800

Financial.—*First Cost*, $40,000. No allowance for public use of water. **Management.**—Water Com. appointed every year by burgess. Supt., Wm. Kuhn, elected by town council.

125. **MOUNT PLEASANT**, *Westmoreland Co.* (Pop., 1,197—3,652.) Built in '86 by Mount Pleasant Water Co.; franchise neither controls rates nor gives city right to buy works. **Supply.**—Jacob's Creek, pumping to reservoir and stand-pipe. Fiscal year closes Feb. 28.

Year.	Mains.	Taps.	Meters.	Hyd'ts.	Cons'pt'n.	Cost.	Exp.	Rev.
'89-90	8.5	159	0	74	1,053,248	$132,303	$6,815	$16,276
'90-91	8.5	238	0	74	1,036,000	150,000	6,753	15,651

Financial.—*Cap. Stock:* authorized, $150,000, all paid up; paid 2% div'd in '89, and 6% in '90. *No Debt*. *Expenses:* operating, in '89, $6,500, in '90, $6,753; taxes, '89, $315. in '90, $457; total, in '89, $6,815, in '90, $6,209. *Revenue:* consumers, in '89, $2,200, in '90, $2,123; city, fire protection, in '89, $1,650, in '90, $1,625; other purposes, in '89, $12,-426, in '90, $11,903; total, in '89, $16,276, in '90, $15,651. Rev. from "other purposes" was from supply furnished to coke ovens, the number having been 2,037 in '90. **Management.**—Prest., H. C. Frick. Secy., G. B. Bosworth. Treas., H. M. Curry. all of Pittsburg. Supt., Hugh Cole, Broad Ford.

126. **MUNCY**, *Lycoming Co.* (Pop., 1,174—1,295.) Built in '89 by Muncy Water Co.; put in operation Jan. 1, '90. Engr. and Contr., J. P. Herdic, Williamsport. **Supply.**—Springs, gathered in mountain stream, by gravity from reservoir. Dam and Reservoir.—Cap , 1,500,000 galls. Dam of stone in hydraulic cement, 120 ft. long, 8 ft. high, 16 ft. wide at bottom, 4 ft. at top; 250 ft. above town. **Distribution.**—*Mains*, 10 to 4-in. c. i., 7½ miles; from Mellert Foundry & Machine Co., Reading. *Taps*, 24. No *Meters*. *Hydrants*, Mellert, 20. *Consumption*, 30,000 galls. *Pressure*, 110 lbs. **Financial.**—*Cost*, $35,000. *Cap. Stock:* authorized, $36,000, all paid up. *Debt*, none; $35,000 bonds authorized. *Operating expenses*, $250. *Revenue:* consumers, $1,710; city, $600; 5-years contract, self-renewing unless borough buys. **Management.**—Prest., W. S. Steuger, 439 Chestnut St., Philadelphia. Secy. and

PENNSYLVANIA.

Treas., R. L. Brown, Seventh National Bank, Philadelphia. Supt., J. Shoemaker. Gen. Mgr., J P. Herdic, Williamsport. Report, Jan. 1, '91.

127. NANTICOKE, *Luzerne Co.* (Pop., 3,884—10,014.) No sewers. Has electric lights. *History.*—Built in '85 by Nanticoke Water Co. Engr., J. H. Bowden, Wilkes-Barre. *Supply.*—Honeoye Creek, by gravity to pumping station, thence pumping to stand-pipe. *Pumping Machinery.*—Dy. cap., 3,600,000 galls.; two Gordon & Maxwell dup., 22-in. steam and 12¼-in. water cylinder, with 18-in. stroke. *Boiler:* from Vulcan Iron Wks., Wilkes-Barre. *Stand-Pipe.*—Is 5 × 54 ft.; from Vulcan Iron Wks., Wilkes-Barre. *Distribution.*—*Mains*, 11 to 4-in. 1., 15 miles. *Services.* ½-in. Pipe and specials from Warren Foundry, Phillipsburgh, N. J.; trenching and pipe-laying done by Coon & Mooney, Kingston. *Taps,* 1,200. *Meters,* none. *Hydrants,* Ludlow, 47. *Valves,* Ludlow. *Consumption,* 2,000,000 galls. *Pressure:* ordinary, 20 to 95 lbs.; fire, 27 to 105 lbs. *Financial.*—Withheld. *Management.*—Prest., I. J. Wister. Secy., Horace Whiteman. Treas., Arthur Haviland. All of 233 South 4th st., Philadelphia. Supt., G. T. Morgan.

128. NAZARETH, *Northampton Co.* (Pop., 984—1,318.) Built in '89 by Nazareth Water Co.; franchise neither controls rates nor gives city right to buy works. *Supply.*—Springs, by gravity and pumping to reservoir. Fiscal year closes Oct. 31. In '90 an artesian well was sunk near reservoir, to increase supply. *Distribution.*—*Mains.* 2 miles. *Taps,* 110. *No Meters. Hydrants,* 19. *Consumption,* 18,000 galls. *Financial.*—*Cost,* $15,500. *Cap. Stock:* authorized, $10,000; paid up, $7,350; paid 5% dividend in 1890. *Bonded Debt,* $3,800, at 4½%. *Expenses:* operating, $300; int., $171; taxes, $33; total, $504. *Revenue:* consumers, $650; city, fire protection, $285; total, $935. *Management.*—Prest., H. F. Kingkinger. Secy. and Treas., Frank Kunkel. Supt., Jno. F. Bardill.

129. NEW BETHLEHEM, *Clarion Co.* (Pop., 773—1,026.) Built in '83 by Citizens' Water Co.: franchise neither controls rates nor gives city right to buy works. *Supply.*—Red Bank Creek, pumping to reservoir; filtered. Fiscal year closes May 31. *Distribution.*—*Mains,* 1.5 miles. *Taps,* 53. *No Meters. Hydrants,* 13. *Consumption,* 30,000 galls. *Financial.*—*Cost,* $11,592. *Cap. Stock:* authorized, $10,000, all paid up; paid 4% dividend in '89-90 and same in '90-1. *No Debt. Expenses:* operating, $558 in '89-90 and $600 in '90-1; taxes, in '89-90, $43. *Revenue:* consumers, $778 in '89 and $1,200 in '90; fire protection, $325 in both '89-50 and '90-1; other purposes, $452 in '89-90. *Management.*—Prest., Wm. M. Moore. Secy., E. R. Marsh. Treas., J. R. Foster. Supt., Geo. S. Thomas.

130. NEW BRIGHTON, *Beaver Co.* (Pop., 3,653—5,616.) Built in '78-9 by New Brighton Water-Works Co. *Supply.*—Beaver River, pumping to reservoir and direct; filtered. Fiscal year closes Jan. 31.

Yr.	Mains.	Taps.	Meters.	Hyd'ts.	Cons'pt'n.	Cost.	B'd'd d't.	Int.ch'g's.	Exp.	Rev.
'84-5	7	220	0	50	259,200	$60.000	$23,000	$1,380	$2,500	$5,600
'86-7	7.5	300		51	270,000	63,000	23,000	1,380	2,400	5,650
'87-8	11	350	4	55	300,000	65,000	25,000	1,500	2,800	6,500
'90-1	13	459	4	55	......	65,000	25,000	1,500	4,000	8,800

Financial.—Int., 6%. *Revenue:* consumers, $6,300; city, $2,500; total, $8,800. *Management.*—Prest., R. S. Kennedy. Secy. and Treas., H. S. McConnel. Supt., J. D. King.

131. NEW CASTLE, *Lawrence Co.* (Pop., 8,418—11,600.) Built in '81 by city of New Castle Water Co. *Supply.*—Shenango River, pumping to reservoir; filtered. Fiscal year closes March 31.

Year.	Mains.	Taps.	Meters.	Hyd'ts.	Cons'pt'n.	Cost.	B'd'd d't.	Int.ch'g's.	Exp.	Rev.
'82-3	12.8	200	3	104	140,000	....	......	......	....	....
'86-7	16	456	0	113	400,000	....	......	......	$10,988	$14,984
'87-8	17	652	3	117	500,000	....	$100,000	$6,000	5,000	17,900
'88-9	17	700	0	117	600,000	....	100,000	6,000	4,8:0	19,600
'90-1	17.6	909	5	117	1,000,000	$325,488	100,000	6,000	6,220	23,350

Financial.—*Capital Stock:* authorized, $200,000, all paid up; paid 2% dividend in '90. *Bonded Debt,* $100,000; at 6%, due in 1931. *No Sinking Fund. Expenses:* operating, $5,863; int., $6,000; taxes, $358; total, $6,220. *Revenue:* consumers, $17,590; fire protection, $5,680; other purposes, $180; total, $23,350. *Management.*—Prest., J. B. Finley, Monongahela City. Secy., F. L. Stephenson; Treas., L. H. Williams; both of Pittsburgh. Supt., J. W. Taylor. NOTE.—In '91 there was talk of purchase of works by city.

FOR FURTHER DESCRIPTIVE MATTER see Manual for 1889-90. CHANGES since 1890 are here given. POPULATIONS are for 1880 and 1890, respectively.

38. NEW HAVEN, *Fayette Co.* (Pop., 442—1,221.) Supply from Connellsville.

132. NEWTOWN, *Bucks Co.* (Pop., 1,001—1,213.) Built in '88-9 by Newtown Artesian Water Co.; franchise neither fixes rates nor provides for purchase of works by city. **Supply.**—Two artesian wells, pumping to reservoir. Fiscal year closes Jan. 10.

Year.	Mains.	Taps.	Meters.	Hydrants.	Consumption.	Cost.	Exp.	Rev.
'89-0	4	85	0	22	36,000	$23,000	......	......
'90-1	4	95	0	23	40,000	$24,000	$1,301	$1,487

Financial.—*Cap. Stock.* authorized $20,000; all paid up. No dividend in 1890. *Floating Debt,* $3,000 at 5%. *Expenses,* operating, $1,235; int., $150; taxes, $66; total, $1,551. *Revenue:* consumers, $1,200; city, fire protection, $287; total, $1,487. **Management.**—Prest., Geo. C. Worstall. Secy., Paul Blaker. Treas., Harry Smith. Supt., Robt. Shields.

81. NOBLE, *Montgomery Co.,* see Jenkintown.

133. NORRISTOWN, *Montgomery Co.* (Pop., 13,063—19,791.) Built in '47 by Norristown Insurance & Water Co. **Supply.**—Schuylkill River, pumping to reservoir. Fiscal year closes Sept. 10. In '90 a 5,000,000-gall. Worthington high duty comp. cond. engine was added.

Year.	Mains.	Taps.	Meters.	Hyd'ts.	Cons'pt'n.	Cost.	B'd'd d't.	Int. ch'g's.
'81-2	27	2,500	0	85	800,000	$275,000	$90,000	$4,500
'84-5	27	unk'n	0	130	800,000	275,000	19,000	5,950
'87-8	37	"	0	150	1,500,000	325,000	90,000	4,500
'89-90	20	"	0	175	2,250,000	310,027	80,000	2,500

Financial.—(In '89-90.) *Cap. Stock:* authorized, $200,000, all paid up; paid 8% div'd in 1890, *Bonded Debt,* $50,000 at 5%, due in 1900. Undivided profits, $116,132. *Expenses:* operating, $9,597; int., $2,500; taxes, $800; total, $12,897. *Revenue:* consumers, $40,513; fire protection, none. **Management.**—Prest., Jno. Slingluff. Secy., H. C. Crawford. Treas., W. H. Slingluff. Supt., Jno. Straight.

134. NORTH CLARENDON, *Warren Co.* (Pop., 295—not given.) Owned by Elston estate; built in '81 by Brown Bros. **Supply.**—Artesian well, pumping direct and to tanks for domestic use and to tank for fire protection. **Distribution.**—*Mains,* 1.5 miles. *Taps,* 90. *No Meters. Hydrants,* 9. **Financial.**—*Cost,* $4,300. *Operating Expenses,* $1,300. *Revenue:* consumers, about $1,500. **Management.**—Supt., A. F. Crossman. Chn. of Fire and Water Committee, Thos. McNally.

135. NORTH EAST, *Erie Co.* (Pop., 1,396—1,528.) Built in '86 by town. **Supply.**—Springs, by gravity. Fiscal year closes May 1. **Distribution.**—(In '88.) *Mains,* 6 miles. *Taps,* 100. *Meters,* 9. *Hydrants,* 34. **Financial.**—*Cost,* $30,000. *Bonded Debt,* $20,000 at 5%, due in '95, or 1905, at option of town. *Expenses:* operating, none; int., $1,000; total, $1,000. *Revenue,* consumers, $1,700. **Management.**—Council. Prest., R. A. Davidson. Secy. and Supt., E. W. Merrill.

136. NORTH WALES, *Montgomery Co.* (Pop., 678—1,060.) Built in '86-8 by North Wales Water Co.; franchise neither controls rates nor gives city right to buy works. **Supply.**—Two artesian wells, pumping through main to tanks. Fiscal year closes third Saturday in November.

Year.	Mains.	Taps.	Meters.	Hydrants.	Cost.	Exp.	Rev.
88-89	2.5	76	0	20	$15,000	$290	$1,110
89-90	2.5	107	0	21	17,274	795	1,460

Financial.—*Cap Stock:* authorized, $16,000; paid-up, $15,000. Paid 3% dividend in 1890. *Floating Debt,* $1,500, at 5%. *Expenses:* operating, $750; int., $39; taxes, $15; total, $804. *Revenue:* consumers, $1,307; fire protection, $153; total, $1,460. **Management.**—Prest., Elias K. Freed. Secy., Henry S. Kriebel. Treas., Abel K. Shearer. Supt., Franklin T. Kriebel.

137. OIL CITY, *Venango Co.* (Pop., 7,315—10,932.) Built in '72 by city. **Supply.**—Allegheny River, filtered, pumping to reservoir, and springs, by gravity. River water pumped through sand filter. Fiscal year closes Mar. 31. In '90 a 1,500,000-gall. pumping engine and two 125,000-gall. steel tanks were added. Summer of '91 comrs. recommended new 7,400,000-gall. reservoir; est. cost, $38,600.

Year.	Mains.	Taps.	Meters.	Hydrants.	Cost.	B'd'd debt.	Exp.	Rev.
'80	10	1,000	..	72	$160,000	......	$6,000	......
'82	12	..	3	90	160,000	$93,000	......	......
'87	..	1,000	30	..	160,000	75,000	4,000	$17,789
'90-1	15	1,270	37	113	210,000	69,000	18,000	21,608

Financial.—*Total Debt,* $79,000: floating, $10,000 at 4%; bonded, $69,000 at 5%; due $3,000 on Sept. 1 of each year '91 to 1902, inclusive, and $36,000 in 1902. *No Sinking Fund,* bonds redeemed from taxes. *Expenses:* operating, $18,000; int., about $3,500;

total, about $21,500. *Revenue,* total, $21,608. No allowance for water for public purposes. **Management.**—Board. Prest., W. H. Loots. Clerk, A. M. Breckenridge. Supt., M. F. Johnson.

138. OLYPHANT, *Lackawanna Co.* (Pop., 2,604—4,083.) Neither sewers nor electric lights. **History.**—Built in '89 by Olyphant Water Co.; put in operation Jan. 1, 1890. Engr., W. M. Maple. **Supply.**—Mountain stream, by gravity to reservoir. **Reservoir.**—Cap., 2,000,000 galls. **Distribution.**—*Mains,* 10 to 6-in., 6 miles. *Taps,* 300. *Meters,* 20. *Hydrants,* Ludlow, 20. *Valves,* Ludlow. *Pressure:* ordinary, 140 lbs.; fire, 175 lbs. **Financial.**—*Cost,* withheld. *Bonded Debt,* $37,500 at 6%. *Ann. op. exp.,* $100. *Revenue:* consumers, $3,500; city, $500. **Management.** —Prest., Jas. Jordon. Secy., Jas. J. Lynch. Treas., Jno. T. Richards.

139. ORWIGSBURGH, *Schuylkill Co.* (Pop., 796—1,290.) Built in '85 by town. **Supply.**—By gravity. **Distribution.**—(In '87.) *Mains,* 2.5 miles. *Taps,* 250. *Hydrants* (in '86), 17. **Financial.**—(In '87.) *Cost,* $13,000. *Bonded Debt,* $13,000, at 4%. *Operating Expenses,* $400. *Revenue,* total, $1,100. **Management.**—Council. Supt., T. J. Reed.

140. OXFORD, *Chester Co.* (Pop., 1,502—1,711.) Owned by borough. Built in '68 by Oxford Gas and Water Co.; bought in '70 by borough for $30,000. **Supply.**— Two bored wells, pumping to reservoir. Fiscal year closes March 3. **Distribution.**—(Mar. 3, '90.) *Mains,* 4.11 miles. *Taps,* 131. *No Meters. Hydrants,* 25.

Year.	Mains.	Taps.	Meters.	Hyd'ts.	Cost.	B'd'd debt.	Int. ch'g's.	Exp.	Rev.
'30	2.5	110		18				600	
'82	4	300	0	35		28,000	1,400		
'87	3.87	141	0		44,283	37,400		950	1,575
'88	3.87		0	22	44,652	40,000	1,969	1,753	1,612
89-90	4.11	131	0	25	45,437	39,700	1,954	1,195	1,727

Financial.—(In '89-90.)—*Cost,* $45,437. *Debt, bonded,* $39,700; $12,200 at 7% due in '90; $27,500 at 4% due in 1906. *No Sinking Fund.* May 1, '90, $12,000 of 4½% bonds were issued, and the 7% bonds paid off. *Expenses:* operating, $1,195; int., $1,954; total, $3,149. *Revenue,* $1,727. **Management.**—Com. of three. Chn., Jeremiah King. Secy. Council, J. C. Kerr.

154. PALO ALTO, *Schulykill Co.* (Pop., 1,588-1,424.) Supply from Pottsville.

141. PARKERSBURG, *Chester Co.* (Pop., 817—1,514.) Partially sewered. No electric lights. Built in '89-90 by Parkersburgh Water Co.; put in operation June 1. **Supply.**—Artesian well, pumping to reservoir. **Pumping Machinery.**— Windmill pump. **Reservoir.**—Cap., about 290,000 galls.; is 106 × 56 × 7½ ft.; built by Co. **Distribution.**—*Mains,* 8 to 6-in., from R. D. Wood & Co., Philadelphia. *Taps,* 14. *Hydrants,* 12. **Financial.**—*Cost,* about $21,000. *Cap. Stock:* authorized, $20,000; all paid up. *Debt,* $1,300, floating, at 5%. *No Bonds.* Ann. op. exp., about $375. *Revenue:* consumers, $100; city, $1,300. **Management.**—Prest., Horace A. Beale. Secy., Chas. C. Owens. Supt., Lewis Reynolds.

142. PARKERS CITY, *Armstrong Co.* (Pop., 1,835—1,317.) Built in '72 by Parker's Landing & Laurenceburgh Water Co. **Supply.**—Allegheny River, pumping to tanks.

Year.	Mains.	Taps.	Meters.	Hydrants.	Exp.	Rev.
'80	2	200		37	$300	
'84	8	240	0	28	1,800	$4,000
'88	8	250	0	34	1,000	

Financial.—*First Cost,* $14,000; in '87, $20,000. *Cap. Stock* (in '88), $10,000. *Revenue,* city (in '87), $225. **Management.**—(In '89.) Prest., O. Tinsman. Secy., R. W. Moore. Supt., John Walker.

XX. PECKVILLE, *Lackawanna Co.* (Pop., est., 2,500.) Now owned by Peckville Water Supply Co. First supplied in '70; in '89-90 mains were laid by both Archibald and Olyphant Water Cos. In '90 reorganization of old Co. took place since which mains of Archibald and Olyphant Cos. have been bought and Olyphant pipes taken up. **Supply.**—Grassy Island Creek, by gravity. **Distribution.**—*Mains,* 10 to 4-in. c. i., 3 miles. *Taps* (Aug. 11, '91), 250. *Hydrants,* none, expect to and later. *Pressure,* 100 lbs. **Financial.**—(Aug. 11, '91.) *Cost,* $15,000. *Cap. Stock:* authorized, $20,000; paid up, $15,000. *Bonded Debt,* $7,500 at 6%. *Revenue:* consumers, $1,500; city, none. **Management.**—Prest., Thos. Law. Secy. and Treas., Jno. Carroll. Supt., S. W. Arnold.

FOR FURTHER DESCRIPTIVE MATTER see Manual for 1889-90. CHANGES since 1890 are here given. POPULATIONS are for 1880 and 1890, respectively.

136   MANUAL OF AMERICAN WATER-WORKS.

143. **PEN ARGYL,** *Northampton Co.* (Pop., 572—2,108.) Built in '89 by Pen Argyl Water Co.; city has no control over rates; franchise does not provide for purchase by city. Supply.—Spring, by gravity. Distribution.—(March 1, '90.) *Mains,* 3½ miles. *Hydrants,* 20. Financial.—*Hydrant Rental,* $25 each. Management.—(March 1, '90.) Secy., Treas. and Supt., W. H. Sanger.

66. **PENN** *Westmoreland Co.* (Pop., 604—931.) See Greensburgh.

213. **PENN TOWNSHIP,** *Westmoreland Co.* (Pop., 2,798—3,811.) Supply for part or all of township from Wilkinsburg.

144. **PETROLIA,** *Butler Co.* (Pop., 1,186—546.) Owned by T. Keighron; built in '74 by Co.; bought in '80 by Petrolia Water-Works Co., Ltd. Supply.—Artesian well, pumping to tank for domestic use and direct for fire protection. Distribution—(In '89.) *Mains,* 1 mile. *Taps,* 30. *No Meters. Hydrants,* 20. Financial.—*Cost* (in '82), $3,000. *No Debt* in '87. *Operating Expenses* (in '87), $1,700. *Revenue,* $1,000. Management.—Owner, T. Keighron. Address W. C. Foster.

145. **PHILADELPHIA,** *Philadelphia Co.* (Pop., 847,170—1,046,964.) Built in 1800-1 by city. Supply.—Schuylkill and Delaware rivers and springs, pumping to reservoirs, and direct; main supply from Schuylkill River. Fiscal year closes Dec. 31. Since '51 all turbines, pumps, etc., connected with Fairmount works have been furnished by Emile Geyelin, Philadelphia. Feb. 2, '91, bids were received for new 20,000,000-gall. pumping engine. In '89 the third section of the East Fairmount Park reservoir was completed. Close of '90 total reservoir cap. was 860,288,814 galls., and theoretical dy. cap. of pumps was 185,290,000, of which cap. 33,290,000 galls. is water power.

Year.	Mains.	Meters.	Hyd'ts.	Cons'ption.	Exp.	Rev.
'60	325	....	2,845	20,398,197	......	$552,532
'70	488	....	3,118	36,720,030	....	935,371
'80	716	....	5,833	57,707,082	$386,062	1,484,357
'81	754	....	6,014	62,249,355	450,963	1,509,541
'86	853	284	6,490	77,649,411	552,454	1,933,328
'87	876	253	6,919	87,949,972	590,339	2,030,034
'88	902	267	6,929	100,364,422	634,499	2,114,927
'89	930	301	7,433	115,557,209	708,847	2,242,000
'90	960	522	7,749	137,736,703	712,497	2,381,038

Distribution.—*Taps,* 170,911. Financial.—*Cost,* about $18,750,000. *Debt* (in '87), $6,401,800. Management.—Bureau of Water. Chief of Bureau, Jno. L. Ogden. Gen. Supt., F. L. Hand. *References.*—See '88 and '89-90 Manuals. Also "Civil Engineering of North America," pp. 189-195. By David Stevenson, in Weale's Rudimentary Series. London, 1859.

146. **PHILLIPSBURGH,** *Centre Co.* (Pop., 1,779—3,245.) Built in '81 by Phillipsburgh Water Co.; franchise neither controls rates nor gives city right to buy works. Supply.—Cold Stream, pumping to reservoir. Fiscal year closes June 1.

Year.	Mains.	Taps.	Meters.	Hyd'ts.	Consumption.	Cost.	Expenses.	Revenue.
'86	..	400	0	44	200,000	$40,000	$2,400	$5,000
'88	6.5	450	..	48	400,000	40,000	......	7,395
'90	7	475	3	47	500,000	50,000	2,000	9,025

Financial.—*Cap. Stock:* authorized, $40,000; all paid up; paid 6% div'd in '90. *No Debt. Expenses:* operating, $2,000; taxes, $180; total, $2,180. *Revenue,* total, $9,025, including $705 from city for fire protection. Management.—Prest., J. N. Casanova. Secy., W. E. Irwin. Treas., O. Perry Jones. Supt., Henry Southard.

147. **PHŒNIXVILLE,** *Chester Co.* (Pop., 6,682—8,514.) Built in '72-3 by borough. Supply.—Schuylkill River, pumping to reservoir. Fiscal year closes March 1.

Year.	Mains.	Taps.	Meters.	Hyd'ts.	Cons'pt'n.	B'd d't.	Int.ch'g's.	Exp.	Rev.
'74	....	......	.....	....	....	$131,038	$10,460	$3,420	$6,488
'75	....	......	.....	....	....	182,000	13,505	3,342	8,610
'76	....	......	.....	....	....	191,000	14,126	3,034	8,071
'77	....	......	.....	....	312,852	194,000	13,974	3,029	8,589
'78	....	......	.....	....	375,000	194,000	14,240	2,589	8,744
'79	....	......	.....	....	363,900	194,000	14,250	2,868	9,110
'80	....	453	.....	126	354,700	194,000	13,060	2,842	8,685
'81	....	......	.....	....	387,280	194,000	14,107	3,265	9,545
'82	10.2	490	54	362	394,540	194,000	14,135	3,383	9,424
'83	....	......	.....	....	476,687	194,000	15,219	2,859	11,174
'84	10.8	761	5	125	533,753	194,000	9,675	3,191	11,908
'85	11.	802	0	125	552,769	191,000	9,382	2,390	9,186
'86	....	......	.....	....	560,471	....	9,351	3,205	13,963
'87	12.1	......	4	144	621,700	192,200	9,874	2,865	12,938
'88	13.	922	4	144	500,000	192,200	7,688	13,169	12,938
'90	13.	1,000	4	150	800,000	188,200	9,377	7,806	20,589

Financial.—*Cost*, $195,000. *To'al Debt*, $188,200, bonded at 5%. *Sinking Fund*, $760. *Expenses*: operating, $7,806; int., $9,377; State tax, $546; total, $17,730. *Revenue*: consumers, $20,589; city, fire protection, $6 per hydrant. **Management.**—Prest., W. H. Bitting. Supt., A. S. Vanderslice.

148–149. **PITTSBURGH**, *Allegheny Co.* (Pop., 156,389–238,617.)
CITY WORKS.
Built in '26 by city. **Supply.**—Allegheny River, pumping to reservoirs. Fiscal year closes Jan. 31.

Year.	Mains.	Taps.	Meters.	Hyd'ts.	Cons'pt'n.	Cost.	B'd'd	d't.Int.ch'g's.	Exp.	Rev.
'80	111	1	...	...	912	16.021.624	....	......	$85,617	$302,085
'84	134.4	11,000	0	848	20,180,000	....	4,582,500	320,775	140,000	400,000
'86	147	11,600	6	1,051	21,000,000	....	4,500,000	312,000	118,310	132,230
'87	175	16,000	27	1,200	30,000,000	4,500,000	4,500,000	......	198,000	510,000
'88	100	17,081	45	1,612	34,000,000	4,752,000	4,752,000	329,640	100,000	517,008
'89	192	17,744	43	1,532	35,000,000	4,582,000	4,582,500	317,775	78,010	541,909

Financial.—(In '89.) *Bonded Debt*, $4,582,500: $4,282,500 at 7%; $300,000 at 6%. *Sinking Fund*, $1,356,033. **Management.**—Chief Dept. of Pub. Wks., E. M. Bigelow, appointed by Council. Secy., L. A. Denison. Supt., G. H. Brown.

MONONGAHELA WATER CO.
Built in '65. **Supply.**—Monongahela River, by pumping. *Consumers*, about 10,000. **Management.**—Prest., M. W. Watson. Secy. and Gen. Mgr., Martin Prentz. Treas. Henry Stamm.

150. **PITTSTON** *Luzerne Co.* (Pop., 7,472–10,302.) Built in 55-7 by Pittston Water Co., now owned by People's or Citizens' Water Co.; supplies W. Pittston; city has no control over rates; franchise does not provide for purchase by city. **Supply.**, Susquehanna River, pumping to reservoirs.

Year.	Mains.	Taps.	Meters.	Hydrants.	Cost.	B'd'd Debt.	Int Ch'g's.	Exp.	Rev.
'82	15	1,200	1	27	$106,593	$14,000	$840	$7,300	$18,000
'84	11.5	....	3	22	114,000	......	......	5,500	24,028
'87	21	....	...	28*	114,000	......	2,760	8,000	28,350
				*35 in '90.					

Financial.—(In '87.) *Cap. Stock*, $76,400. *Debt*, $69,000 at 4%. *Expenses*: operating, $8,000; int., $2,760. *Revenue*: consumers, about $28,000; city, fire protection, $350 in '87; free in '90. **Management.**—Prest., L. D. Shoemaker. Secy., G. F. Nesbit. Treas., Abram Nesbit. Supt., Thos. Monie.

122. **PLEASANT VALLEY** *(Avoca P. O.), Bucks Co.* Supply from Moosic.

151. **PLYMOUTH**, *Luzerne Co.* (Pop., 6,065–9,344.) Built in '75-6 by Plymouth Water Co.; city does not control rates but has right to purchase works after 20 years. **Supply.**—Surface water and springs, by gravity from impounding reservoir; artesian well and Susquehanna River, pumping to reservoir. Fiscal year closes Dec. 21. Additional supply was introduced in '90 from Spring Brook Water Co., Kingston. Cap. stock was increased to $25,000.

Year.	Mains.	Taps.	Meters.	Hydrants.	Consumption.	Cost.	Exp.	Rev.
'80	12	600	..	1	......	$50,000	$1.20	$3,600
'86	15	850	0	0	450,000	85,000	3,675	....
'87	15	900	4	40	600,000	100,000	4,000	15,000
'88	15	1,000	4	40	650,000	100,000	4,500	15,100
'89	18	1,012	5	41	750,000	121,828	3,300	20,023

Financial.—(In '89.) *Cap. Stock*: authorized, $100,000; all paid up; paid 10% div'd in '89. *Floating Debt*, $8,000 at 6%. *Expenses*: operating, $3,300; int., $480; total, $3,780. *Revenue*: consumers, $19,000; fire protection, $1,023; total, $20,023. **Management.**—Prest., Draper Smith. Secy. and Treas., J. W. Chamberlain. Supt., O. M. Lance.

152. **PORT ALLEGANY**, *McKean Co.* (Pop., 731–1,230.) Built in '85-6 by Port Allegany Water Co.; put in operation in Jan. Engr., W. E. Smith, present Supt. Contr., F. H. Steck. **Supply.**—Streams, by gravity. *Reservoir.*—Cap., about 180,000 galls.; 200 ft. above village; bottom cemented. *Distribution.*—*Mains*, 5 miles. *Services*, galv. i. *Takers*, 55. *Hydrants*, 20. *Valves*, Eddy. *Pressure*, 90 lbs. *Financial.*—*Cost* $20,000. *Capital stock*, authorized and paid up, $10,000. *Bonds*, authorized and issued, $7,500 at 6%. *Revenue*: consumers, $1,000; city, $200. **Management.**—Prest., Thos. McDowell. Secy., S. W. Smith. Treas., J. H. Williams. Supt., E. W. Smith.

154. **PORT CARBON**, *Schuylkill Co.* (Pop., 2,346–1,976.) Supply from Pottsville.

For Further Descriptive Matter see Manual for 1889-90. Changes since 1890 are here given. Populations are for 1880 and 1890, respectively.

138   MANUAL OF AMERICAN WATER-WORKS.

201. **PORT PERRY**, *Allegheny Co.* (Pop., 1,100—1,031.) Supply from Wall.

153. **POTTSTOWN**, *Montgomery Co.* (Pop., 5,305—13,285.) Built in '69-71 by Pottstown Gas & Water Co. Supply.—Schuylkill River, pumping to reservoir. Fiscal year closes May 31. Unofficially reported that in '91 intake was changed from point below to one above river and 5,000,000-gall. reservoir added.

Year.	Mains.	Taps.	Meters.	Hyd'ts.	Cons'pt'n.	Cost.	B'dd	d't.Int.chgs.	Exp.	Rev.
'80	7	490		45					$3,000	
'81	8	623	5	60	155,000	$90,000	$18,000	$900	$3,971	$7,951
'87	15	1,200	6	79	450,000	125,000	10,000	450		
'90-91	19.6			5						

Financial.—(In '87.) *Cap. Stock*, $90,000. *Bonded Debt*, $10,000 at 4½%. Management.—Prest., G. B. Lessig. Secy. and Treas., J. H. Maxwell.

154. **POTTSVILLE**, *Schuylkill Co.* (Pop., 13,252—14,117.) Built in '34-6 by Pottsville Water Co. Supplies St. Clair, Palo Alto and Port Carbon; city has no control over rates, and franchise does not provide for purchase of works by it. Supply.—Mountain streams, by gravity from impounding to distributing reservoir. Fiscal year closes Dec. 31. Distribution.—*Mains*, 38 miles. *Meters*, 15. *Hydrants*, 113. Financial.—*Cap. Stock:* authorized, $200,000; all paid up. *Bonded Debt*, $57,000 at 4%. No hydrant rental. Management.—Prest., D. W. Bland. Secy., W. D. Pollard.

155. **PUNXSUTAWNEY**, *Jefferson Co.* (Pop., 674—2,792.) Built in '87 by Punxsutawney Water Co.; city has no control over rates; franchise does not provide for purchase by city. Supply.—Mahoning Creek, pumping to tank; filtered through stone and charcoal.

Year.	Mains.	Taps.	Meters.	Hydrants.	Consumption.	Cost.	Exp.	Rev.
'87	3	100	0	13	70,000	$15,000	$1,050	$1,795
'88	4	135	2	20	175,000	16,000	1,000	1,575
'90	4	175	2	20	225,000	19,000	1,400	3,250

Financial.—*Cap. Stock:* authorized, $15,000. *Total Debt*, $15,000, floating at 6%. Revenue: consumers, $2,900; city, fire protection, $350; total, $3,250. Hydrant rental, $17.10 until '93, then $25 each. Management.—Secy., G. G. Downes. Prest., J. A. Weber. Supt., C. E. Ratz. Treas., L. Pantall.

156. **READING**, *Berks Co.* (Pop., 43,278—58,661.) Now owned by city; built in '19-21 by Reading City Water Co.; bought by city in '65. Supply.—Springs and creeks, by gravity, with reserve from Maiden Creek, by pumping. Fiscal year closes Apr. 4. Additional supply introduced in '90-91 from Maiden Creek, a branch of the Schuylkill River, 7 miles north of Reading. Pumping Machinery.—Dy. cap., 5,000,-000 galls.; Worthington. Supply Main.—30, 24 and 20-in. pipe; cost, $157,500. Reservoirs now have total storage cap. of about 175,000,000 galls. July 14, $150,000 of bonds were voted to lay larger mains and connect Maiden Creek pumping station with Hampden reservoir. Distribution.—*Mains*, 90 miles; in '65, 12.7 miles; in '70, 28.9 miles; in '74, 36.6 miles; in '85, 39.8 miles. *Taps*, about 10,000. *Meters*, 6. *Hydrants*, 500. *Consumption*, 5,000,000 galls. Financial.—*Cost*, $1,250,000. No floating debt.

Year.	Mains.	Taps.	Mts.	Hydt's.	Cons'pt'n.	Cost.	B'dd	d't. Int.chgs.	Exp.	Rev.
'84	38.8	8,500	30	400	3,000,000	$342,866	$479,125		$39,673	$72,521
'86							475,500			
'87	44		11	379		1,166,405	471,500		24,585	83,421
'88			30						22,120	82,014	108,795
'90-91	90	10,000	6	500	5,000,000	1,250,000	465,000	38,858	32,331	132,057

Bonded Debt, $465,000: $57,500 due Jan. 1, 1900; $20,000 due Jan. 1, 1905; $63,000 due Jan. 1, 1906; $125,000 due Jan. 1, 1919; $200,000 due July 1, 1920; all at 4%. *Sinking Fund*, $57,310: $16,000 invested in 6% bonds; $21,000 do., 5%; $15,000 do., 4% and $5,310 in cash. *Expenses:* operating, $32,331; int., $33,858; total, $66,189. *Revenue:* consumers, $132,057; city, none. Rev. in '65, $11,346; in '70, $37,222; in '75, $61,519; in '80 $67,047; in '85, $73,432. Management.—Four comrs. Prest., A. A. Heizmain. Secy., E. A. Howell. City Treas., Jno. Obold. Supt., W. B. Harper.

157. **RENOVO**, *Clinton Co.* (Pop., 3,708—4,154.) Built in '72-3 by city. Supply. —Peters' run, by gravity, with pumping when water is low. May, '91, new dam was being built; unofficially reported that dam 60 ft. long and 10 ft. high was to be built, and 1,000 ft. 8-in. c. i. pipe laid to connect reservoir formed with new supply. Distribution.—(In '88.) *Mains*, 5 miles. *Taps*, 475. *No Meters*. *Hydrants*, 62. Financial.—(In '88.) *Cost*, $57,000. *Debt*, $16,350 at 4 and 4½%. *Operating Expenses*, $125. *Revenue*, consumers, $2,850. Management.—Address R. N. Martin.

PENNSYLVANIA. 139

158. REYNOLDSVILLE, *Jefferson Co.* (Pop., 1,410—2,789.) Built in '88 by Reynoldsville Water Co.; franchise neither controls rates nor gives city right to buy works. Supply.—Springs, pumping to tank. Fiscal year closes Sept. 30. Distribution.—*Mains,* 3.5 miles. *Taps,* 30. *Hydrants,* 18. Financial.—*Cost,* $13,000. *No Debt. Financial.—Cap. Stock:* authorized, $12,000, all paid up; paid 3% div'd in '90. *Expenses:* operating, $1,000; taxes, $50; total, $1,050. *Revenue,* from city, $150. Management.—Prest., Geo. Millinger. Treas., A. Reynolds. Secy., M. M. Davis. Supt., M. S. Sterley.

159. REYNOLDTON, (*McKeesport P. O.*), *Allegheny Co.* (Pop., in '90, 1,379.) Built in '89 by borough. Supply.—McKeesport Water-Works. Fiscal year closes Jan. 16.

Year.	Mains.	Taps.	Meters.	Hydr'ts.	Cost.	B'd'd debt.	Int. ch'g's.
'89	.75	..	1	21	$10,000	$10,000	$450
'90	2.1	85	0	21	11,000	11,000	495

Financial.—*Bonded Debt,* $11,000 at 4½%, due in 1911, *Expenses:* operating, nominal; interest, $495. *Revenue:* consumers, $700; fire protection, $40 per hydrant. Management.—Burgess, J. A. Geeting. Secy. and Supt., O. H. Osborne, Hero.

160. RIDGWAY, *Elk Co.* (Pop., 1,100—1,903) Built in '68 by city. Supply.—Mountain streams, by gravity. Fiscal year closes March 1. Distribution.—*Mains,* about 5 miles. *Taps,* 125. *Hydrants,* 30. Financial.—(In '88.) *Cost,* $23,879. *Bonded Debt,* $23,879, at 4%. Management.—Secy., W. C. Healy. Supt., T. J. Barry.

161. RIDLEY PARK, *Delaware Co.* (Pop., 2,533—4,529.) Built in '90 by Ridley Park Cold Spring Water Co. Supply.—Artesian wells and lake, pumping direct and to tank. Fiscal year closes May 31. Distribution.—*Mains,* 3 miles. *Taps,* 70. *No Meters. Hydrants,* 19. *Consumption,* 10,000 galls. Financial.—*Cost,* $26,000. *Cap. Stock:* authorized, $25,000; paid up, $8,000. *No debt. Expenses:* operating, $1,600; no int.; taxes withheld. *Revenue:* consumers, $1,525; city, fire protection, $475; total, $2,000. Management.—Prest., H. C. Keyes. Secy., Charles S. Salin. Treas., F. E. Harrison. Supt., J. B. Coleman.

162. ROCHESTER, *Beaver Co.* (Pop., 2,552—3,649.) Built in '90-1 by Valley Water Co.; was to be in operation April 1. Connected with Union Water Co., Beaver Falls. Supply.—Beaver River, direct pumping. Management.—Prest., H. W. Hartman, Beaver Falls.

163. ROYERSFORD and SPRING CITY, *Montgomery* and *Chester Co.* (Pop.: Royersford, 658—1,815; Spring City, 1,112—1,797.) Built in '89-90 by Home Water Co.; city has no control over rates, nor does franchise provide for purchase of works by city. Supply.—Artesian wells, pumping to stand-pipe. Distribution.—*Mains,* 6 miles. *Hydrants,* 15. Financial.—*Cost,* $40,000. *Cap. Stock,* authorized $60,000. *No debt.* Hydrant rental, $20. Management.—Prest., J. A. Buckwater. Secy., U. S. G. Finkbiner. Treas., S. H. Latshow.

154. SAINT CLAIR, *Schuylkill Co.* (Pop., 4,149—3,680.) Supply from Pottsville.

164. SAINT MARY, *Elk Co.* (Pop., 1,501—1,745.) Built in '89 by St. Mary's Water Co. Supply.—Silver Creek, pumping to reservoir. Fiscal year closes July 1. Distribution.—*Mains,* 6.5 miles. *Taps,* 150. *No Meters. Hydrants,* 40. *Consumption,* 20,000 galls. Financial.—*Cost,* $25,000. *Cap. Stock:* authorized, $40,000; paid up, $25,000. *No Debt. Expenses:* operating, $600; taxes, $90; total, $690. *Revenue:* consumers, $1,400; city, fire protection, $1,000; total, $2,400. Management.—Prest., J. K. P. Hall. Secy., Geo. Weidenboemer. Treas., Chas. Luke.

165. SAINT PETERSBURGH, *Clarion Co.* (Pop., 1,044—655.) Built in '75 by borough. Supply.—Spring, 1½ miles distant, by gravity and well, ¼ mile distant. There are three tanks; combined cap., 140,000 galls; two, 50,000, one 40,000. Fiscal year closes March 1. Distribution.—*Mains,* 3 miles; 2 miles 4-in. c. i., 1 mile 2-in. w. i. *Taps,* 150. *Hydrants,* 2-in , 12. *Consumption,* 12,000 galls. *Pressure,* 60 lbs. Financial.—*Cost,* about $10,000. *No Debt. Operating Expenses,* $498. *Revenue:* consumers, $688; city, none. Management.—Com. of three, including J. A. Dittman. Secy., C. O. Davis. Supt., Josiah Fillman.

166. SAYRE and ATHENS, *Bradford Co.* (Pop., Sayre, in '80, 729; Athens 1,592—3,274.) Built in '84 by Sayre Land Co., and leased to Sayre Water Co.;

FOR FURTHER DESCRIPTIVE MATTER see Manual for 1889-90. CHANGES since 1890 are here given. POPULATIONS are for 1880 and 1890, respectively.

140   MANUAL OF AMERICAN WATER-WORKS.

franchise neither controls rates nor gives city right to buy works. Supply.—Susquehanna River, pumping to reservoir. Fiscal year closes May 1.

Year.	Mains.	Taps.	Meters.	Hydrants.	Consumption.
'86-7	7	269	12	38	450,000
'87-8	9	330	25	46	400,000
'88-9	10.4	385	24	48	450,000
'89-90	1.3	510	40	44	500,000
'90-1	13.5	521	43	....	......

Financial.—*No Debt* in '88; no floating debt June 10, '91. Hydrant rental, $20. **Management.**—Prest., Howard Elmer. Secy. and Treas., Wm. Stevenson. Supt., A. B. Shearer.

167. **SCHUYLKILL HAVEN**, *Schuylkill Co.* (Pop.: 3,052–3,088.) Built in '84-5 by Schuylkill Haven Gas & Water Co.; franchise neither controls rates nor gives city right to buy works. **Supply.**—Springs, four miles from town, by gravity. Fiscal year closes Dec. 31. Reservoir built since '89.

Year.	Mains.	Taps.	Hydrants.	Cost.	B'd'd Debt.	Int. ch'g's	Exp.	Rev.
'87	9	300	26	$60,000	......	......	$2,000	......
'88	9	300	27	75,000	$5,000	$360	......	......
'90	13	325	28	......	......	......	......	$4,624

Financial.—*Revenue:* consumers, $2,600; city, fire protection, $224; other purposes, $1,800; total, $4,624. **Management.**—(Dec., '90.) Supt., W. H. Mellon.

168. **SCOTTDALE**, *Westmoreland Co.* (Pop., 1,275–2,693.) Built in '90 by Citizens Water Co. under 25 years franchise; put in operation Jan. 1, '91. Des. Engr., C. W. Knight, Rome. Const. Engr., W. W. Jamison, Greensburg. Contrs., Stark Bros., Greensburg. **Supply.**—Stream, by gravity. **Reservoir.**—Cap., 60,000,000 galls. **Distribution.**—*Mains*, 12 to 4-in. c. i. 13 miles. *Services*, lead. *Taps*, 150. *No Meters. Hydrants*, Mathews, 35. *Valves*, Eddy. *Pressure*, 80 lbs. **Financial.**—*Cost*, $150,000. *Cap. Stock:* authorized, $150,000; paid up, $10,000. *Bonds:* authorized, $150,000; issued, $140,000 at 6%. Hydrant rental, $10. **Management.**—Prest., W. W. Jamison, Greensburg. Secy. and Treas., P. W. Janeway, 243 S. Third St., Philadelphia. Supt., S. F. Potter.

169, 170. **SCRANTON**, *Lackawana Co.* (Pop., 45,850–75,215.)

PROVIDENCE GAS AND WATER CO.

Built in '67-8. **Supply.**—Storage reservoirs on streams, by gravity. Fiscal year closes April 1. Reported in '91 that 3,000,000-gall. reservoir was to be added.

Year.	Mains.	Taps.	Meters.	Hydrants.	Exp.	Rev.
'80	27.5	....	....	115	$8,000	......
'81	29.3	1,047	0	106	......	$3,000*
'87	30	1,324	0	60	2,000	3,500
'88	35	1,514	0	70	2,500	26,000
'90-1	35.5	1,764	0	85	2,750	26,330

*From city.

Financial.—*Revenue:* Consumers, $25,330; fire protection, $1,000; total, 26,330; other information withheld. **Management.**—Prest., W. R. Storrs. Secy., H. F. Atherton. Supt., J. B. Fish.

SCRANTON GAS AND WATER CO.

Built in '57. **Supply.**—Surface water, by gravity from impounding reservoir, with small reservoir and stand-pipe as equalizers of pressure. **Distribution.**—(In '81.) *Mains*, 29.3 miles. *Taps*, 1,047. *Hydrants*, 106. **Financial.**—(In '81.) *Cost*, gas and water-works, $680,000. *Cap. Stock*, $680,000. *No Debt. Revenue:* city, $3,000. **Management.**—(In '88.) Prest., W. W. Scranton. Secy. and Treas., G. B. Hand. NOTE.—No later information can be secured.

171. **SELIN'S GROVE**, *Snyder Co.* (Pop., 1,431–1,315.) Built in '85-6 by Selin's Grove Water-Works Co. **Supply.**—Penn's Creek, pumping to reservoir and direct. **Distribution.**—(In '88.) *Mains*, 5 miles. *Hydrants*, 30. *Consumption*, 100,000 galls. **Financial.**—(In '88.) *Bonded Debt*, $75,000, at 5%. *Operating Expenses*, $600. *Hydrant Rental*, $40 each for 25, $30 each for remainder. **Management.**—(In '88.) Prest., H. B. Melick. Secy. and Treas., H. Hinckley. NOTE.—Later information refused.

172. **SEWICKLEY**, *Allegheny Co.* (Pop., 2,053). Built in '70-1 by borough. **Supply.**—Springs, by gravity to and from reservoir, and, in dry weather, pumping from wells near river.

PENNSYLVANIA. 141

Year	Mains.	Taps.	Meters.	Hydrants.	Cost.	B'd'dDebt.	Int. Ch'g's.	Exp.	Rev.
'82	..	312	0	38	$75,000	$74,000	$5,640	$900	$4,293
'83	..	..	..	..	....	....	....	....	4,007
'84	..	384	0	36	75,000	75,000	5,700	1,000	4,208
'85	..	..	..	..	....	....	....	....	5,168
'86	..	425	1	36	....	21,000	1,380	....	5,829
'87	..	425	1	30	95,000	75,000	5,700	950	6,860
'88	9	443	1	38	100,000	75,000	5,700	2,147	6,802

Financial.—(In '88.) *Total Debt*, $75,000: $60,000 at 8%, $15,000 at 6%. *Revenue:* consumers, $6,602; city, none; ice, $130; total, $6,732. Revenue from '75 to '81, inclusive has been as follows for successive years: $1,401, $2,311, $3,192, $3,025, $3,400, $3,641, $3,522. **Management.**—(In '88.) Five comrs. Secy., Thos. Dickson. Supt., John Patton, Jr.

173. **SHAMOKIN BOROUGH and COAL TOWNSHIP,** *Northumberland Co.* (Pop., Shamokin, 8,124—14,403; Coal, 4,320—8,616.) Built in '73-4 by Shamokin Water Co.; city neither controls rates nor has right to buy works. See '89-90 Manual, p. 272. **Supply.**—Mountain streams, by gravity. Fiscal year closes April 1. In '88-90 Manual, Coal Water Co. should read Anthracite Water Co.; Roaring Run should read Roaring Creek.

Year.	Mains.	Taps.	Mt's.	Hy'ts.	Con'pt'n.	Cost.	B'd't d'bt.	Int. ch's.	Exp.	Rev.
'80-1	9	1,000	0	20	....	....	....	....	$2,000	...
'84-5	10	...	0	25	230,000	$110,000	none.	none.	1,500	$12,000
'87-8	35	1,100	4	43	1,000,000	450,000	$124,000	....	5,000	860*
'88-9	35	1,300	4	43	2,000,000	500,000	124,000	....	6,000	860*
'89-90	35	1,400	5	47	2,250,000	525,000	124,000	....	....	2,000*

'90-1 Report states no change for year. *From city.

Management.—Prest., W. C. McConnell. Secy., Geo. O. Martz. Treas., C. L. McWilliams.

174. **SHARON,** *Mercer Co.* (Pop., 6,684—7,459.) Built in '85 by Sharon Water-Works Co.; rates are fixed by council and water co. jointly. G. H. Pierson, N. Y. City, states that he was Engr. **Supply.**—Chenango River, pumping to reservoir. Fiscal year closes Dec. 31.

Year.	Mains.	Taps.	M't'rs.	Hyd'ts.	Con'tn.	Cost.	B'd'd debt.	Int. ch'gs.	Exp.	Rev.
'86	8.5	312	10	80	300,000	$165,000	$100,000	$6,000	$3,200	$10,215
'87	10	398	4	80	....	140,000	100,000	6,000	3,000	11,150
'88	10	..	..	80	....	300,000	100,000	6,000	....	4,000*
'89	10	590	1	80	750,000	300,000	100,000	6,000	6,140	9,700
'90	11	700	2	89	500,000	300,000	100,000	6,000	....	12,000

*From city.

Financial.—*Cap. Stock*, authorized, $200,000. *Bonded Debt*, $100,000 at 6%, due in 1905. *Expenses*, $10,500. **Management.**—Prest., H. S. Hopper. Secy., W. Owlsey. Treas., Walter Wood. Supt., E. J. Robinson.

175. **SHARPSBURGH,** *Allegheny Co.* (Pop., 3,466—4,898.) Built in '86 by borough. **Supply.**—Mountain springs, by gravity. Fiscal year closes Nov. 31.

Year.	Mains.	Taps.	Meters.	Hydrants.	Cost.	B'd'd debt.	Int. charges.	Exp.	Rev.
'87-8	4	450	..	42	$52,000	$60,000	$3,000	$3,000	$5,500
'88-9	7.5	500	..	41	54,000	50,000	2,500	3,400	6,500
'90-1	7.5	609	..	41	....	....	....	4,200	....

Financial.—(In '88.) Int., 5%. **Management.**—Supt., Robt. Beattie. Report by Wm. Weckbecker.

176. **SHENANDOAH,** *Schuylkill Co.* (Pop., 10,147—15,944.) Built in '70 by Citizens' Water & Gas Co.; franchise neither controls rates nor gives city right to buy works. **Supply.**—Mountain springs, by gravity. Fiscal year closes Oct. 1. **Distribution.**—*Mains*, 10 to 3-in., c. i., 8 miles. *Meters*, 1. *Hydrants*, 29. *Pressure*, 90 lbs. **Financial.**—*Cost*, $154,000. *Cap. Stock*, authorized, $60,000. *No Debt. Operating Expenses*, $154. *Revenue*, withheld. **Management.**—Prest., John Cather. Secy., J. O. Roads. Treas., E. J. Wesley. Supt., S. D. Hess.

XXI. **SHICKSHINNY,** *Luzerne Co.* (Pop., 1,058—1,448.) Built in '84 by Shickshinny Water Co. **Supply.**—Shickshinny Creek, fed by springs, by gravity. Fiscal year closes Nov. 1. **Distribution.**—*Mains*, 2½ miles. *Taps*, 275. *No Meters. Hydrants*, 1. **Financial.**—*Cost and Cap. Stock*, $12,000. *No Debt. Operating Expenses*, $150. *Revenue*, about $1,200. **Management.**—Prest., G. W. Search. Secy. and Supt., M. B. Hughes. Treas., Jesse Beadle.

For Further Descriptive Matter see Manual for 1889-90. Changes since 1890 are here given. Populations are for 1880 and 1890, respectively.

177. SHIPPENSBURGH, *Cumberland Co.* (Pop., 2,213—2,188.) Built in '86 by borough. Supply.—Mountain springs, by gravity. Fiscal year closes March 1.

Years.	Mains.	Taps.	Hydrants.	Cons'ption.	Cost.	B'd'd d't.	Int.ch'gs.	Exp.	Rev.
'86-7	8	100	39	......	$76,000	$26,000	$1,440	...	...
'87-8	8	200	39	300,000	29,000	26,000	1,410	$125	$1,400
'88-9	8	211	42	100,000	29,000	26,000	1,410	150	1,500
'89-90	8	220	45	400,000	29,500	26,000	1,410	250	1,500
'90-1	8	5	260	45	500,000	30,000	26,000	1,410	200	1,650

Financial.—*Bonded Debt*, $26,000, at 4%. Management.—Six members of council, elected every three years. Supt., B. F. Landis, appointed by council.

XXII. SHREWSBURY, *York Co.* (Pop., 565—562.) Built in '91 by Shrewsbury Water Co. Supply.—Well, pumping to tank. Financial.—*Cost*, $2,000. *Cap. Stock*, authorized, $2,500. Management.—Address Jas. Gerry.

178. SLATINGTON, *Lehigh Co.* (Pop., 1,634—2,716.) Now owned by borough; built in '61 by Slatington Water Co., bought in '83 and extended by borough. Supply.—Mountain springs, by gravity. Fiscal closes Mar. 1.

Year.	Mains.	Taps.	Meters.	Hyd'ts.	Cons'pt'n.	Cost.	B'd'd Debt.	Int.ch'gs.	Exp.	Rev.
'81	3	150	0	0	......	$10,000	$3,500	$210	$100	$9,000
'85	4	200	0	10	40,000	......	20,000	......	100	2,000
'87	6	312	0	25	......	26,396	25,000	1,040	100	2,400
'88	6.5	337	1	28	......	26,635	26,635	1,065	150	2,381

'90 report states that there are no changes since '88.

Financial.—*First Cost*, $8,000. All expenses paid from revenue. Management.—Com. appointed every year by prest. of council. Chn., J. F. Berkemeyer. Secy. and Supt., G. McDowell.

179. SMETHPORT, *McKean Co.* (Pop., 872—1,150.) Built in '81 by Smethport Water Co.; franchise neither controls rates nor provides for purchase of works by city. Supply.—Springs, by gravity. Fiscal year closes Sept. 30. Distribution.—*Mains*, 6 miles. *Taps*, 200. *Meters*, 0. *Hydrants*, 30. Financial.—*Cost*, $23,500. *Cap. stock*: authorized, $30,000; paid up, $23,500; 10% div'd in 1890. *No debt. Expenses*: operating, $420; taxes, $187. *Revenue*: city, fire protection, $300. Management.—Prest., Henry Hamlin. Secy., Treas. and Supt., A. B. Armstrong.

180. SOUTH BETHLEHEM and WEST BETHLEHEM, *Northampton Co.* (Pop.: So. Bethlehem, 4,925—10,302; West Bethlehem, not given.) First supplied in '75. Built in '85-6 by South Bethlehem Gas & Water Co. Supply.—Lehigh River, pumping to reservoir. Fiscal year closes Dec. 31. Unofficially reported that reservoir of brick, 69 ft. in diam. by 18½ ft. deep, was completed in Aug., '91. Apr., '91, contract for new 10,000,000-gall. reservoir was let to Thos. F. Kerns, Reading. Engr., J. Marshall Young, Easton.

Year.	Mains.	Taps.	Meters.	Hyd'ts.	Consump'n.	Cost.	Bd'd debt.
'87	12.5	550	1	56	600,000	$147,589	$34,500
'58	14	714	3	72	700,000	150,000	34,500
'90	14.8	900	..	72	844,200	160,954	34,500

Financial.—*Cap. stock*: authorized, $140,000, all paid up; paid 6% div'd in 1890. *Bonded debt*, $34,500, at 7%, due in 1905. Stock and debt include both gas and water plant. *Expenses and revenue*, withheld. Management.—Prest. E. P. Wilbur. Secy., C H. Weisser. Treas., H. S. Goodwin. Supt., B. Lehmen. NOTE.—Aug., '91, there was talk of purchase of works by borough.

181. SOUTH EASTON, *Northampton Co.* (Pop., 4,534—5,616.) Built in '86-8 by South Easton Water Co.; franchise neither controls rates nor gives city right to purchase works. Supply.—Tunnel into mountain, by gravity. Fiscal year closes Dec. 1. Distribution.—*Mains*, 68 miles. *Taps*, 250. *Meters*, 0. *Hydrants*, 19. *Consumption*, 60,000 galls. Financial.—*Cost*, $40,000. *Cap Stock*: authorized, $20,000, all paid up. *Total Debt*, $20,000; floating, $10,000 at 6%; bonded, $10,000 at 5%, due in 1905. *Expenses*: operating, $100, int., $600; taxes, $40; total, $740. *Revenue*: consumers, $1,325; fire protection, $475; total, $1,800. Management.—Prest., W. F. Pascoe. Secy., J. F. Andrews. Treas., Jno. Stewart. Supts., McKeen & Coyle.

163. SPRING CITY, *Chester Co.* (Pop., 112—1,797.) Supply from Boyersford.

182. STROUDSBURG, *Monroe Co.* (Pop., 1,860—2,419.) Built in '76 by Stroudsburg Water Co. Supply.—Springs, by gravity. Fiscal year closes Oct. 1.

Year.	Mains.	Taps.	Meters.	Hyd'ts.	Cost.	Exp.	Rev.
'79-80	1	......	..	15	$21	......	
'81-2	3	75	0	14	$16,000	138	$1,084
'85-6	3	80	0	15	......	270	1,691
'89-90	5	200	0	25	17,000	214	1,400

PENNSYLVANIA. 143

Financial.—*Cap. Stock :* authorized, $15,000; all paid up; paid 6% div'd in 1890. *No Debt. Expenses :* operating, $150; taxes, $64; total, $214. *Revenue :* consumers, $1,400; fire protection, withhold. **Management.**—Prest., J. B. Stone. Secy., J. S. Brown. Treas., A. N. Snoover. Supt., Fred. Phillips.NOTE.—July, '91, reported that people would soon vote on construction of $10,000 works.

213. **STERRIT TOWNSHIP,** *Allegheny Co.* (Pop., 2,356—1,182.) See Wilkinsburgh.

183. **SUMMIT HILL,** *Carbon Co.* (Pop., 1,763—2,816.) Built in '76 by Summit Hill Water Co Supply.—Mountain stream, pumping to reservoir. **Distribution.**—(In '87.) *Mains,* 4 miles. *Taps.* 175. *No Meters. Hydrants,* 18. **Financial.**—*First Cost,* $12,000. *Cap. Stock,* (in '83), $10,000. *No Debt* in '87. *Operating Expenses* (in '84), $1,300. *Revenue* (in '84), consumers, $2,100, **Management.**—Address Jas. McCrady or Samuel Rickard.

184. **SUNBURY,** *Northumberland Co.* (Pop., 4,077—5,930.) Built in '83–84 by Sunbury Water Co. Supply. -Surface and spring water, pumping from impounding to distributing reservoir. **Distribution.**—(In '88.) *Mains,* 11 miles. *Taps,* 450, *Meters,* 2. *Hydrants,* 40. *Consumption,* 750,000 galls. **Financial.**—(In '88.) *Cost,* $410,000. *Cap. Stock,* $33,000. Int. on debt, 6%. *Revenue,* consumers, $0,000 ; city $1,200; total, $10,200. **Management.**—Prest., John Haas. Secy. and Treas., L. T. Roheback. Supt., W. H. Roheback.

185. **SUSQUEHANNA,** *Susquehanna Co.* (Pop., 3,467—3,872.) Built in '74 by Susquehanna Water Co.: town neither controls rates nor has right to purchase works, under franchise. **Supply.**—Springs, by gravity. Fiscal year closes, Oct. 15.

Year.	Mains.	Taps.	Hydrants.	Cost.	Exp.	Rev.
'81	.7	100	5	....	$55	$600
'86	.8	100	5	$5,000	50	660
'89	1	140	5	5,000	162	1,132
'90	2.1	185	7	20,000	352	1,590

Financial.—*First Cost,* $5,000. *Cap. Stock:* authorized, $5,000; all paid up; paid 18% div'd in '89, 6% in '90. *No Debt. Expenses:* operating, $257; taxes, $95; total, $352. *Revenue:* consumers, $1,590; city, none **Management.**—Prest., Gaylord Curtis. Secy., C. F. Curtis. Supt., V. Backburn.

186. **TACONY,** *Philadelphia Co.* (Suburb and branch P. O. of Philadelphia.) Built in '78-9 by Disston Water Co.; city has no control of rates; city of Philadelphia may buy works at any time, price fixed by three arbitrators, under franchise of '88. **Supply.**—Delaware River, pumping direct and to tanks. Fiscal year closes Sept. 30. Nov. 25, '90, following improvements were reported: 1½ miles of 12-in. main, 3 miles of 6-in. main, three additional tanks with cap. of 225,000 galls.; cost, $25,000. **Distribution.**—*Mains,* 7½ miles. *Taps,* 800. *No Meters. Hydrants,* 61. **Financial.**—*No Debt.* No hydrant rental. **Management.**—Prest. Wm. Disston. Secy., Wm. Miller. Treas., J. S. Disston. Supt., Samuel Bevan.

187. **TAMAQUA,** *Schuylkill Co.* (Pop., 5,730—6,054.) Built in '51 by borough. Des. Engr., D. K. Goodwin. **Supply.**—Rabbit Run and Owl Creek, by gravity from impounding reservoir. Rabbit Run reservoir has cap. of 20,000,000 and Owl Creek of 37,000,000 galls. Owl Creek dam is 2,350 ft. long, 600 ft. wide at bottom; raised 6 ft. in '90. **Distribution.**—*Mains,* 10 miles. *Taps* (probably faucets), 1,450. *No Meters. Hydrants,* 97. *Consumption,* unknown. **Financial.**—*Cost* (in '88), about $58,000. *Bonded Debt,* $43,000, $20,000 at 4½%, due in 1905. Debt reduced by calling in bonds. *Expenses:* operating, $920; int., $1,831; total, $2,751. *Revenue,* total, $7,154. **Management.**—Council. Clerk, Samuel Beard. Supt., J. L. Mackey.

188. **TARENTUM,** *Allegheny Co.* (Pop., 1,245—4,627.) Owned and enlarged by Tarentum Water Co.; franchise neither controls rates nor gives city right to buy works. Built in '86 by H. M. Breckenridge. **Supply.**—Springs, pumping to reservoir. Fiscal year closes Jan. 1. A 1,000,000-gall. pump and 75-HP. boiler has been added since '89.

Year.	Mains.	Taps.	Meters.	Hydr'ts.	Cost.	B'd'd debt.	Int. ch'gs.	Exp.	Rev.
'88	7.5	177	0	1	$46,000	$25,000	$1,500	....	$3,240
'90	10.3	369	1	3	79,300	25,000	1,500	$2,075	2,960*

* From consumers.

FOR FURTHER DESCRIPTIVE MATTER see Manual for 1889-90. CHANGES since 1890 are here given. POPULATIONS are for 1880 and 1890, respectively.

144   MANUAL OF AMERICAN WATER-WORKS.

Financial.—*Cap. Stock:* authorized, $50,000; paid up, $15,500; paid no div'd in 1890. *Total Debt,* $29,960; floating, $4,960 at 6%; bonded, $25,000 at 6%. *Expenses:* operating, $2,000; int., $1,800; taxes, $75; total, $3,875. *Revenue:* consumers, $2,960. **Management.**—Prest., J. B. Ford. Secy. and Treas., J. W. Hemphill. Supt., F. W. McDowell.

190. **TAYLORVILLE,** (*Scranton P.O.*) *Lackawanna Co.* (Pop., not given.) Supply from Minooka.

189. **TIDIOUTE,** *Warren Co.* (Pop., 1,255—1,328.) Built in '71 by Tidionte Water Co.; city has no control of rates; franchise does not provide for purchase of works by city. **Supply.**—Mountain stream, by gravity from impounding to distributing reservoirs.

Year.	Mains.	Taps.	Meters.	Hydrants.	Cost.	Bonded debt.	Exp.	Rev.
'80	2.5	90		13	......	......	$850	
'84	7	111	0	23	$40,000	$3,000	600	$500*
'87	9	135	0	27	40,000	1,200	500	4,000
'90	11	167	0	27	......	none	700	....

* From city.

Financial.—*First cost,* $26,000. *Cap. Stock,* $26,000. Hydrant rental, $500 for all. **Management.**—Prest, and Supt, T. W. Irvin. Secy., W. R. Dawson. Treas., H. H. Cummings.

190. **TIOGA,** *Tioga Co.* (Pop., 500—557.) Built in '74 by T. A. & C. H. Wickham; franchise neither controls rates nor gives city right to buy works. **Supply.**—Beatty Creek, by gravity from impounding to distributing reservoir. Fiscal year closes Jan. 1. **Distribution.**—*Mains* (in '88), 3.2 miles. *Taps* (in '82), 85. *No Meters. Hydrants,* 11.* **Financial.**—*Cost* (in '88), $20,000. *Cap. Stock,* $15,000; all paid up; no div'd in '90. *No Debt.* Hydrant rental, $350 per year. **Management.**—Prest., E. G. Schieffelin. Secy. and Supt., T. A. Wickham.

191. **TITUSVILLE,** *Crawford Co.* (Pop., 9,046—8,073.) Built in '73 by city. **Supply.**—Wells, pumping direct. **Distribution.**—(In '87.) *Mains,* 11 miles. *Taps,* 850. *Meters,* 25. *Hydrants,* 66. *Consumption,* 1,000,000 galls. **Financial.**—(In '87.) *Cost,* $139,292. *Bonded Debt,* $112,000. *Expenses,* $5,600. *Revenue,* $12,000. **Management.**—Unknown.

192. **TOWANDA,** *Bradford Co.* (Pop., 3,814—4,169.) Built in '79 by Towanda Water-Works Co.; franchise fixes rates; city may buy works 5 years from Oct. 1, '96. **Supply.**—Surface water, by gravity and river, by pumping. Fiscal year closes Dec. 1. A new 10-in. c. i. supply main, 16 miles long, from Ellenberger Springs, was to be put in use Oct. 1, '91. **Distribution.**—*Mains,* 8 miles. *Taps,* 450. *No Meters. Hydrants,* 50. **Financial.**—*Revenue:* consumers (in '87), $6,000; city, $2,500 in '87, $3,000 under new contract of '91; total, $8,500. **Management.**—Prest. and Supt., J. Griffiths. Secy., F. E. Beers. Treas., N. N. Betts.

193. **TREMONT,** *Schuylkill Co.* (Pop., 1,785—2,046.) Built in '66; operated since '74 by Tremont Gas & Water Co.; franchise neither controls rates nor gives city right to purchase works. **Supply.**—Mountain stream and spring, by gravity; filtered. Fiscal year closes Dec. 31. During '90-1 supply was increased threefold and pressure doubled. Source of new supply not stated, but additional basin was built to receive it.

Year.	Mains.	Taps.	Meters.	Hyd'ts.	Cons'pt'n.	Cost.	B'd'd d't.int.ch'g's.	Exp.	Rev.
'80	2	50	......	3	26,000	......		40	400
'86	2.3	75	0	3	$10,000	26,000	$10,000	$500	40	800
'89	2.5	96	0	3	25,000	26,000	13,000	650	121	967
'90	2.9	116	0	3	35,000	28,000	13,000	650	112	1,127

Financial.—*First cost,* $20,000. *Cap. Stock.*—Common: authorized, $40,000; paid up $10,500; paid no div'd in 1889, 5% in 1890. Pref'd: authorized, $9,000; paid up, $6,000; 5% div'd guaranteed. *Total Debt,* $13,071: floating, $71 at 5%; bonded, $13,000 at 4%, due in '95. *Expenses:* operating, $51; int., $874; taxes, $61; total, $762. *Revenue,* consumers, $1,067; fire protection, $60; total, $1,127. **Management.**—Prest. and Supt., J. P. Bechtel. Secy., L. J. Lewis. Treas., C. B. Sillyman.

194, 195. **TROY,** *Bradford Co.* (Pop., 1,241—1,307.)

EDGEWOOD WATER CO.

Built in '80. **Supply.**—Springs, by gravity. Fiscal year closes Dec. 31. **Distribution.**—*Mains,* 4.5 miles. *Taps* (in '88), 50. *No Meters. Hydrants,* 10. **Financial.**

PENNSYLVANIA. 145

—*Cost* (in '88), $15,000. *No debt. Operating Expense* (in '88), $400. **Management.**
—Prest., Treas. and Supt., R. J. Reddington. Secy., N. S. McReem.

TROY WATER CO.

Built in '80. **Supply.**—Springs, by gravity. **Distribution** (in '88).—*Mains*, 2 miles. *Taps* [or faucets], 200. *No meters. Hydrants*, 3. **Financial.**—*Cost* (in '88)), $4,000. **Management.**—Unknown.

196, 197. **TUNKHANNOCK,** *Wyoming Co.* (Pop., 1.241—1,253.)

SLOCUM SPRINGS WATER CO.

Built in '59 by Slocum Water Co. **Supply.**—Slocum Spring and Susquehanna River, pumping to reservoir; filtered. **Distribution.**—(In '88.) *Mains*, 6 miles. *Taps*, 150. *No Meters. Hydrants*, 27. **Financial.**—*Cost* (in '88)), $30,000. *Debt*, none in '56. **Management.**—(In '88.) Prest., George A. Vail, New York City. Treas. and Supt., W. C. Kittridge. NOTE.—Later information refused.

TUNKHANNOCK WATER CO.

Built in '70. **Supply.**—Springs, by gravity. Water from five springs, at different altitudes. impounded by as many dams. Supply Main,—C. i., 4 and 3-in. for 4,050 ft., 8-in. for 675 ft. **Distribution** —*Mains*, 3-in., 4.75 miles. *Takers*. 200. *Hydrants*, 10. **Management.**—Secy., Felix Ansart.

NOTE.—All information save name of Secy. for '80. No later official report could be secured. Unofficially reported that stockholders were to vote on Sept. 11, '91, on increase of cap. stock to $40,000 to buy property of Slocum Spring Water Co.

198. **TYRONE,** *Blair Co.* (Pop., 2,678—4,705.) Built in '70, by Tyrone Gas & Water Co. **Supply.**—Sinking Spring run, by gravity from impounding reservoir. Fiscal year closes Jan. 1.

Year.	Mains.	Taps.	Meters.	Hydrants.	Cost.	Exp.	Rev.
'84	4	400	0	25	15,000	....	....
'86	6	500	0	27	30,000	....	....
'87	7	520	0	28	30,000	100	84*
'88	7	574	0	31	30,000	100	84*
'90	7.5	580	0	31	35,000	125	87*

*From city.

Financial.—*No Debt. Operating Expenses*, $125. *Revenue*, fire protection, $87. **Management.**—Prest., C. Guyer. Secy., A. R. Stevens. Treas., A. B. Howe. Supt., C. H. Diffenbaugh.

199. **ULYSSES,** *Potter Co.* (Pop., 638—801.) Built in '85 by Lewisville Water Co. **Supply.**—Springs, by gravity for domestic use and pumping for fire protection. **Distribution.**—(In 88.) *Mains*, 3.5 miles. *Taps*, 60. *No Meters. Hydrants*, 7. **Financial.**—(In '88.) *Cost*, $7,000. *Cap. Stock*, $5,000. *Bonded Debt*, $3,000 at 6%. *Operating Expenses*, "light." *Revenue*: consumers, $500; city, $120; total, $620. **Management.**—Prest., E. N. Eaton. Treas. and Supt., Perry Brigham.

Union City, *Erie Co.* (Pop., 2,171—2,261.) Owned by borough. Built in '85 by Wm. Dunmeyer. **Supply.**—Creek, pumping to tank. Works were bought by borough in '89 or '90, and are used for fire protection only. See Manual for '89-90 p. 280; also see projected works.

200. **UNIONTOWN,** *Fayette Co.* (Pop., 3,265—6,359.) Built in '83 by Uniontown Water Co. **Supply.**—Springs, by gravity from impounding reservoir.

Year.	Mains.	Taps.	Meters.	Hydrants.	B'd'd debt.	Int. ch g's
'86	8	400	..	50	$25,000	$1,500
'88	15	600	....	60	25,000	1,500
'90	25	850	..	65	......	......

Financial—Int., 6%. **Management.**—Prest., W. H. Playford. Secy., W. B. McCorm'ck. Supt., G. H. Seaton.

33. **UPLAND,** *Delaware Co.* (Pop., 2,028—2,275.) Supply from Chester.

201. **WALL AND WILLMERDING,** *Allegheny Co.* Pop.: Wall, est., 703; Willmerding, in '90, 419.) Built in '89-90 by Turtle Valley Water-Works Co.; put in operation April 1. Franchise neither controls rates nor gives city right to buy works. **Supply.**—Pumped to reservoir. Fiscal year closes Nov. 1. **Distribution.**—*Mains*, 8 miles. *Taps*, 160. *Hydrants*, 31. **Financial.**—*Cap Stock*, authorized, $150,000, all paid up. *Bonded Debt*, $100,000, at 6%, due in 1920. *Expenses and*

FOR FURTHER DESCRIPTIVE MATTER see Manual for 1889-90. CHANGES since 1890 are here given. POPULATIONS are for 1880 and 1890, respectively.

146   MANUAL OF AMERICAN WATER-WORKS.

Revenue: not determined at time of report. **Management.**—Prest., Geo. Westinghouse, Jr. Secy., Treas. and Mgr., J. R. McGinley.

XXIII. **WALSTON,** *Jefferson Co.* (Pop., est., 2,000. (Built in '82-3 by Rochester & Pittsburg Coal & Iron Co., to supply their coke ovens and town. **Supply.**—Two artesian wells and Mahoning River, pumping to tanks. Cap. of tank, 24,000 galls. **Distribution.**—*Mains*, 1 mile. *Taps*, 25. *No Meters or Hydrants.* Pressure, 83 lbs. **Management.**—Prest., G. E. Merchant, Rochester, N. Y. Gen. Mgr., L. W. Robinson. Supt., G. W. Snyder.

202. **WARREN,** *Warren Co.* (Pop., 2,810—4,332.) Built in '81-2 by Warren Water Co. **Supply.**—Allegheny and Morrison's Rivers, by gravity and pumping to reservoir. Fiscal year closes June 30. **Distribution.**—*Mains*, 15 miles. *Hydrants,* 54. **Financial.**—*Cost,* $132,000. *Cap. Stock* (in '88), $30,000. *Bonded Debt*, $30,000, at 6%. *Revenue,* city, $1,030. **Management.**—Prest., G. C. Orr. Secy. and Supt., W. B. Meredith. Treas., Wm. Pollock.

203. **WASHINGTON,** *Washington Co.* (Pop., 4,292—7,063.) Built in '87-8 by Citizens' Water Co. Controlled by Kittanning Consolidated Water Co., Kittanning. **Supply.**—Chartier's Creek, pumping from filter wells, near impounding reservoir, to distributing reservoir. Fiscal year closes Jan. 1. During '91, Co. were to build a new dam and storage reservoir, covering 3 acres, and put in new filter.

Year.	Mains.	Taps.	Meters.	Hydrants.	Cost.	B'd'd Debt.	Int. Ch'g's.	Rev.
'87	10	...	..	40	$325,000	$75,000	$4,500	$2,000*
'88	10	250	0	40	355,000	75,000	4,500	......
'80	15	320	2	40	355,000	75,000	4,500	18,200
'9)	15	360	2	40	480,000	75,000	4,500	16,000

* From city.

Financial.—*Cap. Stock:* authorized, $250,000, all paid up. *Bonded Debt,* $75,000, at 6%, due in 1907. *Expenses:* int., $4,500; taxes, $75. *Revenue:* consumers, $14,000; fire protection, $2,000; other purposes, $2,200; total, $18,200. **Management.**—Prest., S. L. Hazlett. Secy., G. H. Fox, Kittanning. Treas., Wm. Pollock, Kittanning. Supt., J. MacJones.

204. **WATSONTOWN,** *Northumberland Co.* (Pop., 1,481—2,157.) Built in '87 by Watsontown Water Co. **Supply.**—White Deer Creek, pumping to reservoir. **Distribution.**—*Mains*, 5 miles. *No Meters. Hydrants,* 31. *Consumption,* 300,000 galls. **Financial.**—(In '88.) *Cost,* $40,000. *Bonded debt,* $40,000 at 6 %. *Expenses:* operating, $1,000; int., $2,400. *Revenue:* consumers, $2,500: city, fire protection, $300. **Management.**—(In '88.) Prest., S. B. Morgan. Secy., W. F. Shay. Treas. and Supt., Robt. Buck.

205. **WAYNE,** *Delaware Co.* (Pop., in '90, 997.) Built in '81 by Drexel & Childs, proprietors of Wayne estate; franchise neither controls rates nor gives city right to purchase works; surface well and springs, pumping to reservoir and direct. Water from springs is conveyed into retaining ponds, having cap. of 450,000 galls. Well has dy. cap. of 100,000 galls. In '90 an additional water supply by driven wells having cap. of 250,000 galls. per day was contracted for.

Year.	Mains.	Taps.	Meters.	Hydrants.	Consumption.	Cost.	Exp.	Rev.
'84	2.5	17	0	18	32,500	$20,000	$800	......
'88	4	120	0	35	133,051	40,000	....	......
'89	5	....	....	....	147,126	41,054	....	$4,001
'90	8	....	....	50	242,051	......	1,210	4,450

Financial.—*No capital stock or debt.* **Management.**—Mgr., Frank Smith. Engr., Benj. F. Shaw.

206. **WAYNESBOROUGH,** *Franklin Co.* (Pop., 1,888—3,811.) Built in '82-4 by Waynesborough Water Co. **Supply.**—Mountain springs, by gravity.

Year.	Mains.	Taps.	Meters.	Hd'ts.	Cons'pt'n.	Cost.	B'd'd't.	Int.Ch'g's.	Exp.	Rev.
'87	9	500	0	26	1,000,000	$75,000	$39,000	......	$500	$5,990
'90	9	560	0	34	....	75,000	38,000	$2,880	730	6,537

Financial.—*Cap. Stock,* authorized, $80,000. *Bonded Debt,* $38,000, bonded at 6%. **Management.**—Prest., C. H. Horne. Secy. and Supt., G. S. Meyer. Treas., G. J. Beecher.

207. **WAYNESBURGH,** *Greene Co.* (Pop., 1,208—2,101.) Built in '87-8 by Waynesburgh Water Co.; city has no control over rates; State law provides city may buy works after 20 years, or in 1907 or 1908. **Supply.**—Ten Mile run, pumping to reservoir; filtered. Fiscal year closes Dec. 31. **Distribution.**—*Mains,* 4 miles. *Taps,* 50. *No Meters. Hydrants,* 27. *Consumption,* 30,000 galls. **Financial**—*Cost,* $35,000. *Cap. Stock,* $50,000. *Bonded Debt* (in '88): authorized, $25,000; issued, $12,000

PENNSYLVANIA. 147

at 6%. *Expenses*, operating, $1,200. *Revenue:* consumers, $1,285; city, $690; total, $1,885. **Management.**—Secy. and Treas., I. U. Singley, Uniontown. Agent, J. A. F. Randolph.

208. **WEATHERLY,** *Carbon Co.* (Pop., 1,977–2,931.) Built in '81-2 by Weatherly Water Co. **Supply.**—Penrose creek and springs, by gravity. **Distribution.** —(In '88.) *Mains,* 6 miles. *Taps,* 325. *No Meters. Hydrants,* said to be 150. **Financial.**—(In '88.) *Cost,* $25,000. *Cap. Stock,* $12,000. *Debt,* $3,000, at 5%. *Revenue:* consumers, $2,000. **Management.**—Prest., W. W. Blakeslee, Sr. Treas., A. J. Landerburn. Secy., S. G. Eby.

209. **WELLSBOROUGH,** *Tioga Co.* (Pop., 2,228–2,931.) Built in '86 by Wellsborough Water Co.; franchise neither controls rates nor gives city right to buy works. **Supply.**—Spring, by gravity to and from reservoir; filtered through white sand-stone, crushed to the size of peas. Fiscal year closes second Monday in January.

Year.	Mains.	Taps.	Meters.	Hydrants.	Cost.	B'd d't. Int. Ch'g's.		Exp.	Rev.
'86	14	49	10	40	$35,000				
'88	16	241	0	43	95,000	$20,000	$1,200	$1,200	$3,950
'89	16	251	10	44	95,000	20,000	1,200	1,300	2,200*
'90	16	258	4	45	95,000	20,000	1,200	1,500	2,250*

*From city.

Financial.—*Cap. Stock:* authorized, $75,000; all paid up; paid 6% div'd in '90. *Bonded Debt,* $20,000, at 6%, due in 20 yrs. *Expenses:* operating, $1,200; int., $1,200; taxes, $300; total, $2,700. *Revenue:* consumers, withheld; fire protection, $2,250. **Management.**—Prest., Chn., Wm. Bache. Secy., J. Harrison. Supt , W. C. Kress.

WEST BETHLEHEM, *Southampton Co.* Supply from South Bethlehem.

210. **WEST CHESTER,** *Chester Co.* (Pop., 7,046–8,028.) Built in '45-6 by borough. **Supply.**—Chester Creek, fed by springs, pumping to reservoir. Fiscal year closes March 4. *Distribution.*—(In '89.) *Mains,* 14.5 miles. *Taps,* 1,550. *Meters,* 3. *Hydrants,* 144. *Consumption,* 400,000 galls. **Financial.**—(In '88.) *Cost,* $56,000, *Operating Exp'nses,* $3,123. *Revenue,* $16,014. **Management.**—(In '88.) Com. of three. Chn., C. B. Lear. Supt., E. P. Mercer.

79. **WEST INDIANA,** *Indiana Co.* (Pop., 1,077–1,534.) Supply from Indiana.

150. **WEST PITTSTON,** *Luzerne Co.* (Pop., 2,544–3,906.) Supply from Pittston.

XXIV. **WEST READING** *(Reading P. O.), Berks Co.* (Pop., est., 650.) Built by West Reading Water Co. in '86. Engr., P. W. Alexander, Reading. **Supply.**—Schuylkill River, pumping to tanks. **Pumping Machinery.**—No. 5 Guild & Garrison. **Tanks.**—Cap., 50,000 galls.; two 25,000 galls.; from Smith & Sons. Philadelphia. *Mains,* 4-in. c. i., about 3 miles; from Mclbert Foundry & Machine Co., Reading. *Taps,* about 150. *Consumption,* about 30,000 galls. *Pressure,* 50 lbs. **Financial.**—*Cost,* $10,000. *Cap. Stock:* authorized, $10,000; "free." *No debt.* **Management.**—Prest., G. R. Frill. Secy., C. H. Schaefer. Treas., G. W. Alexander.

211. **WHITE HAVEN,** *Luzerne Co.* (Pop., 1,408–1,634.) Built in '70 by White Haven Water Co. No Engr. **Supply.**—Springs, by gravity. **Distribution.**— *Mains,* iron, 4 miles. *Service,* lead, iron. *Taps,* 190. *No Meters. Hydrants,* 6. **Financial.**—*Cost,* $19,000. *Operating Expenses,* $350. **Management.**—Prest., C. L. Kirk. Secy. and Treas., Stephen Maguire.

103. **WICONISCO,** *Dauphin Co.* (Pop., 2,130–2,280.) Supply from Lykens.

212. **WILKES-BARRE,** *Luzerne Co.* (Pop., 23,339–37,718.)

WILKES-BARRE WATER CO.

Built in '57-59; city has no control over rates; franchise does not provide for purchase of works by city. **Supply.**—Laurel River, by gravity and pumping to reservoirs. **Dams and Reservoirs.**—Combined cap., about 2,046,000,000 galls. *Laurel Run Intake, No.* 1: Cap., about 10,000,000 galls.; area flooded, about 5 acres. Dam of masonry, faced on upper side with cut rubble, broken range and filled against with earth puddling, unpaved; curved in plan, 125 ft. long, 20 ft. high, about 5 ft. wide on top. Inner batter, 1 in.; outer, 2 ins. to 1 ft. Spillway in west end of dam, 30 ft. from end, 66 ft. long and 3 ft. deep; masonry, controlled by two gates. A 30-in. discharge pipe branches into two pipes outside of dam. Dam does

FOR FURTHER DESCRIPTIVE MATTER see Manual for 1889-90. CHANGES since 1890 are here given. POPULATIONS are for 1880 and 1890, respectively.

not leak. Built in '59 in connection with first supply, and since raised twice, in all about 8 ft., to present height. Engr., C. F. Ingham. No contr.

Laurel Run Storage. No. 2: Cap., about 30,000,000 galls.; area flooded, about 5 acres, about ½ mile above No. 1. Dam of rubble masonry, founded on conglomerate rock; curved, 280 ft. long, 30 ft. high, 6 ft. wide at top. Inner batter, 1 in., outer, 2 ins. to 1 ft. Spillway, 80 ft. long, 2½ ft. deep, of masonry; one gate culvert 2½ × 3 ft., but no discharge pipes, as water passes through stream to lower reservoir. Dam does not leak. Built in '86. Engr., W. H. Sturdevant; Asst. Engr., A. E. Hess; Contr., Jos. Hendler; all of Wilkes-Barre.

Intake, No. 3: Cap., 6,000,000 galls.; area flooded, 1½ acres. Dam of cut rubble masonry, broken range, with loam puddle against inner face; founded on rock and concrete; 85 ft. long, 16 ft. high, 4.5 ft. wide on top, with 13 ft. water. Inner batter, 1 in., outer, 2 ins. to 1 ft. Spillway, 60 ft. long, 3 ft. deep; masonry, in dam, 10 ft. from east end; two gates. A 30-in. discharge pipe and 2 ft. 8-in. × 3 ft 8-in. culvert. Built '90-1. Engr., W. H. Sturdevant; Asst. Engr., T. A. Wright; both of Wilkes-Barre. Contrs., A. H. Coon & Co., Kingston.

Toby's Creek Storage Reservoir, No. 4: Cap., 2,000,000,000 galls.; area flooded, 396 acres; drainage area, about 8 sq. miles. About 3 miles above No. 3. Dam of cut rubble masonry, broken range, with clay and loam puddle against inner face; founded on rock; curved, 252 ft. long, 25 ft. high, 6 ft. wide on top, 23 ft. water. Inner batter 1 in., outer 3 ins. to 1 ft.; slope of puddled embankment, 1¼ to 1. Spillway, 70 ft. long, 2 ft. deep, in two sections separated by gate base of 10 ft.; masonry, in center of dam; two gates. Two culverts, 2 ft. 8 ins. × 3 ft. 8 ins. Engr., W. H. Sturdevant; Asst. Engr., T. H. A. Wright, both of Wilkes-Barre. Contr., R. C. Mitchells, Plainsville, Pa.

Distribution.—(In '88.) *Mains*, 45 miles. *Taps*, 4,100. *Meters*, 4. *Hydrants*, 100. Financial.—Withheld. Management.—Prest., Geo. R. Wright. Mgr., Sam'l L. Marshall.

CRYSTAL SPRINGS WATER CO.

Prest., J. R. Maxwell. Secy. and Treas., Walter Gaston. Supt., E. H. Lauall. Supplies part of city and Ashley. No further information.

213. **WILKINSBURGH,** *Allegheny Co.* (Pop., 1,529—4,662.) Built in '88-9 by Pennsylvania Water Co. to supply boroughs of Wilkinsburgh, Edgewood and Brushton, and townships of Sterrett, Penn, Wilkins and Braddock. Figures given for '90-1 are for July 1. Distribution.—*Mains*, 21 miles. *Taps*, 674. *Meters*, 110. *Hydrants*, 20. Financial.—*Cost*, $300,000. *Cap. Stock*, $300,000; 6% stock div'd in '91. *Bonds* authorized, $10,000; issued $50,000 at 6%. Management.—Prest., J. D. Cherry, Secy., E. J. Harlow. Treas., Sam'l Doubt. Supt., W. A. Alexander.

213. **WILKINS TOWNSHIP,** *Allegheny Co.* (Pop., 4,426—2,304.) See Wilkinsburgh.

214. **WILLIAMSPORT,** *Lycoming Co.* (Pop., 18,931—27,932.) Built in '53 by Co., owned and operated by Williamsport Water Co. and Citizens' Water & Gas Co.; city has right to buy works, terms specified in franchise; has no control of rates. Supply—Two mountain s'reams, by gravity, filtered. Fiscal year closes April 1. In '90 a new reservoir was built by Citizens' Water & Gas Co., see below. In '91 a 5,000,000-gall. reservoir was being constructed by Williamsport Water Co. New reservoir gives additional head of 51 ft. *New Dam and Reservoir:* Cap., 20,000,000 galls.; area flooded, 6 acres. Dam of earth with heart-wall, inner section being of loam, clay and stones, and outer, next to heart-wall, of same material, but with large stone; founded on clay and boulders; 1,250 ft. long, 16 ft. high, 10 ft. wide on top, 66 at bottom, 13 ft. water. Inner slope, 2 to 1, 18 ins. riprap at bottom, 9 at top; outer slope, 1½ to 1, covered with "pitched," or loose stone, to depth of 4 to 6 ft. Heart-wall, blue clay puddle, well tamped, 5 ft. wide at top, 6 at natural surface, 7 at bottom; carried within 3 ft. top of dam and 7 ft. below bottom. Spillway, 25 ft. long, 3 ft. deep; lined with masonry, free. Discharge pipes, two 24-in., supply and blow-off. Engr. and contr., co.

Year.	Mains.	Taps.	M't'rs.	Hyd'ts.	Cons'pt'n.	Cost	B'd'd d't.	Int.ch'g's.	Exp.	Rev.
'86-7	41	3,000	3	239	1,000,000	$500,000	$80,500	$4,830	.....	$40,000
'88 9	53	.....	1	174	.....	.....	.....	.....	.....	.....
'89-90	60	..	0	183	4,000.000	681,531*	225,000 †	11,250	$9,000	2.305¶
'90-1	60	.....	0	184	4.500.000	.....	.....	.....	.....	2,310¶

*$421,581 to Williamsport and $260,000 to Citizens' Water Co. †$125,000 Williamsport, and $100,000 Citizens' Water Co. ¶ From city.

Financial.—*First Cost*, $250,000. *Cap. Stock*, authorized, Williamsport Water Co., $255,750; Citizens Water Co., $100,000; all paid up. Paid 8% dividend in 1890. Bonded

Debt, $225,000; bonded, Williamsport Water Co., $125,000 at 5%, due '89—1904; Citizens' Water Co., $100,000 at 5%, due '91—1906. *No Sinking Fund. Expenses,* operating, $9,000. *Interest,* $11,250. *Revenue,* consumers, withheld; fire protection, $1,830; other purposes, $475. **Management.**—Prest., J. V. Brown. Secy. and Treas., W. H. Bloom.

XXV. **WILLIAMSTOWN,** *Dauphin Co.* (Pop., 1,771—2,324.) Gravity works built by Williamstown Water Co., date not given. About 2 miles mains. Address G. H. Ronk or John Lynch.

201. **WILMERDING,** *Allegheny Co.* (Pop., in '90, 419.) See Wall.

81. **WYNCOTE,** *Montgomery Co.* See Jenkintown.

215. **YORK,** *York Co.* (Pop., 13,940—20,793.) Built in '16 by York Water Co.; town neither controls rates nor has right to buy works, under franchise. **Supply.**—Springs and surface water, pumping to reservoir. Fiscal year closes April 1.

Year.	Mains.	Taps.	Meters.	Hyd'ts.	Cons'p'tn.	Cost.	B'd'd Debt.	Int. Ch'g's.	Rev.
'86-7	..	3,800	..	58	600,000	$200,000	none	none	$31,030
'88-9	22	4,000	2	70	600,000	200,000	none	none	39,000
'89-0	24	4,040	9	80	1,250,000	200,000	$23,000	$1,240	40,000
'90-91	26	4,500	11	85	1,400,000	200,000	......	....	...

Financial.—*Cost,* $165,000. *Cap. Stock:* authorized, $200,000, all paid up; paid 10% div'd in 1889, and 12% in 1890. *Total Debt* (in '80), $23,000, floating, at 1½%. *Expenses,* withheld. *Revenue,* consumers, $40,000; fire protection, free. **Management.**—Prest., J. Carl. Secy., Snyser Williams. Treas., W. H. Griffith. Supt., J. L. Kuehn.

216. **YORKVILLE** (*Pottsville P. O*), *Schuylkill Co.* (Pop., 610—915.) Distributing system put in in '91 by borough. **Supply.**—Pottsville Water Co. **Management.**—Above Co will have supervision of system; W. D. Pollard, Secy. Borough Prest., Martin Otterbein.

Water-Works Projected with Fair Prospects of Construction.

BLOSSBURG, *Tioga Co.* (Pop., 2,140—2,568.) Blossburg Water Co. let contract for works in Sept., '91, to Wm. Monie, Scranton. Supply, Taylor Run, 4 miles distant. Cap. stock, authorized, $60,000. Address L. A. Waters, or L. Amerman, Scranton.

BRYN MAWR, *Montgomery Co.* (Pop., est., 2,400.) Lower Merion Water Co. chartered in '85 and began construction of works with S. H. Lockett, N. Y. City, as Dec., and J. M. Walker as Chf. Engr. Charter covered territory from Wayne station, Pennsylvania R. R., to Philadelphia limits. Supply was to be from Indian Brock, by pumping to 15,000,000-gall. reservoir. Construction was started, but works had not been completed, June 20, '91. By terms of charter same would expire Jan., '91, as works were not finished. Co. applied to Court of Common Pleas for extension of charter, but remonstrances were filed by parties who wished to start new company and obtain supply from another source. About March, '91, application was made for incorporation of Bryn Mawr Water Co. by C. A. Chipley, C. S. Farnum and others. June 20, '91, nothing definite had been accomplished. Address H. W. Barrett or C. A. Chipley.

CHARLEROI, *Washington Co.* (Pop., 2,188—1,041.) Charleroi Water Co. in. corp. in '90. Proposed supply, Monongahela River, pumping to reservoir. Nothing had been done, Aug. 8, '91.

BUTLER, *Butler Co.* (Pop., 3,163—8,734.) **Present Supply.**—Works of Butler Water Co. See p. 113. **Proposed Supply.**—In '91 right to lay pipe was granted to Mutual Water Association and to H. W. Christie. Midsummer, '91, latter had one well drilled, from which windmill was to pump. Aug. 15, '91, Mr. Christie had nothing definite to report. Mutual Water Association consists of individuals who propose to supply south side, or "Springdale," from artesian wells and tanks on hill 260 ft. above town. July 9, '91, an 8-in. well, 500 ft. deep, had been successfully sunk. Estimated cost, $13,000, for part of works first proposed. Trustees: H. H. Boyd, L. C. Wick and F. J. Klingler.

CARBONDALE, *Lackawanna Co.* (Pop., 7,714—10,833.) **Present Supply.**—Crystal Springs Water Co. See p. 114. **Proposed Supply.**—July 28, '91. Carbondale Water Co. was incorp., with C. E. Spencer, Treas. Cap. Stock, $10,000; paid up, $1,000.

For Further Descriptive Matter see Manual for 1889-90. Changes since 1890 are here given. Populations are for 1880 and 1890, respectively.

CORAOPOLIS, *Allegheny Co.* (Pop., in '90, 962.) Coraopolis Water Co. incorp. in '90, or earlier, but had done nothing July 13, '91. Hopes to build soon. Proposed supply, wells near Ohio River, by pumping. Address C. E. Cornelius, 406 Grant St., Pittsburg.

EAST FRANKLIN TOWNSHIP (*P. O. not given*), *Armstrong Co.* (Pop., 1,695—1,575.) Franklin Water Co. incorp. Aug. 4, '91. Cap. Stock, $5,000. Address Joseph Buffington, Kittaning.

EVERETT, *Bedford Co.* (Pop., 1,247—1,679.) Present Supply—See Mt. Pleasant, under completed works. Proposed Supply.—Aug 10, '91, there was talk of works being built by Everett Water Co., with supply pumped from springs to reservoir. Report by C. G. Masters.

FACTORYVILLE, *Wyoming Co.* (Pop., 462—577.) In '90, or earlier, Nokomis Water Co. was incorp. Address Prof. D. Brown.

FELL TOWNSHIP, (*P. O. not given*), *Lackawanna Co.* (Pop., 441—1,154.) Panther Creek Water Co. incorp. July 24, '91. Authorized cap. stock, $20,000; paid up $100. Treas., T. H. Walters, Scranton.

FLEETWOOD, *Berks Co.* (Pop., 802—878.) June, '91, works were projected by council, but action was deferred owing to legal complications.

HYNDMAN, *Bedford Co.* (Pop., 323—1,056.) Hyndman Water Co. incorp. Aug. 19, '91. Proposed supply, spring, by gravity. Est. cost, $30,000. Authorized cap. stock, $20,000; paid up, $10,000. Prest., J. K. White. Secy., L J. McGregor.

MANSFIELD, *Tioga Co.* (Pop., 1,611—1,762.) Mansfield Water Co. incorp. Sept. 11, '91, by L. A. Waters and L. Amerman, Scranton. About 7 miles of pipe and 40 hydrants proposed. Authorized cap. stock, $60,000. Address above.

MONROETON, *Bradford Co.* (Pop., est. 600.) May be supplied from new main of Towanda Water Co., which was laid through village in '91.

MONTROSE, *Susquehanna Co.* (Pop., 1,722—1,735.) Construction was started in '91, but owing to high price demanded for water privileges new source was sought and work stopped while proposed supply was being tested. Address F. Pecke, Watertown, N. Y.

MOSCOW, *Lackawanna Co.* (Pop., 320—582) Works projected in '91 to be owned by Moscow Water Co. Proposed supply, springs, by gravity. Est. cost, $20,000. Prest., J. A. Mean. Engr., J. M. Rittenhouse. Report by J. T. Richards, Scranton; Aug. 1, '91.

MORRISVILLE, *Bucks Co.* (Pop., 968—1,203.) June 23, '91. Committee of Board of Trade were investigating. Proposed supply, Delaware River, by pumping. Address W. H. Moon, Secy. Board of Trade.

MOUNT JEWETT, *McKean Co.* (Pop., not given.) Mount Jewett Water Co. incorp. in '90. Nothing done to July, '91. Address W. W. Brewer.

NESCOPECK, *Luzerne Co.* (Pop., 1,205—1,456.) Nescopeck Water Co. incorp. July 10, '91. Authorized cap. stock, $5,000; paid up, $2,500. Prest., M. F. Williams. Secy., S. P. Hanley. Treas., F. H. Eaton, Berwick, Pa.

NEWMANSTOWN, *Lebanon Co.* (Pop., 511—612.) Charter for Co. to supply this place and Sheridan has been secured, but nothing had been done June 20, '91. Address W. W. Stewart.

NEWPORT, *Perry Co.* (Pop., 1,319—1,417.) Newport Water Co. chartered ate in '91, and proposes to complete works by July 1, '91. Supply, creek, by pumping. Address W. H. Norris, Treas., 3237 Powelton Ave., Philadelphia, or C. Jesse Young, 1,404 S. Penn. Sq., Philadelphia.

NORTHUMBERLAND, *Northumberland Co.* (Pop., 2,293—2,744.) July, '91, Council appointed committee to investigate and report on works. Town Clerk, C. H. Pete.

PINE GROVE, *Schuylkill Co.* (Pop., 957—1,103.) Pine Grove Water Co. incorp. in '90. Nothing done, June 20, '91. Address N. T. Felty.

RED LION, *York Co.* (Pop., 241—524.) Red Lion Water Co. projected. Supply, springs, pumping to reservoir 225 ft. above, by hydraulic ram. Est. cost, $5,000. Address P. W. Snyder or B. F. Zarfos.

SALTSBURG, *Indiana Co.* (Pop., 855—1,088.) Saltsburg Water Co. projected. Supply, river, pumping to stand-pipe. Address J. C. Moore.

PENNSYLVANIA.

SHARPSVILLE, *Mercer Co.* (Pop., 1,821—2,330.) Sharpsville Water Co. incorp. in '90. Proposed supply, artesian wells. Address J. J. Pierce.

SOMERSET, *Somerset Co.* (Pop., 1,197—1,713.) Works talked of in '91 Address W. H. Welfley.

STEELTON, *Dauphin Co.* (Pop., 2,447—9,250.) Under construction July 21, '91. Prest., Jas. Gamble, 68 Wall St., N. Y. City.

THOMPSON, *Susquehanna Co.* (Pop., 249—302.) Thompson Borough Water Co. incorp. July 28, '91. Cap. stock, authorized and paid up, $5,000. Treas., C. E. Gelatt.

UNION CITY, *Erie Co.* (Pop., 2,171—2,261.) In '90 or '91 borough bought fire protection works (see Manual for '89-90, p. 230). June 16 borough voted to extend works with proceeds of $15,000 of bonds. It is expected that water will be filtered and used for domestic purposes. Address J. C. McLean.

WESTFIELD, *Tioga Co.* (Pop., 579—1,128.) Westfield Water Co. hopes to build in '92. Supply, springs, by gravity. Est. cost, $31,000. Secy., T. M. Leonard.

Water-Works Projected in '90 and '91, but Now Indefinite.

Clifton Heights, Delaware Co. *McSherrytown,* Adams Co. *New Holland,* Lancaster Co.

DELAWARE.

Water-Works Completed or in Process of Construction.

1. DOVER, *Kent Co.* (Pop., 2,811—3,061.) Built in '82 by town. Supply.—Wells pumping direct. Fiscal year closes Feb. 28.

Year.	Mains.	Taps.	M't'rs.	Hd'ts.	Cons'pt'n.	Cost.	B'd'd d't.	Int. ch'g's.	Exp.	Rev.
'82	2.5	...	0	28	......	$20,000	$20,000	$80	.....	.....
'88	2.8	237	1	29	59,102	23,000	21,500	800	$2,409	$3,168
'90	5	283	0	32	70,710	26,800	20,000	780	2,700	3,565

Financial.—*Bonded Debt,* $20,000, at 4%, due in 1900. *Expenses,* total, $3,180. *Revenue:* consumers, $3,565; city, none. Management.—Com. of three appointed by city council every year. Chn., H. H. Reedy. Secy., W. W. Boggs. Supt., J. T. Hofecker. Chf. Engr., T. F. Cooke.

2. MIDDLETOWN, *New Castle Co.* (Pop., 1,280—1,454.) Built in '88 by town. Supply.—Wells, pumping to stand-pipe. Fiscal year closes March 1. Distribution.—*Mains,* 3 miles. *Taps,* 111. *Hydrants,* 10. Financial.—*Cost,* $17,000. *Debt,* $17,000 at 5%. *Expenses,* $1,100. *Revenue,* $700. Management.—Prest., Henry Clayton. Secy., W. S. Leatherberry. Supt., J. M. Leatherberry.

3. NEWARK, *New Castle Co.* (Pop., 1,148—1,191.) Built in '88-9 by town. Supply.—Well, pumping to stand-pipe and direct. Fiscal year closes Feb. 28.

Year.	Mains.	Taps.	Meters.	Hyd'ts.	Cons'pt'n.	Cost.	B'd'd d't.	Int. Ch'g's.	Exp.	Rev.
'88	2.9	27	0	35	50,000	$57,000	$57,000	$1,850	$1,000	$1,900
'90	2.9	59	0	36	53,226	37,000	37,000	1,850	900	1,847

Financial.—*Revenue:* consumers, $1,847; city, $900. Management.—Supt., J. H. Haye.

4. NEW CASTLE, *New Castle Co.* (Pop., 3,700—4,010.) Built in '70 by New Castle Water-Works Co. Supply.—None Such Creek, pumping to reservoirs. Distribution.—(In '88.) *Mains,* 8.5 miles. *Taps,* 507. *Meters,* 0. *Hydrants,* 40. *Consumption,* 450,000 galls. Financial.—(In '83.) *Cost,* $112,000. *Bonded Debt,* $54,000 at 5%. *Expenses:* operating, $4,500; int., $2,700; total, $7,200. *Revenue:* consumers, $3,000. Management.—Prest., J. G. Shaw. Secy., Wm. H. Clark.

NOTE.—No statistics have been received since '88.

5. SMYRNA, *Kent Co.* (Pop., 2,423—2,455.) Built in '86 by town. Supply.—Surface well, pumping to stand-pipe and direct. Fiscal year closes Feb.1. Distribution.—(In '89.) *Mains,* 3.5 miles. *Taps,* 288. *Meters,* 0. *Hydrants,* 34. *Consumption,* 28,000 galls. Financial.—(In '89.) *Cost,* $24,000. *Bonded Debt,* $21,000 at 4%. *Expenses:* operating, $1,212; int., $960; total, $2,172. *Revenue:* consumers, $2,082. Management.—Comrs. Secy., J. B. Cooper. Supt., A. Taylor.

FOR FURTHER DESCRIPTIVE MATTER see Manual for 1889-90. CHANGES since 1890 are here given POPULATIONS are for 1880 and 1890, respectively.

152   MANUAL OF AMERICAN WATER-WORKS.

6. WILMINGTON, *New Castle Co.* (Pop., 42,478—61,431.) Owned by borough. Built in '04 by Wilmington Spring Water Co. Supply.—Brandywine Creek, pumping to low-service reservoir and repumping to high-service reservoir. Fiscal year closes Dec. 31.

Year	Mains	Taps	M'trs	H'd'ts	C'ns'pt'n	B'd'd Debt	Int. Ch'g's	Exp.	Rev.
'77	51.65	4,344	..	470	......	......	......	26,193	61,146
'82	58	8,658	3	520	4,671,553	......	......	35,499	75,517
'85	61	9,540	..	535	4,625,042	......	......	27,205	81,352
'87	69.5	10,869	7	625	5,272,334	632,000	......	50,000	104,536
'88	71.34	11,205	10	627	5,814,105	622,500	33,456	52,000	114,461
'89	73.75	71,751	14	633	5,931,068	586,000	33,443	37,236	116,901
'90	75.5	12,238	28	636	6,934,912	532,000	30,629	39,549	126,706

Financial.—*First Cost,* $874,566. *Bonded Debt,* $532,000: $237,000 at 6%, $150,000 at 5%, $60,000 at 4½%, $25,000 at 4%, $60,000 at 3½%; city debt, including the above, is subject to a sinking fund scheme which cancels it in 1925. *Expenses,* total, $70,178. *Revenue,* consumers, $126,706. An act creating the Water Commission provides that rates to consumers must be large enough to pay all expenses. Net revenue is paid into city treasury. **Management.**—Three com'rs, appointed every six years by mayor. Chn., Wm. J. Porter. Secy., John S. Grohe. Supt., Jos. A. Bond; appointed by com'rs.

Water-Works Projected With Prospects of Construction.

MILFORD, *Kent Co.* (Pop., 1,240—1,226.) Contracts were let on Sept. 2, '91, for parts of the works, to be owned by town. Engr., I. C. Cassin, Philadelphia. Supply.—Well, pumping to stand-pipe. About 4 miles of mains and 50 hydrants are projected at est. cost of $25,000. Chn., Nathan Pratt.

MARYLAND.

Water-Works Completed or in Process of Construction.

1. ANNAPOLIS, *Anne Arundel Co.* (Pop., 6,642—7,604.) Built in '65-6 by Annapolis Water Co.; city neither controls rates nor has right to buy works. Supply.—Springs and surface water, pumping from impounding to distributing reservoir; filtered. Fiscal year closes Feb. 23. Distribution.—*Mains,* 8 miles. *Taps,* 525. *Hydrants,* 42. **Financial.**—*Cost,* $61,771. *Cap. Stock:* authorized, $6,450; all paid up; paid 7% div'd in 1890. *Bonded Debt,* $10,000 at 6%, due in '09. *Expenses:* operating, $10,467; int., $693; total, $11,160. *Revenue:* consumers, $12,906; city, $815; total, $13,721. **Management.**—Prest., Jas. H. Brown. Secy. and Treas., J. Harwood Igehart. Supt., Edward Powers.

2. BALTIMORE, *City of Baltimore Co.* (Pop., 332,313—434,439.) Owned by city since '54. Built in '07 by Co. Supply.—Impounding reservoirs, pumping to distributing reservoirs. Fiscal year closes Dec. 31.

Year	Mains	Taps	Mtrs	Hd'ts	C'n'p'n	Cost	B'd't d't	Int. ch.	Exp.	Rev.
'60	278	48,609	..	825	......	......	......	......	......	......
'82	290	52,876	677	892	25,000,000	$10,197,867	$9,000,000	......	$93,012	637,212
'83	301.3	57,238	...	913	28,071,312	10,375,649	9,500,000	......	98,917	611,112
'86	327	61,276	717	971	26,878,715	10,738,018	9,500,000	......	101,952	692,614
'87	358	66,089	784	1,042	34,000,000	10,896,979	9,500,000	472,630	108,091	741,679
'88	380	68,000	791	1,169	37,069,250	11,212,218	9,900,000	480,630	106,658	598,168
'89	406.2	73,017	849	1,215	36,838,451	11,539,820	10,100,000	488,630	200,266	589,212
'90	422	74,728	913	1,815	40,978,229	12,018,377	12,100,000	496,630	144,121	607,803

Financial.—*Bonded Debt,* $10,109,000: $263,000 at 6%, due Ir. '91; $8,737,000 at 5%, due in '94; $500,000 at 4%, due in 1922; $400,000 at 4%, due in 1926. *Expenses,* total, $640,751. *Sinking Fund* in hands of financial committee. *Revenue:* consumers, $607,803; city, $51,588; total, $659,391. **Management.**—Board of which Mayor is ex-officio president, appointed every two years by Mayor. Prest., Robert C. Davidson. Secy., Jno. F. Hunter. Supt., Chf. Engr., Robt. K. Martin.

NOTE.—For description of aqueduct bridge truss see *Engineering Record*, Mar. 14, '91.

3. BEL AIR, *Hartford Co.* (Pop., in 90, 1,116.) History.—Built in '90 by Belair Water & Light Co.; put in operation Oct. 1. Des. Engr. A. R. T. Lackie. Const. Engr. and Contr., Jas. H. Harlow & Co., Pittsburgh, Pa. Supply.—Spring, by gravity to reservoir. Reservoir.—Cap., 1,700,000 galls. Distribution.—*Mains,* 8

MARYLAND. 153

to 4-in. c. i., 6 miles. *Taps,* 60. *Meters,* 2. *Hydrants,* 40. *Pressure,* 40 lbs. **Financial.**—*Cost.* $10,000. *Int.*, 6%. *Revenue:* city, $800. **Management.**—Prest., Otho S. Lee. Secy., Geo. R. Cairnes. Treas., Richard Dallam. Supt., Jno. S. Dallam.

4. BRUNSWICK, *Frederick Co.* (Pop., not given.) **General Description.**—Construction of works was begun during summer of '91 by Brunswick Water Co. Engr., R. A. Rager, Frederick. **Supply.**—Mountain stream, by gravity to and from reservoir. Prest., Chas. W. Wenner. Secy., W. L. Gross. Address Thos. Cannon.

5. CATONSVILLE, *Baltimore Co.* (Pop., 1,712–2,115.) Built in '87 by Catonsville Water Co. **Supply.**—Artesian well and stream, pumping to tank. Fiscal year closes May 1. **Distribution.**—*Mains,* 10 miles. *Taps,* 150. *Hydrants,* 10. *Consumption,* 200,000 galls. **Financial.**—*Cost,* $93,000. *Cap. Stock,* $33,000. No debt. *Revenue,* $7,000. **Management.**—Prest., Jos. M. Cone. Secy., Miles W. Ross. Supt., F. A. McCrea.

6. CENTREVILLE, *Queen Anne Co.* (Pop., 1,193–1,309.) Built in '89 by town. **Supply.**—Stream, pumping to stand-pipe. Fiscal year closes April 1. Filter described in '89-90 Manual has been abandoned. **Distribution.**—*Mains,* 25 miles. *Hydrants,* 40. **Financial.**—*Total Debt,* $16,000; floating, $1,000; bonded, $15,000 at 5%, due in 1937. **Management.**—Chn., W. W. Busterd. Secy., Geo. A Whitely. Supt., Horace Vasey, appointed by comrs.

7. CHESTERTOWN, *Kent Co.* (Pop., 2,350–2,632.) Built in '85 by Chester town Water Co; city neither controls rates nor has right to purchase works. **Supply.**—Springs, pumping from collecting to distributing reservoir. **Distribution.**—*Mains,* 6 miles. *Taps,* 210. *Meters,* 6. *Hydrants,* 29. *Consumption,* 50,000 galls. **Financial.**—*Cost,* $25,000. *Cap. Stock:* common, $10,000; preferred, $5,000. *No Debt. Expenses:—* operating, $1,150; taxes, $150; total, $1,300. *Revenue,* city, $550. **Management.**—Prest., Chas. T. Westcott. Secy., M. W. Thomas.

8. CUMBERLAND, *Allegheny Co.* (Pop., 10,693–12,729.) Built in '80-1 by city. **Supply.**—Potomac River, pumping direct. To secure pure water, supply pipes have been extended 1,300 ft. farther up Potomac. **Supply Main.**—A 21-in. vitrified terra cotta pipe; brings water to deep well, at right angles with river; conduit contains two screens and a gate. Nearly all the 3 and 4-in. mains have been replaced with 6-in. mains. **Distribution.**—(In '87.) *Mains,* 30 miles. *Taps,* 1,800. *Meters,* 0. *Hydrants,* 200. **Financial.**—*Cost,* $150,000. *Bonded Debt,* $150,000 at 5%. *Expenses,* $7,000. *Revenue,* $10,000. **Management.**—City Council. Address Sam'l P. Gaffrey.

9. EASTON, *Talbot Co.* (Pop., 3,005–2,939.) Built in '86 by Easton Water Co.; town may buy works after 10 yrs.; has no control of rates. **Supply.**—Artesian well-, pumping to stand-pipe. Fiscal year closes June 30. **Distribution.**—*Mains,* 6.5 miles. *Taps,* 260. *Meters,* 1. *Hydrants,* 10. *Consumption,* 40,000 galls. **Financial.**—*Cost.* $10,000. *Cap. Stock,* $50,000. *No Debt. Expenses,* $1,300. *Revenue:* city, $1,100. **Management.**—Prest., Jos. H. White. Secy., Wm. E. Shannahan. Treas., Robert B. Dixon.

10. ELKTON, *Cecil Co.* (Pop., in '90, 2,318.) **General Description.**—During summer of '91, Elkton Water Co. was granted franchise and let contracts for construction of works, which were to be completed Oct. 1. Des. Engr., I. S. Cassin, Philadelphia, Pa. Const. Engrs., Young Bros., Philadelphia, Pa. **Supply,** wells, pumping to 1,000,000-gall reservoir. *Mains,* 5 miles. *Cost,* $30,000. Capital secured. Prest., W. T. Warburton. Comr., Geo. A. Blake.

11. FREDERICK, *Frederick Co.* (Pop., 8,659–8,193.) Built in '44 by city; previously supplied by Co. **Supply.**—Springs, artesian wells, and mountain stream, by gravity. In '91 new supply was projected. **Distribution.**—(In '88.) *Mains,* 10 miles. *Meters,* 1. **Financial.**—*Cost* (in '83), $125,091. *Bonded Debt* (in '82), $70,000 at 5%. **Management.**—Supt., C. N. Hahn. NOTE.—Enquiries sent since '89 have been ignored.

12. FROSTBURGH, *Allegheny Co.* (Pop., 4,057–3,801.) Built in '54 by Frostburgh Water Co.; town neither controls rates nor has right to buy works. **Supply.**—Springs and four artesian wells, by gravity from collecting reservoir, with tank for fire protection. Fiscal year closes Dec. 31. In '90 contract was let for new reser-

FOR FURTHER DESCRIPTIVE MATTER see Manual for 1889-90. CHANGES since 1890 are here given. POPULATIONS are for 1880 and 1890, respectively.

voir. Distribution.—*Mains.* 4.5 miles. *Taps*, 315. *Meters*, 0. *Hydrants*, 33. Consumption, 30,000 galls. Financial.—*Cost*, $27,500. *Cap. Stock:* authorized, $40,000; all paid up; no div'd, net revenue used in extension. *Total Debt*, $4,100: floating, $1,600; bonded, $2,800 at 6%. *Expenses:* operating, $450; int., $223; taxes, $86; total, $759. *Revenue:* consumers, $2,312; city, $150; total, $2,463. Management.—Pres., W. O. Sprigg. Secy., Carroll Sprigg. Treas., Hazlehurst Sprigg. Supt., D. J. Williams.

13. **HAGERSTOWN,** *Washington Co.* (Pop., 6.627—10,118.) Built in '81-2 by Washington County Water Co.; rates are subject to approval of city, but it has no right to buy works. Supply.—Beaver Creek and wells, by gravity. Fiscal year closes June 30. During '90-1 an open timbered well 80 ft. deep was dug, in which a 500 000 gall. Cameron pump was placed, which forces water into reservoir a few feet distant. In '91 a 1,000,000-gall. Smith & Vaile was also put in.

Year.	Mains.	Taps.	Meters.	Hydrants.	Cost.	B'd'd Debt.	Int. ch'g's.	Exp.	Rev.
'84	15.3	325	5	56	$123,500	$40,000	$2.000	$1,000	$9,000
'87	18	640	17	56	140,000	40,000	1,600	1,200	9.000
'89	13.5	710	20	58	150,000	40,000	1,600	3,527	10,919
'90	19	778	20	58	150,000	40,000	1,600	3,536	11,312

Financial.—*Cap. Stock:* authorized, $80,000; all paid up; paid 6% div'd in '90. *Bonded Debt*, $40,000 at 4% *Expenses:* operating $2,100; int., $1,600; taxes, $1,133; total, $5,133. *Revenue:* consumers, $10,312; city, $1,000; total, $11,312. Management.—Prest., H. H. Keedy. Secy., Edward Stake. Treas and Supt., C. F. Manning.

14. **HAVRE DE GRACE,** *Hartford Co.* (Pop., 2,816—3,244.) Built in '83-4 by Havre de Grace Water Co. Supply.—Susquehanna River, pumping to reservoir. Fiscal year closes Dec. 31. Distribution.—*Mains*, 2 miles. *Taps*, 130. *Meters*, 2. *Hydrants*, 32. Financial.—*Cost*, $20,000. *Debt*, $15,000 at 6%. *Expenses:* operating $1.100; int. $900; total, $2,000. *Revenue:* consumers, $1,700; city, $500; total, $2,200. Management.—Prest., A. P. McCombs. Secy., Chas. Zeitler. Treas., R. K. Vaunemaux. Supt. J. W. Maslin.

15. **MECHANICSTOWN,** *Frederick Co.* (Pop., 730—930.) Built in '87 by Mechanicstown Water Co.; city neither controls rates nor has right to buy works. Supply.—Mountain springs, by gravity. Fiscal year closes in July. Distribution.—*Mains*, 3 miles. *Hydrants*, 18. Financial.—*Cost*, $10,000. *Cap. Stock:* authorized, $10,000; all paid up; paid 6% dividend in 1890. *Expenses*, $100. Management.—Prest., Van B. Osler. Secy. and Supt., S. M. Birely.

16. **SALISBURY,** *Wicomico Co.* (Pop., 2,581—2,905.) Built in '88 by Salisbury Water Co.; Des. Engr., L. S. Bell, not Isaac Cassin, as stated in 1889-90 Manual. Supply.—Wells for domestic use and Wicomico River for fire protection, pumping to stand-pipe and direct. Supply was increased in '90 by eight 6-in artesian wells. Distribution.—*Mains*, 5 miles. *Taps*, 170. *Meters*, 3. *Hydrants*, 7. *Consumption*, 50,000 galls. Financial.—*Cost*, $27,500. *Cap. Stock:* authorized, $50,000; paid-up, $8,100. *Bonded Debt*, $15,000 at 5%, due in 1910. *Expenses:* operating, $1,300; int., $250. *Revenue*, not determined. Management.—Prest., S. P. Dennis. Secy. and Supt., L. S. Bell. Treas., Wm. H. Jackson.

17. **UNION BRIDGE,** *Carrol Co.* (Pop., 579—743.) Built in '86-7 by Union Bridge Water Co.; city has no control of rates; may purchase works at any time by paying actual cost and 6% int. on investment. Supply.—Springs, pumping from receiving to distributing reservoir. Fiscal year closes Dec. 31. Distribution.—*Mains*, 2.8 miles. *Taps*, 110. *Meters*, 115. *Hydrants*, 18. *Consumption*, 40,000 galls. Financial.—*Cost*, $17,000. *Cap. Stock:* authorized, $12,000, all paid up. *Bonded Debt*, $5,000 at 5 and 6%. *Expenses:* operating, $300; int., $290; total, $590. *Revenue:* consumers, $510; city, $340; total, $850. Management.—Prest., M. C. McKinstry. Secy., Jno. S. Repp. Treas., Sol. T. Shepherd. Supt., J. M. Hollenberger.

18. **WESTMINSTER,** *Carroll Co.* (Pop., 2,507—2,903.) Built in '83 by Westminster Water Co.; town neither controls rates nor has right to buy works. Supply.—Three springs, pumping from receiving to distributing reservoir. Fiscal year closes May 1.

Year.	Mains.	Taps.	Meters.	Hyd'ts.	Cons'pt'n.	Cost.	D'd'd d't.	Int.chgs.	Exp.	Rev.
'84-5	5	175	0	40	.....	$25,000	$20,000	$1,200	$1,000	$2,700
'89-90	5.1	200	10	45	12,000	.....	20,000	1,200	1,030	3,400
'90-1	5.3	315	11	45	13,000	46,000	20,000	1,200	2,030	3,500

Financial.—*Cap. Stock*.—Common, authorized, $20,000; all paid up; paid 6% div'd. Pref'd; authorized, $5,000; all paid up; guaranteed div'd, 7%. *Total Debt*, $20,500: floating, $500; bonded, $20,000 at 6%. *Expenses:* operating, $1,000; int., $1,550; taxes,

MARYLAND. 155

$30; total, $2,580. *Revenue:* consumers, $2,500; city, $900; total, $3,400. **Management.**—Prest., Wm. A. McKellip, Secy., A. H. Huber. Treas., Wm. B. Thomas. Supt., F. A. Meyers.

Water-Works Projects Reported in '90 and '91, but Now Indefinite.

Cambridge, Dorchester Co.

DISTRICT OF COLUMBIA.

Water-Works Completed.

1. WASHINGTON and GEORGETOWN. (Pop.: Washington, 147,293; Georgetown, 12.578; total, 159,871)—figures for 1890 are evidently for whole district, 230,392. Built in '53-9 by U. S. Government. Fiscal year closes June 30.

Year.	Mains.	Taps.	M't'rs.	Hyd'ts.	Cons'pt'n.	Int. and S'k'g Fund.	Exp.	Rev.
'79-80	...	...	...	...	...	$74.025	$95,551	$165,611
'80-1	175.6	...	...	825	26 000.0 0	74,124	66,623	109,738
'81 2	...	...	...	...	...	43,796	51,173	101,021
'82-3	176.3	20,534	...	833	26,525 991	44,610	66,841	95,752
'83-4	...	...	...	...	...	44,575	62,994	119,610
'84-5	165	22,400	0	848	26,500,000	58 296	111,739	118,528
85 6	170	26,280	4	925	23,000,000	86,532	80,739	121,896
'86 7	128	22,095	...	1,013	26,147,052	57,735	86,939	138,559
'87-8	186	27,639	14	1,017	26,518,914	88,724	111,936	171,892
'88 9	201.9	31,219	62	1,062	...	121,266	147,836	189,407
'89-90	215.1	33,270	87	1,080	...	125,893	104,336	197,053

Financial.—*Expenses,* total, $230,229. *Revenue:* rents, $197,053; taxes, $45,387; taps, $5,311; permits, $6,328; total. $254,083. **Management.**—Engr. Comr. of D. C., Col. Henry M. Robert. Supt., H. F. Hayden.

GROUP 3.—South Atlantic States.—Water-Works Comp'eted, in Process of Construction or Projected, in Virginia, West Virginia, North Carolina, South Carolina, Georgia and Florida.

VIRGINIA.

Water-Works Completed or in Process of Construction.

1. ABINGDON, *Washington Co.* (Pop., 1,061–1,674.) Built in '80-'90 by Abingdon Water-Works Co. Fiscal year closes Dec. 31. **Distribution.**—*Mains,* 5 miles. *Taps,* 42. *Meters,* 0. *Hydrants,* 19. *Consumption,* 30 000 galls. **Financial.**—*Cost,* $18,000. *Cap. Stock:* authorized, $25,000, including electric lighting plant; paid-up, $19,000; paid 6% div'd in 1890. *No Debt. Expenses,* about $1,500. *Revenue:* consumers, $4,000; city, $950; total, $4,950. **Management.**—Prest., T. P. Trigg. Secy. and Treas., R. M. Page. Supt., T. L. Pickle.

2. ALEXANDRIA, *Alexandria Co.* (Pop., 13,659—14,339.) Built in '58 by Alexandria Water Co.; franchise provides neither for control of rates nor purchase of works by city. **Supply.**—Cameron's run, pumping to reservoir. **Distribution.** —(In '81.) *Mains,* 15.9 miles. *Taps,* 573. *Hydrants,* 147. **Financial.**—(In '79.) *Expenses,* $1,241. *Revenue,* $14,631. **Management.**—Prest., Benona Wheal. **NOTE.**—Co. has ignored all inquiries since '81.

I. BASIC CITY, *Augusta Co.* (Pop., not given.) Built in '90-1 by Basic City Mining, Mfg. & Land Co.; put in operation Feb. 1, '91. Engr. and Contr., W. B. Holtzclaw. **Supply.**—Spring, pumping to reservoir. **Pumping Machinery.**- Dy. cap., 1,000,000 galls.; Deane. **Reservoir.**—Cap., 330,000 galls. **Distribution.**— There are 1¼ miles of 8-in. c. i. pipes, supplying two hotels and a few private families. **Management.**—See above.

FOR FURTHER DESCRIPTIVE MATTER see Manual for 1889-90 CHANGES since 1890 are here given. POPULATIONS are for 1880 and 1890, respectively.

156   MANUAL OF AMERICAN WATER-WORKS.

3. BEDFORD CITY (*formerly known as Liberty*), *Bedford Co.* (Pop., 2,191, —2,897.) Built in '85 by town. Supply.—Spring, by gravity. Distribution.—(In '88.) *Mains,* 12 miles. *Taps,* 350. *Meters,* 0. *Hydrants,* 32. Financial.—(In '88.) *Cost,* $45,000. *Bonded Debt,* $42,400 at 6%. *Expenses:* $500. *Revenue:* consumers, $3,200. Management.—City Council. Supt., Jno. S. Hancock.

4. BERRYVILLE, *Clarke Co.* (Pop., 2,629–2,610.) History.—Built in '91 by Berryville Water Co., under 20-yrs. franchise; put in operation Sept. 1. Des. Engrs., A. Y. Lee, Pittsburg, Pa., and Robt. Martin, Baltimoro, Md. Const. Engr., Jas. P. Herdic, Williamsport, Pa. Supply.—Impounding reservoir, supplied by springs in Blue Ridge Mts., by gravity. Reservoir.—Cap., 1,600,000 galls.; dam is 80 ft. high. Distribution.—*Mains,* 12 to 4-in. c. i., 12 miles; pipes and specials from R. D. Wood & Co., Philadelphia, Pa. *Meters,* Crown. *Hydrants,* Eddy, 30. *Valves,* Eddy. *Consumption,* 60,000 galls. *Pressure,* 96 lbs. Financial.—*Cost,* $52,000. *Cap. Stock:* authorized, $100,000; paid up, $30,000. *Bonds:* authorized, $25,000 at 6%; issued, none. *Operating Expenses,* $250. *Revenue,* city, $775. Management.—Prest., Geo. C. Thomas. Secy. and Treas., Thos. S. Thompson.

5. BIG STONE GAP, *Wise Co.* (Pop., 1,333–2,484.) History.—Built in '90-1 by Big Stone Gap Water Co.; put in operation June 1, '91. Engr., C. A. Alden. Contrs., Glamorgan Co., Lynchburg. Supply.—Howell's River, by gravity to reservoir. Reservoir.—Cap., 3,500,000 galls. Distribution.—*Mains,* c. i., 13 miles. *Hydrants,* Mathews, 30. *Valves,* Ludlow. *Pressure,* 180 lbs. Financial.—*Cost,* $65,000. *Cap. Stock:* authorized, $200,000; paid up, $200,000. *Bonded Debt:* authorized $100,000; issued, $65,000. Management.—Prest., Wm. McGeorge, Jr. Secy. and Treas., W. A. McDowell, Jr.

6. BRISTOL (*Bristol, Tenn., P. O*), *Washington Co.* (Pop., 1,562–2,902.) Built in '80 by Co. Supply.—Mountain stream, by gravity. Fiscal year closes Dec. 31. Distribution.—*Mains* (in '89), 7 miles. *Taps,* 290. *Meters,* 0. *Hydrants* (in '89), 40. Financial.—*Cost,* $33,000. *Bonded Debt,* $33,000 at 6%. *Expenses:* operating, $300; int., $1,990; taxes, 0; total, $2,290. *Revenue:* consumers, $1,000; fire protection, none; other purposes, $900; total, $1,900. Management.—Secy., J. H. Winston, Jr. Treas., J. L. Smith. Supt., A. R. Littlejohn.

7. BUCHANAN, *Botetourt Co.* (Pop., 5,694–5,867.) History.—Built in '91 by Central Land Co., to aid in development of town; put in operation Sept. 10. Engr., W. L. Gordon. Contr., Standard Construction Co., Newport, Ky. Supply.—James River, pumping to stand-pipe. Pumping Machinery.—D'y. cap., 500,000 galls.; Worthington dup. Stand-Pipe.—Cap., 50,000 galls.; 18 × 26 ft.; of steel. Distribution.—*Mains,* 6-in. c. i., 1¾ miles. *Taps,* 25. *Hydrants,* Bourbon, 20. *Pressure:* ordinary, 100 lbs.; fire, 130 lbs. Financial.—*Cost,* $75,000. *No Debt.* Management.—Prest., C. M. Clark, Philadelphia, Pa. Vice-Prest., Mosby H. Payne. Secy. and Treas., Jno. M. Miller.

8. BUENA VISTA, *Rockbridge Co.* (Pop., in '90, 1,044.) History.—Built in '90 by town; put in operation in Sept. Supply.—Mountain streams, by gravity. Reservoir.—Cap., 2,000,000 galls. Distribution.—*Mains,* 15 to 4-in. c. i. *Taps,* 95. *Hydrants,* Ludlow, 20. *Valves,* Ludlow. *Consumption,* 10,000 galls. *Pressure,* 100 lbs. Financial.—*Cost,* $20,000. *No Debt.* *Revenue,* about $2,000. Management. —C. Bargainn.

9. CHARLOTTESVILLE, *Albemarle Co.* (Pop., 2,676–5,591.) Built in '86 by city. Supply.—Surface water and springs, by gravity. Distribution.—(In '88.) *Mains,* 15 miles. *Taps,* 450. *Meters,* 0. *Hydrants,* 60. *Consumption,* 10,000 galls. Financial.—(In '88.) *Cost,* $115,000. *Bonded Debt,* $70,000 at 6%. *Expenses:* operating, $900; int., $4,200; total, $5,100. *Revenue:* consumers, $5,000; city, none. Management.—Com.: H. T. Nelson, C. D. Carter, A. M. Peyton. Supt., T. J. Williams.

10. CHATHAM, *Pittsylvania Co.* (Pop., 513–757.) Built in '88 by town. Supply.—Springs, by gravity. Distribution.—(In '88.) *Mains,* 2 miles. *Taps,* 4. *Hydrants,* 4. *Consumption,* 1,000 galls. Financial.—(In '88.) *Cost,* $7,000. *Expenses,* $25. Management.—Comrs., Chas. E. Holt, J. Q. De Nutt.

11. CLIFTON FORGE, *Alleghany Co.* (Pop., '90, 1,792.) Built in '88-'90 by Clifton Forge Water Co. Supply.—Smith's creek, pumping to reservoir. Fiscal year closes April 1. Distribution.—*Mains,* 3.8 miles. *Taps,* 90. *Meters,* 0. *Hydrants,* 10. *Consumption,* 20,000 galls. Financial.—*Cost,* $14,000. *Cap. Stock:* authorized, $100,000; paid up, $70,000. *No debt.* *Expenses:* operating, $1,235. *Revenue:* consumers, $1,205, including $800 profit on plumbing; city, $208; total, $1,413. Manage

VIRGINIA.

ment.—Prest., Chas. E. Whitlock; Vice-Prest., B. W. Branch;] both of Richmond. Secy. and Supt., Edw. W. Scott.

12. **COVINGTON,** *Allegheny Co.* (Pop., 436—704.) Built in '89 by Covington Water Co.; franchise fixes maximum rates and provides for purchase of works by city in 1914. Supply.—Spring branch, by gravity. Fiscal year closes Dec. 31. Distribution.—*Mains*, 4 miles. *Taps*, 65. *Meters*, 0. *Hydrants*, 11. *Consumption*, unk'n. Finan ial.—*Cost*, $8,000. *No debt. Expenses*, $50. *Revenue*, $765. Management.—Prest. and Mgr., E. M. Nettleton.

13. **CREWE,** *Nottoway Co.* (Pop., in '90, 887.) Built in '88 by Virginia Water Co., now owned by Crewe Land & Improvement Co. Supply.—Four springs, pumping to tank. Distribution.—*Mains*, 4 miles. *Taps*, 40. *Meters*, 1. *Hydrants*, 21. Financial.—*Cost*, $25,000. Management.—Pres., J. H. Dingee, 330 Walnut St., Philadelphia, Pa. Secy. and Treas., Malcolm Bryan. Supt.· C. D. Epes.

14. **DANVILLE,** *Pittsylvania Co.* (Pop., 7,526—10,305.) Built in '75, by co., and purchased by city in '76, including gas works. Supply.—Dan river, pumping to reservoir and direct. Distribution.—*Mains* (in '80), 8.2 miles. *Taps* (in '88), 760. *Meters*, 0. *Hydrants* (in 88), 87. *Consumption*, 1,000,000 galls. Financial.—Withheld. Management.—Water committee, appointed by city council every year. Chn., J. G. Covington. Secy., Frank Smith. Supt., C. A. Ballou.

11. **ELKTON,** *Rockingham Co.* (Pop., not given.) Elkton Lithia Water Co. supplies water through system of pipes to all residences and stores desiring it. No further information could be obtained.

15. **FREDERICKSBURGH,** *Spottsylvania Co.* (Pop., 5,010—4,528.)
FREDERICKSBURGH AQUEDUCT CO.
Built in '32. Supply.—Poplar Springs. Address P. V. Conway, Secy.

CITY'S SYSTEM.

Built in '84-5. Supply.—Rappahannock River, pumping to reservoir. Fiscal year closes June 30.

Year.	Mains.	Taps.	Meters.	Hydt's.	Cost.	B'd'd d't.	Int. c'g's.	Exp.	Rev.
'88-9	9.5	693	1	50	$30,000	$30,000	$1,800	$1,263	$4,079
'89-90	10	769	1	50	35,000	30,000	1,800	.....	2,671*

* Works were inoperative for 4 months, on account of destruction of dam by flood.

Financial.—*Bonded Debt*, $50,000, at 6%, due '95—1915. *Sinking Fund*, 10% of debt, invested in city bonds. *Revenue*, consumers, $2,671. Policy of city council is to place rates as low as possible, and pay expenses and sinking fund. Management.—Committee of three, appointed by city council every two years. Chn., M. G. Willis. Secy. and Supt., S. J. Quinn.

16. **GATE CITY,** *Scott Co.* (Pop,, not given.) Estillville Water Co. built system in '88-89, under 25-yrs'. franchise. Contr., E. A. Hague. Supply.—Impounding reservoir, by gravity. Reservoir.—Cap., 100,000 galls. Distribution.—*Mains*, 4-in., 2 miles. *Taps*, 40. *Hydrants*, 15. Financial.—*Cost*, $1,500. *Cap. Stock*, $1,500. *No Debt. Expenses*, $150. *Hydrant rental*, $25. Management.—Piest., W. F. Edmonds. Secy., W. F. Hoekam. Supt., John A. Quillin.

17. **GLASGOW,** *Rockbridge Co.* History.—Built in '96 by Rockbridge Co.; put in operation July 1. Engr., A. D. Exall. Supply.—Springs in mountains, by gravity. Reservoir.—Cap., 3,000,000 galls. Dam of masonry, backed by puddled clay, riprapped; 12 ft. high, 75 ft. long at top, 45 ft. long at bottom, 10 ft. below bed of stream. Distribution.—*Mains*, c. 1., 5¼ miles. *Services*, 2 in. w. i. Pipes and specials from Glamorgan Works, Lynchburg. *Taps*, 45. *Hydrants*, Galvin, 12. *Valves*, Galvin. *Pressure*, 80 lbs. Financial.—*Cost*, $15,000. Management.—Prest., Fitzhugh Lee. Secy. and Treas., T. Williams.

6. **GOODSON** (*Bristol, Tenn., P. O.*), *Washington Co.* (Pop., not given.) Supply from Bristol.

18 **GORDONSVILLE,** *Orange Co.* (Pop., 919—932.) Built in '89 by town. Supply.—Springs, by gravity. Fiscal year closes June 30. Distribution.—*Mains*, 3 miles. *Taps*, 120. *Meters*, 0. *Hydrants*, 14. Financial.—(In '89.) *Cost*, $1,000. *Debt*, $1,200 at 6%. Management.—Chn., C. F. Ayer. Supt., W. H. Porter.

FOR FURTHER DESCRIPTIVE MATTER see Manual for 1889-90. CHANGES since 1890 are here given. POPULATIONS are for 1880 and 1890, respectively.

158   MANUAL OF AMERICAN WATER-WORKS.

HAMPTON, *Elizabeth Co.* (Pop., 2,684—2,513.) April 30, 1891. H. R. Booker, Secy. of Hampton Water Co., reported that works had suspended operation, and co, was searching for a better water supply. See 1889-90 Manual, p. 302.

19. **HARRISONBURGH**, *Rockingham Co.* (Pop., 2,831—2,792.) Built in '88-9 by town. **Supply.**—Artesian wells, pumping to reservoir. Fiscal year closes Dec. 21. **Distribution.**—*Mains*, 4.5 miles. *Taps*, 76. *Meters*, 2. *Hydrants*, 22. *Consumption*, 42,485 galls. **Financial.**—*Cost*, $25,000. *Bonded Debt*, $25,000 at 6%. *Expenses:* operating, $1,410; int., $1,500; total, $2,910. *Revenue:* consumers, $1,528; city, none. **Management.**—Committee appointed every two years by mayor. Chn., R. E. Sullivan. Supt., P. S. Thomas.

37. **JEFFERSONVILLE** (*Tazewell P. O.*), *Tazewell Co.* (Pop., 598—601. Supply from Tazewell.

III. **LEESBURGH**, *Loudoun Co.* (Pop., 1,726—1,650.) Built in '45 by village. **Supply.**—Rock spring, by gravity. Fiscal year closes June 30. **Distribution.**—*Mains*, 4 miles. *Hydrants*, 0. **Financial.**—*Cost*, $20,000. No *Debt*. *Expenses*, $250. *Revenue*, $300. **Financial.**—See above. **Management.**—Committee of three appointed every year by mayor. Chn., W. J. Harrison. Secy., B. F. Sheetz. Supt., Geo. R. Head.

20. **LEXINGTON**, *Rockbridge Co.* (Pop., 2,771—3,059.) Built in '33 by town. **Supply.**—Springs, by gravity to and from receiving well and reservoir. Fiscal year closes June 30. **Distribution.**—*Mains*, 10 miles. *Taps*, 400. *Meters*, 3. *Hydrants*, 29. *Consumption*, 75,000 galls. **Financial.**—*Cost*, $33,600. *Bonded Debt*, $37,500 at 6%. *Expenses:* operating, $300; int., $2,250; total, $2,850. *Revenue:* consumers, $4,000. **Management.**—City Council. Supt., Jos. B. Holmes.

3. **LIBERTY**, *Bedford Co.* See Bedford City.

21. **LYNCHBURGH**, *Campbell Co.* (Pop., 15,950-19,709.) Built in '28 by city; first supplied in 1790 by Co. **Supply.**—James River, pumping to low and high service reservoirs. Fiscal year closes Dec. 31. Steam pumps and filters were to be introduced in '91. **Distribution.**—*Mains*, 21 miles. *Taps*, 1,853. *Meters*, 18. *Hydrants*, 160. *Consumption*, 2,462,042 galls. **Financial.**—*Cost*, $374,051. *Bonded Debt*, not separate from general debt of city. *Expenses:* operating, $10,031. *Revenue:* consumers, $22,320; city, $8 500; total, $30,910. **Management.**—Committee of three. Chn., W. D. Adams. Clerk, T. W. Green. Supt., J. B. Page.

22. **MANCHESTER**, *Chesterfield Co.* (Pop., 5,729—9,243.) Built in '88-9 by White Bros.; first supplied in '40 by town, but this system has been abandoned. **Supply.**—James River, pumping to reservoir and direct in case of fire. Fiscal year closes April 1. **Distribution.**—*Mains*, 9.5 miles. *Taps*, 420. *Meters*, 4. *Hydrants*, 52. **Financial.**—Withheld. **Management.**—Prest. and Treas., W. H. White. Secy. and Supt., H. S. Bradley.

23. **MARION**, *Smyth Co.* (Pop., 919—1,651.) Built in '83 by town **Supply.**—Springs, by gravity to stand-pipe. Fiscal year closes July 1. **Distribution.**—*Mains*, 3.5 miles. *Taps*, 91. *Hydrants*, 13. **Financial.**—*Cost*, $11,750. *Bonded Debt*, $11,750 at 6% due in 1903. *Expenses:* operating, $25; int, $705; total, $730. *Revenue*, consumers, $800. **Management.**—Chn., J. P. Sheffey. Secy., J. H. Gallehan. Supt., F. S. Snavley.

24. **MAX MEADOWS**, *Wythe Co.* (Pop., not given.) **History.**—Built in '90-1 by Max Meadows Land & Improvement Co.; to be in operation Oct. 1, '91. Engr., Geo. P. Maury. Contrs., S. B. Mosby & Co., Bedford Cy. **Supply.**—Reed Creek, pumping to reservoir. **Pumping Machinery.**—Not determined. **Reservoir.**—Cap., 1,000,000 galls. **Distribution.**—*Mains*, 10-in. c. i., 4.5 miles; pipes and specials from Glamorgan Co., Lynchburgh. *Hydrants*, Ludlow, 12. *Valves*, Ludlow. *Pressure*, 100 lbs. **Financial.**—Est. *Cost*, $300,000. **Management.**—Prest., Clarence M. Clark, Vice-Prest., E. C. Pechin. Secy. and Treas., Edward Isey, Bullitt Building, Philadelphia, Pa. Mgr., H. C. Baker.

25. **NORFOLK**, *Norfolk Co.* (Pop., 21,966—34,871.) Built in '72 by city. **Supply.**—First Moores' Bridges, Lakes Lawson and Bradford, pumping direct. Fiscal year closes June 30. Contracts were let in '91 for 10,000,000 and 5,000,000-gall. Worthington pumping engines.

VIRGINIA. 159

Year	Mains	Taps	M't'rs	H'd'ts	Cons'pt'n	Cost	B'd'd d't	Int. ch'g's	Exp	Rev
'80-1	22	2,100	..	115	......	......	......	......	$12,500	......
'82-3	26	2,455	77	122	810,000	$531,618	$500,000	......	15,023	$33,513
'83-4	27.4	2,758	87	123	920,316	531,618	500,000	......	16,175	37,643
'94	..	..	..	123	......	......	......	......	......	......
'85-6	28	3,676	70	124	1,400,000	590,000	572,500	$43,825	15,078	41,660
'86-7	30	3,925	70	126	2,377,848	795,030	640,000	47,000	16,049	50,825
'87-8	32	4,177	85	126	1,861,522	837,000	610,000	47,000	18,155	55,278
'88-9	..	4,378	..	...	3,154,000	855,459	640,000	49,217	19,850	60,312

Management.—(In '83.) Three comrs. Secy., D. S. Burwell. Registrar, Jno. R. Todd. Supt., R. Y. Zachery.

26. **PETERSBURGH,** *Dinwiddie Co.* (Pop., 21,656—22,680.) Built in '56 by city. **Supply.**—Small brook, pumping from impounding to distributing reservoir. **Distribution.**—(In '87.) *Mains*, 11 miles. *Taps*, 1,930. *Meters*, 0. *Hydrants*, 110. *Consumption*, 750,000 galls. **Financial.**—(In '87.) *Cost*, $150,000. *Bonded debt*, $125,000. *Operating Expenses*, $9,000. *Revenue*, $12,000. **Management.**—Supt., Jno. S. Cooke

27. **PORTSMOUTH,** *Norfolk Co.* (Pop., 11,390—13,268.) Built in '83 by Portsmouth Water Co. **Supply.**—Kilby's Lake, pumping through stand-pipe to reservoir; thence pumped directly to mains. Fiscal year closes Dec. 31. During '89-90 one Knowles' steam and two Holly water-power pumps were added.

Year	Mains	Taps	Meters	Hd'ts	Cons,pt'n	B'd'd d't	Int. ch'g's	Exp.	Rev.
'88	30.5	124	4	100	330,000	$175,000	$10,500	$9,000	$13,500
'89	31	349	13	100	257,909	......	......	10,000	25,500
'90	31.5	632	19	100	295,000	......	......	10,500	27,100

Management.—Prest., J. R. Planten. New York City. Secy., L. B. Ward, Jersey City, N. J. Supt., G. H. Coleman.

28. **PULASKI CITY,** *Pulaski Co* (Pop. in '90, 2,112.) **History.**—Built in '91 by Pulaski Land & Improvement Co.; to be in operation in October. Contrs., Glamorgan Iron Co., Lynchburgh. **Supply.**—Artesian wells, pumping to stand-pipe. **Pumping Machinery.**—Cap., not given; Gordon. **Stand-Pipe.**—Cap., 250,000 galls.: 35x35 ft.; of steel; from Camden, (N. J.) Iron Works. **Distribution.**—*Mains*, 12-in. 7 miles. *Pressure*, 100 lbs. **Financial.**—*Cost*, $50,000. **Management.**—Supt., Henry Taylor.

29. **RADFORD,** *Montgomery Co.* (Pop., in '90, 2,060.) **History.**—Built in '90 by Radford Land & Improvement Co. Des. Engr., Howard Murphy, Philadelphia, Pa. Contr., John B. Foley. **Supply.**—Connelly's branch of New River, pumping to subsiding reservoir, thence to stand-pipe. **Pumping Machinery.**—Dy. cap., 1,000,000 galls.; one Gaskill comp., cond., dup., direct-acting. **Reservoir.**—Cap., not given; 142x277x14½ ft. deep; slope, 2½ to 1. Conduit.—Diam, 12 ins.; from Donaldson Iron Co., Emaus, Pa. **Stand-Pipe.**—Cap., 183,589 galls.; 25x50 ft. on elevation of 200 ft.; steel; from American Bridge and Iron Co., Roanoke. **Distribution.**—*Mains*, 12 to 4-in., i., 7½ miles. **Management.**—Address Radford Land and Improvement Co.

NOTE.—March 19, '91, City Clerk reported that town would build water-works to supply eastern portion of town as soon as necessary bonds could be negotiated. Aug. 18, Arthur Roberts, Clerk, reported that system would not be built till '92.

30. **RICHLANDS,** *Tazewell Co.* (Pop., not given.) **History.**—Built in '90-1 by Richlands Water, Gas & Electric Light Co., to be in operation Aug. 1, '91. Engr., Howard Murphy, 326 Walnut St., Philadelphia, Pa, **Supply.**—Clinch River, pumping to stand-pipe. **Pumping Machinery.**—Dy. cap., 1,000,000 gallv.; Gaskill comp., cond., dup., direct acting. **Stand Pipe.**—Is 25 × 50 ft., from American Bridge Co., Roanoke. **Distribution**—*Mains*, 16 to 4-in., 3 miles. *Services*, ½-in. galv. i. Pipe from Gloucester Pipe Fdry., specials from I. S. Cassin & Co.; trenching and pipe-laying done by Fred C. Dunlap, 2,900 Poplar St., Philadelphia, Pa. *Taps*, 100. *Hydrants*, Cassin, 35. *Valves*, Cassin. *Pressure:* ordinary, 80 lbs.; fire, 80 to 125 lbs. *Cost*, $45,000. **Management.**—Prest., Geo. R. Dunn. Vice-Prest., Geo. McCall. Secy., C. Graham. Treas., W. K. Hewson. Supt., Hubert Raven.

31. **RICHMOND,** *Henrico Co,* (Pop., 63,600—81,388.) Built in '30 by city. **Supply.**—James River, pumping to low and high service reservoirs. Fiscal year closes Dec. 31.

FOR FURTHER DESCRIPTIVE MATTER see Manual for 1889-90. CHANGES since 1890 are here given. POPULATIONS are for 1880 and 1890, respectively.

Year.	Mains.	Taps.	Meters.	Hydrants.	Consumption.	Exp.	Rev.
'82	53.1	6,200	0	324	5,599,315	$32,569	$32,748
'84	55	6,500	40	400	8,000,000	50,000	85,000
'85	57.9	7,062	38	402	9,906,122	21,629	92,632
'86	61.46	7,608	38	429	12,329,778	29,084	104,805
'87	73	8,900	66	430	11,133,017	30,368	111,024
'88	73	9,606	85	475	11,311,013	36,222	116,999
'89	73	10,315	115	486	10,852,842	35,893	114,697
'90	76.05	10,983	113	497	10,507,102	35,635	126,111

Management.—Committee of seven chosen from aldermen and council every two years. Chn., J. W. White. Secy., E. E. Davis. Supt., Chas. E. Bolling. Appointed by both branches of council in joint session. NOTE.—For descriptive matter see *Engineering Record*, May 23, '91, et seq.

32. **ROANOKE,** *Roanoke Co.* (Pop., 669—16,159.) Built in '84 by Roanoke Land & Improvement Co.; franchise provides neither for control of rates nor purchase of works by city. Supply.—Spring, pumping to reservoir and direct when necessary. Fiscal year closes Dec. 31. During '90 two 2,100,000-gall. Gaskill pumps were added.

Year.	Mains.	Taps.	M't'rs.	Hy'ts.	Cost.	B'd't debt.	Int. ch'gs.	Exp.	Rev.
'89	17.8	1,827	0	53	$251,400	$100,000	$5,000	$11,533	$25,376
'90	26.5	1,887	0	64	302,149	100,000	5,000	10,252	33,318

Financial.—*Cap. Stock*, $298,510; paid 4% dividend in 1890. *Bonded Debt*, $100,000, at 5%. *Expenses*, total, $15,253. *Revenue*, total, $33,318. **Management.**—Prest., F. J. Kimball, Bullitt Bldg., Philadelphia, Pa. Secy., H. E. Gerhard. Mgr., Jno. C. Rawn.

33. **SALEM,** *Roanoke Co.* (Pop., 1,752—3,279.) Built in '75-6 by town. Supply. —Springs, pumping to reservoir. Fiscal year closes July 1. Since close of fiscal year '90 $4,000 have been expended in new pump and pump house. Distribution.—*Mains*, 8 miles. *Taps* (in '88), 250. *Meters*, 0. *Hydrants*, 50. Financial.—*Cost*, $13,000. *Bonded Debt* (July 1, '90), $10,000; since that date $13,000 have been issued. *Expenses:* operating, $668; int., $600; total, $11,268. *Revenue,* consumers, $1,386. **Management.**—Chf., Thos. R. Boon. Supt., Alexander Johnson.

South Boston, *Halifax Co.* (Pop., 328—1,789.) Built in '88-9 by town for fire protection.

34. **STANLEY,** *Page Co.* (Pop., not given.) System was built in '89 by Stanley Furnace & Land Co. Supply.—Impounding reservoir, by gravity. *Cost*, $10,000 No further information.

35. **STAUNTON,** *Augusta Co..* (Pop., 6,664—6,975.) Built in '76-'77 by city. Supply.—Ten springs, pumping from receiving to distributing reservoir. Fiscal year closes March 31. Distribution.—*Mains*, 10 miles. *Taps*, 325. *Meters*, 5. *Hydrants*, 53. *Consumption.* 57,806 galls. Financial.—*Cost*, $200,000. *No Debt. Expenses,* $4,000. *Revenue:* consumers, $12,000; city, none. **Management.**—Chn., Andrew J. Butts. Secy., Newton Argenbright. Supt., Jas. A. Moore.

36. **SUFFOLK,** *Nansemond Co.* (Pop., 1,963—3,354.) Built in '89 by Nansemond Water Co. Supply.—Portsmouth water-works. Fiscal year closes Dec. 31.

Year.	Mains.	Taps.	Meters.	Hydrants.	Consumption.	Exp.	Rev.
'89	6	85	5	35	62,621	$800	$4,779
'90	6.5	156	12	40	152,596	1,000	7,728

Financial.—*Cap. Stock* (in '83), $10,000. *Revenue:* consumers, $5,728; city, $2,000 total, $7,728. **Management.**—Supt., Geo. H. Colman, Portsmouth.

37. **TAZEWELL,** *Tazewell Co.* (Pop., not given.) Built in '86-8 by village ; Supply.—Spring, by gravity. Distribution.—(In '88.) *Mains*, 1.3 miles. *Taps*, 20. *Meters*, 0. *Hydrants*, 7. Financial.—(In '88.) *Cost*, $5,000. *Bonded Debt*, $4,000 at 6%. *Expenses:* operating, $150; int., $240; total, $390. *Revenue:* consumers, $350. **Management.**—Supt., J. B. Candill.

38. **VINTON,** *Roanoke Co.* (Pop., in '90, 1,057.) History.—Built in '90 by Roanoke & Vinton Light & Water Co.; put in operation Jan. 1, '91. Engr. and Contr., S. B. Mosby, Bedford City. Supply.—Spring and subsiding reservoir, pumping to stand-pipe. Pumping Machinery.—Dy, cap., 500,000 galls.; Deane. Stand-Pipe. —Cap., 300,000 galls.; from Simpkin & Hyllier, Richmond. Distribution.—*Mains*, 10 to 4-in., 2.5 miles. Pipe and specials from Glamorgan Pipe Co., Lynchburgh. *Hydrants*, Eddy. *Valves*, Eddy. *Consumption,* 200,000 galls. Pressure: ordinary, 125 lbs.; fire, 150 lbs. Financial.—*Cost*, $32,000. *Cap. Stock*, $50,000. *No Debt.* **Management.**—Prest., W. S. Goech; Secy., W. M. Yager; Treas., W. T. Claxton; all of Roanoke. Supt., C. O. Dunn.

39. WARRENTON, *Fauquier Co.* (Pop., 1,464—1,316.) **History.**—Built·in 90 by town; put in operation Nov. 1. Contrs., Glamorgan Co. **Supply.**—Springs, by gravity. **Reservoir.**—Cap., 1,250,000 galls. **Distribution.**—*Mains*, 8 to 4-in. c. l., 4 miles. *Hydrants*, Ludlow, 20. *Valves*, Bourbon. *Pressure*, 50 lbs. **Financial.**— *Cost*, $21,000. **Management.**—Not given.

40. WINCHESTER, *Frederick Co.* (Pop., 4.958—5,196.) Built in last century by town. **Supply.**—Spring, by gravity. Fiscal year closes May 1. **Distribution.** —*Mains*, 8.5 miles. *Taps*, 1,000. *Meters*, 0. *Hydrants*, 29. **Financial.**—*Expenses.* $800. *Revenue*, $3,000. **Management.**—City Council, City Clerk, R. L. Gray, Supt., Jno. N. Nelton, appointed by council.

IV., 41. WYTHEVILLE, *Wythe Co.* (Pop., 1,885—2,570.)

WYTHEVILLE WATER CO.

Built in '38. For further description see 1889-'90 Manual, p., 310.

VILLAGE WORKS.

Built in '89. **Supply.**—Spring, pumping from storage to high-service reservoir. Fiscal year closed July 1. In '91 pipes were extended five miles to spring in mountains for additional supply. **Distribution.**—(July 1, '91.) *Mains*, 3 miles. *Taps*, 60. *Meters*, 1. *Hydrants*, 21. **Financial.**—(July 1, '91.) *Cost*, $25,162. *Debt*, $23,000 at 6%. *Sinking Fund*, on total bonded debt of town, $1,720. *Revenue*, consumers, $1,230. **Management.**—Mayor, W. L. Yost. Chn., C. W. Pike. Supt., W. H. Burch.

Water-Works Projected with Fair Prospects of Construction.

ELLISTON, *Montgomery Co.* (Pop., not given.) Elliston Development Co. contemplates constructing system at early date. Supply, springs, by gravity. Mains, 1½ miles. Hydrants, 2. Const. Engr., B. D. Bachelder.

FRONT ROYAL and RIVERTON, *Warren Co.* (Pop., of town, 820—868.) Royal Water Co. was to begin construction of system by Aug. 1, '91. Supply, impounding reservoir on mountain stream, by gravity. Reservoir, cap, 1,000,000 galls.; 218 ft. above town. Supply Main, 8,880 ft. long. Cost., from $30,000 to $40,000. Cap. Stock, $20,000. Bonded Debt, to cover balance of cost. Prest., H. L. Cook, Baltimore. Secy., C. A. Macatee.

GRAHAM, *Tazewell Co.* (Pop., in '90, 1,021.) Franchise has been granted co. and preliminary surveys and estimates made. Latest report, July 22, '91. Address Walter Graham, Engr.

NEWPORT NEWS, *Warwick Co.* (Pop., in '90, 4,449.) Newport News Water & Light Co. has been organized, and water rights have been procured. Prest., C. B. Orcutt, No. 1 Broadway, N. Y. City. Address Theo. Livezey.

SHENANDOAH, *Page Co.* (Pop., 197—751.) Co. was organized during summer of '91, and franchise granted. Works to be erected at once. Des. Engr., Howard Murphy, 326 Walnut St., Philadelphia, Pa. Address Wm. H. Reese.

WAYNESBOROUGH, *Augusta Co.* (Pop., 184—646.) Projected in '91 by W. W. Fishburne under name of Waynesborough Street Railway, Light & Water Co. Supply, springs, pumping by water power direct or to stand-pipe. Mains, 8 to 4-in., about 2 miles. Hydrants, 20. Cost, $7,000. Address W. W. Fishburne. Latest report, Oct. 9, '91.

WEST VIRGINIA.

Water-Works Completed or in Process of Construction.

1. BLUEFIELD, *Mercer Co.* (Pop. in '90, ',775.) Built in '88 by Norfolk & Western R. R. Co.; bought Sept. 24, '90, by Bluefield Water-Works & Improvement Co. **Supply.**—Springs, by gravity. Fiscal year closes Sept. 24. **Distribution.**— *Mains*, 8 miles. *Meters*, 1. *Hydrants*, 21. *Consumption*, 150,000 galls. **Financial.**—*Cost*, $25,000. **Management.**—Prest., J. H. Dingee, 330 Walnut St., Philadelphia, Pa., Secy. and Supt., Malcolm W. Bryan.

FOR FURTHER DESCRIPTIVE MATTER see Manual for 1889-90. CHANGES since 1890 are here given. POPULATIONS are for 1880, and 1890, respectively.

2. **CHARLESTON,** *Kanawha Co.* (Pop., 4,192–6,742.) Built in '86 by Charleston Water-Works Co., under 20-yrs. franchise. Rates are fixed in franchise; city may buy works at intervals of five yrs. Supply.—Elk River, pumping direct; filtered. Fiscal year closes Dec. 31.

Year.	Mains.	Taps.	Meters.	Hyd'ts.	Cons'ption.	Cost.	B'd'd debt.	Int. ch'g's.	Rev.
'87	12	700	2	30	450,000	$126,500	$84,500	$5,070	$9,090
'89	12.75	851	1	51	500,000	126,765	88,000	5,280	2,550*
'90	14.75	852	1	51	800,000	134,322	88,000	5,280	2,550*

* From city.

Financial.—*Cap. Stock:* paid up, $42,000; paid 8% div'd in 1890. *Bonded Debt,* $88,000 at 6%, due in 1907. *Expenses:* int., $5,280; taxes, $903. **Management.**—Prest., E. B. Knight. Secy., W. S. Laidly. Treas. and Supt., Frank Woodman.

3. **CLARKSBURGH,** *Harrison Co.* (Pop., 2,037–3,003.) Built in '88 by town. Supply.—West fork of Monongahela River, pumping to tanks. Distribution.—(In '88.) *Mains,* 6.5 miles. *Hydrants,* 51. Financial.—(In '88.) *Cost* $50,000. *Bonded Debt,* $50,000, at 5%. **Management.**—(In '88.) Three comrs. Supt., W. R. Alexander.

5. **GUYANDOTTE,** *Cabell Co.* (Pop., 819–1,500.) Supply from Huntington.

4. **HINTON,** *Summers Co.* (Pop., 879–2,570.) **History.**—Built in '90-1 by Hinton Water-Works Co.; put in operation May 1, '91; under 99-yrs. franchise. Des. Engr , Jas. H. Harlow. Const. Engr., F. R. Van Antwerp. Contrs., Glamorgan Co., Lynchburgh, Va. Supply —Greenbrier River, pumping to reservoir, **Pumping Machinery.**—Dy. cap., 1,000,000 galls.; two 500,000 Smith & Vaile. **Reservoir.**—Cap., 1,000,000 galls. Distribution.—*Mains,* 8 to 4-in. c. i., 2.9 miles. *Services,* galv. i. Trenching and pipe-laying done by S. B. Mosby & Co., Bedford Cy., Va. *Taps,* 185. *Meters,* 0. *Hydrants,* Mathews, 30. *Valves,* Kennedy. *Consumption,* 325,000 galls. *Pressure,* 152 lbs. **Financial.**—*Cost,* $75,000. *Cap. Stock:* authorized, $75,000; paid up, $20,000. *No bonds. Floating debt,* $5,000 at 6%. *Expenses,* $1,500. *Revenue:* consumers, $600; Ry., $900; city, $900; total, $2,700. *Hydrant Rental,* $30. **Management.**—Prest. and Supt., F. R. Van Antwerp. Secy., H. Evart. Supt., J. A. Riffe.

5. **HUNTINGTON and GUYANDOTTE,** *Cabell Co.* (Pop., 3,174–10,108.) Built in '87 by Huntington Water Co.; controlled by American Water-Works Guarantee Co., Ltd., Pittsburgh, Pa.; maximum rates fixed in franchise, which provides for purchase of works by city in '98, and at intervals of 5 yrs. thereafter, price to be determined by arbitration. Supply.—Ohio River, pumping to reservoir and direct; filtered. **Pumping Machinery.**—Dy. cap., 2,500,000 galls.; two Worthington, one 1,500,000 high-pressure, 20 × 12 × 15, and one comp.; 12 × 18½ × 9 × 10. Pumps are set in a pit 20 ft. in diam., 50 ft. deep. *Boilers:* three 80 HP., horizontal, tubular. **Reservoir.**—Cap., 2,000,000 galls.; 261 ft. above main streets of city.

Year.	Mains.	Taps.	Meters.	Hydrants.	Cons'pt'n.	Cost.	B'd'd debt.	Int. charges.
'87	8.5	326	0	84	300,000	$90,000	$75,000	$4,500
'90	14	1,050	0	108	700,000	.....	......	......

Financial.—Withheld. **Management.**—Prest., J. F. Cockburn. Vice-Prest., A. E. Clancy. Treas., W. S. Kuhn, Pittsburgh, Pa. Secy., A. B. Darragh. Supt., D. W. Immel.

6. **MARTINSBURGH,** *Berkeley Co.* (Pop., 6,335–7,226.) Built in '73-4 by city. Supply.—Springs, pumping direct. Fiscal year closes May 15. Distribution.—*Mains,* 7 miles. *Taps,* 800. *Meters,* 0. *Hydrants,* 61. *Consumption,* 375,000 galls. **Financial.**—*First Cost,* $80,000. *Bonded Debt,* $75,000 at 5%. *Expenses:* operating, $1,800; int., $3,750; total, $5,500. *Revenue:* consumers, $3,000; city, none. **Management.**—Board, consisting of one member of council from each ward, appointed every year by mayor. Chn., H. C. Berry. Secy., C. A. Young. Supt., J. M. Shaffer.

7. **MORGANTOWN,** *Monongalia Co.* (Pop., 745–1,011.) Built in '89 by Union Improvement Co.; city neither controls rates nor has right to buy works. Supply.—Mountain stream, by gravity. Fiscal year closes Jan. 1. Distribution.—*Mains,* 11 miles. *Taps,* 128. *Meters,* 0. *Hydrants,* 15. *Consumption,* 40,000 galls. **Financial.**—*Cost,* $30,000. *Cap. Stock,* including both water and gas plants: authorized, $100,000; paid-up, $60,000; paid no div'd in 1890. *Expenses,* $960. *Revenue:* consumers, $1,900; city, $600; total, $2,500. **Management.**—Prest., J. W. Rowland. Secy., O. M. Sloan. Both of Emlenton, Pa. Supt., E. M. Grant.

8. **PARKERSBURGH,** *Wood Co.* (Pop., 6,582–8,408.) Built in '85 by city. Supply.—Ohio River, pumping to tanks.

WEST VIRGINIA. 163

Yr.	Mains.	Taps.	Meters.	Hyd'ts.	Cons'pt'n.	Cost.	B'd'd d't.	Int.ch'g's.	Exp.	Rev.
'87	12	680	0	60	365,000	$100,000	$80,000	$4,800	$3,000	$9,500
'86	13	815	4	63	365,000	107,000	80,000	4,800	4,000	11,649

Distribution.—(In '88.) *Mains,* 13 miles. *Taps,* 815. *Meters,* 4. *Hydrants,* 63. **Consumption,** 365,000 galls. **Financial.**—*Cost,* $107,000. *Bonded Debt,* $80,000, at 6%. **Expenses:** operating, $4,000; int., $4,800; total, $8,800. *Revenue:* consumers, $11,649. **Management.**—Mayor, H. S. Wilson. Clerk, Lee W. Hughes. Supt., J. V. Mayhall.

NOTE.—In '91, Spring Water Supply Co. was organized to supply the suburban districts of Elberon and Riverside. *Supply,* springs. *Cost,* about $20,000. Supt. and Engr., S. F. Shaw. Treas., Z. T. Taylor.

8. **ROMNEY,** *Hampshire Co.* (Pop., 371—451.) Built in '70-1 by town for domestic use only. **Supply.**—Springs, by gravity. *Mains,* 3 in. 1., 3 miles. *Cost,* $6,500. *No Debt. Expenses,* $75. *No Revenue.* Supt., W. F. Davis.

9. **WELLSBURG,** *Brooke Co.* (Pop., 1,815—2,235.) Built in 86 by town. **Supply.**—Ohio River, pumping to reservoir and direct. Fiscal year closes May 1. **Distribution.**—*Mains,* 3.5 miles. *Taps,* 250. *Meters,* 0. *Hydrants,* 24. **Financial.**—*Cost,* $27,000. *Total Debt,* $23,250: floating, $750 at 7%; bonded, $22,500 at 6%, due in 1915. *Expenses:* operating, $1,200; int., $1,350; total, $2,250. *Revenue,* consumers, $1,800. **Management.**—Board of three, appointed every three years by council. Chn., J. E. Montgomery. Secy.; C. W. Windsor. Supt., Nath. Nelson. Appointed by board.

10. **WHEELING,** *Ohio Co.* (Pop., 30,737—34,522.) Built in '31-4 by town. **Supply.**—Ohio River, pumping to reservoir. During '90-91 an additional reservoir was built. In '91 new supply was proposed. **Distribution.**—*Mains* (in '87), 41 miles. *Hydrants* (in '86), 350. *Consumption,* 5,000,000 galls. **Financial.**—*No Debt. Expenses,* $65,000. **Management.**—Chn., M. Pollock. Secy., Geo. Baird. Supt., Jno. W. Cummings.

Water-Works Projected with Fair Prospects of Construction.

ELKINS, *Randolph Co.* (Pop., in '90, 737.) Village contemplates constructing works or granting franchise to Co. Supply, creek, 3 miles distant. June 23, no definite plans had been formed.

FAIRMONT, *Marion Co.* (Pop., 900—1,023.) In '91 city voted to issue $40,000 worth of bonds at 5% to pay construction of works. Supply, Monongahela River, pumping to reservoir or tank. June 26, preliminary surveys were being made by J. W. Prickett and Perry Thompson. Address T. A. Fleming, Recorder.

GRAFTON, *Taylor Co.* (Pop., 3,030—3,159.) In '91 City Council employed J. B. Strawn, Salem, O., to prepare plans for system of water-works. Supply, Valley River, pumping to reservoir. Ownership not determined. Hydrants, about 50. Cost, about $40,000. Address, J. J. Gilligan, Town Clerk.

MANNINGTON, *Marion Co.* (Pop., in '90, 908.) Mannington Water Supply Co. was organized July 20, '91, to construct system for fire protection only. Supply, Buffalo Creek, pumping to reservoir. Cap. stock, $25,000. Cost, $5,000. Prest., F. R. Stewart. Secy., A. L. Prichard.

RONCEVERTE, *Greenbrier Co.* (Pop., 395—481.) In '91 Ronceverte Water Co. secured franchise for construction of system. Supply, Greenbrier River, pumping to reservoir or stand-pipe. Cost, about $30,000. Address E. M. Nettleton, Covington, Va.

SAINT ALBANS, *Kanawha Co.* (Pop., of town, 2,515—3,270.) R. T. Owey, Charleston, has been granted right to construct system of water-works.

Water-Works Projects Reported in '91, but Now Indefinite.

Keyser, Mineral Co. *Weston,* Lewis Co.

FOR FURTHER DESCRIPTIVE MATTER see Manual for 1889-90. CHANGES since 1890 are here given. POPULATIONS are for 1880 and 1890, respectively.

NORTH CAROLINA.

Water-Works Completed or In Process of Construction.

1. ASHEVILLE, *Buncombe Co.* (Pop., 2,616—10,235.) Built in '83 by city. Supply.—Springs and surface water, filtered, by gravity from receiving reservoir; and Swananoa River, pumping to stand-pipe; filtered. Fiscal year closes June 30. Summer of '91 new pump and 12,000,000-gall. reservoir were talked of.

Year.	Mains.	Taps.	M't's.	Hy'ts.	Con'pt'n.	Cost.	B'd't d'bt.	Int. ch's.	Exp.	Rev.
'87	10	300	0	45	300,000	$100,000	......	......	$600	$3,000
'88	11	250	0	50	300,000	150,000	$120,000	$7,260	1,000	3,000
'90	15	500	0	80	350,000	150,000	120,000	7,200	4,000	3,660

Management.—Supt., J. L. Murray, appointed by aldermen.

Bryson City, *Swain Co.* (Pop., est., 400.) Built in '89 by town, for fire protection.

2. CHARLOTTE, *Mecklenburgh Co.* (Pop., 7,094—11,557.) Built in '81-2 by Charlotte City Water-Works Co. Supply.—Springs, pumping from impounding reservoir to stand-pipe. Fiscal year closes Dec. 31. Distribution.—*Mains*, 15.5 miles. *Hydrants*, 137. *Consumption*, 500,000 galls. Financial.—*Bonded Debt*, $140,000 at 6%, due in 1917. *Expenses:* operating, $6,068; int., $8,423; total, $14,491. *Revenue:* consumers, $15,997; city, $4,448; total, $20,415. Management.—Prest., A. J. Breward. Secy. and Supt., D. B. Hutchison. Treas., E. B. Spring.

3. CONCORD, *Cabarrus Co.* (Pop., 1,264—4,339.) History.—Built in ('84-5 by P. B. Fetzer under 25-yrs. exclusive franchise. Contract with city provides for 2½-in. double hydrants, not over 300 ft. apart. Adding one or more hydrants, at any time, by resolution of the board of councilmen, will constitute a renewal of the franchise. Supply.—Eno River, pumping to tanks. Distribution.—(In '88.) *Mains*, 700 ft. *Hydrants*, 8. Separate one of pipe distributes domestic supply, which is also stored in separate tanks. Supply of 30,000 galls. for every 15 plugs is required to be constantly stored for fire protection. Financial.—*Hydrant Rental*, $40. Management.—Owner, P. B. Fetzer.

4. DURHAM, *Durham Co.* (Pop., 3,014—5,485.) Built in '86-7 by Durham Water Co. under exclusive franchise for 20 years, when city may buy works, or at end of each 10 years thereafter. Supply.—Eno River, pumping to reservoir. Fiscal year closes Dec. 31. Distribution.—*Mains*, 14 miles. *Taps*, 268. *Meters*, 8. *Hydrants*, 86. *Consumption*, 220,000 galls. Financial.—*Cost*, $251,273. *Cap. Stock:* authorized, $125,000; all paid up. *Total Debt*, $139,220: floating, $14,220 at 6%; bonded, $125,000 at 6%. *Expenses:* operating, $1,767; int., $2,893; taxes, $649; total, $5,314. *Revenue:* consumers, $3,445; city, fire protection, $3,950; total, $7,395. Management.—Prest. and Supt., S. W. Homan. Secy., W. W. Fuller. Treas., A. H. Howland.

5. FAYETTEVILLE, *Cumberland Co.* (Pop., 3,485—4,222.) Built in '29 by Fayetteville Water-Works Co.; leased since '75 by S. W. Tillinghast; city neither controls rates nor has right to purchase works. Supply.—Springs, by gravity, to *Hydrants* (in '88), 9. Fiscal year closes July 1. Distribution.—*Mains*, 4.3 miles. and from reservoir. *Consumption*, 80,000 galls. Financial.—*Cap. Stock:* authorized, $8,000, all paid up; paid 8% dividend in '90. *No Debt. Expenses:* operating, $500; taxes, $25; total, $525. *Revenue:* consumers, $800; city, none. Lessee, S. W. Tillinghast. NOTE.—See projected works, p. 166.

6. GOLDSBOROUGH, *Wayne Co.* (Pop., 3,286—4,017.) Built in '89 by Goldsborough Water Co., under 30-yrs. franchise which fixes rates and provides for purchase of works by city at end of 10 yrs. Supply.—Little River, pumping to stand-pipe. Distribution.—*Mains*, 8.5 miles. *Taps*, 150. *Meters*, 1. *Hydrants*, 62. *Consumption*, 100,000 galls. Financial.—(In '89.) *Cost*, $30,000. *Bonded Debt* (in '89), $80,000 at 6%. *Expenses:* operating, $2,580; int., $4,800; total, $7,380. *Revenue:* consumers, $7,000; city, $2,790; total, $9,790. Management.—Prest., Frank R. Dudley. Secy., Nath P. Hobart. Treas., Jos. E. Lopey. Supt., H. P. Dortch.

7. GREENSBOROUGH, *Guilford Co.* (Pop., 2,105—3,317.) Built in '87-8 by Greensborough Water Co. Supply.—Creek and springs, pumping to tank. Distribution.—*Mains*, 6.5 miles. *Taps*, 150. *Hydrants*, 80. *Consumption*, 150,000 galls. Financial.—*Cost*, $75,000. *Cap. Stock*, $75,000. *Bonded Debt*, $50,000 at 6%. *Revenue*, $8,000. Management.—Prest., D. R. Schenk. Secy., Chas. Vance. Treas., Robt. Smith. Supt., Hugh B. Peters.

NORTH CAROLINA.

8. HENDERSON, *Vance Co.* (Pop., in '90, 4,191.) History,—Built in '90-1 by Henderson Water Supply Co., under 40-yrs. franchise; put in operation July 15, '91. Des. Engr., Geo. F. Keith. Const Engr., J. W. Nelson. Contrs., Woltmann, Keith & Co., 11 Wall St., N. Y. City. Supply.—Lake, fed by springs, pumping to standpipe and direct. **Pumping Machinery.**—Dy, cap., 3,000,000 galls.; two 1,500,000-gall. Deane. *Boilers:* from Tippett & Wood, Phillipsburgh, N. J. **Stand-Pipe.**—Cap., 195,500 galls., 16x130 ft.; from Tippett & Wood, Phillipsburgh, N. J. **Distribution.**—*Mains,* c. 1., 8 miles; pipes and specials from Anniston, (Ala.) Pipe Co. and McNeal Pipe Co., Burlington, N. J. *Services,* w. i. *Hydrants,* Ludlow, 68. *Valves,* Ludlow. **Financial.**—*Cost,* $107,000. *Cap. Stock,* authorized, $100,000; all paid up. *Bonded Debt:* authorized, $100,000 at 6%; issued, $90,000. *Revenue,* city, $2,750. **Management.**—Not given.

9. HENDERSONVILLE, *Henderson Co.* (Pop., 551—1,216.) Built in '89-90 by town. **Supply.**—Spring, by gravity. Fiscal year closes May 1. **Distribution.**—*Mains,* 3 miles. *Taps,* 100. *Meters,* 0. *Hydrants,* 18. *Consumption,* 20,000 galls. **Financial.**—*Cost,* $20,000. *Bonded Debt,* $20,000 at 6%, due in 1813. *Expenses and Revenue,* not determined. **Management.**—Supt., T. E. Broswell. Report by V. L. Hyman.

NEW BERNE, *Craven Co.* (Pop., 6,433—7,843.) Built in '85 by A. K. Denison. Supply,—See 1889-90 Manual, page 316. NOTE.—See Projected Works, p. 166.

10. RALEIGH, *Wake Co.* (Pop., 9,265—12,678.) Built in '87 by Raleigh Water Co.; franchise provides that city may purchase works after 10 years; rates are fixed. **Supply.**—Walnut Creek, pumping to tank and direct for fire protection; filtered.

Year.	Mains.	Taps.	Meters.	H'd'ts.	C'ns'pt'n.	Cost.	B'd'd d'bt.	Int. ch'g's.	Rev.
'88	15	270	18	122	300,000	$184,000	$160,000	$9,600	$4,480*
'90	17.5	492	27	129	375,000	175,000	175,000	10.500	14,270

*From city.

Financial.—*Cap. Stock* (in '88), $100,000. *Bonded Debt,* $175,000 at 6%. *Expenses:* int., $10,500; taxes, $500. *Revenue:* consumers, $9,000; city, $5,270; total, $14,270. **Management.**—Prest., Julius Lewis. Secy., M. M. Moon, N. Y. City. Treas., W. W. Smith, Dayton, O. Supt., E. B. Englehard. NOTE.—In '91 there was talk of the purchase of the works by the city.

11. SALEM, *Forsyth Co.* (Pop., 1,340—2,771.) Built in '78 by Salem Water Supply Co.; franchise provides neither for control of rates nor purchase of works by city. **Supply.**—Springs and two wells, pumping to reservoir. Fiscal year closes Dec. 31. Additional supply was obtained in 1890 by extending main to spring 1½ miles from town. *Cost,* $5,000. **Distribution.**—*Mains,* 3.5 miles. *Taps,* 124. *Meters,* 105. *Hydrants,* 43 **Financial.**—*Cost,* $21,000. *Cap. Stock:* authorized, $250,000; paid up, $16,620; paid no div'd in 1890. *Floating debt,* $11,000 at 6% and 8%. *No bonds.* Bonds will be issued during '91. *Expenses:* operating, $500; int., $560; taxes, $166; total, $1,226. *Revenue:* consumers, $2,200; city, none. Company closed contract with town to pay hydrant rental of $25 each after Jan, 1, '91. **Management.**—Prest., F. H. Fries. Secy. and Treas., H. F. Shaffner. Supt., C. H. Fogle.

12. SALISBURY, *Rowan Co.* (Pop., 2,723—4,418.) Built in '87 by Moffett, Hodgkins & Clarke, Syracuse, N. Y. **Supply.**—Springs, pumping to tank. **Distribution.**—(In '88.) *Mains,* 8.5 miles. *Taps,* 80. *Meters,* 2. *Hydrants,* 62. *Consumption,* 84,400 galls. **Financial.**—(In '88.) *Cost,* $100,000. *Debt,* $85,000 at 6%. *Expenses:* operating, $1,200; int., $5,100; total, $6,300. *Revenue:* consumers, $4,800; city, $2,540; total, $7,340. **Management.**—(In '88.) Prest., J. V. Clarke. Supt., E. B. Neave.

13. TARBOROUGH, *Edgecombe Co.* (Pop., 1,600—1,924.) History.—Built in '89 by town. Engr., A. L. Spandor, Norfolk, Va. **Supply.**—Hendricks Creek, pumping to tank and direct. **Pumping Machinery.**—Dy, cap., not given; Worthington dup. Tank.—Cap., 20,000 galls. **Distribution.**—*Mains,* 5-in. w. i., 1.5 miles; pipe and specials from R. D. Wood & Co., Philadelphia, Pa. *Hydrants,* Mathews, 14. *Pressure:* ordinary, 40 lbs.; fire, 110 lbs. **Financial.**—No Debt. *Expenses.* $55. *No Revenue.* **Management.**—Supt., E. Haywood. Report by A. Fountain, Mayor.

14. WILMINGTON, *New Hanover Co.* (Pop., 17,350—20,056.) Built in '81-2 by Clarendon Water-Works Co. **Supply.**—Cape Fear River, pumping to stand-pipe, and direct for fire protection. Fiscal year closes Sept. 30.

FOR FURTHER DESCRIPTIVE MATTER see Manual for 1889-90. CHANGES since 1890 are here given. POPULATIONS are for 1880 and 1890, respectively.

MANUAL OF AMERICAN WATER-WORKS.

Year.	Mains.	Taps.	Meters.	Hyd'ts.	Cons'pt'n.	Cost.	B'd'd	d't.Int.ch'g's.	Exp.	Rev.
'82	11.5	254	32	105	100,000	$137,000	$80,000	$5,600	$10,000	$13,750
'88	13.1	527	142	115	506,785	195,000	80,000	4,800	7,350	16,650
'90	13.3	600	240	137	431,374	198,000	80,000	4,800	8,800	17,450

Financial.—*Cap. Stock*, authorized, $100,000. *Bonded Debt*, $80,000 at 6%, due in 1911. *Expenses:* operating $7,700; int., $4,800; taxes, $1,100; total, $9,600. *Revenue:* consumers, $10,100; city, $5,750; other purposes, $650; total, $17,450. *Management.*—Prest., J. F. Divine. Treas., Geo. W. Kidder. Supt., J. C. Chase.

15. **WINSTON**, *Forsyth Co.* (Pop., 2,854—8,018) Built in '81-2 by Winston Water Co., under 99-years franchise. *Supply.*—Two wells, pumping to reservoir. Fiscal year closes March 31. At '91 session of the General Assembly authorized stock of co. was increased to $500,000. Stockholders have fixed the capital and (April 27) were about to issue $75,000 of 5 and 6% bonds. With this sum old 8% bonds will be retired and balance be expended in following improvements: additional wells; a 3,000,000-gall. reservoir; 1,000,000-gall. pumping plant; four or five miles of 8- and 6- in. mains.

Year.	Mains.	Taps.	Meters.	Hyd'ts.	Cons'pt'n.	Cost.	B'd'd	d't.Int.ch'g's.	Exp.	Rev.
'84-5	4	92	3	40	75,000	$42,500	$24,000	$1,920	$400	$2,800
'86-7	4.5	120	28	43	113,000	46,939	24,000	1,920	895	4,372
'88-9	4.75	150	100	45	120,000	52,000	24,000	1,920	1,209	4,441
'90-1	4.75	155	100	45	150,000	54,416	24,000	1,920	1,579	5,582

Financial.—*Cap. Stock:* authorized, $500,000; paid up, $24,000; paid 6% div'd in 1890. *Surplus*, $404. *Total Debt*, $26,327; floating, $2,327 at 8%; bonded, $24,000 at 8%, due in '92. *Expenses*, total, $3,741. *Revenue:* consumers, $4,457; city, $1,125; total, $5,582. **Management.**—Prest., J. Wilson. Secy., Geo. W. Henshaw.

Water-Works Projected with a Fair Prospects of Construction.

ELIZABETH CITY, *Pasquotank Co.* (Pop. '90, 3,251.) Preliminary survey was made in '91. Probably driven-well system. Address Henry E. Knox, Engr., Norfolk, Va.

FAYETTEVILLE, *Cumberland Co.* (Pop., 3,482—4,222.) Present supply.—See p. 164. Proposed Works.—Summer of '91 franchise was granted National Water Supply & Guarantee Co., J. M. Gardiner, Ashland, Ky., Gen. Mgr. Construction was to be started July 10 and works to be in operation Dec. 31, '91. Supply, head waters of Little Cross Creek, water power pumping to stand-pipe. Mains, 7 miles. Hydrants, 60. Est. cost, $40,000 to $45,000. Address above, or J. D. McNeill, Chn. Water-Works Com.

HICKORY, *Catawba Co.* (Pop., '90, 2,025.) Town contemplates building works. Supply, artesian well, pumping to stand-pipe. Mains, 4 miles. Hydrants, 35. Hartford, Hebert & Co., Chattanooga, Tenn., have been employed to prepare plans. Address L. R. Whitener, Mayor. Latest report, June 23, '91.

NEW BERNE, *Craven Co.* (Pop., 6,443—7,843.) Present Supply.—See '89-'90 Manual, p. 316. Proposed Supply.—In '91 New Berne Water Co. was organized. Construction was to begin June 15, and be finished Dec. 31. Supply, springs, pumping to stand-pipe ; 1,000,000-gall. pump and 200,000-gall. stand-pipe. Address J. A. Bryan.

SHELBY, *Cleaveland Co.* Pop., '90, 1,394.) July 6, '91, it was reported that Shelby Improvement Co. hoped to have system of water supply in operation soon.

WILSON, *Wilson Co.* (Pop., '90, 2,099.) Decided in '91, by popular vote, to introduce works. Nothing further had been done Aug. 18.

Water-Works Projects Reported in '90 and '91, but Now Indefinite.

Lincolnton, Lincoln Co. *Morganton,* Burke Co. *Oxford,* Granville Co. *Rutherfordton,* Rutherford Co. *Statesville,* Iredell Co.

SOUTH CAROLINA.

Water-Works Completed or in Process of Construction.

1. **ANDERSON,** *Anderson Co.* (Pop., 1,850—3,018. **History.**—Built in '90 by Anderson Water Supply Co. under 25-yrs. franchise; put in operation in Sept. Engr., W. C. Whitner, Rock Hill. **Supply.**—Springs, pumping to stand-pipe and direct.

SOUTH CAROLINA.

Storage Reservoir.—Cap., 1,000,000 galls.; about 300 × 60 × 6 ft. deep, one mile from city; in excavation, lined with gravel; slope, 2 to 1. *Dam*, of masonry, 30 ft. long on crest, 6 ft. high. *Spillway*, 8 ft. opening in center, with gate. **Conduit.**—Is 12 ins. in diameter, 2,500 ft. long, leading to pumps. Thence water flows through 10-in. force main. **Pumping Machinery.**—Dy. cap., 2,000,000 galls.; two 1,000,000 Smith & Vaile comp. *Boilers*, from Whittier Machine Co., Boston. **Stand-pipe.**—Cap., 200,000 galls.; 18 × 100 ft.; from Tippert & Wood, Phillipsburgh, N. J. **Distribution.**—*Mains*, 10 to 4-in. c. i., 9 miles. *Services*, galv. i. Pipes and specials from So. Pittsburgh (Tenn.) Pipe Co. *Taps*, 250. *Meters*, Crown, 5. *Hydrants*, Ludlow, 68. *Valves*, Ludlow. *Consumption*, 50,000 galls. *Pressure*, ordinary, 80 lbs.; fire, 110 lbs. **Financial.** *Cost*, $65,000. *Cap. Stock*, paid up, $10,000. *Bonds*, authorized, $75,000. *Expenses*, $3,000. *Revenue* : consumers, $3,000; city, $3,500; total, $6,500. *Hydrant Rental*, $50. **Management.**—Prest., W. L. Roddey.

2. **CHARLESTON**, *Charleston Co.* (Pop., 49,984—54,952.) Built in '76-8; owned by City of Charleston Water-Works Co.; franchise provides that city may purchase works at end of 10 years time has expired; rates established in franchise. **Supply.**—Three artesian wells, flowing into two reservoirs, thence pumping to stand-pipe. Wells are in order of sinking, 1,974, 1,950 and 1,945 ft. deep; 2⅛, 3¼ and 5 ins. in diam.; dy. flow, 170,000, 250,000, and 1,200,000 galls.; cost, $25,000, $48,000, and $38,000 respectively. May 28, '91, contract was made with J. C. Miller, Chicago, for fourth well; depth, 2,000 ft.; contract price, $44,000. New reservoir w s built to hold 10,000,000 galls., but owing to leakage depth of water has been le-sened, decreasing cap. to 5,000,000 galls. Fiscal year closes Jan 31.

Year.	Mains.	Taps.	Meters.	Hydrants.	Revenue.*
'80	10	500	..	208	$10,000
'81	11	500	..	208	10,000
'82	16	500	..	208	10,000
'84	17.5	600	20	275	10,200
'86	18	800	20	310	11,500
'87	18	900	20	300	11,800
'88	18	1,000	20	300	11,800
'89	26	1,300	20	398	12,700
'90	26.8	1,150	20	409	13,800

*From city.

Financial.—Withheld. **Management.**—Prest., Geo. I. Cunningham. Sec., Zimmerman Davis. Supt., Caspar A. Chisholm.

3. **COLUMBIA**, *Richland Co.* (Pop., 10,036—15,353.) Built in (about) '85 by city. **Supply.**—Congaree River and springs, pumping to reservoir. In '87 there were 16 miles of pipe and 110 hydrants. May 23, '91, Supt. reported that new plant, including 7 miles of mains would be constructed at cost of $75,000. Unofficially reported that 3,000,000-gall. Worthington pump and Victor turbine had been bought. New supply will be from Congaree River, filtered. Address J. A. Fetner, Supt.

4. **DARLINGTON**, *Darlington Co.* (Pop., 940—2,389.) **History**—Built in '91 by Darlington Light, Water, & Power Co., under 30-yrs. franchise; put in operation June 30. Engr., S. S. Farman. **Supply.**—Artesian well, pumping to tank and direct for fire protection. **Pumping Machinery.**—Cap., 144,000 galls.; Smith & Vaile. **Tank.**—Cap., 40,000 galls. **Distribution.**—*Mains*, 3 miles. *Hydrants*, 7. *Pressure:* ordinary, 65 lbs.; fire, 120 lbs. **Financial.** *Cost*, $7,000. *Cap. Stock:* authorized, $10,000; paid up, $7,000. *Bonds:* authorized, $3,000; issued, none. *Revenue*, city, $350. **Management.**—Prest. and Treas., W. F. Dargan. Secy., W. H. Early. Treas., B. B. Farman.

5. **GREENVILLE**, *Greenville Co.* (Pop., 6,160-8,607.) **History.**—Built in '90-1, by Paris Mountain Water Co., under 25-yrs.' franchise; put in operation June 1, '91. Engr., J. W. Ledoux. Rome, N. Y. Contr., American Pipe Mfg. Co. **Supply.**—Mountain Creek, by gravity. **Reservoir.**—Cap., 25,000,000 galls.; formed by masonry dam 45 ft. high and 250 ft. above town. **Supply Main.**—Is 12 ins. in diam.; 6½ miles long. **Distribution.**—*Mains*, 10.5 miles. *Services*, w. i. *Taps*, 115. *Meters*, Crown, Worthington, 6. *Hydrants*, Ludlow, 113. *Valves*, Ludlow. *Pressure*, 60 to 110 lbs. **Financial.**—*Cap. Stock:* authorized, $200,000; all paid up. *Bonds:* authorized, $200,000; issued, $185,000 at 6%. *Hydrant Rental*, $40. **Management.**—Prest., Wm. S. Perot, Jr. Secy. and Treas., H. Bayard Hodge. Mgr., Jos. S. Keen. Jr. All of 1,326 Chestnut St., Philadelphia, Pa.

For Further Descriptive Matter see Manual for 1889-90. Changes since 1891 are here given. Populations are for 1880 and 1890, respectively.

6. ORANGEBURGH, *Orangeburgh Co.* (Pop., 2,140-2,964.) Owned by Geo. H. Cornelson; franchise provides that city may purchase works after 10 yrs. Built in '87 by town, Mr. Cornelson owning wells and in '89 buying distribution. **Supply.**—Two artesian wells, pumping to stand-pipe. Fiscal year closes Aug. 1. **Distribution.**—*Mains*, 6 miles. *Meters*, 0. *Hydrants*. 10½. **Financial.**—*Cost*, $24,085. *No Debt. Expenses*, $600. *Revenue*, $2,580. **Management.**—Geo. H. Cornelson, owner.

7. SPARTANBURGH, *Spartanburgh Co.* (Pop., 3,253—5,544.) Built in '88-9 by Spartanburgh Water-Works Co.; maximum rates fixed in franchises which provide, for purchase of works by city in 1907. **Supply.**—Chinquapis Creek, pumping to stand-pipe. **Distribution.**—*Mains* (in '89), 7 miles. *Taps* (in '89), 67. *Meters* (in '89), 2. *Hydrants*, 51. *Consumption* (in '89), 29,930 galls. **Financial.**—*Cost* in '89), $100,000. *Cap. Stock* (in '89), $100,000. *Bonded Debt* (in '89), $85,000. *Revenue*, city, $3,420. **Management**—Prest., H. C. Hodgkins. Supt.. E D. Ball.

Water-Works Projected with Fair Prospects of Construction.

BLACKSBURGH, *York Co.* (Pop., 145—1,245.) July 10, '91, Blacksburgh Land & Improvement Co. contemplated building system of water-works at once. Prest., J. W. F. Jones.

FLORENCE, *Florence Co.* (Pop, 1,914—3.395.) City intends to build works and began sinking artesian well March 1, '91. Address E. W. Lloyd, City Clerk.

NEWBERRY, *Newberry Co.* (Pop., 2,342—3,020) At citizens' meeting held March 19, '91, city council was instructed to procure propositions from companies for construction of system of water-works.

ROCK HILL, *York Co.* (Pop., 869—2,744.) City has decided to construct water-works, probably in '92, and was to petition '91 session of Legislature (in November) for authority to issue $50,000 worth of bonds. Address W. B. Wilson, Jr.

SUMTER, *Sumter Co.* (Pop., 2,111—3,865.) In '91 franchise was granted to H. O. Reid. 206 Broadway, N. Y. City, providing for artesian water, 5 miles of mains, 55 hydrants. Hydrant rental, $50 each for first 55, $45 each for additional and $40 each for all over 100.

Water-Works Projects Reported in '90 and '91, but now indefinite.

Gaffney City. Spartanburgh Co.

GEORGIA.

Water-Works Completed or in Process of Construction.

1. ALBANY, *Dougherty Co.* (Pop., 3,216—4,008.) City is supplied with water from 10 artesian wells, one of which is owned by city and others by private parties. Wells are from 2 to 5 ins. in diam., from 630 to 730 ft. deep, throw water 30 ft. above surface, and yield from 50 to 120 galls. per minute. Pipes extend from these wells to supply town. June 23, '91, city was boring a 12-in. well from which water is to be pumped to stand pipe, for fire protection and domestic supply. As expense is met from ordinary revenue of city, without issuing bonds, two or three years may be required for its completion. Address Nelson Tift.

1. AMERICUS, *Sumter Co.* (Pop., 3,635—6,398.) Built in '88 by city. **Supply.**—Springs and artesian wells, pumping to tank. Fiscal year closes Dec. 31. During '90 supply was increased by two artesian wells and a 1,000,000-gall Worthington comp. pump was added. **Distribution.**—*Mains*, 11 miles. *Taps*, 500. *Meters*, 15. *Hydrants*, 72. *Consumption*. 250,000 galls. **Financial.**—*Cost*, $55,000. *Bonded Debt*, $51,000, at 6%. *Expenses:* operating, $7,800; int., $3,060 ; total, $10,860 *Revenue:* consumers, $8,700; city, $3,700; total, $12,400. **Management.**—Chn., G. W. Glover. Clerk, D. K. Binder. Supt., G. W. Edridge.

2. ATHENS, *Clarke Co.* (Pop., 6,099—8,639.) Built in '82-3 by Athens City Water Works Co.; maximum rates are fixed in franchise, which provides for purchase of works in '94, and at intervals of 10 yrs. thereafter. **Supply.**—Two creeks, pumping to stand-pipe. Fiscal year closes Dec. 31.

GEORGIA.

Year.	Mains.	Taps.	Met'rs.	H'd'ts.	Cons'n.	Cost.	B'd'd d'bt.	Int. ch'gs.	Exp.	Rev.
'81	7	65	2	50	50,000	.....	$75,000	$3,750	....	$4,884
'87	7	250	8	53	150,000	$86,000	75,000	3,750	$2,300	8,180
'89	8.1	288	16	64	78,000	123,000	75,000	5,000	3,143	11,234
'90	8.5	313	16	68	89,500	128,000	100,000	5,000	3,495	11,945

Financial.—*Cap. Stock:* authorized, $75,000; all paid up. *Bonded Debt,* $100,000: $75,000 at 5%, due in 1912; $25,000 at 6%, due in 1914. *Expenses:* operating, $2,800; int., $6,380; taxes, $495; total, $9,675. *Revenue:* consumers, $7,800; city, $4,145; total, $11,945. **Management.**—Prest. and Treas., A. H. Howell, N. Y. City. Secy., G. M. Church. Providence, R. I. Supt., W. L. Wood.

3. **ATLANTA,** *Fulton Co.* (Pop., 37,409—65.533.) Built in '74-75 by city. Supply.—South River, pumping direct from impounding reservoir. Fiscal year closes Dec. 31. In '91 plans by R. Herring, N. Y. City. for a new supply from Chattahoochee River were accepted by council. The plan included a pumping and repumping stations, and a filtering plant. Water will be pumped about 7 miles. July, '91, it was reported that Comrs. had awarded contracts for three 10,000,000-gall. Gaskill pumps, and in Sept. that contract for 48, 30 and 20-in. pipe had been awarded to the Howard-Harrison Iron Co., Bessemer, Ala.

Year.	Mains.	Taps.	Meters.	Hyd'ts.	Cons'pt'n.	B'd't d't.	Int. ch'gs.	Exp.	Rev.
'76	....	393	....	....	607,312	......	......	......	......
'80	14	1,153	....	221	1,673,386	......	......	$3,500	......
'82	17.6	1,431	74	229	2,200,974	......	......	17,940	$27,414
'84	19.6	1,832	550	255	3,137,761	$440,000	$30,800	21,400	40,000
'85	21.8	1,973	1,392	273	2,400,152	440,000	30,800	25,116	39,283
'86	25	2,077	1,500	315	1,575,157	440,000	30,800	17,886	32,751
'87	27	2,192	2,000	342	1,721,858	440,000	30,800	21,000	......
'88	31.2	2,464	2,286	440	1,718,391	440,000	30,800	21,500	43,347
'89	35.8	2,883	....	444	2,073,322	440,090	30,800	30,332	56,370
'90	40.5	3,273	2,934	503	2,359,564	440,000	30,800	28,363	63,439

Financial.—*Cost,* $532,744. **Management.**—Comrs. Chn., H. C. Hutchison. Secy., Geo. W. Terry, Jr. Supt., W. G. Richards.

4. **AUGUSTA,** *Richmond Co.* (Pop., 21,891—33,300.) Present works built in '59-61 by city; first supplied in '28 by Turknett Springs Water Co. Supply.—Savannah river, through Augusta canal and settling basins. pumping direct and to tank. Fiscal year closes Dec. 31.

Year.	Mains.	Taps.	Meters.	Hydrants.	Consumption.	Exp.	Rev.
'82	21.6	1,324	0	175	2,247,085	$8,312	$17,333
'84	22.6	1,567	0	186	2,250,000	6,500	23,560
'86	27.8	1,728	0	213	2,177,595	8,816	28,030
'88	35.5	...	0	296	2,572,193	8,523	26,551
'89	35.8	....	0	302	3,385,182	5,907	24,089*
'90	36.2	....	0	306	3,385,484	5,640	35,133

* For 8 months, owing to change in fiscal year.

Financial.—*Floating Debt,* $13,000, at 6%. *No Bonds. Expenses:* operating, $6,640; int., $780; total, $6,420. *Revenue:* consumers, $35,133; city, none. **Management.**—Com. appointed by mayor every three years from council. Chn., D. Kerr. Supt., Wm. Bennett.

5. **BAINBRIDGE,** *Decatur Co.* (Pop., 1,433—1,663.) Built in '82-6 by town. Supply.—Artesian wells, pumping to tank. Distribution.—*Mains,* 6-in. w. i. *Services,* w. i. *Taps,* 200. *Meters,* 0. *Hydrants,* Holyoke, 20. *Consumption,* 25,000 galls. **Financial.**—*Cost,* $15,000. *Bonded Debt,* $5,000 at 7%. *Expenses:* operating, $800; int., $350; total, $1,150. *No Revenue.* **Management.**—City Council. Mayor, G. F. Westmoreland. Clerk, J. S. Bradwell.

6. **BRUNSWICK,** *Glynn Co.* (Pop., 2,891—8.459.) Built in '86 by Brunswick Light & Water Co., under 20 yrs.' franchise, after expiration of which city may purchase works, price to be settled by arbitration; city does not control rates. Supply. —Artesian well, pumping to tank. Fiscal year closes third Tuesday in Jan. A 2,500,000-gall. Worthington pump was added in '60.

Year.	Mains.	Taps.	Mtrs.	Hyd'ts.	Cons'pt'n.	Cost.	B'd'd d't.	Int. ch'g's.	Exp.	Rev.
'87*	7	285	5	55	250,000	$96,000	$50,000	$3,500	....	$3,750*
'88	10	411	5	56	400,000	96,000	50,000	3,500	....	10,725
'89	10.5	550	2	69	750,000	97,000	100,000	7,000	6,125	12,200
'90	14	800	0	91	800,000	125,000	200,000	12,000	....	17,062

* From city.

Financial.—All except revenue applied to water, gas and electric lights. *Cap. Stock:* authorized, $50,000; all paid up; paid 10% div'd in 1890. *Total Debt,* $106 000: floating, $6,000; bonded, $50,000 at 7%, due in 1907; $50,000 at 7%, due in 1909. *Expenses:*

FOR FURTHER DESCRIPTIVE MATTER see Manual for 1889-90. CHANGES since 1890 are here given. POPULATIONS are for 1880 and 1890, respectively.

int., $12,000; taxes, $1,125. *Revenue*, total, $17,062. **Management.**—Prest., W. E. Burbage. Secy. and Supt., W. E. Sutton.

7. **CARTERSVILLE,** *Bartow Co.* (Pop., 2,037—3,171.) Built in '89 by Cartersville Water Co., under 30-yrs'. franchise. **Supply.**—Springs, on bank of Etowah River, pumping to tank, Fiscal year closes Dec. 31. **Distribution.**—*Mains.* 8 miles. *Taps*, 150. *Meters*, 1. *Hydrants*, 50. **Financial.**—*Cost*, $67,000. **Management.**—Prest., J. S. Baxter, Macon. Secy., Douglas Wilke. Treas., A. E. Boardman, Brevard, N. C. Supt., J. A. Campbell.

8. **COLUMBUS,** *Muscogee Co.* (Pop., 10,123—17,303.) Built in '82 by Columbus Water-Works Co. **Supply.**—Holland's Creek, by gravity from impounding reservoirs to stand-pipe. **Distribution.**—*Mains*. 18 miles. *Taps*, 818. *Meters*, 722. *Hydrants*, 97. **Financial.**—Withheld. **Management.**—Prest., H. H. Epping. Secy. and Treas., J. G. Beasley. Supt., M. H. Tuggle.

NOTE.—Corporation known as Rose Hill Water Co. was chartered in '90, but has not begun construction of works.

9. **CORDELE,** *Dooly Co.,* (Pop., in 90, 1,578.) **History.**—Built in '90 by city; to be in operation Oct. 10. Engr., J. A. Raiford. Contrs., Wilcox & Ellis. Macon. **Supply.**—Seven mineral springs, pumping direct. **Pumping Machinery.**—Dy. cap., 300,000 galls.: Knowles. **Distribution.**—*Mains*. 8 to 4-in., c. i., 4¼ miles. Pipes and specials from So. Pittsburgh (Tenn.) Pipe & Iron Foundry. *Taps*, 40. *Meters*, 0. *Hydrants*, 12. *Valves*, Eddy. *Pressure*, 80 lbs. **Financial.**—*Cost*, $11,800. *Bonded Debt*, $14,000 at 6%, due in 20 years. *Expenses*, $200. **Management**—Prest., G. M. McMillan. Secy., J. B. Scott. Supt., J. A. Raiford.

10. **DALTON,** *Whitfield Co.* (Pop., 2,516—3,046.) Built in '88 by town. **Supply.**—Spring, pumping from settling reservoir to stand-pipe. Fiscal year closes Dec 31. **Distribution.**—*Mains*, 5.8 miles. *Taps*, 225. *Meters*, 0. *Hydrants*, 50. *Consumption*, 135,000 galls. **Financial.**—*Cost*, $33,600. *Bonded Debt*, $15,000 at 5%, due in 1918. *Expenses:* operating. $993; int., $750; total, $1,743. *Revenue:* consumers. $1,659; city, none. **Management.**—Committee of three appointed by mayor every year. Chn., J. H. Kenner. Secy. and Supt., C. G. Spencer.

11. **DAWSON,** *Terrell Co.* (Pop., 1,576—2,284.) **History.**—Built in '91 by Dawson Water-Works Co., under 90-yrs'. franchise; put in operation Sept. 1. Engrs. and Contra., American Pipe Mfg. Co., Philadelphia. **Supply.**—Creek, pumping to tank. **Pumping Machinery.**—Dy. cap., 750,000 galls.; one 500,000 and one 250,000 Worthington. *Boilers,* from Sterns Mfg. Co., Erie, Pa. Tank.—Cap., 75,000 galls. **Distribution.**—*Mains*, Phipps Hydraulic cem., 5¼ miles, *Services*, w. i. *Hydrants*, Ludlow, 50. *Valves*, Ludlow. **Financial.**—*Cap. Stock:* authorized, $100,000; issued, $10,000. *Bonded Debt:* authorized, $40,000; issued, $38,000. **Management.**—Prest., Wm. S. Perot, Jr.; Secy. and Treas., H. Bayard Hodge; both of Philadelphia. Mgr., Jas. S. Keen, Jr.

12. **FORT GAINES,** *Clay Co.* (Pop., 867—1,097,) Built in '77-8 by town. **Supply.**—Artesian well, pumping to tank. **Distribution.**—(In '88.) *Mains*, 2,000 ft. *Hydrants*, 16 *Consumption*, 12,000 galls. **Financial.**—(In '88.)*Cost*, $2,000. *Debt*, $300. *Expenses*, $550. *Revenue*, $300. **Management.**—(In '88.) Mayor, W. M. Speight.

13. **GAINESVILLE,** *Hall Co.* (Pop., 1,919—3,202.) **History.**—Built in '90-1 by city; put in operation Jan. 10. Engr., N. W. Davis, Atlanta, Ga. **Supply.**—Mountain stream, pumping to stand-pipe; filtered through gravel and sand. **Pumping Machinery**—Dy. Cap., 500,000 galls.; Dean dup., 14 × 8½ × 10. *Boiler:* 65 H. P ; Dean. **Stand-Pipe.**—Cap., 120,000 galls.; 16 × 80 ft.; of iron, enclosed in brick wall; cost $7,000; from Tippett & Wood, Phillipsburg, N. J. **Distribution.**—*Mains*, 10 to 4-in. c. i., 5¼ miles. *Services*, ¾-in. galv. iron. Pipe and specials from So. Pittsburgh (Tenn.) Pipe Works; trenching and pipe-laying done by T Wm. Harris & Co., N. Y. City, *Hydrants*, Ludlow, 30. *Valves*, Ludlow. *Pressure:* ordinary, 70 lbs.; fire, 90 to 100 lbs. **Financial.**—*Cost*, $31,822. *Bonded Debt*, $30,000 at 5 and 6%. *Operating Expense*, $600. **Management.**—Supt., G. W. Walker.

14. **GRIFFIN,** *Spalding Co.* (Pop., 3,620—4,503.) Works were put in operation in April, '91, by Griffin Water-Works Co. Attempt was soon afterwards made to float bonds to pay liabilities, but owing to financial depression bonds could not be sold at fair prices, and were withdrawn from the market. H. C. Burr was temporary Receiver. It was hoped in September that matters would soon be settled. J. A. Stewart was Mayor in May, '91.

15. **KENSINGTON,** *Walker Co.* (Pop. not given.) Construction was begun May 1, '90, by Kensington Water & Electric Light Co.; system to be in operation by Dec. 1. Engineer, G. L. Hogan, Nashville, Tenn. **Supply.**—Spring and creek, pumping to 100,000 gall. stand-pipe. *Mains*, a 10-in supply and 8-in. delivery main with 5 miles of distribution pipe. Prest., F. R. Pemberton. Secy. and Treas., Floyd S. Paterson. Supt., G. L. Hogan. Latest report Aug. 10, '90.

16. **MACON,** *Bibb Co.* (Pop., 12,749—22,746.) First supplied in '67; present works built in '81 by Macon Gaslight & Water Co.; franchise fixes maximum rates but does not provide for purchase of works by city. **Supply.**—Springs, pumping to reservoir and tank. Fiscal year closes April 1. Summer of '91 reported that company would increase cap. stock, put in new pumps and a filtering plant.

Year.	Mains.	Taps.	Meters.	Hd'ts.	Cons'pt'n.	Cost.	B'd' d't.	Int. ch'g's.	Exp.	Rev.
'82-3	9.6	216	104	53	175,000	$78,000	$75,000	$4,500	$2,460	$6,320
'84-5	15.5	524	143	81	376,000	108,500	110,500	6,600	5,496	12,503
'86-7	17.5	818	...	104	620,000	215,000	115,000	6,900	9,532	22,500
'88-9	32.5	1.500	248	200	1,600,000	287,000	115,000	6,900	9,850	29,600
'90-1	32.5	2.273	65	200	1,647,000	356,473	......	......	......	......

Financial.—*(Including Gas and Electric Plants.) Cap. Stock;* authorized, $300,000; all paid up. Total Debt, $513,000; floating, $13,000; bonded, $500,000 at 6%, due in 1917. **Management.**—Prest. and Treas., A. E. Boardman, Brevard. N. C. Vice-Prest., W. A. Jeter, Brunswick. Secy., J. B. Hall. Supt., J. W. Wilcox.
NOTE.—Summer of '91 city engaged B. F. Church, N. Y. City, to report on city works, and secured authority to issue about $100,000 of bonds to pay for works.

17. **MILLEDGEVILLE,** *Baldwin Co.* (Pop., 3,800—3,322.) System was constructed in '91 by Sam'l Walker; to be in operation Oct. 1. **Supply.**—Springs, pumping direct. Works designed and constructed by owner. No further information could be obtained.

11. **RISING FAWN,** *Dade Co.* (Pop., 1,198—927.) Built in '90 by G. W. Cureton. **Supply.**—Springs, by gravity from reservoir. **Reservoir.**—Cap., 250,000 galls. **Distribution.**— *Mains,* 3-in., 3 miles. *Pressure,* 125 lbs. **Financial.**—*Cost,* $10,000. **Management.**—Owner, Geo. W. Cureton.

18. **ROME,** *Floyd Co.* (Pop., 3,877—6,957.) Built in '71 by city. **Supply.**—Wells, pumping to stand-pipe. Fiscal year closes March 31. A well about 375 ft. deep was added in '91.

Year.	Mains.	Taps.	Meters.	Hydrants.	Cost.	B'd'd d't.	Int. ch'g's.	Exp.	Rev.
'80-1	5	350	..	55	......	......	...	$2,500	.....
'84-5	6	187	..	50	......	......	...	3,700	$5,500
'88-9	10	625	45	80	$96,000	$85,000	$5,100	4,000	8,000
'89 90	10	625	45	...	......	86,000	5,110	...	...
'90-1	12	719	50	102	150,000	86,000	5,031	4,594	9,626

Management.—Com. of three, one appointed by Mayor every year. Chn., W. W. Cain. Supt., A. J. Wagner, appointed by city council.

19. **SAVANNAH,** *Chatham Co.* (Pop., 30,709—43,189.) Built in '53-4 by city. **Supply.**—Artesian wells, pumping direct. Fiscal year closes Dec. 31. During '91 supply was increased by two 10-in. artesian wells, and contract was let for new conduit, 3,664 ft. of which is of masonry. Contrs.: for pipe and specials, Howard-Harrison Cc., Bessemer, Ala.; for trenching and pipe-laying, Robertson & Weaver, Baltimore, Md. In '91 contract was let for two 10,000,000-gall. Gaskill pumping engines.

Year.	Mains.	Taps.	Hydrants.	Cons'ption.	Exp.	Rev.
'80	22	....	257	......	$11,000	.....
'83	28.5	....	264	3,180,090	17,899	$45,225
'86	32.6	2,600	312	5,715,000	24,000	47,000
'89	33.25	4,000	320	......	43,521	54,921
'90	34.9	....	331	5,851,610	38,445	51,975

Financial.—*First Cost,* $225,000. *No Debt. Expenses,* total, $8,446. *Revenue,* consumers, $51,975. **Management.**—Com. of five, appointed every two years by Mayor. Chn., J. J. McDonough. Clerk, I. H. Connell. Supt., James Manning.

20. **TALLAPOOSA,** *Haralson Co.* (Pop., 52—1,609) **History.**—Built in '90-1 by Tallapoosa Water Co., under 30-yrs. franchise; put in operation Feb. 19. Engr., Geo. P. Keith. Contrs., Woltman, Keith & Co., 11 Wall St., N. Y. City. **Supply.**—Springs and stream, pumping to stand-pipe. *Pumping Machinery.*—Dy. cap., 1,500,000 galls.; two 750,000 Worthington horizontal, comp., non-cond. *Boilers:* from Annis-

ton (Ala.) Boiler Wks. **Stand-Pipe.**—Cap., 115,000 galls.; 15 × 100 ft.; from Tippett & Wood, Phillipsburgh, N. J. **Distribution.**—*Mains*, 12 to 4-in. c. i., 4½ miles. **Hydrants**, Ludlow, 32. *Valves*, Ludlow. **Financial.**—*Cost*, $15,000. *Bonded Debt*: authorized, $75,000; issued, $40,000. *Revenue:* city, $1,600; hydrant rental, $50. **Management.**—Supt., A. M. Greene.

21. **THOMASVILLE**, *Thomas Co.* (Pop., 2,555—5,514.) Built in '88 by town. **Supply.**—Two artesian wells, pumping to stand-pipe. Fiscal year closes April 1. **Distribution.**—*Mains*, 4 miles. *Taps*, 200. *Meters*, 1. *Hydrants*, 87. *Consumption*, 80,000 galls. **Financial.**—*Cost* (in '88), $30,000. *Expenses*, $6,000. **Management.**—Supt., E. O. Thompson.

III. **TYBEE** (*Savannah P. O.*). *Chatham Co.* (Pop., est., 500.) Built in '89 by Artesian Water, Ice & Lighting Co., to supply private families and bathing establishments, chiefly during the summer months. **Supply.**—Artesian well, pumping to tank. Fiscal year closes April 30. *Takers*, 25. *Cost*, $7,000. Address W. P. Lovell, Secy. NOTE.—In '91 it was reported that new works were to be built.

Water Works Projected with Fair Prospects of Construction.

BUENA VISTA, *Marion Co.* (Pop., 529—783.) City contemplated building works, Aug. 24, '91. Address W. D. Crawford, Mayor.

CARROLLTON, *Carroll Co.* (Pop., 926—1,451.) July 10, '91, it was reported that city would grant 40-yrs. franchise and pay liberal hydrant rental for 15 hydrants, to induce co. to build system.

CEDARTOWN, *Polk Co.* (Pop., 843—1,625.) June 26, '91, it was reported that committee was considering plans for construction of work by city. Later bill was introduced in legislature to authorize city to issue bonds. Address A. Richardson or Cedartown Land Improvement Co.

EAST ROME (*Rome, P. O.*), *Floyd Co.* (Pop., in '90, 514.) East Rome Water Co. was granted charter in '91, and was to construct system at once. Address J. P. Cooper, Secy.

FORT VALLEY, *Houston Co.* (Pop., 1,277—1,752.) In Jan., '91, citizens voted to bond town to construct system of water-works. Board of comrs. was organized with H. C. Harris, Prest. System described in 1883—90 Manual, p. 327, was not constructed.

HAWKINSVILLE, *Pulaski Co.* (Pop., 1,512—1,755.) Legislature was petitioned in '91 by town for authority to issue $15,000 of bonds to construct water-works. Town already owns well which furnishes water for domestic use.

LA GRANGE, *Troup Co.* (Pop., 2,295—3,090.) July 7, '91, O. A. Bull, City Clerk stated that one or two companies had submitted propositions to construct works, but no action had been taken. Projects described in 1883-90 Manual, p. 327, were not carried out.

MACON, *Bibb Co.* (Pop., not given.) Highland Park Land & Improvement Co. proposed to let contracts Oct. 1, '91, to supply suburb of Macon; works to be in operation by April 1, '92. Supply, springs, pumping to tank. Cost, $50,000. Capital secured. Address, A. M. Rodgers, Prest.

MADISON, *Morgan Co.* (Pop., 1,974—2,131.) Early in '91, town voted to have water-works, to be owned by a private co. Contract was awarded W. A. Robinson, Atlanta; construction to be begun Aug. 1 and completed by Dec. 31. Hydrants, 50. July 13, E. W. Butler, Mayor, reported that construction had not begun.

NEWMAN, *Coweta Co.* (Pop., 2,006—2,859.) Fall of '91 it was reported that Sinclair & Leavenworth, Chattanooga, Tenn., had contract or franchise for works.

SMITHVILLE, *Lee Co.* (Pop., est., 750.) Small system projected. Supply, two artesian wells, pumping to tank. Cost, $1,000. Address, R. F. Johnson.

TENNILLE, *Washington Co.* (Pop., 99—953.) Contract was let in '91 to Hartford, Hebert & Co. for construction of system, to be owned by city. Supply, artesian well, pumping to 30,000-gall. stand-pipe. Mains, 1 mile. Hydrants, 7. Cost, $6,750. Work had not begun June 25.

VIENNA, *Dooly Co.* (Pop., in '90, 536.) System projected June 25, '91. Supply, artesian well. Address J. O. Hamilton, Mayor, or C. F. Stovall, Alderman.

GEORGIA.

WASHINGTON, *Wilkes Co.* (Pop., 2,199—2,651.) See 1889-90 Manual, p. 328. Legislature was petitioned again in '91 for authority to issue bonds and construct works. Address T. B. Greene.

WAYCROSS, *Ware Co.* (Pop., 1,231—3,364.) In '90 town was authorized to issue bonds for $30,000 at 6%, to construct works. Des. Engr., G. H. Craft, Atlanta. Supply, source not determined, pumping to stand-pipe. System had not been built June 25, '91. Chn., H. Murphy.

WRIGHTSVILLE, *Johnson Co.* (Pop., 272—910.) July 10, '91, town voted to issue $5,000 in bonds for construction of water-works, contracts to be let at once. Town Clerk, J. W. Brinson.

Water-Works Projects Reported in '90 and '91, but Now Indefinite.

Trenton, Dade Co. *West End,* Fulton Co.

FLORIDA.

Water-Works Completed or in Process of Construction.

1. FERNANDINA, *Nassau Co.* (Pop., 2,562—2,803.) Built in '89 by city. Supply.—Artesian well, pumping to tank. Fiscal year closes July 1. Distribution. —*Mains,* 4 miles. *Taps,* 128. *Hydrants,* 53. *Consumption,* 60.000 galls. Financial. —*Bonded Debt,* $40,000, at 6%; due in 1909, may be paid in '94. *Expenses:* operating, $2,600; int., $2,400; total, $5,000. *Revenue:* consumers, $1,200. Revenue is turned into general fund of city, from which all expenses are paid. **Management.**—Board of Public Works, of three, elected or appointed every three years by Common Council. Chn., Geo. R. Fairbanks; Secy, Jno A. Edwards; Supt., Chas. Williams; appointed by City Council.

2. JACKSONVILLE, *Duval Co.* (Pop., 7,650—17,201). Built in '79-80 by city Supply.—Artesian wells, pumping direct. Fiscal year closes July 1.

Year.	Mains.	Taps.	Meters.	Hyd'ts.	Cons'pt'n.	Cost.	B'd' d't.	Exp.	Rev.
'80-1	6	150	...	87	186,037	$95,000	......	......	......
'81-2	7.8	150	...	92	309,334	......	......	$4,820	$1,324
'82-3	8.5	284	160	99	362,317	98,000	......	4,170	1,536
'83-4	8.5	382	166	99	400,146	......	$100,000	6,928	7,618
'84-5	8.5	450	226	99	573,502	100,000	100,000	9,326	10,593
'85-6	8.5	583	240	100	782,500	......	100,000	9,497	13,108
'86-7	8.5	707	287	100	918,305	112,000	100,000	9,168	16,521
'87-8	8.8	820	385	105	1,122,766	112,000	100,000	10,309	18,785
'88-9	8.9	823	385	106	878,207	117,000	100,000	10,992	14,441
'89 90	9.1	1,081	391	107	991,194	119,179	100,000	10,910	15,160

Management.—Chn., A. S. Baldwin. Secy., Geo. R. Foster.

3. LAKE CITY, *Columbia Co.* (Pop., 1,379—2,020.) Built in '87-8 by Lake City Water & Light Co.; franchise fixes maximum rates and provides for purchase of works by city. Supply.—Lake Isabella, pumping direct. Tank formerly used has been discontinued. Distribution.—*Mains,* 2 miles. *Taps,* 40. *Hydrants,* 32. Financial.—*Cost,* $20,000. *Bonded Debt,* $25,000 at 6%, $15,000 issued in '90. *Expenses,* operating, $250. *Revenue:* consumers, $720; city, $640; total, $1,360. **Management.** —Prest. and Treas., Noyes S. Collins. Secy., L. Harrison.

Melbourne, *Brevard Co.* (Pop., in '90, 99.) Built in '89 by individuals, chiefly for fire protection. Supply.—Artesian well, by gravity, owned by parties whose property is protected thereby. Address E. P. Branch.

4. OCALA, *Marion Co.* (Pop., 803—2,904.) Built in '87-8 by Ocala Water Co., under 25-yrs. franchise, in which maximum rates are fixed; at expiration of franchise city may buy works, price determined by arbitration, or may renew franchise for 5 yrs. Supply.—Artesian wells, pumping to tank. Distribution.—*Mains,* 7.5 miles. *Taps,* 266. *Meters,* 3. *Hydrants,* 85. *Consumption,* 125,000 galls. Financial.—*Cost,* $95,599. *Cap. Stock,* $50,000. *Bonded Debt,* $60,000 at 6%. *Expenses:* operating, $3,600; int., $3,600; total, $7,200. *Revenue:* consumers, $4,500; city, $3,825; total, $3,325. **Management.**—Prest., A. D. Schofield. Secy., R. L. Anderson. Treas., A. E. Boardman, Brevard, N. C. Supt., C. H. Campbell.

FOR FURTHER DESCRIPTIVE MATTER see Manual for 1889-90. CHANGES since 1890 are here given. POPULATIONS are for 1880 and 1890, respectively.

5. ORLANDO, *Orange Co.* (Pop., in '90, 2,856.) Built in '87 by Orlando Water Co.; city does not control rates but has right to buy works at end of every 10 years. Supply.—Highland Lake, pumping direct and to stand-pipe. Fiscal year closes Dec, 31. Distribution.—*Mains*, 13 miles *Taps*, 380. *Meters*, 7. *Hydrants*, 55. *Consumption*, 150,000 galls. **Financial**—*Cost*, $150,000. *Cap. Stock:* authorized, $50,000; all paid up; paid no div'd. in 1890. *Bonded Debt*, $100,000 at 6%. *Expenses:* operating, $4,500; int., $6,000; total, $10,500. *Revenue:* consumers, $5,500; city, $3,200; total, $8,700. **Management.**—Prest., Geo. W. Lintz. Williamsport, Pa. Secy., Andrew Johnson, Chattanooga, Tenn. Treas., H. Hinckley, Williamsport, Pa. Supt., H. W. Guetham.

6. PALATKA, *Putnam Co.* (Pop., 1,616—3,039.) Built in '86-7 by Palatka Water Works Co., under 20-yrs. franchise, in which maximum rates are fixed; city may buy works at any time after '97, price determined by three disinterested persons. Supply.—White Water branch, pumping to stand-pipe. Fiscal year closes Dec. 31.

Year.	Mains.	Taps.	Meters.	Hyd'ts.	Cons'pt'n.	Cost.	B'd d't.	Int.ch'gs.	Exp.	Rev.
'87	6	130	4	65	....	$65,000	$60,000	$3,600	$2,000	$5,500
'88	8.8	210	6	70	....	140,000	60,000	3,600	1,600	7,170
'90	11.5	281	7	75	115,000	146,664	60,000	3,600	2,090	9,933

Financial.—*Cap. Stock*, $75,000. *Total Debt*, $70,000: floating, $10,000; bonded, $60,000 at 6%, due in 1912. *Expenses:* operating, $2,690; int., $4,266; taxes, $119. total, $7,075. *Revenue:* fixtures, $5,261; meters, $1,514; hydrants, $3,158; total, $9,933. **Management.**—Prest., Harry Wheeler; Secy., J. F. J. Mulhall; Treas., Elbert Wheeler; all of Boston, Mass. Supt., Jno. H. Yeaton.

7. PENSACOLA, *Escambia Co.* (Pop., 6,845—11,750.) Built in '85-6 by Pensacola Water Co., under 50-yrs. franchise which fixes maximum rates and provides that the city may purchase works at any time after 10 yrs., price determined by arbitration. Fiscal year closes July 1. During '90 supply was increased by driven well 4 ins. in diam., 120 ft. deep. Stand-pipe is stated to be 20 × 137 ft., instead of 25 × 125 ft., as in last Manual. Distribution.—*Mains*, 12 miles. *Taps*, 650. *Meters*, 34. *Hydrants*, 105. *Consumption*, 350,000 galls. **Financial.**—*Cap. Stock:* authorized, $150,000; all paid up. *Bonded Debt*, $150,000 at 6%. *Expenses:* operating, $7,500; int., $9,000; taxes, $750; total, $17,250. *Revenue:* consumers, $10,960; city, $6,350; total, $17,310. **Management.**—Prest. and Supt., B. B. Pitt. Secy., H. G. DeSilva.

1. RIVERSIDE *(Jacksonville, P. O.), Duval Co.* (Pop., probably included in Jacksonville.) Built in '90 by Artesian Well Co.; put in operation Dec. 1. Supply, Artesian well, by natural pressure. *Mains*, 1 mile. *Taps*, 40. *Cost*, $5,000 *Cap. Stock*, $5,000. *No Debt*. Prest., B. R. Powell. Secy., Stephen E. Foster. Treas., A. J. Hendrick.

8. ST. AUGUSTINE, *St. John's Co.* (Pop., 2,293—4,742.) **History.**—Built in '84 by St. Augustine Water Co. under 99-yrs. exclusive franchise, which provides neither for control of rates nor purchase of works by city. Des. Engr., E. L. Parker. Const. Engr., E. F. Joyce. Supply.—Artesian wells, by gravity. Distribution.—*Mains*, w. i., 5 miles. *Services*, galv. i. *Taps*, 210. *Meters*, 0. *Hydrants*, 37. **Financial.**—*Cost*, $33,000. *Cap. Stock:* authorized, $50,000; paid up, $33,000. *No Debt. Expenses*, none. *Revenue:* consumers, $2,000; city, $1,480. **Management.**—Supt., E. F. Joyce.

9. SANFORD, *Orange Co.* (Pop., 1,300—2,016.) Built in '83 by Sanford Water-Works Co. under 50-yrs. franchise. Supply, lake, by gravity to reservoir, thence pumping direct. Fiscal year closes Feb. 3. March 16, '91, Co. had just completed a 12-in. supply main from lake to reservoir in town, having changed source of supply to a larger lake of purer spring water, about three miles south of works. Distribution.—*Mains*, 9 miles. *Taps*, 350. *Meters*, 12. *Hydrants*, 60. *Consumption*, 125,000 galls. **Financial.**—*Cap. Stock*, $40,000. *Bonded Debt*, $20,000 at 8%. *Revenue*, city, $3,000. **Management.**—Prest., H. S. Sanford. Secy. and Supt., W. Beardall. Treas., F. H. Rand.

10. SEVILLE, *Volusia Co.* (Pop., est., 500.) Built in '86 by Seville Co. Supply.—Lake Louise, pumping to tanks. Distribution.—(In '87.) *Mains*, 1.5 miles. *Hydrants*, 8. **Financial.**—*Cap. Stock* (in '87), $50,000. **Management.**—Prest., Mason Young, 35 Wall St., N. Y. City. Supt., Geo. R. Scott.

11. TALLAHASSEE, *Leon Co.* (Pop., 2,494—2,934.) **History.**—Built in '89-90 Tallahassee Water-Works Co., under 25-yrs. franchise; put in operation July 30. Engrs. and contrs., American Pipe Co. Supply.—Springs, pumping to tank.

FLORIDA. 175

Pumping Machinery.—Barr pumping engine. Tank.—Cap., 75,000 galls. Distribution.—*Mains*, Phipps Hydraulic cem., 6 miles. *Services*, w.i. *Taps*, 100. *Meters*, Crown. *Hydrants*, Ludlow, 60. *Pressure*, 70 lbs. Financial.—*Cost*, $10,000. *Cap. Stock:* authorized, $10,000; all paid up. *Bonded Debt*, $40,000 at 6%. *Expenses:* operating, $900; int., $2,400; total, $3,300. *Revenue:* consumers, $1,200; city, $2,400; total, $3,600. Management.—Prest., Jos. S. Kean; Secy., H. Bayard Hodge; Treas., Wm. S. Perot, Jr.; all of 1,326 Chestnut St., Philadelphia. Supt., Geo. W. Saxon.

12. **TAMPA**, *Hillsborough Co.* (Pop., 720—5,532.) Built in '88-9 by Tampa Water-Works Co.; franchise does not fix rates, but provides for purchase of works by city at any time within 30 yrs., price determined by arbitration. Supply.— Spring, pumping to tank. Distribution.—*Mains*. 14 miles. *Taps*, 569. *Hydrants*, 147. *Consumption*, 320,000 galls. Financial.—*Cost*, $152,000. *Cap. Stock*, $100,000. *Bonded Debt*, $82,000 at 6%. *Sinking Fund*, 10% of net earnings. *Revenue:* consumers, $12,500; city, $6,615; total, $19,115. Management.—Prest., R. L. Henry. Secy. and Supt., W. A. Campbell. Treas., A. E. Boardman, Brevard, N. C.

Water-Works Projected with Fair Prospects of Construction.

DE LAND, *Volusia Co.* (Pop., est., 2,500.) Works have been projected since '86. Fall of '91 it was reported that city was to pay $1,000 yearly for hydrants and exempt works from taxation.

GAINESVILLE, *Alachua Co.* (Pop., in '90, 2,790.) Bids for works to be owned by city were received in September. Unofficially reported that contract was awarded to Hartfor, Hebert & Co., Chattanooga, Tenn. Supply, springs, pumping to stand-pipe. Cost, $50,000. Chn., A. J. McArthur.

GREEN COVE SPRINGS, *Clay Co.* (Pop. 320—est., 1,200.) Mutual Water-Works Co. was to begin construction of water-works Nov. 1, '91. Supply, artesian well. Est. cost, $8,000. Prest., C. E. Garner. Secy., J. F. Gruir.

JASPER, *Hamilton Co.* (Pop., 311—est., 1,800.) City proposes to build works. July 7, '91, committee had been appointed to let contracts. Supply, artesian well, pumping direct. Cost, $3,000. Address Jno. M. Caldwell.

KEY WEST, *Monroe Co.* (Pop., 9,890—est., 18,000.) Works have been projected for some time. Sept., '91, there was talk of granting a franchise for works.

LEESBURG, *Lake Co.* (Pop., in '90, .722.) Company was organized in '91 to construct water-works. Supply, well, pumping to tank. Cost, $10,000. Prest., G. C. Staplyton. Secy., B. R. Milan.

SOUTH JACKSONVILLE, *(Jacksonville P. O.), Duval Co.* (Pop. in '90, 378) In '91 South Jacksonville Water-Works Co. was incorporated; construction began Aug. 1. Supply, artesian well. Cost, $5,000. Cap. Stock, authorized, $5,000. Prest , C. E. Garner. Secy. and Treas., A. J. Henduck.

GROUP 4.—Gulf and Mississippi Valley States.—Water-Works Completed, in Process of Construction or Projected, in Alabama, Mississippi, Louisiana, Tennessee and Kentucky.

ALABAMA.

Water-Works Completed or n Process of Construction.

1. **ANNISTON,** *Calhoun Co.* (Pop., 942—9,876.) First supplied in '81 by Woodstock Iron Co. Later works were transferred to Anniston City Land Co., and in Oct., '89, to Anniston Water Supply Co., which improved works and obtained new supply. Fiscal year closes Sept. 30. Supply.—Springs, pumping to reservoir. Springs have a cap. stated as 22,000,000 galls. dy. Pumping Machinery.—Cap., 8,000,000 galls.; two 4,000,000-gall. comp. dup. Gaskill. *Boilers*, three 80 HP. Force Main.—Diam., 20 ins. Reservoir.—Cap., 6,000,000 galls. Supply. Mains.—One 16 and two 12-in. Distribution.—*Mains*, 21 miles. *Taps*, 433. *Meters*, 37. *Hydrants*,

FOR FURTHER DESCRIPTIVE MATTER see Manual for 1889-90. CHANGES since 1890 are here given. POPULATIONS are for 1880 and 1890, respectively.

143. *Consumption*, 600,000 galls. **Financial.**—*Cap. Stock:* authorized, $500,000; all paid up. *Bonded Debt*, $300,000. *Revenue*, $29,000. **Management.**—Prest., T. G. Bush. Secy. and Treas., Osborne Parker. Supt., H. B. Rudisell.

2. **BESSEMER,** *Jefferson Co.* (Pop. in '90, 4,544.) Built in '87-8 by Bessemer Land & Improvement Co. **Supply.**—Streams, by gravity to pump pond, thence by pumping to stand-pipe. Fiscal year closes March 31. **Distribution.**—*Mains*, 11 miles. *Taps*, 383. *Hydrants*, 45. *Consumption*, 120,000 galls. **Financial.**—*Cost*, $100,950. *No Debt*. *Expenses*, $7,000. **Management.**—Prest., H. F. De Bardeleben. Secy., H. M. McNutt. Supt., G. H. Stevenson.

3. **BIRMINGHAM,** *Jefferson Co.* (Pop., 3,086–26,178.) Built in '72-3 by Elyton Land Co., now owned by Birmingham Water-Works Co.; franchise fixes rates and provides for purchase of works by city in 1917, or at end of any ten years thereafter, price to be determined by arbitration, if necessary. **Supply.**—Five Mile Creek, through canal to settling basin, thence by pumping to reservoir. Fiscal year closes April 1. Cahaba River extension completed in Dec., '90.

Year.	Mains.	Taps.	M't'rs.	H'd'ts.	C'ns'pt'n.	Cost.	B'd'd d't.	Int. ch'g's.	Exp.	Rev.
'80-1	8	450		15					$500	
'84-5	17	938	50	34	1,500,000	$150,000	None.	None.	8,000	$30,000
'86-7	25,32	1,547	80	40	2,000,000	400,000	None.	None.	12,000	40,029
'87-8	30,38	2,041	250	57	3,800,000	606,487	$100,0.0	$6,000	20,000	71,879
'88-9	40	2,361	334	60	4,250,000	781,170	300,000	18,000	21,000	80,829
'89-90	50	3,000	500	65	4,250,000	1,128,020	400,000	22,703	24,748	85,387
'90-1	58	3,281	500	209	5,072,966	1,276,732	455,000	27,300	30,951	

Financial.—*First Cost*, $90,000. *Cap. Stock:* authorized, $500,000; all paid-up: profits used for extensions. *Total Debt*, $628,353; floating, $173,353 at 7%; bonded, $155,000 at 6%, due in 1917. *Expenses:* operating, $29,056; int., $27,300; taxes, $1,895; total, $58,251. **Management.**—Prest., H. M. Caldwell. Treas., Jno. London. Secy. and Supt., W. J. Milner.

4. **BRIDGEPORT,** *Jackson Co.* (Pop., est., 1,000.) **History.**—Built in '90-1 by Bridgeport Water-Works Co. Engr., David Giles. **Supply.**—Mountain spring, by gravity. **Stand-Pipe.**—Cap., as projected, 1,000,000 galls. **Distribution.**—*Mains*, c. 1., 9 miles. *Services*, galv. i. *Taps*, 40. *Meters*. Hersey, 25. *Hydrants*, Ludlow, 50. *Valves*, Ludlow. *Pressure*, 125 lbs. **Financial.**—Withheld. **Management.**—Prest., David Giles. Supt., Wm. H. Brundige. NOTE.—Above description is subject to correction as latest report was sent Jan. 11, before completion of system.

5. **DECATUR and NEW DECATUR,** *Morgan Co.* (Pop., 1,063–2,765.) Built in '83-9 by Decatur Water Co., under 25-yrs. franchise. **Supply.**—Tennessee River, pumping to stand-pipe or direct. Fiscal year closes Dec. 31. **Distribution.** —*Mains*, 18.5 miles. *Taps*, 262. *Meters*, 3. *Hydrants*, 151. *Consumption*, 300,000 galls. **Financial.**—*Cap. Stock*, authorized, $350,000, all paid up. *Bonded Debt*, $300,000 at 6%, due in 1907. *Expenses and Revenue*, not determined. **Management.** Prest., John D. Roquemon. Secy., George A. Ellis, Boston, Mass. Treas., C. C. Harris. Supt., L. B. Wyatt.

6. **EUFAULA,** *Barbour Co.* (Pop., 3,836–4,394.) Built in '87-8 by Eufaula Water Co.; franchise provides that rates must not be higher than in other cities similarly situated and that works may be bought by city in 1906. **Supply.**—Little Chewalla Creek and Shorter Springs branch, pumping from storage reservoir to stand-pipe. Oct. 5, '91, Supt. Wulker wrote that two wells would be started soon as contracts could be let. **Distribution.**—*Mains*, 7 miles. *Hydrants*, 52. **Financial.** —*Cost* (in '88), $60,000. *Bonded Debt*, $60,000 at 6%. *Revenue*, city, $3,120. **Management.**—Prest., S. H Dent. Secy. and Treas., E. B. Young. Supt., R. H. Walker.

7. **FLORENCE,** *Lauderdale Co.* (Pop., 1,359–6,012.) First supplied in '87 by Cypress Water-Works Co. In '89 Jeter & Boardman Water & Gas Assn. bought plant and combined it with new system. Rates are fixed in franchise, which provides for purchase of works by city in 1912. **Supply.**—Cypress Creek, pumping to stand-pipe. **Distribution.**—*Mains*, 13 miles. *Hydrants*, 103. **Management.**—Prest., W. A. Jeter. Secy., H. C. Fairman. Treas., A. E. Boardman, Brevard, N. C. Supt., T. A. Howell.

8. **FORT PAYNE,** *De Kalb Co.* (Pop., in '90, 2,598.) Built in '89 by Fort Payne Coal & Iron Co. **Supply.**—Two springs, pumping to stand-pipe. Fiscal year closes March 1. **Distribution.**—*Mains*, 9 miles. *Taps*, 304. *Meters*, 65. *Hydrants*, 52. *Consumption*, 600,000 galls. **Financial.**—*Cost*, $68,000. *Cap. Stock:* authorized,

ALABAMA.

$400,000; paid-up, $80,000. *Bonded Debt*, $80,000, at 6%. *Expenses:* operating, $500; int., $4,800. *Revenue:* consumers, $5,200; city, $2,600; total, $7,800. **Management.**—Prest., J. W. Spaulding. Secy., J. S. Spaulding.

9. **GADSDEN**, *Etowah Co.* (Pop., 1,697—2,901.) Built in '85-6 by Gadsden Water-Works Co., under 20-yrs. franchise. **Supply.**—Coosa River, pumping through filter to reservoir. Fiscal year closes Feb. 1. **Distribution.**—*Mains*, 7 miles. *Taps*, 253. *Meters*, 0. *Hydrants*, 58. **Financial.**—*Cost*, $62,000. *Cap. Stock*, $75,000. *Bonded Debt*, $62,000 at 8%. *Expenses:* operating, $1,500; int., $4,900; total, $6,460. *Revenue:* consumers, $2,100; city, $3,750; total, $5,850. **Management.**—Prest., A. L. Glenn. Secy., W. G. Brockway. Supt., M. E. Jones.

10. **GREENVILLE**, *Butler Co.* (Pop., 2,171—2,806.) **History.**—Built in '90 by Greenville Water-Works Co., under 25-yrs. franchise; accepted in Dec. Engrs. and Contrs., American Pipe Mfg. Co., Philadelphia. **Supply.**—Creeks, pumping to tank. **Pumping Machinery.**—Dy. cap., 500,000 galls.; Barr. *Boilers:* from Sterns Mfg. Co., Erie, Pa. **Tank.**—Cap., 75,000 galls.; on elevation of 73 ft. above street. **Distribution.**—*Mains*, Phipps Hydraulic cem., 5½ miles. *Taps*, 102. *Meters*, Crown. *Hydrants*, Ludlow, 48. *Valves*, Ludlow. *Pressure*, 65 lbs. **Financial.**—*Cost*, $28,000. *Cap. Stock*, $28,000. *Bonded Debt*, $28,000 at 6%. *Operating Expenses*, $900. *Revenue:* consumers, $1,200; city, $2,150; total, $3,450. *Hydrant rental*, 35 at $50 each, balance at $40 each. **Management.**—Prest., F. M. Brooke; Secy., H. Bayard Hodge; Treas., W. S. Perot, Jr.; Gen. Mgr., J. S. Keen, Jr.; all of 1326 Chestnut St., Philadelphia.

11. **HUNTSVILLE**, *Madison Co.* (Pop., 4,977—7,995.) Now owned by city. First supplied in '25; ownership not stated. In '36 Thos. Fearn built works, city owning reservoir jointly with him, which were bought by city in '58. **Supply.**—Artesian wells, discharging into reservoirs, thence by pumping through distributing mains to stand-pipe. Fiscal year closes April 1. Sept. 30, '91, bids were received for 3,000,000-gall. pump to work against 175-ft. head. E. R. Hartford, Chattanooga, Tenn., recommended addition of stand-pipe. **Distribution.**—*Mains*, 4.5 miles. *Taps*, 470. *Meters*, 0. *Hydrants*, 22. *Consumption*, 75,000 galls. **Financial.**—*Bonded Debt*, $15,000 at 5%. *Expenses*, $2,733. *Revenue*, $4,000. **Management.**—Committee of three, appointed every two years by mayor. Chn., S. J. Mayhew, elected by aldermen. Secy., E. R. Matthews. Supt., B. M. Blake.

12. **JACKSONVILLE**, *Calhoun Co.* (Pop., 822—1,237.) Now owned by Martin & Gaboway, who bought works in '89. Built in '83 by town. **Supply.**—Springs, by gravity, to and from reservoir. System was extended in '90. **Distribution.**—(In '87.) *Mains*, 4 miles. *Taps*, 200. *Meters*, 0. *Hydrants*, 30. *Consumption*, 500,000 galls. **Financial.**—(In '87.) *Cost*, $40,000. *Bonded Debt*, $20,000 at 5%. *Expenses:* operating, $1,800; int., $1,000; total, $2,800. *Revenue:* consumers, $3,000; city, $3,000; total, $6,000. **Management.**—Owners, Martin & Gaboway.

13, 14. **MOBILE**, *Mobile Co.* (Pop., 29,132—31,076.)

MOBILE WATER-WORKS CO.

Built in '40 by individual, and owned by Louis Stein in '85. **Supply.**—Creek, pumping to stand-pipe. **Distribution.**—(In '85.) *Mains*, 22. *Taps*, 1,200. *Meters*, 1. *Hydrants*, 150. **Financial.**—(In '85.) *Cost*, $200,000. *Expenses:* operating, $8,000. **Management.**—See above.

BIENVILLE WATER SUPPLY CO.

Built in '86-8. **Supply.**—Creek pumping to reservoir or direct. Fiscal year closes Dec. 31.

Year.	Mains.	Taps.	M't'rs.	H'd'ts.	C'ns'n.	Cost.	B'd'd d.	Int.ch'g's.	Exp.	Rev.
'59	32	900	3	260	900,000	$1,250,000	$750,000	$45,000	$26,000	*$20,000
'90	36	1,500	20	260	2,500,000	1,250,000	750,000	45,000	25,200	35,000
										*from city.

Financial.—*Cost*, $1,250,000. *Cap. ¦Stock*, $500,000. *Bonded Debt*, $750,000 at 6%. *Expenses:* operating, $20,000; int., $45,000; taxes, $5,200; total, $70,200. *Revenue:* total, $35,000. **Management.**—Prest., Geo. A. Ketchum. Secy., Pope St. John. Treas., Walter Wood, 400 Chestnut St., Philadelphia, Pa. Supt., Jno. H. Turner. Gen. Supt., A. M. McCallum.

FOR FURTHER DESCRIPTIVE MATTER see Manual for 1889-90. CHANGES since 1890 are here given. POPULATIONS are for 1880 and 1890, respectively.

15. MONTGOMERY, *Montgomery Co.* (Pop., 16,723—21,883.) Present works built in '87 by Capital City Water Co. First works built in '73 by Montgomery Water-Works Co., which was reorganized in '80. **Supply.**—Artesian wells, discharging into reservoirs, thence pumping through distributing mains to stand-pipe. Fiscal year, closes Dec. 31. **Distribution.**—*Mains,* 38 miles. *Taps,* 1,830. *Meters,* 652. *Hydrants,* 206. *Consumption,* 800,000 galls. **Financial.**—*Cost,* $668,306. *Cap. Stock:* authorized $350,000; all paid up. *Total Debt,* $356,556: floating, $11,556 at 6%; bonded, $345,000 at 6%, due in 1907. *Expenses:* operating, $13,447; int., $21,191; taxes, $3,417; total, $38,-055. *Revenue:* consumers, $28,771; city, $14,918; total, $13,689. **Management.**—Prest., J. W. Dimmick. Secy. and Supt., Fred Crosby. Treas., A. H. Howland, Boston, Mass.

NOTE.—Under franchise city had right to buy works Oct. 7, '90, and in '91 secured authority from legislature to buy or build works. Arbitrators were appointed in '91 to determine price.

5. NEW DECATUR, *Morgan Co.* (Pop., in '90, 3,565.) Supply from Decatur.

Pell City, *St. Clair Co.* (Pop., est., 500.) Built in '91 by Hercules Foundry Co. to furnish water and fire protection for their shops and, incidentally, for city. **Supply,** springs, pumping to reservoir. Reservoir is on hills, 140 ft. above town. *Cost,* $3,000. Prest., Avery L. Rand. Treas. and Mgr., O. B. Cole. Supt., Jno. Foran.

16. SELMA, *Dallas Co.* (Pop., 7,529—7,622.) Built in '86 by Selma Water Co. **Supply.**—Artesian wells, pumping to stand-pipe and direct. In '91 co. bought two wells at Mullen's lake and was to lay mains therefrom. Fiscal year closes May 1. **Distribution.**—*Mains,* 11.5 miles. *Taps,* 450. *Meters,* 1. *Hydrants,* 112. **Financial.**—*Bonded Debt,* $100,000 at 6%, due in 1906. *Expenses,* $7,000. **Management.**—Secy. and Supt., Thos. H. Gatchell.

17. SHEFFIELD, *Colbert Co.* (Pop., in '90, 2,731.) First supplied in '85 from temporary works built by Sheffield Land, Iron & Coal Co., which then owned townsite. Present works built in '89-90 by Sheffield Water Co., under 30-years' franchise. **Supply.**—Tennessee River, pumping to stand pipe. Fiscal year closes April 15. Aug. 6, '91, contract for improvements was let to Scovel & Irwin Construction Co., Nashville, Tenn., for $100,000. There will be a 2,000,000-gall. comp. dup. cond. and two 1,000,000-gall. dup. high-pressure pumps, located in new station about ¾ of a mile above present station. A 16 in. force main will lead to a 20 × 80-ft. stand-pipe. **Distribution.**—*Mains,* 10 miles. *Consumption,* 1,000,000 galls. **Financial.**—*Cost,* $32,000. *Cap. Stock,* $250,000. *No Debt:* in April, '91, company proposed to issue bonds for enlargement of works. *Expenses,* $4,000. *Revenue,* $8,600. **Management.**—Prest., W. L. Chambers. Secy., J. I. Allen. Supt., Thos. Gothard.

18. SHELBY, *Shelby Co.* (Pop., 567—753.) **History.**—Built in '90-1 by Alabama Coal & Iron Co. to supply their furnaces and, incidentally, the town; put in operation Jan. 15. Des. Engr., J. H. Turner, Mobile. Const. Engr., E. S. Safford. Contr., H. Merigold, Lima, O. **Supply.**—Springs, pumping direct. Reservoir.—Cap., 1,000,000 galls. Pumping Machinery.—Dy. cap., 2,500,000 galls.; Dean. **Distribution.**—*Mains,* 12-in., 4.5 miles. *Hydrants,* Bourbon, 10. *Valves,* Bourbon. *Pressure:* ordinary, 70 lbs.; fire, 150 lbs. **Financial.**—*Cost,* $40,000. **Management.**—Prest., T. G. Bush; Secy., J. W. Lapsey; both of Anniston. Treas., W. S. Gurnee, New York City.

I. SPRING GARDEN, *Cherokee Co.* (Pop., est., 400.) Built in '89 by Spring Garden Water-Works Co.; franchise neither regulates rates nor provides for purchase of works by city. **Supply.**—Glade Springs, pumping to reservoir. Fiscal year closes April 1. **Distribution.**—*Mains,* ¾ miles. *Taps,* 4. *Hydrants,* 1. **Financial.**—*Cost,* $1,500. *Cap. Stock:* authorized, $2,500; paid up, $1,300. *Floating Debt,* $200. *Expenses,* $10. *No Revenue.* **Management.**—Prest., Jas. W. Tucker. Secy., W. M. Graham. Treas., W. F. Clark.

19. TALLADEGA, *Talladega Co.* (Pop., 1,233—2,063.) Built in '86 by Talladega Gas Light & Water Co., under 25-yrs. franchise. **Supply.**—Spring, pumping to stand-pipe. **Distribution.**—(In '87.) *Mains,* 4 miles. *Taps,* 163. *Meters,* 0. *Consumption,* 40,000 galls. **Financial.**—(In '87.) *Cost,* $40,000. *Cap. Stock,* $60,000 *Bonded Debt,* $40,000. *Revenue:* city, $2,580. **Management.**—Mgrs., A. H. Glenn, D. M. Rogers, W. H. Skegg, R. P. Anderson.

TROY, *Pike Co.* (Pop., 2,294—3,449.) Temporary system was built by town in '86. Permanent system projected in '91. Address A. C. Worthy, City Clerk.

ALABAMA.

20. **TUSCALOOSA,** *Tuscaloosa Co.* (Pop., 2,481—1,215.) Built in '88 by Tuscaloosa Water Works Co. under 30-yrs. franchise. Supply.—Warrior river, pumping to tank. Distribution.—*Mains*, 8 miles. *Taps*, 237. *Meters*, 0. *Hydrants*, 53. *Consumption* 250,000 galls. Financial.—*Cost*, $75,000. *C p. Stock*, $75,000. *Bonded Debt*, $50,000 at 6%. *Sinking Fund*, $2,000 per annum. Management.—Prest., J. W. Wilcox, Secy., W. C. Fitts, Treas., A. E. Boardman, Brevard, N. C. Supt., W. T. McCormick.

Water-Works Projected with Fair Prospects of Construction.

ATTALLA, *Etowah Co.* (Pop., 351—1,251.) Construction of city works was to begin by Aug. 1, '91. Des. Engrs., Hartford, Hebert & Co., Chattanooga, Tenn. Const. Engr., Phil. S. Fitzgerald, Gadsden. Supply, spring. Mains, 3½ miles. Hydrants, 31. Pressure, 40 to 60 lbs. Cost, $38,000. Bonded debt, $50,000 at 6%.

AVONDALE, *Jefferson Co.* (Pop., est., 2,000.) Fall of '91 Robt. Jamison, Prest. East Lake Land Co., proposed to supply city.

COLUMBIA, *Henry Co.* (Pop., 290—960.) Small system projected, ownership not determined. Supply, artesian well, pumping to tank. Cost, about $5,000. Mayor, W. W. Sanders. Latest report, July 14, '91. See 1889-90 Manual, p. 324.

DOTHEN, *Henry Co.* (Pop., in '90, 247.) Town has voted to build works, and obtained authority from legislature to issue bonds. Supply, springs or artesian well, pumping to stand-pipe or direct. City Clerk, W. Pilcher. Latest report, July 13, '91.

EAST LAKE, *Jefferson Co.* (Pop., not given.) System was projected early in '91 by East Lake Water & Electric Light Co. Address Robt. Jennison, Birmingham.

GREENSBORO, *Hale Co.* (Pop., 1,833—1,759.) July 25, '91, city was boring artesian well. If supply of water was obtained, water-works were to be constructed. Mayor, W. W. Powers.

LUVERNE, *Crenshaw Co.* (Pop., in '90, 451.) July 9, '91, town contemplated building works. Address Mayor.

MARION, *Perry Co.* (Pop., 2,074—1,982.) July 14, '91, system was contemplated, contracts to be let as soon as possible. Supply, wells or river, pumping to stand-pipe. Chn., A. F. Redd.

NORTHPORT, *Tuscaloosa Co.* (Pop., 564—413.) System contemplated, committee appointed, but no definite action had been taken Aug. 24, '91. City Clerk, R. S. Cox.

ONEONTA, *Blount Co.* (Pop., est., 450.) July 17, '91, it was reported that Oneonta Land & Improvement Co. had bought water rights and proposed putting in water-works in the near future. Address F. H. Eaton.

OPELIKA, *Lee Co.* (Pop., 3,245—3,703.) In '91, 20-yrs. franchise was granted to American Pipe Mfg. Co., Philadelphia. Franchise fixes rates and provides for 6 miles of pipe, not more than 60 hydrants; hydrant rental, $40. Construction was to begin Aug. 1. Address H. Bayard Hodge, Philadelphia.

OZARK, *Dale Co.* (Pop., 512—1,195.) City Council received bids till Oct. 1, '91, for construction of water-works. City Clerk, T. J. Fain.

PIEDMONT, *Calhoun Co.* (Pop., in '90, 711.) In '91, City Council voted to grant 20-yrs. franchise to co. June 25, '91, franchise had not been taken. Mayor, A. McAllister.

PRATT MINES, *Jefferson Co.* (Pop., in '90, 1,946.) July 11, '91, city had granted contract to Drennen & Co. to put in water-works, to be completed within two yrs. from March 1. Mayor, J. H. Russell.

UNIONTOWN, *Perry Co.* (Pop., 810—854.) City will issue $25,000 of 6% bonds, to construct water-works. Contracts to be let as soon as bonds are sold. Artesian well already bored. Latest report July 21, '91. City Clerk, J. H. Hudson.

WARRIOR, *Jefferson Co.* (Pop., est., 2,000.) June 25, '91, B. F. Reuse & Co. were making estimates for construction of water-works.

For Further Descriptive Matter see Manual for 1889-90. Changes since 1890 are here given. Populations are for 1880 and 1890, respectively.

MISSISSIPPI.

Water-Works Completed or in Process of Construction.

1. **BILOXI,** *Harrison C.* (Pop., 1,540—3,234.) Built in '89 by Biloxi Artesian Water-Works Co. under 25-yrs. franchise, which neither regulates rates nor provides for purchase of works by city. Supply.—Artesian wells, by gravity. Fiscal year closes June 1. Distribution.—*Mains,* 3 miles. *Taps,* 183. *Meters,* 1 *Hydrants,* 32. *Consumption,* 36,000 galls. Financial.—*Cost,* $9,000. *Cap. Stock:* authorized $8,000: all paid up; no div'd in 1890, net revenue of $1,000 being used for extensions. *No Debt. Expenses,* $150. *Revenue,* $1,073. Management.—Prest. and Supt., John A. Walker. Secy., E. Theobold. Engr., W. L. Covel.

2. **COLUMBUS,** *Lowndes Co.* (Pop., 3,995—4,559.) Built in '67 by city. Supply.—Two artesian wells, pumping to reservoirs. Fiscal year closes Oct. 31. In '90 Knowles pump was replaced by a Deane dup. Apr., '91, increase storage cap. was projected. Distribution.—*Mains,* 4 miles. *Taps,* 300. *Meters,* 0. *Hydrants.* 30. *Consumption,* 150,000 galls. Financial.—*Cost,* $30,000. *No Debt. Expenses,* $1,500. *Revenue.* $1,500. Management.—Committee of two, appointed by mayor every two years. Chn., J. M. McDorin. Supt., A. S. Frierson, appointed by city council.

GREENVILLE, *Washington Co.* (Pop., 2,191—6,658.) Under construction '88-91 by Greenville Water-Works Co. Supply.—Artesian well, with Mississippi river as reserve, pumping to reservoir. In '91 works were not completed.

3. **JACKSON,** *Hinds Co.* (Pop., 5,201—5,920.) Built in '88-9 by Jackson Light, Heat & Water Co. Franchise fixes maximum rates and provides for purchase of works by city in 1930. Supply.—River, pumping to stand-pipe. Distribution.—*Mains,* (in '89), 13 miles. *Hydrants,* 80. Financial.—*Cost,* $100,000. *Revenue,* city, $4,800. Management.—(In '89.) Prest., A. L. Saunders. Supt., Chas. Dudley.

4. **MERIDIAN,** *Lauderdale Co.* (Pop., 4,008—10,624.) Built in '86-7 by Meridian Water-Works Co., now controlled by American Water-Works & Guarantee Co., Ltd., Pittsburgh, Pa.; maximum rates fixed in franchise which provides for purchase of works by city in '97 and at intervals of 5 yrs thereafter. Supply.—Springs and surface water, pumping from collecting and storage reservoirs. Fiscal year closes Dec. 31. Pumping Machinery.—Dy. Cap., 4,000,000 galls.; one 1,750,000 Worthington comp., 14 × 20 × 12 × 15 and one 2,250,000 Gordon comp., 14 × 26 × 14 × 24. Boilers, two 150 HP., Hazleton. Distribution.—*Mains,* 16.5 miles. *Taps,* 631. *Meters,* 8. *Hydrants,* 123. *Consumption,* 880,000 galls. Financial.—Withheld. Management.—Prest., J. H. Purdy. Secy., A. E. Clairey. Supt., Chas. Zeigler.

5. **NATCHEZ,** *Adams Co.* (Pop., 7,058—10,101.) Built in '89 by Natchez Water & Sewer Co., under 30-yrs. franchise which provides that rates shall be those of St. Louis. Supply.—Mississippi River, pumping to stand-pipe. June 10, '91, system was in hands of a receiver. Distribution.—*Mains* (in '89), 8 miles. *Hydrants,* 110. Financial.—*Cost* (in '89), including sewers, $160,000. *Revenue,* city, $5,450. Management.—Receiver, Geo. W. Koontz. Supt., C. D. Shaw. City Clerk, Thos. P. Quartennam.

I., II., III. **PASS CHRISTIAN,** *Harrison Co.* (Pop., 1,410—1,705.) General Description.—Three companies supply water for private consumption. Pass Christian Artesian Well Ass'n. put down artesian well and began furnishing water in '88, East End Co. in '89, and West End Co. in '90. Wells are 2½ ins. in diam., from 435 to 610 ft. deep, furnish about 65 galls. of water a minute, and throw it to an elevation of from 25 to 35 ft. Stand-pipes, 1-in. in diam., are so placed as to insure a constant flow, even when all plugs and faucets are closed. Water is furnished by first and last named companies at $7.50 per year and $2.50 for each additional faucet. Wells cost from $500 to $700. Prost. of first and last named companies, Jno. H. Lang.

IV. **SCRANTON,** *Jackson Co.* (Pop., 1,052—1,353.) History.—Built in '90-1 by F. H. Lewis for domestic use only; put in operation July 1. Supply.—Artesian wells, by natural pressure, which raises water 33 ft. Dy. cap., 80,000 galls. Distribution.—*Mains,* c. i., 3.4 miles. *Taps,* 60. *Hydrants,* 0. Financial.—*Cost,* $3,200. *No Debt. Revenue,* $800. Management.—Owner, F. H. Lewis. Supt., Jno. Sutter.

6. **VICKSBURG,** *Warren Co.* (Pop., 11,814—13,373.) Built in '87-8 by Vicksburg Water Supply Co. Supply, Mississippi river, pumping from settling basin direct or to stand-pipe. Fiscal year closes Dec. 31.

MISSISSIPPI. 181

Year.	Mains.	Taps.	Meters.	Hyd'ts.	Cons'pt'n.	B'd'd d't.	Int. ch'g's.	Rev.
'89	14.3	305	1	86	160,000	$250,000	$15,000	$3,250*
'90	14.3	416	20	95	312,625	250,000	15,000	5,900*

*From city.

Financial.—Withheld, except data above. **Management.**—Secy., W. G. Snow, 2 Wall St., N. Y. City. Supt., Chester R. McFarland.

7. **YAZOO CITY,** *Yazoo Co.* (Pop., 2,542—3,286.) **History.**—Built in '91 by Yazoo Improvement Co.; put in operation July 1. Engr., J. A. Jones. Contrs., Michigan Pipe Co., Bay City, Mich. **Supply.**—Artesian wells, pumping to stand-pipe. **Pumping Machinery.**—Dy. cap., 1,500,000 galls.; comp., high pressure, from Laidlow & Dunn, Cincinnati, O. **Stand-Pipe.**—Cap., 70,000 galls.; 20 × 30 ft., on elevation 200 ft. above city; from Dan'l Snea & Co., Memphis, Tenn. **Distribution.**—*Mains*, 10 to 4-in. wood, 6 miles. *Scruwes*, iron. *Taps*, 20. *Hydrants*, Galvin, 50. *Valves*, Galvin. *Pressure*, 100 lbs. **Financial.**—*Cost.*, $100,000. *Cap. Stock:* authorized, $100,000; all paid up. *Bonded Debt*, $75,000 at 6%. *Revenue*, city, $1,650. **Management.**—Prest., Wm. Hamel. Secy., D. R. Barnett. Treas., J. H. D. Haverkamp.

Water-Works Projected with Fair Prospects of Construction.

BAY ST. LOUIS, *Hancock Co.* (Pop., 1,978—1,974.) July 28, '91, it was reported that Bay St. Louis Wells Co. contemplated building water-works. Address E. J. Bowers.

CLARKSDALE, *Coahoma Co.* (Pop., in '90, 761.) June 26, '91 it was reported that Clarksdale Electric Light & Water Co. would build works. Supply, artesian well, pumping to tank. Address, J. H. Clark.

PORT GIBSON, *Claiborne Co.* (Pop., in '90, 1,594.) City granted 21-yrs. franchise to M. S. Hasie who will construct water-works and then transfer them to Port-Gibson Water-Works Co., organized in '91. Supply, Bayou Pierre, pumping direct. A 500,000-gall. Holly pump. Cost, $20,000.

WATER VALLEY, *Yalobusha Co.* (Pop., 2,220—est., 4,000.) A. F. Benson & Co., Batesville, Ark., were granted contract to construct water-works. Supply, wells, by natural pressure. Cost, $20,000. June 25, '91, construction had not begun.

Water-Works Projects Reported in '90 and '91, but Now Indefinite.

Macon, Noxubee Co.

LOUISIANA.

Water-Works Completed or In Process of Construction.

1. **BATON ROUGE,** *East Baton Rouge Co.* (Pop., 7,197—10.478.) Built in '88 by Baton Rouge Water Co.; maximum rates fixed in franchise, which provides for purchase of works by city in 1908 at actual value, determined by board of experts. **Supply.**—Well, pumping to stand-pipe and direct. Well is 8 ins. in diam., 758 ft. deep, constructed in 1890; previously water was taken from Mississippi River.

Year.	Mains.	Taps.	Meters.	Hyd'ts.	Cons'pt'n.	Cost.	B'd'd Debt.	Int. Ch'g's.	Exp.	Rev.
'88	6.5	100	2	83	125,000	....	$60,000	$3,600		$7,000
'90	8	180	5	96	200,000	....	60,000	3,600	$3,750	10,150

Financial.—*Cap. Stock* (in '88), 100,000. *Bonded Debt*, $60,000 at 6%. *Expenses*, total, $7,350. *Revenue:* consumers, $5,000; city, $5,150; total, $10,150. **Management.**—Prest., Wm. Garig. Secy., Jno. H. Wood, Dubuque, Ia. Treas., C. H. Meyer. Supt., L. P. Amiss.

2. **DONALDSONVILLE,** *Ascension Co.* (Pop., 1,907—3,121.) **History.**—Built in '91 by city. Engr., S. A. Poche. **Supply.**—River, pumping direct. **Pumping Machinery.**—Dy. cap., 1,700,000 galls.; Barr comp. dup. *Boiler*, steel, horizontal tubular. **Distribution.**—*Mains*, 10-in., 2½ miles. *Pressure:* ordinary, 125 lbs.; fire, 150 lbs. **Financial.**—Withheld. **Management.**—Mayor, S. Goette, Jr.

NOTE.—At time of last report, July 16, distribution system was not laid and water used chiefly for flushing sewers. System was to be completed before close of '91.

FOR FURTHER DESCRIPTIVE MATTER see Manual for 1889-90. CHANGES since 1890 are here given. POPULATIONS are for 1880 and 1890, respectively.

3. NEW ORLEANS, *Orleans Co.* (Pop., 216,090—242,039.) Now owned by New Orleans Water Co.; built in '53 by co., sold in '68 to city, and in '78 by city to New Orleans Water Co. Franchise for 50 yrs. from '78, at expiration of which time city may buy works at price fixed by five arbitrators; if city does not buy, franchise continues 50 yrs. longer. Rates regulated by franchise to keep net profits at or below 10%. Supply.—Mississippi river, pumping to reservoirs, stand-pipe, and direct. Fiscal year closes April 1.

Yr.	M'ns.	Taps.	M't'rs.	Hydts.	C'n'tion.	Cost.	B'd'd debt.	Int. ch'g's.	Exp.	Rev.
'80-1	61	6,000	..	1,200	9,000,000	$2,100,000	......	......	$45,000	$91,371
'82-3	68.7	5,000	0	790	9,000,000	2,300,000	......	......	41,000	100,000
'84-5	70	4,250	0	1,135	8,500,000	2,300,000	$350,000	$21,000	40,000	130,000
'86-7	71	4,400	0	1,191	6,000,000	2,297,000	395,000	23,700	39,640	255,840
'87-8	72	4,400	8	1,191	6,000,000	2,397,000	305,000	23,700	......	......
'88-9	75	4,174	2	1,195	6,038,776	2,414,270	395,000	23,700	84,363	173,889
'89-90	76.5	4,334	20	1,208	7,758,000	2,420,783	395,000	23,700	86,613	218,245
'90-1	76.6	4,450*	20	1,208	8,976,715	2,420,783	395,000	23,525	98,481	200,694

* For some time a large additional number of taps have been available but not in use. April 1, '91, number not in use was 5,850.

Financial.—*First Cost,* $2,000,000. *Cap. Stock,* $2,000,000; paid 4% div'd in 1891. *Bonded Debt,* $395,000 at 6%, due in 1909. *Expenses:* operating, $58,697; int., $23,525; taxes, $39,784; total, $122,006. *Revenue:* consumers, $200,694. **Management.**—Prest., R. E. Craig. Secy. and Treas., R. S. Macmurdo. Supt., A. H. Gardner.

4. SHREVEPORT, *Caddo Co.* (Pop., 8,009—11,979.) Built in '86-7 by Shreveport Water-Works Co., under 30-yrs. franchise. **Supply.**—Cross Bayou, pumping to settling basin and stand-pipe; filtered. Fiscal year closes Dec. 31. Statement in 1889-90 Manual that system had been sold to First National Bank of N. Y. City was incorrect.

Year.	Mains.	Taps.	Meters.	Hyd'ts.	Consumption.	Cost.	B'd'd Debt.	Int. Ch'g's	Rev.
'87	8.5	434	0	106	500,000	$250,000	$250,000	$15,000	$17,800
'89	10	382	1	106	816,295	......	250,000	15,000	8,800*
'90	10	535	14	114	1,159,115	......	250,000	15,000	9,120*

*From city.

Financial.—Withheld except data above. **Management.**—Secy., W. G. Snow, 2 Wall St., New York City. Supt., Jno. B. Crawley.

Water-Works Projected with Fair Prospects of Construction.

LAKE CHARLES, *Calcasieu Co.* (Pop., (838—3,442.) July 17, '91, it was reported that town was to have system of water-works.

PLAQUEMINE, *Iberville Co.* (Pop., 2,513—3,222.) July 8, '91, Mayor reported that Water-Works Com. would advertise for bids in about three weeks. Stand pipe, 12 × 100 ft. Mains, 8 to 4 ins. Cost, about $15,000. Chn., J. McWilliams. Unofficially reported in Sept., '91, that Plaquemine Water-Works Co. had been organized; Prest., J. A. Grace. Vice-Prest., A. N. Stine. Secy., David Altemus.

TENNESSEE.

Water-Works Completed or In Process of Construction.

1. ATHENS, *McMinn Co.* (Pop., 1,100—2,224.) Built in '88 by Athens Mining & Mfg. Co. **Supply.**—Spring, pumping to stand-pipe. *Mains,* 4 miles. *Cost,* $25,000. No Debt. Prest., F. F. McElwee. Mgr., R. J. Fisher.

2. BRISTOL, *Sullivan Co.* (Pop., 1,617—3,324.) Built in '88-9 by Bristol-Goodson Water Co.; franchise fixes maximum rates and provides for purchase of works by city in '98. **Supply.**—Spring, pumping to tank. **Distribution.**—*Mains,* 7.5 miles. *Taps,* 300. *Meters,* 4. *Hydrants,* 34. *Consumption,* 100,000 galls. **Financial.**—*Cap. Stock:* authorized, $75,000; all paid up; paid no dividend in 1890. *Net revenue used for extensions. Bonded Debt,* $40,000, at 6%, due in 1909. *Expenses:* int., $2,400; taxes, $246. **Management.**—Prest., Sam'l L. King. Secy., B. G. McDonell. Treas., F. B. Hubbell, Baltimore. Md. Supt., Geo. F. Crush.

3. CARDIFF, *Roane Co.* (Pop., in '90, 430.) Cardiff Coal & Iron Co. owns and operates system of water-works. No description has been received.

TENNESSEE. 183

4, 5. CHATTANOOGA, *Hamilton Co.* (Pop., 12,892—29,100.) Owned by American Water-Works & Guarantee Co., Ltd., Pittsburgh, Pa.; maximum rates fixed in franchise, but no provision made for purchase of works by city. Built in '61 by U. S. Government. Supply.—Tennessee River, pumping to reservoir; filtered. Fiscal year closes Dec. 31. Pumping Machinery.—Dy. cap., 17,500,000 galls.; one 4,000,000 Knowles, high-pressure 3 × 20½ × 21, one 6,000,000 Deane,comp. cond., 30 × 40 × 22½ × 36; one 8,500,000 Worthington comp. cond., high-duty, 25 × 50 × 23 × 36. Pumps are in a pit 50 ft. in diam , 49 ft. deep, supported by 40-in. brick walls. A brick-lined tunnel, 8½ × 6½ × 9) ft., leads from pit to river, in which are two suction and other pipes. Boilers: 500 H. P., seven horizontal tubular and one Heine.

Year.	Mains.	Taps.	Meters.	Hydrants.	Consumption.
'82	15	1,000	0	53	1,000,000
'86	23	1,900	5	88	3,000,000
'88	54	3,550	25	150	5,250,000
'90	69	3,028	30	164	6,000,000

Financial.—Withheld. Management.—Prest., J. S. Kuhn. Secy., A. B. Darragh. Treas. and Supt., N. Wingfield.

CHATTANOOGA WATER & POWER CO.

History.—Built in '89 and enlarged in '90 by Debenture Guarantee & Assurance Co., Chicago, Ill.; co. also controls electric railway and incline. Engr., Geo. H. Pierson. Supply.—City Water Co.'s mains pumping to stand-pipe on Cameron Hill. Pumping Machinery.—Dy. cap . 500,000 galls.; Worthington. Stand-Pipe.—Cap.. 150,000 galls.; 15 × 100 ft. Distribution.—*Mains*, 2.5 miles. Financial.—*Cost*, $100,000. *Bonded Debt*, $75,000, at 6%. Management.—Supt., D. W. Hughes.

6. CLARKSVILLE, *Montgomery Co.* (Pop., 3,880—7,924.) Owned by Clarksville Water Co.; rates are fixed in franchise which provides for purchase of works by city at any time after '89, price determined by three disinterested parties. Built in '79 by Travers Daniels. Supply.—Cumberland River, pumping to stand-pipe. Distribution.—*Mains*, 8.5 miles. *Taps*, 395. *Meters*, 0. *Hydrants*, 48. *Consumption*, 500,000 galls. Financial.—*First Cost*, $50,000. *Revenue*, city, $2,400. Management.—Prest., F. P. Gracey. Secy., B. W. Macrae. Supt., R. E. McCulloch.

7. COLUMBIA, *Maury Co* (Pop., 3,400—5,370.) Built in '40 by city; now owned by Columbia Improvement Co.; franchise does not regulate rates, but provides for purchase of works by city after 10 yrs. Supply.—Duck River, pumping to reservoir. Distribution.—(In '88.) *Mains*, 7.5 miles. *Taps*, 242. *Meters*, 11. *Hydrants*, 56. *Consumption*, 125,000 galls. Financial.—(In '88.) *Cap Stock*, $50,000. *Bonded Debt*, $60,000 at 6%. *Expenses*: operating, $4,200; int., $3,600; total, $7,800. *Revenue*: consumers, $5,200; city, $3,000; total, $8,200. Management.—Prest., F. J. Ewing. Secy., C. T. Jones. Mgr., W. F. Tucker.

8. DYERSBURG, *Dyer Co.* (Pop., 1,010—2,009.) Built in '89 by Frank Shepard; now owned by Dyersburg Water & Electric Light Co.; put in operation Oct. 1. Supply.—River, pumping to tank. Pumping Machinery.—Dy. cap., 150,000 galls.; Laidlaw & Dunn, dup. Tank.—Cap.. 49,000 galls. Distribution.—*Mains*, 4 and 6-in. w. i., 3 miles. *Taps*, 131. *Meters*, Gersey, 6. *Hydrants*, White, 10. *Valves*, Powell. *Consumption*, 100,000 galls. *Pressure:* ordinary, 20 lbs ; fire, 100 lbs. Financial.—*Cost*, $10,000. *No Debt*. *Expenses*, $580. *Revenue:* consumers, $4,000; city, $500; total, $4,500. Management.—Prest., Frark Shepard. Secy., H. A. Klyce. Supt., F. H. Shepard, Jr.

9. HARRIMAN, *Roane Co.* (Pop., in '90, 716.) Temporary works were built in '91 by East Tennessee Land Co. for domestic use and fire protection. Supply.—Emory River, pumping to tank. Pumping Machinery.—Dy. cap , 75,000 galls.; Worthington. Tank.—Cap., 15,000 galls. Distribution.—*Mains*, 3 to 1½-in. w. i., 2 miles. *Taps*, 50. *Hydrants*, 5. Water is furnished free. Management.—Prest., A. W. Wagnalls. Gen. Mgr., W. H. Russell.

NOTE.—Co. own franchise of streets, and are ready to treat with parties wishing to put in permanent works.

10. JACKSON, *Madison Co.* (Pop., 5,377—10,039.) Built in '85 by city. Supply. —Tennessee river, pumping to reservoir. Fiscal year closes June 30.

FOR FURTHER DESCRIPTIVE MATTER see Manual for 1889-90. CHANGES since 1890 are here given. POPULATIONS are for 1880 and 1890, respectively.

184   MANUAL OF AMERICAN WATER-WORKS.

Year.	Mains.	Taps.	Meters.	Hyd'ts.	Cons'pt'n	Cost.	B'd'd d't	Int. chgs.	Exp.	Rev.
'87-8	11	450	3	132	500,000	$100,000	$100,000	$6,000	$4,800	$8,000
'88-9	12	649	4	141	500,000	100,000	100,000	6,000	4,700	8,650
'89-90	14.23	707	5	157	725,000	110,111	100,000	6,000	6,450	9,175

Financial.—*Bonded Debt*, $100,000 at 6%, due in 1915. *Expenses:* total, $12,450. *Revenue:* consumers, $9,175; city, none. **Management.**—Board of directors elected by city council every three years. Chn., W. F. Alexander. Secy. and Supt., A. D. Muse.

11. **JOHNSON CITY,** *Washington Co.* (Pop., 1,505—4,161.) **History.**—Built in '90-1 by Watauga Water Co. under 20-yrs. franchise.; put in operation July 1. Engr., A. E. Boardman, Brevard, N. C. Contrs., Jeter & Boardman, Gas & Water Ass'n, Macon, Ga. **Supply.**—Mountain springs, by gravity. **Reservoir.**—Cap., 15,-000,000 galls. **Distribution.**—*Mains*, 14 to 4 in. c. i., 9 miles. *Hydrants*, 50. *Valves*, Eddy. *Pressure*, 75 lbs. **Financial.**—*Cost*, $135,000. *Cap. Stock:* authorized, $200,000; paid-up, $120,000. *Bonded Debt*, $135.000 at 6%. *Hydrant rental*, $10. **Management.**—Prest., W. A. Jeter. Secy. and Supt., Jno. Evans. Treas., A. E. Boardman.

12. **KNOXVILLE, WEST KNOXVILLE and NORTH KNOXVILLE,** *Knox Co.* (Pop.: Knoxville, 9,693—22,535; N. Knoxville, in '90, 2,297; W. Knoxville, in '90, 2,114.) Built in '83 by Knoxville Water Co., under 30-yrs. franchise, which does not regulate rates, but provides that works may be purchased by city in 15 yrs. from June, '83. **Supply.**—Tennessee River, pumping to reservoir. Fiscal year closes Dec. 31. Mar. 3, '91, co. stated that franchise had been secured for extension to North Knoxville, and that franchise from West Knoxville was soon expected. See note under West Knoxville. Previously mains extended into each town, but evidently without contracts with towns.

Year.	Mains.	Taps.	Meters.	Hyd'ts.	Cost.	B'd'd d'bt.	Int. ch'g's.	Exp.	Rev.
'84	9	391	3	82	$130,741	$125,000	$7,500	......	$10,500
'86	13	680	6	93	......	200,000	12,000	......	14,470
'88	21	1,288	18	122	......	235,000	14,100	11,168	24,593
,90	21	1,500	80	131	......	......	......	......	5,595*

*From city.

Financial.—Withheld. **Management.**—Prest., R. N. Hood. Secy., E. T. Sanford. Treas., E. M. McClung. Supt., C. F. Wood.

I. **MANCHESTER,** *Coffee Co.* (Pop., 438—621.) Built in '89 by J. E. Thompson; purchased in '91 by R. L. Wolfe, Evansville, Ind. **Supply.**—Spring, pumping direct. Fiscal year closes Aug. 15. During summer of '91, steam pump was to be added, to be used in case of accident to water-power pump. **Distribution.**—*Mains*, 2 miles. *Taps*, 80. *Hydrants*, 0. *Consumption*, 35,000 galls. **Financial.**—*Purchase Price*, $7,000. *Cap. Stock*, $5,000; paid 20% div'd. in 1890. *No Debt*. *Expenses*, $290. *Revenue*, $1,680. **Management.**—Owner, R. L. Wolfe, Evansville, Ind.

13. **MEMPHIS,** *Shelby Co.* (Pop., 33,592—64,495.) Now owned by Artesian Water Co. City has right to purchase works after 9 yrs.; rates to be no higher than average of Nashville, Atlanta and Kansas City. Built in '72 by Co. **Supply.**—Artesian wells, pumping direct and to stand-pipe. In '89-91 42 wells were added, making 80 in all; connecting tunnel for wells ¾ miles long was driven; three 10,000,000 Worthington vertical, comp., cond., high service pumps and new engine and boiler rooms and six new boilers were added; also 500,000-gall. stand-pipe.

Year.	Mains.	Taps.	Meters.	Hyd'ts.	Cons'pt'n.	Cost.	B'd'd d'bt.	Int. ch's.	Exp.	Rev.
'80	28	2,759	20	184	4,000,000	$325,000	$100,000	$6,000	$25,000	$50,000
'82	35	3,700	70	200	4,500,000	350,000	100,000	6,000	30,000	70,000
'84	35	4,000	150	225	6,000,000	500,000	100,000	6,000	40,000	90,000
'85	37	....	150	250	6,000,000	500,000	100,000	6,000	89,000	106,000
'86	40	4,300	140	250	6,000,000	500,000	100,000	6,000	85,000	112,000
'87	60	4,500	210	300	6,000,000	850,000	100,000	6,000	92,000	130,000
'88	65	4,750	350	300	6,000,000	900,000	100,000	6,000	100,000	140,000
'89	62	5,035	260	300	7,000,000	1,500,000	750,000	37,500	160,000	150,000
'90	70	5,400	200	329	8,000,000	1,750,000	750,000	37,500	100,000	160,000

Financial.—*Cap. Stock:* authorized $750,000; all paid up; paid 8% div'd in 1890. Feb. 1, '91, authorized stock was increased to $1,000,000, all paid up. *Bonded Debt*, $750,000 at 5%. *Expenses*, total, $137,500. *Revenue*, total, $160,000. **Management.**—Prest., T. J. Latham, Secy., Lawrence Simpson. Treas., G. W. Macrae.

14. **MURFREESBORO,** *Rutherford Co.* (Pop., 3,800—3,739.) Built in '91 by Murfreesboro Water-Works Co. under 33-yrs. franchise. (In '88 a fire-protection system was built by town.) **Supply.**—Springs, pumping to stand-pipe and direct. **Stand-Pipe.**—Is 20 × 20 ft. **Distribution.**—*Mains*, 6-in. l. *Hydrants*, 20. **Financial.**—*Cost*, $7,000. *Cap. Stock*, $20,000. *Revenue:* consumers, not determined; city, $1,000.

TENNESSEE.

Management.—Prest., Hickman Weakley. Secy. and Supt., J. E. Thompson. Report by F. B. Selphi.

15. **NASHVILLE,** *Davidson Co.* (Pop., 43,350—76,168.) Built in '32 by city. Supply.—Cumberland River, pumping to stand-pipe and reservoir. Fiscal year closes Oct. 1. A 10,000,000-gall. Worthington pumping engine was added in '90 or '91, and a new intake pipe was laid in the bed of the Cumberland River. The pipe is 48 ins. in diameter and 1,750 ft. long. At the river end it terminates in a crib 8 ft. inside diameter, about 60 ft. high. See note below.

Year.	Mains.	Taps.	Mtrs.	H'd'ts.	Cons'pt'n.	B'd'd d't.	Int. ch'g's.	Exp.	Rev.
'80	42.2	3,000	...	205	3,736,000	...	...	26,000	...
'84	37	3,000	11	250	6,000,000	75,000	4,500	79,018	73,009
'85	39	3,861	3	261	...	525,000	...	46,848	71,364
'86	42.5	4,141	2	300	7,772,369	600,000	36,000	38,258	80,552
'87	42.5	4,321	8	318	8,522,762	600,000	35,156	43,128	80,773
'88	44.7	4,496	15	550	8,970,506	900,000	39,000	45,182	83,885
'89	47.4	4,767	...	563	10,400,000	1,075,000	...	47,916	91,313
'80	49.6	5,098	...	571	11,153,885	1,200,000	...	62,956	95,376

Financial.—Cost, about $2,700,000. **Management.**—Supt., Geo. Reyer. City Engr., J. A. Jowett.

NOTE.—For illustrated description of laying submerged main see ENGINEERING NEWS, vol. xxvi., p. 127.

12. **NORTH KNOXVILLE,** *Knox Co.* See Knoxville.

11. **SEQUACHIE,** *Marion Co.* (Pop., not given.) In '90 Sequachie Town and Improvement Co. made contract with G. Sherman to put in water-works. Temporary supply, for domestic use only, was introduced probably in '93, consisting of 1¼ miles of 1-in. pipe. **Supply.**—Springs, by gravity to tank, affording pressure of 130 lbs. Owner, G. Sherman.

16. **SOUTH PITTSBURG,** *Marion Co.* (Pop., 1,045—1,479.) Built in '78 by Southern Coal, Iron & Land Co.; franchise regulates rates. July 6, '90, works were transferred to South Pittsburg City Water Co. **Supply.**—Springs, by gravity, and Tennessee River, pumping. Fiscal year closes Dec. 31. **Distribution.**—*Mains,* 4.5 miles. *Hydrants,* 12. *Consumption,* 250,000 galls. **Financial.**—*Cap. Stock,* $50,000. *Bonded debt* (probably) $30,000. **Management**—Prest., J. H. Curry. Treas. H. M. Hyde. Secy. and Supt., F. P. Clute.

12. **WEST KNOXVILLE,** (*Knoxville, P. O.*) *Knox Co.* See Knoxville.

NOTE.—Aug. 18, '91, reported that survey was being made and contracts were about to be let for new works. Engr., G. W. Sturdevant. Supply, springs, pumping to stand-pipe. Prest., H. P. Cole. Secy. and Treas., D. S. McIntyre.

Water-Works Projected with Fair Prospects of Construction.

CLEVELAND, *Bradley Co.* (Pop., 1,874—1,863.) Franchise was granted early in '91 to W. H. Whitcomb & Co., but June 26, '91, construction had not begun. Sept., '91, reported that contract for works had been let to Sturdevant Bros. Address J. H. Hardwick.

CLINTON, *Anderson Co.* (Pop., 263—1,198.) Co. secured franchise in '90 for constructing water and electric lighting plant. Prest., Henry Clear, Jr. Secy., C J. Sawyer.

DAYTON, *Rhea Co.* (Pop., 200-2,719) March 9, '91, it was reported that franchise had been granted to Geo. J. Morris, Boston, Mass. Nothing had been done June 26.

DEPTFORD, *Marion Co.* (Pop., not given.) Address Deptford Land & Improvement Co.

FAYETTEVILLE, *Lincoln Co.* (Pop., 2,104—2,410.) Franchise was granted in '90 to Fayetteville Water-Works Co., providing for 30 hydrants at $50 each. Nothing further had been done June 6. 91. Electric lighting plant and ice factory were for sale and water-works could be run in connection with these. Address H. H. Bryson, Mayor.

GREENEVILLE, *Greene Co.* (Pop., 1,066—1,779.) Aug. 15, '91, it was reported that town needed water-works and would grant perpetual franchise. Address H. Reaves or H. R. Brown, Secy. Water-Works Com.

"FOR FURTHER DESCRIPTIVE MATTER see Manual for 1889-90. CHANGES since 1890 are here given. POPULATIONS are for 1880 and 1890, respectively.

HARRIMAN, *Roane Co.* (Pop., in '90, 716.) See note under Harriman in Water-Works Completed, p. 183.

McMINNVILLE, *Warren Co.* (Pop., 1,244–1,577.) June 26, '91, it was reported that preliminary surveys had been made, but nothing further had been done.

MORRISTOWN, *Hamblen Co.* (Pop., 1,350–1,999.) June 26, '91, it was reported that Morristown Water-Works Co. would construct water-works. Engr., Edw. D. Bolton, 44 Broadway, N. Y. City. Supply, springs, pumping to 30 × 80-ft. tank. Mains, 5 miles. Hydrants, 40. Prest., Jas. G. Martin.

NEWPORT, *Cocke Co.* (Pop. 347–658.) July 31, '91, it was reported that Newport Water Co. had been chartered and contracts would be let during fall of '91. Engr., C. D. Moss. Supply, springs, by gravity. Cost, about $25,000. Prest., P. T. South. Secy. and Treas., Guy M. Betton.

TRENTON, *Gibson Co.* (Pop., 1,383–1,693.) City intended to build water works during '91, provided its bonds could be sold at par. Engr. and Contr., Geo. Cadagan Morgan. Supply, driven wells, pumping to stand-pipe. Mains, 5 miles Hydrants, 33. Cost, $20,000. Address W. B. Stafford. Latest report, July 11.

Water-Works Projects Reported in '90 and '91, but Now Indefinite.
Covington, Tipton Co. *Newbern,* Dyer Co. *Shelbyville,* Bedford Co.

KENTUCKY.

Water-Works Completed or in Process of Construction.

1. **ASHLAND,** *Boyd Co.* (Pop., 3,280–4,195.) **History.**—Built in '90-1 by Ashland Water Supply Co. under 20-yrs. franchise; to be put in operation Oct. 1. Des. Engr., M. M. Tidd, Boston, Mass. Const. Engr., S. A. Forbes. Contrs., Gardiner & McGlasson, St. Louis, Mo. **Supply.**—Ohio River and driven wells on its bank, pumping to reservoir. **Pumping Machinery.**—Dy. cap , 4,000,000 galls.; four 1,000,000 Deane, vertical, high-pressure. **Reservoir.**—Cap., 1,500,000 galls.; of stone. Cost, $20,375. Contr., J. J. Newman, Providence, R. I. (See note, below.) **Distribution.**—*Mains,* 16 to 4-in. c. i., 9½ miles. *Services,* lead. Pipes and specials from Union Storage Co., Pittsburgh, Pa.; trenching and pipe-laying done by Jno. T. Foley, Nashville, Tenn. *Taps,* 600. *Hydrants;* Chapman. 90. *Valves.* Eddy. *Consumption,* 300,000 galls. *Pressure,* 100 lbs. **Financial.**—*Cost,* $110,000. *Cap. Stock:* authorized, $250,000; all paid-up. *Bonded Debt.* $75,000 at 6%. *Revenue:* city, $4,000. *Hydrant rental,* $4,000 for 90, $40 each for additional. **Management.**—Prest., Jno. M. Gardiner. Secy., F. D. McGlasson. Treas., R. D. Davis.

NOTE.—For illustrated description of reservoir see ENGINEERING NEWS, vol. xxv., p. 342.

2. **BELLEVIEW and DAYTON,** *Christian and Campbell Cos.* (Pop.: Belleview, 1,400–3,163; Dayton, 3,210–4,264.) Built in '87 by Belleview Water & Gas Co. **Supply.**—Newport water-works Fiscal year closes May 1. **Distribution.**—*Mains,* 12 miles. *Taps,* 550. *Meters,* 2. *Hydrants,* 110. **Financial.**—*Cost,* $80,000. *Cap. Stock,* authorized, $250,000. *Total Debt* (including gas-plant), $84,000: floating, $5,000 at 6%; bonded, $79,000. *Expenses,* operating, $2,500. *Revenue:* consumers, $6,000; city, $2,930; total, $8,930. **Management**—Prest., Chas. W. Baker. Secy. and Supt., G. R. Harmes.

3. **BOWLING GREEN,** *Warren Co.* (Pop., 5,114–7,803.) Built in '68 by town. **Supply.**—Barren River, pumping to reservoir. Fiscal year closes Dec. 31.

Year.	Mains.	Taps.	Meters.	Hyd'ts.	Cons'pt'n.	B'd'd d't.	Int. Ch'g's.	Exp.	Rev.
'82	7.5	503	0	100	316,443	$85,000	$5,100	$4,000	$7,000
'84	9.1	578	22	103	331,480	80,000	4,800	4,000	9,500
'86	12.2	778	28	109	400,358	90,000	5,400	4,000	10,500
'88	14.2	886	45	109	400,000	90,000	5,400	4,000	13,500
'89	15.8	1,092	34	109	426,302	85,000	4,250	.....	15,000
'90	17.8	1,100	30	117	264,125	85,000	4,250	5,221	11,062

Management.—Comrs., appointed every two years by city council. Chn. Jno Demuth. Secy., G. S. Hollingsworth. Supt., Jas. H. Wilkerson.

KENTUCKY.

4. CATTLETTSBURGH, *Boyd Co.* (Pop., 1,225—1,374.) **History.**—Built '91 by Cattlettsburgh Water Co., under 30 yrs. franchise; to be in operation about Nov. 1. Des. Engr., C. E. Cook, Syracuse, N. Y. **Supply.**—Big Sandy River, pumping to reservoir and direct. **Pumping Machinery.**—Dy. cap., 1,500,000 galls.; two 750,000 Smith & Vaile comp. cond. **Reservoir.**—Cap., 1,003,000 galls. **Distribution.**—*Mains*, 8 to 4-in. c. i., 3.3 miles. *Services*, lead. Pipe and specials from Addyston Pipe & Steel Co., Cincinnati. *Hydrants*, Eddy, 40. *Valves*, Eddy. *Pressure*, 120 lbs. **Financial.**—*Cost*, about $15,000. *Cap. Stock*, $15,000. *Hydrant Rental*, $14.75. **Management.**—Prest., W. A. Patton. Secy. and Treas., Thos. R. Brown. Supt., Thos. L. Rosser, Jr.

5. CENTRAL CITY, *Muhlenburg Co.* (Pop., in '90, 1,144.) Built in '89 by Central City Water Co; city neither controls rates nor has right to buy works. **Supply.**—Green River, pumping to reservoir. Fiscal year closes Dec. 31. **Distribution.**—*Mains*, 10.7 miles. *Taps*, 100. *Me'ers*, 0. *Hydrants*, 15. **Financial.**—*Cap. Stock*, $10,000; paid no div'd in 1890. *Bonded Debt*, $30,000; $13,000 at 6%, due in '99; $20,000 at 6% due in 1909. *No Sinking Fund*. *Expenses*: operating, $1,800; int., $1,800; taxes, $5; total, $3,605. *Revenue*: consumers, $2,700; city, fire protection, $750; total, $3,450. **Management.**—Prest., Chn. Thos. Harris, Louisville. Secy. and Treas., B. Coleman. Supt., C. Du Pont.

6. COVINGTON, *Kenton Co.* (Pop., 29,700—37,371.) Built in '68-'70 by city. Rebuilt in '88-'91. **Supply.**—Ohio River, pumping direct. Fiscal year closes Dec 31. New works described in '88 Manual were completed in '90. See '88 Manual, p. 293, also note below.

Year.	Mains.	Taps.	Meters.	Hyd'ts.	Cons'pt'n.	Cost.	Exp.	Rev.
'87	28	2,500	0	220	....	$100,000	$25,000	$30,000
'88	30	2,946	0	230	2,506 000	500,000	24,000	35,000
'89	31	3,195	0	242	4,468,020	....	30,000	30,000
'90	32.5	3,399	162	251	912,500	1,226,000*	28,000	33,000

* Including cost of new works.

Financial.—See above. **Management.**—Three comrs. appointed every year by common council. Chn., Chas. J. Fronter. Secy., N. F. Lindsay. Supt., W. H. Glore.

NOTE.—For illustrated description of new supply see ENGINEERING NEWS, vol. XXIII., pp. 439 and 458.

2. DAYTON, *Campbell Co.* See Belleview.

7. FRANKFORT, *Franklin Co.* (Pop., 6,958—7,892.) First supplied in '41 by town. Present works built in '84-6 by Frankfort Water Co., under perpetual charter. **Supply.**—Kentucky River, pumping to reservoir. Fiscal year closes Sept. 1. **Distribution.**—*Mains*, 11 miles. *Taps*, 670. *Meters*, 20. *Hydrants*, 103. *Consumption*, 800,000 galls. **Financial**—*Cost*, $200,000. *Cap. Stock*: authorized, $150,000; paid-up, $145,000. *Bonded Debt*, $125,000 at 6%. *Expenses*, $9,000. *Revenue*: consumers, $14,000; city, $6,350; total, $20,350. **Management.**—Prest., D. W. Lindsey. Secy., A. H. McClure. Supt., Jno. D. Griffin.

8. GEORGETOWN, *Scott Co.* (Pop., 2,061—not given.) Built in '89 by Georgetown Water Co., which also operates gas and electric lighting plants; franchise provides neither for control of rates nor purchase works by city. **Supply.**—Spring, pumping direct. Fiscal year closes Oct. 1. **Distribution.**—*Mains*, 5.5 miles. *Taps*, 218. *Hydrants*, 60. **Financial.**—*Cost*, $54,000. *Cap. Stock* (for entire plant): authorized, $100,000; paid-up, $22,000. *Bonded Debt*, $35,000 at 6%, due in 1910. No separate account of expenses and revenue for water-works. **Management.**—Prest., R. A. Mitchell. Treas., E. H. Patterson. Secy. and Supt., Jno. Nichols.

9. GLASGOW, *Barren Co.* (Pop., 1,510—2,051.) Built in '82 by Glasgow Water Co.; franchise neither regulates rates nor provides for purchase of works by city. **Supply.**—Spring, pumping to reservoir. Fiscal year closes July 15. *Distribution.* —*Mains*, 4 miles. *Hydrants*, 6. *Consumption*, 38,400 galls. **Financial** —*Cost*, $15,000. *Cap. Stock*: authorized, $10,000; paid-up, $5,000. No Debt. *Expenses*, operating, $750. *Revenue*, consumers, $1,836. **Management.**—Prest, and Treas., T. P. Dickinson.

10. HENDERSON, *Henderson Co.* (Pop., 5,365—8,835.) Built in '74-76 by city. **Supply.**—Ohio River, pumping to reservoir and and stand-pip . Fiscal year closes Dec. 31.

For FURTHER DESCRIPTIVE MATTER see Manual for 1889-90. CHANGES since 1890 are here given. POPULATIONS are for 1880 and 1890, respectively.

Year.	Mains.	Taps.	Meters.	H'd'ts.	Cons'pt'n.	Cost.	B'd'd Debt.	Exp.	Rev
'82	7.5	347	0	47	150,000	$125,000	$113,500	2,500	$5,50
'86	12.	674	10	64	300,000	140,900	113,000	4,500	9,000
'88	12.5	859	10	70	500,000	175,000	113,000	4.000	12,175
'89	15.1	980	13	75	500,000	....	....	3,595	14,513
'90	17.5	1,092	16	76	500,000	200,000	125,000	4,105	15,705

Financial.—See above. Bonds are redeemed by taxation. Net revenue is used for extensions. In '90 a surplus of $6,506 was paid into general fund of city. **Management.**—Comrs. Chn., W. I. Marshall. Secy., R. C. Blackwell. Supt., Jas. P. Wigal.

11. **LEBANON**, *Marion Co.* (Pop., 2,051—2,816.) Built in '88-9 by city. **Supply.**—Rolling Fork Creek, pumping either from filter-gallery or direct from stream to reservoir. Fiscal year closes June 30. **Distribution.**—*Mains*, 9.1 miles. *Taps*, 145. *Meters*, 3. *Hydrants*, 43. *Consumption*, 100,000 galls. **Financial.**—*Cost*, $55,649. *Bonded Debt*, $58,550 at 6%, due in 1920. *Sinking Fund*, $1,500 at 6%. *Expenses*: operating, $2,500; int., $3,513; total, $6,013. *Revenue*: consumers, $2,900; city, none. **Management.**—Board of Directors, appointed by town trustees for two years. Prest., Andrew Offut. Secy., Jno. Rubel. Supt., Harry Davis.

12. **LEXINGTON**, *Fayette Co.* (Pop., 16,656—21,567.) First supplied in '32 by city. Built in '81 by Lexington Hydraulic & M'f'g. Co.; franchise regulates rates and provides for purchase of works by city, within 15 yrs., price determined by three trustees. **Supply.**—Surface water, pumping direct from impounding reservoir. Fiscal year closes Dec. 31. In '91, B. S. Church, N. Y. City, was engaged to report on the proposed enlargement of the impounding reservoir.

Year.	Mains.	Taps.	Meters.	H'd'ts.	Cons'pt'n.	B'd'd D't.	Int. Ch'g's.	Exp.	Rev.
'86	16	250	10	212	200,000				
'58	18	630	357	230	750,000	$235,500	$14,380	$8,000	$23,950
'90	18	800	558	215	800,000	235,500	....	....	12,250*

*From city.

Financial.—*Cap. Stock*, $250,000. Other data, see above. **Management.**—Prest., C. G. Hildreth, Lockport, N. Y. Secy., L. C. Kinney, N. Y. City. Treas. and Supt., T. A. Charles.

13. **LOUISVILLE**, *Jefferson Co.* (Pop., 123,758—161,129.) Built in '57-59 by Louisville Water Co., of which city owns all stock but two $100 shares. **Supply.**—Ohio river, pumping to reservoir, stand-pipe and direct. Fiscal year closes Dec. 31. March 27, '90, stand-pipe was wrecked during cyclone, and contract has since been let for new one. A new 16,000,000-gall. pump was put in operation in '90 or '91; designed by E. D. Leavitt, Cambridgeport, Mass., and Chas. Hermany, Engr. of water-works; built by the I. P. Morris Co., Philadelphia.

Year.	Mains.	Taps.	Meters.	Hyd'ts.	Cons'pt'n.	Cost.†	Int. ch's.	Exp.*	Rev.
'61	26.	512	....	0	$640,627	$837,477	$57,360	$63,693	$12,146
'65	30.8	1,766	....	0	1,706,835	1,095,689	74,469	109,873	71,149
'70	58.	3,436	....	0	2,517,290	1,883,039	92,843	132,996	104,279
'75	93.9	6,234	....	0	3,610,835	2,659,555	95,357	130,475	174,164
'80	111.7	7,506	....	361	6,507,141	4,045,476	137,090	198,513	189,821
'81	114.3	7,947	....	368	8,653,826	1,210,017	137,180	208,013	216,803
'82	116.4	8,293	....	388	7,396,441	4,369,452	135,390	226,078	227,052
'83	119.8	8,730	....	408	8,202,206	4,182,145	134,220	190,728	208,720
'84	123.1	9,261	....	450	8,892,907	4,510,333	127,980	200,378	215,995
'85	126.3	9,700	....	541	9,920,340	4,659,791	126,720	196,550	228,670
'86	133.3	10,243	579	554	10,391,897	4,759,790	124,380	195,091	246,348
'87	137.4	11,001	678	570	10,307,056	5,008,809	108,613	178,007	262,548
'88	142.4	11,521	674	612	10,267,269	5,223,801	96,054	166,615	2:2,942
'89	150.7	12,569	707	620	11,175,654	5,681,299	97,320	178,027	291,493

* Including int. chgs. † Including op. exp.

Financial.—*Capital Stock*, common, $1,275,100; paid 8% dividend in 1890. No allowance for public uses of water. **Management.**—Prest., Chas. R. Long. Secy., J. B. Collins. Treas., W. P. McDowell. Supt., Chas. Hermany.

NOTE.—For view of stand-pipe wrecked by cyclone of March 27, '90, see *Scientific American*, Apr. 12, 90. For illustrated description new 16,000,000-gall. pump see *Iron Age*, Sept. 5, '89.

MAYFIELD, *Graves Co.* (Pop., 1,839—2,909.) Works reported in '89-90 Manual were not built, on account of failure of contractors. See projected works.

14. **MAYSVILLE**, *Mason Co.* (Pop., 5,220—5,358.) Built in '60 by Co. **Supply.**—Ohio River, pumping to reservoir. Fiscal year closes April 1.

Year.	Mains.	Taps.	Meters.	Hyd'ts.	Cons'p'tn.	B'd'd debt.	Int. ch'g's.	Exp.	Rev.
'83-4	9	402	30	50	1,0,000	$50,000	$3,000	$6,000	$7,000
'89-90	11	660	40	51	300,000	50,000	3,000	5,300	12,500
'90-1	11	691	55	51	350,000	50,000	3,000	5,500	13,000

KENTUCKY.

Financial.—*First Cost*, $150,000. *Cap. Stock*, $100,000. *Bonded Debt*, $50,000 at 6%, due in '99. *Expenses*: total, $3,500. *Revenue*: consumers, $13,000; city paid $50,000 for perpetual fire protection. **Management.**—Prest., Jacob Krieger. Secy., A. R. Cooper. Supt., A. Schaeffer.

15. MIDDLESBOROUGH, *Bell Co.* (Pop., in '90 3,271.)

TEMPORARY WORKS.

Built in '89 by C. W. Loverin. Water is brought from Bennets fork of Yellow Creek, through a 3-in. pipe with 2-in. and 1-in. laterals. Upon completion of permanent works this plant will be abandoned.

NEW WORKS.

History.—Built in '90-1 by Middlesborough Water Co., under 25-yrs. franchise. Engr., A. H. Martin. Conts., Allison, Shafer & Co. **Supply.**—Little Yellow Creek, pumping to stand-pipe. **Pumping Machinery.**—Dy. cap., 6,000,000 galls.; united triple expansion, from E. P. Allis & Co., Milwaukee, Wis. **Stand-Pipe.**—Cap., 1,000,000 galls.: 37 × 75 ft.; from Cunningham Iron Works, Charleston, Mass. **Distribution.**—*Mains*, 2¼ to 4-in. c. i., 12 miles. *Services*, galv. i. Pipe and specials from So. Pittsburgh (Tenn.) Pipe Wks. and Anniston (Ala.) Pipe Wks. *Meters*, Crown. *Hydrants*, Ludlow, 50. *Valves*, Ludlow. *Pressure*. 110 lbs. **Financial.**—*Cost*, not determined. *Cap. Stock*, authorized, $750,000, all paid up. Bonds, $100,000. Hydrant rental, $65. **Management.**—Prest., C. M. Woodbury. Secy., E. E. Malcolm Treas., J. B. Carey. Supt., A. H. Martin.

16. NEWPORT, *Campbell Co.* (Pop., 20,443—24,918.) Built in '72 by city. Supply, Ohio River, pumping to reservoir. Fiscal year closes Dec. 31.

Year.	Mains.	Taps.	Mtrs.	Hd'ts.	Cons'pt'n.	Cost.	B'd'd d't.	Int.ch'g's.	Exp.	Rev.
'75	....	....	..	....	......	......	$800,000	$58,400	$11,535	$10,669
'80	20.45	1,582	16	175	......	$696,342	800,000	58,400	8,719	17,759
'81	17.7	1,778	25	110	unk'n	780,781	800,000	58,400	9,850	18,172
'83	23.5	2,328	40	130	700,000	745,845	800,000	58,400	19,325	21,050
'86	24.7	2,915	59	139	800,000	782,497	800,000	58,400	11,161	29,628
'87	26.6	3,256	78	155	900,000	775,051	800,000	58,400	9,916	33,037
'88	27	3,473	89	160	900,000	781,093	800,000	58,400	11,275	34,979
'89	27.6	3,671	101	....	1,300,000	785,229	800,000	58,400	10,927	36,267
'90	....	....	....	....	......	......	800,000	58,400	......	38,66

Financial.—*First Cost*, $615,000. *Bonded Debt*, $800,000, at 7½%. **Management.**—Prest., A. S. Berry.

17. OWENSBOROUGH, *Daviess Co.* (Pop., 6,231—9,837.) Built in '79 by Owensboro Water Co. **Supply.**—Ohio River, pumping direct. Fiscal year closes March 31. Holly pumps have been remodelled from piston to vertical, double-acting plunger. Additional 3,500,000-gall. plunger pump was erected in early part of '91. **Distribution.**—*Mains*, 17 miles. *Taps*, 800. *Meters*, 33. *Hydrants*, 126. *Consumption*, 1,700,000 galls. **Financial.**—Withheld. **Management.**—Prest., Sam'l A. Nutter, P. O. Box 9, Louisville. Secy., Wm. Reinecke, 446 West Market St., Louisville. Supt., H. P. Martin.

18. PADUCAH, *McCracken Co.* (Pop., 8,036—13,076.) Built in '86 by Paducah Water Supply Co.; now owned by Paducah Water Co., under 20-yrs. franchise. **Supply.**—Ohio River, pumping to stand-pipe. Fiscal year closes Dec. 31.

Year.	Mains.	Taps.	M't'rs.	Hd'ts.	Cons'pt'n.	Cost.	B'd'd Debt.	Int.ch'g's.	Exp.	Rev.
'87	13	620	2	16?	280,000	$182,030	$150,000	$9,000	$9 400	$18,000
'88	14.5	1,000	4	164	500,000	192,138	150,000	9,000	8,500	21,400
'89	15	1,020	7	169	300 000	......	150,000	9,000	9,089	22,218
'90	15	1,150	7	169	400,000	......	150,000	9,000	9,867	24,177

Financial.—*Cap. Stock*, $150,000. *Bonded Debt*, $150,000 at 6%. *Expenses*, total, $18,867. *Revenue*: consumers, $17,907; city, $6 570; total, $24,477. **Management.**—Prest., Henry Burnett. Secy., S. M. Rieke. Treas. and Supt, Muscoe Burnett.

19. PARIS, *Bourbon Co.* (Pop, 3,204—4,218.) **History.**—Built in '90-1 by Paris Water Co., under perpetual franchise. Put in operation March 1, '91. Engrs. and Contrs., Wheeler & Parks, Boston, Mass. **Supply.**—Homer River, pumping to stand-pipe. **Pumping Machinery.**—Dy. cap., 2,500,000 galls.; two 1,250,000 Deane, one comp. dup. and one high-pressure. *Boilers*, from E. Hodge & Co., E. Boston, Mass. **Stand-Pipe.**—Cap., 295,000 galls.; 25 × 80 ft.; from E. Hodge, E. Boston, Mass. **Distribution.**—*Mains*, 12 to 4-in. o. i., 7.5 miles. *Services*, galv. i. Pipe

---

FOR FURTHER DESCRIPTIVE MATTER see Manual for 1889-90. CHANGES since 1890 are here given. POPULATIONS are for 1880 and 1890, respectively.

from Chattanooga (Tenn.) Foundry & Pipe Works; specials from Builder Iron Foundry, Providence, R. I. *Taps*, 96. *Meters*, Crown, Hersey, 2. *Hydrants*, Coffin, 9v. *Valves* Coffin. *Pressure*, 60 lbs. Financial.—*Cost*, $175,000. *Cap. Stock*, $100,000. *Bonded Debt*, $75,000 at 6%. *Expenses*, not determined. *Revenue:* consumers, $1,300; city, $2,800; total, $4,100. Hydrant rental, $2,500 for 80 hydrants, $30 each for additional hydrants. Management.—Prest., J. M. Thomas, Paris. Vice-Prest. and Treas., Wm. A. Bowker. Secy., J. F. J. Mullhall, both of Boston, Mass. Supt., H. R. Oldson.

20. **PINEVILLE,** *Bell Co.* (Pop., '83—est., 1,200.) Built in '88 by Pineville Water Co., under perpetual franchise which provides neither for control of rates nor purchase of works by city. Supply.—Mountain streams by gravity. Fiscal year closes March 1. Distribution.—*Mains*, 6 miles. *Taps*, 200. *Meters*, 0. *Hydrants*, 13. *Consumption*, 100,000 galls. Financial.—*Cost*, $33,000. *Cap Stock:* authorized, $100,000; paid up, $12,500; new stock to be issued in '91, to pay floating debt. *Total Debt*, $20,500; floating, $2,000; bonded, $12,500 at 6%, due in 1901. *Expenses:* operating, $1,500; int., $1,200; taxes, $150; total, $2,850. *Revenue:* consumers, $2,000; city, $195; other purposes, $500; total, $2,695. Management.—Prest. and Treas., F. A. Hull, Danbury, Conn. Secy.; H. C. Lee. Supt., Thos. A. Hull.

I. **PLEASANT HILL,** *Mercer Co.* (Pop., est., 200.) Built in '33 by Mercer Water Co. Supply.—Spring, pumping to tank. About 1½ miles of pipe supply water to 15 "large families." *Cost*, $1,500. *Expenses*, $150. Supt., Jas. A. Vale.

21. **RICHMOND,** *Madison Co.* (Pop, 2,909—4,753.) History.—Built in '90-1 by Richmond Water & Light Co., under 20-yrs. franchise; put in operation May 1. Des. Engr., A. E. Boardman, Brevard, N. C. Supply.—Creek, pumping to stand-pipe and direct. Pumping Machinery.—Dy. cap., 2,500,000 galls.; two 1,250,000 Worthington comp. *Boilers:* from J. S. Schofield & Son, Macon, Ga. Stand-Pipe.—Cap., 150,000 galls.; from R. D. Wood & Co. Distribution.—*Mains*, 12 to 4-in. c. i., 10½ miles. Pipe and specials from Addyston Pipe & Steel Co., Cincinnati, O. *Taps*, (June 29) 52. *Meters*, 2. *Hydrants*, Bourbon, 62. *Valves*, Kennedy. *Pressure:* ordinary, 62 lbs. Financial.—*Cost*, $110,000. *Cap. Stock*, $200,000. *Bonded Debt*, $175,000 at 6%. *Hydrant rental*, $50. Management.—Prest., A. E. Boardman, Brevard, N. C. Secy., Chas. A. Powell, Richmond, Va. Treas., Edw. H. York, Portland, Me. Supt., M. N. Driggars.

22. **SOMERSET,** *Pulaski Co.* (Pop. 805—2,625.) History.—Built in 90-1 by Somerset Water Co. under 20-yrs. franchise; put in operation March 1. Engr., Jno. A. Geary. Contrs., Water, Heating & Illuminating Co. Supply.—Pitman Creek pumping to reservoir. Pumping Machinery.—Dy. cap, 700,000 galls.; from Laidlow & Dunn Mfg. Co., Cincinnati. Reservoir.—Cap., 1,000,000 galls. Distribution.—*Mains*, 10 to 4-in. c. i., 6 miles. *Services*, w. i. and lead. Pipe and specials from Anniston (Ala.) Pipe Co. and from Addyston Pipe & Steel Co., Cincinnati, O. *Taps*, 120. *Meters*, Hersey, 6. *Hydrants*, 55. *Consumption*, 200,000 galls. *Pressure*, 115 lbs. Financial.—*Cost*, $50,000. *Cap. Stock:* authorized, $100,000; paid-up, $20,000. *Bonded Debt*, $45,000 at 0%. *Revenue:* consumers, $1,600; Ry., $1,300; city, $2,750; total, $5,650. *Hydrant rental*, $50. Management.—Prest., Jno. A. Geary. Secy., J. M. Roche. Treas., Wm. Bright. Supt., H. C. Miles.

23. **WINCHESTER,** *Clark Co.* (Pop., 1,260—4,519.) History.—Built in '90-1 by Winchester Water Co., under perpetual franchise. Put in operation July 1, '91. Engrs. and Contrs., Wheeler & Parks, Boston, Mass. Supply.—Storage reservoir on Boone Creek, pumping to stand-pipe. Pumping Machinery.—Dy. cap., 2,000,000 gal's.; two Deane, one comp. dup. and one high-pressure. *Boiler:* from E. Hodge & Co., E. Boston, Mass. Stand-Pipe.—Cap., 275,400 galls.; 25 × 75 ft.; from E. Hodge & Co., E. Boston. Distribution.—*Mains*, 12 to 4-in. c. i., 9 miles. *Services*, galv. i. Pipes and specials from Addyston Pipe & Steel Co., Cincinnati, O. *Meters*, Hersey. *Hydrants*, Coffin, 72. *Valves*, Coffin. *Pressure*, 50 to 100 lbs. Financial. *Cost*, $180,000. *Cap. Stock*, $100,000. *Bonded Debt*, $80,000, at 6%. *Revenue*, city, $3,280. Management.—Prest., Lucius M. Fuller, Putnam, Conn. Secy., J. F. J. Mulhall. Treas., Jno. H. Chapman. Both of Boston, Mass.

Water-Works Projected with Fair Prospects of Construction.

FRANKLIN, *Simpson Co.* (Pop., 1,633—2,324.) City proposes to erect water-works, but no definite action had been taken July 25, '91. In Sept., '91, it was reported that works would be built following spring. Town Clerk, S. M. Forline.

KENTUCKY.

FULTON, *Fulton Co.* (Pop., 826—1,818.) Jan. 25, '91, it was reported that city would build works, to be in operation July 1. Engr. and Contr., Geo. Cadogan Morgan, Chicago. Supply, large open wells, pumping to stand-pipe. A 1,250,000-gall. pump and 300,000-gall. stand-pipe. Mains, 3½ miles. Hydrants, Morgan, 43. Valves, Peet. Cost, $23,000. No later report could be obtained.

HARRODSBURG, *Mercer Co.* (Pop., 2,202—3,230.) In '91 J. A. Geary and J. F. Roche, Lexington, Ky., were granted franchise. Sept.. '91, it was expected that construction would be started at once; works must be in operation by June 19, '92. Supply, Salt River, one mile distant, probably by pumping to stand-pipe. Des. Engr., Jas. T. Herdic, Williamsport, Pa. Hydrant rental, $50.

HICKMAN, *Fulton Co.* (Pop., 1,264—1,652.) In '91 city issued $20,000 worth of bonds to construct works; to be in operation March 1, '92. Engrs. and Contrs., Hartford, Hebert & Co., Chattanooga, Tenn. Supply, wells, pumping to 10×60-ft. stand-pipe. Mains, 4 miles. Cost, $20,000. Mayor, J. W. Cowgill.

MAYFIELD, *Graves Co.* (Pop., 1,839—2,909.) Graves County Water & Light Co. let contract for works summer or fall of '91; to be completed Jan. 1, '92. Supply, two 8-in. wells, pumping to stand-pipe 140 ft. high. Est. cost, $15,000. Secy.. and Supt., B. A. Neale.

MOUNT STERLING, *Montgomery Co.* (Pop., 2,087—3,629.) Mt. Sterling Water-Works Co. was preparing to build works July 14, '91. Address J. W. Hedden or M. S. Taylor.

ROWLAND, *Lincoln Co.* (Pop. in '90, 512.) June 22, '91, it was reported that J. W. Peyton, Stanford, was pushing a water-works project.

SMITH'S GROVE, *Warren Co.* (Pop., 388—est., 800.) Aug. 4, '91, J. R. Kirby intended to build water-works. Supply, springs, pumping to reservoir or stand-pipe.

STAMFORD, *Lincoln Co.* (Pop., 1,213—1,385.) Address J. F. Peyton, Stamford, or M. H. Miller.

GROUP 5.—Central Northern States.—Works Completed, in Process of Construction, or Projected, in Ohio, Indiana, Michigan, Illinois, and Wisconsin.

OHIO.

Water-Works Completed or in Process of Construction.

1. **AKRON,** *Summit Co.* (Pop., 16,512—27,601.) Built in '80-1 by Akron Water-works Co.; franchise provided for purchase of works by city, July 1, 1891; also provides that rates shall be governed by those of Cleveland. Supply.—Well in gravel, pumping to reservoir. Fiscal year closes May 31. In '91 supply was increased by driving wells.

Years.	Mains.	Taps.	Meters.	Hydrants.
'82-3	22	925	10	157
'84-5	23	950	25	157
'86-7	24	1,200	55	175
'88-9	27	1,643	65	197
'90-1	30	1,811	110	214

Financial.—Bonded Debt, $200,000 at 6%, due in 1900. *Revenue*, city, $9,560. *Management.*—Prest., Frank Adams. Secy. and Treas., H. C. Starr. Supt., E. A. Lawton.

2. **ALLIANCE,** *Stark Co.* (Pop., 4,635—7,607.) Built in '83 by Alliance Water-works Co. Supply.—Mahoning River, pumping to stand-pipe, and direct. *Distribution.*—*Mains* (In '88), 6 miles. *Taps,* 488. *Meters,* 41. *Hydrants,* 66. *Consumption,* 280,000 galls. *Financial.*—*Cost,* $155,615. *Cap. Stock:* authorized, $75,000; all paid up; paid 2% dividend in 1890. *Bonded Debt,* $70,000 at 6%. **Management.**—Prest. and Treas., Wm. Runkle, 160 Broadway, N. Y. City. Secy., F. W. Campbell. Supt., Simon Johnson.

FOR FURTHER DESCRIPTIVE MATTER see Manual for 1889-90. CHANGES since 1890 are here given. POPULATIONS are for 1880 and 1890, respectively.

3. **ASHTABULA,** *Ashtabula Co.* (Pop., 4,445—8,838.) Built in '87 by Ashtabula Water Supply Co., under 20-yrs. franchise; now controlled by Debenture Guarantee & Assurance Co., Chicago. Ill. **Supply.**—Series of wells and tubes on sandy shore of Lake Erie, pumping to stand-pipe. Fiscal year closes Dec. 31. During 1890 pumping station was enlarged and a 3,000,000-gall. reservoir built. Reservoir is lined with brick and cem., and has puddled banks. Engr., Geo. H. Pierson, N. Y. City. **Distribution.**—*Mains,* 15 miles. *Taps,* 916. *Meters,* 20. *Hydrants,* 100. *Consumption,* 800,000 galls. **Financial.**—*Cost,* $320,000. *Cap. Stock,* paid-up, $250,000. *Bonded Debt,* $250,000 at 6%. *Expenses:* operating and taxes, $7,920; int., $15,000; total $22,920. *Revenue:* consumers, $18,000; Ry., $2,560; city, $4,500; total, $25,060. **Management.**—Prest., Theo. Hall. Secy., Chas. B. Ludwig, 72 Broadway, N. Y. City. Treas., J. S. Blyth. Supt., O. B. Clark.

4. **BELLAIRE,** *Belmont Co.* (Pop., 8,025—9,931.) Built in '72 by city. **Supply,**—Ohio River, pumping to reservoir. Fiscal year closes March 1.

Year.	Mains.	Taps.	M't'rs.	H'd'ts.	Cost.	B'd'd debt.	Int. ch'g's.	Exp.	Rev.
'82-3	9	900	0	76	$112,000	......	....	$4,000	$11,000
'88-9	10	1,143	6	82	164,443	$130,000	$7,800	11,000	18,000
'90-1	30	1,246	7	86	167.664	125,000	7,500	12,000	18,000

Management.—Prest., Geo. Walters. Secy. and Supt., H. G. Wilson.

5. **BELLEFONTAINE,** *Logan Co.* (Pop., 3,998—4,245.) Built in '83 by town. **Supply.**—Artesian wells, by pumping. New water supply consists of two artesian wells cased with 8-in. pipe; drilled into coniferous lime-stone; water is slightly sulpho-saline; dy. cap., 1,000,000 galls. **Pumping Machinery.**—Dy. cap., 2,000,000 galls.; two Gordon, comp , cond., dup., direct-acting. **Force Main.**—Of 12-in. c. i., 3,400 ft. long, connecting pumping station with old distribution system. Des. and Const. Engr., Jno. W. Hill, Cincinnati. **Distribution.**—(In '88.) *Mains,* 7.5 miles. *Taps,* 175. *Meters,* 3. *Hydrants,* 72. *Consumption,* 500 galls. **Financial.**—*Cost,* $80,000. *Bonded Debt,* $80,000 at 6%. **Management.**—Secy., I. N Zearing. Supt., C. F. Martin.

6. **BELLEVUE,** *Huron and Sandusky Cos.* (Pop., 2,169—3,052.) Built in '73 by city. **Supply.**—Surface water, collected in reservoir by ditch, pumping to standpipe. Fiscal year closes April 1. New stand-pipe, 20 × 100 ft., and new pump were contemplated, Oct. 5, '90. **Distribution.**—*Mains,* 10 miles. *Taps,* 390. *Meters,* 0. *Hydrants,* 33. **Financial.**—*Cost,* $40,000. *Bonded Debt,* $22,000 at 5%. *No Sinking Fund. Expenses:* operating, $1,617; int., $1,100; total, $2,717. *Revenue:* consumers, $2,440. **Management.**—Trustees, elected every three years. Chn., W. H. Erdrick. Secy., Jno. L. Painter. Supt., Jno. A. Dampsey.

7. **BROOKLYN,** *Cuyahoga Co.* (Pop., 1,295—4,585.) Built in '88 by village. **Supply.**—Cleveland water-works, through meter. **Distribution.**—(In '88.) *Mains,* 7 miles. *Taps,* 200. *Meters,* 47. *Hydrants,* 59. **Financial.**—(In '88.) *Cost,* $35,000. *Bonded Debt,* $20,000 at 6%. **Management.**—(In '89.) Prest., C. Trowbridge. Secy. H M. Farnsworth, 219 Supsrior St., Cleveland.

8. **BUCYRUS,** *Crawford Co.* (Pop., 3,835—5,974.) Built in '83-4 by private co. under 30-yrs. franchise; city has option of buying works in '91; city and co. each select a person, and these two a third to appraise the works; city controls rates according to law of state. **Supply.**—Well, by direct pumping; filtered. Fiscal year closes Nov. 1. During '90, a 1,500,000-gall. Gaskill pump was added. **Distribution.**—*Mains,* 12 miles. *Taps,* 350. *Meters,* 30. *Hydrants,* 102. *Consumption,* 360,000 galls. **Financial.**—*Cap. Stock,* $60,000, *Bonded Debt,* $60,000 at 6%. *Expenses:* int., $3,600; taxes, $450. *Revenue,* city, $5,180. **Management.**—Prest., J. B. Gormley. Secy. and Treas., G. P. Smith. Supt., Dorsey L. Beall.

9. **CANTON,** *Stark Co.* (Pop., 12,258—26,189.) Built in '69 by city. **Supply.**—Myers' lake, Nimishillen creek, and wells, by direct pumping. Fiscal year closes March 1. During '90, supply was increased by 18 driven wells, 275 ft. deep; entire supply is to be from wells. **Distribution.**—*Mains,* 31 miles. *Taps,* 2,024. *Meters,* 6. *Hydrants,* 250. *Consumption,* 1,200,000 galls. **Financial.**—*Cost,* $300,000. *Bonded Debt,* $130,000 at 5%. *Expenses:* operating, $7,500; int., $6,500; total, $14,000. *Revenue,* consumers, $20,000. **Management.**—Trustees elected every three years. Chn., Martin J. Hogan. Secy., Paul Field. Supt., Louis B. Ohliger.

10. **CARTHAGE,** *Hamilton Co.* (Pop., in '90, 2,257.) **History.**—Built in '90-1 by village; put in operation about April 1, '91. Des. Engr., Thomas Birch. Contrs., Laidlaw & Dunn Co., Cincinnati. **Supply.**—Deep wells, pumping direct. **Pumping**

OHIO.

Machinery.—Cap., not given; from contrs. Distribution.—*Mains*, 8 to 4 in, c. l., 4.5 miles. Pipe and specials from Jackson & Woodin Mfg Co., Berwick, Pa.; trenching and pipe-laying done by W. A. Cain, Dayton. *Hydrants*, Galvin, 30. *Valves*, Galvin. Financial.—*Cost*, $20,000. *Bonded Debt*, $20,000 at 5%. Management.—Prest., Charles E. McCammon. Secy., Harry Simms.

11. **CHILLICOTHE**, *Ross Co.* (Pop., 10,938—11,288.) Built in '81-2 by Chillicothe Gaslight & Water Co.; franchise fixes rates but does not provide for purchase of works by city. Supply.—Well on bank of Scioto River, pumping to reservoir; filtered through natural bed of gravel and sand. Fiscal year closes Dec. 31. Distribution.—*Mains*, 23 miles. *Taps*, 900. *Meters*, 25. *Hydrants*, 125. *Consumption*, 750,000 galls. Financial.—*Cost*, $260,000. *Cap. Stock:* authorized $225,000; paid up, $100,000. *Bonded Debt*, $100,000 at 5%. *Expenses:* int., $5,000; taxes, $1,825. *Revenue*, city, $8,500. *Hydrant rental*, $55 each; new 10-yrs. contract made in March, '91. Management.—Prest. and Supt., F. A. Stacey. Secy., A. C. Kopp. Treas., Thomas G. McKell.

12. **CINCINNATI**, *Hamilton Co.* (Pop., 255,139—296,908.) Owned by city since '39; built in '20 by Sam'l Davis. Supply.—Ohio River, pumping to reservoir for low and middle service, to tank for high service. Fiscal year closes Dec. 31. In '91 contract was let for four Wilson-Snyder 5,000,000-gall. pumps, and two Gordon pumps with combined cap. of 3,350,000 galls.

Year.	Mains.	Taps.	Meters.	Cons'pt'n.	Cost.	B'd'd d'bt.	Int. ch'g's.	*Exp.*	Rev.*
'40	24.2	3,118	.....	1,080,000	.....	$300,000	$29,281	$15,314	$58,440
'45	27.6	4,200	.....	1,400,000	.....	300,000	27,452	6,873	43,929
'50	45.8	6,135	.....	2,400,000	.....	700,000	53,259	23,476	78,016
'55	63.12	9,013	.....	3,500,000	.....	875,000	53,634	38,534	154,897
'60	83.66	12,00	36	4,817,013	.....	875,000	50,590	52,959	160,713
'65	93.26	14,920	.....	5,433,990	.....	575,000	.....	120,538	269,067
'70	133.08	18,629	127	10,442,028	.....	875,000	.....	139,445	380,812
'75	157.78	21,'36	133	14,314,010	.....	1,625,000	4,908	156,588	439,364
'80	188.86	23,627	545	19,476,739	6,727,838	1,625,000	102,768	186,528	499,857
'81	195.14	24,256	619	23,037,272	.....	1,625,000	102,235	217,529	478,991
'82	201.62	24,858	792	19,514,848	7,140,029	1,625,000	101,430	212,183	507,501
'83	209.33	25,870	972	19,401,200	7,236,395	1,625,000	100,052	207,311	534,281
'84	222.33	27,117	1,096	22,404,122	.....	1,625,000	93,896	218,501	518,931
'85	227.41	28,522	1,176	19,800,404	.....	1,525,000	103,376	263,261	569,989
'86	232.31	30,219	1,177	24,171,526	.....	1,375,000	101,900	272,424	597,477
'87	248.06	32,930	1,221	29,906,589	.....	1,375,000	91,400	297,525	629,732
'88	263.99	33,082	1,304	32,271,426	7,847,395	1,375,000	91,400	271,731	651,024
'89	282.31	34,400	1,428	32,679,082	.....	1,075,000	63,282	288,385	682,319
'90	287.85	35,439	1,450	33,997,007	8,697,597	975,000	21,452	491,691	605,018

* Given in annual report as "net."

Financial.—*Bonded Debt*, $975,000 : $575,000 at 6%; $400,000 at 7%. *Sinking Fund*, $286,089. *Expenses*, total, $513,146. *Revenue*, consumers, $605,018. Management. Committee of three, from Board of Public Works. Chn., Henry Lewis. Supt., Willis P. Sharpe.

13. **CIRCLEVILLE**, *Pickaway Co.* (Pop., 6,046—6,556.) Built in '86-7 by Circleville Water Supply Co., under 20-yrs. franchise, which fixes rates and provides for purchase of work by city, price determined by three appraisers. Supply.—Darby creek, pumping to stand-pipe. Distribution.—*Mains* (in '88), 12 miles. *Taps* (in '88), 250. *Hydrants*, 102. Financial.—(In '88.) *Cost*, $180,000. *Bonded Debt*, $150,000 at 6%. *Expense:* operating, $3,000; int., $9,000; total, $12,000. *Revenue:* consumers, $4,100; city, $3,500; total, $7,600. Management.—(In '89.) Prest., C. C. Wolcott; Secy. and Treas., C. E. Kimball; both of N. Y. City. Supt., W. E. Bolin.

14. **CLEVELAND**, *Cuyahoga Co.* (Pop., 160,146—261,353.) Built in '54-6 by city. Supply.—Lake Erie, pumping to reservoir. Fiscal year closes Dec. 31. New intake tunnel was completed in '91; see 1889-90 Manual, p. 365. In '91 plans were completed for extension of tunnel to a total length of 4½ miles. In '90 tunnel was driven beneath Cuyahoga River to carry a 40-in. wrought-iron pipe connecting at each end with cast iron pipe. Tunnel has horizontal diameter of 7.5 and vertical of 7.66 ft. Bottom is 70 ft. below river surface, and length between centre of shafts is 511 ft. Tunnel lined with 12½ ins. of brick.

For Further Descriptive Matter see Manual for 1889-90. Changes since 1890 are here given. Populations are for 1880 and 1890, respectively.

Yr.	M'ns.	Taps.	Meters.	Hyd'ts.	Cons'pt'n.	Cost.	B'd'd d't.	Int.ch'g's.	Exp.	Rev.
'69	25.5	1,000	0	136	710,931	$518,732	......	......	......	$10,794
'70	56.3	3,227	0	258	3,095,558	798,244	......	......	......	62,870
'80	115	11,564	411	908	10,179,461	2,640,829	......	......	$55,915	202,378
'82	152.9	14,192	704	1,264	12,513,804	3,074,502	$1,725,000	......	65,693	248,879
'83	173.9	17,371	913	1,571	14,212,144	3,581,107	1,150,000	......	73,211	274,614
'84	....	....	....	....	15,307,153	......	......	......	75,687	296,392
'85	212	18,441	1.175	1,983	17,050,634	4,560,374	1,775,000	$108,150	120,649	317,661
'86	....	....	....	....	15,930,611	......	1,775,000	108,150	85,617	353,192
'87	252	22,658	1,525	2,517	22,266,155	5,053,091	1,775,000	108,150	89,916	392,376
'88	275.8	23,477	1,411	2,894	23,199,703	5,639,433	1,775,000	108,150	106,396	401,775
'89	301	28,287	1,725	3,257	24,875,326	6,684,273	1,775,000	108,150	101,378	448,498
'90	324	30,993	1.794	3,561	27,757,158	5,989,255	1,775,000	108,150	108,347	502,954

Management.—Secy., W. A. Madison. Supt., Jno. Whitelaw.

15. **CLIFTON,** *Hamilton Co.* (Pop., est., 1,200.) Built in '83 by village. **Supply.** —Cincinnati water-works. Engr., Jno. M. Hill, Cincinnati. For further description, see 1889-90 Manual, pp. 366.

16. **CLYDE,** *Sandusky Co.* (Pop., 2,380–2,327.) Built in '83-4 by city. **Supply.** —Wells in settling reservoir, pumping direct. Fiscal year closes Dec. 31. **Distribution.**—*Mains,* 6 miles. *Taps,* 240. *Meters,* 0. *Hydrants,* 50. **Financial.**—*First Cost,* $15,000. *Bonded Debt,* $26,000 at 6%. *Expenses:* operating, $1,388; int., $1,568; total, $3,256. *Revenue,* consumers, $1,500. **Management.**—J. P. McConnell. Secy., W. E. Gillett. Treas., H. B. Tiffany.

17. **COLUMBUS,** *Franklin Co.* (Pop., 51,647–88,150.) Built in '70 by city. **Supply.**—Surface water, pumping direct; filtered. Fiscal year closes March 31. New pumping station was completed in '91; located in Alum Creek valley,; buildings are of brick, with stone trimmings; engine-room is 61 × 75 ft.; boiler room, 91 × 50 ft.; smokestack 8 ft. in diam. × 135 ft. high. Water is taken from well 25 ft. in diam., 55 ft. deep, having dy. cap. of 10,000,000 galls. Pumping machinery consists of two 7,500,000-gall. Holly comp. triple vertical engines, and four tubular steel boilers, each 5 × 18 ft.

Year.	Mains.	Taps.	Meters.	Hyd'ts.	Cons'pt'n.	Cost.	B'd'd d't.	Exp.	Rev.
'71-2	6.2	42	0	30	475,955	$128,146	....	$11,967	$11,219
'75-6	...	...	...	...	1,333,950	623,301	......	23,751	28,136
'80-1	39.5	2,115	534	319	2,500,000	718,489	......	20,891	48,274
'82-3	57.9	2,288	546	363	2,060,964	872,778	$692,000	27,361	60,122
'84-5	60	2,781	557	448	3,030,000	1,031,128	772,000	31,863	73,128
'85-6	67	4,087	449	519	5,086,957	1,086,771	......	25,833	78,160
'86-7	69.2	4,531	449	555	5,297,833	1,125,979	772,000	26,626	80,412
'87-8	70.2	5,207	393	567	5,840,061	1,160,175	772,000	30,919	101,434
'88-9	90.1	6,080	503	710	6,188,708	1,329,122	894,908	33,180	120,957
'89-90	98.4	6,816	468	762	6,959,100	1,550,729	900,000	35,715	125,296
'90-1	103.5	7,649	...	790	6,882,333	1,709,076	1,047,000	42,888	116,175

Financial.—*Bonded Debt,* $922,000: $535,000 at 6%, due '91-5; $387,000 at 4%, due '93-1909. Bonds due in '91 were refunded. Revenue pays all expenses except redemption of bonds. **Management.**—Board of Public Works. Chn., Jas. M. Loren. Secy., D. A. Filler. Supt., A. H. McAlpine.

18. **CONNEAUT,** *Ashtabula Co.* (Pop., 1,256–3,241.) **History.**—Built in '90 by Conneaut Water-Works & Supply Co., under 20-years' franchise; put in operation Jan. 1, '91. Engrs. and Contrs., Ferris, Halladay & Richards, Equitable Building, N. Y. City. **Supply.**—Wells and Lake Erie, pumping to stand-pipe. **Pumping Machinery.**—Dy. cap., 2,500,000 galls.; two Hughes. **Stand-Pipe.**—Cap., 280,000 galls.; 20 × 120 ft.; from Porter Mfg. Co. **Distribution.** —*Mains,* 12 to 4-in. c. i., 10 miles. *Services,* galv. i. Pipe and specials from Jackson & Woodin Co., Berwick, Pa. *Meters,* Thompson, 6. *Hydrants,* Eddy, 90. *Valves,* Eddy. *Pressure:* ordinary, 60 lbs.; fire, 135 lbs. **Financial.**—*Cost,* $150,000, *Cap. Stock:* authorized, $125,000; all paid-up. *Bonded Debt,* $100,000 at 6%. *Revenue,* city, $3,275. *Hydrant Rental,* $37.50 each for first 75; $35 each for additional. **Management.**—Prest., D. N. Sill. Secy., Harry Parfield. Treas. and Supt., H. D. C. Richards.

19. **COSHOCTON,** *Coshocton Co.* (Pop., 2,044–3,672.) **History.**—Built in 90-1 by village; put in operation June 1. Engrs., Dunham & Paine, 150 Broadway, N. Y. City. **Supply.**—Driven wells and Walnounding River, pumping to reservoir. **Pumping Machinery.**—Dy. cap., 1,000,000 galls.; Hughes. Other engines were to be added later. **Reservoir.**—Cap., 330,000 galls.; circular, 50 ft. in diam., 250 ft. above and 9,400 ft. from pumping station. Cost, $4,310. **Distribution.**—*Mains,* c. i., 8½ miles. Pipe and specials from Erie City Iron Works, Erie, Pa.; trenching and pipe-laying done by Snyder & Williams, Dayton, O. *Taps,* 110. *Meters,* 1.

Hydrants, 93. *Pressure*, 105 lbs. **Financial.**—*Cost*, $60,000. *Bonded Debt*, $60,000 at 6%. **Management.**—Prest., Geo. W. Ames. Secy., E. J. Kuntz.

20. **CRESTLINE,** *Crawford Co.* (Pop., 2,848—2,911.) Built in '69-70 by city. **Supply.**—Springs, by gravity from collecting reservoir. Fiscal year closes May 1. **Distribution.**—*Mains*, 10 miles. *Taps*, 241. *Meters*, 1. *Hydrants*, 40. **Financial.**—*Cost*, $120,000. *Bonded Debt*, $10,000 at 5%; $5,000 paid annually. *Sinking Fund*, $900. *Expenses*: operating, $600; int., $750: total, $1,350. **Management.**—Prest., W. Robinson. Secy., Jas. Losenbeimer. Treas., Jos. Schill. Supt., M. C. Archer.

Cuyahoga Falls, *Summit Co.* (Pop., 2,284—2,611.) Built for fire protection by town.

NOTE.—Unofficially reported,[Mar. '91, that authority was being secured to issue $15,000 in bonds for works.

21. **DAYTON,** *Montgomery Co.* (Pop., 38,678—61,220.) Built in '69-70 by city. **Supply.**—Tube wells, by direct pumping. Fiscal year closes Dec. 31. During '90, supply was increased by ten 8-in. tube wells, and an additional 20-in. suction pipe connecting them with pump.

Year	Mains	Taps	Meters	Hyd'ts	C'ns'pt'n	Cost	B'd'd d'bt	Int. ch'gs	Exp.	Rev.
'70	20	562	0	198	502,829	$324,450				$3,149
'75	31.4	1,021			967,803					20,726
'80	32.8	1,211	18	286	1,058,737	705,918			$16,000	18.277
'81	35.8	1,293			1,367,314	786,728	$555,000			21,308
'82	35.9	1,564	58	315	1,182,474	815,031	575,000		18,663	23,214
'83	36.1	1,137	69	319	1,182,589	878,584	573,000		17,559	21,404
'86	37	1.685	82	321	1,643,000	971.526	562,000	$22,480	17,000	23,876
'87	37.2	1,799	96	326	1,851,015	979,459	602,000	21,900	19,000	23,430
'88	49.6	2,005	96	473	1,982,535	1,064,469	617,000		31,718	32,629
'89	51.5	2,536	125	496	2,353,305	1,161,051	612,000		24,000	40,638
'90	54.5	3,014		525	2,848,926	1,186,691	612,000		21,223	46,746

Financial.—See above. **Management.**—Trustees. Prest., Jno. Tesseyman. Secy., Chas. Rowe. Chf. Engr., Martin L. Weaver.

22. **DEFIANCE,** *Defiance Co.* (Pop., 5,907—7,694.) Built in '87-8 by Defiance Water Co.; city has no control of rates. **Supply.**—Maumee River, pumping to stand-pipe. Fiscal year closes Dec. 3. Early in '91 stand-pipe described in a previous issue burst. **Distribution.**—(In '88.) *Mains*, 13 miles. *Taps*, 296. *Meters*, 0. *Hydrants*, 139. *Consumption*, 225,000 galls. **Financial.**—*Cap. Stock*: authorized, $200,000; all paid up. *Bonded Debt*, $200,000 at 6%. **Management.**—Prest., David A. Boody, N. Y. City. Secy., Edward Squires. Treas., Chas. W. McClellan. Supt., S. M. Stevens.

NOTE.—For description of failure of stand-pipe, Mar. 29, '91. see ENGINEERING NEWS, vol. xxv., pp. 313 and 343. Accident occurred in connection with ice in the pipe.

23. **DELAWARE,** *Delaware Co.* (Pop., 6,895—8,224.) Built in '88-9 by Delaware Water Co.; controlled by Moffett, Hodgkins & Clarke, Watertown, N. Y. **Supply.**—Wells and springs, pumping to stand-pipe and direct. **Distribution.**—(In '89.) *Mains*, 18 miles. *Hydrants*, 206. **Financial.**—(In '89.) *Cost*, $200,000. *Bonded Debt*, $180,000 at 6%. **Management.**—(In '89.) Prest., J. F. Moffet.

24. **DENNISON AND URICHSVILLE,** *Tuscarawas Co.* (Pop.: Dennison, 2,518—2,925; Urichsville. 2.700—3.842.) Built in '88 by Dennison Water Supply Co.; controlled by Delaware Co., Philadelphia, Pa., under 15-yrs. franchise; rates must not exceed average rates of other companies in Ohio; Dennison has right to buy works in 1903 and Urichsville in 1908; price to be determined by arbitration. **Supply.**—Big Stillwater Creek, pumping to reservoir; filtered. Fiscal year closes Dec. 31.

Year	Mains	Taps	Meters	Hyd'ts	C'ns'mpt'n	Cost	B'd Debt	Int.	Exp.	Rev.
'88	10.75			103	600,000	$250,000	$125,000	$7,500	$1,500	$9,100
'89	10.75	125	1	103	425,000	250,000	125,000	7,500	3,165	9,414
'90	10.75	200	1	112	480,000	253,311	125,0.0	7,500	3,965	10,645

Financial.—*Cap Stock*: authorized, $125,000, all paid up. *Bonded Debt*, $125,000 at 6%, due in 1908. *Expenses*: operating $3,500; int., $7,500; taxes, $165; total, $11,165. *Revenue*: consumers, $6,945; city, $3,700; total. $10,645. **Management.**—Prest., Clifford Stanley Sims, 242 So. Third St., Philadelphia, Pa., to whom all inquiries should be addressed. Treas., P. W. Janeway, Philadelphia.

FOR FURTHER DESCRIPTIVE MATTER see Manual for 1889-90. CHANGES since 1890 are here given. POPULATIONS are for 1880 and 1890. respectively.

25. EAST LIVERPOOL, *Columbiana Co.* (Pop., 5,568—10,956.) Built in '79 by village. **Supply.**—Ohio River, pumping to reservoir, and direct if desired. Reported in '90 that 3,000,000-gall. Gordon pump was to be added, and a 1,100,000-gall. high service reservoir built. **Distribution.**—(In '88.) *Mains,* 16 miles. *Meters,* 12. *Hydrants,* 75. *Consumption,* 1,000,000. **Financial.**—(In '88.) *Cost,* $51,000. *Bonded Debt,* $25,000. *Expenses,* $5,000. *Revenue,* consumers, $15,000. **Management.**—Trustees, N. A. Frederick, J. Chetwind, Hugh McNickol. Supt., Philip Morley.

26. EAST PALESTINE, *Columbiana Co.* (Pop., 1047-1,816.) Built in '80-90 by village. **Supply.**—Four drilled wells, pumping through distribution to reservoir. **Distribution.**—*Mains,* 3.7 miles. *Hydrants,* 36. **Financial,**—*Cost,* $23,500. *Bonded Debt,* $25,000 at 5%. **Management.**—Trustees, H. J. Frazer, Thos. Atchison, Supt., A. C. Billingsley.

27. EATON, *Preble Co.* (Pop., 2,143—2,934.) **History.**—Built in '91 by city Engr., Jno. W. Hill, Cincinnati. **Supply.**—System of 6-in. driven wells, for domestic use, creek for fire protection, pumping to stand-pipe and direct. **Pumping Machinery.**—Dy. cap., 1,500,000 galls.; two 18-in Gordon comp., cond., dup., direct-acting. *Boilers,* two return tubular. **Force-Main.**—C. i., 12 inches in diam. **Stand-Pipe.**—Cap., 188,000 galls.; 20 × 80 ft., on masonry pier 50 ft. high; from Variety Iron Wks., Cleveland. **Distribution.**—*Mains,* 12 to 4-in. c. i., 9.81 miles. *Hydrants,* 73. *Valves,* Galvin. *Pressure,* 60 lbs. **Financial,**—*Est. Cost,* $67,000. *Bonded Debt,* $67,000 at 6%. **Management.**—Mayor, A. C. Risinger. Chn.; J. C. Acton,

28. ELYRIA, *Lorain Co.* (Pop., 4,777—5,611.) Built in '79 by Elyria Gas & Water Co. **Supply.**—Black River, pumping to tank. Fiscal year closes Dec. 31. **Distribution.**—*Mains,* 11 miles. *Taps,* 400. *Meters,* 0. *Hydrants,* 61. *Consumption,* 1,000,000 galls. **Financial.**—*Cost,* $160,000. *Cap Stock:* authorized, $100,000; all paid up; paid no div'd in 1890. *Bonded Debt,* $75,000 at 5%, due in 1920. **Management.**—Prest., Geo. H. Ely. Secy. and Supt., D. M. Clark.

NOTE.—Franchise was granted in '90 to second co. to procure supply from driven wells, but adequate supply had not been found June 26, '91.

62. FAIRPORT, *Lake Co.* (Pop., 296—1,171.) Supply from Painesville.

29. FINDLAY, *Hancock Co.* (Pop., 4,633—18,553.) Built in '88-9 by city. **Supply.**—Blanchard River, pumping direct; filtered. Fiscal year closes April 1. **Distribution.**—*Mains,* 18 miles. *Taps,* 200. *Meters,* 2. *Hydrants,* 146. **Financial,**—*Cost,* $275,000. *Bonded Debt,* $300,000 at 5%. **Management.**—Three trustees. Chn., B. F. Kimmons. Secy., D. L. Grable. Supt., D. A. B. Bailey.

30. FOSTORIA, *Seneca and Hancock Cos.* (Pop., 3,569—7,070.) Built in '90-1 by city. **Supply.**—Surface water, pumping from reservoirs to stand-pipe; filtered. Improvements were to be completed early in '91 consisting of 30,000,000 and 110,000,000-gall. impounding reservoirs, connected by a brick conduit ¾ mile long, and a 25 × 100 ft. steel stand-pipe on a 60-ft. masonry base; base encloses pumping station and offices. Engr., J. M. Howells, Chicago. **Distribution.**—*Mains,* 15 miles. *Taps,* 132. **Financial.**—*Cost.* $165,000. *Bonded Debt,* $200,000 at 5%. *Sinking Fund,* $35,000 at 4%. **Management.**—Prest., C. Herman. Secy., C. W. Wickey. NOTE.—Plant was not complete at time of last report, March, 3, '91.

31. FRANKLIN, *Warren Co.* (Pop., 2,385—2,729.) Built in '87-8 by town. **Supply.**—Driven wells, pumping direct. **Distribution.**—(In '88.) *Mains,* 5.8 miles. *Taps,* 186. *Meters,* 1. *Hydrants,* 61. **Financial,**—(In '88.) *Cost,* $43,000. *Bonded Debt,* $45,000 at 5%. *Expenses:* operating, $2,400; int., $2,250; total, $4,650. *Revenue,* consumers, $1,500. **Management.**—Prest., W. W. Earivant. Secy., R. B. Moodie.

NOTE.—Secy. refused to furnish statistics for 1890.

32. FREMONT, *Sandusky Co.* (Pop., 8,446—7,141.) Built in '82-3 by city. **Supply.**—Mainly from springs, pumping from storage reservoir to stand-pipe. Fiscal year closes second Monday in May. **Distribution.**—*Mains,* 12 miles. *Taps,* 514. *Meters,* 7. *Hydrants,* 110. *Consumption,* 676,317 galls. **Financial.**—*Cost.* $153,554. *Bonded Debt,* $131,000 at 5%. *Expenses:* operating, $4,540; int., $6,550; total, $11,090. *Revenue:* consumers, $7,256. **Management.**—Three trustees, one elected each year. Prest. and Supt., C. F. Reiff. Secy., Jos. Schwartz.

33. GALION, *Crawford Co.* (Pop., 5,635—6,326.) Built in '82-4 by Galion Water-Works Co., under 30-yrs. franchise. **Supply.**—Wells, pumping to stand-pipe. **Distribution.**—(In '87.) *Mains,* 10 miles. *Taps,* 250. *Meters,* 0. *Hydrants,* 95. *Con-*

sumption, 350,000 galls. **Financial.**—(In '87.) *Cost*, $100,000. *Cap. Stock*, $100,000. *Bonded Debt*, $70,000 at 6%. *Expenses:* operating, $3,000; Int., $4,200; total, $7,200. *Revenue:* consumers, $3,000; city, $5,000; total, $8,000. **Management.**—(In '88.) Prest., J. H. Green. Secy., and Treas., O. L. Hays.

34. **GRANVILLE**, *Licking Co.* (Pop., 1,127–1,306.) Built in '85 by town. Supply.—Driven wells, pumping to tank. **Distribution.**—*Mains*, 5.5 miles. *Taps*, 144. *Meters*, 3. *Hydrants*, 37. *Consumption*, 25,000 galls. **Financial.**—*Cost*, $18,000. *Debt*, $9,000 at 6%. *Expenses:* operating, $900; Int., $540; total, $1,440. *Revenue*, consumers, $1,008. **Management.**—Prest., J. L. Gilpatrick. Supt., J. K. Black.

35. **GREENFIELD**, *Highland Co.* (Pop., 2,104–2,460.) **History.**—Built in '91 by city; put in operation July 10. Engr. and Contr., Jno. P. Martin, Xenia. **Supply.**—Paint Creek, pumping to stand-pipe and direct. **Pumping Machinery.**—Dy. cap., 1,500,000 galls.; Gordon. **Stand-Pipe.**—Cap., 85,000 galls.; 14 × 75 ft. **Distribution.**—*Mains*, 6 miles. *Hydrants*, Bourbon, 57. **Financial.**—*Cost*, $40,000. Int., 5%. **Management.**—Chn., H. D. Drekey.

36. **HAMILTON**, *Butler Co.* (Pop., 12,122–17,565). Built in '84 by city. **Supply.** —Wagner steamed filter wells, pumping to reservoir and direct. Fiscal year closes May 31.

Year.	Mains.	Taps.	Meters.	Hyd'ts.	C'n'ption.	Cost.	B'd D'bt.	Int.Ch'g's.	Exp.	Rev.
'86-7	27	800	35	212	1,000,000	$354,000	$350,000	$17,500	$7,000	$7,000
'88-9	27	1,345	75	2'5	800,000	367,440	365,000	18,250	12,279	43,544
'89-90	28	1,347	70	220	751,082	369,226	......	......	12,279	15,544
'90-1	29	1,496	90	225	981,041	372,256	......	......	16,165	19,331

Financial.—See above. No allowance for water for public uses. Int. is paid and bonds redeemed by direct taxation. **Management.**—Trustees, elected every three years. Chn., Asa Shuler. Secy., R. N. Andrews.

37. **HARRISON**, *Hamilton Co.* (Pop., including part of Harrison, Dearborn Co., Ind., 1,850-1,690.) **History.**—Built in '91 by Harrison Water, Light & Power Co., under 10-yrs. franchise; put in operation May 23. Engr., Frank C. Horne, 32 Exchange Bldg., Chicago. Contrs., Heidenrich Co., 99 Metropolitan Block, Chicago. **Supply.**—Whitewater River and wells, pumping to stand-pipe and direct. **Pumping Machinery.**—Dy. cap., 1,000,000 galls.; Gordon dup. **Stand Pipe.**—Cap., 93,000 galls.; 12 × 110 ft. **Distribution.**—*Mains*, 2 miles. Pipe and specials from Addyston Pipe & Steel Co. *Hydrants*, Bourbon, 21. *Pressure:* ordinary, 50 lbs.; fire, 150 lbs. **Financial.**—*Cost*, $54,284. *Cap Stock:* authorized, $50,000; all paid up. *Bonded Debt*, $40,000. **Management.**—Prest., F. C. Horne, 32 Exchange B'l'd'g, Chicago. Vice-Prest., E. L. Heidenrich, 99 Metropolitan Block, Chicago. Secy. and Supt., J. P Mittler.

38. **HICKSVILLE**, *Defiance Co.* (Pop., 1,212–2,141.) **History.**—Built in '90 by village; put in operation Jan. 1. '91. Engr., J. D. Cook. **Supply.**—Gang-wells, pumping direct. **Pumping Machinery.**—Dy. cap., 1,500,000 galls.; Hughes. **Distribution.**—*Mains*, 10 to 4 in. c. i., 4 miles. Pipe and specials from Lake Shore Foundry Co. *Hydrants*, Bourbon, 31. *Valves*, Rensselaer. **Financial.**—*Cost*, $20,000. *Bonded Debt*, $20,000 at 6%. **Management.**—Prest., T. C. Kimmons. Secy., W. D. Wilson. Treas., Jas. Fisher.

39. **IRONTON**, *Lawrence Co.* (Pop., 8,857–10,939.) Built in '70-1 by city. Supply.—Ohio River, pumping to tank. Fiscal year closes March 31. Two pumps and engines were added during '89-90.

Year.	Mains.	Taps.	M't'rs.	Hyd'ts.	Consumpt'n	B'd'd d't.	Int. ch'g's.	Exp.	Rev.
'82-3	8	965	0	114	1,088,000	$165,000	$13,200	$5,000	$9,000
'86-7	9.5	1,120	0	126	1,600,000	150,000	7,500	7,000	12,000
'89-90	12	1,270	0	163	1,700,000	155,000	6,380	14,990	11,429
'90-1	12	1,309	0	163	1,700,000	155,000	6,380	7,893	13,646

Financial.—*Cost*, $170,437. *Bonded Debt*, $155,000; $18,000 at 8%, due in 1890; $90,000 at 5%, due in 1905; $27,000 at 4%, due in 1906. *Expenses*, total, $14,276. *Revenue;* consumers. $13,644. **Management.**—Three trustees, elected every three years, Prest., W. C. Frailey. Clerk, T. N. Davey. Supt., E. Lawton.

40. **KENT**, *Portage Co.* (Pop., 3,309–3,501.) Built in '86-7 by Kent Water Co., under 20 yrs. franchise. **Supply.**—Plum Creek, pumping to stand-pipe and direct.

FOR FURTHER DESCRIPTIVE MATTER see Manual for 1889-90. CHANGES since 1890 are here given POPULATIONS are for 1880 and 1890, respectively).

Fiscal year closes Dec. 31. **Distribution.**—*Mains*, 10 miles. *Taps*, 270. *Meters*, 0. *Hydrants*, 79. *Consumption*, 265,000 galls. **Financial.**—*Cap. Stock*, $100,000. *Revenue:* consumers, $2,796; city, $3,950. Total, $6,746. **Management.**—Prest., Archer N. Martin; Secy., Chas. C. Kimball; Treas., Chas. C. Pomeroy; all of New York City· Supt., A. B. Young.

41. KENTON, *Hardin Co.* (Pop., 3,940—5,557.) Owned by city; built in '81-2. **Supply.**—Well, by direct pumping. Fiscal year closes Dec. 31. **Distribution.**—*Mains*, 7 miles. *Taps*, 400. *Meters*, 5. *Hydrants*, 73. *Consumption*, 600,000 galls. **Financial.**—*Cost*, $83,000. *Bonded Debt*, $60,000 at 6%. *Sinking Fund*, $30,000. *Expenses:* operating, $3,800; int., $3,600; total, $7,400, *Revenue:* consumers, $4,000; city, none; county, $400; total, $4,400. **Management.**—Prest., W. W. Young. Clerk and Supt., Louis Fecker.

42. LANCASTER, *Fairfield Co.* (Pop., 6,803—7,555.) Built in '80-2 by town. **Supply.**—Gallery and well, pumping to tank. Fiscal year closes March 31. In '93 new reservoir was projected. **Distribution.**—(In '87.) *Mains*, 6 miles. *Taps*, 300 *Meters*, 0. **Financial.**—(In '87.) *Cost* (in '82), $16,801. *Bonded Debt*, $25,000. *Expenses*, $2,000. *Revenue*, $3,000. **Management.**—Trustees. Prest., J. B. Orman. Secy. and Supt., Louis Snyder.

43. LEETONIA, *Columbiana Co.* (Pop., 2,552—2,826.) Built in '89 by village. **Supply.**—Spring, by gravity to collecting basin, thence by gravity to supply reservoir at pumps, thence by direct pumping, surplus going to small storage reservoir. **Distribution.**—(In '89.) *Mains*, 6 miles. *Taps*, 130. *Meters*, 55. **Financial.**—(In '89.) *Cost*, $41,000. *Bonded Debt*, $45,000 at 5%. **Management.**—Trustees, Jno. O. Hoffert, Jno. Scully, Jno. H. Crane.

44. LIMA, *Allen Co.* (Pop., 7,567—15,967. Built in '85-7 by city. **Supply.**—Surface water, by gravity from impounding to storage reservoir, thence by direct pumping. Fiscal year closes Dec. 31. In '91, R. H. Gamble, as Engr., and a joint com. reported in favor of increasing storage cap.

Year.	Mains.	Taps.	Met'rs.	H'd'ts.	Cons'n.	Cost.	B'd'd d'bt.	Int. ch'gs.	Exp.	Rev.
'87	2.0	525	2	159	400,000	$338,000	$338,000	$16,900		
'84	20.3	832	6	164	411,193	338,000	338,000	16,900	$5,169	$9,117
'89	24.5	1,032	12	173	464,000	358,205	338,000	16,900	5,877	14,282
'90	26.9	1,346	..	183	609,371	366,368	338,000	16,900	6,024	16,926

Financial.—*Total Debt*, $342 000; floating, $4,000; bonded, $338,000 at 5%, due 1900-16. *Expenses:* total, $22,924. *Revenue:* consumers, $16,926; city, none. **Management.**—Three trustees, elected every three years. Chn., W. H. Lamberton. Secy. and Supt., D. E. Fritz.

45. LONDON, *Madison Co.* (Pop., 3,067—3,316.) Built in '89-90 by London Water Co. **Supply.**—Three 6-in. wells, pumping direct and to stand-pipe; wells are 56 ft., 156 ft. and 172 ft. deep respectively, and have cap. of 600,000 galls.; two of the wells flow. Fiscal year closes Dec. 31. **Pumping Machinery.**—Dy. cap., 2,000,000 galls.; two Gordon, one comp. and one dup. **Distribution.**—*Mains*, 11 miles. *Taps*, 126. *Meters*, 0. *Hydrants*, 101. **Financial.**—*Cost*, about $110,000. *Expenses*, $2,500. *Revenue:* consumers (for 5 mos.), $1,500; city, $5,035. **Management.**—Prest., A. B. Voorhees, Cincinnati. Secy. and Treas., Epes Randolph, Lexington, Ky. Supt., Frank O. Smith.

46. LORAIN, *Lorain Co.* (Pop., 1,595—4,863.) Built in '84 by town. **Supply.**—Cook wells, pumping direct and to stand-pipe. Contract was let fall of '91 for new 2,000,000-gall. Gordon pump. **Distribution.**—*Mains*, 13 miles. *Taps*, 652. *Meters*, 0. *Hydrants*, 16. *Consumption*, 550,000 galls. **Financial.**—*Cost*, $60,000. *Bonded Debt* (in '88), $50,000 at 6%. **Management.**—Three comrs., elected every three years. Prest., C. H. Washburton. Clerks, E. C. Bartell and Thos. Tidman. Supt., T. C. Norcross. Appointed by comrs.

47. MANSFIELD, *Richland Co.* (Pop., 9,859—13,473.) Built in '70-2 by city. **Supply.**—Flowing artesian wells and two springs, by direct pumping. Fiscal year closes April 30. During '90-1, 12 wells were added to supply, and high service system was introduced. **Distribution.**—*Mains*, about 25 miles. *Hydrants*, 170. **Financial.**—*Expenses*, about $6,000. *Revenue*, about $11,000. **Management.**—Supt., Wm. Ritter.

48. MARIETTA, *Washington Co.* (Pop., 5,444—8,273.) Built in '89-'90 by city. **Supply.**—Well, near Ohio River, pumping direct and to stand-pipe. **Distribution**—*Mains*, 12 miles. *Meters*, 10. *Hydrants*, 101. **Financial.**—*Cost*, $90,000. *Bonded*

Debt, $9,000 at 5%. Management.—Committee of Council: S. M. McWilliam, D. R. Greene, Jacob Gephar.

49. **MARION,** *Marion Co.* (Pop., 3,899–8,327.) Built in '89–'90 by Scioto Water Co., under 25-yrs. exclusive franchise. Engr., D. C. Henny. Supply.—Water, bearing gravel, pumping to stand-pipe. Fiscal year closes Dec. 31. Distribution.—*Mains*, 11.5 miles. *Hydrants*, 143. Financial.—*Bonded Debt*, $200,000 at 6%. Management.—Prest., C. C. Reid. Secy., W. E. Schofield. Treas., D. C. Henny, Denver, Col.

50. **MARTIN'S FERRY,** *Belmont Co.* (Pop., 3,819–6,250.) First supplied in '54; present works built in '86-7 by city. Supply.—Ohio River, pumping through distribution to reservoir. Fiscal year closes Dec. 31. During summer of '90 additional 12-in. force main was laid to reservoir.

Year.	Mains.	Taps.	Meters.	Hyd'ts.	Cons'ption.	Cost.	B'd'd d't.	Int. c'g's.	Exp.	Rev.
'87	7	140	0	47	20,000	.....	$120,000	$6,000	$2,000	$1,200
'88	7.5	206	0	49	181,330	.....	120,000	6,000	2,000	1,200
'89	7.5	400	2	49	181,330	$111,022	120,000	6,000	2,500	4,304
'90	13	679	6	83	206,127	133,173	145,000	6,000	3,222	6,424

Financial.—See above. Management.—Three trustees, one elected every year. Prest., S. Hipkins, Sr. Secy., J. S. Mitchell, Supt., C. S. Morse, appointed by board.

51. **MARYSVILLE,** *Union Co.* (Pop., 2,061–2,810.) History.—Built in '90-1 by Marysville Light & Water Co., under 30-yrs. franchise; put in operation Jan. 28, '91. Engr., Jno. W. Hill. Supply.—System of 6-in. driven wells for domestic use, impounding reservoir on Mill Creek for fire protection, pumping direct. Impounding Reservoir.—Cap., "several" million galls., formed by low masonry dam, provided with spillway, 85 ft. long, 4 ft. deep, and 12-in. waste pipe. Contrs., Wm. Patrick & Co., Phœnix, N. Y. Pumping Machinery.—Dy. cap., 1,500,000 galls ; two Hughes horizontal, comp., non-cond., dup., direct-acting. *Boilers:* three return tubular. Distribution.—*Mains*, 12 to 4-in. c. i., 7 miles. *Services*, galv. i. Pipe and specials from Ohio Pipe Co., Columbus; trenching and pipe-laying, Wm. Patrick & Co., Phœnix, N. Y. *Taps*, 197. *Meters*, 4. *Hydrants*, Holly, 72. *Valves*, Galvin, *Consumption*, 160,000 galls. *Pressure:* ordinary, 60 lbs.; fire, 100 lbs. Financial.—*Cost*, $70,000. *Cap. Stock*, authorized, $100,000. *Bonded Debt*, $60,000 at 6%. *Expenses:* operating, $1,800; int., $3,600; total, $5,400. *Revenue:* consumers, $2,000; city, $3,076; total, $5,070. Management.—Prest. and Supt., Jno. F. Zwerner. Secy., Geo. M. McPeck. Treas., Walter C. Fullington.

52. **MASSILLON,** *Stark Co.* (Pop., 6,836–10,092.) First supplied in '54. Present system built in '85-6 by Massillon Water Co. under 20-yrs. franchise. Supply, artesian well, pumping to stand-pipe for domestic use; Sippo Lake, by gravity, for other purposes. In '91 plans were to be made by Dunham & Paine, N. Y. City, for moving pumping station. Distribution.—(In July,'91.) *Mains*, 19 miles. *Taps*, 700. *Meters*, 9. *Hydrants*, 212. Financial.—Withheld. Management.—Prest., Jno. G. Warwick. Secy. and Treas., R. B. Dodson, 2 Wall St., N. Y. City. Supt., W. S. Wade.

Medina, *Medina Co.* (Pop., 1,484–2,073.) Built by town for fire protection.

53. **MIDDLETOWN,** *Butler Co.* (Pop., 4,538–7,681.) Built in '73-4 by town. Supply.—Well, by direct pumping. Fiscal year closes April 1. Distribution.—*Mains*, 12 miles. *Hydrants*, 85. *Consumption*, 1,000,000 galls. Financial.—*Cost*, $84,500. *Bonded Debt*, $78,500. *Expenses:* operating, $6,852; int., $3,925; total, $10,767. *Revenue:* consumers, $5,502; from county treasury, $1,732; total, $7,234. Management.—Prest., Daniel Bowman. Secy., B. J. Niederlander.

54. **MILAN,** *Erie Co.* (Pop., 797–627.) Built in '88 by village. Supply.—Ground water, gathered in infiltration reservoir, thence by gravity. Fiscal year closes Sept. 15. Distribution.—*Mains*, 23 miles. *Taps*, 30. *Meters*, 0. *Hydrants*, 23. *Consumption*, 10,000 galls. Financial. — *Cost*, $15,392. *Bonded Debt*, $8,000. at 5%. *Expenses:* operating, $995; int., $400; total, $1,395. *Revenue*, $147. Management.—Three trustees, Miner Curtis, Levi Reseve, F. C. Smith, one elected every year. Supt., Maltby Smith, appointed by trustees.

55. **MINERVA,** *Stark and Carrol Cos.* (Pop., 565–1,139.) Built in '86 by town. Supply.—Well, on bank of creek, pumping to tank. Pipe connects with creek for use in case of necessity. Fiscal year closes April 1. Distribution.—*Mains*, 3 miles. *Taps*, 125. *Meters*, 0. *Hydrants*, 26. *Consumption*, 40,000 galls. Financial.—*Cost*, $15,000. *Bonded Debt*, $12,000 at 6%, due in 1906. *Expenses:* operating, $393; int.,

FOR FURTHER DESCRIPTIVE MATTER see Manual for 1889-90. CHANGES since 1890 are here given. POPULATIONS are for 1880 and 1890, respectively.

$726; total, $1,113. *Revenue:* consumers, $836; city, $390; total, $1,226. **Management.**—Three trustees, one elected every year. Chn., Wm. W. Hooper. Secy., S. Morehead. Supt., Jno. Young, appointed by trustees.

MOUNT GILEAD, *Warren Co.* (Pop., 1,216—1,329.) Built in '87 by town. Supply.—Two large springs, pumped by two hydraulic rams to five public watering places. Address M. B. Talmage.

56. **MOUNT VERNON,** *Knox Co.* (Pop., 5,249—6,027.) Built in '82 by town. Supply.—Surface and artesian well, pumping to stand-pipe. Fiscal year closes April 30. Distribution.—*Mains*, 21 miles. *Taps*, 700. *Meters*, 0. *Hydrants*, 90. *Consumption*, 375,000 galls. Financial.—*Cost*, $100,000. *Bonded Debt*, $56,000 at 6%. *Expenses:* operating, $3,000; int., $3,360; total, $6,360. *Revenue:* consumers, $5,640. **Management.**—Three trustees, elected every three years Chn., W. A. Bounds. Secy., P. B. Chase. Supt., C. W. Koons.

57. **NEWARK,** *Licking Co.* (Pop., 9,600—14,270.) Built in '85-6 by Moffett, Hodgkins & Clark, Syracuse, N. Y., under 10-yrs. franchise, which fixes maximum rates and provides for purchase of works by town in 9 yrs. Supply.—Two surface wells, in gravel, pumping to reservoir. Distribution.—*Mains* (in '88), 25 miles. *Taps* (in '88), 622. *Meters* (in '88), 18. *Hydrants* (in '91), 242. *Consumption* (in '88), 589,900 galls. Financial.—(In '88.) *Cost*, $300,000. *Bonded Debt*, $260,000 at 6%. *Expenses:* operating, $6,000; int., $15,600; total, $21,600. *Revenue:* consumers, $12,500; city (in '90), $10,890. **Management.**—(In '89.) Prest., Jno. F. Moffett. Supt., Wm. A. Veach.

58. **NEW LISBON,** *Columbiana Co.* (Pop., 2,028—2,278.) Built in '40; new works built in '88 by town. Supply—Four artesian wells, pumping to reservoir. Fiscal year closes Oct. 1. Distribution.—*Mains*, 3.3 miles. *Taps*, 154. *Meters*, 4. *Hydrants*, 40. *Consumption*, 35,000 galls. Financial.—*Cost* (in '88), $25,000. *Bonded Debt* (in '88), $16,000 at 6%. *Expenses*, operating, $1,200. *Revenue*, $850. **Management.**—Prest. Jas. Farrell. Secy., L. F. Briggs.

59. **NEW PHILADELPHIA,** *Tuscarawas Co.* (Pop., 3,070—4,456.) Built in '86 by New Philadelphia Water Co.; controlled by American Water-Works & Guarantee Co., Ltd., Pittsburgh, Pa., under 15-yrs franchise; maximum rates are fixed in franchise which provides for purchase of works by city in '96, price determined by arbitration. Supply.—Wells, pumping to reservoir. There are ten 6-in. open wells and one large well which furnishes water by filtration from the river. Fiscal year closes Dec. 31. Pumping Machinery.—Dy. cap., 2,000,000 galls.; two 1,000,000 Blake, 18¼ × 10½ × 12. Boilers.—Two 50 HP., horizontal tubular. Reservoir.—Cap., 2,000,000 galls.; of earth, 225 ft. above city. Distribution.—*Mains*, 9.5 miles. *Taps*, 261. *Meters*, 7. *Hydrants*, 9. *Consumption*, 100,000 galls. Financial.—Withheld. **Management.**—Prest., O. P. Taylor. Vice-Prest. J. H. Purdy. Secy. and Supt., D. B. Ludwick. Treas., W. S. Kuhn, Pittsburg, Pa.

60. **NORWALK,** *Huron Co.* (Pop., 5,704—7,175.) Built in '71-2, and reconstructed in '78 by town. Supply.—Branch of Huron River, pumping direct. Source of supply was changed in '89 and impounding reservoir constructed. Engr., Jno. W. Hill, Cincinnati. In '91 further improvements were proposed and it was reported that C. A. Judson, Sandusky, was to make plans. Fiscal year closes March 3. Distribution.—*Mains*, 14 miles. *Taps*, 700. *Meters*, 0. *Hydrants*, 111. *Consumption*, 100,000 galls. Financial.—*Cost*, $150,000. *No debt*. *Expenses*, $5,553. *Revenue:* consumers, $8,616; city, none. **Management**—Chn., D. McIntaffer. Secy. and Supt., O. A. White.

61. **OBERLIN,** *Lorain Co.* (Pop., 3,242—4,376.) Built in '87 by town. Supply.—Springs, by direct pumping; water is filtered through sand and gravel at the intake, and is brought through vitrified pipe, 5 miles by gravity, to pumping station. Fiscal year closes March 31. Distribution.—*Mains*, 7.5 miles. *Taps*, 192. *Meters*, 5. *Hydrants*, 60. Financial.—*Cost*, $56,000. *Bonded Debt*, $50,000 at 4½%, due in 1916. *Expenses:* operating, $1,300; int., $2,250; total, $4,550. *Revenue*, $1,550. **Management.**—Three trustees elected every three years. Chn., A. W. Mitchell. Secy., O. A. Wright.

62. **PAINESVILLE,** *Lake Co.* (Pop., 3,841—4,775. Pop. supplied, 4,028—6,390.) Built in '82 by Painesville Water-Works Co. New system built in '91 to supply Painesville, Richmond and Fairport. Supply.—Wells and springs and Lake Erie

pumping to stand-pipe and direct. **Pumping Machinery.**—Dy. cap., 5,500,000 galls.; two 2,000,000 and one 1,500,000 Knowles, comp., cond. *Boilers*: two steel tubular. **Stand-Pipe.**—20 × 75 ft., 200 ft. above pumping engines. **Distribution.**—*Mains,* c. i., 15 miles. *Taps,* 600. *Hydrants,* 70. *Consumption,* 1,500,000 galls. *Pressure:* ordinary, 90 lbs.; fire, 125 lbs. **Financial.**—*Cost,* $250,000. *Cap. Stock,* $250,000. *Bonded Debt.,* $250.000 at 5. *Expenses:* operating, $3,700; int., $12,500; total, $16,200. *Revenue:* consumers, $13,000; city, $5,000; total, $18,000. *Hydrant rental* $60. **Management**—Prest., Benj. F. Stephens, Flatbush, N. Y. Secy. and Treas., Jno. A. Brainard.

63. **PIQUA,** *Miami Co.* (Pop., 6,031—9,000.) Built in '75 by city, **Supply.**—Surface water, by direct pumping. **Distribution.**—*Mains,* 16 miles. *Taps,* 880. *Meters,* 0. *Hydrants,* 148. *Consumption,* 2,000,000 galls. **Financial.**—*Cost,* $380,000. *Bonded Debt,* $290,000 at 6 and 8%. *Expenses,* operating, $2,400. *Revenue,* consumers, $7,200. **Management.**—Trustees. Prest., L. Kiefer. Secy., J. C. Smiley. Supt., Adam Conoris.

64. **PORTSMOUTH,** *Scioto Co.* (Pop., 11,391—12,361.) **History.**—Built in 71 by city. **Supply.**—Ohio River, by direct pumping. Fiscal year closes May 31. In 90 contract was let to Portsmouth Foundry & Machine Co. for new pumping engines; price. $20,000.

Yr.	Mains.	Taps.	M't'rs.	H'd'ts.	Cons'pt'n.	Cost.	B'd'd d't.	Int. ch'g's.	Exp.	Rev.
'82	18	1,120	0	90	1,900,000	$200,000	$150,000	$6,000	$11,596	$13,881
'84	18	6 1,135	0	90	....	200,000	150,000	6,000	11,985	12,941
'89	20	1,245	0	108	1,500,060	225,000	150,000	6,000	......	......
'90	21	1,300	0	110	1,500,000	227,000	150,000	6,000	10,367	17,956

Financial.—*Bonded Debt,* $150,000 at 4%. *Expenses:* operating, $10,367; int., $6,000; total, $16,367. *Revenue:* consumers, $17,956; city, none. **Management.**—Three trustees elected every three years. Chn., L. B. Smith. Secy., J. W. Kicker. Supt., Job Phillip.

65. **RAVENNA,** *Portage Co.* (Pop., 3,255—3,417.) Built in '85 by town. **Supply.**—Crystal Lake, by direct pumping. Fiscal year closes Dec. 31.

Year.	Mains.	Taps.	Hyd'ts.	Cons'pt'n.	Cost.	B'd'd d't.	Int. ch'g's.	Exp.	Rev.
'86	10	350	84	125,000	$75,000	$75,000	$1,500	......	......
'87	12.3	550	91	250,000	81,000	81,000	4,350	$2,600	$5,100
'88	13	655	96	200,000	90,000	90,000	4,650	2,700	6,000
'89	13	678	98	200,000	90,000	85,000	4,350	2,600	5,747
'90	13	721	98	225,000	90,000	84,250	......	2,700	5,583

Financial.—*Bonded Debt,* $84,250; $75,000 at 5%, due in annual payments of $3,500 after '95; $10,000 at 6%, due in annual payments of $2,000 after '91; $7.50 was paid in '90. Revenue pays expenses, new construction, and int. on $10,000 of bonds. **Management.**—Three trustees, elected every three years. Chn., D. C. Coolman. Secy., Wm. Grinnell. Supt., Wm. H. Linton.

62. **RICHMOND,** *Lake Co* (Pop., 491—444.) Supply from Painesville.

65. **SALEM,** *Columbiana Co.* (Pop., 4,011—5,780.) First supplied in '62 by A. Phillips. Present system built in '87 by Salem Water Co., under 20-yrs. franchise. **Supply.**—Springs, pumping from settling reservoir to stand-pipe. Fiscal year closes Dec. 31. **Distribution.**—*Mains,* 16 miles. *Taps,* 923. *Meters.* 6. *Hydrants.* 173. *Consumption,* 179,945 galls. **Financial.**—*Cap. Stock:* authorized, $250,000; all paid up. *Bonded Debt,* $135,000 at 6%. **Management.**—Prest., B. S. Ambler. Secy. and Supt., F. W. Allison. Treas., A. B. Turner. Boston, Mass.

67. **SANDUSKY,** *Erie Co.* (Pop, 15,838—18,471.) Built '75-6 by city. **Supply.**—Sandusky Bay, pumping to stand-pipe. Fiscal year closes Dec. 31.

Year.	Mains.	Taps.	Meters.	Hyd'ts.	Cons'pt'n.	Cost.	B'd'd d't.	Exp.	Rev.
'77	18.5	567	0	135	466,200	$371,296	$375,000	$6,523	$6,872
'80	21.1	969	0	146	1,128,300	396,637	347,000	8,212	11,129
'81	21.1	1,103	0	148	1,487,275	391,657	347,000	9,853	13,162
'82	21.3	1,059	0	148		395,437	347,000	10,504	14,533
'83	21.4	1,077	0	149	1,860,000	395,289	347,000	11,367	14,857
'84	21.8	1,260	4	152	2,186,882	396,442	347,000	13,901	15,582
'85	21.9	1,497	23	152	1,800,000	397,473	347,000	14,067	16,711
'86	23.8	1,544	52	161	1,747,000	401,704	347,000	12,793	16,449
'87	24.8	1,618	61	164	1,759,000	409,481	347,000	12,859	17,738
'88	26.5	1,692	63	179	2,089,630	433,142	379,000	11,956	18,591
'89	28.5	1,824	74	197	2,283,106	453,630	379,000	14,318	19,061
'90	30.8	1,981	72	223	2,475,823	470,575	217,000	16,539	22,519

For Further Descriptive Matter see Manual for 1889–90. Changes since 1890 are here given. Populations are for 1880 and 1890, respectively.

Financial.—*Bonded Debt*, $217,000 at 4½ and 5%. *Sinking Fund*, $3,186. *Revenue:* consumers, $22,519; city, $3,704; total, $25,223. **Management.**—Three trustees. Prest., C. F. Schoepfle. Secy., Wm. J. Affleck. Supt., Chas. A. Judson.

68. SIDNEY, *Shelby Co.* (Pop., 3,823—4,850.) First supplied in '73 by city; new works built in '88, by city. **Supply.**—Great Miami River and wells, pumping to stand-pipe; American filter.

Year	Mains	Taps	Meters	Hyd'ts	Cons'pt'n	Cost	B'd'd d't	Int	ch'g's	Exp	Rev
'82	5	150	0	40	8,000	$56,000				$300	$500
'89	12	207	1	55	300,000		$97,000			2,000	1,700
'90	13	267	?	76	300,000	120,000	97,000			2,500	3,000

Financial.—*Bonded Debt*, $97,000: $17,000 at 6%; $80,000 at 5%. *No Sinking Fund*. *Expenses:* operating $2,500; int, $5,020; total, $7,520. *Revenue:* consumers, $3,000; city, $900; total, $3,900. **Management.**—Three trustees, one elected every year. Prest., Jas. N. Anderson. Secy., L. M. Studevant. Supt., J. Wagner.

69. SPRINGFIELD, *Clarke Co.* (Pop., 20,730—31,895.) Built in '81-2 by city. **Supply.**—Springs, in gravel, pumping from collecting reservoir to stand-pipe. Fiscal year closes March 31. Pumping cap. was increased in '90 to 8,000,000 galls. by change of 2,000,000 Cope & Maxwell to 5,000,000 Gaskill horizontal.

Year	Mains	Taps	Meters	Hyd'ts	Cons'pt'n	Cost	B'd'd d't	Int	ch'g's	Exp	Rev
'82-3	30	250	0	250	362,055	$380,000	$340,000				
'84-5	30	1,200	25	279	1,071,810	400,000	400,000	$20,000		$11,000	$14,000
'86-7	31	1,422	30	283	1,235,024	419,596	440,000	22,000		13,012	21,088
'87-8	32.5	1,690	37	285	1,493.224	438,500	400,000	20,000		13,000	23,000
'88-9	33.5	1,856	60	295	1,519,077	419,622	410,000	20,500		16,600	24,361
'89-90	37.3	2,043	46	314	1,704,060	458,503	398,000	19,500		17,140	26,600
'90-1	39.1	2,293	46	322	2,000,000	507,505				15,911	28,480

Management.—Board of three members, one elected every year. Prest., E. C. Guyn. Secy. and Supt., Wm. R. Smith.

70. STEUBENVILLE, *Jefferson Co.* (Pop., 12,093—13,391.) Built in '35 by town. **Supply.**—Ohio River, pumping to reservoir. Fiscal year closes Feb. 15. During fall of '91, a Worthington pump was to be introduced. **Distribution.**—*Mains*, 23 miles. *Taps*, 3,000. *Meters*, 1. *Hydrants*, 140. *Consumption*, 1,500,000 galls. **Financial.**—Withheld. **Management.**—Chn., David McGowan. Secy., R. E. Blinn. Supt., F. B. Ford.

71. TIFFIN, *Seneca Co.* (Pop., 7,879—10,801.) Built in 78-9 by Tiffin Water-Works Co. under 25-yrs. franchise. **Supply.**—Sandusky River, by direct pumping. Fiscal year closes March 1. **Distribution.**—*Mains*, 15 miles. *Taps*, 810. *Meters*, 32. *Hydrants*, 114. **Financial.**—Withheld. **Management.**—Prest. and Supt., M. Scannell.

72. TOLEDO, *Lucas Co.* (Pop., 50,137—81,434.) Built in '73-4 by city. **Supply.**—Maumee River, pumping to stand-pipe. Fall of '91 filter bed was taken up; see Manual for '89-90, p. 383. Fiscal year closes Dec. 31. Since Jan. 1, '91, all new services entering a building are so arranged that meters may be set at any time without alteration, and meters have been set at expense of city for nearly or quite all large consumers.

Year	Mains	Taps	M't'rs	H'd'ts	Cons'pt'n	Cost	B'd'd't	Int	ch'g's	Exp	Rev
'75		1,166			1,818,580	$886,550					$14,373
'80	16.8	1,732	41	349	3,270,873	967,257				$20,777	26,124
'82	18.5	1,907	110	361	3,105,718	998,108				13,770	39,435
'84	49	2,250	230	374	3,618,738	1,015,791				20,335	45,751
'85	50.6	2,417	279	376	3,740,128	1,018,539	$1,000,000	$80,000		24,989	50,171
'86	51.8	2,592	304	383	3,798,313	1,024,517	1,000,000	80,000		21,548	53,478
'87	56	2,879	322	404	4,250,051	1,029,272	1,000,000	80,000		23,782	58,597
'88	62.9	3,210	349	435	4,774,829	1,053,390	1,000,000	80,000		21,683	65,398
'89	69.65	3,671	377	464	5,633,929	1,066,940	998,000	79,840		28,015	74,553
'90	78.5	4,374	441	507	5,842,768	1,080,907	998,000	79,840		22,530	78,868

Financial.—*Bonded Debt*, $998,000. *Sinking Fund*, $10,000. *Expenses*, total, $102,379. *Revenue:* consumers, $78,868; city, none. **Management.**—Three trustees, one elected every year. Prest., Milton S. Huntley. Secy., Daniel Segur. Supt, Harry C. Cotter.

73. TORONTO, *Jefferson Co.* (Pop., in '90, 2,536.) **History.**—Built in '90-1 by village; put in operation July 20. Engr., J. D. Cook, Toledo. Contr., C. H. Conwell, Youngstown. **Supply.**—Ohio River, pumping to reservoir. **Pumping Machinery.**—Dy. Cap. 1,500,000 galls.; Gordon. **Reservoir.**—Cap., 3,000,000 galls. **Distribution.**—*Mains*, 8 and 6-in. c. i., 5 miles. Pipe and specials from Addyston Pipe

& Steel Co., Cincinnati. *Services*, lead. *Hydrants*, Bourbon, 45. *Valves*, Rensselaer. *Pressure*, 133 lbs. **Financial.**—*Cost*, $75,000. *Bonded Debt*, $75,000 at 5%. **Management.**—Prest., A. H. Taylor. Secy., O. Y. Crawford. Supt., M. Edwards.

74. **TROY,** *Miami Co.* (Pop., 3,803—4,594.) Built in '86-8 by town. **Supply.**—Well, 1 y direct pumping. Fiscal year closes May 1. **Distribution.**—*Mains*, 8.5 miles. *Taps*, 461. *Meters*, 0. *Hydrants*, 95. *Consumption*, 21,000 galls. **Financial.**—*Cost*, $30,000. *Bonded Debt*, $80,000 at 6%. *Expenses:* operating, $2,325; int., $4,800; total, $7,125. *Revenue:* consumers. $2,421; city, none. **Management.**—Three trust es, one elected every year. Prest., J. L. Barry. Secy. and Supt., I. N. Price.

75. **UPPER SANDUSKY,** *Wyandot Co.* (Pop., 3,510—3,572.) Built in '89'-'90 by Upper Sandusky Water Co., under 20-yrs. franchise, which does not regulate rates, but provides for purchase of works by city in 1900 and at intervals of 10 yrs. thereafter. **Supply.**—Springs, pumping from two collecting wells to stand-pipe, and direct. Fiscal year closes July 1. **Distribution.**—*Mains*, 11 miles. *Taps*, 300. *Meters*, 5. *Hydrants*, 125. *Consumption*, 150,000 galls. **Financial.**—*Cost*, $60,000. *Cap. Stock*, $60,000. *No Debt. Expenses*, $3,000. *Revenue:* consumers, $750; city, $5,000; total, $5,750. **Management.**—Prest., N. McConnell. Secy. and Treas., C. E. Coon. Supt., W. S. Coon & Son.

76. **URBANA,** *Champaign Co.* (Pop., 6,252—6,510.) Built in '78 by Urbana Water-Works Co., under 10-yrs. franchise. **Supply.**—Well, by direct pumping. Fiscal year closes Oct. 15. **Distribution.**—*Mains*, 15.5 miles. *Taps*, 630. *Meters*, 11. *Hydrants*, 120. *Consumption*, 750,000 galls **Financial.**—*Cost*, $125,000. *Cap. Stock* (in '88), $40,000. *Bonded Debt*, $100,000 at 6%. **Management.**—Prest., A. F. Vance, Jr. Secy. and Supt., F. W. Ambrose. Treas., W. W. Wilson.

24. **URICHSVILLE,** *Tuscarawas Co.* (Pop., 2,790—3,842.) Supply from Dennison.

77. **VAN WERT,** *Van Wert Co.* (Pop., 4,709—5,512.) **History.**—Built in '90-1 by city; put in operation in Sept. Engr., Jno. W. Hill, Cincinnati. **Supply.**—System of 6-in. driven wells, pumping direct, and to reservoir for fire protection. Wells are from 140 to 228 ft. deep, connected with each other and with pumping station at depth of 20 ft., and cased to shut out ground water. **Pumping Machinery.**—Dv, cap., 3,000,000 galls.; two 1,500,000 Holly, comp., cond., dup., direct-acting. *Boilers*, two return tubular. *Reservoir*.-Cap.,3,000,000 galls.; in excavation and embankment; depth 17 ft., 9 ft. of excavation, 8 ft. of embankment; 160 x 264 ft. at top, 75 x 179 ft. at bottom; lined with puddle 16 ins. thick; bottom and lower portion of sides paved with broken stone, remainder paved with dressed blocks laid in cem. mortar. Contr., H. Merigold, Lima. *Cost*, $8,142. **Distribution.**—*Mains*, 14 to 4-in. c. i., 16 miles. Pipe and specials from Addyston Pipe & Steel Co., Cincinnati; trenching and pipe-laying done by R. B. Carothers. *Hydrants* Mathews, 147. *Valves*, Rensselaer. **Financial.**—*Est. Cost*, $120,000. **Management.**—Board of five: O. A. Balyeat, A. B. Gleason, W. C. Brooks, H. Butler, W. B. Jones Clerk, W. B. Jones.

78. **WADSWORTH,** *Medina Co.* (Pop., 1,219—1,574.) Owned by R. F. Weaver. Built in '80 by Wadsworth Water-Works Co, under 10-yrs. franchise. **Supply.**—Four flowing springs, by gravity. **Distribution.**—*Mains*, 6.3 miles. *Takers* (in '88), 800. *Hydrants*, 0. **Financial.**—*Cost*, $20,000. *No Debt.* **Management.**—Owner, R. F. Weaver.

79. **WARREN,** *Trumbull Co.* (Pop., 4,928—5,973.) Built in '86-7 by Sam'l R. Bullock & Co., New York City, under 20-yrs. franchise, which does not regulate rates but provides for purchase of works by city at intervals of 5 yrs. **Supply.**—Mahoning River, pumping from filter gallery to stand-pipe. **Distribution.**—(in '88.) *Mains*, 12 miles. *Taps*, 309. *Meters*, 5. *Hydrants* ('91), 105. *Consumption*, 225,000 galls. **Financial.**—*Cost* (in '87), $180,000. *Bonded Debt* (in '88), $150,000 at 6%. *Expenses:* operating, $3,500; int., $9,000; total, $12,500. *Revenue*, consumers (in '88), $7,500; city (in '90), $4,926. **Management.**—(In '89.) Prest, C. C. Pomeroy. Secy., C. E. Kimball. Both of N. Y. City. Supt., Geo. H. Quimby.

80. **WASHINGTON,** *Fayette Co.* (Pop., 3,798—5,742.) Built in '89 by Washington Water Co., under 20-yrs. franchise. **Supply.**—Battery of wells, pumping direct and to stand-pipe. Fiscal year closes Dec. 31. **Distribution.**—*Mains*, 17 miles. *Taps*, 385. *Meters*, 2. *Hydrants*, 131. **Financial.**—*Cost*, $150,000. *Bonded Debt*,

FOR FURTHER DESCRIPTIVE MATTER see Manual for 1889-90. CHANGES since 1890 are here given. POPULATIONS are for 1880 and 1890, respectively.

$125,000 at 6%. *Expenses:* operating, $2,500; int., $7,500; total, $10,000. *Revenue:* consumers, $3,000; city, $5,270; other purposes, $500; total, $8,770. **Management.**—Prest., E. Randolph Lexington, Ky. Secy., Harvey Meyers, Covington, Ky. Supt., J. P. Maynard.

81. **WELLSVILLE,** *Columbiana Co.* (Pop., 3,773—5,247.) Built in '81 by city. Supply.—Ohio River, pumping to reservoir. Fiscal year closes Dec. 31. Distribution.—*Mains.* 7.5 miles. *Taps,* 700. *Hydrants,* 63. **Financial.**—*Cost,* $64,890. *Bonded Debt,* $50,000 at 5 and 6%. *Sinking Fund,* $210. *Expenses:* operating, $5,806; int., $3,420; total, $9,226. *Revenue:* consumers, $6,350; city, none. **Management.**—Trustees, Chn., John Syth. Secy., Sam'l Stevenson. Acting Supt., Jno. W. Quinn.

82. **WEST CLEVELAND,** *Cuyahoga Co.* (Pop., 1,781—4,117.) Built in '78 by town. Supply.—Cleveland water-works. Distribution.—(In '88.) *Mains,* 4 miles. *Taps,* 170. *Meters,* 20. *Hydrants,* 45. *Consumption,* 44,957 galls. **Financial.**—*Cost,* $19,000. *Bonded Debt,* $12,000 at 6%. *Expenses:* operating, $200; int., $720; total, $920. *Revenue:* consumers, $2,460. **Management.**—Trustees: E. R. Doty, J. W. Greene. J. D. Sinn. Secy., F. J. Phelan.

83. **WOOSTER,** *Wayne Co.* (Pop., 5,840—5,901.) Built in '69 by town. Supply.—Springs and well, by gravity from former, by pumping from latter. Fiscal year closes March 16. Distribution.—*Mains,* 11.1 miles. *Taps,* 250. *Meters,* 7. *Hydrants,* 105. *Consumption,* 150,000 galls. **Financial.**—*Cost,* $115,000. *Bonded Debt,* $12,500: $2,500 at 8%, due in '93; $10,000 at 6%, due in 1900. *Expenses:* operating, $1,600; int., $800; total, $2,400. *Revenue,* consumers, $2,840. **Management.**—Chn., A. B. Babb. Secy., A. B. Snyder. Supt., Edmund Keyser.

84. **XENIA,** *Greene Co.* (Pop., 7,026—7,301.) Built in '87 by Xenia Water-Works Co., under 2½-yrs. franchise; city has right to purchase works in '97 or 1907; rates are determined in franchise. Supply—Springs, pumping direct and to stand-pipe. Fiscal year closes Aug. 1. Distribution.—*Mains.* 19.5 miles. *Taps,* 500. *Meters,* 9. *Hydrants,* 180. *Consumption,* 200,000 galls. **Financial.**—*Cost,* $205,000. *Cap. Stock,* $200,000. *Bonded Debt,* $200,000 at 6%, due in 1907. *Expenses:* int., $12,000; taxes, $900. *Revenue,* city, $7,500. **Management.**—Prest., Jno. Little. Secy. and Supt., G. F. Cooper. Treas., Alfred Birnic, Springfield, Mass.

85. **YOUNGSTOWN,** *Mahoning Co.* (Pop., 15,435—33,220.) Built in '72 by city. Supply.—Mahoning River, by direct pumping. Fiscal year closes March 31. Summer of '91 J. D. Cook, Toledo, reported on new supply from impounding reservoirs on Mill Creek and adding of stand-pipe; est. cost, $300,000.

Year.	Mains.	Taps.	Meters.	Hyd'ts.	Cons'pt'n.	Cost.	B'd d't.	Int.ch'g's.	Exp.	Rev.
'82-3	15	822	15	111	800,000	$175,000	....	....	$9,383	$14,746
'84-5	17	1,122	53	147	882,068	175,000	$155,600	....	11,507	20,222
'86-7	17	1,322	75	146	907,000	208,225	150,000	....	10,000	25,089
'87-8	22.2	1,423	85	185	1,417,911	247,221	140,000	$7,000	10,506	27,159
'88-9	25	1,522	90	195	1,717,274	261,890	125,000	6,400	12,000	29,231
'89-90	25.8	1,627	..	208	1,634,687	274,018	115,000	5,000	12,125	32,147
'90-1	27.4	1,816	..	230	1,799,486	290,625	105,000	5,300	12,933	35,016

Financial.—*Bonded Debt,* $105,000: $100,000, at 5%, due in 1901; $5,000 at 6%, due in '91. *Expenses,* total, $18,233. *Revenue:* consumers, $35,016; city, none. **Management.** —Three trustees. Secy., D. N. Simpkins. Supt., W. S. Hamilton.

86. **ZANESVILLE,** *Muskingum Co.* (Pop., 18,133—21,009.) Built in '42 by town. Supply.—Muskingum River, pumping to reservoir. Fiscal year closes March 31. System was generally reconstructed during '90-1. Engr., Jno. W. Hill, Cincinnati. Sept., '91, 5,000,000-gall. high duty Gaskill pump was contracted for.

Year.	Mains.	Taps.	M't'rs.	Hydt's.	Cons'pt'n.	Cost.	B'd d't.	Int.chgs.	Exp.	Rev.
'82-3	35	2,500	5	170	2,000,000	$500,000	$300,000	....	$11,500	$28,600
'84-5	37.6	2,663	..	212	2,420,751	517,473	300,000	$21,000	16,006	28,515
'86-7	42	3,219	3	202	3,420,854	....	300,000	21,000	19,000	29,168
'88-9	44.1	3,374	3	207	....	....	310,000	21,000	19,016	31,054
'89-90	46.8	3,614	3	218	2,411,095	....	310,000	21,000	25,301	31,825
'90-1	48.1	3,861	3	228	3,191,036	....	305,000	21,000	23,647	32,782

Management.—Three trustees. Secy., Pius Padgitt. Supt., R. M. Saup.

Water-Works Projected with Fair Prospects of Construction.

BRIDGEPORT, *Belmont Co.* (Pop., 2,395—3,369.) City will build works during '91. Supply, water-works at Martin's Ferry. Cost, $25,000, to be paid by bonds. Contracts were let June 29, '91: for reservoir, to Moore & Staub; for other work, to Hibbard & Son, Wheeling, W. Va.

OHIO.

CAMBRIDGE, *Guernsey Co.* (Pop., 2,883—4,361.) Proposals for construction of works were received July 15, '91. Pumping to stand-pipe. Village Clerk, Jno. S. Black. City Engr., O. M. Hoge.

FERNBANK, *Hamilton Co.* (Pop. in '90, 367.) Address Ohio Water & Light Co., or H. W. Woodruff, 29 West 4th St.. Cincinnati.

GLENDALE, *Hamilton Co.* (Pop., 1,400—1,414.) Works projected by village. Address R. D. Mussey.

HILLSBOROUGH, *Highland Co.* (Pop., 3,234—3,620.) Water-works projected but no definite action taken July 6, '91.

NELSONVILLE, *Athens Co.* (Pop., 3,095—4,558.) City has secured authority from Legislature to issue $60,000 worth of bonds to build works. No further steps had been taken July 20, '91. Address C. S. Cable or Jos. Slater.

NILES, *Trumbull Co.* (Pop., 3,879—4,289.) Contracts were let during summer of '91 and village proposed to build works at once. Engr., J. B. Strawn, Salem. Contr., C. H. Connell, Youngstown. Village Clerk, M. J. Flaherty.

NORTH BALTIMORE, *Wood Co.* (Pop., 701—2,857.) North Baltimore Water Co. incorp. and secured franchise and proposed to build system during '91; cap. stock, $100,000. Supply, Portage River, filtered, pumping to stand-pipe. Contrs., W. S. Coon & Son, Upper Sandusky.

PLAIN CITY, *Madison and Union Cos.* (Pop., 665—1,245.) Well was drilled in '89 by Gas and Oil Co. In '90 two hydraulic rams were procured, 2,000 ft. of pipe laid and public drinking place erected. Co. hopes to develop water-supply system.

WAPAKONETA, *Auglaize Co.* (Pop., 2,765—3,616.) Town had voted to build works, but no further action had been taken July 20, '91. Address Jno. B. Walsh, County Surveyor.

WELLSTON, *Jackson Co.* (Pop., 952—4,377.) City was to begin construction of works Aug. 1. Pumping to stand-pipe. Mayor, J. M. Ervin.

WILLOUGHBY, *Lake Co.* (Pop., 1,001—1,219.) System projected but no definite action Sept. 2, '91. Address E. J. Dickey.

Water-Works Projects Reported in '90 and '91, but Now Indefinite.

Napoleon, Henry Co.

INDIANA.

Water-Works Completed, in Process of Construction, or Projected.

1. **ANDERSON,** *Madison Co.* (Pop., 4,126—10,741.) Built in '86 by town. Supply.—Well and White River, by direct pumping; filtered through sand and gravel. Fiscal year closes April 30.

Year.	Mains.	Taps.	Meters.	H'd'ts.	C'na'pt'n.	Cost.	B'd'd d'bt.	Int. ch'g'e.	Exp.	Rev.
'87-8	4	100	0	39	100,000	$30,000	$9,000	$540	$1,500	$1,500
'88-9	7	300	1	70	450,000	45,000	6,000	540	1,800	2,100
'90-1	7.5	400	2	74	1,000,000	57,000	21,000	1,260	2,507	3,590

Management.—Chn., L. J. Burr. Secy., B. McMahon. Supt., Frank Davis.

2. **ATTICA,** *Fountain Co.* (Pop., 2,150—2,320.) Built in '73-4 by town. **Supply.**—Springs, pumping to reservoir. **Distribution.**—*Mains,* 7 miles. *Taps,* 330. *Meters,* 1. *Hydrants,* 7. *Consumption,* 114,000 galls. **Financial.**—*Cost,* $50,000. *Bonded Debt,* $10,000 at 6%. *Expenses:* operating, $1,500; int., $600; total, $2,100. *Revenue:* consumers, $1,650; city, none. **Management.**—Com. of three, appointed by mayor every three years. Chn., Sam'l. Clark. Supt., S. D. Mentzer, appointed by town council.

3. **AVILLA,** *Noble Co.* (Pop., 446—576.) **History.**—Built in '90 by town; put in operation Nov. 20. **Supply.**—Source not given, pumping to tank. **Pumping Machinery.**—Dy. cap., 18,720 galls.; from Henion & Hubbell, Chicago, Ill. **Tank.**—Cap., 16,000 galls. **Distribution.**—*Mains,* 6-in. galv. i., ½ mile. *Taps,* 6. *Hydrants,* 6. *Consumption,* 3,000 galls. *Pressure,* 120 lbs. **Financial.**—*Cost,* $2,800. *Bonded*

FOR FURTHER DESCRIPTIVE MATTER see Manual for 1889-90. CHANGES since 1890 are here given. POPULATIONS are for 1880, and 1890, respectively.

Debt, $1,800 at 0%. **Management.**—Prest. and Supt., G. S. Henry. Secy., J. N. DeCamp.

4. **BLUFFTON**, *Wells Co.* (Pop., 2,354—3,569.) Built in '86 by town. **Supply.** —Wells, in rock near Wabash River, by direct pumping. In '90, connection was made with river, for fire protection. Fiscal year closes Sept. 1.

Year.	Mains.	Taps.	Hydrants.	Cost.	B'd debt.	Int. ch'g's.	Exp.	Rev.
'86	2	....	32	18,000	16,000	800	....	....
'88	2.8	220	36	20,000	16,000	800	1,350	1,700
'90	3.6	190	42	22,000	16,000	800	1,285	1,500

Distribution.—*Meters*, 0. Financial.—Int., 5%. *Expenses*, total, $2,085. *Revenue*. city, none. **Management.**—Three comrs., one appointed every year. Chn., G. F. McFarren. Secy., W. H. Bennett. Supt., W. B. Dehaven.

5. **BRAZIL**, *Clay Co.* (Pop., 3,441—5,905.) Built in '75 by town. **Supply.**—Surface water, by direct pumping from well, sunk in coarse sand and gravel stratum. Fiscal year closes Dec. 31. **Distribution.**—*Mains*, 4.8 miles. *Taps*, 195 *Meters*, 0. *Hydrants*, 47. Financial.—*Cost*, $50,000. *Bonded Debt*, $28,500. *Expenses*, $5,300. *Revenue*: consumers, $2,700; city, none. **Management.**—Committee of three elected every year. Chn., R. S. Stewart. Clerk, Oscar Thomas. Supt., Thos. Henderson.

6. **COLUMBUS**, *Bartholomew Co.* (Pop., 4,813—6,719.) Built in '71 by town. Supply.—Driftwood River, by direct pumping. Distribution.—(In '85) *Mains*, 11 miles. *Taps*, 400. *Meters*, 1. *Hydrants*, 70. *Consumption*, 1,000,000 galls. Financial.—(In '88.) *Cost*, $52,000. *Expenses*, $5,000. *Revenue*, $5,300. **Management.**—Trustees. Prest., Henry Lang. Secy., Jno. R. Gent. Supt., E P. Smith.

7. **CONNERSVILLE**, *Fayette Co.* (Pop., 3,228—4,548.) Built in '69 by city. Supply.—Hydraulic canal, by direct pumping. Distribution.—*Mains* (in '88), 8 miles. *Hydrants* (in '87), 75. *Consumption* (in '88). 400,250 galls. Financial.—(In '87.) *Cost*, $40,000. No *Debt*. *Expenses*, $1,500. **Management.**—Trustees. Peter Lavel. Geo. Keller, W. J. Cain.

8. **CRAWFORDSVILLE**, *Montgomery Co.* (Pop., 5,251—6,089.) Built in '86 by Crawfordsville Water-Works Co., under 20-yrs. franchise; rebuilt in '90 by Crawfordsville Water & Light Co.; now controlled by Debenture Guarantee and Assurance Co., Chicago, Ill. Electric light and gas plants are consolidated with water-works. Supply.—Van Cleve Springs, by direct pumping and pumping to reservoir and stand-pipe. Fiscal year closes Dec. 31. During '90, an impounding reservoir with cap. of 3,000,000 galls. was built. Engr., Geo. H. Pierson, N.Y. City. Distribution.—*Mains*, 13 miles. *Taps*, 600. *Meters*, 3. *Hydrants*, 137. *Consumption*, 400,000 galls. Financial.—*Cost*, $425,000. *Cap. Stock*, paid-up, $300,000. *Bonded Debt*, $300,000 at 6%. *Sinking Fund*, 2% after 10 years. *Expenses*: operating and taxes, $12,600; int., $18,000; total, $30,600. *Revenue*: consumers, $23,000; Ry., $4,100; city, $10,830; total, $13,930. **Management.**—Prest., B. A. May. Secy. and Treas., Jno. T. Martindale. Supt., F. H. Sheetz.

9. **ELKHART**, *Elkhart Co.* (Pop., 6,953—11,360.) Built in '84 by Elkhart Water Co., under 20-yrs. franchise. Maximum rates are fixed in franchise, which provides for purchase of works by city in '99 at cost and 10 % additional. Supply.—Surface well, pumping to tank, and direct. May 16,'91 it was reported that two 2,000,000 gall. pumps would be added at once. Distribution.—(In '88.) *Mains*, 12 miles. *Taps*, 700. *Meters*, 25. *Hydrants*, 125. *Consumption*, 450,000 galls. Financial.—(In '88.) *Cost*, $115,000. *Bonded Debt*, $100,000 at 6 %. *Expenses*, operating. $4,500; int., $6,000; total, $10,500. *Revenue*: consumers $5,500; city, $3,975; total, $11,475. **Management.**—Prest., E H. Bucklie. Secy., C. A. Sharpe.

10. **EVANSVILLE**, *Vanderburgh Co.* (Pop., 29,280—50,756.) Built in '70-3 by city. Supply.—Ohio River, pumping direct. Fiscal year closes Dec. 31.

Year.	Mains.	Taps.	Hyd'ts.	Cons'pt'n.	B'd'd debt.	Int. chg's.	Exp.	Rev.
'82	32	1,589	238	3,514,082	$300,000	$21,900	$36,068	$50,519
'86	31.4	2,012	255	4,000,000	300,000	21,900	24,179	33,857
'87	34	2,230	274	4,000,000	300,000	21,900	27,691	39,850
'88	34	2,460	280	4,200,000	400,000	20,000	21,357	39,576

Management.—(In '89.) Three trustees. Secy., N. A. Riggs. Supt., Jno. Fetz.

11. **FORT WAYNE**, *Allen Co.* (Pop., 26,880—35,393.) Built in '79-'90 by city. Supply.—Spy Run Creek, pumping direct; filtered. Fiscal year closes Dec. 31. A 6,000,000-gall. Gaskill triple comp. pump has been added: cost, $32,500.

INDIANA.

Yr.	Mains.	Taps.	Meters.	Hyd'ts.	Cons'p.'n.	Cost.	B'd'd d't.	Int chgs.	Exp.	Rev.
'81	26	765	....	212	685,081	$271,215	.....	....	$8,023	$10,002
'82	27.6	901	27	218	956,960	278,159	$270,000	......	9,026	15,110
'84	27.8	1,247	80	220	1,293,730	271,263	270,000	......	8,896	18,836
'86	28.5	1,460	103	220	1,297,639	280,010	26,000	$15,800	8,700	21,000
'87	30	1,607	133	252	1,415,314	295,702	260,000	15,800	9,000	25,000
'88	31	1,830	160	260	1,540,417	310,061	270,000	16,200	11,753	26,000
'89	32	2,140	212	267	....	321,567	270,000	.....	14,535	23,595
'90	35	2,713	250	317	....	....	270,000	.....	......	......

Management.—Committee of three, one elected every year. Chn., Chas. McCulloch. Secy., P. J. McDonald.

12. **FRANKFORT,** *Clinton Co.* (Pop., 2,803—5,919.) **History.**—Built in '85 by Frankfort Water-Works Co., under 20 yrs. franchise **Supply.**—Artesian well, pumping to tank and direct. **Pumping Machinery.**—Dy. cap., 3,000,000 galls.; two 1,500,000-gall. Knowles. **Distribution.**—*Mains.* 10 miles. *Hydrants,* 75. *Pressure:* ordinary, 65 lbs; fire, 65 to 100 lbs. **Financial.**—*Cap. Stock:* authorized. $100,000; all paid up. *Bonded Debt,* $60,000 at 6%. *Expenses:* operating, $3,000; int., $3,600; total, $6,600. *Revenue:* consumers, $6,500; city, $3,680; total, $10,180. *Hydrant Rental,* $60 and $40. **Management**—Prest. and Supt., Eli Marvin. Secy., J. B. Marvin. Supt., G. Norman Weaver, Newport, R. I.

13. **FRANKLIN,** *Johnson Co.* (Pop., 3,116—3,781.) **History.**—Built in '90-1 by Franklin Water, Light & Power Co.; put in operation Sept. 4. Engr., Jno. Hainsworth, Peru, Ill. **Supply.**—Driven wells, pumping to stand-pipe. **Pumping Machinery**—Dy. cap., 2,000,000 galls.; two 1,000,000 Gordon. **Stand-Pipe.**-Cap., 120,000 galls.: 12½ × 40 ft.; of steel; from Porter Boiler Mfg. Co., Chicago. **Distribution.**—*Mains.* c. i., 7 miles. *Services,* galv. i. Pipe and specials from Addyston Pipe & Steel Co.. Cincinnati, O *Taps,* 30). *Meters.* Worthington, 10. *Hydrants,* Beaumont, 61. *Valves,* Chapman. *Consumption.* 200,000 galls. *Pressure:* ordinary. 45 lbs.; fire, 130 lbs. **Financial.**-*Cap. Stock,* $160,000; *Bonded Debt,* $100,000 at 6%. *Revenue,* city, $3,000. **Management.**—Prest. and Treas., W. G. Rieve; Vice-Prest. and Supt., Jno. J. Hainsworth; both of Peru, Ill. Secy., M. L. Johnson.

14. **GOSHEN,** *Elkhart Co.* (Pop., 4,123—6,033.) Built in '81 by town. **Supply.**—Six artesian wells, pumping direct. Fiscal year closes June 1. **Distribution.**—*Mains,* 9 miles. *Taps,* 690. *Meters,* 0. *Hydrants,* 89. *Consumption,* 360,000 galls. **Financial.**—*Cost* $80,000. *Bonded Debt,* $15,000 at 5%. *Expenses:* operating, $3,500; int., $750; total. $4,250. *Revenue:* consumers, $3,600. **Management.**—Three trustees elected every three years. Chn., Wm. R. Ellis. Secy., I. D. Wolfe. Supt., Lew Wanner.

15. **GREENCASTLE,** *Putnam Co.* (Pop., 3,644—4,390.) Built in '86 by Greencastle Water-Works Co. **Supply.**—Six artesian wells, pumping direct. Fiscal year closes Dec. 31.

Year.	Mains.	Taps.	M'tr's.	Hy'd'ts.	C'ns'n.	Cost.	B'd'd D't.	Int. Cb'g's.	Exp.	Rev.
'88	8.5	280	2	85	......	$300,000	$150,000	$9,000	$3,800	$4,700
'89	9	306	2	85	200,000	300,000	150,000	9,000	5,900	7,500
'90	9	375	3	85	300,000	300,000	150,000	9,000	4,000	8,500

Financial.—*Cap. Stock,* $150,000. *Bonded Debt,* $150,000 at 6%. *Expenses:* operating, $4,000; int., $9,000; taxes, $1,500; total, $14,500. *Revenue,* total, $3,500. **Management.**—Pres., D. E. Williamson. Secy., Walter Wood, Philadelphia. Supt., Z. Potter. Gen. Supt., A. W. MacCullam, 400 Chestnut St., Philadelphia.

16. **GREENSBURGH,** *Decatur Co.* (Pop., 3,138—3,596.) Built in '89 by Greensburgh Water Co., under 30 yrs. franchise; city may purchase works at end of 10 years; does not con'rol rates. **Supply.**—Tube wells, pumping direct. Fiscal year closes April 25. **Distribution.**—*Mains,* 7.5 miles. *Taps,* 75. *Meters,* 0. *Hydrants,* 101. **Financial.**—*Cost,* $100,000. *Cap. Stock,* $100,000. *Bonded Debt,* $100,000. *Revenue* (for 6 mos.), $4,700. **Management.**—Secy., J. H. Christie. Supt., R. Thompson.

17. **HAMMOND,** *Lake Co.* (Pop., 699—5,428.) Built in '88 by town. **Supply.**—Well, pumping to stand-pipe. **Distribution.**—(In '88.) *Mains,* 5 miles. *Taps,* 300. *Meters,* 6. *Hydrants,* 70. *Consumption,* 200,000 galls. **Financial.**—(In '88.) *Cost,* $60,000. *Bonded Debt,* $36,000, at 6%. *Expenses:* operating, $3,000; int., $2,160; total, $5,160. *Revenue,* consumers, $3,000. **Management.**—Trustees. J. F. Krost, Chas. Larone, J. H. Kasper.

FOR FURTHER DESCRIPTIVE MATTER see Manual for 1888-9). CHANGES since 1890 are here given. POPULATIONS are for 1880 and 1890, respectively.

18. **HUNTINGTON,** *Huntington Co.* (Pop., 3,863—7,328.) History.—Owned by city. Built in '90-1 by Huntington Water-Works Co.; put in operation May 1. On completion of works Co. assigned its stock to city, which is now sole owner. Des. Engr., Jno. W. Hill, Cincinnati, O. **Supply.**—Five 6-in. driven wells for domestic use. Wabash River for fire protection, pumping direct. Wells are on bank of Wabash River, two miles from city, and a timber intake-pier in channel of stream is connected with pump-well by 16-in. intake-pipe for fire emergencies. **Pumping Machinery.**—Dy. cap., 3,000,000 galls.; two 1,500,000 Gordon comp., cond., dup., direct acting. *Boilers,* two return tubular. **Distribution.**—*Mains,* 14 to 4-in. c. i., 13.4 miles. *Services,* lead. Pipe and specials from Addyston Pipe & Steel Co., Cincinnati. *Taps,* 175. *Hydrants,* Bourbon, 125. *Valves,* Galvin. *Consumption,* 200,000 galls. *Pressure;* ordinary, 60 lbs.; fire, 90 lbs. **Financial.**—*Cost,* $110,000. *Bonded Debt,* $100,000 at 6%. Aug. 24, '91, city issued $12,000 additional bonds, proceeds to be used for 24 × 34 ft. well and 2 miles of pipe. **Management.**—Trustees, James D. Culp., H. L. Elvin, Sam'l Buchanan.

19. **INDIANAPOLIS,** *Marion Co.* (Pop., 75,056—105,436.) Owned by Indianapolis Water Co.; city has right to purchase works in '95; rates are limited. Built in '70 by co., and sold at sheriff's sale to present co. in '80 or '81. **Supply.**—White River, pumping direct; filtered. Fiscal year closes April 1. In '91 co. contracted to extend mains to supply West Indianapolis. Pumping station was built in '90, in which was placed a new 12,000,000-gall. Gaskill horizontal, comp., pump. A force main was also laid.

Year.	Mains.	Taps.	M't's.	Hy'ts.	Con'pt'n.	Cost.	B'd't d'bt.	Int. ch's.	Rev.
'82-3	43	1,100	25	600	4,000,000	$1,200,000	$500,000	$30,000	$70,940
'84-5	56	1,554	128	550	3,500,000	1,500,000	550,000	33,000	53,565
'86-7	68	1,679	135	608	4,241,000		550,000	33,000	37,147
'88-9	78	2,245	350	660	5,500,000	2,000,000	550 000	33,000	111,229
'89-90	78	2,807	210	730					
'90-1	84	2,903	226	747	7,500,000	2,200,000	690,000	43,060	149,338

Financial.—*Cap. Stock;* authorized, $500,000; all paid-up. *Expenses,* taxes, $5,980. *Revenue;* consumers, $111,420; city, $38,418. **Management.**—Prest., T. A. Morris. Secy., M. A. Morris. Treas., F. A. W. Davis.

20. **JEFFERSONVILLE,** *Clark Co.* (Pop., 9,307—10,666.) Built in '87-8 by Jefferson Water Supply Co., under 50 years' franchise. **Supply.**—Infiltration well near Ohio River, pumping to stand-pipe. Until '91 supply was direct from river. Engr. for well, Jno. W. Hill, Cincinnati. Fiscal year closes June 30. **Distribution.**—*Mains,* 12 miles. *Taps,* 348. *Hydrants,* 118. *Consumption,* 150,000 galls. **Financial.**—*Cap. Stock,* $200,000. *Bonded Debt,* $171,000 at 6%, due 1918-21. *Expenses;* operating, $4 500; int., $10,260; taxes, $1,200; total, $15,960. *Revenue;* consumers, $10,200; city, $6,680, total, $16,880. **Management.**—Prest., T. C. Woodbury; Secy., E. H. York; Treas., F. F. Talbot; all of Portland, Me. Supt., D. M. Alien.

21. **KENDALLVILLE,** *Noble Co.* (Pop., 2,287—2,960.) Built in '87 by city. **Supply.**—Wells, pumping direct. Fiscal year closes Dec. 31. **Distribution.**—*Mains,* 4.5 miles. *Taps,* 204. *Meters,* 0. *Hydrants,* 36. **Financial.**—*Cost,* $27,000. *Total Debt,* $16,000: floating, $3,000 at 6%; bonded, $13,000 at 6%. *Expenses;* operating, $2,200; int., $780; total, $2,980. *Revenue;* consumers, $1,600; city, $2,500; total, $4,100. **Management.**—Three trustees elected every three years. Chn., D. C. Welling. Secy., Wm. Osborn. Supt., Herman Kruezer.

22. **KOKOMO,** *Howard Co.* (Pop., 4,042—8,261.) Built in '87-8 by Kokomo Water Works Co.; controlled by American Work-Works & Guarantee Co., Ltd., Pittsburgh, Pa., under 15-yrs. franchise, which fixes maximum rates and provides for purchase of works by city in '97 or 1902, price to be determined by arbitration. **Supply.**—Wells, pumping direct. There are two large open wells and in the bottom of each are drilled three 6-in. artesian wells 100 ft. deep; additional wells are drilled in vicinity of pump-house and supply conducted to pump-wells. Supply may be increased whenever required by sinking more wells. **Pumping Machinery.**—Dy. cap., 3,000,000 galls.; two high-pressure Deane, each 15½ × 12 × 18. *Boilers,* two 75 H P. horizontal tubular. **Distribution.**—*Mains,* 12.3 miles. *Taps,* 436. *Meters,* 0. *Hydrants,* 169. *Consumption,* 300,000 galls. **Financial.**—Withheld. **Management.**—Prest., J. H. Purdy. Vice-President, W. S. Kuhn, Pittsburgh, Pa. Secy., A. E. Claney. Supt., C. H. Taylor.

23. **LA FAYETTE,** *Tippecanoe Co.* (Pop., 14,860—16,243.) Built in '75-6 by city. **Supply.**—Wabash River, pumping to reservoir and direct, and to stand-pipe for fire pressure in highest part of city. Fiscal year closes April 30. Aug., '91, wells were

INDIANA. 209

being sunk in the river by the National Water Supply Co., Cincinnati, O., and connected with the intake for future supply. **Distribution.**—*Mains*, 21 miles. *Taps*, 1,217. *Meters*, 11. *Hydrants*, 211. *Consumption*, 1,500,000 galls. **Financial.**—*First Cost*, $300,000, *Bonded Debt*, $300,000 at 8%, due in '95. *Expenses:* operating, $9,617; int., $24,000; total, $33,617. *Revenue:* consumers, $19,215; city, none. **Management.**— Three Trustees appointed every year by Mayor. Prest., T. J. Levering. Secy. and Supt., Ed. Cunningham.

24. **LA PORTE,** *La Porte Co.* (Pop., 6,195—7,126.) Built in '70 by city. **Supply.**—Clear Lake, pumping direct. **Distribution.**—*Mains*, 18 miles. *Taps*, 700. *Meters*, 0. *Hydrants*, 110. *Consumption*, 800,000 galls. **Financial.**—*Cost*, $150,000. *Bonded Debt*, $50,000 at 7%. *Expenses:* operating. $6,000; int., $3,500; total, $9,500. *Revenue*, consumers, $5,000. **Management.**—Trustees, Jno. Ball, J. B. Silliman, J. Shumm. Supt., Geo. Storey.

25. **LIGONIER,** *Noble Co.* (Pop., 2,010—2,195.) Built in '88-9 by village. **Supply.**—Well, pumping to tank and direct. Fiscal year closes April 21. **Distribution.** —*Mains*, 2.3 miles. *Taps*, 144. *Meters*, 1. *Hydrants*, 19. *Consumption*, 7,000 galls. **Financial.**—*Cost*, $14,600. *Total Debt*, $15,800; floating, $1,800 at 6%; bonded, $14,000 at 6%. *Expenses:* operating, $1,200; int., $950; total, $2,150. *Revenue:* consumers, $760; city, none. **Management.**—Prest., H. R. Cornell. Secy., E. B. Gerber. Supt., W. W. Noe.

26. **LOGANSPORT,** *Cass Co.* (Pop., 11,198—13,328.) Built in '75-6 by city. **Supply.**—Eel River, pumping direct. During summer of '90 a system of wells was put in, and as soon as these were connected with works supply was to be wholly derived from them. Fiscal year closes June 1.

Year.	Mains.	Taps.	Meters.	Hydrants.	Cost.	B'd'd debt.	Int. ch'g's.	Exp.	Rev
'82	9.5	400	6	89	$132,000	$172,000	$13,760	$3,000	$6,900
'86	12	531	0	92	175,000	170,000	13,600	3,000	10,000
'90	15	893	0	107	225,000	160,000	8,000	4,617	12,451

Financial.—*Bonded Debt:* $160,000, at 5%; due, '96-1906. *Expenses*, total, $12,617. *Revenue:* consumers, $12,451; city, $5,500; total, $17,951. **Management.**—Prest., Thos. J. Obenchain. Sec., Terrence McGovern. Supt., Patrick Walsh.

27. **MADISON,** *Jefferson Co.* (Pop., 8,945—8,936.) Built in '49 by city. **Supply.**— Infiltration well near Ohio River, pumping to reservoir. During '90-91 supply was changed to well 18 ft. in diam., 35 ft. deep, bricked. Engr., Jno. W. Hill, Cincinnati. **Distribution.**—(In '87.) *Mains*, 12 miles, *Taps*, 700. *Meters*, 0. *Hydrants*, 240, *Consumption*, 465,000 galls. **Financial.**—(In '86.) *Cost*, $135,000. *Bonded Debt*, $100,000, at 8%. *Expenses:* operating, $6,500; int., $8,000; total, $14,500. *Revenue*, consumers, $7,000. **Management.**—Clerk, Jno. C. Roberts. Supt., Chas. Godman.

28. **MARION,** *Grant Co.* (Pop., 3,182—8,769.) Built in '77 by city. **Supply.**— Well, pumping direct, Fiscal year closes April 30. **Distribution.**—*Mains*, 11 miles. *Taps*, 420. *Meters*, 0. *Hydrants*, 75. *Consumption*, 350,000 galls. **Financial** —*Cost*, $50,500. *Bonded Debt*, $20,000, at 6%. *Expenses:* operating, $2,500; int., $1,200; total $3,700. *Revenue*, consumers, $4,000. **Management.**—Three trustees, elected every. year. Chn. and Supt., Chas. W. Hamilton. Secy., H. F. Clunk.

29. **MICHIGAN CITY,** *La Porte Co.* (Pop., 7,366—10,776.) First supplied in '75, present system built in '88 by city. **Supply.**—Crystal Springs and Lake Michigan, pumping to tank and direct. New supply from Lake Michigan introduced in '90. Contrs., Dowagiac (Mich.) Construction Co., Sept., '91, reported that new pumping station was projected. Fiscal year closes in May.

Year.	Mains.	Taps.	Mtrs.	Hyd'ts.	Cons'n.	Cost.	B'd'd. d't.	Int.	Exp.	Rev.
'84	7	80	3	78	180,000	$10,000	$35,000	$2,450	$5,700	$1,638
'86	8	165	0	101	200,000	40,000			2,000	2,326
'88	12	260	3	102	800,000	60,000	15,000	900	4,000	4,400
'90	13	320	3	106	1,000,000	85,000	15,000	900	5,000	5,000

Management.—Board of three, one elected every year. Secy., Henry Fish. Supt., Wm. Blinks.

30. **MISHAWAKA,** *St. Joseph Co.* (Pop., 2,640—3,371.) **History.**—Built in '90-1 by Mishawaka Water-Works Co.; put in operation in July. Engr., W. B. Hosford. Fire protection system was built in '73 by Perkins Windmill & Axe Co. See 1889-90 Manual, p. 395. **Supply.**—Well, with emergency supply from river, pumping direct. **Pumping Machinery.**—Dy. cap., 3,000,000 galls.; two 1,500,000 comp. cond.

FOR FURTHER DESCRIPTIVE MATTER see Manual for 1889-90. CHANGES since 1890 are here given. POPULATIONS are for 1880 and 1890, respectively.

Distribution.—*Mains*, 12 to 4-in. c. i., 3.8 miles. *Hydrants*, 38. Financial.—Withheld. Management.—Prest., W. H. Dodge. Vice-Prest., H. G. Niles. Secy., Jno. J. Schindler. Consult. Engr., J. R. Maxwell.

31. **MOUNT VERNON,** *Posey Co.* (Pop., 2,730—4,705.) Built in '60 by Mount Vernon Water Co.; controlled by American Water-Works & Guarantee Co., Ltd., Pittsburg, Pa., under 25-yrs. franchise, which fixes maximum rates and provides for purchase of works by city in '97, price determined by arbitration. Supply.—Ohio River, pumping to stand-pipe. Fiscal year closes Dec. 31. Pumping Machinery.—Dy. cap., 2,000,000 galls.; two 1,000,000 Blake, one high-pressure 18½ × 10 × 12, and one comp., 12 × 18½ × 10 × 12. *Boilers*, two 80-HP., horizontal tubular. Distribution.—*Mains*, 8 miles. *Taps*, 385. *Meters*, 1. *Hydrants*, 75. *Consumption*, 200,000 galls. Financial.—Withheld. Management.—Prest., J. H. Purdy. Secy., A. B. Danagh. Supt., W. B. Dunbar.

32. **MUNCIE,** *Delaware Co.* (Pop., 5,219—11,345.) Built in '85 by Muncie Water-Works Co., controlled by American Water-Works & Guarantee Co., Ltd., Pittsburg, Pa., under 25-yrs. franchise, which fixes maximum rates and provides for purchase of works by city in 1935, price determined by arbitration. Fiscal year closes Dec. 31. Supply.—Four artesian wells, pumping direct. Water rises 10 ft. above surface of ground, and is connected by pipes with pump-well. Three 8-in. wells were drilled in '90, giving daily flow of 2,000,000 galls. A filter gallery along bank of river is also connected with pump-well, but only the artesian water is used. Pumping Machinery.—Dy. cap., 4,500,000 galls.; three Worthington, one 12½ × 18½ × 12 × 10, other two each 14 × 20 × 12 × 15. *Boilers*, 200 HP., three tubular. In '90, new engine house was built three miles from town in which an additional 2,000,000-gall. pump and one boiler were placed. A gas well was also sunk for fuel for boilers. Distribution.—*Mains*, 13 miles. *Taps*, 553. *Meters*, 1. *Hydrants*, 85. Financial.—Withheld. Management.—Prest., W. S. Kuhn, Pittsburg, Pa. Vice-Prest., J. H. Purdy. Secy., A. B. Danagh. Supt., Quince Walling.

33. **NEW ALBANY,** *Floyd Co.* (Pop., 16,423—21,059.) Built in '75 by New Albany Water-Works Co. Supply.—Ohio River, pumping to settling reservoir. In '90-1 new reservoir having cap. of 20,000,000 or 30,000,000 galls. was built. Distribution.—(In '88.) *Mains*, 21 miles. *Taps*, 1,200. *Meters*, 60. *Hydrants*, 165. *Consumption*, 1,000,000 galls. Financial.—(In '87.) *Cost*, $265,000. *Cap. Stock*, $150,000. *Bonded Debt*, $150,000 at 7%. *Expenses:* operating, $7,000; int., $10,500; total, $17,500. *Revenue:* consumers, $33,000; city, $13,975; total, $43,975. Management.—Prest., W. H. Dillingham. Secy., E. J. Brooks.

34. **NEW CARLISLE,** *St. Joseph Co.* (Pop., 530—607.) Built in '79 by town. Supply.—Well, pumping to tank, and direct for fire protection. Distribution.—(In '87.) *Mains*, 2 miles. *Taps*, 100. *Meters*, 0. *Hydrants*, 17. *Consumption*, 20,000 galls. Financial.—*Cost*, $10,400. *Bonded Debt*, $4,500 at 7%. *Expenses:* operating, $575; int., $315; total, $890. *Revenue*, consumers, $400. Management.—Three comrs.: M. S. Brunwidt, J. Ackley, S. L. Argalinet.

35. **NEW CASTLE,** *Henry Co.* (Pop., 2,299—2,697.) Built in '89 by city. Supply.—Well, pumping direct. Fiscal year closes March 1. Distribution.—*Mains*, 5 miles. *Taps*, 125. *Hydrants*, 45. Financial.—*Cost*, $27,000. *Debt*, $20,000. Management.—Supt., Geo. T. Melle.

36. **PERU,** *Miami Co.* (Pop., 5,280—7,028.) Built in '79 by city. Supply.—Wabash River, pumping to reservoir. Fiscal year closes May 1.

Year.	Mains.	Taps.	Meters.	Hyd'ts.	Cons'pt'n	Cost.	B'd'd d't.	Int.chgs.	Exp.	Rev.
'80	11.8	130	0	101	213,000	$111,069	......	$4.873	$3,845
.86	12.7	376	5	109	500,000	122,600	$110,000	$8,800	3,500	7,570
'88	12.7	466	2	111	631,674	127,503	100,000	8,000	4,182	8,529
'90	12.8	548	3	113	685,482	128,361	100,000	8,000	4,216	8,583

Financial.—*First Cost*, $109,550. *Bonded Debt*, $100,000 at 8%, due in '98. *Expenses*, total, $12,216. *Revenue:* consumers, $4,603; other purposes, $3,980; total, $8,583. Management.—Prest., Jno. K. Brown. Supt., Jno. Stewart.

37. **PLYMOUTH,** *Marshall Co.* (Pop., 2,570—2,723.) Built in '88 by village. Supply.—Four driven wells from which water flows to cistern, thence by direct pumping. Fiscal year closes May 10.

Year.	Mains.	Taps.	M't'rs.	H'd'ts.	Cons'pt'n.	Cost.	B'd'd debt.	Int.ch'g's.	Exp.	Rev.
'88-9	3.3	93	0	30	18,000	$17,800	$16,000	$800	$1,600	$475
'90-1	4	118	0	35	25,000	17,800	16,000	800	1,600	725

Management.—Prest., Jos. Elch. Supt., A. R. Underwood.

38. RICHMOND, *Wayne Co.* (Pop., 12,742—16,608.) Built in '86 by Richmond City Water-Works Co.; franchise fixes maximum rates and provides for purchase of works by city in '93, price determined by appraisers. **Supply.**—Springs, pumping to reservoir and direct; filtered. Fiscal year closes July 16.

Year.	Mains.	Taps.	Mtrs.	Hd'ts.	Cons'pt'n.	Cost.	B'd d d't.	Int.ch'g's.	Exp.	Rev.
'86	23	100	30	170	200,000	$325,000	$200,000	$12,000		$10,200*
'88	25	1,050	24	180	1,000,000	400,000	2 0,000	12,000	6,000	18,690
'90	25	1,400	50	200	1,000,000		200,000	12,000	10,500	27,000

*From city.

Financial.—*Cap. Stock,* $250,000; paid no div'd in 1890. *Total Debt,* $250,000: floating, $50,000 at 6%; bonded, $200,000 at 6%. *Expenses:* operating, $9,000; int., $15,000, taxes, $1,500; total, $25,500. *Revenue:* consumers, $15,000; city, $12,000; total, $27,000. **Management.**—Prest., D. K. Zeller, Secy., J. D. Craighead. Treas., W. P. Hutton, Supt., F. M. Curtis.

39. ROCKPORT, *Spencer Co.* (Pop., 2,382—2,314.) Built in '86 by Rockport Water-Works Co., under 50-yrs. franchise; city neither controls rates nor has right to buy works. Electric lighting system run in connection with water-works. **Supply.**—Ohio River, pumping direct and to stand-pipe. Fiscal year closes Jan. 1. **Distribution.**—*Mains,* 3.5 miles. *Taps,* 125. *Meters,* 4. *Hydrants,* 17. *Consumption,* 75,000 galls. **Financial.**—*Cap. Stock,* $25,000. *Total Debt,* $17,700: floating, $1,700; bonded $16,000 at 6%. *Expenses:* operating, $1,600; int., $960; taxes, $100; total, $2,660. *Revenue,* total, $1,700. **Management.**—Prest., S. W. Stocking. Treas., Secy. and Supt., A. W. Kennedy.

40. SALEM, *Washington Co.* (Pop., 1,615—1,975.) Built in '86 by town. **Supply.**—Spring, pumping to reservoir, with direct pumping for fire protection. **Distribution.**—*Mains,* 4 miles. *Taps,* 211. *Hydrants,* 19. *Consumption,* 60,000 galls. **Financial.**—*Cost,* $16,000. *Bonded Debt,* $15,000 at 5%. *Expenses:* operating, $700; int., $750; total, $1,450. **Management.**—Three trustees, one elected every year. Chn., T. F. Wilson. Secy., J. B. Berkey. Supt., E. Craycroft.

41. SEYMOUR, *Jackson Co.* (Pop., 4,250—5,337.) Built in '89 by Seymour Water Co.; city has right to purchase works at end of 10 yrs., or of any subsequent 5. yrs.; franchise fixes rates. **Supply.**—East fork of White River, pumping direct and to stand-pipe; filtered. Fiscal year closes Sept. 12. **Filter Crib.**—Width, 6 ft.; timber, filled with sand, one end abuts upon old mill; sluiceway is 18 in. sq., of 2-in. oak plank, passes through center of intake-well. Gates regulate supply, so it can be passed directly to intake-well without filtration. **Intake-Well.**—Old wheel pit of mill. **Distribution.**—*Mains,* 10 miles. *Taps,* 107. *Meters,* 5. *Hydrants,* 101. **Financial.**—*Cap. Stock,* $80,000. *Bonded Debt,* $80,000 at 6%. *Expenses:* operating, $1,775; int., $4,800; taxes, $850; total, $7,425. *Revenue:* consumers, $1,726; city, $4,535; total, $6,261. **Management.**—Prest., E. C. McMillan, Paris, Ill. Secy. and Supt., Judson A. Lamon. Treas., Benj. F. Price.

42. SHELBYVILLE, *Shelby Co.* (Pop., 3,745—5,451.)
INDIANA WATER AND LIGHT CO.

Organized in '90 to operate Water and Lighting plants at Shelbyville and Warsaw; controlled by Debenture Guarantee & Assurance Co., Chicago, Ill. Fiscal year closes Dec. 31. **Distribution.**—Given separately below and under Warsaw. **Financial,**—*Cost,* $225,000. *Cap. Stock,* paid-up, $300,000. *Bonded Debt,* $180,000 at 6% *Expenses:* operating and taxes, $5,300; int., $10,800; total, $16,100. *Revenue:* consumers, $8,362; Ry., $1,400; city, $12,862; total, $22,624. **Management.**—Prest., Chas. Whann. Secy., Jas. Greehey. Treas., Chas. B. Ludwig,172 Broadway, N. Y. City.

LOCAL PLANT.

Built in '86 by Shelbyville Water-Works Co., now operated by Indiana Water & Light Co., which see above. **Supply.**—Well, pumping to stand-pipe. **Distribution.**—*Mains,* 9½ miles. *Taps,* 480. *Meters,*1. *Hydrants,* 76. *Consumption,* 400,000 galls. Financial and **Management.**—See Indiana Water & Light Co., above. Supt., F. M. Nuckolls.

43. SOUTH BEND, *St. Joseph Co.* (Pop., 13,280—21,819.) Built in '73 by town. **Supply.**—Artesian wells, pumping from receiving reservoir to stand-pipe. Fiscal year closes Dec. 31. During summer of '90 new flume to water-wheels was built and old flume abandoned. March 1, '91, an additional 6,000,000-gall. Deane pump was completed. Six wells added since last report.

FOR FURTHER DESCRIPTIVE MATTER see Manual for 1889-90. CHANGES since 1890 are here given. POPULATIONS are for 1880 and 1890, respectively.

Year.	Mains.	Taps.	Meters.	Hyd'ts.	Cons'pt'n.	Cost.	B'd'd d'bt.	Int. ch's.	Exp.	Rev.
'80	16	500	....	200	....	....	....	......	$3,000	......
'82	17	600	30	300	800,000	$225,000	....	......	4,500	$6,000
'84	18	700	75	275	2,500,000	225,000	$200,000	$10,000	6,000	8,000
'86	19	1,200	75	275	1,500,000	250,000	230,000	11,500	5,000	8,000
'88	25	1,435	85	305	1,750,000	250,000	178,000	9,215	6,000	12,000
'90	29	1,460	130	305	1,873,653	....	....	....	......	......

Management.—Trustees, elected every three years. Chn., C. H. Pavey. Secy., Jno. E. Callahan. Supt., Ira S. Schrop.

44. **TERRE HAUTE,** *Vigo Co.* (Pop., 26,042—30,217.) Built in '73 by Terre Haute Water-Works Co. **Supply.**—Wabash River, pumping direct; filtered. Fiscal year closes Dec. 31. Following additions have been made since '88: Pump and boiler-house has been built, in which a 6,000,000-gall. Gaskill comp., vertical engine and three 60 × 16-ft. horizontal tubular boilers were placed. National filtering plant (see note below) and a 1,000,000-gall. storage reservoir have been added.

Year.	Mains.	Taps.	Meters.	Hyd'ts.	Cons'p'tn.	Cost.	B'd'd debt.	Int. ch's.	Exp.	Rev.
'80	17.8	453	......	274	....	....	....	......	$6,500	......
'82	20.1	590	3	339	900,000	$301,169	....	......	8,989	$30,123
'84	22	620	6	339	1,000,000	330,000	$100,000	$8,000	7,000	32,000
'86	26	700	3	416	1,200,000	345,000	100,000	8,000	8,000	34,000
'88	30	900	3	479	2,000,000	380,000	100,000	8,000	5,000	53,000
'90	38	1,225	40	573	2,500,000	800,000	100,000	8,000	15,000	57,000

Financial.—*First Cost,* $300,951. *Cap. Stock* (in '88), $220,000. *Bonded Debt,* $100,000 at 8%. *Expenses:* operating, $15,000; int., $8,000; total, $23,000. *Revenue:* consumers, $35,000; city, $22,000; total, $57,000. **Management.**—Prest., J. C. Kolsem. Secy., H. K. Sease. Treas., N. W. Harris. Supt., S. S. Williamson.

NOTE.—For illustrated description of filtering plant, see ENGINEERING NEWS, vol. XXV., p. 127.

45. **UNION CITY,** *Randolph Co.* (Pop., 2,478—2,681.) Built in '73 by town. **Supply.**—Surface wells, pumping direct. Late in '90 second well was completed, which furnishes an additional daily supply of 2,500,000 galls. Fiscal year closes June 1. **Distribution.**—*Mains,* 7 miles. *Taps,* 300. *Meters,* 31. *Hydrants,* 29. *Consumption,* 156,000 galls. **Financial.**—*Cost* (in '88), $60,000. *Expenses,* $3,372. *Revenue,* $2,641. **Management.**—Three trustees elected every three years. Chn., J. M. Shank. Secy., J. L. Heck. Supt., E. S. Pettis. Appointed by trustees.

46. **VALPARAISO,** *Porter Co.* (Pop., 4,461—5,090.) Built in '85-6 by Valparaiso City Water Supply Co. **Supply.**—Flint Lakes, pumping direct. **Distribution.**—(In '88.) *Mains,* 12.5 miles. *Taps,* 304. *Meters,* 39. *Hydrants,* 55. *Consumption,* 500,000 galls. **Financial.**—*Cost,* $100,000. *Cap. Stock,* $15,000. *Bonded Debt,* $55,000 at 6%. *Expenses:* operating, $5,500; int., $3,300; total, $8,800. *Revenue:* consumers, $8,000; city, $4,250; total, $12,250. **Management.**—Prest., S. V. Saleno; Secy. and Treas., H. B. Smith; Supt., Don A. Salver; all of Bay City, Mich.

47. **VINCENNES,** *Knox Co.* (Pop., 7,680—8,853.) Built in '85 by Vincennes Water Supply Co., under 20-yrs. franchise; city has right to purchase works at end of 10 yrs.; rates as fixed in franchise. **Supply.**—Wabash River, pumping to standpipe. Fiscal year closes Dec. 31.

Year.	Mains.	Taps.	M'trs.	Hd'ts.	Cons'pt'n.	Cost.	B'd'd d'bt.	Int. ch'g's.	Exp.	Rev.
'87	14	356	8	150	......	$220,000	$175,000	$10,500	$3,850	$14,640
'89	14	375	9	150	230,000	375,000	175,000	10,500	6,180	10,000*
'90	15	400	12	153	500,000	375,000	175,000	10,500	4,800	12,000*

*From city.

Financial.—*Cap. Stock,* authorized, $200,000. *Bonded Debt,* $175,000 at 6%. *Expenses,* total, $15,300. **Management.**—Prest., W. G. Hopper. Secy. and Treas., Walter Wood. Supt., L. J. Weisentroger. Gen. Supt., A. W. MacCullum, 400 Chestnut St., Philadelphia, Pa.

48. **WABASH,** *Wabash Co.* (Pop., 3,800—5,105.) Built in '86-7 by Wabash Water Co., under 20-yrs. franchise; city may buy works any time after five yrs.; maximum rates fixed in franchise. **Supply.**—Treaty Creek, springs and driven well, pumping to stand-pipe from impounding reservoir. Fiscal year closes July 1. **Distribution.**—*Mains,* 13 miles. *Taps,* 400. *Meters,* 1. *Hydrants,* 134. *Consumption,* 150,000 galls. **Financial.**—*Cap. Stock,* authorized, $100,000; all paid up; paid no div'd in 1590. *Bonded Debt,* $130,000 at 6%. *Expenses:* operating, $2,500; int., $7,800; taxes, $1,260; total, $11,560. *Revenue:* consumers, $5,500; city, $5,000; total, $10,500. **Management.**—Prest., Chas. C. Pomeroy. Secy. and Treas., C. E. Kimball, 45 Wall St., N. Y. City.

INDIANA.

49. WARSAW, *Kosciusko Co.* (Pop., 3,123—3,574). Built in '86 by Warsaw Water-Works Co.; rebuilt in '90; now controlled by Debenture Guarantee & Assurance Co., Chicago, Ill. Engr., Geo. H. Pierson, N. Y. City. **Supply.**—Artesian well pumping to stand-pipe and direct. Well is 8 ins. in diam., 200 ft. deep. Fiscal year closes Dec. 31. **Distribution.**—*Mains*, 6 miles. *Taps*, 320. *Hydrants*, 54. *Consumption*, 300,000 galls. **Financial and Management.**—See Indiana Water & Light Co., Shelbyville. Supt., Jno. G. Sutton.

50. WASHINGTON, *Daviess Co.* (Pop., 4,323—6,064.) Built in '88 by Washington Water-Works Co., controlled by Moffett, Hodgkins & Clarke, Syracuse, N. Y. **Supply.**—White River, pumping to stand-pipe and direct. **Distribution.**—(In '89.) *Mains*, 7.5 miles. *Taps*, 96. *Meters*, 10. *Hydrants*, 90. *Consumption*, 111,-325 galls. **Financial.**—(In '88.) *Cost*, $130,000. *Bonded Debt*, $115,000 at 6%. *Expenses:* operating, $2,000; int., $6,000; total, $8,900. **Management.**—(In '80.) Prest., H. C. Hodgkins, Syracuse, N. Y. Supt., R. B. Boyce.

WOLCOTTVILLE, *Lagrange Co.* (Pop., est., 500.) Built in '89 by Wolcottville Water-Works Co. for fire protection.

Water-Works Projected with Fair Prospects of Construction.

CANNELTON, *Perry Co.* (Pop., 1,834—1,901.) Reported Oct., '91, that 25-yrs. franchise had been granted to American Pipe Co., Philadelphia, and that construction would be started at once.

BLOOMINGTON, *Monroe Co.* (Pop., 2,756—4,018.) Franchise was granted March 3, '91, to Jesse W. Starr. Construction had not begun July 8, '91.

BROOKVILLE, *Franklin Co.* (Pop., 1,813—2,028.) About Sept. 17, '91, town let contract for works to T. A. Hardman, Olney, Ill.; to be completed by May 1, '92. Laidlaw & Dunn Co., Cincinnati, will furnish pumping machinery.

DANVILLE, *Hendricks Co.* (Pop., 1,598—1,569.) In Sept., '91, J. B. Homon was chn. of com. on proposed works. For earlier project see Manual for 1889-90, p. 402.

DELPHI, *Carroll Co.* (Pop., 2,040—1,923.) City expected to begin construction of works July 10, '91, and to have them in operation by Dec. Engr. and contr., Geo. C. Morgan, Chicago. Supply, springs, by gravity to 28,000-gall. tank. Mains, 3½ miles. Hydrants, 40. Pressure, 80 lbs. Cost, $38,000.

NAPPANEE, *Elkhart Co.* (Pop., 517—1,493.) Bids for building town works were to be received until Sept. 21, '91. Supply, tubular wells, pumping to stand-pipe. Town Clerk, Perry A. Early.

NOBLESVILLE, *Hamilton Co.* (Pop., 4,550—5,274.) Noblesville Water & Light Co. let contracts for construction of water-works July 24, '91, under 25-yrs. franchise; city reserving right to purchase works at end of 5 years and at any time thereafter, by giving 6 months notice. Construction to begin Aug. 15 and be completed Nov. 1. Supply, wells, pumping to reservoir. Two 1,500,000-gall. Deane pumps. Mains, 12 to 4-in. c. i., 6.4 miles. Hydrants, 75. Engr., J. D. Cook, 47 Produce Exchange, Toledo, O. Prest., Jas. R. Christian.

PRINCETON, *Gibson Co.* (Pop., 2,566—3,076.) Oct, 5, '91 it was reported that two companies had made propositions to City Council. Address, C. W. Benton Clerk.

Water-Works Projects Reported in '90 and '91 but Now Indefinite.

Lebanon, Boone Co. *Martinsville*, Morgan Co. *Portland*, Jay Co. *Rushville*, Rush Co.

MICHIGAN.

Water-Works Completed or in Process of Construction.

1. ADRIAN, *Lenawee Co.* (Pop., 7,819—8,756.) Built in '83-4 by Adrian Water-Works Co., under 30-yrs. franchise which does not regulate rates but provides for purchase of works by city. **Supply.**—Surface and artesian wells, by direct pumping.

For FURTHER DESCRIPTIVE MATTER see Manual for 1889-90. CHANGES since 1890 are here given. POPULATIONS are for 1880 and 1890, respectively.

Jan. 6, '91, works were sold at auction to R. A. Watts, representing C. H. Venner & Co., N. Y., for $80,000. Distribution.—(In '88.) *Mains*, 15 miles. *Taps*, 400. *Meters*. 5. *Hydrants*, 120. *Consumption*, 450,000 galls. Financial,—(In '88.) *Bonded debt*, $145,000 at 7%. *Expenses:* operating, $4,500; int., $10,150; total, $14,650. *Revenue:* consumers, $4,000; city, 10,000; total, $14,000. Management.—Address W. A. Underwood, 33 Wall St., N. Y, City, or M. G. Gippert.

2. **ALBION**, *Calhoun Co.*, (Pop., 2,716—3,763.) Has sewers and electric lights. History.—Built in '89 by city; put in operation Jan. 1, '90. Des. Engr., Geo. C. Morgan, Chicago, Ill. Contr., T. C. Burks, Jackson. Supply.—Two 6-in wells, pumping to stand-pipe. Pumping Machinery.—Dy. cap., 3,000,000 galls., two 1,500,000 Hughes. *Boilers* from Lansing Engine & Iron Works. Stand-Pipe.—Cap., not given; 12x128 ft.; from Lansing Engine & Iron Works. Distribution.—*Mains*, c. 1, 9 miles; pipe and special, from Detroit Foundry. *Services*, leads. *Taps*, 300. *Hydrants*, Mathews, 68. *Valves*, Eddy. *Pressure*, 75 lbs. Financial,—*Cost*, $50,000. *Debt*, $50,- 000. Int., 6%. Ann. op. exp., $1,700. Ann. rev.: consumers, $1,500; no allowance for public uses of water. Management.—City Council, Supt., R. J. Frost.

3. **ALLEGAN**, *Allegan Co.*, (Pop., 2,305—2,669.) Built in '71 by village. Supply.— Well, on bank of Kalamazoo river, by direct pumping. Fiscal year closes March 1. Distribution.—*Mains*, 6 miles. *Taps*, 325. *Meters*, 0. *Hydrants*, 44. *Consumption*, 350,000 galls. Financial,—*Cost*, $45,000. *No debt*, *Expenses*, $1,500. *Revenue*, $1,950. Management.—Chn., Phil Padghan. Supt., Morris Dyer.

4. **ALMA**, *Gratiot Co.* (Pop., 925—1,655). Built in '82 by village. Supply, Pine River, pumping direct. Distribution.—(In '88.) *Mains*, 4 miles. *Taps*, 50. *Meters*, 0. *Hydrants*, 35. Financial,—(In '88.) *Cost*, $15,000. *Bonded Debt*, $8,000 at 6%. *Expenses:* operating, $1,000; int. $480; total, $1,480. Management.—Supt., P. M. Smith.

5. **ALPENA**, *Alpena Co.* (Pop., 6,153—11,283.) Built in '79 by Alpena City Water Co.; city may buy works in '94, price determined by disinterested non-resident persons according to law of state; rates are to be on average of Detroit, Port Huron and Bay City. Supply.—Thunder Bay and River, pumping direct. Fiscal year closes Oct. 1. "Co. has purchased a water-power on stream 1¼ miles distant, and proposes to utilize this power by transmission, electric or otherwise, in running the pumps. *Cost*, $29,000. This power is also rented to an electric lighting co. The issue of $60,000 in bonds mentioned below is purchase price and improvement fund of this investment."

Year.	Mains.	Taps.	Meters.	Hydrants.	Cost.	B'd'd d't.	Int. ch'g's.	Exp.	Rev.
'85	12.5	771	0	106	$100,000	$57,500	$4,025	$3,000	$10,965
'88	17.5	1,261	0	113	200,000	47,500	3,325	5,000	16,250
'90	22.5	1,500	0	133	200,000	67,500	3,578	7,521	17,089

Financial,—*First Cost*, $50,000. *Cap Stock*, $75,000. *Bonded Debt*, $67,500 at 7%. *Expenses:* operating, $7,290; int., $3,598; taxes, $231; total, $11,099, *Revenue:* consumers, $9,363; city, $7,726; total, $17,089. Management.—Prest. and Supt., Wm. H. Johnson, Secy., F. W. Gilchirst. Treas., A. W. Comstock.

6. **ANN ARBOR**, *Washtenaw Co.* (Pop., 8,061—9,431.) Built in '85-6 by Ann Arbor Water Co., under 30-yrs franchise; rates must be reasonable and works may be purchased by city at any time by paying their full value. Supply.—Springs and surface water, pumping to reservoir. Supply was increased in '90 by three additional wells. Fiscal year closes Dec. 31.

Year.	Mains.	Taps.	M't'rs.	H'd'ts.	C'ns'n.	Cost.	B'd'd d.	Int.ch'g's.	Exp.	Rev.
'87	20.5	900	3	105	350,000	$215,000	$150,000	$9,000	$3,345	$14,745
'88	20.5	1,000		112	500,000	215,000	150,000	9,000	3,334	14,798
'89	24.6	1,136	2	112	441,282	227,213	150,000	9,000	4,491	16,254
'90	24.6	1,240	2	117	521,089	233,129	150,000	9,000	5,623	17,789

Financial,—*Cap. Stock:* authorized, $75,000; all paid up; paid 4% div'd in 1889, 5% in 1890. Pref'd: authorized, $100,000; paid up, $93,450; 5% div'd. *Total Debt*, $173,685; floating, $13,685 at 6%; bonded, $150,000 at 6%. *Expenses:* operating, $4,608; int., $10,049; taxes, $1,015; total, $15,672. *Revenue:* consumers, $12,674; city, $5,115; total, $17,789; Management.—Prest. and Supt., Alex. W. Hamilton. Secy., Chas. E. Hiscock. Treas., A. K. Hale, Adams, N. Y.

7. **AU SABLE**, *Iosco Co.* (Pop., 1,328—4,328.) Built in '85 by town. Supply.— Au Sable River, pumping direct. Fiscal year closes March 5. Distribution.— *Mains*, 6 miles. *Taps*, 225. *Hydrants*, 70. Financial,—*Cost*, $35,000. *Floating Debt*, $7,000. *Expenses*, $1,800. *Revenue*, $1,500. Management.—Supt., J. S. Duncan.

MICHIGAN. 215

8. BATTLE CREEK, *Calhoun Co.* (Pop., 7,063—13,197.) Built in '87 by city. Supply.—Goguac Lake, pumping to stand-pipe. Fiscal year closes Dec. 31.

Year.	Mains.	Taps.	Meters.	H'd'ts.	Cons'pt'n.	Cost.	B'd'd d't.	Int. ch's.	Exp.	Rev.
'87	12.3	193	10	110	160,000	$127,000	$100,000	$1,500	$5,000	$5,000
'88	15	528	22	135	230,000	138,315	100,000	4,530	4,168	5,939
'89	1.7	779	28	159	321,000	151,116	100,000	4,500	5,047	8,423
'90	19.4	921	57	179	410,000	163,906	100,000	4,500	5,730	11,191

Financial.—*Bonded Debt,* $100,000 at 4½%, due in annual payments of $10,000 after '96. *Expenses,* total, $10,230. *Revenue:* consumers, $11,191; city, none. Net revenue is used for extensions, int. on bonds being paid by taxation. **Management.**—Board of Public Works, of five, one elected every year. Prest., E. C. Nichols. Secy., W. C. Gage. Supt., W. W. Bridgen.

9. BAY CITY, *Bay Co.* (Pop., 20,693—27,839). Built in '72 by city. Supply.—Saginaw Bay, arm of Lake Huron, by gravity to pumps, thence by direct pumping. Fiscal year closes Dec. 31. Changes or additions since 1889-90 Manual.—Nov. 4, '90, town voted to issue bonds to the amount of $50,000, May 1, '91, for new 20-in. main.

Year.	Mains.	Taps.	Meters.	Hyd'ts.	Cons'pt'n.	Cost.	B'd'd d't.	Int. ch'g's.	Exp.	Rev.
'72	3.75			33		$141,408	$177,030	$14,160	$612	
'73	8.5	139		81	390,350	197,326	227,000	18,160	7,000	$4,500
'75	18	441		127	1,141,800	349,801	377,000	30,160	9,500	9,800
'80	19.25	819		138	2,266,000	367,113	377,000	30,160	10,800	13,900
'81	19.25	942	6	138	2,587,350	367,113	377,000	30,160	11,000	15,350
'82	20	1,046	28	140	2,125,048	572,739	377,000	30,160	10,509	17,900
'83	21	1,158	37	150	2,120,391	379,074	377,000	30,160	11,700	18,750
'84	22.25	1,236	44	160	2,120,390	386,732	377,000	30,160	14,600	20,000
'85	24.25	1,294	151	160	2,293,069	390,012	377,000	30,160	13,100	20,200
'86	22.2	1,245	151	160	2,136,334	426,777	377,000	30,160	13,080	19,825
'87	25.5	1,332	190	190	2,266,288	130,363	367,000	30,080	11,740	19,380
'88	27.6	1,389	215	215	2,385,219	440,078	357,500	29,280	10,249	21,332
'89	29.75	1,546	249	229	2,589,597	447,584	347,500	27.800	9,780	22,220
'90	31.5	1,565	267	289	2,708,983	466,226	307,000	27,960	11,321	23,912

Financial.—*Total Debt* (in '89), $361,000: floating, $13,500 at 4%; bonded, $347,500 at 8%, due '90 to 1913. *No Sinking Fund. Expenses,* total, $39,231. *Revenue:* consumers, $23,942; city none. Int. on bonds is paid by taxation, net revenue being used for extensions. **Management.**—Board of seven members, one appointed every year by common council, on nomination of Mayor. Chn., Andrew Walton. Secy. and Supt., E. L. Dunbar, appointed by Board.

10. BELLAIRE, *Antrim Co.* (Pop., 65—est., 500.) Built in '88 by Robt. Rickardi; city does not control rates. Supply.—Cedar River, pumping to reservoir. Fiscal year closes Dec. 31. Distribution.—*Mains,* 2 miles. *Taps.* 80. *Meters,* 2. *Hydrants,* 7. Financial.—*Cost,* $5,000. *Expenses,* $190. **Management.**—Owner, Robt. Rickardi. Secy. and Supt., H. C. Adamson.

11. BENTON HARBOR, *Berrien Co.* (Pop., 1,230—3,692.) **History.**—Built in '91 by Benton Water Supply Co., under 20-yrs franchise; to be in operation Aug. 1. Engr., Peter English, Supply.—Cook well system, pumping to stand-pipe. Pumping Machinery.—Dy. cap., 2,500,000 galls.; Hughes comp. dup. *Boilers,* from C. C. Warner Machine Co., Detroit. Stand-Pipe.—Cap., 84,600 galls.; 12 × 100 ft.; from Porter Boiler Mfg. Co., Chicago. Distribution—*Mains,* c. 1., 6.5 miles. *Services,* lead. Pipe and specials from Ohio Pipe Co., Columbus. *Hydrants,* Bourbon, 75. *Valves,* Bourbon. *Pressure,* 50 to 75 lbs. Financial.—*Cost,* $65,000. *Cap. Stock,* $100,000. *Bonded Debt,* $40,000. *Hydrant rental,* $35. **Management.**—Prest. and Supt., Peter English. Secy. and Treas., Frank English.

12. BESSEMER, *Gogebic Co.* (Pop., in '90, 2,566.) Now owned by city. Built in '87-8 by Gogebic Water & Gas Supply Co. Supply.—Springs and wells, by direct pumping from collecting reservoir. Distribution.—*Mains,* 2 miles. *Hydrants,* 24. Financial.—*Cost,* $50,000. *Bonded Debt* $40,000 at 6%. **Management.**—City Council.

13. BIG RAPIDS, *Mecosta Co,* (Pop., 3,552—5,303.) Built in '71 by town. Supply.—Driven wells pumping direct from collecting reservoir. Distribution.—*Mains,* 7 miles. *Taps,* 400. *Meters,* 0. *Hydrants,* 73. *Cosumption,* 800,000 galls. Financial.—Withheld. **Management.**—Mayor and aldermen. Supt., Willard Jefts.

FOR FURTHER DESCRIPTIVE MATTER see Manual for 1889-90. CHANGES since 1890 are here given. POPULATIONS are for 1880 and 1890, respectively.

216   MANUAL OF AMERICAN WATER-WORKS.

14. BIRMINGHAM, *Oakland Co.* (Pop., 733-899.) History.—Built in '90 by village; put in operation Sept. 1. Engr., Geo. Blakeslee. Supply.—Flowing wells, pumping direct. Pumping Machinery.—Dy. cap., 1,500,000 galls.; Hughes. *Boilers*, from Lansing Iron & Engine Wks. Distribution.—*Mains,* 4 miles. Pipe and specials from Lake Shore Foundry, Cleveland, O. *Taps,* 70. *Hydrants,* Galvin, 40. *Consumption,* 75,000 galls. *Pressure:* ordinary, 70 lbs.; fire, 120 lbs. Financial.—*Cost,* $20,000. *Bonded Debt,* $17,000: $15,000 at 5%; $2,000 at 5½%. *Expenses:* operating, $1,200; int., $860; total, $2,060. *Revenue,* consumers, $500. Management.—Prest., Lyman B. Peabody. Secy., Geo. H. Mitchell. Supt., Geo. Blakeslee.

15. CADILLAC, *Wexford Co.* (Pop., 2,213—4,161.) Built in '78 by individual or Co. Supply.—Clam Lake, pumping direct. Reported in '91 that works were to be improved by Western Construction Co., Chicago, which held 30-yrs. franchise. Distribution—(In '88.) *Mains*, 5.5 miles. *Taps*, 500. *Meters,* 0. *Hydrants,* 45. *Consumption,* 1,000,000 galls. Financial.—*Cost,* $35,000. *No Debt. Revenue:* consumers $9,500; city, $3,250; total, $12,750. Management.—Owner (June 9, '91), Holden N. Greene.

93. CALUMET, *Houghton Co.* (Pop., in '90, 1,159.) Supply from Red Jacket.

Capac, *St. Clair Co.* (Pop., est, 500.) Built in '87-8 by village, for fire protection only.

16. CARO, *Tuscalo Co.* (Pop., 1,282—1,701.) Built in '88 by Caro Water Co.; franchise prescribes rates and provides for purchase of works by city in '98. Supply.—Springs, pumping to stand-pipe and direct. Fiscal year closes Sept. 1. Distribution.—*Mains,* 5.5 miles. *Taps,* 180. *Meters,* 15. *Hydrants,* 28. *Consumption,* 29,102 galls. Financial.—*Cost* (in '88), $40,000. *Cap. Stock:* authorized, $25,000; all paid up. [*Total Debt,* $25,000, at 6%, due in 30 yrs. *Expenses:* int. $1,540; taxes, $154. *Revenue,* city, $1,500. Management.—Prest., Geo. N. Phelps, Mt. Morris, N. Y. Secy. and Supt., C. O. Thomas. Treas., A. E. Carrier, Detroit.

17. CARSON CITY, *Montcalm Co.* (Pop., 709—921.) Built in '88 by village. Supply.—Fish Creek, pumping to tank, and direct in case of fire. Fiscal year closes April 22. Distribution.—*Mains,* 1 mile. *Taps,* 8. *Hydrants,* 7. *Consumption,* 12,-800 galls. Financial.—*Cost,* $10,000. *Debt,* $13,000. *Expenses,* $300. Management.—Supt., Byron Abbott.

18. CEDAR SPRINGS, *Kent Co.* (Pop., 1,141—1,035.) Built in '89 by village. Supply.—Well and creek, pumping direct. Fiscal year closes Feb. 28. Distribution.—*Mains,* 1 mile. *Taps,* 10. *Hydrants,* 10. Financial.—*Bonded Debt,* $11,000: $5,000 at 4%; $6,000 at 6%. *Expenses,* $650. *Revenue,* $745. Management.—Chn., S. H. Chapman. Secy., C. C. Shay.

19. CHARLOTTE, *Eaton Co.* (Pop., 2,910—3,867.) Built in '86-7 by town. Supply.—Wells, pumping direct. Fiscal year closes first Monday in March.

Year.	Mains.	Taps.	Meters.	Hyd'ts.	Cons'pt'n.	Cost.	B'd'd debt.	Int.	ch'g's.	Exp.	Rev.
'88	9.3	438	0	61	266,666	$45,000	$40,000	$1,600	$2,500	$3,000
'90-1	9.5	506	1	62	233,333	45,000	40,000	1,600	2,600	3,300

Financial.—*Bonded Debt,* $40,000 at 4%. *Expenses,* total, $4,200. *Revenue:* consumers, $3,300; city, $1,240; total, $4,540. Management.—Secy., L. C. Holbrook. Supt., E. Shepherd.

20. CHASE, *Lake Co.* (Pop., 273—338.) Built in '83 by village. Supply.—Pere Marquette River, pumping direct. Fiscal year closes March 31. Distribution.—(In '88.) *Mains,* 1 mile. *Taps,* 20. *Meters,* 0. *Hydrants,* 4. Financial.—(In '83.) *Cost,* $11,000. *Debt,* $1,500 at %. *Expenses:* operating, $500; int., $105; total, $605. Management.—(In '88.) Comrs., appointed by council. Prest., J. H. Kneville. Secy., J. C. Brace.

21. CHEBOYGAN, *Cheboygan Co.* (Pop., 2,629—6,235.) Built in '82 by village. Supply.—Two artesian wells, pumping direct. Fiscal year closes first Monday in March. Distribution.—*Mains,* 5 miles. *Taps,* 212. *Meters,* 24. *Hydrants,* 34. *Consumption,* 200,000 galls. Financial.—*Cost,* $20,000. *Bonded Debt,* $25,000 at 5½%. *Expenses:* operating, $2,700; int., $1,375; total, $4,075. *Revenue:* consumers, $2,000; city, $1,900; total, $3,900. Management.—Chn., F. E Rich. Secy., A. J. Finn. Supt., Chas. Adams.

22. CLARE, *Clare Co.* (Pop., 502—1,174) Built in '86 by town. Supply.—Well, pumping direct. Fiscal year closes March 31. Distribution.—*Mains,* 2.

miles. *Taps,* 100. *Hydrants,* 15. Financial.—*Cost,* unk'n. *Expenses,* $1,300. *Revenue,* $740. Management. — Comrs., appointed every three years by town council. Chn., Henry Hart. Secy., A. J. Doherty.

23. **COLDWATER,** *Branch Co.* (Pop., 4,681—5,247.) History.—Built in '90-1 by city; put in operation March 1, '91. Engr., J. D. Cook, Toledo, O. Supply.—Wells, pumping direct. Pumping Machinery.—Dy. cap., 3,000,000 galls.; two 1,500,000 Hughes comp. *Boilers,* from E. Mansell. Distribution.—*Mains,* 16 to 4-in c. i., 13 miles. *Services,* lead and i. Pipe and specials from Lake Shore Foundry Co., Cleveland, O.; trenching and pipe-laying done by Snyder & Williams, Dayton, O. *Taps,* 125. *Hydrants,* Mathews, 100. *Valves.* Rensselaer. *Pressure:* ordinary, 45 to 55 lbs.; fire, 100 to 120 lbs. Financial.—*Cost,* $70,000. *Bonded Debt,* $70,000 at 5%. Management.—Trustees, O. R. Root, L, O. Rose, L. C. Dillingham.

24. **CONSTANTINE,** *St. Joseph Co.* (Pop., 1,405—1,316.) Built in '78 by town. Supply.—Fawn River, pumping direct. Fiscal year closes March 3. Distribution. —*Mains,* 5.3 miles. *Taps,* 188. *Hydrants,* 51. Financial.—*Cost,* $22,800. *Bonded Debt,* $12,500: $2,500 at 6%, $10,000 at 5%. *Expenses:* operating, $320; int., $650; total, $870. *Revenue:* consumers, $752; city, none. Management.—Common Council, elected every year. Prest., Chas. H. Barry. Secy., Thomas Harrison.

25. **CRYSTAL FALLS,** *Iron Co.* (Pop., in '90, 1,176.) Built in '89-90 by village. Supply.—Collecting reservoir and small stream, pumping direct. Distribution. —*Mains,* 1 mile. *Taps.* 10. *Hydrants,* 15, *Consumption.* 150,000 galls. Financial.—*Cost,* $20,000. *Bonded Debt,* $18,000 at 6%. Management.—Clerk, A. Lustfield. Supt., F. Klyersteuber.

26. **DETROIT,** *Wayne Co.* (Pop., 116,340—205,876.) Owned by city. Built in '27 by Rufus Wells. Supply.—Detroit River, pumping to stand-pipe. Fiscal year closes Dec. 31.

Year.	Mains.	Taps.	M'tr's.	H'd'ts.	Cons'p'n.	Cost.	B'd'd d't.Int.ch'g'r.	Exp.	Rev.
'81	220.5	14,317	....	780	17,926,377	$3,082,708	$1,652,000 $106,000	$52,769	$241,844
'82	232.1	15,781	....	872	17,261,440	3,200,967	1,651,000 106,260	49,000	261,726
'83	241.4	16,978	....	918	20,217,334	3,315,989	1,555,000 114,323	54,451	280,049
'84	252	18,324	....	975	23,253,044	3,478,565	1,551,000 97,733	60,321	300,467
'85	267.4	20,874	....	1,041	27,317,341	3,618,489	1,451,000 100,850	62,376	313,205
'86	301.1	23,297	....	1,227	23,976,907	3,885,240	1,447,000 91,862	71,176	314,954
'87	322.8	28,259	....	1,379	26,079,166	4,091,453	1,377,000 86,890	106,614	332,835
'88	343.1	31,821	48	1,571	39,397,716	4,267,149	1,327,000 83,400	101,019	344,315
'89	343	37,725	194	1,818	35,294,888	3,410,910*	1,326,000 83,330	102,587	367,925
'90	362	40,351	856	1,828	33,208,067	3,631,599	1,229,000 76,540	102,391	387,877

* Previous figures estimated.

Financial.—*Bonded Debt,* $1,229,000 at 7, 6 and 4%. Expenses, total, $178,931. *Revenue,* consumers, $377,877. Management.—Five comrs., one appointed every year by mayor and confirmed by city council. Prest., H. M. Duffield. Secy., L. N. Case.

27. **DOWAGIAC,** *Cass Co.* (Pop., 2,100—2,806.) Built in '89 by city. Supply.— Artesian wells, pumping direct from receiving reservoir. Fiscal year closes April 30.

Year.	M'ns.	Taps.	Meters.	Hy'ts.	C'ns'pt'n.	Cost.	B'd'd debt.	Int. ch'g's.	Exp.	Rev.
89-90	9	205	0	73	90,000	$40,072	$40,000		$2,000	$2,600 $1,592
90-1	9.5	360	0	73	150,000	......	51,000		.....	2,950 2,840

Bonded Debt, $51,000: $40,000 at 5%; $11,000 at 7%. *Revenue:* consumers, $2,840; city, $1,000. Management.—Chn., W. M. Vrooman. Secy., A. E. Rudolphi. Supt., E. Barlow Jewell.

96. **EAST SAGINAW,** *Saginaw Co.* Consolidated with Saginaw Apr. 1, '90. See Saginaw.

28. **EAST TAWAS,** *Iosco Co.* (Pop., 1,086—2,226.) Built in '87-8 by village. Supply.—Lake Huron, pumping direct. Fiscal year closes Feb. 1. Distribution.—(In '88.) *Mains,* 5 miles. *Taps,* 220. *Meters,* 0. *Hydrants,* 76. *Consumption,* 375,000 galls. Financial.—(In '88.) *Cost,* $30,000. *Bonded Debt,* $25,000 at 6%. *Expenses:* operating, $2,000; int., $1,500; total, $3,500. *Revenue:* consumers, $1,000; city, $1,500; total, $2,500. Management.—(In '89.) Secy., C. H. Hubbell. Supt., J. J. Alguire.

29. **EDMORE,** *Montcalm Co.* (Pop., 704—735.) Built in '81 by village. Supply. —Tube wells, pumping to reservoir and direct. Fiscal year closes April 1. Dry season of '89 compelled addition of two driven wells and pumping machinery having

FOR FURTHER DESCRIPTIVE MATTER see Manual for 1889-90. CHANGES since 1890 are here given. POPULATIONS are for 1880 and 1890, respectively.

dy, cap of 25,000 galls. Distribution.—*Mains*, 1 mile. *Taps*, 50. *Hydrants*, 8. *Consumption*, 4,000 galls. Financial.—*Cost*. $8,500. *No Debt. Expenses*, $350. *Revenue*, $150. Management.—Committee of three, appointed every year by mayor. Prest., Wm. H. Gardner. Secy., Hiram Morehead. Supt., Jas. S. Slemons.

30. **ESCANABA,** *Delta Co.* (Pop., 3,036—6.808.) Built in '87 by Escanaba Water-Works Co.; controlled by Moffett, Hodgkins & Clarke, Syracuse, N. Y. Supply.—Green Bay, filtered, by direct pumping.

Year.	Mains.	Taps.	M'rs.	Hyd'ts.	Cons'n.	Cost.	B'd'd debt.	Int. ch'g's.	Exp.	Rev.
'87	8	113	1	70	123,677	$125,000	$100,000	$6,000	......	......
'88	10	287	1	83	122,000	125,000	100,000	6,000	$3,000	$10,000

Financial.—(In '88.) *Int.*, 6%. *Expenses:* operating, $3,000; int., $6,000; total, $9,000. *Revenue:* consumers, $6,000; city, $4,000; total, $10,000. Management.—(In '89.) Prest., J. V. Clarke. Supt., W. N. Lafleur.

31. **EVART,** *Osceola Co.* (Pop., 1,302—1,269.) Built in '79-80 by village. Supply. —Spring and two wells, pumping direct. Fiscal year closes March 31. During 1890 works were entirely reconstructed. New pumping station was erected and new pumps and boiler put in. Distribution.—*Mains*, 2.2 miles. *Taps*, 150. *Hydrants*, 17. *Consumption*, 150,000 galls. Financial.—*Cost*, $30,000. *Bonded Debt*, $8,000 at 6%, due in annual payments of $1,000, after '91. *Expenses:* operating, $1,600; int., $480; total, $2,080. *Revenue:* consumers, $1,300; city, none. Management.—Prest., Ellery E. Canton. Secy., Wm. E. Davis. Supt., Edward Seath.

32. **FARWELL,** *Clare Co.* (Pop., 521—584.) Built in '87-8 by village. Supply. —Spring brook, pumping direct. Fiscal year closes in May. Distribution.—*Mains*, 2.5 miles. *Taps*, 40. *Hydrants*, 35. Financial.—*Cost*, $13,000. *Bonded Debt*, $13,000 at 7%. *Expenses*, $1,050. *Revenue*, $200. Management.—Board of three, appointed every three years by village council. Chn., W. H. Safford. Supt., J. N. Brown.

33. **FENTON,** *Genesee Co.* (Pop., 3,807—3,667.) Built in '87 by village. Supply. —Artesian wells, flowing into reservoir, 20 × 20 ft., thence pumping direct. Fiscal year closes Oct. 1. Distribution.—*Mains*, 6 miles. *Taps*, 140. *Hydrants*, 65. *Consumption*, 75,000 galls. Financial.—*Cost*, $25,000. *Bonded Debt*, $25,000 at 5%. *Expenses:* operating, $1,700; int., $1,250; total, $3,950. *Revenue:* consumers, $700. Management.—Three comrs., appointed every three years by council. Prest., Josiah Buckbee. Secy., Jno. Jennings. Supt., M. Walker.

34. **FLINT,** *Genesee Co.* (Pop., 8,409—9,803.) Built in '83 by Flint City Water-Works Co., under 30-yrs. franchise; city has option of buying works at end of 15 years; does not control rates. Supply.—Flint River, pumping direct. Fiscal year closes June 1. In '90 company moved plant to point above city, and takes its supply direct from Flint River. A 4,000,000-gall. pump was bought, and $30,000 in bonds were issued to defray expenses of these improvements.

Year.	Mains.	Taps.	M't'rs.	H'd'nts.	C'ns'pt'n.	Cost.	B'd'd d't.	Int.ch'g's.	Exp.	Rev.
'86	11.6	396	2	100	468,000	$140,000	$100,000	$6,000	$6,000	$13,471
'87	12.3	642	2	107	528,000	140,000	100,000	6,000	6,500	17,200
'88	13.75	627	5	109	762,316	134,026	100,000	6,000	4,225	14,579
'90	14	704	4	110	936,016	148,507	130,000	6,000	6,178	15,629

Financial.—*Cap. Stock*, paid up, $150,000; paid no dividend in 1890. *Bonded Debt*, $130,000 at 6%. *Expenses:* operating, $5,216; int., $6,000; taxes, $962; total, $12,178. *Revenue:* consumers, $5,898; city, $6,963; other purposes, $2,768; total, $15,629. Management.—Prest., Wm. Hamilton. Secy., D. Anderson. Treas., C. S. Brown. Supt., Wm. L. Fisher.

35. **FORT GRATIOT,** *St. Clair Co.* (Pop., 1,280—2,832.) Built in '86 by village. Supply.—St. Clair River, pumping direct. Fiscal year closes April 1. Distribution.—*Mains*, 6.5 miles. *Taps*, 350. *Meters*, 0. *Hydrants*, 45. *Consumption*, 130,000 galls. Financial.—*Cost*,$30,000. *Bonded Debt*, $25,000 at 5%. *Expenses:* operating, $2,200; int., $1,250; total, $3,450. *Revenue:* consumers, $3,000; city, $700; total, $3,700. Management.—Com. of three from common council appointed every year by mayor. Supt., W. V. Burger.

36. **FRANKFORT,** *Benzie Co.*(Pop., 782—1,175.) History.—Built in '90 by Frankfort City Water Co., under 5-yrs. franchise; put in operation Dec. 1. Engr., S. V. Saleno, Bay City, Mich. Contrs. Michigan Pipe Co., Bay City. Supply.—Well, pumping to tank. Pumping Machinery.—Dy. cap., 84,000 galls; Smith & Vaile. Tank.—Cap., 97,000 galls. Distribution.—*Mains*, Wyckoff wood, 3.8 miles. *Hydrants*, Galvin, 34. *Valves*, Galvin. *Pressure:* ordinary, 60 lbs.; fire, 100 lbs.

MICHIGAN. 219

Financial.—*Cost*, $25,000. *Cap. Stock,* authorized, $30,000. *Bonded Debt* $20,000 at 7%. *Revenue,* city, $1,800. *Hydrant rental*, $35 **Management.**—Prest., Geo. L. Davis. Vice-Prest., Fred. A. Aldrich. Secy., and Supt., D. E. McIntyre. Treas., E. R. Chandler.

37. **FREMONT,** *Newaygo Co.* (Pop., 902—1,097.) Built in '83 by village. **Supply.**—Springs and four flowing wells, pumping direct. Fiscal year closes Dec. 1. **Distribution.**—*Mains,* 1 mile. *Taps,* 20. *Hydrants,* 8. **Financial.**—*Cost,* $6,000. *No. Debt. Expenses,* $400. *Revenue:* consumers, $150; R'y., $250; total, $400. **Management.**—Village Board elected every two years. Prest. and Supt., J. Gerber. Secy., B. E. Tanner.

38. **GAYLORD,** *Otsego Co.* (Pop., 292—661.) Built in '88 by village. **Supply.**—Well, pumping to tank. **Distribution.**—*Mains,* 2 miles. *Taps,* 75. *Hydrants,* 18. *Consumption,* 4,500 galls. **Financial.**—*Cost,* $8,500. *Bonded Debt,* $8,500 at 6%. *Expenses:* operating, $605; int., $510; total, $575. *Revenue,* $400. **Management.**—Secy., W. S. Filmore. Supt., E. B. Bolton.

39. **GLADSTONE,** *Delta Co.* (Pop., in '90, 1,337.) Built in '89 by city. **Supply.**—Little Bay de Noc, at head of Green bay. Fiscal year closes third Tuesday in March. **Distribution.**—*Mains,* 4 miles. *Taps,* 150. *Hydrants,* 32. **Financial.**—*Cost,* $34,000. *Total Debt,* $35,500: floating, $500 at 7%; bonded, $35,000 at 6%, due in 1909. *Expenses:* operating, $2,461; int., $2,100; total, $4,561. *Revenue,* consumers, $150. **Management.**—Prest., J. W. Collins. Secy., J. H. Le Clair. Supt., Jas. W. Williams.

40. **GLADWIN,** *Gladwin Co.* (Pop., 467—903.) Built in '88 by village. **Supply.**—Artesian wells, pumping direct; river water is used for fire protection. Fiscal year closes March 1. **Distribution.**—*Mains,* 1.3 miles. *Taps,* 34. *Hydrants,* 9. **Financial.**—*Cost,* $7,000. *Total Debt,* $7,700: floating, $700; bonded, $7,000 at 7%. *Expenses:* operating, $539; int., $900; total, $1,439. *Revenue,* $300. **Management.**—Village Council. Engr., M. F. Pike.

41, 42. **GRAND HAVEN,** *Ottawa Co.* (Pop., 4,862—5,023.)

CITY'S WORKS.

Built in '80 and extended in '88. **Supply.**—Driven wells and one open well, pumping direct. Fiscal year closes March 15.

Year.	Mains.	Taps.	Meters.	Hydrants.	Cost.	B'd'd debt.	Int. ch'g's	Exp.	Rev.
'82	1	30	0	8				$1,500	$350
'89	5.5	200	0	47	$21,600	$50,000	$2,500	2,990	1,139
'90	6	235	0	48	21,300	47,000	2,350	2,987	1,200

Financial.—*Bonded Debt,* $57,000 at 5%, due in '98. *Expenses,* total, $5,337. *Revenue:* consumers, $1,200; city, $1,500; total, $2,700. **Management.**—Recorder, C. Reynolds. Supt., Jos. Palmer, appointed by council.

GRAND HAVEN WATER-WORKS CO.

Built in '84. City has no control of rates. **Supply.**—Driven wells, pumping direct. Fiscal year closes April 1. **Distribution.**—*Mains,* 7 miles. *Taps,* 200. *Meters,* 4. *Hydrants,* 60. *Consumption,* 200,000 galls. **Financial.**—*Cost,* $200,000. *Bonded Debt,* $70,000 at 6%. *Expenses:* operating, $4,300; int., $4,200. *Revenue:* not determined, city and co. being in litigation over payment of hydrant rental. **Management.**—Prest., S. L. Wiley, Omaha, Neb.

43, 44. **GRAND RAPIDS,** *Kent Co.* (Pop., 32,016—60,278.) Spring of '91, large territory was annexed to city.

GRAND RAPIDS HYDRAULIC CO.

Built in '40. **Supply.**—Two receiving wells, supplied by springs, pumping to standpipe. First well is 20 ft. in diam., 25 ft. deep; cap., 1,500,000 galls.; of brick, lined with cem., built on an iron shoe sunk in the earth, by digging on the interior and dredging down into the gravel. Second well is 20 ft. square, 20 ft. deep, sunk in lower strata of gravel. These two wells are supplemented by lines of pipe leading from various springs, at or below the gravel line. Reported in May, '91, that co. was to duplicate force main to stand-pipe. **Distribution.**—*Mains,* 12 to 6-in. c. i., 8 miles. In '88, a 12-in. main, 900 ft. long, was laid across Grand River at cost of $6,000. *Taps,* 1,000. *Meters,* 150. *Hydrants,* 20. *Consumption,* 130,000 galls. **Financial.**— (In '89.) *Cost,* $500,000. *Bonded Debt,* $450,000 at 6%. *Expenses:* operating, $9,000;

FOR FURTHER DESCRIPTIVE MATTER see Manual for 1889-90. CHANGES since 1890 are here given. POPULATIONS are for 1880 and 1890, respectively.

int., $27,000. *Revenue:* consumers, $30,000; city, $6,500; total, $36,500. **Management.** —Address M. R. Crow.

CITY WORKS.

Built in '73-5. **Supply.**—Grand River, filtered, pumping to reservoir. Fiscal year closes May 1. See note below. **Filter Galleries.**—Three; 14 × 16 ft., 12 × 110 ft., 40 × 480 ft., respectively; gravel beds, built in excavation in solid rock in center of river and below river bottom. In event of trouble with filter an intake crib, having copper wire screen, is built in the river to take water direct. **Conduit.**—Is 3 × 6 ft., leading from galleries to pump-well on bark of river. **Pumping Machinery.**—Dy. cap., 12,000,000 galls. *Low service:* two pumps; one 5,000,000-gall., horizontal, direct-acting, condensing, with 33-in. steam and 20-in. water cylinder of 72-in. stroke, from Eagle Iron Works, Detroit; one 3,000,000-gall. horizontal direct acting condensing, with 33-in. steam and 15-in. water cylinders of 72-in. stroke, from Butterworth & Lorne, Grand Rapids. *High service:* two pumps; one 2,500,000, Smith & Vaile, comp., cond., dup., with 13-in. and 30-in. steam and 16-in. water cylinders of 24-in. stroke; one 1,500,000 Gordon & Maxwell, dup., com., cond., with 20-in. steam cylinders and 12-in. water cylinders of 18-in. stroke, to be used as an auxiliary. Pumping machinery and force mains of high and low service are interchangeable. **Force Mains.**—To reservoirs for low service, 24 ins. in diam.; to standpipe for high service, 16 ins. in diam. **Distributing Reservoir.**—Cap., 6,000,000 galls.; 190 ft. in diam. at bottom, 271 ft. at top, 25 ft. deep, with 20 ft. of water, 177 ft. above river, in excavation and embankment. Bank is of sand, with puddle in center, along the natural surface and over the bottom; is 12 ft. wide on top, with inner slope of 1½ to 1 and outer slope of 2 to 1. Inner slope is paved with 1 ft. of cobblestones, laid on 1 ft. of concrete, except the upper 2 ft. of slope, which has 6 ins. of cobbles set in gravel. Bottom has an 8-in. cobblestone pavement, on 8 ins. of concrete. Reservoir does not leak. The 16-in. force main and 20-in. effluent pipe are carried under the bank, the joints being inclosed in masonry piers; the force main ends in a mass of masonry and the effluent begins in a small masonry well. **Stand-Pipe.**—Cap., 397,000 galls.; 30 ft. in diam., 75 ft. high; of iron on masonry foundation. **Distribution.**—*Mains*, all of c. i., except 3 miles of wood. Two sub merged mains of c. i., across the river, one 12 and one 16 ins. in diam.

Year.	Mains.	Taps.	M't's.	H'd'ts.	C'ns'p'n.	Cost.	B'd'd d't.	Int. ch'g's.	Exp.	Rev.
'78-9	22.3	730	90	235		897,024	$401,590	$382,000	$29,560	......	$13,734
'79-80	22.3	819	90	235	1,107,427	404,296	382,000	29,560	$9,417	16,174
'80-1	22.3	919	183	235	1,335,655	420,975	382,000	29,560	10,798	15,847
'81-2	24.7	962	260	260	1,157,230	443,319	382,000	29,560	12,071	20,239
'82-3	24.8	1,057	328	260	1,224,366	446,510	382,000	29,500	12,988	20,347
'83-4	26.4	1,176	338	275	1,533,731	456,890	382,000	29,560	16,873	24,637
'84-5	26.6	1,319	350	278	1,937,994	464,608	382,000	29,560	17,243	28,676
'85-6	27.3	1,453	364	285	2,046,080	485,066	382,000	29,560	16,102	34,915
'86-7	27.6	1,518	467	290	2,396,367	503,165	382,000	29,560	21,218	34,273
'87-8	28	1,696	492	296	2,769,333	536,177	389,000	29,560	22,481	41,696
'88-9	48.4	2,493	471	526	2,716,736	715,836	612,000	42,080	22,081	41,839
'89-90	65.9	2,866	452	636	3,315,119	841,804	612,000	42,080	26,606	43,830
'90-1	67	3,819	459	646	4,392,193	880,060	812,000	42,080	30,776	50,666

Financial.—*First Cost*, $404,296. *Bonded Debt*, $612,000; $250,000 at 8%, due in '93; $100,000 at 8%, due in '95; $32,000 at 8%, due in '96; $150,000 at 5%, due in 1908; $80,000 at 5%, due in 1909. *No Sinking Fund. Expenses*, total, $72,836. *Revenue:* consumers, $50,666; city, none. Net revenue is used for new construction; bonds and int. are paid by taxation. No allowance for public uses of water. **Management.**—Board of Public Works, appointed every three years by Mayor. Prest., Jas. W. Davis. Secy., F. A. Twamley. Supt., J. Van Amberg.

NOTE.—June 30, '91, city voted to issue $300,000 of bonds for supply from driven wells and extension of mains 33 miles.

45. GRAYLING, *Crawford Co.* (Pop., 245—1,558.) Built in '85 by Grayling, Hanson & Co. **Supply**—An Sable River by gravity. Fiscal year closes Dec. 31. **Distribution.**—*Mains*, 3.3 miles. *Taps*, 56. *Hydrants*, 12. **Financial.**—*Cost*, $3,000. *No Debt. Expenses*, $175. *Revenue*, $1,460. **Management.**—Owners, Grayling & Hanson.

46. GREENVILLE, *Montcalm Co.* (Pop., 3,144—3,053.) Built in '88 by city. **Supply.**—Driven wells and river, pumping direct; river for fire protection. Fiscal year closes April 1. Apr., '91, well 24 ft. in diam., 20 ft. deep was being sunk. **Distribution.**—*Mains*, 5.8 miles. *Taps*, 201. *Hydrants*, 46. *Consumption*, 250,000 galls. **Financial.**—*Cost*, $46,116. *Bonded Debt*, $45,000 at 3%, due in annual payments of $3,000. *Expenses:* operating, $2,576; int., $1,400; total, $3,976. *Revenue:*

consumers, $1,331; city, $2,700; total, $5,031. **Management.**—Chn., Geo. G. Douglas. Secy. and Supt., Jas. Gracey.
NOTE.—For description of leak in submerged main see *The Technic* for 1890, and abstract of article in ENGINEERING NEWS, vol. XXIV., p. 173.

47. **GROSSE POINT,** *Wayne Co.* (Pop., 189–293.) **History.**—Built in '90 by Newberry & McMillen, under 25-yrs. franchise; put in operation April 1. Engr. and Contr., M. Walker, Port Huron. **Supply.**—Lake St. Clair, pumping direct from settling basin. **Pumping Machinery.**—Dy. cap., 1,500,000 galls.; Walker. **Distribution.**—*Mains*, 8-in. i., 3 miles. *Taps,* 90. *Hydrants,* Ludlow, 27. *Valves,* Flower. *Consumption,* 60,000 galls. *Pressure:* ordinary, 35 lbs.; fire, 90 lbs. **Financial.**—*Cost,* $35,000. *No Debt. Hydrant rental, free.* **Management.**—Prest., Hugh McMillen. Secy., T. H. Newberry. Supt., Geo. W. Wortman.

HANCOCK, *Houghton Co.* (Pop., 1.783–1,772.) Built in '65 by village. There is no regular system of works and no revenue except from parties outside village. Water is taken from cisterns, supplied by springs. Reported in May, '91, that plans for works were to be made by C. F. Loweth, St. Paul.

48. **HARRISON,** *Clare Co.* (Pop., 129–752.) Built in '85 by village. **Supply,** —Budd Lake, pumping direct. Fiscal year closes Dec. 31. **Distribution.**—*Mains.* 1.5 miles. *Taps,* 100. *Hydrants,* 18. *Consumption,* 45,000 galls. **Financial.**—*Cost,* $6,000. *Bonded Debt,* $1,000 at 5%, due in '91. *Expenses:* operating, $990; int., $50; total, $950. *Revenue:* consumers, $325. **Management.**—Supt., Wm. H. Brown.

49. **HART,** *Oceana Co.* (Pop., 464–757.) Built in '86 by village. **Supply.**—River, pumping direct. **Distribution.**—*Mains,* 0.8 mile. *Taps,* 50. *Hydrants,* 5. **Financial.**—*Cost,* $4,500. *Debt,* $1,500 at 8%. *Expenses,* $175. *Revenue,* $122. **Management.**—Prest., Jno. F. Widoe. Secy. and Supt., Frank H. Edwards.

50. **HASTINGS,** *Harry Co.* (Pop., 1,284–1,187.) Built in '86-7 by village. **Supply.**—Well, pumping direct. **Distribution.**—(In '88.). *Mains,* 4 miles. *Taps,* 200. *Meters,* 0. *Hydrants,* 40. *Consumption,* 300,000 galls. **Financial.**—(In '88.) *Cost,* $30,000. *Bonded Debt,* $25,000 at 4½%. *Expenses:* operating, $2,000; int., $1,125; total, $3,125. *Revenue* consumers, $1,500. **Management.**—(In '89.) Three comrs.: W. L. Wilkins, W. D. Hayes, W. C. Kelly.

51. **HILLSDALE,** *Hillsdale Co.* (Pop., 3,441–3,915.) Built in '85-86 by village. **Supply.**—Bawbeese Lake, pumping direct. Fiscal year closes March 1. **Distribution.**—*Mains,* 10 miles. *Taps,* 540. *Meters,* 4. *Hydrants,* 75. **Financial.**—*Cost,* $55,000. *Bonded Debt,* $45,000. *Expenses:* operating, $2,509; int., $2,025; total, $4,534. *Revenue,* consumers, $3,158. **Management.**—Three comrs., one appointed every year by city council. Prest., C. F. Cook. Secy., M. J. Davis.

52. **HOLLAND,** *Ottawa Co.* (Pop., 2,620–3,945.) Built in '85 by village. **Supply.**—Springs, pumping direct from collecting reservoir. Fiscal year closes third Monday in March. Apr., '91, $17,000 of bonds were voted to extend mains about 4½ miles. **Distribution.**—(In '88.) *Mains,* 4.5 miles. *Taps,* 123. *Meters,* 0. *Hydrants,* 30. *Consumption,* $150,000 galls. **Financial.**—(In '88.) *Cost,* $28,868. *Bonded Debt,* $23,600 at 5%. *Expenses:* operating, $2,700; int., $1,180; total, $3,880. *Revenue:* consumers, $1,600. **Management.**—(In '89.) Secy., Geo. H. Sipp. Supt., Jno. Kramer.

53. **HOLLY,** *Oakland Co.* (Pop., 1,443–1,266.) Built in '79 by city. **Supply.**—Undriven wells, pumping direct. Until '90 supply was taken from river. Fiscal year closes June 1. **Distribution.**—*Mains,* 3 miles. *Taps,* 125. *Hydrants,* 22. **Financial.**—*Cost* (in '83), $10,099. *Expenses,* $1,300. *Revenue,* $1,400. **Management.**—Committee of three appointed every three years by council. Secy., Chas. Barrell. Supt., L. J. Striggon.

54. **HOWARD CITY,** *Montcalm Co.* (Pop., 921–1,137.) Built in '90 by village. **Supply.**—River, pumping direct. Fiscal year closes May 1. **Distribution.**—*Mains,* 1.3 miles. *Hydrants,* 13. **Financial.**—*Cost,* $8,500. *Bonded Debt,* $8,000 at 6%. **Management.**—Three comrs., one appointed every year by council. Chn., Albert O'Donald. Secy., Warren Fisk.

55. **HUDSON,** *Lenawee Co.* (Pop., 2,254–2,178.) **History.**—Built in '90 by village; put in operation Oct. 1. Engr., J. D. Cook, Toledo. **Supply.**—Wells, pump-

FOR FURTHER DESCRIPTIVE MATTER see Manual for 1889-90. CHANGES since 1890 are here given. POPULATIONS are for 1880 and 1890, respectively.

ing direct. **Pumping Machinery.**—Dy, cap., 2,000,000 galls.; two 1,000,000 Hughes, *Boilers*, from Lansing Iron & Engine Wks. **Distribution.**—*Mains*, 12 to 4 in., 5 miles. Pipes and specials from Addyston Pipe Co., Cincinnati, O.; trenching and pipe laying done by Snyder & Williams, Dayton. *Hydrants*, Bourbon, 51. *Valves*, Rensselaer. *Pressure*: ordinary, 40 lbs.; fire, 75 lbs. **Financial**—*Cost*, $30,000. *Bonded Debt*, $30,000 at 5%. **Management.**—Prest., G. I. Thompson. Clerk, Ira C. Wykoff.

56. **IMLAY CITY,** *Lapeer Co.* (Pop., 971-1,251.) **History.**—Built in '90-1 by city; put in operation Jan. 24. Engr. and Contr., Geo. Cadogan Morgan, 15 Major Block, Chicago. **Supply.**—Deep well, flowing into reservoir, thence pumping to tower. **Reservoir.**—Cap., 40,000 galls. **Pumping Machinery.**—Dy cap., 1,000,000 galls. Deane. **Tower.**—Cap., 30,000 galls. **Distribution.**—*Mains*, Wyckoff wood, 2.5 miles. *Hydrants*, Morgan, 31. *Valves*, Peet. **Financial.**—*Cost*, $18,000. **Management.**—Not given.

57. **IONIA,** *Ionia Co.* (Pop., 4,190-4,482.) Built in '76 by village. **Supply.**—Springs, surface water, and artesian wells, pumping direct. Fiscal year closes March 31.

Yr.	M'ns.	Taps.	M't'rs.	Hydts.	C'n'tion.	Cost.	B'd'd debt.	Int. ch'g's.	Exp.	Rev.
'84-5	8.5	325	0	74	35,000	$80,000	$30,000	$1,501	$3,500	$1,800
'87-8	12	550	0	99	300,000	93,000	30,000	1,500	1,600	3,700
'89-90	14.6	568	0	107	350,600	107,369	30,000	1,500	4,286	4,229
'90-1	15.1	585	0	109	340,000	108,750	30,000	1,500	4,250	4,287

Financial.—*Total Debt*, $39,500; floating, $9,500 at 7%; bonded, $30,000 at 5%, due in 1903. *No Sinking Fund*. *Expenses*, total, $2,000. *Revenue*: consumers, $4,287; city, none. **Management.**—Board of Public Works elected every three years. Prest., H. H. Hearsey. Clerk and Supt., Frank C. Sibley, appointed by Board.

58. **IRON MOUNTAIN,** *Menominee Co.* (Pop., in '90, 8,599.) **History.**—Built in '90 by Iron Mountain Water-Works Co.; put in operation Dec. 1. Engr., W. R. Coats, Kalamazoo. **Supply.**—Wells, pumping to elevated reservoir. Aug., '91, filter gallery was being put in, and main 8,000 ft. long being laid from it beneath lake. Engr., as above. **Pumping Machinery.**—Dy. cap., 4,000,000 galls; Gordon comp. dup. *Boilers*, from Boynton Engineering Co., Cincinnati. **Reservoir.**—Cap., 500,000 galls.; 60 ft. in diam., 25 ft. deep. **Distribution.**—*Mains*, 16 to 4-in. c. i., 13 miles. *Services*, w. i. Pipe and specials from Detroit Pipe & Foundry Co. *Taps*, 229. *Hydrants*, Chapman, 163. *Valves*, Chapman. *Consumption*, 250,000 galls. *Pressure*. 90 lbs. **Financial.**—*Cost*, $150,000. *Cap. Stock*, $125,000. *Bonded Debt*: authorized, $125,000; issued, $100,000. *Revenue*: city, $9,000. **Management.**—Prest., F. A. Todd. Secy., Robt. W. Murphy. Treas., S. L. Quirk. All of Ypsilanti.

59. **IRONWOOD,** *Gogebic Co.* (Pop., in '90, 7,715.) **History.**—Construction was begun in '87 by Gogebic Gas & Water Supply Co. and completed in March '91 by Ironwood Water-Works Co., under 50 yr. franchise. Const. Engr., Saml. A. Souther. **Water Supply.**—Montreal River, pumping to stand-pipe. **Pumping Machinery.**—Dy. cap., 2,000,000 galls.; two Deane, 10 × 20 × 19½ × 18. *Boilers*, from Parish Mfg. Co.. Ashland. **Filters.**—Dy. cap., 500,000 galls ; Jewell. **Stand-Pipe.**—Cap., 264,000 galls.; 30 × 50 ft.; from Tippett & Wood, Phillipsburgh, N. J. **Distribution.**—*Mains*, 12 to 4-in. kal. i., 5 miles. Pipe and specials from National Tube Wks., Chicago. *Taps*, 132. *Meters*, 4. *Hydrants*, Ludlow. 60. *Valves*, Ludlow. *Consumption*, 300,000 galls. *Pressure*, 80 lbs. **Financial.**—See Hurley, Wis. **Management.**—Prest., C. L. Hyde. Secy., Frank Lynde. Treas., C. H. Jackson. Supt., S. A. Souther.

60. **ISHPEMING,** *Marquette Co.* (Pop., 6,039—11,197.) Built in '82 by city. **Supply.**—Lake Sally, pumping direct. Fiscal year closes Dec. 31. Source of supply was changed in '90 from Lake Angeline to Lake Sally. Water is conveyed through 14-in. pipe under 84-ft. head. In case of fire, pressure is increased by three large steam pumps.

Year.	Mains.	Taps.	Meters	Hydrants.	B'd'd d't.	Int. ch'g's.	Exp.	Rev.
'86	6.5	311	0	34	$63,000	$3,150	$5,746	$5,867
'87	9.2	...	0	54	45,000	2,250	6,292	8,383
'89	16.3	808	0	69	......	....	6,890	9,346
'90	18.7	909	0	81	......	....	5,411	11,254

Management.—Three comrs., F. P. Mills, P. H. Devine, Thomas Walters.

Ithaca, *Gratiot Co.* (Pop., 600–1,627.) Built in '84 by village, for fire protection.

MICHIGAN. 223

61. JACKSON, *Jackson Co.* (Pop., 16,105—20,798.) Bought by city in '71. Built in '70 by Co. **Supply.**—Artesian wells, pumping direct. Oct. 19, '91, city sold $100,000 of 5% bonds. About same time W. S. Parker was engaged as consult. engr. to remodel works. W. R. Coats reported on improvements earlier in the year.

Year.	Mains.	Taps.	Mtrs.	Hy't's.	Cons't'n.	Cost.	Bdd. Debt.	Int. C'gs.	Exp.	Rev.
'80	12	600	....	85	782,203	....	....	....	$7,000	$8,250
'82	13	807	4	86	8,500,000	$180,000	$163,500	$13,550	8,000	9,000
'84	14	1,000	7	108	1,000,000	160,000	160,000	12,800	9,000	9,000
'87	21	3,000	30	218	1,800,000	250,000	160,000	....	10,000	18,000
'88	28.5	1,940	56	270	1,800,000	400,000	160,000	11,300	9,500	18,000

Management.—Board of Public Works. Secy., J. K. Rogent.

62. KALAMAZOO, *Kalamazoo Co.* (Pop., 11,937—17,853.) Built in '60 by city. **Supply.**—Two surface wells, pumping direct. Fiscal year closes March 31. Oct., '91, W. R. Coats reported in favor of adding a 2,000,000-gall. elevated reservoir and making necessary connections therewith.

Year.	Mains.	Taps.	Meters.	Hyd'ts.	Consumption.	Cost.	B'd'd debt.	Exp.	Rev.
'82	16.7	1,237	..	155	741,575	....	none.	$6,701	$5,889
'84	18.5	1,780	0	174	909,331	$192,412	"	7,198	7,566
'86-7	20	1,197	7	203	1,526,495	322,847	"	9,732	13,529
'88-89	24.3	1,269	..	232	1,790,660	350,000	"	9,832	13,173
'89-90	27.5	1,307	..	247	1,896,600	367,327	"	8,942	14,092

Management.—Comr., Hugh Beggs. Supt., Geo. H. Chandler.

63. KALKASKA, *Kalkaska Co.* (Pop., 496—1,161.) **History.**—Built in '90-1 by village; put in operation Jan. 3. Engr. and Contr., M. Walker, Port Huron. **Supply.**—River, pumping direct. **Pumping Machinery.**—Dy. cap., 1,500,000 galls.; Walker. **Distribution.**—*Mains*, 10 and 8-in. wood, 3.5 miles. *Services*, galv. i. Pipe, specials, trenching and pipe-laying, Michigan Pipe Co., Bay City. *Hydrants*, Ludlow, 36. *Pressure:* ordinary, 30 lbs.; fire, 90 lbs. **Financial.**—*Cost*, $18,000. *Bonded Debt*, $25,000, at 6%. **Management.**—Prest., Geo. R. Colton. Secy., Cassius M. Phelps. Treas., Oscar Watson. Supt., M. Helmer.

64. LAKE CITY, *Missaukee Co.* (Pop., 61—663.) Built in '88 by Dan'l Reeder & Co.; city has right to buy works in '94, each party to appoint an appraiser and these two a third; does not control rates. **Supply.**—Muskrat Lake, pumping direct. Fiscal year closes Aug. 18. **Distribution.**—*Mains*, 1 mile. *Taps*, 65. *Hydrants*, 7. **Financial.**—*Cost*, $8,053. *No Debt. Expenses*, $620. *Revenue*, $1,600. **Management.**—Dan'l Reeder, Owner.

65. LAKE LINDEN, *Houghton Co.* (Pop., 2,600—est., 3,500.) Built in '87 by village. **Supply.**—Artesian well and creek, by direct pumping from former for domestic use, by gravity from latter for fire protection. **Distribution.**—*Mains*, 4 miles. *Takers*, 278. *Hydrants*, 36. **Financial.**—*Cost* (in '87), $40,000. *Bonded Debt* (in '87), $10,000. *Int.*, 6%. *Expenses*, [$]200. *Revenue*, $1,500. **Management.**—Supt., F. O. Mayotte.

66. LANSING, *Ingham Co.* (Pop., 8,319—13,102.) Built in '86 by city. **Supply.** —Eight tile-lined wells, pumping to stand-pipe. Fiscal year closes May 1.

Year.	Mains.	Taps.	M't'rs.	Hyd'ts.	Cons'pt'n.	Cost.	B'd'd d't.	Int. ch'g's.	Exp.	Rev.
'87-8	17.5	600	15	179	800,000	$135,000	$100,000	$4,500	$4,000	$7,300
'88-9	22	950	30	200	900,000	145,000	100,000	4,500	6,000	8,150
'90-1	27	1,412	50	257	1,070,831	197,657	125,000	5,375	6,600	9,945

Financial.—*Bonded Debt*, $125,000: $75,000 at 4½%; $50,000 at 4%. *Expenses*, total, $11,975. *Revenue:* consumers, $9,945; city, $9,380; total, $19,325. **Management.**— Prest., Chas. Connell. Supt., Wm. S. Wright.

67. LAPEER, *Lapeer Co.* (Pop., 2,911—2,753.) Built in '88 by village. **Supply.** —Three flowing artesian wells, pumping direct. Fiscal year closes April 1. **Distribution.**—*Mains*, 11 miles. *Taps*, 275. *Hydrants*, 70. **Financial.**—*Cost*, $43,000. *Bonded Debt*, $40,000 at 5%. *Expenses:* operating, $2,000; int., $2,500; total, $4,500. *Revenue:* consumers, $2,800; city, none. **Management.**—Chn., W. S. Johnson, Supt., S. D. Brown.

68. LOWELL, *Kent Co.* (Pop., 1,538—1,829.) Built in '87 by Lowell Water Co. **Supply.**—Twenty driven wells, pumping to reservoir. **Distribution.**—(In '87.) *Mains*, 3.6 miles. *Hydrants*, 50. **Financial.**—(In '87.) *Cost*, $22,000. *Cap. Stock*, $30,000. **Management.**—Secy., Jno. E. More, 107 Ottawa St., Grand Rapids.

FOR FURTHER DESCRIPTIVE MATTER see Manual for 1889-90. CHANGES since 1890 are here given. POPULATIONS are for 1880 and 1890, respectively.

69. LUDINGTON, *Mason Co.* (Pop., 4,190—7,517.) Built in '82 by Ludington Water Co.; franchise fixes maximum rates, but does not provide for purchase of works by city. **Supply.**—Lake Michigan and Pere Marquette River, pumping direct. Fiscal year closes Jan. 1. Two new boilers, each 5 × 16 ft., were put in in '90. **Distribution.**—*Mains*, 8.5 miles. *Taps*, 300. *Meters*, 0. *Hydrants*, 90. *Consumption*, 1,500,000 galls. **Financial.**—*Cap. Stock:* authorized, $30,000, all paid up; paid 15% div'd in 1890. *Total Debt,* $11,085; floating, $1,085 at 6%; bonded, $10,000 at 6%, due in 1903. *Expenses:* operating, $1,600; int., $348; taxes, $712; total, $5,660. *Revenue:* consumers, $5,230; city, $5,770; total, $11,000. **Management**—Prest., G. G. Wing. Secy. and Supt., N. J. Gaylord. Treas., G. N. Stray.

Lyons, *Ionia Co.* (Pop., 747—612.) Built in '79 by village] for fire protection and stree sprinkling.

70. MANISTEE, *Manistee Co.* (Pop., 6,930—12,812.) Built in '82-3 by Manistee Water Co.; city does not control rates, but has right to buy works at any time, paying cost and 10% premium. **Supply.**—Surface and driven wells, pumping direct. Fiscal year closes May 1. **Distribution.**—*Mains*, 13.5 miles. *Taps*, 890. *Hydrants*, 83. *Consumption*, 625,000 galls. **Financial.**—*Cost*, $172,047. *Cap. Stock:* authorized, $200,000; paid up, $80,000; paid no div'd in 1890. *Bonded Debt,* $80,000 at 6%. *Expenses:* operating, $5,849; int., $4,833; taxes, $275; total, $10,959. *Revenue:* consumers, $9,118; fire protection, $8,326; other purposes, $33; total $17,507. **Management.**—Prest., Simeon Babcock. Secy., Jas. Dempsey. Treas., T. J. Ramsdell. Supt., S. A. Cabill.

71. MARINE CITY, *St. Clair Co.* (Pop., 1,673—3,238.) Built in '84 by town. **Supply.**—St. Clair River, pumping direct. Fiscal year closes March 31. **Distribution.**—*Mains*, 9 miles. *Taps*, 500. *Hydrants*, 96. *Consumption*, 110,584 galls. **Financial.**—*Cost*, $12,000. *Bonded Debt,* $32,000 at 6%. *Expenses:* operating, $1,892; int., $1,920; total, $3,812. *Revenue,* consumers, $3,321. **Management.**—Comrs. Secy., Wm. Caswell. Supt., S. W. Hadden.

72. MARQUETTE, *Marquette Co.* (Pop., 4,690—9,093.) Built in '70 by town. **Supply.**—Lake Superior, pumping direct. Fiscal year closes Dec. 31. In '91 one 3,000,000 Worthington comp. dup. and one 3,000,000 Gaskill horizontal pump were to be added.

Year.	Mains.	Taps.	Meters.	Hyd'ts.	C'n'ption.	Cost.	B'd d'bt.	Int.ch'g's.	Exp.	Rev.
'70	5	300	0	40	......	$74,887	......	....	....	$2,107
'80	5.5	500	4	55	......	118,009	......	....	$3,500	6,973
'85	7	623	2	55	......	......	$90,000	....	7,500	10,000
'88	13	950	4	116	......	171,838	75,000	....	7,500	10,800
'90	15	1,100	3	134	878,282	192,213	75,000	$4,496	10,053	18,386

Financial.—*Bonded Debt,* $75,000: $35,000 at 6%, due in '91; $40,000 at 5%, due in '96 '08, 1900. Bonds to be redeemed by taxation. *Expenses,* total, $14,549. *Revenue:* consumers, $18,386; city, none. **Management.**—Five comrs. appointed by council every five years. Chn., J. W. Wilkinson. Supt., Jno. P. Kern.

73. MARSHALL, *Calhoun Co.* (Pop., 3,795—3,908.) Built in '89 by Marshall Water-Works Co.; controlled by Moffett, Hodgkins & Clarke, Syracuse, N. Y. **Supply.**—Well, pumping to stand-pipe. **Distribution.**—(In '89.) *Mains,* 8 miles. *Taps,* 50. *Hydrants,* 85. **Financial.**—(In '89.) *Cost,* $100,000. *Bonded Debt,* $80,000 at 6%. *Revenue,* city, $4,000. **Management.**—(In '89.) Prest., J. L. Hotchkin. Supt., R. H. Sanderson. NOTE.—Latest report Sept. 1, '89.

74. MENOMINEE, *Menominee Co.* (Pop., 3,288—10,630.) Built in '84 by Menominee Water Co.; city may purchase works in 1905, price settled by agreement or arbitration; franchise fixes rates. **Supply.**—Green Bay, pumping direct. Fiscal year closes Dec. 31. Extensive improvements were contemplated in '91, but had not been authorized Mar. 7.

Year.	Mains.	Taps.	M't's.	Hy'ts.	Con'pt'n.	Cost.	B'd d'bt.	Int. ch's.	Exp.	Rev.
'86	9.3	275	1	94	280,000	$210,000	$100,000	$6,000	$1,900	$12,830
'88	14	760	15	150	750,000	235,000	135,000	8,100	5,500	23,192
'90	15.8	897	..	150	565,000	252,262	132,500	7,950	9,873	23,413

Financial.—*Cap. Stock,* $100,000. *Total Debt,* $132,750; floating, $250 at 6%; bonded, $132,500 at 0%. *Expenses,* total, $17,668. *Revenue:* consumers, $14,023; fire protection, $9,390; total, $23,413. **Management.**—Prest., Wm. A. Lynch. Secy., R. J. Brown. Treas., A. H. Howland, Boston, Mass. Supt., K. H. Adams.

75. MIDLAND, *Midland Co.* (Pop., 1,529—2,277.) Built in '82 by ci'.y. **Supply.**—Chippewa River, pumping direct. **Distribution.**—(In '88.)*Mains,* 4.6 miles. *Taps,*

MICHIGAN.

150. *Meters*, 0. *Hydrants*, 32. *Consumption*, 500,000 galls. **Financial.**—(In '88.) *Cost*, $22,000 *Bonded Debt*, $20,000 at 5%. *Expenses:* operating, $2,600; int., $1,000, total, $3,600. *Revenue:* consumers, $2,637; city, $133; total, $2,770. **Management,**—(In '89.) Secy., Wm. Reardon. Supt., Chas. Fournia.

76. **MIO,** *Oscoda Co.* (Pop., not given.) Built in '89 by Mio Water Co.; organized under general State Law|which provides that town may purchase works after 25 yrs.; franchise does not fix rates. **Supply.**—Spring Creek, pumping to reservoir. Fiscal year closes Nov. 30. **Distribution.**—*Mains*, 0.6 miles. *Taps*, 12. *Hydrants*, 4. *Consumption*, 2,000 galls. **Financial.**—*Cost*, $2,841. *Cap. Stock:* authorized, $10,-000: paid up, $1,550; paid no div'd in 1890. Difference between cost of works and paid-up stock was mainly made up by municipal aid granted to co. in form of advanced hydrant rental. *No Debt.* *Expenses:* operating, $33; taxes, $3; total, $36. *Revenue:* consumers, $95; fire protection. $1,000; total, $1,095. **Management.**—Prest., Alex. McKay. Secy. and Supt., Jno. Randall. Treas., J. J. McCarthy.

77. **MONROE,** *Monroe Co.* (Pop., 4,930—5,258.) Built in '89 by Monroe Water Co. **Supply.**—Lake Erie, pumping direct. **Distribution.**—*Mains*, 15 miles. *Hydrants*, 128. **Financial.**—*Cost*, $50,000. *Cap. Stock*, $50,000. *Hydrant rental*, $46.90. **Management.**—Prest., Geo. Spaulding. Secy., C. W. Scott. Treas., E. H. York, Portland, Me.

Montague, *Muskegon Co.* (Pop., 1,297—1,623.) Built for fire protection.

78. **MOUNT CLEMENS,** *Macomb Co.* (Pop., 3,057—4,748.) Built in '88-9 by city. **Supply.**—Clinton River, pumping direct; filtered.

Year.	Mains.	Taps.	M'rs.	Hy'ts.	Cons'pt'n.	Cost.	B'd'd d't.	Int.ch'g's.	Exp.	Rev.
'89	4	129	1	99	175,000	$49,358	$50,000	$2,500	$3,000	$6,450
'90	7	249	4	105	400,000	49,358	48,000	2,400	3,000	......

Management.—Prest., Paul Ulrich. Secy., Chas. S. Fitch.

79. **MOUNT PLEASANT,** *Isabella Co.* (Pop., 1,115—2,701.) Built in '82-3 by village. **Supply.**—Well, pumping to tank and, in case of fire, direct. **Distribution.**—*Mains*, 4.5 miles. *Taps*, 200. *Hydrants*, 28. **Financial.**—(In '88.) *Cost*, $15,500. *Debt*, $15,000 at 5%. **Management.**—City Marshal, Wm. Dixon.

80. **MUSKEGON,** *Muskegon Co.* (Pop., 11,262—22,702.) Built in '74-5 by co. **Supply.**—Well and branch of Muskegon River, pumping direct. Fiscal year closes July 1. Works for supply from Lake Michigan was designed in '90 by Chester B. Davis, of Chicago, Ill. Contracts were awarded as follows: Building crib and laying pipe in Lake Michigan, Thacher & Breyman, Toledo, O. Pipe-laying in city. James Wallace, Chicago, Ill. Pumping machinery, one 2,000,000 and one 3,000,000 gall, vertical pump, Holly Mfg. Co. July 1, '90, city issued $250,000 worth of bonds at 5% to defray expensenses of improvements.

Year.	Mains.	Taps.	M'tr's.	Hy'd'ts.	C'ns'n.	Cost.	B'd'd d't.	Int. ch'g's.	Exp.	Rev.
'80	16.8	625	0	200	......	$114,736	......	......	$4,900	$7,147
'86	22	1,000		165	$1,000,000	200,000	$140,000	$11,200	9,509	32,044
'89	25.8	1,300	6	302	1,300,000	216,000	110,000	8,800	12,590	15,000
'90	28.6	1,140	0	350	1,385,188	221,286	100,000	8,000	17,192	16,856

Financial.—*Bonded Debt*, $100,000 at 8%. *Expenses*, total, $25,192. *Revenue:* consumers, $16,856; city, none. **Management.**—Board of Public Works, seven members, appointed every three years by mayor, approved by common council. Prest., mayor of city. Secy., Jno. Kuppenheimer. Supt., Wm. Dixon.

81. **NEGAUNEE,** *Marquette Co.* (Pop., 3,931—6,078.) Built in '82-3 by city. **Supply.**—Teal Lake, pumping direct. Fiscal year closes March 31.

Year.	Mains.	Taps.	Meters.	H'd'ts.	B'd'd d'bt.	Int. ch'g's.	Exp.	Rev.
'88-9	9	306	0	33	$20,000	......	$5,837	$5,518
'90-1	10.9	539	0	42	20,000	$1,046	6,101	6,973

Financial.—*Cost* (in '89), $50,000. **Management.**—Supt., A. Benishek.

82. **NEWAYGO,** *Newaygo Co.* (Pop., 1,097—1,330.) Built in '84 by village. **Supply.**—Well, pumping direct. In case of fire, water is pumped from river, by water-power. Fiscal year closes April 1. **Distribution.**—*Mains*, 1.5 miles. *Taps*, 45. *Hydrants*, 14. **Financial.**—*Cost*, $13,000. *Bonded Debt*, $10,000 at 6%. *Sinking Fund*, $2,000. *Expenses:* operating, $450; int., $600; total, $1,050. *Revenue*, consumers, $250. **Management.**—Committee of three, appointed every year by village president. Prest., Geo. E. Taylor. Secy., C. K. Porter. Supt., E. S. Bennett.

For Further Descriptive Matter see Manual for 1889-90. Changes since 1890 are here given. Populations are for 1880 and 1890, respectively.

83. NILES, *Berrien Co.* (Pop., 4,197–4,197.) Built in '78-9 by Niles Water-Works Co.; city neither controls rates nor has right to purchase works. **Supply.**—Barren Lake, by gravity to distributing reservoir. **Distribution.**—*Mains,* 9 miles. *Taps,* 300. *Hydrants,* 50. **Financial.**—Withheld. **Management.**—Prest., S. L. Wiley, Omaha, Neb. Secy. and Supt., Ernest Bacon.

84. NORTH MUSKEGON, *Muskegon Co.* (Pop., in '90, 1,580.) Built in '85-6 by village. **Supply.**—Rain Lake and driven wells, pumping direct. **Distribution.** —(In '88.) *Mains,* 3.5 miles. *Taps,* 150. *Meters,* 0. *Hydrants,* 60. *Consumption,* 250,000 galls. **Financial.**—(In '88.) *Cost,* $30,000. *Bonded Debt,* $23,000: $15,000 at 6%, $8,000 at 5%. *Expenses:* operating, $2,000; int., $1,300; total, $3,300. *Revenue,* consumers, $1,000. **Management.**—(In '89.) Supt., Wm. Collins.

85. OSCODA, *Iosco Co.* (Pop., 1,951–3,593.) Built in '86 by village. **Supply.**—Lake Huron, pumping direct. Fiscal year closes April 30. **Distribution.**—*Mains,* 4 miles. *Taps,* 310. *Hydrants,* 34. *Consumption,* 150,000 galls. **Financial.**—*Cost,* $28,300. *Bonded Debt,* $7,500. *Expenses:* operating, $2,505; int., $505; total, $3,010. *Revenue,* consumers, $1,000. **Management.**—Three comrs., one appointed every year by village president. Prest., Robt. Thompson. Secy., Geo. H. Cosgrove.

86. OTSEGO, *Allegan Co.* (Pop., 1,000–1,626.) Built in '88 by village. **Supply.**—Kalamazoo River, pumping direct. Fiscal year closes March 31.

Year.	Mains.	Taps.	Meters.	Hyd'ts.	Cost.	B'd'd d'bt.	Int. ch'g's.	Exp.	Rev.
'89-90	4.5	8	0	20	$15,000	$9,000	$540	$100	$36
'90-91	4.5	67	0	22	15,000	8,000	480	100	268

Management.—Chn., C. R. Conrad. Supt., C. T. Clapp.

87. OVID, *Clinton Co.* (Pop., 1,479–1,423.) Built in '88 by village. **Supply.**—Wells, pumping direct. Fiscal year closes May 1. **Distribution.**—*Mains,* 4.5 miles. *Taps,* 125. *Hydrants,* 34. **Financial.**—*Cost,* $15,000. *Bonded Debt,* $15,000 at 5%. *Expenses,* $3,658. *Revenue:* consumers, $411; Ry., $500; Mill Co., $300; total, $1,211. **Management.**—Prest., C. C. Taylor. Clerk, Geo. Faxon. Supt., E. F. Luthy.

88. OWOSSO, *Shiawassee Co.* (Pop., 2,501–6,561.) Built in '89 by village. **Supply.**—Wells and river, pumping direct. Reported in '90 that open well was to be sunk, with W. R. Coats, Kalamazoo, Engr. and Contr. Fiscal year closes Dec. 31. **Distribution.**—*Mains,* 15.1 miles. *Taps,* 407. *Hydrants,* 119. *Consumption,* 106,- 849 galls. **Financial.**—*Cost,* $80,000. *Bonded Debt,* $75,000. *Expenses,* operating, $3,300. *Revenue:* consumers, $2,200; city, none. **Management.**—Chn., E. M. Estey. Secy., W. R. Smith. Supt., R. Waters.

89. PETOSKEY, *Emmet Co.* (Pop., 1,815–2,872.) Built in '81 by village. **Supply.**—Well, pumping direct and to tank for fire protection. Fiscal year closes March 14. Dec. 19, 1890, village contemplated building new reservoir 265 ft. above pumping station; also to put in dup. pump and extend mains. A new conduit was completed in '90. **Distribution.**—*Mains,* 2.8 miles. *Taps,* 225. *Hydrants,* 20. *Consumption,* 150,000 galls. **Financial.**—*Cost,* $25,500. *Bonded Debt,* $16,000 at 5%. *Expenses:* operating, $700; int., $800; total, $1,500. *Revenue,* consumers, $1,500. **Management.**—Committee, composed of three members of village council, appointed every year by mayor. Prest., W. W. Price. Secy., A. C. Huntley. Supt., G. S. Richmond, elected by council.

90. PINCONNING, *Bay Co.* (Pop., 300–885.) Built in '88 by village. **Supply.**—Pond, pumping direct. Water-works are not a success, owing to the poorness of water. Fiscal year closes May 1. **Distribution.**—*Mains,* 1.3 miles. *Taps,* 13. *Hydrants,* 15. *Consumption,* 50,000 galls. **Financial.**—*Cost,* $8,500. *Total Debt,* $9,300: floating, $2,300 at 6%; bonded, $7,000 at 7%. *Expenses,* operating, $1,300. *Revenue,* $175. **Management.**—Committee of public works, appointed every two years by village president. Secy., D. A. McDowell. Supt., C. L. Bingham. Report by Geo. Barrie.

Plainwell, *Allegan Co.* (Pop., 1,356–1,414.) Built in 70 by village for fire protection.

91. PONTIAC, *Oakland Co.* (Pop., 4,509–6,200.) Built in '88 by town. **Supply.**—Clinton River and wells, pumping direct. Fiscal year closes Dec. 31.

Year.	M'ns.	Taps.	Hyd'ts	C'ns'pt'n.	Cost.	B'd'd d't.	Int. ch's.	Ex.	Rev.
'88	13	175	115	65,000	$85,000	$85,000	$3,825	$3,300	$1,475
'90	13	285	116	215,143	85,000	85,000	3,825	4,593	2,512

MICHIGAN.

Financial.—*Bonded Debt,* $85,000 at 4½%. *Expenses,* total, $8,118. *Revenue*: consumers, $2,512; city, $5,700; total, $8,212. **Management.**—Chn., D. S. Hammond, Mayor. Secy., J. D. Hammond. Supt., H. G. Monroe.

92. PORT HURON, *St. Clair Co.* (Pop., 8,883—13,543.) Built in '72 by city. **Supply.**—Lake Huron, pumping direct. Fiscal year closes March 31.

Year.	Mains.	Taps.	M'trs.	H'd'ts.	C'ns't'n.	Cost.	Bd'd d'bt.	Int.ch's.	Exp.	Rev.
'82-3	18.5	1,050	0	109	575,907	$180,000	$150,000	......	$7,135	$13,146
'84-5	21.8	1,305	1	119	738,093	189,569	150,000	$7,500	8,286	15,280
'86-7	25.4	1,591	2	129	1,329,216	203,155	110,000	7,000	8,407	18,212
'87-8	27.5	1,818	2	136	1,672,095	222,751	120,000	6,000	9,256	19,056
'88-9	29	1,930	0	147	1,834,751	239,786	120,000	6,000	9,313	21,512
'89-90	30.6	2,139	0	156	1,992,567	254,530	120,000	6,000	9,056	24,547
'90-1	34.6	2,391	0	174	2,001,671	281,698	131,000	6,550	12,196	27,696

Financial.—*Bonded Debt,* $131,000 at 5%, due in 10 yrs. *Expenses,* total, $18,746. *Revenue:* consumers, $27,686; city, none. **Management.**—Three comrs. Secy., Lewis Atkins. Supt., Hugh F. Doran.

Portland, *Ionia Co.* (Pop., 1,670—1,678.) Built in '79 by village for fire protection only.

Quinnesec, *Menominee Co.* (Pop., in '90, 322.) Built in '86 by town. See 1889-90 Manual, p. 426.

93. RED JACKET, *Houghton Co.* (Pop., 2,140—3,073.) Built in '89 by Calumet & Hecla Mining Co., to supply Red Jacket and Calumet and also mining camp. **Supply.**—Lake Superior, pumping direct. A 1,560,000-gall. pump. *Mains,* 5 miles Gen. Mgr., M. Whiting.

94. REED CITY, *Osceola Co.* (Pop., 1,091—1,776.) Built in '82 by village. **Supply.**—Well, spring and Hersey Creek, pumping direct. Fiscal year closes April 1. **Distribution.**—*Mains,* 2 miles. *Taps,* 111. *Hydrants,* 21. **Financial.**—*Cost* (in '88), $24,000. *Bonded Debt,* $2,000 at 6%. *Expenses:* operating, $1,832; int., $120; total, $1,952. *Revenue,* consumers, $1,400. Revenue from private consumers goes into water fund and deficiency is made up from general fund, the proceeds of direct taxation. **Management.**—Three comrs., one appointed every year by common council. Chn., Geo. W. Morris. Secy., Low B. Winsor. Supt., Benj. F. Edwards.

95. ROSCOMMON, *Roscommon Co.* (Pop., 114—511.) Built in '83 by town. **Supply.**—Spring Creek, pumping to tank. **Distribution.**—*Mains,* 2 miles. *Taps,* 30. *Hydrants,* 11. *Consumption,* 20,000 galls. **Financial.**—*Cost,* $5,000. No debt. *Expenses,* $500. *Revenue,* 100. **Management.**—Chn., Jas. B. McCrea. Secy., W. L. Dildine. Supt., Jno. O'Brien.

96. SAGINAW, *Saginaw Co.* (Pop., 29,541—46,322. Apr. 1, '90, East Saginaw consolidated with Saginaw. Pop. of Saginaw in '80, 10,525; of East Saginaw, 19,016.) Saginaw and East Saginaw plants now under same management.

SAGINAW PLANT.

Built in '72 by city. **Supply.**—Saginaw River, pumping direct. In '91 J. D. Cook, Toledo, was engaged as engr. of proposed improvements. Fiscal year closes Dec. 31.

Yr.	Mains.	Taps.	M't's.	H'd'ts.	C'ns'pt'n.	Cost.	B'd'd d't.	Int. ch'gs.	Exp.	Rev.
'82	9	325	0	133	1,000,000	$175,000	$150,000	......	$6,029	$3,618
'86	14.5	353	0	191	2,000,000	200,000	135,000	......	6,546	5,001
'87	15.6	370	0	199	2,000,000	205,000	135,000	......	7,473	5,020
'88	16.5	450	0	207	2,800,000	205,000	135,000	......	8,602	5,537
'89	19.3	562	4	242	2,533,650	229,117	125,000	$7,375	11,184	7,304

Financial.—*Total Debt,* $125,000; *Bonded,* $40,000 at 8%, $25,000 at 5½%, $20,000 at 5%, $40,000 at 4½%; all due '91 to 1903.

EAST SAGINAW PLANT.

Built in '73 by East Saginaw. **Supply.**—Saginaw River, pumping direct; filtered. A 12,000,000-gall. Gaskill hor. comp. cond. was put in operation Feb. 13, '90. Fiscal year closes Dec. 31.

Yr.	Mains.	Taps.	Meters.	Hyd'ts.	Cons'pt'n.	Cost.	B'd'd't.	Int. c'g's.	Exp.	Rev.
'83	25.4	1,093	72	172	2,218,336	$360,314	$400,000	......	$13,919	$22,763
'86	36.5	1,633	82	247	2,136,764	......	414,500	......	13,847	25,052
'87	40	1,911	120	266	2,812,912	486,379	410,000	......	15,967	26,971
'88	40.3	1,995	134	270	3,075,456	503,090	410,000	$22,805	16,387	28,892
'89	41	2,083	144	276	3,152,237	500,857	410,000	21,803	16,070	31,400
'90	43.5	2,291	154	287	3,352,329	533,985	410,000	21,073	16,693	32,198

For Further Descriptive Matter see Manual for 1889-90. Changes since 1890 are here given. Populations are for 1880 and 1890, respectively.

Financial.—*Total Debt*, $439,160: *Floating*, $29,160 at 5%; due, $10,180, Nov. 1, '91; $9,720, Nov. 1. '92: $9,260, Nov. 1, '93. *Bonded*, $410,000, of which $18,500 is at 8%, due May 1, '93; $10,000 at 7%, due May 1, '91; $50,000 at 6%, due May 1, '92; $247,500 at 5%, due from '94 to 1905; $31,000 at 4½%, due in 1903; $52,500 at 4%, due in '97 and '99. **Management.** Comrs. Secy , J. F. Maher.

97. **SAINT CLAIR**, *St. Clair Co.* (Pop., 1,923—2,353.) Built in '86 by village. **Supply.**—St. Clair River, pumping direct. **Distribution.**—*Mains*, 5.3 miles. *Taps*, 507. *Hydrants*, 38. **Financial.**—*Cost* (in '87), $32,500. *Expenses*, $2,700. *Revenue*, $2,900. **Management.**—Supt., Frank Schrepperman.

98. **SAINT IGNACE**, *Mackinac Co.* (Pop., 934—2,704.) Built in '89 by city. **Supply.**—Lake Huron, pumping to stand-pipe. Fiscal year closes March 1. **Distribution.**—*Mains*, 4.8 miles. *Taps*, 120. *Hydrants*, 39. *Consumption*, 40,000 galls. **Financial.**—*Cost*, $38,707. *Bonded Debt*, $40,000 at 5%, *Expenses:* operating, $1,500; int., $2,000; total, $3,500. *Revenue*, consumers, $800. **Management.**—Chn., Wm.. Saulson. Secy., P. W. Murray.

99. **SAINT JOHN'S**, *Clinton Co.* (Pop., 2,370—3,127.) Built in '89 by village. **Supply.**—Artesian wells, pumping direct. Fiscal year closes April 1. During '90 two dup. pumps were added. **Distribution**—*Mains*, 11.5 miles. *Taps*, 243. *Hydrants*, 88. *Consumption*, $71,768 galls. **Financial.**—*Cost*, $57,444. *Bonded Debt*, $60,000 at 5%, due in annual payments of $1,000 after 1900. *Expenses:* operating, $3,400; int., $2,625; total, $6,025. *Revenue:* consumers, $1,473; city, $3,656; total, $5,129. **Management.**—Prest., Jno. D. Henderson. Secy., De Witt H. Hunt. Supt., E. King.

100. **SAINT LOUIS**, *Gratiot Co.* (Pop., 1,975—2,246.) Built in '80 by village. **Supply.**—Two flowing wells, pumping direct. **Distribution.**—*Mains*, 2 miles. *Hydrants*, 32. **Financial.**—*Cost*, $8,000. *Revenue*, $800. **Management.**—Chn., B. S. Nelson. Supt., L. A. Thompson.

Sand Beach, *Huron Co.* (Pop., 534—1,046.) Built in '81 by J. Jenks & Co., for fire protection only.

101. **SAULT DE STE MARIE**, *Chippewa Co.* (Pop., 1,947—5,760.) Built in '68 by city. **Supply.**—Sault Ste Marie River, pumping direct. Fiscal year closes Feb. 28.

Year.	Mains.	Taps.	Meters.	Hydrants.	Cost.	B'd'd d't.	Int. chgs.	Exp.	Rev.
'89-90	6	450	0	50	$30,000	$50,000	$3,240	$4,500	$5,500
'90-1	8.8	480	0	59	58,871	50,000	3,210	4,841	5,270

Financial.—*Bonded Debt*, $50,000: $40,000 at 6½%; $10,000 at 6%. *Expenses*, total, $3,081. *Revenue:* consumers, $5,270; city, none. **Management.**—Board of Public Works, appointed every three years by Mayor, approved by Council. Mayor, A. M. Fowle. Chn., B. Disemberg. Secy., G. S. Trempe. Supt., D. McGregor.

Sheridan, *Montcalm Co.* (Pop., 661—447.) Built in '82 by village for fire protection.

Sparta, *Kent Co.* (Pop., 507—904.) Sparta Water Co. built system in '91 for fire protection only. Cost, $10,000.

102. **STAMBAUGH**, *Iron Co.* (Pop., in '90, 711.) **History.**—Built in '90 by village; put in operation Nov. 1. **Supply.**—Iron River, pumping direct. **Pumping Machinery.**—Dy. cap., not given; from National Tube Wks., Chicago. **Distribution.**—*Mains*, 6-in. i., 1.5 miles. *Services*, ¾-in. i. *Taps*, 6. *Hydrants*, 10. *Pressure:* ordinary, 100 lbs.; fire 175 lbs. **Financial.**—*Cost*, $5,000. *Bonded Debt*, $5,000 at 7%. **Management.**—Prest., J. N. Porter. Treas., F. C. Vilas.

103. **STANTON**, *Montcalm Co.* (Pop., 1,760—1,352.) Built in '85 by village. **Supply.**—Flowing well, pumping to reservoir. **Distribution.**—(In '88.) *Mains*, 2.2 miles. *Taps*, 113. *Meters*, 0. *Hydrants*, 32. **Financial.**—(In'88.) *Cost*, $15,094. *Bonded Debt*, $7,000 at 6%. *Expenses:* operating, $1,500; int., $420; total, $1,920. *Revenue*, consumers, $950. **Management.**—Comr., J. B. Smith.

104. **STURGIS**, *S. Joseph Co.* (Pop., 2,060—2,489.) Built in '89 by city. **Supply.**—Well, pumping direct. Fiscal year closes March 9. **Distribution.**—*Mains*, 7.5 miles. *Taps*, 232. *Hydrants*, 54. *Consumption*, 42,874 galls. **Financial.**—*Cost*, $30,358. *Bonded Debt*, $30,000 at 5%. *Expenses:* operating, $1,478; int., $1,500; total, $2,978. *Revenue*, consumers, $2,637. **Management.**—Chn., Dennis Kane. Supt., Jas. W. Beck.

MICHIGAN.   229

105. **TECUMSEH,** *Lenawee Co.* (Pop., 2,111—2,310.) Built in '77 by village.
Supply.—Well, pumping to tank. Distribution.—*Mains,* 1 mile. *Hydrants,* 19.
Financial.—*Cost,* $7,000. *No Debt. Expenses,* $150. *Revenue,* $50. Management.—Supt., S. W. Bordine.

106. **THREE RIVERS,** *St. Joseph Co.* (Pop., 2,525—3,131.) Built in '82 by
village. Supply.—Nine driven wells, pumping direct. Fiscal year closes in Feb.
Distribution.—*Mains,* 5 miles. *Taps,* 195. *Hydrants,* 56. *Consumption,* 2,000,000
galls. Financial.—*Cost,* $31,000. *Bonded Debt,* $18,000 at 5%. *Expenses :* operating, $100; int., $900; total, $1,300. *Revenue:* consumers, $1,300. Management.—Chn., J. W. French. Sec.. H. D. Cushman. Supt., E. H. Burgert.

107. **TRAVERSE CITY,** *Grand Traverse Co.* (Pop., 1,897—4,353.) Built in '81-2
by H. D. Campbell. Franchise provides for purchase of works by city in '96, price
to be determined by three disinterested persons. Rates must be approved by city
council. Supply.—Grand Traverse Bay, pumping direct from settling basin.
Fiscal year closes Dec. 31.

Year.	Mains.	Taps.	Meters.	Hyd'ts.	Consumption.	Cost.	Exp.	Rev.
'86	5.5	196	3	45	200,000	$28,000	$1,800	$4,800
'89	6.	248	6	45	200,000	35,000	2,000	5,654
'90	7.5	283	6	50	205.000	36,100	2,000	6,287

Financial.—*No Debt. Revenue :* consumers, $2.935; fire protection. $3.312; other
purposes, $71; total, $6,287. Management —Owner, H. S. Campbell. Supt., D. F.
Campbell.

108. **VASSAR,** *Tuscola Co.* (Pop., 670—1,632.) Built in '89 by village. Supply.
—Four artesian wells and Cass River, pumping to stand-pipe and direct. Reported
June '91, that wells were to be added. Fiscal year closes July 1. Distribution.—
Mains, 5.5 miles. *Taps,* 85. *Hydrants,* 55. *Consumption,* 33,500 galls. Financial.—
Cost, $32,000. *Bonded Debt,* $30,000 at 5%. *Expenses:* operating, $950; int., $1,500; total,
$2,450. *Revenue:* consumers, $600; city, none. Management.—Three comrs., one appointed every year by council. Chn., Frank L. Fales. Clerk, Jno. Loranger. Supt.,
Wm. J. Bennett.

109. **WEST BAY CITY,** *Bay Co.* (Pop., 6,397—12,981.) Built in '80-3 by city.
Supply.—Saginaw River, pumping direct. Fiscal year closes Dec. 31. Late in '90,
test wells were being sunk for purpose of securing purer supply of water.

Year.	Mains.	Taps.	Meters.	Hydrants.	Consumption.	Cost.	Exp.	Rev.
'84	8	114	0	74	400,000	$75,000	$3,100	.....
'86	9	146	0	83	600,000	78,500	2,975	$5.612
'89	9.5	267	1	121	810,063	79,734	4,923	3,034
'90	11	380	1	122	1,100,000	......	6,000	3,500

Financial.—*No Debt. Revenue:* consumers, $3,500 ; city, $3,000 ; total, $6,500.
Expenses : operating, $14,923 ; int., $938; total, $15,861. *Revenue :* consumers,
$3,034 ; city, fire protection, $12,000. Management.—Board of Water-Works,
five members, appointed every five years, by Common Council. Chn., Thos. Tookey.
Secy. and Supt., W. H. Phillips.

110. **WHITE CLOUD,** *Newaygo Co.* (Pop., 440—743.) Built in '86 by village.
Supply.—White River, pumping direct. Distribution.—*Mains,* 1.5 miles. *Taps,*
180. *Hydrants,* 11. Financial.—*Cost,* $10,500. *Expenses,* $15. *Revenue,* $650. Management.—Secy., L. D. Sidebottom, Supt., Geo. W. Oaks.

111. **WHITEHALL,** *Muskegon Co.* (Pop., 1,724—1,903.) History.—Built in '91
by village. Engr. and Contr., M. Walker, Port Huron. Supply.—Wells, pumping
direct. Pumping Machinery.—Dy, capacity, 1,500,000 galls.; Walker. Distribution.—*Mains,* 10 to 4-in. c. i., 3½ miles. *Services,* lead and galv. i. *Hydrants,*
Ludlow, 32. *Valves,* Walker. *Pressure:* ordinary, 40 lbs.; fire, 70 to 120 lbs.
Financial.—*Cost,* $17,000. *Bonded Debt,* $15,000 at 5%. Management —Prest., J.
H. Williams. Secy., D. E. Staples. Supt., F. Hinman.

112. **WYANDOTTE,** *Wayne Co.* (Pop., 3,631—3,817.) Built in '89-90 by city.
Supply.—Detroit River, pumping direct. Fiscal year closes third Monday in
March. Distribution.—*Mains,* 8 miles, *Taps,* 256. *Hydrants,* 81. *Consumption,*
60,595 galls. Financial.—*Cost,* $61,371. *Bonded Debt,* $50,000 at 7%, due 1900-30. *Expenses :* operating, $6.701; int., $3,500; total, $10,201. *Revenue :* consumers, $1,902;
city, $4,785; total, $6,687. Management.—Chn., T. J. Langlois. Secy., John McG.
Laughlin. Supt., H. N. Osobock.

For Further Descriptive Matter see Manual for 1889-90. Changes since 1890
are here given. Populations are for 1880 and 1890, respectively.

230   MANUAL OF AMERICAN WATER-WORKS.

113. **YPSILANTI**, *Washtenaw Co.* (Pop., 1,459—6,129.) Built in '89 by city. Supply.—Well; pumping to elevated tank. Fiscal year closes April 30.

Year.	Mains.	Taps.	Meter.	Hyd'ts.	Cons'ption.	Cost.	B'd'd d't.	Int. c'g's.	Exp.	Rev.
'89-90	17.5	471	1	125	245,238	$146,285	$125,000	$5,000	$1,606	$1,167

Management. Five comrs. Prest., Daniel L. Quirk. Secy., Frank Joslyn.

Water-Works Projected with Fair Prospects of Construction.

BUCHANAN, *Berrien Co* (Pop., 1,894—1,994.) Aug. 13, '91, it was reported that city had voted to issue $50,000 of bonds, to construct water-works, that contract for building dam had been let to E. R. Beardsley, Waldron, Ill., and that construction of system would begin at once. Chn., C. H. Baker.

ELK RAPIDS, *Antrim Co.* (Pop. of township, 741—1,486.) Sept. 12, '91, Elk Rapids Water Supply Co. was incorp.; cap., $100,000. About same time people voted in favor of works.

L'ANSE, *Barago Co.* (Pop., 1,014—855.) Oct. 26, '90, S. Kinney reported that village had let conditional contract to G. C. Morgan, Chicago, to be completed June 1, '92. Supply, Spring Creek, by gravity. Est. cost, $15,000. Engr. for surveys, H. G. Rothwell, L'Anse.

MASON, *Ingham Co.* (Pop., 1,809—1,875.) In '91 city voted to issue $35,000 of bonds. In Sept. contract was let to Lansing Iron & Engine Works for $31,740. Consult. Engr., W. S. Parker. Mayor, M. J. Murray.

MILFORD, *Oakland Co.* (Pop., 1,251—1,138.) June 6, works were talked of but no definite steps had been taken.

NASHVILLE, *Barry Co.* (Pop., 978—1,029.) System voted by village, pumping to stand-pipe and direct. Prest., L. E. Lentz. Clerk. H. C. Zuschnitt. Latest report Aug. 27, '91.

NORTHVILLE, *Wayne Co.* (Pop., 634—1,573.) Aug. 13, '91, village voted to have water-works; further details not decided. Village Clerk, C. A. Downs.

PLYMOUTH, *Wayne Co.* (Pop., 1,025—1,172.) Sept. 14, '91, it was reported that supply had been secured from spring, 4 miles distant, and a 12- and 10-in. pipe laid to village. Bids had been advertised for and construction was to begin at once. Engr., W. S. Parker. Chn., W. F. Markham.

ROMEO, *Macomb Co.* (Pop., 1,629—1,637.) Aug., '91, village voted to issue $26,000 of bonds to construct works, 2¼ miles of pipe, 40 hydrants, 2 pumps. Bids were asked in Sept. Unofficially reported that all bids were rejected and nothing would be done till spring of '92. Clerk, B. N. Seaman.

ST. JOSEPH, *Berrien Co.* (Pop., 2,603—3,733.) Bids were to be received by city until Nov. 21, '91, for building works, including two non-cond. dup. 1,500,000-gall pumping engines, 7 miles 12 to 4-in. iron mains, with hydrants, etc. City Clerk, C. A. Keeler.

SOUTH HAVEN, *Van Buren Co.* (Pop., 1,442—1,924.) Water-works talked of but no definite steps taken. Address Geo. S. Seaver.

Water-Works Projects Reported in '90 and '91, but Now Indefinite.

Fowlerville, Livingston Co.   *West Branch*, Ogemaw Co.

ILLINOIS.

Water-Works Completed or in Process of Construction.

1. **ALEDO**, *Mercer Co.* (Pop., 1,192—1,601.) Built in '89 by city. Supply.—Artesian well, pumping to tank. Distribution.—*Mains*, 1.5 miles. *Hydrants*, 8. Financial.—*Cost*, $22,000. *Debt*, $16,000 at 5%. Management.—Mayor, J. A. Cummins. City Clerk, R. B. Talcaferro.

ILLINOIS.   231

2. ALTON, *Madison Co.* (Pop., 8,975—10,294.) Built in '75 by Alton Water-Works Co.; franchise fixes rates and provides for purchase of works by city. Supply.—Mississippi River, pumping to tank and direct; filtered. Reported early in '90 that new filter had been placed in river.

Year.	Mains.	Taps.	Meters.	Hydrants.	Consumption.	Cost.	Exp.	Rev.
'80	12	106	51	87	......	......	$3,600	......
'86	12	230		90	......	$110,000	9,413	*$8,850
'88	16	250	46	97	200,000	110,000	10,000	12,000
'90	23	335	55	113	100,000	150,000	9,500	*9,500

* From city.

Financial.—Withheld, except data above. **Management.**—Prest., T. A. Taylor. Secy., W. A. Underwood. Supt., W. G. Curtis.

3. AURORA, *Kane Co.* (Pop., 11,873—19,688.) Built in '85 by city. Supply.—Fox River, pumping to stand-pipe and direct; filtered. Fiscal year closes Dec. 31. Sept. 26 artesian well had been sunk to depth of 1,370 ft. and est. dy. flow of 500,000 galls. struck.

Yr.	M'ns.	Taps.	M't'rs.	Hyd'ts.	Cons'pt'n.	Cost.	B'd'd d't.	Int.ch'g's.	Exp.	Rev.
'86-7	20.5	485	2	215	....	$170,213	$138,000	$3,467	$5,404	$5,674
'87-8	20.8	738	125	221	713,499	172,336	138,000	6,937	6,764	9,535
'89	25	1,118	215	247	799,389	188,702	138,000	6,934	6,576	10,521

Financial.—*Total Debt*, $138,000. bonded at 5%; $46,000 due in '95, $46,000 in 1900, $46,000 in 1905. *Expenses and Revenue,* see above. Bonds are redeemed by taxation. No allowance for public uses of water. **Management.**—Board of Public Works, five members, appointed every two years by mayor and council. Prest., C, C. Earle. Secy., J. M. Kennedy. Supt., M. W. Corbett.

72. AUSTIN, *Cook Co.* (Pop., 1,359—4,051.) Supply from Oak Park.

Barry, *Pike Co.* (Pop, 1,392—1,354.) Built in '80-1 by village for fire protection, stock watering and street sprinkling.

4. BELLEVILLE, *St. Clair Co.* (Pop., 10,683—15,361.) Built in '85 by City Water Co. Supply.—Richland Creek and branch of same, pumping from impounding reservoir and stand-pipe; filtered. Fiscal year closes June 23, In '89 works were put into hands of receiver, by whom they were sold to bondholders June 23, '90. In '90, reservoir No. 3 was built similar to the one at Lake Christine; cap., 80,000,000 galls ; area, 15 acres; height of dam, 25 ft. Distribution.—*Mains,* 14 miles. *Taps,* 450. *Hydrants.* 105. *Consumption,* 800,000 galls. Financial.—Withheld. *Cost* (in '87), $175,000. **Management.**—Mgr., M. T. Stookey.

5. BELVIDERE, *Boon Co.* (Pop., 2,951—3,867.) History.—Built in '91 by city, to be in operation Oct. 15. Engr. and contr., Geo. Cadogan Morgan, 49 Major Block, Chicago. Supply.—Artesian wells, pumping to Cadogan water tower and to reservoir. Pumping Machinery.—Dy. cap., 2,000,000 galls.; two 1,000,000 Deane. Boilers, two 50 HP. Tower.—Cap., 42,000 galls.; 124 ft. high, on elevation of 48 ft. above pump. Reservoir.—Cap., 60,000 galls. Distribution.—*Mains,* 10 to 4-in. c. i., 5 miles. *Hydrants.* Mathews, 36. *Valves,* Eddy. *Pressure,* ordinary, 50 lbs.; fire, 90 to 100 lbs. Financial.—*Cost,* $35,000. **Management.**—Com. of three, Chn., F. R. Smiley. Supt., E. E. Spooner.

6. BLOOMINGTON, *McLean Co.* (Pop., 17,180—20,048.) Built in '75 by city. Supply.—Surface and tube-wells, pumping to stand-pipe and direct. During winter of '90-1 seven additional 8-in. tube wells were sunk. Fiscal year closes May 30.

Year.	Mains.	Taps.	M't'rs.	H'd'ts.	Cons'pt'n.	B'd'd d't.	Int.ch'g's.	Exp.	Rev.
'82-3	11	600	0	150	500,000	$75,000	$6,000	$7,203	$8,502
'86-7	15	1,100	50	175	....	75,000	6,000	9,000	12,000
'88-9	18	1,136	210	205	700,000	75,000	6,000	8,340	13,635
'89-90	18	1,126	180	205	799,730	50,000	3,500	9,343	13,927
'90-1	22	1,240	300	220	1,100,000	55,000	7,100	......	14,899

Financial.—*Bonded Debt*, $55,000 at 8%, due in annual payments of $10,000. *Revenue,* consumers, $14,899; city, none. **Management.**—Board of three, appointed from City Council each year. Prest., Cheney Moulton. Secy., Rolla N. Evans. Supt., H. W. Schmidt.

7. BLUE ISLAND, *Cook Co.* (Pop., 1,542—2,521.) Built in '76 by village. Supply.—Well, pumping to tank. Distribution.—(In '88.) *Mains,* 4 miles. *Taps,* 175.

FOR FURTHER DESCRIPTIVE MATTER see Manual for 1889-90. CHANGES since 1890 are here given. POPULATIONS are for 1880 and 1890, respectively.

Meters, 4. Hydrants, 50. Consumption, 60,000 galls. Financial,—(In '88.) Cost, $30,000. Bonded Debt, $13,000 at 6%. Management.—(In '89.) Clerk, F. Hohman. Supt., A. Reiner.

Blue Mound, Macon Co. (Pop., in '90, 696.) Built in '86 by village for fire protection.

Buckley, Iroquois Co. (Pop., 324—433.) Built in '89—90 by village, for fire protection.

BUDA, Bureau Co. (Pop., 778—990.) System described in 1889-'90 Manual, p. 434, is not in operation. April 27, '91, T. J. Lockwood, Prest. Buda Water-Works Co. wrote: "At present we are without water-works, but hope soon to have business revival."

8. **BUSHNELL,** McDonough Co. (Pop., 2,316—2,314.) History.—Built in 89-90 by city. Supply.—Tube wells, pumping to stand-pipe. Wells are 6 ins. in diam and 110 ft. deep. Pumping Machinery.—Dy. cap., 500,000 galls.; Deane, dup. Boiler, 54 × 14 ins.; from Frost Mfg. Co., Galesburg, Ill. Stand-Pipe.—Cap., not given; 12 × 52 ft., on tower 52 ft. high. Distribution.—Mains, 8 to 4-in. c. i., 2 miles. Pipes and specials from Chattanooga (Tenn.) Foundry Co. Taps, 200. Meters 1. Hydrants, 25. Financial.—Cost, $21,000. Bonded Debt, $21,000 at 5%. Management.—City Council. Prest., H. M. Harrison, Mayor. City Clerk, O. C. Hicks, Supt., Wm. Vanderver.

☞9. **CABERY,** Ford Co. (Pop., 114—342.) Built in '87 by village. Supply.—Artesian well, pumping to tank. Distribution.—Mains, 4,000 ft. Taps, 23. Hydrants, 14. Financial.—Cost, $5,800. Debt, $1,000 at 6%. Expenses: operating, $35; int., $60; total, $95. Revenue, consumers. $115. Management.—Village Prest., W. Willis.

10. **CAIRO,** Alexander Co. (Pop., 9.611—10,324.) Built in '85 by Cairo Water Co. Supply.—Ohio River, pumping to stand-pipe, or direct; American filter. Fiscal year closes Dec. 31.

Year.	Mains.	Taps.	Met's.	Hyd's.	Consump'n.	Cost.	B'd'd dt.	Int.	ch's	Exp.	Rev.
'88	15	345	0	158	500,000		$100,000	$6,000			$8,300
'89	15	659	0	157	700,000	$150,000	100,000	6,000		$10,000	21,203
'90	16	691	0	160	800,000	152,500	100,000	6,000		9,000	26,404

Financial.—Cap Stock, $100,000. Bonded Debt, $100,000 at 6%. Expenses, total, $15,000. Revenue: consumers, $20,134; city, $6,270; total, $26,404. Management.—Prest., H. B. Mellick, Williamsport, Pa. Sec., H. Hinkley, Supt., Thos. W. Gaines.

11. **CANTON,** Fulton Co. (Pop., 3,762—5,604.) Built in '84 by town. Supply.—Surface wells, pumping to stand-pipe and direct. Distribution.—(In '88.) Mains, 1.8 miles. Taps, 125. Meters, 6. Hydrants, 30. Consumption, 100,000 galls. Financial.—(In '88.) Cost, $28,000. Bonded Debt, $25,000 at 5%. Expenses, operating, $1,000; int., $1,250; total, $2,250. Revenue: consumers, $1,100; city, $3,000; total, $4,100. Management.—Supt., Geo. W. Seaton.

12. **CARLINVILLE,** Macoupin Co. (Pop., 3,117—3,293.) Built in '89-90 by Carlinville Water Co. Supply.—Macoupin Creek, pumping to stand-pipe or direct; American filter. Distribution.—Mains, 5.5 miles. Taps, 76. Meters, 1. Hydrants, 52. Financial.—Cost, $50,000. Cap. Stock, authorized, $75,000. No stocks or bonds had been sold Feb. 18, '91. Expenses: operating, $2,000; taxes, $700; total, $2,700. Revenue: consumers, $2,500; city, $3,000; total, 5,500. Management.—Prest. and Treas., W. G. Reeve. Secy., F. Wm. Raeder. Supt, Geo. Siegel.

13. **CARLYLE,** Clinton Co. (Pop., 2,017—1,784.) Built in '87 by Carlyle Water, Light & Power Co.; city regulates rates and may buy works in 1902. Supply.—Kaskaskia River, pumping to stand-pipe: Hyatt filter. Fiscal year closes Dec. 31. Distribution.—Mains, 3.5 miles. Taps, 50. Meters. 2. Hydrants, 43. Consumption, 10,000 galls. Financial.—Cap. Stock, $20,000. Bonded Debt, $35,000 at 6%, due in 1907. Expenses: operating, $750; int., $2,100; taxes, $130; total, $2,980. Revenue: consumers, $540; fire protectirn, $2,100; total, $2,640. Management.—Prest. and Supt., J. W. Randall. Secy., T. Wm. Raeder.

14. **CARROLLTON,** Greene Co. (Pop., 1,934—2,258.) History.—Built in '90 by city; put in operation Jan. 1, '91. Engr. and Contr., Geo. Codogan Morgan, 49 Major Block, Chicago. Supply.—Artesian well, flowing to storage reservoir, thence pumping to tank and direct. Well is 1,321 ft. deep and was bored in '89-90 by W. H.

ILLINOIS. 233

Gray & Co., Chicago. **Pumping Machinery.**—Dy. cap., 1,000,000 galls.; Deane dup. **Reservoir.**—Cap., 56,520 galls. **Tank.**—Cap., 35,000 galls. **Distribution.**—*Mains*, 8 to 4-in., c. i., 4 miles. *Services*, galv. i. *Taps*, 78. *Hydrants*. Morgan 35. *Valves*, Peet. *Consumption*, 50,000 galls. *Pressure:* ordinary, 50 lbs.; fire, 125 lbs. **Financial.**—*Cost*, $23,000. *Bonded Debt*, $20,000 at 5%. *Expenses:* op-rating, $1,200; int., $1,000; total, $2,200. *Revenue*, consumers, $500. **Management.**—Mayor, Ed. Miner. City Clerk, D. E. Fox.

15. **CARTHAGE,** *Hancock Co.* (Pop., 1,594—1,654.) Built in '87 by town. **Supply.**—Artesian well, pumping from receiving reservoir to tank and direct. Fiscal year closes April 1. **Distribution.**—*Mains*, 2 miles. *Taps*, 12. *Hydrants*, 32. **Financial.**—*Cost*, $15,300. *Total Debt*, $13,300 : floating, $1,000 ; bonded, $12,300 at 6%, due in annual payments of $1,000. *Expenses:* operating, $675 ; int., $738; total, $1,413. *Revenue*, consumers, $309. **Management.**—Chn., U. P. M'Kee.

16. **CHAMPAIGN AND URBANA,** *Champaign Co.* (Pop.: Champaign, 5,103—5,839; Urbana, 2,942—3,880.) Built in '85 by Union Water Supply Co. City has right to purchase works at end of every five years from time of construction. **Supply.**—Surface well by direct pumping and pumping to reservoir; National filter. Fiscal year closes May 1. During '91, an additional pump was put in.

Year.	Mains.	Taps.	M't'rs.	Hyd'ts.	C's pt'n.	Cost.	B'd debt.	Int.ch'g's.	Exp.	Rev.
'87	13	....	....	100	250,000	$200,000	$100,000	$6,000	$10,000	$12,006
'89	13	476	20	100	300,000	140,000	100,000	6,000	9,000	15,320
'90-1	14	600	15	99	400,000	200,000	100,000	6,000	7,000	16,520

Financial.—*Cap. Stock*, $50,000. *Bonded Debt*, $100,000 at 6%. *Expenses:* operating, $6,000; int. $6,000; taxes, $1,000; total, $13,000. *Revenue:* consumers, $4,000; fire protection, $5,320; other purposes, $7,200; total, $15,320. **Management.**—Prest., Chas. Zilly. Secy. and Supt., W. B. McKinley.

17. **CHARLESTON,** *Coles Co.* (Pop., 2,867—4,135.) Built in '76 by city. **Supply.**—Embarras River, pumping direct. **Distribution.**—(In '87.) *Mains*, 10 miles. *Hydrants*, 35. **Financial.**—(In '87.) *Cost*, $10,000. *Bonded Debt*, $21,000 at 10%. *Expenses:* operating, $3,300; int., $2,100; total, $5,400. *Revenue*, consumers, $3,500. **Management.**—(In '88.) Secy., J. Trainer. Supt., Frank Cooper.

18. **CHEBANSE,** *Iroquois Co.* (Pop., 728—616.) Built in '76 by Chebanse Water-Works Co. Supply, source not given, pumping to tank. **Distribution.**—(In '88.) *Mains*, 1,500 ft. *Taps*, 20. *Meters*, 0. *Hydrants*, 16. **Financial.**—(In '88.) *Cost*, $3,500. *Cap. Stock*, $3,500. *No Debt*. *Expenses*, $50. **Management.**—Mgr., Geo. Spies.

19, 20, 21. **CHICAGO,** *Cook Co.* (Pop., 503,185—1,099,850.) Present works built by city in in '52-4. First supplied by hydraulic co. in '42; co. evidently ceased to operate works after city plant was built. Hyde Park and Lake (see below). First supplied in '74, Lake View in '76. **Supply.**—Lake Michigan, pumping direct. Fiscal year closes Dec. 31.

In '89 and '90 area of city was increased by annexation from area of 43.56 to 180.5 sq. miles. Annexed territory includes former city of Lake View, towns of Lake and Jefferson, and part of Cicero; villages of Hyde Park, Washington Heights, Gano and West Roseland, and Pullman. Previous to annexation Hyde Park and Lake were supplied jointly, and Lake View, Washington Heights and Pullman each had independent works, all of which, save Pullman plant, came under control of Chicago. All plants but Washington Heights were described in '89-90 Manual, pp. 438, 439. See Washington Heights and Pullman plants below. New tunnels in connection with the main works and Lake View division are still under construction, a contract having been let in '91 to extend latter some 5,000 ft. further. For illustrated description of former, by Asst. City Engr. Bernhard Feind, see ENGINEERING NEWS, vol. xxv., p. 2. At Lake View pumping station new w. i. intake pipe, 30-in., 1,100 ft. long, was added in '90; in addition there are now two 20-in. w. i. intakes, 2,000 ft. and 1,400 ft. in length, and a 16-in. w. i., 1,700 ft. in length. These pipes have total dy. cap. of 28,000,000 galls. A measure to guard against formation of anchor ice at mouth of 30-in. intake by compressed air was illustrated and described in ENGINEERING NEWS, vol. xxv., p. 622.

FOR FURTHER DESCRIPTIVE MATTER see Manual for 1889-90. CHANGES since 1890 are here given. POPULATIONS are for 1880 and 1890, respectively.

234   MANUAL OF AMERICAN WATER-WORKS.

Total pumping cap. close of '90, is given in annual report for '90 as 260,000,000 galls., but owing to unfinished state of tunnels this cap. was really not available and in addition some of the pumps had not been tested. Pumping engines, as reported at various stations, are as follows: *North Side*, 36,000,000, 18,000,000, 12,000,000, 8,000,000 condensing beam and two 12,000,000 Gaskill; total, 98,000,000 galls.; the twelve and eight-million beams have suctions often out of water. *West Side*, four 15,000,000 comp. cond. beam; total, 60,000,000 galls. *Central*, two 15,000,000 Allis; total, 30,000,000 galls.; neither tested time of report, and water for one only. *Lake View*, 12,000,000 Gaskill, 5,000,000 Worthington, 2,000,000 Vergennes; total, 19,000,000 galls. *Hyde Park*, 12,000,000 Worthington, tested but not accepted, time of report; two 12,000,000 Gaskill, 8,000,000 Holly vertical, two 3,000,000 Gordon & Maxwell, one 3,000,000 Knowles; total, 53,000,000 galls. In '91 contract was let for two 12,000,000 Gaskill high duty, crank and fly wheel, pumping engines to be placed at Hyde Park station. An independent main will be laid from these to supply Columbian Exposition under 90 lbs. pressure. After exposition pumps will be used for regular supply.

Y'r.	M's.	Taps.	M't's.	H'd's.	C'p'n.	Cost.	B'd'd d't.	Int. c'gs.	Exp.	Rev.
'60	91	6,350	....	....	4,703,525	$1,020,060	....	....	....	$131,163
'70	272	35,318	....	....	21,776,260		....	....	....	539,318
'80	455	67,949	2,113	3,361	57,384,376	8,802,725	$3,955,000	$282,873	$220,000	860,369
'81	472	73,627	2,163	3,553	63,922,709	8,979,523	3,955,000	274,133	289,000	936,802
'82	497	78,840	2,310	3,825	66,166,939	9,353,315	3,955,000	269,535	317,132	1,049,033
'83	519	85,496	2,503	4,144	73,265,592	9,695,956	3,955,000	264,924	371,860	1,142,868
'84	543	92,133	2,685	4,616	80,017,990	10,099,658	3,955,000	264,929	391,000	1,204,339
'85	565	98,688	2,897	4,943	91,650,190	10,416 344	3,955,000	264,955	345,000	1,339,038
'86	596	106,771	3,085	5,350	97,789,706	10,931,397	3,955,000	264,814	480,000	1,374,837
'87	638	115,578	3,278	5,899	101,937,842	11,398,422	3,955,000	264,374	504,000	1,490,000
'88	677	125,867	3,122	6,367	104,315,624	12,390,464	3,955,000	262,525	520,100	1,577,116
'89	730	133,284	3,287	6,983	110.895,707	13,772,562	3,955,000	255,937	556,000	1,621,750
'90*	1,205	155,096*	3,924	11,836	152,372,288	16,902,190	1,938,900	309,625	912,893	2,109,508

* Figures for '90 are for the whole city including all works in annexed district, save those in Washington Heights described separately below. Annexation in '89 and '90 added 346 miles of mains, an unknown number of taps, 291 meters, and 3,465 hydrants. Taps for '90 are too small by all in annexed districts prior to annexation. Cost of works in annexed district, $1,160,164.

Financial.—Total Bonded Debt, $4,938,900, as follows: Main works, $3,955,000; Lake bonds, $401,900; Hyde Park bonds, $434,000; Lake View bonds, $148,000. *Chicago Bonds:* $2,847,000 at 7%, due $821,000 July 1, '92, $541,000 due July 1, '94, $1,485,000 due July 1, '95; $132,000 at 6%, due $50,000 July 1, '97, $82,000 due July 1, '98; $150,000 at 4%, due July 1, 1908; $333,000 at 3.65%, due July 1, 1902; $493,000 at 3½%, due July 1, 1909. *Lake Bonds:* $222,000 at 7%, due $107,000 Jan. 1, '91, $115,000 due July 1, '91; $179,000 at 5%, due $15,000 April 1, '91, $14,000 April 1, '92, and $15,000 on April 1 of each year, 1893 to 1903 inclusive. *Hyde Park:* $384,000 at 7%, due $30,000 July 1, '93, $23,000 Jan. 1, '94, $322,000 Jan. 1, '95, $9,000 Jan. 1, '96; $50,000 at 5%, due Jan. 1, 1904. *Lake View Bonds:* $75,000 at 7%, due July 1, '95; $23,000 at 5%, due July 1, 1904; $50,000 at 4%, due July 1, 1907. Summarized by rates of interest the debt is as follows: $3,528,000 at 7%; $132,000 at 6%; $252,000 at 5%; $200,000 at 4%; $333,000 at 3.65%; $493,000 at 3.5%; total debt, $4,938,000.

WASHINGTON HEIGHTS PLANT.

History.—Built in '89 by village and came under contract of Chicago by annexation latter part of '90. Water Supply.—Artesian well, pumping by deep-well pump to reservoir, repumping to tank. Pumping Machinery.—Deep-well pump has cap. of 430,000 and main pump of 720,000 galls. Reservoir.—Cap., 60,000 galls.; 22 ft. above deep-well pump. Tank.—Cap., not given; 115-ft. lift for main pump. Distribution. —*Mains*, about 17 miles 6 and 8-in. Financial.—*Cost*, about $90,000.

Management.—(All works, save Pullman plant, below.) Dept. Pub. Works. Comr., J. Frank Aldrich. City Engr., A. W. Cooko. Asst. City Engr., Bernard Feind.

PULLMAN PLANT.

Built in '80-'82 by Pullman Palace Car Co. Supply.—Hyde Park division of Chicago Water-Works, pumping direct, and in case of fire, to tank. Fiscal year closes July 13. Distribution.—*Mains*, 15 miles. *Taps*, 808. *Meters*, 50. *Hydrants*, 130. *Consumption*, 1,476,553 galls. Financial.—*Cost*, $176,000. *No Debt.* Management.— Supt., D. Doty.

22. **CLINTON,** *De Witt Co.* (Pop., 2,709–2,598.) Built in '87-8 by city. Supply.—Well, pumping direct. Fiscal year closes May 1. Distribution.—*Mains*, 6 miles. *Taps*, 106. *Meters*, 3. *Hydrants*, 45. *Consumption*, 70,000 galls. Financial.—*Cost*, $37,000. *Bonded Debt*, $17,000 at 5%. *Expenses*, operating, $1,600; int..

ILLINOIS.

$850; total, $2,450. *Revenue:* consumers, $1,100; city, none. **Management.**—City Council. Chn., J. H. Conklin. Secy., J. W. Bowen. Supt., J. E. Moffett.

23. **DANFORTH,** *Iroquois Co.* (Pop., 207—est., 500.) **History.**—Built in '91 by village; put in operation June 15. Engrs. and Contrs., United States Wind Engine & Pump Co., Batavia. **Supply.**—Artesian wells, pumping to tank. **Pumping Machinery.**—Dy. cap., not given; Eureka. **Tank.**—Cap., 28,800 galls. **Distribution.**—*Mains,* 6-in. c. i. **Financial.**—*Cost,* $2,700. *Bonded Debt,* $1,500 at 6%. **Management.**—Chn., Jno. Overacker, Supt., J. V. Ballard.

24. **DANVILLE,** *Vermilion Co.* (Pop., 7,733—11,491.) Built in '82-3 by Danville Water Co., under 30 yrs. franchise. **Supply.**—North fork of Vermilion River and Clear Lake reservoir, pumping to stand-pipe. **Distribution.**—(In '88.) *Mains,* 14 miles. *Taps,* 561. *Meters,* 29. *Hydrants,* 107. *Consumption,* 300,000 galls. **Financial.**—(In '88.) *Cost,* $250,000. *Cap Stock,* $200,000 *Bonded Debt,* $150,000 at 7%. *Revenue* (in '87), consumers, $10,500; city, $7,937; total, $18,437. **Management.**—Prest., S. A. Miller, Louisville, Ky.

25. **DECATUR,** *Macon Co.* (Pop., 9,547—16,841.) Built in '71 by town. **Supply.**—Sangamon River, pumping direct. Fiscal year closes April 30.

Year.	Mains.	Taps.	Meters.	Hyd'ts.	C'ns'mpt'n.	B'd debt.	Int. ch'g's.	Exp.	Rev.
'82-3	9.7	385	..	100	250,000	......	....	$8,620	$6,256
'86-7	12	630	11	124	1,150,000	......	......	9,050	16,000
'89-90	14.5	723	8	140	1,650,000	$35,000	$3,500	8,550	17,250
'90-1	15	819	8	157	2,000,000	35,000	3,500	8,500	18,500

Financial.—*Bonded Debt,* $35,000 at 10%, due in '91. *Expenses,* total, $12,000. *Revenue:* consumers, $18,500; city, none. **Management.**—City Clerk, F. C Betzer.

26. **DE KALB,** *De Kalb Co.* (Pop., 1,598—2,579.) Built in '74 by city. **Supply.**—Artesian well, pumping to stand-pipe. Fiscal year closes May 1. **Distribution.**—*Mains,* 4 miles. *Taps,* 300. *Hydrants,* 75. *Consumption,* 67,935 galls. **Financial.**—*No Debt. Expenses,* $2,041. *Revenue,* consumers, $2,112. **Management.**—City Clerk, E. A. Porter.

27. **DIXON,** *Lee Co.* (Pop., 3,658—5,161.) Owned by Dixon Water Co. Built in '84 by A. H. McNeal and S. S. Murphy, Burlington, N. J.; afterwards sold to Co. **Supply.**—Rock River and springs, pumping to reservoir and direct. About Jan. 1, '90, court decreed that there was due American Loan & Trust Co., from Dixon Water Co., $90,401, and ordered works sold unless money was paid within five days **Distribution.**—(In '89). *Mains,* 7 miles. *Taps,* 200. *Meters,* 5. *Hydrants,* 65. *Consumption,* 200,000 galls. **Financial.**—(In '88.) *Cost,* $80,000. *Bonded Debt,* $80,000 at 6%. *Expenses:* operating, $2,000; int., $4,800; total, $6,800. *Revenue,* city, $4,875. **Management.**—(In '91.) Prest., Jno. D. Crabtree. Treas., Jas. A. Hawley. Supt., Jno. Masley.

28. **EAST DUBUQUE,** *Jo Daviess Co.* (Pop., 1,037—1,069.) Built in '83 by city. **Supply.**—Artesian well, pumping to reservoir. **Distribution.**—(In '88.) *Mains,* 1.5 miles. *Taps,* 20. *Meters,* none. *Hydrants,* 10. **Financial.**—(In '88.) *Cost,* $8,000. *No Revenue.* **Management.**—(In '89.) Pumping Engr., T. Dames.

29. **EAST SAINT LOUIS,** *Saint Clair Co.* (Pop., 9,185—15,169.) **History.**—Built in '86 by City Water Co.; controlled by American Water-Works & Guarantee Co., Ltd., Pittsburgh, Pa.; under 20-yrs. franchise, which fixes rates and provides for purchase of works by city in '95 and at intervals of 10 yrs. thereafter, price determined by arbitration. **Supply.**—Mississippi River: low service, pumping to settling reservoir and stand-pipe, thence to city; high service, pumping direct. **New Intake Tunnel.**—Water is taken from river through a tunnel $8 \times 8 \times 350$ ft., in which a 30-in. suction and other pipes are laid, leading to pump well 30 ft. deep, from which water is pumped for high service system; this system was built in '89 or '90. **Pumping Machinery.**—Combined dy cap., 17,000,000 galls. *Low Service:* dy. cap., 12,000,000 galls.; two 3,000,000-gall. Worthington comp., located in elliptical shared pit $36\frac{1}{2} \times 25 \times 24$ ft., for pumping water to settling reservoir, and one 3,000,000 Deane, $16 \times 39 \times 17\frac{1}{2} \times 24$ in. and one 3,000,000 Worthington for pumping into distribution mains. *High Service:* dy. cap., 5,000,000 galls.; one Gaskill, built in '90. *Boilers:* four 60-HP. horizontal tubular. **Low Service Reservoir.**—Cap., 7,500,000 galls.: $14 \times 160 \times 440$ ft.; divided into two compartments by cross-wall so constructed that water is allowed to settle in one while filling other; water level in this reservoir is about 24 ft.

FOR FURTHER DESCRIPTIVE MATTER see Manual for 1889-90. CHANGES since 1890 are here given. POPULATIONS are for 1880 and 1890, respectively.

above river. Stand-Pipe.—Cap., 132,500 galls.; 15 × 100 ft. Contr., T. Wagner. Fiscal year closes Dec 31. Distribution.—*Mains*, 32 miles. *Taps*, 1,034. *Meters*, 65. *Hydrants*, 192. Financial—Withheld. [Management.—Prest., J. H. Purdy. Secy. and Treas., A. E. Claney. Supt., F. M. Horner.

30. **ELGIN,** *Kane Co.* (Pop., 8,787 – 17,823.) Built in '87-8 by city. Supply.—Well and Fox River, pumping to stand-pipe or direct; American filters. Fiscal year closes Dec. 31.

Yr.	Mains.	Taps.	Met'rs.	Hyd'ts.	Cons'pt'n.	Cost. B'd'd debt.	Int ch'g's.	Exp.	Rev.
'88	16	287	37	196	232,000	$162,000 $86,000	$4,360	$5,060	$8,000
'89	17	591	35	196	254,372	182,700 1·8,000	5,960	5,481	8,278
'90	20.5	999	35	218	....	186,482 118,000	7,080	6,042	11,861

Financial.—*Bonded Debt*, $118,000 at 6%. *Expenses*, total, $13,122. *Revenue:* consumers, $11,861; city, none. **Management.**—Three comrs., appointed by mayor every two years. Chn., S. F. Buckley. Secy., G. R. Wilson. Supt., R. R. Parkin, appointed by comrs.

31. **ELMHURST,** *Du Page Co.* (Pop., 723–1,050.) History.—Built in '90-1 by Elmhurst Spring Water Co., under 30-yrs. franchise; put in operation July 1. Engr. and Contr., Geo. Cadogan Morgan, 49 Major Block, Chicago. Supply —Mammoth spring, pumping to water tower. Spring is 3¼ miles distant from town. **Pumping Machinery.**—Dy. cap., 350,000 galls; Worthington. **Water Tower.**—Cap., 30,000 galls.; 100 ft. height. Distribution.—*Mains*, Wyckoff wood, 7 miles. *Hydrants*, Morgan. *Valves*, Peet. Financial.—*Cost*, $25,000. *Cap. Stock:* authorized, $50,000; all paid up. *Bonded Debt*, $20,000. **Management.**—Prest., Thos. B. Bryan. Secy. and Treas., Wilbur E. Hagans.

32. **EL PASO,** *Woodford Co.* (Pop., 1,390–1,353.) Built in '82 by city. Supply.—Well, pumping to tank, and direct. Distribution.—(In '87.) *Mains*, 2.5 miles. *Taps*, 35. *Meters*, 0. *Hydrants*, 20. *Consumption*, 6,000 galls. Financial.— (In '87.) *Cost*, $12,000. No Debt. *Expenses*, $50. **Management.**—(In '91.) Chn., Richard Hebden.

33. **EUREKA,** *Woodford Co.* (Pop., 1,185–1,481.) Built in '86 by village. Supply.—Well, pumping to tank or direct. Fiscal year closes April 21. Distribution.—*Mains*, 0.8 miles. *Taps*, 45. *Hydrants*, 9. *Consumption*, 15,000 galls. Financial.—*Cost*, $6,500. *Bonded Debt*, $5,000 at 6%. *Expenses*: operating, $200; int., $300; total, $500. *Revenue*, consumers, $280. **Management.**—Committee, apointed every year by village president from village trustees. Chn., G. W. Anthony. Clerk, A. V. S. Barnard.

34. **EVANSTON,** *Cook Co.* (Pop., 4,400–est., 16,000.) Built in '74 by town. Supply.—Lake Michigan, pumping direct. During summer of '90 new inlet pipe was laid on and under the bed of Lake Michigan. Pipe is 30 ins. in diam, 2,600 ft, long, on gravel bottom, with depth of 26 ft. of water. Pipe was laid by Jos. G. Falcon. Old intake is 1,600 instead of 1,000 ft. long, as in '89-90 Manual. Distribution.—(In '88.) *Mains*, 15 miles. *Taps* (in '81), 400. *Meters*, 0. *Hydrants*, 100. *Consumption*, 1,200,000 galls. Financial.—(In '87.) *Cost*, $118,000. *Bonded Debt*, $32,000 at 6%. *Expenses:* operating, $6,000; int., $1,920; total, $7,920. *Revenue:* total, $12,000. **Management.**—Supt., Sam'l Peeney.

35. **FAIRBURY,** *Livingston Co.* (Pop., 2,140–2,324.) Built in '87 by village. Supply.—Source not given, pumping to tank. Nov. 5, '91, people voted to issue $15,000 of bonds for artesian well and extensions to mains. Distribution.—(In '88.) *Mains*, 1 mile. *Taps*, 18. *Meters*, 0. *Hydrants*, 12. *Consumption*, 1,500 galls. Financial.—(In '88.) *Cost*, $3,000. *Bonded Debt*, $5,000 at 6%. **Management.**—Town Board. Prest., S. M. Barnes. Secy., Z. T. Hanna.

36. **FREEPORT,** *Stephenson Co.* (Pop., 8,516–10,189.) Built in '82 by Freeport Water Co. Supply.—Artesian wells and Pecatonica River, pumping to stand-pipe and direct; filtered. Distribution.—*Mains*, 13.8 miles. *Taps*, 669. *Meters*, 44. *Hydrants*, 110. *Consumption*, 700,000 galls. Financial.—*Cost* (in '88), $251,000. *Cap. Stock*, $150,000. *Bonded Debt*, $130,000 at 6%, due in 1915. **Management.**—Prest., Nathan Shelton. Secy., D. S. Brewster. Treas., O. B. Bidwell. Supt., F. E. Josel.

Fulton, *Whiteside Co.* (Pop., 1,733–2,099.) Built in '87 by village for fire protection.

37. **GALENA,** *Jo Daviess Co.* (Pop., 6,451–5,635.) Built in '86-7 by Galena Water Co. Supply.—Artesian well, pumping to stand-pipe and direct. Distribution.—*Mains*, 6 miles. *Hydrants*, 74. *Consumption*, 400,000 galls. Financial.—

Cost (in '88), $75,000. *Cap. Stock*: authorized, $65,000; all paid up. *Bonded Debt*, $65,000 at 6%. *Expenses:* operating, $2,500; int., $3,900; taxes, $300; total, $6,700. *Revenue:* consumers, $4,000; fire protection, $2,920; other purposes, $400; total, $7,320. **Management.**—Prest., R. Barrett. Secy., H. S. Raymond, Chicago. Supt., Geo. R. Merill.

38. **GALESBURCH,** *Knox Co.* (Pop., 11,437—15,264). **History.**—First supplied in '83 by Galesburgh Water Co.. whose system has not been in operation for three or four years, and whose franchise was declared void in '90; unofficially reported that city bought company's works summer of '91 for $6,500. Present works built in '90-1 by city. Engr., C. B. Davis, Chicago. Fiscal year closes March 31. **Supply.**—Tube wells, pumping direct. Wells are 4 ins. in diam. and from 65 to 85 ft. deep, yielding from 50 to 300 galls. each, per minute. **Distribution.**—*Mains*, 15 miles. *Hydrants*, 194. **Financial.**—*Cost*, $125,000. *Bonded Debt*, $86000 at 5%, due 1900-10. **Management.**—Chn., J. B. Holland. Secy., J. C. Stevens.

39. **GENESEO,** *Henry Co.* (Pop., 3,518—3,182.) Built in '79-82 by city. **Supply.** —Artesian wells, pumping direct from reservoir. Fiscal year closes March 31. **Distribution.**—*Mains* (in '88), 8 miles. *Takers*, 110. *Hydrants*, 90. **Financial.**—*Cost*, $20,000. *Debt*, $6,000. *Expenses*, $1,800. *Revenue*, $600. **Management.**—City Clerk, J. F. Leoberknecht.

40. **GREENVILLE,** *Bond Co.* (Pop., 1,886—1868.) Built in '81 by town. **Supply.**—Well, pumping direct. Fiscal year closes April 16. Unofficially reported in '91 that stand-pipe was to be added. **Distribution.**—*Mains*, 2 miles. *Taps*, 120. *Hydrants*, 23. *Consumption*, 100,000 galls. **Financial.**—*Cost*, $24,000. *Bonded Debt*, $16.000 at 6%. *Expenses:* operating, $1,300; int., $960; total, $2,260. *Revenue*, consumers, $1,200. **Management.**—Comrs. appointed every year by mayor. Chn., J. C. Sanderson. Secy., Ward Reid. Supt., Vance McLain.

41. **HAVANA,** *Mason Co.* (Pop., 2,118—2,525.) Built in '89-'90 by city. **Supply.**— Six driven wells, pumping to stand-pipe. Wells are 6 ins. in diam, 60 ft. deep. **Stand-Pipe.**—Cap., 53,000 galls; 15×40 ft. on brick tower, 50 ft. high; from Titusville (Pa.) Iron Co. **Distribution.**—*Mains*, 10 to 4-in. c. i., 3.5 miles. *Services*, galv. i. *Taps*, 185. *Hydrants*, Chapman, 32. *Valves*, Chapman. *Pressure:* ordinary, 45 lbs.; fire, 80 lbs. **Financial.**—*Cost*, $25,000. *Bonded Debt*, $15,009 at 5%. *Expenses:* operating, $1,500,; int., $750; total, $8,250. **Management.**—Not given.

42. **HILLSBORO,** *Montgomery Co.* (Pop., 1,803—est., 2,100.) Built in '88 by village. **Supply.**—Well pumping to tank; pumping direct in case of fire. **Distribution.**—*Mains*, 1.8 miles. *Taps*, 81. *Hydrants*, 18. **Financial.**—*Cost*, $15,600. *Bonded Debt*, $16,600 at 6%. *Sinking Fund*, $540. *Expenses*, $600. *Revenue*, $684. **Management.**—City Clerk, Frank P. Winchester. Secy., Geo. T. Seward.

43. **HINSDALE,** *Du Page Co.* (Pop., 819—1,584.) **History.**—Built in '91 by village; put in operation Oct. 1. Engr. and contr. Geo. Cadogan Morgan, 49 Major Block, Chicago. **Supply.**—Artesian well, pumping from reservoir to tank and water tower. **Pumping Machinery**—Dy. cap., 1,000.000 galls.; one 500,000 Dean vertical, deep-well, and one 500,000 Worthington dup. **Reservoir**—Cap., 60,000 galls. **Tank.**—45,000 galls. **Tower**—Cap., 75,000 galls. **Distribution**—*Mains*, c. i. 8 miles. *Hydrants*, 40. *Pressure*, ordinary, 45 lbs., fire, 100 lbs. **Financial**—*Cost*, $37,000. *Debt*, $11,100 at 5%.

44. **HOOPESTON,** *Vermilion Co.* (Pop., 1,272—1,911.) Built in '88 by village. **Supply.**—Deep well and reservoir, pumping to tank. **Distribution.**—(In '88.) *Mains*, 3 miles. *Taps*, 90. *Meters*, 0. *Hydrants*, 24. **Financial.**—(In '88.) *Cost*, $16,500. *Bonded Debt*, $12,900 at 6%. *Expenses:* operating, $700; int., $645; total, $1.345. *Revenue:* consumers, $700. **Management.**—(In '89.) Com. Chn., A. T. Catherwood.

Hopedale, *Tazewell Co.* (Pop., 362—471.) Built in '88-9 by village for fire protection.

45. **IPAVA,** *Fulton Co.* (Pop., 675—667.) **History.**—Built in '90 by Ipava Water Co. under 10-yrs. franchise with yearly option of purchase by village; put in operation Nov. 1. Village apparently helped pay for works. Engr. and Contr., Geo. W. Sturtevant, Jr., Bushnell. **Supply.**—Tube wells, pumping to stand-pipe. Wells are 4½ ins. in diam., 1,550 ft. deep. **Pumping Machinery.**—Dy. cap., 500,000 galls.;

FOR FURTHER DESCRIPTIVE MATTER see Manual for 1889-90. CHANGES since 1890 are here given. POPULATIONS are for 1880 and 1890, respectively.

Barr comp., dup. **Stand-Pipe.**—Cap., 42,3C0 galls.; 12 × 50 ft., on 50-ft. tower. **Distribution.**—*Mains,* 6 and 4-in. c. i., 1.8 miles. *Services,* galv. i. *Hydrants,* Empire, 91. *Valves,* Scott. **Financial.**—*Cost,* $12,500. *Cap. Stock,* $5,690. *Bonded Debt :* Co., $5,690 ; village, $5,000 ; total, $10,690. **Management.**—Prest., J. L. McCune.

46. **JACKSONVILLE,** *Morgan Co.* (Pop., 10,927—10,740.) Built in '73-4 by city. **Supply.**—Surface water and artesian wells, pumping from impounding to distributing reservoir, or direct. Fiscal year closes March 1. During '90 supply was increased by addition of artesian well. Another well was to be bored during '91.

Year.	Mains.	Taps.	M't'rs.	H'd'nts.	Cons'pt'n.	B'd'd debt.	Int. ch'ges.	Exp.	Rev.
'80	9	350	..	75	212,000	....	....	$3,500	....
'82	11	380	124	78	252,000			5,432	$9,000
'84	12.5	285	135	84	300,000	$150,000	$9,000	5,000	8,000
'87	13.5	316	186	88	400,000	105,500	6,330	4,600	8,812
'89	14	503	250	104	408,427	105,500	4,725	6,074	9,519
'90	15.8	550	290	116	500,000	105,5 0	4,725	6,000	11,000

Financial.—*First Cost,* $150,000. *Total Debt,* $105,500, bonded, at 4½%, due 1889-1909. *No Sinking Fund. Bonded Debt,* $105,500 at 4½%, due '09-1909. *Expenses,* total, $10,725. *Revenue :* consumers, $11,000; city, none. **Management.**—Three comrs., appointed every three years by mayor and city council. Prest., F. G. Farrell. Secy., W. E. Capps. Supt., D. C. Fry.

47. **JERSEYVILLE,** *Jersey Co.* (Pop., 2,891—3,207.) Built in '89 by city. **Supply.**—Artesian well, pumping to reservoir, then pumped to tank. Fiscal year closes Oct. 1. **Distribution.**—*Mains,* 4.5 miles. *Taps,* 75. *Hydrants,* 31. *Consumption,* 30,000 galls. **Financial.**—*Cost,* $28,000. *Bonded Debt,* $21,000 at 5%. *Expenses:* operating, $1,670; int., $1,052; total, $2,722. **Management.**—Committee appointed by mayor. Chn. J. S. Daniells.

48. **JOLIET,** *Will Co.* (Pop., 11,657—23,264.) Owned by city since '88. Built in '80-1 by Joliet Water-Works Co. **Supply.**—Pond and artesian wells, pumping to stand-pipe and direct. Fiscal year closes April 30.

Yr.	Mains.	Taps.	M't'rs.	H'd'ts.	Cons'pt'n.	Cost.	B'd'd d't.	Int. ch'g's.	Exp.	Rev.
'83	10	175	0	60	600,000	$75,000			$6,000	$3,000
'86	11	365	365	60	700,000	205,000	$80,000	$4,800	4,000	7,000
'88	16	650	..	65	1,700,000	186,620	126,000	6,300	....	....
'89-90	15.3	...	...	..	1,747,134	..	....	....	9,296	8,814

Management.—Secy., Thos. B. Supplee, Jr. Supt., J. F. Gleason.

49. **KANKAKEE,** *Kankakee Co.* (Pop., 5,651—9,025.) Built in '86-87 by Kankakee Water Co., controlled by Moffett, Hodgkins & Clarke, Syracuse. N. Y., under 20-yrs. franchise. **Supply.**—Kankakee River, pumping to stand-pipe; filtered. **Distribution.**—(In '88.) *Mains,* 13 miles. *Taps,* 526. *Meters,* 0. *Hydrants,* 111. *Consumption,* 250,000 galls. **Financial.**—(In '88.) *Cost,* $160,000. *Bonded Debt,* $125,000 at 6%. *Expenses :* operating, $3,000; int., $7,500; total, $10,500. *Revenue:* consumers, $8,400; city, $5,000; total, $13,400. **Management.**—(In '89.) Prest., H. C. Hodgkins

50. **KENILWORTH,** *Cook Co.* Part or suburb of Chicago. (Pop., not given.) **History.**—Built in '90 by Kenilworth Co.; put in operation Sept. 15. Engr. and Contr., Geo. Cadogan Morgan, 49 Major Block, Chicago. **Supply.**—Lake Michigan, pumping to water tower; filtered. **Pumping Machinery.**—Dy. cap., 350,000 galls.; Deane. **Water Tower.**—Cap., 350,000 galls. **Distribution.**—*Mains,* 8 to 4-in. c. i., 3.5 miles. *Services,* lead. *Taps,* about 90. *Hydrants,* Morgan. *Valves,* Peet. *Pressure:* ordinary, 45 lbs.; fire, 110 lbs. **Financial.**—*Cost,* $30,000. **Management.**—Prest., Jos. Sears.

51. **KEWANEE,** *Henry Co.* (Pop., 2,704—4,569.) Built in '87-8 by village. **Supply.**—Wells, pumping direct and to storage reservoir for fire protection. Fiscal year closes April 30. **Distribution.**—*Mains,* 2.5 miles. *Taps,* 20. *Hydrants,* 25. *Consumption,* 100,000 galls. **Financial.**—*Cost,* $40,000. *Bonded Debt,* $16,000 at 6%. *Expenses:* operating, $3,000; int., $960; total, $3,960. *Revenue :* consumers, $700. **Management.**—Trustees. Prest., E. E. Baker, Clerk, Hugh Hill. Supt., A. W. Erretti.

52. **LA GRANGE,** *Cook Co.* (Pop., 531—2,314.) **History.**—Built in '90 by Moffett, Hodgkins & Clarke, Syracuse, N. Y., under 30-yrs. franchise; put in operation Dec. 15. **Supply.**—Artesian wells, pumping to stand-pipe. **Pumping Machinery.** —Dy. cap., 3,000,000 galls.; Worthington. **Stand-Pipe.**—Cap., 180,000 galls.; 16 × 120 ft. **Distribution.**—*Mains,* c. i., 6.5 miles. *Taps,* 100. *Hydrants,* Eddy, 77.

Valves, Eddy. *Pressure:* ordinary, 60 lbs.; fire, 120 lbs. **Financial.**—*Cost*, $160,000. *Cap. Stock*, $100,000. Int., 6%. *Revenue*, city, $3,360. **Management.**—Prest. and Secy., J. T. Moffett; Treas., J. V. Clarke; both of Syracuse. Supt., W. S. Reed.

53. LAKE FOREST, *Lake Co.* (Pop., 877—1,203.) **History.**—Built in '91 by Lake Forest Water Co., under 15-yrs. franchise; put in operation Oct. 1. Engr., Jno. A. Cole. **Supply.**—Lake Michigan, pumping to stand-pipe. **Pumping Machinery.**—Dy. cap., 1,250,000 galls.; Gordon. *Boilers:* from Porter Mfg. Co., Chicago. Stand-Pipe.—Cap., 125,000 galls.; 16½ × 80 ft.; from Tippett & Wood, Phillipsburg, N. J. **Distribution.**—*Mains*, 10 and 8-in., c. i., 4.5 miles. Pipe and specials from Addyston Pipe & Steel Co., Cincinnati, O.; trenching and pipe-laying done by M. Philbin & Co., Chicago. *Hydrants*, 40. **Financial.**—*Cap. Stock*, authorized, $50,000; all paid up. **Management.**—Prest., Jos. B. Durand. Secy., Edmund F. Chapin. Treas., Jos. T. Bowen.

Lanark, *Carroll Co.* (Pop., 1,198—1,295.) Built in '88 by city, probably for fire protection. See '89-'90 Manual, p. 447.

54. LA SALLE, *La Salle Co.* (Pop., 7,847—9,855.) **History.**—Built in '88-9 by town. **Supply.**—Springs, pumping direct. **Pumping Machinery.**—Dy. cap., not given; Holly comp. cond. dup. **Distribution.**—*Mains*, 12 to 4-in., c. i., 11 miles. *Taps*, 600. *Meters*, 50. *Hydrants*, Ludlow, Galvin, 65. *Consumption*, 500,000 galls. *Pressure*, 90 lbs. **Financial.**—*Cost*, $90,000. *Debt*, $30,000. *Revenue*, $4,500. **Management.**—Supt., Wm. P. Rounds.

55. LEMONT, *Cook Co.* (Pop., 2,108—est., 4,000.) Built in '84 by village. **Supply.**—Artesian wells, pumping to reservoir, with direct pumping from pond for fire protection. During '90 supply was increased by boring second well 2,100 ft. from first. Fiscal year closes April 30. **Distribution.**—*Mains*, 4.5 miles. *Taps*, 215. *Hydrants*, 27. **Financial.**—*Cost*, $27,475. *Bonded Debt*, $12,000 at 6%. Bonds are payable $2,000 each year after '93; to be redeemed from saloon licenses. Water is furnished to consumers at net cost. *Expenses:* operating, $2,100; int., $720; total, $2,820. *Revenue:* consumers, $1,865; city, none. **Management.**—Prest., D. C. Norton. Clerk, Geo. Wiemer, Jr.

56. LEWISTON, *Fulton Co.* (Pop., 1,771—2,166.) Now owned by city; built in '88 by Lewiston Water Co., in which town had interest from the start. **Supply.**—Driven well, pumping to stand-pipe. Fiscal year closes April 1. **Distribution.**—*Mains*, 5.3 miles. *Taps*, 165. *Hydrants*, 28. *Consumption*, 80,000 galls. **Financial.**—*Cost*, $32,000. *Bonded Debt*, $31,000. *Int.*, $2,481. **Management.**—Committee of three, appointed by Mayor every two years. Chn., P. C. Ross.

57. LINCOLN, *Logan Co.* (Pop., 5,639—6,200.) Built in '86 by Lincoln Water-Works Co., controlled by Moffett, Hodgkins & Clarke, Syracuse, N. Y., under 30-yrs. franchise. **Supply.**—Wells, pumping to stand-pipe. **Distribution.**—(In '88.) *Mains*, 9 miles. *Taps*, 250. *Meters*, 2. *Hydrants*, 51. *Consumption*, 223,856 galls. **Financial.**—(In '88.) *Cost*, $110,000. *Bonded Debt*, $85,000 at 6%. *Expenses:* operating, $2,500; int., $5,100; total, $7,600 *Revenue:* consumers, $7,200; city, $3,000. **Management.**—Prest., J. F. Moffett. Supt., A. J. Aldrich.

58. LITCHFIELD, *Montgomery Co.* (Pop., 4,326—5,811.) Built in '74 by town. **Supply.**—Shoal Creek, pumping direct from impounding reservoir. Fiscal year closes March 20. **Distribution.**—*Mains*, 5.5 miles. *Taps*, 125. *Hydrants*, 53. *Consumption*, 125,000 galls. **Financial.**—*Cost*, $85,000. *No Water Bonds*. *Expenses*, $2,500. *Revenue:* consumers, $2,000; city, $2,120; total, $4,120. **Management.**—Comrs. Prest., Jno. Lange. Secy., Dennis Corrigan. Supt., Alfred Reeves.

NOTE.—In '91, Litchfield Water Supply Co. was incorporated. Cap. Stock, $15,000. System to be built for use of several factories.

59. LOCKPORT, *Will Co.* (Pop., 1,679—2,449.) Built in '77 by Lockport Artesian Well Co. **Supply.**—Artesian well, pumping to reservoir. **Distribution.**—*Mains*, 1 mile. *Taps*, 60. *Hydrants*, 20. **Financial.**—*Cost*, $5,000. *Cap. Stock*, $3,000. *No Debt*. *Expenses*, $500. *Revenue*, $500. **Management.**—Prest., J. L. Norton. Secy., W. G. Norton. Supt., W. O. Fisher.

Mason City, *Mason Co.* (Pop., 1,714—1,869.) Built by city in '90-1 for fire protection, only. Engr., Geo. W. Sturtevant. Bushnell. **Supply.**—Tubular well, pump-

FOR FURTHER DESCRIPTIVE MATTER see Manual for 1889-90. CHANGES since 1890 are here given. POPULATIONS are for 1880 and 1890, respectively.

ing to 12 × 40-ft. stand-pipe. *Mains*, 1 mile. *Hydrants*, 12. *Pressure*, 60 lbs. *Cost*, $12,000. *Debt*, $9,000 at 6%.

60. MATTOON, *Coles Co.* (Pop., 5,737—6,833.) Built in '85 by Mattoon Clear Water Co.; rates are fixed in franchise, which provides for purchase of works in 1900, price, if not agreed upon, to be determined by arbitration. **Supply.**—Kickapoo Creek and wells alongside, pumping to reservoir, thence by repumping direct or to stand-pipe. Fiscal year closes Dec. 31. **Distribution.** *Mains,* 7.5 miles. *Taps,* 150. *Hydrants* (in '86), 73. *Consumption,* 200,000 galls. **Financial.**—*Cap. Stock,* authorized, $100,000; all paid up. *Bonded Debt,* $73,000 at 6%: $11,000 due in 1900; $65,000 due in 1905. *Revenue,* total, $5,930. **Management.**—Prest., H. S. Clark. Secy., Jno. F. Scott. Treas., Russell S. Clark. Supt., Geo. Gray.

61. MENDOTA, *La Salle Co.* (Pop., 4,142—3,542.) Built in '88 by town. **Supply.** —Artesian well, pumping to storage reservoir, thence to tank. Fiscal year closes May 1. **Distribution.**—*Mains,* 6 miles. *Taps,* 200. *Meters,* 4. *Hydrants,* 70. *Consumption,* 30,000 galls. **Financial.**—*Bonded Debt,* $15,500 at 6%. **Management.**— Supt., D. L. Harris.

62. MINONK, *Woodford Co.* (Pop., 1,913—2,316.) Built in '87-9 by village. **Supply.**—Well, pumping to tank. Fiscal year closes June 30. **Distribution.**—*Mains,* 1.5 miles. *Hydrants,* 17. **Financial.**—*Cost,* $12,000. *Bonded Debt,* $10,000 at 5%, due in annual payments of $1,000, after 10 years. *Expenses:* operating, $250; int., $500; total, $750. *Revenue,* withheld, not sufficient to pay operating expenses. **Management.**—Comrs., elected every year. Prest., Dan'l Neilinger. Supt., appointed by Mayor, Josiah Kerrick.

63. MOLINE, *Rock Island Co.* (Pop., 7,800—12,000.) Owned by city since '86. Built in '83 by Davis & Co. **Supply.**—Mississippi River, pumping direct. Fiscal year closes May 31.

Year.	Mains.	Taps.	Meters.	Hyd'ts.	Cons'pt'n.	Cost.	B'd't d't.	Int.ch'gs.	Exp.	Rev.
'84	9	300	20	120	700,000	....	$50,000	$2,500	$3,000	$5,600
'88	10	350	20	152	750,000	$105,000	50,000	2,500	5,200	7,200
'90	11.5	450	20	166	850,000	130,000	50,000	2,500	6,000	8,500

Financial.—*Bonded Debt,* $50,000 at 5%, payable at option of city. *Sinking Fund,* $8,000. *Expenses,* total, $8,500, *Revenue:* consumers, $8,500; city, none. **Management.**—Committee of three, appointed by mayor from Council for three years. Prest., J. M. Donahay. Supt., Fred Alsterlund.

64. MONMOUTH, *Warren Co.* (Pop., 5,000—5,936.) Built in '88-9 by city. **Supply.**—Artesian well, pumping to reservoir, thence repumped directly to mains. Fiscal year closes April 1.

Year.	Mains.	Taps.	M't'rs.	Hyd'ts.	C'ns'pt'n.	Cost.	B'd'd d't.	Int.ch'gs.	Exp.	Rev.
'89-90	5	75	0	45	55,000	$40,000	$35,000	$1,750	$2,160	.....
'90-1	11	300	1	110	150,000	70,000	50,000	2,500	3,000	$3,000

Financial.—*Bonded Debt,* $50,000 at 5%, due 1905-10. *Sinking Fund,* $2,500. *Expenses:* total, $5,500. *Revenue:* consumers, $3,000; fire protection, $3,400; total, $6,400. **Management.**—Supt., M. G. Browning.

65. MORGAN PARK, *Cook Co.* (Pop., 187—1,027.) Built in '74 by village. **Supply.**—Artesian well, pumping to tank. **Distribution.**—(In '87.) *Mains,* 8 miles. *Taps,* 80. *Meters,* 0. *Hydrants,* 4. *Consumption,* 30,000 galls. **Financial.**—(In '87.) *Cost,* $35,000. *Expenses,* $1,500. **Management.**—(In '88.) Comrs. Prest., T. W. Goodspeed.

66. MORRISON, *Whiteside Co.* (Pop., 1,951—2,088.) Built in '69 by town. **Supply.**—Spring, pumping to tank and direct in case of fire. Fiscal year closes April 30. During summer of '90, steel tank, 35 ft. in diam. and having cap. of 216,-000 galls., was built, cost, $5,000. Reservoir affords ordinary pressure of 33 lbs. **Distribution.**—*Mains,* 4.3 miles. *Taps,* 300. *Hydrants,* 40. *Consumption,* 200,000 galls. **Financial.**—*Cost,* $23,000. *Bonded Debt,* $11,000: $6,000 at 5%; $5,000 at 6%. *Expenses:* operating, $2,000; int., $600; total, $2,600. *Revenue:* consumers, $2,500; city, none. **Management.**—Committee appointed by mayor from city council every two years. City Clerk, J. N. Baird. Supt., H. G. Parrish.

67. MORRISONVILLE, *Christian Co.* (Pop., 748—844.) Built in '88-9 by village. **Supply.**—Well, pumping to tank and direct. **Distribution.**—*Mains,* 1 mile. *Taps,* 10. *Meters,* 1. *Hydrants,* 5. *Consumption,* 50,000 galls. **Financial.**—

ILLINOIS.

Cost, $8.000. *Bonded Debt*, $5,000 at 6%. **Management.**—Secy., W. E. Forbes. Supt. and Engr., A. L. McAuley. NOTE.—List report, May 14, '89.

68. MOUNT CARROLL, *Carroll Co.* (Pop., in '90, 1,836.) Built in '88 by village. **Supply.**—Well, pumping to tank. **Distribution.**—*Mains*, 2 miles. *Taps*, 80. *Hydrants*, 22. **Financial.**—*Cost* (in '88), $18,000. *Bonded Debt* (in '85), $13,000 at 6%. *Expenses*, $800. *Revenue*, $750. **Management.**—Supt., H. M. Melendy.

69. MOUNT PULASKI, *Logan Co.* (Pop., 1,125—1,357.) Built in '82 by village. **Supply** —Well, pumping to tank. **Distribution.**—(In '88.) *Mains*, 4 miles. *Taps*, 200. *Meters*, 0. *Hydrants*, 36. *Consumption*, 150,000 galls. **Financial.**—(In '88.) *Cost*, $6,000. *No Debt*. *Expenses*, $1,500. *Revenue*, consumers, $1,200. **Management.**—Supt., J. M. Hopkins.

70. MURPHYSBOROUGH, *Jackson Co.* (Pop., 2,196—3,880.) Built in '90 by Murphysborough Water-Works, Electric & Gaslight Co. **Supply.**—Big Muddy River, pumping to stand-pipe; Hyatt filter added in '90. Fiscal year closes Dec. 31. **Distribution.**—*Mains*, 4.5 miles. *Taps*, 200. *Meters*, 1. *Hydrants*, 60. **Financial.** —*Cost*, $75,000. *Cap. Stock* (entire plant) authorized, $60,000; all paid-up. *Total Debt*, $62,000: floating $2,000; bonded, $60,000 at 6%, due in 1910. *Expenses:* operating, $10,000; int., $3,600; total, $13,600. *Revenue:* consumers, $7,200; fire protection $3,000; other purposes, $3,000; total, $13,200. **Management.**—Prest., W. K. Murphy. Secy. and Treas., Willard Wall. Supt., W. Henderson.

71. NORWOOD PARK, *Cook Co.* (Pop., est., 750.) Built in '89 by village. **Supply.**—Artesian well, pumping from reservoir to tank. *Mains* (in '88), 2.5 miles. Supt., Jno. Stockwell.

72. OAK PARK, *Cook Co.* (Pop., 1,888—4,771.) Owned by Cicero Water, Gas, & Electric Light Co. Built in '85 by Jas. W. Scoville; sold, latter part of '87, to above co.; city neither controls rates nor has right to purchase works. Supplies Austin. **Supply.**—Four artesian wells, pumping to stand-pipe. Fiscal year closes Sept. 1. During '90, supply was increased by two artesian wells, 2,180 ft. deep; pumping machinery increased by one comp. Dean pump.

Year.	Mains.	Taps.	Met's.	Hyd's.	Cons'pt'n.	Cost.	B'd'd dt.	In. c'g's.	Exp.	Rev.
'87	9	200	70	85	30,000	$35,000	$30,000	$1,800	$4,000	$4,000
'98	15.5	370	98	165	110,000	100,000	30,000	1,800	5,000	10,890
80	27	650	70	197	100,000	100,000	100,000	6,000	....	....
'90	33	1,275	125	277	150,000	150,000	100,000	6,000	13,000	25,000

Financial.—*Cap. Stock:* authorized, $100,000; all paid up, paid 6% dividend in 1890. *Bonded Debt*, $100,000, at 6%; 5-20s. *Expenses*, total, $19,000. *Revenue:* consumers, $12,000; fire protection, $7,000; other purposes, $6,000; total, $25,000. **Management.**—Prest., J. W. Scoville. Secy., C. E. Simmons. Treas., O. W. Herrick. Supt., F. W. Jackson.

73. OREGON, *Ogle Co.* (Pop., 1,088—1,566.) Built in '78 by village. **Supply.**—Rock River, pumping to reservoir. Fiscal year closes May 1st. **Distribution.**—*Mains*, 1.5 miles. *Taps*, 60. *Meters*, 4. *Hydrants*, 18. **Financial.**—*Cost*, $14,000, *No debt*. *Expenses*, $500. *Revenue*, $575. **Management.**—Chn., F. M. Gilbert. Secy., S. G. Mason.

74. OTTAWA, *La Salle Co.* (Pop., 7,834—9,985.) Built in '60 by Co.; franchise neither regulates rates nor provides for purchase of works by city. **Supply.**—Springs by gravity. **Distribution.**—*Mains*, 4 miles. *Taps*, 125. *Meters*, 1. *Hydrants*, 3. *Consumption*, 180,000 galls. **Financial.**—Withheld. **Management.**—Prest. and Supt., J. D. Caton. Secy., C. A. Caton.

75. PARIS, *Edgar Co.* (Pop., 4,373—4,996.) Built in '79 by city. **Supply.**—Ground water, pumping direct from reservoir. Fiscal year closes first Monday in May. Unofficially reported. Mar., '91, that Cook wells were being sunk. **Distribution.**—*Mains*, 10 miles. *Taps* (in '88), 200. *Hydrants*, 60. **Financial.**—*Cost* (in '88), $53,277. *Bonded Debt*, $30,000 at 7%. **Management.**—Prest., W. R. Swinford, Mayor. Clerk, O. H. Hodge. Supt., W. S. Hopkins.

76. PARK RIDGE, *Cook Co.* (Pop., 457—987.) Built in '87 by village. **Supply.**—Artesian well, pumping to tank. **Distribution.**—*Mains*, 3,600 ft. *Taps*, 36. *Meters*, 11. *Hydrants*, 2. **Financial.**—*Cost*, $6,000. **Management.**—Prest., Sam'l Cochran. Clerk, S. E. Cummings.

FOR FURTHER DESCRIPTIVE MATTER see Manual for 1889-90. CHANGES since 1890 are here given. POPULATIONS are for 1880 and 1890, respectively.

77. **PAXTON,** *Ford Co.* (Pop., 1,725—2,187.) Built in '87-8 by village. Supply —Artesian well, pumping to tank. Fiscal year closes April 30. Distribution.— *Mains,* 2.5 miles. *Taps,* 200. *Hydrants,* 15. *Consumption,* 43,000 galls. Financial. —*Cost,* $10,000. *Debt,* $1,500 at 6%, due in annual payments of $500. *Expenses,* $1,200, *Revenue,* consumers, $700. Management.—Com. Chn., F. L. Coon. City Clerk, H. H. Kerr.

78. **PECATONICA,** *Winnebago Co.* (Pop., 1,029—1,059.) History.—Built in '89-90 by village; put in operation June 1. Eng. and Contr., Geo. Cadogan Morgan, 49 Major Block, Chicago. Supply.—Springs, pumping to water tower. Pumping Machinery.—*Dy. Cap.,* 250,000 galls ; Deane. Water Tower.—*Cap.,* 64,000 galls, Distribution.—*Mains,* c. i., 1 mile. *Taps,* 50. *Hydrants,* Morgan, 12. *Valves,* Peet. Financial.—*Cost,* $15,000. *Debt,* $8,000 at 5%. Management.—Prest., C. E. Hallock, Secy., B. J. Weaver. Treas., S. Niles. Supt., P. Colvin.

79. **PEKIN,** *Tazewell Co.* (Pop., 5,993—6,347.) Built in '86 by Pekin Water-Works Co. Supply.—Driven well, pumping to tank. Distribution.—*Mains,* 11 miles. *Taps,* 300. *Meters,* 2. *Hydrants,* 111. *Consumption,* 350,000 galls. Financial.— *Cost,* $100,000. *Cap. Stock,* $100,000. *Bonded Debt,* $100,000 at 6%. Management.— Prest., C. A. Lamb, Chicago, Secy. and Supt., Henry Lautz.

80. **PEORIA,** *Peoria Co.* (Pop., 29,259—41,024.) Built in '67-8 by city; bought in '89 by Moffett, Hodgkins & Clarke, Syracuse, N. Y., and improved '89-91. Franchise regulates rates and provides for purchase of works by city in '09, and at intervals of 10 years thereafter. Supply.—Illinois River, pumping direct. Distribution.—(In '88.) *Mains,* 45 miles. *Taps* (in '87), 3,000. *Meters,* 12. *Hydrants,* 367. *Consumption,* 4,000,000 galls. Financial.—(In '88.) *Cost* (in '87), $450,000. *Bonded Debt,* $450,000 at 6%. *Revenue,* consumers, $34,000. Management.—(In '89.) Moffett, Hodgkins & Clarke, Syracuse, N. Y.

81. **PETERSBURGH,** *Menard Co.* (Pop., 2,332—2,342.) Built in '67-8 by village. Supply.—Four 8-in. tube wells, pumping to tank. Fiscal year closes in July. Distribution.—*Mains,* 2.5 miles. *Taps,* 102. *Hydrants,* 25. *Consumption,* 700,000 galls. Financial.—*Cost,* $26,500. *Bonded Debt,* $19,000 at 5%, due '97-1907. *Expenses:* operating, $1,200; int., $950; total, $2,150. *Revenue:* consumers, $1,600; city, none. Management.—Chn., Geo. F. Luthrunger. Supt., J. C. Bishop.

Pinckneyville, *Perry Co.* (Pop., 964—1,298.) Small system is owned jointly by town, and C. S. L. Ry. Co. Supply.—Impounding reservoir on stream. Mayor, J. L. Murphy.

Pittsfield, *Pike Co.* (Pop., 2,104—2,295.) Built in '87-8 by town for fire protection and street sprinkling.

Polo, *Ogle Co.* (Pop., 1,819—1,728.) Water for fire protection supplied from tank of Ill. C. R. R. Co.

NOTE.—In '91 city was drilling artesian well. Contrs., J. P. Miller & Co., Milwaukee. Upon completion of well system of water-works was to be constructed. Oct., '91 reported that construction of tank 36 ft. high on 80-ft. masonry structure had been started. Contr., Geo. Cadogan Morgan, 49 Major Block, Chicago. City Clerk, G. H. Butts.

82. **PRINCETON,** *Bureau Co.* (Pop., 3,439—3,396.) History.—Built in '89-90 by city ; put in operation March 1. Engrs., Sturtevant Bros., Bushnell. Supply.— Tube wells, pumping to stand-pipe. Wells are 8 ins. in diam., 2,515 ft. deep. Pumping Machinery.—*Dy,* cap., 925,000 galls.; one 175,000 Cook deep-well, and one 750,000 Barr comp., dup. Stand-Pipe.—Cap., 56,000 galls.; 14 × 50 ft., on 50-ft. tower. Distribution.—*Mains,* 10 to 4-in. c. i., 3.3 miles. *Services,* galv. i., lead. *Taps,* 210. *Hydrants,* Matthews, 32. *Valves,* Scott. *Consumption,* 65,000 galls. *Pressure:* ordinary, 50 lbs.; fire, 80 lbs. Financial.—*Cost,* $27,500. *Bonded Debt,* $26,000 at 5%. *Expenses,* $1,750. Management.—Not given.

21. **PULLMAN,** *Cook Co.* Now part of Chicago, but has independent system. See Chicago.

83. **QUINCY,** *Adams Co.* (Pop., 27,268—31,494.) Owned by L. & W. B. Bull; city neither controls rates nor has right to buy works. Built in '73 by co. Supply.— Mississippi River, pumping to reservoir. Fiscal year closes Jan. 1.

Year.	Mains.	Taps.	Meters.	Hydrants.	Consumption.
'84	18	500	20	120	671,000
'88	22	1,400	190	160	750,000
'90	28	1,600	235	169	632,000

ILLINOIS. 243

Financial.—*First Cost*, $115,000. *No Cap. Stock or Bonded Debt. Expenses:* operating, $7,954; taxes, $3,166; total, $11,120. *Revenue:* consumers, $34,581; fire protection, $10,000; other purposes, $2,516; total, $45,100. **Management.**—Supt, Wm. B. Bull.

NOTE.—April 11, '91, City Clerk wrote that ordinance had been passed by Board of Aldermen, providing for building public works, but that people voted against it. As City Council has right to go ahead without consent of people, it was uncertain what will be done. In '91 MacRitobie & Nichol, Chicago, reported on works; est. cost, $620,000. Nov. 17, '91, people again voted against building works.

84. **RIVERSIDE**, *Cook Co.* (Pop., 450—est., 1,100.) Built in '70 by village. Supply.—Artesian well, pumping to tank. *Cost,* $25,000. *Bonded Debt,* $5,000. Chn., C. F. Wardell.

85. **ROCHELLE**, *Ogle Co.* (Pop., 1,893—1,789.) Built in '77 by village. Supply.—Spring, pumping direct and to stand-pipe. Stand-pipe is 12 × 40 ft., on brick tower 70 ft. high; built in '90. Fiscal year closes May 31. **Distribution.**—*Mains,* 2.5 miles. *Taps,* 88. *Hydrants,* 38. *Consumption,* unk'n. **Financial.**—*Cost,* $30,000. *No Debt. Expenses,* $1,800. *Revenue,* consumers, $723. **Management.**—Comrs., appointed every year by mayor. Chn., Wm. Stocking. Secy., E. E. Ogden. Supt., E. F. Spooner.

95. **ROCK FALLS**, *Whiteside Co.* See Sterling.

86. **ROCKFORD**, *Winnebago Co.* (Pop., 13,121—23,584.) Built in '74-5 by city. Supply —Five artesian wells, pumping direct. Fiscal year closes Dec. 31. During '90 a 6 000 000-gall. Gaskill pump was added.

Year	Mains	Taps	Meters	Hy'd'ts	Cons'p't'n	Cost	B'd'd d't	Int.	Exp.	Rev.
'82	29	1,080	0	189	917,238	$280,070			$8,792	$10,319
'86	29.6	1,533		225	1,650,000	322,218	$94,000		10 835	
'87	29.6	1,784	0	235	1,452,000	342,689	94,000		11,500	15,850
'88	37.5	1,969	0	263	1,963,942	381,942	92,500	$5,087	11,310	22,096
'89	42.1	2,253	8	285	1,904,960	399,755	92,500	5,087	11,050	22,368
'90	44	2,509	8	296	2,373,100	422,400	92,500	5,087	12,791	25,539

Financial.—*Bonded Debt,* $92,500 at 5% and 6%. *Expenses,* total, $17,878. *Revenue:* consumers, $25,539; city none. **Management.**—Committee of five elected every year. Chn., T. W. Cole, Secy. and Supt., Jno. T. Lakin, appointed by Mayor.

87. **ROCK ISLAND**, *Rock Island Co.* (Pop., 11,659—13,634.) Built in '71-2 by city. Supply.—Mississippi River, pumping direct. Apr., '91, a 2,000,000-gall. Jewell filter was put in. Fiscal year closes third Monday in April. **Distribution.**—*Mains,* 19 miles. *Taps,* 1,110. *Meters,* 13. *Hydrants,* 140. *Consumption,* 1,750,000 galls. **Financial.**—*Cost,* about $300,000. *Bonded Debt:* $50,000; $25,000 at 8% due in '92; $25,000 at 5% due in 1902. *Sinking Fund,* $16,000 at 5%. *Expenses:* operating, $9,500; int., $3,450; total, $12,950. *Revenue:* consumers, $15,400; city, $7,000; total, $22,400. **Management.**—Committee of three appointed every year by town council. Chn., W. W. P. Tindall. Secy., Robert Kennedy. Supt. Jno. A. Murrin.

88. **ROGERS PARK**, *Cook Co.* (Pop., 529—1,708.) Built in '89 by Rogers Park Water Co., under 30-yrs. franchise. Des. Engr., C. B. Davis, Chicago. Const. Engr., A. W. Stevens. Spring of '91, 25 ft. was being added to height of stand-pipe. Supply.—Lake Michigan, pumping to stand-pipe and direct. **Distribution.**—*Mains,* 12 miles. *Taps,* 230. *Meters,* 40. *Hydrants,* 110. **Financial.**—*Cost* (in '88), $70,000. *Capital Stock:* authorized, $50,000, all paid up. *Bonded Debt,* $57,000 at 6%. *Expenses:* operating, $2,400; int., $3,420; total, $5,820. *Revenue,* city, $5,675. **Management.**—Prest., H. E. Keeler. Secy., Chas. Howard; Treas., W. A. Partridge; both of Chicago. Supt., Wm. Burnham.

89. **SANDWICH**, *De Kalb Co.* (Pop., 2,352—2,516.) Built in '83 by city. Supply.—Well, pumping to stand-pipe and direct in case of fire. Fiscal year closes May 1. **Distribution.**—*Mains,* 3.5 miles. *Taps,* 125. *Meters,* 1. *Hydrants,* 34. *Consumption,* 40,000 galls. **Financial.**—*Cost,* $30,000. *Bonded Debt,* $5,000 at 6%. *Expenses:* operating, $1,400; int., $300; total, $1,700. *Revenue:* consumers, $1,000; city, none. **Management.**—Chn., H. A. Marcy. Secy., E. S. Johnson. Supt., L. D. Smith.

90. **SAVANNA**, *Carroll Co.* (Pop., 1,000—3,097.) Built in '86 by town. Supply.—Well on island in Mississippi River, pumping to reservoir; filtered. Artesian well was to be sunk in '91; upon its completion, island well was to be abandoned. Fiscal

FOR FURTHER DESCRIPTIVE MATTER see Manual for 1889-90. CHANGES since 1890 are here given. POPULATIONS are for 1880 and 1890, respectively.

year closes May 1. Distribution.—*Mains*, 4 miles. *Taps*, 90. *Hydrants*, 21. *Consumption*, 105,000 galls. Financial.—*Cost*, $26,000. *Bonded Debt*, $15 000 at 5%. *Expenses:* operating, $1,000; int., $775; total, $1,775. *Revenue:* consumers, $927; city, $950; total, $1,877. Management.—Three comrs., one appointed every year by council. Chn., Wm. H. Griffith. Secy., F. S. Greenleaf. Supt., J. S. Hunt.

91. **SHELBYVILLE**, *Shelby Co.* (Pop., 2,939–3,162.) Built in '85-6 by Shelbyville Water Co.; controlled by Debenture Guarantee and Assurance Co., Chicago, Ill. Co. operates an electric lighting plant. Supply.—Kaskaskia River, by infiltration well, thence pumping to stand-pipe. Fiscal year closes Dec. 31. Distribution.—*Mains*, 7 miles. *Taps*, 367. *Meters*, 1. *Hydrants*, 50. *Consumption*, 175,000 galls. Financial.—*Cap. Stock*, paid up, $75,000. *Bonded Debt*, $100,000 at 6%. *Expenses:* operating and taxes, $2,000; int., $6,000. *Revenue:* consumers, $6,482; city, $3,000; total, $9,482. Management.—Prest., Chas. Whann. Secy. and Treas., C. B. Ludwig, 72 Broadway, N. Y. City. Supt., Chas. H. Marshall.

92. **SOMONAUK**, *De Kalb Co.* (Pop., 587—est., 800.) Built in '89-'90 by village. Supply.—Well, pumping to tank. Fiscal year closes May 1. Distribution.—*Mains*, 1.5 miles. *Taps*, 25. *Hydrants*, 12. Financial.—*Cost*, $7,500. *Expenses*, $25. Management.—Chn., H. Wright.

93. **SOUTH EVANSTON**, *Cook Co.* (Pop., 1,517—est., 3,800.) Built in '84 by village. Supply.—Lake Michigan, through inlet pipe, pumping to tank. Fiscal year closes April 15. Plant to be enlarged during '91. Distribution.—*Mains*, 12 miles. *Taps*, 695. *Meters*, 4. *Hydrants*, 102. Financial.—*Cost*, (in '88), $55,000. *Bonded Debt*, $14,000 at 4%. *Expenses:* operating, $5,000; int., $560; total, $5,560. *Revenue*, consumers, $4,757. Management.—Engr., W. F. Seiber.

94. **SPRINGFIELD**, *Sangamon Co.* (Pop., 19,743–24,963.) Built in '66 by city. Supply.—Sangamon River, pumping to stand-pipe and reservoir; filtered. Fiscal year closes Feb. 28. During '90 a second filter gallery was constructed and a dam built to raise the water in the river. New gallery is 986 ft. long, in sections 12 ft. long and 5 ft. wide and 4 ft. deep, formed by plank with ½-in. interstices. Space for about 1 ft. each side of gallery filled with broken stone. Fall of '91 contract for additional force main, 24 ins. in diam., 16,000 ft. long, was let to Geo. McIntosh, Bloomington.

Year.	Mains.	Taps.	M't'rs.	H'd'ts	Cons'pt'n.	B'd' d't.	Int. ch'g's.	Exp.	Rev.
'80	10.5	700		91	1,500,000			$15,000	$23,868
'84	21.5		36	96	1,975,151	$450,000	$36,000	22,062	26,642
'85	23.5	1,240	34	109	2,330,585	450,000	36,000	17,975	28,864
'86	27.4	1,395	41	125	2,265,755	450,000	36,000	14,271	28,145
'87	29.6	1,504	47	148	2,140,780	450,000	36,000	16,652	34,240
'88	33	1,580	70	147	2,598,478	450,000	36,000	15,805	29,795
'89	34	1,658	84	148	2,411,357	450,000	36,000	16,811	35,137
'90	35.1	1,774	86	153	2,571,223	450,000	36,000	21,809	37,549

Financial.—*First Cost*, $450,000. *Bonded Debt*, $450,000. *Expenses*, total, $57,609. *Revenue:* consumers, $37,649; city, none. Management.—Comrs., elected every three years. Chn., H. C. Bolles. Secy., H. H. Beecher. Supt., Peter Berriman.

95. **STERLING AND ROCK FALLS**, *Whiteside Co.* (Pop., Sterling, 5,087–5,824; Rock Falls, 894–1,900.) Built in '86 by Sterling Water Co.; rates are specified in franchise, which provides that city may purchase works in '95, and at close of every 5 years thereafter. Supply.—Artesian well and Rock River, pumping to stand-pipe. Fiscal year closes Dec. 31. Distribution.—*Mains*, 14 miles. *Taps*, 389. *Meters*, 6. *Hydrants*, 142. *Consumption*, 325,000 galls. Financial.—*Cost*, $258,745. *Cap. Stock:* authorized, $150,000; paid up, $126,900. *Total debt*, $143,887: floating, $53,887 at 6%; bonded, $90,000 at 6%, due in 1906. *Expenses:* operating, $3,267; int., $5,736; taxes, $1,752; total, $10,655. *Revenue:* consumers, $3,729; city; $6,270; total, $9,999. Management.—Prest., A. H. Howland, Boston. Secy., H. C. Ward. Treas., H. Butler. Supt., W. J. Watson.

96. **STREATOR**, *La Salle Co.* (Pop., 5,137–11,414.) Built in '86-7 by Streator Aqueduct Co. under 30 yrs. franchise, which fixes maximum rates and provides for purchase of works by city. Supply.—Tube wells and Ottar Creek, pumping to stand-pipe. Distribution.—*Mains*, 20 miles. *Taps*, 525. *Meters*, 21. *Hydrants*, 110. *Consumption*, 500,000 galls. Financial.—*Cost* (in '87), $275,000. *Revenue:* city, $4,400. Management.—Prest., L. M. Bronson, N. Y. City. Secy., T. M. Birnie; Treas., Wm. Birnie; both of Springfield, Mass. Supt., F. D. Ferguson.

Sullivan, *Moultrie Co.* (Pop., 1,305–1,468.) Built in '88 by city, probably for fire protection. See '89-'90 Manual, p. 458.

97. SYCAMORE, *De Kalb Co.* (Pop., 3,028—2,987.) Built in '76 by village. Supply.—Surface well, pumping to stand-pipe and direct. Fiscal year closes May 1. Extensive improvements were to be made during '91.

Y'r.	M'ns.	Taps.	M't's.	H'd'ts.	C'ns'pt'n.	Cost.	B'd'd d't.	Int. ch'g's.	Exp.	Rev.
'88	2	60	3	25	35,000	$40,000	$25,000	$1,250	$1,900	$900
'89	3	90	10	36	27,126	40,000	25,000	1,250	2,100	961
'90	3	100	40	36	29,017	40,000	25,000	1,200	2,100	1,200

Financial.—*Bonded Debt*, $25,000, at 5%, due in 10, 15 and 20 yrs. *Expenses*, total, $3,350. Revenue goes into general fund of city, from which all expenses are paid. **Management.**—Committee of three, appointed every year by mayor from aldermen. Chn., Walter Waterman. City Clerk, F. O. Van Galder. Supt., S. Othello Pike.

98. TAYLORVILLE, *Christian Co.* (Pop., 2,237 - 2,829.) Built in '86 by town. Supply.—Open and tube wells, pumping to tank and direct. Fiscal year closes May 1. During '90 supply was increased by addition of three 8-in. tubular wells, 52 ft. deep; a 750,000-gall. Gaskill dup. pump was also added. **Distribution.**—*Mains*, 6 miles. *Taps*, 310. *Hydrants*, 40. *Consumption*, 350,000 galls. **Financial.**—*Cost*, $36,000. *Bonded Debt*, $18,000, at 6%. *Expenses:* operating, $650; int., $1,080. **Management.**—Supt., T. B. Kraft.

16. URBANA, *Champaign Co.* See Champaign.

99. VENICE, *Madison Co.* (Pop., 612—532.) Built in '89 by village. Supply.—Mississippi River, pumping to tank. Fiscal year closes May 1. **Distribution.**—*Mains*, 2 miles. *Taps*, 30. *Hydrants*, 10. **Financial.**—*Cost*, $12,000. *Bonded Debt*, $7,000 at 6%, due in 1900. *Expenses:* operating, $1,000; int., $420; total, $1,420. **Management.**—Chn, Trustees, Theo. Selb. Secy., Jas. Brownell.

Warsaw, *Hancock Co.* (Pop., 3,105—2,721.) Built in '87 by city, for fire protection. See 1889-90 Manual, p. 459.

100. WASHINGTON, *Tazewell Co.* (Pop., 1,397—1,301.) Built in '88-9 by city. Supply.—Driven wells, pumping to stand-pipe. Fiscal year closes first Monday in May. **Distribution.**—*Mains*, 2.3 miles. *Taps*, 56. *Hydrants*, 22. *Consumption*, 12,000 galls. **Financial.**—*Cost*, $15,000. *Bonded Debt*, $12,090 at 6%, due in annual payments of $1,000. *Expenses:* operating, $500; int., $780; total, $1,280. *Revenue:* consumers, $365; city, none. **Management.**—City Clerk, Chas. Koker.

20. WASHINGTON HEIGHTS, *Cook Co.* Annexed to Chicago in '90; independent description under Chicago.

101. WAUKEGAN, *Lake Co.* (Pop., 4,012—4,915.) Built in '74 by town. Supply.—Three flowing artesian wells, by gravity. **Distribution.**—*Mains*, 7 miles. *Taps*, 250. *Consumers*, 900. *Hydrants*, 40. **Financial.**—*Cost*, $25,000. *No Debt. Revenue*, $900. **Management.**—City Clerk, Thos. Carney.

102. YORKVILLE, *Kendall Co.* (Pop., 365—375.) Built in '87 by village. Supply.—Springs, by gravity. Fiscal year closes April 7. **Distribution.**—*Mains*, 2 miles. *Taps*, 48. *Hydrants*, 15. **Financial.**—*Cost*, $6,000. *Bonded Debt*, $1,500 at 6%, due in annual payments of $500. *Expenses:* operating, $15; int., $270; total, $285. *Revenue:* consumers, $366. **Management.**—Chn., W. R. Newton. Clerk, R. M. Newton.

Water-Works Projected with Fair Prospects of Construction.

BEARDSTOWN, *Cass Co.* (Pop., 3,135—4,226.) Aug. 12, '91, city voted to build works. Bids for construction were to be received until Oct. 14. Mayor, H. M. Schmoldt.

CENTRALIA, *Marion Co.* (Pop., 3,621—4,763.) System projected, to be built by city. Est. cost, $65,000. Mayor, H. L. Rhodes.

DOWNERS GROVE, *Du Page Co.* (Pop., 586—900.) Works projected by city. Address Messrs. Heckman & Straub. Latest report Aug. 12.

WEST DUNDEE, (*Dundee P. O.*) *Kane Co.* (Pop., 585—873.) May 11, '91, J. G. Quick, Clerk, reported that village wished to grant 20-yrs. franchise to a co. to construct water and electric lighting plants. Supply, springs. Hydrants, 25.

FOR FURTHER DESCRIPTIVE MATTER see Manual for 1889-90. CHANGES since 1890 are here given. POPULATIONS are for 1880 and 1890, respectively.

DUQUOIN, *Perry Co.* (Pop., 2,807—4,052.) See 1889-90 Manual. lp. 460. System still contemplated by Duquoin Water-Works Co., June 26, '91. Address J. W. Winters.

DWIGHT, *Livingston Co.* (Pop., 1,295—1,354.) July 30, '91. A. T. Doherty, Town Clerk, wrote that town had decided to build works but plans were not perfected. Engr., Geo. Cadogan Morgan, 49 Major Block, Chicago.

EFFINGHAM, *Effingham Co.* (Pop., 3,065—3,260.) Aug. 23, '91, city intended to build works and would advertise for bids at once. Address L. H. Bissell or J. B. Walker, Mayor.

FLANAGAN, *Livingston Co.* (Pop., in '90, 384.) July 25, '91, it was reported that city would begin construction of works Aug 1, and complete it by Nov. 1. Contrs., Fairbanks, Morse & Co., Chicago. Supply, artesian wells, pumping to tank. Contract price, $5,436. Prest., E. Litehfield. Clerk, H. F. Mette.

GLENCOE, *Cook Co.* (Pop., 387—est., 500.) System projected by Moffett, Hodgkins & Clarke, Syracuse, N. Y., to supply Glencoe, Highland Park, Wilmette and Winnetka. Est. cost, $200,000. Latest report in Oct., when contract had not been closed.

HARVEY, *Cook Co.* (Pop., not given.) Suburb of Chicago. Harvey Transit Co. may build works. Gen. Mgr., W. S. Reed. Latest report, June 6, '91.

LA HARPE, *Hancock Co.* (Pop., 958—1,113.) Works talked of but no definite action taken June 26, '91. Mayor, G. F. Otto.

HIGHLAND PARK, *Cook Co.* (Pop., est., 1,200.) See Glencoe, above.

METROPOLIS, *Massac Co.* (Pop., 2,068—3,573.) July 7, '91, it was reported that Geo. Cadogan Morgan, 49 Major Block, Chicago, was drawing plans for construction of water-works and electric lighting plants. Bids were opened Sept. 10. Mayor, A. Quante. City Clerk, Bart. Kerr.

MONTICELLO, *Piatt Co.* (Pop., 1,337—1,643.) In '91, city voted to build water-works. June 30, an artesian well was being drilled. City Clerk, W. P. Smith.

PAW PAW, *Lee Co.* (Pop., 52—est., 625.) Small system projected by village. Supply, well, pumping by wind-mill to tank. Cost, about $8,000. Construction to begin Sept. 15. Prest., C. F. Peribou.

PERU, *La Salle Co.* (Pop., 4,632—5,550.) Peru Water-Works Co. was to begin construction of works about Aug. 1, '91; following Oct. reported that stand-pipe was being erected. Supply, artesian well, pumping to reservoir and stand-pipe. Contracts were awarded in June. Engr., C. H. Ledlie, St. Louis. Contr., E. Sutphin, St. Louis. Prest., H. Bellingham. Secy., G. Lassig.

PONTIAC, *Livingston Co.* Pop., 2,242—2,784.) A 20-yrs. franchise was granted in '91. Supply, well, pumping to 15 × 125-ft. stand-pipe. Mains, 5 miles. Hydrant rental, $2,500. Address Jno. J. Hainsworth, Peru, Ill.

ROODHOUSE, *Greene Co.* (Pop., in '90, 2,360.) Water-works talked of, but no definite action, June 29, '91. Address W. C. Roodhouse.

TUSCOLA, *Douglas Co.* (Pop., 1,457—est., 1,800.) Town voted in Oct., '91, to have works.

WILMETTE, *Cook Co.* (Pop., 419—1,458.) July 6, '91, F. H. Drury, Village Clerk, wrote that engr. had been engaged to make surveys and draw plans for water-works. See Glencoe above.

WINNETKA, *Cook Co.* (Pop., 584—1,079.) Co. made proposition in '91 to supply town from plant about to be erected at Glencoe and controlled by Moffett Hodgkins & Clarke, Syracuse, N. Y.; Aug. 28 it was not decided whether this proposition would be accepted or independent plant erected. See Glencoe above.

Water-Works Projects Reported in '90 and '91, but Now Indefinite.

Keithsburg, Mercer Co. *Morris,* Grundy Co. *Olney,* Richland Co. *Pana,* Christian Co. *Wheaton,* Du Page Co.

WISCONSIN.

Water-Works Completed or In Process of Construction.

1. **ANTIGO,** *Langlade Co.* (Pop., in '90, 4,424.) **History.**—Built in '90-1 by Antigo Water-Works Co; put in operation Jan. 20. Des. Engr., R. C. Pearsons, Oshkosh. Const. Engr.– A. W. Stevens, Logansport, Ind. Conts., C. E. Gray, Jr., & Co., Whitewater. **Supply.**—Well, pumping to stand-pipe and direct. **Pumping Machinery.**—Dy. cap., 3,000,000 galls.; two 1,500,000-Gordon dup. boilers, from Ft. Wayne Iron Wks. **Stand-Pipe.**—Cap.,150,366 galls. **Distribution.**—*Mains*, 12 to 4-in.1, 5 miles. Pipe and specials from Lake Shore Foundry, Cleveland. *Services*, Kal. i. *Taps*, 112. *Hydrants.* Ludlow, 66. *Valves*, Ludlow. *Pressure:* ordinary, 50 lbs.; fire, 150 lbs. **Financial.**—*Cost*, $60,000. *Hydrant Rental*, $3,750. **Management.**—Prest., H. C. Humphrey. Treas., W. G. Maxey, Oshkosh. Supt., Edmund La Lande.

NOTE.—In March, '91, it was reported that city council would not accept works on ground that they were not completed during contract time.

2. **APPLETON.** *Outagamie Co.* (Pop., 8,005—11,860.) Built in '82 by Wiley Construction Co., under 20-yrs. franchise. **Supply.**—Fox River and three artesian wells, pumping to reservoir; Jewell filter. During '90 a 2,500,000-gall. reservoir was constructed, 59 ft. above pumps. **Distribution.**—*Mains*, 21 miles. *Taps*, 800, *Meters*, 60. *Hydrants*, 187. **Financial.**—*Revenue*, city, $8,000. **Management.**—Prest., C. H. Venner, 33 Wall St., N. Y. City. Secy., J. C. Rogers. Supt., J. H. Hayes.

3. **ASHLAND,** *Ashland Co.* (Pop., 951—9,956.) Built in '84-5 by Ashland Water Co., under 50 yrs. franchise. **Supply.**—Chequamegon Bay, Lake Superior, pumping direct. Fiscal year closes Aug. 1.

Year.	Mains.	Taps.	Meters.	Hyd'ts.	C'ns'pt'n.	Cost.	B'd'd d'bt.	Int.	Exp.	Rev.
'86-7	5.5	225	6	42	350,000	$165,000	$70,000	$4,200	.....	$4,200*
'87-8	8	435	10	72	600,000	165,000	70,000	4,200	$5,500	20,480
'84-9	10	611	16	90	700,000	205,935	100,000	6,000	6,577	18,875
'89-90	13	687	19	90	900,000	258,318	150,000	9,000	7,000	23,490

*From city.

Financial.—*Cap. Stock:* authorized, $100,000; all paid up. Oct., 91, co. increased cap. stock to $300,000, of which $125,000 was 6% cumulative preferred. *Total Debt*, $157,000: floating, $7,000 at 6%; bonded, $150,000 at 6%. *Expenses*, total, $16,000. No taxes. *Revenue:* consumers, $16,920; fire protection, $6,570; total, $23,490. **Management.**—Prest., T. T. Flagler, Lockport, N. Y. Secy., P. H. Linneen, 301 Home Insurance Bldg., Chicago, Ill. Treas., H. H. Flagler, Lockport, N. Y. Supt., J. I. Jones.

NOTE.—Unofficially reported summer of '91 that works had been sold to Boston parties represented by Wheeler & Parks.

4. **BARABOO,** *Sauk Co.* (Pop., 3.266—4,605.) Built in '86-7 by Baraboo City Water-Works Co. under 30-yrs. franchise; now controlled by Moffett, Hodgkins & Clarke, Syracuse, N. Y. **Supply.**—Springs, pumping to stand-pipe.

Year.	Mains.	Taps.	Meters.	Hyd'ts.	C'ns'pt'n.	Cost.	B'd'd d'bt.	Int. ch'g's.	Exp.	Rev.
'87	8.5	184	1	81	90,000	$130,000	$100,000	$6,000	.....	.....
'88	9.5	292	4	84	60,730	130,000	100,000	6,000	$1,000	$4,260

Management.—(In '89.) Prest., J. F. Moffett. Secy., Chas. T. Moffett.

5. **BAYFIELD,** *Bayfield Co.* (Pop., 495—1.373.) Built in '86 by Bayfield Hydraulic Co. under 20-yrs. franchise. **Supply.**—Springs, by gravity. Fiscal year closes Dec. 31. **Distribution.**—*Mains*, 2 miles. *Taps*, 60. *Hydrants*, 14. *Consumption*, 15,000 galls. **Financial**—*Cost*, $23,034. *Cap. Stock*, $50,000; paid 10%, div'd. in '90. *Floating Debt*, $1,200 at 8%. *Expenses:* operating, $500; int., $14; taxes, $55; total, $599. *Revenue*, $2,612. **Management.**—Prest., A. Tate. Secy., F. W. Denison. Treas., Fred Fisher. Supt., C. G. Bell.

6. **BEAVER DAM,** *Dodge Co.* (Pop., 3,416—4,222.) Built in '88 by Moffett, Hodgkins & Clarke, Syracuse, N. Y. **Supply.**—Springs, pumping to stand-pipe. **Distribution.**—(In '88.) *Mains*, 8 miles. *Hydrants*, 78. *Consumption*, 40,000 galls. **Financial.**—(In '88.) *Cost*, $125,000. *Bonded Debt*, $100,000 at 6%. *Expenses:* operating, $1,500; int., $6,000; total, $7,500. *Revenue*, city, $3,400. **Management.**—(In '89.) Prest., E. G. Ferris. Supt., Geo. L. Camp.

FOR FURTHER DESCRIPTIVE MATTER see Manual for 1889-90. CHANGES since 1890 are here given. POPULATIONS are for 1880 and 1890, respectively.

248   MANUAL OF AMERICAN WATER-WORKS.

7. BELOIT, *Rock Co.* (Pop., 4,790–6,315.) Built in '85 by Beloit Water-Works Co. Supply.—Well, pumping to tank and direct. Distribution.—*Mains*, 9 miles. *Hydrants*, 75. Financial.—*Cost* (in '88), $95,000. *Revenue*, $12,000. Management.—Prest., J. B. Peet. Secy., C. B. Salmon. Supt., W. H. Wheeler.

8. BLACK RIVER FALLS, *Jackson Co.* (Pop., 1,427–2,261.) Built in '86-7 by town. Supply.—Forty-eight driven wells, pumping to stand-pipe. Fiscal year closes April 1. Distribution.—*Mains*, 3 miles. *Taps*, 117. *Meters*, 7. *Hydrants*, 13. *Consumption*, 45,000 galls. Financial.—*Cost*, $27,000. *Bonded Debt*, $20,000 at 5%, due in 1906. *Sinking Fund*, $2,500. *Expenses:* operating, $110; int., $1,000; total, $1,110. *Revenue*, consumers, $1,247. Management.—Three comrs., elected for three years. Chn., A. Meinhold, Secy., G. M. Hull. Supt., E. J. Greenlee. Appointed by comrs.

9. BURLINGTON, *Racine Co.* (Pop., 1,611–2,043.) Built in '90 by town. Supply.—Artesian well, pumping to tank. Fiscal year closes May 1. Reported Nov., '91, that bonds had been voted for 2.5 miles additional mains, pumping station and reservoir. Reported July, '91, that "new works" had been accepted. *Mains*, 2.6 miles. *Hydrants*, 23. Address H. E. Zimmerman, Village Clerk.

10. CHIPPEWA FALLS, *Chippewa Co.* (Pop., 3,982–8,670.) Built in '85 by Moffett, Hodgkins & Clarke, Syracuse, N. Y. Supply.—Chippewa River, pumping to stand-pipe, filtered.

Year.	Mains.	Taps.	Meters.	Hydrants.	Consumption.	Cost.	Exp.	Rev.*
'86	10	180	12	115	75,000	$150,000	......	$6,325
'88	11	650	43	128	120,000	300,000	$1,000	14,000

* From city,

Management.—(In '89.) Prest., H. W. Phelps, Minneapolis, Minn. Secy., L. M. Newman. Supt., H. B. Shamp.

11. DARLINGTON, *LaFayette Co.* (Pop., 1,372–1,589.) Built in '86-7 by town. Supply.—Well, pumping to reservoir. Fiscal year closes first Monday in March. Distribution.—*Mains*, 3.5 miles. *Taps*, 118. *Meters*, 4. *Hydrants*, 20. *Consumption*, unknown. Financial.—*Cost*, $17,000. *Bonded Debt*, $15,000 at 5%, due in annual payments of $1,000 after Feb. 1, '92. *Expenses:* operating, $100; int., $750; total, $1,150. *Revenue:* consumers, $1,200. Management.—Committee of three, appointed by mayor every year. Chn., Jno. V. Swift. Secy., Edw. Schreiter. Supt., C. C. H. Carey. City Engr., J. H. Martin.

12. DE PERE, *Brown Co.* (Pop., 817–969.) Owned by Kimberly & Clarke Co., Neenah, known as Shattuck & Babcock Co. Built '86 by Artesian Water Supply Co., under 25-yrs. franchise. Supply.—Flowing artesian well, by gravity. Fiscal year closes Dec. 1. Distribution.—*Mains*, 6 miles. *Taps*, 120. *Hydrants*, 20. Financial.—*Cost* (in '88), $14,000. *Cap. Stock:* authorized, $25,000; all paid-up; paid 5½% div'd in '90. *No Debt. Taxes*, $70. *Revenue:* consumers, $1,000; city, $535; total, $1,535. Management.—Prest., B. F. Smith. Secy., E. F. Parker. Supt., Robert Jackson.

Drummond, *Bayfield Co.* (Pop., in '90, 696.) Built for fire protection only.

Eagle River, *Oneida Co.* (Pop., in '90, 1,154.) Built in '89 for fire protection only.

13. EAU CLAIRE, *Eau Claire Co.* (Pop., 10,119–17,415.) Built in '85 by Eau Claire Water-works Co. Supply.—Wheaton springs, by gravity from impounding reservoir to pump-well, thence by direct pumping. Fiscal year closes Dec 31.

Year.	Mains.	Taps.	Meters.	Hydrants.	Cons'pt'n.
'86	24	170	0	269	300,000
'88	26.5	600	0	331	8 0,000
'90	28.5	944	16	312	1,000,000

Financial.—*Expenses*, operating, $7,500. *Revenue:* consumers, $13,000; fire protection, $15,800; other purposes, $400; total, $29,200. Management.—Prest., O. H. Ingram Secy. and Treas., C. A. Sharpe, Lake and La Salle Sts., Chicago, Ill. Supt., W. H. Willard.

14. FOND DU LAC, *Fond Du Lac Co.* (Pop., 13,094–12,024.) Built in '35 by Fond Du Lac Water Co. under 20-yrs. franchise, which does not regulate rates, but provides for purchase of works by city in '95 and at intervals of 5 yrs. thereafter, price determined by appraisers. Supply.—Four artesian wells, by direct pumping, with storage reservoir for fire protection.

WISCONSIN.

Year.	Mains.	Taps.	Meters.	Hydrants.	Cost.	B'd'd debt.	Int. ch'g's.	Exp.	Rev.
'86	13	203	0	140	$200,000	$150,000	$9,000	....	$9,000*
'88	14	484	5	150	......	150,000	9,000	$6,000	15,800
'90	15	606	10	165	205,290	158,500	9,540	6,150	17,300

*From city.

Financial.—*Cap Stock*, paid-up, $41,800. Net income used for extensions. Total *Debt*, $150,500: floating, $1,000 at 6%, bonded, $158,500 at 6%, due in 1915 with option of redeeming in '95. *Expenses:* operating, $4,400; int., $9,540; taxes, $1,650; total, $15,090. *Revenue:* consumers, $7,500; city, $9,800; total, $17,300. **Management.**—Prest., T. T. Flagler, Lockport, N. Y. Secy., P. H. Linneen, 301 Home Insurance Building, Chicago, Ill. Treas., L. M. Winter. Supt., W. H. Masson.

16. **FORT HOWARD,** *Brown Co.* (Pop., 3,083—4,754.) Supply from Green Bay.

15. **GLENWOOD,** *St. Croix Co.* (Pop., in '90, 1,656.) **History.**—Built in '91 by town; put in operation Aug. 1. Engrs. and Contrs., Fairbanks, Morse & Co., Chicago. **Supply.**—Creek, pumping direct. **Pumping Machinery.**—Dy. cap., 2,500,-000 galls.; Smith & Valle dup. **Distribution.**—*Mains*, 8 to 4-in. c. i., 1.5 miles. *Services,* w. i. *Hydrants*, Mathews, 27. *Valves,* Eddy. *Pressure:* ordinary, 40 lbs.; fire, 175 lbs. **Financial.**—*Cost,* about $8,500. *Bonded Debt*, $10,000. **Management.**—Chn., H. J. Baldwin.

16. **GREEN BAY AND FORT HOWARD,** *Brown Co.* (Pop.: Green Bay, 7,464—9,069; Fort Howard, 3,083—4,754.) Built in '85-6 by Green Bay & Fort Howard Water-Works Co. Franchises establishes rates and provide for purchase of works by city after 20 yrs. on usual conditions. **Supply.**—Flowing artesian wells, direct pumping; two small storage reservoirs used only in case of fire. **Distribution.**—*Mains,* 25 miles, *Taps,* 1,000. *Hydrants,* 227. **Financial.**—*Cost* (in '88), $300,000. *Cap. Stock:* authorized, $300,000; paid up, $150,000; paid 1% div'd in 1890. *Total Debt,* $150,000; bonds at 6%. *Bonded Debt,* $150,000 at 6%. *Revenue,* city, $10,000. **Management.**—Prest., Eli Marvin. Secy., A. C. Neville. Supt. and Treas., W. K. Harrington.

17. **HAYWARD,** *Sawyer Co.* (Pop., in '90, 1,349.) Owned by city. Built in '87-8 by co. **Supply.**—Namakagon River, direct pumping. Fiscal year closes Jan. 1.

Year.	Mains.	Taps.	Meters.	Hydrants.	Consumption.	Cost.	Exp.
'89	4.3	338	0	57	155,500	27,000	$2,800
'99	4.3	334	0	60	237,440	27,000	3,100

Financial.—Withheld, except data above. **Management.**—Secy., H. P. Faley. Treas., H. Bigler. Supt., P. Oleson.

18. **HUDSON,** *St. Croix Co.* (Pop., 2,298—2,885.) Built in '87-8 by village. **Supply.**—Artesian well, pumping to underground reservoir. Fiscal year closes April 30.

Year.	Mains.	Taps.	Meters.	Hyd'ts.	Cost.	B'd'd debt.	Int. ch'g's.	Exp.	Rev.
'88-9	3.3	56	0	40	$34,183	$24,000	$960	$1,500	$1,200
'89-90	4	120	0	45	40,000	24,000	960	1,200	2,500
'90-1	5.3	156	0	48	44,000	24,000	960	1,800	2,600

Financial.—*Bonded Debt,* $24,000 at 4½%, due in annual payments of $5,000 after '91. *No Sinking Fund;* money for redemption of bonds raised by direct taxation. *Expenses,* total, $2,760. *Revenue:* consumers, $2,60); city, none. **Management.**—Board of five, appointed by common council every year. Chn., W. H. Phipps. Secy., Jos. Bench. Supt., C. Wangsgard.

19. **HURLEY,** *Ashland Co.* (Pop., in '90, 2,267.) **History.**—Construction was begun in '87 by Gogebic Water & Gas Supply Co., and completed early in Feb., '91, by Hurley Water Co., under 30-yrs. franchise. Const. Engr., Sam'l A. Souther. **Supply:**—Montreal River, pumping to stand-pipe. **Pumping Machinery.**—Dy. cap., 1,500,000 galls.; one Worthington dup., high-pressure, and one Barr dup. *Boilers,* from Garr Co., Minneapolis, Minn. **Stand-Pipe.**—Cap., 264,000 galls.; 30 × 50 ft., from Tippett & Wood, Phillipsburgh, N. J. **Filters.**—Jewell, improved. **Distribution.**—*Mains,* 12 to 4-in. kal. i., 3 miles. Pipes and specials from National Tube Wks., Chicago, Ill. *Taps,* 95. *Hydrants,* Ludlow, 35. *Valves,* Ludlow. *Consumption,* 300,000 galls. *Pressure,* 80 lbs. **Financial.**—*Cost,* $175,000. *Cap. Stock,* $200,000. *Bonded Debt,* $250,000 authorized, $150,000 issued. *Expenses:* operating, $4,800. *Revenue:* consumers, $12,500; city, $9,250, *Hydrant rental,* $100 for each first 50, $75 each for second 50, $50 each for remainder. **Management**—Prest., C. H.

For Further Descriptive Matter see Manual for 1889-90. Changes since 1890 are here given. Populations are for 1880 and 1890, respectively.

Jackson. Secy., Frank Linde. Treas., C. L. Hyde, 50 Broadway, N. Y. City. Supt., S. A. Souther, Ironwood,
NOTE.—Data under Financial and Management include both Hurley, Wis., and Ironwood, Mich.

20. **JANESVILLE,** Rock Co. (Pop., 9,018—10,836.) Built in '87-8 by Janesville Water Co. Supply.—Flowing artesian well, pumping to stand-pipe and direct. Fiscal year closes Dec. 31.

Year.	Mains.	Taps.	Meters.	Hydrants.	Consumption.
'88	13	480	12	180	
'90	15	640	111	197	202,705

Financial.—Cap. Stock, $200,000. Bonded Debt, $164,000 at 0%. Management.—Prest., B. R. Clarke, Boston, Mass. Treas., A. B. Turner, Boston, Mass. Secy. and Supt., W. C. Mitchell.

21. **KENOSHA,** Kenosha Co. (Pop., 5,039—6,532.) Built in 78-9 by Park City Water Co., under 20-yrs. exclusive franchise; city neither controls rates nor has right to buy works. Supply.—Four flowing artesian wells, by gravity. Fiscal year closes Dec. 1. Distribution.—Mains, 5.5 miles. Taps, 509. Meters, 22. Hydrants, 34. Financial.—Cost, $30,000. Cap. Stock: authorized, $17,000 all paid up; paid 8% div'd in 1890. Expenses: operating, $600; taxes, $300; total, $900. Revenue: consumers, $3,800; city, $850; total, $4,650. Management.—Prest., Lewis Bain. Treas., N. J. Lewis. Sec. and Supt., E. G. Hazelton.

22. **KILBOURN CITY,** Columbia Co. (Pop., 915—951.) Built in '89-90 by village; put in operation April 1, '90. Supply.—Surface well near river, pumping to tanks; filtered. Fiscal year closes April 1. In '90 supply for fire purposes was increased by sinking a second well. Further extensions of water supply were contemplated Jan. 27, '91. Distribution.—Mains, 3,500 ft. Taps, 12. Hydrants, 8. Consumption, 1,000 galls. Financial.—Cost, $6,500. Debt, in notes, $4,000 at 7%, due in '94. Expenses: operating, $550; int., $280; total, $830. Revenue, consumers, $100. Private consumers lay their own service pipes, and are paid in service at prescribed rates. Management.—Village Board. Contr. and Supt., G. M. Marshall.

23. **LA CROSSE,** La Crosse Co. (Pop., 14,505—25,090.) Built in '77 by city. Supply.—Mississippi River, pumping direct. Fiscal year closes Dec. 31. Improvements were under discussion in '91.

Year.	Mains.	Taps.	Meters.	Hyd'ts.	Cons'pt'n.	Cost	B'd'd d't.	Int. ch'g's.	Exp.	Rev.
'80	8.5	325		90		$90,000	none		$6,540	$1,000
'84	16	550	0	105	950,000	95,400			6,000	6,500
'87	16.5	1,069	2	143	2,197.628	169,964			9,977	13.813
'88	18.1	1,220	4	172	2.219,789	186,409			11.025	15.561
'89	19.5	1,364	4	182	2,284,402	191,164	$75,000	$3,750	13,117	16,779
'90	20.7	1,468	6	193	2,162,196	197,060	75,000	3,750	9,759	13,412

Financial,—Bonded Debt, $75,000 at 5%; $10,000 due in 1904; $40,000 in 1915; $5,000 in 1916; $20,000 in 1907. Sinking Fund, $7,850. Expenses, total, $13,509. Revenue: consumers, $12,412; city, none. Management.—Board of Public Works, three members, elected every two years. Chn., J. E. Parker. Secy., T. S. Vickers.

24. **LAKE GENEVA,** Walworth Co. (Pop., 1,999—2,297.) History.—Built in '90 by Lake Geneva Water & Light Co., under 25-yrs. franchise; put in operation Aug. 1. Const. Engr., E. P. Wheeler, Beloit. Contrs., Hinchliffe & Wheeler. Supply.—Well, pumping to tank. Pumping Machinery.—Dy. cap., 1,000,000 galls.; Smith & Vaile dup. Tank.—Cap., 100,000 galls. Distribution.—Mains, 10 to 4 in. c. i., 4 miles. Services, lead. Trenching and pipe-laying done by C. E. Gray & Co., Whitewater. Hydrants Mathews, 50. Pressure, 60 lbs. Financial.—Cost (including lighting plant), $75,000. Hydrant rental, $2,500 for 50 hydrants. Management.-Prest., Geo. Hinchliffe, Chicago. Secy. and Treas., E. P. Wheeler, Beloit. Supt., S. D. Ross, Beloit.

25. **MADISON,** Dane Co. (Pop., 10,324—13,426.) Built in '82 by city. Supply.—Eight artesian wells, stand-pipe, direct pumping. Fiscal year closes Sept. 30.

Year.	Mains.	Taps.	Meters.	Hyd'ts.	C'ns'mpt'n.	Cost.	B'd'd d't.	Int. ch'gs.	Exp.	Rev.
'83-4	12.5	224	0	84		$100,000	$76,000	$3,800	$3,765	$3,176
'84-5	14	307		84	381,615	130,085	76,000	3,800	3,765	3,176
'85-6	16	699	2	94	546,120	138,934	76,000	3,800	8,789	9,060
'86-7	18.3	876		106	648,355	156,641	76,000	3,800	8,335	11,023
'87-8	21	980	7	118	715,885	173,102	76,000	3,800	8,740	13,480
'88-9	22	1,099	210	125	701,050	183,397	76,000	3,800	9,485	14,017
'89-90	22	1,229	385	127	535,180	206,522	76,000	3,800	11,726	13,680

Management.—Three comrs., one elected every year. Prest., Jno. Lamont. Supt., E. M. Nichols.

WISCONSIN.

26. MANITOWOC, *Manitowoc Co.* (Pop.. 6,367—7,710.) Built in '88 by Moffett, Hodgkins & Clarke, Syracuse, N. Y., under 20-yrs. franchise. Supply.—Lake Michigan, through filter galleries, pumping to stand-pipe. Distribution.—(In '88.) *Mains,* 15 miles. *Taps,* 150. *Hydrants,* 132. *Consumption,* 1,200,000 galls. Financial.—(In '88.) *Cost,* $200,000. *Bonded Debt,* $165,000 at 6%. *Expenses:* operating. $2,500; int., $9,900; total, $12,400. *Revenue:* consumers, $8,000; city, $6,500; total, $14,- 500. Management.—(In '89.) Prest., J. L. Hotchkin. Supt., E. O. Green.

27. MARINETTE, *Marinette Co.* (Pop., 2,750—11,523.) Built in '88 by American Water-Works & Guarantee Co., Ltd., Pittsburgh, Pa.; under 20-yrs. franchise, which fixes rates and provides for purchase of works by city in '97 and at intervals of 5 yrs. thereafter. Supply.—Green Bay, pumping direct. A 20-in. pipe extends from suction well, 10 × 18 ft., for 6,000 ft. to where the water is 30 ft. deep. Fiscal year closes Dec. 31.

Year.	Mains.	Taps.	Metres.	Hyd'ts.	C'ns'pt'n.	Cost.	B'd'd d't.	Int. ch'g's.	Rev.
'88	17	950	10	150	450,000	$155,000	$125,000	$7,500	$12,700
'90	18.5	1,069	3	153	600,000				

Financial.--Withheld. Management.—Prest., J. H. Purdy. Secy. and Treas., A. E. Claney. Supt., M. Kirkpatrick.

Marshfield, *Wood Co.* (Pop., 669—3,450.) Built by village, for fire protection. See '89-'90 MANUAL, p. 469.

NOTE.—In '91. Moffett, Hodgkins & Clarke, Syracuse, N. Y., made a proposition to build works with supply from Yellow River or artesian wells.

Mazo Manie, *Dane Co.* (Pop., in '90, 1,034.) Built in '88 by village, for fire protection and street sprinkling. See '89-90 Manual, p. 469.

28. MENOMINEE, *Dunn Co.* (Pop., 2,589—5,491.) Built in '84-5 by Moffett. Hodgkins & Clarke, Syracuse, N. Y., under 20-yrs. franchise. Supply.—Springs and Red Cedar River, pumping to stand-pipe and direct.

Year.	Mains.	Taps.	Meters.	Hyd'ts.	Cons'p't'n.	Cost.	B'd'd d'bt.	Int. ch'g's.	Exp.	Rev.
'86	9	80	10	101	20,000	$125,000	$75,000	$4,500	$1,900	$6,400
'88	10.6	190	...	106	75,853	125,000	90,000	4,500	2,400	10,336

Management.—(In '89.) Prest., Jno. F. Moffett. Supt., E. H. Weber.

29. MERRILL, *Lincoln Co.* (Pop., in '90, 6,803.) Built in '87 by City Water Works Co.; controlled by American Water-Works & Guarantee Co., Ltd., Pittsburgh, Pa., under 30-yrs franchise, which fixes maximum rates and provides for purchase of works by city in '97 and at close of each succeeding 5 yrs. Supply.— Filter gallery on bank of Prairie River, pumping direct. Gallery is 250 ft. long and 6 ft. below bed of river, in sand and gravel. Pumping Machinery.—Dy. cap., 3,000,- 000 galls.; two 1,500,000 Deane, one being 12 × 24 × 12 × 18 comp., the other 18½ × 12 × 18 high-pressure. *Boilers,* two 80-HP. horizontal tubular. Distribution.—*Mains,* 9.5 miles. *Taps,* 163. *Hydrants,* 104. *Consumption,* 300,000 galls. Financial.— Withheld. Management.—Prest., J. H. Purdy. Secy. and Treas., A. E. Claney.

30. MILWAUKEE, *Milwaukee Co.* (Pop., 115,587—204,468.) Built in 72-3 by city. Supply.—Lake Michigan, pumping to stand-pipe and reservoir, with repumping to stand-pipe for high service. Fiscal year closes Dec. 31. In '90 contracts were let for new intake, extending to 60 ft. of water and consisting of brick-lined tunnel 7.5 ft. in diam., about 3,000 ft. long, terminating in crib, from which will extend two lines 5-ft. c. i. pipe laid in trench 8 ft. deep and 18 ft. wide, and ending in submerged cribs about 15 ft. high. Contrs.: shafts and tunnels, Shailer & Shinglau, Chicago, for $122,000; crib, J. O'Neill, Milwaukee, for $34,000; c. i. pipe, Addystone Pipe & Steel Co., Cincianati, O., $126,000; dredging trench, C. H. Starke, Milwaukee, $65,- 000; pipe-laying, Shailer & Shinglau, Chicago, $53,500; total contract price, not including submerged crib, $400,500. In '90 contract was also let for 18,000,000-gall. Allis triple expansion pumping engine and 5 steel boilers, for $76,000.

Yr.	Mains.	Taps.	M't'rs.	Hyd's.	Cons'p't'n.	Cost.	Bd. d'bt.	In. ch's.	Exp.	Rev.
'74	37.6	1,787	...	409	1,250,000	1,855,401	1,600,000	3,670	33,923	27,156
'80	91	7,524	30	718	12,269,000	2,171,195	....	48,494	60,000	137,333
'82	99.2	7,974	91	794	14,690,413	1,851,591	1,600,000	50,000	86,700	177,190
'84	106	10,010	350	856	14,800,000	2,400,000	1,485,000	103,959	93,000	213,573
'85	120	11,942	572	1,013	16,062,475	2,589,841	....	97,085	94,610	223,854
'86	133.9	12,980	741	1,113	17,878,436	2,828,325	1,505,000	106,680	103,011	249,286
'87	147.4	14,072	1,728	1,208	20,089,014	3,012,877	1,709,000	......	119,176	246,738
'88	162.5	15,738	3,523	1,321	19,159,432	3,136,920	1,477,003	88,760	140,851	263,829
'89	176.48	16,986	4,749	1,434	19,745,408	3,229,856	1,591,500	92,130	139,706	279,523
'90	190.1	18,422	5,376	1,532	22,380,783	3,368,432	2,057,000	124,403	106,789	300,792

FOR FURTHER DESCRIPTIVE MATTER see Manual for 1889-90. CHANGES since 1890 are here given. POPULATIONS are for 1880 and 1890, respectively.

Financial.—*Bonded Debt*, $2,057,000: $1,055,000 at 7%. due in 1903; balance at 4%, due 1903-7; 5% of above bonds are redeemed every year. *Expenses*, total, $231,192 *Revenue:* consumers, $300.792; city, $11.271; total, $312,063. Op. exp , int., and large part of bonds are paid from revenue from consumers; the remainder is paid from direct taxation. New construction is met by fresh issue of bonds **Management.**—Board of Public Works appointed every year by mayor. Prest. and City Engr., G. H. Benzenbug. Secy., Geo. P. Traeumer.

31. **MONROE,** *Green Co.* (Pop., 3.293—3,768) **History.**—Built in '89-90 by Monroe Water-Works Co., under 15-yrs. franchise; put in operation in Feb Des. Engr., W. C. Wheeler. Const. Engr., E. P. Wheeler. Both of Beloit. **Supply.**—Well, pumping to reservoir and direct. Well is 40 ft. in diam., 40 ft. deep, with drill holes in rock bottom, 120 ft. deep. **Pumping Machinery.**—Dy. cap., 1.500.000 galls. *Reservoir*, cap., 100,000 galls. **Distribution.**—*Mains*, 10 to 4-in. c. i., 7 miles. *Services*, lead. *Meters*, Hersey. *Hydrants*, Wood. 72. *Valves*, Ludlow. **Financial.**—*Revenue*, city, $5,000. **Management.**—Prest., W. E. Wheeler, Beloit. Secy., E. P. Wheeler, Ashland. Supt., S. D. Ross, Beloit.

32. **NEILLSVILLE,** *Clark Co.* (Pop., 1,050—1,936.) Built in '85 by city; operated jointly by city and Electric & Water Supply Co. **Supply.**—Creek, pumping to stand-pipe and direct. Stand-pipe is 16 × 100 ft. **Distribution.**—*Mains*, 5 miles. *Taps*, 60. *Hydrants*, 16. **Financial.**—*Revenue*, $650. **Management.**—Prest., Tom Lowe. Supt., R. F. Kountz.

33. **OCONTO,** *Oconto Co.* (Pop., 4,171—5,219.)

OCONTO WATER-WORKS CO.

Built in '83. **Supply.**—Oconto River, pumping to tank. *Mains*, 1.4 miles, *Taps*, 20. *Hydrants*, 6. *Cost*, $1,000. Secy., W. K. Smith.

OCONTO WATER CO.

History.—Built in '90-1; put in operation June 1. Des Engr., H. J. Hillert, Milwaukee. Const. Engr., F. H. Ladd, St. Paul, Minn. **Supply.**—Well, pumping direct. Well is 100 ft. from Oconto River. **Pumping Machinery.**—Dy. cap., 3,500,000 galls.; two 1,500,000 Deane cond., direct-acting; one 500,000 Barr. *Boilers*, from Atlas Boiler Wks., Cincinnati. **Distribution.**—*Mains*, 16 to 6-in., 8½ miles. *Services*, galv. i. Pipe and specials from National Foundry & Pipe Wks.. Pa.; trenching and pipe laying by C. E. Gray, Jr.. & Co., Chicago, Ill. *Taps*, 60. *Hydrants*, Mathews, 100. *Valves*, Chapman. *Pressure:* ordinary, 40 lbs.; fire, 100 lbs. **Financial.**—*Cost*, $100,000. *Cap. Stock*., $100,000. *Bonded Debt*. $100,000. *Revenue*, city, $6,500, *Hydrant rental*, $65. **Management.**—Prest., G. W. Sturtevant, Bushnell, Ill.. Secy. and Treas., S. W. Ford.

34. **OSHKOSH,** *Winnebago Co.* (Pop., 15,748—22,836.) Built in '84 by Oshkosh Water Co., under 30 yrs. franchise which fixes rates and provides for purchase of works by city, price determined by arbitration. **Supply.**—Lake Winnebago, filtered, pumping direct, Intake extends 1,000 ft. into lake. Attempt at filtration was made before present plant was adopted; original supply was from three artesian wells. In '91 a 2,000,000-gall. Warren filter was put in, water being lifted to filters by 8-in. Webber centrifugal driven by 15 HP. Westinghouse engines. Water level in filters is about 12 ft. above lake surface. Filtered water passes to original pump-well, and small amount may be stored in 500,000-gall. covered reservoir formerly used to receive artesian supply. See note below. **Distribution.**—*Mains*, 21 miles, *Taps*, 1,008. *Meters*, 11. *Hydrants*, 239. *Consumption*, 1,700,-000 galls. **Financial.**—*Revenue*, city, $16 340. **Management.**—Prest., W. G. Maxcy. Secy., C. C. Clausen, Washburn. Treas., J. S. Maxcy. Gardiner, Me. Supt., Ben Edgarton.

NOTE.—For reproduction of photographic view of filters in place and some details concerning them see ENGINEERING NEWS, vol. 26, p. 421; *Fire and Water*, for July 8, 1891.

35. **PORTAGE,** *Columbia Co.* (Pop., 4,346—5,143.) Built in '87 by Portage Water Co.; controlled by Moffett, Hodgkins & Clarke, Syracuse, N. Y., under 20-yrs. franchise. **Supply.**—Wisconsin River, pumping to stand-pipe and direct; filtered.

Yr.	Mains.	Taps.	Met's.	Hyd'ts.	C'ns'pt'n.	Cost.	B'd'd debt.	Int. ch'g's.	Exp.	Rev.
'87	8	150	9	75	....	125,000	$100,000	$6,000	.....	.....
'88	9	181	8	75	140,000	125.000	100.000	6,000	$2.500	$9,000

Management.—(In '89.) Prest., E. G. Ferriss. Supt., Jas. W. Reade.

36. **PRAIRIE DU CHIEN,** *Crawford Co.* (Pop., 3,777—3,131.) Built in '75-6 by Prairie du Chien Artesian Water Co.; town neither controls rates nor has right to

purchase works. Supply.—Flowing artesian well, by gravity. Distribution.—
Mains, 1 mile. *Taps*, 22. *Hydrants*, 7. Financial.—*Cost*, $10,000. *No Debt. Expenses*, $100. Management.—Prest., M. Menger. Treas., B. F. Fay. Secy. and Supt., F. H. Poebler.

37. **RACINE,** *Racine Co.* (Pop., 16,031—21,014.) Built in '86-7 by Racine Water Co., under 25-yrs. franchise, which does not regulate rates, but provides for purchase of works by city in 1903 and at intervals of 5 yrs. thereafter, price determined by three comrs. Supply.—Lake Michigan, pumping to stand-pipe. Fiscal year closes Nov. 1. Distribution.—*Mains*, 33.5 miles. *Taps*, 1,838. *Meters*, 144. *Hydrants*, 322. *Consumption*, 484,360 galls. Financial —*Cost* (in '88), $750,000. *Bonded Debt*, $400,000 at 6%, due in 1906. Management.—Prest., Henry E. Cobb, Boston, Mass. Treas., Chas. E. Eddy. Boston, Mass. Secy., J. E. Dodge, Supt., Wm. H. Laing.

38. **RHINELANDER,** *Oneida Co.* (Pop., in '90, 2,658.) Built in '87 by town. Jan. 17, '91, improvements to works were tested; contrs., Moffett, Hodgkins & Clarke, Syracuse, N. Y. Two 1,000,000-gall. Deane pumping engines, two new boilers and steel stand-pipe were included in improvements. Supply.—Wisconsin River, pumping to stand-pipe. Distribution.—(In '88.) *Mains*, 1.5 miles. *Taps*, 20. *Hydrants*, 12. Financial.—(In '88.) *Cost*, $12,000. *No Debt. Expenses.* $2,000. *Revenue*, $400. Management.—(In '89) Chn., W. E. Brown. Supt., Jno. Doherty.

39. **RICHLAND CENTRE,** *Richland Co.* (Pop., 1,227—1,819.) Built in '89 by city. Supply.—Artesian wells, pumping to reservoir. Fiscal year closes Oct. 1. Distribution.—*Mains*, 4 miles. *Taps*, 84. *Hydrants*, 39. Financial.—*Cost*, $19,500 *Bonded Debt*, $18,000, at 5%, due in annual payments of $1,000. *Expenses*: operating, $1,000; int., $900; total, $1,900. *Revenue*, consumers, $500. Management.—City Council. Mayor, Geo. H. Strang. City Clerk, K. W. Eastland.

40. **RIPON,** *Fond du Lac Co.* (Pop., 3,117—3,358.) Built in '90 by Ripon Water Works Co.; city has right to purchase works in 1900 and at close of each 5 yrs. thereafter, appraisers selected by both parties; city has no control over rates. Supply. —Filter gallery, pumping to stand-pipe and direct in case of fire. Fiscal year closes Jan. 1. Contr. for original works, A. W. Stevens, Logansport, Ind. Pumps, Gordon. Distribution.—*Mains*, 8 miles. *Taps*, 85. *Meters*, ?. *Hydrants*, 75. *Consumption*, 40,000 galls. Financial.—*Cost*, $100 000. *Cap. Stock*, $80,000. *Bonded Debt*, $75,000, at 6%. Management.—Prest., T. W. Gray, Chicago. Secy. and Supt., I. R. Gray.

41. **SHEBOYGAN,** *Sheboygan Co.* (Pop., 7,314—16,359.) Built in '87 by City Water Co., now controlled by American Water-Works and Guarantee Co., Ltd., Pittsburgh, Pa.; franchise fixes maximum rates and provides for purchase of works by city in '97 and at intervals of 5 yrs. thereafter, price determined by arbitration. Supply.—Lake Michigan, pumping to stand-pipe. About June, '91, reported that contract had been let for extension of intake 600 ft. further into lake. Pumping Machinery.—Dy. cap., 4,500,000 galls.; two 1,500,000 Worthington, each 15 × 26 × 12 × 15, and one 1,500,000 Dean high-pressure, 18½ × 12 × 10. Fiscal year closes Dec. 31. Distribution.—*Mains*, 22 miles. *Taps*, 728. *Meters*, 13. *Hydrants*, 298. *Consumption*, 450,000 galls. Financial.—Withheld. Management.—Prest. and Supt., W. D. Cockburn. Secy., W. S. Kuhn, Pittsburg, Pa. Treas., A. E. Claney.

42. **SHELL LAKE,** *Washburn Co.* (Pop., in '90, 1,535.) History.—Built in '90 by city. Des. Engr., Jas. Devereaux. Const. Engr., L. E. Thomas. Supply.—Lake, pumping to stand-pipe. Pumping Machinery.—Dy. cap., 720,000 galls.; Wells. Stand-Pipe.—Cap., 20,000 galls. Distribution.—*Mains*, 8-in. c. i., 2.3 miles. *Taps*, 50. *Hydrants*, 12. *Consumption*, 50,000 galls. *Pressure*: ordinary, 40 lbs.; fire, 120 lbs. Financial.—*Cost*, $10,000. *Expenses*, $2,000. Management.—Prest. and Supt., A. H. Earle. Secy., A. Ryan. Treas., Jas. Devereaux.

45. **SOUTH SUPERIOR,** *Douglas Co.* (Pop., not given). Supply from Superior.

43. **STEVENS POINT,** *Portage Co.* (Pop., 4,149—7.896.) Built in 87-'8 by Stevens Point Water Co. Supply.—Wisconsin River, pumping to stand-pipe; Hyatt filters added in '90. Fiscal year closes Dec. 31.

Year.	Mains.	Taps.	Meteis.	H'd'ts.	Consp't'n.	B'd'd debt.	Int. ch'g's.	Rev.*
'89	9.5	74	22	128	63,000	$175.000.	$10,500	$5,000
'90	10.3	100	35	133	177,889	175,000	10.500	5,032

*From city.

FOR FURTHER DESCRIPTIVE MATTER see Manual for 1889-90. CHANGES since 1991 are here given. POPULATIONS are for 1880 and 1890, respectively.

Financial.—*Cap. Stock*, $175,000. *Bonded Debt*, $175,000, at 6%. **Management.**—Secy., W. G. Snow, 2 Wall St., N. Y. City. Supt., W. O. Lamoreux.

44. STOUGHTON, *Dane Co.* (Pop., 1,353—2,170.) Built in '85 by village. **Supply.**—Surface well, pumping to tank. Fiscal year closes April 7. **Distribution.**—(In '88.) *Mains*, 3.8 miles. *Taps*, 110. *Meters*, 1. *Hydrants*, 34. *Consumption*, 10,000 galls. **Financial.**—*Cost*, $30,000. *Bonded Debt*, $30,000 at 5%, due in 25 yrs. *Sinking Fund*, $2,500. **Management.**—City Clerk, C. J. Rolter. Supt., Julius Fix.

45. SUPERIOR AND WEST SUPERIOR, *Douglas Co.* (Pop.: Superior. 655—11,983; West Superior, not given.) Original works built in '87-8 by Superior Water-Works Co., under 20-yrs. franchise, which fixes maximum rates, and provides for purchase of works by city in 1903, and at intervals of 5 yrs thereafter. Aug. '89, Superior Water, Light & Power Co. was organized, and has since bought out above Co., also Superior & Duluth Electric Light Co., Superior Light & Fuel Co., Superior Arc Light & Power Co. **Supply.**—Superior Bay, pumping direct. During '90-1 a c. i. intake pipe 8,000 ft. long was extended 2,600 ft. into Lake Superior. New pumping station was also completed, costing $15,000. Purchase of 5.000,000-gall. pump was under consideration April 28, '91. Fiscal year closes Oct. 31. **Distribution.**—*Mains*, 32 miles. *Taps*, 298. *Meters*, 47. *Hydrants*, 432. *Consumption*, 750,000 galls. **Financial.**—*Cost*, $678,000. *Cap. Stock*, paid up, $100,000. *Bonded Debt*, $850,000 at 6%. Cap. stock and bonded debt include electric lighting plant. *Hydrant rental*, $75 each for first 80, $60 each for second 80, $50 each for additional. **Management**—Prest., A. H. Wilder, St. Paul, Minn. Secy., F. A. Ross, West Superior. Treas., T. M. Watkins, St. Paul, Minn. Supt., Jno, Mather.

46. WASHBURN, *Bayfield Co.* (Pop., in '90, 3,039.) Built in '89 by Washburn Water-Works Co. **Supply.**—Chequamegon Bay, Lake Superior, pumping to reservoir and direct. Fiscal year closes May 13. **Distribution.**—*Mains*, 4 miles. *Hydrants*, 50. *Consumption*, 30,000 galls. **Financial.**—*Cap. Stock*, paid-up, $100,000. *Bonded Debt*, $75,000 at 6%. **Management.**—Prest., W. G. Maxcy, Oshkosh. Secy., C. C. Chausen. Treas., D. M. Maxcy.

47. WAUKESHA, *Waukesha Co.* (Pop., 4,613—6,321.) Built in '88 by Waukesha Water-Works Co.: rates are fixed in franchise; city may buy works after 10 yrs. **Supply.**—Two surface wells and artesian well, pumping to stand-pipe and direct. **Distribution.**—*Mains*, 13 miles. *Taps* (in '88), 225. *Meters* (in '88), 0. *Hydrants* (in '88), 92. **Financial.**—*Cost*, $135,000. *Cap. Stock:* authorized, $200,000; paid up, $100,000. *Bonded Debt*, $60,000 at 6%. *Expenses:* operating, $2,200; int., $3,600; taxes, $300; total, $6,100. *Revenue:* consumers, $5,000; fire protection, $4,080; total, $9,080. **Management.**—Prest., Eli Marion. Secy., T. W. Haight. Treas., A. J. Frame. Supt., Geo. F. Deakin.

48. WAUSAU, *Marathon Co.* (Pop., 4.277—9,253.) Built in '85 by city. **Supply.**—Well, pumping direct. Fiscal year closes Nov. 30.

Year	Mains	Taps	Meters	Hyd'ts	C'ns'pt'n	Cost	B'd'd d't	Int. ch'g's	Exp.	Rev.
'86	12.6	299	4	104	523,919	$128,687	$90,000	$4,500	$5,563	$4,739
'88	14.5	502	0	111	438,249	132,126	90,000	4,500	4,852	6,480
'89	15	599	0	112	507,480	132,898	90,000	4,500	4,678	6,230
'90	15.5	675	0	113	631,321	135,290	90,000	4,500	4,458	6,750

Financial.—*Bonded Debt*, $90,000 at 5%, due in 1905. *Expenses*, total, $3,953. *Revenue:* consumers, $6,750; city, none. Net revenue goes into general city fund, from which interest on bonds is paid. **Management.**—Three comrs. appointed every three years by city council. Chn., J. W. Smith. Secy., Jno. Ringle. Supt., Philip Ringle.

49. WEST DE PERE, *Brown Co.* (Pop., 1,870—est., 2,100.) Built in '87-8 by West De Pere Artesian Water Supply Co., under 25-yrs. franchise. **Supply.**—Two artesian wells, by gravity. **Distribution.**—*Mains*, 3.5 miles. *Taps*, 100. **Financial.**—*Cap. Stock*, $20,000. *No Bonds*. *Floating Debt*, $1,200 at 6%. **Management.**—Prest., Paul Scheuring. Secy., E. A. Lange. Supt., B. Carpenter.

45. WEST SUPERIOR, *Douglas Co.* See Superior.

50. WHITE WATER, *Walworth Co.* (Pop., 3,617—4,359.) Built in '89 by White Water Water-Works Co., under 25-yrs. franchise; city has right to purchase in '99, and at end of every five years thereafter; has no control over rates. **Supply.**—Artesian well, pumping direct and to tank and reservoir. Fiscal year closes Jan. 1. **Distribution.**—*Mains*, 8.5 miles. *Taps*, 150. *Meters*, 7. *Hydrants*, 96. *Consumption*, 15,000 galls. **Financial.**—*Cost*, $100,000. *Cap. Stock*, authorized, $90,000; all paid up. *Bonded Debt*, $72,500 at 6%, due in 1909. *Expenses:* operating, $2,100; int.,

$4,350; taxes, $200; total, $6,650. *Revenue:* consumers, $3,500; fire protection, $4,000; total, $7,500. **Management.**—Prest., C. E. Gray, Chicago, Ill. Secy., S. W. Gray, Chicago, Ill. Supt., H. A. Sanford.

Water-Works Projected with Fair Prospects of Construction.

BRODHEAD, *Green Co.* (Pop., 1,254—1,461.) July 1, '91, it was reported that city would build small system during '91-2, construction to begin July 10. Supply, artesian well, pumping to stand-pipe. Chn., Wm. Roantree.

ELROY, *Juneau Co.* (Pop., 663—1,413.) City intended to begin construction of water-works April 1, '91. Supply, well, pumping to reservoir. Cost, $10,000. Latest report, Aug. 13, '91. Supt., E. S. Willey.

MARSHFIELD, *Wood Co.* (Pop., 669—3,450.) About Sept. 1, '91, it was reported that Moffett, Hodgkins & Clarke, Syracuse, N.Y., had refusal of franchise till Oct. 1.

TOMAHAWK, *Lincoln Co.* (Pop., in '90, 1,816.) Bids for construction of water-works were received by City Council till Sept. 2, '91; unofficially reported that contract was afterward let for $18,586. Pumping to stand-pipe. City Clerk, A. J. Olson.

WAUPACA, *Waupaca Co.* (Pop., 1,392—2,127.) Franchise was granted to R. N. Roberts and J. H. Woodworth. Supply, Spring Lake. July 10, '91, it was unofficially reported that construction had begun.

Water-Works Projects Reported in '90 and '91, but Now Indefinite.

Delavan, Walworth Co. *Watertown,* Jefferson Co.

GROUP 6.—Northwestern States.—Water-Works Completed, in Process of Construction or Projected, in Iowa, Minnesota, Nebraska, South Dakota, North Dakota, Wyoming and Montana.

IOWA.

Water-Works Completed or in Process of Construction.

Algona, *Kossuth Co.* (Pop., 1,359—2,063.) System for fire protection only, constructed in '90. Well, pumping to stand-pipe. *Mains,* 1,600 ft. *Hydrants,* 5. *Cost,* $9,000.

1. **ANAMOSA,** *Jones Co.* (Pop., 2,083—2,078.) Built in '73-4 by Anamosa Water-Works Co.; city neither controls rates nor has right to buy works. **Supply.**—Wapsipinicon River, pumping to reservoir and direct. Fiscal year closes April 1. Distribution.—*Mains,* 2.5 miles. *Taps,* 110. *Hydrants,* 11. *Consumption,* 80,000 galls. **Financial.**—*First Cost,* $15,000. *Cap. Stock:* authorized, $30,000; paid-up, $6,200. *Bonded Debt,* $9,000, at 8%, $500 due in 1900 and every six months thereafter. *Expenses:* operating, $1,000; int., $750; taxes, $64; total, $1,784. *Revenue:* consumers, $836; fire protection, $1,325; other purposes, $1,040; total, $3,201. **Management.**—Prest., B. Huggins. Secy., W. R. Parsons. Treas., T. W. Shapley. Supt., J. C. Griffith.

2. **ATLANTIC,** *Cass Co.* (Pop., 3,662—4,351.) Built in '82-3 by Atlantic City Water Co. under 25-yrs. franchise. **Supply.**—Cook system of wells, pumping direct. Supply was increased in '90 by eight 6-in. wells. Fiscal year closes March 14.

Yr.	M'ns.	Taps.	M't'rs.	H'd'ts.	C'n'tion.	Cost.	B'd'd debt.	Int. ch'g's.	Exp.	Rev.
'86-7	5.5	200	18	63	192,345	$70,000	$50,000	$3,000	$3,000	$6,500
'88-9	6	403	0	72	250,000	75,000	50,000	3,000	4,374	7,542
'90-91	9	643	0	104	350,000	93,416	50,000	3,000	4,320	9,461

FOR FURTHER DESCRIPTIVE MATTER see Manual for 1889-90. CHANGES since 1890 are here given. POPULATIONS are for 1880 and 1890, respectively.

Financial.—*Cap. Stock*, $21,000; paid 6% div'd in 1891. *Total Debt*, $67,103; floating, $17,103; bonded, $50,000 at 6%; due in '93. *Expenses:* operating, $1,320; int., $3,864; total, $5,184. *Revenue:* consumers, $5,297; fire protection, $3,894; other purprses, $270; total, $9,461. **Management.**—Secy. and Supt., W. A. Wilken. Treas., H. J. Cavenaugh.

3. **AUDUBON,** *Audubon Co.* (Pop., 794—1,310.) Built in '89 by city. **Supply.** —Well, pumping to tank. Fiscal year closes Jan. 1. In Nov., '90, engine-house was built in which was placed a 750,000-gall. Barr pump and a 50 HP. boiler; total cost, $2,000. **Distribution.**—*Mains,* ½ mile. *Taps,* 45. *Hydrants,* 7. *Consumption,* 50,000 galls. **Financial.**—*Cost,* $14,000. *Bonded Debt,* $5,000 at 6%, due in '92. *Expenses:* operating, $800; int., $250; total, $1,050. *Revenue:* consumers, $700; city, none. Revenue goes into general fund of city. **Management.**—Com. of two, appointed every year by mayor from council. Chn., A. L. Armstrong, Mayor. Secy., D. L. Freeman. Supt., E. J. Freeman, appointed by council.

4. **AVOCA,** *Pottawattamie Co.* (Pop. of town, 2,344—2,231.) **History.**—Built in '91 by town; put in operation July 30. Engr., F. L. Burrell. Contrs., Godfrey & Meals All of Fremont. **Supply.**—Driven wells, pumping to stand-pipe and direct. **Pumping Machinery.**—D; cap., 750,000 galls. Smedley horizontal, dup. *Boilers,* from Fairbanks, Morse & Co., Chicago. **Stand-Pipe.**—Cap., 59,227 galls.; 12 × 70 ft.; furnished by Contr. **Distribution.**—*Mains,* 8 to 4-in. c. i., 2.3 miles. *Services,* lead. Pipe and specials from Shickle, Harrison & Howard, St. Louis, Mo. *Taps,* 90. *Hydrants,* Galvin, 26. *Valves,* Eddy. *Consumption,* 60,000 galls. *Pressure:* ordinary, 68 lbs.; fire, 100 lbs. **Financial.**—*Cost,* $13,689. *Bonded Debt,* $15,000 at 6%. **Management.**—Not given.

5. **BOONE,** *Boone Co.* (Pop., 3,330—6,520.) Built in '84 by city. **Supply.**—Artesian well, pumping to tank and direct. Fiscal year closes Dec. 31. Well reported as in progress in 1889-90 Manual has been completed and deep-well pump been placed in it. Well is 3,011 ft. deep, with pump head down 260 ft., and furnishes abundant supply of good water; former supply was not potable. **Distribution.**—*Mains,* 2 miles. *Taps,* (in '87), 35. *Meters,* 3. *Hydrants,* 21. *Consumption,* 80,000 galls. **Financial.**—*Cost,* $38,500. *Total Debt,* $29,826: floating, $1,326 at 6%; bonded, $28,500 at 6%. Bonds are optional and sinking fund is applied at once to their redemption. *Expenses:* operating, $2,526; int., $1,789; total, $4,315. *Revenue:* consumers, $800; fire protection, $2,700; total, $3,500. **Management.**—Com. of three appointed every year by mayor. Chn., C. Deeming. Secy., F. D. Gay.

6. **BURLINGTON,** *Des Moines Co.* (Pop., 19,450—22,565.) Built in '77-8 by Burlington Water Co., under 25-yrs. franchise. **Supply.**—Mississippi River, pumping direct. Fiscal year closes Dec. 31.

Year.	Mains.	Taps.	Meters.	Hydrants.	Consumption.	B'd'd d'bt.	Int. ch'g's.
'84	19.5	1,027	86	209	600,000	$200,000	$12,000
'86	19.5	1,122	90	211	850,000	186,000	11,180
'88	19.5	1,542	90	215	1,250,000	178,000	10,680
'90	19.5	1,719	80	218	1,592,509	174,000	10,440

Financial.—*First Cost,* $185,000. *Cap. Stock:* authorized, $300,000; paid up, $30,000; paid 12% guaranteed dividend in 1890. *Bonded Debt,* $174,000 at 6%. *Expenses:* operating, about $18,000; int., $10,440; total, $120,440. *Revenue:* consumers, $18,000; city "collects 5-mill tax levy of $13,000 and pays int. on co.'s bonds. **Management.** —Prest., Geo. D. Rand. Supt., Chas. Hood.

7. **CARROLL,** *Carroll Co.* (Pop., 1,385—2,448.) Built in '83-4 by town. **Supply.**—Surface well, pumping to tank. In '91, a 750,000-gall. pump was put in. **Distribution.**—*Mains,* 1 mile. *Takers,* 21. *Hydrants,* 11. **Financial.**—*Cost,* $10,500. *Bonded Debt,* $10,000 at 6%. *Expenses:* operating, $300; int., $600; total, $900. *Revenue,* consumers, $225. **Management.**—Supt., J. H. Giesing.

8. **CEDAR FALLS,** *Linn Co.* (Pop., est., 3,500.) Built in '87-8 by town. **Supply.**—Springs, pumping to tank. **Distribution.**—(May '89.) *Mains,* 3.7. *Taps,* 92. *Meters,* 2. *Consumption,* 50,000 galls. **Financial.**—(May, '89.) *Cost,* $33,100. *Bonded Debt,* $30,000 at 6 %. *Expenses:* operating, $1,000; int., $1,800; total, $2,800. *Revenue,* consumers, $900. **Management.**—(In '89.) Secy., Geo. A. Newman. Supt., Chas. Boehmer.

9. **CEDAR RAPIDS,** *Linn Co.* (Pop., 10,104—18,020.) Built in '75-76 by Cedar Rapids Water Co.; city has right to purchase works in 1900, at appraised valuation; franchise prescribes that rates must not exceed those of Dubuque, Davenport,

Des Moines and Burlington. **Supply.**—Cedar River for fire protection, and artesian well for domestic use, 1 umping direct. Fiscal year closes March 31.

Year.	Mains.	Taps.	Meters.	Hydrants.	Cost.	B'd'd debt.	Int. ch'g's.	Rev.
'82-3	8	375	0	76	.....	$65,000	.....	$15,877
'84-5	12	550	0	110	$135,000	65,000	$5,200	21,000
'86-7	16	700	0	112	175,000	80,000	6,400	10,000*
'88-89	20	946	4	153	200,000	85,000	6,800	13,000*
'90-91	20	1,163	12	160	215,000	71,000	5,560	14,000*

* From city.

Financial.—*First Cost*, $119,245. *Cap. Stock :* authorized, $200,000; paid up, $145,-000. *Bonded Debt*, $71,000 ; $65,000 at 8%, due in '95; $6,000 at 6%, due on call. **Management.** - Prest., C. H. Souther. Supt., Chas. J. Fox.

10. **CHARLES CITY,** *Floyd Co.* (Pop., 2,421—2,802.) Built in '87 by town. **Supply.**—Spring, impounded, pumping direct. Fiscal year closes May 1. **Distribution.**—*Mains,* 2.5 miles. *Taps*, 70. *Hydrants*, 25. **Financial.**—*Cost*, $21,000. *Bonded Debt*, $20,000 at 5%. *Expenses:* operating, $1,450; Int., $1,000; total, $2,450. *Revenue*, consumers, $854. Revenue goes into general fund of city, from which all expenses are paid. **Management.**—Committee of five, appointed every year by council. Chn., H. C. Baldwin. Clerk, Fred Miner. Supt., Jule C. Sterns.

11. **CHEROKEE,** *Cherokee Co.* (Pop., 1,523—3,441.) Built in '89 by city. **Supply.** —Artesian wells, pumping to stand-pipe. Fiscal year closes Feb. 9. **Distribution.**—*Mains,* 2.5 miles. *Taps,* 150. *Meters,* 2. *Hydrants,* 18. *Consumption*, 65,000 galls. **Financial.**—*Cost*, $26,000. *Bonded Debt*, $20,000 at 5%. *Expenses:* operating, $1,200 ; Int., $1,000; total, $2,200. *Revenue,* consumers, $1,200. **Management.**—Chn., J. C. Hall. Secy. and Supt., J. H. Unhoefer.

12. **CLARENCE,** *Cedar Co.* (Pop., 607—629.) **History.**—Built in '91 by village for general supply; put in operation July 10. Engrs. and contrs., Fairbanks, Morse & Co., Chicago, Ill. **Supply.**—Artes'an well, pumping to reservoir. **Pumping Machinery.**—Dy. cap., 50,000 galls.; Smith & Vaile. **Reservoir.**—Cap., 25.000 galls. **Distribution.**—*Mains,* 6 and 4 in., 1 mile. *Hydrants,* Beaumont, 7. *Pressure,* 40 lbs. **Financial.**—*Cost*, $6,000. *Bonded Debt*, $5,000 at 6%. **Management.**—Supt., S. W. Spaulding.

13. **CLARINDA,** *Page Co.* (Pop., 2,011—3,262). Built in '86 by town. **Supply.** —Driven wells, pumping direct. Fiscal year closes March 30. In March, '91, standpipe was projected. **Distribution.**—*Mains,* 5 miles. *Taps*, 150. *Meters*, 30. *Hydrants,* 36. *Consumption,* 100,000 galls. **Financial.**—*Cost*, $41,000. *Bonded Debt,* $35,000 at 5%, due in 1906. *Expenses:* operating, $2,500; Int., $1,750; total, $4,250. *Revenue:* consumers, $1,450; city, $2,500; total, $3,950. **Management.**—City Council. Supt., T. J. Bracken.

14. **CLINTON,** *Clinton Co.* (Pop., 9,052—13,619.) Built in '74-5 by Clinton Water-Works Co.; now controlled by American Water-Works & Guarantee Co., Ltd., Pittsburg, Pa.; franchise fixes maximum rates and provides for purchase of works by city in 1914, price determined by arbitration. See Lyons. **Supply.**—Three 6-in. artesian wells, pumping to tank, and discharge in case of fire. Wells are 1,000 ft. deep and discharge directly into earth reservoir having cap. of 5,000,000 galls. **Pumping Machinery.** — Dy. cap., 8,500,000 galls.; two 2,250,000 Gordon, one comp. cond., 16 × 28 × 14 × 24, one comp. non-cond., 18½ × 28 × 14 × 24, and two 2,000,000 Cope & Maxwell, high pressure, each 16 × 10 × 30. *Boilers*, four 320-HP. **Tank.**—Is 25 ft. high, on top of brick tower 85 ft. high. Fiscal year closes Dec. 31.

Year.	Mains.	Taps.	Meters.	Hydrants.
'86	10	300	12	78
'88	12	720	84	150
'90	17.5	1,056	125	169

Financial.—Withheld. **Management.**—Prest., J. H. Purdy. Secy. and Treas., A. E. Claney. Supt., S. M. Highlands.

15. **COLFAX,** *Jasper Co.* (Pop., 620—957.) Built in '87 by town. **Supply.**— Well, pumping to reservoir. Supply was formerly taken from three 3-in. driven wells. June 11, '91, began pumping from well 6 ft. 8 ins. in diam., 22 ft. deep, thereby increasing supply from 57,000 to 216,000 galls. daily. Fiscal year closes March 1. **Distribution.**—*Mains,* 8,200 ft. *Taps*, 103. *Meters*, 5. *Hydrants*, 9. *Consumption,* 30,000 galls. **Financial.**—*Cost,* $7,500. *Bonded Debt,* $6,000 at 6%, due in 1902.

FOR FURTHER DESCRIPTIVE MATTER see Manual for 1889-90. CHANGES since 1890 are here given. POPULATIONS are for 1880 and 1890, respectively.

368   MANUAL OF AMERICAN WATER-WORKS.

Sinking Fund, $1,200. *Expenses:* operating, $300; Int., $380; total, $680. *Revenue:* consumers, $800; city, none. **Management.**—Comr., Geo. A. Goodrich, appointed every year by council.

16. **CORNING**, *Adams Co.* (Pop., 1,525—1,531.) **History**.—Built in '88 by city. **Supply**.—Wells, pumping to tank. **Pumping Machinery**. -Cap., about 70,000 galls. **Tank**.—Cap., 61,000 galls. **Distrbution**.—*Mains*, 2 miles. *Taps*, 100. *Hydrants*, 29. *Consumption*, 12,800 galls. *Pressure*, 80 lbs. **Financial.**—*Cost*, $20,000. *Bonded Debt*, $15,000 at 6%. *Operating Expenses*, $600. *Revenue*, consumers, $750. **Management.**—Supt., Jno. W. Bixby.

17. **COUNCIL BLUFFS,** *Pottawattamie Co.* (Pop., 18,063—21,477.) Built in '82-3 by Council Bluffs City Water-Works Co., under 20-yrs. franchise; rates are fixed by city ordinance. **Supply**.—Missouri River, pumping to two subsiding reservoirs and repumping to distributing reservoir; 10,000,000-gall. subsiding reservoir added in '91; water flows from first to second, then second to third. **Distribution**— *Mains*, 31 miles. *Taps*, 1,910. *Meters*, 217. *Hydrants*, 227. *Consumption*, 2,000,000 galls. **Financial.**—*Cost*, $798,000. *Cap S'ock*, $1,100,000. *Bonde 1 Debt*, $650,000 at 6%. *Expenses:* operating, $20,350; int., $39,000; total, $59,350. *Revenue:* consumers, $51,975; city, $22,025 ; total, $77,000. **Management.**—Prest., Robt. Sewell; Secy., S. W. Hopkins, Jr.; Treas., Geo. P. Sheldon; all of Western Union Bldg., N. Y. City.

18. **CRESTON,** *Union Co.* (P)p., 5,031—7,200. **History**. - Built in '90-1 by Creston Water-Works Co. under 25-yrs. franchise, which fixes rates and provides for purchase of works by city at intervals of 10 yrs.; put in operation Aug. 1, '91. Eng., E. A. Rudiger, Nebraska City, Neb. Contrs., Union Construction Co. **Supply**.—Impounding and clear water reservoirs on West Fork of Platte River, pumping to stand pipe; Jewell filter. **Impounding Reservoir**.—Cap., not given. Dam, 900 ft. leng. 133 ft. wide at bottom, 10 ft. wide at top, 32 ft. high. Contr., P. D. Watson & Co., Cameron, Mo. **Pumping Machinery**.—Dy. cap., 2,500,000 galls.; Barr comp., 14 × 20 × 12 × 12 and Smith & Valle 14 × 20 × 12 × 15. *Boilers*, two, steel. **Stand-Pipe**.—Cap., 132,200 galls.; 15 × 100. **Distribution**.—*Mains*, c. i.. 13 miles. *Services*, galv. i., lead. Pipe and specials from Shickle, Harrison & Howard, St. Louis, and New Philadelphia (O) Pipe Co. *Hydrants*, Mathews, 140. *Valves*, Eddy. *Pressure:* ordinary, 85 lbs.; fire, 125 lbs. **Financial.**—*Cost*, $150,000. *Cap. Stock:* authorized, 125,000; paid up. $30,000. *Debt*, $83,000 at 5%. *Revenue*, city, $7,800. *Hydrant Rental*, $50. **Management.**—Prest., H. S. Clarke. Secy., Jno. Gibson. Treas., J. B. Harsh. Supt., J. W. Kendall.

19. **DAVENPORT,** *Scott Co.* (Pop., 21,831—26,872.) Built in '74 by Davenport Water Co., under 25-yrs. franchise, which regulates rates. In '89 city renewed franchise for another 25 yrs., reserving right to purchase works at end of that time. **Supply**.—Mississippi River, pumping direct and to reservoir; American filters, having dy. cap. of 6,000,000 galls., added in '91. Fiscal year closes April 1.

Year.	Mains.	Taps.	Meters.	Hydrants.	Cost.
'82	22	780	178	251	$550,000
'84	26	881	300	257	550,000
'86	28	1,337	45	261	830,000
'88	33	1,710	50	320	920,000
'90	34	1,750	50	320	980,000

Financial.—*First Cost*, $500,000. *Cap. Stock:* authorized, $1,000,000; all paid up. *No Debt*. *Expenses:* operating, $26,800; co. is exempt from taxation. *Revenue:* consumers, $40,000; fire protection, $12,800; total, $52,800. **Management.**—Prest., Nicholas Kuhner. Vice-Prest., F. H. Griggs. Secy. and Treas., J. P. Donahue. Supt., Thos. N. Hooper.

20. **DECORAH,** *Winneshiek Co.* (Pop., 2,951—2,801.) Built in '81 by city. **Supply**.—Surface wells, pumping to reservoir, with provision for direct pumping. Extension of 6,400 ft. 6-in. pipe was made in '90, to Luther College; cost, $5,000; contrs., Harrison & Hawley. **Distribution**.—*Mains*, 2 miles. *Taps*, 75. *Hydrants*, 23. **Financial**—*Cost*, $28,000. *Expenses*, operating, $1,372. *Revenue*, $1,566. **Management.**—Supt., O. Hoffas.

21. **DENISON,** *Crawford Co.* (Pop., 1,411—1,782.) Built in '83 by town. **Supply**.—Wells, pumping to tank. Supply was increased in '90 by twelve 2-in. driven wells. A 100,000-gall. dup. steam pump was also added. Fiscal year closes March 15. **Distribution**.—*Mains*, 2 miles. *Taps*, 95. *Hydrants*, 12. *Consumption*, 32,000 galls. **Financial**—*Cost*, $11,800. *Bonded Debt*, $7.500 at 6%, due in '91. *Sinking Fund*, $1,302. *Expenses:* operating, $1,300; int., $450; total. $1,750. *Rev-

enuc, consumers, $600. **Management.**—Three members of city council appointed every year by mayor. Chn., J. B. Romans. Sec., W. J. McAhren. Supt., N. J. Wheeler.

22. DES MOINES, *Polk Co.* (Pop., 22,408—50,093.) Built in '71 by Des Moines Water-Works Co., under perpetual franchise, which regulates rates and provides for purchase of works by the city. Supply.—Grour d water on Raccoon River pumping direct, filtered through sand. Fiscal year closes Dec. 31. During '90, steel stand-pipe was erected, 30 × 100 ft., on elevation of 176 ft., ¾ miles from pumping station.

Year.	Mains.	Taps.	Meters.	Hydrants.	Cons'tion.	Cost.	B'd'd debt.	Int. ch'g's.
'82	20	1.037	30	192	1,500,000	$388,618	$350,000	$21,000
'86	30	2,500	500	300	1,703,997	750,000	350,000	21,000
'90	60	...	1,500	650	2,750,000	1,590,000	800,000	48,000

Financial.—*First Cost,* $350,000. *Cap. Stock,* $1,500,000. *Total Debt,* $825,000: floating, $25,000; bonded, $800,000 at 6%. *Expenses:* operating, $20,000; int., $18,000; total, $68,000. **Management.**—Prest., F. M. Hubbell. Secy. and Mgr., A. N. Denman. Treas., H. K. Love.

23. DUBUQUE, *Dubuque Co.* (Pop., 22,254—30,311.) Built in '71 by Dubuque Water Co.; maximum rates fixed in franchise, which provides for purchase of works by city at any time at appraised value. Supply.—Abandoned tunnel for draining mines, by gravity, to and from reservoir. Fiscal year closes May 1.

Year.	Mains.	Taps.	Meters.	Hyd'ts.	Rev.*
'82	13	517	10	121	......
'88	20	750	10	174	$11,400
'90	23	931	12	209	11,944

*From city.

Financial.—*No Debt.* *Taxes,* $5,250. **Management.**—Prest., E. R. Perkins. Cleveland, O. Secy. and Supt., N. W. Kimball. NOTE.—In '90, Mayor wrote that city contemplated buying works.

24. ELDORA, *Hardin Co.* (Pop., 1,584—1,577.) Built in '83 by city. Supply.—Combined surface and driven well, pumping to tank. Fiscal year closes March 1. Distribution.—(In '88.) *Mains.* ½ mile. *Taps,* 7. *Meters,* 0. *Hydrants,* 5. Financial.—(In '88.) *Cost,* $7,800. *Bonded Debt,* $6,000 at 7%. *Expenses:* operating, $100; int., $400; total, $500. *Revenue,* consumers, $100. **Management.**—(In '89.) Secy., S. R. Edington. Supt., H. E. Gardner.

25. EMMETSBURG, *Palo Alto Co.* (Pop., 879—1,584.) **History.**—Built in '90 by town; put in operation in Oct. Engr. and Contr., Fremont Turner, Nevada. Supply.—Well, pumping to stand-pipe. Well is 8 ins. in diam., 260 ft. deep. Pumping Machinery.—Not described. Stand-Pipe.—Cap., 70,000 galls.; 12½ × 80 ft. Distribution.—*Mains,* 6-in., 1,790 ft. *Services,* "only a few." *Hydrants,* Ludlow, 6. *Valves,* Ludlow. Financial.—Cost, $9,500. *Bonded Debt,* $9,500 at 5%. **Management.**—Address Mayor.

26. FAIRFIELD, *Jefferson Co.* (Pop., 3,086—3,391.) Built in '83-4 by city and Geo. B. Inman. Supply.—Surface water, impounded by dam, pumping to stand-pipe. Fiscal year closes March 1.

Year.	Mains.	Taps.	Meters.	Hydrants.	Con'pt'n.	B'd'd debt.	Int. ch'g's.	Exp.	Rev.
'87-8	4.5	155	1	33	53,000	$30,000	$1,500	$8,000	$1,632
'88-9	5.5	165	1	33	45,424	25,000	1,250	2,155	1,410
'89-90	5.9	176	1	33	50,826	25,000	1,250	2,238	1,580
'90-1	6.2	220	1	33	51,104	25,000	1,250	3,631	2,035

Financial.—*Cost,* $55,646. *Total Debt,* $25,420: floating, $420 at 7%; bonded, $25,000 at 6%, due in 1903. *S. nking Fund,* $1,990. *Expenses,* total, $4,881. *Revenue:* consumers, $2,033; city, none. **Management.**—Committee of three. Supt., T. F. Higley.

27. FORT DODGE, *Webster Co.* (Pop., 3,586—1,871.) Built in '80-1 by city. Supply.—Des Moines River, pumping direct; filtered. Fiscal year closes Jan. 31. Distribution.—*Mains,* 5.5 miles. *Taps,* 212. *Hydrants,* 12. Financial.—*Cost,* $40,000. *Bonded Debt,* $36,000 at 6%. *Expenses:* operating, $5,085; int., $2,160; total, $7,245. *Revenue:* consumers, $4,072; fire protection, $3,001: total, $7,673. **Management.**—City Clerk, C. M. Hamilton. Supt., W. Cann.

28. FORT MADISON, *Lee Co.* (Pop., 4,679—7,901.) Built in '85 by Ft. Madison Water Co.; rebuilt in '90; now controlled by Debenture Guarantee & Assurance Co., Chicago, Ill. Supply.—Artesian well, pumping to tank. Fiscal year closes Dec. 31. During '90, a 10,000,000-gall reservoir was built, pumping station enlarged, 3,000,000-

FOR FURTHER DESCRIPTIVE MATTER see Manual for 1889-90. CHANGES since 1890 are here given. POPULATIONS are for 1880 and 1890, respectively.

gall. pump added. Engr., Geo. H. Pierson, N. Y. City. Distribution.—*Mains*, 12 miles. *Taps*, 230. *Meters*, 14. *Hydrants*, 87. *Consumption*, 400,000 galls. Financial.— *Cost* $225,000. *Cap. stock*, paid up, $150,000. *Bonded Debt*, $150,000 at 6%. *Expenses:* operating and taxes, $3,600; int., $9,000; total, $12,600. *Revenue:* consumers, $8,136; city, $5,820; other purposes, $3,500; total, $17,756. Management.—Prest., Chas. B. Ludwig, 72 Broadway, N. Y. City. Secy., W. R. Adams. Treas. H. Van C. Homans. Supt., W. J. McCray.

29. **GLADBROOK,** *Tama Co.* (Pop., in '90, 556.) Built in '83 by town. Supply. —Artesian well, pumping to tank. Fiscal year closes Nov. 1. Distribution.— *Mains*, 2.9 miles. *Taps*, 9. *Hydrants*, 3. Financial.—*Cost*, $3,600. *Debt*, $2,500 at 6%. *Expenses:* operating, $100; int., $150; total, $250. Management.—Supt., L. Wieland.

30. **GUTHRIE CENTER,** *Guthrie Co.* (Pop., 571—1,027) History.—Built in '90 by town; put in operation Oct. 1. Engr., F. L. Burrell, Fremont, Neb. Contrs.— Eagle Iron Wks., Des Moines. Supply.—Driven wells, pumping to stand-pipe. Pumping Machinery.—Cap., 500,000 galls.; Smedley horizontal, dup. Stand-Pipe.—Cap., 25,000 galls.; 13 × 25 ft. Distribution.—*Mains*, 8 to 4-in. c. i., 1 mile. *Services*, lead. *Taps*, 130. *Hydrants*, Empire, 12. *Valves*, Galvin. *Consumption*, 13,000 galls. *Pressure:* ordinary, 84 lbs.; fire, 120 lbs. Financial.—*Cost*, $7,625. *Bonded Debt*, $9,000 at 6%. *Expenses:* operating, $510; int., $665; total, $1,205. Management.—Supt, W. F. Brown.

31. **HARLAN,** *Shelby Co.* (Pop., 1,304—1,765.) History.—Built in '91 by town; to be in operation Oct. 1. Engr., F. L. Burrell, Fremont, Neb. Contrs., Fremont Foundry & Machine Co. Supply.—Tube wells, pumping to stand-pipe. Pumping Machinery.—Dy. cap., 500,000 galls.; Barr dup. Stand-Pipe.—Cap., 30,000.; 10 × 50 ft. Distribution.—*Mains*, 1.5 miles. *Consumers*, 50. *Hydrants*, Ludlow, 11. *Pressure:* ordinary, 50 lbs.; fire, 80 lbs. Financial.—*Cost*, $14,327. Management.—Chn., O. F. Graves.

32. **HAWARDEN,** *Sioux Co.* (Pop., in '90, 744.) History.—Built in '90 by town; put in operation Sept. 15. Engr and Contr., H. L. Wood. Supply.—Driven wells, pumping to tank. Pumping Machinery.—Dy. cap., 500,000 galls.; from National Iron & Brass Wks., Dubuque. Tank.—Cap., 60,000 galls. Distribution.—*Mains*, 6 and 4 in. c. i., 1 mile. Pipes and specials from So. Pittsburg, Tenn. *Services*, tarred iron. *Taps*, 20. *Hydrants*, Mathews, 6. *Valves*, Galvin. *Pressure:* ordinary, 50 lbs.; fire, 112 lbs. Financial.—*Cost*, $8,000. *Debt*, $3,000 at 6%; $5,000 bonded. Management.—Prest., J. A. Ashley. Secy., H. L. Wood. Treas., L. L. Harlow. Supt., R. J. O'Brien.

Holstein, *Ida Co.* (Pop., in '90, 539.) Built in '86 for fire protection, by town.

33. **HUMBOLDT,** *Humboldt Co.* (Pop., 655—998.) History.—Built in '90 by city; put in operation Dec. 1. Engr. and contr., Fremont Turner, Ames. Supply. —Spring and river, pumping to reservoir and direct. Pumping Machinery.—Dy. cap., 266,000 galls.; Deane dup. Reservoir.—Cap., 112,000 galls.; of stone masonry, 35 ft. in diam., 15 ft. deep, on hill 70 ft. above town. Distribution.—*Mains*, 6-in., c. i. and w. i., 1 mile. Pipe and specials from Schickle, Harrison & Howard, St. Louis, Mo., and from National Tube Wks., Chicago. *Taps*, 27. *Hydrants*, Ludlow, 8. *Valves*, Ludlow. Financial.—*Cost*, $10,000. *Bonded Debt*, $10,000 at 5½%. Management.—Not given.

34. **IDA GROVE,** *Ida Co.* (Pop., 759—1,503.) Built in '82 by town. Supply.— driven wells, pumping to tank. Fiscal year closes March 1. Distribution.—*Mains*, 2 miles. *Hydrants*, 13. *Consumption*, 13,000 galls. Financial.—*Cost*, $13,000. *Bonded Debt*, $7,500 at 5%. *Expenses:* operating, $1,000; int., $375; total, $1,375. *Revenue*, consumers, $1,100. Management.—Chn., F. A. Lusk. Secy., W. J. Scott.

35. **INDEPENDENCE,** *Buchanan Co.* (Pop., 3,123—3,163.) Built in '86-7 by town. Supply.—Driven wells and Wapsipinicon River, pumping direct. Fiscal year closes March 31. Distribution.—*Mains*, 5 miles. *Taps*, 200. *Meters*, 89. *Hydrants*, 40, *Consumption*, 250,000 galls. Financial.—*Cost*, $40,000. *Bonded Debt* $40,000 at 5%. *Sinking Fund*, $5,332. *Expenses:* operating, $3,120; int., $2,000; total, $55,120. *Revenue:* consumers, $4,517; city, $1,400; total, $5,917. Management.— Comrs., appointed every year by common council. Chn., H. Marvins. Secy., Rufus Brewer. Supt., A. D. Guernsey.

36. IOWA CITY, *Johnson Co.* (Pop., 7,123–7,016.) Built in '82 by Iowa City Water Co., under 20-yrs. franchise, which fixes rates but does not provide for purchase of works by city. **Supply.**—Wells, pumping direct; brick filter. During '90 supply was increased by several wells. Fiscal year closes Dec. 31.

Year.	Mains.	Taps.	M'rs.	Hyd'ts.	Cons'n.	Cost.	B'd'd debt.	Int. ch'g's.	Exp.	Rev.
'83	5.5	110	10	70	150,000	$85,000	$70,000	$4,200	......	....
'86	5.5	120	3	74	160,000	...	70,000	4,200	......	*$4,400
'88	9.5	285	1	93	350,000	125,000	70,000	4,200	$3,000	9,350
'90	9.5	321	1	93	325,000	125,000	70,000	4,200	3,537	9,556

* From city.

Financial.—*Cap. Stock*, $55,000. *Bonded debt*, $70,000 at 6%. *Expenses:* operating, $2,638; int., $4,200; taxes, $899; total, $7,737. *Revenue:* consumers, $4,200; fire protection, $5,350; total, $9,556. **Management.**—Prest., A. B. Brown. Secy., T. Cowell. Treas., Robt. L. Thompson. Supt., Jay Chatham.

37. KEOKUK, *Lee Co.* (Pop., 12,117–14,101.) Built in '77-8 by Keokuk Water-Works Co.; now controlled by American Water-Works & Guarantee Co., Ltd., Pittsburg, Pa., under 20-yrs. franchise, which fixes maximum rates but does not provide for purchase of works by city. **Supply.**—Mississippi River, pumping direct; filtered. **Pumping Machinery.**—Cap., 3,750,000 galls.; one 1,500,000 Holly-Gaskill, having four water cylinders of 8-in. diameter; four steam cylinders of 14½-in. diam., 22-in. stroke, and one 2,250,000 Gordon, 14 × 26 × 14 × 24. *Boilers:* two 16 HP. horizontal, tubular. **Distribution.**—*Mains*, 17 miles. *Taps*, 790. *Meters*, 15. *Hydrants*, 100. *Consumption*, 1,100,000 galls. **Financial.**—Withheld. **Management.**—Prest., J. H. Purdy. Secy., A. E. Claney. Supt., D. W. Swartz.

38. KNOXVILLE, *Marion Co.* (Pop, 2,577–2,632.) Built in '87 by town. **Supply.**—White Breast Creek, pumping from reservoir to stand-pipe. Fiscal year closes March 1. Previous to summer of '90 supply was taken from a series of springs; this source failing a 6-in. main was extended 2½ miles to White Breast Creek. Well, 10 ft. in diam. was dug 75 ft. from and 9½ ft. below bed of creek, and connected with it by gallery, 4 ft. square, a part of which is filled with filtering material. Well is lined with brick and cement. Water is pumped from well to reservoir by a 7,000,000 galls Knowles pump, or it can be pumped direct to stand-pipe. Cost, $15,000. **Distribution.**—*Mains*, 5 miles. *Taps*, 30. *Hydrants*, 30. **Financial.** -*Bonded Debt*, $33,500: $18,000 at 5%, due in 1907; $5,500 at 6%, due in 1908; $15,000 at 5%, due in 1910. *Sinking Fund*, $500. *Expenses:* operating, nominal; int., $1,230. Revenue pays about one-half of expenses; balance met by taxation. **Management.**—Two members of council, appointed every year by mayor. Chn., H. Cunningham. Secy., T. G. Gilson.

39. LANSING, *Allamakee Co.* (Pop., 1,811–1,683.) Built in '77 by Lansing Artesian Well Co. **Supply.**—Two flowing wells, by gravity. **Distribution.**—*Mains*, ½ mile. *Hydrants*, 7. **Financial.**—*Cost*, $7,500. *Profits*, 10%. *No Debt.* **Management.**—Prest., M. Kerndt. Treas., J. W. Thomas. Supt., Ed. Boeckh.

40. LE MARS, *Plymouth Co.* (Pop., 1,895–4,036.) Built in '88-9 by Le Mars Water & Light Co.; bought May 1, '90, by J. H. Winchel. **Supply.**—Driven wells, pumping direct; in case of fire, water is pumped from reservoir. Fiscal year closes Dec. 31. Extensions of considerable importance were to be made during '91. **Distribution.**—*Mains*, 8 miles. *Taps*. 446. *Meters*, 4. *Hydrants*, 54. **Financial.**—*Cost*, $33,446. *No Cap. Stock*, *Bonded Debt*, $40,000, at 6%, due in 1903. *Expenses:* operating, $3,000; int., $2,400; no taxes; total, $5,400. *Revenue:* consumers, $3,549; fire protection, $2,560; other purposes, $375; total, $6,484. **Management.**—Owned and operated by J. H. Winchel.

41. LOGAN, *Harrison Co.* (Pop., 644–827.) Built in '84 by village. **Supply.**—Two wells, pumping to reservoir. **Distribution.**—(In '88.) *Mains*, 1½ miles. *Taps*, 10. *Hydrants*, 16. **Financial.**—(In '88.) *Cost*, $7,400. *Debt*, $3,800 at 6%. *Expenses:* operating, $240; int., $228, total, $468. *Revenue*, consumers. $175. **Management.**—Supt., Almon Stern.

42. LYONS, *Clinton Co.* (Pop., 4,095–5,799.) Built in '76 by Lyons Water-Works Co.; in '91 American Water-Works & Guarantee Co., Pittsburg, Pa., secured controlling interest. **Supply.**—Artesian wells, probably by pumping to reservoir and direct; previous to '91 supply was from Mississippi River, but connection has been made with Clinton works. Fiscal year closes April 1. **Distribution.**—*Mains*, 9

For Further Descriptive Matter see Manual for 1889-90. Changes since 1890 are here given. Populations are for 1880 and 1890, respectively.

miles. *Taps*, 307. *Meters*, 2. *Hydrants*, 76. **Financial.**—*Cost*, $74,865. *Cap. Stock*, authorized, $75,000; paid up, $46,000; paid 7% div'd. in '91. *Surplus*, $18,048. *Total Debt*, $20,500: floating,$500; bonded, $20,000 at 5½%. *Expenses:* operating, $2,073; int., $1,-160; taxes, $420; total, $3,753. *Revenue:* consumers, $3,814; city, $3,000; total, $6,814. **Management.**—Address American Water-Works & Guarantee Co., Pittsburg, Pa.

McGregor, *Clayton Co.* (Pop., 1,002–1,160.) Built by McGregor Artesian Well Co. for fire protection only.

43. MANCHESTER, *Delaware Co.* (Pop., 2,275–2,344.) **History.**—Built in '90 by city; put in operation Dec. 23. Engr. and contr., E. Smedley Dubuque. **Supply.**—Springs for domestic use, river for fire protection, pumping to stand pipe. **Pumping Machinery.**—Dy. cap., 1,500,000 galls.; two 750,000 Smedley. **Stand-Pipe.** —Cap., 105,000 galls.; 15 × 80 ft.; from D. H. Young. **Distribution** —*Mains*, 8 to 4-in. c. i., 3 miles. *Taps*, 20. *Hydrants*, Chapman, 26. *Valves*, Chapman. *Pressure:* ordinary 55 lbs.; fire, 90 to 100 lbs. **Financial.**—*Cost*, $21,000. *Bonded Debt*, $20,000 at 5%. **Management.**—Chn., W. B. Sherman. Secy., C. H. Day. Supt., D. B. Allen.

44. MAQUOKETA, *Jackson Co.* (Pop., 2,467–3,077.) Built in '83–84 by city. **Supply.**—Well, pumping to stand-pipe and direct. Fiscal year closes March 1. **Distribution.**—*Mains*, 3 miles. *Taps*, 250. *Hydrants*, 3. **Financial.**—*Cost*, $31,653. *Bonded Debt*, $15,500 at 6%. *Expenses:* operating, $1,046; int., $880; total, $2,576. *Revenue:* consumers, $1,600; city, none. **Management.**—Three comrs., appointed every year by council. Chn., Harvey Reid, Secy. and Supt., Fred Fischer.

Marengo, *Iowa Co.* (Pop., 1,738–1,710.) Built in '75 by village for fire protection only. See 1889-90 Manual, p. 486.

45. MARION, *Linn Co.* (Pop., 1,939–3,094.) Built in '84–5 by Jesse W. Starr, Jr., Philadelphia, Pa.; reorganized in '87, under 25-yrs. franchise by Marion City Water Co. **Supply.**—Springs, pumping to stand-pipe and direct; filtered. Connection with Indian Creek is made to each pump for reserve. Water is conducted through 12-in. w. i. pipe to filtering gallery, thence pumped direct into mains, surplus water returning to reservoir, and overflow from reservoir to creek. Fiscal year closes Dec. 31. **Distribution.**—*Mains*, 9 miles. *Taps*, 166. *Meters*, 3. *Hydrants*, 58. *Consumption*, 150,000 galls. **Financial.**—*Cost* (in '83), $127,000. *Revenue:* city, $3,240. **Management.**—Prest., Sam'l Taylor, Philadelphia, Pa. Vice-Prest., W. S. Twogood. Secy. and Gen. Mgr., L. M. Rupert. Treas., A. L. Daniels.

46. MARSHALLTOWN, *Marshall Co.* (Pop., 6,240–8,914.) Built in '76 by city. **Supply.**—Gravel bed near Iowa River, pumping direct. Supply was changed in '91 and filter-bed abandoned. Vitrified, perforated sewer pipe is laid in bed of gravel on opposite side of river from pumping station; water is conveyed under river-bed in 24-in. c. i. pipe. A 2,500,000-gall. Gordon pump was added in '91. Fiscal year closes March 9.

Yr.	Mains.	Taps.	Met'rs.	Hyd'ts.	Cons'p'tn.	Cost.	B'd'd d't.	Int. ch'g's.	Exp.	Rev.
'80–1	7.9	425		69					$3,380	
'83–4	9.9	454	1	79	282,593	$86,302	$47,000		5,208	$4,790
'85–6	10.2	516	0	70	300,934	86,587	17,000		5,513	4,651
'88–9	12.5	475	1	90	644,131	105,049	37,000	$2,960	5,292	7,000
'90 $1	14	750	1	110	800,000	120,000	37,000	2,960	6,400	9,575

Financial.—*First Cost*, $46,500. *Bonded Debt*, $37,000 at 8%, due in '96. *Expenses:* total, $9,380. *Revenue:* consumers, $9,575; fire protection, none. **Management.**—Secy., J. G. Trotter.

47. MASON CITY, *Cerro Gordo Co.* (Pop., 2,510–4,007.) Built in '85 by town. **Supply.**—Springs and Lime River, pumping direct; reported in '91 that contract had been let for artesian well. Fiscal year closes March 1. **Distribution.**—(In '88.) *Mains*, 2.8 miles. *Taps*, 125. *Meters*, 4. *Hydrants*, 26. *Consumption*, 185,000 galls. **Financial.**—(In '89.) *Cost*, $28,000. *Bonded Debt*, $26,000 at 6%. *Expenses:* operating, $2,657; int., $1,560; total, $4,217. *Revenue:* consumers, $3,407; city, $1,518; total, $4,925. **Management.**—(In '89.) Chn., C. H. Tondro, City Clerk, A. R. Sale.

48. MISSOURI VALLEY, *Garrison Co.* (Pop., 1,154–2,797.) Built in '89 by city. **Supply.**—Ten driven wells, pumping to reservoir. Fiscal year closes March 17. **Distribution.**—*Mains*, 1.5 miles. *Taps*, 28. *Meters*, 2. *Hydrants*, 10. **Financial.**—*Cost*, $16,000. *Bonded Debt*, $15,000 at 5%, due in 19 years. *Expenses:* operating,

$060; int., $750; total, $1,410. *Revenue:* consumers, $335. **Management.**—Comr., D. P. Baker. Chn., W. M. Harman. Clerk, O. B. Walker.

49. **MONTICELLO,** *Jones Co.* (Pop., 1.877—1,938.) Built in '78 by city. **Supply.**—Artesian well, pumping to reservoir. **Distribution.**—(In '88.) *Mains,* 3 miles. Taps, 215 *Meters,* 0. *Hydrants,* 21. *Consumption,* 60,000 galls. **Financial.** —(In '88.) *Cost,* $20,000. *Bonded Debt,* $7,000 at 5½%. *Expenses:* operating, $1,200 ; int., $335; total, $1.185. *Revenue:* consumers, $2,000; city, $1,500; total, $3,500. **Management.**—(In '89.) Prest., A. Kemp. Secy., F. Coop. Supt., J. D. Graves.

50. **MOUNT PLEASANT,** *Henry Co.* (Pop., 4,410—3,937.) Built in '81 by Mt. Pleasant Water-Works Co. **Supply.**—Big Creek, pumping to stand-pipe; two Hyatt filters. **Distribution.**—*Mains,* (in '88), 6 miles. *Taps,* 65. *Meters,* 7, *Hydrants,* 50. **Financial.**—*Cap. Stock* (in '88), $130,000. *No Debt.* **Management.** —Prest. and Supt., H. M. Phillips, Springfield, Mass. Secy., J. H. Hewson, 39 Broad St., New York City. Supt. C. M. Glover. NOTE.—Reported, Aug. '91 that citizens voted to buy works for $32,500.

51. **MUSCATINE,** *Muscatine Co,* (Pop., 8,295—11,454.) Built in 75-6 by Muscatine Water-Works Co.; franchise provides for purchase of works by city in 10 years. Rates are to be no higher than those of other cities of same size. **Supply.**—Mississippi River, pumping to reservoir and direct; filtered. Fiscal year closes April 30, Two 100-HP. boilers, costing $3,500, and an additional pump house, costing $5,000, were in course of construction, Feby. 9, '91. Reported June, '91, that contract had been let for 3,000,000-gall. Gordon pump. **Distribution.**—*Mains,* 5 miles. *Taps,* 400. *Meters,* 30. *Hydrants,* 65. *Consumption,* 400,000 galls. **Financial.**—*Cost,* $95,000. *Cap. Stock,* $41,000. *Bonded Debt,* $30,000, at 5%. *Expenses:* operating, $5,500; int., $1,500; taxes, $2,000; total, $10,000. *Revenue:* consumers, $5,000; city, $5,470; total, $10,740. **Management.**—Prest., H. G. Jackson. Secy., J. Carscaddan. Treas., H. W. Moore. Supt., W. Molis.

52. **NASHUA,** *Chickasaw Co.* (Pop., 1,116—1,240.) **History.**—Built in '85 by city. Engr., A. Y. MacDonald, Dubuque. **Supply.**—Cedar River. pumping direct, **Pumping Machinery.**—Dy. cap. not given; rotary. **Distribution.**—*Mains,* 8-in., 2 miles. *Taps,* 50. *Hydrants,* Eclipse. 24. *Pressure:* ordinary, 40 lbs.; fire, 110 lbs. **Financial.**—*Cost,* $13,000. *Debt,* $11,000 at 6%. *Expenses:* operating, $220; int., $660; total, $880. *Revenue,* consumers, $300. **Management.**—Secy., E. H. Barnes. Supt., W. A. Lytle.

Neola, *Pottawattamie Co.* (Pop., 286—917,) Built in '87 by village for fire protection. See 1889-90 Manual, p. 439.

53. **NEVADA,** *Story Co.* (Pop., 1,541—1,662.) Built in '88 by village. **Supply.**— Well, pumping to tank. **Distribution.**—*Mains,* 1.3 miles. *Taps,* 35. *Hydrants,* 18. **Financial.**—*Cost,* $12,000. *Bonded Debt,* $12,000 at 5½%. *Expenses:* operating, $100; int., $660; total, $750. **Management.**—Supt., H. C. Boardman.

54. **NEW HAMPTON,** *Chickasaw Co.* (Pop., 1,105—1,314.) **History.**—Built in '91 by town; put in operation June 20. Engrs. and Contrs, Fairbanks, Morse & Co., St. Paul, Minn. **Supply.**—Driven wells, pumping to tank. **Pumping Machinery,**—Dy. cap., 500,000 galls.; Smith & Vaile dup. Steam supplied by boilers in flour mill. **Tank.**—Cap., 55,000 galls.; on 80 ft. wooden substructure. **Distribution.**—*Mains,* 8 to 4-in. c. i., ¾ miles. *Services,* w. i. *Hydrants,* Mathews, 18. *Valves,* Eddy. *Pressure:* ordinary, 45 lbs.; fire, 125 lbs. **Financial.**—*Cost,* $6,827. *Bonded Debt,* $7,000. **Management.**—Prest., Jno. Foley.

55. **NEWTON,** *Jasper Co.* (Pop., 2,607—2,564.) Built in '84 by town, chiefly for fire protection. **Supply.**—Artesian well, pumping to reservoir. There are 7,000 ft. of pipe, supplying 25 private consumers and 17 hydrants. *Cost,* $20,000. *Revenue,* $250. Address Jos. Stevens, Mayor.

22. **NORTH DES MOINES,** *Polk Co.* (Pop., not given.) In '90 franchise was granted to Co. to supply North Des Moines, on account of dissatisfaction with service of Des Moines Water Co. There was to be a 3,000,000-gall. pump, and 8 miles of 16 to 4-in. mains. Supply, Des Moines River. Contr., D. M. Carey. Engrs., Dunham & Paine, N. Y. City. City Council afterward entertained propositions from the Des Moines Water Co. which led to sale of franchise by new co. to the old.

FOR FURTHER DESCRIPTIVE MATTER see Manual for 1889-90. CHANGES since 1890 are here given. POPULATIONS are for 1880 and 1890, respectively.

56. **ODEBOLT,** *Sac Co.* (Pop., 637–1,122.) Built in '84 by Odebolt Water-Works Co., under 20-yrs. franchise; at its expiration city may buy works at cost; city does not control rates. **Supply.**—Driven wells, pumping to tank. Tank has cap. of 50,000 galls; is 24 × 18 ft. Fiscal year closes March 1. **Distribution.**—*Mains*, 1 mile. *Taps*, 30. *Hydrants*, 10. *Consumption*, 5,000 galls. **Financial.**—*Cost*, $9,000. *Cap. Stock*, $8,000. *No Debt. Sinking Fund*, $1,000. *Expenses*, $200. *Revenue:* consumers, $325; city, $500; total, $825. **Management.**—Prest., H. C. Wheeler. Secy., W. A. Helsell. Treas., J. R. Mattes.

57. **OSKALOOSA,** *Mahaska Co.* (Pop., 4.598–6,558.) Built in '80 by Oskaloosa Water Co., under 20-yrs. franchise. **Supply.**—Surface water, pumping to stand-pipe and direct; Hyatt filter. Sept., '91, 20-ft. well was being sunk on steel shoe in bottom of Skunk River. **Distribution.**—(In '88.) *Mains*, 14 miles. *Taps*, 452. *Meters*, 16. *Hydrants*, 60. *Consumption*, 500,000 galls. **Financial.**—Withheld. **Management.**—Prest., Jas. Gamble, N. Y. City. Treas., H. L. Spencer. NOTE.—Statistics for '90 refused.

58. **OTTUMWA,** *Wapello Co.* (Pop., 9.004–14,001.) Built in '77-8 by Iowa Water Co., under 20-yrs. franchise, which neither regulates rates nor provides for purchase of works by city. **Supply.**—Des Moines River, water passing through canal to settling basin, thence to pump-well, from which it is pumped to reservoir; filtered. Contrs. for filters, Albany (N. Y.) Steam Trap Co.

Year.	Mains.	Taps.	Meters.	Hydrants.
'80	8	340	..	76
'84	12	460	11	84
'88	15	615	15	100
'90	16	758	64	142

Financial.—*Cost*, $600,000. *Expenses:* operating, $8,500. *Revenue:* city, $10,000. *Hydrant rental*, $67. **Management.**—Prest. and Treas., S. L. Wiley. Secy., W. H. Sherwin. Supt., Frank Fuller.

59. **PERRY,** *Dallas Co.* (Pop., 952–2,880.) June 9, '91, contract was let to A. L. Strang & Son, Omaha, to build works, consisting of about 4 miles of pipe and 30 hydrants, to be completed within 90 days. Oct. 6, '91, reported works were to be tested at once. Mains to be owned by C., M. & St. P. R, R. Co.; hydrants owned by city. City already had benefit of 1 mile of pipe and 3 hydrants for fire protection. **Supply.**—Artesian well, pumping to stand-pipe. *Cost*, $20,000. City Clerk, I. L. Townsend.

60. **RED OAK,** *Montgomery Co.* (Pop., 3,755–3,321.) Built in '80-1 by village **Supply.**—Eighteen drilled wells for domestic supply and two open wells for fire protection, pumping direct. There are eight 2-in. and ten 3-in. wells connected with an 8-in. supply pipe. Former supply from well 20 ft. in diam. failed, and new supply has been introduced since 1888. Fiscal year closes third Monday of March. **Distribution.**—*Mains*, 7 miles. *Hydrants*, 47, *Consumption*, 250,000 galls. **Financial.** —*Cost*, $50,000. *Bonded Debt*, $21,000 at 5%. *Expenses:* operating, $3,400; int., $1,050; total, $4,450. **Management.**—Chn., Chas. Kelley.

Reinbeck, *Grundy Co.* (Pop., 482–731.) Built in '83 by village for fire protection. See 1889-90 Manual, p. 491.

61. **ROCKFORD,** *Floyd Co.* (Pop., 739–1,010.) **History.**—Built in '91 by city; put in operation in Sept. Engr. and Contr., Fremont Turner. **Supply.**—Shell Rock River, pumping direct, filtered. **Reservoir.**—Not described. **Pumping Machinery.**—Dy. Cap., 130,000 galls.; Blake. **Distribution.**—*Mains*, 8 to 3-in., 1 mile. *Hydrants*, 10. *Consumption*, 75,000 galls. *Pressure:* ordinary, 100 lbs.; fire, 180 lbs. **Financial.**—*Cost*, $6,750. *Bonded Debt*, $7,000 at 6%. *Revenue:* ry., $800: city, $400. **Management.**—Mayor, C. Yorker. Clerk, N. D. Bowles.

62. **ROCK RAPIDS,** *Lyon Co.* (Pop., in '90, 1,394.) Built in '83-1 by Rock Rapids Water Co., under 21-yrs. franchise; now owned by W. C. Reeves, Peru, Ill. **Supply.**—50 "drive points." pumping to stand-pipe. **Pumping Machinery.**—Dy. cap., 1,500,000 galls.; Knowles comp. **Distribution.**—*Mains*, 3.3 miles. *Hydrants*, 25. **Financial.**—*Cap. Stock*, $60,000. *Revenue*, city, $1,500. **Management.**—Prest., W. G. Reeve, Peru, Ill. Gen. Mgr., H. B. Pierce.

63. **SAC CITY,** *Sac Co.* (Pop., 595–1,249.) Built in '84 by city for general supply. **Supply.**—Wells, pumping by wind-mill pump. Comr., J. Hanger.

64. **SIOUX CITY,** *Woodbury Co.* (Pop., 7,366–37,806.) Owned by city. Built in '84-5 by City Water-Works Co. **Supply.**—Driven wells, pumping to reservoir. Fiscal

year closes March 16. In '91 Wagner steamed filter wells and pumping station were added. Contr., National Water Supply Co., Cincinnati, O. Two 1,250,000-gall. Dean pumps were put in, new station built and main laid to old station. Contract cap. of wells was 2,500,000 galls. Contract called for completion of new supply in Mar., '91; Oct., '91 contract cap. had not been developed. Cost of improvements, about $45,000. Wells are intended for temporary additional supply, permanent supply from Missouri River being projected. Distribution.—*Mains*, 28 miles. *Taps*, 1,500. *Meters*, 90. *Hydrants*, 147. Financial.—*Cost*, $269,771. *Expenses*, operating, $11,014. *Revenue*, consumers, $41,800. Management.—Secy., V. A. Swan. Supt., Phil Carlin.

65. **SIOUX RAPIDS,** *Buena Vista Co.* (Pop., 181—650.) Built in '90 by town. Supply.—Well, pumping to tank. Fiscal year closes Feb. 5. Distribution.—*Mains*, 1.5 miles. *Taps*, 20. *Hydrants*, 7. Financial.—*Cost*, $8,400. *Total Debt*, $7,000; floating, $1,000 at 6%; bonded, $6,000 at 3½%. *Revenue*, $75. Management,—Secy., C. Colewell. Report by James M. Hoskins, Mayor.

66. **STATE CENTRE,** *Marshall Co.* (Pop., 880—354.) Built in 77 by city. Supply.—Two wells, pumping to tank. Fiscal year closes March 2. Distribution. —*Mains*, ½ mile. *Taps*, 3. *Hydrants*, 11. Financial.—*Cost*, $10,000. *Debt*, $5,000 at 6%. *Expenses:* operating, $186; int., $300; total, $186. *Revenue*, $25. Management.—Chn., O. Johnson. Secy.. W. M. Palmer, Town Clerk. Supt., Jas. F. Cowan.

67. **STORM LAKE,** *Buena Vista Co.* (Pop., 1,031—1,682.) History.—Built in '91 by city; put in operation Aug. 15. Engr., F. L. Burrell. Fremont, Neb. Contrs., U. S. Wind Engine & Pump Co.. Batavia, Ill. Supply.—Storm Lake. pumping to stand-pipe; filtered. Filter Bed.—About 18 ft. in diameter, inclosed with brick wall and roofed; water passes from lake beneath bed and up through 1 ft., each, broken stone, charcoal, coarse sand, small gravel and coarse gravel. Suction draws from water above bed. Pumping Machinery.—Dy. cap., 500,000 galls, Smith & Vaile comp. dup. *Boiler*, 6) HP., Brownell. Stand-Pipe.—Cap., 58.750 galls.; 10 × 100 ft. from Porter Mfg. Co., Chicago. Distribution.—*Mains*, 10 to 4-in. c. i., 2.15 miles. *Services*, lead pipe and specials from Shickle, Harrison & Howard, St. Louis, Mo. *Hydrants*, Mathews, 23. *Valves*, Rensselaer. *Pressure:* ordinary, 50 lbs.; fire, 100 lbs. Financial.—*Cost*, $15,484. *Bonded Debt*, $15,000 at 6%. Management.—Not given.

68. **TIPTON,** *Cedar Co.* (Pop., 1,299—1,599.) Built in '89 by city. Supply.—Artesian well, pumping to stand-pipe. Fiscal year closes July 15. Distribution.— *Mains*, ½ mile. *Taps*, 20. *Hydrants*, 12. *Consumption*, 5,000 galls. Financial.— *Cost*, $18,400. *Debt*, $18,000 at 6%. *Expenses:* operating, $1,066; int., $1,080; total, $2,146. *Revenue*, consumers, $230. Management.—Secy. and Supt., A. Shaw.

69. **TOLEDO,** *Tama Co.* (Pop., 1,026—1,836.) History.—Built in '91 by city; put in operation Sept. 15. Engr. and Contr., G. C. Morgan, 49 Major Block, Chicago. Supply.—Well, pumping from reservoir to stand-pipe. Reservoir.—Cap., 25,000 galls. Pumping Machinery.- Dy. cap., 1,000.000 galls.; two Deane dup. Stand-pipe.—Cap., 27,000 galls.; 103 ft. high; "Morgan's Special." Distribution.—*Mains*, c. i., 3.8 miles, *Hydrants*, Morgan, 24. *Valves*, Poet. *Pressure*, 150 lbs. Financial. —*Cost*, $16,648 *Bonded Debt*, $18,000; $14,000 at 5%; $4,000 at 6%. Management,— Not given.

70. **VILLISCA,** *Montgomery Co.* (Pop., 1.299—1,744.) Built in'86 by town. Supply.—Wells, supplied by springs, pumping to tank. Fiscal year closes April 4. Distribution.—*Mains*, 1.3 miles. *Taps*, 75. *Meters*, 2. *Hydrants*, 12. *Consumption*, 20,000 galls. Financial.—*Cost*, $11,500. *Bonded Debt*, $8,000 at 6%. *Expenses:* operating, $500; int., $480; total, $980. *Revenue*, consumers. $300. Management.—Chn., F. F. Jones. Secy.. F. L. Ingram. Supt., M. Cowgill.

71. **VINTON,** *Benton Co.* (Pop., 2,906—2,865.) Built in '88 by town. Supply.— Artesian well for domestic use, Cedar River for fire protection, direct pumping. Fiscal year closes March 1.

Year.	Mains.	Taps.	Meters.	Hyd'ts.	Cons'pt'n.	Cost.	B'd'd d'bt.	Int.	ch's.	Exp.	Rev.
'89	3.5	110	0	41	unk'n	$30,000	$29,000		$1,520	$1,600	$2,800
'90-1	4	208	2	46	......	30,400	28,000		1,400	1,737	2,980

Financial.—*Bonded Debt*, $28,000 at 5% due in 1908. *Expenses*, total, $3,137, *Reve-

FOR FURTHER DESCRIPTIVE MATTER see Manual for 1889-90. CHANGES since 1890 are here given. POPULATIONS are for 1880 and 1890, respectively.

nue; consumers, $2,980; city, none. **Management.**—Com. on Public Improvements appointed every year by mayor from council. Chn., J. W. Barr. Clerk, C. S. Bennett.

72. **WASHINGTON,** *Washington Co.* (Pop., 2,919—3,235.) **History.**—Construction begun during summer of '91 by city. Contrs., Heidenrich Co., Chicago. Works consist of two 85,000-gall. dup. pumps, two 50 HP. boilers, 30,000-gall. storage reservoir, 14 × 110 ft. stand-pipe, one artesian well pump, 3½ miles of mains, 38 hydrants. *Cost,* about $25,000. **Management.**—Com. F. E. Lampherc, W. S. Reisler, Frank Stewart, Aaron Hise, W. W. Hood.

73. **WATERLOO,** *Black Hawk Co.* (Pop., 5,630—6,674.) Built in '86 by Waterloo Water-Works Co., under 20-yrs franchise. **Supply.**—Cedar River, pumping direct to reservoir and to stand-pipe. Since '88, stand-pipe having cap. of 111,090 galls. has been built; Jewell gravity filter has been introduced; a 500,000-gall. Gordon pump has been added; an 83,000-gall. reservoir has been built. **Distribution.**—*Mains,* 9.8 mile. *Taps,* 125. *Meters,* 8. *Hydrants,* 111. **Financial.**—(In '83) *Cost,* $89,000. *Bonded Debt,* $70,000 at 6%. **Management.**—Supt., G. F. Harris.

74. **WAVERLY,** *Bremer Co.* (Pop., 2,315—2,316.) **History.**—Built in '90-1 by city; put in operation in April. Engr., Chas. F. Loweth. Contrs.. Fairbanks, Morse & Co., St. Paul, Minn. **Supply.**—Well and river, pumping direct, **Pumping Machinery.** -Dy. cap., 2,500,000 galls.; one 1,000,000 and one 1,500.000 Smith & Vaile dup. **Distribution.**—*Mains,* 8 to 4-in. c. i., 3 miles. *Taps,* 190. *Hydrants,* Mathews, 31. *Valves,* Galvin. *Pressure:* ordinary, 50 lbs.; fire, 100 lbs. **Financial.**—*Cost,* $19,000. *Bonded Debt,* $10,000 at 6%. **Management.**—Mayor, H. C. Holt.

75. **WEBSTER CITY,** *Hamilton Co.* (Pop., 1,818—2,829.) Built in '83 by village. **Supply.**—Artesian well, pumping to tank and direct. **Distribution.**—(In 88.) *Mains,* 1.5 miles. *Taps,* 160. *Hydrants,* 15. *Consumption.* 75,000 galls. **Financial.**—(In '88.) *Cost,* $21,000. *Bonded Debt,* $13,000 at 6%. *Expenses:* operating, $1,200; int., $780; total, $1,930. *Revenue,* consumers, $1,000. **Management.**—Comrs. Chn., A. Austin. Engr., R. H. Stevenson.

West Burlington, *Des Moines Co.* (Pop., in '90, 836.) Built in '87-8 by village for fire protection. See 1889-90 Manual, p. 493.

76. **WEST LIBERTY,** *Muscatine Co.* (Pop., 1,141—1,268.) Built in '87-9 by town. **Supply.**—Artesian well, by gravity. Fiscal year closes March 15. Storage reservoir and other extensions contemplated in '91.

Year.	Mains.	Taps.	Meters.	Hyd'ts.	Cost.	B'd'd debt.	Int. ch'g's.	Exp.	Rev.
'89-90	1.75	86	0	..	$10,000	$10,000	$600	....	$700
'90-1	3.3	153	0	18	13,620	9,500	590	$75	1,208

Financial.—*Bonded Debt,* $9,500: $5,000 at 6%, due in '93; $2,500 at 6%, due in '91; $2,750 at 7%, due in '91-2; $3,500 will be refunded in '91. *Expenses,* total, $765. *Revenue:* consumers, $1,208; city, none. **Management.**—Mayor, T. W. Swan. Clerk, F. K. Chase. Supt., C. E. Cheeseborough.

77. **WILTON JUNCTION,** *Muscatine Co.* (Pop., of township, 2,517—2,007.) **History.**—Built in '90-1 by village; put in operation Feb. 1. **Supply.**—Artesian well, pumping to tanks. **Pumping Machinery.**—Dy. cap., 432,000 galls.; Hughes. **Tank.**—Cap., 35,000 galls. **Distribution.**—*Mains,* 6 and 4-in., 2 miles. *Hydrants,* 15. **Financial.**—*Cost,* $9,000. *Bonded Debt,* $11,700 at 6%. **Management.**—Recorder, J. E. Park.

Water-Works Projected with Fair Prospects of Construction.

ALBIA, *Monroe Co,* (Pop., 2,435—2,359.) Aug. 15, '91, it was reported that franchise had been granted to D. W. Hunt & Co. to construct water, electric lighting and gas plants. Work to begin at once.

CRESCO, *Howard Co.* (Pop., 1,875—2,018.) Aug. 29, '91, it was reported that city would build works at once. Wells, pumping to stand-pipe. Cost, $15,000. Contr., Fremont Turner, Nevada. Mayor, D. A. Lyons.

DE WITT, *Clinton Co.* (Pop., 1,608—1,359.) System projected by city in '91. Well, pumping to tank. Cost, $7,000. Contracts to be let Oct. 25. Mayor, D. Armentrout. Address F. Small.

ESTHERVILLE, *Emmet Co.* (Pop., 138—1,475.) July 14, 91, it was reported that city wished to grant 20-yrs. franchise to co.; 1½ mile of pipe, 10 hydrants; hydrant rental, $60. Mayor, A. O. Paterson.

IOWA.

GARNER, *Hancock Co.* (Pop., 321—679.) Town voted in '91 to build works, consisting of wind mill, tank, 8 hydrants, 2,600 ft. of 6-in. pipe. Engr., C. S. Hall.

GRINNELL, *Poweshiek Co.* (Pop., 2,415—3,332.) April 18, '91, it was reported that town intended to build works. Well, pumping to stand-pipe. Cost, $15,000. Address J. R. Lewis.

HULL, *Sioux Co.* (Pop., in '90, 566.) Aug. 13. '91, it was reported that town was drilling well, and if sufficient water was obtained works would be built. City Recorder, B. H. Tamplin.

MALVERN, *Mills Co.* (Pop., 743—1,003.) Sept. 25, '91, Jno. R. Foulks, Mayor, wrote that city would build works soon. Supply, well, pumping to tank. Cost, $7,000.

MANSON, *Calhoun Co.* (Pop., 377—822.) Aug. 13, '91, it was reported that city would begin construction of works within 30 days. Oct., '91, reported works would probably be completed during month. Engr. and Contr., H. L. Wood, Hawarden. Bonds, authorized, $5,000. Chn., R. A. Horton.

MAPLETON, *Monona Co.* (Pop., 379—782.) July 1, '91, it was reported that works were talked of and town would vote on question soon.

ORANGE CITY, *Sioux Co.* (Pop., 320—1,246.) July 2, '91, it was reported that town had decided to build works. Address Mayor.

PAULLINA, *O'Brien Co.* (Pop., in '90, 510.) In '91 town voted to build works. Mayor, L. N. B. La Rue.

SHELDON, *O'Brien Co.* (Pop., 730—1,478.) In '91 franchise was granted to co., works to be completed in May, '92. City Clerk, Homer W. Conant.

TAMA *Tama Co.* (Pop., 1,289—1,741.) Oct. 5, '91, it was reported that city was receiving bids for construction of works. Est. cost, $15,000. Wells, pumping to stand-pipe. Mayor, C. A. Hilton.

WALL LAKE, *Sac Co.* (Pop., 208—439.) Wall Lake Water-Works Co. granted contract for construction of works Sept., '91, to U. S. Wind Engine & Pump Co., Batavia, Ill., to be completed Nov. 1. Supply, well, pumping to 40,000-gall. tank. About 3,000 ft. of 4 and 6-in pipe. Director, C. A. Johnston. Secy., R. M. Hunter.

WEST UNION, *Fayette Co.* (Pop., 1,551—1,676.) In June, '91, city voted to build works at est. cost of $16,500. Bids received till Sept. 3, work to begin Oct. 1. Chf. Engr., Frank Hobson.

Water-Works Projects Reported in '90 and '91 but Now Indefinite.

Centreville, *Appanoose Co.*

MINNESOTA.

Water-Works Completed or In Process of Construction.

1. **ALBERT LEA,** *Freeborn Co.* (Pop., 1,966—3,305.) History.—Built in '91 by city; put in operation Oct. 15. Des. Engr., C. Curtis, St. Paul, Const. Engr., D. F. Stacy. Contr., E. T. Sykes, Minneapolis. Supply.—Artesian wells, pumping, to tank and direct. Pumping Machinery.—Dy, cap., 1,000,000 galls. Tank.—Cap. 1,000,000 galls. Distribution.—*Mains,* 1.3 miles. *Taps,* 15. Financial.—*Cost,* $35,000. *Bonded Debt,* $20,000 at 5%. Management.—Cem.; A. Wigand, W. A. Morin, C. W. Ransom, T. E. Richards.

2. **ALEXANDRIA,** *Douglas Co.* (Pop., 1,335—2,118.) Built in '89 by city. Supply.—Thirty driven wells, pumping direct. Fiscal year closes Dec. 31. It is proposed to take water from lake instead of wells. Distribution.—*Mains,* 2 miles. *Taps,* 50. *Hydrants,* 19. *Consumption,* 50,000 galls. Financial.—*Cost.* $16,000. *Bonded Debt,* $10,000 at 6%. *Expenses:* operating, $2,500; int., $600; total, $3,100. *Revenue:* consumers. $1,100; city, none. Management.—Prest., N. P. Ward. Secy., A. G. Sexton. Supt., D. R. Dimond.

FOR FURTHER DESCRIPTIVE MATTER see Manual for 1889-90. CHANGES since 1890 are here given. POPULATIONS are for 1880 and 1890, respectively.

3. ANOKA, *Anoka Co.* (Pop , 2,706—4,252.) Built in '89 by city. Supply.—Rum River, pumping direct. Fiscal year closes April 30. Distribution.—*Mains.* 4.3 miles. *Taps,* 163. *Hydrants,* 128. *Consumption,* 80,000 galls. Financial.—*Cost,* $32,000. *Bonded Debt,* $32,000 at 6%, due in '98. *Expenses:* operating, $1,807; int., $1,920; total, $3,727. *Revenue:* consumers, $2,762; city, $2,240; total, $5,002. Management.—Mayor, Jno. Shune. Clerk, H. Metzgan. Supt., H. G. Cass.

4. AUSTIN, *Mower Co.* (Pop., 2,305—3,901.) Built in '87 by town. Supply.—Artesian well for domestic, Cedar River for fire purposes, pumping direct. Fiscal year closes Feb. 28. Distribution.—*Mains,* 4.5 miles. *Taps,* 160. *Hydrants,* 48. *Consumption,* 50,000 galls. Financial.—*Cost,* $10,000. *Bonded Debt,* $30,000 at 6%, due in 1909. *Expenses:* operating, $2,573; int., $1,800; total, $4,373. *Revenue,* consumers, $1,500. Management.—M. Greenman,.

5. BRAINERD, *Crow Wing Co.* (Pop., 1,865—5,703.) Built in '83 by Brainerd Water Co. Supply.—Mississippi River, pumping direct. Jan 14, '90, system was put in hands of Receiver. Distribution.—*Mains,* 14.5 miles. *Taps* (in '28), 700. *Hydrants,* 79. Financial.—*Revenue,* city, $3,950. Management.—Receiver, Ambrose Tighe.

6. BRECKENRIDGE, *Wilkin Co.* (Pop., 436—655.) History.—Built in '91 by village; to be put in operation Sept. 1. Engrs. and Contrs., Harrison & Hawley, St Paul. Supply.—River, pumping to tank. Pumping Machinery.—Dy. cap., 500,000 galls.; Smith & Vaile dup. Tank.—Cap., 55,000 galls.; on 80-ft. wooden tower. Distribution.—*Mains,* 8 to 4-in. c. i., 1 mile. *Hydrants,* Mathews, 12. *Valves,* Eddy. Financial.—*Cost,* $10,600. *Bonded Debt,* $10,000. Management.—Prest., W. E. Truax. Recorder, Robt. J. Wells.

7. CANNON FALLS, *Goodhue Co.* (Pop., 942—1,078.) Built in '87-8 by village. Supply.—Well, pumping to reservoir. *Mains,* ½ mile. *Taps,* 15. *Hydrants,* 2. *Cost,* $5,500. *Debt,* $5,500 at 6%. Supt., W. H. Scofield.

I. CLOQUET, *Carlton Co.* (Pop., in '90, 2,531.) Built in '86 by Cloquet Water Co. for domestic use. Supply.—St. Louis River, pumping to tank. *Cost,* $1,000. *Cap. Stock,* authoriz:d, $7,500. Prest., Aug. Peterson. Secy.-Supt., A. M. Sheldon.

8. CROOKSTON, *Polk Co.* (Pop., 1,227—3,457.) Built in '82 by Crookston Water-Works Co.; now controlled by Moffett, Hodgkins & Clarke, Syracuse, N. Y. Supply.—Red Lake River, pumping direct. Distribution.—(In '87.) *Mains,* 6.5 miles. *Taps,* 125. *Meters,* 1. *Hydrants,* 41. *Consumption,* 500,000 galls. Financial.—(In '87.) *Cost,* $150,000. *Bonded Debt,* $105,000 at 7%. *Expenses:* operating, $2,000; int., $7,350; total, $9,350. *Revenue:* consumers, $9,000; city, $4,000; total, $13,-000. Management.—Moffett, Hodgkins & Clarke, Syracuse, N. Y.

9. DULUTH, *St. Louis Co.* (Pop., 3,483—33,115.) Built in '83 by Duluth Gas & Water Co., under 30-yrs. exclusive franchise, which provides for purchase of works by city at intervals of 5 yrs., price determined by appraisers. Co. owns or controls works of Lakeside Light & Water Co., described under Lakeside, which is a suburb of Duluth and will become part of the city Jan. 1, '93. Co. also supplies water to West Duluth Gas & Water Co.; see West Duluth. Supply.—Lake Superior, pumping to reservoir. Fiscal year closes April 30.

Year.	Mains.	Taps.	M't'rs.	H'd'ts.	C'ns'n.	Cost.	B'd d'bt.	Int.ch'g's.	Exp.	Rev.
'84	8	200	4	72						
'87	25	700	28	107	1,000,000	$587,000	$282,000	$16,920	$36,000	$45,000
'88	25	950	82	118	1,000,000	732,430	426 00	25,170	39,005	55,162
'89 '90	30	1,200	93	199	2,247,000	1,079,046	751,000	29,746	19,813	67,737

Financial.—*Cap. Stock:* authorized, $500,000; all paid up; paid 3% div'd in 1889. *Total Debt,* $75',000, bonded; $177,000 at 6%; $574,000 at 5%. *Expenses:* operating $16,-516; int., $29,746; taxes and insurance, $3,297; total, $49,559. *Rev·nue:* consumers, $47,198; fire protection, $20,534; total, $67,732. Management.—Prest., R. L. Belknap, N. Y. City. Secy., Geo. C. Squires; Treas., R. B. C. Bement; both of St. Paul. Supt, Wm. Craig.

NOTE.—Summer of '91 city voted $300,000 in bonds for public works. Proposed bond issue declared legal by lower court. Reported case would be carried to Supreme Court. Oct. 21. '91, C. B. Davis, Chicago, reported on value of company's gas and water-works, including plant of Lakeside Water Co., which above co. owns. On revenue basis value water and gas plants was est. as $1,756,857; cost of duplicating Duluth Co.'s water-works est. at about $1,000,000; neither est. includes real estate. Nov. 14, '91, City Engr. E. H. Keating submitted a report estimating the cost of constructing the gas and water plants, including stock on hand, at $1,091,107, of which $283,665 was for the gas plant, leaving est. cost of water-works about $1,000,000. Nov. 19, '91, people voted on issuing $700,000 additional bonds and on the purchase of co.'s property for $1,950,000, under an option offered by W. S. Sargeant. Vote was adverse.

Ely, *St. Louis Co.* (Pop., in '90, 901.) Built in '89 by village for fire protection only. Domestic supply will probably be introduced during '91. Report by D. H. Martin, Engr.

10. **FARIBAULT**, *Rice Co.* (Pop., 5,415—6,520.) Built in '83-4 by Faribault Water-Works Co., under 25-yrs. franchise. Supply.—Surface wells, pumping to reservoir. Distribution.—(In '87.) *Mains*, 7 miles. *Taps*, 106. *Meters*, 8. *Hydrants*, 59. *Consumption*, 100,855 galls. Financial.—(In '87.) *Cost*, $140,000. *Bonded Debt*, $80,000 at 6%. *Expenses:* operating, $2,800; int., $4,800; total, $7,600. *Revenue:* consumers, $3,000; city, $5,010; total, $8,010. Management.—Address A. C. Miller.

11. **FERGUS FALLS**, *Otter Tail Co.* (Pop., 1,635—3,772.) Built in '83 by Fergus Falls Water Co., under 30-yrs. franchise. Supply.—Red River of the North, pumping direct. Distribution.—(In '88.) *Mains*, 8 miles. *Taps*, 327. *Meters*, 1. *Hydrants*, 52. Financial.—(In '88.) *Cost*, $168,000. *Cap. Stock*, $82,000. *Bonded Debt*, $88,000 at 6%. *Sinking Fund*, $3,016. *Expenses:* operating, $583; int., $5,280; total, $5,833. *Revenue:* consumers, $7,150; city, $6,000; total, $13,150. Management.—(In '89.) Prest., C. E. Gray, Chicago, Ill. Secy., F. W. Burnham. Treas., C. D. Wright. Supt., O. Moore.

12. **GLENWOOD**, *Pope Co.* (Pop., 464-627.) Built in '83 by village. Supply.—Spring Creek, by gravity from impounding reservoir. Fiscal year closes Jan. 1. Distribution.—*Mains*, 1.5 miles. *Taps*, 46. *Hydrants*, 7. Financial.—*Cost*, $8,000. *Debt*, $4,000 at 8%, due in annual payments of $500. *Expenses*, operating, $200; int., $320; total, $520. *Revenue*, consumers, $300. Management.—Prest., C. T. Wallen. Secy., C. L. Peterson. Supt., C. Brown.

II. **HUTCHINSON**, *McLeod Co.* (Pop., 580—1,414.) Water, Light & Power Co. built water-works in '89-91. Supply.—Artesian well pumping to three 11,000-gall. cisterns. Financial.—*Est. Cost*, $1,500. Management.—Prest., Isaac Konwe. Secy., E. J. Stearns. Treas., Wm. E. Harrington.

13. **LAKESIDE** (*Duluth P. O.*), *St. Louis Co.* (Pop. in '90, 897.) Will become part of Duluth Jan. 1, '93. History.—Built in '90 by Lakeside Light & Water Co., controlled by Duluth Gas & Water Co., under 30-years franchise; put in operation in Nov. Des. Engr., Robt. Bement & Co. Const. Engr., Wm. Craig. Contr., H. E. Stevens, St. Paul. Supply.—Lake Superior, pumping direct. Intake.—C. i., 12-in., about 600 ft. long, extending to 27 ft. of water and at other end terminating in well in solid rock. Pumping Machinery.—Dy. cap., 1,100,000 galls.; Blake dup. *Boilers*, from Robinson & Cary, St. Paul, and National Water Tube Co. *Force-Main*, diam., 12-in., branching into 10-in. Distribution.—*Mains*, 10 to 6-in., 3 miles. *Services*, lead. Pipe and specials from West Superior Iron & Steel Co. *Taps*, 67. *Meters*, Hersey, 1. *Hydrants*, Galvin, 31. *Valves*, Galvin. *Consumption*, 40,000 galls. *Pressure:* ordinary, 75 lbs.; fire, 100 lbs. Financial.—*Cost*, $50,000. *Bonded Debt*, $50,000 at 5%. *Hydrant Rental*, $100 each for first 80, $80 each for next. Management.—Same as Duluth.

Lanesboro, *Fillmore Co.* (Pop., 1,032-898.) Built in '85-6 by village for fire protection. See 1889-90 Manual, p. 493.

14. **LITCHFIELD**, *Meeker Co.* (Pop., 1,250—1,899.) History.—Built in '90 by village. Des. Engr., Geo. Cadogan Morgan, 49 Major Block, Chicago. Const. Engrs. and Contrs., Fairbanks, Morse & Co., St. Paul. Supply.—Driven wells, pumping to stand-pipe. Pumping Machinery.—Dy. cap., not given; two Smith & Vaile dup. Stand Pipe.—Cap., 67,000 galls. Distribution.—*Mains*, 1, 2.8 miles. *Services*, galv. 1. *Taps*, 22. *Hydrants*, not given. Financial.—*Cost*, $20,000. *Bonded Debt*, $20,000. Management.—City Clerk, G. B. Phelps.

15. **LITTLE FALLS**, *Morrison Co.* (Pop., 508—2,354.) Built in '90 by Little Falls Electric and Water Co. Supply, Mississippi River, pumping direct. Fiscal year closes Dec. 31. Distribution.—*Mains*, 3.5 miles. *Taps*, 30. *Hydrants*, 50.

For further descriptive matter see Manual for 1889-90. Changes since 1890 are here given. Populations are for 1880 and 1890, respectively.

Consumption, 150,000 galls. Financial.—Withheld. Management.—Prest. and Treas., S. Stoll. Supt., Russell Baker.

16 **MANKATO**, *Blue Earth Co.* (Pop., 5,550—8,838.) Fire protection introduced in '79. Built in '88-9 by city. Supply, two flowing artesian wells, pumping to reservoir. Fiscal year closes April 1.

Year.	Mains.	Taps.	Meters.	Hydrants.	Cost.	B'd'd debt.	Exp.	Rev.
'88-9	4	150	21	45	$45,000	....	$2,500	$1,000
'89-90	6.5	165	21	45	78,000	$77,000	3,500	1,200
'90-1	8.5	283	21	74	85,000	77,000	3,500	3,000

Management.—City Engr., M. B. Haynes.

17. **MINNEAPOLIS**, *Hennepin Co.* (Pop., 46,887—164,738.) Built in '67 by city. Supply.—Mississippi River, pumping direct. Fiscal year closes Dec. 31. During '90 two 15,000,000-gall. Worthington pumps were added.

Year.	Mains.	Taps.	Mtrs.	H'd'ts.	C'ns'pt'n.	Cost.	B'd'd d't.	Int.ch'g's.	Exp.	Rev.
'78	22	264	....	268	3,500,000	....	....	....	$11,000	$12,402
'82	20.5	1,703	6	252	3,000,000	$404,237	....	....	15,000	39,770
'84	31.6	2,539	....	365	7,360,964	653,162	$507,000	....	51,177	81,115
'86	81.2	4,650	126	1,032	8,228,805	953,119	735,000	$26,138	61,515	115,269
'87	114	5,598	306	1,410	10,814,005	1,422,376	1,235,000	49,400	68,089	133,171
'89	157	8,473	372	1,757	12,933,072	....	1,235,000	46,400	62,631	157,511
'90	166.1	9,990	633	1,996	12,416,117	....	1,285,000	....	86,087	172,510

Management.—Com. of five, appointed by president and council every two years. Clerk, T. F. Moody. Supt., J. H. McConnell.

18. **MOORHEAD**, *Clay Co.* (Pop., in '90, 2,088.) Built in '81 by village. Supply.—Red River of the North, pumping direct. Fiscal year closes Jan. 1. Distribution.—*Mains*, 4 miles. *Taps*, 60. *Hydrants*, 55. *Consumption*, 180,000 galls. Financial.—*Cost*, $45,000. *Bonded Debt*, $30,000 at 8%. *Expenses:* operating, $3,635; int., $2,400; total, $6,035. *Revenue*, consumers, $2,000. Management.—Supt., L. M. Anderson, appointed by Mayor.

19. **NEW ULM**, *Brown Co.* (Pop., 2,471—3,741.) Built in '89-90 by city; put in operation Nov., '90. Contrs., Fairbanks, Morse & Co., St. Paul. Supply.—Well, pumping to reservoir, thence pumping to tank. Fiscal year closes April 1. 11,000 ft. of mains and 15 hydrants to be added in '91. Distribution.—*Mains*, ½ mile. *Taps*, 11. *Hydrants*, 14. Financial.—*Cost*, $20,000. *Bonded Debt*, $10,000 at 6%, due in 1900. Management.—Chn., F. G. Koch. Secy., Louis Schilling. Supt., Chas. Stall.

20. **NORTH SAINT PAUL**, *Ramsey Co.* (Pop., in '90, 1,099.) Built in '89 by North St. Paul Water Co. Supply.—Artesian wells, for domestic use; Silver Lake for fire protection; pumping direct from lake and reservoir, and pumping to standpipe. Distribution.—*Mains*, 2.3 miles. *Hydrants*, 20. Financial.—Withheld. *Cost* (in '89), $29,286. Management.—Prest., W. S. Morton. Secy., Wm. H. Lightner. Treas., F. Driscoll, Jr. Supt., C. R. McKinney.

21. **OWATONNA**, *Steele Co.* (Pop., 3,161—3,849.) History.—Built in '90 by city; put in operation Nov. 1. Engr., W. W. Curtis. Contrs., Harrison & Hawley, St. Paul. Supply.—Artesian well, pumping to stand-pipe. Water rises to 5 ft. below surface. Pumping Machinery.—Dy. cap., 1,000,000 galls.; Smedley comp. dup. Stand-Pipe.—Cap., 132,000 galls. 15 × 100 ft. Distribution.—*Mains*, c. 1., 3.5 miles. *Taps*, 25. *Hydrants*, Mathews, 37. *Valves*, Galvin, Eddy Pressure: ordinary, 70 lbs.; fire, 125 lbs. Financial.—*Cost*, $30,000. *Bonded Debt*, $30,000, at 5%. Management.—Board: Jo. Hoffman, B. Casper, J. Brown.

22. **PIPE STONE**, *Pipe Stone Co.* (Pop., 222—1,232.) Built in '88 by village. Supply.—Bored wells, pumping to stand-pipe. Distribution.—*Mains*, 4.5 miles. *Taps*, 113. *Hydrants*, 21. *Consumption*, 25,000 galls. *Pressure*, 43 lbs. Financial.—*Cost*, $35,000. *Bonded Debt*, $20,000 at 6%. *Expenses:* operating, $980; int., $1,200; total, $2,-180. *Revenue*, consumers, $1,200. Management.—Comr., Clarence Walkup. Supt., Frank Raymond.

23. **RED WING**, *Goodhue Co.* (Pop., 5,876—6,294.) Built in '84 by city. Supply.—Mississippi River, pumping to reservoir; filtered. Fiscal year closes March 31.

Year.	Mains.	Taps.	Meters.	H'd'ts.	C'n'p'n.	Cost.	B'd'd debt.	Int. ch'g's.	Exp.	Rev.
'85-6	6.5	73	....	56	77,583	$81,568	$80,000	$4,000	$3,176	$626
'87-8	7.2	169	32	61	81,307	....	80,000	4,000	6,123	2,901
'89-90	7.5	208	59	65	92,195	....	80,000	4,000	4,740	3,998

Management.—Chn., F. W. Howe. Secy., S. J. Willard. Supt., W. I. Miller.

MINNESOTA. 271

24. ROCHESTER, *Olmstead Co.* (Pop., 5,103—5,321.) Built in '87 by Rochester Water-Works Co.; controlled by Moffett, Hodgkins & Clarke, Syracuse, N. Y., under 30-yrs. franchise. **Supply.**—Springs, pumping to stand-pipe. **Distribution.**—(In '88.) *Mains,* 10 miles. *Taps,* 150. *Meters,* 23. *Hydrants,* 139. *Consumption,* 40,000 galls. **Financial**—(In '88.) *Cost,* $150,000. *Bonded Debt,* $115,000 at 6%. *Expenses:* operating, $2,000; int., $6,900; total, $8,900. *Revenue:* consumers, $4,600; city, $4,536; total, $9,136. **Management.**—(In '89.) Prest., J. F. Moffett. Supt., C. W. Streator.

25. SAINT CLOUD, *Sherbourne and Benton Cos.* (Pop., 2,462—7,686.) Built in '83 by city. **Supply.**—Mississippi River, pumping to stand-pipe and direct **Distribution.**—*Mains,* 10.8 miles. *Taps,* 210. *Hydrants,* 132. **Financial.**—Withheld. *Cost* (in '83), $175,000. **Management.**—Prest., E. T. Sykes. Secy., J. A. Clarke.

26. SAINT PAUL, *Ramsey Co.* (Pop., 41,173—133,156.) Bought since '82 by city. Built in '70 by St. Paul Water Co. **Supply.**—Phalen, Vadnais, Pleasant, and others of chain of lakes, by gravity, and by direct pumping for high service. Fiscal year closes Nov. 1. A contract was let, fall of '90, for a 6,000,000-gall. pump.

Year.	Mains.	Taps.	Meters.	Hydrants.	Cost.	B'd't d'bt.	Int.	ch's.	Exp.	Rev.
'80	17.5	1,817	0	188	$510,000					
'82	24.2	2,130	130	197	523,802	$510,000	$7,009	$1,858		$7,538
'83	28.8	2,431	135	236	671,025	510,000	28,371	8,693		69,872
'84	45.1	2,931	144	315	1,451,768	1,250,000	50,337	14,022		95,241
'85	67.8	3,940	166	491	1,699,534	1,660,000	79,319	21,516		116,195
'86	85.9	4,947	200	673	1,857,856	1,760,000	85,009	22,331		130,024
'87	104	6,017	248	868	2,116,174	1,860,000	84,395	27,063		166,396
'88	131.6	7,235	284	1,134	2,483,936	2,000,000	100,280	32,238		156,713
'89	166.8	9,052	354	1,471	2,858,400	2,360,000	108,733	30,682		211,672
'90	193.08	10,458	440	1,748	3,095,145	2,460,000	113,431	39,934		187,008

Financial.—*First Cost,* $290,000. *Bonded Debt,* $2,460,000: $1,150,000 at 5%; $860,000 at 4½%; $340,000 at 4%; $100,000 at 4%. Bonds are all issued for 30 yrs., from '83-90. *No Sinking Fund. Expenses,* total, $153,365. *Revenue:* consumers, $187,098; frontage tax for fire protection, $73,080; in addition, $60,000 were raised by taxation in '89 and $67,000 in '90. It is thought that in '91 revenue will cover all expense. **Management.**—Five comrs. appointed every four years by mayor. Secy., Jno. Caulfield. Supt., J. H. Overton.

27. SAINT PETER, *Nicollet Co.* (Pop., 3,436—3,671.) Built in '88-9 by city. **Supply.**—Well, near river, pumping to stand-pipe. Fiscal year closes April 1. **Distribution.**—*Mains,* 3 miles. *Taps,* 40. *Meter,* 1. *Hydrants,* 35. **Financial.**—*Cost,* $38,000. *Bonded Debt,* $32,000 at 6%. *Expenses:* operating, $700; int., $1,920; total, $2,620. *Revenue,* consumers, $300. **Management.**—Com., appointed by city council from its members every year. Engr., J. M. Peterson.

28. SANDSTONE, *Pine Co.* (Pop., in '90, 517.) Owned by Sandstone Water Co. Built in '88 by W. H. Grant. **Supply.**—Springs, impounded by dam, pumping to reservoir, *Mains,* 2.2 miles. *Hydrants,* 7. *Cost,* $15,090. Prest., W. H. Grant. Secy., W. H. Grant, Jr.

29. SAUK CENTRE, *Stearns Co.* (Pop., 1,201—1,695.) Bought by city since '88. Built in '84. **Supply.**—Sand Lake, pumping direct. Fiscal year closes April 1. Filter to be built during '91. **Distribution.**—*Mains,* 1.5 miles. *Taps,* 100. *Hydrants,* 15. *Consumption,* 60,000 galls. **Financial.**—*Cost* (purchase price paid by city).—$14,800. *Bonded Debt,* $14,000 at 7%. *Expenses:* operating, $1,900; int., $980; total, $2,880. *Revenue:* consumers, $1,100; city, $1,125; other purposes, $200; total, $2,425. **Management.**—City Council. Secy., F. Cooper. Supt., L. E. Coe.

30. STILLWATER, *Washington Co.* (Pop., 9,055—11,260.) Built in '80 by Stillwater Water Co., under 30-yrs. franchise. **Supply.**—McKusick's Lake, by gravity for low, and pumping to tanks for high-service; filtered. A 1,000,000-gall. pump and a steel stand-pipe, 15 × 65 ft., were to be added during '91. **Distribution.**—*Mains,* 9 miles. *Taps,* 600. *Hydrants,* 126. *Consumption,* 1,000,000 galls. **Financial.**—*Cost,* $215,000. *Bonded Debt,* $100,000 at 7%. *Expenses:* operating, $3,150; int., $7,000; total, $10,450. *Revenue:* consumers, $9,500; city, $7,250; total, $16,750. **Management.**—Prest., E. W. Durant. Secy. and Supt., H. H. Harrison. Treas., R. S. Davis.

31. TWO HARBORS, *Lake Co.* (Pop., est., 1,200.) Built in '89 by village. Contrs., Fairbanks, Morse & Co., St. Paul. **Supply.**—Lake Superior, pumping direct and to tank. Fiscal year closes April 3. **Distribution.**—*Mains,* 2 miles. *Taps,* 160.

For Further Descriptive Matter see Manual for 1889-90. Changes since 1890 are here given. Populations are for 1880 and 1890, respectively.

Meters, 2. *Hydrants*, 13. *Consumption*, 40,000 galls. **Financial.**—*Cost*, $30,000. *Total Debt*, $20,300: floating, $300 at 5%; bonded, $20,000 at 5%. *Expenses:* operating $1,500; int., $1,000; total, $2,500. *Revenue*, consumers, $1,200. **Management.**—Secy., S. C. Holden. Supt., F. E. Duncan.

32. **WEST DULUTH**, *St. Louis Co.* (Pop., in '90, 3,368.) West Duluth Gas & Water Co. put system in operation Aug. 1, '91. **Supply.**—From Duluth Gas & Water Co., through 16-in. main 4,854 ft. long, owned by latter co. West Duluth Co. pays 7% on cost of main, est. as $16,500, and stated sum for water. Prest., R. C. Elliott.

33. **WILMAR**, *Kandiyohi Co.* (Pop., in '90, 1,825.) **History.**—Built in '91 by village; put in operation Sept. 1. Contrs., Harrison & Hawley, St. Paul. **Supply.**—Wells, pumping to tank. Wells are 4½ ins. in diam., 120 ft. deep. **Pumping Machinery.**—Dy. cap., 1,000,000 galls.; two 500,000 Smith & Vaile. **Tank.**—Cap., 65,000 galls.; 22 × 20 ft. **Distribution.**—*Mains*, 8 to 4-in. c. i., 3 miles. *Services*, w. i. *Hydrants*, Mathews, 25. *Valves*, Eddy. *Pressure :* ordinary, 45 lbs.; fire, 125 lbs. **Financial.**—*Cost*, $20,900. *Bonded Debt*, $20,000 at 6%. **Management.**—Prest., Jno. Williams. Recorder, H. J. Ramsett.

34. **WINONA**, *Winona Co.* (Pop., 10,208—18,208.) Built in '76 by city. **Supply.** Two wells, near Mississippi River, pumping to stand-pipe. Fiscal year closes March 31. March 21, '91, it was reported that city would issue $50,000 of water bonds in order to put in a 5,000,000-gall. pump and a large discharge pipe, and add about 2 miles of mains.

Year.	Mains.	Taps.	Meters.	Hyd'ts.	Cons'pt'n.	Cost.	B'd't d't.	Int. chgs.	Exp.	Rev.
'81	8	300	0	90	400,000	$160,000-$40,000	$2,800	$6,500	$1,900
'87	16	1,000	0	175	1,184,634	200,000	100,000	4,500	14,000	8,000
'89	21	1,200	0	228	1,500,000	211,663	323,000	16,150	6,000	8,594
'90	21.1	1,300	0	229	1,600,000	211,953	317,000	15,850	6,500	9,863

Financial.—*Bonded Debt*, $317,000 at 5%. *Expenses*, total, $22,350. *Revenue:* consumers, $9,863 ; city, $14,740 ; total, $21,603. **Management.**—Comr., Jno. B. Fellows.

Water-Works Projected with Fair Prospects of Construction.

BLOOMING PRAIRIE, *Steele Co.* (Pop., 338—308.) Village talks of building works. President, G. H. Johnson.

DETROIT CITY, *Becker Co.* (Pop., 554—1,510.) Aug. 13, '91, it was reported that village would build works and that contract had been let to Fairbanks, Morse & Co., St. Paul. Clerk, O. D. Holmes.

EXCELSIOR, *Hennepin Co.* (Pop., 417—619.) Village voted in '91 to build works. Supply, Lake Minnetonka, pumping to reservoir. Est. cost, $20,000. Mayor, W. T. Milnor.

FAIRMONT, *Martin Co.* (Pop., 541—1,205.) It was reported in '91 that village had voted to issue $15,000 worth of bonds to construct water-works and electric lighting plant.

LU VERNE, *Rock Co.* (Pop., 679—1,466.) In '91 village let contract for construction of water-works to Harrison & Hawley, St. Paul. Village Recorder, Albert Barck.

MADELIA, *Watonwan Co.* (Pop., 489—852.) Construction began Aug. 10, to be completed Oct. 15 by city. Contrs., Fairbanks, Morse & Co., St. Paul. Well, pumping to tank. Cost, about $9,000. Village Recorder, C. Cooley.

NORTHFIELD, *Rice Co.* (Pop., 2,296—2,659.) July 2, '91, it was reported that preliminary plans were being made for construction of system of water-works by city. Address Wm. Clark.

TRACY, *Lyon Co.* (Pop., 322—1,400.) May 27, '91, village voted to issue bonds to construct water-works. Recorder, F. S. Brown.

WINNEBAGO, *Faribault Co.* (Pop., 933—1,108.) Aug. 29, '91, it was reported that construction of water-works would be begun at once by city. Des. Eng., F. Turner. Const. Engrs. and Contrs., Smedley Mfg. Co., Dubuque, Ia. Well, pumping direct. Mains, 2 miles. Hydrants, 22. Cost, $10,000. Mayor, W. I. Haigly. Secy., S. T. Conklin.

WORTHINGTON, *Nobles Co.* (Pop., 636—1,164.) City was to begin construction of Works July 1, '91, to be completed Oct. 1. Contrs., Harrison & Hawley, St. Paul. Well, pumping to tank. A 750,000-gall. pump, 2¼ miles of pipe, 16 hydrants. Cost, $15,000. Recorder, Frank Lewis.

KANSAS.

Water-Works Completed or in Process of Construction.

1. **ABILENE,** *Dickinson Co.* (Pop., 2,360—3,547.) Built in '82 by Abilene Water & Electric Light Co., rebuilt in '90; now owned by Debenture Guarantee & Assurance Co., Chicago, Ill. Supply.—Sands springs, pumping direct. Fiscal year closes Dec. 31. During '90 new pumping station was built and stand-pipe, 15 × 125 ft. Engr., Geo. H. Pierson, New York City. Distribution and Financial.—Withheld. Management.—Supt., Geo. A. Rogers.

2. **ANTHONY,** *Harper Co.* (Pop., 345—1,806.) Built in '87 by Anthony Improvement Co. Supply.—Eighteen driven wells, adjacent to Bluff Creek, pumping to stand-pipe. Distribution.—(In '88). *Mains*, 8.5 miles, *Taps*, 250. *Hydrants*, 65. *Consumption*, 160,000 galls. Financial.—(In '83.) *Cap. Stock*, $100,000. *Bonded Debt*, $60,000. *Expenses*, operating, $7,000. *Revenue:* consumers, $4,500; city, $3,500; total, $8,000. Management.—(In '89.) Prest., C. C. Pomeroy. Secy. and Treas., Chas. E. Kimball, 3 Pine St., N. Y. City. Supt., S. A. Darrough.

3. **ARGENTINE,** *Wyandotte Co.* (Pop., in '90, 4,732.) Built in '89-90 by Argentine Water & Electric Works Co., under 20-yrs. franchise. Supply.—Six Wagner wells, pumping to tank. Fiscal year closes March 31. Distribution.—*Mains*, 4 miles. *Taps*, 90. *Hydrants*, 45. *Consumption*, 50,000 galls. Financial.—*Cost*, $25,000. *Cap Stock:* authorized, $90,000; paid up, $75,000. *Bonded Debt*, $100,000 at 6%, due in 1909. *Expenses:* operating, $4,000; int., $4,500; taxes, $800; total, $9,300. *Revenue:* consumers, $1,000; city, $4,500; total, $5,500. Management.—Prest., J. D. Cruise; Secy., E. D. Cruise; both of Kansas City. Supt., Fred. Payne.

4. **ARKANSAS CITY,** *Cowley Co.* (Pop., 1,012—8,347.) Built in '83 by Arkansas City Water Co. Supply.—Arkansas River, pumping to stand-pipe.

Year.	Mains.	Taps.	Meters.	Hydrants.	Cost.	Rev.*
'86	7	150	0	100	$100,000	$6,000
'87	13	300	5	150	150,000	9,000
'88	13	500	18	151	150,000	9,240
'90	22	700	60	184	150,000	11,040

*From city.

Financial.—*Cap. Stock*, $150,000. *No Debt*. Management.—Prest., J. B. Quigley, St. Louis, Mo. Secy., L. P. Andrews. Treas. and Supt., M. L. Andrews.

NOTE.—Reported Sept., '91, that city would buy works for $190,000.

5. **ATCHISON,** *Atchison Co.* (Pop., 15,105—13,963.) Built in '80 by Atchison Water Co.; franchise fixes rates and provides for purchase of works by city in 1909, price determined by arbitration. Supply.—Missouri River, pumping to reservoir. Fiscal year closes Dec. 31. Distribution.—*Mains*, 12 miles. *Taps*, 800. *Meters*, 65. *Hydrants*, 102. *Consumption*, 600,000 galls. Financial.—*Bonded Debt*, $80,000 at 7%, due in 1900. *Revenue*, city, $6,120. Management.—Prest., Jno. R. Lionberger; Secy., Jno. D. Davis; both of St. Louis, Mo. Supt., E. S. Wells.

6. **BELLEVILLE,** *Republic Co.* (Pop., 238—1,868.) Built in '89 by Belleville Water, Light & Power Co., under 30-yrs. franchise. Supply.—Wells and impounding reservoir on creek; filtered. Distribution.—(In '89.) *Mains*, 5.3 miles. *Hydrants*, 43. Financial.—(In '89.) *Cost*, $55,000. *Bonded Debt*, $35,000 at 6%. *Revenue*, city, $3,440.

NOTE.—In July, '91, it was unofficially reported that co. was no longer in existence and works were in litigation.

7. **BELOIT,** *Mitchell Co.* (Pop., 1,835—2,455.) Built in '87 by city; now owned by A. T. Rogers; 25-yrs. franchise. Supply.—Well, on bank of Solomon river, pumping to tank. Distribution.—*Mains*, 5.5 miles. *Taps*, 120. *Meters*, 3. *Hydrants*, 70. *Consumption*, 70,000 galls. Financial.—*Cost*, $54,000. *Bonded Debt*, $51,000 at 6%. *Revenue*, $5,290. Management.—Owner and Mgr., A. T. Rogers.

FOR FURTHER DESCRIPTIVE MATTER see Manual for 1889-90. CHANGES since 1890 are here given. POPULATIONS are for 1880 and 1890, respectively.

8. **BIRD CITY,** *Cheyenne Co.* (Pop., in '90, 145.) Built in '87-8 by village. Supply.—From tube wells, pumping to tank. **Distribution.**—*Mains,* 1 mile. *Consumers,* reported as 200. *Hydrants,* 6. **Financial.**—*Cost,* about $5,000. *Expenses,* $500. *Revenue,* $590. **Management.**—Address City Clerk.

9. **BLUE RAPIDS,** *Marshall Co.* (Pop., 829—905.) Built in '89-'91 by city; works were put in operation by city in '74, but were probably for fire protection only. See '89-'90 Manual, p. 507. **Supply.**—Wells, pumping to reservoir. **Distribution.**—*Mains,* 2.5 miles. *Hydrants,* 17. **Financial.**—*Cost,* $20,000. *Bonded Debt,* $20,000; $8,000 at 6%; $12,000 at 5%. **Management.**—City Clerk. A. J. Loomis.

NOTE.—Feb. 19, '91, City Clerk reported that system was not yet in operation, City Council refusing to accept plant when completed on ground that construction had not fully conformed to contract.

10. **BUNKER HILL,** *Russell Co.* (Pop., 135—157.) Now owned by city. Built in '76 by Bunker Hill Water Co. **Supply.**—Springs, pumping to tank and direct. Fiscal year closes April 6. In '91 a 100,000-gall. Deane dup. pump was added. **Distribution**—*Mains,* 3.8 miles. *Taps,* 35. *Hydrants,* 12. *Consumption,* 10,000 galls. **Financial.**—*Cost* (to city), $6,500. *Bonded Debt,* $6,500 at 7%, due in 1909. *Sinking Fund,* $16 at 7%. *Expenses:* operating, $120; int., $435; total, $855. *Revenue,* consumers, $120. **Management.**—Mayor, J. W. Shaffer. Supt., W. H. Blays.

11. **BURLINGTON,** *Coffey Co.* (Pop., 2,011—2,239.) Built in '86-7 by Burlington Water-Works Co., under 20-yrs. franchise. **Supply.**—Neosho River, pumping to stand-pipe; filtered. Fiscal year closes Jan. 1. **Distribution.**—*Mains,* 5.3 miles. *Taps,* 80. *Hydrants,* 70. *Consumption,* 500,000 galls. **Financial.**—*bonded Debt,* $60,000 at 6%. *Expenses:* operating, $1,800; int., $1,600; total, $5,400. *Revenue:* consumers, $1,910; fire protection, $3,500; total, $5,410. **Management.**—Prest., H. L. Jarboe. Secy., C. H. Race.

12. **CALDWELL,** *Sumner Co.* (Pop., 1,005—1,612.) Built in '86 by Co. under 20-yrs. franchise. Rates were established by city. New co. was to be organized in '91. **Supply.**—Creek, pumping from impounding reservoir to stand-pipe; filtered. Fiscal year closes Dec. 31. During '89-90 works were completely overhauled. Two 8 × 10-ft. National filters were introduced. A 10-in. main was extended to another creek and an 8-ft. stone dam built on it to secure water in dry weather. Cost of improvements, $8,282. New co. was about to be organized at the time of report, Feb., '91. **Distribution.**—*Mains,* 6.8 miles. *Taps,* 180 *Hydrants,* 55. *Consumption,* 85,000 galls. **Financial.**—*Cost,* $73,282. *Revenue,* city, $3,000. **Management.**—Address Post, Martin & Co., 3 Wall St., N. Y. City. Supt., S. H. Horner.

13. **CHERRY VALE,** *Montgomery Co.* (Pop., 620—2,140.) Built in '85 by Cherry Vale Water & Mfg. Co., under 21-yrs. franchise, which regulates rates. **Supply.**—Drum Creek, pumping from impounding reservoir to stand-pipe; filtered. **Distribution.**—*Mains,* 5 miles. *Taps,* 250. *Hydrants,* 45. *Consumption,* 50,000 galls. **Financial**—*Cost,* $100,000. *Cap. Stock:* authorized, $200,000; paid-up, $100,000. Paid no div'd in 1890. *Bonded Debt,* $60,000 at 7%. *Expenses:* operating, $2,100; int., $4,200; taxes, $200; total, $6,500. *Revenue:* consumers, $1,200; Ry., $900; fire protection, $2,700; total, $4,800. **Management.**—Prest., S. P. White; Secy., C. T. Ewing; both of Thayer, Kans. Supt., J. C. Paxson.

14. **CLAY CENTRE,** *Clay Co.* (Pop., 1,753—2,802.) Built in '85 by Clay Centre Water Co., under 20-yrs. franchise; furnished fire protection Dec., '84. **Supply.**—Surface water, pumping direct. **Distribution.**—*Mains,* 4 miles. *Taps,* 222. *Meters,* 1. *Hydrants,* 40 **Financial.**—*Cost,* $80,000. *Cap. Stock,* $100,000. *Bonded Debt,* $60,000 at 7%. *Expenses:* operating, $2,300; int., $4,200; total, $6,500. *Revenue:* consumers, $2,000; city, $4,000; total, $6,600. During '91 city and co. were in litigation over payment of hydrant rental. **Management.**—Prest., C. A. Beebe, Hutchinson. Secy. and Supt., Jas. A. Hanna.

15. **CLYDE,** *Cloud Co.* (Pop., 956—1,137.) Built in 87 by village. **Supply.**—Steamed wells, pumping to stand-pipe. **Distribution.**—(In '88.) *Mains,* 4 miles. *Taps,* 137. *Meters,* 0. *Hydrants,* 40. *Consumption,* 80,000 galls. **Financial.**—(In '88.) *Cost,* $30,000. Int., 6%. *Expenses,* operating, $1,240. *Revenue,* consumers, $1,452. **Management.**—Supt., Wm. L. Brandon.

16. **COLUMBUS,** *Cherokee Co.* (Pop., 1,164—2,160.) **History.**—Built in '87 by Long & Doubleday, under 21-yrs. franchise. **Supply.**—Artesian wells, pumping to stand-pipe, reservoirs and direct. **Stand-Pipe.**—Cap., 126,900 galls. **Reservoir.**—

Cap., 300,000 galls. **Distribution.**—*Mains, 1.5* miles. Pipe and specials from Chattanooga (Tenn.) Pipe Foundry. *Taps,* 110. *Hydrants,* White, 50 **Financial.**—*Cost,* $60,000. *Expenses,* $1,100. *Revenue:* consumers, $4,000; city, $3,000; total, $5,000. *Hydrant Rental,* $60. **Management.**—Prest. and Secy., R. A. Long. Treas., L. L. Doubleday. Supt., C. McNulty.

17. **CONCORDIA**, *Cloud Co.* (Pop., 1,853—3,181.) Built in '87-8 by town. **Supply.**—Cook well system, near Platte River, pumping to stand-pipe. Fiscal year closes Aug. 25, '91. **Distribution.**—*Mains,* 6.3 miles. *Taps,* 193. *Meters,* 1. *Hydrants,* 50. **Financial.**—*Cost,* $15,000. *Bonded Debt,* $15,000 at 5%, due in 1907. *Expenses:* operating, $1,800; int.. $2,250; total, $4,050. *Revenue,* consumers, $1,800. **Management.**—Chn. Homer Kennett. Clerk, Ed. E. Chapman.

18. **CONWAY SPRINGS,** *Sumner Co.* (Pop., 258—681.) Built in '88 by Conway Springs Water Co. **Supply.**—Well, pumping to stand-pipe. **Distribution.**—(In '88.) *Mains,* 2.8 miles. *Taps,* 27. *Hydrants,* 8. **Financial.**—(In '88.) *Cost,* $12,000. *Expenses,* $7,000. *Revenue,* $1,200. **Management.**—Receiver, J. M. Wilson.

NOTE.—Co. became insolvent in Nov. '89, and part of plant is in litigation, hence it was impossible to give statistics for 1890. Works are now being operated by Receiver.

19. **COOLIDGE,** *Hamilton Co.* (Pop., in '90, 472.) Built in '88-9 by Peck Water Works Co; franchise provides neither for control of rates nor purchase of works by city. **Supply.**—Two artesian and one surface well, pumping from receiving reservoir direct and to stand-pipe. Supply was increased in '90 by one artesian well, 500 ft. deep, and one surface well, 17 ft. in diam., 30 ft. deep. Fiscal year closes July 25. **Distribution.**—*Mains,* 3 miles. *Consumption,* 100,000 galls. **Financial.**—*Cost,* $13,000. *Cap. Stock:* anthorized, $60,000; paid up, $13,000; paid 3½% div'd in 1890. Reserved for extensions, $17,000. *No debt. Sinking Fund,* $259. *Expenses:* operating, $1,500; taxes, $135; total, $1,635. *Revenue:* consumers, $900; Ry., $1,500; fire protection, $720; school, $120; other purposes, $24; total, $3,264. **Management.**—Vice-Prest., C. O. Miller. Secy. and Supt., E. H. Peck.

20. **COUNCIL GROVE,** *Morris Co.* (Pop., 1,041—2,211.) Built in '87-'88 by Council Grove Water Co., under 20-yrs. franchise. Const. Engr., F. H. Crosby. **Supply.**—Well, near river, pumping to stand-pipe and direct. Fiscal year closes Feb. 1. **Distribution.**—*Mains,* 4 miles. *Taps,* 75. *Meters,* 2. *Hydrants,* 43. **Financial.**—Withheld. **Management.**—Prest., Wm. Runkle. Secy. and Treas., W. E. Sliger. Supt., Preston Little.

21. **CULLISON,** *Pratt Co.* (Pop., est., 500.) Built in '88 by village and Wichita & Western R. R. Co., village owning distribution only. **Supply.**—Well, pumping to tank. **Distribution.**—*Mains,* 1.5 miles. *Taps,* 40. *Hydrants,* 12. **Financial.**—(*For City.*) *Cost,* $6,500. *Bonded Debt,* $6,500 at 7%. *Expenses,* paid by Ry. *Revenue,* $300. **Management.**—Clerk, J. M. McLane.

22. **DODGE CITY,** *Ford Co.* (Pop., 996—1,763.) Built in '87 by Dodge City Water Supply Co. **Supply.**—Source not given, pumping to tank. *Mains,* 5 miles. *Hydrants,* 50. *Revenue,* city, $2,500. **Management.**—(In '88.) Prest. and Treas., F. W. Gilbert. Secy., E. E. Soule. Supt. F. A. Heineke.

23. **DOWNS,** *Osborne Co.* (Pop., 465—938.) Built in '88 by village. **Supply.**—Eight Wagner wells, pumping to stand-pipe. Fiscal year closes Dec. 31. **Distribution.**—*Mains,* 4 miles. *Taps,* 80. *Hydrants,* 20. *Consumption,* 14,000 galls. **Financial.**—*Cost,* $35,000. *Bonded Debt,* $30,000 at 6%. *Expenses:* operating, $1,100; int., $1,800; total, $2,900. *Revenue,* consumers, $275. **Management.**—Supt., Alfred Jackson.

24. **EL DORADO,** *Butler Co.* (Pop., 1,411—3,339.) Built in '87 by El Dorado Water Supply Co. **Supply.**—Well, pumping to stand-pipe and direct. **Distribution.**—(In '88.) *Mains,* 8 miles. *Taps,* 200. *Meters,* 6. *Hydrants,* 37. **Financial.**—(In '88.) *Cap. Stock,* $100,000. *Bonded Debt,* $60,000 at 6%. *Revenue:* consumers, $3,600; city, $3,350; total, $6,950. **Management.**—(In '89.) Prest. J. F. Thompson. Secy., C. H. Baldwin. Treas. and Supt., C. L. Turner.

25. **ELLSWORTH,** *Ellsworth Co.* (Pop., 929—1,620.) Built in '86 by town. **Supply.**—Driven wells, pumping to tank. **Distribution.**—(In '88.) *Mains,* 7 miles. *Taps,* 160. *Meters,* 0. *Hydrants,* 62. *Consumption,* 250,000 galls. **Financial.**—(In

FOR FURTHER DESCRIPTIVE MATTER see Manual for 1889-90. CHANGES since 1890 are here given. POPULATIONS are for 1880 and 1890, respectively.

'26.) *Cost*, $50,000. *Bonded Debt*, $50,000 at 6%. *Expenses*. operating, $1,500; int.,
$3,000; total. $4,500. *Revenue*, consumers, $1,200. **Management.**—Secy., J. M.
Champion, Supt., C. F. Pohlman.

26. EMPORIA, *Lyon Co.* (Pop., 4,631—7,551.) Built in '80-2 by city. **Supply.**—
Neosho River, pumping to reservoir and direct. Fiscal year closes April 1.

Year	Mains.	Taps.	M't'rs.	H'd'ts.	C'n'p'n.	Cost.	B'd'd d't.	Int. ch'g's.	Exp.	Rev.
'80	11.5	104		47		$62,500			$4,500	
'84	14	400	0	52	unk'n	75,000	$60,000	$4,200	5,000	$6,000
'88	22	740	0	123	$150,000	255,000	225,000	12,350	5,000	7,000
'89	23.75	767	0	123	41,000	268,000	225,000	12,350	6,800	7,000
'90	24.5	850	0	125	50,000	270,000	225,000	12,350	6,800	7,500

Financial.—*Bonded Debt*, $225,000: $50,000 at 7%, due in '99; $160,000 at 5%, due in 1907;
$15,000 at 5%, due in 1910. *Sinking Fund*, to be started in '91. *Expenses*, total, $19,-
150. *Revenue:* consumers, $7,500; city, $6,000; total, $13,500. **Management.**—Com.
of three appointed every year from city council by its members. Chn., D. M. Davis.
Secy., H. S. Alexander. Supt., W. O. Phillips.

27. EUREKA, *Greenwood Co.* (Pop., 1,127—2,259.) Built in '88-9 by city. **Supply.**—Fall River, pumping to stand-pipe; filtered through sand, charcoal and gravel.
Fiscal year closes April 1.

Year	M'ns.	Taps.	Met'rs.	Hyd'ts.	Cons'pt'n.	Cost.	B'd'd debt.	Int. ch'g's.	Exp.	Rev.
'89-90	6	60	0	32	100,000	$50,000	$40,000	$2,000	$793	$800
'90-1	6	110	0	36	500,000	52,000	40,000	2,000	900	904

Financial.—*Total Debt*, $42,000: floating, $2,000 at 6%; bonded, $40,000 at 5%, due in
1920 or 1905 at option of city. *Expenses*, total, $2,900. *Revenue:* consumers, $804;
city, none. **Management.**—City Council. Prest., A. R. Ingalls. Secy. and Supt.,
F. A. Watson.

28. FLORENCE, *Marion Co.* (Pop., 954—1,229.) Built in '87-8 by Florence Wate
Supply Co.; now owned by E. H. Queen, St. Louis, Mo. **Supply.**—Cotton River,
pumping to stand-pipe and direct; filtered. Fiscal year closes April 1. **Distribution.**
—*Mains*, 3.3 miles. *Taps*, 28. *Hydrants*, 25. **Financial.**—*Cost*, $42,000. *Expenses:*
operating, $1,100; taxes, $173. *Revenue:* consumers, $325; fire protection, $1,200; other
purposes, $1,050; total, $2,575. **Management.**—Supt., D. Waterman.

29. FORT SCOTT, *Bourbon Co.* (Pop., 5,372—11,946.) Built in '82 by Fort Scott
Water Co. **Supply.**—Marmaton River, pumping from two impounding reservoirs
to stand-pipe and direct. **Distribution.**—*Mains*, 24 miles. *Taps*, 1,500. *Meters*, 50.
Hydrants, 80. *Consumption*, 500,000 galls. **Financial.**—*Expenses*, about $4,200.
Revenue, about $18,000. **Management.**—Prest., Alger G. Black. Secy., J. W. Maguire. Both of New York City. Supt., R. S. Worthington.

30. FREDONIA, *Wilson Co.* (Pop., 923—1,515.) Built in '87-8 by Jas. O'Neil under name of Fredonia Water Co. **Supply.**—Wells, near Fall River, pumping to
reservoir. Fiscal year closes July 15. **Distribution.**—*Mains*, 7 miles. *Hydrants*,
50. *Consumption*, 230,000 galls. **Financial.**—*Cost*, $88,000. *Bonded Debt*, $75,000 at
6%. *Expenses:* operating, $1,800; int., $4,500; total, $6,300. *Revenue:* consumers,
$6,300; city, $3,120; total, $9,420. **Management.**—Prest., R. T. Race. Treas. and
Supt., A. S. Race. Secy., B. B. Tuttle.

31. GALENA, *Cherokee Co.* (Pop., 1,463—2,496.) Built in '89 by Galena Water
Co., under 20-yrs. franchise. **Supply.**—Shoal River, pumping to stand-pipe and
direct. Fiscal year closes first Tuesday in March. **Distribution.**—*Mains*, 6 miles.
Taps, 500. *Meters*, 6. *Hydrants*, 40. **Financial.**—*Cost*, $65,589. *Bonded Debt*,
$40,000 at 6%. *Expenses :* operating, $1,897; int., $2,400; total, $4,297. *Revenue :* consumers, $3,414; city, $2,500; total, $6,914. **Management.**—Prest., W. B. Stone.
Secy., A. H. McCormick. Supt., W. J. Land.

32. GARDEN CITY, *Finney Co.* (Pop., in '90, 1,490.) Built in '87 by city.
Supply.—Surface well, pumping to stand-pipe. Fiscal year closes April 15. **Distribution.**—*Mains*, 4 miles. *Taps*, 150. *Hydrants*, 42. *Consumption*, 250,000 galls.
Financial.—*Cost*, $46,000. *Bonded Debt*, $20,000 at 6%. *Expenses:* operating, $2,000;
int., $1,800; total, $3,800. *Revenue*, consumers, $1,600. **Management.**—Chn., Andrew Sabine. Secy., Arthur Powers. Supt., W. H. Trull.

33. GARNETT, *Anderson Co.* (Pop., 1,389—2,191). History.—Built in '90 by
city; put in operation Jan. 1, '91. Engr. and Contr., Geo. W. Sturtevant, Bushnell,
Ill. **Supply.**—Crystal Lake, pumping to stand-pipe; filtered through gravel, charcoal and sponge. **Pumping Machinery.**—Dy. cap., 750,000 galls.; Barr dup.

Stand-Pipe.—Cap., 12,000 galls., 12 X 50 ft., on 72-ft. brick tower. Distribution.—*Mains*, 12 to 4-in. c. l., 4.5 miles. *Services*, galv. i., lead. *Hydrants*, Empire, 30. *Valves*, Scott. *Pressure:* ordinary, 50 lbs; fire, 70 to 80 lbs. **Financial.**—*Cost*, $32,000. *Bonded Debt*, $30,000 at 6%. **Management.**—Supt., H. A. Cleavland.

34. GIRARD, *Crawford Co.* (Pop., 1,289—2,541.) Built in '82 by city. **Supply.**—Artesian well, pumping to tank. **Distribution.**—(In '84.) *Mains*, 1 mile. *Taps*, 35. *Meters*, 0. *Hydrants*, 4. **Financial.**—(In '84.) *Cost*, $10,000. *Debt*, $8,000. Int., 7%. *Expenses*, operating, $150. *Revenue*, consumers, $600. **Management.**—City Clerk, Austin Hawley.
NOTE.—Works are in dilapidated condition, though still in operation. Montauk Construction Co. failed to construct system mentioned in 1889-90 Manual. City is in need of new works and is ready to grant franchise on favorable terms.

35. GREAT BEND, *Barton Co.* (Pop., 1,071—2,450.) Built in '87-8 by Great Bend Water Supply Co.; city has right to purchase works in '97 and must purchase them in 1907 at a value determined by disinterested parties: maximum rates are fixed in franchise. **Supply.**—Wells, pumping to stand-pipe. Co. is to be reorganized in '91 and extensive improvements made. **Distribution.**—*Mains*, 4.8 miles. *Taps*, 118. *Hydrants*, 50. *Consumption*, 100,000 galls. **Financial.**—*Cap. Stock*, $100,000. *Bonded Debt*, $40,0.0 at 6%. *Expenses:* operating, $2,000; Int., $2,400; taxes, $400; total, $4,800. *Revenue:* consumers, $1,700; fire protection, $2,500; total, $4,200. **Management.**—Prest., E. B. Smith. Secy., F. E. Smith. Treas., A. G. Farr, Chicago. Supt., J. A. Pritchard.

36. GREENLEAF, *Washington Co.* (Pop., 316—316.) Built in '86 by city. **Supply.**—Eight Wagner wells, pumping to stand-pipe. Two miles of pipe supply about 100 families, Ry. and 13 hydrants. Cost, $25,000. Address City Clerk.

37. GREENSBURGH, *Kiowa Co.* (Pop., in '90, 515.) Built in '87-8 by Greensburgh Water Supply & Hydraulic Power Co., un ler 30-yrs. franchise; now owned by Kansas National Bank, Wichita, which purchased system at foreclosure sale. **Supply.**—Wells, pumping to tank and direct. Cost, $33.000. Address H. W. Lewis, Prest. National Bank, Wichita.

38. HARPER, *Harper Co.* (Pop., in '90, 1,579.) Built in '83-8 by Harper Water Co., under 21-yrs. franchise. **Supply.**—Artesian wells, pumping to stand-pipe. Fiscal year closes June 11. **Distribution.**—(June 11, '89.) *Mains*, 5.5 miles. *Taps*, 50. *Meters*, 1. *Hydrants*, 64. *Consumption*, 260,000 galls. **Financial.**—(In '89.) *Cost*, $100,000. *Cap. Stock*, $150,000. *Bonded Debt*, $100,000. *Operating Expenses*, $2,000. *Revenue*, consumers, $100. **Management.**—Supt., W. J. Foster.

39. HIAWATHA, *Brown Co.* (Pop., 1,375—2,486.) Built in '87-8 by town. **Supply.**—System of small surface wells, pumping to reservoir. Fiscal year closes Jan. 1. Improvements costing $7,000 made in '90-1. Previously works were unsatisfactory, owing to failure of water supply. **Distribution.**—*Mains*, 5.5 miles. *Taps*, 75. *Meters*, 8. *Hydrants*, 42. **Financial.**—*Cost*, $57,000. *Bonded Debt*, $50,000 at 6%, due in 1917. **Management.**—Com. of three, appointed from council by mayor every year. Chn., M. L. Guelich. Report by S. H. Lawrence.

40. HUTCHINSON, *Reno Co.* (Pop., 1,540—8,682.) Built in '86 by Water, Light & Telephone Co.; franchise provides that city may purchase works in '95 at its appraised value; if not purchased, franchise is extended indefinitely. Maximum rates are fixed in franchise. **Supply.**—Driven wells, in conjunction with large dug well, pumping direct. Fiscal year closes Dec. 31.

Year.	Mains.	Taps.	Meters.	Hydrants.	Cons'pt'n.	B'd'd d'bt.	Int. ch'g's.	Rev.
'88	9.5	347	11	111	425,000	$200.00	$12,000	$13,521
'89	14.6	124	35	174	......	200,000	12.000	18,055
'90	16	457	50	186	714,848	200,000	12,000	20,367

Financial.—*Cap. Stock*, $200,000. *Revenue:* consumers, $8,367; city, $12,000; total, $20,367. **Management.**—Prest., L. A. Beebe. Secy., J. R. Swigart. Treas., L. A. Bigger. Supt., S. C. Bennett.

41. INDEPENDENCE, *Montgomery Co.* (Pop., 2,915—3,127.) Built in '85 by Independence Water Co., under 20-yrs. franchise. **Supply.**—Verdigris River, pumping to tank and direct. L. C. Mason reported in July, '90, that co. intended to put

---

FOR FURTHER DESCRIPTIVE MATTER see Manual for 1889-90. CHANGES since 1890 are here given. POPULATIONS are for 1880 and 1890, respectively.

in mechanical filter, also new pumps. In '91 reported that contract for 500,000-gall. American filter had been let. Feb. 10, '91, it was reported that works had passed into hands of bondholders. Distribution.—(In '88.) *Mains*, 6.8 mile٠. *Hydrants*, 13. Financial.—(In '88.) *Cost*, $35,000. Management.—Not given.

42. **JUNCTION CITY,** *Geary Co.* (Pop., 2,684—4,502.) Built in '87 by town. Supply.—Wells, pumping to reservoir. Fiscal year closes March 15. Distribution. —*Mains*, 7.6 miles. *Taps*, 226. *Meters*, 1. *Hydrants*, 64. *Consumption*, 75,000 galls. Financial.—*Cost*, $65,000. *Total Debt*, $64,000: floating, $14,000 at 10%; bonded, $50,000 at 6%, due in 1917. *Expenses:* operating, $1,500; int., $3,000; total, $4,500. *Revenue*, consumers, $1,160. Management.—Committee, appointed by mayor and council. Chn., A. L. Barnes. Supt., C. T. Wisler.

KANSAS CITY, *Wyandotte Co.* (Pop., 3,200—38,316; since '80, Wyandotte City with pop., in '80, of 6 149, has been annexed to Kansas City.) Supply from Kansas City, Mo.

43. **KINGMAN,** *Kingman Co.* (Pop., 970—2,390.) Built in '86-7 by Kingman Water & Power Co., under 30-yrs. franchise. Supply.—Sixty 2-in. driven wells, pumping to tank. Fiscal year closes Jan. 1. Distribution.—*Mains*, 4.5 miles. *Taps*, 125. *Hydrants*, 47. *Consumption*, 50,000 galls. Financial.—*Cost*, $42,500. *Revenue*, $4,500. Management.—Supt., R. A. Munson.
NOTE.—During '90, works were in the hands of receiver. Feb. 10, '90, co. was being organized with paid-up cap. stock of $150,000, known as Kingman Water & Light Co. Officers had not been elected at time of report.

44. **KINSLEY,** *Edwards Co.* (Pop., 457—771.) Built in '87-8 by town. Supply. —Well, pumping to stand-pipe. Distribution.—(In '88.) *Mains*. 5 miles. *Taps*, 75. *Meters*, 0. *Hydrants*, 25. Financial.—(In '88.) *Cost*, $40,000. *Bonded Debt*, $40,000 at 6%. *Revenue*, consumers, $700. Management.—Supt., R. G. Stearns.

45. **LARNED,** *Pawnee Co.* (Pop., 1,036—1,861.) Built in '87 by R. L. Walker & Co., Topeka, Kan., under 21 yrs. franchise, which establishes rates. City has right to purchase works in '96, and at close of every 5 yrs. thereafter, one appraiser to be appointed by city, one by co., and third by these two. Supply.—Driven wells, pumping to stand-pipe. In '90, pumping station was removed from Pawnee to Arkansas River, and system of driven wells put in, to secure better water supply. Distribution.—*Mains*, 7.5 miles. *Taps*. 200. *Meters*, 7. *Hydrants*, 80. *Consumption*, 700,000 galls. Financial.—Withheld. Management.—Prest., Thos. G. Fitch, Wichita. Supt., J. J. Nesbit.

46. **LAWRENCE,** *Douglas Co.* (Pop., 8,510-9,937.)

KANSAS WATER & LIGHT CO.

Organized in '89 to operate works at Lawrence and Paola; controlled by Debenture Guarantee & Assurance Co., Chicago, Ill. Fiscal year closes Dec. 31. Distribution.—Given separately under Lawrence and Paola. Financial.—*Cost*, $350,000. *Cap. Stock*, paid up, $300,000. *Bonded Debt*, $255,000 at 6%. *Sinking Fund*, 2% after 10 yrs. *Expenses:* operating and taxes, $5,490; int., $15,300; total, $20 70). *Revenue:* consumers, $15,135; Ry., $3,844; city, $9,112; total, $28,091. Management.—Prest., Chas. Whann. Secy. and Treas., Chas. B. Ludwig, 74 Broadway, N. Y. City.

LOCAL PLANT.

Built in '86 by City Water Co., owned and operated by Kansas Water & Light Co., which see above. Supply.—Kansas River, pumping to stand-pipe. During '90 three settling basins were built, having cap. of 6,000,000 galls.; Engr., Geo, H. Pierson, N. Y. City. Distribution.—*Mains*, 18 miles. *Taps*, 516. *Meters*, 21, *Hydrants*, 150. *Consumption*, 250,000 galls. Financial and Management.—See Kansas Water & Light Co., above. Supt., O. J. Woodard.

47. **LEAVENWORTH,** *Leavenworth Co.* (Pop., 16,546—19,768.) Built in '82 by Leavenworth City & Fort Leavenworth Water Co., under 20-yrs. franchise. Supply.—Missouri River, pumping to settling basins and repumping to reservoir. A 2,000,000-gall. settling reservoir was built in '90.

1Year.	Mains.	Taps.	M't'rs.	H'd'ts.	B'd' d't.	Int. ch'g's.	Exp.	Rev.
'84	23	817	208	114	$324,000	$19,440	$11,200	$33,783
'88	25	1,210	337	116	600,000	36 000	16,000	50,825
'90	25	1,250	300	116	......	......	......	*8,700

*From city.

Financial.—Withheld for '90, except hydrant rental, $75. Management.—Prest., L. T. Smith. Secy.; W. T. Hewitt. Supt., S. Hastings.

48. McPHERSON, *McPherson Co.* (Pop., 1,590—3,172.) Purchased and enlarged in '90 by Western Water & Electric Co.; built in '83 by city **Supply.**—8-in. driven wells, pumping to stand-pipe. **Pumping Machinery.**—Dy. cap., 690,000 galls.; Cook deep-well. **Stand-Pipe.**—Cap., 93,000 galls.; 12 × 110 ft.; from Porter Mfg. Co. **Distribution.**—*Mains,* 10 to 4-in., c. i., 7 miles. *Services,* w. i. Pipe and specials from Ripley & Bronson. *Taps,* 250. *Meters,* 1. *Hydrants,* 55. *Consumption,* 67,000 galls. *Pressure:* ordinary, 30 lbs.; fire, 80 lbs. **Financial.**—*Cost,* $125,000. *Cap. Stock,* authorized, $75,000; all paid-up. *Bonded Debt,* $50,000 at 6%. *Expenses:* operating, $5,000; int., $3,000; total, $8,000. Preceding figures include both water and lighting plants. *Revenue,* consumers, $3,000; city, $3,500; total, $6,500. **Management.**—Prest. and Supt., Jas. B. Danah. Treas., B. A. Allison. Secy., C. R. Wallace.

49. MANHATTAN, *Riley Co.* (Pop., 2,105—3,004.) Built in '87-8 by city; first customer supplied Feb. 15, '88. **Supply.**—Twelve 4-in. Cook wells, driven 20 ft. in bed of Blue River, pumping to reservoir; wells distributed over area of 40 × 120 ft. During summer of '90, pumping engines were lowered 12 ft. and wells were sunk. A 10-in. c. i. suction pipe was laid from wells to pump, 418 ft., at 4 ft. below low water level. Cost of improvements, $2,350.

Year.	Mains.	Taps.	Meters.	Hyd'ts.	Cons'pt'n.	Cost.	B'd't d't.	Int. ch'gs.	Exp.	Rev.
'89	8	143	1	49	101,000	$68,173	$70,000	$4,200	$2,140	$2,041
'90	8	154	1	49	209,315	70,620	70,000	4,200	2,435	2,860

Financi.l.—*Bonded Debt,* $70,000, at 6%: $50,000 due in 1908; $20,000 due in 1909. Bonds are sold at sufficient premium to cover cost of construction. *Expenses,* total, $6,635. *Revenue:* consumers, $2,860; city, none. **Management.**—Com. of three, appointed by City Council from its members for one year. Chm., W. C. Johnston. City Clerk, C. E. Bowen. Supt., Geo. E. Hopper.

50. MARION, *Marion Co.* (Pop., 857—2,047.) Built in '87 by Marion Water Supply, Electric Light & Power Co., under 21-yrs. franchise, when city has right to purchase works at appraised value, giving city bonds at 5% for 20 yrs.; city does not control rates. **Supply.**—Well, pumping to stand-pipe and direct. Fiscal year closes Aug. 1. **Distribution.**—*Mains,* 5 miles. *Taps,* 65. *Hydrants,* 50. *Consumption,* 10,000 galls. **Financial.**—*Cost* (in '8s), $80,000. *Cap. Stock,* authorized, $100,000; paid no div'd in 1890. *Total Debt,* $72,500: floating, $2,500; bonded, $70,000 at 6%. *Expenses:* operating, $3,500; int., $4,200; taxes, $530; total, $8,230. *Revenue:* consumers, $1,800; fire protection, $2,500; other purposes, $1,700; total, $6,000. **Management.**—Prest., J. F. Thompson. Secy. and Treas., F. E. Palmer. Supt., Taylor Riddle.

51. MARYSVILLE, *Marshall Co.* (Pop., 1 249—1,913.) Built in '89 by Marysville Water Co., under 20-yrs. franchise. **Supply.**—Thirty-six driven wells, near Blue River, pumping to stand-pipe. **Distribution.**—*Mains,* 4.5 miles. *Meters,* 45. *Hydrants,* 50. **Financial.**—*Cost,* $50,000. *Bonded Debt,* $42,000 at 6%. *Revenue,* consumers, $2,500. **Management.**—Prest., W. J. Sherman, Dallas, Tex. Supt., E. L. Miller.

52. MEDICINE LODGE, *Barber Co.* (Pop., 373—1,095.) Built in 89-'90 by city. Operated by Eldred & Hinman. **Supply.**—Spring, pumping direct; filtered. **Pumping Machinery.**—Dy. cap., 1,250,000 galls.; two Knowles dup. **Distribution.**—*Mains,* 8 to 4 in. c. i., 3 miles. *Taps,* 68. *Hydrants,* Ludlow, 21. *Pressure:* ordinary, 40 lbs.; fire, 120 lbs. **Financial.**—*Cost,* $35,000. *Bonded Debt,* $35,000. *Expenses,* operating, $1,500. *Revenue:* consumers, $1,000; city, none. **Management.** —Lessees, Eldred & Hinman. Engr., D. J. Aber.

53. MINNEAPOLIS, *Ottawa Co.* (Pop., 1,084—1,756.) Built in '87 by Minneapolis Water Co., under 20-yrs. franchise. **Supply.**—Wells, connected by tunnel, pumping direct. Fiscal year closes Dec. 31 **Distribution.**—*Mains,* 4.5 miles. *Taps,* 197. *Meters,* 28. *Hydrants,* 51. *Consumption,* 52,460 galls. **Financial.**—*Cap. Stock,* authorized, $100,000. *Bonded Debt,* $63,000 at 6%. These figures include both water and electric lighting plants. **Management.**—Prest., A. B. Turner, Boston, Mass. Secy., Adolph Gilbert. Treas., J. W. Brown, Boston, Mass. Supt., W. F. Grover.

54. NEWTON, *Harvey Co.* (Pop., 2,601—5,605.) Built in '84 by Newton Water Co., under perpetual franchise, unless city buys works. **Supply.**—Two batteries

---

FOR FURTHER DESCRIPTIVE MATTER see Manual for 1889-90 CHANGES since 1890 are here given. POPULATIONS are for 1880 and 1890, respectively.

Wagner wells, pumping to stand-pipe. Distribution,—(In '88.) *Mains,* 13 miles. *Hydrants,* 112. *Consumption,* 500,000 galls. Financial.—(In '83.) *Cost,* $210,000. Management—Secy., W. J. Graham, Supt., John E. Ford. NOTE.—Co. declined to give '90 statistics.

55. **NORTON,** *Norton Co.* (Pop., 634—1,074.) Built in '88 by town. Supply.— Eight Wagner wells, pumping to stand-pipe and direct. Fiscal year closes March 1. Distribution.—*Mains,* 5 miles. *Taps,* 200. *Meter,* 1. *Hydrants,* 15. *Consumption,* 120,000 galls. Financial.—*Cost,* $38,000, *Bonded Debt,* $38,000: $35,000 at 6%; $3,000 at $10. *Expenses;* operating, $1,426; Int., $2,400; total, $3,826. *Revenue:* consumers, $1,263; city, $262; total, $1,525. Management.—Supt., L. C. Hawlin.

56. **OBERLIN,** *Decatur Co.* (Pop. of township, 1,196—1,243.) Built in '88 by town. Supply.—Tube wells, pumping to stand-pipe and direct. Fiscal year closes June 30. Distribution.—*Mains,* 3.5 miles. *Taps,* 40. *Hydrants,* 23. *Consumption,* 60,000 galls. Financial.—*Cost,* $32,000. *Total Debt,* $32,000: floating, $2,030 at 6%; bonded, $30,000 at 6%, due in 1908. *Expenses;* operating, $2,000; int., $1,800; total, $3,800. *Revenue:* consumers, $450; city, none Management.—Mayor and Council elected every year. Mayor, S. W. McElroy. City Clerk, J. C. Lathrop. Report by Geo. W. Keys.

57. **OLATHE,** *Johnson Co.* (Pop., 2,285—3,294.) Built in '81 by town. Supply. —Surface water, gathered in impounding reservoir; Hyatt filter. Fiscal year closes May 1. Distribution.—*Mains,* 5 miles. *Meter,* 1. *Hydrants,* 49. Financial.— *Bonded Debt,* $37,000 at 7%. *Expenses,* operating $1,200. *Revenue,* consumers, $1.500. Management.—Com. of three, appointed every two years by mayor. Chn., Geo. H. Beach. Engr., Jno. Pitts.

58. **OSWEGO,** *Labette Co.* (Pop., 2,351—2.574.) Built in '87 by Oswego Water Supply Co., under 21-yrs. franchise. Supply.—Neosho River, pumping to stand-pipe; filtered through charcoal and gravel. In '88 system was in hands of receiver, but in '90 co. was re-organized. Distribution.—*Mains,* 6 miles. *Taps,* 125. *Hydrants,* 51. Financial.—*Cost,* $60,000. *Cap. Stock,* $100,000. *Bonded Debt,* $40,000 at 6%. *Expenses:* operating, $2,000; int., $2,400; total, $4,400. *Revenue,* city, $2,530. Management.—Prest., J. F. Thompson, Pine Bluff, Ark. Secy, Geo. Rogers. Supt., Carpenter Harrison.

59. **OTTAWA,** *Franklin Co.* (Pop., 4,052—6,248.) Built in '86 by C. T. Ewing, Thayer. Rates are subject to approval of city, but it has no right to buy works. Supply.—Marais de Cygnes River and well, pumping to stand-pipe. Fiscal year closes Dec. 31.

Year.	Mains.	Taps.	Mtrs.	H'd'ts.	C'ns'pt'n.	Cost.	B'd'd d't.	Int. ch'g's.	Exp.	Rev.
'86	9.5	50	1	80	.....	$100,000	$80,000	$4,800	.....	.....
'89	9.9	358	5	78	213,000	110,519	80,000	4,800	$429	$7,424
'90	9.9	382	9	79	249,000	110,577	80,000	4,800	516	8,099

Financial.—*No Cap. Stock. Bonded Debt,* $80,000 at 6%, due in 1906. *Expenses,* total, $5,316. *Revenue:* consumers. $5,084; fire protection, $3,015; total, $8,099. Management.—Owner, C. T. Ewing. Supt., W. Mitchell.

60. **PAOLA,** *Miami Co.* (Pop., 2,312—2,943.) Built in '85-6 by Paola Water Co., under 20 yrs. franchise, which provides for purchase of works by city at expiration of this period. Works are operated by Kansas Water & Light Co., Lawrence, which see. Supply.—Bull Creek, pumping and direct. Fiscal year closes Dec. 31. During 1890 a masonry dam was built across Mud Creek; Engr., Geo. H. Pierson, N. Y. City. Distribution.—*Mains,* 6 miles. *Taps,* 381, *Hydrants,* 57. *Consumption,* 160,000 galls. Financial and Management.—See Kansas Water & Light Co., Lawrence. Supt., E. W. Robinson.

61. **PARSONS,** *Labette Co.* (Pop., 4,199—6,736.) Built in '83 by Parsons Water Co. City has right to purchase works at intervals of five years ; rates are regulated by city ordinance. Supply.—Big Labette Creek, pumping to tank and direct; filtered. Fiscal year closes Dec. 31.

Year.	Mains.	Taps.	Mt'rs.	H'd'ts.	C'ns'pt'n.	Cost.	B'd debt.	Int. ch'g's.	Exp.	Rev.
'87	7	200	6	32	300,000	$100,000	$65,000	$4,200	$2,000	$7,500
'89	7.9	260	14	54	300,000	135,000	65,000	4,200	2,990	8,276
*'90	8.1	282	15	56	320,000	137,000	65,000	4,200	3,310	9,172

Financial.—*Cap. Stock:* authorized, $100,000; all paid up. *Total Debt,* $65,400: floating, $400; bonded, $50,000 at 6%; $15,000 at 8%. *Expenses:* operating, $3,145; int., $4,- 200; taxes, $165; total, $7,510. *Revenue:* consumers, $6,007; fire protection, $3,175;

total, $9,172. Management.–Prest., C. H. Kimball. Treas., E. H. Edwards. Secy. and Supt., W. K. Hayes.

NOTE.—In '91 co. was reorganized under name of Parsons Water Supply Co., with intention of rebuilding works. City refused to make necessary concessions in franchise, and nothing further had been done Sept. 25.

62. **PEABODY,** *Marion Co.* (Pop., 1,087—1,474.) Built in '86 by town. Supply.—Wells, pumping to tank and direct. Distribution.—*Mains,* 3 miles. *Taps,* 42. *Hydrants,* 25. *Consumption,* 18,000 galls. Financial.—*Cost,* $19,000. *Bonded Debt,* $15,000 at 7%. *Expenses:* operating, $200; int, $1,050; total, $1,250. *Revenue,* consumers, $300. Management.—Supt., E. F. Davidson. Engr, S. L. Morrell.

63. **PHILLIPSBURGH,** *Phillips Co.* (Pop., 309—992.) History.—Built in '90 by city; put in operation Oct. 12. Engr., S. K. Felton, Omaha. Supply.—Eigth wells, pumping to tank and direct. Pumping Machinery.—Dy. cap., 250,000 galls.; two Deane. Tank.—Cap., 58,600 galls. Distribution.—*Mains,* Kal. 1., 5 miles. *Services,* lead and I. Pipe and specials from National Tube Co., St. Louis, Mo. *Taps,* 50. *Hydrants,* Holly, 25. *Valves,* Ludlow. *Consumption,* 28,800 galls. *Pressure:* ordinary, 40 lbs.; fire, 100 lbs. Financial.—*Cost,* $23,000. *Bonded Debt,* $19,000 at 6%. *Expenses,* operating, $800; int., $1,140; total, $1,940. *Revenue,* consumers, $400. Management.—Supt., E. E. Brainerd.

64. **PITTSBURGH,** *Crawford Co.* (Pop., 624—6,697.) Built in '85 by Pittsburgh Water & Irrigating Co., under 21-yrs. franchise, which regulates rates. Supply.—Artesian wells, pumping to tank and direct. Fiscal year closes Dec. 31. Following additions were made in '90 : Supply, two wells. *Pumping Machinery,* one 1,000,000-gall., and one 1,500,000-gall. pump. *Boilers,* three. Distribution.—*Mains,* 13 miles. *Meters,* 30. *Hydrants,* 100. Financial.—Withheld. Management.—Prest., Frank Playter. Treas. and Supt , J. A. Bendure.

65. **PLEASANTON,** *Linn Co.* (Pop., 709—1,139.) Built in '84 by town. Supply.—Artificial lake, pumping to reservoir; filtered through sand and gravel. Fiscal year closes April 1. Distribution.—*Mains,* 3 miles. *Taps,* 65. *Hydrants,* 28. *Consumption,* 50,000 galls. Financial.—*Cost,* $17,000. *Bonded Debt,* $14,000: $9,000 at 7%; $5,'00 at 6%; due in annual payments of $1,000. *Expenses;* operating, $100; int , $930; total, $1,030. *Revenue,* consumers, $700. Management.—Supt. and Marshall under direction of com. of two appointed by mayor from council. Report by S. H. Allen. City Clerk, B. F. Blakie.

66. **PRATT,** *Pratt Co.* (Pop., in '90, 1,418.) Built in '86 by city. Address S. D. Gephart.

67. **SALINA,** *Saline Co.* (Pop., 3,111—6,149.) Built in '83 by Salina Water-Works Co., under 20-yrs. franchise which fixes rates. Supply.—Wells, pumping direct. Fiscal year closes Dec. 31. In winter of '89-90, four Wagner steamed wells were added.

Year.	Mains.	Taps.	Meters.	Hydrants.	Consumption	B'd'd debt.	Int. ch'g's.
'84	5	100	2	48	....	$60,000	$3,600
'86	5.5	205	0	55	....	80,000	4,800
'88	8.5	357	17	64	150,000	92,000	5,520
'90	8.8	395	33	66	265,000	92,000	5,520

Financial.—*Cap. Stock:* authorized, $100,000; paid up, $35,000. *Bonded Debt,* $92,-000, at 6%, due in 1,903. Management.—Prest., C. H. Payson.

68. **SMITH CENTRE,** *Smith Co.* (Pop., 254—767.) Built in '89-90 by city. Supply.—Beaver Creek, pumping direct and to stand-pipe; filtered. Fiscal year closes April 30. In '90 cap. of reservoir was increased. Distribution.—*Mains,* 9 miles. *Taps,* 73. *Hydrants,* 20. *Consumption,* 60,000 galls. Financial.—*Cost,* $28,000. *Total Debt,* $29,000; floating, $4,000; bonded, $25,000 at 6%, due in 1,909. *Expenses,* total, $4,260. *Revenue:* consumers, $144; city, none. Management.—Mayor, D. M. Relihan. City Clerk, A. S. Kingsbury. Supt., Paul Arnold.

69. **SPRINGFIELD,** *Seward Co.* (Pop., in '90, 347.) Bought in '89 by city; built in '88-9 by Springfield Light, Heat & Water Co. Address W. S. Allen.

70. **STERLING,** *Rice Co.* (Pop., 1,011—1,641.) Built in '87 by city. Supply.—38 driven wells, pumping direct. Fiscal year closes Aug. 1. Financial.—*Bonded Debt,* $42,000 at 6%. *Sinking Fund,* $7,000. *Expenses:* total, $4,620. *Revenue:* consumers, $950; city, none. Management.—Comrs., appointed every year by city council. Chn., J. T. Gaskill, mayor. Secy., Thos. Powers.

FOR FURTHER DESCRIPTIVE MATTER see Manual for 1889-90. CHANGES since 1890 are here given. POPULATIONS are for 1880 and 1890, respectively.

71. STOCKTON, *Rooks Co.* (Pop., 111—880.) Built in '87 by town. Supply.—Wells, pumping to stand-pipe and direct. Supply was improved in '90, by sinking one well and by lowering pumps 10 ft. Distribution.—*Mains*, 3 miles. *Taps*, 85. *Hydrants*, 75. Financial.—*Cost*, $40,000. *Bonded Debt*, $33,000 at 5%. Management.—Mayor, Jacob Hendricks. City Clerk, F. A Chipman.

72. TOPEKA, *Shawnee Co.* (Pop., 15,452—31,007.) Built in '81-2 by Topeka Water Co; rebuilt in '90: now controlled by Debenture Guarantee & Assurance Co., Chicago. Ill.; franchise provides for purchase of works by city in 1911. Supply.—Twenty-eight 6-in. wells, driven into a sandbar in Kansas River, pumping direct. Fiscal year closes Dec. 31. In '91, steel stand-pipe, 30 × 180 ft., was under contract. New wells and pumping station and other improvements were under construction, including laying of 20 miles of pipe, and removal of works to point 4 miles further up river. Total cost of improvements, $300,000. Engr., Geo. H. Pierson, N. Y. City.

Year.	Mains.	Taps.	Meters.	Hyd'ts.	Cons'pt'n.	Cost.	B'd d't.Int.ch'gs.Exp.		Rev.
'82	15	360	4	152	unk'n.	$162,000	$120,000	$7,200	......
'84	18	734	8	171	400,000	200,000	120,000	7,200	9,500 23,550
'86	28	1,450	60	258	800,000	200,000	120,000	7,200	...... 12,160*
'90	40	1,585	97	309	1,600,000	......	750,000	45,000	11,111 70,960

*From city.

Financial.—*Capital Stock*, authorized, $1,000,000. Bonded debt, $750,000 at 6%. *Expenses:* operating and taxes, $11,111; int., $45,000; total, $56,111. *Revenue:* consumers, $55,999; city, $14,961; total, $70,960. Management.—Prest., J. R. Mulvane. Secy. and Treas., B. A. May. Supt., Jesse Shaw.

73. UDALL, *Cowley Co.* (Pop., 85—338.) Built in '88 by village. Supply.—Well, pumping to tank. Distribution.—*Mains*, 2.5 miles. *Taps*, 16. *Meters*, 4. *Hydrants*, 7. Financial.—*Cost*, $7,000 *Bonded Debt*, $6,500 at 6%. Management.—Chn., J. H. Hildebrand. Secy., B. F. Baker. Report by F. M. Ammon.

74. VALLEY FALLS, *Jefferson Co.* (Pop., 1,016—1,180.) Built in '88 by city. Supply.—Wells, pumping to tank. Distribution.—*Mains*, 2 miles. *Taps*, 25. *Hydrants*, 15. Financial.—*Cost*, $15,800. *Bonded Debt*, $15,000 at 5%. Management.—Address City Clerk.

75. WEIR, *Cherokee Co.* (Pop., 376—2,138.) Built in '89 by Weir City Water-Works Co. Supply.—Two artesian wells, pumping to reservoir and direct. Reservoir is 75 ft. in diam., 16 ft. deep, of brick and cem. Distribution.—*Mains*, 2.5 miles. *Taps*, 78. *Hydrants*, 18. Financial.—Withheld. Management.—Prest., B. F. Hobart, Bank of Commerce Building, St. Louis. Mo. Secy., W. P. Heath. Treas., A. B. Cockerill.

76. WELLINGTON, *Sumner Co.* (Pop., 2,694—4,391.) Built in '84-5 by Wellington Water Co., under 20-yrs. franchise. Now owned by former bondholders under name of Wellington Water-Works Co. Supply.—Slate Creek, pumping from settling reservoir to tank, stand-pipe and direct; filtered. During '90 stone dam was built across Slate Creek and a tank erected. Fiscal year closes Dec. 31. Distribution.—*Mains*, 10 to 4-in. c. i., 10.3 miles. *Services*, 1. *Taps*, 340. *Meters*, Crown, Hersey, 33. *Hydrants*, Ludlow, Chapman, 71. *Valves*, Ludlow, Chapman. *Consumption*, 265,331 galls. Financial.—*Cap. Stock:* authorized, $100,000; all paid up. *Expenses:* operating, $3,300; taxes, $784. *Revenue:* consumers, $6,800; fire protection, $4,050, *Hydrant rental*, $60 each for 50, $50 each for additional. Management.—Prest. Chas. Perry, Jr.; Secy., A. Chester; both of Westerly, R. I. Supt., J. M. Thralls.

77. WICHITA, *Sedgwick Co.* (Pop., 4,911—23,853.) Built in '82 by Wichita Water Co.; controlled by American Water-Works & Guarantee Co., Ltd., Pittsburg, Pa., under 21-yrs. franchise, which fixes maximum rates but does not provide for purchase of works by city. Supply.—Ninety-six driven wells, pumping to stand-pipe and direct. Fiscal year closes Dec. 31. During 89-90 a 5,000,000 Holly high duty pump and a 125-HP. boiler were added.

Year.	Mains.	Taps.	Meters.	Hydrants.	Consumption.
'84	8	400	0	60	......
'86	11	630	0	80	......
'88	28	1,500	0	200	2,000,000
'90	34	1,211	33	217	2,400,000

Financial.—Withheld. Management.—Prest., J. H. Purdy. Vice Prest. and Supt., F. C. Amsbary. Secy., A. E. Clancy. Gen. Mgr., W. S. Kuhn, Pittsburg, Pa.

78. WINDOM, *McPherson Co.* (Pop., est., 500.) Built in '88 by city. Supply.—Well, pumping to tank. Distribution.—*Mains*, 1 mile. *Taps*, 9. *Hydrants*, 3.

Consumption, 1,000 galls. **Financial.**—*Cost*, $5,000. *Debt*, $5,900 at 6%. *Revenue*, consumers, $200. **Management.**—Mayor, Royal Matthews. Secy., C. C. Bliss.

79. **WINFIELD,** *Cowley Co.* (Pop., 2,844—5,184.) Built in '83 by Winfield Water Co.; city may purchase works in '93 and at intervals of five yrs. thereafter; rates must be approved by City Council. **Supply.**—Walnut River, pumping to reservoir. Fiscal year closes Feb. 28. In '90 supply was obtained further up river, 1,500 ft. above station, through 16-in. southern pine conduit. Intake can be cleaned by water from reservoir. See note below. **Distribution.**—*Mains*, 9 miles. *Taps*, 400. *Meters*, 12. *Hydrants*, 73. **Financial.**—*Cap, Stock:* authorized, $150,400; paid up, $35,000. *Bonded Debt*, $75,000 at 6%, due in 1903. **Management.**—Prest., Chas. H. Payson. Portland, Me.

NOTE.—For view showing new intake and crib in position to sink to river bottom see *Fire & Water*, March 8, '90.

WYANDOTTE, *Wyandotte Co.* Now part of Kansas City, Kan. Supply from Kansas City, Mo.

NEBRASKA.

Water-Works Completed or in Process of Construction.

1. **AINSWORTH,** *Brown Co.* (Pop., in '90, 733.) **History.**—Built in '90 by village; put in operation Oct. 15. Engrs., Andrews & Burroll ; contrs., Godfrey & Meals; all of Fremont. **Supply.**—Wells, pumping to stand-pipe and direct. **Pumping Machinery.**—Dy. cap., 400,000 galls.; Smedley dup. **Stand-Pipe.**—Cap., 41,000 galls.; 10 × 70 ft. **Distribution**—*Mains*, 6 and 4-in. c. i., 1.8 miles. *Hydrants*, Alaska, 11. *Valves*, Lorain. *Consumption*, 20,000 galls. *Pressure:* ordinary, 37 lbs.; fire, 150 lbs. **Financial.**—*Cost*, $8,692. *Bonded Debt*, $8,000 at 6%.

2. **ALBION,** *Boone Co.* (Pop., 330—926.) Built in '87 by town. **Supply.**—Not given, pumping from reservoir, tank and direct. Fiscal year closes April 30. **Distribution.**—*Mains*, 1.3 miles. *Taps*, 20. *Hydrants*, 9. **Financial.**—*Cost*, $9,500. *Bonded Debt*, $6,500 at 7%. *Expenses*, $100. *Revenue*, $200. **Management**—Chn., C. S. Barnes. Clerk, W. A. Hosford. Supt., E. M. Morehead.

3. **ASHLAND,** *Saunders Co.* (Pop., 978—1,601.) Built in '87-8 by village. Des. Engr., A. A. Richardson, Lincoln. **Supply.**—Two wells, pumping to tank. Fiscal year closes May 1. **Distribution.**—*Mains*, 3.5 miles. *Taps*, 110. *Meters*, 0. *Hydrants*, 32. **Financial.**—*Cost*, $19,000. *Bonded Debt*, $19,000 at 6%. *Expenses:* operating, $850; int., $1,140; total, $1,996. *Revenue*, consumers, $1,000. **Management.**—Comrs., appointed every year by mayor. Supt., A. H. Gould.

4. **ATKINSON,** *Holt Co.* (Pop., in '90, 701.) **History.**—Built in '90 by city; put in operation Jan. 1, '91. Engr., Chas. H. Godfrey. Contrs., Godfrey & Meals, Fremont. **Supply.**—Godfrey system of wells, pumping to stand-pipe. **Pumping Machinery.**—Dy. cap., 500,000 galls.; Smedley. **Stand-Pipe.**—Cap., 44,070 galls.; 10 × 75 ft. **Distribution.**—*Mains*, 8 to 4-in. c. i., ½ mile. *Hydrants*, Alaska, 7. *Valves*, Lorain. *Consumption*, 75,000 galls. *Pressure:* ordinary, 30 lbs.; fire, 100 lbs. **Financial.**—*Cost*, $7,000. **Management.**—Not given.

5. **AURORA,** *Hamilton Co.* (Pop., in '90, 1,862.) Built in '88-9 by town. **Supply.**—Wells, pumping to stand-pipe, reservoir, and direct. Fiscal year closes April 30. During '90-1, a Cook dup. well pump and 50-HP. boiler were added. **Distribution.**—*Mains*, 4.3 miles. *Taps*, 166. *Meters*, 0. *Hydrants*, 28. **Financial.**—*Cost*, $31,000. *Bonded Debt*, $31,000 at 6%, due in 20 yrs. *Expenses:* operating; $1,839; int., $2,040; total, $3,879. *Revenue*, consumers, $2,186. **Management.**—Supt., H. G. Cass.

6. **BEATRICE,** *Gage Co.* (Pop., 2,447—13,836.) Built in '86 by city. **Supply.**—Well and Big Blue River, pumping direct from reservoir. Fiscal year closes March 31. Jewell filter has been discarded. Sept., '91, city contracted with Godfrey & Meals. Fremont, Neb., for a daily supply of 2,000,000 galls. from wells, and for

FOR FURTHER DESCRIPTIVE MATTER see Manual for 1889-90. CHANGES since 1890 are here given. POPULATIONS are for 1880 and 1890, respectively.

moving of pumping station, new 1,500,0.0-gall pump and necessary changes and connections; price, $26,512.

Year.	Mains.	Taps.	M't'rs.	H'd'nts.	C'ns'pt'n.	Cost.	B'd'd d'bt.	Int.ch'gs.	Exp.	Rev.
'87-8	7	200	3	77	500,000	$80,000	$90,000	$4,800	$3,000	$1,200
'90-1	11.2	440	3	101	533,286	93,000	90,000	4,800	6,706	6,385

Financial,—*Total Debt*, $96,300: floating, $6,300 at 7%; bonded, $90,000 at 6%. *Expenses*' total, $11,506. *Revenue:* consumers, $6,385; fire protection, $6,875; total, $13,260. Management.—Comr., E. D. Wheelock, and committee of three: R. S. Bibb, Wm. Gillespie, I. N. Caspers,

7. **BLAIR,** *Washington Co.* (Pop., 1,317—2,069.) Built in '86 by town. Supply.—Thirty tube wells, pumping to reservoir. Fiscal year closes May 1. Distribution—*Mains*, 1 mile. *Taps*, 20. *Meters*, 1. *Hydrants*, 23. Financial.—*Cost*, $21,-000. *Bonded Debt*, $20,000 at 7%. Management.—Comr., E. J. Farr, appointed every year by mayor.

8. **BLUE HILL,** *Webster Co.* (Pop., 138—796.) Built by city in '90-1; put in operation March 1. Des. Engr., A. A. Richardson, Lincoln. Supply.—Well owned by ry. co., pumping to 12 × 110-ft. stand-pipe. *Hydrants*, Adams, 19. *Valves* Lorain. *Cost*, $10,000.

9. **BROKEN BOW,** *Custer Co.* (Pop., in '90, 1,647.) Built in '88-9 by Broken Bow Water-Works Co., under 20-yrs. franchise. Supply.—Ten wells, pumping to reservoir. Fiscal year closes Dec. 31. Distribution.—*Mains*, 3.5 miles. *Taps*, 100. *Hydrants*, 34. *Consumption*, 60,000 galls. Financial.—*Cost*, $40,170 *Cap. Stock:* authorized, $50,000; paid up, $14,500; paid 5% div'd in 1890. *Bonded Debt*, $25,000 at 7%, due 1908. *Expenses:* operating, $2,000; int., $1,750; no taxes; total, $3,750. *Revenue:* consumers, $1,500; fire protection, $2,830; total, $4,330. Management.—Prest., Jno. Ruse. Secy. and Supt., Frank A. Young.

10. **CENTRAL CITY,** *Merrick Co.* (Pop., 648—1,368.) History.—Built in '88-9 by village. Des. Eng., A. A. Richardson, Lincoln. Contr., Ben S. Clarke, Ashland. Supply.—Four wells, pumping to stand-pipe and direct. Wells are 7 ins. in diam , 40 ft.deep. Pumping Machinery.—Dy. cap., 1,000,000 galls.; Gordon & Maxwell. Stand-Pipe.—Cap., 100,000 galls.; 12 × 110 ft. Distribution.—*Mains*, c. i., 3½ miles. *Hydrants*, Adams, 30. *Valves*, Detroit. *Pressure:* ordinary, 60 lbs.; fire, 125 lbs. Financial.—*Cost*, $30,000. *Bonds*, authorized, $20,000 at 6%. Management.—Report by A. A. Richardson, Engr.

11. **CHADRON,** *Dawes Co.* (Pop., in '90, 1,867.) Built in '88-9 by town. Des. and Consult. Engr., A. A. Richardson, Lincoln. Supply.—Chadron Creek, pumping to reservoir. Fiscal year closes May 1. July 6, '91, bonds were voted to pay for cost of securing a gravity supply, which would require about 4 miles of mains. Distribution.—*Mains*, 7 miles. *Taps*, 200. *Meters*, 0. *Hydrants*, 2. Financial.—*Cost*, $45,000. *Bonded Debt*, $45,000 at 6%. Management.-Mayor, Geo. Harper. Secy., L. G. F. Jaeger. Comr., C. E. Foster.

12. **COLUMBUS,** *Platte Co.* (Pop., 2,131—3,134.) Built in '86-7 by town. Supply.—Sixteen driven wells, pumping to stand-pipe and direct. Fiscal year closes May 1. Distribution.—*Mains*, 6 miles. *Taps*, 150. *Meters*, 2. *Hydrants*, 36. *Consumption*, 12,500 galls. Financial.—*Bonded Debt*, $35,000: $25,000 at 7%, due in 1906; $10,000 at 6%, 1908. *Expenses:* operating, $2,716; int., $2,100, total, $4,816. *Revenue*, consumers, $2,029. Management.—Comr., O. L. Baker.

13. **CRAWFORD,** *Dawes Co.* (Pop., in '90, 571.) History.—Built in '90 by city; put in operation Nov. 20. Engr., L. Harper. Contrs., U. S. Wind Engine & Pump Co., all of Omaha. Supply.—Spring, by gravity to reservoir. Reservoir.—Cap., 300,000 galls. Distribution.—*Mains*, c. i., 3.5 miles. *Services*, galv i. *Taps*, 20. *Hydrants*, Eclipse, 10. *Valves*, Lorain. *Pressure*, 60 lbs. Financial.—*Cost*, $20,-000. *Bonded debt*, $17,000 at 6%. Management.—Clerk, E. W. Darley.

14. **CRETE,** *Saline Co.* (Pop., 886—2,310.) Built in '88-9 by A. L. Strang, Omaha. Supply.—System of wells, pumping to reservoir. March 8, '91, Supt. reported that the works had not been accepted, on account of controversy beeween contractors and city. Distribution.—*Mains*, 4.4 miles. *Taps*, 32. *Hydrants*, 50. Financial.—Withheld. Management.—Supt., Frank McConnell.

15. **CULBERTSON,** *Hitchcock Co.* (Pop., 108—460.) Built in '88-9 by town. Supply.—Tube well, pumping to tank, chiefly for fire protection. Fiscal year closes

May 5. Distribution.—*Mains*, ½ mile. *Taps*, 7. *Hydrants*, 6. **Financial.**—*Cost*, $5,965. *Debt*, $6,000 at 7%. *Expenses:* operating, $85; int , $420; total, $505. *Revenue*, $40. **Management.**—Village Council.

16. **DAVID CITY,** *Butler Co.* (Pop., 1,000—2,028.) Built in '88-9 by city. **Supply.**—Two Cook wells, pumping to stand-pipe and direct. Fiscal year closes April 30. Distribution.—*Mains*, 3.5 miles. *Taps*, 95. *Meters*, 0. *Hydrants*, 32. **Financial.**—*Cost*, $24,000. *Expenses*, $1,800. *Revenue*, $1,200. **Management.**—Mayor, S. Clingman. Comr., I. J. West.

17. **EDGAR,** *Clay Co.* (Pop., 577—1,105.) Built in '88-9 by city. **Supply.**—Source not given, pumping to stand-pipe. Distribution.—*Mains*, 1 mile. *Taps*, 125. *Meters*, 1. *Hydrants*, 14. *Consumption*, 4,800 galls. **Financial.**—*Cost*, $14,000 *Bonded Debt*, $13,000 at 6%. *Expenses:* operating, $720; int., $780; total, $1,500! *Revenue*, $700. **Management.**—Mayor, T. B. McClellan. Clerk, J. W. Boder. Supt.. J. J. Rogers.

18. **FAIRBURY,** *Jefferson Co.* (Pop., 1,251—2.630.) Built in '88-9 by Fairbury Water-Works Co., under 25-yrs. franchise. **Supply.**—Wells, pumping to stand-pipe: Blue River in case of fire. Distribution.—*Mains*, 5.7 miles. *Hydrants*, 63. **Financial.**—*Cap. Stock*, $75.000. *Bonded Debt*, $30.000. *Hydrant rental*, $50. **Management.**—Prest., W. H. Burnham. Secy., W. S. Derby, Batavia, Ill. Supt., R. D. Russell.

19. **FAIRMONT,** *Fillmore Co.* (Pop., 600—1,029.) **History.**—Built in '85-6 by city; Engrs., Peabody & Williams. **Supply.**—Driven wells, pumping to tank. **Pumping Machinery.**—Not described. **Tanks.**—Cap., 35,000 galls.: two, one 40 and one 100 ft. high. Distribution.—*Mains*, 8 and 6-in. wood, 2 miles. *Services*, galv. 1. *Hydrants*. 25. **Financial.**—*Cost*, $17,500. *Bonded Debt*, $10,000 at 7%. **Management.**—Report by Elmer E. Lesh.

20. **FALLS CITY,** *Richardson Co* (Pop., 1,583—2,102.) Built in '83 by city. **Supply.**—Ground water, pumping to stand-pipe. Fiscal year closes May 1. Distribution.—*Mains*, 3.8 miles. *Taps*, 108. *Meters*, 6. *Hydrants*, 27. *Consumption*, 55,000 galls. **Financial.**—*Cost*, $40,800. *Bonded Debt*, $34,000; $21,000 at 5%, $10,000 at 6%. *Expenses:* operating. $3,100; int., $1,800; total, $1,900. *Revenue*, consumers, $3,300. **Management.**—Comrs., appointed every two years by mayor. Report by N. T. Van Winkle. Mayor.

21. **FREMONT,** *Dodge Co.* (Pop., 3,013—6.747.) **History.**—Built in '85-6. remodeled in '91, by city. Engr. F. L. Burrell, Fremont, Neb **Supply.**—Godfrey system of driven wells, pumping to stand-pipe and direct. Fiscal year closes last Tuesday in April. **Pumping Machinery.**—Dy, cap., 2,000,000 galls.; one 500,000 Deane and one 1,500,000 Smedley, added in '91. *Boilers:* two 30-HP.; from Fremont Foundry & Machine Co. *Stand-Pipe.*—Cap., 30,000 galls. Distribution.—*Mains*, 12 to 4-in. c. i., 10.5 miles. *Services*. lead. Pipe and specials from Nat'l Iron & Brass Wks., Dubuque, Ia.; trenching and pipe-laying done by Fremont Foundry & Machine Co. *Taps*, 178. *Hydrants*, Morgan, Ludlow, 74. *Consumption*, 400,000 galls. *Pressure:* ordinary, 43 lbs.; fire, 90 lbs. **Financial.**—*Cost*, $59,50?. *Bonded Debt*, $60,000 at 6% *Expenses:* operating. $2,120; int., $3,600; total, $5,720. *Revenue*, consumers, $1.970. **Management.**—Comr., Geo. Marshall, appointed every two years by mayor.

22. **GENEVA,** *Fillmore Co.* (Pop.. 376—1,580.) Built in '89 by city. **Supply.**—Two Cook wells, pumping to stand-pipe. Distribution.—*Mains*, 3 miles *Taps*, 101. *Hydrants*, 33. **Financial.**—*Cost*, $25,000. *Bonded Debt*, $27,000. *Expenses:* operating, $1,200. *Revenue*, consumers, $1,400. **Management.**—City Clerk, J. D. Hamilton.

23. **GENOA,** *Nance Co.* (Pop., 137—793.) **History.**—Bui't in '89-90 by city: put in operation Jan. 15. Engr, M. V. Mondy. Contr., Jonas Welch, Columbus. **Supply.**—Two wells, pumping to reservoir. **Pumping Machinery.**—Dy. cap., not given; one Cook steam and one 10,000-gall. wind-mill pump. **Reservoir.**—Cap., 100,000 galls. Distribution.—*Mains*, 10 to 4-in., c. i., 1.5 miles. *Taps*, 15. *Hydrants*, 11. *Consumption*, 3,000 galls. *Pressure*, 75 lbs. **Financial.**—*Cost*, $9,000. *Bonded Debt*, $5,000 at 7%. *Expenses:* operating. $350; int.. $200; total, $550. *Revenue:* consumers, $575. **Management.**—Comr., M. V. Mondy.

---

*For Further Descriptive Matter see Manual for 1889-91. Changes since 1890 are here given. Populations are for 1880 and 1890, respectively.

24. GRAND ISLAND, *Hall Co.* (Pop., 2,963—7,536.) Built in '86 by city. Supply.—Driven wells, pumping to stand-pipe and direct. Distribution.—*Mains*, 13.3 miles. *Taps*, 400. *Meters.* 6. *Hydrants*, 75. Financial.—*Cost*, about $75,000. *Bonded Debt*, $70,000 at 6%, due in 1907-9. *Sinking Fund.* $3,000. *Expenses :* operating, $5,674, Int., $4,200; total, $9,674. *Revenue :* consumers, $6,394. Management.— Supt., Geo. Loan, appointed by mayor.

25. GRANT, *Perkins Co.* (Pop., in '90, 315.) Built in '89 by Lincoln Land & Town Site Co., Lincoln; franchise provides that city may buy works after 3 yrs. at cost; does not control rates. Supply.—Wells, pumping to stand-pipe and direct. Water is furnished by B. & M. Ry. Co. Fiscal year closes Dec. 31. Distribution.— *Mains*, 3.3 miles. *Taps*, 42. *Meters*, 0. *Hydrants*, 15. *Consumption.* 8,000 galls. Financial.—*Revenue*, $1,400. Management.—See above. Report by C. H. Meeker, McCook.

26. HASTINGS, *Adams Co.* (Pop., 2,817—13,584.) Built in '86-7 by city. Des. Engr., A. A. Richardson, Lincoln. Supply.—Tubular wells, pumping to stand-pipe. Fiscal year closes July 31.

Year.	Mains.	Taps.	Meters.	Hydrants.	Cost.	Bonded debt.	Int. ch'gs.	Exp.	Rev.
'88-9	16	680	5	84	$100,000	$100,000	$6,700	$6,500	$7,500
'89-90	18	826	3	89	115,000	110,000	5,800	6,500	8,200

Financial.—*Bonded Debt*, $110,000: $85,000 at 5%, due in 1906; $10,000 at 6%, due in '97; $15,000 at 5%, due in 1908. *Sinking Fund*, $10,000. *Expenses*, total, $12,100. *Revenue*, consumers, $8,200. Management.—Comrs. and committee appointed by mayor from council every two years. Report by Aug. Rice. *

27. HEBRON, *Thayer Co.* (Pop., 466—1 502.) History.—Built in '90-1 by city; put in operation May 1. Des. Engr., A. A. Richardson, Lincoln. Supply.—Dug wells, pumping to stand-pipe. Pumping Machinery.—Dy. cap., 1,000,000 galls.; two 500,000 Buffalo dup. Stand-Pipe.—Cap., 84,000 galls.; 12 × 100 ft.; from Porter Boiler Mfg. Co. Distribution.—*Mains* c. f., 2.5 miles. *Hydrants*, Adam, 34. *Valves*, Lorain. *Pressure:* ordinary, 60 lbs.; fire, 125 lbs. Financial.—*Cost*, $25,000. Management.—Town Clerk, Edwin Kirby.

28. HOLDREDGE, *Phelps Co.* (Pop., in '90, 2,601.) Built in '88 by Holdredge Water Co., under 20-yrs. franchise. Supply.—Two wells, pumping to reservoir and repumping to stand-pipe and direct. Distribution.—(In '88.) *Mains*, 4 miles. *Taps*, 50. *Hydrants*, 52. Financial.—(In '88.) *Cost*, $62,000. *Cap. Stock*, $32,000. *Bonded Debt*, $45,000 at 5%. *Revenue*, city, $3,380. Management.—H. Brown, Omaha.

NOTE.—In July, '89, co. defaulted, and system was put in hands of receiver. Feb 28, '91, works were sold to city of Holdredge for $35,000, but city failing to make payment, sale was set aside July 6. Sept. 29, system was sold to A. Paxton, Omaha, for $25,005.

29. HOOPER, *Dodge Co.* (Pop., 204—670.) History.—Built in '90 by village; put in operation June 18. Engrs., Andrews & Burrell, Fremont. Contrs., Fremont Foundry & Machine Co. Supply.—Driven wells, pumping to tank. Pumping Machinery.—Dy. cap., 240,000 galls.; Smedley comp. Distribution.—*Mains*, 6 and 4-in. c. i., 1 mile. *Consumption*, 5,000 galls. *Pressure:* ordinary, 52 lbs., fire, 100 lbs. Financial.—*Cost*, $5,245. *Bonded Debt*, $5,100 at 6%. Management.—Comr., Fred. F. Heine.

30. KEARNEY, *Buffalo Co.* (Pop., 1,782—3,074.) Built in '87 by City Water Co., under 25-yrs. franchise; now controlled by American Water-Works & Guarantee Co., Ltd., Pittsburg, Pa.; franchise provides for purchase of works by city in '96 and at intervals of 5 yrs. thereafter, price to be determined by arbitration. Supply.—Wells, pumping direct. There are 48 3-in. driven wells and one large well 32 ft. in diam., 30 ft. deep Pumps are in well 23 ft. sq. and 11 ft. deep. Reported in '91 that a 1,000-HP. turbine wheel had been added to replace a 750-HP. Fiscal year closes Dec. 31. Distribution.—*Mains*, 13 miles. *Taps*, 400. *Meters*, 0. *Hydrants*, 05. *Consumption*, 600,000 galls. Financial.—Withheld. Management.—Prest., J. H. Purdy. Secy., A. E. Claney. Supt., W. W. Cunningham.

31. LINCOLN, *Lancaster Co.* (Pop., 13,003—55,154.) Built in '85 by city. Des. Engr., A. A. Richardson, Lincoln. Supply.—One surface and about 160 driven wells pumping to stand-pipe. Fiscal year closes Sept. 1.

NEBRASKA. 287

Year.	M'ns.	Taps.	M't'rs.	Hyd'ts	C'ns'pt'n.	Cost.	B'd'd d't.	Int. ch's.	Exp.	Rev.
'86	11	543	24	77	600,000	$110,000	$110,000	$4,600	$7,000	$10,000
'87	15	1,070	40	90	8,000,000	130,000	120,000	7,200	8,000	16,000
'88	19	1,230	60	153	1,500,000	150,000	110,000	6,600	10,673	20,000
'89	19.5	1,251	54	168	2,000,000	194,0.0	130,000	7,800	11,004	20,964
'90	28.2	1,492	95	246	2,500,000	233,744	130,000	7,800	17,320	25,475

Financial.—*Bonded Debt*, $130,000 at 6%. *Expenses*, total, $25,120. *No Sinking Fund.* Revenue from private consumers pays all expenses except new construction, which is paid for by direct taxation. **Management.**—Committee and comrs., former elected, latter appointed by Mayor every two years. Chn., H. M. Rice. Supt., I. L. Lyman, appointed by Mayor.

32. LONG PINE, *Brown Co.* (Pop., in '90, 562.) Built in '88 by village. Supply.—Silver Springs, pumping to tank. **Distribution.**—*Mains*, 3 miles. *Taps*, 150. *Meters*, 0. *Hydrants*, 13. **Financial.**—*Cost*, $15,000. *Expenses*, $600. *Revenue*, $625. **Management.**—Clerk, J. S. Davidson.

33. LOUP CITY, *Sherman Co.* (Pop., 74—671.) Built in '87-8 by city. Supply. Four driven wells, pumping to reservoir and direct. Fiscal year closes first Monday in May. **Distribution.**—*Mains*, 1.3 miles. *Taps*, 12. *Hydrants*, 12. **Financial.**—*Cost*, $16,000. *Debt*, $14,000. *Expenses*, $600. *Revenue*, $200. **Management.**—Mayor, G. W. Hunter. Comr., S. G. Wookey.

34. McCOOK, *Red Willow Co.*(Pop., in '90, 2,346.)**History.**—Built in '83 by Lincoln Land & Town Site Co. Engr., G. W. Irving. **Supply.**—Three wells, pumping to tank. Wells are 20 ins. in diam. **Pumping Machinery.**—Dy. cap., 2,000,000 galls.; two Deane dup. **Tank.**—Cap., 50,000 galls. **Distribution.**—*Mains*, 12 to 3-in. c. i., 10 miles. *Services*, lead. *Taps*, 308. *Meters*, Union, 1. *Hydrants*, Eclipse, 20. *Valves*, Ludlow. *Consumption*, 3,000,000 galls. *Pressure:* ordinary, 60 lbs.; fire, 120 lbs. **Financial.**—*Cost*, $75,000. *No Debt*. *Hydrant Rental*, $60. **Management**—Supt., C. H. Meeker.

35. MADISON, *Madison Co.* (Pop., 417—930.) Built in '85 by city. **Supply.**—Driven wells, pumping direct and to tank. Fiscal year closes June 1. Blake deep-well pump was replaced in '90 by Knowles. **Distribution**—*Mains*, "42 blocks. *Taps*, 67. *Meters*, 0. *Hydrants*. 23. *Consumption*, 8,000 galls. **Financial** – *Cost*, $13,800. *Bonded Debt*, $13 000 at 5%. *Expenses:* operating, $100; int., $245; total, $615. *Revenue*, consumers, $570. **Management.**—Supt., A. G. Cleveland.

36. MASON CITY, *Custer Co.* (Pop., est., 200.) **History.**—Built in '91 by village; put in operation May 1. Const. Engr., T. J. Wood. Contrs., W. J. Cooper and Cole Bros., Lincoln. **Supply.**—Well, pumping to tank. **Pumping Machinery.**—A 5,000-gall. deep-well and an Althouse wind-mill pump. **Tank.**—Cap., 36,000 galls. **Distribution.**—*Mains*, 1,890 ft. *Taps*, 12. *Hydrants*, 6. **Financial.**—*Cost*, $2,000. *Bonded Debt*, $2,000 at 7%. *Expenses:* operating, $100; int., $140; total, $240. *Revenue*, consumers, $30. **Management.**—Village Clerk, Henry M. Kidder.

37. MINDEN, *Kearney Co.* (Pop., 98—1,380.) **History.**—Built in '90 by city; put in operation Dec. 20. Des. Engr., A. A. Richardson, Lincoln. Const. Engr., Jno. Cummings. **Supply.**—Tube wells, pumping to stand-pipe. **Pumping Machinery.**—Dy. cap., not given; Cook. **Stand Pipe.**—Cap., 93,000 galls.; 12 × 110 ft.; from Porter Mfg. Co. **Distribution.**—*Mains*, c. i., 4 miles. *Hydrants*, Adams. 53. *Valves*, Detroit. *Pressure:* ordinary, 50 lbs.; fire, 120 lbs. **Financial.**—*Cost*, $25,000. **Management.**—Supt., J. P. Dunsmore.

38. NEBRASKA CITY, *Otoe Co.* (Pop, 4,183—11,494.) Built in '87 by City Water Co. under 20-yrs. franchise; rebuilt in '90; now controlled by Debenture Guarantee & Assurance Co., Chicago, Ill. Company owns and operates in connection with works an electric lighting plant. **Supply.**—Missouri River, pumping to settling basins thence to stand-pipe; filtered. Fiscal year closes Dec. 31. During '90, a 2,000,000-gall. Worthington pump was added and two settling basins built; basins have cap. of 6,000,000 galls each, have puddled banks, and are lined with concrete. Engr., Geo. H. Pierson, N. Y. City. **Distribution.**—*Mains*, 8 miles. *Taps*, 521. *Meters*, 37. *Hydrants*, 81. *Consumption*, 200,000 galls. **Financial.**—*Cap. Stock*, paid up, $375,000. *Bonded Debt*, $150,000 at 6%. *Expenses:* operating and taxes, $5,500; int., $9,000; total, $14,500. *Revenue:* consumers, $18,116; ry., $1,634; city, $1,050; total $23,800. **Management.**—Prest., Chas. Whann, Secy. and Treas., Chas. B. Ludwig, 72 Broadway, N. Y. City. Supt., D. P. Rolf.

FOR FURTHER DESCRIPTIVE MATTER see Manual for 1889-90. CHANGES since 1890 are here given. POPULATIONS are for 1880 and 1890, respectively.

MANUAL OF AMERICAN WATER-WORKS.

39. NELIGH, *Antelope Co.* (Pop., 326—1,209.) Built in '87 by city. **Supply.**—Elkhorn River, pumping to tank; filtered. Fiscal year closes May 1. Dy. cap. of pump is 250,000 galls., not 500,000 galls., as stated in '89-90 Manual. In '90 an 80-HP. boiler was added. **Distribution.**—*Mains*, 4 miles. *Taps*, 100. *Meters*, 20. *Hydrants*, *Consumption*, 55,000 galls. **Financial.**—*Cost*, $20,000. *Total Debt*, $16,900: floating, 32-$2,000 at 7%; bonded, $14,900 at 6%. *Expenses:* operating, $1,600; int., $840; total, $1,840. *Revenue:* consumers, $500; city, $500; total, $1,000. **Management.**—Chn., T. H. Brenton. Secy., W. H. Karls. Supt., E. C. Million, appointed by mayor.

40. NORFOLK, *Madison Co.* (Pop., 547—3,038.) Built in '88 by Norfolk Water-Works Co. **Supply.**—Two artesian wells, pumping to stand-pipe. New stand-pipe, 12 X 110 ft., on elevation of 96 ft. above city, was built in '90. Fiscal year closes July 1.

Year.	Mains.	Taps.	Meters.	Hydrants.	Cost	Exp.	Rev.
'88	5	80	3	60	$35,000	$1,800	$4,200
'90	7	108	5	77	....	2,200	6,320

Management.—Sept. 1, '90, system passed into hands of bondholders. Receiver, Jas. Daniels, St. Louis, Mo. Supt., C. C. Voorhees.

41. NORTH PLATTE, *Lincoln Co.* (Pop., 363—3,055.) Built in '87 by North Platte Water-Works Co. under 20-yrs. franchise. Managed by American Water-Works & Guarantee Co., Ltd., Pittsburg, Pa. Maximum rates fixed in franchise, which provides for purchase of works by city in '97, and at intervals of 5 yrs. thereafter, price determined by arbitration. **Supply.**—Twenty-two driven wells, pumping direct. Wells are 6 ins. in diam., 70 ft. deep. **Distribution—***Mains*, 7 miles. *Taps*, 279. *Meters*, 0. *Hydrants*, 51. *Consumption*, 350,000 galls. **Financial.**—Withheld. **Management.**—Prest. J. H. Purdy. Vice-President, J. W. Bixler. Secy., A. E. Claney. Supt., Jas. Wilson.

42. OMAHA, *Douglas Co.* (Pop., 30,518—140,452.) Owned by American Water Works Co. Built in '80-1 by City Water-Works Co., under perpetual franchise. Supplies South Omaha. **Supply.**—Missouri River, pumping to settling basins, thence repumped to city and to reservoir; repumped from latter for high service. Fiscal year closes Dec. 1. June, '91, pumping machinery was as follows: River plant, two 15,000,000 Allis, one 15,000,000 Gaskill, making total cap. of 45,000,000 galls.; general service, one 18,000,000 Allis, one 7,000,000 Gaskill; high service, one 7,000,000 and one 3,000,000 Gaskill. **Distribution.**—*Meters*, about 1,200. *Consumption*, about 14,000,000 galls.

Year.	Mains.	Taps.	Hydrants.	B'd'd debt.	Int. chgs.	Exp.	Rev.
'87	75	2,575	502	....	....	....	$168,983
'88	111	3,673	773	$2,000,000	$120,000	$68,000	230,053
'89	...	4,862	1,026	3,000,000	170,000	71,668	231,559
'90	157	6,193	1,252	....	....	.....	....

Financial.—(Jan. 1, '89.) *Cap. Stock:* common, $4,000,000; preferred, authorized, $1,000,000, 6% cumulative. *Hydrant rental* (in '89), 250 at $84, 31 at $10, balance at $60. **Management.**—Prest., W. A. Underwood, N. Y. City. Secy., W. H. Hall. Supt., A. B. Hunt.

43. ORD, *Valley Co.* (Pop., 181—1,208.) Built in '87 by town. **Supply.**—Driven wells, pumping to reservoir and direct. Fiscal year closes May 1. In '90 pump-house was moved to spring, in bottom of which two 5-in. Cook wells have been driven 10 ft. deep. **Distribution.**—*Mains*, 3.5 miles. *Taps*, 63. *Hydrants*, 21. **Financial.**—*Cost*, $19,500. *Total Debt*, $26,000: floating, $12,000 at 6%, due in '97; bonded, $4,000 at 7%, due in '99. *Expenses:* operating, $874; int., $720; total, $1,594. *Revenue*, consumers, $823. **Management.**—Comr., J. G. Sharpe.

ORLEANS, *Harlan Co.* (Pop., 409 - 812.) Built in '88-91 by city. **Supply.**—Tube wells, pumping to tank. May 25, '91, City Clerk reported that system described in 1880-90 Manual was not yet completed.

44. PAWNEE CITY, *Pawnee Co.* (Pop., 763—1,550.) Built in '87-8 by city. **Supply.**—Two wells, pumping to stand-pipe and direct. **Distribution.**—*Mains*, 4 miles. *Taps*, 100. *Meters*, 1. *Hydrants*, 35. *Consumption*, 15,000 galls. **Financial.**—*Cost*, $28,000. *Expenses*, $1,100. **Management**—Mayor, W. S. Storey. Supt., Wm. Headley.

45. PLATTSMOUTH, *Cass Co.* (Pop., 4,175—8,392.) Built in '86-7 by Wisconsin Construction Co., Boston, Mass., under 20-yrs. franchise. **Supply.**—Missouri River, pumping to stand-pipe. Fiscal year closes Dec. 31. **Distribution.**—*Mains*,

NEBRASKA.

6.5 miles. *Taps*, 261. *Meters*, 38. *Hydrants*, 71. *Consumption*, 250,000 galls. **Financial.**—*Cap. Stock:* authorized, $200,000; all paid up. *Total Debt*, $137,000: floating, $12,000; bonded, $125,000 at 6%, due in 1906. *Expenses:* int., $7,500; taxes, $1,100. *Revenue:* consumers, $7,925; fire protection and other purposes, $5,050; total, $12,-975. **Management.**—Prest., A. B. Turner, Boston, Mass. Secy., M. S. Polk. Treas., B. R. Clarke, Boston, Mass. Supt., T. F. Coursey.

46. RANDOLPH, *Cedar Co.* (Pop., in '90, 374.) Built in '89 by village. **Supply.**—Tubular well, pumping to tank. **Distribution.**—*Mains*, 1 mile. *Taps*, 80. *Hydrants*, 3. **Financial.**—*Cost*, about $2,800. *Revenue*, $630. **Management.**—Supt., J. C. Failey.

47. RED CLOUD, *Webster Co.* (Pop., 677—1,839.) Built in '87 by city. **Supply.**—Two surface wells, pumping to stand-pipe, and direct. Fiscal year closes Jan. 1. **Distribution.**—*Mains*, 4.3 miles. *Taps*, 100. *Meters*, 3. *Hydrants*, 32. *Consumption*, 100,000 galls. **Financial.**—*Cost*, $35,000. *Bonded Debt*, $25,000 at 7%. *Expenses:* operating, $1,700; int., $1,750; total, $3,450. *Revenue*, consumers, $800. **Management.**—Mayor and council, elected every two years. Prest., A. S. Marsh. Secy. and Supt., J. Ward.

48. RIVERTON, *Franklin Co.* (Pop., 426—389.) Built in '88 by town. **Supply.**—Thompson Creek, pumping to tank. Fiscal year closes Jan. 1. **Distribution.**—*Mains*, 2 miles. *Taps*, 47. *Meters*, 0. *Hydrants*, 8. *Consumption*, 30,000 galls. **Financial.**—*Cost*, $6,000. *Bonded Debt*, $4,000 at 8%, due in 1908. *Expenses:* operating, $50; int., $320; total, $370. *Revenue*, consumers, $160. **Management.**—Water com., appointed every year by mayor and council. Chn., D. Eastwood. Secy., L. H. Rollins.

Rushville, *Sheridan Co.* (Pop., in '90, 484.) Built in '88 by village for fire protection. See 1889-90 Manual, p. 536.

49. SAINT PAUL, *Howard Co.* (Pop., 482—1,263.) Built in '87-8 by town. **Supply.**—Thirty-two driven wells, pumping to stand-pipe and direct. **Distribution.**—(In '88.) *Mains*, 3.5 miles. *Hydrants*, 26. **Financial.**—(In '88.) *Cost*, $22,000. *Bonded Debt*, $20,000 at 6%. **Management.**—Comr., J. L. Johnson. Supt., E. T. Frame.

50. SEWARD, *Seward Co.* (Pop., 1,525—2,108.) Built in '88-9 by city. **Supply.**—Eighteen driven wells, pumping to stand-pipe. In '89-90 Manual it was erroneously stated that Cook wells and pumps are used. Pump is a 1,000,000-gall. from Pond Engineering Co., St. Louis, Mo. **Distribution.**—*Mains*, 3.3 miles. *Hydrants*, 48. **Financial.**—*Cost*, $30,000. *Bonded Debt*, $30,000 at 6%. **Management.**—Mayor, G. W. Fuller. Supt., Wm. Allen.

42. SOUTH OMAHA, *Douglas Co.* (Pop., in '90, 8,026.) Supply from Omaha.

51. STROMSBURGH, *Polk Co.* (Pop., est., 1,000.) Built in '88-9 by town. **Supply.**—Cook wells, pumping to stand-pipe and direct. Fiscal year closes April 30. **Distribution.**—*Mains*, 2 miles. *Taps*, 52. *Meters*, 0. *Hydrants*, 15. **Financial.**—*Cost*, $12,000. *Bonded Debt*, $12,000 at 6%. *Expenses:* operating, $1,000; int., $720; total, $1,720. *Revenue*, consumers, $500. **Management.**—Mayor and Council.

52. SUPERIOR, *Nuckolls Co.* (Pop., 458—1,614.) Built in '88-9 by town. **Supply.**—Well, pumping direct. Fiscal year closes May 1. **Distribution.**—*Mains*, 3.3 miles. *Taps*, 130. *Hydrants*, 21. **Financial.**—*Cost*, $27,500. *Bonded Debt*, $25,000 at 5%. *Expenses:* operating, $2,000; int., $1,250; total, $3,250. *Revenue*, consumers, $800. **Management.**—Comrs. Mayor, C. E. Williams.

53. SUTTON, *Clay Co.* (Pop., in '90, 1,544.) **History.**—Built in '91 by city; put in operation May 23. Engr., A. A. Richardson, Lincoln. **Supply.**—Wells, pumping to stand-pipe. **Pumping Machinery.**— Dy. cap., 160,000 galls.; **Cook. Stand-Pipe.**—Cap., 62,500 galls.; 12 × 75 ft.; from Porter Boiler Mfg. Co., Chicago. **Distribution.**—*Mains*, 8-in. c. i., 2.5 miles. Pipe and specials from So. St. Louis Foundry, St. Louis. *Taps*, 12. *Hydrants*, Adams, 27. *Valves*, Bourbon. *Pressure:* ordinary, 50 lbs.; fire, 60 lbs. **Financial.**—*Cost*, $16,500. *Bonded Debt*, $16,500 at 6%. **Management.**—Mayor, E. W. Woodruff.

For FURTHER DESCRIPTIVE MATTER see Manual for 1889-90. CHANGES since 1890 are here given. POPULATIONS are for 1880 and 1890, respectively.

290   MANUAL OF AMERICAN WATER-WORKS.

54. TECUMSEH, *Johnson Co.* (Pop., 1,268—1,654.) Built in '88-9 by town. Supply.—Wells, pumping to tank. Godfrey system of wells was introduced in '90, at a cost of $2,275. Fiscal year closes first Tuesday in May. Distribution.—*Mains* (in '89), 4 miles. *Taps* (in '89), 40. *Meters*, 6. *Hydrants* (in '89), 30. *Consumption*, 12,700 galls. Financial.—*Cost*, $25,000. *Total Debt*, $21,950: floating, $950; bonded, $21,000 at 6%. *Expenses*, $1,725. *Revenue*, consumers, $2,700. Management.—Mayor, Geo. Hill, Supt., C. K. Chubbuck.

55. VALENTINE, *Cherry Co.* (Pop., est., 400.) History.—Built in '89 by village. Contr., U. S. Wind Engine & Pump Co., Omaha. Supply.—Minnehaduza River, pumping to tank. Pumping Machinery.—Dy. cap., 62,000 galls. Tank.—Cap., 58,000 galls. Distribution.—*Mains*, 6 and 4-in. c. i., 1¼ miles. *Taps*, 30. *Hydrants*, 12. Financial.—*Cost*, $8,400. *Bonded Debt*, $8,400 at 7%. *Expenses*, $1,200. *Revenue*, $1,200. Management.—Clerk, E. W. Harvey.

56. VALPARAISO, *Saunders Co.* (Pop., 300—515.) History.—Built in '87-8 by town. Des. Engr., A. A. Richardson, Lincoln. Const. Engr., E. Ashland. Supply.—Well, pumping to tank. Pumping Machinery.—Dy. cap., 50,000 galls. Tank.—Cap., 40,000 galls.; from Nebraska Planing Mill Co. Distribution.—*Mains*, 8 and 4-in. c. i., 1 mile. Pipe and specials from Chattanooga (Tenn.) Foundry & Pipe Wks.; trenching and pipe-laying done by Ben Clarke. No *Meters*. *Hydrants*, 12. *Pressure*, 45 lbs. Financial.—*Cost*, $7,000. *Bonded Debt*, $7,000 at 6%. *Expenses:* operating, $300; int., $420; total, $720. *Revenue*, consumers, $50. Management.—Clerk, J. C. Stephen.

57. WAHOO, *Saunders Co.* (Pop., 1,064—2,006.) Built in '87 by Wahoo Water-Works Co., now controlled by American Water-Works & Guarantee Co., Ltd., Pittsburg, Pa., under 20-yrs. franchise which fixes maximum rates and provides for purchase of works by city in '99, price determined by arbitration. Supply.—Driven wells and Wahoo River, pumping direct. Fiscal year closes Dec. 31. Distribution.—*Mains*, 6 miles. *Taps*, 114. *Hydrants*, 41. *Consumption*, 140,000 galls. Financial.—Withheld. Management.—Prest., J. H. Purdy. Secy., A. E. Claney. Supt., C. H. Adams.

58. WEEPING WATER, *Cass Co.* (Pop., 317—1,350.) Built in '88 by town Supply.—Wells, pumping to reservoir. Fiscal year closes first Tuesday in May. Distribution.—*Mains*, 4 miles. Financial.—*Cost*, $15,000. *Bonded Debt*, $9,000 at 6%. Management.—Comr., A. L. Timbliss.

59. WEST POINT, *Cuming Co.* (Pop., 1,009—1,842.) Built in '86 by town. Supply.—Wells, pumping to reservoir and direct. Distribution.—*Mains*, 3.5 miles. *Taps*, 117. *Hydrants* (in '88), 34. Financial.—*Cost*, $16,000. *Bonded Debt*, $16,000 at 7%. *Expenses:* operating, $1,000; int., $1,120; total, $2,220. Revenue is sufficient to pay operating expenses only. Management.—Comr., appointed every year by Mayor and Council. City Clerk, F. E. Krause.

60. WISNER, *Cuming Co.* (Pop., 282—310.) Built in '84 by town. Supply.—Well, pumping to tank. Distribution.—*Mains*, 8,500 ft. *Taps*, 40. *Hydrants*, 18. Financial.—*Cost*, about $5,000. *Expenses:* operating, $200; int., $280; total, $480. *Revenue*, consumers, $330. Management.—Comr., J. C. Grogan.

61. WYMORE, *Gage Co.* (Pop., in '90, 2,420.) Built in '89—'90 by city. Supply.—Wells, pumping to stand-pipe; filtering contemplated. Fiscal year closes April 30. Distribution.—*Mains*, 4.5 miles. *Taps*, 63. *Hydrants*, 51. *Consumption*, 15,000 galls. Financial.—*Cost*, $38,000. *Total Debt*, $38,000: floating, $8,000 at 7%; bonded, $30,000 at 6%, due In 1900. *Expenses*, operating, $1,200. *Revenue*, consumers, $250. Management.—Mayor, E. P. Reynolds. Chn., Julius Newman. Supt., Frank Acton.

62. YORK, *York Co.* (Pop., 1,259—3,405.) Built in '87-8 by York Water-Works Co. Supply.—Cook wells and connection with Beaver Creek, pumping to stand-pipe and direct. Fiscal year closes Dec. 31. Distribution.—*Mains*, 6.5 miles. *Taps*, 160. *Meters*, 8. *Hydrants*, 60. *Consumption*, 34,729 galls. Financial.—*Cost* (in '89), $75,000. *Bonded Debt*, $60,000 at 6%. *Revenue:* consumers, $2,397; city, $3,600; total, $5,997. Management.—Receiver, Jas. Daniels. Supt., J. R. Strasser.

NOTE.—Reported in latter part '90 that works had been sold to T. A. Post, St. Louis, Mo.

NEBRASKA.

Water-Works Projected with Fair Prospects of Construction.

AUBURN, *Nemaha Co.* (Pop., in '90, 1,537.) City talks of building waterworks, and surveys and plans had been made July 2, '91. Address, Albert D. Gilmore.

BENKLEMAN, *Dundy Co.* (Pop., in '90. 357.) May 1, '91, it was reported that attempts were being made to obtain supply of water through irrigating ditches. July 2 it was reported that construction had not begun. City Clerk, J. F. Haskin.

BLUE SPRINGS, *Gage Co.* (Pop., 513—963.) Construction was to be begun Oct. 1 by city. Contrs., Shepard & Co., Beatrice. Est. cost, $10,000. City Clerk, Jas. H. Casebeer.

NIOBRARA, *Knox Co.* (Pop., 475—633.) In '91, co. was organized and work on artesian well was begun about Sept. 1. If sufficient water was obtained, distribution system was to be constructed in '92. Address G. G. Bayha, or S. Draper.

OAKLAND, *Burt Co.* (Pop., 345—633.) City was to begin construction of water-works Aug. 17, '91. Engr., S. K. Felton. Contrs., U. S. Wind Engine & Pump Co. Well, pumping to 12 × 50-ft. stand-pipe. Est. cost, $10,000. Latest report Aug. 13.

O'NEILL, *Holt Co.* (Pop., 57—1,226.) July 3, '91, it was reported that construction of water-works would begin at once. Address M. Hardoker.

PONCA, *Dixon Co.* (Pop., 591—1,009.) During fall of '90, city voted to issue bonds to construct water-works. July 10, '91. bonds had not been sold nor contracts let. City Engr., E. B. Stough.

SCHUYLER, *Colfax Co.* (Pop., 1,017—2,160.) In Feb., 91, city voted to issue $25,000 worth of 5% bonds to build water-works. Wells, pumping to stand-pipe. Engrs., Andrews & Burrell, Fremont. Contrs., Michigan Pipe Co., Bay City. Clerk, W. T. Howard.

SOUTH SIOUX CITY, *Dakota Co.* (Pop., in '90, 603.) Water-works talked of but no definite steps taken Aug. 13, '91. Address C. S. Smiley.

TEKAMAH, *Burt Co.* (Pop., 776—1,244.) City was to begin construction of works July 25. Des. Engr., Andrew Rosewater, Omaha. Wells, pumping to 500,000-gall. reservoir. Est. cost, $13,000. City Clerk, E. B. Atkinson.

WAYNE, *Wayne Co.* (Pop. in '90, 1,178.) City expected to begin construction of water-works July 10, '91. Only a portion of contracts had been let July 17. Wells, pumping to 12 × 75-ft. stand-pipe. Mains, 3¼ miles. Hydrants, 33. Est. Cost, $18,000. Mayor, Frank Fuller.

Water-Works Projects Reported in '90 and '91, but Now Indefinite.

Alliance, Boxbutte Co. *Lexington*, Dawson Co. *Liberty*, Gage Co. *Ravenna*, Buffalo Co.

SOUTH DAKOTA.

Water-Works Completed or in Process of Construction.

1. **ABERDEEN,** *Brown Co.* (Pop., in '90, 3,182.) Built in '85 by town. Supply.—Two flowing wells, by gravity. Fiscal year closes May 1. Distribution.—Mains, 1.5 miles. Taps, 160. Meters, 0. Hydrants, 26. Financial.—*Cost*, $25,000. Bonded Debt, $8,000 at 7%. Revenue, $2,300. Management.—Com., appointed from council by mayor every year. Engr., W. P. Butler.

2. **ANDOVER,** *Day Co.* (Pop., in '90. 262.) Built in '88 by village. Supply.—Artesian well, by gravity. Fiscal year closes May 6. Distribution.—*Mains*,¾ miles. Taps, 15. Meters, 0. Hydrants, 4. Financial.—*Cost*, $3,000. Debt, $2,000 at 7%. Expenses, nominal. Revenue, $180. Management—Clerk, Wm. Carpenter.

FOR FURTHER DESCRIPTIVE MATTER see Manual for 1889-90. CHANGES since 1890 are here given. POPULATIONS are for 1880 and 1890, respectively.

MANUAL OF AMERICAN WATER-WORKS.

3. **BROOKINGS**, *Brookings Co.* (Pop., in '90, 1,518.) **History.**—Built in '89-90 by city; put in operation March 10, '90. Engr., E. E. Gaylord. **Supply.**—Wells, pumping to tank. Wells are 7 ins. in diam., 430 ft. deep. **Pumping Machinery.**—Dy. cap., 48,330 galls.; Hughes dup., 12 × 12 × 7, for fire; pump for domestic supply not described. **Tank.**—Cap., 600,000 galls.; 40 ft. high; from Eclipse Wind Mill Co. **Distribution.**—*Mains*, 9 and 7-in. w. i., 1,360 ft. *Hydrants*, Eclipse, 5. *Pressure*, ordinary, 20 lbs.; 120 to 160 lbs. **Financial.**—*Cost*, $6,000. *Debt*, $5,000. *Expenses*, $900. **Management.**—City Engr., E. E. Gaylord.

4. **CANTON**, *Lincoln Co.* (Pop., 675—1,101.) **History.**—Built in '91 by city; put in operation July 25. Des. Engr., A. A. Richardson. Const. Engr., J. W. Percival. Both of Lincoln, Neb. **Supply.**—Driven wells, pumping to stand-pipe. **Pumping Machinery.**—Dy. cap., 500,000 galls.; Smedley dup. **Stand-Pipe.**—Cap., 86,400 galls.; 12 × 100 ft.; from Porter Boiler Mfg. Co., Chicago. **Distribution.**—*Mains*, 8 to 4-in., c. i., 3.5 miles. *Services*, w. i. *Taps*, 75. *Hydrants*, Adams, 40. *Valves*, Lorain. *Consumption*, 40,000 galls. *Pressure*, 80 lbs. **Financial.**—*Cost*, $20,000. *Bonded Debt*, $17,000 at 6%. **Management.**—Com.: N. Noble, J. W. Hewitt, C. S. Kartvedr, J. P. Hann.

5. **CHAMBERLAIN**, *Brule Co.* (Pop., in '90, 939.) Built in '84 by city. **Supply.**—Artesian well, flowing into reservoir. Well was completed in '91; 6 ins. in diam., 800 ft. deep, with dy. flow of 700,000 galls. **Distribution.**—*Mains*, 2.5 miles. *Taps*, 75. *Meters*, 35. *Hydrants*, 25. **Financial.**—*Cost*, $25,000. *Bonded Debt*, $21,000 at 8%. **Management.**—Mayor, Jno. F. Anderson. Comr., Jas. D. Farrell. NOTE.—In '91 there was talk of city selling works.

1. **COLUMBIA**, *Brown Co.* (Pop., 133—400.) Built in '86 by city. **Supply.**—Artesian well, 1,000 ft. deep, 6 ins. in diam. at top, 4 at bottom. **Distribution.**—*Mains*, 6-in., c. i., 1,000 ft. *Hydrants*, 4. *Pressure*, 180 lbs. No revenue had been derived from works at time of last report. Address A. Loomis, mayor.

6. **DEADWOOD**, *Lawrence Co.* (Pop., 3,777—2,366.) Built in '79 by Black Hills Canal & Water Co., under 20-yrs. exclusive franchise. **Supply.**—Whitewood Creek, by gravity, to tanks. Fiscal year closes Dec. 31. **Distribution.**—*Mains*, 4.3 miles. *Taps*, 180. *Meters*, 0. *Hydrants*, 19. *Consumption*, 90,000 galls. **Financial.**—*Cost*, $73,000. *No Debt*. *Expenses*, $6,207. **Management.**—Supt., T. J. Grier.

NOTE.—June 22, '91, $40,000 worth of 6% bonds were offered for sale by city, proceeds to be used for purchasing and extending works and building sewers.

DOLAND, *Spink Co.* (Pop., in '90, 216.) Built in '89 by village. See 1889-90 Manual, p. 540.

7. **HURON**, *Beadle Co.* (Pop., 164—3,038.) Built in '84 by town. **Supply.**—Flowing artesian well, by gravity, James River, pumping direct.

Year	Mains	Taps	Meters	Hyd'ts	C'ns'pt'n	Cost	B'd'd d't	Int.	ch'g's	Exp.	Rev.
'84	2.3	70	0	16	200,000	$24,500	$25,000			....	$2,000
'88	2.3	125	0	17	200,000	29,000	29,000	$2,030		$700	2,500
'90	4	200	0	25	......	60,000	60,000	3,600		...	....

Management.—Mgr., Chas. C. Dunlap. Supt., Alex. McIntosh. NOTE.—Reported that Huron Water-Works Co. was incorporated in '90 or '91.

8. **MELLETTE**, *Spink Co.* (Pop., in '90, 341.) **History.**—Built in '89-90 by village; put in operation in Jan. Contrs., Swan & Stacey, Aberdeen. **Supply.**—Artesian well. **Distribution.**—*Mains*, Kal., ½ mile. *Taps*, 35. *Hydrants*, 4. *Pressure:* ordinary, 85 lbs.; fire, 165 lbs. **Financial.**—*Cost*, $3,200. *Debt*, $3,200 at 7%. **Management.**—Town Clerk, W. I. Leitch.

Millbank, *Grant Co.* (Pop., in '90, 1,207.) Built in '86 by C. M. & St. P. Ry. Co. and city jointly, for fire protection only. Address Jno. W. Bell, city clerk.

9. **MILLER**, *Hand Co.* (Pop., in '90, 536.) Built in '86 by town. **Supply.**—Flowing artesian well, by gravity. **Distribution.**—*Mains*, 3 miles. *Hydrants*, 26. **Financial.**—*Cost*, $16,000. *Bonded Debt*, $16,000 at 6%. *Expenses:* operating, $250; int., $960; total, $1,210. *Revenue*, $250. **Management.**—Mayor, M. F. Cahalan. Clerk, Fred Alber. Supt., A. S. Sales.

10. **MITCHELL**, *Davison Co.* (Pop., 320—2,217.) Built in '87 by town. Engr., S. K. Felton, Omaha. **Supply.**—Two artesian wells, pumping to reservoir and direct.

SOUTH DAKOTA. 293

Fiscal year closes Aug. 31. **Distribution.**—*Mains*, 4.5 miles. *Taps*, 115. *Meters*, 0. *Hydrants*, 27. **Financial.**—*Cost*, $65,000. *Bonded Debt*, $65,000. *Sinking Fund*, $3,000. *Expenses:* operating, $1,500; int., $4,550; total, $6,050. *Revenue*, consumers, $850. **Management.**—Treas., H. R. Kibbee. Supt., W. S. Crouse.

11. **PIERRE,** *Hughes Co.* (Pop., in '90, 3,235.) Built in '85 by Pierre Water-Works Co., under 25-yrs. exclusive franchise; city may purchase works after giving 6 months' notice in 1900, '05, or '10; value to be determined by appraisers. **Supply.**—Missouri River, pumping from settling well to reservoir; filtered through sand and gravel from river to large well, 300 ft. from bank. In '90, pumping capacity increased by new 1,250,000-gall. pump and three new boilers. Fiscal year closes Sept. 2.

Year.	Mains.	Taps.	M't'rs.	H'd'ts.	C'ns'pt'n.	Cost.	B'd'd d't.	Int.ch'g's.	Exp.	Rev.
'86	4.5	..	45	45	.....	$100,000	none.	none.	$1,500	$6,000
'88	7.5	60	8	60	30,000	129,500	$75,000	$4,500	3,000	8,500
'90	9.25	386	28	84	150,000	134,820	100,000	6,000	....	13,225

Financial.—*Cap. Stock,* $125,000; paid no div'd in 1890. *Total Debt,* $104,000: floating, $4,000 at 8%; bonded, $100,000 at 6%, due in 1905. *Sinking Fund,* "just started." *Expenses,* withheld. Co. is exempt from taxation. *Revenue:* consumers, $1,800; fire protection, $7,675; other purposes, 750; total, $13,225. **Management.**—Prest., James Bixby. Secy. and Supt., T. W. Pratt.

12. **PLANKINGTON,** *Aurora Co.* (Pop., in '90, 604.) **History.**—Built in '88-9 by city. Engr., J. F. Barton. **Supply.**—Flowing artesian well, by gravity. **Distribution.**—*Mains,* 6-in., l., 1 mile. *Taps,* 25. *Hydrants,* Chapman, 12. *Pressure:* ordinary, 50 lbs.; fire, 75 lbs. **Financial.**—*Cost,* $7,000. *Debt,* $10,000 at 7%. *Expenses:* operating, $200; int., $700; total, $900. *Revenue,* consumers, $200. **Management.**—Supt., Chas. A. Griswold.

13. **RAPID CITY,** *Pennington Co.* (Pop., 292—2,128.) Built in '85-6 by city. **Supply.**—Spring, by gravity to reservoir. Fiscal year closes March 1. **Distribution.**—*Mains,* 5 miles. *Taps,* 210. *Meters,* 0. *Hydrants,* 23. *Consumption,* 200,000 galls. **Financial.**—*Cost,* $65,000. *Bonded Debt,* $51,000 at 7%, due in 1906. *Revenue,* consumers, $4,100. **Management.**—Chn., James M. Woods. Secy., W. H. Tompkins. Supt., H. J. McMahan.

14. **REDFIELD,** *Spink Co.* (Pop., in '90, 796.) Built in '86 by town. **Supply.**—Flowing artesian well, by gravity. **Distribution.**—(In '88.) *Mains,* 3 miles. *Taps,* 300. *Meters,* 0. *Hydrants,* 4. **Financial.**—(In '88.) *Cost,* $5,000. *Expenses,* $251. **Management.**—Supt., W. D. Beebe.

15. **SALEM,** *McCook Co.* (Pop., in '90, 429.) Built in '87-8 by village. **Supply.**—Well, pumping to tank; 2-in. pump has been replaced by a 5-in. pump. Fiscal year closes April 1. **Distribution.**—*Mains,* ½ mile. *Taps,* 9. *Meters,* 0. *Hydrants,* 3. *Consumption,* 2,000 galls. **Financial.**—*Cost,* $3,949. *No Debt. Expenses,* $200. *Revenue,* $50. **Management.**—Prest., L. S. Tyler. Auditor, W. H. Cross. Supt., M. A. Blackburn.

Scotland, *Bon Homme Co.* (Pop., 150—1,083.) Built in '89 by city. Contrs., Fairbanks, Morse & Co., St. Paul, Minn. **Supply.**—Well, pumping direct and to tank. Fiscal year closes Dec. 31. Water has thus far been used for fire protection and watering stock. Mains will be extended during '91 to residence portion of town to supply water for domestic consumption. Chn., Jno. C. Clark. Supt., A. D. Ardery.

16. **SIOUX FALLS,** *Minnehaha Co.* (Pop., 2,164—10,177.) Built in '84 by Sioux Falls Water Co.; now controlled by American Water-Works & Guarantee Co., Ltd., Pittsburg., Pa., under 25-yrs. exclusive franchise which fixes maximum rates and provides for purchase of works by city in '94, price determined by arbitration. Fiscal year closes Dec. 31. **Supply.**—Wells, pumping to stand-pipe. There are twelve 6-in. Cook wells, 40 ft. deep, having combined dy. cap. of 10,000,000 galls. **Pumping Machinery.**—Dy. cap., 5,000,000 galls.; one 3,000,000 and two 1,000,000 Worthington. *Boilers,* three 80-HP., horizontal tubular. **Distribution.**—*Mains,* 16 miles. *Taps,* 416. *Meters,* 0. *Hydrants,* 76. *Consumption,* 600,000 galls. **Financial.**—Withheld. **Management.**—Prest., J. H. Purdy. Secy., A. E. Clancy. Supt., R. W. Barnes.

FOR FURTHER DESCRIPTIVE MATTER see Manual for 1889-90. CHANGES since 1890 are here given. POPULATIONS are for 1880 and 1890, respectively.

17. SPEARFISH, *Lawrence Co.* (Pop., 170—678.) Built in '87 by town. Supply.—Springs, by gravity to tank. Fiscal year closes March 1. Distribution.— *Mains*, 5 miles. *Taps*. 75. *Meters*, 0. Financial.—*Cost*, $23.550. *Bonded Debt*, $20,000 at 6%, due in 1907. *Expenses :* operating, $200; int., $1,200; total, $1,400. *Revenue*, consumers, $1,150. Management.—Chn., J. F. Summers. Secy., G. W. Mitchell. Supt., M. Berg.

18. TYNDALL, *Bon Homme Co.* (Pop., in '90, 509.) Built latter part of '89 by village. *Mains*, 6 miles. *Taps*, 60. *Hydrants*, 27. *Cost*, $7,000. *Expenses*, nominal. *Revenue*, $200. City Auditor, W. P. Dunlap. Report by Henry B. Phœnix.

19. WATERTOWN, *Codington Co.* (Pop., 746—2,672.) Built in '88 by Watertown Water-Works Co., under 20-yrs. franchise. Const. Engrs., H. F. Dunham, Cleveland, O., and D. C. Henny. Denver. Col. Supply.—Lake Kampeska, pumping to stand-pipe. Distribution.—(In '89.) *Mains*, 8.8 miles. *Taps*, 140. *Meters*, 15 *Hydrants*, 80. *Consumption*, 220,000 galls. Financial.—(In '89.) *Cost*, $276,000. *Cap. Stock*, $150,000. *Bonded Debt*, $126,000 at 6%. *Expenses:* operating, $3,600; int., $7,560 ; total, $11,160. *Revenue*, $12,000. Management.—Prest., W. E. Scarritt. Treas., E. W. Thomas. Secy., Ed. Scarritt. Supt., F. Smith.

NOTE.—In '89, system came into possession of A. H. Howland, Boston, Mass. Statistics above were furnished by old co. Press dispatch, Nov. 20, '90, stated that Eastern directors had resigned and works had passed into hands of "McIntyre Syndicate."

20. WHITEWOOD, *Lawrence Co.* (Pop., in '90, 443.) History.—Built in '90 by Whitewood Water-Works Co., under 20-yrs. franchise. Contrs., U. S. Wind Engine & Pump Co., Omaha, Neb. Supply.—Springs, by gravity to tanks. Tanks. —Cap., 100,000 galls.; two 50,000-gall. Distribution.—*Mains*, 6-in., kal. i., 1¼ miles, *Taps*, 24. *Hydrants*, 7. *Consumption*, 40,000 galls. *Pressure*, 54 lbs. Financial. —*Cost*, $10,000. *Cap. Stock*, $10,000; all paid up. *Revenue:* consumers, $840; city, $300 ; total, $1,140. *Hydrant Rental*, $50. Management.—Prest., William Sellie.

21. WOONSOCKET, *Sanborn Co.* (Pop., in '90, 687.) Built in '89-90 by town. Supply.—Flowing artesian well, by gravity. *Mains*, 1 mile. *Consumers*, 40. *Hydrants*, 13. *Cost*, $6,500. *Expenses*, $100. *Revenue*, $200. Report by C. E. Hinds, Mayor.

22. YANKTON, *Yankton Co.* (Pop., 3,431—3,670.) Built in '84 by city. Supply. —Flowing artesian well, by gravity to tanks. Fiscal year closes July 1.

Year.	Mains.	Taps.	Meters.	Hyd'ts.	Cost.	B'd'd debt.	Int. ch'g's.	Exp.	Rev.
'86-7	1.9	40	0	14	$20,000	$18,000	$1,260	$175	....
'88-9	3.5	140	0	21	21,000	18,000	1,680	500	$1,750
'90-1	3.5	176	0	32	21.000	4,000	240	195	1,635

Management.—Committee: C. H. Bruce, W. B. Valentine, Jacob Huber. Supt., Jno. Martin. City Engr., E. D. Palmer.

Water-Works Projected with Fair Prospects of Construction.

FLANDRAU, *Moody Co.* (Pop., 471—569.) Aug. 31, '91, it was reported that $10,000 worth of bonds were for sale, to construct water-works. City wished either to build works or grant franchise to Co. Clerk, Alf. I. Whitman.

VERMILLION, *Clay Co.* (Pop., 714—1,496.) In '91 franchise was granted. Vermillion Artesian Well, Electric Light, Mining, Industrial & Improvement Co. were building Oct. 6, '91. Cap. stock, $30,000, of which S. V. Saleno, of Bay City, Mich., held $20,000. Direct pumping. Hydrants, 40. Hydrant rental, $50. Wells, pumping to stand-pipe.

WEBSTER, *Day Co.* (Pop., in '90, 618.) July 6, '91, it was reported that city would build water-works. Contrs. for artesian well, Swan Bros., Andover. Est. cost, $10,000. Prest., L. G. Oehsmint.

NORTH DAKOTA.

Water-Works Completed or In Process of Construction.

1. **BISMARCK**, *Burleigh Co.* (Pop., 1,758–2,186.) Built in '87 by Bismarck Water Co. Supply.—Missouri River, pumping to reservoirs. Fiscal year closes May 1. Distribution.—*Mains*, 6 miles, *Taps*, 110. *Meters*, 21. *Hydrants*, 43. Financial.—(In '88.) *Cost*, $110,000. *Cap. Stock*, $100,000. *Bonded Debt*, $100,000 at 6%. Management.—Prest., Ebor H. Bly. Secy. and Treas., C. B. Little.

2. **ELLENDALE**, *Dickey Co.* (Pop., in '90, 761.) Built in '86 by town. Supply.—Artesian well, by gravity. Distribution.—*Mains*, ¾ miles. *Taps*, 37. *Meters*, 0. *Hydrants*, 10. Financial.—*Cost*, $7,075. *Bonded Debt*, $7,950. *Expenses*: operating, nominal; int., $556. *Revenue*, consumers, $300. Management.—Com. of three, appointed every year by City Council. City Auditor, Chas. Ackley.

3. **FARGO**, *Cass Co.* (Pop., 2,693–5,664.) Bought by city Oct 1, '90, at a valuation of $65,748. Built in '89 by Fargo Water & Steam Co., under 30-yrs. franchise. Supply.—Red River of the North, pumping direct from settling basins. Distribution.—*Mains*, 13 miles. *Taps*, 460. *Meters*, 160. *Hydrants*, 65. Financial.—*Cost*, to city, $65,748. *Bonded Debt*, $30,000 at 5%. due in 1910. *Sinking Fund*, $1,000 annually. *Revenue*, $11,000. Management.—Mayor, Wilbur F. Ball. Supt., Wm. Hart.

4. **GRAFTON**, *Walsh Co.* (Pop., in 90, 1,594.) Built in '86 by city. Supply.—Artesian well, pumping direct. Distribution.—*Mains*, 1.5 miles. *Taps*,' 118. *Meters*, 0. *Hydrants*, 12. Financial.—*Cost*, $29,460. *Debt*, $10,000 at 6%. *Sinking Fund*, $8,000. *Expenses*, $2,500. *Revenue*, $350. Management.—Mayor, Jos. Deschenes. City Recorder, J. H. Kelly.

5. **GRAND FORKS**, *Grand Forks Co.* (Pop., 1,705–4,979.) Built in '82 by city. Supply.—Red Lake River, pumping direct. Fiscal year closes third Tuesday in April.

Y'r.	M'ns.	Taps.	M't's.	H'd'ts.	C'ns'pt'n.	Cost.	B'd'd d't.	Int. ch'g's.	Exp.	Rev.
'86-7	5	68	1	33	500,000	$55,000	$48,000	$6,360	$5,600	$4,200
'90-1	6.2	300	21	46	300,000	78,000	48,000	3,360	3,781	7,566

Financial.—*Bonded Debt*, $48,000 at 7%, due in '97-99 and 1905. *Sinking Fund*, $11,500, invested at 7%. *Expenses*, total, $7,141. *Revenue*, consumers, $7,556. Management.—Com. of five, chosen from council by its president. Chn., C. B. Ingalls. Supt., Hugh P. Ryan, appointed by Mayor.

6. **JAMESTOWN**, *Stutsman Co.* (Pop., 393–2,296.) Built in '87-8 by town. Supply.—Flowing artesian well, by gravity. Fiscal year closes Dec. 31. Distribution.—*Mains*, 1 mile. *Taps*, 16. *Meters*, 0. *Hydrants*, 20. Financial.—*Cost*, $12,000. *Bonded Debt*, $10,000 at 7%. *Expenses*, nominal. Management.—Secy., Andrew Blewitt. Supt., Jno. F. Venner.

Mandan, *Morton Co.* (Pop., 239–1,328.) Built in '84 by town for fire protection. See 1889-90 Manual, p. 548.

OAKES, *Dickey Co.* (Pop., in '90, 379.) Works reported in '89-90 Manual, p. 548. were not built. System will probably be constructed during '91.

7. **WAHPETON**, *Richland Co.* (Pop., 400–1,510.) Built in '85 by Wahpeton Water Co.; later passed to control of New Hampshire Trust Co., Manchester. In '90, city leased works, and by making semi-annual payments will own them at end of 20 yrs. Fiscal year closes Oct. 1. Supply.—Otter Tail Creek, by gravity to settling well, thence pumping direct. Distribution.—*Mains*, 5 miles. *Taps*, 360. *Meters*, 2. *Hydrants*, 25. *Consumption*, 100,000 galls. Financial.—*Cost*, $92,000. *Expenses*, operating, $5,000. *Revenue*: consumers, $7,000; city, $700; total, $7,700. Management.—Chn., Jno. Nelson. Secy., A. L. Roberts. Supt, E. H. Bishop.

Water-Works Projects Reported In '90 and '91, but Now Indefinite.

Devil's Lake, Ramsey Co.

FOR FURTHER DESCRIPTIVE MATTER see Manual for 1889-90. CHANGES since 1890 are here given. POPULATIONS are for 1880 and 1890, respectively.

WYOMING.

Water-Works Completed or in Process of Construction.

1. CHEYENNE, *Laramie Co.* (Pop., 3,456—11,690.) Built in '82 by town. Supply.—Infiltration wells in bed of Crow Creek, by gravity, with direct pumping in case of fire; storage reservoirs for emergency. Fiscal year closes Dec. 31. Reservoir of stone having cap. of 1,100.000 galls. constructed in '90; cost, $14,000. Also submerged dam of masonry across Crow Creek, just below filter galleries; cost, $6,000.

Year.	Mains.	Taps.	Meters.	Hy'd'ts.	Cons'p't'n.	B'd'd d't.	Int. ch'g's.	Exp.	Rev.*
'86	11	600	0	70	700,000	$130,000	$9.100	$3,500	$12,500
'88	14	750	2	90	1,500.000	....	......	7,000	24,000
'90	20	800	2	101	1,500,000	100,000	7,000	....	26,600

* Including allowance for hydrant rental.

Financial.—*Bonded Debt*, $100,000 at 7%. *Revenue:* consumers, $15,000; city, $11,600; total, $26,600. **Management.**—Water and Sewerage Committee of City Council. Chn., Louis G. Jenks. Supt., P. S. Cook.

2. DANA AND HANNA, *Carbon Co.* (Pop., in '90, Dana, 253; Hanna, 260.) **History.**—Built in '90 by Rattle Snake Creek Water Co., controlled by U. P. Ry. Co.; put in operation Oct. 10. Des. Engr., E. C. Kinney. Const. Engr., T. J. Wyche. **Supply.**—Rattle Snake Creek, by gravity to reservoir (probably including supply main). *Distribution.*—*Mains*, 12 to 4-in. c. i., 30 miles. *Consumption*, 300,000 galls. *Pressure:* ordinary, 85 lbs.; fire, 100 lbs. **Financial.**—Withheld. **Management.**—Supt., T. J. Wyche, Rock Springs.

3. DOUGLAS, *Converse Co.* (Pop., in '90, 491.) Built in '89-G0 by town. Supply.—Platte River, pumping, with Knowles dup. pump, to reservoir; filtered. Filter is a box 4 ft. wide, open at bottom, sunk 3 ft. below bed of river; connects with pump-well, 150 ft. from river. Fiscal year closes Dec. 31. *Distribution.*—*Mains*, 2 miles. *Taps*, 168. *Hydrants*, 10. *Consumption*, 5,000 galls. **Financial.**—*Cost*, $12,299. *Bonded Debt*, $7,000 at 7%. *Sinking Fund*, $350. *Expenses:* operating, $1,429; int., $490; total, $1,923. *Revenue*, consumers (for six months), $651. **Management.**—Comrs., appointed from Council every year. Report by J. H. Selkirk.

4. EVANSTON, *Uinta Co.* (Pop., 1,277—1,995.) **History.**—Built in '89-'90 by town; put in operation April 1. **Supply.**—Bear River, pumping to reservoir. **Pumping Machinery.**—Dy. cap., 1,000,000 galls.; Deane. Reservoir is 75 × 100 × 15 ft. *Distribution.*—*Mains*, kal. i.,6 miles. *Services*, kal., galv. i. Pipe and specials from National Tube Wks., Chicago. *Taps*, 110. *Hydrants*, National, 50. *Valves*, Ludlow. *Pressure:* ordinary, 75 lbs.; fire, 100 lbs. **Financial.**—*Cost*, $24,000. *Bonded Debt*, $20,000 at 6%. *Expenses:* operating, $1,500; int., $1,200; total, $2,700. *Revenue:* consumers, $2,000; city, $2,500; total, $4,500. **Management.**—Mayor and Council. Report by A. C. Sloan.

8. GREEN RIVER, *Sweetwater Co.* See Rock Springs.

2. HANNA, *Carbon Co.* See Dana.

5. LARAMIE, *Albany Co.* (Pop., 2,696—6,388.) Built in '70 by U. P. Ry. Co. **Supply.**—Springs, by gravity from reservoir or pond. Town paid ry. co. $18,000 toward construction of works, and has use of water free of charge, for fire protection and for private consumption to all who will pipe it from mains at their own expense. *Mains* (in '88), 5 miles. *Hydrants* (in '88), 24.

6. NEW CASTLE, *Weston Co.* (Pop., in '90, 1,715.) Built in '90 by New Castle & Cambria Water Supply Co.; city may buy works in 20 yrs., or the time may then be extended; rates are regulated by city ordinance. **Supply.**—Sweet Water Creek, by gravity. *Distribution.*—*Mains*, 3 miles. *Taps*, 50. *Meters*, 3. *Hydrants*, 11. *Consumption*, 150,000 galls. **Financial.**—*Cost*, $95,000. *Revenue,* city, $15,000. **Management.**—Prest., W. H. Kilpatrick. Secy. and Treas., R. A. Weston. Supt., F. W. Mondell.

7. RAWLINS, *Carbon Co.* (Pop., 1,451—2,235.) U. P. Ry. Co. supplied town from '82 till '84, supply being derived from artesian wells. In '84 well was drilled by Carbon Co., 500 ft. deep, water being within 40 ft. of surface. This well supplied town till '90, when break in casing, about 200 ft. below surface, ruined well. In '90 ry. co. drilled second well, which now supplies town. April 14, '91, voters authorized issuance of $35,000 of bonds to sink new well, pipe town, and erect 54 hydrants. Report by H. Rosmusson, Mayor. City Clerk, C. P. Hill.

WYOMING.

8. ROCK SPRINGS AND GREEN RIVER, *Sweetwater Co.* (Pop.: Rock Springs, 763—3,406; Green River, 327--723.) Built in '86-8 by U. P. Ry.

GREEN RIVER PLANT.

Supply.—Green River, pumping through mains to tanks. Fiscal year closes Dec. 31. **Distribution.**—*Mains*, 4 miles. *Taps*, 110 *Hydrants*. 7. *Consumption*, 250,000 galls. **Financial.**—Withheld. **Management.**—See Rock Springs Plant.

ROCK SPRINGS PLANT.

Supply.—Green River, pumping direct, surplus going to reservoir. Fiscal year closes Dec. 31. **Distribution.**—*Mains*, 31 miles. *Taps*, 310. *Meters*, 190. *Hydrants*. 15. *Consumption*, 500,000 galls. **Financial.**—Withheld. **Management.**—Genl. Supt., T. J. Wyche. Asst. Supt., W. S. Avery.

9. SUNDANCE, *Crook Co.* (Pop., in '90, 515.) Built in '88 by Sundance Water Supply Co., under 20-yrs. franchise; city may purchase works in 1903 or 1908, price determined by arbitration. **Supply.**—Springs, by gravity to and from reservoir. Fiscal year closes Dec. 31. During '91 supply was to be increased. reservoir built, and 5 miles of mains laid. **Distribution.**—*Mains*, 1.5 miles. *Taps*, 50. *Hydrants*, 10. **Financial.**—*Cost*, $20,000. *Cap. Stock:* authorized, $40,000; paid-up, $20,000; paid 10% div'd in '90. *No Bonds*. *Floating Debt*, $7,000 at 12%. *Expenses:* operating, "nominal"; int., $840; taxes, $60. *Revenue:* consumers, $1,350; fire protection, $750; other purposes, $300; total, $2,400. **Management.**—Prest., W. S. Metz. Secy. and Treas., T. M. Pettigrew.

Water-Works Projected With Fair Prospects of Construction.

CASPER, *Natrona Co.* (Pop., in '90, 544.) Town voted to build works, and July, 1, '91, it was reported that construction would begin Aug. 1. River, pumping to tank. *Est. Cost.*, $10,000. Address O. K. Garry.

LANDER, *Fremont Co.* (Pop., 193--525.) Oct. 18, '91, town wished to grant 20-yrs. franchise for works taking gravity supply from Big Popo-Agie River, which flows through town with fall of 60 ft. per mile. Water-Works Com., F. T. Wright and N. H. Brown. Town Engineer, J. E. Hill.

MONTANA.

Water-Works Completed or in Process of Construction.

1. ANACONDA, *Deer Lodge Co.* (Pop., in '90, 3,975.) **History.**—Built in '88 by Anaconda Water-Works Co., under perpetual franchise. Des. Engr., Jos. H. Harper, Butte. **Supply.**—Mountain springs, by gravity to reservoir. **Reservoir.**—Cap., 8,500,000 galls. **Distribution.**—*Mains*, 10, 8 and 6-in. c. i., 9.5 miles. *Services*. ¾-in. *Taps*, 650. *Hydrants*, Ludlow, 27. *Valves*, Ludlow. *Consumption*, 225,000 galls. *Pressure:* ordinary, 105 lbs.; fire, 120 lbs. **Financial.**—*Cost*, $100,000. *Cap. Stock:* authorized, $100,000 ; all paid up. *Hydrant Rental* $100. **Management.**—Prest., Marius Daly. Secy., and Supt., D. F. Hallahan.

2. BILLINGS, *Yellowstone Co.* (Pop., in '90, 836.) Built in '86-7 by Billings Water-Power Co., under 20-yrs. franchise; city may purchase works in 1906 at price fixed by three comrs. chosen by city and co. **Supply.**—Yellowstone River, through canal, pumping direct. Fiscal year closes Nov. 1. In '90, Buffalo pump, 10 × 16 × 12 × 12, was added, to use in emergency, when troubled with ice.

Year.	Mains.	Taps.	Meters.	Hyd'ts.	C'ns'pt'n.	Cost.	B'd'd d't.	Int.	ch'g's.	Exp.	Rev.
'87	5.3	120	0	15	150,000	$60,000	$30,000		$2,730	$5,725	$5,500
'90	6	75	0	20	250,000	80,000	50,000		3,500	4,600	12,700

Financial.—All data include both water and lighting plants. *Cap. Stock:* authorized, $50,000; paid up, $30,000. *Bonded Debt*, $50,000 at 7%, due in 1906. *Expenses:* operating, $4,000; int., $3,500; taxes, $600; total, $8,100. *Revenue:* consumers, $10,700; fire protection, $2 000; total, $12,700. **Management.**—Prest., Henry Belknap, N. Y. City. Secy. and Supt., H. W. Rowley. Treas., H. H. Mund.

FOR FURTHER DESCRIPTIVE MATTER see Manual for 1889-90. CHANGES since 1890 are here given. POPULATIONS are for 1880 and 1890, respectively.

3. **BOZEMAN,** *Gallatin Co.* (Pop., 894—2,143.) Built in '89 by Bozeman Water-Works Co. **Supply.**—Lyman Creek, spring-fed mountain stream, by gravity. Fiscal year closes Dec. 31. **Distribution.**—*Mains,* 7.7 miles. *Taps,* 155. *Meters,* 5. *Hydrants,* 30. **Financial.**—*Cost.* $110,000. *Cap. Stock:* authorized, $100,000; all paid up. *Bonded Debt,* $25,000 at 6%, due in 1910. **Management.**—Prest., S. M. Cary. Secy., Tracy Lyon. Treas., R. B. C. Bement. All of St. Paul, Minn. Supt., E. P. Whipple.

4. **BUTTE CITY,** *Silver Bow Co.* (Pop., 3,363—10,723.)

OLD WORKS.

History.—Built in '82-3 by Silver Bow Water Co., which in '90 was merged with Summit Valley Co., evidently under original name. Franchise does not provide for purchase of works by city; city has no control of rates. **Supply.**—Creeks, pumping to reservoir and direct. First supply was four springs at head of Bull Run Gulch. About '86 water was first pumped from Silver Bow Creek, near southern end of city, by means of 800,000-gall. Knowles pump. Complaint of water being contaminated led to building of 20,000,000 gall. reservoir on Basin Creek, 12 miles south of town; reservoir formed by earth dam. Water was conveyed from reservoir in 10 × 12-in. wooden flume. In '90 preparations were made to bring water from German Gulch, 20 miles west of city. In '91, 50,000,000-gall. reservoir was being added on Basin Creek, and 2,000,000 Allis pump was added. Fiscal year closes Aug. 20. **Distribution.**—*Mains,* about 15 miles. *Taps,* withheld. *Meters,* 120. *Hydrants,* about 47. *Consumption,* 850,000 galls. **Financial.**—*Cost,* $260,000. *Cap. Stock,* authorized and paid up, $60,000; no div'd since Apr., '87, all earnings going into plant. *Total Debt,* $285,000: floating, $60,000 at 12%; bonded, $225,000 at 7%. *Expenses:* operating, $25,441; licenses and taxes, $28,368. *Revenue:* consumers, $59,905; city, $6,000; total, $65,905.

CHANGES AND PROPOSED NEW WORKS.

New works will be built in '92 by Butte City Water-Works Co. In '89 Silver Bow Hydraulic Mining Co. was organized to carry on hydraulic mining and supply water; Mar. 5, '90, this co. received franchise to supply Butte. Its sources of supply were Jerry and Divide creeks, 20 and 27 miles southwest of city. During '89-90 this co. built canal 20 miles long from source toward city at cost of $60,660; Engrs., Ray & Leonard, Butte. During '90 Silver Bow Hydraulic Mining Co. was bought out, or bargains made, by parties who, in Apr., '91, incorp. Butte City Water-Works Co., with authorized cap. stock of $2,000,000. Apr. 10, '91, latter co. bought, or bargained for, property of Silver Bow Water Co., described above. May 2, '91, Butte City Water-Works Co. received franchise from city. Owing, evidently, to stringency in money market, new co. was unable to complete transactions with two old cos. until Oct., '91, when it was reported that arrangements were being perfected by old cos. to transfer water-works property to new, the considerations named in deed being $300,000 to Silver Bow Water Co. and $100,000 to Silver Bow Hydraulic Mining Co.

May 3, '91, N. C. Ray, Butte, temporary trustee of new co., reported that a new supply would be introduced as follows: Waters of Jerry and Divide creeks, tributaries to Missouri River, would be stored in 1,000,000,000-gall. reservoir at head of Jerry Creek, some 20 miles from city and about 7,600 ft. above sea. From reservoir 24-in. pipe would carry water across ravine to summit of ridge between creeks and deliver water into channel of Jerry Creek; max. head on this pipe line, 1,100 ft. On Divide Creek two reservoirs with combined cap. of 400,000,000 galls. were proposed. Thence a 30-in. pipe about 20 miles long would lead to city, crossing main range of Rocky Mts. in Deer Lodge Pass, 5,700 ft. above sea. Reservoirs at different elevations, and with combined cap. of 20,000,000 galls., were proposed near city. About 20 miles of mains and 85 hydrants were projected for addition to old works. Consult. Engr., C. B. Davis, Chicago, Ill. Const. Engrs., Ray & Leonard, Butte. Moffett Hodgkins & Clarke, Syracuse, N. Y., are interested.

1. **CASCADE,** *Cascade Co.* (Pop. of precinct, in '90, 134.) Built in '89 by Perkins & Taylor. **Supply.**—Flowing artesian well, by gravity. *Mains,* 3.5 miles. *Taps,* 35. *Hydrants,* 1. *Cost,* $7,000. Owners, Perkins & Taylor.

5. **DEER LODGE CITY,** *Deer Lodge Co.* (Pop., 941—1,463.) **History.**—Built in '87 by Deer Lodge Water Co. Engr., H. B. Davis. **Supply.**—Mountain stream, by gravity. **Reservoirs.**—Two ; one 20 × 100 × 12 ft. deep; one 25 × 25 × 12 ft.

Distribution.—*Mains*, 10 to 4-in., 12 miles. *Taps*, 200. *Hydrants*, Ludlow, 16. *Valves*, Ludlow. Consumption, about 70,000 galls. *Pressure*, 45 lbs. Financial.—*Cost*, $54,000. *Cap. Stock:* authorized, $60,000; paid up, $54,000. *No Bonds. Floating Debt*, $6,000 at 10%. *Expenses*, $1,200. *Revenue:* consumers, $5,500; city, $800; total, $6,300. Management.—Prest., N. G. Bielenberg. Secy., Willard Bennett. Treas., Wm. Coleman. Supt., Frank S. Davey.

6. **FORT BENTON,** *Choteau Co.* (Pop., 1,618—621.) Owned by National Tube Works Co., McKeesport, Pa. Built in '87-8 by Fort Benton Water Co., under 20-yrs. franchise; rates fixed in franchise. Supply.—Missouri River, pumping direct. Distribution.—*Mains* (in '88), 5.5 miles. *Taps*, 225. *Meters*, 0. *Hydrants*, 25. Financial.—*Cost* (in '88), $42,000. *Expenses*, operating, $1,800. *Revenue*, city, $2,500. Management.—Supt., H. K. Holmes.

7. **GREAT FALLS,** *Cascade Co.* (Pop., in '90, 3,979.) Built in '89 by Great Falls Water Co. Supply.—Missouri River, pumping direct. Fiscal year closes Dec. 31. In '91 city limits were extended and council ordered 6½ miles new mains, of which 800 ft. of 12-in pipe will be laid beneath Missouri River, deepest water being 21 ft. Filter and stand-pipe projected for spring of '92.

Year.	Mains.	Taps.	Meters.	Hyd'ts.	C'ns'pt'n.	Cost.	B'd'd debt.	Int. ch'g's.	Rev.
'89	7.1	114	0	71	68,000	$85,000	$75,000	$5,250	......
'90	10.1	294	2	101	285,000	119,906	92,000	6,440	$14,325

Financial.—*Cap. Stock*, $150,000. *Bonded Debt*, $92,000 at 7%. *Expenses:* operating, cannot be determined, as works are run in connection with an electric lighting plant; int., $6,440; taxes and licenses, $510. *Revenue:* consumers, $6,240; fire protection, $6,690; other purposes, $1,395; total, $14,325. Management.—Prest., E. G. Maclay. Secy., Jno. G. Maclay. Supt., E. W. King.

8. **HELENA,** *Lewis and Clarke Co.* (Pop., 3,624—13,834.) Now owned by Helena Consolidated Water Cos. In '87-8 Helena Water Co. built works under 20-yrs franchise. Three other cos. built works prior to '89, and in that year consolidated. Supply.—(In '88.) Combined surface and driven wells, pumping to reservoir. Early in '91 gravity supply was introduced from reservoir in mountain 9 miles from city. Distribution.—*Mains* (in '89), 40 miles. *Taps* (in '88), 550. *Meters* (in '88), 20. *Hydrants* (in '89), 200. Financial.—Withheld. Management.—In '90, Frank J. Stever.

9. **LIVINGSTON,** *Park Co.* (Pop., in '90, 2,850.) History.—Built in '90-1 by Livingston Water-Works & Electric Light Co., under 20-yrs. franchise, which fixes rates and provides for purchase of works by city. Put in operation June 1. Engr., Sam'l Burdock. Contr., E. Sutphin, St. Louis, Mo. Supply.—Yellowstone River, pumping to reservoir and direct; filtered through gravel into large well on bank of river. Pumping Machinery.—Dy. cap., 1,250,000 galls.; one 500,000 and one 750,000 Deane. Reservoir.—Cap., 350,000 galls; 140 ft. above city. Distribution.—*Mains*, 10 to 4-in., c. i., 4.5 miles. *Taps*, 175. *Meters*, 2. *Hydrants*, 35. *Consumption*, 200,000 galls. *Pressure*, 70 lbs. Financial.—*Cost*, $60,000. *Cap. Stock:* authorized, $50,000; all paid up. *Bonded Debt*, $30,000 at 6%. *Revenue*, city, $3,375. Management.—Prest., W. J. Anderson. Vice-Prest., J. A. Lavy. Secy. and Treas., E. C Day. Supt., E. C. Ross.

10. **MILES CITY,** *Custer Co.* (Pop., 629—956.) History.—Built in '90 by Miles City Water & Electric Light Co., under 15-yrs. franchise. Put in operation July 19. Des. Engr., B. Ullman. Const. Engr., J. M. Buckner. Supply.—Artesian and surface wells, pumping direct. Two surface wells, each 30 ft. in diam., 20 ft. deep; six 4½-in. artesian wells, from 179 to 480 ft. deep. Pumping Mahinery.—Dy. cap., 1,250,000 galls.; one 1,000,000 Worthington, one 250,000 Knowles. Distribution.—*Mains*, 6 and 4-in. c. i., 2 miles. *Taps*, 16. *Hydrants*, Eclipse, 15. *Valves*, Ludlow. *Pressure:* ordinary, 15 lbs.; fire, 100 to 150 lbs. Financial.—*Cost*, $16,000. *Hydrant Rental*, $100. Management.—Prest., E. R. Gillman, Chicago. Secy. and Treas., H. P. Batchelor. Supt., B. Ullman.

11. **MISSOULA,** *Missoula Co.* (Pop., in '90, 3,426.) Built in '83 by Missoula Water-Works & Mining Co.; franchise provides neither for control of rates nor purchase of works by city. Supply.—Rattle Snake Creek, by gravity. Fiscal year closes Dec. 31. Distribution.—*Mains*, 11.5 miles. *Taps*, 300. *Hydrants*, 32. Financial.—*Cost*, $75,000. *Cap. Stock*, $70,000; paid no div'd. in '90. *No Bonds. Floating Debt*,

For Further Descriptive Matter see Manual for 1889-90. Changes since 1890 are here given. Populations are for 1880 and 1890, respectively.

$30,000. *Expenses:* operating, $2,400; int., $3,000; taxes, $270; total, $5,670. *Revenue,* city, $7,200. In '90 hydrant rental was increased from $400 to $600 per month **Management.**—Prest. F. G. Higgins. Treas., J. M. Reitz. Secy. and Supt., Alfred Cave.

II. **WALKERVILLE,** *Silver Bow Co.* (Pop., 444—1,743.) Mountain Water Co. supplies city. Address manager, at Butte City. It is expected that Butte City Water Co. will eventually supply city.

Water-Works Projected With Fair Prospects of Construction.

CASTLE, *Meagher Co.* (Pop., in '90, 383.) Franchise for 20 yrs. granted in '91 to A. M. Holter and Jas. King, Helena, who organized Castle Water & Power Co., with Mr. King, Secy. and Treas. In '91 ditch one mile long, 3 ft. wide, 2 ft. deep, with fall of ¼-in. per rod was built to town from Boulder and Aliebaugh creeks. A 17,000,000-gall. reservoir will be built at head of creek. Mains must be laid by July 10, '91. Est cost, $60,000.

DILLON, *Beaver Head Co.* (Pop., in '90, 1,012.) Aug. 18, '91, it was reported that J. H. Lawrence and others had been granted a 20-yrs. franchise, which would expire Sept. 1, and as yet nothing further had been done.

PHILLIPSBURGH, *Deer Lodge Co.* (Pop., 299—1,058.) Water-works talked of. Address K. J. Pardee.

Water-Works Projects Reported In '90 and '91, but Now Indefinite.

White Sulpur Springs, Meagher Co.

GROUP 7.—Southwestern States.—Water-Works Completed, In Process of Construction or Projected, In Missouri, Arkansas, Texas, Colorado, New Mexico and Oklahoma.

MISSOURI.

Water-Works Completed or In Process of Construction.

1. **BONNE TERRE,** *St. Francois Co.* (Pop., in '90, 3,719.) Built in '87 by St. Joe Lead Co. Supply.—Big Run, pumping direct. A number of 6 to 3-in. mains run through portion of town for free use of people. About 15 fire plugs afford fire protection and are fed by pump in engine house.

2. **BOONVILLE,** *Cooper Co.* (Pop., 3,854—4,141.) Built in '83 by Boonville Water Co., under 20-yrs. franchise. Supply.—Missouri River, pumping to reservoir and tank. Fiscal year closes Dec. 31. Boiler was added in '90; additional reservoir was in course of construction in '91.

Year.	Mains.	Taps.	Meters.	Hydrants.	Consumption.	Cost.	B'd't	d'bt.	Int.	ch's.
'84	6	187	0	50	200,000	$83,000	$50,000		$3,000	
'86	6.5	245	2	50	200,000	84,000	25,000		3,000	
'88	7.5	260	2	52	250,000	84,000	50,000		3,000	
'90	8	200	1	52	250,000		50,000		3,000	

Financial.—*Cap. Stock,* authorized, $100,000. *Bonded Debt,* $50,000 at 6%. **Management.**—Prest., Jno. S. Elliott. Secy., W. C. Culverhouse. Treas., Jno. Cosgrove. Supt., R. Hodelich.

3. **BUTLER,** *Bates Co.* (Pop., 2,162—2,812.) History.—Built in '91 by Butler Water & Light Co., under 20-yrs. franchise, which fixes rates and provides for purchase of works by city at end of 10 yrs. or at intervals of 5 yrs. thereafter; put in operation Sept. 1. Des. Engr., J. W. Neir, St. Louis. Const. Engr., C. B. Ingel, Nevada. Supply.—Miami River, pumping to reservoir, thence to stand-pipe; filtered. Pumping Machinery.—Dy. cap., 1,750,000 galls.; Deane. Reservoirs.—There are two settling and one storage reservoir. Stand-Pipe,—Cap., 115,000 galls.; 14 × 100 ft.; from Riter & Conley, Pittsburg, Pa. Distribution.—*Mains,* 8-in., 9.5 miles. Pipe and specials from Shickle, Harrison & Howard, St. Louis. *Hydrants,* Empire, 75. Financial.—*Cap. Stock:* authorized, $100,000; paid up, $50,000. *Revenue,* city, $1,000. Management.—Prest., F. J. Tygard. Secy., Wm. E. Walton. Treas., J. E. Clark. Supt., C. B. Ingels.

4. CARROLLTON, *Carroll Co.* (Pop., 2,313—3,878.) Built in '85 by Carroll Water-Works Co., under 20-yrs. franchise, when city has right to purchase works, price to be determined by arbitration; city does not control rates; they are same as those of Kansas Cy. **Supply.**—Driven wells, pumping direct and to stand-pipe. During '90 a 12 × 115-ft. stand-pipe was built. **Distribution.**—*Mains*, 3 miles. *Taps*, 132. *Hydrants*, 216. **Financial.**—*Cap. Stock*, $60,000. *Bonded Debt*, $65,890: $40,000 at 6%, due in 1910; $25,000 at 7%, due in 1904; $390, rate not given. *Expenses:* operating (in '89), $2,500; int., $2,400; total, $4,900. Co. is exempt from taxation. *Revenue:* consumers, $1,000; fire protection, $2,415; other purposes, $600; total, $4,015. **Management.**—Prest., B. F. Jones. Secy. and Treas., F. C. Gunn, both of Kansas Cy. Supt., S. K. Turner.

37. CARTERVILLE, *Jasper Co.* (Pop., 483—2,881.) Supply from Webb City.

5. CARTHAGE, *Jasper Co.* (Pop., 4,167—7,981.) Built in '81-2 by Carthage Water-Works Co. **Supply.**—Spring River, pumping to stand-pipe and direct; filtered.

Year.	Mains.	Taps.	M't'rs.	H'd'ts.	C'ns'pt'n.	Cost.	B'd'd d't.	Int. c'g's.	Exp.	Rev.
'81	8	430	4	55	300,000	$65,000	$50,000	$3,000	$4,000	$9,000
'86	10	600	10	55	350,000	75,000	50,000	3,000	3,800	10,725
'90	14	750	18	68	750,000					

Financial.—Withheld. **Management.**—Prest., C. F. Martin. Secy. and Supt., H. C. Messenger.

6. CHILLICOTHE, *Livingston Co.* (Pop., 4,078—5,717.) Built in '87 by Chillicothe Water, Gas & Electric Light Co., under franchise which provides for purchase of works by city in '97 and at intervals of 5 years thereafter; rebuilt in '90; now controlled by Debenture Guarantee & Assurance Co., Chicago, Ill. Engr. for improvements, Geo. H. Pierson, N.Y. City. **Supply.**—Grand River, pumping through filters to clear water reservoir, thence by repumping to stand-pipe and direct. Fiscal year closes Dec. 31. **Distribution.**—*Mains*, 11 miles. *Taps*, 470. *Meters*, 1. *Hydrants*, 115. *Consumption*, 160,000 galls. **Financial.**—*Cap. Stock*, paid up, $250,000. *Bonded Debt*, $145,000 at 6%. *Expenses:* operating and taxes, $5,000; int., $8,700; total, $13,700. *Revenue:* consumers, $9,115; ry., $3,000; city, $6,900; total, $19,015. **Management.**—Prest., B. A. May. Secy. and Treas., Jno. T. Martingale. Supt., H. G. Seymour.

7. CLINTON, *Henry Co.* (Pop., 2,868—4,737.) Built in '86 by Clinton Water Co. **Supply.**—Artesian wells, pumping to stand-pipe, reservoir and direct. During '90 Deane pump and 3,000,000-gall. reservoir were added. **Distribution.**—*Mains*, 10 miles. *Taps*, 350. *Hydrants*, 55. **Financial.**—Withheld. **Management.**—Prest., B. F. Hobart, St. Louis. Secy. and Supt., W. H. Allen.

8. FULTON, *Callaway Co.* (Pop., 2,409—4,314.) Built in '88-9 by city. **Supply.**—Artesian well, by gravity to reservoir for domestic use, and by direct pumping for fire protection. Fiscal year closes June 1. During '90 stand-pipe and reservoir described below were added. **Stand-Pipe.**—*Cap.*, 100,000 galls.; 15 × 75 ft.; of boiler plate steel. **Reservoir.**—*Cap.*, 132,000 galls.; 40 × 40 × 11ft. **Distribution.**—*Mains*, 3.2 miles. *Taps*, 111. *Hydrants*, 47. *Consumption*, 30,000 galls. *Pressure:* ordinary 75 lbs.; fire, 150 lbs. **Financial.**—*Bonded Debt*, $20,000 at 6%; $8,000 due in 1900, $12,000 due in 1910. *Sinking Fund*, $3,000, not yet invested. *Expenses:* operating, $1,500; int., $1,200; total, $2,700. *Revenue:* consumers, $1,200; city, none. **Management.**—Prest., W. H. Wilkerson, Mayor. Secy., T. M. Bolton. Supt., W. H. Dawson.

9. HANNIBAL, *Marion Co.* (Pop., 11,074—12,857.) Built in '79-80 by Hannibal Water Co., under 20-yrs. franchise, which fixes rates and provides for purchase of works by city at its expiration. **Supply.**—Mississippi River, pumping to reservoir; filtered through charcoal and gravel. Fiscal year closes Dec. 31. A 25,000,000-gall. reservoir was to be constructed in '91.

Year.	Mains.	Taps.	M't'rs.	Hy'd'ts.	Cons'p't'n.	Cost.	B'd'd d't.	Int.ch'g's.	Exp.	Rev.
'80	11	500	..	75	130,000	$100,000				
'82	13.2	660	12	88	301,210	102,714	$90,000	$6,300	$6,813	$19,036
'84	14	760	20	88	50',000	112,000	90,000	6,200		
'86	14	750	31	92	500,000	128,000	105,000	8,400	5,700	5,000*
'88	14.5	800	24	92	750,000					
'90	15.5	1,164	44	97	1,130,000	172,000			9,600	33,740

*From city.

For Further Descriptive Matter see Manual for 1889-90. Changes since 1890 are here given. Populations are for 1880 and 1890, respectively.

302   MANUAL OF AMERICAN WATER-WORKS.

Financial.—*Expenses,* operating, $9,600. *Revenue:* consumers, $17,780; fire protection, $5,660; total, $33,740. **Management.**—Prest., E. Whitaker. Secy., H. D. Wood. Supt., Chas. J. Lewis.

10. **HOLDEN**, *Johnson Co.* (Pop., 2,014—2,520.) Built in '87 by Holden Water-Works Co., under 20 yrs. franchise which fixes rates and provides for purchase of works by city at intervals of 10 yrs. **Supply.**—Impounding reservoir, pumping through filter to stand-pipe. **Distribution.**—(In '88.) *Mains,* 6.3 miles. *Taps,* 46. *Hydrants,* 50. *Consumption,* 23,500 galls. **Financial.**—(In '88.) *Cost,* $65,000. *Bonded Debt,* $75,000 at 6%. *Expenses:* operating, $3,000; int., $4,500; total, $7,500. *Revenue,* city, $3,000. **Management.**—Prest., Theo. Plate; Secy., A. Schenk; both at 52 and 53 Gay Building, St. Louis, Mo. Supt., R. P. Adams.

11. **INDEPENDENCE**, *Jackson Co.*—(Pop., 3,146—6,380.) Built in '83-4 by Independence Water Co., under 20-yrs. franchise which regulates rates; city has right to purchase works in '94. **Supply.**—Missouri River, pumping to reservoir; filtered. Fiscal year closes Jan. 1. Sept. 4, '90, stockholders voted to increase bonded indebtedness by issuing $40,000 more of bonds. As soon as bonds are placed, following improvements will be made: Stand-pipe, 15 x 150 ft.; two 750,000-gall. dup. pumps, to pump to elevation of 535 ft.; 2 miles of 14-in. and several miles of 4-in. mains.

Year.	Mains.	Taps.	Mtrs.	H'd'ts.	C'ns'pt'n.	Cost.	B'd'd d't.	Int.ch'g's.	Exp.	Rev.
'84	8.5	200	0	50	250,000	$90,000	$50,000	$3,500	$2,500	$8,000
'90	9	350	9	50	250,000	100,000	60,000	4,900	5,910	9,031

Financial.—*Cap. Stock:* authorized, $100,000; all paid-up; paid no div'd in 1890. *Bonded Debt,* $60,000 at 7%; $50,000 due in 1904; $10,000 due in '96. *Expenses:* operating, $5,810; int., $4,900; taxes, $100; total, $10,810. *Revenue:* consumers, $5,031; fire protection, $4,000; total, $9,031. **Management.**—Prest., W. E. Winner. Secy., C. S. Crysler. Treas., W. W. Anderson. Supt., R. D. Wirt. NOTE.—Reported Sept., '91, that R. D. Wirt had been appointed Receiver.

12. **JEFFERSON CITY**, *Cole Co,* (Pop., 5,271—6,742.) Built in '88 by Jefferson City Water-Works Co., under 20-yrs. franchise which provides that rates must not be higher than in other places on Missouri River. City has right to purchase works in 20 years and every 5 years thereafter. **Supply.**—Missouri River, pumping to aerating and settling basins, from which water flows to clear water basin; thence repumped to stand-pipe or direct. Fiscal year closes Oct. 1. System of æration, described in '89-90 Manual, was designed by J. B. Johnson, St. Louis. Plant supplies prison, breweries and railway, in addition to regular consumers.

Year.	Mains.	Taps.	Meters.	Hyd'ts.	C'ns'pt'n.	Cost.	B'd'd d'bt.	Int. ch'g's.	Exp.	Rev.
'88	8.5	150	12	42	385,000	$149,000	$100,000	$5,000	$3,200	$10,560*
'90	8.5	..	..	72	400,000	153,000	100,000	5,000	10,200	13,550

*For 7 months.

Financial.—*Cap. Stock,* $100,000. *Bonded Debt,* $100,000 at 5%, due in 1908. *Sinking Fund,* $1,000 annually; used to pay off bonds. **Management.**—Prest., F. H. Bindler. Secy., W. A. Dallmeyer. Treas., W. W. Wagner.

13. **JOPLIN**, *Jasper Co.* (Pop., 7,038—9,943.) Built in '80 by Joplin Water Co., under 20 yrs. franchise; rates fixed by city ordinance. **Supply.**—Shoal Creek, pumping to reservoir. In '91 two 1,500,000-gall. Deane pumps, new boilers and pumping station were added. Engr., J. W. Nier, Kansas City. Fiscal year closes Dec. 31.

Year.	Mains.	Taps.	M'rs.	Hyd'ts.	B'd'd debt.	Int. ch'g's.	Exp.	Rev.
'84	13	580	6	65	$75,000	$5,250	$2,700	$8,500
'86	13	650	6	65	75,000	5,250	2,700	5,300*
'88	14	674	10	60	75,000	5,250	4,750	15,504
'89	14	...	13	60	75,000	5,250	5,000	18,257
'90	17	...	18	91	75,000	5,250	6,010	23,725

* From city.

Financial.—*First Cost,* $124,000. *Bonded Debt,* $75,000 at 7%, due in 1901. *Expenses:* operating, $5,450; int., $5,250; taxes, $650; total, $11,350. *Revenue:* consumers, $18,434; fire protection, $5,291; total, $23,725. **Management.**—Prest., Thos. Connor. Secy. and Supt., C. W. Glover. Treas., Jno. F. Wise.

14. **KANSAS CITY**, *Jackson Co.* (Pop., 55,785—132,716.) Built in '73 by National Water-Works Co.; city may buy in '93 at price to be mutually agreed upon or determined by courts. **Supply.**—Missouri River, pumping direct. Fiscal year closes Dec. 31. A 10,000,000-gall. Gaskill pump was added in '90.

MISSOURI. 303

Year.	Mains.	Taps.	M'trs.	H'd'ts.	Cons'p'n.	Cost.	B'd'd dbt.	Int.chgs.	Exp.	Rev.
'80	25	1,700		265	2,070,000				$43,000	
'82	42	2,763	140	303	3,450,000				30,000	
'84	53	3,200	300	495	4,500,000	$1,150,000	$1,150,000	$90,000	32,750	$191,561
'86	75	6,080	615	726	6,957,660	4,300,000	3,000,000	180,000	45,425	262,630
'87	85	7,786	950	852	11,000,000	4,500,000	3,000,000	180,000	56,283	392,764
'90	138	11,198	1,971	1,417	12,000,000					

Financial.—Figures for '90 withheld. **Management.**—Prest., Giles E. Taintor, N. Y. City. Secy., Robt. N. Weems. Supt., B. F. Jones.

NOTE.—Summer of '91 city voted $2,000,000 in bonds for public works. Board of Public Works. G. W. Pearsons, Supt. of construction, engaged G. H. Benzenberg, Milwaukee; Chas. Hermany, Louisville; M. L. Holman, St. Louis, to report on plan for, and cost of new works. Preliminary work was done by Mr. Pearsons. Oct., '91, experts reported est. cost of works to supply 200,000 people at $2,425,620, and that works for present population could be built for $2,002,434. Plan included intake crib in Missouri River at Harlem bend, pumping to settling basins for low service, repumping direct for high service.

15. **LEBANON,** *Laclede Co.* (Pop., 1,419—2,218.) **History.**—Built in '90 by Lebanon Light & Water Co., under 20-yrs. franchise; put in operation in Dec. Engr., W. H. Allen, Clinton. **Supply.**—Artesian well, pumping to stand pipe. Well is 1,000 ft. deep. **Pumping Machinery.**—Not described. **Stand-Pipe.**—Cap., 85,000 galls.; 12½ × 100 ft.; from Shickle, Harrison & Howard, St. Louis. **Distribution.**—*Mains*, 10 to 4-in. *Hydrants,* Perkins, 12. *Valves,* Scott. **Financial.**—Withheld. **Management.**—Prest., B. F. Hobart; Secy. and Treas., W. P. Heath; both of St. Louis.

16. **LEXINGTON,** *La Fayette Co.* (Pop., 3,995—4,537.) Built in '84-5 by Lexington Water Co., under 20-yrs. franchise; city may purchase works at intervals of 5 yrs. after '94. **Supply.**—Missouri, pumping to settling reservoirs, then repumping to tank. Fiscal year closes June 1.

Year.	M'ns.	Taps.	M't'rs.	Hy'd's.	C'ns'tn.	Cost.	B'd'd debt.	Int. ch'gs.	Exp.	Rev.
'86	7	75	1	50	50,000	$125,000	$60,000	$4,200		*$4,000
'88	8	342	15	50	160,000	93,000	75,000	4,500	$3,600	8,500
'90	8	375		50	110,000		75,000	4,500		

*From city.

Financial.—*Cap. Stock* (in '88), $100,000. *Bonded Debt,* $75,000 at 6%. **Management.**—Prest., Wm. Morrison. Vice-Prest., E. W. Abendroth. Secy., Jas. Wentworth. Supt., Louis C. Yates.

17. **LOUISIANA,** *Pike Co.* (Pop., 4,325—5,090.) Built in '87-8 by Louisiana Water Co. Controlled by American Water-Works & Guaranteee Co., Ltd., Pittsburg, Pa.; franchise fixes rates and provides for purchase of works by city in '97 and at intervals of 5 yrs. thereafter; price determined by arbitration. **Supply.**—Mississippi River, pumping through mains to reservoir; filtered. Fiscal year closes Dec. 31. **Distribution.**—*Mains,* 6 miles. *Taps,* 231. *Meters,* 3. *Hydrants,* 55. *Consumption,* 200,000 galls. **Financial.**—Withheld. **Management.**—Prest., J. H. Purdy. Secy., A. B. Darragh. Supt., S. B. Culver.

18. **MASON CITY,** *Macon Co.* (Pop., 3,046—3,371.) **History.**—Built in '90-1 by city; put in operation Feb. 1. Des. Engr., Edgar B. Kay; Const. Engrs., Western Engineering Co.; Contr., Jno. Maxwell; all of Kansas City. **Supply.**—Impounding reservoir on east fork of Sharitan River, pumping direct; gravity filter through, sand and gravel, from dam to reservoir. **Reservoir.**—Impounding. *Cap.,* 60,000,- 000 galls. **Pumping Machinery.**—Dy. cap., 1,750,000 galls.; two Deane dup. **Distribution.**—*Mains,* c. i., 3.5 miles. *Services,* galv. i. and lead. *Hydrants,* Mathews, 10. *Valves,* Eddy. *Pressure:* ordinary, 60 lbs.; fire, 125 lbs. **Financial.**—*Cost,* $30,600. *Bonded Debt,* $25,000 at 6%. **Management.**—City Clerk, Chas. L. Farrar.

19. **MARSHALL,** *Saline Co.* (Pop., 2,701—4,297.) Built in '85 by Marshall Water Co.; under 20-yrs. franchise, which fixes rates and provides that works may be purchased by city at any time. **Supply.**—Well, pumping direct. **Distribution.** —*Mains,* 5 miles. *Taps,* 120. *Hydrants,* 67. *Consumption,* 225,000 galls. **Financial.**—*Cost* in ('88), $75,000. *Cap. Stock,* $75,000. *Bonded Debt,* $75,000 at 6%. *Expenses :* operating, $2,200; int., $4,500; total, $6,700. *Revenue :* fire protection, $5,025; other purposes, $2,600. **Management.**—Prest., Jno. R. Hall. Secy. and Supt., E. R. Page. Treas., J. H. Ordell.

FOR FURTHER DESCRIPTIVE MATTER see Manual for 1889-90. CHANGES since 1890 . are here given. POPULATIONS are for 1880 and 1890, respectively.

20. **MARYVILLE,** *Nodaway Co.* (Pop., 3,485–4,037.) Built in '86 by Municipal Investmen Co., under exclusive 20-yrs. franchise, which fixes maximum rates and provides for purchase of works by city in '95, and at intervals of 5 yrs. thereafter. **Supply.**—One Hundred and Two River, pumping to stand-pipe and direct. **Distribution.**—(In '88.) *Mains,* 8.3 miles. *Taps,* 135. *Meters,* 27. *Hydrants* (in '91), 56. **Financial.**—(In '88.) *Cost,* $85,000. *Bonded Debt,* $75,000 at 6%. *Expenses:* operating, $1,440; int., $4,500; total, $5,940. *Revenue:* city, in '91, $3,700. **Management.**—Prest., W. O. Cole. Vice-Prest., Chas. H. Cottln. Secy. and Treas., C. P. Denny. NOTE.—Co. refused to furnish statistics for '90.

21. **MEXICO,** *Audrain Co.* (Pop., 3,835–4,789.) Built in '84–5 by Mexico, Mo., Water-Works Co., under 20-yrs. franchise. **Supply.**—Springs and surface water, pumping direct; filtered. Fiscal year closes Nov. 1. **Distribution.**—*Mains,* 4.8 miles. *Taps,* 125. *Meters,* 4. *Hydrants,* 46. *Consumption,* 126,215 galls. **Financial.**—*Cost* (in '87), $80,000. *Cap. Stock,* $60,000. *Total Debt,* $61,422: floating, $1,422; bonded, $60,000 at 4.16%. *Expenses:* operating, $3,407; repairs, $349; int., $2,496; ex'ensions, $636; total, $7,388. *Revenue:* consumers, $4,106; fire protection, $2,500; total, $6,600. **Management.**—Prest., G. B. Macfarlane. Secy., J. F. Llewellyn. Supt., J. J. Steele.

22. **MOBERLY,** *Randolph Co.* (Pop., 6,070–8,215.) Built in '85 by Moberly Water-Works Co., under 20-yrs. franchise; June 10, '91, works were to be sold for bondholders. **Supply.**—Surface water, pumping direct from reservoir. Fiscal year closes Dec. 31. **Dam and Reservoir.**—Cap., 140,000,000 galls.; area of reservoir, 42 acres. Dam of earth, with heart-wall, on rock foundation; 383 ft. long, 40 ft. high, 18 ft. wide on top. Inner slope, 3 to 1, riprapped from 3 ft. below water line to top; outer, 2 to 1, sodded. Heart-wall, masonry in Louisville hydraulic cement, 10 ft thick at bottom, 3 at top, 35 ft. high; 2 ft. of tamped clay against each side. Spillway, 20 ft. from end of dam, in rock, lined with rubble and plank, free. Dam does not leak. Built in '86. Cost, $65,000. Des. Engr., C. S. Masten, Phœnix, Ariz. Const. Engr., W. C. Masten, Kansas City. Contrs., Locke & Masten.

Yr.	Mains.	Taps.	Met'rs.	Hyd'ts.	Cons'p'tn.	Cost.	B'd'd d't.	Int.	ch'g's.	Exp.	Rev.
'87	6.5	260	4	70	200,000	$125,000	$100,000	$6,000		$3,758	$10,528
'90	6.5	387	33	71	355,000	....	100,000	6,000		5,692	14,514

Financial.—*Cap. Stock:* authorized, $20,000; all paid up; paid no div'd in 1890. *Bonded Debt,* $100,000 at 6%. *Expenses:* operating, $5,051; int., $6,000; taxes, $642; total, $11,692. *Revenue:* consumers, $9,014; fire protection, $5,500; total, $14,514. **Management.**—Feb. 6, '91, works were put in hands of Receiver, H. V. Estello.

23. **MOUND CITY,** *Holt Co.* (Pop., 678–1,193.) **History.**—Built in '90-1 by city; put in operation Feb. 1. Engr., W. Kiersted. Contrs., W. Kiersted & Co., Kansas City. **Supply.**—Four 5-in. tube wells, pumping to stand-pipe. Wells are 6 ins. in diam., 88 ft. deep, 14 ft. apart; Contrs., Cook Well Co. **Pumping Machinery.**—Dy. cap., 500,000 galls.; Deane. **Stand-Pipe.**—Cap., 35,000 galls.; 10 × 60 ft.; on hill 165 ft. above business portion of city. **Distribution.**—*Mains,* 8 to 4-in., 2 miles. *Services,* w. i. *Hydrants,* Eddy, 16. *Valves,* Eddy. *Pressure,* 90 lbs. **Financial.**—*Cost,* $14,000. *Bonded Debt,* $14,000 at 6%. **Management.**—Mayor, D. W. Porter. Clerk, J. M. Hasness.

24. **NEVADA,** *Vernon Co.* (Pop., 1,913–7,262.) Built in '85 by Nevada Water Co., under 20-yrs. franchise; put in hands of receiver May, '91. **Supply.**—Marmaton River, pumping to reservoir. Fiscal year closes Dec. 31. **Distribution.**—*Mains,* 15 miles. *Taps,* 515. *Meters,* 7. *Hydrants,* 70. **Financial.**—Withheld. *Cost* (in '87), $77,000. *Bonded Debt* (in '87), $50,000. **Management.**—Receiver, C. F. Strohm.

25. **NEOSHO,** *Newton Co.* (Pop., 1,631–2,198.) **History.**—Built in '91 by Neosho Water-Works Co., under 20-yrs. franchise; put in operation Aug. 1. Engr. and Contr., A. L. Holmes, Grand Rapids. Mich. **Supply.**—Springs, by gravity. **Reservoir.**—Cap., 1,000,000 galls. **Distribution.**—*Mains,* 10 to 4-in., Wyckoff, wood, 5.5 miles. *Hydrants,* Galvin, 50. *Valves,* Flower. *Pressure,* 55 lbs. **Financial.**—Withheld. **Management.**—Prest., S. V. Saleno, Bay City, Mich.

26. **PALMYRA,** *Marion Co.* (Pop., 2,479–2,515.) **History.**—Built in '90-1 by Palmyra Light & Water Co., under 20-yrs. franchise; put in operation Feb. 1. Engr., W. H. Allen, Clinton. **Supply.**—Springs, collected in reservoir, pumping to stand-pipe. **Pumping Machinery.**—Dy. cap., 2,000,000 galls.; from Springfield Car & Foundry Co. **Stand-Pipe.**—Cap., not given; from Shickle, Harrison & Howard, St. Louis. **Distribution.**—*Mains,* 10 to 4-in. c. i., 4 miles. *Hydrants,* Perkins, 40.

MISSOURI. 305

Valves, Ludlow, Scott. *Pressure:* ordinary, 90 lbs.; fire, 140 lbs. **Financial.**—*Cap. Stock*, paid up, $50,000. *Bonds*, authorized, $50,000. *Revenue*, city, $2,500. **Management.**—Prest., B. F. Hobart; Secy. and Treas., W. P. Heath; both of St. Louis.

27. **PIERCE CITY,** *Lawrence Co.* (Pop., 1,350—2,511.) Built in '88 by J. Guinney. Supply.—Guinney's Lake, by gravity to stand-pipe. Distribution.— (In '88.) *Mains*, 6 miles. *Hydrants* 25. *Consumption*, 50,000 galls. **Financial.**—(In '88.) *Cost*, $45,230, *Bonded Debt*, $25,000 at 6%. *Expenses:* operating, $900; int., $1,500; total, $2,460. *Revenue:* consumers, $1,540; city, $1,750; total, $3,290. **Management.** —(In '89.) Prest. and Mgr., J. Guinney. Secy. and Treas., G. N. Bennett.

28. **RICH HILL,** *Bates Co.* (Pop., in '90, 4,008.) Built in '83; owned since July 1, '90, by Rich Hill Water, Light & Fuel Co.; city has right to buy works in '93, '98, and 1903; rates fixed in franchise. Supply.—Marais des Cygnes River to well, thence pumping direct; filtered. Fiscal year closes July 1. Reported in Sept., '91, that new pump had been ordered. Distribution.—*Mains*, 8 5 miles. *Taps*, 343. *Meters*, 15. *Hydrants*, 48. *Consumption*, 100,000 galls. **Financial.**—*Cost* (in '88), $90,000. *Cap. Stock:* authorized, $80,000; all paid up. *Bonded Debt*, $75.000 at 6%, due in 20 yrs. *Expenses:* int., $4,500; taxes, $600. *Revenue*, city. $1,500. **Management.**—Prest., Thos. Irish. Secy. and Treas., R. Walters. Supt., F. J. Pilgrim.

29. **SAINT CHARLES,** *St. Charles Co.* (Pop., 5,014—6,161.) Built in '81 by St. Charles Water & Heating Co., under 20-yrs. franchise, which provides for purchase of works by city in 10 yrs.; city regulates rates. Supply.—Missouri River, pumping to settling tanks, thence by repumping to tank and direct. Fiscal year closes Feb. 4. Distribution.—*Mains*, 7.5 miles. *Taps*, 340. *Hydrants* (in '89), 40. **Financial.**—*Cost* (in '88), $80,000. *Cap. Stock*, $80,000. *No Debt.* **Management.**— Prest., F. L. Deming. Treas., E. B. Deming. Supt., W. L. Vick.

30. **SAINT JOSEPH,** *Buchanan Co.* (Pop., 32,431—52,321.) Built in '80 by St. Joseph Water Co.; now controlled by American Water-Works & Guarantee Co., Ltd., Pittsburg, Pa., under 20-yrs. franchise which fixes maximum rates and provides for purchase of works by city in 1900, price determined by arbitration. Supply.— Missouri River, pumping to reservoir. Fiscal year closes Dec. 31. A 6,000,000-gall. Gaskill pump was put in operation in Oct., '91.

Year.	Mains.	Taps.	Meters.	Hydrants.	Consumption.	Cost.	Bonded debt.	Int. ch'g's.
'82	24	553	47	212	1,500,000	$850,000	$400,000	$24,000
'90	40.5	1,900	250	265	2,500,000	......	......	......

Financial—*Withheld.* **Management.**—Prest., W. S. Kuhn, Pittsburg, Pa. Secy., A. E. Claney. Supt., L. C. Burnes.

31. **SAINT LOUIS,** *St. Louis Co.* (Pop., 350,518—451,770.) Built in '30 by city. Supply.—Mississippi River, pumping to settling basin, thence repumped through stand-pipes and distributing system to reservoir. Fiscal year closes April 9. In '90-1 low service extension was in progress, including inlet tower. Two 20,000,000-gall. Worthington engines were contracted for in '91. See note, below.

Year.	M'ns.	Taps.	M't'rs.	Hyd'ts.	Cons'pt'n.	Cost.	B'd't d't.	Int. chgs.	Exp.	Rev.
'80-1	212	20,204	573	1,842	25,000,600	$12,000,000	......	......	......	$660,024
'82 3	224 8	21,745	905	1,980	27,500,000	......	......	$253,629	......	......
'84-5	238	25,500	1,700	2,200	30,000,000	13,000,000	$5,250,000	$315,000	300,000	750,000
'86-7	279	40,193	2,143	2,650	26,900,000	13,000,000	5,200,000	312,000	310,000	800,326
'87-8	314.4	31,022	....	3,197	30,457,000	13,000,000	5,200,000	312,000	270,000	920,000
'88-9	336.1	30,082	2,840	2,366	31,984,000	13,315,000	5,200,000	312,000	287,956	952,689
'89-90	353.7	38,183	3,115	3,515	32,500,000	......	5,200,000	312,000	251,258	1,017,016

Financial.—See above. Revenue covers all expenses, including extensions, except int. on bonds. No sinking fund. **Management.**—Comr., M. L. Holman, appointed by mayor every four years.

NOTE.—For illustrated description of low service extension see ENGINEERING NEWS, vol. xxv., p. 380, vol. xxvi., p. 4; for illustrated description of cement testing machine, by S. Bent Russell, see ENGINEERING NEWS, vol. xxv., p. 2.

32. **SEDALIA,** *Pettis Co.* (Pop., 9,561—14,068.) Built in '71 by city; bought in '87 by Sedalia Water-Works Co.; rates must be St. Louis rates of '8"; city may purchase works in 1907 and at intervals of 20 yrs. thereafter. Supply.—Flat River and Spring Creek pumping to stand-pipes; filtered.

FOR FURTHER DESCRIPTIVE MATTER see Manual for 1889-90. CHANGES since 1890 are here given. POPULATIONS are for 1880 and 1890, respectively.

306   MANUAL OF AMERICAN WATER-WORKS.

Year.	Mains.	Taps.	Meters.	Hydrants.	Cost.	B'd'd debt.	Int. ch'g'e.	Rev.*
'82	8	400		4	38	$135,000	$125,000	$6,250	$9,445
'84	13	575	0	38	135,000	125,000	6,250	10,500
'88	25	696	63	150	400,000*	200,000	12,000	.....
'89	30	750	218	150	400,000	200,000	12,000	9,000
'90	30	812	218	150	400,000	200,000	12,000	9,000

*From city.

Financial.—*First Cost,* $125,000. *Cap. Stock,* $200,000. *Bonded Debt,* $200,000 at 6%, due 1907. **Management.**—Prest. and Treas., J. B. Quigley. Secy., L. P. Andrews. Supt., S. F. Resse.

33. SPRINGFIELD, *Green Co.* (Pop., 6,522—21,850.) Built in '83 by Springfield Water Co., under 20 yrs. franchise, which fixes rates and provides for purchase of works in '93 and at intervals of 5 yrs. thereafter. **Supply.**—Fullbright spring, pumping to reservoir, tank and direct. Fiscal year closes Dec. 31. Pumping station was built in '90 at Fullbright spring, in which were placed a 4,500,000 Deane comp., dup. pump and a 100-HP. boiler. New 18-in. force main, 3½ miles long, connects pumping station with city reservoir.

Year.	Mains.	Taps.	Meters.	Hyd'ts.	Con'pt'n.	Cost.	B'd'd debt.	Int.	Exp.	Rev.
'84	17.5	570	3	86	500,000	$162,000	$100,000	$6,000	$3,000	$7,000
'87	30	1,200	6	90	1,000,000	375,107	200,000	12,000	9,200	31,900
'89	33	1,397	7	155	1,000,000	......	......	......	......	8,510*
'90	42.6	1,495	21	211	1,217,000	......	......	......	......	8,510*

*From city.

Financial.—*Cap. Stock,* $500,000. **Management.**—Prest., G. P. Westcott. Secy., G. F. West. Both of Portland, Me. Mgr., W. C. Hornbeak.

34. TRENTON, *Grundy Co.* (Pop., 3,312—5,039.) Built in '86 by Trenton Water Co., under 20 yrs. franchise. **Supply.**—Grand River and well close by, pumping to stand-pipe and direct; Hyatt filter. Fiscal year closes Dec. 31. **Distribution.**—*Mains,* 6 miles. *Taps,* 150. *Meters,* 45. *Hydrants,* 60. *Consumption,* 150,000 galls. **Financial.**—*Cost,* $78,200. *Cap. Stock:* authorized, $75,000; paid up, $55,000. Paid no div'd in 1890. Earnings used for new construction. *Total Debt,* $81,750: floating, $1,750 at 6%; bonded, $80,000 at 6%, due in 1908. *Expenses:* operating, $2,300; int., $4,'00; taxes, 300; total, $7,400. *Revenue:* consumers, $2,500; fire protection, $2,000; ry., $3,000; total, $8,100. **Management.**—Prest., F. H. Snyder. Treas. and Supt., W. W. Hubbell.

35. WASHINGTON, *Franklin Co.* (Pop., 2,121—2,725.) Built in '88-9 by Washington Water & Light Co., under 20-yrs. franchise. **Supply.**—Missouri River, pumping to filter and clear water reservoir, thence by repumping to stand-pipe and direct. Fiscal year closes Dec. 31. **Distribution.**—*Mains,* 5 miles. *Hydrants,* 50. *Consumption,* 45,000 galls. **Financial.**—*Cost,* $60,000. *Cap. Stock,* $80,000; paid no dividend in 1890. *Bonded Debt,* $50,000 at 6%. *Expenses:* operating, $1,200; int., $3,000; taxes, $300; total, $4,500. *Revenue:* consumers, $1,72.; fire protection, $3,000; total, $4,720. **Management.**—Prest. and Treas., Theo. Plate. Secy., A. Schenck.

36. WEBB CITY and CARTERVILLE, *Jasper Co.* (Pop.: Webb City, 1,588—5.043; Carterville, 483—2,884.) Built in '90 by co., under 20-yrs. franchise, which states maximum rates; city has right to purchase works in 15 yrs., and every 5 yrs. thereafter. **Supply.**—Springs, and Centre Creek in case of emergency, pumping to stand-pipe. Works were accepted May 21, '90. **Distribution.**—*Mains,* 11.3 miles. *Taps,* 250. *Meters,* 1. *Hydrants,* 102. *Consumption,* 175,000 galls. **Financial.**—*Cost,* $91,400. *Bonded Debt,* $75,000 at 6%, due in 20 yrs. *Expenses:* operating (for 6 months), $2,597. *Revenue* (for 6 months), $6,970. *Hydrant Rental,* $5,100. **Management.**—Prest., James O'Neil. Secy., J. W. Aylor. Supt., Geo. Bruen.

I. WEBSTER GROVES, *St. Louis Co.* (Pop., in '90, 1,783.) In '91 Webster Groves Water Co. laid about 1,500 ft. of pipe to supply ten stockholders with water. **Supply.**—Well, pumping to tank. *Cost,* $3,500. *No Revenue.* Secy., E. W. Pike.

Water-Works Projected with Fair Prospects of Construction.

ALBANY, *Gentry Co.* (Pop., 979—1,334.) July 6, '91, it was reported that city had voted to issue $19,000 worth of bonds to put in water-works and electric lights, and that contracts had been let. Address Mayor.

AURORA, *Lawrence Co.* (Pop., in '90, 3,482.) Aug. 1, '91, it was reported that 20-yrs. franchise had been granted to J. Guinney, Pierce City, construction to begin

MISSOURI.

at once. Des. Engr., J. W. Neir, St. Louis. Deane pumps, **stand-pipe**, 12½ miles of pipe, 63 hydrants. Hydrant rental, $37.35. Est. cost, $85,000.

BROOKFIELD, *Linn Co.* (Pop., 2,264—4,547.) In '91, city voted to issue $25,000 worth of bonds to construct water works and awarded contract to Cook Well Co., St. Louis. Secy., D. F. Howard.

BRUNSWICK, *Chariton Co.* (Pop., 1,801—1,748.) See 1889-90 Manual, p. 366. July 6, '91, it was reported that artesian well had been sunk, affording sufficient water supply. Construction to begin in fall. Prest., L. Bonecke.

STANBERRY, *Gentry Co.* (Pop., 1,207—2,035.) April 14, '91, city voted to issue $17,000 worth of bonds to construct water-works. City Clerk, J. A. Moon.

TARKIO, *Atchison Co.* (Pop., in '90, 1,156.) Rankin Water Co. was to begin construction of works Sept. 20, '91. Wells pumping to stand-pipe. Est. cost, $15,000. Latest report Sept. 9.

Water-Works Projects Reported in '90 and '91 but Now Indefinite.

Columbia, Boone Co.

ARKANSAS.

Water-Works Completed or In Process of Construction.

I. **ALEXANDER,** *Pulaski Co.* (Pop., in '90, 146.) Built in '88 by Alexander Water Co. **Supply.**—Springs, by gravity to tank. Fiscal year closes Jan. 1. There are 1,400 ft. of pipe and 10 taps; no hydrants. Water is used chiefly by stockholders. Cost, $160. Cap. stock all paid up. No debt. Op. exp., $10. Taxes, $6. Rev., withheld. Prest., Esten Peloubet. Secy., A. B. Holland. Supt., Ernest Koch.

1. **ARKADELPHIA,** *Clark Co.* (Pop., 1,506—2,455.) History.—Built in '90-1 by Arkadelphia Water & Light Co., under 30-yrs. franchise; put in operation Sept. 1. -Engr., E. N. Maxwell, Texarkana. **Supply.**—River, pumping to stand-pipe. Pumping Machinery,—Dy. cap., 1,500,000 galls.; comp. dup. **Stand-Pipe.**—Cap., 235,000 galls.; from Porter Boiler Mfg. Co., Chicago. **Distribution.**—*Mains,* steel, 7.5 miles. *Services,* galv. i. Pipe and specials from National Tube Works. *Taps,* 100. *Meters,* 2. *Hydrants,* Ludlow, 30. *Valves,* Ludlow. *Consumption,* 200,000 galls. *Pressure:* ordinary, 50 lbs.; fire, 100 lbs. **Financial.**—*Cost,* $70,000. *Cap. Stock:* authorized, $100,000; paid up, $60,000. *Bonded Debt,* $60,000 at 6%. *Revenue,* city, $2,000. **Management.**—Prest., M. P. Hillyer, Topeka, Kan. Secy., E. N. Maxwell, Texarkana. Treas., H. E. Corder.

2. **ARKANSAS CITY,** *Desha Co.* (Pop., 503—est., 750.) Built in '89 by Arkansas City Water Co. **Supply.**—Mississippi River, pumping to tank. Fiscal year closes Dec. 31. **Distribution.**—*Mains,* ¼ miles. *Taps,* 150. *Hydrants,* 10. **Financial.**—*Cost,* $3,000. *Cap. Stock:* authorized, $10,000; paid up, $4,000. *Expenses,* $600. *Revenue,* $1,000. **Management.**—Prest., J. M. Whitehill. Secy., Sam'l. Marks. Supt., W. M. Ogburn.

II. **BRINKLEY,** *Monroe Co.* (Pop., 325—1,510.) Built in '86-7 by Brinkley Car Wks. & Mfg. Co. **Supply.**—Well, pumping to tank. About 3,000 ft. of pipe supply consumers in town. Secy., Harry H. Meyers.

NOTE.—Reported in '90 that Brinkley Water-Works, Electric Light & Power Co. had been incorporated with authorized cap. stock of $25,000, and with W. S. McCullough, President, and T. H. Jackson, Secretary; also that Cook wells were being sunk.

3. **CAMDEN,** *Ouachita Co.* (Pop., 1,503—2,571.) Built in '89 by Camden Water Works Co. **Supply.**—Ouachita River, pumping to stand-pipe. **Distribution.**—(July, '89.) *Mains,* 2.3 miles. *Taps,* 125. *Meters,* 0. *Hydrants,* 10. *Consumption,* 175,000 galls. **Financial.**—(July, '89.) *Cost,* $17,500. *Cap. Stock,* authorized, $30,000; paid-up, $16,000. *Revenue:* consumers, $3,000; city, $750; total, $3,750. **Management.**—Prest. and Supt., S. Q. Sevier. Secy. and Treas., J. B. Friedheim. NOTE.—Co. refused to furnish statistics for '90.

FOR FURTHER DESCRIPTIVE MATTER see Manual for 1889-90. CHANGES since 1890 are here given. POPULATIONS are for 1880 and 1890, respectively.

4. **DARDANELLE,** *Yell Co.* (Pop., 748—1,456.) **History.**—Built in '85 by W. E. De Long, under 20-yrs. franchise. **Supply.**—Driven wells, pumping to tank. **Pumping Machinery.**—Dy. cap., 360,000 galls.; Smith & Valle. **Tank.**—Cap., 48,000 galls. **Distribution.**—*Mains*, 1-in., 1 mile. *Taps*, 50. *Hydrants*, 6. *Consumption*, 8,000 galls. **Financial.**—*Cost*, $7,500. *No Debt. Revenue:* consumers. $1,000; city, $600; total, $1,600. **Management.**—Owner, W. E. De Long.

5. **FORT SMITH,** *Sebastian Co.* (Pop., 3,099 —11,311.) Built in '84 by Fort Smith Water Co., under 20-yrs. franchise. **Supply.**—Poteau River, pumping to reservoir; filtered. Nov., '90, plans for improvements were made or submitted by M. M. Tidd, Boston. Reported, Jan., '91, that contract for stand-pipe had been let to Sharon (Pa.) Boiler Works; also that 2,500,000-gall. low-pressure Blake pump had been contracted for. **Distribution.**—(In '88.) *Mains*, 16 miles. *Taps*, 860. *Meters*, 8. *Hydrants*, 72. *Consumption*, 500,000 galls. **Financial.**—(In '88.) *Cost*, $230,000. *Cap. Stock*, $200,000. *Bonded Debt*, $125,000 at 6%. *Expenses:* operating, $5,005; int., $7,500; total, $12,505. *Revenue:* consumers, $15,000; city, $4,000; total, $19,000. **Management.**—Prest., T. B. Sweet. Secy. and Treas., Harry E. Kelley. Supt., Clark Kellogg.

NOTE.—Oct. 19, '91, attorneys for Washington Trust Co., New York City, notified co. that it would apply for receiver, on account of alleged unsatisfactory management.

6. **HOPE,** *Hempstead Co.* (Pop., 1,233—1,937.) Built in '85 by town. **Supply.**—Artesian well, pumping to tank. Pump has dy. cap. of 100,000 galls. Tank has cap. of 72,000 galls. **Distribution.**—*Mains*, 6 miles. *Taps*, 360. *Hydrants*, 4. *Consumption*, 100,000 galls. **Financial.**—*Cost*, $15,000. *No Debt. Expenses*, $1,800. **Management.**—Com. appointed by City Council, from its members. Chn., G. H. Andrews. NOTE.—In '91 city talked of building.

7. **HOT SPRINGS,** *Garland Co.* (Pop., 3,554—8,066.) Built in '82 by Hot Springs Water Co., under 50 yrs. exclusive franchise. **Supply.**—Bull Bayou and springs, pumping to distributing reservoir. **Distribution.**—*Mains*, 10 miles. *Hydrants*, 73. **Financial.**—Withheld. **Management.**—Prest., A. B. Gaines. Secy., J. W. Corrington. Supt., A. N. Sill.

8. **LITTLE ROCK,** *Pulaski Co.* (Pop., 13,138—25,874.) Built in '78 by Little Rock Water Co , now controlled by American Water-Works & Guarantee Co., Ltd., under 50-yrs. franchise, which fixes maximum rates and provides for purchase of works by city in 1905, price determined by arbitration. **Supply.**—Arkansas River, pumping to stand-pipe and reservoir; filtered. Fiscal year closes Dec. 31. Since '88 reported that a 6,000,000-gall. high duty pump and three 60 × 16 boilers have been placed in a new engine house. An "American" filter plant was in course of construction. Dec. 31, '90.

Yr.	Mains.	Taps.	Meters.	H'd'ts.	C'n'p'n.	Cost.	B'd'd d't.	Int. ch'g's.	Exp.	Rev.
'82	8.5			85						$7,225*
'87	12	1,044	15	104	2,000,000	$430,000	$80,000	$4,800	$9,000	36,820
'88	19		33	117	2,151,172	430,000	250,000	15,000	12,000	32,583
'90	26	1,125	31	130	3,000,000					

*From city.

Financial.—Withheld. **Management.**—Prest., J. H. Purdy. Secy., A. E. Clancy. Supt., W. A. Bixby.

9. **NEWPORT,** *Jackson Co.* (Pop., 683—1,571.) Built in '86-7 by Newport Irrigation & Water Co. **Supply.**—White River, pumping to tank. Fiscal year closes Dec. 31. **Distribution.**—*Mains*, 3.5 miles. *Meters*, 15. *Hydrants*, 54. **Financial.**—*Cost* (in '88), $52,000. *Bonded Debt*, $40,000 at 7%. **Management.**—Prest., Adam H. McCormick. Secy., J. M. Stayton. Treas., L. Hirsch. Supt., W. M. Knowles.

10. **PINE BLUFF,** *Jefferson Co.* (Pop., 3,203—9,952.) Owned by Pine Bluff Water & Light Co. Built in '88 by Water, Light & Power Co.; city has right to purchase works in 15 yrs., and at close of every 5 yrs. thereafter; rates based on those of St. Louis. **Supply.**—Cook wells, on bank of Arkansas River, pumping direct and to stand-pipe. Fiscal year closes May 1. **Distribution.**—*Mains*, 15 miles. *Taps*, 631. *Meters*, 5. *Hydrants*, 184. *Consumption*, 1,250,000 galls. **Financial.**—*Cap. Stock*, $300,000. **Management.**—Prest., J. F. Thompson. Secy., N. J. Thompson. Supt., W. P. Thompson.

11. **ROGERS,** *Benton Co.* (Pop., in '90, 1,265.) Built in '88 by Rogers Lime & Water Co. **Supply.**—Spring, pumping to tank. **Distribution.**—*Mains*, 4.5 miles

ARKANSAS.

Taps, 113. *Hydrants,* 8. *Consumption,* 30,000 galls. **Financial.**—*Cost,* $17,500. *Cap. Stock,* $12,000. *No Bonds. Floating Debt,* $2,090 at 10%. *Expenses:* operating, $1,200; int., $200; taxes, $50; total, $1,450. *Revenue:* consumers, $2,301; city, $300; total, $2,601. **Management.**—Prest., J. A. C. Blackburn. Secy., B. F. Dyer. Supt., J. R. Bredin.

12. **TEXARKANA,** *Miller Co.* (Pop.: in Arkansas, 1,390—3,528; in Arkansas and Texas, 3,223—6,380.) Built in '87 by Texarkana Water Co., under 30-yr. franchise; consolidated in '91 with Texarkana (Texas) Water Co. **Supply.**—Twelve flowing Wagner wells, pumping. Fiscal year closes Dec. 31. Pumping station on Texas side has been abandoned.

Year.	Mains.	Taps.	Meters.	Hyd'ts.	B'd'd d't.	Cost.	Int. ch'gs.	Exp.	Rev.
'87	4.5	110	1	28	$64,000	$60,000	$3,600	$800	$11,400
'88	7	194	5	30	76,000	60,000	3,600	4,000	14,025
'90	12	510	4	62	170,000	250,000	15,000	5,850	22,560

Financial.—*Cap. Stock:* authorized, $100,000; all paid up; paid no div'd in '90. *Bonded Debt,* $250,000 at 6%, due in 1910. *Sinking Fund* to be started in '95. *Expenses:* operating $5,100; int., $15,000; taxes, $750. *Revenue:* consumers, $7,500; fire protection, $5,000; ry. and other large consumers, $10,000; total, $22,560. **Management.**—Prest., M. P. Hillyer, Topeka, Kan. Secy., E. N. Maxwell. Treas., Ben Collins. Supt., J. H. McLaughlin.

Water-Works Projected with Fair Prospects of Construction.

EUREKA SPRINGS, *Carroll Co.* (Pop., 3,984—3,706.) City expected to begin construction of water-works in '91. Aug. 14 contracts had been let. Engr., Jno. W. Riley. Springs, by gravity and pumping to tanks. Mains, 7.5 miles.

HELENA, *Phillips Co.* (Pop., 3,652—5,183.) In '91, a 30-yrs. franchise was granted to W. P. Hillyer, Topeka, Kan., and E. N. Maxwell, Texarkana. Mayor, N. J. Fritzton. Clerk, J. O. Bagwell.

MORRILLTON, *Conway Co.* (Pop., 770—1,644.) Aug. 14, '91, W. E. DeLong reported he had secured 30-yrs. franchise and would begin construction of waterworks in '92. Direct pumping, two miles of pipe, 12 hydrants. Est. cost, $10,000.

PRESCOTT, *Nevada Co.* (Pop., 1,253—1,287.) Water-works talked of. Ownership not decided. Mayor, Jno. H. Arnold.

STUTTGART, *Arkansas Co.* (Pop., in '90, 1,165.) Water-works projected. No definite steps had been taken July 6, '91. Address, T. H. Leslie.

TEXAS.

Water-Works Completed or in Process of Construction.

1. **ABILENE,** *Taylor Co.* (Pop., in '90, 3,194.) Owned by town. Built in '85-6 by co. Bought by town in '87. **Supply.**—Wells and Lytle Creek, pumping to stand-pipe and direct. Fiscal year closes April 1. **Distribution.**—*Mains,* 5 miles. *Taps,* 175. *Meters,* 6. *Hydrants,* 37. *Consumption,* 50,000 galls. **Financial.**—*Cost,* $43,500. *Bonded Debt,* $31,000 at 7%. *Sinking Fund,* $45,000, loaned at 10%. *Expenses:* operating, $2,000; int., $2,480; total, $4,480. *Revenue:* consumers, $1,800; city, none. **Management.**—Mayor, D. W. Wister. Supt., J. R. Spaulding.

2. **ARLINGTON HEIGHTS** *(Fort Worth P. O.), Tarrant Co.* (Pop., not given.) **History.**—Built in '90 by Chamberlain Investment Co.; put in operation in Dec. Engrs., O'Brian, Rudiger & Temple. **Supply.**—Artesian wells, pumping to tank. Pumping Machinery.—Dy. cap., 260,000 galls.; one 80,000 and one 130,000 Cook. *Boilers,* from Heine Safety Boiler Co., St. Louis. **Tank.**—Cap., 165,000 galls.; 30 × 20 ft.; from Jas. McEvan, Ft. Worth. **Distribution.**—*Mains,* 8 to 4-in. 1., 3.2 miles. Pipe and specials from Chattanooga Iron Wks. *Hydrants* Ludlow, 40. **Financial.**—*Cost,* $60,000. **Management.**—Supt., H. W. Tallant, Ft. Worth.

3. **AUSTIN,** *Travis Co.* (Pop., 11,013—14,476.) Owned by Austin Water, Light & Power Co. Franchise fixes rates and provides for purchase of works in '92. Built in '75 by City Water Co. **Supply.**—Colorado River, pumping direct; filtered.

FOR FURTHER DESCRIPTIVE MATTER see Manual for 1889-90. CHANGES since 1890 are here given. POPULATIONS are for 1880 and 1890, respectively.

Pumping Machinery.—Dy. cap., 8,000,000 galls.; one 5,000,000-Gaskill horizontal and one 3,000,000-Holly quadruplex. **Distribution.**—*Mains*, 40 miles. *Taps*, 2,000. *Meters*, 984. *Hydrants*, Holly, 165. *Consumption*, 2,500,000 galls. **Financial.**—*Revenue*, city, $12,050. **Management.**--Prest., M. D. Mather. Secy., D. L. Wickes. Supt., R. H. Felter.

NOTE.—In '90 city voted $1,400,000 of 4% bonds for city works and to develop power, with supply from Colorado River. In '91 masonry dam 1,150 ft. long, 50 ft. high above summer level, was being built across Colorado River to develop power for pumping, electrical and manufacturing purposes.

4. **BEAUMONT**, *Jefferson Co.* (Pop., in '90, 3,296.) Built in '88 by city; leased by Beaumont Ice, Light & Refrigerating Co. for 30 yrs. City has no control over rates. **Supply.**—Neches River, pumping to stand-pipe. Fiscal year closes April 30. **Distribution.**—*Mains*, 3 miles. *Hydrants*, 42. **Financial.**—(In '88.) *Cost*, $20,000. *Cap. Stock*, $40,000. *Revenue*: consumers, $1,800; city, none. Works were leased by city for consideration of free hydrant rental for 42 hydrants. **Management.**—Prest., J. F. Knit. Secy. and Supt., E. L. Bacon. Treas., H. C. Herring.

5. **BELTON**, *Bell Co.* (Pop., 1,797—3,000.) Leased for 10 yrs. by Belton Water, Light & Power Co. Built in '81 by town. **Supply.**—Artesian wells, having daily flow of 2,000,000 galls. Pumping station, used formerly by city, has been abandoned. **Distribution.**—(In '88.) *Mains*, 6 miles. *Taps*, 450. *Meter*, 1. *Hydrants*, 28. *Consumption*, 300,000 galls. **Financial.**—(In '88.) *Cost*, $45,000. *Bonded Debt*, $40,000 at 6%. *Expenses:* operating, $3,500; int., $2,400; total, $5,900. *Revenue:* consumers, $5,500; city, none. **Management.**—Prest., N. C. Denny. Vice-Prest., Silas Baggett. Secy., J. H. Wear. Treas., J. Z. Miller. NOTE—No statistics have been furnished since '89.

6. **BRENHAM**, *Washington Co.* (Pop., 4,101—5,209.) Built in '85 by company, under 25-yrs. franchise. **Supply.**—Springs, pumping to stand-pipe. Fiscal year closes Aug. 1. **Distribution.**—*Mains*, 5 miles. *Taps*, 150. *Hydrants*, 35. *Consumption*, 60,000 galls. **Financial.**—*Cost*, $74,000. *Bonded Debt*, $60,000 at 6%, due in 1905. *Expenses:* operating, $2,200; int., $3,600; total, $5,800. *Revenue:* consumers, $4,900; city, $1,750; total, $6,650. **Management.**—Prest., W. A. Wood. Secy., D. C. Giddings, Jr. Supt., J. H. Ledlie.

7. **BROWNWOOD**, *Brown Co.* (Pop., 725—3,361.) Built in '86-7 by town. **Supply.**—Springs, pumping to stand-pipe. Fiscal year closes Dec. 31. Feb., '91, dam 50 ft. high, 500 ft. long, flooding 100 acres was projected. **Distribution.**—*Mains*, 4.8 miles. *Taps*, 200. *Hydrants*, 37. *Consumption*, 40,000 galls. **Financial.**—*Cost*, $47,000. *Bonded Debt*, $45,000 at 6%; $19,000 due '92-94. $26,000 due in 1935. *Expenses:* operating, $1,500; int., $2,700; total, $4,200. **Management.**—Mayor, Jno. T. Mayo. Secy., Ed. F. Smith, Supt., Jno. Kennedy.

8. **BRYAN**, *Brazos Co.* (Pop., in '90, 3,079.) Built in '89-90 by Bryan Water, Ice & Electric Light Co., under 25-yrs. franchise. **Supply**—Wells, pumping to stand-pipe. **Distribution.**—*Mains*, 4 miles. *Hydrants*, 16. **Financial.**—*Cost*, $18,000. *Cap. Stock:* authorized, $50,000; paid-up, $15,000. *Bonded Debt*, $18,000. **Management.**—Prest., J. N. Henderson. Secy., A. D. Counice. Treas., J. P. Borough. Supt., T. M. Carey. NOTE.—Latest report Feb. 10, '90.

9. **CALVERT**, *Robertson Co.* (Pop., 2,280—2,632.) Built in '86-7 by Calvert Water, Ice & Electric Light Co., under 25-yrs. franchise; city has no control over rates but may purchase works at any time, by giving one year's notice and paying cash; value determined by arbitration. **Supply.**—Artesian well, pumping to stand-pipe and reservoir. Fiscal year closes May 10. *Mains*, 5.5 miles. *Taps*, 180. *Meters*, 36. *Hydrants*, 12. *Consumption*, 50,000 galls. **Financial.**—*Cost*, including ice factory, $29,000. *Cap. Stock*, $21,500; paid 10% div'd, in '91. *Bonded Debt*, $7,500 at 8%, due in 1902. *Surplus*, $1,125. *Expenses:* operating, $1,500; int., $600; taxes, $300; total, $2,400. *Revenue*, $2,600. **Management.**—Prest., L. T. Fuller. Secy., J. Adone. Engr., A. Day.

10. **CASTROVILLE**, *Medina Co.* (Pop., 731-679.) Built in '87 by J. Courand; put in operation in May. **Supply.**—Stream, pumping to tank. **Distribution.**—*Mains*, 2 miles. *Taps*, 30. *Consumption*, 15,000 galls. **Financial.**—*Cost*, $2,500. *No Debt*. **Management.**—Owner, J. Courand.

11. **CHANNING,** *Hartley Co.* (Pop., est., 200.) Construction begun in June, '91, by J. H. Hamlin & Co. Wells, pumping to 16,000-gall. reservoir. *Mains,* 3,000 ft. *Cost,* $2,000.

12. **CISCO,** *Eastland Co.* (Pop., in '90, 1,063.) **History.**—Built in '91 by Cisco Water-Works Co., under 25-yrs. franchise; put in operation Sept. 1. Engr., J. S. Thatcher, Dallas. Contr., Z. T. Williams, Ft. Worth. **Supply.**—Surface water, pumping from reservoir to stand-pipe. **Reservoir**—Cap., 8,500,000 galls.; area, 18 acres; embankment of earth, 20 ft. high, faced with masonry. **Pumping Machinery,**—Dy. cap., 500,000 galls.; comp., non-cond., from Laidlaw-Dunn Co., Cincinnati, O. **Stand-Pipe.**—Cap., 50,000 galls.; 12 × 60 ft.; on hill 60 ft. above business portion of city. **Distribution.**—*Mains,* 6 and 4-in. c. i., 3.5 miles. Pipe and specials from Chattanooga Pipe & Foundry Co. *Hydrants,* Ludlow, 12. *Valves,* Ludlow. *Pressure:* ordinary, 55 lbs.; fire, 125 lbs. **Financial.**—*Cost,* $25,000. *Hydrant Rental,* $100. **Management.**—Jno. F. Patterson. Secy. and Treas., F. C. LeVeaux. Supt., J. H. Holcomb.

13. **CLEBURNE,** *Johnson Co.* (Pop., 1,855–3,278.) Owned by individual or co. Built in '83-4 by town; early in '91 plant was sold to S. E. Moss. Later it was reported that Mr. Moss sold works to S. W. Lovelady. In '91 well was bored for additional supply and pump was to be placed in it. **Supply.**—Springs, pumping to tank and direct. Fiscal year closes Dec. 15. **Distribution.**—*Mains,* 9 miles. *Taps,* 296. *Meter,* 1. *Hydrants,* 40. *Consumption.* 260,000 galls. **Financial.**—*Cost,* $51,000. *No Debt.* **Management.**—See above.

14. **COLORADO,** *Mitchell Co.* (Pop., in '90, 1,532.) Built in '85 by town. Leased by Caldwell & Fletcher. **Supply.**—Wells, pumping to stand-pipe. During '90, well was drilled and Halladay wind-mill pump put up. Cost $1,500. In '90, suit was brought against city by tax-payers. Owing to legal defect, charter was cancelled, and "city left without government." Works are thus left on hands of lessees who continue to operate them. **Distribution.**—*Mains,* 7.5 miles. *Taps,* 150. *Meters,* 36. *Hydrants,* 16. *Consumption.* 40,000 galls. **Financial.**—*Cost,* $56,000. *Bonded Debt,* $10,000 at 7%, due in 35 yrs. *Expenses:* operating, $2,000; int., $2,800; total, $4,800. *Revenue,* $3,000. **Management.**—Lessees, Caldwell & Fletcher.

15. **COLUMBUS,** *Colorado Co.* (Pop., 1,959–2,199.) Built in '84 by town. **Supply.**—Colorado River, pumping to tank and direct. In '90, artesian well was being sunk. **Distribution.**—*Mains,* 3.5 miles. *Taps,* 85. *Meters,* 0. *Hydrants,* 9. *Consumption,* 20,000 galls. **Financial.**—*Cost,* $30,000. *Bonded Debt,* $25,000 at 8%. *Expenses:* operating, $800; int., $2,000; total, $2,800. *Revenue,* consumers, $1,000. **Management.**—Supt., Chas. De Lany. Pumping Engr., J. I. Dick.

16. **CORSICANA,** *Navarro Co.* (Pop., 3,373–6,285.) Built in '84 by Corsicana Water Co., under 25-yrs. franchise. **Supply.**—Lake, pumping to stand-pipe. **Distribution.**—(In '88.) *Mains,* 7 miles. *Taps,* 406. *Hydrants,* 61. *Consumption,* 500,000 galls. **Financial.**—(In '88.) *Cost,* $250,000. *Cap. Stock,* $75,000. *Bonded Debt,* $130,000 at 7%. *Expenses:* operating, $4,000; int., $9,100; total, $13,100. *Revenue:* consumers, $10,000; city, $5,300; total, $15,300. **Management.**—(In '89.) Prest., Sam'l R. Frost. Secy. and Treas., Bryan P. Barry. Supt., Chas. Seevir.

17. **CUERO,** *De Witt Co.* (Pop., 1,333–2,442.) Built in '89-90 by city; put in operation June 1, '90. Prior to this time, fire protection was afforded to small part of town. **Supply.**—Guadalupe River, pumping direct to stand-pipe. Fiscal year closes March 31. Following is list of contrs.: *Pumping Machinery,* Knowles; *stand-pipe,* Ripley & Bronson, 16×110 ft.; *hydrants,* Empire; *valves,* Scott. **Distribution.**—*Mains,* 6 miles. *Taps,* 135. *Hydrants,* 45. *Consumption,* 70,000 galls. **Financial.**—*Cost,* $38,396. *Bonded Debt,* $32,500 at 6%, due 1939. *Sinking Fund,* $1,829 at 6%. *Expenses,* 1,030. *Revenue:* consumers (for 10 months), $1,107; city none. **Management.**—Mayor, H. F. Hill. Secy., Geo. H. Law. Supt., G. Dietze.

18. **DALLAS,** *Dallas Co.* (Pop., 10,358–38,067.) Built in '76 by city. **Supply.**—Trinity River and springs, pumping to reservoir and stand pipe. Fiscal year closes April 15. Unofficially reported Oct. '91, that contract had been let for 10,000,000 Gaskill pump. In '90 East Dallas was annexed and works of East Dallas Water Supply Co. were bought by city.

FOR FURTHER DESCRIPTIVE MATTER see Manual for 1889-90. CHANGES since 1890 are here given. POPULATIONS are for 1880 and 1890, respectively.

Year.	Mains.	Taps.	Met's.	Hyd's.	Cons'pt'n.	Cost	B'd'd	d't. Int. ch'g's.	Ex.	Rev.
'82	10	500	3	73	145,000	$85,000	$70,000		$4,200	$5,000 $12,000
87	25	1.500	0	140	2,225,000	350,000	300,000		.....	 30,000
'89	25.3	1.764	0	177	2,548,132	350,000	295,000	16,700	29,327	14,956

Management.—Not given.

NOTE.—For brief description of failure of new Turtle Creek reservoir embankment see ENGINEERING NEWS, vol. xxv., p. 555, of Oct., '91. R. Hering, F. Harris and D. A. Poyner reported on repairs to reservoir and improvements to works, recommending investigation of artesian wells as supply and changes in distribution.

19. **DECATUR,** *Wise Co.* (Pop., 579—1,746.) **History.**—Built in '83 by A. R. Whitehead and enlarged in '89. **Supply.**—Deep wells, pumping to tank and standpipe. **Pumping Machinery.**—Dy. cap., 375,000 galls.; one 125,000 Cook deep well, and one 250,000 Smith & Vaile. **Tank.**—Cap., 80,000 galls. **Stand-Pipe.**—Cap., 36,000 galls. **Distribution.**—*Mains*, 6 and 4-inc., . 1., 1.5 miles. *Taps*, 200. *Meters*, 0. *Hydrants*, 9. *Consumption*, 30,000 galls. *Pressure:* ordinary, 30 lbs.; fire, 150 lbs. **Financial.**—*Cost*, $15,000. *Expenses*, operating, $800. *Revenue:* consumers, $3,600; city, $1,800. **Management**—Owner, A. R. Whitehead.

20. **DEL RIO,** *Val Verde Co.* (Pop., 50—1,980.) Built in '83 by S. P. R. R. Co. **Supply.**—Springs, pumping to tank. **Distribution.**—*Mains*, 10 miles. *Taps*, 150. *Meters*, 0. *Hydrants*, 30. *Consumption*, 75,000 galls. **Financial.**—*Expenses*, $3,500. *Revenue:* consumers, $2,600; city, $1,050; total, $3,650. **Management.**—Supt., Paul Flato.

21. **DENISON,** *Grayson Co.* (Pop., 3,975—10,958.) Built in '86-7 by Denison City Water Co., under 20-yrs. franchise. City regulates rates and has right to purchase works. **Supply.**—Wells and surface water, latter impounded in reservoir, pumping to stand pipe; filtered. Fiscal year closes Dec. 31. Aug. 29, '91, 8 additional wells had been sunk.

Year.	Mains.	Taps.	M't'rs.	H'd'ts.	C'ns'pt'n.	Cost.	B'd debt.	Int.ch'g's.	Exp.	Rev.
'87	10	375	38	75	......	$240,000	$200,000	$12,000	$1,100	$15,620
'89	11.1	500	265	81	275,000	400,000	200,000	12,000	13,525	8,100*
'90	13	700	270	90	450,000	400,000	200,000	12,000	9,500	21,000

* From city.

Financial.—*Cap. Stock*, $200,000. *Bonded Debt*, $200,000 at 6%. *Expenses:* operating, $9,000; int., $12,000; taxes, $500; total, $21,500. *Revenue*, total, $21,000. **Management.**—Prest., H. S. Hopper, So. 3rd St., Philadelphia, Pa. Secy., Walter Wood, 400 Chestnut St., Philadelphia, Pa. Supt., M. J. Fitzgerald. Gen. Supt, A. W. MacCullum, 400 Chestnut St., Philadelphia. Pa.

22. **EAGLE PASS,** *Maverick Co.* (Pop., 1,627—est., 2,500.) Built in '84 by Eagle Pass Water Supply Co. City neither controls rates nor has right to buy works. **Supply.**—Rio Grande River, pumping to reservoir; filtered. Fiscal year closes Sept. 1. In '90, 8 and 10-in. mains were laid 1½ miles to point 120 ft. above where a 2,000,000-gall. reservoir is to be built. **Distribution.**—*Mains*, 6.5 miles. *Taps*, 310. *Hydrants*, 2. *Consumption*, 500,000 galls. **Financial.**—*Cost*, $41,993. *Cap. Stock*, $20,000. *Total Debt*, $1,690, floating. Bonds were issued in '89 to the amount of $20,000, but had not been sold Nov. 8, '90; int., 8%; due Jan. 1, '99. *Expenses:* operating, $2,622; taxes, $390; total, $3,012. *Revenue*, total, $6,843. **Management.**—Prest., Fred H. Hartz. Secy., M. Hartz. Supt., Ed. Schmidt.

EAST DALLAS, *Dallas Co.* Built in '86 by East Dallas Water Supply Co., under 50-yrs. franchise. In '90 Dallas and East Dallas were consolidated, and waterworks of East Dallas were purchased by city. See Dallas.

23. **EL PASO,** *El Paso Co.* (Pop., 736—10,338.) Built in '82 by El Paso Water Co., under 25-yrs. franchise. **Supply.**—Rio Grande River, pumping to reservoirs. **Distribution.**—*Mains*, 10 miles. *Taps*, 369. *Meters*, 450. *Hydrants*, 44. *Consumption*, 600,000 galls. **Financial.**—*Cost*, $120,000. *Cap. Stock* (in '88), $20,000. *Bonded Debt*, $80,000 at 6%. *Expenses:* operating, $15,000; int., $4,800; total, $19,800. *Revenue*, $27,000. **Management.**—Prest., J. W. Parker. Secy., H. Lawton, Supt., W. H. Watts.

NOTE.—Reported in '90 that city was to issue $100,000 of bonds to sink well and build works.

24. **ENNIS,** *Ellis Co.* (Pop., 1,350—2,171.) **History.**—Built in '91 by Ennis Ice, Light & Water Co., under 30-yrs. franchise; put in operation Sept. 1. Engrs., O. J. Gorman and J. S. Thatcher, Dallas. Contrs., Dallas Construction Co. **Supply.**—Lake, pumping to stand-pipe. **Pumping Machinery.**—Dy. cap., 1,000,000 galls; two

dup. **Stand-Pipe.**—Cap.. 85,000 galls.; 12 × 100 ft. Contr., M. J. Williams. **Distribution.**—*Mains*, 8 to 4-in., c. I., 4.8 miles. *Hydrants*, Empire, 32. *Pressure:* ordinary, 50 lbs.; fire, 100 lbs. **Financial.**—*Cost*, $64,000. *Cap. Stock:* authorized, $75,000; paid up, $34,000. *Bonded Debt*, $30,000 at 7%. Hydrant rental, $65. **Management.**—Prest., J. Baldridge. Secy. and Treas., M, Lattimer. Supt., C. Dankerly.

25. **FORT WORTH,** *Tarrant Co.* (Pop., 6,66?—23,076.) Owned by city. Built in '82-3 by Co. **Supply.**—Trinity Creek, Clear Creek and gang wells, pumping direct. Fiscal year closes March 31. **Distribution.**—*Mains*, 28 miles. *Hydrants*, 191. *Consumption*, 3,000,000 galls. **Financial.**—*Cost* (in '84), $206,223. *Bonded Debt*, $200,000 at 6%. *Revenue*, $54,550. **Management.**—Committee of three, appointed by mayor. Chn.. Geo. F. Nics. Secy., E. W. Amentrout. Supt., A. W. Scoble, appointed by committee.

NOTE.—Nov. 17, '91, bids were to be received for new works including 12 8-in. wells 1,000 ft. deep, or a daily flow of 8,000,000 galls; two 8,000,000 gall. vertical triple comp. pumps, 4 boilers, stand-pipe; 55,000,000-gall. storage reservoir; about 48 miles 48 to 4-in. pipe; hydrants, valves and other material.

26. **GAINESVILLE,** *Cook Co.* (Pop., 2,667—6,594.) Built in '83-4 by Gainesville Water Co., under 50-yrs. franchise. **Supply.**—Elm River, pumping direct. Fiscal year closes March 31. **Distribution.**—*Mains*, 9.4 miles. *Taps*, 369. *Meters*, 202. *Hydrants*, 84. **Financial.**—Withheld. **Management.**—Vice-Prest., F. L. Cleaves. Secy. and Supt., H. P. Fletcher.

27. **GALVESTON,** *Galveston Co.* (Pop., 22,248—29,084. First supplied in '84 by Galveston Water Co. Present works built in '88-9 by city. **Supply.**—13 artesian wells, pumping to stand-pipe and storage tank. Fiscal year closes Feb. 28. In '91, city had been authorized to issue $900,000 5% bonds for purpose of securing purer water supply. March 31, '91, contract had been awarded and work was about to be begun on deep well, to be sunk 3,000 ft., if necessary; price $75,000. If adequate supply cannot be obtained from this source, it will be obtained from some watershed on mainland. **Distribution.**—*Mains*, 34 miles. *Taps*, 415. *Hydrants*, 359. *Consumption*, 905,753 galls. **Financial.**—*Cost*, $450,000. *Total Debt*, $452,187; floating, $2,480; bonded, $450,000 at 5%, due in 1928, subject to call. *Sinking Fund*, $20,69); invested in 5% bonds. *Expenses:* operating, $16,448; int. and sinking fund, $31,500; total, $47,948. *Revenue:* consumers, $12,213; fire protection, $31,500; total, $43,713. **Management.**—Prest., Albert Weis. Secy., Dan'l J. Buckley. Supt., Geo. L. Parker.

28. **GEORGETOWN,** *Williamson Co.* •(Pop., 1,354—2,447.) Built in '84 by city. **Supply.**—San Gabriel River, pumping to stand-pipe and direct. Fiscal year closes July 1. During '91, additional pump and boiler were to be put in. **Distribution.**—*Mains*, 8 miles. *Taps*, 300. *Meters*, 5. *Hydrants*, 21. *Consumption*, 150,000 galls. **Financial.**—*Bonded Debt*, $14,000 at 8%. *Sinking Fund*, $1,120. *Expenses:* operating, $2,360; int., $1,120; total, $3,480. *Revenue:* consumers, $5,000; city, none. **Management.**—Chn., Emzy Taylor. Secy. and Supt., W. J. Thomas.

29. **GONZALES,** *Gonzales Co.* (Pop., 1,581—1,641.) Built in '84 by Gonzales Water-Works Co. **Supply,** Guadalupe River, pumping to stand-pipe, tank and direct. Fiscal year closes Dec. 31. **Distribution.**—*Mains*, 8 miles. *Taps*, 300. *Meters*, 0. *Hydrants*, 16. **Financial.**—*Cost*, $22,000. *Cap. Stock*, $12,000. *Bonded Debt*, $5,000 at 8%, due in 1900. *Expenses:* operating, $800; int., $400; total, $1,200. *Revenue:* consumers, $3,600; city, none. **Management.**—Prest., T. H. Harwood. Secy., R. H. Walker. Supt., T. S. Walker.

30. **GREENVILLE,** *Hunt Co.* (Pop., in '90, 4,330.) Built in '88-9 by Greenville Water & Electric Light Co. **Supply,** Cowleach fork of Sabine River, pumping from reservoir through filter to stand-pipe; American filter. Fiscal year closes Sept. 4. **Distribution.**—*Mains*, 5.8 miles. *Taps*, 74. *Meters*, 8. *Hydrants*, 60. **Financial.**—*Cost*, $104,772. *Bonded Debt*, $100,000 at 6%. *Expenses:* operating, $1,900; int., $6,000; total, $10,500. *Revenue*, $9,860. **Management.**—Prest., E. B. Perkins. Vice-Prest. and Gen. Mgr., Chas. H. Ledlie, 417 Olive St., St. Louis, Mo. Secy., L. D. Lasater.

31. **HALLETTSVILLE,** *Lavaca Co.* (Pop., 588—1,011.) **History.**—Built in '91 by city; put in operation July 1. Engr. and Contr., G. Jaeger, Rich Hill, Mo. **Supply.**—Artesian well, pumping from reservoir to stand pipe. **Reservoir.**—Cap..

FOR FURTHER DESCRIPTIVE MATTER see Manual for 1889-90. CHANGES since 1890 are here given. POPULATIONS are for 1880 and 1890, respectively.

120,000 galls.; 18 ft. in diam., 16 ft. deep. **Pumping Machinery.**—Dy. cap., 500,000 galls.; Knowles comp., dup. **Stand-Pipe.**—Cap., 85,000 galls.; 12 × 100 ft.; from Tippett & Wood, Phillipsburgh, N. J. **Distribution.**—*Mains*, 8 to 4-in., c. i., 2 miles. Pipe and specials from Ripley & Bronson, St. Louis. *Hydrants*, Empire, 25. *Valves*, Scott. *Pressure:* ordinary, 45 lbs.; fire, 125 lbs. **Financial.**—*Cost*, $14,750. *Bonded Debt*, $15,050 at 6%. **Management.**—Mayor, M. Y. Townsend.

32. **HOUSTON,** *Harris Co.* (Pop., 16,513—27,557.) Built in '78-9 by Houston Water-Works Co.; city has right to buy works after 25 yrs.; has no control of rates. **Supply.**—Flowing artesian wells, pumping to stand-pipe and direct. Reported in '90 that well No. 14 had been completed, and that it is 8 ins. in diam. and 590 ft. deep. Fiscal year closes Dec 31.

Year.	Mains.	Taps.	Meters.	Hydrants.	Consumption.	Cost.	B'd'd d'bt.	Int. ch'g's.
'80	13	350	....	......	....	$75,000	....	....
'85	18	600	12	91	....	175,000	$60,000	$1,266
'87	28	900	12	150	1,250,000	300,000	150,000	9.000
'88	29	1,100	22	185	1,750,000	325,000	150,000	9,000
'90	38	1,200	22	198	2,000,000	......	......	....
				258	3,750,000			

Financial.—Withheld. **Management.**—Prest., T. H. Scanlon. Secy., C. H. Sprong. Treas., T. W. House. Supt., F. J. Smith.

33. **KYLE,** *Hays Co.* (Pop., in '90, 779.) Built in '87-8 by Kyle Water Co. **Supply.**—Rio Blanco River, pumping to tanks. Fiscal year closes Oct. 27. **Distribution.**—*Mains*, 7.9 miles. *Meters*, 2. *Hydrants*, 11. *Consumption*, 25,000 galls. **Financial.**—*Cost*, $12,000. *Bonded Debt*, $7,000. *Expenses*, $800. *Revenue*, $1,300. **Management.**—Prest., H. Hillman. Secy., C. L. Sledge. Supt., J. L. Ray.

34. **LAMPASAS,** *Lampasas Co.* (Pop., 613—2,408.) Built in '85 by town. **Supply.**—Sulphur Creek, pumping to stand-pipe and direct. **Distribution.**—*Mains*, 6.5 mile. *Taps*, 182. *Meters*, 9. *Hydrants*, 25. *Consumption*, 50,000 galls. **Financial.**—*Cost*, $45,000. *Bonded Debt*, $40,000 at 7%. *Expenses:* operating, $2,000; int., $2,800; total, $4,800. *Revenue*, $3,000. **Management.**—Com. of three. Chf. Engr., H. E. Hedeman.

NOTE.—June, '91, it was reported that town had collapsed and water works-system was in litigation.

35. **LAREDO,** *Webb Co.* (Pop., 3,521—11,319.) Built in '83-4, by Laredo Water Co.; franchise fixes rates and provides for purchase of works by city in 1900; price determined by three disinterested persons. **Supply.**—Wells and Rio Grande River, pumping direct; filtered. Fiscal year closes Dec. 31. During '90 a 1,500,000-gall. Worthington was introduced. It is proposed to build an 8,500,000-gall. reservoir.

Year.	Mains.	Taps.	Hydrants.	Cons'p'n.	Cost.	B'd'd debt.	Int. ch'g's.	Exp.	Rev.
'84	7	100	100	20,000	$120,000	$90,000	$7 200	....	$7,500*
'89	10	410	100	600,000	120,000	90,000	7,200	6,600	21,500
'90	11	520	109	700,000	129,045	97,000	7,200	13,306	28,511

*From city.

Financial.—*Cap. Stock:* authorized, $140,000; paid up, $36,000. Paid no div'd in 1890. *Total Debt*, $152,377: floating, $5,377; bonded, $90,000 at 8%; $7,000 at 7%, all due in 1903. *Expenses:* operating, $12,547; int., $7,200; taxes, $759; total, $20,506. *Revenue:* consumers, $9,526; fire protection, $9,385; other purposes, $8,800; total $28,511. Hydrant rental has been changed, by mutual agreement, from $67.50 to $45 each for first 200 hydrants and $25 instead of $30 each for additional. **Management.**—Prest. and Supt., A. L. McLane. Secy., Raymond Martin.

I. **LLANO,** *Llano Co.* (Pop., 213—est., 1,000). Built in '86 by W. W. Knowles & Son. **Supply.**—Llano River, pumping to tank. Fiscal year closes Dec. 31. **Distribution.**—*Mains*, 3 miles. *Taps*, 60. *Meter*, 1. *Hydrants*, 0. **Financial.**—*Cost*, (in '88), $4,000. *Revenue*, $1,000. **Management.**—Owners, W. W. Knowles & Son.

II. **LOCKHART,** *Caldwell Co.* (Pop., 718—1,233.) Owned by Dr. Bennett, Galveston. Built in '87 by Lockhart Water Supply Co. Water is forced by two hydraulic rams into 8,000-gall tank, 70 ft high. Mains extend around public square and one or two blocks in each direction. In '88 there were 24 taps. Cost, $5,000. No debt. Report by A. R. Chew, former supt.

36. **LULING,** *Caldwell Co.* (Pop., 1,114—1,792.) History.—Built in '91-1 by Luling Water-Works Co., under 50-yrs. franchise; put in operation Feb. 1. Des. Engr., W. A. Freeman, San Antonio. Contr., W. J. Williams, Dallas. **Supply.**—San Marcos River, pumping to stand-pipe. **Pumping Machinery.**—Dy. cap., 750,000 galls.; Deane water-power. **Stand-Pipe.**—Cap., 85,000 galls.; 12 × 100 ft. **Distribution.**—*Mains*, 8 to 2-in. *Services*, w. i. *Taps*, 75. *Hydrants*, 10. *Pressure:*

ordinary, 45 lbs.; fire, 200 lbs. Financial.—*Cos'*. $23.000. Cap. Stock: authorized, $50,000; paid up, $7.500. Bonded Debt, $16,500 at 6%. Hydrant Rental, $60. Management.—Prest., W. A. Johnston. Secy., Otis McGaffey. Treas., W. W. Lipscomb.

37. **MARSHALL**, *Harrison Co.* (Pop., 5,624—7,207.) Built in '88-9 by city. Supply.—Twenty-four Wagner steamed wells, pumping to stand-pipe and direct. Fiscal year closes April 1.

Year.	Mains.	Taps.	Meters.	Hydrants.	Cost.	B'd'd debt.	Int. ch'gs.	Exp.	Rev.
'89-'90	8	170	2	52	$67,000	$02,000	.....	.....	$2,442
'90 1	8	330	3	52	67,000	62,000	.....	$3,600	4,127

Financial.—*Bonded Debt*, $62,000 at 6%, due in 1929. *Expenses*, total, $7,300. *Revenue:* consumers, $4,127; city, none. Management.—Com. of three from City Council. Supt., A. F. McAllister.

38. **MEXIA**, *Limestone Co.* (Pop., 1,298—1,674.) Built in '87-8 by Mexia Water, Ice & Light Co.; franchise provides that rates must not be greater than those of other cities similarly situated; no provision is made for purchase of works. Supply.—Artificial lake, pumping to stand-pipe and direct. Fiscal year closes April 1. Distribution.—*Mains*, 5 miles. *Hydrants*, 16. *Consumption*, 50,000 galls. Financial.—*Cost*, $33,000. *Cap. Stock.*, authorized, $30,000; all paid up; paid 2% div'd in 1891. *Bonded Debt*, $18,500 at 5%, due in 1927. *Sinking Fund*, $1,520, loaned at 10%. Management.—Prest., D. M. Prendergast. Secy. and Supt., Jno. R. Corley. Treas., R. E. Map.

III. **MORGAN**, *Bosque Co.* (Pop., 347—426.) Built in '89 by J. Muirhead, under 15-yrs. exclusive franchise; city controls rates. Supply.—Flowing artesian well, by gravity to tank. Fiscal year closes Dec. 31. Distribution.—*Mains*, 2 miles. *Taps*, 41. *Hydrants*, 0. Financial.—*Cost*, $4,000. *No Debt. Expenses*, nominal. Management.—Owner, J. Muirhead.

39. **NAVASOTA**, *Grimes Co.* (Pop., 1,611—2,997.) Built in '86-8 by city; leased to E. L. Bridge for 25 yrs. Supply.—Artesian wells, pumping to stand-pipe and direct. Fiscal year closes Nov. 30. During '90 additional well was sunk. 810 ft. deep, and new pump added.

Year.	Mains.	Taps.	Meters.	Hydrants.	Consumption.	Cost.	B'd'd debt.	Exp.	Rev.
'88	4.5	110	3	12	60 000	$20,500	$15,000	$2,500	$3,900
'90	6.8	190	3	12	100,000	27,500	15,000	1,900	4,600

Financial.—*Bonded Debt*, $15,000 at 6 and 8%, *Sinking Fund*, $1,200. *Revenue:* consumers, $3,400; city, $1,200; total, $4,690. Management.—Lessee, E. L. Bridge. Supt., Wm. Bassett.

40. **NEW BRAUNFELS**, *Comal Co.* (Pop., 1,938—1,603.) Built in '86 by town. Supply.—Comal River, pumping to reservoir and direct. Apr., '91, bids were asked for steel stand-pipe, 30 × 60 ft. Fiscal year closes July 1. Distribution.—*Mains*, 6 miles. *Taps*, 161. *Hydrants*, 28. *Consumption*, 100,000 galls. Financial.—*Cost* (in '88), $21,000. *Total Debt*, $12,200: floating, $1,200; bonded, $11,000 at 6%. *Expenses:* operating, $200; int., $660; total, $860. *Revenue*, consumers, $2,400. Revenue covers all but 3% of expenses, including int. on bonds. Management.—Committee, appointed by mayor every two years. Chn., C. Jahn. Secy., C. Rudorf. Supt., C. Ulzes.

41. **PALESTINE**, *Anderson Co.* (Pop., 2.997—5,838.) Built in '81-2 by Palestine Water Co., under 30-yrs. franchise. Supply.—Springs, pumping from receiving reservoir to stand-pipe and direct; brick filter. In '91 pumping cap. was increased by 1,000,000 galls. Reported that an 18 × 150-ft. stand-pipe was to be added in '91. Fiscal year closes Dec. 31. Distribution.—*Mains*, 8 miles. *Taps*, 250. *Meters*, 10. *Hydrants*, 40. *Consumption*, 310,450 galls. Financial.—*Cost*, $120,000. *Bonded Debt*, $80,000 at 6%. Management.—Prest., M. E. Gray. Mgr., J. W. Ozment. Treas., Dwight Treadway. Supt., Ed. McCleery.

42. **PARIS**, *Lamar Co.* (Pop., 3,980—8,254.) Built in '88 by city and Paris Water Co.; leased to J. D. Thomas. Rates are fixed at average of five similar cities in Texas. Supply.—Well, pumping to stand-pipe. Fiscal year closes Dec. 31. Distribution.—*Mains*, 11 miles. *Taps*, 196. *Meters*, 121. *Hydrants*, 85. *Consumption*, 150,000 galls. Financial.—*Cost*, $110,000. *Bonded Debt*, $83,000 at 6%. *Expenses:* operating, $2,000; int., $4,980; total, $6,980. *Revenue*, $7,200. *Hydrant Rental:* first 81 free; additional, $100 each. Management.—Lessee and Mgr., J. D. Thomas.

For Further Descriptive Matter see Manual for 1889-90. Changes since 1890 are here given. Populations are for 1880 and 1890, respectively.

43. SAN ANGELO, *Tom Greene Co.* (Pop. in '90, 2,613.) Built in '84 by San Angelo Water-Works Co., under 50-yrs. franchise. Supply—North Concho River pumping to tank and direct. In '90, capital stock was increased to $120,000. During '91 system was completely remodeled, work to be completed about June 1; standpipe or tank with cap. of 70,000 galls., 120 ft. high, was added. Distribution.—(In '88.) *Mains,* 4 miles. *Taps,* 138. *Meters,* 6. *Hydrants,* 44. *Consumption,* 60,000 galls. Financial.—Withheld. Management.—Prest. and Mgr., J. L. Millpaugh.

44. SAN ANTONIO, *Bexar Co.* (Pop., 20,550—37,673.) Built in '78-9 by San Antonio Water-Works Co., under 50-yrs. franchise which does not regulate rates, but provides for purchase of works by city in 1903 and every 5 yrs. thereafter, price determined by arbitration. Supply.—San Antonio Springs, pumping to reservoir and direct.

Year.	Mains.	Taps.	Meters.	Hyd'ts.	Rev.*
'80	12	565	....	100	$5,000
'82	15	1,020	5	130	6,950
'88	75	3,600	25	420	21,000
'90	90	5,050	35	561	28,050

* From city.

Financial.—Withheld. Management.—Prest., George W. Breckenridge. Secy., J. Ulrich.

45. SAN MARCOS, *Hays Co.* (Pop, 1,232—2,335.) Built in '83 by San Marcos Water Co. Supply.—San Marcos River, pumping to reservoir and direct. Distribution.—(In '88.) *Mains,* 7 miles. *Taps,* 350; *Meters,* 5. *Hydrants,* 21. *Consumption,* 500,000 galls. Financial.—(In '88) *Cost,* $350,000. *Cap. Stock.* $25,000. *Bonded Debt,* $25,500 at 7%. *Expenses:* operating $1,400; int., $1,785; total, $3,175. *Revenue:* consumers, $5,000; city, $1,300; total, $6,300. Management.—(In '89.) Prest., W. O. Hutchinson. Secy., D. Hopkins. Treas., D. A. Glover. Supt., R. Petitt.

46. SEGUIN, *Guadalupe Co.* (Pop., 1,363—1,716.) Built in '86-7 by Seguin Water & Ice Co., under 20-yrs. franchise. Supply.—Guadalupe River, pumping to stand-pipe and direct. Distribution.—(In 88.) *Mains,* 6 miles. *Taps,* 150. *Hydrants,* 15. *Consumption,* 50,000 galls. Financial.—(In '88.) *Cost,* $27,900. *Cap. Stock,* $25,000. *Bonded Debt,* $19,000 at 7%. Management.—Prest., Jas. P. Holmes. Secy., C. M. Holmes. Supt., J. M. Blanks. NOTE.—Co. refused to furnish statistics for '90.

47. SHERMAN, *Grayson Co.* (Pop., 6,093—7,335.) Built in '87-8 by city. Supply—Gang-wells, pumping to stand-pipe. Fiscal year closes first Monday in May. March 23, '91, artesian well was being drilled for additional supply; depth, 2,500 ft.; cost, $13,500; contrs., Miller & Co., Chicago, Ill.

Year.	Mains.	Taps.	Meters.	Hydrants.	Cons'pt'n.	Cost.	Bonded debt.	Int. ch'g's.
'88-0	9	250	30	70	100,000	$70,000	$70,000	$5,500
'90-1	10	350	30	82	100,000	70,000	70,000	5,600

Financial.—*Bonded Debt,* $70,000, at 8%, due in 1936. *Expenses and Revenue.—Withheld,* on account of litigation. Management.—Mayor, J. S. Porter. Secy, C. S. Dustin. Supt., Robt. Harvey.

48. SONORA, *Sutton Co.* (Pop., est., 200.) History.—Built in '89 by Sonora Water-Works Co. Leased for one year from Oct. 21, '91, to Jesse W. Taylor. Supply.—Drilled well, pumping to tank. Pumping Machinery.—Cap., about 40,000 galls. Tank.—Cap., 40,000 galls. Distribution.—*Mains,* 4-in., 2 miles. *Taps,* 15. *Consumption,* 15,000 galls. Financial.—*Cost,* $3,000. *No Debt.* Management.—Prest., C. F. Adams. Lessee, J. W. Taylor.

49. TAYLOR, *Williamson Co.* (Pop., in '90, 2,584.) Built in '82-3 by Taylor Water Co., under 50-yrs franchise. Supply.—Springs and San Gabriel River, pumping to tank, stand-pipe and direct. Fiscal year closes Dec. 31.

Year.	Mains.	Taps.	Meters.	Hydrants.	Cost.	Rev.
'87	7.4	140	0	21	....	$7,500
'90	16	300	5	36	$80,000	7,900

Financial.—*Cap. Stock,* $72,000. *No Debt. Taxes,* $350. *Revenue:* consumers, $4,800; fire protection, $1,500; ry., $1,600; total, $7,900. Management.—Prest., G. W. Binkett. Secy., A. A. Stephens, both of Palestine. Supt., W. H. Riley.

50. TEMPLE, *Bell Co.* (Pop., in '90, 4,017.) Built in '84 by Temple Water-Works Co., under 50-years franchise, which does not provide for purchase of works by city. Supply.—Surface wells, pumping to stand-pipe. Fiscal year closes in Sept. Oct. 26,

'90, stand-pipe failed. It was 120 ft. high and had cap. of 55,000 galls.; steel on masonry foundation; from Ripley & Bronson, St. Louis, Mo.; cost $10,000; cause of disaster, unknown. Distribution.—*Mains*, 13 miles. *Taps*, 247. *Meters*, 0. *Hydrants*, 27. Financial.—*Cost*, $97,000. *Cap. Stock*, authorized, $100,000; paid-up, $33,000; paid no dividend in 1890. *Total Debt*, $15,600: floating, $3,600; bonded, $10,000 at 6%, due in 20 yrs.; $60,000 of bonds have been issued, of which $20,000 are unsold. *Expenses:* operating, $2,000; int., $1,200; taxes, $360; total $3,560. *Revenue:* consumers, $4,000; fire protection, $2,065; other purposes, $4,900; total, $10,965. **Management.**—Prest., Geo. E. Wilcox. Treas., A. Bentley. Secy., Otto K. Bentley.

51. **TERRELL,** *Kaufman Co.* (Pop., 2,003—2,988.) Now operated by Terrell Ice, Light & Water Co., which leased works in '91 for 30 yrs., with privilege of buying after first year. Built in '84 by town. Supply.—Two wells, pumping to stand-pipe and direct. Fiscal year closes March 31. Distribution—*Mains*, 5 miles. *Taps*, 95. *Meter*, 1. *Hydrants*, 32. *Consumption*, 50,000 galls. Financial.—*Cost* (in '88), $37,200. *Bonded Debt*, $20,000 at 7%. *Sinking Fund*, $14,000. **Management.**—Supt., V. Reinhardt.

52. **TEXARKANA,** *Bowie Co.* (Pop., 1,333—2,852.) See Texarkana, Ark.

53. **TYLER,** *Smith Co.* (Pop., 2,423—6,908.) Built in '83 by Tyler Water Co. under 50-yrs. franchise. Supply.—Springs, pumping from impounding reservoir to stand-pipe. Fiscal year closes Jan. 31. During '91, pumping station, with Worthington pump, was constructed on Indian Creek, 3½ miles from city; cost, $30,000. Additional stand-pipe was projected. Distribution.—*Mains*, 16 miles. *Taps*, 200. *Meters*, 4. *Hydrants*, 100. *Consumption*, 300,000 galls. Financial.—*Cost*, $114,000. *Cap. Stock*, $100,000. *Bonded Debt*, $100,000 at 7%, due in 1908. *Expenses:* operating, $4,700; int., $7,000; taxes, $240; total, $11,940. *Revenue:* consumers, $7,000; fire protection, $5,000; total, $12,000. **Management.**—Prest., F. B. Fish, Secy. and Supt., B. W. Rowland. Treas., H. G. Askew.

54. **UVALDE,** *Uvalde Co.* (Pop., 794—1,265.) Uvalde Water Supply & Power Co. completed works about Oct. 20, '91. Livara River, pumping to stand-pipe. Est. cost, $19,000. Prest., N. D. Townes. Secy. and Treas., W. W. Collier. Const. Engr., R. B. Copeland. Contr., M. S. Kelley.

55. **VERNON,** *Wilbarger Co.* (Pop., in '90, 2,857.) History,—Built in '90 by Vernon Ice, Light & Power Co.; bought by city in '91. Engrs. and Contrs., Thomas & Gorman, Houston. Supply.—Wells, pumping to stand-pipe. Pumping Machinery.—Dy. cap., 1,000,000 galls.; Deane. Stand-Pipe.—Cap., 115,000 galls.; 14 × 100 ft. Distribution.—*Mains*, 8 to 2 in. c. i., 2.5 miles. *Taps*. 40. *Hydrants*, 41. *Consumption*, 35,000 galls. *Pressure:* ordinary, 40 lbs.; fire, 125 lbs. Financial.—*Cost*, $25,000. *Bonded Debt*, $23,500 at 6%. *Expenses*, $3,000. *Revenue:* consumers, $1,500; city, $1,500; total, $3,000. **Management.**—Supt., M. F. Thompson. NOTE.—Unofficially reported in '91 that system had been purchased by city.

56. **VICTORIA,** *Victoria Co.* (Pop., in '90, 3,046.) Built in '85 by city; Supply.—Guadaloupe River, pumping to stand-pipe and direct. Fiscal year closes Jan. 1.

Year.	Mains.	Taps.	H'd'ts.	C'ns't'n.	Cost.	Bd'd d'bt.	Int. ch'g's.	Exp.	Rev.
'86	4	150	40	25,000	$11,250	$15,000	$2,700	$1,200	$1,200
'90	5	200	43	45,000	48,000	45,000	2,700	1,450	2,800

Financial.—*Bonded Debt*, $15,000 at 6%, due in 1915, option, 1925. *Sinking Fund*, 2½% per annum in '90, $2,500 invested in county bonds, bearing 6% interest. *Expenses*, total, $4,150. *Revenue:* consumers, $2,800; city, none. **Management.**—Committee of three, appointed every two years by council, from its members. Chn., Theo. Bubler. Supt., Wm. Wheeler.

57, 58. **WACO,** *McLennan Co.* (Pop., 7,295—14,445.)
WACO LIGHT & POWER CO.
Built in '78 by Waco Water Co.; rebuilt in '90; now controlled by Debenture Guarantee & Assurance Co.,Chicago, Ill. Supply.—Bosque River, pumping to reservoir. Fiscal year closes Dec. 31. During '90, supply was changed from Brazos to impounding basin on Bosque River; cap. of reservoir, 100,000,000 galls.; pumping station was built in which a new 3,000,000-gall. Knowles pump was placed. Engr., Geo. H. Pierson, N. Y. City. Distribution.—*Mains*, 25 miles. *Taps*, 1,200. *Meters*, 75. *Hydrants*, 150. *Consumption*, 1,000,000 galls. Financial.—*Cap. Stock*, paid up,

FOR FURTHER DESCRIPTIVE MATTER see Manual for 1889-90. CHANGES since 1890 are here given. POPULATIONS are for 1880 and 1890, respectively.

318   MANUAL OF AMERICAN WATER-WORKS.

$500,000. *Bonded Debt*, $300,000 at 6%. *Expenses and Revenue*, withheld. **Management.**—Prest., Geo. Clark. Secy. and Treas., Geo. H. Pierson, N. Y. City. Supt., Jno. B. Tschamer.

BELL WATER CO.

Built in '89. **Supply.**—Five artesian wells, by gravity to stand-pipe; two 8-in., two 6-in. and one 4-in well, all 1,842 ft. deep. **Distribution.**—*Mains*, 25 miles. *Taps*, 1,000. *Hydrants*, 160. **Financial.**—*Cost*, $200,000. *Cap. Stock*, $200,000. *No Bonds*. *Hydrant Rental*, $40 each for 125, $35 each for additional. **Management.**—Prest., J. D. Bell. Secy. and Supt., A. M. Prescott. Treas., C. H. Higginson.

59. **WAXAHACHIE,** *Ellis Co.* (1,354—3,076.) **History.**—Built in '91 by Waxahachie Improvement & Water Co., under 25-yrs. franchise. Des. Engr., W. R. Freeman. Const. Engr., F. Harris. **Supply.**—Reservoir, pumping to stand-pipe; filtered through stone and gravel. **Pumping Machinery.**—Not described. Stand-Pipe.—Cap., 115,000 galls.; 14 × 100 ft. **Distribution.**—*Mains*, 8 to 4-in. c. i., 5.5 miles. *Services*, w. i. **Financial.**—*Cost*, $45,000. *Cap. Stock*: authorized, $30,000; paid up, $15,000. *Bonded Debt*, $30,000 at 6%. *Revenue*, city, $2,700. *Hydrant Rental*, $75 each for 20, $60 each for additional. **Management.**—Prest., T. R. Anderson. Secy., M. B. Templeton. Treas., H. W. Trippett.

60. **WEATHERFORD,** *Parker Co.* (Pop., 2,046—3,369.) Built in '88 by Weatherford Water, Light & Ice Co., under 25-yrs. franchise, in which rates are specified but no provision is made for purchase of works by city. **Supply.**—Large well with radiating tunnels, pumping to elevated reservoir, thence by gravity. Fiscal year closes Jan. 1. **Pumping Machinery.**—Dy. cap., 1,000,000 galls.; two 500,000 Deane, comp. dup. **Reservoir.**—Engr., Jno. M. Bassett.

Year.	Mains.	Taps.	Mtrs.	H'd'ts.	B'd'd d't.	Int. Ch'g's.	Exp.	Rev.
'88	3.5	42	0	25	$50,000	$3,500	$4,800	$2,895
'89	3.5	65	1	25	15,000	1,050	2,589	3,731
'90	4	123	4	26	15,000	1,050	1,597	3,734

Financial.—*Cap. Stock*, $50,000. *Bonded Debt*, $15,000 at 7%. *Expenses*, total, $2,647. *Revenue*: consumers, $1,846; city, $1,888; total, $3,734. **Management.**—Prest., W. J. Sherman. Secy., R. W. Kindle. Supt., J. M. Bassett.

Water-Works Projected with Fair Prospects of Construction.

AMARILLO, *Potter Co.* (Pop., in '90, 482.) In Aug., '91, it was unofficially reported that system was in course of construction.

BAIRD, *Callahan Co.* (Pop., in '90, 850.) Aug. 29, 91, it was reported Baird Water & Power Co. had been incorporated and contracts for construction of works would be let in about 60 days. Wells, pumping to tank. Cap. Stock, $20,000. Prest., J. L. Lea. Secy., T. E. Powell. Treas., W. C. Powell.

BALLINGER, *Runnells Co.* (Pop., est., 1,500.) In '91 city let contract for construction of works to Delamer & Lee, Waco, for $14,500. Mayor, B. M. Brooks.

BASTROP, *Bastrop Co.* (Pop., 1,516—1,634.) July 29, '91, A. B. McLavy, Secy. of Bastrop Improvement & Investment Co., reported that contracts for construction of works would be let Aug. 1. Supply, artesian wells. Est. cost, $15,000. Prest., J. C. Buchanan.

BONHAM, *Fannin Co.* (Pop., 1,880—3,361.) See 1889-90 Manual, p. 586. July 6, '91, it was reported that contractors were still at work on artesian wells and hoped oon to obtain adequate water supply.

BOWIE, *Montague Co.* (Pop., est., 1,900.) Reported Oct., '91, that works would be completed by Nov. 1 following.

CAMERON. *Milam Co.* (Pop., 441—1,608.) Oct. 6, '91, it was reported that city was sinking artesian well, but no further plans had been made. Address B. I. Arnold, Mayor.

COMANCHE, *Comanche Co.* (Pop., 704—1,226.) July 11, '91, City Council had decided to build works. Wells, pumping to stand-pipe. Est. cost, $30,000. Address J. F. McCarty or H. Dunaway.

CORPUS CHRISTI, *Nueces Co.* (Pop., 3,257—4,387.) Aug. 15, '91, contract to build works for city was let to Frank P. McCullum, N. Y. City. Contract price, $82,000. Wells, pumping to stand-pipe. Mains, 6½ miles. Construction to begin Sept. 20. Mayor, H. Keller.

DENTON, *Denton Co.* (Pop., 1,194—2,558.) Denton Water, Light & Power Co. were putting down well, Aug. 24, '91. If adequate water supply is obtained, works will be built at once. Prest., J. Roark. Vice-Prest., W. J. Williams. Secy., D. Head.

DUBLIN, *Erath Co.* (Pop., 264—2,025.) May 21, '91, Dublin Water-Works Co. were boring for artesian water. Prest., R. H. McCain.

ELGIN, *Bastrop Co.* (Pop., 164—est., 700.) Contracts for city works to be let about Dec. 1. Supply, wells, pumping to stand-pipe. Est. cost, $10,000. Mayor, C. A. King.

GATESVILLE, *Coryell Co.* (Pop., 434—1,375.) Oct. 2, '91, Gatesville Water Supply Co. let contracts for construction of works. Artesian well, pumping to stand-pipe. Well is 6 ins. in diam., 700 ft. deep, having daily flow of 150,000 galls. A 350,000-gall. pump and 12 × 60-ft. stand-pipe. Mains, 20,000 ft. Hydrants, 8. Cap. Stock, $25,000. Prest., W. E. Brown. Secy. and Treas., R. O. Potts.

HARROLD, *Wilbarger Co.* (Pop., est., 200.) Co. was organized in '91, to build works. Gang-wells, pumping by windmill to tank.

HARTLEY, *Hartley Co.* (Pop., est., 150.) Co. began construction of small system Sept. 15. Contr., C. E. Williams.

HONDO CITY, *Medina Co.* Hondo City Water Works Co. began to lay mains in Oct., '91. Address C. A. Jackson or A. J. Battle.

HUBBARD CITY, *Hill Co.* (Pop., in '90, 894.) July 11, '91, it was reported that artesian well was being bored and works would be built by Co. Address, T. C. Mo.'gan.

LADONIA, *Fannin Co.* (Pop., 223—765.) May 18, '91, Ladonia Artesian Water-Works Co. was boring artesian well and expected to begin construction of works about July 1. Prest., C. W. T. Weldon. Secy., W. A. Anderson.

LA GRANGE, *Fayette Co.* (Pop., 1,325—1,626.) Construction of works was to be started late in Oct., '91, by City. Engrs. and Contrs., M. P. Kelley & Co., Dallas. Colorado River, pumping to stand-pipe. Mains, 2½ miles. Hydrants, 20. Cost, $27,000. Bonds, $20,000 at 6%.

LONGVIEW, *Gregg Co.* (Pop., 1,525—2,034.) Works projected by city. Artesian well system. Address, M. Emerson. Latest report, Aug. 14.

MACGREGOR, *McLennon Co.* (Pop. in '90, 771.) MacGregor Artesian Water Co. has been granted 50-yrs. franchise and began boring artesian wells in Feb., '91. Small system to be put in at once. Cap. stock, authorized, $25,000; paid-up, $6,100. Prest., W. R. Blailoch. Secy., Chas. F. Smith. Treas., A. J. Sewell.

MARBLE FALLS, *Burnet Co.* (Pop. in '90, 587.) A. R. Johnson wrote Sept. 2, '91, that he proposed putting in works at est. cost of $12,000. Engr., Louis G. Hester. Colorado River, pumping to reservoir or tank.

QUANAH, *Hardeman Co.* (Pop., in '90, 1,477.) Aug. 14, '91, it was reported that franchise had been granted to Co. Supply.—Artesian well.

ROCKDALE, *Milam Co.* (Pop., 1,185—1,505.) Rockdale Water-works Co. were to begin construction of works about June 1, '91, under 25-yrs. franchise. Well, pumping to stand-pipe. Est. cost, $43,000. Latest report May 11.

SULPHUR SPRINGS, *Hopkins Co.* (Pop., 1,854—3,038.) Bids for franchise open till Sept. 15, '91. Wells, pumping to stand-pipe. Est. cost, $30,000. Address O. M. Patl.

YOAKUM, *De Witt Co.* (Pop. in '90, 767.) Yoakum Water & Power Co. were granted 25-yrs. franchise, and were to begin construction of works Sept. 14, '90. Contr., M. P. Kelley, San Antonio. Artesian wells, pumping to 16 × 110-ft. stand-pipe. Mains, 6 miles. Hydrants, 30. Est. cost, $60,000. Latest report, Aug. 14 '91, when co. were still boring for water supply. Secy., J. G. Blanks.

For Further Descriptive Matter see Manual for 1889-90. Changes since 1890 are here given. Populations are for 1880 and 1890, respectively.

OKLAHOMA.

Water-Works Projected with Fair Prospects of Construction.

GUTHRIE, *Oklahoma Co.* (Pop., in '90, 2,788.) Fire protection system built '89 by Jno. E. Ford. In '91 city voted to issue $50,000 worth of 7% bonds to construct water-works. July 6, contracts had not been let. Chn., R. R. Carlin. City Clerk, F. Seixall.

OKLAHOMA, *Oklahoma Co.* (Pop., in '90, 4,151.) In '91 franchise was granted to J. F. Thompson, Pine Bluffs, Ark. July 21, '91, it was reported that contracts had been let to Heidenrich Co., Metropolitan Block, Chicago, and that construction had begun and was to be completed within nine months.

COLORADO.

Water-Works Completed or in Process of Construction.

1. **AKRON,** *Washington Co.* (Pop., in '90, 359.) History.—Built in '90-1 by town; put in operation Feb. 1. Engr. and Contr., S. ,T Wicks, Colorado Springs. Supply.—Well, pumping to stand-pipe. Pumping Machinery.—Dy. cap., not given; Deane. Stand-Pipe.—Cap., 86,500 galls.; 14×75 ft.; from Riter & Conley, Pittsburg, Pa. Distribution.—*Mains,* 8 to 4-in. w. i., 3 miles. Pipe and specials from C. W. Bagley & Co., Pittsburg, Pa. *Hydrants,* Ludlow, 15. *Consumption,* 20,000, *Pressure:* ordinary, 35 lbs.; fire, 135 lbs. Financial.—*Cost,* $25,000. Bonded *Debt,* $25,000 at 8%. Management.—Mayor, R. H. Northcott. Supt., S. T. Wicks, Colorado Springs.

I. **ANTONITO,** *Conejos Co.* (Pop., in '90, 315.) First supplied from irrigating ditches of Taos Valley Co. Jan. 3, '91, it was reported that supply would be introduced that year by town, as bonds had already been voted for. Address, S. W. Hatch, Clerk and Recorder.

2. **ASPEN,** *Pitkin Co.* (Pop., in '90, 5,108.) Built in '86 by Castle Creek Water Co. Supply.—Castle and Hunter creeks, by gravity. Fiscal year closes Dec. 31. Distribution.—*Mains,* 6 miles. *Hydrants,* 55. *Consumption,* 600,000 galls. Financial,—*Cost,* $110,000. *Cap. Stock:* authorized, $50,000; all paid-up. Bonded Debt, $50,000 at 8%, due in 1903. *Expenses:* int., $4,000; taxes, $1,350. *Revenue,* city, $3,500. Management.—Prest., H. P. Cowenhover. Treas., D. R. C. Brown. Secy., J. H. Devereux. Supt., H. G. Koch.

3. **ASPEN JUNCTION,** *Eagle Co.* (Pop., not given.) August 28, '91,it was reported that system of works was owned by Gabe Lucksinger. Supply.—Spring, by gravity to reservoir. Cost, about $3,000.

16. **BARNUM AND VILLA PARK (WEST DENVER),** *Arapahoe Co.* (Pop., not given.) First supplied in '89 by Mountain Water Co.; now consolidated with Denver City Water Co., which see.

30. **BERKELEY** *(Highlands P. O.), Arapahoe Co.* (Pop., not given.) Supply from Highlands.

4. **BERTHOUD,** *Larimer Co.* (Pop., in '90, 228.) Owned by town. Built in '86 by Berthond Ditch, Reservoir & Water Co.; bought by town in '89. Supply.—Surface water, gathered in reservoir by gravity. Fiscal year closes April 1. Distribution—*Mains* (in '88), 3 miles. *Hydrants,* 5. Financial.—*Cost* (in '88), $12,000. *Expenses,* $150. *Revenue,* $1,200. Management.—Mayor, F. A. Crane. Secy., J. H. Newell. Supt., C. R. Blackwell.

5. **BLACK HAWK,** *Gilpin Co.* (Pop., 1,540—1,067.) Built in '83-4 by town. Supply.—Mountain springs, by gravity. Fiscal year closes April 1. Distribution, —*Mains,* 5 miles. *Taps,* 200. *Meter,* 1. *Hydrants,* 15. Financial.—*Cost,* $15,000: Bonded *Debt,* $15,000 at 8%. *Expenses:* operating, $1,000; int., $1,200; total, $2,200. *Revenue:* consumers, $4,000; city, none. Management.—City Marshal, Wm. R. Backus, appointed every year by council.

6. **BOULDER,** *Boulder Co.* (Pop., 3,069—3,330.) History.—Built in '75 and extended in '90 by town. Engr. for extensions, E. S. Snell. Supply.—Boulder

Creek, by gravity to and from storage reservoir. **Supply Main.**—An 3-in. pipe, 3 miles long. Larger pipe was recommended by engr., but was not acted on by city council; 8-in. pipe is found to be inadequate. *D· n and Reservoir.*—Cap., 6,000,000 galls.; area, 1¼ acres. Dam of sand, gravel and clay, mostly clay, without heartwall; 320 ft. long, 32 ft. high, 12 ft. wide on top. Inner slope, 2½ to 1, puddled with 4 ft. of clay, gravel and sand, and paved; outer, 2 to 1. Puddling and paving dispensed with by council, but afterward ordered. Spillway, at end of dam, consisting of two 30-in. w. i. pipes laid in concrete in rock trench; free; slope, 10%; bottom is 7 ft. below top of dam. Discharge-pipe, 20-in.c. i. Concrete gate-houce, with intake pipe extending to screen at bottom of reservoir, and another drawing water same distance above bottom. Dam does not leak. Built in '90. Cost, $10,000. Des. Engrs., Henry Fulton and E. L. Snell, Boulder. Const. Engr., E. L. Snell. Contr., T. O'Connell, Denver. **Distribution.**—*Mains*, 20 to 4-in. c. i., 15 miles. *Services,* w. i. Pipe and specials from Ripley & Bronson, St. Louis, Mo.; trenching and pipe laying done by Igo & Howard, Greeley. *Taps,* 650. *Hydrants,* Ludlow, Mathews, 87. *Valves,* Ludlow, Holly. *Consumption,* 100,000 galls. *Pressure,* 125 lbs. **Financial.** —*Cost,* $140,000. *Total Debt,* $125,000: floating, $25,000 at 5%; bonded, $100,000 at 5%. *Expenses,* operating, $500. *Revenue:* consumers, $8,200; city, none. **Management**— City Engr., E. S. Snell.

7. **BRECKINRIDGE,** *Summit Co.* (Pop., 1,657—not given.) Built in '86 by city. **Supply.**—Small stream, by gravity to reservoir. Reservoir is 28×100 ft. **Distribution.** -*Mains*, gas-pipe, 5,500 ft. *Services*, ¾-in. *Hydrants,* 35. *Pressure,* 100 lbs. **Financial.**—*Cost,* $8,000. *Bonded Debt,* $8,000 at 8%. **Management.**— Mayor, M. B. Corbin.

8. **BUENA VISTA,** *Chaffee Co.* (Pop., est., 1,500.) Built in '83 by town. **Supply.**—Cottonwood Creek, by gravity Fiscal year closes March 1. **Distribution.**—*Mains*, 3.5 miles. *Taps*, 135. *Meters*, 0. *Hydrants,* 16. *Consumption,* 100,000 galls. **Financial.**—*Cost,* $35,000. *Bonded Debt,* $30,000 at 8%. *Expenses:* operating, $80; int., $2,400; total, $2,480. *Revenue,* consumers, $1,379. **Management.**— Mayor, M. L. Mason. Secy., S. A. Safford. Supt., E. M. Kemble.

9. **CANON CITY,** *Fremont Co.* (Pop., 1,501—2,825.) Owned by city. Built in '80 by Cañon City Water Co. **Supply.**—Ground water, pumping direct. Fiscal year closes March 31. **Distribution.**—*Mains,* 5.8 miles. *Hydrants,* 35. **Financial.**— *Cost,* $60,000. *Bonded Debt,* $60,000 at 7%. *Expenses:* extensions, $511; operating, $3,307; int., $4,221; total, $8,039. *Revenue:* consumers, $6,817 ; city, $1,353 ; other purposes, $284; total, $8,454. **Management.**—Mayor, J. M. Bradbury. Clerk, H. S. Conoway.

10. **CARIBOU,** *Boulder Co.* (Pop., 549—not given.) Built in '78 by city. **Supply.**—Ditch, by gravity from reservoir. Ditch is 1½ miles long, and is fed by springs and melting snow. **Reservoir.**—Is 20 × 20 ft., × 12 ft. deep. In '80 there were 1,400 ft. of mains, 15 water-takers, and five fire plugs. No later information.

11. **CASCADE,** *El Paso Co.* (Pop., est., 600.) Built '87 by Cascade Town & Improvement Co., which has been succeeded by Cascade Town Co. **Supply.**— Reservoir on Cascade Creek, by gravity. **Distribution.**—*Mains,* 3 miles. *Taps,* 22. *Meters,* 0. *Hydrants,* 7. **Financial.**—*Cost,* $10,000. *No Debt.* **Management.** —Prest., D. N. Herzer. Secy. and Treas., W. W. Miller. Supt., E. E. Doane.

12. **CENTRAL CITY,** *Gilpin Co.* (Pop., 2,626—2,480.) First supplied in '73 by Wilson Bros.; present works built in '86-9 by city. **Supply.**—Springs, by gravity. **Distribution.**—(In '88.) *Mains,* 4 miles. *Taps,* 250. *Meters,* 12. *Hydrants,* 35. *Consumption,* 26,400 galls. **Financial.**—(In '88.) *Cost,* $80,000. *Bonded Debt,* $45,000; $15,000 at 10 %; $30,000 at 7 %. *Expenses:* operating, $2,500; int., $3,600; total, $1,600. *Revenue:* consumers, $8,400. **Management.**—(In '89.) City Clerk, R. S. Haight. Supt., Jas. Davidson.

13. **COLORADO CITY,** *El Paso Co.* (Pop., 347—1,788.) Built in '88 by Colorado City Water Co. **Supply.**—Sutherland Creek, by gravity to reservoir. **Distribution.**—*Mains,* 4 miles. *Taps,* 150. *Meters,* 0. *Hydrants,* 24. *Pressure,* 126 lbs. **Financial.**—*Cost,* $30,000. *Cap. Stock,* $20,000. *No Debt. Revenue:* consumers, $1,500; city, none. **Management.**—Prest. and Supt. A. Bott. Secy., Jas. M. Jackson.

FOR FURTHER DESCRIPTIVE MATTER see Manual for 1889-90. CHANGES since 1890 are here given. POPULATIONS are for 1880 and 1890, respectively.

14. **COLORADO SPRINGS,** *El Paso Co.* (Pop., 4,226—11,140.) Built in '79 by town. Supply.—Ruxton's Creek, by gravity to and from distributing reser voirs. Fiscal year closes April 1. Two 2,500,000 gall. reservoirs were in course of construc. tion in '90. Sept., '91, $70,000 of 6% extension bonds were sold.

Year.	Mains.	Taps.	Meters.	Hyd'ts.	Cost.	B'd'd debt.	Int. ch's.	Exp.	Rev.
'80-1	13	450	0	10	$80,000	......	......	$2,000	$7,000
'86-7	25	1,100	0	40	125,000	$82,000	$5,740	......	......
'87-8	27	1,500	2	56	250,000	117,000	7,020	3,000	20,000
'88-9	40	1,800	1	57	300,000	202,000	12,370	3,000	22,600
'89-90	52	2,100	3	90	100,000	336,000	20,410	3,000	26,165
'90-1	61.5	2,066	3	99	450,000	407,000	18,110	3,000	32,346

Financial.—*Bonded Debt*, $407,000: $25,000 at 7%, due in '95; $282,000 at 6%, due 1900-5; $100,000 at 5%, due in 1905-6. *Sinking Fund*, $12,219. *Expenses*, total, $21,110. *Revenue*: consumers, $32,346; city, none. Management.—City Council. Chn., W. R. Roby. Clerk, A. H. Corman. Supt., E. W. Frost.

15. **CRESTED BUTTE,** *Gunnison Co.* (Pop., in '90, 857.) Built in '89 by Crested Butte Light & Water Co.; accepted Oct. 27, '90. Supply.—Coal Creek, by gravity to reservoir. Distribution.—*Mains*, 3.5 miles. *Taps*, 27. *Hydrants* (in '89), 112. Financial.—*Cost*, $35,000. *Expenses*, $2,600. *Revenue*, $6,600. Management. —Prest., C. W. Badgley. Mgr., H. C. Wright.

16, 17. **DENVER,** *Arapahoe Co.* (Pop., 35,629—106,713).

DENVER CITY WATER CO.

Built in '72 by Denver Water Co. Supply.—Underflow of So. Platte River, by gravity through conduit to reservoir, thence by direct pumping; also wells, by gravity. Fiscal year closes Nov. 1. In '90, Denver Water Co., Beaver Brook Water Co. and Mountain Water Co. consolidated under name of Denver City Water Co. New Co. supplies Denver, Highlands, West Denver, Barnum, Villa Park and West Colfax Assn.

Year.	Mains.	Taps.	Meters.	Hyd'nts.	Cons'ption.	Cost.	B'd'd dbt.	Int. ch'gs.
'80	35	2,500	..	25	......	......	......	......
'84	40	3,500	0	300	3,000,000	$800,000	$350,000	$21,000
'86	45	4,000	20	350	4,500,000	600,000	350,000	......
'88	60	6,311	20	350	5,500,000	1,200,000	500,000	......
'90	225	10,797	85	1,970	15,000,000	3,707,000	2,500,000	175,000

Financial.—*Cap. Stock*, authorized, $7,000,000. "*Earnings*": In '87, $202,720; '88, $265,859; '89, $308,444; '90, $382,000. Management.—Prest., W. P. Robinson. Secy., M. A. Wheeler. Treas., Chas. B. Rhode.

CITIZENS' WATER CO.

Began construction in June, '88; Sept. 29, '90, began to supply water from temporary source. Permanent supply works still under construction, Nov., '91. See '89-90 Manual, p. 591. Engr., C. P. Allen, Denver. Consult. Engr., J. D. Schuyler, San Diego. Supply.—*Temporary*: Jan. 22, '90, co. contracted with town of South Denver to operate pumping station of latter free of cost, in return for which it was to have all water above needs of town. Supply is pumped from well and filter galleries. *Permanent supply*: South Platte River, from point in the mountains, by gravity to and from storage reservoir; water collected in underground galleries. As planned, two high earth dams will form 10,000,000,000-gall. reservoir. Main conduit to and from reservoir will be of wood. A tank, 50 × 40 ft., has been built; one 50 × 40 ft. and another 100 × 400 ft. are under construction. A 23,000,000-gall. reservoir has been completed and a 22,000,000-gall. is under construction. Distribution. —(Nov. 24, '91.) *Mains*, c. i., 100 miles; from Shickle, Harrison & Howard, St. Louis and Chattanooga (Tenn.) Pipe Works. *Services*, lead. *Taps*, 4,500. *Hydrants*, Ludlow, 50. *Valves*, Ludlow. *Pressure*, business part of city, 110 lbs. Financial.— (Nov. 24, '91.) *Cost*, $1,300,000. *Cap Stock*: authorized, $3,000,000; paid-up, $1,500,000. *Expenses*: operating, $12,000. *Revenue*, consumers, $175,000. Management.— Prest., D. H. Moffatt. Vice-Prest., E. F. Hallack. Secy., Richard Holme. Chf. Engr., C. P. Allen. Report, Nov. 24, '90.

18. **DURANGO,** *La Plata Co.* (Pop., in '90, 2,726.) Built in '83-4 by town. Supply.—Animas River, pumping to reservoir. Fiscal year closes April 14.

Year.	Mains.	Taps.	M't'rs.	Hyd'ts.	C'n'p't'n.	Cost.	B'd't debt.	Int. ch'g's.	Exp.	Rev.
'87	3.5	150	0	25	150,000	$80,000	$80,000	$6,400	$1,000	$3,500
'89	5.3	250	0	25	250,000	80,000	80,000	6,400	4,885	2,680
'90	5.3	300	0	25	300,000	80,000	80,000	6,400	3,917	6,000

Financial.—*Bonded Debt*, $85,000: $5,000 at 10%, $80,000 at 8%, due in '97. *Expenses*, total, $10,347. *Revenue*: consumers, $6,000; city, none. Management.—Committee of three, appointed by Mayor every year from City Council. Chn., Harry Jackson. Supt., Blair Burwell, appointed by committee.

COLORADO.

19. EAST COLFAX (P. O. Montclair), *Arapahoe Co.* (Pop., not given.) Supply.—Subterranean wells and Sand Creek, by gravity through 3 miles of wood pipe to reservoir. *Mains*, c. 1. *Services*, w. 1. and lead.

20. ELYRIA, *Arapahoe Co.* (Pop., not given.) History.—Built in '91 by Denver Water Co., under 20-yrs. franchise, which provides for purchase of works by city; put in operation in June. Supply.—Source not given, by gravity and pumping. *Mains*, 6 miles. *Taps*, 100. *Hydrants*, 60. *Pressure*, 60 lbs. *Hydrant Rental*, $35.

21. FAIRPLAY, *Park Co.* (Pop., 450—301.) Built in '82 by village. Supply.—Spring, by gravity to and from reservoir. Fiscal year closes March 31. Distribution.—*Mains*, 1.5 miles. *Taps*, 100. *Hydrants*, 12. Financial.—*Cost*, $10 000. *Bonded Debt*, $8,000 at 8%, due in '92. *Sinking Fund*, $450. *Expenses:* operating, $150; int., $720; total, $870. *Revenue*, consumers, $450. Management.—Supt., Chas A. Wilkin.

22. FORT COLLINS, *Larimer Co.* (Pop., 1,356—2,011.) Built in '82-3 by city. Supply.—Cache la Poudre River, by gravity to settling reservoir, thence by direct pumping. Fiscal year closes April 1. Distribution.—*Mains*, 12.5 miles. *Taps*, 443. *Hydrants*, 47. Financial.—*Bonded Debt*, $105,000: $85,000 at 7%, due in '92; $20,000 at 8%, due in '93. *Expenses:* operating, $1,559; int., $7,550; total, $9,109. *Revenue:* consumers, $8,201; city, 0. Management.—Mayor, W. B. Minor. Supt., C. B. Rosenow.

23. GEORGETOWN, *Clear Creek Co.* (Pop., 3,291—1,927.) Now owned by city; see note, below. Built in '74 by Clear Creek Water Co. Supply.—Clear Creek, by gravity. Distribution.—*Mains*, 3 miles. *Hydrants*, 19. Financial.—*Cost*, $35,000. *No Debt. Expenses*, $2,500. Management.—Prest., A. R. Forbes. Secy., O. C. Warner.

NOTE.—June 11, '91, Chn. of Finance Com. reported that city had bought plant for $27,000 and would take possession July 1.

45. GLEN PARK, *El Paso Co.* (Pop., not given.) Supply from Palmer Lake.

24. GLENWOOD SPRINGS, *Garfield Co.* (Pop., in '91, 920.) Built in '88 by Glenwood Light & Water Co. Supply.—No Name Creek, by gravity. Distribution.—*Mains*, 6.3 miles. *Taps*, 132. *Meters*, 2. *Hydrants*, 21. *Consumption*, 1.500,-000 galls. Financial.—*Cost*, $132,000. *Cap. Stock:* authorized, $250.000; all paid-up. No debt. Management.—Prest., Jas. H. Devereux, Aspen. Gen. Mgr., Mason W Mather.

25. GOLDEN, *Jefferson Co.* (Pop., 2,730—2,383.) Built in '79 by city. Supply —Infiltration from irrigating canal to two reservoirs, pumping direct; filtered. During '91 supply was increased by laying drain-pipe 10 ft. deep under Clear Creek; pipe is 811 ft. long, 348 ft. of wood, 463 ft. of cem. Fiscal year closes March 31. Distribution.—*Mains*, 5 miles. *Taps*, 185. *Meters*, 1. *Hydrants*, 40. Financial.—*Cost* $49,000. *Bonded Debt*, $49,000 at 8%, due in 1900. *Expenses:* operating, $4.500; int., $5,400; total, $9,900. *Revenue:* consumers, $4,734; city, none. Management.—Committee, appointed by mayor every year. Report by E. W. Zerthoud, ex-Mayor. Supt., Wm. Gruever.

26. GRAND JUNCTION, *Mesa Co.* (Pop., in '90, 2,030.) Built in '88-9 by Grand Junction Water Co., under 20-yrs. franchise. Supply.—Grand River, pumping to stand-pipe. Fiscal year closes Aug. 7. Distribution.—*Mains*, 4 miles. *Taps*, 300. *Meters*, 2. *Hydrants*, 55. Financial.—Withheld. Management.—Supt., W. J. Kruser.

27. GREELEY, *Weld Co.* (Pop., 1,297—2,395.) Built in '80 by city. Supply.—Well and galleries, pumping to stand-pipe. Additional gallery 100 ft. long 18 ft. deep, was built in '91. Fiscal year closes March 31.

Year.	Mains.	Taps.	Hyd'ts.	C'ns'pt'n.	Cost.	B'd debt.	Int. ch'g's.	Exp.	Rev.
'89-90	11	135	46	120,000	$63,128	$65,000	$3,900	$2,154	$1,201
'90-1	11.5	179	46	173,414	63,716	65,000	3,900	3,754	3,008

Financial.—*Total Debt*, $82,100: floating, $17,100 at 8%; bonded, $65,000 at 6% due in '94 *Expenses*, total, $7,654. *Revenue:* consumers, $3,008; city, none. Management.—City Council. Mayor, S. B. Wright. Secy. and Supt., E. F. Dauley.

FOR FURTHER DESCRIPTIVE MATTER see Manual for 1889-90. CHANGES since 1890 are here given. POPULATIONS are for 1880 and 1890, respectively.

324   MANUAL OF AMERICAN WATER-WORKS.

28. GREEN MOUNTAIN FALLS AND UTE PARK, El Paso Co. (Pop., not given.) History.—Built in '90 by Ute Pass Land & Water Co.; put in operation July 17. Engr., H. I. Reid, Colorado Springs. Supply.—South fork of Fountain qui Bouille, by gravity. Settling Reservoir.—Cap., 160,000 galls. Distribution.—Mains, 10 to 6-in c. i., 4.2 miles. Takers, 25. Valves, Chapman. Pressure: ordinary, 90 lbs; fire, 200 lbs. Financial—Cost, about $35,000. Expenses, $780. Management.—Prest., Frank R. Ehrich. Secy. and Treas., Frank White. Both of Colorado Springs. Supt., Chas. Bauer.

29. CUNNISON, Gunnison Co. (Pop., 888—1,105.) Built in '82 by Gunnison Gas & Water Co. under 25-yrs. franchise, which does not control rates but provides for purchase of works by city in 1907, 10% being added to actual value. Supply.—Gunnison River. pumping to tanks and direct. Fiscal year closes Aug. 31. Distribution.—Mains, 4.5 miles. Taps, 119. Meter, 1. Hydrants, 52. Consumption, 181,-800 galls. Financial.—Cost, $85.925. Cap. Stock, authorized, $20,000; all paid up; paid no div'd in '99. No Debt. Expenses, $3,508. Revenue, $5.559. Management.—Prest. and Treas., B. H. Lewis. Secy. and Supt., D. J. McCanne.

16. HARMON, Arapahoe Co. (Pop., in '90, 743.) Supply from Denver City Water Co.

30. HIGHLANDS, Arapahoe Co. (Pop., in '90, 5,161.) Owned by Denver City Water Co.; see Denver; city does not have right to buy works; rates must be same as at Denver. Built in '87 by Beaver Brook Water Co. Supply.—Artesian wells, pumping to tank. In '90 two Deane pumps were added. Fiscal year closes Nov. 1.

Year.	Mains.	Taps.	Mtrs.	H'd'ts.	C'ns'pt'n.	Cost.	B'd'd d't.	Int. ch'g's.	Exp.	Rev.
'88	12.5	520	..	7	1,000,000	$89,000	$48,000	......	$6,000	$11,500
'90	30	1,100	7	160	1,000,000	357,000	75,000	$1,900	13,200	28,000

Financial.—Cap. Stock, authorized, $300,000, all paid up. Expenses: operating, $12,000; int., $4,900; taxes, $1,200; total, $18,100. Revenue: consumers, $21,000; fire protection, $6,000; other purposes, $1,000; total, $28,000. Management.—Prest., W. A. Underwood. Vice-Prest. and Treas., Dennis Sullivan. Secy. and Gen. Mgr., Frank P. Arbuckle. Asst. Gen. Mgr., R. H. Porter.

NOTE.—System is now operated in connection with plant of Denver City Water Co., at Denver, which see.

31. HOLYOKE, Phillips Co. (Pop., in '90. 649.) Built in '89 by Lincoln Land & Town Site Co.; franchise does not provide for purchase of works by city. Supply.—Well, pumping to stand-pipe. Water is furnished by Burlington & Missouri Ry. Co. Fiscal year closes Dec. 31. Distribution.—Mains, 5.5 miles. Taps, 60. Meters, 4. Hydrants, 16. Consumption, 5,000 galls. Financial.—Cost, $30.000. Expenses, nominal. Revenue: consumers, $835; city, $1,045; total, $1,880. Management.—Lincoln Land Co.

32. IDAHO SPRINGS, Clear Creek Co. (Pop., 733—1,338.) Built in '80 by city. Supply.—Mountain stream, by gravity to reservoirs. In '90 the 4-in. wooden supply pipe was replaced by 6 and 8-in. cem. pipe; new pipe being laid at perfect grade has increased supply five times. Fiscal year closes March 31. Distribution.—Mains, 2 miles. Taps, 165. Hydrants, 22. Financial.—Cost (in '88). $30,000. Bonded Debt. $20,000 at 8%, due in '91. Sinking Fund, $7,000. Expenses, $3,099. Revenue: consumers, $2,956; city, none. Management.—Committee of three appointed by mayor every year. Prest., R. B. Griswold. Supt., H. O. Walker.

33. LA JUNTA, Otero Co. (Pop., in '90, 1,439.) Owned by La Junta Water Co. Built in '82-3 by A. T. & S. F. R. R. Co. Franchise neither regulates rates nor provides for purchase of works by city. Supply.—Well on bank of river, pumping to tank and direct. Fiscal year closes June 30.

Year.	Mains.	Taps.	Meters.	Hydrants.	Consumption.	Cost.	Exp.	Rev.
'89-90	3.5	200	0	15	250,000	$32,405	$3,776	$3,894
'90-1	3.5	220	0	15	300,000	32,897	3,328	4.317

Financial.—Cap. Stock: authorized, $55,000; paid-up, $22,000. Paid no div'd in 1891. No debt. Expenses: operating, $3,013; taxes, $315; total, $3,328. Revenue: consumers, $3,387; fire protection and other purposes, $960; total, $4,347. Management.—Prest., Allen Mannell, Chicago, Ill. Secy., Edw. Wilder. Topeka, Kan. Mgr., A. A. Robinson.

34. LAKE CITY, Hinsdale Co. (Pop., in '90, 607.) Built in '90 by town; accepted Sept. 10. Supply.—Lake fork of Gunnison River, by gravity from impounding reservoir. Distribution.—*Mains*, 3 miles. *Hydrants*, 17. Financial.—*Cost*, $25,000. *Bonded Debt*, $25,000 at 8%, due in 1904. Management.—Prest., C. D. Peck, Mayor. Secy., A. M. Wilson. Supt., Geo. Boyd.

35. **LAMAR,** *Prowers Co.* (Pop., in '90, 566.) Owned by city. Built in '87 by Lamar Water Co. Supply.—Nine driven wells, pumping to tank and direct. Distribution.—*Mains*, 4 miles. *Taps*, 100. *Meters*, 0. *Hydrants*, 20. Financial.—*Cost*, $22,000. *Bonded Debt*, $22,000 at 6%. *Expenses:* operating, $2,000; int., $1,320; total, $3,320. Management.—Mayor, W. H. Vanorsdale.

36. **LAS ANIMAS,** *Bent Co.* (Pop., 52—611.) Built in '89 by city. Supply.—Open well, near river, pumping to tank and direct. Fiscal year closes April 1. Distribution.—*Mains*, 3 miles. *Taps*, 100. *Meters*, 0. *Hydrants*, 30. *Consumption*, 80,000 galls. Financial.—*Cost*, $31,000. *Bonded Debt*, $28,000 at 8 %. *Sinking Fund*, $2,000. *Expenses:* operating $1,100; int., $2,240; total, $3,340. *Revenue*, consumers, $500. Management.—City Council.

37. **LEADVILLE,** *Lake Co.* (Pop., 14,820—11,212) Built in '79 by Leadville Water Co. Supply.—Mountain streams, by gravity and pumping to reservoir. Fiscal year closes Dec. 31. Distribution.—*Mains*, 14 miles. *Taps*, 1,000. *Meters*, 4. *Hydrants*, 140. *Consumption*, 1,000,000 galls. Financial.—*Cost*, $600,000. *Cap. Stock* (in '88), $300,000. *Bonded Debt*, $300,000 at 6%. *Revenue:* consumers, $35,000; city, $12.000; total, $47,000. Management.—Prest., Wm. E. Hawks. Secy. and Supt., C. N. Priddy.

38. **LONGMONT,** *Boulder Co.* (Pop., 773—1,543.) Built in '82 by town. Supply.—South St. Vrain River, by gravity to and from reservoirs; filtered. Fiscal year closes March 31.

Year.	Mains.	Taps.	Meters.	Hyd'ts.	Cost.	B'd't d't.	Int. ch'gs.	Exp.	Rev.
'86-7	14.5	324	0	35	$72,00	$70,000	$5,600	$250	$5,000
'88-9	15.8	3-9	1	35	75,000	70,000	5,600	600	6,250
'90-1	16	427	1	35	75,500	70,000	5,600	1,000	7,000

Financial.—*Bonded Debt*, $70,000 at 8%, due in '97. *Expenses*, total, $6,600. *Revenue:* consumers, $7,000; city, none. Management.—Prest., Geo. W. Brown.

39. **LOVELAND,** *Larimer Co.* (Pop., 236—698.) Built in '87 by city. Supply.—Little Cañon of Big Thompson River, by gravity. Fiscal year closes April 1. Distribution.—*Mains*, 12.5 miles. *Taps*, 167. *Hydrants*, 12. Financial.—*Cost*, $44,-000. *Bonded Debt*, $41,000. *Expenses:* operating, $300; int., $2,840; total, $3,140. *Revenue:* consumers, about $2,200. Management.—Supt., David James.

40. **MANITOU SPRINGS,** *El Paso Co.* (Pop., 422—1,439.)—Built in '87-8 by town. Supply.—French's Creek, by gravity to reservoir. Fiscal year closes March 31.

Year.	Mains.	Taps.	Hydrants.	Cost.	B'd'd debt.	Int. ch'gs.	Exp.	Rev.
'88-9	2.8	100	18	$40,000	$40,000	$2,400	$500	$1,000
'90-1	5	220	26	65,000	65,000	3,900	500	5,000

Financial.—*Bonded Debt*, $65,000 at 6%; due in 1902-5. *Expenses*, total, $4,400. *Revenue*, consumers, $5,000. Management.—Prest., Geo. W. Snider. Secy., C. H. Frownie. Supt., M. F. Bowers.

16. **MONTCLAIR,** *Arapahoe Co.* (Pop., in '90, 380.) Supply from Denver City Water Co.

41. **MONTROSE,** *Montrose Co.* (Pop., in '90, 1,330.) Built in '88 by town. Supply.—River, pumping direct. Fiscal year closes March 12. Distribution.—*Mains*, 5 miles. *Taps* (in '88), 98. *Meters* (in '88), 0. *Hydrants* (in '88), 11. *Consumption*, 500,000 galls. Financial.—*Cost*, $25,000. *Bonded Debt*, $20,000 at 6%, due in 20 yrs. *Expenses:* operating, $3,000; int., $1,200; total, $4,200. *Revenue*, consumers, $1,000. Management.—Chn., J. E. McClure. Clerk, W. C. Redding. Supt., W. H. Franklin.

II. **MORRISON,** *Jefferson Co.* (Pop., 186—251.) Built in '85-6 by Morrison Water Co., under 20-yrs. franchise; town neither controls rates nor has right to buy works. Supply.—Bear Creek, by gravity. Fiscal year closes Jan. 31. Distribution.—*Mains*, ¾ miles. *Taps*, 25. *Hydrants*, 0. Financial.—*Cost*, $3,500. *Cap. Stock:* authorized, $5,000; paid up, $3,500; paid 9¼% div'd in '91. No *Debt*. *Expenses*, $19. *Revenue*, $348. Management.—Prest., P. Christensen. Secy. and Treas., W. S. Smith. Supt., J. H. Pratt.

FOR FURTHER DESCRIPTIVE MATTER see Manual for 1889-90. CHANGES since 1890 are here given. POPULATIONS are for 1880 and 1890, respectively.

326   MANUAL OF AMERICAN WATER-WORKS.

42. NEW CASTLE, *Garfield Co.* (Pop., in '90, 311.) History.—Built in '80 by town; put in operation July 1. Engr., W. H. Trumbore. Contr., Mason W. Mather Glenwood Springs. Supply.—East fork of Elk Creek, by gravity. Conduit.—A wooden flume, 5½ miles long. Distribution.—*Mains.* kal. i., 4 miles *Taps*, 34. *Hydrants*, 7. *Pressure*, 120 lbs. Financial.—*Cost*, $16,500. *Bonded Debt*, $15,000 at 8%. *Expenses:* op:rating, $500; int., $1,200; total, $1,700. *Revenue*, $2,700. Management.—Secy., Chauncey Sheldon. Supt., Jno. F. Ritter.

16. NORTH DENVER (*Denver P. O.*), *Arapahoe Co.* (Pop., not given.) Supply from Denver City Water Co.

43. ORDWAY, *Otero Co.* (Pop., est., 50; new town.) Built in '90-1 by co.; put in operation June 1. A 1,500,000-gall. reservoir. *Mains*, 1½ miles. *Hydrants*, 3. *Cost*, $5,000. *Cap. Stock:* authorized, $10,000; paid up, $7,000. Prest., G. N. Ordway. Secy. and Supt., J C. Mosher. Treas., C. N. Clinton, Denver.

44. OURAY, *Ouray Co.* (Pop., 864-2,531.) Built in '81 by town: Supply.— Cañon Creek and spring, by gravity. Fiscal year closes May 1. Distribution.— *Mains*, 3.2 miles. *Taps*, 1,000. *Meters*, 0. *Hydrants*, 17. *Consumption*, 600,000 galls. Financial.—*Cost*, $42,000. *Bonded Debt*, $40,000 at 8%. *Expenses:* operating, $1,500; int., $3,200; total, $4,700. *Revenue*, withheld, is sufficient to pay all expenses. Management.—City Council. Mayor, W. W. Roan. Secy., Geo. M. Stebbins. Supt., G. P. Jaggers. Report by M. S. Corbett.

45. PALMER LAKE, *El Paso Co.* (Pop., est., 1,000.) Built in '87 by Palmer Lake Water Co., under 20-yrs. franchise. Supply.—Monument Creek, by gravity. Distribution.—(In '88.) *Mains*, 3 miles. *Taps*, 87, *Hydrants*, 5. Financial.— (In '88.) *Cost*, $7,000. *No Debt.* *Expenses*, $50. *Revenue*, $1,100. Management.— Prest., W. Failey Thompson. Secy. and Supt., J. W. Drinkwater. NOTE.—Co. declined to give statistics for '90.

54. PETERSBURGH, *Arapahoe Co.* (Pop., not given.) See South Denver.

46, 47. PUEBLO, *Pueblo Co.* (Pop., 3,217--24,558.)

CITIZENS' WORKS.

Built in '74-5 by city; now owned by citizens of portion of city, under management of Board of Trustees. See Pueblo Water Co. Supply.—Arkansas River, by gravity to settling reservoirs, thence by direct pumping.

Year.	Mains.	Taps.	Meters.	Hyd'ts.	Cons'pt'n.	Cost.	B'd d't.	Int. ch'gs.	Exp.	Rev.
'90	6.5	350		60					$6,590	
'82	'8	485	0	75	367,0 0	$160,000	$140,000	$11,200	0,000	$11,000
'88	23		12	75	2,030.000					

Management.—Supt., Frank Barndollar.

PUEBLO WATER CO.

Built in '81 by South Pueblo Water Co.; now owned by Pueblo Water Co.; maximum rates fixed by city ordinance; no provision for purchase of works by city. Upon consolidation of cities of Pueblo and South Pueblo in '86 it was agreed that citizens of Pueblo, which city owned its works, should retain their ownership, while that part formerly known as South Pueblo should continue their contract with the co. Fiscal year closes Dec. 31. In '90 a 5,000,000-gall. Worthington pump was added, and franchise renewed for 25 yrs., giving co. exclusive right to supply water to territory south of Arkansas River. Reported later that Pueblo Water Co. would succeed above co. and double cap. of works. Distribution.—*Mains*, 35 miles. *Taps*, 1,350. *Meters*, 0. *Hydrants*, 125. *Consumption*, 2,000,000 galls. Financial.—*Cap. Stock*, authorized, $300,000; all paid up; income used in improvements. *Bonded Debt*, $250,- 000 at 6%, issued April 1, '91. Management.—Gen. Mgr., Geo. J. Dunbaugh. Secy. and Treas., J. A. Joy. Supt., G. J. Shields.

NOTE —In '90 works were projected by the Pueblo Gravity Water Supply Co., the supply to be developed from the underflow of the Fontain qui Bouille, by means of drain-pipe. D. C. Henny, Denver, was Engr. for co. and J. D. Schuyler, San Diego. Cal., Consult. Engr. For illustrated abstract of latter's report see ENGINEERING NEWS, vol. xxv., p. 53.

48. RED CLIFF, *Eagle Co.* (Pop., in '90, 383.) History.—Built in '87 by city. Supply.—Mountain stream, by gravity to reservoir. Reservoir.—Is 10 × 20 × 40 ft. Distribution.—*Mains*, 3½-in., 1 mile. *Taps*, 50. *Meters*, 0. *Hydrants*, 4. *Pressure*, 50 lbs. Financial.—*Cost*, $4,000. *Debt*, $1,300 at 8%. *Revenue*, consumers $1,200. Management.—Report by H. W. Smith.

COLORADO. 327

49. RICO, *Dolores Co.* (Pop., 891—1,131.) Bu'lt in '87 by town. Silver Creek, by gravity. *Mains*, 2,000 ft. *Takers*, number unknown. *Hydrants*, 5. *Cost*, $8,000. *Bonded Debt*, $5,000 at 8%. Supt., W. B. Hess.

50. RIDGWAY, *Ouray Co.* (Pop., not given.) Built in '90-1 by co.; put in operation in July. Contr., Gordon Kimball, Ouray.

51. ROCKY FORD, *Otero Co.* (Pop., 47—468.) **History.**—Built in '90 by city. Contr., Jno. J. Brown. **Supply.**—Irrigating ditch, by gravity; brick filter. **Distribution.**—*Mains*, 6-in. c. i., 1 mile. *Taps*, 8. *Hydrants*, Torrent, 7. *Valves*, Lud low. *Pressure*, 55 lbs. **Financial.**—*Cost*, $6,450. *Bonded Debt*, $7,000 at 10%. *Revenue*, consumers, $100. **Management.**—Supt., C. D. Steward.

52. SALIDA, *Chaffee Co.* (Pop., est., 3,000.) Built in '82 by town. **Supply.**— South Arkansas River, by gravity from reservoir. **Distribution.**—(In '88.) *Mains*, 3 miles. *Taps*, 250. *Meters*, 0. *Hydrants*, 0. *Consumption*, 300,000 galls. **Financial.**—(In '88.) *Cost*, $25,000. *Bonded Debt*, $24,000 at 8%. *Expenses:* operating, $600; int., $1,920; total, $2,520. *Revenue*, consumers, $2,400. **Management.**—(In '89.) Mayor, J. R. Eddy. Chn., J. W. Deen.

SILVER CLIFF, *Custer Co.* (Pop, 5,040—546.) Now owned by Silver Cliff Water Supply Co.; city built in '80 and turned works over to bondholders in '86; not in operation since early part of '89. See 1889-90 Manual, p. 508.

53. SILVERTON, *San Juan Co.* (Pop., 264—est., 2,000.) Built in '82-3 by Silverton Water-Works Co.; now owned by Silverton Water Supply Co. **Supply.**— Springs, by gravity from reservoir. During '90 reservoir was lined with cem. at cost of $3,000. **Distribution.**—*Mains*, 5 miles. *Taps*, 200. *Meters*, 0. *Hydrants*, 15. **Financial.**—*Cost*, $53,000. *No Debt*. *Expenses*, $1,200. *Revenue*, $7,900. **Management.**—Prest. B. L. Carr; Secy. and Treas., F. P. Secor; both of Longmont, Mgr., H. O. Montague.

54. SOUTH DENVER, *Arapahoe Co.* (Pop..in '90, 1,491.) **History.**—Built in '89 by town; put in operation Nov. 25. Jan. 22, '90, town contracted with Citizens' Water Co., Denver, to operate pumping station free of cost to town, in return for which co. has all water above town's needs. Supplies Petersburg. **Supply.**—Well and galleries, by pumping. Pumping Machinery.—Cap., 2,000,000 galls.; hor. compr. cond. high duty Gaskill. Fiscal year closes, Apr. 1. **Distribution.**—*Mains*. 12 to 4-in. c. i., about 19 miles. *Taps*, Mar. 17, '91, 416; Apr. 1, '90, 112. *Hydrants*, 25 or more. **Financial.**—*Cost*, $165,207. *Bonded Debt*, $160,000 at 6%, due $10,000 year from Jan. 1, '95. *Expenses:* operating (as near as can be determined from town report), $3 711; int., $8,600; total, $13,311. *Revenue*, consumers, $4,943. **Management.**—Comr., Henry Domire. Town Clerk, C. H. Peters.

SOUTH PUEBLO, *Pueblo Co.* See Pueblo.

55. TELLURIDE, *San Miguel Co.* (Pop., in '9), 766.) Built in '85 by town. **Supply.**—Creek, by gravity. *Bonded Debt*, $22,000 at 8%. Supt., Geo. Phillips.

56. TRINIDAD, *Las Animas Co.* (Pop., 2.226—5,523.) Built in '79 by Trinidad Water-works Co.; city has right to buy works in '90 and every 5 yrs. thereafter; does not control rates. **Supply.**—Las Animas River, by gravity to settling reservoir, thence to city. In '90 supply main was extended 8 miles up river. Head, 320 ft. Fiscal year closes Oct. 1.

Year.	M'ns.	Taps.	M't'rs.	Hyd'ts	C'ns'pt'n.	Cost.	B'd'd d't.	Int. ch's.	Exp.	Rev.
'80	5	250	....	30	90,000	$80,000	.....	.....	$1,500	.....
'82	5.5	265	150	32	100,000	100,000	None.	None.	6,000	$10,000
'84	8	375	150	40	250,000	100,000	$50,000	$3,500	..	10,500
'87	8	450	....	50	500,000	110,000	50,000	3,500	4,000	14,500
'90	20	700	150	79	750,000	240,330	200,000	8,168	8.353	*6,320

*From city.

Financial.—*Cap. Stock*, $50,000; paid no div'd in 1890. *Bonded Debt*, $200,000 at 6%, due in 1909; authorized bonds, $250,000. *Expenses:* operating, $7,013; int., $8,168; taxes, $1,340; total, $16,521. *Revenue*, city, $6,320. **Management.**—Prest., D. A. Chappell, Secy. and Supt., Wm. B. Cunningham.

NOTE.—In '90 or early in '91 W. Kiersted, Kansas City, Mo., reported on gravity supply for city, which proposed to build works, but was enjoined from doing so on ground that co. charter was exclusive.

28. UTE PARK, *El Paso Co.* (Fop. not given.) **Supply.**—From Green Mountain Falls.

FOR FURTHER DESCRIPTIVE MATTER see Manual for 1889-90. CHANGES since 1890 are here given. POPULATIONS are for 1880 and 1890. respectively.

16. **VILLA PARK,** *Arapahoe Co.* Part of Denver. Supply from Denver City Water Co.

57. **WALSENBURG,** *Huerfano Co.* (Pop., 377—923.) Built in '88 by Walsenburg Water Co., under exclusive franchise. Supply.—Lake Miriam, by gravity. Distribution.—(In '88.) *Mains,* 6.8 miles. *Taps,* 160. *Meters,* 0. *Hydrants,* 16. Financial.—(In '88.) *Cost,* $70,000. *Cap. Stock,* $75,000. *Bonded Debt,* $30,000 at 8%. *Expenses:* operating, $400; int., $2,400; total, $2,800. *Revenue:* consumers, $3,400; city, $3,200; total, $8,600. **Management.**—(In '89.) Prest., Fred Walsen. Secy. and Treas., T. F. Martin. Supt., F. O. Roof.

16. **WEST DENVER,** *Arapahoe Co.* (Pop., not given.) Supply from Denver.

Water Works Projected with Fair Prospects of Construction.

BRIGHTON, *Arapahoe Co.* (Pop., in '90, 306.) Contracts let by Co. in '91 to D. F. Cormichael, Brighton, for works to be completed March 1, '92. Supply, wells, pumping to tank. Est. cost, $10,600. Prest., W. H. Malone. Secy., Jos. MacKean. Treas., W. C Kidder.

CASTLE ROCK, *Douglas Co.* (Pop., 88—315.) Aug. 13, '91, town advertized for bids from contractors for construction of works. Mayor, J. H. Craig.

COAL CREEK, *Fremont Co.* (Pop., est., 1,100.) Coal Creek Water & Light Co. were to begin construction of water-works Sept. 15, '91, under 20-yrs. franchise. Spring, by gravity to reservoir. Mains, 6 miles. Hydrants, 22. Pressure, 75 lbs. Cap. Stock: authorized, $50,000; paid up. $10,000. Prest., B. F. Rockfeller, Cañon City. Secy., Jno. Young, Colebrook. Treas., Geo. A. Baker, Rockvale.

DEL NORTE, *Rio Grande Co.* (Pop., 729—736.) July 6, '91, it was reported that city would begin construction of water-works as soon as $25,000 worth of bonds had been floated. Rio Grande River, pumping to reservoir. Des. Engr., Geo. D. Nickel. Contrs., U. S. Wind Engine & Pump Co., Kansas City, Mo. Aug. 15 it was reported that bonds had been sold and construction would begin at once.

ERIE, *Weld Co.* (Pop., 358—662.) Aug. 6, '91, it was reported that city would build water works as soon as bonds were sold. Supply, reservoir, by gravity. Chn., Robt. Lawly. Engr., E. C. Kinney, Denver. Address Osborn & Taylor, 27 Tabor Block, Denver.

IRONTON, *Arapahoe Co.* (Pop., not given.) July 15, '91, it was reported that water-works would soon be built. Address A. McL. Hawks, 23 Essex B'ld'g., Denver.

LOUISVILLE, *Boulder Co.* (Pop., 450-590.) City voted to build water-works, contracts to be let July 15, '91. July 13 bonds had not been sold. Supply, mountain stream, by gravity from reservoir. Est. cost, $25,000. Mayor, John M. Wilson. City Clerk, Wm. Condery.

LYONS, *Boulder Co.* (Pop., in '90, 574.) Water-works talked of and badly needed. No definite steps had been taken July 6. Address E. S. Lyons.

SAINT ELMO, *Chaffee Co.* (Pop., est., 500.) Town was to begin construction of works July 1, '91. Mountain stream, by gravity. Est. cost, $4,000. Chn., P. Hurley. Clerk, A. W. Root.

WRAY, *Yuma Co.* (Pop. in '90, 125.) Reported Aug. 14, '91, that town had let contracts and construction would begin Sept. 1. Republican River, pumping to tank. Contrs., Wray Milling Co. Est. cost, $5,000. Chn., E. Hitchings. Engr., E. G. Howard.

NEW MEXICO.

Water-Works Completed or in Process of Construction.

1. **ALBUQUERQUE,** *Bernalillo Co.* (Pop., 2,315—5,515; figures include old and new towns; pop. of old town for '90 was 1,733.) Built in '86 by Albuquerque Water Co.; name changed since to Water Supply Co. Supply.—Surface and driven wells, pumping to reservoir. Fiscal year closes Sept. 30.

NEW MEXICO.

Year.	Mains.	Taps.	Meters.	Hyd'ts.	C'ns'p'tn.	Cost.	B'd'd debt.	Int.ch's.	Exp.	Rev.
'87	7.5	160	1	62	160,000	$118,000	$110,000	$7,700	$5,000	$11,030
'89	8	320	18	64	285,000	......	110,000	7,700	......	......
'90	8	349	22	64	285,000	123,000	110,000	7,700	6,150	14,160

Financial.—*Cap. Stock*, $200,000. *Bonded Debt*, $110,000 at 7%, due in 1915. *Expenses:* operating, $5,800; int., $7,700; taxes, $650; total, $14,150. *Revenue:* consumers, $9,000; fire protection, $5,160; total, $14,160. **Management.**—Prest., A. A. Grant. Secy., B. F. Davis. Supt., C. J. Stetson.

2. **CLAYTON,** *Colfax Co.* (Pop., in '90, 140.) Built in '86 by U. P. Ry. Co.; town has no control over rates. **Supply.**—Spring, pumping to tank. **Distribution.** —*Mains*, 2 miles. *Hydrants*, 4. *Consumption*, 18,000 galls. *Cost*, $3,000. **Management.**—Pumping Engr., H. S. Simpson.

3. **LAS VEGAS,** *San Miguel Co.* (Pop., in '90, 2,385.) Built in '81-2 by Aqua Pura Co. under 50-yrs. franchise; city neither controls rates nor reserves right to purchase works. **Supply.**—Spring, pumping to tank; filtered. Fiscal year closes Aug. 31.

Year.	Mains.	Taps.	Meters.	Hydrants.	Cost.	Bonded debt.	Int. ch'gs.	Exp.	Rev.
'82	12	300	0	30	$110,000	......	......	$3,000	$9,000
'87	14	646	0	34	......	$120,000	$7,200	4,000	17,700
'90	20	722	1	34	135,678	120,000	7,200	......	......

Financial.—*Cap. Stock:* authorized, $200,000; paid up, $187,500; paid no div'd in 1890. *Bonded Debt*, $120,000 at 6%. *Sinking Fund*, $7,000, invested in Aqua Pura Co.'s bonds at 6%. **Management.**—Prest., Wm. A. Vincent. Secy., W. E. Baker. Treas., Jefferson Reynolds. Supt., A. K. Nones.

4. **MAXWELL CITY,** *Colfax Co.* (Pop., not given.) **History.**—Built in '90 by Maxwell City, Land & Improvement Co. Des. Engr., L. S. Preston. Const. Engr., E. S. Warren. Contr., S. T. Wicks, Colorado Springs, Col. **Supply.**— Springs, by gravity. **Storage Reservoir.**—Cap., 50,000 galls.; lined with cem. **Distribution.**—*Mains*, w. i., 2.3 miles. *Hydrant*, 1. *Pressure*, 65 lbs. **Financial.** —*Cost*, $8,000. **Management.**—Mgr., E. S. Warren.

5. **RATON,** *Colfax Co.* (Pop., in '90, 1,255.) Built in '80-1 by A. T. & S. F. R. R. Co.; see note below. Town neither controls rates nor has right to buy works. **Supply.**—Springs and wells, by gravity from former and by pumping to reservoir from latter. Supply was increased in '90 by sinking additional wells. See note below. Fiscal year closes June 30. **Distribution.**—*Mains*, 3.5 miles. *Taps*, 165. *Hydrants* (in '87), 10. **Financial.**—*Cost*, $55,370. *Cap. Stock:* authorized, $100,000; paid-up, $50,000; paid no div'd in '90. *No Debt. Expenses:* operating, $3,870; taxes, $203; total, $4,133. *Revenue:* consumers, $2,983; other purposes, $1,200; total, $4,183. **Management.**—See note below.

NOTE.—In '91 plant was transferred to Raton Water-Works Co., and contracts for improvements let to Gate City Construction Co. Engr., Wm. Kingman, Topeka, Kan. Oct. 24, '91, co. was building a dam across narrow valley 7 miles from town to impound water from drainage area of 29.5 sq. miles. Dam of earth, with heart-wall 48 ft. high, 360 ft. long, forming 52,000,000-gall. reservoir 458 ft. above town. Dam 20 ft. wide at top, with highest part on rock. Slopes, 2 to 1; inner, riprapped. Heart-wall, concrete, 6 ft. thick at bottom, 4 at top. Spillway, at end of dam, in rock; 100 ft. long, 6 ft. deep. As designed, there will be filter bed in reservoir consisting of V-shaped trenches made water-tight by concrete lining. Trench 10 ft. wide, about 4 ft. deep. Water passes down through 3 ft. of sand, then through felt, then through broken stone, and is collected by 6-in. sewer pipes and conveyed to gate of reservoir discharge pipes. Supply main will consist of 2 miles 10-, and 5 miles 8-in. c. i. pipe, and will have dy. cap. of 1,000,000 galls. Prest., J. W. Dwyer. Secy., E. D. Somer. Treas., Marcy Weaver. Supt., C. A. Fox.

6. **SANTA FE,** *Santa Fe Co.* (Pop., 6,635–6,185.) Built in '81-3 by Water & Improvement Co. **Supply.**—Santa Fe River, by gravity from impounding reservoir. **Distribution.**—(In '87.) *Mains*, 10 miles. *Hydrants*, 25. **Financial.**—(In '87.) *Cost* (in '83), $175,000. *Cap. Stock*, $180,000. *Bonded Debt*, $110,000 at 6%. *Expenses:* operating, $6,000; int., $6,600; total, $12,600. **Management.**—(In '88.) Prest. and Treas., Robt. E. Carr, St. Louis, Mo. Secy. and Supt., E. B. Seward.

FOR FURTHER DESCRIPTIVE MATTER see Manual for 1889-90. CHANGES since 1890 are here given. POPULATIONS are for 1880 and 1890, respectively.

MANUAL OF AMERICAN WATER-WORKS.

7. **SILVER CITY,** *Grant Co.* (Pop., 1,800—2,102.) Built in '86-7 by Silver City Water Co., under 50 years franchise. Supply.—Wells, pumping to reservoir and direct. Fiscal year closes Dec. 3¹. Distribution.—*Mains.* over 6 miles. *Taps,* over 200. *Meters,* 46. *Hydrants,* 23. *Consumption,* 65,000 galls. Financial.—*Cap. Stock,* authorized, $100,000; all paid up. *Total Debt,* $77,500; floating, $2,500; bonded, $75,000 at 6%. *Expenses,* operating, $3,200; int., $4,500; taxes. $600; total, $8,300. *Revenue:* consumers, over $7,000; city, $3,375 was due Apr., 28, '91, but city was attempting to annul franchise and had not paid hydrant rental. Management.—Prest., Geo. H. Utter. Secy. and Supt , Tom Foster.

8. **SOCORRO,** *Socorro Co.* (Pop., 1,272—1,601) Built in '87 by city. Supply.—Large mountain springs, by gravity. Distribution.—(In '88.) *Mains,* 5.5 miles. *Hydrants,* 25. *Consumption,* 200,000 galls. Financial.—(In '88.) *Cost,* $30,000. *Bonded Debt,* $30,000 at 6%. *Expenses:* operating, $1,500; int., $1,800; total, $3,300. Management.—(In '89.) W. Com.

9. **SPRINGER,** *Colfax Co.* (Pop., 34—600.) History.—Built in '90 by Aqua Crystilina Co. Contr., I. W. Wicks, Colorado Spgs., Colo. Supply.—Wells, pumping direct. Pumping Machinery.—Dy, cap., not given; Knowles. Distribution.—*Mains,* 6 and 4-in., 2 miles. Pipe and specials from Bendner, Bollhoff & Co., Denver, Colo. Financial.—*Cost,* $13,000. *Bonds:* authorized, $8,000; issued. $6,000. *Expenses,* $1,000. *Revenue:* consumers, $1,800; city, none. Management.—Prest., M W. Mills. Secy., G. W. Abbott. Supt., W. W. Jacobs.

Water-Works Projected with Fair Prospects of Construction.

GALLUP, *Bernalillo Co.* (Pop., in '90, 1,208.) Water-works talked of; ownership not determined. Well, pumping to reservoir. Est. cost, $12,000. Address E. Hart.

GROUP 8.—PACIFIC STATES.—Water-Works Completed, in Process of Construction or Projected in Washington, Oregon, California, Arizona, Nevada, Utah and Idaho.

WASHINGTON.

Water-Works Completed or in Process of Construction.

1. **ABERDEEN,** *Chehalis Co.* (Pop., in '90, 1,638.) History.—Built in '91 by city; put in operation Sept. 1. Small system, chiefly for fire-protection, had previously been built, and was bought by city. Engrs. and Contrs., Cummings & Cook. Supply.—Brook, pumping direct. Pumping Machinery.—Dy. cap., 2,000,000 galls.; one 1,500,000 Worthington comp. dup. and one 500,000 Dow. Distribution.—*Mains,* 12-in. steel and 12 to 4-in. kal. i., 3 miles. *Services,* w. i. Pipe and specials from Francis Smith & Co., San Francisco. *Hydrants,* Golden Gate, 15. *Valves,* 15. *Pressure:* ordinary, 50 lbs.; fire, 90 lbs. Financial.—*Cost,* $45,000. *Bonded Debt,* $40,000. Sept., '91, $60,000 in bonds are voted to pay for old works bought by city and construction of new works. Management.—City Clerk, J. H. White.

2. **ANACORTES,** *Skapit Co.* (Pop., in '90, 1,131.) Built in '90 by Anacortes Water Co; controlled by Oregon Improvement Co; franchis e provides neither for control of rates nor purchase of works by city. Supply.—Sprints, by gravity to and from reservoir. *Mains,* 3.6 miles. *Hydrants,* 8. *Cost,* $50,000. Hydrant rental, free. Agents, Hogan & Hogan. Gen. Mgr., C. J. Smith. Seattle.

3. **BALLARD,** *King Co.* (Pop., in '90, 1,173.) History.—Built in '88 by West Coast Water Co. Des. Engr., R. H. Thompson. Supply.—Springs, by gravity. *Reservoir.*—Cap., 500,000 galls. Distribution.—*Mains,* 6-in., 4 miles. *Pressure,* 110 lbs. Financial.—*Cost,* $20,000. Management.—Prest., W. R. Ballard. Secy., C. De Cesta. Treas., R. W. Graves.

BLAINE, *Whatcom Co.* (Pop., in '90, 1,563.) Built in '90-1 by Northwest Water Co. Supply.—Springs, by gravity to 6,000,000-gall. reservoir. *Cost,* $30,000. Prest., J. M. Yestling. Seattle. Secy., E. A. Curtis.

WASHINGTON. 331

5. CENTRALIA, *Lewis Co.* (Pop., in '90, 2,026.) **History.**—Built in '90-1 by Centralia Water Co., under 20-yrs. franchise; put in operation March 1. Des. and Const. Eng., J. B. Wood. **Supply.**—Well, near Skorkum Chuck River, pumping to reservoir. **Pumping Machinery,**—Dy. cap., 750,000 galls.; Worthington. **Reservoir.** - Cap., 1,200,000 galls. **Distribution.**—*Mains*, kal. 1., 6.5 miles. *Services.*—galv. 1. Pipe and specials from National Tube Works, McKeesport, Pa. *Hydrants*, National, 25. *Valves*, Eddy. *Pressure*, 95 lbs. **Financial.**—*Cost*, $1,000. *Cap. Stock*, authorized, $50,000. *Operating Expenses*, $2,500. *Revenue*, city, $2,100. *Hydrant Rental*, $84. **Management.**—Prest., R. J. Miller. Secy. and Treas., A. A. Miller.

6. CHENEY, *Spokane Co.* (Pop., in '90, 647.) **History.**—Built in '90 by Cheney Water & Land Co., under 30-yrs. franchise, which provides for regulation of rates by city. Des. Engr., F. J. Carroll, Spokane. Const. Engr. and Contr., Jno. Corkish, Portland, Ore. **Supply.**—Lake, pumping to reservoir. **Pumping Machinery.**—Dy. cap., 1,000,000 galls.; Knowles. **Reservoir,**—Cap., 250,000 galls. **Distribution.**—*Mains.* 8-in., 3.5 miles. *Services*, galv. 1. *Hydrants*, 11. *Consumption*, 50,000 galls. *Pressure:* ordinary, 50 lbs.; fire, 80 lbs. **Financial.**—*Cost*, $11,000, *Cap. Stock:*, authorized, $100,000; paid-up, $25,000. *Bonded Debt*, $35,000 at 7%. *Expenses;* operating, $1,920; int., $2,450; total, $4,370. *Revenue:* consumers, $2,640; city, $2,100; total, $4,740. *Hydrant Rental*, $96. **Management.**—Prest., D. F. Percival. Secy., J. Melleville. Treas., W. E. Weygant.

7. COLFAX, *Whitman Co.* (Pop., 447-1,649.) **History.**—Built in '88 by Spring Mountain Co. **Supply.**—Springs, by gravity to 600,000-gall. reservoir. Dec. 8, '90, attempt was being made to obtain artesian water. *Mains*, 4 miles. *Cost*, $10,000. Owner, H. W. Livingstone. NOTE.—Aug. 11, '91, it was reported that city expected to put in works at once. Address Mayor J. D. Chadwick.

I. COUPEVILLE, *Island Co.* (Pop., 90—513.) Built in '89 by Coupeville Water Co; city neither controls rates nor has right to buy works. **Supply.**—Spring, pumping to tank and direct. Fiscal year closes Dec. 31. **Distribution.**—*Mains*, 2 miles. *Taps*, 25. *Hydrants*, 0. *Consumption*, 3,500 galls. **Financial.**—*Cost*, $1,500. *Cap. Stock*, $4,000; paid no div'd in '90. *No Debt.* Expenses, $865. **Management.**—Prest., J. B. Libbey. Treas., Jos. Goodwin. Secy., J. W. Clapp.

8. DAYTON, *Columbia Co.* (Pop., 996—1,880.) **History.**—Built in '91 by city; put in operation Sept. 15. Engr., A. L. Adams, Pendleton, Ore. Contr., Geo. A Sutherland, Walla Walla. **Supply.**—Infiltration gallery near stream, by gravity to reservoir. **Reservoir.**—Cap., 60,000 galls.; Contrs.. Oregon Bridge Co., Portland. **Distribution.**—*Mains*, 10 to 4-in. kal 1. *Services*, galv. 1. *Taps*, 16). *Hydrants*, Ludlow, 156. *Valves*, Ludlow. *Pressure*, 66 lbs. **Financial.**—*Cost*, $75,000. *Cap. Stock*, $75,000. *Bonded Debt*, $75,000 at 6%. *Expenses*, $6,500. *Revenue:* consumers, $5,400; city, $1,800; total, $7,200. **Management.**—Chn., J. H. Day.

9. ELLENSBURGH, *Kittitas Co.* (Pop., in '90, 2,768.) Built in '88 by Capital Hill Water Co. **Supply.**—Wilson Creek, by gravity to reservoir. (In '88.) *Mains*, 3.5 miles. *Taps*, 200. *Meters*, 0. *Hydrants*, 8. *Consumption*, 200,000 galls. **Financial.**—(In '88.) *Cost*, $50,000. *Cap. Stock*, $50,000. *No Debt. Hydrant Rental*, $75. **Management.**—(In '89.) Prest., C. A. Sanders. Secy. and Supt., R. E. Craig.

NOTE.—Summer of '90 city sold $150,000 of bonds, proceeds of which were to be used to put in public works and sewers. Later city requested bids for sinking 1,500-ft. well.

10. ELMA, *Chehalis Co.* (Pop., in '90, 315.) **History.**—Built in '90 by Elma Water Co., under 25-yrs. franchise; put in operation Jan. 1, '91. Engr., Jno J. Carney. **Supply.**—Impounding reservoir on Spring Brook, by gravity. **Reservoir.**—Area, ¾ acres. **Distribution.**—*Mains*, 8 to 2½-in. wood, banded with steel, 2.5 miles. *Services*, galv. 1. *Taps*, 115. *Hydrants*, Chapman, 6. *Valves*, Chapman. *Pressure:* ordinary, 20 lbs.; fire, 30 lbs. **Financial.**—*Cost*, $12,000. *Cap. Stock :* authorized, $25,000; paid up, $12,000. No bonds issued. *Expenses*, $300. *Revenue :* consumers. $1,800; city, $576; total, $2,376. *Hydrant Rental*, $96. **Management.**—Prest. and Supt., John J. Carney. Secy., C. E. Sackett. Treas., Q. Z. Carney.

FOR FURTHER DESCRIPTIVE MATTER see Manual for 1889-90. CHANGES since 1890 re here given. POPULATIONS are for 1880 and 1890, respectively.

11. **FAIRHAVEN,** *Whatcom Co.* (Pop., in '90, 4,076.) Built in '90 by Fairhaven City Water & Power Co. Rates are fixed in franchise; city has right to purchase works. **Supply.**—Lake Padden, by gravity. Fiscal year closes Oct. 15. **Distribution.** *Mains,* 10 miles. *Taps,* 250. *Hydrants,* 23. *Consumption,* 800,000 galls. **Financial.**—*Cost,* $75,000. *Cap. Stock:* authorized, $100,000; paid-up, $65,000. *No Bonds. Floating Debt,* $4,400. *Expenses:* operating, $1,600; taxes, $300; total, $1,900. *Revenue:* consumers, $3,480; fire protection, $920; total, $4,400. **Management.**—Prest., G. A. Black. Secy., Jno. H. Ware. Supt., T. W. Gillette.
NOTE.—May 2, '91, election was to be held to vote on issuance of $250,000 of bonds for city works.

12. **GOLDENDALE,** *Klickitat Co.* (Pop., 545—702.) **History.**—Built in '90 by town. Engr. and Contr., Jno. Corkish, Portland, Ore. **Supply.**—Spring, pumping to reservoir. **Pumping Machinery,**—Cap., 750,000 galls.; Hall. **Reservoir.**—Cap., 100,000 galls. **Distribution.**—*Mains,* 6-in., wood, banded with steel, 3 miles. *Takers,* 100. *Hydrants,* Chapman, 12. *Valves,* Chapman. *Consumption,* 100,000 galls. *Pressure,* 95 lbs. **Financial.**—*Cost,* $12,500. **Management.**—City Clerk, W. B. Presley.

13. **HOQUIAM,** *Chehalis Co.* (Pop., in '90, 1,302.) **History.**—Built in '82 by Northwestern Lumber Co. Engr. and Contr., Geo. H. Emerson. **Supply.**—Small streams, collected in reservoir by gravity; also artesian well, pumping direct. **Reservoir.**—Cap., 1,000 galls. **Distribution.**—*Mains,* 6 to 2-in. wood, 3 miles. *Takers,* 275. *Hydrants,* 10. **Financial.**—*Cost,* $10,000. *Expenses,* $3,000. *Revenue:* consumers, $6,000; city, none. **Management.**—Prest., A. M. Simpson. Secy., E. J. Holt. Treas., Sam'l Perkins. Supt., Geo. H. Emerson.
NOTE.—Jan. 22, '91, it was reported that City Council was considering question of introducing more adequate supply. Mayor, F. D. Arnold. Chn., Wm. H. Pedler.

14. **KENT,** *King Co.* (Pop., in '90, 853.) **General Description.**—Built in '89 by Kent Water Supply Co. **Supply.**—Springs, by gravity. Prest. (in '89), J. J. Crow. Secy. and Treas. (in '89), C. E. Guiberson.
NOTE.—First part of '91, reported that city was inviting proposals for franchise.

15. **KIRKLAND,** *King Co.* (Pop., est., 1.000.) Kirkland Land & Improvement Co. began construction of water-works June 1, '91. Gordon pump. A 2,000,000-gall. reservoir. *Mains,* c. i., 2 miles. *Services,* w. i. Engr., T. A. Moble.

16. **LA CONNER,** *Skagit Co.* (Pop., in '90, 398.) Built in '88 by La Conner Water-Works Co. **Supply.**—Spring, pumping from receiving to distributing tank. **Distribution.**—(In '88.) *Mains,* 3 miles. *Hydrants,* 60. *Consumption,* 3,000 galls. **Financial.**—(In '89.) *Cost,* $6,000. *Cap. Stock,* $6,000. *No Debt. Expenses,* $300. *Revenue,* $1,200. **Management.**—(In '90.) Prest., G. V. Calhoun. Secy., L. L. Andrews. Treas., W. E. Schricker. Supt., Jno. S. Church.

17. **NEW WHATCOM,** *Whatcom Co.* (Pop., in '90, 4,059.) **History.**—Built in '89-90 by Bellingham Bay Water Co., under 50-yrs. franchise. Engr., M. L. Stangroom. **Supply.**—Lake Whatcom, by gravity. Lake has area of 7½ sq. miles, and affords head of 315 ft. **Distribution.**—*Mains,* riveted steel and c. i., 18.3 miles. *Services,* galv. i. *Taps,* 432. *Hydrants,* 40. *Consumption,* 2,000,000 galls. *Pressure,* 125 lbs. **Financial.**—Withheld. *Hydrant Rental,* free. **Management.**—Prest., E. Eldridge. Secy., E. Fischer. Treas. and Supt., M. L. Stangroom.

18. **NORTH YAKIMA,** *Yakima Co.* (Pop., in '90, 1,535.) Built in '90-1 by Yakima Water Co.; franchise regulates rates but does not provide for purchase of works by city. **Supply.**—Natchez River, by gravity, with direct pumping to increase pressure, when necessary. Fiscal year closes April 30. **Distribution.**—*Mains,* 6.5 miles. *Hydrants,* 30. **Financial.**—*Cost,* $150,000. *Cap. Stock,* $150,000. *No Debt.* **Management.**—Prest., E. S. Whitson. Secy. and Supt., F. S. Woodward. Treas., W. L. Steinweg.

19. **OLYMPIA,** *Thurston Co.* (Pop., 1,232—1,698.) **History.**—Built in '66 by Olympia Water Co.; reconstructed in '90 by Olympia Water-Works Co.; 25-yrs' franchise, which neither regulates rates nor provides for purchase of works by city. Fiscal year closes March 15. **Supply.**—Moxlie Creek through wooden flume, 1,000 ft. long. pumping to reservoir. **Pumping Machinery.**—Dy. cap., 2,000,000 galls.; two Worthington, com. **Reservoir.**—Cap., 1,750,000 galls. **Distribution.**—*Mains,* 12 to 4-in., c. i. and wood, 15 miles. *Services,* w. i. *Taps,* 357. *Meters,* 3. *Hydrants,* Mathews, 42. *Valves,* Eddy. *Consumption,* 40,000 galls. *Pressure,* 50 to 90 lbs. **Financial.**—*Cost,* $245,202. *Cap. Stock,* $150,000. *Bonded Debt,* $75,000 at 7%, due in

20 yrs. **Hydrant Rental**, $100 each for first 20, $75 each for additional. **Management.**—Prest., E. W. Andrews. Secy., W. F. Newell. Treas., C. I. Eaton. Supt., Chas. Otis.

20. **ORTING,** *Pierce Co.* (Pop., in '90, 623.) **History.**—Built in '89 by Orting Light & Water Co. **Supply.**—Spring, by gravity. **Tank.**—Is 14 × 14 × 10 ft. **Distribution.**—*Mains*, wood, i., 5 miles. *Services*, galv. i. *Taps*, 120. *Consumption*, 70,000 galls. *Pressure*, 100 lbs. **Financial.**—*Cost*, $5,000. *Cap. Stock*, $5,000. *No Debt. Expenses*, $100. *Revenue*, $1,400. **Management.**—Prest., F. E. Eldredge. Secy., Mrs. E. A. Boatman. Treas., Mrs. F. E. Eldredge. Supt., E. A. Boatman.

21. **PALOUSE,** *Whitman Co.* (Pop., 148—1,119.) Built in '88 by village. **Supply.**—Palouse River, pumping to reservoir. **Distribution.**—*Mains*, 2 miles. *Taps*, 43. *Hydrants*, 8 *Consumption*, 10,000 galls. **Financial.**—*Cost*, $9,000. *Debt*, $3,800. *Expenses:* operating, $1,000; int., $285; total, $1,285. **Management.**—City Council. Supt., F. M. Martin.

22. **POMEROY,** *Garfield Co.* (Pop., in '90, 661.) Built in '87 by city. **Supply.**—Pataha Creek, through mill race, pumping to reservoir. Fiscal year closes first Tuesday in Aug. **Distribution.**—*Mains*, 1.7 miles. *Taps*, 60. *Hydrants*, 12. **Financial.** - *Cost*, $10,000. *No Bonds. Floating Debt*, $6,000 at 10%. Revenue is sufficient to pay all expenses. **Management.**—Report by E. Burlingame.

23. **PORT ANGELES,** *Clallam Co* (Pop., est., 3,000.) **History.**—Built in '91 by Port Angeles Gas & Water Co., under 25-yrs. franchise; put in operation June 12. Engr., Jas. L. Stanford. **Supply.**—Mountain stream, by gravity. **Reservoir.**—Cap., 3,000,000 galls. **Distribution.**—*Mains*, 12-in. steel, 8 and 6-in. c. i., 6 miles. *Mains*, 12-in. steel, 8 and 6-in. c. i. *Meters*, "rotary piston," 7. *Hydrants*, Ideal, 10. *Valves*, Galvin. *Pressure*, 115 lbs. **Financial.**—*Cost*, $80,000. *Cap. Stock:* authorized, $300.000; paid up, $80,000. *No Bonds. Revenue:* consumers, $300; city, $63; total, $363. *Hydrant rental*, $84. **Management.**—Prest., Henry Croft, Victoria, B. C. Secy. and Supt., C. E. Mallette.

24. **PORT TOWNSEND,** *Jefferson Co.* (Pop., 917—4,558.) **History.**—Temporary works (evidently) built in '89-90 by Mt. Olympia Water Co., under 25-yrs franchise. In '90 contracts were reported as let for gravity supply from Quilcene River, some 18 miles distant. Engr., new works, H. B. Smith, Victoria, B. C. Contrs., new works, U. S. Construction & Pipe Mfg. Co., San Francisco. **Supply.**—Wells, pumping to reservoir. Wells are 20 ft. in diam., 40 ft. deep, **Pumping Machinery.**—Cap., 400,000 galls.; two Dow. *Boilers*, a 15 and a 25-HP. **Reservoir.**—Cap., 275,000; brick and cem. **Distribution.**—(Sept., '91.) *Mains*, 6 to 2½-in. w. i., riveted steel, wood, about 2.5 miles. *Hydrants*, 17. **Financial.**—(Sept., '91.) *Cost*, $73,975. *Cap. Stock*, authorized, $750,000. *Bonded Debt*, authorized, $500,000 at 6%. *Revenue*, consumers, Apr. to Sept., '91, inclusive, $7,179. **Management.**—Prest., A. W. Bash. Secy., N. Q. Hill. Treas., R. C. Hill.

NOTE.—Oct. 31, '91, people were to vote on issuance of about $300,000 of bonds to buy co.'s works and complete same by bringing water from Little Quilcene River. Co. had offered to sell for $62,500.

25. **PULLMAN,** *Whitman Co.* (Pop., in '90, 868.) **History.**—Built in '90 by city; put in operation Dec. 1. Engr. and Contr., G. H. Sutherland, Walla Walla. **Supply.**—Artesian well, pumping to reservoir and direct. Well raises water 27 ft. above surface of ground. **Pumping Machinery.**—Dy. cap., 600,000 galls.; dup. **Reservoir.**—Cap., 150,000 galls. **Distribution.**—*Mains*, 8 to 4-in., ¾ miles. *Services*, galv. i. Pipe and specials from American Tube & Iron Co., Middletown, Pa. *Taps*, 50. *Hydrants*, Eclipse, 7. *Valves*, Ludlow. *Consumption*, 10,000 galls. *Pressure*, 65 lbs. **Financial.**—*Cost*, $8,500. *Expenses*, $2,000. *Revenue*, $2,000. **Management.**—Supt., Edward Barber.

II. **PUYALLUP,** *Pierce Co.* (Pop., 297—1,732.) No sewers. Has electric lights. **History.**—Bought June, '91, by Tacoma Light & Water Co., of Tacoma. Built in '76 by individuals or co.; Puyallup Water & Light Co, incorporated in '86; city may regulate rates, but made no provision in franchise for purchase of works. **Supply.**—Springs, pumping to reservoir. **Pumping Machinery.**—Dy. cap., not given; two Worthington. **Reservoir.**—Cap., 400,000 galls. **Distribution.**—*Mains*, wood, with special bands, 10 miles. *Services*, galv. i. *Meters*, 2. *Hydrants*, 0. **Financial.**—

FOR FURTHER DESCRIPTIVE MATTER see Manual for 1889-90. CHANGES since 1890 are here given. POPULATIONS are for 1880 and 1890, respectively.

MANUAL OF AMERICAN WATER-WORKS.

Cost, $75,000. *Cap. Stock*, $30,000. *Bonded Debt*, $31,000 at 8%. *Expenses*, $3,000. *Revenue*, $9,200. **Management.**—Prest., R. F. Radebaugh, Tacoma. Secy., Robt. Wilson. Treas., Frank O. Meeker. Supt., Wm. Slythe.

26. **ROSLYN,** *Kittitas Co.* (Pop., in '90, 1,484.) **History.**—Built in '89 by city; put in operation Nov. 1. Const. Engr., Wm. McKay. **Supply.**—Springs, by gravity. **Reservoir.**—Cap., 900,000 galls. **Distribution.**—*Mains* (in '89), 1 mile. *Hydrants*, 17. *Pressure*, 55 lbs. **Financial.**—*Cost*, $10,000. *Debt*, $3,000 at 10%. *Expenses*, $200. *Revenue*, $1,400. **Management.**—City Clerk, L. F. McConihe. Oct., '91, surveys for new supply from Cle-Elum River were made by D. B. Ogden, of Ogden & Bosworth, Tacoma. Pumping to present reservoir was recommended.

III. **SEA HAVEN,** *Pacific Co.* (Pop., not given. Built in '90 by Thomas Potter. **Supply.**—Springs, collected in reservoir, by gravity. Reservoir is 150 ft. above and ¾ miles from town. *Mains*, 6-in., wood, bound with strap iron, 1.5 miles. *Taps*, 10. *Consumption*, 10,000 galls. *Pressure*, 30 lbs. *Cost*, $2,000. *Owner*, Thos. Potter.

27, 28. **SEATTLE,** *King Co.* (Pop., 3,533—42,837.)

CITY'S PLANT.

Built in '84 by Spring Hill Water Co.; bought by city in '89, which began to operate works late in '90; price, $352,266. **Supply.**—Lake Washington, pumping to reservoir and direct. Fiscal year closes Jan. 1. Two Deane pumps were added in '90, and a 5,000,000-gall. Gaskill high duty was contracted for in '91. A gravity supply is projected.

Year.	M'ns.	Taps.	M't'rs.	Hyd's.	C'n's'pt'n.	Cost.	B'd'd d't.	Int. ch'g's.	Exp.	Rev.
'84	5	....	3	20	200,000	$250,000			......	$150*
'87	12	....	30	43	800,000	250,000	$50,000	$3,000	....	3,612*
'89	32	1,603	60	73	2,000,000	352,126[1]	50,000	3,000	$24,000	....
'90	36	3,000	150	100	4,000,000	552,126	845,000	......	4,253	10,525[2]

[1] Price paid by city for plant. [2] For month of Dec., '90. * From city.

Management.—Board of Public Works. Chn., Jesse Cochran. Secy., F. W. D. Holbrook.

LAKE UNION WATER CO.

Built in '82. **Supply.**—Springs, by gravity, and Lake Union, by pumping; filtered. **Distribution.**—*Mains*, 13 miles. *Taps*, 400. *Meters*, 16. *Hydrants*, none. *Consumption*, 150,000 galls. **Financial.**—*Cost*, $113,000. *Cap. Stock:* authorized, $120,000; paid up, $97,500; paid no dividend in 1890. *Debt*, $22,000. *Expenses:* operating, $5,000; int., $1,600, taxes, $710; total, $7,310. *Revenue*, consumers, $8,000; city, none. **Management.**—Prest., David L. Denny. Secy. and Treas., Edgar Bryan. Supt., Frank Maess.

NOTE.—At time of last report, April 18, '91, city was considering proposition to buy works. May, '91, co. unofficially reported that it offered to sell for $27,500; co. valued its plant at $89,000; but city did not want all of it.

17. **SEHOME,** *Whatcom Co.* See New Whatcom.

29. **SHELTON,** *Mason Co.* (Pop., est., 900.) Built in '91 by city; put in operation in Oct. Contrs., Olympia Wood Pipe Co. **Supply.**—Springs, by gravity to and from reservoir. **Tank.**—Cap., 100,000 galls.; 75 ft. above town. **Distribution.**—*Mains*, wood. *Hydrants*, 10. **Financial.**—*Cost*, $10,000. *Bonded Debt*, $10,000. **Management.**—Not given.

IV. **SOUTH ABERDEEN** (*Aberdeen P. O.*), *Chehalis Co.* (Pop., evidently included under Aberdeen.) **History.**—Built in '91 by Sunset Light & Water Co.; put in operation July 1. Engrs. and Contrs., Cummings & Cook, Aberdeen. **Supply.**—Brook, by gravity. **Reservoir.**—Cap., 500,000 galls **Distribution.**—*Mains*, 10 to 4-in. riveted steel, 3 miles. *Services*, w. 1. *Taps*, 12. *Hydrants*, none. *Valves*, Scott. *Pressure*, 30 lbs. **Financial.**—*Cost*, $15,000. *Cap. Stock*, $50,000. **Management.**— Prest., W. P. Rice. Secy., J. M. Ashton, Tacoma. Supt., B. F. Johnston.

30. **SOUTH PRAIRIE,** *Pierce Co.* (Pop., est., 300.) South Prairie Water Co. have built works. **Supply.**—Spring, by gravity to reservoir. Address Hadder & Bissan.

31, 32. **SPOKANE FALLS,** *Spokane Co.* (Pop., 350—19,922.)

CITY'S PLANT.

Built in '84. **Supply.**—Spokane River, pumping direct. Fiscal year closes April 30.

Year.	Mains.	Taps.	Meters.	H'd'ts.	C'ns't'n.	Cost.	B'd'd d't.	Int. chgs.	Exp.	Rev[3]
'86-7	4	250	0	35	300,000	$50,000	$50,000	$4,000	$2,500	$5,000
'87-8	6	450	6	44	500,000	50,000	50,000	4,000	2,500	13,000
'88-9	9	600	6	84	2,000,000	177,000	110,000	7,600	4,000	20,000
'90-1	10 6	...	60	99	2,013,000	258,216	120,000	8,200	12,067	38,201

Financial.—*Bonded Debt*, $120,000: $50,000 at 8%, due in '95; $70,000 at 6% due in 1908. *Expenses*, total, $21,167. *Revenue:* consumers, $38,201; city, none. **Management.**—Board of Public Works: G. G. Smith, W. H. Wiscomb, Jos. Monahan.

NOTE.—In '91 about $500,000 of bonds were voted for a new supply. Capt. T. W. Symonds, City Engr. Oscar Huber and others had reported on new supply, but nothing had been decided Sept., '91.

LIDGERWOOD WATER-WORKS CO.

History.—Built in '90 to supply suburb; put in operation April 1. Dis. Eng., F. D. Stanley. Contrs., Puget Sound Pipe Co., Olympia. **Supply.**—Artesian well, pumping to reservoir. **Pumping Machinery.**—Dy. cap., 450,000 galls.; Knowles. **Reservoir,**—Cap , 50,000 galls. **Distribution.**—*Mains*, 8 to 4 in. banded wood, 5.5 miles. *Services.* galv. i. *Taps*, 95. *Hydrants*, Portland, 12. *Valves*, Crane. *Consumption*, 25,000 galls. *Pressure:* ordinary, 30 lbs.; fire, 70 lbs. **Financial.**—*Cost*, $21,000. No Debt. **Management.**—Owners: J. H. Lidgewood, 93 Liberty St , N. Y. City; Chester Glass, P. S. Byrne.

33. **SPRAGUE,** *Lincoln Co.* (Pop., '91—1,689.) Built in '89 by Sprague Water Co., under 25-yrs. franchise, which does not regulate rates, but provides for purchase of works by city. **Supply.**—Well, pumping to reservoir and direct. Fiscal year closes June 28. **Distribution.**—*Mains*, 2.5 miles. *Taps*, 55. *Hydrants*, 25. *Consumption*, 50,000 galls. **Financial.**—*Cost*, $16,153. *Cap. Stock:* authorized, $20,000; paid-up, $14,500. *No Bonds. Floating Debt*, $1,649 at 12%. *Expenses*, $985. *Revenue:* consumers, $590; fire protection, $942; total, $1,532. **Management.**—Prest., Geo. I. Brook. Secy. and Supt., W. B. Lottman.

34. **SUMNER,** *Pierce Co.* (Pop., in '90, 580.) Built in '87 by Spring Branch Water Co. **Supply.**—Springs, by gravity to and from reservoir. Fiscal year closes Aug. 1. **Distribution.**—*Mains*, 6 miles. *Taps*, 87. *Hydrants*, 2. *Consumption*. 43,000 galls. **Financial.**—*Cost*, $7,300. *Cap. Stock*, authorized, $10,000; paid up. $4,000. *Total Debt*, $3,250: floating, $250 at 10%; bonded, $3,000 at 10%, due Nov. 7, '90. *Expenses:* operating, $125: int., $310; total, $735. *Revenue*, total, $1,004. **Management.**— Prest , E. C. Meade. Secy., F. E. Thompson. Supt., F. Sweinson.

35. **TACOMA,** *Pierce Co.* (Pop., 1,098—36,006.) Built in '85 by Tacoma Light & Water Co.; no provision for purchase of works; rates must be reasonable. **Supply.**—Spanaway Lake and Clover Creek, by gravity to and from reservoir for low, springs by pumping for high service; portion of supply is filtered through sand and gravel. Fiscal year closes Dec. 31. Additional supply of 3,000,000 galls. was obtained in '90 from springs near city. Two 1,000,000-gall. Worthington pumps were added. Standpipe, 25 × 125 ft., was constructed. Early in '91, contract was let for two 2,500,000 Gordon pumps. June, '91 co. bought entire works of Puyallup Water & Light Co., of Puyallup, for $112,500. Purchase included Maplewood Springs, with est. dy. cp. of 11,000,000 galls., part of which co. proposes to pipe to Tacoma. Co. also bought Clear Creek Springs between Puyallup and Tacoma, with est. dy. cap. of 3,500,000 galls. It is contemplated to build entire new system, taking its supply from Green River, at point 32 miles east of city. Proposed system includes construction of 25,-000,000-gall. storage reservoir. City will be divided into three services; high service to be supplied by stand-pipe built in '90; middle service, having elevation of from 200 to 300 ft., by new reservoir.

Year.	Mains.	Taps.	M't'rs.	Hy'd'ts.	Cost.	H'd' d't.	Rev.
'86	11.3	250	0	41	$250,000	......	......
'87	13	600	20	46	250,000	......	$11,000
'88	16	1,350	35	50	300,000	......	53,500
'89	38.5	2,548	84	140	......	......	......
'90	40	3,300	90	160	......	$1,700,000	......

Financial.—Figures include both water and lighting plants. *Cap. Stock*, $1,500,000. *Bonded Debt*, $1,700,000 at 7%. **Management.**—Prez., Theodore Hosmer. Secy. and Treas., J. H. Houghton. Gen'l. Supt., Chas. R. Hurley. Asst. Supt., R. H. Lloyd.

NOTE.—Aug. 8, '91, com. of 11. Nelson Bennett, Chn., was appointed to report on works to be owned by city. Oct., '91, Engr. for Com., F. G. Plummer, made preliminary report on gravity supply and recommended investigation of impounding dam on Mashell River, 32 miles from, and 700 ft. above city. Investigation of artesian wells as a source also recommended. In '91, North End Water Co. was incor-

FOR FURTHER DESCRIPTIVE MATTER see Manual for 1889-90. CHANGES since 1890 are here given. POPULATIONS are for 1880 and 1890, respectively.

porated by R. B. Mullen, Anna Mullen and F. M. Harsberger to develop the "Mullen Water-Works" in section not supplied by Tacoma Light & Water Co. Cap. stock, $40,000.

V., 36. **VANCOUVER**, *Clarke Co.* (Pop., 1,722—3,545.)

COLUMBIA LAND AND IMPROVEMENT CO.

VANCOUVER WATER CO.

Built in '68, under 25-yrs. exclusive franchise, which determines maximum rates; city does not have right to buy works. **Supply.**—Springs, by gravity from impounding to distributing reservoir. Fiscal year closes May 1.

Year.	Mains.	Taps.	Meters.	Hydrants.	Cost	Exp.	Rev.
'82	10	175	0	6	$50,000	$2,000	.....
'84	13	350	1	10	75,000	......	.....
'88	13.5	350	1	..	......	2,000	$11,100
'89	16	450	0	none	......	......	.....
'90	17	500	0	none	......	......	.....

Financial.—*Cap. Stock:* authorized, $45,000; all paid-up; paid 10% div'd in 1890. *Floating Debt,* $6,500 at 7%; was to be paid May 1, '91, from surplus funds. **Management.**—Prest., S. W. Brown. Secy., G. H. Steward. Supt., J. B. Wintler.

COLUMBIA LAND AND IMPROVEMENT CO.

History.—Built in '89-91. **Supply.**—Deep well, pumping to tank and reservoir. **Pumping Machinery.**—Dy. cap., 500,000 galls; Dow. **Tank.**—Cap., 300,000 galls. **Reservoir.**—Of stone masonry. **Distribution.**—*Mains,* 16 miles. *Hydrants,* 3, "on trial." *Pressure:* ordinary, 100 lbs.; fire, 180 lbs. **Financial.**—*Cost,* $50,000. *Bonded Debt,* $25,000 at 8%. *Expenses* (July 10), $90 per month. *Revenue* (July 10), $120 per month. **Management.**—Prest., J. Gibbon. Vice-Prest. and Mgr., L. M. Hidden. Secy., Chas. Brown.

37. **WALLA WALLA**, *Walla Walla Co.* (Pop., 3,558—4,709.) Built in '65 by Walla Walla Water Co. **Supply.**—Springs, by gravity from collecting reservoir. **Distribution.**—(In '87.) *Mains,* 14 miles. *Taps,* 500. *Hydrants,* 25. *Consumption,* 600,000 galls. **Financial.**—(In '87.) *Cost,* $125,000. *Cap. Stock,* $100,000, *Bonded Debt,* $50,000 at 8%. **Management.**—(In '89.) Prest., R. R. Rees. Supt., Jerome F. Bowman. NOTE.—In '90, city talked of buying the works.

38. **WEST SEATTLE**, *King Co.* (Pop., est., 1,400.) **History.**—Built in '90 by West Seattle Land & Improvement Co. Engr., R. H. Stretch. **Supply.**—Springs, pumping to tank. **Pumping Machinery.**—Cameron. **Reservoir.**—Cap., 50,000 galls. **Distribution.**—*Mains,* 4 miles. *Services,* ¾-in. **Financial.**—*Cost,* $17,500. *Cap. Stock,* $100,000. *No Debt.* **Management.**—Prest., Thos. Ewing. Secy., M. S. Bates. Supt., Geo. W. Walker.

Water-Works Projected with Fair Prospects of Construction.

KETTLE FALLS, *Stevens Co.* (Pop., est., 800.) Rochester & Kettle Falls Land Co. had begun construction of water-works Aug. 22, '91. Springs, by gravity to and from reservoir. Mains, wood, 4 miles. Hydrants, 15. Est. cost, $13,000. Engr. Geo. C. Mills.

MONTESANO, *Chehalis Co.* (Pop., in '90, 1,632.) City was to begin construction of water-works Oct. 1, '91. Engr., C. E. Fenner. Lake and creek, pumping to reservoir. Est. cost, $18,000. Mayor, C. E. Jameson. Address H. R. Ballamy.

MOUNT VERNON, *Skagit Co.* (Pop., in '90, 770.) Mount Vernon Water Co. was granted franchise Aug. 12, '91, and was to begin construction of works Aug. 25. River and spring, pumping to stand-pipe. Est. cost, $35,000. Address J. N. Turner.

OAKESDALE, *Whitman Co.* (Pop., in '90, 528.) City wanted to build works, Aug. 14, '91. Address W. G. Gilstrap.

SNOHOMISH, *Snohomish Co.* (Pop., 149—est., 600.) In '91, city voted $60,000 worth of bonds to construct water-works.

SOUTH BEND, *Pacific Co.* (Pop., est., 2,500.) A 30-yrs. franchise granted June 16, '91, to W. R. Forrest, Jno. H. McGraw and H. K. Owen. Hydrants, 50. Hydrant rental, $90. Est. cost, $100,000.

WAITSBURGH, *Walla Walla Co.* (Pop., 248—817.) July 9, '91, it was reported that preliminary surveys had been made by A. L. Adams, Pendleton, Ore., and city was about to vote on issuing of $30,000 worth of bonds to construct water-works.

WASHINGTON.

WATERVILLE, *Douglas Co.* (Pop., in '90, 293.) July 10, '91, it was reported that Waterville Improvement Co. had let contract for construction of water-works to H. K. Owens, Seattle, for $20,000, and that construction would begin Aug. 1. Prest. W. R. Ballard, Seattle. Address R. S. Steiner.

OREGON.

Water-Works Completed or in Process of Construction.

1. **ALBANY,** *Linn Co.* (Pop., 1,867—3,079.) Built in '80 by J. A. Crawford; sold in '89 to Albany Canal, Water, Transportation & Lighting Co.; franchise neither regulates rates nor provides for purchase of works by city. Supply.— Santiam Canal and Calapooya River, pumping direct. Distribution.- (In '88.) *Mains*, 3.6 miles. *Taps*, 200. *Meters*, 0. *Hydrants*, 5. Financial.—*Revenue*, city, $100. Management.—Gen. Mgr., Wm. M. Hoag, Corvallis. NOTE.—Statistics for '90 refused.

21. **ALBINA,** *Multnomah Co.* Consolidated with Portland July 1, '91. See Albina Light & Water Co., under Portland.

2. **ARLINGTON,** *Gilliam Co.* (Pop., in '90, 356.) Built in '88 by Arlington Water Co.; franchise does not regulate rates, but provides for purchase of works by city. Supply.—Columbia River, pumping to reservoir. Distribution.—(In '88.) *Mains*, 2 miles. *Taps*, 70. *Meters*, 0. *Hydrants* (In '90), 8. *Consumption*, 100,000 galls. Financial.— (In '88.) *Cost*, $7,000. *Cap. Stock*, $10,000. *No Debt. Expenses*, $1,200. *Revenue:* consumers, $3,200; city, $960; total, $4,130. Management.—(In '89.) Prest., J. E. Frick. Secy., A. C. Fry. Supt. and Engr., E. D. Parrot.

3. **ASHLAND,** *Jackson Co.* (Pop., 842—1,784.) History.—Built in '90 by city; put in operation July 20. Engr., Jno. O'Connor. Supply.—Mountain stream, by gravity. Distribution.—*Mains*, 10-in. spiral steel welded, 8-in. riveted steel, 6 to 2-in. converse lock joint, 15 miles. *Services*, galv. i. *Taps*, 166. *Hydrants*, 34. *Pressure:* ordinary, 95 lbs.; fire, 95 to 120 lb . Financial.—*Cost*, $56,000. *Bonded Debt*, $56,000 at 6%. *Operating expenses*, $1,500. Management.—Treas., E. V. Carter. Supt., Eugene Walrod.

1. **ASTORIA,** *Clatsop Co.* (Pop., 2,803—6,184.)

FIRST SUPPLY.

Small works built in '70. No further information.

COLUMBIA WATER CO.

Built in '83-4. Supply.—Mountain spring, by gravity. Distribution.—(In '87.) *Mains*, 6-in., 13 miles. *Meters*, 6. *Hydrants*, 1. *Consumption*, 300,000 galls. Financial.—(In '87.) *Cost*, $80,000. *Cap. Stock* $100,000. *No Debt.* Management—(In '88.) Prest., F. De Kum. Supt., Jas. W. Welch.

NOTE.—April 19, '91, W. E. Dement, Clerk, wrote that last session of legislature passed bill appointing water commission of 7 members to buy or build works.

4. **BAKER CITY,** *Baker Co.* (Pop., 1,253—2,604.) Built in '88-9 by city. Supply.—Artesian wells, pumping to reservoir and direct. Distribution.—(In '89.) *Mains*, 2 miles. *Taps*, 75. *Meters*, 0. *Hydrants*, 16. *Consumption*, 150,000 galls. Financial.—(In '89.) *Cost*, $20,000. *Bonded Debt*, $20,000, at 6%. Management.—(In 89.) Supt., Wm. O. Reynolds.

5. **COQUILLE,** *Coos Co.* (Pop., 176—194.) Built in '85-6 by Occident Water Co.; city has neither right to buy works nor control over rates. Supply.—Mountain stream, by gravity. Fiscal year closes Dec. 31. Distribution.—*Mains*, 1 mile. *Taps*, 85. *Meter*, 1. *Hydrants*, 3. Financial.—*Cost*, $3,500. *No Debt. Expenses*, $26. *Revenue*, $600. Management.—Prest. and Supt., R. E. Buck. Secy., A. J. Sherwood.

6. **CORVALLIS,** *Benton Co.* (Pop., 1,128—1,527.) Built in '87 by Corvallis Water Co. Supply.—Willamette River, pumping to tank and direct. Fiscal year closes Jan. 1. Early in '90 contract was let for a 1,270,000-gall. Worthington pump.

FOR FURTHER DESCRIPTIVE MATTER see Manual for 1889-90. CHANGES since 1890 are here given. POPULATIONS are for 1880 and 1890, respectively.

Year.	Mains.	Taps.	Meters.	Hydrants.	Consumption.	Cost.	Exp.	Rev.
'88	4	200	1	8	200,000	$16,000	$600	$1,150
'89	4	250	1	9	100,000	......	1,400	600*
'90	5	300	1	9	150,000	......	1,700	400*

* From city.

Financial.—*Cap. Stock*, $100,000. *No Bonds*. *Floating Debt*, $3,000. **Management.**—Prest., G. R. Farrer. Secy., Wm. Graves.

19, 20. EAST PORTLAND, *Multnomah Co.* Consolidated with Portland July 1, '91. See East Portland and East Side Water cos. under Portland.

7. ENTERPRISE, *Wallowa Co.* (Pop., in '90, 242.) **History.**—Built in '90 by Enterprise Water Co., under 20-yrs. franchise; put in operation Oct. 1. Engr., H. Feldnur, St. Johns. **Supply.**—Wallowa River, pumping to reservoir and direct. **Pumping Machinery.**—Dy, cap., 750,000 galls.; Gaskill. **Reservoir.**—Cap., 1,500,000 galls. **Distribution.**—*Mains*, kal. i. Pipe and specials from National Tube Wks., Chicago. *Hydrants*, 6. **Financial.**—*Cost*, $8,500. *Cap. Stock* : authorized, $10,000; paid up, $5,000. *No Bonds*. *Debt*, $3,000. *Expenses*, nominal. *Hydrant Rental*, $100. **Management.**—Prest. and Supt., J. M. Church. Secy., J. P. Gardner.

8. EUGENE CITY, *Lane Co.* (Pop., est., 4,000.) Built in '86 by co., under exclusive franchise. **Supply.**—Willamette River, pumping to reservoir and direct. **Distribution.**—(In '88.) *Mains*, 12 miles. *Hydrants*, 27. *Consumption*. 170,-000 galls. **Financial.**—(In '88.) *No Debt*. *Expenses*, $1,700. *Revenue*, city, $,1000. **Management.**—Secy., James F. Robinson. NOTE.—Co. declined to furnish statistics for '90.

9. GRANT'S PASS, *Josephine Co.* (Pop., in '90, 1,432.) **History.**—Built in '90-1 by Grant's Pass Water, Light & Power Co., under 50-yrs. franchise; put in operation Feb. 1. Des. Eng. and Contr., Wm. Knox, Haywards, Cal. Const. Engr., F. M. Johnson. **Supply.**—Rogue River, pumping to reservoir. **Pumping Machinery.**—Dy. cap., 1,000,000 galls.; Dow water-power, with steam reserve. **Reservoir.**—Cap., 500,000 galls. **Distribution.**—*Mains*, 8 to 4-in., 7.5 miles. Pipe and specials from W. Montague Co., San Francisco. *Taps*, unk'n, *Hydrants*, National. 15, *Valves*, Eddy. *Pressure*, 100 lbs. **Financial.**—*Cost*, $75,000. *Cap. Stock*: authorized, $30,000, all paid up. *Total Debt*, $45,000: floating, $15,000; bonded, $30,000 at 7%, *Hydrant Rental*, $96. **Management.**—Prest. and Treas. J. W. Howard. Secy. and Supt., F. M. Johnson.

II. HOOD RIVER, *Waco Co.* (Pop., in '90, 201.) Built in '81 by E. F. & H. C. Coe, under name of Hood River Hydrant Co. Town neither controls rates nor has right to purchase works. **Supply.**—Spring, by gravity. **Distribution.**—*Mains*, 1.5 miles. *Taps*, 40. *Hydrants*, none. **Financial.**—*Cost*, $2,350. *No Debt*. *Expenses*, nominal. *Revenue*, $750. **Management.**—See above.

III. INDEPENDENCE, *Polk Co.* (Pop., 691—est., 850.) **History.**—Built in '88 by an individual, under 20-yrs. franchise, for domestic use only. **Supply.**—Driven wells, pumping to tank. **Pumping Machinery.**—Dy. cap., 144,000 galls.; Worthington. **Tank.**—Cap., 30,000 galls. **Distribution.**—*Mains*, 6 to 2-in. w. i., ¾ miles. *Services*, galv. i. Pipes and specials from Jno. Bassett Co., Portland. *Taps*, 45. *Hydrants*, 0. *Consumption*, 20,000 galls. *Pressure*, 25 lbs. **Financial.**—*Cost*, $5,000. *No Debt*. *Expenses*. $800. *Revenue*, $1,510. **Management.**—Owner, L. C. Gilmore.

10. JOSEPH, *Walla Walla Co.* (Pop., in '90, 249.) Built in '89 by Joseph Water Co.; city has no control of rates, but is to have first option of buying works. **Supply.**—Lake, by gravity. Fiscal year closes in May. **Distribution.**—*Mains*, 2 miles. *Taps*, 40. *Hydrants*, 7. **Financial.**—*Cost*, $1,200. *Cap. Stock*, $3,000; paid 18% div'd in 1890. *No Debt*. *Expenses:* operating, $200; taxes, $30; total, $230. *Revenue*, city, $250. **Management.**—Prest., F. D. McCully. Secy., J. D McCully. Supt., Jas. Amey.

11. LA GRANDE, *Union Co.* (Pop., 836—2,583.) Built in '87 by La Grande Water Co. **Supply.**—Mill Creek, by gravity to reservoir. **Distribution.**—(In '88.) *Mains*, 4 miles. *Taps*, 200. *Meters*, 0. *Hydrants*, 10. *Consumption*, 200,000 galls. **Financial.**—(In '88.) *Cost*, $15,000. *Cap. Stock*, $20,000. *No Debt*. *Revenue:* consumers, $4,-000; city, $600; total, $4,600. **Management.**—(In '89.) Prest. and Treas., Jno. Corkish, Portland. Secy., Jas. Dwight.

NOTE.—In '91 La Grande Water & Light Co. was incorporated under 25-yrs. franchise to supply water for domestic use, fire protection and irrigation. Des. Engr., A. L. Adams, Portland. Hydrants, 25. Est. cost, $55,000. Cap. Stock, authorized,

OREGON.   339

$100,000; paid-up, $15,000. Hydrant rental, $50 each for first 40, $36 each for additional. Prest., B. W. Grandy. Secy., H. G. Romig. Treas., Jas. Palmer.

12. **McMINNVILLE,** *Yam Hill Co.* (Pop., 670—3,368.) Built in '89 by city. Supply.—Yam Hill River, pumping direct. Fiscal year closes Dec. 2. Distribution.—*Mains*, 2 miles. *Taps*, 120. *Hydrants*, 16. Financial.—*Cost*, $23,000. *Bonded Debt*, $20,000 at 6%, due in 1909. *Expenses*, $1,000. *Revenue*, $840. Management.—Mayor, C. D. Johnson. Recorder, J. J. Spencer. Supt., F. E. Griffith.

·13. **MEDFORD,** *Jackson Co.* (Pop., in '90, 867.) Built in '89 by city. Supply.—Bear Creek, by gravity through canal and by pumping from canal to reservoir. Fiscal year closes second Tuesday in Jan. Distribution.—*Mains*, 1 mile. *Taps*, 40. *Hydrants*, 12. *Consumption*, 50,000 galls. Financial.—*Cost*, $18.000. *Bonded Debt*, $20,000. *Expenses*, $700. *Revenue*, $200. Management.—Trustees elected every year. Mayor, G. W. Howard. Recorder, D. T. Sears.

14. **MILTON,** *Umatilla Co.* (Pop., est., 500.) Built in '88 by town. Supply.—Walla Walla River, by gravity. Distribution.—*Mains*, 2 miles. *Taps*, 47. *Hydrants*, 4. Financial.—*Cost*, $5,000. *Bonded Debt*, $5,000 at 8%. *Expenses*: operating, $75; int., $400; total, $475. *Revenue*, consumers, $500. Management.—Chn., J. W. O'Connell. Supt., W. W. Miller.

IV., V. **MOUNT TABOR,** *Multnomah Co.* (Pop., est., 300.)

MOUNT TABOR ELECTRIC LIGHT & WATER CO.

Built in '91. Supply.—Wells. Supt., Mr. Hart.

MOUNT TABOR VILLA WATER CO.

History.—Built in '91, under perpetual franchise; put in operation July 1. Supplies suburb of Mount Tabor. Joint stock co. Stock held by propertyowners, no one person to have more than two $50 shares. Supply.—Well, pumping to tank. Well is 4 ft. in diam., 222 ft. deep. Pumping Machinery.—Dy. cap., 50,000 galls. Tank.—Cap., 10,000 galls. Distribution.—*Mains*, 2-in. wood, 2 miles. *Services*, galv. i. *Taps*, 53. *Hydrants*, 0. *Consumption*, 5,000 galls. *Pressure*, 60 lbs. Financial.—*Cost*, $5,000. *Cap. Stock*: authorized, $5,000; paid-up, $2,000. *No Bonds*. *Debt*, $3,000. *Expenses*, $500. *Revenue*, $1,000. Management.—Prest., Geo. Reichwein. Secy. and Supt., J. L. Ivers.

15. **OREGON CITY,** *Clackamas Co.* (Pop., 1,263—3,062.) Built in '86 by town. Supply.—Willamette River, pumping to stand-pipe and direct. Fiscal year closes Dec. 31. During '90 3,000,000-gall. pumping plant was added.

Year.	Mains.	Taps.	Hydrants.	Consumption.	Cost.	Exp.	Rev.
'87	2	250	15	......	......	$500	$3,500
'89	2.25	250	18	250,000	$12,000	900	2,467
'90	4.5	290	30	350,000	13,000	1,500	3,676

Management.—Committee of three appointed every year by mayor from city council. Chn., J. W. O'Connell. Supt., W. H. Howell, appointed by mayor and council.

16. **PENDLETON,** *Umatilla Co.* (Pop., 919—2,506.) Built in '87 by town. Supply.—Well, pumping to reservoir.

Year.	Mains.	Taps.	Meters.	Hydrants.	Cost.	B'd'd debt.	Int. ch'g's.	Exp.
'88	6.5	300	0	21	$32,000	$30,000	$2,100	......
'89	10	378	0	21	39,000	39,000	2,340	......
'90	11	448	1	27	40,000	40,000	2,400	$7,200

Management.—Com. of three appointed every year by mayor from council. Chn., J. H. Raley. Secy., Geo. R. Lash. Supt., Geo. T. Schaffner.

17, 18, 19, 20, 21. **PORTLAND,** *Multnomah Co.* (Pop.: Portland, 17,577—46,385; East Portland, 2,934—10,532; Albina, 143—5,129. See below.) July 1, '91, consolidation act went into effect by which Portland, East Portland and Albina became one municipality. All contracts unfulfilled by annexed cities were assumed by Portland, to be completed by common council of consolidated city. Prior to consolidation there was a water company in operation in both East Portland and Albina and a second plant, to be bought by city, was under construction in East Portland. These plants are described separately below.

MAIN CITY WORKS.

Now owned by city. Built in '53 by Portland Water Co.; bought by city, Jan 1, '87. Supply.—Willamette River, pumping direct. Fiscal year closes Dec. 31. Feb.

FOR FURTHER DESCRIPTIVE MATTER see Manual for 1889-90. CHANGES since 1890 are here given. POPULATIONS are for 1880 and 1890, respectively.

13, 91, authority had been secured previously to issue $2,500,000 of bonds for gravity supply from Bull Run, 30 miles distant, in Cascade Range. Wrought iron conduit, with dy. cap. of 20,000.000 to 25,000,000 galls., three bridges over Bull Run and Sandy River and 2,000 ft. submerged pipe beneath Willamette River, were projected for conduit. At same time, it was proposed to expend within city $500,000 for new mains and reservoirs.

Year.	Mains.	Taps.	Meters.	Hyd'ts.	C'ns'pt'n.	Cost.	B'd'd d't.	Int. ch's.	Exp.	Rev.†
'87	18	5,444*	2	75	4,716,000	$500,000	$500,000	$25,000	$47,001	$97,500
'88	33.51	5,765	89	197	5,900,000	593,544	600,000	30,000	54,792	113,692
'89	51.84	6,665	55	197	7,056,000	752,897	650,000	30,000	51,407	148,716
'90	55.72	7,575	72	...	9,415,000	951,407	700,000	33,750	71,497	181,310

*Est. by Supt.   †All sources; see below.

Financial.—*Total Debt:* Bonded, $700,000; floating, $10,000. *Revenue:* in '90 city paid $2,400 for municipal rents, and $2,920 for street sprinkling. **Management.**—Works were rebuilt by a committee of 15; when completed are to be turned over to a commission of five. Chn., Henry Failing. Secy., Clerk, Frank T. Dodge. Treas., C. H. Lewis. Supt. and Engr., Isaac W. Smith.

EAST SIDE PLANT.

Now owned by Portland. Built in '90-1 by East Side Water Co., and transferred to East Portland May 20, '91, city paying back to each stockholder the money paid in by him and 10% additional, according to provisions of franchise. July 1, '91, works came under control of Portland common council; when stand-pipe is completed works will come under management of Portland Water Com., probably about Jan. 1, '92. **Supply.**—Five 12-in. wells sunk in bottom of pump-well, pumping direct and to stand-pipe. Pump-well is 38 ft. in diameter by 22 ft. deep, excavated to "cem. gravel"; 12-in. wells are 75 ft. 4 ins., 84 ft. 5 ins., 90 ft., 108 ft. and 163.5 ft. deep, respectively, below bottom of pump-well, which is 8 ft. above low water in Willamette River. **Pumping Machinery.**—Cap., 4,000,000 galls.; Worthington; 1,000,000-gall. simple. connected with first two wells and 3,000,000-gall. comp. connected with last three. *Boilers,* two 100-HP., from Willamette Iron Works. **Stand-Pipe.**—Cap., 357,000 galls.; 25 × 100 ft.; from Tippett & Wood. Phillipsburg, N. J. **Distribution.**—(July, '91.) *Mains,* 16-in. w. i., 14 to 4-in. c. i., 16.5 miles; about 8.5 miles had been previously ordered laid. *Hydrants,* Ludlow, 94. **Financial.**—East Portland paid $95,889 for unfinished works; July, '91, additional sums had been paid, bringing total to $134,396, and contracts to the amount of $65,148 were outstanding. *Bonded Debt,* $250,000. **Management.**—Common Council till all contracts are completed, then as given under "main works." above.

EAST PORTLAND WATER CO.

Built in '81-2; franchise regulates rates and provides for purchase of works by city, price to be determined by joint commission. **Distribution.**—*Mains,* 25 miles. *Hydrants,* 30. *Consumption,* 2,000,000 galls. **Financial.**-*Cost,* $100,000. *Cap. Stock* (in '88). $80,000. *No Debt.* **Management.**—Prest. and Supt., H. P. McGuire. Secy., W. S. Chapman. Treas., W. W. McGuire.

ALDINA LIGHT & WATER CO.

Built in '87-'90, under 10-yrs. franchise. **Supply.**—Driven wells, pumping to tank and direct. There are four wells driven 130 ft. from bottom of pump-well 25 ft. in diam., 30 ft. deep; latter is walled and cemented **Pumping Machinery.**—Dy. cap., 1,200,000 galls.; two 600,000 Worthington. **Tank.**—Cap., 100,000 galls. **Distribution.**—*Mains,* 10 to 2-in. Converse patent, 15 miles. Pipe and specials from Dunham, Cairrigan & Hayden Co. *Taps,* 600. *Meters,* Crown, 8. *Hydrants,* Ludlow, 23. *Valves,* Chapman. *Consumption,* 400,000 galls. *Pressure,* 95 lbs. **Financial.**—Purchase price, $20,000. *Cap. Stock:* authorized, $100,000.; paid-up. $75,000. *Hydrant Rental,* $60. **Management.**—Prest., G. W. Bates. Secy., C. F. Sweigert.

KINZEL PARK WATER CO.

Built in '91 to supply that portion of Portland known as Kinzel Park. Put in operation July 1. Well, 200 ft. deep, pumping direct. *Taps,* about 290. *Cost,* $3,000. Address Enterprise Real Estate Co., 88 Morrison St., Portland.

22. ROSEBURGH, *Douglas Co.* (Pop., 852—1,472.) Built in '89 by Roseburgh Water Co., under 30-yrs. exclusive franchise; city has neither right to purchase works nor control of rates. **Supply.**—Umpqua River, pumping direct and to reservoirs. Fiscal year closes Dec. 31. In '90 steam pump was put in to be used in, case of disability of water-power pump; Hall, comp., dup., 7 × 12 × 8 × 12; boiler, 40-HP. **Distribution.**—*Mains,* 7.3 miles. *Taps,* 110. *Hydrants,* 10. Con

OREGON. 341

sumption, 50,000 galls. **Financial**.—*Cost*, $28,500. *Cap. Stock:* $20,000; paid no div'd in 1890. *Floating Debt*. $1,000 at 10%. *Expenses:* operating, $1,200; taxes, $200; total, $1,400. *Revenue:* consumers, $2,600; fire protection, $1,000; total, $3,600. **Management**.—Prest., T. R. Sheridan. Secy., W. S. Hamilton. Treas., S. C. Flint. Supt., O. L. Willis.

23. **SALEM**, *Marion Co.* (Pop., 2,538; est., 10,000.) Built in '71 by Salem Water Co., under 17-yrs. franchise. **Supply**.—Willamette River, pumping direct. Fiscal year closes second Monday in January. Dec. 27, '90, co. reported that early following spring they would build a 2,000,000-gall. reservoir, to be located about 200 ft. above business portion of city and connected with present system of mains. Reservoir to be 200 × 155 ft. by 10 ft. deep; walled with brick laid in cem., 20 ins. thick, backed by stone. Mains are to be extended beyond South Salem to Fair Mount Park. Feb. 2 it was reported that reservoir was in course of construction.

Year.	Mains.	Taps.	Hydrants.	Cost.	Exp.	Rev.
'82	7.7	250	34	$120,000	$4,001	$5,595
'87	10.5	487	17	135,000	3,000	12,000
'89	15	750	47	153,164		12,200
'90	18	800	50	196,173	5,560	15,120

Financial.—*Floating Debt*, $14,000 at 8%. *Expenses:* operating, $5,000; int., $1,000; taxes, $560; total, $6,560. *Revenue:* consumers, $12,600; fire protection, $1,920; other purposes, $1,200; total, $15,120. **Management**.—Prest. and Secy., J. H. Martin, Treas., J. H. Albert.

24. **THE DALLES**, *Wasco Co.* (Pop., 2,232; est., 3,000.) **History**.—Now owned by city. Built in '62 by Robt. Pentland, afterward owned by Dalles Mill & Water Co., bought in '90 by city for $17,600, and extended in '91. Des. Engr. for extension, Gorman Lawe. Const. Engr., J. B. Wallace. **Supply**.—Hood River, by gravity. **Reservoir**.—Cap., 3,000,000 galls.; Contr., Lee Hoffman, Portland. **Distribution**.—*Mains*, 12 to 4-in. c. i., 6 miles. Pipe and specials from Wolf & Ziegler, Durgan Bros. and Oregon Iron Works; trenching and pipe-laying done by Leo Hoffman, Portland. *Consumption*, 500,000 galls. *Pressure*, 100 lbs. **Financial**.—*Cost*, $120,000. *Bonded Debt*, $100,000 at 6%. *Expenses:* operating, $1,800; int., $6,000; total, $7,800. *Revenue*, consumers, $10,000. **Management**.—Secy., C. L. Phillips. Supt., I. J. Norman.

25. **WESTON**, *Umatilla Co.* (Pop., 116—568.) Built in '89 by town. **Supply**.—Pine Creek, by gravity. Fiscal year closes first Wednesday in Dec. **Distribution** —*Mains*, 1.5 miles. *Taps*, 75. *Hydrants*, 10. **Financial**.—*Cost*, $10,000. *Bonds Debt*, $10,000 at 8%, due in '93. *Expenses:* operating, $100; int., $800; total, $900. *Revenue*, $750. **Management**.—Com. of three, chosen from council by mayor each year. Supt., J. W. Young. Report by P. A. Worthington.

26. **WOODBURN**, *Marion Co.* (Pop., in '90, 105.) **History**.—Built in '90-1 by D. L. Remington & Son, under 50-yrs. franchise; put in operation in March. **Supply**. —Wells, pumping to 20,000-gall. tank. **Distribution**.—*Mains*, 1 mile. *Taps*, 100 *Hydrants*, 8. *Consumption*, 50,000 galls. *Pressure*, 40 lbs. **Financial**.—*Cap. Stock*, $10,000. **Management**.—Proprs., D. L. Remington & Son.

VI. **WOODLAWN**, *(Portland P. O.) Multnomah Co.* **History**.—Built in '90 by Woodlawn Land Co. for domestic use only. Des. Engr., H. Grimes. Const. Engr. and Contr., Jno. Corkish. **Supply**.—Well, pumping to 50,000-gall. tank with Knowles pump. **Distribution**—*Mains*, banded wood, 15,000 ft. *Services*, galv. i, *Taps*, 50, *Hydrants*, none. *Pressure*, 25 lbs. **Financial**.—*Cost*, $10,000. **Management**.— Prest., F. De Kum. Secy., H. Shalton. Supt., H. Grimes.

Water-Works Projected with Fair Prospects of Construction.

CORNELIUS, *Washington Co.* (Pop., not given.) See Forest Grove, below.

DALLAS, *Polk Co.* (Pop., 670—848.) In '91 a 30-yrs. franchise was granted to Chas. P. Shore & Co., Palace Hotel, San Francisco, Cal. Supply, La Creole River, by gravity. Hydrants, Ludlow. 25. Valves, Ludlow, 25. Hydrant rental, $100. Est. cost, $55,000. Secy., C. L. Phillips.

FOR FURTHER DESCRIPTIVE MATTER see Manual for 1889-90. CHANGES since 1890 are here given. POPULATIONS are for 1880 and 1890, respectively.

FOREST GROVE, *Washington Co.* (Pop., 517—863.) Gales Peak Water Co. was to begin construction of works Aug. 1, '91, to supply Forest Grove, Cornelius and Hillsboro. Engr., R. A. Habersharer, Portland. Supply, Gales Creek, by gravity. Est. cost., $50,000. Prest., C. M. Keep. Vice-Prest., J. G. Boos. Secy., P. O. Chilstrom. Treas., E. W. Haines. Latest report Dec. 26, '90.

HILLSBOROUGH, *Washington Co.* (Pop., 402—est., 400.) See Forest Grove above. Aug. 20, '91 it was reported that Hillsborough Electric Light & Power Co. began construction of water-works Aug. 1, under 10-yrs. franchise. Cook well, pumping to 60,000-gall. tank and direct. Mains, 4.5 miles. Hydrants, 20. Pressure: ordinary, 60 lbs.; fire, 100 lbs. Engr. and Contr. H. V. Gates, Portland.

MYRTLE POINT, *Coos Co.* (Pop., 52—354.) Myrtle Point Water Supply & Mfg. Co. were granted 99-yrs. franchise, but had not begun construction of works Aug. 20. Des. Engr., B. M. Brower. Const. Engr., L. S. Rogers. Supply, branch of Coquille River, by gravity. Cap. stock, authorized, $500,000. Prest., E. Binder. Secy., Orvil Dodge. Treas., J. H. Roberts.

PORTLAND HEIGHTS (*Portland P. O.*), *Multnomah Co.* Portland Heights Water Co. were to begin construction of system July 15, '91. Supply to be taken from a 10-in. main of Portland water-works and pumped through 6-in. to 50,000 gall. tank, 600 ft. elevation for low service and 30,000-gall. tank, 720 ft. elevation for high service. Cost, $18,000. Engr., F. I. Fuller. Prest., A. T. Smith. Secy. and Treas., G. P. Lamb.

UNION, *Union Co.* (Pop., 416—604.) Sept. 30, '91, unofficially reported that construction of city works began that day and would be completed in 60 days. Contr., Dean Reefe, Walla Walla, Wash. Supply, creek, by gravity. Est. cost, $18,000. City Recorder, J. B. Thompson.

CALIFORNIA.*

Water-Works Completed or in Process of Construction.

1 **ALAMEDA and FITCHBURG,** *Alameda Co.* (Pop.: Alameda, 5,708—11,165; Fitchburg, not given.) Built in '75 by R. R. Thompson. Supply.—Four artesian wells, gathered in receiving well, pumping to tank and direct. Fiscal year closes Dec. 31.

Year.	Mains.	Taps.	Meters.	Hydrants.	Cost.	Exp.	Rev.
'82	30	300	0	56	....	....	....
'89	45	1,300	1,300	85	$383,596	$24,430	$51,604
'90	45	1,600	1,600	85	389,236	28,004	49,967

Financial.—*No Cap. Stock;* earnings paid 7% div'd in 1889 and 6% in '90. *No Debt.* Expenses: operating, $26,169; taxes, $1,835; total, $28,004. Revenue: consumers, $37,-759; fire protection, $2,040; other purposes, $10,168; total, $49,967. Management.—Owner, R. R. Thompson. Supt., I. L. Borden.

2. **ANAHEIM,** *Los Angeles Co.* (Pop., 833—1,273.) Built about '78 by town. Supply.—Wells, pumping to tank. Mains, 1.25 miles. Hydrants, 8. Supt. (in '89), A. Schneider.

3. **ANDERSON,** *Shasta Co.* (Pop., in '90, 508.) Built in '87-8 by Anderson Water Co. Supply.—Mountain stream, by gravity from impounding to distributing reservoir. Fiscal year closes July 24. Distribution.—*Mains,* 1.5 miles. *Taps,* 39. *Hydrants,* 5. Financial.—*Cost,* $10,000. *Cap. Stock:* authorized, $10,000: paid up, $7,530. *No Debt. Expenses,* operating, $261. Management.—Prest., J. H. Sumner. Secy., E. F. Buss. Treas., J. F. Bedford.

4. **ANTIOCH,** *Contra Costa Co.* (Pop., 626—635.) Built in '76 by M. C. Belshaw. Supply.—San Joaquin River, pumping to reservoir. Pumping Machinery.—Cap., 40,000 galls. Reservoir.—Cap., 30,000 galls. Distribution.—*Mains,* 3 miles. *Taps,* 200. *Hydrants,* 10. Financial.—*Est. Cost,* $8,000. Owner, M. C. Belshaw.

5. **ARCATA,** *Humboldt Co.* (Pop., 702—962.) Built by Union Water Co., date not given. Supply.—Springs, by gravity. *Mains,* ¾ miles. *Hydrants,* 16. Pressure, 50 to 60 lbs. *Cost,* $12,000. *Revenue,* city, $300. Supt., O. H. Spring.

*Constitution of State provides for regulation of rates. Only three companies in state, two at Los Angeles and one at San Luis Obispo, report provisions in franchise for regulation of rates and purchase of works.

6. ARCH BEACH and LAGUNA, *Orange and San Diego Cos.* (Pop., not given.) Built in '89-'90 by Arch Beach Land & Water Co. **Supply.**—Lakes in Laguna Cañon, by gravity. Fiscal year closes Dec. 31. **Distribution.**—*Mains,* 2.8 miles. *Taps,* 89. *Hydrants,* 2. *Consumption,* 3,000 galls. **Financial.**—*Cost,* $10,300. *Cap. Stock:* authorized, $60,000; paid up, $18,000. *No Debt. Expenses,* $222. *Revenue:* consumers, $346; other purposes, $820; total, $1,166. **Management.**—Prest., H. S. Goff. Secy., P. I. Cook. Treas., A. B. Judkins, Supt., F. R. Ludlow.

7. AUBURN, *Placer Co.* (Pop., 1,229—1,595.) Built in '87 by F. W. Birdsall, and known as Auburn Water-Works Co. **Supply.**—Bear River Canal, by gravity to reservoir, through inverted syphon. **Distribution.**—*Mains,* 15.5 miles. *Hydrants,* 33. *Consumption,* 200,000 galls. **Financial.**—*Cost,* $10,000. *No debt. Revenue,* city, $1,200. **Management.**—Owner and Supt., F. W. Birdsall. Secy., Jno. Scott.

8. AZUSA, *Los Angeles Co.* (Pop., est., 500.) Built in '87 by Azusa Land & Water Co. **Supply.**—San Gabriel River by gravity to reservoir. **Distribution.**—(In '88.) *Mains,* 6 miles. *Taps,* 85. **Financial.**—(In '88.) *Cost,* $35,000. *Cap. Stock,* $500,000. *No Debt.* **Management.**—(In '89.) Prest., J. S. Slauson. Secy., James Slauson. Treas., F. C. Howe. Supt., D. A. Shaw.

9. BAKERSFIELD, *Kern Co.* (Pop., 801—2,626.) Built in '83-4 by Bakersfield Water Co. **Supply.**—Four artesian wells, pumping to tank. Reported in summer of '91 that tank would be added. Fiscal year closes Feb. 15. **Distribution.**—*Mains,* 7.5 miles. *Taps,* 325. *Hydrants,* 11. *Consumption,* 100,000 galls. **Financial.**—*Cost,* $30,000. *Cap. Stock,* $30,000; paid 15% div'd in 1890. *No Debt.* **Management.**—Prest., S. W. Wible. Secy., W. E. Houghton. Treas. and Supt., W. H. Scribner.

10. BENICIA, *Solano Co.* (Pop., 1,794—2,361.) Built in '80 by Benicia Water Co., under 20-yrs. franchise. **Supply.**—Paddy Ranch and Sulphur Springs Creek, by gravity from impounding to distributing reservoir. **Distribution.**—*Mains,* 7 miles. *Taps,* 398. *Meters,* 47. *Hydrants* 24. *Consumption,* 150,000 galls. **Financial.**—*Cost,* $172,000. *Cap. Stock,* $180,000. *Bonded Debt,* $30,000 at 7%. *Expenses:* operating, $3,000; int., $2,100; total, $10,100. *Revenue:* consumers, $12,900; city, none. **Management.**—Prest., E. H. Nelson. Secy., D. M. Hart. Supt., Alex. Robinson.

96. BERKELEY, *Alameda Co.* (Pop., in '90, 5,101.) Supply from West Berkeley. NOTE—In July it was reported that franchise had been granted to Percy Morgan Later, it was reported that Berkeley & Lorain Water & Light Co. had been incorporated. Directors, W. L. Sheldon, W. F. Martin.

11. BODIE, *Mono Co.* (Pop., in '90, 595.) Built in '83 by town and Prescott & Co., **Supply.**—Shaft of mine, pumping to reservoir. **Distribution.**—(In '88.) *Mains,* ¼ miles. *Taps,* 12. *Hydrants,* 9. **Management.**—(In '89.) Fire Comrs. Chf. Engr., H. M. Hartley.

12. CENTINELA, *Los Angeles Co.* (Pop., not given.) Built in '87-8 by Centinela-Inglewood Land Co. **Supply.**—Inglewood Springs, developed by excavation and artesian wells, pumping to reservoir. Fiscal year closes Feb. 25. **Distribution.** —*Mains,* 18 miles. *Taps,* 105. *Hydrants,* 3. *Consumption,* 6,000 galls. **Financial.** —*Cost,* unk'n. *Expenses,* $750. *Revenue:* consumers, $900; other purposes, $800; total, $1,700. **Management.**—Supt., A. C. Freeman.

13. CHICO, *Butte Co.* (Pop., 3,300—2,691.) Built in '75 by Chico Water Co., under 50-yrs. franchise. **Supply.**—Surface wells, pumping to tank and direct. Fiscal year closes Dec. 31. **Distribution.**—*Mains,* 7 miles. *Taps,* 380. *Hydrants,* 43. **Financial.**—*Cap. Stock:* authorized, $100,000; paid up, $50,000; paid 8% div'd in '90. *No Debt. Hydrant Rental,* $12. **Management.**—Secy. and Supt., W. J. Costar. Treas., J. R. Robinson.

14. CLOVERDALE, *Sonoma Co.* (Pop., 430—763.) Built in '85-6 by Riverside Water Co., under 50-yrs. franchise. **Supply.**—Well, pumping to reservoir. Fiscal year closes first Wednesday in Sept. **Distribution.**—*Mains,* 5.3 miles. *Taps,* 150. *Meters,* Thompson. *Hydrants,* 45. *Consumption,* 75,000 galls. **Management.**— Prest., J. A. Kleiser. Secy. and Supt., J. L. Sedgeley.

15. COLTON, *San Bernardino Co.* (Pop., 878—1,315.) Owned since '89 by Colton City Water Co. Built by Colton Terrace Land & Water Co., under 20-yrs. franchise. **Supply.**— "Cienza" and 8 artesian wells. **Distribution.**— *Mains,* 10.3 miles. *Taps,* 265. *Hydrants,* 20. **Financial.**—*Cost,* $50,000. *Cap. Stock,* $250,000. *Bonded*

FOR FURTHER DESCRIPTIVE MATTER see Manual for 1889-90. CHANGES since 1890 are here given. POPULATIONS are for 1880 and 1890, respectively.

344   MANUAL OF AMERICAN WATER-WORKS.

Debt, $25,000. Management.—Prest., A. B. Miner. Secy., Frank A. Miner. Report Aug. 4, '91.

NOTE.—Oct. '91, G. E. Slaughter, city clerk, reported that city had bought two acres artesian well land and employed engineer to make estimates for city works. It was expected that bonds would be voted and work begun soon.

I. COLUMBIA, *Tuolumne Co.* (Pop., 650—est., 750.) Built in '53 by New England Water Co., under 50-yrs. franchise. Supply.—Springs and Stanislaus River, by gravity. Distribution.—(In '82.) *Mains*, 5 miles. *Taps*, 75. *Hydrants*, 0. *Consumption*, 30,000 galls. Financial.—(In '88.) *Cost*, $24,000. *No Debt*. *Expenses*, $800. *Revenue*, $1,100. Management.—(In '91.) Owner, Mrs. R. S. Fraser, Mgr., A. Tarleton.

16. COLUSA, *Colusa Co.* (Pop., 1,779—1,336.) Built in '69 by J. B. Cooke. Supply.—Sacramento River, pumping to tanks.

Year.	Mains.	Taps.	Meters.	Hydrants.	Cost.	Exp.	Rev.
'82	5	75	0	1	$24,000	$900	$1,200
'87	5.5	175	1	15	34,000	4,000	7,373
'90	5.5	264	1	16	......	......	800*

*From city.

Financial.—*No Cap. Stock.* *No Debt*. Management.—Owner, J. B. Cooke.

17. CUCAMONGA, *San Bernardino Co.* (Pop., est., 300.) Built in '86-7 by Cucamonga Water Co. Supply.—Tunnels into volcanic hill and artesian wells, by gravity through reservoir. Fiscal year closes Oct. 31. Distribution.—*Mains*, iron and cem., 48 miles. *Consumers:* for domestic use, 21; for irrigation, 59. *Hydrants*, none. Financial.—*Cost*, $103,000. *No Debt*. *Expenses*, $1,200. *Revenue:* domestic consumers, $500. Management.—Prest., N. W. Stowell. Secy., O. C. Matthay. Treas., E. T. Wright. Supt., T. G. Graham.

17. DIXON, *Solano Co.* (Pop., in '90, 1,082.) June 1. '90, Dixon Light & Water Co. put works in operation. Engr., Jno. Graves. Supply.—Wells, pumping to reservoir. *Cost*, $30,000. Address A. A. Osborn.

18. DUTCH FLAT, *Placer Co.* (Pop., 939—682.) Owned by Isaac T. Coffin. Built in '59. Supply.—Springs, by gravity from collecting reservoir. Fiscal year closes Dec. 31. Distribution.—*Mains*, 1 mile. *Hydrants*, 15. *Consumption*, 50,000 galls. Financial.—*Cost*, $5,000. *No Debt*. *Expenses*, $374. *Revenue*, city, $150. Management.—Owner, Isaac T. Coffin.

19. ESCONDIDO, *San Diego Co.* (Pop., in '90, 541.) Built in '87-8 by Escondido Land & Water Co. Supply.—Three wells, sunk in bed of river, and 21 driven wells, pumping to reservoir. Water is used chiefly for irrigation. During '90 21 2-in. wells were driven 20 ft. into gravel and connected by 8-in. suction pipe. New wells are 550 ft. from pump and 200 ft. from old wells. Water may be pumped from two systems of wells separately or together. Distribution.—*Mains*, 4 miles. *Taps*, 112. *Meters*, 4. *Hydrants*, 5. *Consumption*, 50,000 galls. Financial.—*Cost*, $38,000. *Entire Cap. Stock of Co.*, $100,000. *No Debt*. *Expenses*, $2,500. Management.—Prest., J. Groundike. Secy., W. W. Thomas. Treas., Jerry Yoles. Supt., J. T. Chambers. All of San Diego.

20. EUREKA, *Humboldt Co.* (Pop., 2,639—4,858.) Owned by Ricks Water Co. under 25-yrs. franchise. Built in '79 by C. S. Ricks. Supply.—Elk River, pumping direct; filtered. Fiscal year closes Dec. 31. In Dec., '90, Hegeman & Olyphant filter was introduced; cap., 612,000 galls. Sept., '91, contract for Redwood tank, 54 × 30 ft., with cap. of about 500,000 galls., was let to W. H. McWhinney, Eureka. Tank will be on concrete piers.

Year.	Mains.	Taps.	Meters.	Hydrants.	Consumption.	Cost.	Exp.	Rev.
'80	5.5	325	2	4	180,000	$38,186	......	$6,381
'88	8	450	8	6	225,000	55,147	$3,600	7,576
'90	16.5	650	20	6	275,000	158,820	3,925	13,800

Financial.—*Cap. Stock:* authorized, $150,000; all paid up. *No Debt*. *Expenses:* operating, $3,600; taxes, $325; total, $3,924. *Revenue:* consumers, $13,860; city, none. Management.—Prest., Richard Sweasey. Secy. and Supt., H. L. Ricks.

NOTE.—Summer of '91 city talked of putting in salt water system for fire protection, sewer flushing and street sprinkling. Reported later that co. was putting in hydrants.

21. FELTON, *Santa Cruz Co.* (Pop., 271—259.) Built in '80 by Felton Water Co. Supply.—Bull Creek, by gravity. Distribution.—*Mains*, 1.3 miles. Financial.—*Cost*, $3,000. *Cap. Stock*, $10,000. Officers: C. H. Winter, Wm. Russell, W. A. Walker. Report July 21, '89.

CALIFORNIA. 345

Ill. FERNDALE, *Humboldt Co.* (Pop., 178-763.) Owned by Mrs. Grace Francis. Built in '73. **Supply.**—Spring, by gravity. **Distribution.**—(In '88.) *Mains,* 7 miles. *Taps,* 140. *Hydrants,* 0. *Consumption,* 50,000 galls. **Financial.**—(In '88) *Cost,* $18,500. *No Debt. Expenses,* $500. *Revenue,* $2,000. **Management.**—(In '89.) Lessee, Geo. M. Brice.

1. **FITCHBURG,** *Alameda Co.* See Alameda.

22. **FOLSOM CITY,** *Sacramento Co.* (Pop., in '90, 699.) **History.**—Built in '88 by Natoma Vineyard Co. **Supply.**—American River, through canal, by gravity. **Distribution.**—*Mains,* 7-in. w. i., 1 mile. *Taps,* 50. *Hydrants,* 7. *Consumption,* 100,000 galls. *Pressure,* 50 lbs. *Cost,* $2,000. Supt., C. H. Schussler.

23. **FORT BRAGG,** *Mendocino Co.* (Pop., in '90, 915.) **History.**—Built in '89 by Fort Bragg Water Co. Engr. and Contr., H. F. Millikin. **Supply.**—Springs, by gravity. **Reservoir.**—Cap., 700,000 galls. **Distribution.**—*Mains,* wood and i., 2 miles. *Consumers,* 55. *Hydrants,* 12. *Consumption,* 20,000 galls. *Pressure,* 25 lbs. **Financial.**—*Cost,* $8,000. *Debt,* $2,000 at 10%. **Management.**—Prest., F. A. Whipple. Secy., C. Huggins. Supt., H. F. Milliken.

24, IV. **FORTUNA,** *Humboldt Co.* (Pop., est., 700.)

HENRY ROHNER'S PLANT.

Built in '87. **Supply.**—Springs, by gravity. **Reservoir.**—Cap., 29,000 galls. **Distribution.**—*Mains,* galv. i., 1¼ miles. *Services,* galv. i. *Taps,* 33. *Hydrants,* 6. **Financial.**—*Cost,* $3,500. *No Debt. Expenses,* $200. *Revenue,* $600. Address Henry Rohner.

W. M. MORGAN'S PLANT.

Built in '90. **Supply.**—Springs by gravity. Cap. of tank, 6,000 galls. *Mains,* ½ mile. *Expenses,* $60. *Revenue,* $360. Address M. M. Morgan.

25. **FRESNO,** *Fresno Co.* (Pop., 1,112-10,818.) Built in '76 by Fresno Water Co.; under 50-yrs. franchise. Early in '90 works were reported as sold to Municipal Investment Co., Chicago and London, for $500,000. **Supply.**—Eight 8-in. wells, pumping to tanks. Wells are from 150 to 600 ft. deep. In '91 reported that 4,000,000-gall. Gaskill pump had been added. **Distribution.**—*Mains,* 27 miles. *Taps,* 2,600. *Hydrants,* 35. *Consumption,* 2,500,000 galls. **Financial.**—*Cap. Stock,* $250.000. *No Debt. Expenses,* $15,000. *Revenue,* $30,500. **Management.**—Prest. and Mgr., Jno. J. Seymour. Secy., J. M. Collier.

26. **GILROY,** *Santa Clara Co.* (Pop., 1,6.11-1,894.) Now owned by city. Built in '71-2 by company. **Supply.**—Uvas River, by gravity to reservoir, with pumping from wells in extreme dry weather. Summer of '91 $50,000 bonds were voted for improvements. **Distribution.**—(In '88.) *Mains,* 3 miles. *Taps,* 250. *Meters,* 0. *Hydrants,* 8. **Financial.**—(In '88.) *Cost,* to city, $18,000. *Bonded Debt,* $15,000 at 7%. *Revenue,* consumers, $4,500. **Management.**—(In '89.) Supt., Jno. W. Norris.

27. **GRASS VALLEY,** *Nevada Co.* (Pop., 4,500.) Built in '66-7 by town. **Supply.**—South Yuba River, by gravity through open ditch to reservoir. **Distribution.**—(In '88.) *Mains,* 10 miles. *Taps,* 700. *Hydrants,* 68. *Consumption,* 170,000 galls. **Financial.**—*Cost* (in '87), $80,000. *Expenses,* $6,000. *Revenue,* $9,000. **Management.**—(In '91.) Supt., J. F. Robinson.

28. **HAYWARDS,** *Alameda Co.* (Pop., 1,231-1,419.) Built in '88 by Knox Bros., under 50-yrs. franchise. **Supply.**—Wells, pumping to reservoir and direct. Fiscal year closes Dec. 31. **Distribution.**—*Mains,* 6 miles. *Taps,* 201. *Meters,* 4. *Hydrants,* 20. *Consumption,* 50,000 galls. **Financial.**—*Cost* (in '83), $42,500. *No Cap. Stock. No Debt. Expenses,* $1,634. *Revenue:* consumers, $3,000; fire protection, $864; other purposes, $121; total, $3,985. **Management.**—Supt., Wm. Knox.

29. **HEALDSBURGH,** *Sonoma Co.* (Pop., 1,133-1,485.) Built in '76; bought May 1, '84, by F. & A. Koenig, under 50-yrs. franchise. **Supply.**—Springs, pumping to reservoir. **Distribution.**—*Mains,* 3 miles. *Taps,* 200. *Hydrants,* 12. **Financial.**—*Cost,* $30,000. *No Debt. Expenses,* $500. **Management.**—Supt., F. Koenig.

30. **HOLLISTER,** *San Benito Co.* (Pop., 1,034-1,234.) Built in '76 by co., under 50-yrs. franchise. **Supply.**—Three artesian wells, pumping to reservoir. Fiscal year closes Dec. 31. **Distribution.**—*Mains,* 11 miles. *Taps,* 400. *Hydrants,* 21.

FOR FURTHER DESCRIPTIVE MATTER see Manual for 1889-90. CHANGES since 1890 are here given. POPULATIONS are for 1880 and 1890, respectively.

Consumption, 55,000 galls. **Financial.**—*Cost*, $10,000. *Cap. Stock:* authorized, $30,-000; all paid up: paid 8% div'd in 1890. *No Debt. Expenses:* operating, $4,000; taxes, $300; total, $4,300. *Revenue:* consumers, $6,200; city, fire protection, withheld; other purposes, $200. **Management.**—Prest., T. S. Hawkins. Secy. and Mgr., T. W. Hawkins.

31. **INDEPENDENCE,** *Inyo Co.* (Pop., est., 300.) Built in '87 by Independence Water Co., under 50-yrs. franchise. **Supply.**—Pinyon and Pine creeks, by gravity. Fiscal year closes April 30. **Distribution.**—*Mains*, 2 miles. *Taps*, 25. *Hydrants*, 5. *Consumption*, 20,000 galls. **Financial.**—*Cost*, $7,650. *No Debt. Expenses*, $200. *Revenue*, $370. **Management.**—Prest., J. A. Lank. Secy., J. C. Irwin.

12. **INGLEWOOD,** *Los Angeles Co.* (Pop., not given.) **Supply.**—From Centinela.

32. **JACKSON,** *Amador Co.* (Pop., 1,010—est., 1,500.) Owned by B. F. Richtmeyer. Built in '65. **Supply.**—Springs and Blue Lakes Water Co.'s canal, by gravity, water passing to reservoir from former and to tank from latter. **Distribution.**—*Mains*, 4.5 miles. *Taps*, 140. *Hydrants*, 15. **Financial.**—*Cost*, $25,000. *No Debt.* **Management.**—Owner, B. F. Richtmeyer.

67. **LACONIA,** *San Bernardino Co.* (Pop., not given.) **Supply.**—From Redlands.

6. **LAGUNA,** *San Diego Co.* (Pop., not given.) Supply from Arch Beach.

33. **LIVERMORE,** *Alameda Co.* (Pop., 855—1,391.) Built in '75-6 by Aylward & Mullany, and known as Livermore Spring Water Co. **Supply.**—Livermore Springs, pumping to reservoirs, and Arroyo Mocho River, by gravity. **Distribution.**—*Mains*, 12 miles. *Taps*, 400. *Hydrants*, 15. *Consumption*, 80,000 galls. **Financial.**—*Cost*, $35,000. *Debt*, $3,700 at 8%. *Expenses*, operating, $350. **Management.**—Supt., John Aylward.

V. **LODI,** *San Joaquin Co.* (Pop., 606—1,013.) **History.**—Built in '83-90 by Bay City Gas, Water & Electric Light Co., under 25-yrs. franchise; put in operation June 1. Engr., Geo. G. Buckland. **Supply.**—Five wells, pumping to tank. **Pumping Machinery.**—Dy., cap , 500,000 galls; triple-acting. **Tank.**—Cap., 70,000 galls. **Distribution.**—*Mains*, riveted steel, 5 miles. *Taps*, 100. *Hydrants*, 0. *Consumption*, 80,000 galls. *Pressure*, 20 lbs. **Financial.**—*Cost*, $100,000. *Cap. Stock*, $100,000. *Bonded Debt*, $22,000. *Revenue:* consumers, $1,800; city, none. **Management.**—Prest. and Supt., G. G. Buckland. Secy., J. D. Fish. Treas., J. L. Vaughan.

34, 35, 36. **LOS ANGELES,** *Los Angeles Co.* (Pop., 11,183—50,395.)

LOS ANGELES CITY WATER CO.

Built in '62; city may purchase works in '98, price determined by appraisement; rates are stipulated in franchise. **Supply.**—Water is purchased from Crystal Springs Land & Water Co.; produced from sprigs by sub-drainage. Fiscal year closes Nov. 1. Infiltration pipes were extended in summer of '90, increasing supply.

Year.	Mains.	Taps.	M'trs.	H'd'ts.	Cons'p'n.	Cost.	B'd'd dbt.	Int. chgs.	Exp.	Rev.
'80	23	1,500		60		$387,000	$100,000	$3,000		$256,000
'82	26	1,726	0	60	1,500,000	435,000	95,000	6,650		55,500
'86	70	3,645	26	80	3,500,000	760,000	75,000	5,250		120,000
'87	72	6,600	150	136	5,000,000	940,000	70,000	4,900	$18,000	200,000
'88	73	8,200	60	190	6,000,000	1,226,000	65,000	4,500	24,000	215,000
'89	75	8,255	58	194	6,000,000	1,441,000	300,000	18,000	36,000	220,000
'90	150	8,662	80	210	7,000,000	1,700,000	300,000	18,000	115,400	225,000

Financial.—*Cap. Stock:* authorized, $1,240,000, all paid up; paid, $74,400 div'd in 1890. *Bonded Debt*, $300,000 at 6%. *Expenses:* operating, $59,647; int., $18,000; water, $89,753; taxes, $5,000; total, $172,400. *Revenue:* consumers, $225,000; fire protection, free. **Management.**—Prest., Chn. W. H. Perry. Vice-President, W. J. Brodrick. Secy., S. H. Mott. Treas., I. W. Hellman. Supt., Wm. Mulholland.

CITIZENS' WATER CO.

Built in '73-4. City controls rates and has right to purchase works by giving one year's notice. **Supply.**—Los Angeles River, by gravity and by pumping to reservoirs. Fiscal year closes July 31. During '90, Worthington comp. dup. pump and small pumping reservoir were added.

Year.	Mains.	Taps.	Meters.	Hydrants.	Cost.	Bonded debt.	Int. ch'gs.	Exp.	Rev.
'86-7	75	1,540	10			$200,000	$12,000		
'87-8	75	2,000	15	10	$275,000	200,000	12,000	$41,000	$58,000
'88-9	75	2,100	20	10	275,000	200,000	12,000	25,000	
'89-90	75	2,200	250	35	300,000	200,000	12,000		47,880

Financial.—*Cap. Stock:* authorized, $1,000,000; paid up, $90,000. *Bonded Debt,* $2?0 C00 at $6%. **Management.**—Prest. and Treas.. M. L. Weeks. Secy., L. M. Anderson. Supt., J. I. Lotspeich.

EAST SIDE SPRING WATER CO.

Serves a high portion of city not reached by other co.s. **Supply.**—Springs. pumping to reservoir. *Consumers,* about 100. Principal owner, J. F. Crank, Seventh St. and Grand Ave.

NOTE.—In '91 Sierra Madre Water Co. obtained franchise from city to supply water for domestic use, **Supply.**—Arroyo Seco, by gravity, to be impounded in winter. Prest., E. L. Stern, 105 No. Spring St. Secy., H. Silver.

37. **LOS GATOS,** *Santa Clara Co.* (Pop., 555—1,652.) **History.**—Built in '91 by Cold Spring Water Co., under 25-yrs, franchise; put in operation about Sept. 1. Des. Engrs., Pieper & Brown. Const. Engrs. and Contrs., U. S. Construction & Pipe Mfg. Co., San Francisco. **Supply.**—Impounding reservoir on mountain stream, by gravity. Reservoir,—Cap., 2,500.000 galls.; Contr., Jno. E. Sexton, San Francisco. **Distribution.**—*Mains,* Phipp cem,·lined, about 4.5 miles. *Hydrants,* Ludlow, 20. *Valves,* Ludlow. **Financial.**—*Cost,* $32,000. *Cap. Stock:* authorized, $60,000. *Bonded Debt,* $30,000. **Management**—Prest., T. G. Murphy. Secy., S. Smith. Both of 303 California St., San Francisco.

VI. **MADERA,** *Fresno Co,* (Pop., 217—950.) **History.**—Built in '85 or '86 by Madera Flume & Trading Co. Des. Engr., Return Roberts. **Supply.**—Well, pumping to tank. **Pumping Machinery.**—Not described. **Tank.**—Cap., not given; 1d × 12 ft. **Distribution.**—*Mains,* 1½ miles. *Taps,* 1C0. *Hydrants,* none. *Consumption,* 20,000 galls. **Financial.**—*Cost,* $5.000. *No Debt. Expenses,* $1.000. *Revenue,* $3,000. **Management.**—Prest. and Supt., Return Roberts. Secy. and Treas., E. H. Cook.

38. **MARTINEZ,** *Contra Costa Co.* (Pop., in '90, 1,600.) Built in '88 by Geo. W. McNear, under 20-yrs. franchise. **Supply.**—Wells, gathered in reservoir, pumping to tank. **Distribution.**—*Mains,* 5 miles. *Taps,* 140. *Meters,* 7. *Hydrants,* 11. *Consumption,* 150,000 galls. **Financial.**—*Cost,* $35,000. *Cap. Stock:* authorized. $40,000; all paid-up; paid no div'd. in '90, *No Debt. Expenses,* $3,000. *Revenue,* $3,900. **Management.**—Prest., Geo. W. McNear. Secy., E. B. Cutter. Supt., Ben Sedgwick.

39. **MARYSVILLE,** *Yuba Co.* (Pop., 4,321—3,991.) **History.**—Built in '58 by Marysville Water Co., under 50-yrs. franchise. **Supply.**—Two artesian wells, pumping to tank. **Pumping Machinery.**—Dy. cap., 1,500,000 galls.; one 1,000,000 Worthington and one 500,000-Gaskill. *Boilers:* from McAffee, San Francisco. **Tank.** —Cap., 250,000 galls.; from Schickle, Harrison & Howard, St. Louis, Mo. **Distribution.**—*Mains,* c. i. *Services,* w. i. *Meters,* Gem, Crown. *Hydrants,* 15. *Pressure,* 22 lbs. In '80 there were 8 miles of mains, 700 taps, 12 hydrants. **Financial.**—*First Cost,* $200,000. *Cap. Stock,* $20,000. *Bonded Debt,* $30,000. *Op. Exp.* (in '80), $1,800. Hydrant rental, free. **Management.**—Prest., Chas. E. Simpkins. Secy., Emerson E. Meek.

40. **MENLO PARK,** *San Mateo Co.* (Pop., est., 300.) Built in '64 by Menlo Park Water Co. **Supply.**—Surface water, by gravity. In '90-1, improvements were in course of construction. **Distribution.**—(In '88.) *Mains,* 11 miles. *Taps,* 125. *Meters,* 43. *Hydrants,* 0. *Cost,* as reported. $250,000. **Management.**—Prest., C. N. Felton. Secy., L. C. Fraser. Supt., D. J. Willis.

41. **MERCED,** *Merced Co.* (Pop., 1,406—2,009.) Bought in '89 by Crocker, Huffman Land & Water Co., from W. L. Silliman and J. J. Stevenson, who built works in '80-1 under 25-yrs. franchise. The plant is operated in connection with irrigation works, the main part of which was built in '88 by the Merced Land & Irrigation Co. For earlier supply see Manual for '89-90, p. 623. **Supply.**—Merced River, by gravity from large storage reservoir, fed by canal 21 miles long, and with max. cap. of 2,000 cu. ft. per second. **Dam and Reservoir.**—Cap., 3,000,000,000 galls.; area, 500 acres. Drainage area of reservoir very small. Dam of clay and sand thoroughly mixed, without heart-wall, about 60 ft. high. Inner slope riprapped or paved from top to point below low water. Masonry gate-house in dam, fed by pipe, and by opening at intersection of gate-house and inner slope. Masonry culvert leads from

FOR FURTHER DESCRIPTIVE MATTER see Manual for 1889-90. CHANGES since 1890 are here given. POPULATIONS are for 1880 and 1890, respectively.

gate house to irrigating canal, and in culvert discharge-pipe connecting with city supply main is placed. **Supply Main.**—C. I., 16 in., 4½ miles long. **Distribution.** —*Mains*, 8-in., 15 miles. *Hydrants*, 42. **Financial.**—*Cost*, city works, $120,000. **Management.**—Prest. and Supt., C. H. Huffman. Secy., David Miller.

42. MISSION SAN JOSE, *Alameda Co.* (Pop., 246—not given.) Built in '70 by co. **Supply.**—Springs, by gravity. **General Description.**—(In '87.) *Mains*, ½ mile. *Taps*, 30. *Hydrants*, 12. *Consumption*, 20,000 galls. *Expenses*, $40. *Revenue*, $600. Supt. (in '91). Juan Gallegos.

43. MODESTO, *Stanislaus Co.* (Pop., 1,693–2,402.) Built in '76 by Modesto Water Co., under 50-yrs. franchise. **Supply.**—Artesian wells, pumping to tank. **Distribution.**—(In '88.) *Mains*, 8 miles. *Taps*, 450. *Hydrants*, 15. *Consumption*, 100,000 galls. **Financial.**—(In '88.) *Cost*, $40,000. *Cap. Stock*, $20,000. *No Debt.* *Expenses*, $7,125. *Revenue*, $10,604. **Management.**—(In '89.) Prest. and Supt., Robt. McHenry. Secy., Ora McHenry.

53. MONTEREY, *Monterey Co.* (Pop., 1,396–1,662.) Supply from Pacific Grove.

44, 45. MONROVIA, *Los Angeles Co.* (Pop., in '90, 907.)

SANTA ANITA WATER CO.

History.—Built in '85. **Supply.**—Big Santa Anita and Sawpit cañons, by gravity. **Reservoir**—Cap., 3,000,000 galls. **Distribution.**—*Mains*, 20 to 8-in., 8 miles. *Taps*, 400. *Consumption*, 2,000,000 galls. **Financial.**—*Cost*, $40,000. *Expenses*, $400. **Management.**—Prest. and Treas., E. J. Baldwin. Secy. and Supt., H. A. Unruh.

CITY'S PLANT.

History.—Built in '91 by city. Put in operation Nov. 1. Engr., Jno. Jackson, Los Angeles. Contrs., J. M. Thomas and F. M. Monroe. **Supply.**—Reservoir in Sawpit Cañon, by gravity. **Reservoir.**—Cap., 3,000,000 galls. **Distribution.**—*Mains*, 24-in., cem., 2 miles. *Hydrants*, 24. *Pressure*, 90 lbs. **Financial.**—*Bonded Debt*, $40,000 at 6%. **Management.**—Prest., J. F. Banning. Secy., C. E. Slossen. Treas., Jno. Bartle. Supt., M. L. Sevier.

46. MORENO, *San Bernardino Co.* (Pop., not known.) Supplied in '91 by branch steel pipe in connection with Allessandro Irrigation District, which is supplied from Bear Valley Irrigation Co.'s system. About 200-ft. head.

47. NAPA, *Napa Co.* (Pop., 3,731–4,395.) Built in '83-4 by Napa City Water Co. **Supply.**—Surface well, pumping to reservoir. Fiscal year closes Sept. 18. In '90 new pumping works were put in. *Cost*, $20,000. **Distribution.**—*Mains*, 6.5 miles. *Taps*, 700. *Meters*, 38. *Hydrants*, 41. **Financial.**—*Cost*, $132,000. *Cap. Stock:* authorized, $150,000; all paid up. *No Debt.* Paid 6% div'd in 1890. *Sinking Fund*, $7,000. *Expenses*, taxes, $554. *Revenue:* consumers, $14,400; fire protection and other purposes, $2,600; total, $17,000. **Management.**—Prest., S. E. Holden. Secy., E. S. Churchill. Treas., Geo. E. Goodman. Supt., T. R. Parker.

48. NATIONAL CITY, *San Diego Co.* (Pop., 248–1,353.) Built in '86-8 by San Diego Land & Town Co. **Supply.**—Sweetwater River, by gravity from impounding reservoir. Fiscal year closes Dec. 31. Co. has been in litigation 3 yrs. over portion of land covered by reservoir, compelling water to be lowered to uncover it. Recovery of the land has increased fire pressure from 65 to 80 lbs.

Year.	Mains.	Taps.	Meters.	Hydrants.	C'nsumpt'n.	Cost.	Exp.	Rev.
'89	65	456	12	38	3,500,000	$780,529	$14,366	$11,968
'90	65	574	18	38	3,500,000	797,330	12,528	12,685

Financial.—*Cap. Stock:* paid up, $2,250,000; paid no div'd in '90. *No Debt. Expenses:* operating, $7,991; taxes, $4,357; total, $12,328. *Revenue:* consumers, $11,349; fire protection, $1,330; total, $12,685. After July 1, '91, hydrant rental will be $31.58 instead of $35 per hydrant. **Management.**—Prest., Benj. Kimball, Boston, Mass. Secy., Edw. Parker, Topeka, Kans. Treas., S. W. Reynolds, Boston, Mass. Gen. Mgr. and Supt., Wm. G. Dickinson.

NOTE.—For a full, illustrated description of the Sweetwater dam see Trans. Am. Soc. C. E., vol. XIX., p. 201. Also see ENGINEERING NEWS, vol. XIX., p. 270.

49. NEVADA CITY, *Nevada Co* (Pop., 4,022–2,524.) Built in '61 by co. **Supply.**—South Yuba River, by gravity. **Distribution.**—(In '87.) *Mains*, 4 miles. *Taps*, 400. *Meters*, 4. *Hydrants*, 41. *Consumption*, 680,000 galls. **Financial.**—(In '87.) *Cost*, $38,000. *No Debt. Expenses*, $4,800. *Revenue*, $16,100. **Management.**—Supt., J. E. Brown.

50. **NIPOMO**, *San Luis Obispo Co.* (Pop., in '90, 215.) Built in '87 by Nipomo Water Co. Supply.—Mountain springs, by gravity to reservoir. Reservoirs have cap. of about 200,000 galls. Fiscal year closes Aug. 1. Distribution.—*Mains*, 2 miles. *Taps*, 99. *Hydrants*, not given. *Consumption*, 15,000 galls. Financial.—*Cost*, $23,500. *Cap. Stock*, $25,000. *Debt*, $2,000 at 10%. *Expenses*, $200. *Revenue*, $1,000. *Hydrant Rental*, free. Management.—Prest., T. B. Thayer. Secy., S. A. Dana. Treas., P. Fry. Supt., Geo. Stone.

51. **NORTH SAN JUAN**, *Nevada Co.* (Pop., 656—3,031.) Small system built in '60 by A. N. Crane. Supply.—Irrigating ditch, by gravity. *Hydrants*, 6. Address A. N. Crane.

52. **OAKDALE**, *Stanislaus Co.* (Pop., 376—1,012.) Built in '81 by co.; now owned by Wallace Ferguson Supply.—Artesian well, pumping direct, surplus going to tank. Tank.—Cap., 760,000 galls. Distribution.—*Mains*, 3 miles. *Taps*, 150. *Hydrants*, 8. *Consumption*, 100,000 galls. Financial.—*Cost*, $15,900. *No Debt*. *Expenses*, $2,500. Management.—Lessee and Supt., A. Rand.

53. **OAKLAND**, *Alameda Co.* (Pop., 34,555—48,682.) Built in '66 by Contra Costa Water Co. Supply.—Artificial lakes, by gravity to reservoirs and artesian wells, by pumping to reservoir to supply highest parts of city; 4,000,000-gall. Hyatt filter added in '91. Fiscal year closes March 30. Unofficially reported that there is at Lake Chabot a masonry-lined tunnel 13 ft. high in the clear. Distribution.—*Mains* (in '87), 164 miles. *Taps* (in '87), 7,400. *Meters* (in '87), 150. *Hydrants* (in '91), 249. Financial.—*Cost* (in '87), $5,200,000. *Expenses*: operating, $173,527; construction, $123,172; dividends, $180,000; total, $476,999. *Revenue*: consumers, $440,- 022; city, $36,977; total, $476,999. Management.—Prest., Henry Pierce. Secy., Wm. H. Mead. Treas., Marshall Pierce. Supt., Orestes Pierce. Report (latest figures) by H. D. Rowe, ex-Deputy City Clerk.

NOTE.—For description of aerating and screening plants see ENGINEERING NEWS, vol. XXIV., p. 150.

54. **OCEANSIDE**, *San Diego Co.* (Pop., est. 1,200.) Now owned by city; built in '86 by Oceanside Water Co. Supply —River, pumping to reservoirs and direct. Pumping Ma.hinery.—Dy. cap., 300,000 galls.; Monitor duo. Reservoirs.—Cap., 268,000 galls.; one 178,000-gall., and one 90,000-gall. Distribution.—*Mains*, 5 miles. *Taps*, 200. *Meters*, 10. *Hydrants*, 6. *Pressure*, 65 lbs. Financial.—*Cost*, $15,000. *Bonded Debt*, $14,250 at 7%. *Expenses*, operating, $1,440. *Revenue*, consumers, $2,700. Management.—City Clerk, Geo. W. Wilbur.

VII. **ONTARIO**, *San Bernardino Co.* (Pop., in '90, 633.) Built in '83 by San Antonio Water Co.. controlled by Ontario Land & Improvement Co. Supply.—San Antonio Creek and infiltration tunnel beneath same, by gravity. Fiscal year closes Nov. 23. Distribution.—*Mains*, 94 miles. *Hydrants*, 9. *Consumption*, 3,500,000 galls. Financial.—*Cost*, $320,000. *Cap. Stock*, $1,500,000. No Debt. *Expenses*, $3,800. *Revenue*, $4,100. Management. - Prest., H. H. Morgan. Secy., W. P. Craft. Supt. and Engr., F. E. Trask. NOTE.—Above system evidently includes irrigation plant.

55. **OROVILLE**, *Butte Co.* (Pop., 1,743—1,787.) Built in '78 by Oroville Water Co. Supply.—West Branch of Feather River, by gravity through ditch to reservoir, supply being purchased from Thermalito Colony Co. Fiscal year closes Dec. 31. Distribution.—*Mains*, 5 miles. *Taps*, 400. *Meter*, 1. *Hydrants*, 47. *Consumption*, 400,000 galls. Financial.—*Cost*, $10,000. *Cap. Stock*: authorized, $50,000; paid up, $10,000. *No Debt*. *Expenses*: operating, $3,000; int., $110; total, $3,110. Management.—Prest., Jno. J. Smith. Secy., F. W. Fogg. Treas., R. Parker. Supt., D. N. Frierlebew.

56. **PACIFIC GROVE**, *Monterey Co.* (Pop., in '90, 1.336.) Built in '83 by Pacific Improvement Co. Supply.—Carmel River, impounded, by gravity to reservoirs. Impounding dam is 423 ft. above sea level. From it 20 miles of 18, 15 and 12-in. pipe convey water to a 140,000,000-gall. reservoir 285 ft. above the sea; 2½ miles below there is a 16,000,000 gall. reservoir at an elevation of 208 ft., from which a 12 and 8-in. supply main leads. *Consumption*, 1,500,000 galls. *Cost*, $2,000,000. Supt., J. A. Skinner.

57, 58, 59. **PASADENA**, *Los Angeles Co.* (Pop., 391—4,882.)

NORTH PASADENA LAND & WATER CO.

Built in '63-71. Supply.—Arroyo Seco and branches, by gravity to reservoir. Dis-

---

FOR FURTHER DESCRIPTIVE MATTER see Manual for 1889-90. CHANGES since 1890 are here given. POPULATIONS are for 1880 and 1890, respectively.

tribution.—*Mains*, 20 miles. *Taps*, 300. *Meters*, 300. *Hydrants*, owned by city. *Consumption*, 150,000 galls. **Financial.**—*Cost*, $63,000. *Cap. Stock:* authorized, $00,000; paid up, $03,000. *No Bonds. Floating Debt*, $7,000. *Expenses:* operating; $2,500; int., $500; taxes, $70; total, $3,070. **Management.**—Prest., W. D. Painter. Secy., A. J. Painter. Supt., J. B. Lambert.

PASADENA LAND & WATER CO.

Built in '74, under 50-yrs. franchise, supplying western portion of Pasadena and nearly all of South Pasadena. **Supply.**—Arroyo Seco River, by gravity and pumping to reservoir. **Distribution.**—*Mains*, 10 miles. *Taps*, 750. *Meters*, 3. *Hydrants*, 15. *Consumption*, 500,000 galls. **Financial.**—*Cost*, $100,000. *Cap. Stock*, authorized, $75,000; all paid up, and $10,000 additional in assessments. *Bonded Debt*, $13,000 at 7%. *Expenses*, $1,500. *Revenue*, $9,500. **Management.**—Prest., A. K. McQuilling. Secy., H. S. Bennett. Treas., S. Washburn. Supt., W. J. Barcus.

PASADENA LAKE VINEYARD LAND & WATER CO.

History.—Built in '83 under 50-yrs. franchise, to supply stockholders only. **Supply.**—Arroyo Seco River, by gravity. **Impounding Reservoir.**—Cap., 6,000,000 galls. **Distribution.**—*Mains*, 11 to 4-in. steel. *Meters*, Crown, 30. *Hydrants*, 30. *Valves*, Ludlow, Crane. **Financial.**—*Cap. Stock*, authorized, $250,000. *Revenue*, consumers, $18,000. **Management.**—Prest., Jno. Allen. Secy., Jno. Habbick. Supt., A. H. Hinde.

PASADENA HIGHLAND WATER CO.

System is in operation, but no description has been received.

60. **PASO DE ROBLES,** *San Luis Obispo Co.* (Pop., in '90, 827.) Built in '87-8 by Paso Robles Water Co. **Supply.**—Springs, by gravity to reservoir. Pumping machinery described in 1889-90 Manual has been dispensed with. Fiscal year closes Dec. 31. In '91 artesian well was being sunk. **Distribution**—*Mains*, 5 miles. *Taps*, 150. *Meters*, 15. *Hydrants*, 10. *Consumption*, 150,000 galls. **Financial.**—*Cost*, $25,000. *Cap. Stock:* authorized, $34,000; paid-up, $25,000; paid $3,500 div'd in 1890. *Floating Debt*, $500 *Expenses:* operating; $150; taxes, $52; total, $202. *Revenue:* consumers, $4,000; fire protection, free; other purposes, $800; total, $4,800. **Management.**—Prest., D. W. James. Secy., D. Speyer. Supt., J. W. Graves.

61. **PETALUMA,** *Sonoma Co.* (Pop., 3,326—3,692.) Built in '71 by Sonoma County Water Co. **Supply.**—Mountain streams, by gravity to reservoir. Fiscal year closes Dec. 31. Reservoir is of brick with concrete bottom, lined with cem. Conduit is of c. i., 9 miles. In '90 impounding reservoir on mountain stream was in course of construction, situated about 6 miles from village in valley, having area of 10 acres; cap., 50,000,000 galls.; area, 8 acres; depth, 25 ft. Dams, three; one 388 × 27½ ft., one 321 × 10½ ft., one 149 × 5 ft. **Distribution.**—*Mains*, 4 miles. *Taps*, 800. *Hydrants*, 31. **Financial,**—*Cost* (in '87), $140,000. *Total Debt*, $15,000; *floating*, $5,000 at 5%; *bonded*, $10,000 at 5%. *Expenses:* operating, $6,000; int., $1,050; taxes, $1,300; total, $8,350. *Revenue*, total $21,500. **Management.**—Prest., H. E. Lawrence. Secy. and Supt., Chas. H. Egan.

62. **PLACERVILLE,** *El Dorado Co.* (Pop., 1,951—1,690.) Built in '52 by Placerville Water Co. **Supply.**—Mining canal, by gravity to reservoirs. *Mains* (in '86), 3.5 miles. *Hydrants* (in '86), 18. Supt. (in '91), Geo. W. Barlow.

63. **POMONA,** *Los Angeles Co.* (Pop., in '90, 3,634.) Built in '82-3 by Pomona Water-Works Co. **Supply.**—Tule River, by gravity to pump, thence by pumping to stand-pipe and tank. Fiscal year closes Oct. 1. **Distribution.**—*Mains*, 34.5 miles. *Taps*, 846. *Meters*, 2. *Hydrants*, 21. *Consumption*, 650,000 galls. **Financial,**—*Cost* (in '88), $100,000. *Cap. Stock*, $100,000. *Bonded Debt*, $25,000 at 6%, due in 1909. **Management.**—Prest., B. S. Nichols. Secy. and Treas., F. L. Palmer. Supt., H. J. Nichols.

64. **PORTERVILLE,** *Tulare Co.* (Pop., 202—606.) Built in '88 by Pioneer Land Co., San Francisco; county board of supervisors may regulate rates, but does not. **Supply.**—Tule River, by gravity to pump, thence by pumping to stand-pipe and tank; filtered. Fiscal year closes Dec. 31. During '89-90 30,000-gall. tank and 150,000-gall. National filter, were added. **Distribution.**—*Mains*, 2.3 miles. *Taps*, 61. *Hydrants*, 7. *Consumption*, 60,000 galls. **Financial.**—*Cost*, $15,000. *No Debt*. *Expenses*, $500. *Revenue*, $2,000. **Management.**—Prest., Wm. Thomas, 101 Sandstone St., San Francisco. Secy., Louis Sloss, Jr., 310 Sandstone St., San Francisco. Supt., E. Newman.

CALIFORNIA. 351

65. RED BLUFF, *Tehama Co.* (Pop., 2,106—2,608.) Built in '61; now owned by Antelope Creek & Red Bluff Water Co. **Supply.**—Antelope Creek, by gravity, and Sacramento River by direct pumping. Fiscal year closes last Monday in April. **Distribution.**—*Mains,* 17 miles. *Taps,* 310. *Hydrants,* 35. *Consumption,* 600,000 galls. **Financial.**—*Cost,* $150,000. *Cap. Stock:* authorized, $150,000, all paid up; paid $5,200 div'd in 1890. *No Debt. Expenses:* operating, $7,000; taxes, $1,000; total, $8,000. *Revenue:* consumers, $12,000; fire protection and other purposes, $1,000; total, $13,- 000. **Management**—Prest., J. H. Jewett. Secy., W. B. Mahoney. Supt., Chas. Cadwalader.

66. REDDING, *Shasta Co.* (Pop., 600—1,821.) Built in '86 by Redding Water Co. **Supply.**—Sacramento River, pumping to reservoir. **Distribution.**—(In '88.) *Mains,* 7.8 miles. *Taps,* 310. *Meters,* 2. *Hydrants* (in '91), 21. *Consumption,* 300,- 000 galls. **Financial.**—(In '88.) *Cost,* $12,000. *No Debt. Expenses,* $2,000. *Revenue,* $7,720. **Management.**—(In '89.) Prest., Francis Smith. Secy., S. J. R. Gilbert. Supt., S. P. Fikman.

67. REDLANDS, *San Bernardino Co.* (Pop. in '90, 1,904.) Built in '87-8 by Redlands, Lugonia & Crafton Domestic Water Co. **Supply.**—Bear Valley reservoir, Santa Ana River and Mill Creek, by gravity to and from distributing reservoir. In Sept., '91, 2,000,000-gall. reservoir (No. 3), for increased pressure, was completed. **Distribution.**—*Mains,* 24 miles. *Taps,* 375. *Hydrants,* 8. *Consumption,* 450,000 galls. **Financial.**—*Cost,* $200,000. *Cap. Stock:* authorized, $500,000; paid-up, $350,000. *No Debt. Expenses,* $3,000. *Revenue,* $8,000. **Management.**—Prest., Geo. A. Cook. Secy., F. G. Ferand. Supt., H. W. Allen.

68. REDONDO BEACH, *Los Angeles Co.* (Pop., in '90, 603.) Built in '87-8 by Redondo Beach Co., put in operation in Jan. **Supply.**—Thirty-two artesian wells, pumping to reservoir. **Distribution.**—*Mains,* 7.5 miles. *Taps,* 60. *Hydrants,* 20. **Financial.**—*Cost,* $50,000. *Hydrant Rental,* free. **Management.**—Prest., G. J. Ainsworth. Secy. and Treas., G. R. Brewer. Supt., L. J. Perry.

69. REDWOOD CITY, *San Mateo Co.* (Pop., 1,383—1,572.) Built in '77 by town. **Supply.**—Salinas Gas & Water Co.'s system, pumping to tanks. Fiscal year closes May 15. **Distribution.**—*Mains,* 4 miles. *Taps,* 270. *Meters,* 50. *Hydrants,* 50. *Consumption,* 70,000 galls. **Financial.**—*No Debt. Expenses,* $6,000. *Revenue,* $7,000. **Management.**—Chn., A. Gordon. Clerk, J. W. Glennon. Supt., Jas. Stalter.

70. RIVERSIDE, *San Bernardino Co.* (Pop., 1,358—4,683.) Built in '86-87 by Riverside Water Co. **Supply.**—Artesian wells, by gravity from receiving reservoir. Fiscal year closes Nov. 30. **Distribution.**—*Mains,* 19 miles. *Taps,* 580. *Meters,* 5. *Hydrants,* 25. **Financial.**—*Cost,* $187,895. *Cap. Stock:* authorized, $240,000; paid-up, $234,470. *Expenses:* operating, $1,337; taxes, $122; total, $1,459. *Revenue:* consumers, $9,898; fire protection, $300; other purposes, $1,521; total, $11.720. Co. operates works both for irrigation and for domestic supply, and previous to '90, financial s'atistics were not kept separately. There is a bonded debt of $300,000 on irrigation works, but none on pipe system. **Management.**—Pres., Jos. Jarviss. Secy., W. A. Correll. Treas., Riverside Banking Co. Supt., Francis Cuttle.

71. SACRAMENTO, *Sacramento Co.* (Pop., 21,420—26,386.) Built in '54 by city. **Supply.**—Spring, by gravity. Fiscal year closes April 1.

Year.	Mains.	Taps.	Hydrants.	Expenses.	Revenue.
'81-2	95	4,600	285	$25,000	$68,500
'90-1	38	6,000	381	40,000	90,000

Financial.—*Cost* (in '87), $350,000. *No Debt.* **Management.**—Not given.

72. SAINT HELENA, *Napa Co.* (Pop., 1,339—1,705.) Built in '77-8 by St. Helena Water Co. **Supply.**—Spring, by gravity. Fiscal year closes April 1. **Distribution.**—*Mains,* 10 miles. *Taps,* 200. *Hydrants,* 26. **Financial.**—*Cost,* $40,250. *Cap. Stock:* authorized, $50,000; paid-up, $40,250; paid $2,415 div'd in '90. *No Debt. Expenses:* operating, $1,200; taxes, $175; total, $1,375. *Revenue:* consumers, $4,500; fire protection, $312; total, $4,812. **Management.**—Prest. and Supt., Seneca Ewer. Secy., Fred. S. Ewer.

For Further Descriptive Matter see Manual for 1889-90. Changes since 1890 are here given. Populations are for 1880 and 1890, respectively.

73. **SALINAS,** *Monterey Co.* (Pop., 1,854—2,339.) Built in '74-5 by Salinas Gas & Water Co. Supply.—Four artesian wells, pumping to tanks. Distribution.— *Mains,* 7 miles. *Taps,* 298. *Meters,* 2. *Hydrants,* 24. *Consumption,* 150,000 galls. Financial.—*Cap. Stock,* $60,000. *No Debt.* Expenses and revenue of gas and water plants are not kept separately. Management.—Prest., J. B. Iverson. Secy., R. L. Porter. Treas., Wm. Vanderhurst. Supt., Frank B. Day.

74. **SAN BERNARDINO,** *San Bernardino Co.* (Pop., 1,673—4,012.) History. —Built in '90 by city. Engr., F. C. Finkle. Supply.—Artesian wells, by gravity to reservoir. There are two 11-in. wells 250 and 300 ft. deep, and nine 3-in. wells from 100 to 220 ft. deep. Combined dy. cap., 1,296,000 galls. Reservoir.—Cap., 1,000,000 galls.; 200 × 100 × 8 ft.; of concrete, covered with corrugated iron; 230 ft. abore and 3 miles from city; contr., E. Vieweger. Conduit.— Is 12 ins. in diam.; of steel, riveted and dipped in asphaltum. Distribution.—*Mains,* 12 to 3-in. kal. i., 22.8 miles. Services, ½ and ¾-in. galv. i. Pipe and specials from Geo. M. Cooley, who did trenching and pipe-laying. *Meters,* Crown, 3. *Hydrants,* Mathews, 70. *Valves,* Eddy. *Consumption,* 40,000 galls. *Pressure,* 100 lbs. Financial.—*Cost,* $130,000. *Bonds:* authorized, $160,000; issued, $135,000 at 6%. Management.—Prest., W. E. W. Lightfoot. Secy., B. B. Harris. Supt., F. O. Finkle.

75. **SAN DIEGO,** *San Diego Co.* (Pop., 2,637—16,159.) Built in '73 by San Diego & Coronado Water Co., sold and transferred to San D.ego Water Co. Aug. 29, '89. No provision in franchise for purchase of works by city, but in '91 city leased plant for 20-yrs. from June 1, with option of purchase at any time at actual cost. City pays $110,000 yearly for 3,000,000 galls. daily, and has option of buying, in excess of above amount, water, at 5 cents per 1,000 galls. Co. keeps works in repair, and in case gravity supply is not delivered will pump water, with old plant, from San Diego River. But failure to supply water from flume for 60 days will terminate contract, if city so elects. City may cancel contract at any time by giving one month's notice. City may extend works at its own expense, but in case co. makes extensions city pays 6% on cost of same and 3% per annum for deterioration, latter amounts to be deducted from purchase price if city buys Supply.—Apr. 9, '90, source changed to flume of San Diego Flume Co. Latter co. owns reservoir about 4,500 ft. above sea called Lake Cuyamaca. Reservoir, when full, has cap. of 2,570,320,000 galls., depth of 31 ft. and floods 804 acres; with 26 ft. of water cap. is 1,422,000,000 galls. Water is drawn from reservoir and flows for 12 miles through rocky bed of Boulder Creek to San Diego River. About one-half mile below junction of creek with river masonry diverting dam turns water into flume through which it flows for 35.75 miles and connects with 15-in. w. i. pipe leading to city City's supply passes through 3 and 6-in. meters. At flume pipe is 630 ft. above city, but head is broken to give pressure of 80 to 90 lbs. Fiscal year closes Dec. 31.

Year.	Mains.	Taps.	M't'rs.	H'd'es.	C'ns'pt'n.	Cost.	B'd'd d't.	Int.ch'g's.	Exp.	Rev.
'80	8	225	...	1	80,000	$79,000	none.	...	$14,828	$17,976
'87	28	1,400	215	69	....	....	$250,000	......	15,000	56,117
'88	54.8	1.801	1,000	100	......	710,000	296,000	$18,400	61,000	110,434
'89	57	2,141	1,248	185	650,000	1,925,740	1,000,000	50,000	71,500	89,700
'90	60.3	2,394	1,252	186	651,286	1,938,197	1,000,000	50,000	48,004	91,384

Financial.—*First Cost,* $76,938. *Cap. Stock:* authorized, $1,000,000; all paid up; paid uo div'd in 1890. *Bonded Debt,* $1,000,000 at 5%, due in 1909. *Expenses:* operating, $12,320; int., $50,000; taxes, $5,084; total, $89,004. *Revenue:* consumers, $67,018; fire protection, $18,557; other purposes, $8,809; total, $94,381. Management.—Prest., E. S. Babcock. Secy., Jos. A. Flint. Supt., E. Winsby. ;Prest. of City Fire Comrs., J. P. Burt. City Clerk, W. M. Gassaway.

76. **SAN FRANCISCO,** *San Francisco Co.* (Pop., 233,959—298,997.) Built in 56-9 by Spring Valley Water-Works Co. About same time San Francisco Water-Works Co. built works, putting them in operation after above co. In '85 two cos. consolidated. Supply.—Mountain streams, by gravity to and from reservoirs, and Lobus Creek and Lake Merced, pumping to reservoirs. In '91 new 6.000,000-gall. pumping station, with two 3,000,000-gall. Corliss pumping engines was being built at Lake Merced, and 60-ft. concrete dam, designed to be 100 ft. high eventually, was under construction near Searsville. In connection with new Lake Merced station an aerating plant is being built. In '91 San Mateo dam was raised to height of 145 ft., while Crystal Spring lower dam has been raised from 76 to 114 ft. Fiscal year closes June 30.

CALIFORNIA. 353

Year.	Mains.	Taps.	Meters.	Hydrants.	Consumption.	Exp.	Rev.
'80	175	20,500	......	1,350	12,643,000	$329,000	$1,300,000
'84	203	24,000	6,250	1,418	17,000,000	......	1,200,000
'86	222	24,000	......	1,470	17,500,000	737,737	1,281,500
'87	250	26,000	8,300	1,480	19,112,000	800,000	1,375,000
'88	326	30,000	12,750	1,570	18,241,000	637,000	1,444,483
'90	341	30,200	12,505	1,670	18,359,000	496,500	1,427,483

Financial.—See above. Profit for '90, $195,975. **Management.**—Prest., Chas. Webb Howard. Secy., Wm. Norris. City Supt., Chas. Elliott. Chief Engr., Herman Schussler.

NOTE.—For illustrated description of screening apparatus see ENGINEERING NEWS, vol. xxiv., p. 160. For some time a section of the city between Bush and Pacific, Larkin and Gough Sts. was supplied by the "Bradbury Water-Works." In '90 this plant was abandoned, owing, it was said, to insufficient patronage.

77. **SAN GABRIEL**, *Los Angeles Co.* (Pop., in '90, 737.) **History.**—Built in '87-8 by San Gabriel Valley Land & Water Co. Engr., W. F. Johnson. **Supply.**—Wells, by gravity to and from reservoir. **Distribution.**—*Mains,* 5 miles. **Financial.**—Cost, $20,000. *Cap Stock:* authorized, $1,600,000; paid up, $533,000. Bonded Debt, $75,000. These figures are for entire property of co. *Expenses,* nominal. Water supply system is part of land sale project and rates have not yet been decided upon. **Management.**—Prest., A. Vanderlip. Secy., F. K. Alexander. Treas., E. P. Johnson.

78. **SAN JACINTO**, *San Diego Co.* (Pop., in '90, 661.) San Jacinto Water Co. has small domestic and fire protection works, consisting of pump and tank at elevation of 24 ft. Village may contract, later, with Hemet Land & Water Co. for more adequate supply. Village Clerk, P A. Munn.

79. **SAN JOSE**, *Santa Clara Co.* (Pop., 12,567—18,000.) Built in '85 by San Jose Water Co. **Supply.**—Mountain streams, by gravity from impounding to distributing reservoirs, and seven artesian wells, by pumping, for about five months, to supplement gravity supply. Fiscal year closes June 2.

Year.	Mains.	Taps.	Meters.	Hydrants.	Consumption.	Cost.
'84	65	2,500	13	50	3,000,000	$100,000
'90	66	3,600	12	154	3,500,000	575,000

Financial.—First Cost, $413,500. *Cap. Stock,* $1,000,000. *Taxes,* $7,000. **Management.**—Prest., E. Williams. Secy., A. S. Williams. Supt., W. Wilcox.

53. **SAN LEANDRO**, *Alameda Co.* (Pop., 1,369—est., 1,800.) Supply from Oakland.

80. **SAN LUIS OBISPO**, *San Luis Obispo Co.* (Pop., 2,243—2,995.) Built in '74 by San Luis Water Co.; franchise provides that rates shall not be over 15% of cost of construction; city may purchase works in '95, price determined by five comrs., two appointed by city, two by co., fifth by these four. **Supply.**—San Luis Creek, by gravity. Fiscal year closes Jan. 31.

Year.	Mains.	Taps.	Hydrants.	Consumption.	Cost.	Exp.	Rev.
'84	9	250	22	250,000	$75,000	......	......
'89	9	541	21	......	96,409	$2,902	$13,156
'90	9	580	21	......	98,404	3,013	11,459

Financial.—*Cap. Stock:* authorized, $60,000; all paid-up; paid 12% div'd in 1890. *No Debt. Expenses:* operating, $2,703; no int.; taxes, $309; total, $3,013. *Revenue:* consumers, $11,281; fire protection, free; other purposes, $178; total, $11,459. **Management.**—Prest., P. W. Murphy. Secy. and Supt., Jno. A. Nock. Treas., W. E. Stewart.

76. **SAN MATEO**, *San Mateo Co.* Pop., 932—est., 1,500.) Supplied, in '71, by gravity from Spring Valley Water Co., San Francisco.

VIII. **SAN MIGUEL**, *San Luis Obispo Co.* (Pop., in '90, 458.) **History.**—Built in '89 by L. D. Murphy, under 25-yrs. franchise. **Supply.**—Well, pumping to reservoir. **Pumping Machinery.**—Dy. cap., 360,000 galls.; Hooker. **Reservoir.**—Cap., 150,000 galls. **Distribution.**—*Mains,* 1 mile. *Taps,* 60. *Hydrants,* 0. *Consumption,* 20,000. *Pressure,* 100 lbs. **Financial.**—*Cost,* $5,000. *No Debt. Expenses* $600. *Revenue,* $1,200. **Management.**—Address L. D. Murphy.

99. **SAN PEDRO**, *Los Angeles Co.* (Pop., in '90, 1,240.) Supply from Wilmington.

FOR FURTHER DESCRIPTIVE MATTER see Manual for 1889-90. CHANGES since 1890 are here given. POPULATIONS are for 1880 and 1890, respectively.

354   MANUAL OF AMERICAN WATER-WORKS.

81. SAN RAFAEL, *Marin Co.* (Pop., 2,276—3,290.) Built in '71 by Marin County Water Co. Supply.—Mountain Lake, by gravity. Fiscal year closes Dec. 31.

Year.	Mains.	Taps.	Meters.	Hyd'ts.	Cost.	B'd'd d't.	Int. ch'g's.	Exp.	Rev.
'84	15	450	19	36	$440,000	$50,000	$3,000	$25,000	$34,620
'88	30	630	10	40	435,000	135,000	6,750		
'89	35	640	40	40	400,000	135,000	6,750	13,698	41,802
'90	40	682	40	40	450,000	135,000	6,750	21,792	41,311

Financial.—*First Cost,* $350,000. *Cap. Stock:* authorized, $600,000; paid up, $300,000; paid 6% div'd in '90. *Bonded Debt,* $135,000, at 5%, due in 1907. *Expenses:* operating, $19,585; int., $6,750; taxes, $2,207; total, $28,542. *Revenue,* consumers, $41,311. Management.—Prest. and Supt., Robt. Walker. Secy., A. D. Harrison.

82. SANTA ANA, *Orange Co.* (Pop., 711—3,628.) History.—Built in '91 by city; put in operation Aug. 25. Des. Engrs., J. P. Leslie and E. S. S. Talle. Const. Engr., G. H. Finley. Contrs., Baker Iron Wks., Los Angeles. Supply.—Artesian wells, pumping direct from reservoir. Pumping Machinery.—Dy. cap., 1,500,000 galls.; Worthington. Reservoir.—Is 78 ft. square, 10 ft. deep. Distribution.—*Mains*, 12 and 10-in. c. i., 8.5 miles. Pipe and specials from McNeal Pipe Co., San Francisco; trenching and pipe-laying done by Frick Bros., Los Angeles. *Hydrants,* Eclipse, 60. *Valves,* Crane. Financial.—*Cost,* $57,000. *Bonded Debt,* $57,000 at $6%. Management.—City Clerk, Ed. Tedford.

83. SANTA BARBARA, *Santa Barbara Co.* (Pop., 3,610—5,864.) Mission and De la Guerra Garden Springs Water Cos., consolidated Jan., 91, as Santa Barbara Water Co. First works built in '30 by Franciscan Monks. De la Guerra Garden Springs Co. built works in '88. Supply.—Mission Creek, by gravity and pumping direct. Summer of '91, $150,000 bonds were voted for improvements. Distribution.—*Mains,* 15 miles. *Meters,* 12. *Hydrants,* 47. *Consumption,* 500,000 galls. Financial.—*Cap. Stock:* authorized, $500,000; paid up, $200,000. *No Debt.* Management.—Prest. R. B. Canfield. Secy. and Supt., A. W. Canfield.

79. SANTA CLARA, *Santa Clara Co.* (Pop., 2,416—2,891.) Supply from San Jose.

84. SANTA CRUZ, *Santa Cruz Co.,* (Pop., 3,898—5,596.) History.—Bought by city Jan. 1, '91. Built in '76 by Santa Cruz Water Co. For further description see 1889-90 Manual, p. 636. In June, '90, city began construction of works, of which following is description. Des. Eng., Geo. Schussler, San Francisco. Contrs., Risdon Iron Co., San Francisco. Chas. Purcell, Supt. of Construction. Supply.—Laguna Creek, by gravity. Impounding Reservoir.—Cap., 65,000,000 galls. Dam of clay, 380 ft. long, 45 ft. high. Inner slope, 3½ to 1; rip-rapped; outer slope, 2½ to 1; no heart-wall. Discharge-pipe, 16-in. c. i., laid in pure cem. on bed rock. Supply Main.—A 14-in. sheet-iron pipe, 12 miles long, extends from head-waters to city reservoir, and thence to city limits. Distribution.—*Mains,* w. i. and c. i., 17 miles. *Services,* galv. i., pipe and specials from Risdon Iron Works. *Hydrants,* Garrett, 45. *Valves,* Eddy. *Pressure,* 100 lbs. Financial.—*Cost,* $375,000. *Bonded Debt,* $375,000 at 5%. Management.—Supt., O. J. Lincoln.

85. SANTA ROSA, *Sonoma Co.* (Pop., 3,616—5,220.) Built in '73 by Santa Rosa Water Co. Supply.—Santa Rosa and Alamos creeks, by gravity. Fiscal year closes Dec. 31. Distribution.—*Mains,* 17 miles. *Taps,* 950. *Hydrants,* 34. Financial.—*Cost,* $150,000. *Expenses:* operating, $8,572; int., $883; taxes, $800; total, $10,254. *Revenue:* consumers, $21,026; city, $1,099; total, $22,125. Management.—Prest. and Supt., M. L. McDonald. Secy., T. A. Proctor.

NOTE.—In '91 com. on public works recommended that an engineer be employed to report on buying company's works or building new plant.

86. SELMA, *Fresno Co.* (Pop., in '90, 1,150.) History.—Built in '90 by Selma Water Co.; previously supplied by two small plants, which were destroyed by fire. Engr. for new works, G. G. Buckland. Supply.—Artesian wells, pumping to tank. Wells are 180 and 220 ft. deep, respectively. Pumping Machinery.—Dy. cap., 700,000 galls.; Hooker, Dow. Distribution.—*Mains,* 4.5 miles. *Hydrants,* 16. *Pressure,* fire, 250 lbs. Financial.—*Cost,* $16,000. Management.—Prest., D. S. Snodgrass. Secy., Frank Couch. Supt., Jno. Hall.

87. SONORA, *Tuolumne Co.* (Pop., 1,492—1,441.) Built in '53 by Gold Mount Water Co. Supply.—Springs, by gravity to reservoir, thence repumped to another reservoir. Fiscal year closes Dec. 31. During '90, 7,000 ft. of 4-in. pipe were laid, connecting with mains and affording pressure of 100 lbs.

Year.	Mains.	Taps.	Hydrants.	Cost.	Exp.	Rev.
'87	3	230	25	$25,000	$2,600	$4,000
'88	3	300	31	40,000	1,800	4,200
'90	5	267	34	40,000	2,330	4,926

Financial.—*No Cap. Stock. No Bonds. Floating Debt*, $10,000 at 0%. *Expenses:* operating, $2,100; int., $600; taxes, $230; total, $2,930. *Revenue:* consumers, $4,554; fire protection, $200; other purposes, $172; total, $4,926. **Management.**—Owner, Jno. Ferguson.

87. **SOUTH PASADENA,** *Los Angeles Co.* (Pop., in '90, 623.) Supply from Pasadena Land & Water Co., Pasadena.

88. **STOCKTON,** *San Joaquin Co.* (Pop., 10,282—14,424.) Built in '57 by P. E. Conner, afterwards owned Stockton Water-Works Co., purchased in '90, and rebuilt and enlarged in '91 by Stockton Water Co. **Supply.**—Artesian well, pumping to reservoir, stand-pipe and tanks. Fiscal year closes April 1.

Year.	Mains.	Taps.	Meters.	Hydrants.	Consumption.
'86	4	600	0	0	750,000
'88	8	750	0	0	1,000,000
'90-1	20	1,000	0	0	1,500,000

Financial.—*Cap. Stock:* authorized, $500,000; paid up, $250,000. *No Debt.* Fire cisterns are supplied with water free. Prest., W. S. McMurtry. Secy., John Flournay. Supt., M. S. Thresher.

89. **TEMPLETON,** *San Luis Obispo Co.* (Pop., in '90, 308.) Built in '87 by Templeton Water Co. **Supply.**—Well, pumping to tank. **Distribution.**—*Mains,* 1 mile. *Taps,* 40. *Meter,* 1. *Hydrants,* 6. *Consumption,* 4,000 galls. **Financial.**—*Cost,* $5,700. *No Debt. Expenses,* $375. *Revenue,* $615. **Management.**—Treas., C. M. Steinbeck. Mgr. and Supt., V. E. Donelson.

90. **THERMALITO COLONY,** *Oroville, Butte Co.* (Pop., not given.) Built in '87-8 by Thermalito Colony Co. **Supply.**—West Branch of Feather River, by gravity to reservoir. Fiscal year closes Jan. 1. **Distribution.**—*Mains,* 13.5 miles. *Taps,* 416. *Hydrants,* 40. *Consumption,* 374,000 galls. **Financial.**—*Cost,* $250,000. *Cap. Stock,* authorized, $1,000,000. *Bonded Debt,* $75,000 at 7%. *Expenses:* operating, $3,000; int., $5,250; total, $8,250. *Revenue,* none. Water was furnished free to consumers till Jan. 1, '91, when rates were established. **Management.**—Prest., A. F. Jones. Secy., H. C. Hills. Treas., C. W. Fogg. Supt., Jas. Lafferty.

91. **TRUCKEE,** *Nevada Co.* (Pop., 1,147—1,350.) Built in '68; now owned by Geo. Schaffer. **Supply.**—Springs, by gravity. *Mains,* 1 mile. *Hydrants,* 15. *Cost,* $1,000. *No Debt. Expenses,* $600. *Revenue,* $2,000. Supt., C. Shields.

92. **TULARE,** *Tulare Co.* (Pop., 447—2,697.) Built in '82 by Tulare City Water Co. **Supply.**—Two artesian wells, pumping to tank and direct. Fiscal year closes Dec. 31. **Distribution.**—*Mains,* 13 miles. *Taps,* 400. *Hydrants,* 40. *Consumption,* 200,000 galls. **Financial.**—*Cost,* $75,000. *Cap. Stock:* authorized, $100,000; paid up, $75,000; paid 6% div'd in 1890. *No Debt. Expenses:* operating, $5,000; taxes, $300; total, $5,300. *Revenue:* consumers, $10,000; fire protection, $1,800; total, $11,800. **Management.**—Prest. and Supt., D. W. Madden. Secy., W. L. Blythe.

93. **UKIAH,** *Mendocino Co.* (Pop., 933—1,627.) Small system of water-works is owned and operated by T. H. Jamison. **Supply.**—Gibson Creek. No further description received.

NOTE.—Gravity works to be owned by town were under consideration in '91. Address Jno. McGlashan or F. Brunner.

94. **VALLEJO,** *Solano Co.* (Pop., 5,987—6,343.) Built in '69-70 by Vallejo Water Co. **Supply.**—Artificial lake, by gravity and by pumping from supply mains to reservoir and stand-pipe. Fiscal year closes Dec. 31.

Y'r.	M'ns.	Taps.	M't'rs.	Hyd'ts.	Cons'pt'n.	Cost.	B'd'd debt.	Int. c'g'a.	Exp.	Rev.
'82	18	950	0	40	300,000	$250,000			$10,000	$25,000
'89	24	1,000	30	48	360,000	283,589	$185,000	$9,250	11,686	25,567
'90	25.9	1,054	40	50	380,000	293,857	185,000	9,250	16,553	36,123

Financial.—*First Cost,* $200,000. *Cap. Stock,* $1,000,000. *Bonded Debt,* $185,000 at 5%. *Expenses:* operating, $15,826; int., $9,250; taxes, $727; total, $34,757. *Revenue:* con-

FOR FURTHER DESCRIPTIVE MATTER see Manual for 1889-90. CHANGES since 1890 are here given. POPULATIONS are for 1880 and 1890, respectively.

sumers, $23,829; fire protection, none; other purposes, $1,294; total, $30,123. Management.—Owner and Gen. Mgr., E. J. Wilson. Prest., S. G. Hilborn. Secy., G. W. Wilson.

IX. **VALLEY SPRINGS,** *Calaveras Co.* (Pop., 1,685—not given.) Built in '84-5 by Thos. Hague. Supply.—Canal and Mining Co.'s ditches. *Mains,* ¼ mile. *Hydrants,* none. *Cost*, $3,000. *No Debt. Expenses,* $500. *Revenue,* $700. Supt., H. A. Mensinger.

95. **VENTURA,** *Ventura Co.* (Pop., 1,370—2,320.) Built in '82 by Santa Ana Water Co. Supply.—San Buenaventura River, by gravity through open ditch to reservoirs, with pumping for few houses on high ground. Fiscal year closes Dec. 31. A 1,000,000-gall. reservoir was built in '90. Distribution.—*Mains,* 7.5 miles. *Taps,* 560. *Meters,* 7. *Hydrants,* 14. *Consumption,* 1,326,000 galls. Financial.—*Cost,* $161,815. *Cap. Stock:* authorized, $100,000, all paid up; paid 5½% div'd in '60. *Bonded Debt,* $40,000, at 6%, due in 10 or 20 yrs. at option of co. *Expenses:* operating, $2.882; int., $2,400; taxes, $654; total, $11,688. *Revenue:* consumers, $11,678; fire protection, $10; total, $11,688. Management.—Prest., G. W. Chri man. Secy., E P. Foster. Supt., W. W. Martin.

96. **VISALIA,** *Tulare Co.* (Pop., 1,412-2,885.) Now owned by A. Crowley. Built in '77 by W. H. Hammond, under 20-yrs franchise. Supply.—Artesian wells, pumping to tanks and direct. Distribution.—(In '87.) *Mains,* 5 miles. *Hydrants,* 12. *Consumption,* 200,000 galls. Financial.—(In '87.) *Cost,* $30,000. *Debt,* $10,000 at 7%. *Expenses:* operating, $3,120; int., $700; total, $3,820. *Revenue:* consumers, $6,270; city, $450; total, $6,720. Management.—Owner and Mgr., A. Crowley.

97. **WATSONVILLE,** *Santa Cruz Co.* (Pop., 1,799—2,149.) Built in '76-78 by Watsonville Water Co., afterward sold to Corralitos Water Co. Supply—Corralitos Creek, by gravity to reservoirs. Fiscal year closes Dec. 31. Distribution.—*Mains,* 10 miles. *Taps,* 600. *Hydrants,* 30. *Consumption,* 300,000 galls. Financial.—*Cost,* $130,000. *No Debt. Expenses:* operating, $2,000; taxes, $900; total, $2,900. *Revenue:* consumers, $7,500; fire protection, $600; irrigation, $3,500; total, $11,600. Management.—Prest., Francis Smith. Supt., A. W. White.

98. **WEST BERKELEY,** *Alameda Co.* (Pop., est., 2,500.) Built in '75; now owned by Alameda Water Co. Supply.—Tunnels driven into hills and springs, by gravity. In '91 unofficially reported that 50,000,000-gall. reservoir was under construction. Distribution.—*Mains,* 12 miles. *Taps,* 860. *Meters,* 75. *Hydrants,* 34. Financial.—*Revenue,* city, $200. Management.—Not given.

NOTE.—Summer of '91, city obtained estimates of cost of salt water fire protection system, including pump, boilers, 1,000 ft 10-in., 15,000 ft. 8-in, 11,100 ft. 6 in. and 75,000 ft. 4-in. pipe. Est. cost, $58,751.

99. **WHEATLAND,** *Yuba Co.* (Pop., 635—630.) Built in '89-90 by Wheatland Water Co. Supply.—Well, pumping to tanks and direct. Reported in '90 that tower was being added. Fiscal year closes Feb. 2. Distribution.—*Mains,* 1.5 miles. *Taps,* 80. *Hydrants,* 21. Financial.—*Cost,* $6,000. *No Debt.* Management.—Prest., Jno. Stineman. Secy. and Supt., H. C. Niemeyer. Treas., T. H. Thomas.

100. **WILMINGTON AND SAN PEDRO,** *Los Angeles Co.* (Pop.: Wilmington, 911—687; San Pedro, in '90, 1,240.) Built in '82 by Wilmington Development Co. Supply.—Wells, pumping to tanks and direct. Fiscal year closes Dec. 31. Holly system, affording pressure of 125 lbs., was introduced in '90 to be used in case of fire. Distribution.—*Mains,* 12 miles. *Hydrants,* 12. *Consumption,* 115,000 galls. Financial.—*Cost,* $50,000. *No Debt.* Management.—Mgr., J. B. Banning. Supt., A. Young.

101. **WOODLAND,** *Yolo Co.* (Pop., 2,257—3,069.) First supplied in '73 by co.; in '85 Woodland Water Co. built works. Plant described below owned by W. W. Porter. Supply.—Four artesian wells, pumping to tank and direct. Fiscal year closes Dec. 31. Distribution.—*Mains,* 8.5 miles. *Taps,* 900. *Meters,* 3. *Hydrants,* 23. *Consumption,* 534,000 galls. Financial.—*Cost,* $53,000. *No Cap. Stock. No Bonds. Expenses:* operating, $7,000; taxes, $426; total, $7,426. Management—Owner, W. W. Porter.

NOTE.—In July, '91, contract was let to W. P. Garrett & Co., San Francisco, to construct works consisting of ten 200-ft. wells, two Holly comp. dup. pumps, four $7,000-gall. tanks, about 7 miles of pipe, 30 hydrants. Est. cost. $50,000.

CALIFORNIA. 357

102. **YREKA,** *Siskiyou Co.* (Pop., 1,059—1,100.) Built(in '82 by H. Scheld. Supply.—Springs, by gravity, and by pumping to reservoir during dry season. Fiscal year closes Dec. 31. Distribution.—*Mains,* 2 miles, Taps, 90. *Hydrants* 12. Financial.—*Cost,* $3,600. Management.—Propr., Henry Scheld. Secy., Louis P. Scheld.

103. **YUBA CITY,** *Sutter Co.* (Pop., in '90, 562.) Built in '80 by Yuba City Water-Works Co. Supply.—Artesian wells, pumping to tank. Pumping Machinery.—Cap., 110,000 galls. Tank.—Cap., 65,000 galls. Distribution.—*Mains,* 2 miles. Financial.—*Cost,* $10,000. Management.—Supt., J. E. Orr.

Water-Works Projected With Fair Prospects of Construction.

LAKEPORT, *Lake Co.* (Pop., in '90, 901.) Aug. 14, '91, it was reported that town wished to grant franchise for construction of water-works. Address M. R. Chamblin.

ROCKLIN, *Placer Co.* (Pop., 624—1,056.) System projected by South Yuba Water Co. Address Jro. Spaulding, Auburn.

WHITTIER, *Los Angeles Co.* (Pop., in '90, 585.) East Whittier Land & Water Co. were constructing works Aug. 6, '91, to be used for irrigation and domestic supply. There are 14 artesian wells, from 80 to 200 ft. deep. Main conduit is 11 miles long and is a concreted trench 6 ft. wide at top, 4 ft. at bottom, and 3 ft. 4 ins. deep. About 1.25 miles of the 11 is made up of flumes, that crossing the San Gabriel River being 3,000 ft. long, on piles. The flumes are lined with asphaltum. A 1,500,-000-gall. pumping plant will lift water to a reservoir for the higher levels. Contrs., Gray Bros. & French, Los Angeles. Chf. Engr. and Mgr., A. L. Reed.

ARIZONA.

Water-Works Completed or In Process of Construction.

1. **PHŒNIX,** *Maricopa Co.* (Pop., 1,708—3,152.) Built in '87-8 by Gardiner & Son, now owned by Phœnix Water-Works Co., under 50-yrs. franchise; franchise fixes maximum rates and provides for purchase of works by city in 25 yrs., price determined by three experts, one appointed by city, one by co., third by the two. Supply.—Wells, pumping to stand-pipe, and direct. Fiscal year closes Dec. 1. Distribution.—*Mains,* 15 miles. Taps, 300. *Meters,* 10. *Hydrants,* 50. *Consumption,* 450,000 galls. Financial.—*Cost,* $260,000. *Cap. Stock,* authorized, $1,000,000. *Bonded Debt,* $250,000 at 6%, due in 1910. *Expenses:* operating, $3,600; int., $15,000; taxes, $1,000; total, $19,600. *Revenue:* consumers, $20,000; fire protection, $3,500; total, $23,500. Management.—Prest. and Supt., T. W. Hine. Secy., B. N. Pratt. Treas., Jerry Millay.

2. **PRESCOTT,** *Yavapai Co.* (Pop., 1,836—1,759.) Built in '84 by town. Supply.—Creek, pumping to reservoir; filtered. Distribution.—*Mains,* 3.8 miles. Taps, 238. *Meters,* 45. *Hydrants,* 24. *Consumption,* 250,000 galls. Financial.—*Cost* (in '88), $60,000. *Bonded Debt,* $60,000 at 3%. *Expenses:* operating, $3,000; int., $4,800; total, $7,800. *Revenue:* consumers, $2,650. Management.—Two comrs., members of city council, elected every two years. Mayor, Jno. Howard. Secy. V. B. Foster. Supt., F. G. Barker, appointed by Mayor. City Engr., J. F. Mahoney.

NOTE.—In '91 a co. made proposition to buy or lease city works and introduce new supply from Banning Creek.

3. **TOMBSTONE,** *Cochise Co.* (Pop., 973—1,875.) Owned by Huachuco Water Co. Built in '81-2 by Sycamore Springs Water Co., under 50-yrs. franchise, which fixes rates but does not provide for purchase of works by city. Supply.—Collecting reservoir at head of Miller's Cañon, by gravity to distributing reservoir. Distribution.—*Mains* (in '88), 25 miles. *Hydrants* (in '91), 213. Financial.—*Cost* (in '88), $500,000. *Revenue,* city (in '91), none. Management.—(In '80.) Prest., Chas. H. Stone. Secy. and Treas., Hugh Porter. Supt., J. W. Clark. Main office, No. 11 Pine St., N. Y. City.

FOR FURTHER DESCRIPTIVE MATTER see Manual for 1890-90. CHANGES since 1890 are here given. POPULATIONS are for 1880 and 1890, respectively.

4. **TUCSON**, *Pima Co.* (Pop., 7,007—5,150.) Built in '81 by Tucson Water Co., under 25-yrs. franchise, which does not regulate rates but provides for purchase of works by city in 1906. Supply, Santa Cruz River, by gravity. Distribution.—*Mains* (in '87), 6 miles. *Taps* (in '87), 500. *Meters* (in '87), 500. *Hydrants* (in '91), 56. *Consumption* (in '87), 160,000 galls. **Financial**.—*Cost* (in '87), $80,000. *Revenue*, city (in '01), $8,400. **Management**.—Mgr., J. R. Watts. Secy., A Crowley. Address S. Watts, No. 901 Olive St., St. Louis, Mo.

Water-Works Projected with Fair Prospects of Construction.

YUMA, *Yuma Co.* (Pop., 1,200—1,773.) In July, '91, Yuma Pumping & Irrigating Co. were granted 50-yrs. exclusive franchise. Supply, Colorado River. Address A. A. Dougherty.

NEVADA.

Water-Works Completed or In Process of Construction.

1. **AUSTIN**, *Lander Co.* (Pop., 1,679—1 215.) Built in '65 by Austin City Water Co.; city does not control rates nor have right to buy works. Supply.—Tunnel driven into mountain, by gravity from reservoir. Fiscal year closes Dec. 31. Distribution.—*Mains*, 4.5 miles. *Taps*, 250. *Hydrants*, 17. **Financial**.—*Cost*, $20,000. *No Debt*. *Expenses*, $1,532. **Management**. Prest. W. W. Bishop. Secy., R. W. Heath. Supt., A. W. Foster.

I. **CANDALARIA**, *Esmeralda Co.* (Pop., 756—345.) Built in '82 by Candalaria Water-Works & Milling Co. Supply.—Springs in White Mts., by gravity. Water is brought 25 miles in a 4-in. iron pipe and sold to families at $4 per month; also to mines at from $100 to $250 a month. Mgr., Col. Sutherland.

2. **CARSON CITY**, *Ormsby Co.* (Pop., 4,229—3,950.) **History**.—Built in '75 by Carson Water Co., under 50-yrs. franchise, which provides neither for control of rates nor purchase of works by city. Supply.—Natural springs, by gravity to reservoir. Reservoir.—Cap., 3,000,000 galls. Distribution.—*Mains*, 5 miles. Services, lap-welded gas pipe. *Taps*, 1,000. *Hydrants*, 26. *Consumption*, 1,000,000 galls. **Financial** —*Cost*, $100,000. *Cap Stock:* authorized, $100,000; all paid up. *Floating Debt*, $8,000 at 8½%. **Management**.—Prest., H. M. Yerington. Secy., G. W. Richards. Treas., Wells, Fargo & Co. Supt., E. S. Dougherty.

3. **ELKO**, *Elko Co.* (Pop., 752—est., 903.) First supplied in '74 by HI Lay Water Co. In '90, Elko Water Co. began to operate works. Supply.- Humboldt River, by gravity. *Mains*, 9,000 ft. *Takers*, 90. *Hydrants*, 4. *Cost*, $20,000. Supt., W. G. Smith.

4. **EUREKA**, *Eureka Co.* (Pop., 4,207—1,600.) Owned by Mrs. May J. McCoy. Built in '71 by W. W. McCoy. Supply.—Springs, collected in tanks, by gravity. Distribution.—*Mains*, 20 miles. *Taps*, 500. *Hydrants*, 18. *Consumption*, 200,000 galls. **Financial**.—*Revenue*, city, $1,560. **Management**.—Supt., E. A. McConnell. County Clerk, F. A. Harmon.

8. **GOLD HILL**, *Storey Co.* Supply from Virginia City, to which it has been annexed.

HAMILTON, *White Pine Co.* (Pop., 203—not given.) Built in '69 by Eberhardt Mining Co. In '90 it was reported that system is no longer in operation. Address Alexander Muir.

5. **PIOCHE**, *Lincoln Co.* (Pop., 745—676.) Built in '72 by Pioche Water Co., purchased in '89 by individuals, but new co. had not yet been incorporated, Nov, 10, '90; city does not control rates nor have right to buy works. Supply.—Two springs, by gravity to tanks. Fiscal year closes Oct. 31. Distribution.—*Mains*, 12 miles. *Taps*, 200. *Hydrants*, 30. **Financial**.—*Cost*, $151,400. *No Cap. Stock*. *No Debt*. *Expenses:* operating, $2,950; taxes, $412; total, $3,360. *Revenue:* consumers, $7,500; city pays one-half expense of repairing mains. **Management**.—Prest., W. S. Goodbe, Salt Lake City, Utah. Secy., Frank Goodbe. Supt., W. L. Cook.

NEVADA.

5. **RENO**, *Washoe Co.* (Pop., 1,302—est., 3,000.) Built in '73 by A. A. Evans, now owned by Reno Gas & Water Co. **Supply.**—Truckee River, by gravity through ditch to reservoir. **Distribution.**—*Mains*, 12 miles. *Taps*, 700. *Hydrants*, 20. **Financial.**—Withheld. **Management.**—Prest., C. C. Powning. Secy., H. B. Rule. Supt., G. F. Bliss.

6. **SILVER CITY**, *Lyon Co.* (Pop., 605—342.) Supply from Virginia City.

7. **TUSCARORA**, *Elko Co.* (Pop., 1,361—1,156.) Built in '88 by Tuscarora Water Co. **Supply.**—Twin Bridges Creek. **Distribution.**—*Mains*, 15 miles. *Taps*, 250. *Hydrants*, 13. *Consumption*, 900,000 galls. **Financial.**—Withheld. **Management.**—Prest., P. C. Hyman; Secy., J. W. Pew; both of San Francisco, Cal. Supt., H. W. Coffin.

8. **VIRGINIA CITY**, *Storey Co.* (Pop., 10,917—8,517.) Built in '72 by Co. **Supply.**—Creeks and artificial lake, by gravity from impounding to distributing reservoir. Fiscal year closes July 1. **Distribution.**—*Mains*. 23 miles. *Taps*, 1,980. *Hydrants* 108 *Consumption*, 7,500,000 galls. **Financial.**—*Cost*, $4,428,390. *Cap. Stock*, authorized, $5,000,000. *No Debt*. *Taxes*, $4,792. **Management.**—Pres., W. S. Hobart; Secy., W. W. Stetson; both of San Francisco, Cal. Supt., J. B. Overton.

9. **WINNEMCCA**, *Humboldt Co.* (Pop., 763—1,037.) Built in '77 by co. **Supply.**—Cross cañon, by gravity and pumping. **Distribution.**—(In '88.) *Mains*, 10 miles. *Taps*, 150. *Hydrants*, 18. *Consumption*, 700,000 galls. **Financial.**—(In '88.) *Cost*. $20,000. *No Debt*. *Expenses*, $5,000. **Management.**—(In '89.) Prest. and Secy., Francis Smith.

UTAH.

Water-Works Completed or in Process of Construction.

1. **EUREKA**, *Juab Co.* (Pop., 122—1,733.) **History.**—Built in '90 by Howansville Water Co.; put in operation Aug. 1. Contrs., David James & Co., Salt Lake City. **Supply.**—Springs, pumping to tank. **Pumping Machinery.**—Dy. cap., 120,000 galls ; Knowles. **Tanks.**—Cap., 60,000 galls. **Distribution.**—*Mains*, 1 to 2-in. w. i., 5.5 miles. *Services*, galv. i. *Taps*, 116. *Hydrants*, Douglass, 6. *Valves*, Ludlow. *Consumption*, 40,000 galls. *Pressure*. 110 lbs. **Financial.**—*No Debt*. *Expenses*, $3,120. *Revenue*, consumers, $4,875. *Hydrant Rental*, $3°. **Management.**—Prest., Moses Thatcher, Logan. Secy., W. J. Beattie, Salt Lake City. Supt. and Engr., A. Burch.

2. **LOGAN**, *Cache Co.* (Pop., 3,396—4,565.) Built in '79-80 by city. **Supply.**—Logan River, by gravity, through two settling reservoirs, to reservoir. **Distribution.**—(In '87.) *Mains*, 3 miles. *Taps*, 20. *Hydrants*, 14. *Consumption*, 240,000 galls. **Financial.**—(In '87.) *Cost*, $13,000. *No Debt*. *Expenses*, $700. *Revenue*, $700. **Management.**—(In '91.) Committee, W. M. Hanson, Seth A. Langton, H. Hayball. Supt., G. W. Lufkin.

3. **NEPHI**, *Juab Co.* (Pop., 1,797—2,034.) Built in '89-90 by city. **Supply.**—Springs, by gravity to and from distributing reservoir. **Distribution.**—*Mains*, 5.5 miles. *Taps*, 70. *Hydrants*, 12. **Financial.**—*Cost*, $15,000. *Bonded Debt*, $12,000 at 7%, due in '99. *Sinking Fund* to be started in '94. *Expenses*, nominal. *Revenue*: consumers, $300; city, none. **Management.**—Mayor, Alma Hague. Recorder, E. R. Booth. Supt., Peter Sutton.

4. **OGDEN**, *Weber Co.* (Pop., 6,069—14,889.)

OLD WORKS.

Built in '80-2 by city and sold in '90 to Bear Lake & River Water-Works & Irrigation Co., which took possession Oct. 27; has changed services to new mains laid by it and abandoned old system. For further information see Manual for '89-90, p. 647.

NEW WORKS.

Built in '89-90 by Bear Lake & River Water-Works & Irrigation Co.; tested Oct. 24, '90. City has perpetual right to buy distributing system at cost and pay for same

FOR FURTHER DESCRIPTIVE MATTER see Manual for 1889-90. CHANGES since 1890 are here given. POPULATIONS are for 1880 and 1890, respectively.

in 6% bonds, but in case it buys it must pay $150,000 for perpetual use of supply conduit. **Supply.**—Ogden River, by gravity from impounding reservoir, about 10 miles from city, to distributing reservoir. **Supply Conduit.**—Said to be 2 ft. in diam., of w. i., where under heavy pressure, otherwise of wood. **Distributing Reservoir.**—Cap., about 7,000,000 galls.; built in '91. **Distribution.**—*Mains*, about 35 miles. **Management.**—Prest., Jas. C. Armstrong. Secy. and Treas., Jas. H. Bacon. Gen. Mgr., H. C. Gilbert. Chf., Engr., S. Fortier.

5. **PARK CITY**, *Summit Co*. (Pop., 1,542–2,850.) Built in '80 by private co.; city neither controls rates nor has right to purchase works. **Supply.**—Mountain stream, by gravity. Fiscal year closes Dec. 1. **Distribution.**—*Mains*, 3 miles. *Taps*, 325. *Hydrants*, 14. **Financial.**—*Cost*, $19,500. *Cap. Stock:* authorized, $20,000, all paid up; paid $1 per share div'd in '90. *No Debt*. **Management.**—Prest., Edwin Kimball. Secy., Wm. M. Ferry. Treas., Thomas Cupit. Supt., Walter Willcocks.

6. **SALT LAKE CITY**, *Salt Lake Co*. (Pop., 20,768–44,843.) Built in '74 by city. **Supply.**—City creek, by gravity; filtered. Fiscal year closes Dec. 31. Spring of '91 contract was let for a 36-in. brick conduit, 6 miles long, for a new supply from Parley's Cañon. Contrs: Conduit proper, Du Bois & Williams, $84,384; grading, Hobson & Wilkinson, $25,250; total, $109,634.

Year.	Mains.	Taps.	Hydrants.	Cost.	Expenses.	Rev.
'80	10.5	700	116	......	......	......
'82	12.7	760	121	$250,000	$19,521	$13,047
'84	14	1,200	134	300,000	6,000	15,000
'86	14.8	1,780	148	320,000	5,445	18,145
'88	28.2	2,813	179	411,000	9,000	41,456
'90	57.18	4,073	322	661,890	15,259	41,952

Financial.—Bonds are redeemed from net income. Property holders are taxed 4 mills per sq. ft. for extension of mains, this assessment being refunded in water rates. **Management.**—Six comrs., appointed every two years by mayor. Con., Jas. Anderson. Secy., Anne Madison. Supt., W. H. Ryan. Appointed by mayor and council.

7. **TOOELE**, *Tooele Co*. (Pop., 1,096–1,008.) **History.**—Built in '90 by Tooele City Water Co., city owning about one-third of stock; put in operation July 24. Engr., Francis M. Lyman, Jr. **Supply.**—Spring, by gravity. **Reservoir.**—Cap., 88,000 galls. **Distribution.**—*Mains*, 8 to 2¼-in. Puget Sound wood, 6 miles. *Services*, galv. i. Pipe and specials from David James, Salt Lake City. *Taps*, 80. *Hydrants*, Ludlow, 5. *Valves*, Ludlow. *Consumption*, 50,000 galls. *Pressure*, 70 lbs. **Financial.** —*Cost*, $20,000. *Cap. Stock:* authorized, $25,000; paid-up, $20,000. *Revenue:* consumers, $600; city, $300; total, $900. **Management.**—Prest., Francis M. Lyman, Sr. Secy., Francis M. Lyman, Jr. Treas., Geo. F. Richards. Supt., Hugh S. Gorvans.

Water-Works Projected with Fair Prospects of Construction.

BRIGHAM CITY, *Box Elder Co*. (Pop., 1,877–2,130.) Works have been projected for some time. Sept., 91, Hyram House, Corinne, proposed to secure franchise, or else contract to build for city.

MANTI, *San Pete Co*. (Pop., 1,748–1,950.) July 2, '91, it was reported that city was constructing water-works. Address Ezra Shoemaker.

PROVO CITY, *Utah Co*. (Pop., 3,432–5,159.) City voted to issue $120,000 worth of 6% bonds to construct water-works. Sept. 8, '91, it was reported that contract had been let to Rhodes Bros. Consulting Eng., C. P. Allen. Const. Engr., D. C. Henny, all of Denver, Colo. Supply, Provo River, by gravity; filtered. Mains, 12 to 4-in., about 13 miles. Hydrants, 80.

IDAHO.

Water-Works Completed or in Process of Construction.

1. **BELLEVUE**, *Logan Co*. (Pop., in '90, 892.) Built in '87 by Bellevue Water Co.; city neither controls nor reserves right to purchase works. **Supply.**—Seaman's Creek, by gravity, from impounding reservoir. Fiscal year closes Aug. 7. **Distri-**

IDAHO. 361

bution.—*Mains*, 5 miles. *Taps*, 165. *Hydrants*, 8. **Financial.**—*Cost*, $16,015. *Cap. Stock:* authorized, $30,000; all paid up; paid 10% div'd in '90. *No Debt. Expenses:* operating, $790; taxes, $15; total, $805. *Revenue:* consumers, $2,990; city, $343; total, $3,333. **Management.**—Prest., H. E. Miller. Secy. and Supt., H. F. Baker. Treas., N. C. De Lano.

2. BOISE CITY, *Ada Co.* (Pop., 1,899—2,311.) **History.**—Built in '90 by two cos., afterward consolidated as Artesian Hot & Cold Water Co. Engr., H. B. Grumbling. Contr., A. C. Fover. **Supply.**—Artesian wells, by gravity. **Reservoirs.**—Cap., 650,000 galls.; three. **Distribution.**—*Mains*, 10 to 3-in. kal. l., 10 miles. Pipe and specials from National Tube Wks. *Taps*, 430. *Hydrants*, Ludlow, 134. *Consumption*, 1,000,000 galls. *Pressure*, 65 lbs. **Financial.**—*Cost*, $175,000. *Cap Stock:* authorized, $250,000; paid up, $175,000. *No Debt. Revenue*, city, $1,350. **Management.**—Prest., C. W. Moore. Secy., E. B. Tage. Treas., Alfred Eoff. Gen. Mgr., Peter Sonna.

EAGLE ROCK, *Bingham Co.* See Idaho Falls.

3. HAILEY, *Alturas Co.* (Pop., in '90, 1,073.) Built in '84 by Alturas Water Co; city neither controls rates nor has right to purchase works. **Supply.**—Indian Creek, by gravity. Fiscal year closes Dec. 31. **Distribution.**—*Mains*, 5 miles. *Taps*, 400. *Hydrants*, 7. **Financial.**—*Cost*, $15,000. *Cap. Stock:* authorized, $15,000; paid-up, $15,000; paid 4% div'd in '89. *No Debt. Expenses:* operating, $1,000; taxes, $250; total, $1,250. **Management.**—Pres., E. A. White. Secy., Alex. Willman. Supt., W. I. Riley.

4. IDAHO FALLS (formerly Eagle Rock), *Bingham Co.* (Pop., in '90, 1,588.) Built in '85 by Eagle Rock Water-Works Co.; franchise allows co. to make net income of 15%; there is no provision for purchase of works. **Supply.**—Snake River, pumping to tanks. Fiscal year closes Dec. 31. **Distribution.**—*Mains*, 1.5 miles. *Taps*, 80. *Hydrants*, 5. *Consumption*, 70,000 galls.

Year.	Mains.	Taps.	Meters.	H'd'ts.	C'ns'pt'n.	Cost.	Exp.	Rev.
'87	1.5	20	1	4	35,000	$8,000	$1,700	$1,700
'90	1.5	80	....	5	70,000	10,000	1,750	700

Financial.—*Cost*, $10,000. *Cap. Stock:* authorized, $8,000; all paid up; paid no div'd in '90. *Floating Debt*, $3,000 at 12%. *Expenses:* operating, $1,700; int., $360; taxes, $50; total, $2,610. *Revenue:* consumers, $1,700; fire protection, none; Ity., $1,000; total, $2,700. **Management.**—Pres., J. A. Clark. Secy. and Supt., W. B. Crow.

5. KETCHUM, *Alturas Co.* (Pop., in '90, 450.) Built in '89 by Ketchum Spring Water Co.; city neither controls rates nor reserves right to purchase works. **Supply.**—Springs, by gravity to and from reservoir. Fiscal year closes July 15. **Distribution.**—*Mains*, 2 miles. *Taps*, 100. *Hydrants*, 7. *Consumption*, 20,000 galls. **Financial.**—*Cost*, $13,000. *Cap. Stock:* authorized, $24,000; paid-up, $10,080; p'd no div'd in '90. *Floating Debt*, $2,500. *Expenses*, $560. *Revenue*, $2,100. **Management.**—Prest., I. I. Lewis. Secy. and Supt., Geo. Stevens. Treas., Wm. Greenhow.

6. LEWISTON, *Nez Perces Co.* (Pop., 739—849.) **History.**—Built in '90-1 by Lewiston Water & Light Co., under 25-yrs. franchise; put in operation May 1, '91. Des. Engr., H. Bloomfield, San Francisco. Contr., H. T. Madgwick. **Supply.**—Clearwater River, pumping to reservoir and direct. **Pumping Machinery.**—Dy. cap., 1,500,000 galls.; comp., dup., from Parke & Lacey, Portland, Ore. **Reservoir.**—Cap., 1,500,000 galls.; contrs., Willamette Iron Works, Portland. **Distribution.**—*Mains*, 12 to 8-in. kal. l., 8 miles. Pipe and specials from Dunham, Carrigan & Hayden Co., Portland. *Taps*, 80. *Hydrants*, 17. *Pressure*, 97 lbs. **Financial.**—*Cost*, $90,000. *Cap. Stock:* authorized, $160,000; paid up, $30,000. *Bonded Debt*, $50,000 at 6%. *Expenses*, $4,000. *Revenue*, city, $2,490. **Management.**—Prest., W. F. Kettenbach. Secy., F. W. Kettenbach. Treas., C. C. Bunnell.

7. MOSCOW, *Latah Co.* (Pop., 76—2,861.) **History.**—Built in '90 by town; put in operation July. Engr. and contr., G. H. Sutherland. During '91, mains were to be extended and 14 × 100-ft. stand-pipe erected. **Supply.**—Two flowing artesian wells, pumping direct. **Pumping Machinery.**—Dy. cap., 40,000 galls. **Distribution.**—*Mains*, 8 to 4-in. kal. l., 2 miles. Pipe and specials from Dunham, Carrigan & Hayden Co., Portland. Ore. *Hydrants*, National, 11. *Valves*, Eddy. *Consump-*

FOR FURTHER DESCRIPTIVE MATTER see Manual for 1889-90. CHANGES since 1890 are here given. POPULATIONS are for 1880 and 1890, respectively.

tion, 10,000 galls. *Pressure:* ordinary, 70 lbs.; fire, 120 lbs. **Financial.**—*Cost*, $17,000. *Bonded Debt*, $16,000 at 7%. *Expenses*, $3,000. **Management.**—Supt., Thomas Lamb.

8. **MULLAN,** *Shoshone Co.* (Pop., in '90, 818.) **History.**—Built in '90 by A. M. Strode. Des. Engr., W. W. Hart. Const. Engrs., W. J. Linthern and H. C. Johnson. Contrs, Holley. Mason, Marks & Co. **Supply.**—Mill Creek, by gravity. Reservoir.—Cap., 50,000 galls. **Distribution.**—*Mains*, 8 and 5-in. w. l., 8,000 ft. *Taps*, 200. *Hydrants*, Chapman, 7. *Valves*, Chapman. *Pressure*, 125 lbs. **Financial.**—*Cost*, $13,000. *No Debt*. *Expenses*, $600). *Revenue*: consumers, $1,800; city, $1,500; total, $6,300. **Management.**—Secy., W. J. Linthern. Supt., A. M. Strode.

9. I. **POCATELLO,** *Bingham Co.* (Pop, in '90, 2,330.)

UNION PACIFIC RAILWAY COMPANY'S PLANT.

Built in '82; as co. controls town, there is no franchise or provision for purchase of works. **Supply.**—Mountain springs, by gravity and Portneuf River, by pumping. Fiscal year closes Dec. 31. **Distribution.**—*Mains*, 3 miles. *Taps*, 65. *Hydrants*, 10. *Consumption*, 50,000 galls. **Financial.**—*Cost*, $13,000. *Expenses*, $1,500. *Revenue*, $100. **Management.**—U. P. Ry. Co. Supt., J. M. Bennett.

CENTER STREET WATER CO.

System was put in operation in Dec., '93. **Supply.**—From Union Pacific Railway Company's Plant. *Mains*, 15,000 ft. *Taps*, 50. *Cost*, $2,500. Prest., D. Sweinhart. Secy., J. A. Hoagland. Supt., L. A. West.

10. **WARDNER,** *Shoshone Co.* (Pop., in '90, 858.) **History.**—Built in '84 by Alexander Monk. Contrs., Holley. Mason, Marks & Co. **Supply.**—Springs and brook, ½ mile from and 400 ft. above town, by gravity to tanks. Tanks.—Cap., 70,000 galls.; one 50,000-gall. and one 20,000-gall. **Distribution.**—*Mains*, 8 to 4-in. w. l., 1.5 miles. *Taps*, 150. *Hydrants*, Morgan, 5. *Valves*, Crane. *Consumption*, 20,000 galls. *Pressure*: ordinary, 60 lbs.; fire, 50 to 103 lbs. **Financial.**—*Cost*, $15,000. *No Debt*. *Expenses*, $2,000. *Revenue*, $1,500. **Management.**—Owner, Alex. Monk.

GROUP 9.—Canada: Maritime Provinces. — Water-Works Completed, in Process of Construction, or Projected in New Brunswick, Nova Scotia, Prince Edward Island and Newfoundland

NEW BRUNSWICK.

Water-Works Completed or in Process of Construction.

1. **FREDERICTON,** *York Co.* (Pop., 6,218—6,502.) Built in '82-3 by city. **Supply.**—River St. John, pumping direct; filtered. Fiscal year closes Oct. 31.

Year	Mains	Taps	M'l'rs	H'd'ts	C'ns'pt'n	Cost	B'd debt	Int.ch'g's	Exp.	Rev.
'84	9	202	0	80		$93,614	$200,000	$10,000		
'85	9	446	0	81	357,181	100,000	100,000	5,000	$4,119	$1,250
'88	9.8	651	0	81	491,542	111,000	100,000	5,000	5,881	5,518
'90	9.8	703	3	82	413,675	114,637	100,000	5,000	5,376	6,441

Financial.—*Bonded Debt*, $100,000 at 5%. *Expenses*, total, $10,376. *Revenue*: consumers, $6,441; city, $1,600; total, $8,041. **Management.**—Com. of three, appointed by council every year. Chn., Marshall Richeg. Secy., Chas. W. Beckwith. Supt., Alex. Burchill. Appointed by City Council.

4. **MILLTOWN,** *Charlotte Co.* (Pop., 1,661—2,146.) Supply from St. Stephens.

2. **MONCTON,** *Westmoreland Co.* (Pop., 5,032—8,765.) Built in '78 by Moncton Gas Light & Water Co. **Supply.**—Northwest branch of Hall's Creek by gravity, with pumping as auxiliary, from impounding reservoir. Fiscal year closes June 1. **Distribution.**—*Mains*, 12.3 miles. *Taps*, 650. *Meters*, 2. *Hydrants*, 50. *Consumption*, 650,000 galls. **Financial.**—*Cost*, $126,360. *Cap. Stock:* authorized, $120,000; all

---

* The large number of taps often reported in this group is probably owing to a different usage of the word tap in Canada, meaning number of faucets instead of number of times the street main is tapped.

NEW BRUNSWICK. 363

paid up; paid 8% div'd. in '90. *Expenses:* operating, $2,927; int., $3,600; taxes, $250; total, $6,477. *Revenue,* total, $14,750. **Management.**—Prest., Jno. L. Harris. Secy., R. A. Borden. Supt., Jno. Edington. Treas., C. P. Harris.

3. **SAINT JOHN,** *Saint John Co.* (Pop., 41,353—39,179.) Owned by city. Built in '36 by private co.; bought by city in '85. Supply.—Little River and Lake Latimer by gravity. Fiscal year closes Dec. 31.

Year.	Mains.	Taps.	Meters.	Hyd'ts.	C'n's'p'n.	Cost.	B'd'd d't.	Int. ch'gs.	Exp.	Rev.
'82	32.1	3,733	127	276	4,000,000	$988,000	......	......	$12,844	$77,370
'83	32.9	3,867	124	277	4,970,000	987.472	$987.472	$59,248	13,004	81,116
'84	46.2	3,915	120	277	4,824,250	993,772	987,472	59,364	12,550	82,705
'85	46.5	4,003	132	282	4,828,220	999,180	992.472	56,915	13,726	83,299
'86	47.3	4,106	131	285	4,833,200	999,180	997,473	61,817	13,212	83,327
'87	47.7	4,165	138	288	4,744,600	1,011,797	1,002,471	58,721	12,332	85,948
'88	48.4	4,211	141	292	4,783,500	1,017,301	1,007.472	60,048	14,711	86,374
'89	60.5	4,258	149	292	4,617,250	....	1,122,472	66,279	....	87,289
'90	62	4,857	158	335	4,935,300	1,038,472	1,122,472	66,279	13,700	94,723

Financial,—*Bonded Debt,* $1,122,472; $1,102,472 at 6%, $20,000 at 4%. *Sinking Fund,* $4,191. *Expenses,* total, $79,979. *Revenue:* consumers, $91,723; city, none. There is a general assessment for water on all real and personal property on lines of mains, whether water is used on the premises or not. Surplus, after paying maintenance and interest, is carried to sinking fund. **Management.**—Board of mgrs., appointed by City Council. Chn., W. D. Baskin. Secy., Richard Seeley. Supt., Gilbert Murdoch, appointed by City Council; has held office since '49.

4. **SAINT STEPHENS and MILLTOWN,** *Charlotte Co.* (Pop.: St. Stephens, 2,338—2,680; Milltown, 1,664—2,146.) Built in '86-7 by St. Croix Water Co.; city neither controls rates nor has right to purchase works. Supply. -Calais, Me., Water-Works. Fiscal year closes Dec. 31. Distribution.—*Mains,* 15 miles, *Taps,* 270. *Hydrants,* 60. *Consumption,* 140,000 galls. Financial.—*Cap. Stock:* authorized, $100,000; paid no div'd in 1890. *Bonded Debt,* $100,000 at 6%. *Revenue,* city, $3,000. **Management.**—Prest., Weston Lewis; Secy., Josiah Maxey; both of Gardiner, Me. Supt., W. E. McAllister.

5. **WOODSTOCK,** *Carleton Co.* (Pop., 2,487—3,290.) Built in '84 by town. Supply.—River St. John, pumping direct; filtered. Fiscal year closes Dec. 31. In '90 stand-pipe, 35 ft. in diam., 40 ft. high, of w. i., was constructed; cost of stand-pipe and extensions, $10,000. Hydrants and Valves, Coffin. Distribution.—*Mains,* 6 miles. *Taps,* 475. *Meters,* 23. *Hydrants,* 79. *Consumption,* 105,000 galls. Financial,—*Cost,* $68,250. *Bonded Debt,* $68,250: $67,000 at 4½%, due in 1917; $1,250 at 4½%, due in 1919. *Expenses;* operating, $3,313; int., $3,015; total, $6,318. *Revenue,* $4,500. **Management.**—Com., appointed from council, every year. Chn., A. Henderson. Supt., Donald Munro.

NOVA SCOTIA.

Water-Works Completed or in Process of Construction.

1. **AMHERST,** *Cumberland Co.* (Pop., 2,274—3,781.) Built in '87-8 by Amherst Water Works Co., Ltd. Supply.—Springs, by gravity to reservoir. Distribution. —(In '88.) *Mains,* 4 miles. *Taps,* 50. *Hydrants,* 40. Financial.—(In '88.) *Cost,* $6,000. *No Debt. Expenses,* $300. **Management**—Prest., G. B. Smith.

NOTE.—May 19, '91, town was authorized to issue $60,000 of 20-30 yrs. bonds, and bids for same were to be received until Nov. 5, '91. Town proposes to build works, taking supply from Nappan River.

1. **ANTIGONISHE,** *Antigonishe Co.* (Pop., not given.) History.—Built in '90-1 by town; put in operation Aug. 1. Engr., W. R. Butler. Windsor. Contrs., McDonald, Sutherland, & Vye. New Glasgow. Supply.—Brooks, by gravity. Reservoir—Cap., 500,000 galls. Distribution.—*Mains,* 8 and 6-in. c. i., 4.5 miles. Pipe and specials from Londonderry Iron Co. and J. Matheson & Co.,.New Glasgow. *Hydrants,* 23. *Pressure;* ordinary, 75 lbs.; fire, 156 lbs. Financial.—*Cost,* $40,000. *Bonded Debt,* $35,000 at 4½%. **Management.**—Town Clerk and Treas., Rupert Cunningham.

FOR FURTHER DESCRIPTIVE MATTER see Manual for 1889-90. CHANGES since 1890 are here given. POPULATIONS are for 1881 and 1891, respectively.

2. **ANNAPOLIS,** *Annapolis Co.* (Pop.,'800—est., 1,000.) Built in '89 by town. Supply.—Springs, by gravity to reservoir. Fiscal year closes May 15. Distribution.—*Mains,* 5.3 miles. *Taps,* 220. *Hydrants,* 23. *Consumption,* 25,000 galls. Financial.—*Cost,* $26,500. *Bonded Debt,* $27,500 at 5%. *Expenses;* operating. $400; int., $1,375; total, $1,775. *Revenue;* consumers, $1,250; city, $400; total, $1,650. **Management.**—Three comrs. Supt., R. J. Uniacke.

3. **BRIDGETOWN,** *Annapolis Co.* (Pop., 800—est., 1,000.) Built in '85 by town. Supply.—Springs, by gravity from reservoir. Fiscal year closes Nov. 1. Distribution.—*Mains,* 4.5 miles. *Taps,* 200. *Hydrants,* 22 Financial.—*Cost,* $22,000. *Bonded Debt,* $21,000 at 5%. *Expenses;* operating, $150; int., $1,050; total, $1,200. *Revenue,* $1,000. **Management.**—Three comrs., one elected every year. Chn., Dan'l Palfrey. Supt., Jno. L. Cox.

4. **HALIFAX,** *Halifax Co.* (Pop., 26,100—38,556.) Owned by city. Built in '45-8 by co.; bought by city in '61. Supply.—Long, Upper and Lower Chain lakes for low and Spruce Hill lakes for high service. by gravity. Fiscal year closes April 30. Distribution.—*Mains,* 49.8 miles. *Taps,* 11,244. *Meters,* 31. *Hydrants,* 343. *Consumption,* 6,500,000 galls. Financial.—*Cost,* $820,0 0. *Bonded Debt,* $802,000. *Expenses;* operating, $15,500; int., $45,000; total, $60,500. *Revenue,* consumers, $70,000. **Management.**—Board of public works; six members, one from each ward, appointed by mayor every year. Chn., J. D. Mac Kintosh. Supt. and Engr., F. W. W. Doane.

5. **KENTVILLE,** *Kings Co.* (Pop., 1,285—1,686.) Built in '87 by city. Supply. Mill Brook, by gravity to reservoir. Fiscal year closes Dec. 31. Distribution.—*Mains,* 5½ miles. *Taps,* 151. *Hydrants* (in '88), 29. Financial.—*Cost,* $30,000. *Bonded Debt,* $30,000 at 5%. *Expenses;* operating, $294; int., $1,500; total, $1,794. *Revenue,* consumers, $887. **Management.**—Mayor, H. B. Webster. Town Clerk, Wm. Eaton. Supt., Watson Bishop.

6. **NEW GLASGOW,** *Pictou Co.* (Pop., 2,595—3,777.) Built in '87-8 by town. Supply.—East River, pumping to reservoir; filtered. Filter bed is of sand and gravel about 1,600 sq. ft. Fiscal year closes Dec. 31. One 1,500,000-gall. pump and 112 HP. boiler were added. Distribution.—*Mains* 10 miles. *Taps,* 1,200. *Meters,* 10. *Hydrants,* 47. *Consumption,* 275,600 galls. Financial.—*Cost,* $80,000. *Bonded Debt,* $80,000. *Expenses,* $4,500. *Revenue,* $5,270. **Management.**—Committee, appointed by council every year. Chn., W. W. McIntosh. Secy., A. M. Fraser. Supt., D. Ormiston.

PICTOU, *Pictou Co.* (Pop., 3,403—2,999.) System is not yet under construction but engineer has been appointed to make surveys. See 1889-90 Manual, p. 654. Address Geo. D. Ives, Town Clerk.

7. **TRURO,** *Colchester Co.* (Pop., 3,461—5,102.) Built in '75 by town. Supply —Surface water, by gravity and direct pumping. Fiscal year closes Dec. 31. Distribution.—*Mains,* about 8 miles. *Taps,* about 450. *Hydrants,* 45. Financial.—*Bonded Debt,* $61,500. *Operating Expenses,* $1,578. *Revenue,* consumers, $2,525. **Management.**—Com., Chn. S. T. Chambers, W. E. Bligh, A. E. McKay.

8. **WINDSOR,** *Hants Co.* (Pop., 6,561—10,322.) Built in '83 by town. Supply. —Brook, by gravity, from impounding reservoir. Fiscal year closes Jan. 1.

Year.	Mains.	Taps.	Hydrants.	Cost.	Bonded debt.	Int. cb'g's.	Exp.	Rev.
'84	8	240	44	$50,000	$50,000	$2,500	....	.....
'88	11	500	45	54,000	52,000	2,600	$650	$3,700
'90	11	520	46	50,000	52,000	2,600	....	.....

Financial.—*Bonded Debt,* $52,000 at 5%, due in 1902. *Sinking Fund,* $7,000, invested in debentures, at 4 and 5%. **Management.**—Secy., Fred. W. Drinck. Supt., J. P. Smith, appointed by council.

9. **WOLFVILLE,** *Kings Co.* (Pop., 900—not given.) Built in '89-90 by town. Supply.—Duncan's Brook, by gravity from impounding to distributing reservoir. Fiscal year closes Jan. 31. Distribution.—*Mains,* 5 miles. *Taps,* 50. *Hydrants,* 14. Financial.—*Cost,* $25,000. *Bonded Debt,* $25,000 at 4½% due in 1910. *Annual Sinking Fund,* $200. Other data not determined at time of report. **Management.**—Three Comrs., one elected every year. Chn., Fred Brown. Secy., A. W. Barss.

NOVA SCOTIA.

11. YARMOUTH, *Yarmouth Co.* (Pop., 3,485—6,089.) Built in '81 by Yarmouth Water Co. **Supply.**—Lake George, by gravity. Fiscal year closes Dec. 31. **Distribution.**—*Mains,* 25 miles. *Taps,* 722. *Hydrants,* 0. **Financial.**—*Cost,* $217,000. *Cap. Stock,* $200,000. *Bonded Debt,* $200,000. *Expenses,* $1,182. *Revenue,* $7,881. **Management.**—Prest., Hugh D. Cann. Secy., Geo. Bingay. Supt., Geo. H. Robertson.

Water-Works Projected with Fair Prospects of Construction.

DARTMOUTH, *Halifax Co.* (Pop., 3,786—6,240.) Town contemplated building works. Lakes, by gravity. Proposals open till Sept. 1. Address Mayor Frederick Scarfe.

MIDDLETON, *Annapolis Co.* (Pop., not given.) Middleton Water Supply Co. was organized in '90, preliminary surveys made and water rights secured, under 20-yrs. exclusive franchise. Supply, reservoir, fed by springs, by gravity. Mains, 4 miles. Cost, $17,500. Bonded Debt, $10,000 at 5%, guaranteed by town. Prest., Hector McLean; Secy., Jno. Erwin; both of Bridgetown.

SPRINGHILL, *Cumberland Co.* (Pop., not given.) Bids were received till June 15, '91, for construction of complete system. No later report. Town Clerk, Daniel McLeod.

STELLARTON, *Pictou Co.* (Pop., 2,297—2,410.) In '91 village voted $11,000 for construction of system, deriving supply from New Glasgow works. Town Clerk, Don Gray.

PRINCE EDWARD ISLAND.

Water-Works Completed.

1. CHARLOTTETOWN, *Queens Co.* (Pop., 11,485—11,374.) Built in '88-9 by city. **Supply.**—Surface well, pumping to reservoir. Fiscal year closes Dec. 31.

Year.	Mains.	Taps.	Hy'd'ts.	Cons'p't'n.	Cost.	B'd'd d't.	Int.ch'g's.	Exp.	Rev.
'88	15	136	87	.....	$155,277	$153,000	......	......	......
'89	16	1,018	87	220,000	177,902	165,000	$7,025	$3,031	$4,622
'90	17	1,347	87	215,000	192,521	175,000	9,130	3,172	9,226

Management.—Comrs., David Laird, Jno. Kelley, Alex. McKinnon.

NEWFOUNDLAND.

1. CARBONEAR. (Pop., 3,000—5,300.) Built in '83-4 by Carbonear Water Co. Government has no control of rates but may purchase works; price determined by arbitration. **Supply**—Small lake, by gravity. Fiscal year closes first Monday in July. **Distribution.**—*Mains,* 5 miles. *Taps,* 46. *Meters,* 0. *Hydrants,* 19. **Financial.**—*Cost,* $34,000. *Cap. Stock:* authorized, $50,000; paid up, $34,000. *Bonded Debt,* $34,000 at 4%. *Expenses:* operating, $1,850; int., $1,360; taxes, $17; total, $3,227. *Revenue:* tax of 8% on appraised property, to be paid by property holders within 200 yards of pipes, amounting to $1,360. **Management.**—Prest., Jno. Rorke. Treas., Alfred Penny. Secy. and Supt., Jas. P. Guy.

2. HARBOR GRACE, *District of Harbor Grace.* (Pop., in '80, 6,770.) Built in '63 by co. **Supply.**—Bannerman Lake, by gravity. **General Description.**—(In '84.) *Mains,* 5 miles. *Taps,* 102. *Hydrants,* 38. *Debt,* $52,400 at 5%. Secy., C. Watts.

3. HEARTS CONTENT, (Pop., in '80, 880.) Built in '83 by Anglo American Telegraph Co., who have built up village. No further information can be obtained.

4. SAINT JOHN'S, *District of St. John's.* (Pop., in '80, 21,000—39,000.) Built in '59-'62 by city. **Supply.**—Pond, by gravity. **Distribution**—(In '88.) *Mains,* 20 miles. *Taps,* 1,850. *Hydrants,* 162. *Consumption,* 3,500,000 galls. **Financial.**—(In '88.) *Cost,* $412,000. *Bonded Debt,* $400,000 at 4%. *Expenses:* operating, $7,200; int., $16,000; total $23,200. *Revenue* (in '86), consumers, $20,900. **Management.**—(In '89.) Chn., J. Goodfellow.

FOR FURTHER DESCRIPTIVE MATTER see Manual for 1889-90. CHANGES since 1890 are here given. POPULATIONS are for 1881 and 1891, respectively.

**GROUP 10.—CANADA: Interior and Western Provinces.—
Water-Works Completed, In Procss of Construction, or Projected, in Quebec, Ontario, Manitoba, Northwest Territory,
and British Columbia.**

QUEBEC.

Water-Works Completed or In Process of Construction.

1. **BEAUHARNOIS,** *Beauharnois Co.* (Pop., 1,419—1,590.) No sewers; has electric lights. History.—Built in '90 by town, put in operation in Oct. Engr., J. Emile, Vanier. Supply.—St. Lawrence River, pumping direct. Water is taken 2,000 ft. from shore. Pumping Machinery.—Dy. Cap., 750,000 galls.; Blake comp. cond., dup. Distribution.—*Mains,* 10 to 4-in. c. i., 3.2 miles. *Services,* lead. *Taps,* 320. *Meters.* 0. *Hydrants,* Boaumont, 13. *Valves,* Chapman. *Consumption,* 85,000 galls. *Pressure:* ordinary, 40 lbs.; fire. 90 lbs. Financial.—*Cost,* $32,000. *Debt:* bonded, $30,000; floating, $2,000; int. 5%. Expenses and revenue, not determined. Management.—Secy. and Treas., L. C. Tarsé. No Supt. Report by J. Emile Vanier, Contr.

2. **BERTHIER,** *Berthier Co.* (Pop., 2,156-1,537.) Built in '79 by J. U. Foucher. Supply.—St. Lawrence River, pumping to tank and direct. Distribution.—*Mains,* 3 miles. *Taps,* 300. *Hydrants,* 15. Financial.—*Cost,* $22,000. *Expenses* $1,000. *Revenue,* $2,400. Management.—Owner, J. U. Foncher. Supt., Gordien Coutu.

3. **CHAMBLY,** *Chambly Co.* (Pop., not given.) History.—Built in '91 by co., under 25-yrs. franchise; to be in operation Dec. 1. Engr. and Contr., L. de la Vallee Poussin. P. O. Box 1889, Montreal. Supply.—Lake, pumping to reservoir. Pumping Machinery.—Dy. cap., 20,000 galls.; from A. R. Williams, Toronto, Ont. Reservoir.—Cap., 60,000 galls. Distribution.—*Mains,* 5 to 2½-in. c. i., 3 miles. *Services,* lead. *Taps,* 200. *Hydrants,* 10. *Consumption,* 10,000 galls. *Pressure:* ordinary, 40 lbs.; fire, 75 lbs. Financial.—*Cost,* $10,000. *Cap. Stock:* authorized, $15,000; paid-up, $10,000. *Expenses,* $500. *Revenue;* consumers, $800; city, $650; total, $1,450. *Hydrant Rental,* $65. Management.—Prest., O. O'Hegny. Supt., L. de la Vallee Poussin.

12. **COTE DE LA VISITATION,** *Montreal Co.* (Pop., not given.) Supply from Montreal, through meter.

12. **COTE DE SAINT LOUIS,** *Montreal Co.* (Pop., 1,571—2,972.) Supply from Montreal, through meter.

4. **COTE SAINT ANTOINE,** *Montreal Co.* (Pop., 884—3,076.) Built in '83 by St. Antoine Water Co.; now owned by Montreal Water & Power Co. Supply,—Montreal Water-Works, through meter. Distribution.—(In '88.) *Mains,* 3 miles. *Taps,* 180. *Meters,* 180. *Hydrants,* 4. Financial.—(In '88.) *Cost,* $19,040. *Debt,* $3,000. *Expenses,* $908. *Revenue,* $3,209. Management.—Montreal Water & Power Co.

5. **DRUMMONDVILLE,** *Drummond Co.* (Pop., 900—1,955.) Built in '87-8 by Comtois & Poirier. Supply.—St. Francis River, pumping to reservoir and tanks. Distribution.—(In '88.) *Mains,* 3 miles. *Taps,* 90. *Hydrants.* 12. Financial.—In '88.) *Cost,* $7,000. *Debt,* $5,000 at 8%. *Expenses:* operating, $150; int., $400; total. $550. *Revenue,* $1,000. Management.—(In '89.) Prop'rs., Comtois & Poirier.

6. **FARNHAM,** *Missisquoi Co.* (Pop., not given.) History.—Built in '91 by city. to be in operation Dec. 1. Des. Engr., V. N. Farnham. Const. Engr., L. de la Vallee Poussin, P. O. Box 1889, Montreal. Contrs., Three Rivers Water Pipe F'dry. Supply.—River, pumping direct. Pumping Machinery.—Dy. cap., 1,000,000 galls. Contr., A. Williams, Toronto. Distribution.—*Mains,* 10 to 4-in. c. i., 6 miles. *Services,* lead. *Taps,* 350. *Meters,* 0. *Hydrants,* Montreal, 30. *Valves,* Scott. *Consumption,* 300,000 gals. *Pressure:* ordinary, 50 lbs.; fire, 110 lbs. Financial.—*Cost,* $36,000. Management.—Not given.

QUEBEC.

7. HULL, *Ottawa Co.* (Pop., 6,890—11,265.) Built in '86 by town. **Supply.**—Ottawa River, pumping to tank and direct. Fiscal year closes Nov. 30. Two additional Knowles pumps and a third boiler, put in in '90.

Year.	Mains.	Taps.	Meters.	H'd'ts.	C'ns't'n.	Cost.	Debt.	Int. ch'g's.	Exp.	Rev.
'86	3	30	0	16	......	$40,000	$40,000	$2,000	......	......
'87	7	38	0	12	600,000	95,780	40,000	2,000	$3,140	$5,987
'88	6.5	996	0	47	1,000,000	118,000	110,000	5,500	5,400	11,000
'90	7	1.106	0	63	1,000,000	180,000	180,000	9,000	10,203	11,370

Financial.—*Bonded Debt*, $180,000 at 5%, due in 25 years. Bonds are redeemed from ordinary city revenue. *Sinking Fund*, $4,500 at 3½%. **Management.**—Com. of five, appointed by council every year. Chn. Geo. G. V. Ardouin. Secy., J. O. Laferriere. Supt., L. Genest, appointed by city council.

8. IBERVILLE, ——— *Co.* (Pop., 1,847—1,719.) Built in '81 by town. **Supply.**—Richelieu River, pumping direct. Fiscal year closes Jan. 1. **Distribution.**—*Mains*, 6.3 miles. *Taps*, 525. *Meters*, 0. *Hydrants*, 20. *Consumption*, 220,000 galls. **Financial.**—*Cost*, $33,000. *No Debt. Expenses*, $1,500. *Revenue*, $2,500. **Management.**—Chn., E. N. Chevalier, Mayor. Secy., F. A. Lussier. Engr., Louis Biron.

9. JOLIETTE, *Joliette Co.* (Pop., 3,268—3,347.) Built in '73 by town. **Supply.**—L'Assomption River, pumping direct. **Distribution.**—(In '88.) *Mains*, 7 miles. *Taps*, 2,000. *Hydrants*, 32. *Consumption*, 150,000 galls. **Financial.**—(In '88.) *Cost*, $50,000. *Bonded Debt*, $50,000 at 5%. *Expenses:* operating, $1,000; int., $2,500; total, $3,500. *Revenue*, consumers, $8,000. **Management.**—(In '91.) Supt., Pierre La Forest.

10. LACHINE, *Jacques Cartier Co.* (Pop., not given.) **History.**—Built in '89 by town. Engr., J. Emile Vanier, 107 St. James St., Montreal. **Supply.**—St. Lawrence River, pumping direct. **Pumping Machinery.**—Dy. cap., 2,000,000 galls.; two 1,000,000 Blake comp., cond., dup. **Distribution.**—*Mains*, 14 to 1-in. c. i., 5.5 miles. *Services*, lead. Pipe and specials from Alex. Gartshore, Hamilton. *Taps*, 780. *Meters*, Crown. *Hydrants*, Beaumont, 61. *Valves*, Chapman. *Consumption*, 180,000 galls. *Pressure:* ordinary, 40 lbs.; fire, 93 lbs. **Financial.**—*Cost*, $95,000. *Bonded Debt*, $95,000 at 4½%. *Expenses:* operating, $2,300; int., $4,275; total, $6,575. *Revenue*, total, $9,800. **Management.**—Secy., H. Robert. Supt., H. Joly.

11. LONGUEIL, *Chambly Co.* (Pop., 2,355—2,757.) Built in '76 by town. **Supply.**—St. Lawrence River, pumping direct. Fiscal year closes Jan. 30.

Year.	Mains.	Taps.	Hyd'nts.	C'ns'pt'n.	Cost.	B'd'd d'bt.	Int. ch'g's.	Exp.	Rev.
'84	7	500	39	115,000	$80,000				
'88	7.25	560	43	120,000	80,000	$80 000	$4,800	$2,000	$4,450
'90	7.5	608	47	120,000	80,000	80,000	4,800	3,000	5,420

Management.—Com. chosen from City Council every year. Chn., Leon Vigir. Secy., L. E. Bourgeois. Supt., J. A. Gariepy, appointed by Council.

12. MAISONNEUVE, *Montreal Co.* (Pop., not given.) Supply from Montreal, through meter.

12. MONTREAL, *Montreal Co.* (Pop., 155,237—216,650.) Owned by city. Built in '01 by co., bought by city in '45. **Supply.**—St. Lawrence River, through canals to pumps, thence, by pumping to reservoirs and direct. Fiscal year closes Dec. 31. Pumping machinery is divided into low and high service, as follows—*High Service:* dy. cap. 3,500,000 galls.; one 500,000 Worthington, one 3,000,000 Gilbert. *Low Service:* dy. cap., 13,000,000 imp. galls.; three turbines with seven pumps, one breast wheel with three pumps.

Year.	Mains.	Taps.	Meters.	Hyd'nts.	Cons'ption.	Cost.	Exp.	Rev.
'65	84	11,683	...	558	4,063,000	......	$26,039	......
'75	12.4	21,573	294	780	8,785,217	......	......	$330,185
'80	132.8	25,752	305	865	9,691,900	......	60,351	366,475
'82	133.5	26,508	361	886	9,566,750	......	68,443	414,818
'83	137.4	26,975	413	933	10,552,174	......	69,599	427,783
'84	139.2	27,627	468	941	10,637,037	......	76,415	463,623
'85	141	28,369	508	955	11,970,500	$6,131,888	80,000	443,900
'86	144.5	29,981	549	995	12,643,000	6,294,308	81,686	521,628
'87	149.6	32,176	600	1,043	13,055,000	6,400,000	90,000	545,700
'88	161.1	34,643	621	1,207	13,240,300	6,645,627	91,569	591,873
'89	167.5	37,577	658	1,403	12,860,900	6,874,000	93,535	637,386
'90	178	40,403	693	1,410	14,400,000	7,443,329	121,186	745,207

Financial.—Debt is consolidated and irredeemable. Revenue paid into general fund of city. **Management.**—Com. of seven, chosen from City Council by its members for one year. Chn., Thos. Conroy. Supt., B. D. McConnell.

FOR FURTHER DESCRIPTIVE MATTER see Manual for 1889-90. CHANGES since 1890 are here given. POPULATIONS are for 1881 and 1891, respectively.

13. **QUEBEC,** *Quebec Co.* (Pop., 63,417—83,000.) Built in '51-3 by city. **Supply.**—St. Charles River, by gravity.

Yr.	Mains.	Taps.	Meters.	H'd'ts.	C'n'p'n.	Cost.	B'd'd d't.	Int. ch'g's.	Exp.	Rev.
'80	27.5	5,016	....	222	3,000,000	$1,300,000	$1,300,000	......	$10,000	$90,000
'88	3.3	6,119	0	251	10,000,000	2,000,000	1,750,000	$1,005,000	25,000	143,000

Management.—Supt., C. Balllarge.

14. **RICHMOND,** *Richmond Co.* (Pop., 1,5.1—2,050.) Built in '89 by Lafontaine & Roussin, Boynton Falls; city has right to purchase works in 1900, but does not control rates. **Supply.**—Source not given, by gravity from reservoir; filtered. Fiscal year closes Dec. 31. **Distribution.**—*Mains,* 2.5 miles. *Taps.* 300. *Hydrants.* 15. **Financial.**—*Cost,* $16,500. *Cap. Stock:* authorized, $16,500; all paid up; profits, in '90, $1,800. *No Debt. Expenses,* $25. **Management.**—Lafontaine & Roussin, Boynton Falls.

15. **SAINT CUNEGONDE,** *Hochelaga Co.* (Pop., 4,819—9,293.) Bought by city in '91; built in '79 by Berger & Béique Co. **Supply.**—St. Lawrence River, pumping direct. Fiscal year closes Dec. 81.

Year.	Mains.	Taps.	Meters.	Hyd'ts.	C'ns'pt'n.	Cost.	B'd'd d'bt.	Int. ch'g's.	Exp	Rev
'84	8.3	1,100	7	14	180,000	$125,000		......	......	......
'88	11.5	1,525	8	18	250,000	175,000	......	......	......	$21,132
'90	17	5,000	18	24	1,200,000	425,000	......	......	$30,000	42,000

Financial.—Withheld, except data above. **Management.**—Not given.

16. **SAINT HENRI,** *Montreal Co.* (Pop., 6,415—13,415.) Bought by city in '91. Built in '80 by Berger & Béique. **Supply.**—Water works at St. Cunegonde. **Distribution.**—*Mains,* 16 miles. *Taps,* 3,000. *Meters,* 125. *Hydrants,* 43. *Consumption,* 750,000 galls. **Financial.**—*Cost,* $155,000. *Revenue,* $26,000. **Management.**—Not given.

17. **SAINT HILAIRE,** *Rouville Co.* (Pop., not given.) **History.**—Built in '91 by co., under 25-yrs. franchise. Engr., L. de la Vallée Poussin, P. O. Box 1289, Montreal. Co.,trs.. Three Rivers Pipe Foundry. **Supply.**—Source not given, by gravity. **Distribution.**—*Mains,* 5 to 2½-in. c. i., about 2 miles. *Services,* lead. *Taps,* 227. *Hydrants,* 6. *Valves,* Scott. *Pressure,* 70 lbs. **Financial.**—*Cost,* $15,000. *Cap. Stock:* authorized, $15,000; paid up, $10,000. *Hydrant rental,* $60. **Management.**—Prest., M. Goulet, Mgr., L. de la Vallee Poussin.

18. **SAINT HYACINTHE,** *St. Hyacinthe Co.* (Pop., 5,321—7,016.) Built in '75 by St. Hyacinthe Water-Works Co. **Supply.**—Yamaska River, pumping direct. Fiscal year closes Feb. 1. **Distribution.**—*Mains,* 20 miles. *Taps,* 1,200. *Hydrants,* 70. *Consumption,* 750,000 galls. **Financial.**—*Cap. Stock* (in '88), $50,000. *Expenses,* $4,750. *Revenue;* consumers, $10,000; fire protection, $600; other purposes, $600; total. $11,200. **Management.**—Pres., Louis Cote. Supt., M. A. Connel.

19. **SAINT JEROME,** *Terrebonne Co.* (Pop., 2,032—2,868.) Built in '79-80 by town. **Supply.**—spring, by gravity. *Mains,* 7½ miles. *Cost,* $20,000. *Expenses,* $400. *Revenue,* $4,000. Supt., J. J. H. Le Clair. Latest report in '89.

20. **SAINT JOHNS',** *St. Johns' Co.* (Pop., 4,314—4,772.) Built in '72 by Louis Molleur's Sons. **Supply.**—Richelieu River, pumping direct and to reservoir. Fiscal year closes Nov. 6. **Distribution**—*Mains,* 16 miles. *Taps* (in '88), 950. *Hydrants* 39. *Consumption,* 500,000 galls. **Financial.**—*Cost,* $140,000. *No Debt. Expenses,* $1,800. *Revenue,* $12,000. **Management.**—Owner P. T. L'Heureux. NOTE.- In '91 city talked of buying the works.

12. **SAINT LOUIS DU MILE END,** *Montreal Co.* (Pop., 1,537—3,537.) Supply from Montreal, through meter.

21. **SHERBROOKE,** *Sherbrooke Co.* (Pop., 7,227—10,110.) Built in '80-1 by Sherbrooke Gas & Water Co.; franchise fixes rates and provides for purchase of works by city in 1900, price determined by arbitration. **Supply.**—Magog River, pumping to reservoir; filtered through gravel. Fiscal year closes Sept. 15.

Year.	Mains.	Taps.	Meters.	Hydrants.	Cons'pt'n.	B'd'd d'bt.	Int. ch's.	Exp.	Rev.
'81	5	160	0	38	......	......	......	......	......
'84	7	220	0	38	$100,000	......	......	$1,200	$5,000
'86	7.5	260	1	38	168,000	......	......	......	......
'88	8.5	400	7	57	300,000	$35,000	......	......	8,400
'90	14	700	12	60	500,000	50,000	$3,000	......	11,700

Financial.—*Cost,* $125,000. *Cap. Stock:* authorized, $20,000; all paid up; paid 5% div'd in 1890. *Bonded Debt,* $50,000 at 6%. **Management.**—Pres., R. W. Hall. Secy., E. F. Waterhouse. Supt., Andrew Sangster.

QUEBEC.

22. SOREL, *William Henry Co.* (Pop., 5,792—6,609.) Built in '72-3 by town. **Supply.**—Richelieu River, pumping direct. Fiscal year closes Dec. 31. **Distribution.**—*Mains* (In '88), 12 miles. *Taps*, 1,230. *Hydrants*, 48. *Consumption*, 300,000 galls. **Financial.**—*Cost*, $80,000. *Bonded Debt*, $40,000 at 6%. *Expenses:* operating, $2,808; int., $2,400; total, $5,208. *Revenue*, $7,649. **Management.**—Com. Prest., A. Martin. Secy., J. G. Crebassa. Treas., Ernest Turcott. Supt., M. C. Blais.

23. THREE RIVERS, *St. Maurice Co.* (Pop., 8,670—8,334.) Built in '75 by city. **Supply.**—St. Lawrence River, pumping direct from settling reservoirs. **Distribution.**—(In '87.) *Mains*, 15 miles. *Taps*, 1.400. *Meters*, 3. *Hydrants*, 57. *Consumption*, 250,000 galls. **Financial.**—(In '87.) *Cost*, $130,000. *Bonded Debt*, $115,000 at 6%. *Expenses:* operating, $5,000; int., $6,900; total. $11,900. *Revenue*, $15,000. **Management.**—Supt., O. Z. Harnel.

24. VALLEYFIELD, *Beauharnois Co.* (Pop., 3,906—5,516.) Built in '86 by Valleyfield Water-Works Co.; bought by city in '87. **Supply.**—St. Lawrence River, pumping direct. **Distribution.**—*Mains*, 5.8 miles. *Taps*, 1,160. *Meters*, 0. *Hydrants*, 72. *Consumption*, 230,000 galls. **Financial.**—*Cost*, $74,000. *Bonded Debt*, $70,000 at 5%. *Expenses:* operating, $950; int., $3,500; total, $4,450. *Revenue*, $6,800. **Management.**—Secy., R. S. Jaron. Supt., Chas May. Report by Contr., J. Emile Vanier, 107 St. James St., Montreal.

Water Works Projected with Fair Prospects of Construction.

COTE ST. PAUL, *Hochelega Co.* (Pop., not given.) A 25-yrs. exclusive franchise was granted in Jan., '91, to Montreal Island Water & Electric Co.; construction to begin in following spring. Address Woltmann, Keith & Co., 11 Wall St., N. Y. City.

HUNTINGDON, *Huntingdon Co.* (Pop., not given.) Committee was appointed in '91 to report to City Council on construction of water-works. Chn., D. Boyd. Chateauguay River, pumping direct. In March, '91, J. E. Vanier, 107 St. James St. Montreal, reported that construction had begun.

SAINT LAMBERT, *Chambly Co.* (Pop., not given.) Jan. 22, '91, it was reported that system was projected by village. St. Lawrence River, pumping to standpipe. In March, '91, J. E. Vanier, 107 St. James St., Montreal, wrote that system was under construction.

SOMERSET, *Megantic Co.* (Pop., not given.) City was to begin construction of works Sept. 30, '91. Engr., L. de la Vallee Poussin, P. O. Box 1889, Montreal. Contrs., Three Rivers Pipe Foundry. Springs, by gravity. Mains, 4½ miles. Hydrants, 15. Pressure, 70 lbs. Cost, $23,000.

ONTARIO.

Water Works Completed or in Process of Construction.

1. AMHERSTBURG, *Essex Co.* (Pop., 2,672—2,279.) **History.**—Built in '91 by town; put in operation Sept. 15. Engr., Jno. Galt, Toronto. Contr., Miles Huntiug Hamilton. **Supply.**—Detroit River, pumping to tank. **Pumping Machinery.**—From Kerr Bros., Walkerville, Ont. **Tanks.**—Cap., not given; 16 × 20 ft. steel. **Distribution.**—*Mains*, 10 to 4-in., 22,000 ft. **Financial.**—*Cost*, $29,000. *Bonded Debt*, $27,000 at 5%. **Management.**—Town Clerk, J. H. C. Leggett.

2. AYLMER, *Ottawa Co.* (Pop., 1,540—2,167.) **History.**—Domestic supply introduced in '91 by town. Works for fire protection built in '84 by town. Engr., Alex. Milne. **Supply.**—Creek and artesian wells, pumping direct. **Pumping Machinery.**—Dy. cap., 2,250,000 galls; one 750,000 Worthington high-pressure, one 1,500,000 Northey. **Distribution.**—*Mains*, 4.3 miles. *Taps*, 83. *Hydrants*, Mathews, 24. *Valves*, Ludlow. *Pressure:* ordinary, 55 lbs.; fire, 100 to 120 lbs. **Financial.**—*Cost* $18,000. *Bonded Debt*, $12,000 at 5%. *Expenses:* operating, $1,500; int., $600; total, $2,100. *Revenue*, consumers, $400. **Management.**—Supt., Alex. Milne.

FOR FURTHER DESCRIPTIVE MATTER see Manual for 1889-90. CHANGES since 189 are here given. POPULATIONS are for 1880 and 1890, respectively.

3. BARRIE, *Simcoe Co.* (Pop., 4,854—5,550.) **History.**—Built in '90-1 by Barrie Water-Works Co.; put in operation Jan. 15. Engrs. and Contrs., Hinds & Bond, Watertown, N. Y. **Supply.**—Artesian wells, pumping to tank. **Pumping Machinery.**—Dy. cap., 2,000,000 galls ; two horizontal comp., cond., dup.; from Northey & Co., Toronto. *Boilers,* from Porter Mfg Co., Syracuse, N. Y. **Tank.**—Cap., 180,000 galls ; 25 × 50 ft.; from Dominion Bridge Co., Montreal. **Distribution.**—*Mains,* 12 to 4-in. c. i., 8 miles. *Services,* galv. i. Pipe and specials from Canada Pipe Foundry, Hamilton. *Taps,* 200. *Meters,* Ontario, 1. *Hydrants,* Ludlow, 67. *Valves,* Ludlow. *Consumption,* 125,000 galls. *Pressure:* ordinary, 75 lbs.; fire, 120 lbs. **Financial.**—Withheld. **Management.**—Prest., E. A. Bond. Secy., Frank A. Hinds. Treas., Talcott H. Camp. All of Watertown, N. Y. Supt., Wm. J. Vallence.

4. BELLEVILLE, *Hastings Co.* (Pop., 9,516—9,914.) Built in '87 by Moffett, Hodgkins & Clark, Syracuse, N. Y. **Supply.**—Lake, filtered, pumping to tank. **Distribution.**—(In '88.) *Mains,* 15.3 miles. *Taps,* 268. *Meters,* 23. *Hydrants,* 168. *Consumption,* 305,000 galls. **Financial.**—(In '88.) *Cost,* $200,000. *Debt.* $100,000 at 6%. *Revenue,* city, $7,000. **Management.**—(In '89.) Prest., J. V. Clarke, Syracuse, N. Y. Supt., Geo. A. Pope.

5. BERLIN, *Waterloo Co.* (Pop., 4,054—7,425.) Built in '88 by Moffett, Hodgkins & Clarke, Syracuse, N. Y. **Supply.**—Springs, pumping to stand-pipe from brick pump-well; connection can also be made with lake, in case of emergency. **Distribution.**—(In '88.) *Mains,* 8.69 miles. *Taps,* 189. *Meters,* 16. *Hydrants,* 79. *Consumption,* 250,000 galls. **Financial.**—(In '88.) *Cost,* $100,000. *Bonded Debt,* $73,000 at 6%. *Expenses:* operating, $3,000; int., $4,380; total, $7,380. *Revenue:* consumers, $6,000; city, $3,200; total, $9,200. **Management.**—(In 89.) Prest., C. T. Moffett. Supt., H. J. Bowman.

6. BRAMPTON, *Peel Co.* (Pop., 2,920—3,252.) Built in '81-2 by town. **Supply.**—Snell's Lake, by gravity. **Distribution.**—(In '88.) *Mains,* 4 miles. *Hydrants,* 60. *Consumption,* 300,000 galls. **Financial.**—(In '88.) *Cost,* $65,000. *Bonded Debt,* $65,000 at 5%. *Expenses:* operating, $200; int., $3,250; total, $3,450. *Revenue:* consumers, $4,100; ry., $1,500; total, $3,600. **Management.**—Chn. (In '91), Wm. Peaker.

Brantford, *Brant Co.* (Pop., 9,616—12,753.) Built in '09-70 by Brantford Water Works Co., chiefly for fire protection.

7. BROCKVILLE, *Leeds Co.* (Pop., 7,608—8,793.) Built in '81-4 by Brockville Water Co.; city does not control rates, but has right to purchase works in 1892. **Supply.**—St. Lawrence River, pumping direct. Fiscal year closes Oct. 31.

Year.	Mains.	Taps.	M't'rs.	Hyd'ts.	Con'p'n.	Cost.	B'd'd debt.	Int. ch'g's.	Exp.	Rev.
'87	9	400	4	82	800,000	$185,000	$40,000	$2,400	$5,000	$13,050
'90	11.25	620	6	87	884,728	200,000	38,270	2,296	7,087	16,000

Financial.—*Cap. Stock:* authorized, $200,000; paid-up, $198,000. *Bonded Debt,* $28,-270, at 6%. Co. has paid but one coupon on debenture, all earnings being used for extension of works. *Expenses:* operating, $7,087; int., $2,296; total, $9,383. *Revenue:* consumers, $9,410; fire protection, $4,300; rys., $2,290; total, $16,000. **Management.**—Prest., W. E. C. Cook. Secy. and Supt., B. F. Steben.

Campbellford, *Northumberland Co.* (Pop., 1,418—2,424.) Built in '89-90 by village for fi·e protection only.

8. CHATHAM, *Kent Co.* (Pop., 7,873—9,052.) Built in '89-91 by Chatham Water-Works Co.; rates are fixed in franchise; town has right to buy works after 10 yrs., price to be determined by arbitration. **Supply.**—Driven wells, pumping to stand-pipe. Fiscal year closes Jan. 1. **Distribution.**—*Mains,* 13. *Hydrants,* 168. **Financial.**—*Cost,* est., $200,000. *Cap. Stock:* authorized, $200,000; paid-up, $180,000. Other data not determined as works were not completed at time of report, Jan 22, '91. **Management.**—Prest, Frank A. Hinds; Secy., Edw. A. Bond ; Treas., Geo. B. Phelps ; all of Watertown, N. Y. Supt., Chas. A. McComb.

9. COBOURG, *Northumberland Co.* (Pop., 4,957—4,829.) Built in '88-9 by Cobourg Water-Works Co., controlled by Moffett, Hodgkins & Clarke, Syracuse, N. Y. **Supply.**—Lake Ontario, pumping through canal to stand-pipe and direct. **Distribution.**—(In '89.) *Mains,* 8 miles. *Taps,* 40. *Hydrants,* 70. **Financial.**—(In '89.) *Cost,* $100,000. *Bonded Debt,* $70,000 at 6%. *Expenses:* operating, $1,500; int., $4,200; total, $5,700. *Revenue:* consumers, $5,000; city, $3,000; total, $8,000. **Management.**—(In '89.) Prest., J. F. Moffett, Syracuse, N. Y. Supt., A. J. Howson.

ONTARIO.

10. COLLINGWOOD, *Simcoe Co.* (Pop., 4,445–4,941.) Built in '89-90 by town.
Supply.—Lake Huron, pumping direct. Fiscal year closes Dec. 31. **Distribution.**
—*Mains,* 8.3 miles, *Taps,* 293. *Hydrants,* 64. *Consumption,* 90,000 galls, **Financial.**—*Cost,* $58,315. *Bonded Debt,* $78,000 at 5%, due in 30 yrs. *Expenses:* operating.
$1,826; int., $3,900; total, $5,726. *Revenue:* consumers (for 6 mos.), $803; city, $2,700.
Management.—Supt., Thos. Bennett.

11. CORNWALL, *Stormont Co.* (Pop., 4,468–6,805.) Built in '86 by Cornwall
Water Co., controlled by Moffett, Hodgkins & Clarke, Syracuse, N. Y. **Supply.**—
St. Lawrence River, pumping to stand-pipe. **Distribution.**—(In '88.) *Mains,* 9
miles. *Taps,* 197. *Meters,* 5. *Hydrants,* 62. *Consumption.* 260,000 galls. **Financial.**—(In '88.) *Cost,* $125,000. *Bonded Debt.* $80,000 at 6%. *Expenses:* operating,
$2,500; int., $4,800; total, $7,300. *Revenue:* consumers, $5,000; city, $3,600; total,
$8,600. **Management.**—(In '89.) Prest., J. V. Clarke; Secy., H. C. Hodgkins: both
of Syracuse, N. Y. Supt., Jas. Strickland.

12. DUNDAS, *Wentworth Co.* (Pop., 3,709–3,516.) Built in '83 by town. **Supply.**—Springs, by gravity from receiving reservoir. **Distribution.**—*Mains,* 6 miles.
Taps, 70. *Meters,* 0. *Hydrants,* 47. **Financial.**—*Cost,* $40,000. *Bonded Debt.*
$35,000 at 7%, due in 1912. *Sinking Fund,* $6,000; $3,000 invested in bank at 4%; $3,000
in debentures at 5%. *Expenses:* operating, $300; int., $1,750; total, $2,050. *Revenue,*
consumers, $565. **Management.**—Prest., R. T. Wilson. Secy., Alex. McKenzie.
Supt., Alex. Bertram.

13. GALT, *Waterloo Co.* (Pop., 5,187–7,535.) **History.**—Built in '90-1 by town;
put in operation Dec. 1. Engr., Willis Chipman, Toronto; Contrs., Garson & Purser,
St. Catharines. **Supply.**—Spring, pumping to stand-pipe. **Pumping Machinery.**—
Dy. cap., 1,000,000 galls.; comp. cond.; from Goldie & McCulloch. **Stand-Pipe.**—
Cap., not given; 30 × 50 ft.; from Goldie & McCulloch. **Distribution.**—*Mains,*
c. 1., 10 miles. *Hydrants,* Ludlow, 88. *Valves,* Ludlow. *Pressure:* ordinary, 115
lbs.; fire, 150 lbs. **Financial.**—*Cost,* $105,000. *Bonded Debt,* $105,000 at 4½%. **Management.**—Town Clerk, I. G. Dykes.

GANANOQUE, *Leeds Co.* (Pop., 2,871–3,669.) Built by town for fire protection.
NOTE.—Town was to receive bids for constructing works until Oct. 20, '90.
Engr., Willis Chipman, Toronto. Pump and stand-pipe were projected at est. cost
of $43,000.

14. GODERICH, *Huron Co.* (Pop., 3,444–2,907.) Built in '88 by city. **Supply.**
—Artesian wells, and Lake Huron in case of emergency, pumping direct. Fiscal
year closes Dec. 31. **Distribution.**—*Mains,* 10 miles. *Taps,* 300. *Meters,* 0. *Hydrants.* 52. **Financial.**—*Cost,* $60,000. *Bonded Debt,* $60,000, at 5%, due in 30 yrs.
Bonds are redeemed and $700 of expenses paid by taxation. *Sinking Fund,* $1,626.
Expenses: operating, $2,800; int., $3,000; total, $5,800. *Revenue,* consumers, $3,000.
Management.—Committee of five, appointed by Town Council from its members
for one year. Chn. and Supt., Jno. Butler. Secy., W. Mitchell.

15. GUELPH, *Wellington Co.* (Pop., 9,890–10,539.) Built in '79-80 by city.
Supply.—Springs and surface water, with river connection for emergency, by
direct pumping from reservoirs. **Distribution.**—*Mains,* 16 miles. *Taps,* 520. *Meters,*
4. *Hydrants,* 108. *Consumption,* 500,000 galls. **Financial.**—*Cost,* $125,000. *Bonded
Debt,* $80,000, at 6%. *Expenses:* operating, $5,000; int., $4,800; total, $9,800. *Revenue:*
consumers, $8,000; city, $2,100; total, $10,100. **Management.**—Supt., Jas. Fordyce.

16. HAMILTON, *Wentworth Co.* (Pop., 35,961–47,252.) Built in '56-9 by city.
Supply.—Lake Ontario. pumping direct and to reservoir; filtered. Fiscal year
closes Dec. 31.

Year.	Mains.	Taps.	M'rs.	H'd'ts.	C'ns't'n.	Cost.	B'd'd d't.	Int ch'g's.	Exp.	Rev.
'82	57	8,150	3	373	2,568,000	$1,280,000	$991,578	$50,512	$23,000	$107,730
'86	72	10,000	0	475	2,750,000		900 000	54,000	26,000	134,000
'89	76	23,000	5	573	3,700,000	1,513,258	972,485	57,825	31,083	139,782

Management.—Com. of seven aldermen. Secy., W. Monk. Sup'., Wm. Haskins,
City Engr.

17. INGERSOLL, *Oxford Co.* (Pop., 4,318–4,191.) Built in '90 by Ingersoll
Water-Works Co., controlled by Moffett, Hodgkins & Clarke, Syracuse, N. Y. **Supply.**—Choate's Pond and springs, pumping to stand-pipe. **Distribution.**—*Mains,* 7

FOR FURTHER DESCRIPTIVE MATTER see Manual for 1889-90. CHANGES since 1890
are here given. POPULATIONS are for 1880 and 1890, respectively.

miles. **Hydrants**, 65. **Financial.**—Withheld. **Management.**—Prest., J. F. Moffett; Secy., H. C. Hodgkins; both of Syracuse, N. Y. NOTE.—Latest report, May 29, '89.

18. **IROQUOIS,** *Dundas Co.* (Pop., 1,001—1,047.) Built in '85-6 by R. H. Buchanan and Gordon Serviss. **Supply.**—St. Lawrence River, pumping direct. Fiscal year closes June 30. **Distribution.**—*Mains*, 1.5 miles. *Consumers*, 75. **Hydrants**, 15. **Financial.**—*Cost*, $18,000. *Cap. Stock*, $18,000. *No Debt. Expenses,* $450. *Revenue,* $1,650. **Management.**—Prest., R. H. Buchanan. Supt., Gordon Serviss.

19. **KINCARDINE,** *Bruce Co.* (Pop., 4,506—3,618.) **History.**—Built in '90 by Moffett, Hodgkins & Clarke, Syracuse, N. Y. Des. Eng., C. I. Bowman. **Supply.**—Lake, pumping to stand-pipe. **Pumping Machinery.**—Dy. cap., 150,000 galls.; Worthington. Stand-Pipe,—Is 15 × 100 ft. **Distribution.**—*Mains*, 10 to 4-in., 3.5 miles. **Hydrants**, 32. **Financial.**—Withheld. **Management.**—Address Moffett, Hodgkins & Clarke, Syracuse, N. Y.

20. **KINGSTON,** *Frontenac Co.* (Pop., 21,725—24,269.) Owned by city. Built in '51 by co. **Supply.**—Lake Ontario, pumping to reservoir and stand-pipe. Fiscal year closes Dec. 31. In '90, a 16 in. force main, 2 miles long, was laid. A steel tank, 80 x 40 ft., has been built on foundation 107 ft. above Lake Ontario, affording pressure of 60 lbs. Jan. 1, 90, rates were reduced one-half. In '91, contract for laying new suction pipe was let.

Year	Mains	Taps	M't'rs	Hyd'ts	C'ns'pt'n	Cost	B'd'd d't	Int. ch'g's	Exp.	Rev.
'82	10	816	20	42	750,000				$8,325	$19,030
'84	11	850	25	42	800,000				9,000	20,030
'86	12.5	880	30	45	1,000,000				10,000	21,000
'88	14	1,200	32	100	1,000,000	$215,000	$200,000	$11,700	6,600	24,030
'83	17.5	1,350	38	150	1,000,000	252,000	260,000	11,700	6,480	25,123

Management.—Com. of seven appointed every year by city council. Chn., J. S. Muckleston. Secy., F. Smith. Supt., T. Hewitt.

21. **LEAMINGTON,** *Essex Co.* (Pop., 1,411—1,910.) **History.**—Built in '91 by town; put in operation Aug. 31. Eng., Alex. Baird. Contr., J. H. Armstrong, Toronto. **Supply.**—Artesian wells, pumping to reservoir. **Pumping Machinery.**—Dy. cap., 100,000 galls.; from Kerr Bros., Walkerville. **Reservoir.**—Cap., 40,000 galls. **Distribution.**—*Mains*, 10 to 4-in., 2.5 miles. **Hydrants**, 27. *Pressure:* ordinary, 45 lbs.; fire, 90 lbs. **Financial.**—*Cost,* $16,000. *Bonded Debt,* $15,000 at 5%. **Management.**—Town Clerk, J. F. McKay. Supt., Jas. Robson.

Lindsay, *Victoria Co.* (Pop., 5,080—6,081.) Built in '74 by town for fire protection.
NOTE.—Aug. 14, '91, it was reported that town wished to grant franchise for complete system.

Listowell, *Perth Co.* (Pop., 2,688—2,587.) Built in '84 by town for fire protection.

22. **LONDON,** *Middlesex Co.* (Pop., 19,746—22,281.) Built in '78 by city. **Supply.**—Springs, pumping to reservoir and direct. Fiscal year closes Nov. 30. In '90 pumping cap. was increased to 1,500,000.

Year	Mains	Taps	Meters	Hyd'ts	Cost	B'd't d't	Int. ch'gs	Exp.	Rev
'82	37	2,278	9	210	$390,000			$8,500	$25,000
'83	40	2,708	15	220	398,819			16,000	25,000
'85	..	3,454	..	240	442,259	$421,629	$24,482	11,596	29,539
'86	..	3,919	..	250	500,193	464,628	26,882	12,446	32,915
'87	50	4,317	25	270	514,827	464,628	26,882	16,016	36,130
'88	51	4,688	..	267	519,546	464,628	26,882	14,905	40,767
'89	60	4,936	..	305	526,881	464,628	26,882	10,897	42,813
'90	65	5,436	63	357	585,092	534,629	30,382	12,486	47,873

Management.—Two comrs. and mayor, elected every year. Chn., Geo. C. Davis. Secy., O. Ellwood. Supt. and Engr., Thos. H. Tracy.

Lucknow, *Huron Co.* (Pop., in '81, 718.) Built in '89-90 by town for fire protection.

23. **MARKHAM,** *York Co.* (Pop., 954—1,100.) **History.**—Built in '90-1 by village; put in operation May 1. Engr., Jno. Galt. Contrs., McMillan & Co., Toronto. **Supply.**—Springs, collected in reservoir, pumping direct and eventually to stand-pipe. **Pumping Machinery.**—Dy. cap., 1,000,000 galls.; dup.; from Inglis & Hunter, Toronto. **Distribution.**—*Mains,* 8 to 4-in., 1 mile. **Hydrants,** 15. *Pressure,* 100 lbs. **Financial.**—*Cost,* $10,000. *Bonded Debt,* $10,000. **Management.**—Chn., M. White. Clerk, R. J. Carson.

ONTARIO.

24. MERRITTON, *Lincoln Co.* (Pop., 1,796—1,813.) Built in '88 by town. Supply.—Lake Erie, by gravity to reservoir. Fiscal year year closes Dec. 31.

Year.	Mains.	Taps.	Hydrants.	Cons.	Cost.	B'd'd debt.	Int.	Exp.	Rev.
'88	5	100	38	......	$70,000	$70,000	$3,850	$450	.....
'89	5	120	38	11,200	70,0 0	68,114	......	1,800	$646
'90	5.5	150	48	27,400	71,200	63,514	......	1,602	1,193

Management.—Five comrs., elected every year. Chn., A. R. Thompson. Secy. and Supt., B. Clark.

25. MILTON, *Halton Co.* (Pop., 1,302—1,450.) Built in '87-8 by town. Supply. Lake Erie, by gravity to reservoir. Fiscal year closes Dec. 15. **Distribution.**—*Mains,* 4 miles. *Hydrants,* 17. **Financial.**—*Cost,* $23,000. *Bonded Debt,* $19,000 at 5%. *Expenses:* operating, $100; int., $950; total, $1,050. *Revenue,* $400. **Management**—Chn., G. A. Homstead.

Mitchell, *Perth Co.* (Pop., 2,284—2,101.) Built in 74 by town for fire protection.

26. MORRISBURG, *Dundas Co.* (Pop., 1,719—1,859.) Built in '87 by town. Supply.—St. Lawrence River, pumping direct. Fiscal year closes Dec. 31. **Distribution.**—*Mains,* 3 miles. *Taps,* 200. *Meters,* 0. *Hydrants,* 22. *Consumption,* 250,000 galls. **Financial.**—*Cost,* $30,000. *Bonded Debt,* $25,000 at 5%, payable in 25 equal annual installments. *Expenses:* operating, $600; int., $2,500; total, $1,500. *Revenue:* consumers, $1,100; city, $1,100; total, $2,200. **Management.**—Chn., Thos. McDonald. Secy., Frank Plautz. Supt., Cyrus Casselman.

27. NAPANEE, *Lenox Co.* (Pop., 3,681—3,434.) Built in '90-1 by Napanee Water-Works Co.; franchise fixes rates; town may buy works after 10 yrs., price to be determined by arbitration. **Supply.**—Napanee River, pumping to stand-pipe. Fiscal year closes Dec. 31. **Distribution.**—*Mains,* 3 miles. *Taps,* 120. *Hydrants,* 33. *Consumption,* 120,000 galls. **Financial.**—*Cost,* $60,000. *Cap. Stock:* authorized, $80,000; all paid up. *Bonded Debt,* $40,000 at 6%, due in 1911. *Expenses:* operating, $1,000; int., $2,000; total, $3,000. *Revenue,* not determined. **Management.**—Prest., Chn., Frank A. Hinds. Secy., Edward A. Bond. Treas., Geo. B. Phelps. All of Watertown, N. Y. Supt., J. R. Dafoe.

28. NEWMARKET, *York Co.* (Pop., 2,006—2,143.) Built in '86 by town. Supply.—Artesian well, pumping to reservoir. F scal year closes Dec. 31. **Distribution.**—*Mains,* 4 miles. *Taps,* 113. *Meter,* 1. *Hydrants,* 25. *Consumption,* 52,000 galls. **Financial.**—*Cost,* $20,100. *Bonded Debt,* $20,000 at 5%, payable in equal annual payments for 30 yrs. *Expenses :* operating, $600; int., $1,000; total, $1,600. *Revenue:* consumers, $1,275; city, none. **Management.**—Com., appointed by Town Council from its members every year. Chn. and Supt., H. S. Cane. Secy., David Lloyd.

29. NIAGARA FALLS, *Welland Co.* (Pop., 2,347—3,349.) Bought by town in '84. Built in '56 by co. **Supply.**—Niagara River, pumping direct, surplus going to reservoir. Fiscal year closes Dec. 31. In '89-90, new system was built by town, using such of old pipe as would bear increased pressure. Submerged raceway, 12 × 12 ft. carries water to wooden conduit, 7 ft. in diam. which delivers water to two turbine wheels, thence through tunnel to pumps. There are two sets of pumps which force water through a 12-in. main to town, two miles distant. Engr., Wm. Kennedy, Jr. Contrs.. Kennedy & Sons, Owen Sound. Cost, $77,500. **Distribution.**—*Mains,* 8 miles. *Taps,* 500. *Hydrants,* 41. *Consumption,* 3,000,000 galls. *Pressure,* increased to about 140 lbs. **Financial.**—*Cost,* $94,500. *Bonded Debt,* $61,411 at 5%; $50,411 due in 30 yrs.; $11,000 due in annual payments of $1,000. *Expenses:* operating, $4,464; int., $2,521; total, $6,985. *Revenue:* consumers, $2,550. **Management.**—Mayor and two comrs., elected every year. Chn., Wm. McHattie. Secy. and Supt., Jno. Robinson.

Orangeville, *Wellington Co.* (Pop., 2,847—2,962.) Built in '76 by town for fire protection.

30. OTTAWA, *Carleton Co.* (Pop., 27,412—37,281.) Built in '72-4 by city. Supply.—Ottawa River, pumping direct. Fiscal year closes Oct. 31.

For Further Descriptive Matter see Manual for 1889-90. Changes since 1890 are here given. Populations are for 1880 and 1890, respectively.

874   MANUAL OF AMERICAN WATER-WORKS.

Year.	Mains.	Taps.	Hydrants.	Consumption.	Cost.	Int. chs. and exp.	Rev.
'75	30	3,690	...	917,645	....	$51,598	$47,681
'76	30.4	3,985	...	1,727,816	....	69,767	76,015
'77	31.5	5,095	...	2,187,175	....	75,761	74,961
'78	38.9	5,422	...	2,293,747	....	81,803	91,406
'79	41.6	5,574	347	2,466,547	$260,195	78,969	83,355
'80	43	5,656	327	2,743,285	1,044,682	75,226	91,412
'81	43.7	5,755	...	2,636,216	....	77,493	92,791
'82	44.2	5,833	341	2,713,531	....	80,138	98,106
'83	44.7	5,942	343	3,049,410	1,095,978	86,640	97,934
'84	45.5	6,085	345	3,112,860	1,100,683	89,597	104,225
'85	46.7	6,332	318	3,204,860	1,108,730	111,323	107,171
'86	49.6	6,642	364	3,414,775	....	124,050	113,591
'87	58.7	7,181	418	3,736 000	1,125,104	117,607	119,294
'88	63.6	7,512	434	4,246,813	1,185,007	127,004	133,105
'89	72.3	8,454	531	4,352,826	1,300,366	122,810	110,290
'90	75.3	8,925	558	4,939,521	1,162,152	129,914	151,845

Financial,—*Expenses:* operating. $47,006; int., $79,028; total, $129,914. *Revenue:* consumers, $140,437; street sprinkling, $8,511; total, $148,948. Management.—City Council. Chn., J. C. Roger. Treas., Thos. H. Kirby. Supt., Rob't. Surtees.

31. **OWEN SOUND**, *Grey Co.* (Pop., 4,759—7,497.) Owned by town since July 1, '90. Built in '80 by Parker & Norther Co. Supply.—Springs, by gravity. Fiscal year closes July 1. July 1, '90, works were bought by town for $55,000. Since then 2 miles of supply main have been laid to connect distribution with spring, also 2 miles of 6-in. distribution pipe and 30 hydrants. Hydrant rental has been reduced from $50 to $30, to be paid by property owners whose buildings are protected. Distribution.—*Mains*, 5.5 miles. *Taps*, 485. *Meters*, 7. *Hydrants*, 29. Financial.—*Cost,* $55,000. Management.—Com. of five chosen from town council by its members. Chn., Jno. Armstrong. Town Engr. and Supt., Jas. C. Kennedy.

Paisley, *Bruce Co.* (Pop., 1,154—1,328.) Built by town for fire protection.

32. **PARIS**, *Brant Co.* (Pop., 3,173—3,094.) Built in '83-4 by town. Supply.— Spring on Smith's Creek, pumping to reservoir. Fiscal year closes Dec. 15. Two springs were utilized to increase supply, and new boiler put in during '90 and new pump during '91.

Year.	Mains.	Taps.	Hydrants.	Cons'pt'n.	Cost.	H'd'd debt.	Int. ch'g's.	Exp.	Rev.
'87	1	250	25	......	$40,000	$38,000	$2,289	$1,290	$3,700
'89	:	250	25	180,000	45,952	34,211	480	1,416	2,946
'90	1.5	270	25	200,000	49,574	39,211	480	2,326	3,254

Financial—*Bonded Debt*, $39,211 at 6%. *Sinking Fund*, $500, invested in bank at 4%. Bonds are redeemed by taxation. Net revenue is applied to interest, balance is paid by taxation. Allowance for public uses of water, $1,250. Management.—Com. of five appointed every year by city council from its members. Chn., Thos. McCosh. Secy., S. Dodson. Supt., Jno. Brockbank.

33. **PENTANGUISHINE**, *Simcoe Co.* (Pop., 1,089—2,110.) History.—Built in '90-1 by town; put in operation May 1. Engr., Jno. Galt. Contrs., McQuillan & Co. All of Toronto. Supply.—Spring, pumping to tank. Pumping Machinery. —Dy. cap., 1,000,000 galls.; from Inglis & Son, Toronto. Tank.—Cap., 45,000 galls. Distribution.—*Mains*, 10 to 4-in. 1., 3 miles. *Meters*, none. *Hydrants*, 35. *Pressure:* ordinary, 60 lbs.; fire, 75 to 80 lbs. Financial.—*Cost,* $19,000. *Bonded Debt,* $19,000 at 5%. *Expenses:* operating, $300; int., $950; total, $1,250. *Revenue:* consumers, $500; city, $1,500; total, $2,000. Management.—Not given.

34. **PETERBOROUGH**, *Peterborough Co.* (Pop., 6,812—9,717.) Built in '82 by Peterborough Water Co. Supply.—Otonabee River, pumping direct. Distribution.—(In '88) *Mains*, 8.5 miles. *Taps*, 400. *Meters*, 4. *Hydrants*, 65. *Consumption*, 500,000 galls. Financial.—(In '87.) *Cost,* $80,000. *Bonded Debt,* $65,000 at 6%. *Expenses:* operating, $8,500; int., $3,250; total, $8,750. *Revenue:* consumers $5,500; city, $3,100; total, $8,600. Management.—(In '89.) Prest., Jno. Burnham. Supt., Wilson Henderson.

Pictou, *Prince Edward Co.* (Pop., 2,975—3,287.) Built in '89 by town, for fire protection.

Port Hope, *Durham Co.* (Pop., 5,588—5,042.) Built in '74 by town, for fire protection.

35. **SAINT CATHARINES**, *Lincoln Co.* (Pop., 9,631—9,170.) Built in 76'-7 by city. Supply.—Beaver Dam Creek, by gravity from reservoir; filtered during spring and fall. Fiscal year closes Dec. 31.

ONTARIO.

Year.	Mains.	Taps.	M't rs.	Hyd'ts.	C'n'pt'n.	Cost.	B'd't d'bt.	Int. ch'g's.	Exp.	Rev.
'80	15.5	546	0	123	.....	$263,071	....	.....	$2,868	$4,246
'87	20.5	1,151	6	155	500,000	275,000	$275,000	$16,500	2,995	8,050
'88	22.5	1,255	8	161	500,000	282,321	275,000	16,500	3,000	8,500
'89	22.7	1,433	10	165	680,000	283,371	305,000	17,700	4,376	9,179
'90	24.1	1,583	10	193	750,000	292,612	305,000	17,700	4,232	10,539

Financial.—*Bonded Debt*, $305,000: $200,000 at 6%, due in '96; $75,000 at 6%, due in '98; $30,000 at 4%, due in 1916. Expenses not covered by revenue are paid by frontage tax. **Management.**—Six comrs., one elected every year. Chn., L. S. Oille. Secy., A. Johnston. Supt., J. Albert Mills.

36. SAINT THOMAS, *Elgin Co.* (Pop., 8,367—10,370.) Built in '74 by town.

Supply.—Kettle Creek, pumping direct from impounding reservoir. Fiscal year closes Dec 31. Following is description of new works built in '90. **Water Supply,**—Impounding reservoir on Kettle Creek and storage reservoir having cap. of 5,500,000 galls. **Pumping Machinery.**—Cap., 4,000,000 galls.; two Worthington comp. dup. *Boilers*, three 100-HP. **Filters.**—Cap., 1,000,000 galls.; two Hyatt, each 8 × 21 ft. **Distribution.**—12.5 miles of mains and 84 hydrants to be added in '91. **Financial.**—Debentures for $125,000 were issued in '90 at 4½%, for 40 yrs., to cover cost of works. Of this sum $75,000 was expended in '90, and balance to be expened in '91. One debenture and accruing interest to be redeemed each year. Des. and Const. Engr. of new works, Jas. A. Bell.

Year.	Mains.	Taps.	Meters.	Hyd'ts.	Cons'pt'n.	Cost.	B'd'd debt.	Exp.	Rev.
'82	4	76	0	40	80,000	$45,000	.....	$4,300	$800
'84	4	150	3	55	3:0,000	70,000	.....	3,000	1,900
'86	5	220	4	56	289,000	70,000	.....	3,752	2,169
'88	5	359	4	56	340,000	70,000	.....	3,920	1,678
'90	8	344	4	76	312,400	75,000	$125,000	3,700	2,498

Management.—Committee of six, appointed every year by city council. Supt., Jas. A. Bell.

37. SARNIA, *Lambton Co.* (Pop., 3,583—2,937.) Built in '75-76 by town.

Supply.—St. Clair River, pumping direct. Fiscal year closes Sept. 30.

Year.	Mains.	Taps.	M't'rs.	H'd'ts.	C'n'p'tn.	Cost.	B'd'd d't.	Int. ch'g's.	Exp.	Rev.
'82	9.2	538	0	74	280,000	$59,590	$70,000	$4,200	$2,636	$1,184
'86	14.6	783	2	97	451,294	66,457	.....	.....	3,476	4,469
'90	15	1,011	3	98	650,400	74,400	.....	.....	5,360	8,819

Financial.—Net revenue is applied to payment of debentures and interest; balance is paid by taxation. No allowance for public uses of water. **Management.**—Com. of three. Supt., Robt. Turner.

Seaforth, *Huron Co.* (Pop., 2,480—2,641.) Built in '79-80 by town for fire protection.

38. SHELBURNE, *Guy Co.* (Pop., 733—1,202.) History.—Built in '89-90 by village. Contrs., Ontario Pump Co., Toronto. **Supply.**—Well, pumping to tank. **Pumping Machinery.**—Dy. cap., 25,000 galls.; Curtis. **Tank.**—Cap., 80,000 galls. **Distribution.**—*Mains*, 10-in., 1., 2,200 ft. *Hydrants*, 7. **Financial.**—*Cost*, $1.200.- Int., 5%. **Management.**—Report by I. G. Dunbar.

39. SMITH'S FALLS, *Lanark Co.* (Pop., 2,087—3,864.) Built in '87 by A. Foster; city neither controls rates nor has right to purchase works. **Supply.**—Rideau River pumping to tank. Fiscal year closes Aug. 1. **Distribution.**—*Mains*, 2.5 miles. *Taps*, 120. *Hydrants*, 4. *Consumption*, 100,000 galls. **Financial.**—*Cost*, $15,000. *No Debt*. *Expenses*: operating $300; taxes, $40; total, $340. *Revenue*: consumers, $600; fire protection, $180; ry., $800; total, $1,580. **Management.**—Owner. A. Foster.

40. STRATFORD, *Perth Co.* (Pop., 8,239—9,501.) Built in '80 by Stratford Water Supply Co.; city has no control over rates, but has right to purchase works at any time, by giving two years' notice, price determined by arbitration. **Supply.**—River Avon, pumping direct; filtered. Fiscal year closes Dec. 31. **Distribution.**—*Mains*, 11 miles. *Taps*, 355. *Hydrants*, 72. *Consumption*, 516,370 galls. **Financial.**—*Cost*, $79,295. *Cap. Stock:* authorized, $35,000; all paid; paid 8% div'd in 1890. *Total Debt*, $13,000: floating, $3,000 at 6%; bonded, $10,000 at 5%, due in 1900. *Sinking Fund*, $3,500. *Expenses*: operating, $4,711; int., $2,207; taxes, none; total, $6,918. *Revenue*, fire protection, $3,460. Hydrant rental was reduced in 1890 from $50 to $40 per hydrant. **Management.**—Prest. and Supt., Jno. Corrie. Secy., D. Burritt.

FOR FURTHER DESCRIPTIVE MATTER see Manual for 1889-90. CHANGES since 1890 are here given. POPULATIONS are for 1880 and 1890, respectively.

41. TILBURY CENTRE, *Kent Co.* (Pop., in '91, 720.) Built in '88 by town. Supply,—Lake St. Clair, pumping direct. Fiscal year closes Dec. 31. Distribution. —*Mains*, 1 mile. *Taps*, 60. *Hydrants*, 9. *Consumption*, 100,000 galls. Financial. —*Cost*, $12,000. *Bonded Debt*, $12,000 at 5%; $1,000 payable yearly. *Expenses*, $500. *Revenue:* consumers, $300; ry., $700; total, $1,000. Management.—Town Council; elected every year. Chn., A. H. Nicol Reeve.

Tilsonburg, *Oxford Co.* (Pop., 1,939–2,163.) Built in '75 by town for fire protection.

42. TORONTO, *York Co.* (Pop., 86,415—144,025.) Owned by city. Built in '41 by Furniss; bought by city in '73. Supply.—Lake Ontario, pumping to reservoir. Fiscal year closes Dec. 31. In '90 a 48-in. steel pipe was laid across bay to Hanlan's Pt. as companion to the old c. i. pipe; and a 60-in. steel pipe, 6,027 ft. long, parallel to the old 48-in. wooden conduit, was laid thence to connection of intake pipe on island. Cost, $160,000. It is proposed to abandon wooden conduit and to extend intake pipe 350 ft. further into lake, to point where it is 70 ft. deep. Two 3,000,000-gall. Blake pumps supplanted the 1,500,000-gall. Worthington.

Yr.	Mains	Taps	M't's	Hyd'ts	Cons'pt'n	Cost	H'd'd d'bt	In.ch's*	Exp.	Rev.
'75	49.8	2,769			3,424,000	$1,507,278				$95,224
'80	113.3	9,582	49	1,070	4,879,422					
'82	116.1	14,062	119	1,118	5,777,809				$87,900	210,000
'84	138.3	18,363	130	1,250	9,960,224		$2,000,000	$120,000	100,500	210,000
'85	143	20,707	195		9,691,733					
'86	156	23,613	256	1,414	11,314,337	2,430,000	2,130,000		104,184	285,050
'87	165	26,893	332	1,683	12,000,610	2,643,000	2,643,000	163,337	130,175	355,700
'88	182.6½	29,883	897	9,954	11,069,784	2,971,600	2,643,000	146,350	135,706	316,920
'89	212.83	34,056	1,347	2,340	11,378,982	3,123,439	3,160,786	181,104	161,500	330,575
'90	230	36,192	1,479	2,599	14,434,722	3,150,478		226,273	168,634	409,788

*Including sinking fund.

Management.—Com. appointed every year from city council by its members. Chn., Wm. J. Hill, Secy., Chas. A. Matthews. Supt., Wm. Hamilton; appointed by council.

I. TRENTON, *Hastings Co.* (Pop., 3,012—4,364.) Built in about '70 by Hugh O'Rourke, for domestic supply only. Water is conveyed from springs through wooden logs with about 10 ft. pressure, which is sufficient to supply only a part of town. There are 3 miles of pipe, supplying water to about 100 families. Cost of works, $2,000. Revenue, about $1,000.

Uxbridge, *Ontario Co.* (Pop., 1,824—2,023.) Built in '74 by town for fire protection.

43. WALKERVILLE, *Essex Co.* (Pop., 600.) Built in '76 by Hiram Walker & Son. Supply.—Source not given, pumping direct. *Mains*, 4 miles. *Hydrants*, 30, Cost, $60,000. Owners, Hiram Walker & Sons, who own town.

44. WATERLOO, *Waterloo Co.* (Pop., 2,066—2,941.) Built (probably) in '89 by Moffett, Hodgkins & Clarke, Syracuse, N. Y. Water pumped from creek to stand-pipe.

45. WELLAND, *Welland Co.* (Pop., 1,870–2,035.) Built in '88 by town. Supply.—Welland Canal, pumping direct; filtered through sand and gravel, with area of about 1,600 ft. Engr. for filter, Wm. Kennedy, Jr., Owen Sound. Fiscal year closes Dec. 31.

Year	Mains	Taps	Hyd'ts	Cons'pt'n	Cost	B'd d'bt	Int.ch'g's	Exp.	Rev.
'88	4	129	30	unk'n	$17,000	$48,000	$2,400	$500	$1,700
'89	6	190	30	75,000	18,763	48,000	2,400	520	1,720
'90	6	213	30	90,000	19,144	18,000	2,400	446	1,767

Financial.—*Bonded Debt*, $18,000 at 5%, due in 1918. *Sinking Fund*, $2,000, deposited in bank at 4%. No allowance for public uses of water. Revenue goes into general fund of town, from which all expenses are paid. Management.—Com. of six, appointed every year by town council. Chn., D. Ross. Secy, and Supt., J. F. Gross.

46. WEST TORONTO JUNCTION, *York Co.* (Pop., in '91, 4,518.) Built in '88-9 by McQuillan & Co. Supply.—Lake Ontario. Cost, $100,000. Address Jno. D. Spears or Asst. Supt. A. H. Clemens. Latest report Oct. 22, '89.

Wiarton, *Bruce Co.* (Pop., 796—1,984.) Built in '88-9 by town for fire protection. Note.—July 11, '91, it was reported that complete system was projected.

47. WINDSOR, *Essex Co.* (Pop., 6,561—10,322.) Built in '72-3 by city. Supply. —Detroit River, pumping direct. Fiscal year ends Dec. 31. Report fall of '91 that 3,000,000-gall. pump had been contracted for. Distribution,—(In '89.) *Mains*, 20.6 miles. *Taps* (in '88), 2,500. *Hydrants*, 213. *Consumption*, 1,200,000 galls. Financial.—(In '89.) *Cost* (in '88), $180,000. *Total Debt,* $212,640: *bonded,* $165,484, at 5 and 6%; *floating,* $47,156. *Operating Expenses,* $9,377. *Revenue:* consumers, $23,401; city, $1,500; total, $24,901. **Management.**—Three comrs. Chn., Jno. Curry. Secy., Chas. J. Bird. Chf. Engr., Jas. Hall.

Wingham, *Huron Co.* (Pop., 1,918—2,167.) Built in '89 by town, for fire protection.

48. **WOODSTOCK,** *Oxford Co.* (Pop., 5,373–8,612.) **History.**—Fire protection system was built by town in '80. Aug. 1, '91, town began construction of complete system. Des. Engr., Thos. C. Keefer. Const. Engrs., Davis & Van Buskirk. **Supply.**—Springs, three miles distant, pumping direct. **Pumping Machinery.**— Dy. cap., 3,500,000 galls.; one 2,000,000-Hughes, comp. dup., direct-acting, and two 750,000 Worthington. **Distribution.**—*Mains,* 12 to 4-in. c. i., 9.5 miles. Pipes and specials from Alex. Gartshore, Hamilton. *Meters,* Crown. *Hydrants,* Ludlow, 120. *Consumption,* 500,000 galls. *Pressure:* ordinary, 60 lbs.; fire, 150 lbs. **Financial.**— Hydrant Rental, $50. **Management.**—Three comrs. Chn., Chas. M. Douglas. Secy., Geo. C. Eden. Treas., Jno. D. Hood. Supt., Wm. M. Davis.

Water-Works Projected with Fair Prospects of Construction.

ESSEX CENTRE, *Essex Co.* (Pop., 800—1,709.) Bids for construction of water-works were received till Aug. 12, '91. Des. Engr., Jno. Galt, Toronto. Artesian wells, pumping to elevated tank. Worthington pump. Mains, 4.5 miles. Hydrants, 21. Cost, $26,500. Mayor, Jno. Milne.

GEORGETOWN, *Halton Co.* (Pop., 1,471—1,509.) Town was to begin construction of works Aug. 1, '91. Engr., Jas. Warren. Springs, by gravity. Est. cost, $40,000. Address G. Goodwillie.

KINGSVILLE, *Essex Co.* (Pop., 863—1,335.) July 25, '91, system was projected by town. Lake, pumping to tank. Address S. T. Copus.

NORTH TORONTO (*Toronto P. O.*), *York Co.* (Pop., not given.) System was under consideration by town council, Dec. 26, '90.

PEMBROKE, *Renfrew Co.* (Pop., 683—801.) System was under consideration by town council, July 31, '91. Des. Engr., Willis Chipman, Toronto. Ottawa River, pumping direct. Address W. C. Irving.

SANDWICH, *Essex Co.* (Pop., 1,143—1,352.) Construction of water-works was awarded by town to Miles Hunting, Hamilton, for $11,375. To be completed Dec. 1, '91. Town Clerk, J. A. Stuart.

MANITOBA.

Water-Works Completed.

1. **WINNIPEG,** (Pop., 7,985—25,642.) Built in '92 by Winnipeg Water-Works Co. **Supply.**—Assiniboine River, pumping direct. *Mains* (in '90), about 16 miles. *Hydrants* (in '91), about 102. *Cost,* $130,000. *Cap. Stock* (in '84): authorized, $300,000; paid-up, $82,500. *Bonded Debt* (in '84), $60,000 at 6%. Prest., Alex. Moffett. Secy., A. B. Elford. Supt., J. E. Hannah. Latest report in '88.

NOTE.—April 1, '91, Mayor Alfred Pearson reported that city was taking steps to purchase co.'s system. City Engr., H. W. Ruttan.

Water-Works Projected with Fair Prospects of Construction.

BRANDON, *Selkirk Co.* (Pop., in '91, 3,778.) Bonds for construction of water-works had been voted by town Sept. 26, '91. City Clerk, Jno. C. Kerr.

NEEPAWA, *Marquette Co.* (Pop., not given.) July 2, '91, it was reported that town would let contract for construction of works July 18. Est. cost, $3,000.

FOR FURTHER DESCRIPTIVE MATTER see Manual for 1889-90. CHANGES since 1890 are here given. POPULATIONS are for 1880 and 1890, respectively.

MANUAL OF AMERICAN WATER-WORKS.

NORTHWEST TERRITORY.

Water-Works Completed.

1. **CALGARY,** Co. not given. (Pop., in '90, 3,876.) **History.**—Built in '90 by Calgary Gas & Water Co. Engr., Wm. Kennedy, Jr. Inspector, Geo. S. Kennedy. **Supply.**—Bow River, pumping direct. **Pumping Machinery.**—Dy. cap., 1,500,000 galls.; one 1,000,000 comp. cond. dup. and one 500,000 crank and fly-wheel; Contrs., Osborne, Worswick Co., Hamilton. **Distribution.**—*Mains,* 12 to 6-in. i., 5 miles. *Services,* lead. Pipe from Londonderry Iron Co., Montreal; specials from Osborne, Worswick Co., Hamilton. *Meters,* 0. *Hydrants,* 30. *Pressure,* 50 to 100 lbs. **Financial.**—*Cap. Stock,* $300,000. **Management.**—Prest., Geo. Alexander. Secy., H. B. Alexander.

BRITISH COLUMBIA.

Water-Works Completed or In Process of Construction.

1. **ESQUIMALT,** *Victoria Co.* (Pop., not given.) Built in '88 by Esquimalt Water-Works Co.; put in operation early in year. **Supply.**—Thetis and Upper Thetis lakes, by gravity. **Distribution.**—(In '88.) *Mains,* 11 miles. *Meters,* 89. *Hydrants,* 32. **Financial.**—(In '88.) *Cost,* $125,000. **Management.**—Prest. and Secy., T. Lubke, Victoria.

2. **KAMLOOPS,** *Yale Co.* (Pop., not given.) Kamloops Water Co. owns system, Thompson River, pumping to tank. Prest., Jas. McIntosh.

3. **NANAIMO,** *Vancouver Co.* (Pop., 1,645—4,595.) Built in '86-7 by Nanaimo Water-Works Co. **Supply.**—Spring-fed mountain stream, by gravity from impounding reservoir; strained through gauze. Fiscal year closes July 31. **Distribution.**—*Mains,* 8 miles. *Taps* (in '88), 200. *Meters,* 4. *Hydrants,* 30. *Consumption,* 150,000 galls. **Financial.**—*Cost,* $51,925. *Cap Stock,* $46,875. *Debt,* $5,050. *Revenue:* $300; city, $8,425; total, $8,725. **Management.**—Prest., Jno. Mahar. Treas., Edwin Pembry. Supt., J. W. Stirtan.

4. **NEW WESTMINSTER,** *New Westminster Co.* (Pop., 1,500—6,641.) **History.**—Built in '90-1 by city; put in operation Oct. 15. Engr., Albert Hill. **Supply.**—Coquitlam Lake, by gravity. **Reservoir.**—Cap., 216,000 galls. Relief tank is 20 ft. in diam., 6 ft. deep. **Force Main.**—Of 22 and 14-in. riveted steel, 16.5 miles; Contr., D. McGillivray, Vancouver. **Distribution.**—*Mains,* c. i., 20 miles. Pipe and specials from A. G. Kidston & Co., Glasgow, Scotland. *Services,* c. i. *Hydrants,* Galvin, 76. *Valves,* Galvin. *Pressure:* ordinary, 100 lbs.; fire, 170 lbs. **Financial.**—*Cost,* $380,000. **Management.**—Comrs.

5. **VANCOUVER,** *New Westminster. Co.* (Pop., in '91, 18,685.) Built in '87-9 by Vancouver Water-Works Co.; franchise does not regulate rates but provides for purchase of works by city at any time, by giving two months notice. See note below. **Supply.**—Capilano River, by gravity from impounding reservoir. Fiscal year closes Dec. 31.

Year.	Mains.	Taps.	Meters.	Hydrants.	Cost.	Expenses.	Revenue.
'87	10			75	$250,000		
'88	26.5	500	22	75	250,000	$6,000	$22,000
'90	27.1	1,070	42	75	322,000	9,826	33,167

Financial.—*Cap. Stock and Debt,* withheld. *Expenses:* operating, $9,826; int., $3,568; taxes, withheld. *Revenue:* consumers, $28,433; fire protection, $3,150; other purposes. $1,584; total, $33,167. **Management.**—Prest., Jno. Irving, Victoria. Secy., J. W. McFarland. NOTE:—In '91, arbitrators appraised value of works at $330,031, city proposing to buy.

6. **VICTORIA,** *Victoria Co.* (Pop., 5,925—16,841.) Built in '74-5 by city. **Supply.**—Creek, by gravity from impounding reservoir on stream, for lower levels of city, and by pumping five hours per day for higher parts of city. Fiscal year closes, Dec. 31.

Year.	Mains.	Taps.	Meters.	Hyd'ts.	Cost.	B'd'd debt.	Int. ch's.	Exp.	Rev.
'87	26.2						$12,284		
'88	32.5	1,738	29	60	$261,316	$257,500	15,650	$7,342	$37,764
'89	35.4	1,924	81	82	426,738	387,500	16,110	10,000	40,427
'90	40	2,234	136	100	484,117	387,500	22,668	18,668	41,385

Management.—Comr., Peter Summerfield.

Classification by Sizes, According to the 1890 Census, of Towns Having Works.

NOTE.—Private or company ownership has been designated by an asterisk prefixed to the name of the city; in case there are public and private plants in the same city a dagger and asterisk are prefixed to the name of the place; if more than one company in a city as many asterisks are prefixed as there are companies. For some of the smaller towns and villages the 1890 populations were not available, and for these estimates have been used, such figures being prefixed by the letter e.

City	Pop.
New York, N. Y.	1,515,301
†*Chicago, Ill.	1,099,850
Philadelphia, Pa.	1,046,964
†*Brooklyn, N. Y.	806,313
St. Louis, Mo.	451,770
Boston, Mass.	448,477
Baltimore, Md.	434,439
*San Francisco, Cal.	298,997
Cincinnati, O.	296,908
Cleveland, O.	261,353
Buffalo, N. Y.	255,664
*New Orleans, La.	242,039
†*Pittsburg, Pa.	238,617
Washington, D. C.	230,392
Detroit, Mich.	205,876
Milwaukee, Wis.	204,468
Newark, N. J.	181,830
Minneapolis, Minn.	164,738
Jersey City, N. J.	163,003
‡Louisville, Ky.	161,129
*Omaha, Neb.	140,452
Rochester, N. Y.	133,896
St. Paul, Minn.	133,156
*Kansas City, Mo.	132,716
Providence, R. I.	132,146
**Denver, Col.	106,713
*Indianapolis, Ind.	105,436
Allegheny, Pa.	105,287
Albany, N. Y.	94,923
Columbus, O.	88,150
*Syracuse, N. Y.	88,143
*New Haven, Conn.	86,045
Worcester, Mass.	84,655
Toledo, O.	81,434
Richmond, Va.	81,388
*Paterson, N. J.	78,317
Lowell,*Mass.	77,696
Nashville, Tenn.	76,168
**Scranton, Pa.	75,215
Fall River, Mass.	74,398
Cambridge, Mass.	70,028
Atlanta, Ga.	65,533
*Memphis, Tenn.	64,495
Wilmington, Del.	61,431
Dayton, O.	61,220
Troy, N. Y.	60,967
†*Grand Rapids, Mich.	60,278
Reading, Pa.	58,661
Camden, N. J.	58,313
Trenton, N. J.	57,458
Lynn, Mass.	55,727
Lincoln, Neb.	55,154
*Charleston, S. C.	54,952
Hartford, Conn.	53,230
*St. Joseph, Mo.	52,324
Evansville, Ind.	50,756
***Los Angeles, Cal.	50,395
*Des Moines, Ia.	50,093
*Portland, Me.	48,803
*Bridgeport, Conn.	48,866
*Oakland, Cal.	48,652
†***Portland, Ore.	46,385
Saginaw, Mich.	46,322
Salt Lake City, U.	44,843
Lawrence, Mass.	44,654
Springfield, Mass.	44,179
Manchester, N. H.	44,126
*Utica, N. Y.	44,007
*Hoboken, N. J.	43,648
Savannah, Ga.	43,189
†*Seattle, Wash.	42,837
*Peoria, Ill.	41,024
New Bedford, Mass.	40,733
Erie, Pa.	40,634
Somerville, Mass.	40,152
Harrisburgh, Pa.	39,385
Dallas, Tex.	38,067
Sioux City, Ia.	37,806
*Elizabeth, N. J.	37,764
*Wilkes-Barre, Pa.	37,718
*San Antonio, Tex.	37,673
*Covington, Ky.	37,371
*Tacoma, Wash.	36,006
Holyoke, Mass.	35,637
Ft. Wayne, Ind.	35,392
Binghamton, N. Y.	35,005
Norfolk, Va.	34,871
Wheeling, W. Va.	34,522
Augusta, Ga.	33,300
Youngstown, O.	33,220
*Duluth, Minn.	33,115
Yonkers, N. Y.	32,033
Lancaster, Pa.	32,011
Springfield, O.	31,895
*Quincy, Ill.	31,494
*Mobile, Ala.	31,076
*Topeka, Kan.	31,007
Salem, Mass.	30,801
Long Island City, N.Y.	30,506
Altoona, Pa.	30,337
*Dubuque, Ia.	30,311
*Terre Haute, Ind.	30,217
*Elmira, N. Y.	29,708
**Chattanooga, Tenn.	29,100
Galveston, Tex.	29,084
Waterbury, Conn.	28,646
*Williamsport, Pa.	27,932
Bay City, Mich.	27,839
*Akron, O.	27,601
*Houston, Tex.	27,557
*Haverhill, Mass.	27,412
Brockton, Mass.	27,294
*Knoxville, Tenn.	26,946
*Davenport, Ia.	26,872
Sacramento, Cal.	26,386
Pawtucket, R. I.	26,333
Canton, O.	26,189
*Birmingham, Ala.	26,178
*Little Rock, Ark.	25,874
*Auburn, N. Y.	25,858
Taunton, Mass.	25,448
Allentown, Pa.	25,228
La Crosse, Wis.	25,090
Springfield, Ill.	24,906
Newport, Ky.	24,918
*Gloucester, Mass.	24,661
†*Pueblo, Col.	24,558
Newton, Mass.	24,379
*Wichita, Kan.	23,853
Rockford, Ill.	23,584
Joliet, Ill.	23,264
Newburgh, N. Y.	23,087
Ft. Worth, Tex.	23,076
Oshkosh, Wis.	22,836
*Macon, Ga.	22,746
Muskegon, Mich.	22,702
*Petersburgh, Va.	22,680
*Burlington, Ia.	22,565
Cohoes, N. Y.	22,509
Poughkeepsie, N. Y.	22,206
Fitchburgh, Mass.	22,37
*Montgomery, Ala.	21,883
*Springfield, Mo.	21,850
Oswego, N. Y.	21,842
South Bend, Ind.	21,819
Johnstown, Pa.	21,805
Lewiston, Me.	21,701
Meriden, Conn.	21,652
*Lexington, Ky.	21,567
*Council Bluffs, Ia.	21,474
*Kingston, N. Y.	21,261
*New Albany, Ind.	21,069
*Racine, Wis.	21,014
Zanesville, O.	21,009
Woonsocket, R. I.	20,830
Jackson, Mich.	20,798
*York, Pa.	20,793
McKeesport, Pa.	20,741
*Chester, Pa.	20,226
*Wilmington, N. C.	20,056
Bloomington, Ill.	20,048
†*Spokane F's, Wash.	19,922
Schenectady, N. Y.	19,902
*Norristown, Pa.	19,791
*Leavenworth, Kan.	19,768
Lynchburgh, Va.	19,709
Aurora, Ill.	19,688
Danbury, Conn.	19,473
*Newport, R. I.	19,467
*Nashua, N. H.	19,311
Bangor, Me.	19,103
Bayonne City, N. J.	19,032
New Britain, Conn.	19,007
Orange, N. J.	18,844
Waltham, Mass.	18,707
New Brunswick, N. J.	18,603
Findlay, O.	18,553
Sandusky, O.	18,471
Winona, Minn.	18,208
*San José, Cal.	*8,060
*Cedar Rapids, Ia.	18,020
Kalamazoo, Mich.	17,853
Elgin, Ill.	17,823
*Warwick, R. I.	17,761
Norwalk, Conn.	17,747
Hamilton, O.	17,565
*Eau Claire, Wis.	17,415
Amsterdam, N. Y.	17,366
*Columbus, Ga.	17,303
Pittsfield, Mass.	17,281
Jacksonville, Fla.	17,201
Concord, N. H.	17,004
Decatur, Ill.	16,841
*Quincy, Mass.	16,723
*Richmond, Ind.	16,608
*New Brighton, N. Y.	16,423
*Sheboygan, Wis.	16,359
Lafayette, Ind.	16,243
*Roanoke, Va.	16,159
‡‡San Diego, Cal.	16,159
Norwich, Conn.	16,156
North Adams, Mass.	16,074
Lockport, N. Y.	16,038
*Jamestown, N. Y.	16,038
Lima, O.	15,987
*Shenandoah, Pa.	15,944
*Stamford, Conn.	15,700
*Belleville, Ill.	15,361
Columbia, S. C.	15,353
Galesburg, Ill.	15,264
*East St. Louis, Ill.	15,169
Central Falls, R. I.	e15,000
Rome, N. Y.	14,991
Northampton, Mass.	14,990
*Ogden, Utah	14,889
Watertown, N. Y.	14,725
Lebanon, Pa.	14,664
Burlington, Vt.	14,590
*Easton, Pa.	14,481
*Austin, Tex.	14,476

† The city owns practically all the stock. ‡‡ City has leased works.

MANUAL OF AMERICAN WATER-WORKS.

Place	Value
**Waco, Tex	14,445
*Biddeford, Me.	14,443
*Stockton, Cal	14,424
*Sh'm'kinb'ro',Pa	14,403
*Alexandria, Va.	14,329
*Newark, O	14,270
*Edgewa:er, N.Y	14,265
*Pottsville, Pa.	14,117
Keokuk, Ia	14,101
*Sedalia, Mo	14,068
*Chicopee, Mass.	14,050
Ottumwa, Ia.	14,001
*Atchison, Kans.	13,933
*Newb'yp't,Mass.	13,947
Gloversville, N.Y.	13,864
Beatrice, Neb.	13,836
*Helena, Mont.	13,834
Marlbo'gh,Mass.	13,805
Malden, Mass.	13,805
NewLondon,Conn	13,757
Rock Island, Ill	13,634
*Clinton, Ia.	13,619
Hastings, Neb.	13,584
Port Huron,Mich.	13,534
Woburn, Mass.	13,499
Mansfield, O.	13,473
Madison, Wis.	13,426
Steubenville, O.	13,394
*Vicksburg, Miss.	13,378
Logansport, Ind.	13,328
*Pottstown, Pa.	13,285
*East Orange,N.J.	13,282
Portsmouth, Va.	13,228
BattleCreek, Mich	13,197
Lansing, Mich.	13,102
*Paducah, Ky.	13,076
*Atl'tic Cy., N.J.	13,055
*Passaic, N.J.	13,028
*West Troy, N.Y.	12,967
*Hannibal, Mo.	12,857
*Manistee, Mich.	12,812
Dover, N.H.	12,790
Cumberland, Md.	12,729
*Raleigh, N.C.	12,678
Portsmouth, O.	12,364
*Waterville,Me.	c12,104
Brookline, Mass.	12,103
*Fond du Lac,Wis	12,024
Moline, Ill	12,000
*Superior, Wis.	11,983
Bay City, Mich.	12,981
*Shreveport, La.	11,979
Middletown, N.Y.	11,977
Saratoga Springs, N.Y.	11,975
*Ft. Scott, Kans.	11,943
*Hazleton, Pa.	11,872
*Appleton, Wis.	11,860
*Pensacola, Fla.	11,750
Cheyenne, W.	11,690
Ogdensburgh,N.Y	11,662
*New Castle, Pa.	11,600
*Charlotte, N.C.	11,557
*Marinette, Wis.	11,523
*Neb'ska Cy.,Neb.	11,494
*Danville, Ill.	11,401
*Muscatine, Ia.	11,454
Bridgeton, N.J.	11,424
*Streator, Ill.	11,414
*Elkhart, Ind	11,360
*Muncie, Ind.	11,345
*Llano, Tex.	11,319
*Ft. Smith, Ark.	11,311
Chillicothe, O.	11,288
*Mahanoy Cy.,Pa.	11,286
*Alpena, Mich.	11,283
*Plainfield, N.J.	11,267
*Stillwater,Minn.	11,260
*Auburn, Me.	11,250
*Leadville, Col.	11,212
Ishpeming, Mich.	11,197
*Alameda, Cal.	11,165
Colo. Springs,Col.	11,140
*Ithaca, N.Y.	11,079
Medford, Mass.	11,079
*Hornellsv'le,N.Y	10,996
*Denison, Tex	10,958
East Liverpool,O.	10,956
Ironton, O	10,939
Oil City, Pa.	10,932
Weymouth, Mass.	10,866
*Janesville, Wis	10,836
*Carbondale, Pa.	10,833
*Mt.Vernon,N.Y.	10,830
Beverly, Mass.	10,821
*Fresno, Cal.	10,818
*Tiffin, O	10,801
MichiganCity,Ind	10,776
Anderson, Ind	10,741
Jacksonville, Ill.	10,740
*Butte Cy, Mont.	10,733
*Massillon, O	10,692
*Jeff'rsonv'le, Ind.	10,666
*Menominee,Mich	10,630
*Meridian, Miss.	10,624
*Columbia, Pa.	10,599
Lansingb'gh,N.Y.	10,555
*Augusta, Me	10,527
Bradford, Pa.	10,514
*Baton Rouge,La.	10,478
Clinton, Mass	10,424
**Ansonia, Conn.	10,342
*El Paso, Tex	10,338
*Cairo, Ill.	10,324
Danville, Va.	10,305
*South Bethlehem, Pa.	10,302
*Pittston, Pa.	10,302
*Alton, Ill	10,294
Asheville, N.C.	10,235
*Hyde Park,Mass.	10,193
*Freeport, Ill.	10,189
*Sioux Falls, S.D.	10,177
Peabody, Mass.	10,158
*Greenwich,Conn.	10,131
*Hagerstown,Md.	10,118
*H'tingdon,W.Va.	10,108
*Natchez, Miss.	10,101
*Nanticoke, Pa.	10,044
Jackson, Tenn	10,039
*Millville, N.J.	10,002
Evanston, Ill.	c10,000
*Flatbush, N.Y.	e10,000
Gravesend,N.Y.	e10,000
*Salem, Ore.	c10,000
*Lawrence, Kan.	9,997
*Ottawa, Ill.	9,985
Ashland, Wis	9,956
*Pine Bluff, Ark.	9,952
Hudson, N.Y.	9,970
*Joplin, Mo	9,943
Bellaire, O.	9,931
*Anniston, Ala.	9,876
La Salle, Ill.	9,855
*Owensb'gh, Ky.	9,837
*Portsmouth, N.H.	9,827
Westfield, Mass.	9,805
*Flint, Mich.	9,803
*Amesbury, Mass.	9,798
*Beaver Falls,Pa.	9,735
Peekskill, N.Y.	9,676
*Meadville, Pa.	9,520
*PerthAmboy,N.J.	9,512
Glens Falls, N.Y.	9,509
*Ann Arbor, Mich.	9,431
Dunkirk, N.Y.	9,416
Sing Sing, N.Y.	9,352
*Plymouth, Pa.	9,344
*Port Jervis, N.Y.	9,327
Warsaw, Wis	9,25
*Manchester, Va.	9,246
*SouthFramingham, Mass.	9,239
Adams, Mass.	9,213
Natick, Mass.	9,118
Marquette, Mich.	9,093
Piqua, O.	9,090
*Green Bay, Wis.	9,069
*Kankakee, Ill.	9,025
Middletown, Conn.	9,013
Madison, Ind	8,936
Marshalltown, Ia.	8,914
*Vincennes, Ind.	8,853
Mankato, Minn.	8,838
*Ashtabula, O.	8,838
Henderson, Ky.	8,835
Little Falls, N.Y.	8,783
*Milford, Mass.	8,780
Marion, Ind.;	8,760
*Adrian, Mich.	8,756
Spencer, Mass	8,747
*Butler, Pa.	8,734
*Bath, Me.	8,723
*Hutchinson, Kan.	8,682
*Chip'waF'ls,Wis.	8,670
Montclair, N.J.	8,656
Willimantic,Conn.	8,648
*Phillipsburg,N.J	8,644
*Athens, Ga.	8,639
*Coal Township,Pa	8,616
*Greenville, S.C.	8,607
*Iron Mt., Mich.	8,599
*Cortland, N.Y.	8,590
Braddock, Pa.	8,561
Corning, N.Y.	8,550
Melrose, Mass.	8,519
*Virginia Cy,Nev.	8,519
Phoenixville, Pa.	8,514
*Brunswick, Ga.	8,459
Flushing, N.Y.	8,436
*Gardner, Mass.	8,424
E. Providence,R.I.	8,422
Parkersb'g,W.Va.	8,408
*Plattsmouth,Neb.	8,392
*Arkansas Cy,Kan.	8,347
Harrison, N.J.	8,338
*Marion, O.	8,327
*Dunmore, Pa.	8,315
Marietta, O	8,273
*Kokomo, Ind.	8,261
*Paris, Tex.	8,254
*Mt. Carmel, Pa.	8,254
Rutland, Vt.	8,239
*Delaware, O.	8,224
*NewHoch'le,N.Y.	8,217
*Moberly, Mo.	8,215
Marblehead, Mass.	8,203
Frederick, Md.	8,193
*Rockland,Me.	8,174
*Morristown, N.J.	8,156
*Kearney, Neb.	8,074
Titusville, Pa.	8,073
*Hot Springs, Ark.	8,066
West Chester, Pa.	8,028
*Manchester,Conn	8,022
*Winston, N.C.	8,018
Danville, Pa.	7,998
Huntsville, Ala.	7,995
*Carthage, Mo.	7,981
*Washington, Pa.	7,963
*Clarksville,Tenn.	7,924
Homestead, Pa.	7,911
*Ft. Madison, Ia.	7,901
*Stevens P't, Wis.	7,896
*Frankfort, Ky.	7,892
Chambersburg,Pa.	7,863
BowlingGreen,Ky.	7,803
*Rockville, Conn.	7,772
St. Albans, Vt.	7,771
*Ironwood, Mich.	7,745
*Manitowoc, Wis.	7,710
*Bloomfield, N.J.	7,708
Johnstown, N.Y.	7,708
*Defiance, O.	7,694
St. Cloud, Minn.	7,686
Middletown, O.	7,681
*Southbridge,Mass	7,655
*Selma, Ala.	7,622
*Carlisle, Pa.	7,620
*Alliance, O.	7,607
*Annapolis, Md.	7,604
Attleborough,Mass	7,577
*Geneva, N.Y.	7,577
Lancaster, O.	7,555
*Emporia, Kan.	7,551
Grand Island,Neb.	7,536
*Ludington, Mich.	7,517
*Sharon, Pa.	7,459
Danvers, Mass.	7,454
Keene, N.H.	7,446
Rochester, N.H.	7,396
*Bristol, Conn.	7,382
Lock Haven, Pa.	7,358
Olean, N.Y.	7,358
Ashland, Pa.	7,346
Sherman, Tex.	7,335
Ware, Mass.	7,329
Huntington, Ind.	7,328
Plymouth, Mass.	7,314
*Greenbush, N.Y.	7,301
*Xenia, O.	7,301
*Calais, Me.	7,290
Leominster, Mass.	7,269
Burlington, N.J.	7,264
*Nevada, Mo.	7,262
*Long Branch,N.J.	7,231
Martinsb'g,W.Va.	7,226
Batavia, N.Y.	7,221
Marshall, Tex.	7,207
*Creston, Ia.	7,200
*Stonington Conn	7,184
Norwalk, O.	7,175
*North Tonawanda, N.Y.	7,145
Tremont, O	7,141
La Porte, Ind.	7,126
*Dedham, Mass.	7,123
Hahway, N.J.	7,105
*Watertown,Mass.	7,073
Fostoria, O.	7,070
Kearney, N.J.	7,064
*Webster, Mass.	7,031
Peru, Ind	7,028
*Iowa Cy., Ia	7,061
*Hoosick F'ls.N.Y.	7,014
Plattsburgh,N.Y.	7,010
*Gardiner, Me.	e7,000
*Wakefield, Mass	6,982
Staunton, Va.	6,975
Rome, Ga.	6,959
*Tyler, Tex.	6,908
*Mattoon, Ill.	6,833
*Westerly, R.I.	6,813
*Barre, Vt	6,812
Merrill, Wis	6,809
*Escanaba, Mich.	6,808
Dennison, O	6,767
Bethlehem, Pa.	6,762
*Charleston,W.Va	6,742
*Jefferson Cy., Mo.	6,712
*Parsons, Kan.	6,736
No. Attleborough, Mass.	6,727
Columbus, Ind	6,719
Petersburgh,Kan.	6,697
*Waterloo, Ia.	6,674
*Gainesville, Tex.	6,594
GloucesterCy.,N.J	6,564
Owosso, Mich	6,564
Oskaloosa, Ia.	6,558
Circleville, O.	6,556
*Bristol, Pa.	6,552
Kenosha, Wis.	6,532
Boone, Ia.	6,520
*Faribault, Minn.	6,520
Palmer, Mass.	6,520
*Putnam, Conn.	6,512
Urbana, O.	6,510
Americus, Ga.	6,298
*Bennington, Vt.	6,391
*Laramie Cy.,Wy	6,388
*Independence,Mo	6,380
*Texarkana, Ark.	6,380
Uniontown, Pa.	6,359
*Pekin, Ill.	6,347
*Vallejo, Cal	6,343
*Galion, O.	6,336
*Waukesha, Wis	6,321
*Athol, Mass.	6,319
*Beloit, Wis	6,315
Red Wing, Minn.	6,294
*P't Richm'd,N.Y.	6,290
*Corsicana, Tex.	6,285
*Oneonta, N.Y.	6,273
Martin's Ferry, O.	6,250
*Ottawa, Kan.	6,218
Cheboygan, Mich.	6,235
*Franklin, Pa.	6,221
*Naugatuck,Conn.	6,218
*Great Falls, N.H.	6,207
Pontiac, Mich.	6,200
*Lincoln, Ill.	6,200
*Santa Fé, N.Mex.	6,185
*St. Charles, Mo.	6,161
*Stoneham, Mass.	6,155
*Dubois, Pa.	6,149
*Salina, Kan.	6,149
*Laconia, N.H.	6,143
Andover, Mass.	6,142
Ypsilanti, Mich.	6,129
College Point,N.Y.	6,127
*Craw'fdsv'le,Ind.	6,089
*Oneida, N.Y.	6,083
Negaunee, Mich.	6,078
Middleb'gh, Mass.	6,065
*Washington, Ind	6,064
*Saco, Me.	6,057
Tamaqua, Pa.	6,050
*Torrington, Conn	6,048
Goshen, Ind.	6,033
*Mt.Vernon, O.	6,027

MANUAL OF AMERICAN WATER-WORKS. 381

Place	Pop.	Place	Pop.	Place	Pop.	Place	Pop.
*Brunswick, Me.	6,012	*Webb Cy., Mo	5,043	*Warren, Pa	4,332	*Thompsonville,Ct.	3,794
*Florence, Ala.	6,012	*Trenton, Mo	5,030	*Greenville, Tex.	4,330	*Lee, Mass	3,785
Monmouth, Ill.	5,996	†*G'd Haven,Mich	5,023	Au Sable, Mich.	4,328	*Franklin, Ind	3,781
*Bucyrus, O	5,974	*Grafton, Mass	5,002	Fulton, Mo	4,314	*Fergus F'ls,Minn.	3,772
*Warren, O	5,973	So.Norwalk,Conn.	e5,000	*Marshall, Mo	4,297	*Monroe, Wis	3,768
*Sunbury, Pa.	5,930	*Moosic, Pa	e5,000	*Exeter, N. H	4,291	*Etna, Pa	3,767
*Jackson, Miss.	5,920	Paris, Ill	4,996	*Milton, Mass.	4,278	Albion, Mich.	3,763
*Frankfort, Ind.	5,919	*Malone, N. Y	4,986	*Dayton, Ky	4,264	Lebanon, N. H.	3,763
Brazil, Ind	5,905	Grand F'ls,N. Dak.	4,974	S.Hadley F's,Mass.	4,261	*Hanover Pa	3,746
Wooster, O.	5,901	Catskill, N. Y	4,920	Abington, Mass.	4,260	*Brockport, N. Y.	3,742
*Buchanan, Va.	5,867	Waukegan, Ill.	4,915	Anoka, Minn	4,252	New Ulm, Minn.	3,741
*SantaBarbara,Cal.	5,864	*Eastport, Me	4,908	*B'dgewater, Mass.	4,249	*Mufr'sboro,Tenn	3,739
*Canandaigua,N.Y.	5,858	Sharpsburg, Pa.	5,898	Bellefonte, O	4,245	Norwood, Mass.	3,733
*Champaign, Ill.	5,889	***Pasadena, Cal	4,882	*Saugerties, N Y	4,237	*Bradford, Mass.	3,720
*Palestine, Tex.	5,838	Ft. Dodge, Ia	4,871	*Bordentown, N.J.	4,232	*Ionne Terre, Mo.	3,718
*Sterling, Ill.	5,824	Lorain, O	4,863	Wallingford,Conn.	4,230	*Catasauqua, Pa.	3,704
Litchfield, Ill	5,811	Winchester,Mass.	4,861	*Beaver Dam,Wis.	4,222	*Petaluma, Cal	3,692
*Lyons, Ia	5,799	*Eureka, Cal	4,858	*Fayetteville.N.C.	4,222	*Bent'nH'b'r,Mich.	3,692
*Salem, O	5,780	*Stoughton, Mass	4,852	*W'm.stown, Mass.	4,221	*Salamanca, N. Y.	3,692
SaultdeSte.Marie.Mich.	5,780	Sidney, O	4,850	*Paris, Ky	4,218	*Mechanicsb'h,Pa.	3,691
*Washington, O.	5,742	Braintree, Mass.	4,848	*Tuscaloosa, Ala.	4,215	Potsdam, N. Y.	3,691
*Huntingdon, Pa.	5,729	Winsted, Conn.	4,846	*Fulton, N. Y	4,214	*Gilberton, Pa.	3,687
*Chillicothe, Mo.	5,717	*Franklin, Mass.	4,831	*Greensburgh, Pa	4,202	Coatesville, Pa.	3,680
*Brainerd, Minn.	5,703	*Hempstead,N. Y.	4,831	*Niles, Mich	4,197	*Greenville, Pa.	3,674
*Corry, Pa	5,677	*Ellsworth, Me.	4,804	*Ashland, Ky	4,195	Saugus, Mass.	3,673
*Revere, Mass	5,668	*Mexico, Mo.	4,789	*Brewer, Me	4,193	Coshocton, O	3,672
Fargo, N. Dak.	5,664	*Painesville, O.	4,775	*Henderson, N. C.	4,191	St. Peter, Minn.	3,671
*Galena, Ill	5,635	*Oak Park, Ill	4,771	*Towanda, Pa	4,169	Yankton, S. Dak.	3,670
*Connellsville, Pa	5,629	*Ft. Howard, Wis.	4,754	*JohnsonCy.,Tenn.	4,161	Fenton, Mich	3,667
Arlington, Mass.	5,629	*Richmond, Ky.	4,753	Renovo, Pa	4,154	*Winooski, Vt.	3,659
*NewBrighton,Pa.	5,616	Mt.Clemens,Mich.	4,748	Red Bank, N. J	4,145	*Mt. Pleasant,Pa.	3,652
*SouthEaston,Pa.	5,616	*St.Augustine,Fla.	4,742	*Lambe'tville,N.J	4,142	*Rochester, Pa.	3,649
*Elyria, O	5,611	*Clinton, Mo.	4,737	*Boonville, Mo.	4,141	*Pomona, Cal	3,634
*Newton, Kan	5,605	*Argentine, Kan.	4,732	Charleston, Ill	4,135	Santa Ana, Cal.	3,628
Canton, Ill.	5,604	*WallaWalla,W'sh.	4,709	*Waverly, N. Y	4,123	*Fishkill, N. Y.	3,617
Santa Cruz, Cal	5,596	*Mt. Vernon, Ind	4,705	WestCleveland,O.	4,117	Montpelier, Vt.	3,617
*Claremont, N. H.	5,595	*Tyrone, Pa.	4,705	*Nyack, N. Y	4,111	Wellesley, Mass.	3,600
Charlottesville,Va.	5,591	Olympia, Wash.	4,698	*MauchChunk,Pa.	4,101	*Latrobe, Pa.	3,598
Kenton, O	5,557	*Riverside, Cal.	4,683	*Monongahela Cy.	4,096	*Greensburgh,Ind	3,596
*Spartanburgh,S.C.	5,544	*Portland, Conn.	4,678	Hopkinton, Mass.	4,068	Oscoda, Mich.	3,593
*Tampa, Fla	5,532	Hudson, Mass.	4,670	Reading, Mass.	4,068	Bluffton, Ind	3,589
*Owego, N. Y	5,525	*Wilkinsburgh,Pa.	4,662	*Caribou, Me.	4,087	Girardville, Pa.	3,584
*Trinidad, Col	5,523	*Bloomsburgh,Pa.	4,635	Frankl'n, N. H.	4,085	*Warsaw, Ind	3,574
Salem, N. J	5,516	*Tarentum, Pa.	4,627	*Olyphant, Pa.	4,083	*Up'r Sandusky,O.	3,572
*Albuq'que, N. M.	5,515	*Camden, Me	4,621	*FairHaven,Wash.	4,076	Tarrytown, N. Y.	3,562
Thomasville, Ga.	5,514	**Great Barrington,Mass.	4,612	*New Whatcom,Wash.	4,059	*Abilene, Kan	3,547
Van Wert, O	5,512	*Baraboo, Wis.	4,605	*Temple, Tex.	4,047	*Vancouver,Wash	3,545
*NiagaraF'ls,N.Y.	5,502	Troy, O.	4,504	*WhitePlains,N.Y.	4,012	Mendota, Ill	3,542
*Southington,Conn	5,501	*Albion, N. Y	4,586	*Maryville, Mo.	4,037	Ballston Spa,N. Y	3,527
*Menomonie, Wis.	5,491	Brooklyn, O	4,585	*Le Mars, Ia	4,036	*Ft. Fairfield, Me.	3,526
*Durham, N. C.	5,485	Kewanee, Ill.	4,569	*Archbald, Pa.	4,032	*Minersville, Pa.	3,504
*Bristol, R. I.	5,478	Logan, Utah	4,565	*Goldsboro, N. C.	4,017	*Summit, N. J.	3,502
*Conshohocken,Pa.	5,476	*Hingham, Mass.	4,564	*Houlton, Me.	4,015	Kent, O	3,501
*Brattleboro, Vt.	5,466	Columbus, Miss	4,559	SanBernardino,Cal	4,012	Asbury Park, N.J.	3,500
*Shelbyville, Ind.	5,451	*Port Townsend,Wash.	4,558	*New Castle, Del.	4,010	*Waterford, N. Y.	e3,500
Hammond, Ind.	5,428	Connersville, Ind.	4,548	*Rich Hill, Mo	4,008	*Georgetown, Ky.	e3,500
*Columbia, Tenn.	5,370	*Bessemer, Ala.	4,544	Mason Cy., Ia	4,007	LakeLinden,Mich	e3,500
*Jamaica, N. Y.	5,361	Canton, Mass.	4,538	*Lansford, Pa.	4,004	Cedar Falls, Ia.	3,500
*Maysville, Ky.	5,338	*Lexington, Mo.	4,537	ChicopeeF'ls,Mass.	4,000	Gouverneur, N.Y.	3,458
*Seymour, Ind	5,337	*Ridley Park, Pa.	4,529	TurnersF'ls,Mass.	e4,000	*Crookston, Minn.	3,457
*Rochester, Minn	5,321	Fredericksb'h, Va.	4,528	*Dan'lsonville,Ct.	e4,000	Cherokee, Ia.	3,441
*Milton, Pa.	5,317	*Winchester, Ky.	4,519	*S.Manchester,Ct.	e4,000	*Wellsville, N. Y.	3,435
Big Rapids, Mich	5,303	*Amherst, Mass.	4,512	Lemont, Ill	e4,000	Mansfield, Mass.	3,432
*Mt. Holly, N. J.	e5,300	*Griffin, Ga.	4,503	*Eugene Cy., Ore.	e4,000	*Missoula, Mont.	3,426
*Belfast, Me.	5,294	Junction City,Kan.	4,502	*Reno, Nev	e4,000	*Princeton, N. J.	3,422
*PortChester,N.Y	5,274	*Dover, N. J	e4,500	*Mt. Pleasant, Ia.	3,997	Ravenna, O	3,417
*Monroe, Mich.	5,258	Grass Valley,Cal.	e4,500	*Marysville, Cal	3,904	Uxbridge, Mass.	3,408
Greenfield, Mass.	5,252	*Warren. R. I	4,489	*Great Falls, Mont	3,979	*Rawlins, Wy	3,406
Coldwater, Mich.	5,247	Ionia, Mich	4,482	*Marshall, Mich.	3,968	*York Neb	3,405
Wellsville, O.	5,247	*Lyons, N. Y	4,475	Anaconda, Mont.	3,957	Fredonia, N. Y.	3,399
*Santa Rosa, Cal.	5,220	*Green Island,N.Y.	4,463	*Carson Cy., Nev.	3,950	Princeton, Ill.	3,396
Oconto, Wis.	5,219	*Cadillac, Mich.	4,461	Bellefonte, Pa.	3,946	*Fairfield, Ia.	3,391
Rockland, Mass.	5,213	*NewPh'delphia,O.	4,456	Randolph, Mass.	3,946	Mason Cy., Mo.	3,371
*Norwich, N. Y.	5,212	Whitman, Mass	4,441	Holland, Mich.	3,945	*Mishawaka, Ind.	3,371
*Brenham, Tex.	5,209	Whitehall, N. Y	4,434	*New Milford, Ct.	3,917	*Weatherford,Tex.	3,369
Winchester, Va.	5,196	Concord, Mass.	4,427	Hillsdale, Mich.	3,915	*Duluth, Minn.	3,358
Westboro'h, Mass.	5,195	*Antigo, Wis	4,424	Woodbury, N. J.	3,911	*Littleton, N. H.	3,365
*Winfield, Kan.	5,184	*Salisbury, N. C.	4,418	Austin, Minn.	3,900	Brownwood, Tex.	3,361
*Haverstraw, N.Y.	5,170	*Birmingham, Ct.	4,413	*Murphysboro, Ill.	3,880	*Ripon, Wis	3,358
*Dixon, Ill.	5,161	Suap'n B'ge, N. Y.	4,405	Urbana, Ill	3,880	Suffolk, Va.	3,354
*Highlands, Col.	5,161	Easthampton,Mass	4,395	*Carrollton, Mo.	3,878	Brookfield, Mass.	3,352
*Tucson, Ariz	5,150	*Napa, Cal	4,395	*Susquehanna, Pa	3,872	*El Dorado, Kan.	3,339
*Portage, Wis	5,148	Eufaula, Ala.	4,394	Belvidere, Ill	3,867	Boulder, Colo	3,330
*Aspen, Col	5,108	*Wellington, Kan	4,391	Charlotte, Mich.	3,867	*Bristol, Tenn	3,324
*Wabash, Ind	5,105	*Greencastle, Ind.	4,390	Somerville, N. J.	3,861	Milledgeville, Ga.	3,322
*Louisiana, Mo.	5,090	Oberlin, O	4,376	St. Johnsbury, Vt.	3,857	Red Oak, Ia.	3,321
*Valparaiso, Ind	5,090	*Whitewater,Wis,	4,359	Owatonna, Minn.	3,849	*Greensb'gh,N.C.	3,317
*Middletown, Pa.	5,080	*TraverseCy.,Mich	4,353	*Vineland, N. J.	3,822	*London, O.	3,316
*West Springfield,Mass.	5,077	*Atlantic, Ia	4,351	Wyandotte, Mich.	3,817	Brandon, Vt.	3,310
*Skowhegan, Me.	5,068	*Waterloo, N. Y.	4,350	*Waynesb'gh, Pa.	3,811	Albert Lea, Minn.	3,305
*Coventry, R. I.	5,068	*Concord, N. C.	4,339	*Frostburgh, Md.	3,804	Beaumont, Tex.	3,296
				So. Evanston, Ill.	e3,800	Olathe, Kan.	3,294
						*Carlinville, Ill.	3,293

City	Pop.	City	Pop.	City	Pop.	City	Pop.
*San Rafael, Cal	3,290	Salida, Col	e3,000	*Ft. Payne, Ala	2,609	*Deadw'd, S.Dak.	2,366
*Yazoo City, Miss.	3,286	The Dalles, Ore	e3,000	New Castle, Ind	2,697	*Benicia, Cal	2,361
Salem, Va	3,279	*Thou.Id.Pk.N.Y.	e3,000	*Tulare, Cal	2,697	*Little Falls, Minn.	2,354
*Cleburne, Tex	3,278	Watkins, N. Y	e3,000	*Scottdale, Pa	2,693	St. Clair. Mich	2,353
*Thomaston, Conn.	3,278	*Th'ma'toCol.Cal.	e3,000	*Clyde, N. Y	2,689	*Stafford Spg., Ct.	2,351
*Athens, Pa	3,274	Navasota, Tex	2,997	Union Cy., Ind	2,681	*McCook, Neb	2,346
*Lewistown, Pa	3,273	*SanLuisOb'po,Cal.	2,995	*Watertown,S.Dak.	2,672	Waverly, Ia	2,346
*Middleb'r'gh, Ky.	3,271	Terrell, Tex	2,988	Allezan, Mich	2,669	Manchester, Ia	2,344
Marine Cy., Mich.	3,268	Sycamore, Ill	2,987	Rhinelander, Wis	2,658	Petersburgh, Ill	2,342
*Nantucket, Mass.	3,268	*Bryan, Tex	2,979	*Cooperstown,N.Y	2,657	*Salinas, Cal	2,339
Clarinda, Ia	3,263	Hollidaysb'gh, Pa.	2,975	*Bridgeport, Pa	2,651	Bethel, Conn	2,335
*Bath, N. Y	3,261	Orangeburgh,S.C.	2,964	*Tilton, N. H	2,636	*San Marcos, Tex.	2,335
*Lewisburgh, Pa.	3,248	*Weatherly, Pa	2,961	*Calvert, Tex	2,632	*No. Conway,N.H.	2,331
*Phillipsburgh, Pa.	3,245	*Wellsboro, Pa	2,961	*Chestertown, Md.	2,632	*Conway, N. H	2,331
*Havre de Gr'e,Md	3,244	Kendallville, Ind.	2,960	Knoxville, Ia	2,632	*Pocatello, Ida	2,330
*Conneaut, O	3,241	*Lehighton, Pa	2,959	*Fairbury, Neb	2,630	Clyde, O	2,327
*Pierre, S. Dak	3,235	*Windsor, Conn	2,954	*Bakersfield, Cal	2,626	Fairbury, Ill	2,324
Washington, Ia	3,235	*Kane, Pa	2,944	*Somerset, Ky	2,625	Attica, Ind	2,320
*Biloxi, Miss	3,234	*Paola, Kan	2,943	*Hollister, Mass	2,619	*Ventura, Cal	2,320
Swanton, Vt	3,231	*Easton, Md	2,939	*San Angelo, Tex	2,615	*Elkton, Nev	2,318
*Gettysburgh, Pa.	3,231	Eaton, O	2,934	*Berryville, Va	3,010	Minonk, Ill	2,316
Jerseyville, Ill	3,207	*Tallahassee, Fla.	2,934	*Red Bluffs, Cal	2,609	Bushnell, Ill	2,314
Gainsville, Ga	3,202	Freehold, N. J	2,932	*Pittsfield, N. H	2,605	*La Grange, Ill	2,314
Swampscott,Mass.	3,198	*Bucksport, Me	2,921	Baker Cy., Ore	2,604	*Rockport, Ind	2,314
*Le'ington, Mass.	3,197	*E.Bdgwat'r, Mass	2,911	*Derry, N. H	2,604	*Boisé Cy., Ida	2,311
Abilene, Tex	3,194	Crestline, O	2,911	*Holdredge, Neb	2,601	*Crete, Neb	2,310
Concordia, Kan	3,184	*Salisbury Md	2,905	Clinton, Ill	2,598	Tecumseh, Mich	2,310
Aberdeen, S. Dak.	3,182	*Ocala, Fla	2,904	*Taylor, Tex	2,584	Irvington, N. Y	2,239
Geneseo, Ill	3,182	*Westminster, Md	2,903	*La Grande, Ore	2,583	*Walton, N. Y	2,298
*Hallowell, Me	3,181	*Bristol, Va	2,902	Canton, N. Y	2,580	*Lake Geneva, Wis	2,297
*Farmington, Conn	3,179	*Gadsden, Ala	2,901	De Kalb, Ill	2,579	Jamest'n, N. Dak.	2,296
*No.Tarryt'n, N.Y.	3,179	Bedford Cy, Va	2,897	*Oswego, Kan	2,574	*Far R'kaw'y,N.Y.	2,288
*McPherson, Kan.	3,172	*Sandy Hill, N. Y.	2,895	*Camden, Ark	2,571	Geneseo, N. Y	2,286
*Cartersville, Ga.	3,171	*Chico, Cal	2,894	*Hinton, W. Va	2,570	*Mt. Morris, N. Y.	2,286
*Bellevue, Ky	3,163	*Lenox, Mass	2,889	Wytheville, Va	2,570	*Dawson, Ga	2,284
Independence, Ia.	3,163	Dalton, Mass	2,885	Bessemer, Mich	2,560	New Lisbon, O	2,278
*Shelbyville, Ky.	3,162	Hudson, Wis	2,885	Newton, Ia	2,564	*Hurley, Wis	2,267
*Phoenix, Ariz	3,152	*Visalia, Cal	2,885	*Raritan, N. J	2,556	*Ticonderoga,N.Y.	2,267
Herkimer, N. Y	e3,150	Ellenville, N. Y	2,881	Girard, Kan	2,541	*Birdsborough, Pa.	2,261
Columbus, Neb	3,134	*Perry, Ia	2,880	Toronto, O	2,536	Bl'k Riv. Falls,Wis	2,261
*Pr'e duChien, Wis.	3,131	Petoskey, Mich	2,872	Ouray, Col	2,534	Eureka, Kan	2,259
ThreeRivers,Mich.	3,131	Vinton, Ia	2,835	Havanna, Ill	2,525	Carthage, O	2,259
*Independ'ce, Kan.	3,127	*Ft. Plain, N. Y.	2,864	*Nevada Cy., Cal.	2,524	Carrollton, Ill	2,258
St. Johns, Mich	3,127	Moscow, Ida	2,861	Blue Island, Ill	2,521	*Clearfield, Pa	2,248
Blairsville, Pa	3,126	Vernon, Tex	2,857	*Holden, Mo	2,520	St. Louis, Mich	2,246
Donaldsonville, La.	3,121	Orlando, Fla	2,856	Sandwich, Ill	2,516	Bedford, Pa	2,242
*Warsaw, N. Y	3,120	*Texarkana, Tex	2,852	*Palmyra, Mo	2,515	*Burlington, Kan.	2,239
*E. Greenw'h, R.I.	3,107	*Livingston, Mont.	2,850	Harrisonburg, Va.	2,513	Wellsboro, W. Va.	2,235
So. Orange, N. J	3,106	*Park City, Utah.	2,850	*Pierce City, Mo.	2,511	Avoca, Ia	2,231
*Kittanning, Pa	3,095	*Cartersville, Mo.	2,844	*Bangor, Pa	2,509	E. Tawas, Mich	2,226
*Marion, Ia	3,094	Washington, N. J.	2,834	Pendleton, Ore	2,506	*Athens, Tenn	2,224
Bellows Falls, Vt.	3,092	Ft. Gratiot, Mich.	2,832	*Haddonfield, N. J	2,502	*Lebanon, Mo	2,218
*Schuylkill H'n,Pa.	3,088	Taylorville, Ill	2,829	*Old Orchard, Me.	e2,500	Mitchell, S. Dak.	2,217
*Richmond, Me	3,182	Webster, Cy, Ia	2,829	Sanford, Me	e2,500	*Co'nell Grove,Ks.	2,211
*Albany, Ore	3,079	Leetonia, O	2,826	N. Easton, Mass	e2,500	*Bethel, Pa	2,209
Maquoketa, Ia	3,077	Garden Cy, Col	2,825	*Whitinsv'le,M's	e2,500	*Lancaster, Mass	2,201
*Waxahachie, Tex.	3,076	*Newark, N. Y	2,824	*Islip, N. Y	e2,500	Columbus, Tex	2,199
*Red Jacket, Mich.	3,073	*Honesdale. Pa	2,816	Eagle Pass, Tex.	e2,500	*Neosho, Mo	2,198
*Woodland, Cal	3,069	Lebanon, Ky	2,816	*Elyria, Col	e2,500	Ligonier, Ind	2,195
Farmington, N. H.	3,004	*Summit Hill, Pa.	2,815	*W. Berkeley, Cal.	e2,500	Garnett, Kan	2,191
Oregon Cy., Ore	3,062	*Butler, Mo	2,812	Galena, Kan	2,496	Shippensburgh,Pa.	2,188
Dover, Del	3,061	*Marysville, O	2,810	Sturgis, Mich	2,489	Paxton, Ill	2,187
Lexington, Va	3,059	Dowagiac, Mich	2,806	Hiawatha, Kan	2,486	*Bismarck, S.Dak.	2,186
Greenville, Mich	3,056	Fernandina, Fla	2,803	*Big Stone Gap,Va.	2,484	Hudson, Mich	2,178
*No. Platte, Neb	3,055	*Greenville, Ala	2,806	Freeport, Me	2,482	Ennis, Tex	2,171
Bellevue O	3,052	Charles Cy, Ia	2,802	Central City, Col.	2,480	*Addison, N. Y	2,166
Dalton, Ga	3,046	*Clay Center,Kan.	2,802	*Brookville, Pa	2,478	Lewiston, Ill	2,166
*Victoria, Tex	3,046	Decorah, Ia	2,801	Holbrook, Mass	2,474	*Clarion, Pa	2,164
*Presque Isle, Me.	3,046	Missouri Valley,Ia.	2,797	Stoughton, Wis	2,470	*Columbus, Kan	2,160
Baldwins'e, N. Y.	3,040	*Punxsutawney,Pa	2,792	Greenfield, O	2,460	*Watsontown,Pa.	2,157
*Palatka, Fla	3,039	Fair Haven, Vt	2,791	*Beloit, Kan	2,455	Apollo, Pa	2,156
*Washburn, Wis	3,039	*Reynoldsville, Pa.	2,789	*Arkadelphia, Ark.	2,455	*Jermyn, Pa	2,150
Huron, S. Dak	3,038	Midland, Mich	2,777	Smyrna, Del	2,455	*Sharon, Conn	2,149
*Norfolk, Neb	3,038	Canastota, N. Y	2,774	*Great Bend, Kan.	2,450	*Watsonville, Cal.	2,149
Needham, Mass	3,035	*E Mauch Ch'k,Pa.	2,772	*Lykens, Pa	2,450	Ayer, Mass	2,118
*N. San Juan, Cal.	3,031	*Salem, N. C	2,771	*Lockport, Ill	2,449	*Emporium, Pa	2,147
Poultney, Vt	3,031	*Ellensburg,Wash.	2,768	Carroll, Ia	2,448	*Bozeman, Mont.	2,143
Wolfbor'gh, N. H.	3,020	*Decatur, Ala	2,765	*Cohasset, Mass	2,448	Hicksville, O	2,141
*Anderson, S. C	3,018	*Winds'r L'ksCon.	2,758	Georgetown, Tex.	2,447	*Mercer, Pa	2,138
Milford, N. H	3,014	*Ocean Grove,N. J	2,754	Cuero, Tex	2,442	*Weir, Kan	2,138
*Thomaston, Me	3,009	Lapeer, Mich	2,753	*Irwin, Pa	2,428	Cape May Cy., N. J	2,136
Clarksb'h, W. Va.	3,008	Media, Pa	2,736	Wymore, Neb	2,420	*Stockbridge, Mass	2,132
Savannah, Ill	3,007	*Sheffield, Ala	2,731	*Stroudsburgh, Pa	2,419	*Palmyra, N. Y	2,131
Manhattan, Kan	3,004	Franklin, O	2,729	Hackettsto'n, N. J	2,417	Rapid Cy., S Dak.	2,128
Belton, Tex	3,000	Durango, Col	2,726	Lampasas, Tex	2,108	*Hamburg, Pa	2,127
Audenreid, Pa	e3,000	*Winthrop, Mass	2,726	*Modesto, Cal	2,402	*Hillsb'ro Br.,N.H.	2,190
Goshen, N. Y	e3,000	*Washington, Mo.	2,725	*Freeland, Pa	2,398	Alexan'Iria, Minn	2,118
Homer, N. Y	e3,000	Plymouth, Ind	2,723	Greeley, Col	2,395	*Catonsville, Md	2,115
Lake Village,N.H.	e3,000	Slatington, Pa	2,716	*Kingman, Kan	2,390	*Pulaska Cy., Va	2,112
*W. Haven,Conn.	e3,000	St. Ignace, Mich	2,701	*Darlington, S. C.	2,389	Pen Argyl, Pa	2,108
Madison, N. J	e3,000	*Berwick, Pa	2,701	*Las Vegas, N. M.	2,385	Seward, Neb	2,106
*P't An'les, Wash.	e3,000	Mt. Pleasant,Mich.	2,701	Golden, Col	2,383	*Cherry Vale,Kan	2,104
*Sag Harbor,N.Y.	e3,000	Maynard, Mass	2,700	*Kingston, Pa	2,381	Keeseville, N. Y	2,103

MANUAL OF AMERICAN WATER-WORKS. 383

Falls Cy., Neb....2,102	*Redding, Cal....1,821	Tipton, Ia........1,599	*Hayfield, Wis...1,373
*Silver Cy., N. M..2,102	E. Stroudsb'gh,Pa.1,819	*Cambridge, N. Y.,1,598	McMinnville,Ore..1,368
*W'ynesburgh, Pa.2,101	Richl'd C'nt'o, Wis.1,819	*Auburn, Cal......1,595	Central Cy, Neb...1,368
Hillsboro, Ill......e2,100	E. Palestine, O....1,816	*Kutztown, Pa....1,595	Granville, O......1,366
*W. De Pere,Wis.e2,100	*Royersford,Pa....1,815	Grafton, N. Dak...1,594	*Shelton, Conn....1,362
Belmont, Mass.....2,096	*Anthony, Kans...1,806	N. Muskegon,Mich.1,590	*Adams, N, Y 1,360
*Canajoharie, N.Y.2,089	*Terryville,Conn.el,800	Darlington, Wis...1,589	*Hughesville, Pa..1,358
Morrison, Ill......2,088	*Spring City, Pa...1,797	*Idaho Falls, Ida..1,588	*Sidney, N. Y.....1,358
Moorhead, Minn...2,088	*Clifton Forge,Va.1,792	Emmetsburgh, Ia..1,584	Mt. Pulaski, Ill...1,357
*Anamosa, Ia......2,078	*Luling, Tex......1,792	Hinsdale, Ill......1,584	El Paso, Ill......1,353
*Ashburnh'm,M'ss1,074	Manchester, Mass.1,789	E. Aurora, N.Y..,1,582	*National Cy. Cal.1,353
*Canisteo, N. Y...2,071	Itchelle, Ill.......1,789	Colorado, Tex....1,582	Stanton, Mich....1,352
*Manheim, Pa.....2,070	*Colorado Cy., Col.1,788	Geneva, Neb......1,580	Truckee, Cal.....1,350
Blair, Neb........2,069	*Oroville, Cal.....1,787	*Harper, Kans....1,579	Weep'gW'ter.Neb.1,350
*Talladega, Ala....2,063	Ashland, Ore.....1,784	Cordele, Ga......1,578	Haywards. Wis...1,349
*Lisbon, N. H.....2,060	*Carlyle, Ill......1,784	Eldora, Ia........1,577	Constantine, Mich.1,346
*Radford, Va......2,060	Denison, Ia.......1,782	*Wadsworth, O ..1,574	Warrenton, Va...1,346
Sewickley, Pa.....2,053	Reed City, Mich...1,776	W. Randolph, Vt .1,573	Idaho Spr'gs, Col..1,338
*Glasgow, Ky.....2,051	*Bluefield, W. Va.1,775	Redwood Cy., Cal..1,572	Gladstone. Mich..1,337
*Marion, Kan.....2,047	Vergennes, Vt....1,773	*Newport, Ark....1,571	*Colusa, Cal......1,336
*Tremont, Pa.....2,046	*Belvidere. N. J...1,768	Oregon, Ill........1,566	*Pacific Grove, Cal.1,336
Burlington, Wis ..2,043	Harlan, Ia........1,765	*Delhi, N. Y......1,564	*Phelps, N. Y.....1,336
Nephi, Utah......2,034	*Dodge City, Kan..1,763	*Blaine, Wash....1,563	*Hamburgh, N.Y.1,331
*Grand Junc.,Col..2,030	Middlebury, Vt...1,762	Ida Grove, Ida....1,563	Montrose, Col.....1,330
David City, Neb...2,028	Prescott, Ariz....1,759	*Skaneateles, N.Y.1,559	Newaygo, Mich....1,330
*Centralia, Wash..2,026	*Minneapolis,Kan.1,753	*Grayling, Mich...1,558	Tidioute, Pa......1,329
Waterville, N. Y..2,024	*Decatur, Tex....1,746	Pawnee Cy., Neb..1,550	Kennett Sq're, Pa..1,326
*Lake City, Fla ...2,020	*St. Mary, Pa.....1,745	Sutton, Neb......1,544	*Nazareth, Pa....1,318
*Sanford, Fla......2,016	Villisca, Ia........1,744	Longmont, Col....1,543	*Parker Cy., Pa...1,317
Fort Collins, Col...2,011	Hinsdale, Mass...1,739	North East, Pa....1,538	*Colton, Cal......1,315
*Dyersburg, Tenn.2,009	*Colebrook, N. H..1,736	Warwick, N. Y...1,537	*Selins Grove, Pa..1,315
*Merced, Cal......2,009	Doylestown, Pa...1,732	*North Yakima,	New Hampton, Ia 1,314
Wilton Junct'n,Ia.2,007	*Eureka, Utah ...1,733	Wash............1,535	Audubon, Ia......1,310
*Wahoo, Neb......2,006	Newport, Vt......1,730	Shell Lake, Wis...1,535	Centreville, Md...1,309
*Silverton, Col....7,000	*Foxcroft, Me.....1,726	*Coudersport, Pa..1,530	*Cheshire, Mass...1,308
*Bar Harbor, Me.e2,000	*Seguin, Tex......1,716	*Deposit, N. Y....1,530	*Troy, Pa.........1,307
*Springvale, Me.e2,000	*New Castle, Wy.1,715	*Corvallis, Ore....1,527	*Hoquiam, Wash..1,302
Cochituate, Mass.e2,000	Oxford, Pa........1,711	*Millersburgh, Pa.1,527	Washington, Ill ...1,301
*Flemington, N.J.e2,000	*Gorham, N. H...1,710	*Bristol, N. H.....1,524	*H'satonic,Mass..1,300
*Gowanda, N. Y.e2,000	*WinterHarbor.Me1,709	Brookings, S.Dak..1,518	*Block Isl'd,R.I...el,300
*Mamaron'k,N.Y.e2,000	*Rogers Park, Ill..1,708	Pulaski, N. Y.....1,517	*Conyngham, Pa..1,299
Mohawk, N. Y...e2,000	*St. Helena, Cal...1,705	*Fredonia, Kan....1,515	*Muncy, Pa.......1,295
*Boyertown, Pa.e2,000	*Caro, Mich.......1,701	*Parkersburgh,Pa.1,514	Orwigsburgh, Pa..1,290
*DerryStation, Pa.e2,000	Granville, N. Y..el,700	*Wahpeton, N. D..1,510	*Annville, Pa.....1,283
*Minooka, Pa....e2,000	*Patchogue, N. Y..1,695	Hobron, Neb......1,502	Groton, N. Y.....1,280
Evanston, Wy.....1,995	*SaukCenter, Minn1,695	Buena Vista, Col.el,500	*Hancock, N. Y...1,279
*Attica, N. Y......1,994	Gilroy, Cal.........1,694	Gwynedotte,W.Va.1,500	Anaheim, Cal.....1,273
*Plainville, Conn..1,993	*Harrison, O......1,690	*Jackson, Cal....el,500	*Jordan, N. Y....1,271
Cazenovia, N. Y...1,987	*Placerville, Cal...1,690	*Locust Gap, Pa.el,500	Clinton, N. Y.....1,269
Westfield, N. Y....1,983	*Sprague, Wash...1,689	*Morrisville, Vt.el,500	Evart, Mich......1,269
Gofftown, N. H....1,981	*Lansing, Ia......1,688	*Vineyard Haven,	W.Liberty, Ia....1,268
*Del Rio, Tex......1,980	Storm Lake, Ia....1,682	Mass.........el,500	Bethlehem, N. H..1,267
Salem, Ind........1,975	Vassar, Mich......1,682	*Nutley, N. J....el,500	Holly, Mich......1,266
*Beverly, N. J.....1,957	*Austin, Pa.......1,679	*Round Lake,	*Rogers, Ark.....1,266
*Indiana, Pa......1,953	*Abingdon, Va....1,674	N. Y..........el,500	*Uvalde, Tex.....1,265
Northboro', Mass..1,952	*Mexia, Tex......1,674	So. Denver, Col...1,491	St. Paul, Neb.....1,263
Monticello, Ia.....1,938	*Tallapoosa, Ga...1,669	Garden Cy., Kan..1,490	*Haton, N. Mex..1,255
Hope, Ark.........1,937	Bainbridge, Ga....1,668	*Hum'lstown, Pa..1,486	*Tunkhannock,Pa.1,253
Nelilsville, Wis ...1,936	*Curwensville, Pa.1,664	*Moravia, N. Y...1,486	Imlay Cy., Mich...1,251
Georgetown, Col...1,927	*Greenwich, N. Y.1,664	Healdsburgh,Cal.1,485	Sac City, Ia......1,249
Tarborough, N. C..1,924	*Blairstown, N. J..1,662	Roslyn, Wash....1,184	*Bartlett, N. H...1,247
*Marysville, Kan..1,913	Nevada, Ia........1,632	Eureka, Ill........1,181	Oberlin, Kan......1,243
*Chatham, N. Y...1,912	Kingston, Mass...1,659	*So.Pittsb'b,Tenn.1,479	Nashua, Ia.......1,240
Hoopeston, Ill.....1,911	Glenwood, Wis....1,656	Peabody, Kans...1,471	*Jacksonville, Ala.1,237
*Redlands, Cal....1,904	Alma, Mich.......1,655	*Roseburgh, Ore..1,472	*Hollister, Cal....1,234
Ridgway, Pa......1,903	St. Petersb'gh,Pa.1,655	*Phoenix, N. Y...1,466	Pipe Stone, Minn.1,232
Whitehall, Mich...1,903	Carthage, Ill......1,654	*Deer Lodge,Mont.1,463	*Port Allegany,Pa.1,230
Camden, N. Y.....1,932	Tecumseh, Neb...1,654	*Round Brook,N.J.1,462	*Florence, Kan...1,220
*Rock Falls, Ill...1,900	†*Avon, N. Y......1,653	*Dardanelle, Ark.1,456	*Merchantv'le,N.J.1,225
Litchfield, Minn...1,899	*Los Gatos, Cal...1,652	Middletown, Del..1,454	*Woodstock, Vt...1,218
*Springville, N. Y.1,881	Marion, Va.......1,651	Sonora, Cal......1,441	Hendersonv'e,N.C.1,216
Dayton, Wash.....1,880	*Colfax, Wash....1,649	*La Junta, Col....1,439	*Austin, Nev.....1,215
*Tombstone, Ariz..1,875	*Broken Bow, Neb.1,647	Manitou Springs,	*Newton, Pa......1,213
*Simsbury, Conn..1,874	*Sharon, Mass....1,643	Col.............1,430	Neligh, Neb.......1,209
*Belleville, Kan...1,868	*Caldwell, Kan...1,642	*Grant's Pass, Ore.1,432	Ord, Neb..........1,208
Greenville, Ill.....1,868	Sterling, Kan.....1,641	Ovid, Mich.......1,423	*Lake Forest, Ill..1,203
Chadron, Neb.....1,867	*Gonzales, Tex....1,641	*Haywards, Cal..1,419	Ebensburgh, Pa...1,202
Aurora, Neb.......1,862	Aberdeen, Wash ..1,638	*Fryburgh, Me....1,418	Randolph, N. Y...1,201
*Larned, Kans....1,861	Corning, Ia........1,634	Pratt, Kan.......1,418	*Adrian Mines,Pa.el,200
*Lansdale, Pa.....1,858	*W. Indiana, Pa..1,634	*Brownsville, Pa..1,417	Clifton, O........el,200
Rouse's Point, N.Y.1,856	*White Haven, Pa.1,634	*Canaan, N. H...1,417	*Island Pond, Vt..el,200
*Jersey Shore, Pa..1,853	*Ukiah, Cal.......1,627	*Belair, Md.......1,416	Oceanside, Cal.el,200
*Plymouth, N. H..1,852	Otsego, Mich.....1,626	*Phenix, R. I....el,400	*Pineville, Ky...el,200
Mt. Joy, Pa........1,848	Richf'd Sps., N.Y.1,623	*Rock Rapids, Ia..1,394	TwoHarbors,Min.el,200
Windsor, Vt......1,848	Ellsworth, Kan...1,620	*Canton, Pa.......1,393	Mound Cy., Mo...1,193
*Meyersdale, Pa...1,847	Superior, Neb.....1,614	*Livermore, Cal..1,381	Newark, Del.....1,191
West Point, Neb..1,842	*Eureka, Nev.....1,609	Cuba, N. Y.......1,386	Fonda, N. Y......1,190
Red Cloud, Neb...1,839	*Jenkintown, Pa..1,609	Avon, Mass.......1,384	Hastings, Mich...1,187
Mt. Carroll, Ill....1,836	NewBraunfels,Tex1,606	*Kent, Conn......1,383	Valley Falls, Kan..1,184
Toledo, Ia.........1,836	Aledo, Ill..........1,601	*Holly, N. Y......1,381	Crystal Falls,Mich.1,176
*Lowell, Mich.....1,829	Ashland, Neb.....1,601	Minden, Neb......1,380	*Hopedale, Mass..1,176
Willmar, Minn....1,825	Socorro, N. Mex..1,601	Reynoldton, Pa...1,379	*Frankfort, Mich..1,175
Cobleskill, N.Y....1,822	*Martinez, Cal....1,600	*Catlettsburgh, Ky.1,374	Clare, Mich......1,174

MANUAL OF AMERICAN WATER-WORKS.

*Ballard, Wash .. 1,173	Medford, Ore...*... 667	Ainsworth, Neb.... 733	*Azusa, Cal e500
*Chatonaug'ay,N.Y.1,17	*Miles Cy., Mont... 906	*Langhorne, Pa... 727	*Bellaire, Mich.... e500
*So. Paris, Mo....1,164	*Arcata, Cal........ 932	*Green River,Wyo. 723	*Cullison, Kan.... e500
*Millerstown, Pa..1,162	Gordonsville, Va... 962	*Harriman, Tenn... 716	Danforth, Ill e500
Kalkaska, Mich...1,161	Kilbourn Cy., Wis. 961	Stambaugh, Mich.. 711	*Millville, Pa...... e500
*Unadilla, N. Y...1,157	Sherburne Quarter,	*Jamestown, R. I . 707	*NorthClardendon,
*Tuscarora, Nev...1,156	N. Y............ 960	*Covington, Va.... 704	Pa................ e500
*Selma, Cal......1,150	Colfax, Ia......... 957	Goldendale, Wash. 702	Milton, Ore....... e500
*Smethport, Pa...1,150	*Waterbury, Vt... 965	Atkinson, Neb..... 701	*Roxbury, N. Y... e500
*Castile. N. Y.....1,146	Avoca, N. Y....... 953	*Fortuna, Cal. ... e700	*Richlands, Va.... e500
*Central Cy., Ky..1,144	*Ft. Bragg, Cal.... 945	*M'garot'sv'le,N.Y. e700	*Seville, Fla...... e500
Minerva, O.......1,139	*Manlius, N.Y ... 942	*Folsom Cy., Cal... 693	*Wenonah, N. J... e500
Pleasanton, Kan...1,139	Chamb'rl'n, S.Dak. 939	Little Valley, N.Y. 698	Windom, Kan.... e500
Clyde, Kan.......1,137	Downs, Kan....... 938	Loveland, Col..... 698	*Coquille, Ore..... 494
Howard City,Mich.1,137	Venice, Ill......... 932	*Wilmington, Cal.. 687	Douglass,Wyo.... 491
Rico, Col........1,134	Madison, Neb..... 930	Woons'ket, S. Dak. 687	Barton L'ding, Vt. 482
*Anacortes, Wash..1,131	*M e c h anicstown,	*Dutch Flat, Cal... 682	*Coolidge, Kan... 472
*Emlenton, Pa...1,126	Md.............. 930	*Conway Sp's,Kan. 681	Rocky Ford, Col.. 468
*Fultonville, N. Y.1,122	*Norway, Me...... 928	*Castroville.. Tex. 679	Culbertson, Tex... 460
*Odebolt, Ia......1,122	*Walsenburg, Col.. 928	Spearfish, S. Dak.. 678	*Ketchum, Idaho.. 450
Palouse, Wash....1,119	Albion, Neb 926	*Elizabethville, Pa. 676	*Whitewood,S.Dak 443
Richford, Vt......1,116	Carson Cy., Mich.. 921	*Pioche, Nev...... 676	*Cardiff, Tenn..... 430
Edgar, Neb.......1,105	*Glenwood Spgs.,	Loup Cy., Neb..... 671	Salem, S. Dak..... 429
*Gunnison, Col....1,105	Col.............. 920	Hooper, Neb...... 670	*Wall, Pa......... 419
Port Byron, N. Y..1,105	Greenleaf, Kan..., 916	*Ipava, Ill......... 667	*Andes, N. Y..... 416
Canton, S. Dak....1,101	Yorkville, Pa...... 916	*Schenevus, N. Y.. 665	*Woodburn, Ore... 405
Riverside, Ill.....1,100	Readsborough, Vt. 910	*Lake Cy., Mich... 663	*Croton, N.Y...... e400
*Yreka, Cal.......1,100	†*Monravia, Cal .. 907	Gaylord, Mich..... 661	*Diamond, 2d Me.. e400
*Lakewood, N. J ..el,100	Blue Rapids, Kan. 905	Pomeroy, Wash... 661	*Kensington, Ga.. e400
*No. St. Paul, Min.1,099	Gladwin, Mich ... 903	*San Jacinto, Cal.. 661	*Centinela, Cal.... e400
Fremont, Mich....1,097	*Elko, Nev....... e900	Breckenri'ge,Minn. 655	*New Dorp, N. Y.. e400
Ft. Gaines, Ga....1,097	Shelton, Wash.... e900	Sioux Rapids, Ia... 650	Valentine, Neb... e400
Medicine L'dg,Kan.1,095	Birmingham, Mich 899	*Holyoke, Colo..... 649	*W. Seattle,Wash. e400
*Hudson, N. H....(1,092	*Lakeside, Minn... 807	*Cheney, Wash.... 647	*La Conner, Wash. 398
*Dixon, Cal.... (S..1,082	*Bellevue, Idaho.. 892	*Foxburgh, Pa..... 640	Riverton. Neb.... 389
*Cottage Cy. Mass.1,080	Edenburgh, Pa... 888	*Mattea wan, N. Y. 638	Chase, Mich...... 388
Cannon Falls, Minn.1078	Knox, Pa......... 888	Millerton, N. Y.... 638	*Prattsville, N. Y. 384
Norton, Kan...,...1,074	*Crewe, Va........ 887	*Antioch, Cal...... 635	Red Cliff, Col..... 383
*Hailey, Idaho....1,073	Pincoming, Mich. 885	*Wheatland, Cal... 630	Yorkville, Ill..... 375
E. Dubuque, Ill...1,069	Emaus, Pa........ 883	Clarence, Ia....... 629	Randolph, Neb... 374
Black Hawk, Col..1,067	Nahant, Mass..... 880	Glenwood, Minn.. 627	Akron, Col........ 359
*Cisco, Tex.......1,063	Stockton, Kan..... 880	Milan, O.......... 627	*Arlington, Col... 356
*No. Wales, Pa... 1,060	*Cattaraugus, N.Y. 878	*Ft. Benton, Mont. 624	Springfield, Kan.. 347
Pecatonica, Ill ...1,059	*Landsdowne, Pa.. 875	*Orting, Wash..... 623	*Elma, Wash 345
*Litchfield, Conn..1,058	Pullman, Wash.... 868	*Chebanse, Ill..... 616	Cabery, Ill........ 342
*Vinton, Va......1,057	*Wardner, Idaho.. 858	Las Animas, Col... 611	Udall, Kan....... 338
*Elmhurst, Ill 1,050	Crested Butte, Col. 857	Wisner, Neb...... 610	*Grant, Neb....... 315
*Bainbridge, N. Y..1,049	State Center, Ia... 854	Lake City, Col..... 607	New Castle, Col... 311
Buena Vista, Va' 1,044	*Kent, Wash...... 853	New Carlisle, Ind. 607	*Templeton. Cal... 308
*Winnemucca.Nev.1,038	Ellicottsville,N.Y. 852	*Ephrata, Pa...... 606	Fairplay. Col..... 301
Guthrie Center, Ia.1,037	*Lewiston, Ida.... 849	*Porterville, Cal... 606	*Independence,Cal. e300
Cedar Spr'gs,Mich.1,033	*Billings. Mont.... 831	Pankington, S. D. 604	*Menlo Park, Cal.. e300
Fairmont, Neb ...1,029	Logan, Ia......... 827	*Redondo B'ch, Cal 603	*So. Prairie, Wash. e300
*Scoharie, N. Y...1,028	*Paso Robles, Cal. 827	*Gate City, Va..... e600	*Grosse Point. Mich.298
Morgan Park, Ill..1,027	*Stamford, N. Y.. 819	*Glasgow, Va...... e600	Andover, S. Dak.. 282
*N'w Beth'hem,Pa.1,026	*Mullan, Idaho.... 818	Tazewell, Va...... e600	*Meriden, N. H... e250
Fr'nklinville, N. Y.1,021	Morrisonville, Ill.. 814	*Mio, Mich e600	*Felton, Cal...... 259
*Oakdale, Cal. ...1,012	Ulysses, Pa....... 801	*Cascade, Col e600	*Joseph, Ore...... 249
Hallettsville, Tex.1,011	*Brunswick, Md .. e800	*Springer, N. Mex 600	*Enterprise, Ore.. 242
*Morg'nt'n,W.Va..1,011	*Max M'dows, Va. e800	*Bodie, Calo 595	Mellette, S. Dak... 241
*Nunda, N.Y.....1,010	Somonauk, Ill.... e800	Farwell, Mich 584	Berthoud, Col..... 228
Rockford, Ia.....1,010	Blue Hill......... 796	Franklin, N. Y.... 581	*Nipoma, Cal..... 215
*Tooele, Utah 1,008	Redfield, S. Dak... 796	Sumner, Wash.... 580	*Channing, Tex ... e200
*Bridgeport, Ala. el,000	Genoa, Neb....... 793	Attica, Ind........ 578	Mason Cy., Neb... e200
*Groveton, N. H. el,000	*Milford, Pa...... 793	*Whitinsv'le, N. Y. 575	*Sonora, Tex...... e200
*Woodsville, N.H.el,000	Northville, N. Y... 792	*Eliz'b'tht'n, N. Y. 573	*Aspen Junct., Col. e200
*Wakefield, R. I. el,000	*Kyle, Tex 779	Crawford, Neb.... 571	Bunker Hill, Kan.. 157
E. Randolph, Vt. el,000	Barton, Vt........ 778	Weston,Ore....... 568	Bird Cy., Kan 145
*Garden Cy., N.Y.el,000	Kinsley, Kan...... 771	Lamar, Col........ 566	*Clayton, N. Mex.. 140
*Kirkland, Wash.el,000	Smith Center, Kan. 767	Long Pine. Neb.... 562	*Ordway, Col..... e50
*Larchmont, N.Y.el,000	Telluride, Col..... 766	*Yuba City, Cal... 562	
*Palm'rL'ke, Col..el,000	*Cloverdale, Cal... 763	*Hobart, N. Y 561	UNKNOWN POP'S.
*Riverton, N. J.. el,000	Ellendale, N. Dak. 761	*Tioga, Pa........ 557	*Forrento, Me.
Stromsb'rgh, Neb.el,000	Chatham, N.Y..... 737	Gladbrook, Ia..... 556	*Pelham Manor, N. Y.
*TuxedoP'k, N.Y.el,000	Hart, Mich 757	Petrolia, Pa...... 546	*Stanley, Va.
*Worcester, N.Y. el,060	*Shelby, Ala...... 753	*Escondido, Cal .. 541	Caribou, Col.
Humboldt, Ia..... 998	Harrison, Mich.... 752	Miller, S. Dak..... 536	*E. Colfax, Col.
*Wayne, Pa...... 997	*Arkans'sCy., Ark.e750	Sandstone, Minn. 517	*Green Mountain Falls
Phillipsb'gh, Kan. 992	Norwood Park, Ill. e750	*Greensburgh,Kan 515	& Ute Park Col.
*Hull, Mass....... 989	*Otego, N. Y...... 749	*Sundance, Wyo... 515	*Ridgway, Col.
Park Ridge, Ill... 987	Hawarden, Ia 744	Valparaiso, Neb... 515	*Maxwell City,N. Mex.
Lincoln, Mass..... 987	White Cloud,Mich. 743	Dana, Wyo........ 513	*Arch Reach& Laguna.
New Berlin, N. Y. 979	*Union Bridge, Md. 743	Roscommon, Mich. 511	Cal.
*Canaan, Conn.... 970	*San Gabriel, Cal.. 737	Tyndall, S. Dak.... 509	*Mission San José, Cal.
*De Pere, Wis. ... 909	Edmore, Mich..... 735	*Anderson, Cal.... 508	*Moreno, Cal.

R. D. WOOD & CO.

Engineers,

Iron Founders,

and Machinists.

Office: 400 Chestnut Street, Philadelphia.

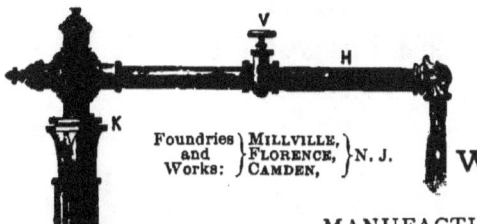

Foundries and Works: MILLVILLE, FLORENCE, CAMDEN, N. J.

Constructors of Water and Gas Works.

MANUFACTURERS OF

CAST IRON PIPE,

OF ALL KINDS AND SIZES.

TURBINE WHEELS.

Pumping Machinery,

Hydraulic Cranes,

Lifts and Machinery,

Heavy Loam Castings,

"Eddy" Valves,

"Mathews" Hydrants.

Iron Floors, Roofs and Girders.

MANUAL OF AMERICAN WATER-WORKS.

THE LUDLOW
VALVE MANUFACTURING CO.

OFFICE AND WORKS:
938 to 954 River St. and 67 to 83 Vail Ave.,
TROY, N. Y.

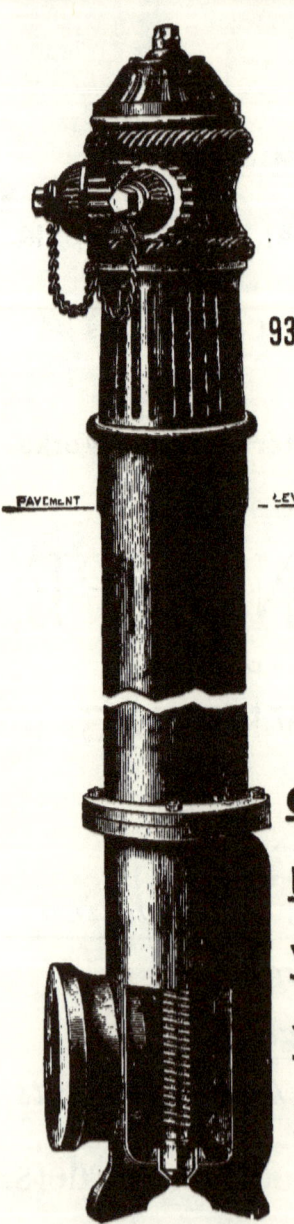

VALVES

Double and Single Gate, 1-2 in. to 48 in., outside and inside Screws, Indicator, etc., for Gas, Water, Steam and Oil.

Check Valves,

Foot Valves,

Valve Boxes,

Yard, Wash &
Fire Hydrants.

Send for Circular.

WATER AND GAS WORKS
SUPPLIES.

READING IRON COMPANY'S
WROUGHT IRON STEAM, GAS AND WATER PIPE, BOILER TUBES, ARTESIAN WELL PIPE AND CASING, AND SPECIAL WORK OF EVERY CHARACTER.

Malleable and Gray Iron Fittings for Gas, Steam and Water.

Cast-Iron Water, Gas and Flange Pipe, Special Castings, Etc.

Chapman Valve Manufacturing Co.'s
GATES AND HYDRANTS.

Specifications for bids solicited and estimates of cost given on Heavy Machinery, Boilers, Tanks, Stand-Pipes, and all engineering work in iron.

PANCOAST & ROGERS,

OFFICE AND WAREHOUSE:

28 PLATT & 15 GOLD STS., NEW YORK.

BOURBON COPPER AND BRASS WORKS.

PATENTED

Fire Hydrant

—AND—

Stop Valves,

ALSO MANUFACTURERS OF

Copper and Brass Goods

Of Every Description.

198, 200, 202, 204
E. FRONT ST.,
CINCINNATI, O.

CARSON TRENCH MACHINE CO., 16 Dorrance St., Boston, Mass.

Narrow Streets kept open.
UTMOST ECONOMY in cost of **excavation**.

Machines to let for digging and back-filling.
Sewer and large Water Trenches. Some simple, some complex. Many sizes. Eleven years' use. Used this season in Boston, Providence, Worcester, Springfield, Montreal, Toronto, Cleveland, Chicago, Cincinnati, St. Louis and many other places.

WELL MADE GOODS AND LOW PRICES ARE BOUND TO BE RESPECTED.

H. MUELLER MFG. CO.
MANUFACTURERS OF
WATER TAPPING MACHINES

Gas Pipe Tapping Machines,

Water Pressure Regulators,

Corporation Stop Cocks,

Round Way Stop Cocks,

Stop and Waste Cocks,

Brass Water Connections,

Check and Waste Cocks, Etc., Etc.

WATER TAPPING MACHINE.

GAS TAPPING MACHINE.

— UNEQUALLED IN EFFICIENCY. —

— SOLD UPON MERIT. —

Our **TAPPING MACHINE** is acknowledged by everybody as the most complete and best machine in the market, and we have over 900 machines sold in the United States. We have on file in our office testimonials from all parts of the Union.

Mr. C. E. Gray, of the Gas and Water Works at Duluth, Minn., writes us as follows:

"GENTLEMEN: Your later improvements in tapping machines I very much like; nevertheless I have no fault to find with any one of the machines of the many I have purchased of you to supply the several Gas and Water Works that I have contracted for during the past seven years. Yours truly,
"C. E. GRAY."

WHAT IS SAID OF OUR IMPROVED GAS TAPPING MACHINE.

"DECATUR, ILL., March 1, 1890.
"H. Mueller Mfg. Co., City.

"GENTLEMEN: I have been using one of your improved gas tapping machines for the past year, and must say that I am so well pleased with it that I would not use an old style—Crow Hook—if I could buy it for $1.
"Yours very respectfully,
"R. J. STRATTON.
"Supt. Decatur Gas Light and Coke Co."

WATER PRESSURE REGULATOR.

We have the most encouraging reports from our **WATER PRESSURE REGULATOR**, which is the best in the market. By the use of this Regulator a steady and uniform pressure from 20 to 30 pounds, as may be desired, can be obtained in any building, notwithstanding a fire-pressure of 120 pounds or more on the mains. In a city of Pennsylvania it is successfully used under a steady pressure of 165 pounds.

We are now prepared to furnish the Mueller Corporation Stop Cock for use with Hexagon Plug, the same as they have always been made up to this season. Also the new style used with screw plugs. All cocks manufactured by us are guaranteed to be perfect in every respect.

NO. 1 CORPORATION STOP COCK.

CORRESPONDENCE SOLICITED.
Prices furnished on application and a liberal discount to the trade.
WRITE FOR CIRCULARS.

H. MUELLER MFG. CO.
Office, 222 E. Main St.
Factory, 111 to 117 State St., Decatur, Ill.

——— MADE UPON HONOR. ———

HAYS MANUFACTURING CO.,

MANUFACTURERS OF

WATER AND GAS COMPANIES' SUPPLIES,

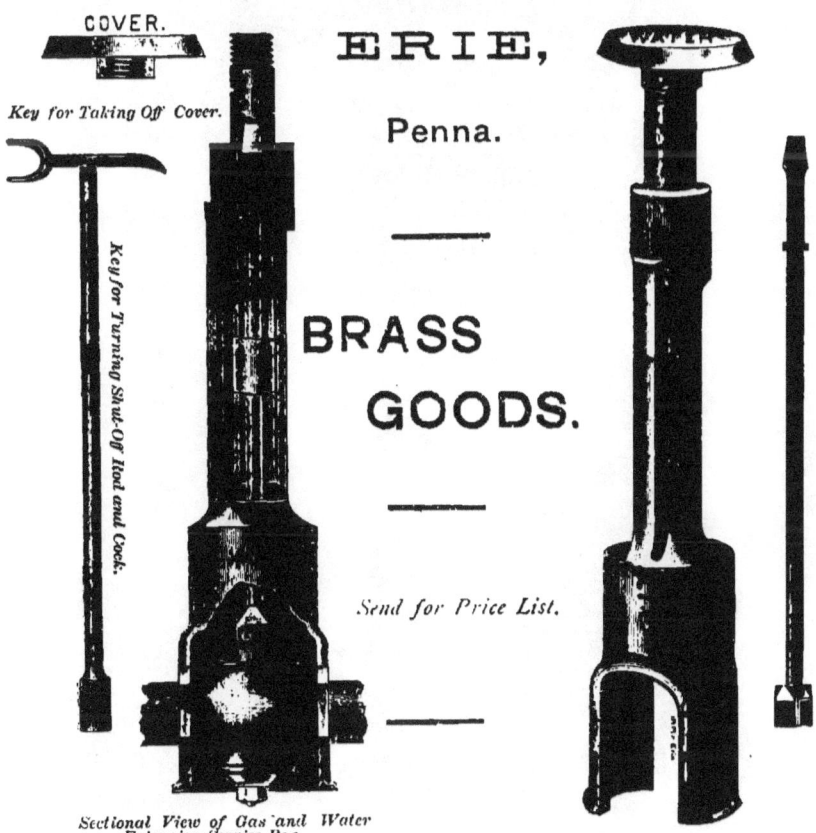

ERIE,

Penna.

BRASS GOODS.

Send for Price List.

SPECIALTIES:

Extension Service Boxes,
Corporation Stop Cocks
for All Tapping Machines.

Service Cocks,
McNamara Hydrants
and Street Washers.

MALLEABLE AND CAST-IRON FITTINGS.

MICHIGAN BRASS AND IRON WORKS,

SUCCESSORS TO
GALVIN BRASS AND IRON WORKS,
DETROIT, MICH.

MANUFACTURERS OF
GALVIN'S COMPOUND WEDGE GATE VALVES,
For Steam, Water and Gas.
GALVIN'S CONICAL CASE GATE FIRE HYDRANT,
GALVIN'S IMPROVED MATHEW FIRE HYDRANTS.
ALSO
GENERAL BRASS AND IRON GOODS,
For Steam, Water and Gas.

CARR'S PATENT HYDRANTS
WITH OR WITHOUT
Supplementary or Double Valves,

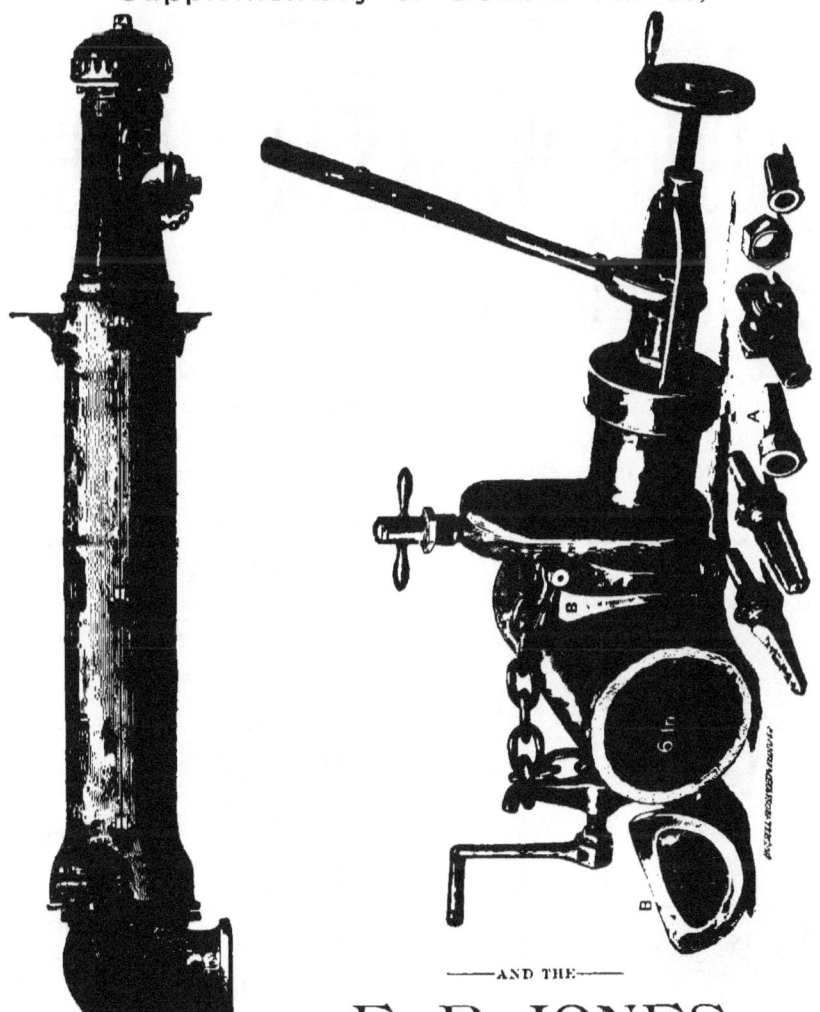

———AND THE———

E. R. JONES
Tapping Machines,

For Inserting service stops from ½ to 1½ inch. Used in all the larger cities.

MANUFACTURED ONLY BY **CHARLES CARR,** CONSULTING MECHANICAL ENGINEER.

Office, 7 Exchange Place, Boston, Mass.

Formerly Vice-President and Manager of Boston Machine Co. and Boston Machine Mfg. Co.

MACHINERY OF ALL KINDS and WATER-WORKS SUPPLIES DESIGNED AND DETAILED DRAWINGS FURNISHED.

Construction Superintended and Contracted for.   Attention given to office and outside work.

RIPLEY & BRONSON,
800 NORTH SECOND STREET,
ST. LOUIS.

WATER WORKS SUPPLIES.

Cast Iron Water and Gas Pipe,

Fire Hydrants, Gate Valves, Pig Lead, Packing Yarn, Yard Hydrants and Street Washers, Service and Valve Boxes, Lawn and Drinking Fountains, Fire Hose and Fire Department Supplies.

Write for Catalogue and Prices.

HOLYOKE | PORTER MFG. CO.,
LIMITED,
Manufacturers and Builders of

Unexcelled for simplicity of construction, durability and low cost of maintenance. All wearing surfaces of composition, drips below the main and cannot freeze. Stones or gravel will not clog it.

HOLYOKE HYDRANT
AND
IRON WORKS.
HOLYOKE, MASS.

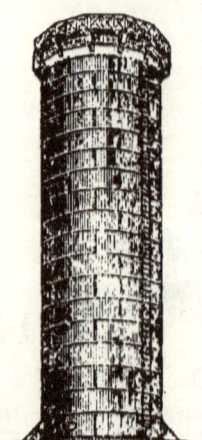

Tanks,
Stand-Pipes,
Engines,
Boilers
and Stone
Crushers.

Specifications Solicited.

ESTIMATES CHEERFULLY GIVEN.

SYRACUSE, N. Y.

COFFIN & STANTON,

BANKERS,

NEW YORK:
72 Broadway.

LONDON:
43 Threadneedle St.

DEALERS IN

Bonds of States,

Municipalities,

—AND—

Corporations.

Globe Special Castings
FOR WATER-WORKS.

PATENTED.

Preferable as regards strength, first cost and convenience in handling.

Sold by the piece.
Carried in stock.

Send for catalogue.

Any form of special made without charge for pattern, and without delay.

BUILDERS' IRON FOUNDRY,
PROVIDENCE, R. I.

M. J. DRUMMOND, No. 192 BROADWAY,
NEW YORK SALES AGENT.

BOILERS.

All Rivet Holes Drilled after Sheets are Rolled and Fitted. Most Complete Boiler Plant in the Southern States.

HYDRAULIC RIVETER, POWER ROLLS, PLATE PLANER, Etc., Etc.

Also Tanks, Stacks, Flume Pipes and Sheet Iron Work.

MACHINERY AND CASTINGS

Valk & Murdoch Iron Works,

14-20 HASELL STREET,

CHARLESTON, S. C.

Please Write for Specifications and Prices.

GET THE BEST.

THE E. STEBBINS MANUFACTURING CO.,

BRIGHTWOOD P. O., SPRINGFIELD, MASS.,

MANUFACTURERS OF

BRASS GOODS,

ROUGH STOPS AND
CORPORATION COCKS.

BROUGHTON PATENT SELF-CLOSING COCKS.

Particular Attention Paid to First-Class Ground Key Work for Water-Works.

ALL WORK GUARANTEED.

Rubber Suction Hose.

Established 1828.

"Smooth-Bore" Rubber Suction Hose.

BOSTON BELTING CO.,
ORIGINAL MANUFACTURERS OF EVERY VARIETY OF

RUBBER BELTING, HOSE, STEAM PACKING, &c.

Also Manufacturers of all other articles of VULCANIZED INDIA RUBBER for Mechanical and Manufacturing Purposes. RUBBER VALVES of all descriptions for Hydrants, Pumps, Engines, Steamships, etc.

256, 258, 260 Devonshire Street,
BOSTON, MASS.

100 Chambers Street,
NEW YORK CITY.

And Agencies in the Principal Cities in the United States and Europe.

NATIONAL FILTER.

24 CITIES
Have Adopted Our Process.

Standard Filter of the World.

GRAVITY
AND
PRESSURE TYPES.

Send for Illustrated Catalogues to

NATIONAL WATER PURIFYING COMPANY,
Refer by permission to Henry R. Worthington. **145 Broadway and 86 Liberty St., New York.**

Has an Unparalleled Record of More than Fifty Years.

"HOFFMAN" ROSENDALE CEMENT.

FROM THE
BEST ROCK.
ADAPTED FOR
Heavy Masonry.
ALWAYS
RELIABLE.

STANDS EVERY
HIGH TEST.
UNIFORM IN
QUALITY
AND
Full Weight.

TAKES THE LEAD OF AMERICAN CEMENTS.
Extensively Used by the U. S. Government.
Receives the approbation of the most eminent
ENGINEERS AND ARCHITECTS.
Send for pamphlet containing valuable information to

THE LAWRENCE CEMENT COMPANY,
GENERAL SALES AGENTS, {M. ALBERT SCULL, ERNEST R. ACKERMAN,} 67 WILLIAM STREET, NEW YORK.

WARREN FOUNDRY & MACHINE CO.,

ESTABLISHED 1856.

Works at Phillipsburg, N. J.

SALES OFFICE:

No. 160 BROADWAY, NEW YORK.

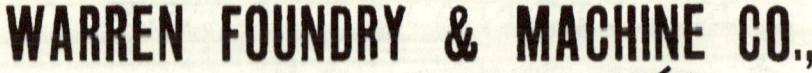

ALL SIZES OF

FLANGE PIPE

—FOR—

MINES, SUGAR HOUSES, ETC.,

Branches, Bends, Retorts, Etc.

M. J. DRUMMOND,

Cast Iron Gas and Water Pipe.

2 to 48 in. diameter.

FLANGE PIPE, LAMP POSTS, STOP VALVES, FIRE HYDRANTS.

BUILDERS' IRON FOUNDRY

"GLOBE" SPECIAL CASTINGS.   PROVIDENCE, R. I.

OFFICE: CORBIN BUILDING, 192 BROADWAY, NEW YORK.

UTICA PIPE FOUNDRY CO.

CAST IRON PIPE AND SPECIALS FOR WATER AND GAS.
Also Flanged Pipe and Fittings, Hydrants, Gates, Pig Lead, etc.

CHAS. MILLAR & SON, Selling Agts., Utica, N. Y.
Manufacturers of Lead Pipe and Plumbers' Materials.
Wholesale Eastern Agents for Akron Vitrified Sewer Pipe.
Also Wrought-Iron Pipe and Fittings.

CAST IRON PIPE
FOR WATER AND GAS.
BELL AND SPIGOT, OR FLANGED
WITH FITTINGS,

Fire Hydrants, Stop Valves, Lamp Posts, Canada Turbines, &c.

GENERAL FOUNDRY AND MACHINE WORK.

MELLERT FOUNDRY & MACHINE CO., Limited,
READING, PA.

JAMES N. GAMBLE, President.　ARCHER BROWN, Sec'y and Treas.　J. K. DIMMICK, Gen'l Manager.

THE RADFORD PIPE AND FOUNDRY CO.
GENERAL OFFICE: CINCINNATI, O.

WORKS AT { ANNISTON, ALA., 200 Tons Daily Capacity.
　　　　　{ RADFORD, VA., 150　　"　　　"　　　"

THE JACKSON & WOODIN MANF'G CO., Berwick, Columbia Co., Pa.

FORGINGS,
MINE WHEELS
STEEL AXLES.
FREIGHT CARS, CAR WHEELS, BAR IRON, Special Castings.

Geo. Ormrod, Man. & Treas., Emaus, Pa.　**EMAUS PIPE FOUNDRY.**　John Donaldson, Pres., 136 S. 4th St., Phila. Pa.

DONALDSON IRON COMPANY,
AND

MANUFACTURERS OF  **SPECIAL CASTINGS**
FOR WATER AND GAS.

ALL PIPES CAST VERTICALLY.　　**Emaus, Lehigh Co., Pa.**

VITRIFIED PIPE
FOR
WATER-WORKS CONDUITS

For Water-Works Conduits, where the supply can be obtained by gravity, there is no material so desirable as our

DOUBLE STRENGTH CULVERT PIPE

for the following reasons:

First : It is clean.

Second : Its cost as compared with Iron is about one half.

Third : It will not corrode nor rust. It is not in any way affected by the action of acids or moisture, but when once laid is indestructible.

Fourth : The thickness of the pipe is uniformly one-twelfth its internal diameter, which greatly increases its strength as compared with common vitrified **Sewer Pipe**.

Fifth : With the **Improved Socket and Spigot** absolute water-tight joints may be made, which cannot be done with the old style sockets.

Sixth : Being made in two and one-half foot sections (the usual length being two feet) one joint is avoided in every ten feet of conduit, thereby increasing its strength and lessening the liability to leakage.

Seventh : The increased length of the sections reduces the cost of labor and cement in laying the pipe about twenty per cent.

VITRIFIED PIPE is now in successful use for **Water-Works** at Little Falls and New Amsterdam, N. Y.; La Salle, Ill., and many other places.

READ THE FOLLOWING LETTER:

LA SALLE, Ill., Jan. 28, 1889.

Messrs. Blackmer & Post, St. Louis, Mo.:

GENTLEMEN : Returning from Bloomington, where I have been attending the meeting of the Association of Illinois Civil Engineers and Surveyors, I find your letter of 21st inst. You owe me nothing; I feel it my duty to answer your questions or send you any blue prints you may require, because you have fulfilled your contract with the city of La Salle in an honorable way, and I take pleasure in stating,

First : The 10,400 feet of 24-inch Double Strength Culvert Pipe, furnished by you to the city of La Salle for a Water-Works Conduit, proved satisfactory in quality, and well adapted for a gravitation water conduit, where the maximum water pressure is 8 feet.

Second : Where tight joints have to be made, as in this case, the four-inch socket, instead of the three-inch, is a great advantage.

Third : The working length of two and one half feet, instead of two feet, is another advantage, as the number of joints is reduced twenty-five per cent.

Fourth : For gravitation water-works, where the pressure of the water does not exceed ten feet head, these pipes are more practical than iron ones, especially where the pipes are partly or temporarily surrounded with water. Iron pipes would sooner or later become corroded, but good vitrified pipes stand for an unlimited time, as the terra-cotta works at Babylon and other ancient cities prove.

Very truly yours, FRED ROTTMANN, *City Engineer.*

Send for Mr. Stephen E. Babcock's Treatise on the use of Vitrified Pipe for Water-Works ; also, our Pamphlet, " Something About Culvert Pipe."

CORRESPONDENCE INVITED.

BLACKMER & POST, St. Louis, Mo.

JOHN F. MOFFETT, President.
JOHN V. CLARKE, C. E., Treasurer.

HENRY C. HODGKINS, C. E., Vice-President.
CHARLES T. MOFFETT, Secretary.

CAPITAL, $500,000.00.

Moffett, Hodgkins & Clarke Co.,
CONTRACTING ENGINEERS,
SYRACUSE, N. Y.

WATER-WORKS.

Within the past eight years have made maps, plans and estimates for Water-Works and Sewerage in over two hundred cities and villages in the United States and Canada. Have constructed and now own and operate water-works in the following places, viz:

West Troy, N. Y.	Spartanburg, S. C.	Manitowoc, Wis.
Watertown, N. Y.	Canandaigua, N. Y.	Portage, Wis.
Adams, N. Y.	Waterford, N. Y.	Baraboo, Wis.
Fulton, N. Y.	White Plains, N. Y.	Escanaba, Mich.
Oswego Falls, N. Y.	Newark, O.	Kankakee, Ill.
Homer, N. Y.	Chippewa Falls, Wis.	Lincoln, Ill.
Cortland, N. Y.	Menominee, Wis.	Cornwall, Ont.
Green Island, N. Y.	Belleville, Ont.	Rochester, Minn.
Greenbush, N. Y.	Beaver Dam, Wis.	Duluth, Minn.
Newark, N. Y.	Berlin, Ont.	Washington, Ind.
Delaware, O.	Salisbury N. C.	Jackson, Miss.
Marshall, Mich.	Kingston, Ont.	Cobourg, Ont.
Waterloo, Ont.	Canton, N. Y.	Ingersoll, Ont.
Peoria, Ill.	La Grange, Ill.	Kincardine, Ont.

Have Water-Works now in process of construction at the following places, viz.:

Montreal, P. Q.	Harvey, Ill.	Butte, Mont.

Irrigation Systems in Southern California.

The construction of Water-Works, Gas Works and Sewerage for Cities and Towns a specialty. Plans, specifications and estimates prepared by careful engineers and receive the benefit of our wide experience. Also general railroad and electrical engineering and construction. Information concerning any proposed work respectfully solicited.

Address

MOFFETT, HODGKINS & CLARKE CO.,
SYRACUSE, N. Y.

New York Office:
NO. 34½ PINE STREET.

Chicago Office:
No. 720 OPERA HOUSE BLOCK.

—A. D. 1891.—

THOMSON METERS

SOLD AND DELIVERED.

APRIL,	-	- 1,008	AUGUST,	-	931
MAY,	-	- 1,252	SEPTEMBER,	-	907
JUNE,	-	- 1,138	OCTOBER,	-	1,007
JULY,	-	- 1,136	NOVEMBER,	-	956

TOTAL FOR EIGHT MONTHS, 8,335.

Average 1,042 a Month.

10,000 Meters Sold from Nov. 1, 1890, to Nov. 1, 1891.

"WHAT DOES IT MEAN?"

These Monthly Sales Explain It.

CATALOGUES UPON APPLICATION.

THOMSON METER CO.,
212 TEMPLE COURT BUILDING,
NASSAU and BEEKMAN STS., NEW YORK.

We make the Largest, Most Durable and Accurate Positive Measuring Water Meter in the World.

SIZES, 5-8 IN. TO 12 IN.

Union Rotary Piston Water Meter, Extra Heavy Pattern.

Meters Sent for Trial.

WRITE FOR LITHOTYPE OF THE

12-in. Water Meters.

LARGEST EVER MADE.

★

THE ONLY RELIABLE AND EFFICIENT AUTOMATIC

Water Pressure Regulator

IN USE.

HAS NEVER FAILED.

★

UNION WATER METER CO.
WORCESTER, MASS., U. S. A.

Union Water Pressure Regulator.

Send for Catalogue containing useful tables for hydraulic engineers.

THACHER & BREYMANN,
Submarine Engineers and Divers.

Contractors for submarine work of all kinds where divers are required.
Recovering sunken property.
Laying Submerged Pipes for Water, Gas and Sewerage a Specialty.
Estimates and Consultation promptly given. Correspondence solicited

306 WATER ST, TOLEDO, O.

Some of the Submerged Pipes Laid:

Diam'r	Where and when laid.	Length.	Diam'r	Where and when laid.	Length.
50 inch.	Milwaukee, Wis., special sewerage, 1884	530 feet	20 inch.	Norristown, Pa., 1876	1,750 feet
48 "	Toledo, O., water-works, 1874	400 "	20 "	Racine, Wis., water-works, river crossing, 1887	400 "
40 "	Milwaukee, Wis., water-works, 1888	250 "	20 "	Nashville, Tenn., water-works, laid in rapid current and bolted to rock	1,200 "
36 "	Sandusky, O., water-works, 1876	1,800 "	20 "	Duluth Minn., extended intake into 50 ft. of water, 1888	300 "
36 "	Saratoga Springs, N. Y., water-works, 1881	202 "	20 "	Sheboygan, Wis., water-works, 1887	1,700 "
33 "	Milwaukee, Wis., water-works, 1886	320 "	20 "	Monroe, Mich., in Lake Erie, 1889	6,000 "
30 "	Pittston, Pa., water-works, 1886	1,075 "	20 "	Wyandotte, Mich., water-works, 1889	300 "
30 "	Decatur, Ala., water-works, 1888	400 "	20 "	Wyandotte, Mich., soda ash works, 1891	500 "
30 "	Milwaukee, Wis., water-works, 1888	450 "	16 "	Milwaukee, Wis., 1888	250 "
30 "	Syracuse, N. Y., Onondaga Lake to 65 ft. of water, 1890	2,000 "	16 "	West Superior, Wis., 1888	4,000 "
30 "	Muskegon, Mich., in trench 7 ft. deep to 27 ft. of water, 1891	1,890 "	14 "	Dunkirk, N.Y., water-works,1881	1,000 "
21 "	Pittston, Pa., water-works, 1886	426 "	14 "	Monroe, Mich., river crossing rock trench, 1889	470 "
24 "	Pittston, Pa., Penn'a Coal Co., in rock trench, 1886	196 "	12 "	Pittston, Pa., water-works, 1886	700 "
24 "	Racine, Wis., water-works, laid in trench 10 ft. deep to 50 ft. of water in Lake Michigan, 1887	7,160 "	12 "	Sheboygan, Wis., two lines, 1887	800 "
			12 "	Janesville, Wis., water-works.	400 "
			12 "	Dunkirk, N.Y., water-works, 1890	650 "
24 "	Racine, Wis., 1887	400 "	8 "	Racine, Wis., water-works, 1887	350 "

We have put in Submerged Cribs for water supply at Racine, Wis.; Muskegon and Wyandotte, Mich.; Dunkirk and Saratoga Springs, N. Y.; a SUBMERGED FILTER at Pittston, Pa.

THE PELTON WATER MOTOR

IS presented as a machine thoroughly scientific in principle, of admirable mechanical design and possessing in the same degree the wonderful energy and power of the Pelton wheels, which has given them a reputation for economy and high efficiency that commands the trade in all parts of the world.

THE PELTON MOTOR will be found to be so superior to all others as regards strength, durability, economy of water, in fact, in all that constitutes a high-class water motor, as to admit of no comparison. Made of sizes varying from the fraction of 1 up to 100 HP.

Recent scientific tests made in the Mechanical Department of the University of Michigan, in connection with several other motors claimed to be the best on the market, showed 45% higher efficiency in favor of the PELTON, while the relative cost per horse power to buy was only one-third to one-half that of others.

Such facts make it apparent that water companies, desiring to make the most of their water supply, should discriminate in favor of the PELTON; also that no other would be cheap to consumers at any price. Deliveries made from New York or San Francisco, as may afford the most favorable freight rates. Write for circulars.

THE PELTON WATER-WHEEL CO.,

143 Liberty Street, } OR { 121 Main Street,
NEW YORK CITY.   SAN FRANCISCO, CAL.

PEET VALVE CO.,

MANUFACTURERS OF THE

GENUINE

DOUBLE GATE

Peet Valves

—FOR—

Steam, Water, Gas, etc.

1-4 to 36 in.

163 ALBANY ST., BOSTON, MASS.

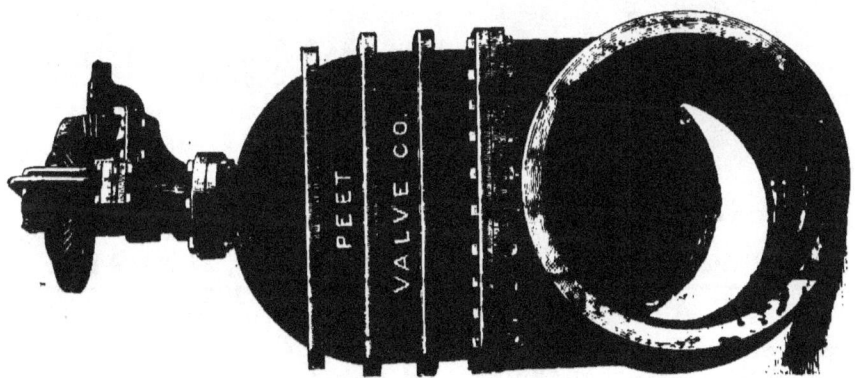

AGENCIES:

Chicago	John Davis & Co.	Chattanooga	C. E. James & Co.
St. Louis	N. O. Nelson Mfg. Co.	New Orleans	M. Schwartz.
Denver	Davis-Creswell Mfg. Co.	Baltimore	Crook, Horner & Co.
Omaha	A. L. Strang & Sons.	Philadelphia	John Mancely.
Cleveland	The W. Bingham Co.	Kansas City, English Supply & Engine Co.	
Buffalo	W. A. Case & Son.	Detroit	A. Harvey & Son.
New York	Colwell Lead Co.	Milwaukee	Rundle, Spence & Co.
St. Paul	Rogers & Ordway.	San Francisco	W. T. Garratt & Co.

STEPHEN E. BABCOCK,
Civil and Hydraulic Engineer.

SEWERAGE
—AND—
WATER=WORKS
A SPECIALTY.

LITTLE FALLS, N. Y.

J. James R. Croes,
M. Am. Soc. C. E., M. Inst. C. E.

13 William Street, New York

Examinations and Reports Made on Projects for Water Supply and Sewerage of Towns.

Plans and Specifications Prepared and Work of Construction Superintended.

JOHN W. HILL,
M. Am. Soc. C. E.,

Consulting and Constructing Engineer

For Water-Works, Sewerage, etc.

21 & 22 Glenn Building,

CINCINNATI, O.

JACOB A. HARMAN,
Member Western Society of Engineers; Member Illinois Society of Engineers and Surveyors,

CIVIL ENGINEER,
Room 100, Y. M. C. A. Bldg.,

PEORIA, ILL.

Designs for and Inspection of Water-Works, Sewerage Systems, Bridges. Paving, Land Drainage, Topographical Surveys, Construction Superintended.

E. WOLTMANN.   GEO. T. KEITH,
M. Am. Soc. C. E.

WOLTMANN, KEITH & CO.,
Engineers and Contractors
For WATER-WORKS, etc.

11 WALL ST., NEW YORK.

RUDOLPH HERING,
MEM. AM. SOC. C. E., M. INST. C. E.,

Civil and Sanitary Engineer
277 PEARL STREET,
Near Fulton Street,   NEW YORK.

Designs for Water Supply and Sewerage. Construction Superintended.

ALAN MACDOUGALL,
M. Can. Soc. C. E.,   M. Inst. C. E.,

Civil and Sanitary Engineer.
Plans, Estimates and Specifications prepared for
Water-Works and Sewerage Systems.
Sewage Disposal, Irrigation and Domestic Sanitation. Construction Superintended.

32 Adelaide St., East,
TORONTO, CANADA.

GEO. W. WYNN,
Civil and Hydraulic Engineer,
CEDAR RAPIDS, IOWA.

Water-Works, Sewerage, Drainage and City Improvements.

Plans, Specifications and Estimates Prepared and Construction Superintended.

J. T. FANNING,

Memb. Am. Soc. C. E.,

CONSULTING ENGINEER

**KASOTA BLOCK,
MINNEAPOLIS, MINN.**

REPORTS on Public Water Supplies, Water Powers, Mills and Sewerage.
DESIGNS for Steam and Hydraulic Pumping Plants, Patent Tubular Triplex Boilers and for Power Transmissions.

Price $2.50. Cloth, 8vo., pp. 281. Cuts, 222, and Folding Plate.

Sent post paid on receipt of price, by the

**ENGINEERING NEWS PUB. CO.,
TRIBUNE BUILDING, NEW YORK.**

CONTENTS:
INTRODUCTORY. PART I.—SEWERAGE.
PART II.—THE PURIFICATION OF SEWAGE.
PART III.—GENERAL MUNICIPAL AND DOMESTIC SANITATION.
American Practice in Street Cleaning and Sewerage.

G. W. PEARSONS,
Civil and Mechanical Engineer,

Member Am. Soc. C. E. (1876).

1611 BALTIMORE AVENUE, - - KANSAS CITY, MO

Record: Over 30 years' experience in building Vessels, Mills, Elevators, etc., and in Civil and Mechanical Engineering.

In the design and construction of Water Works, Sewerage, Water Supply, Dams, Bridges, etc., and in consultation have been employed by cities and corporations in the following places:
* Bangor and § Rockland, Maine; * Marquette and * Kalamazoo, Mich.; * Ogdensburg, * Potsdam and ' Rockaway, N. Y.; † Austin, Minn.; * Memphis, Tenn.; * Canton, Ohio; ‡ New Orleans, La.; ‡ Galveston, Tex.; ‡ Pueblo and † Las Animas, Colo.; ‡ Vicksburg, Miss.; ‡ Birmingham, Ala.; ‡ Little Rock, Ark.; * Kansas City, * Marshall and Holden, Mo.; ‡ Kansas City, * Leavenworth, † Paola, † Minneapolis, † Concordia, † Osborne, § Iathe, * Fort Riley and ‡ Winfield, Kansas; § Oskaloosa, Iowa; § Hiawatha, Kansas.

* Concerned in designing and building.
† Concerned in designing.
‡ Concerned in consulting.
§ Concerned in additions.

Do Not Act as Agent for any Manufacturer.

ARTESIAN
Wells, Oil and Gas Wells, drilled by contract to any depth, from 50 to 4000 feet. We also manufacture and furnish everything required to drill and complete same. Portable Drilling Machines for wells from 100 to 1000 ft. Send 25c. for illustrated catalogue.
Pierce Artesian & Oil Well Supply Co., 80 Beaver St., New York

WATER SUPPLY

In large quantities, from Thousands to Millions of Gallons per day, for Manufacturers, Corporations, Towns, or Cities, by new Patented System. Test Borings made for Water and Estimates of cost furnished. Contracts taken for complete plant for Water-Works, Water supply Stations, Pipes, Valves, Fittings, etc., etc. Address, giving full particulars, requirements, etc.

CHARLES D. PIERCE,
Consulting Engineer and Contractor,
127 Pearl St., NEW YORK.

A. L. HOLMES,
CONTRACTING ENGINEER AND BUILDER OF

Water-Works and Sewerage.

Patentee and Manufacturer of

HOLMES' FLEXIBLE JOINT.

Plans and Specifications furnished.

Submerged Work a Specialty.

CORRESPONDENCE SOLICITED.

GRAND RAPIDS, MICH.

Barr Pumping Engine Company,

Germantown Junction, Philadelphia, Pa.,

MAKERS OF

High Class Pumping Machinery

For Water-Works and Other Services.

All Work Made to Gauge and Thoroughly Interchangeable.

Designs and Estimates Furnished on Application.

Heine Safety Boiler Company.

—PATENT—
SAFETY WATER-TUBE STEAM BOILERS

Economy in Fuel and Space. Freedom from Scaling. Positive Circulation.
Equally adapted for power or heating purposes, for clear or muddy water, and any kind of fuel. Send for circular to

Heine Safety Boiler Co.
706, 707 and 708 Bank of Commerce Bldg.,
ST. LOUIS, MO.

OR TO OUR AGENTS.

Risdon Iron Works............San Francisco, Cal.	Jas. H. Harris ......82 Madison St., Chicago, Ill.
Stearns, Roger Mfg. Co., 4 Duff Blk., Denver, Col.	L. Metzger...31 St. Charles St., New Orleans, La.
Jas. K. Hugg & Co., Room 23 Chamber of Commerce Bldg., Cincinnati, O.	R. M. Huston............45 Broadway, New York.
F. E. Siebenmann...401 Lewis Blk., Pittsburg, Pa.	John MacCormack.......45 Broadway, New York.

G. E. DOWNING, Prest. JAMES BOWRON, Vice-Prest. W. R. GARRETT, Sec'y and Treas.

SOUTH PITTSBURG PIPE WORKS, 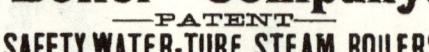 SOUTH PITTSBURG, TENN.,

Manufacturers of  Also

Flange Pipe, Specials, Loam Work.
GENERAL FOUNDERS AND MACHINISTS.
Correspondence Solicited and Prices Furnished on Application.

JAMES H. HARLOW & CO.,
Engineers and Contractors,
FOR

Water Supply and Drainage,

108 Fourth Ave., Pittsburg, Pa.

Have been engaged on the following during construction: Lowell, Lawrence, Boston, Mass.; New Castle, Mercer, Latrobe, Waynesburg, Etna, Tarentum, Wilkinsburg, Millvale, Brushton and Homestead, Pa.; Aurora and Mattoon, Ill.; Eau Claire, Wis.; Minerva, O.; Canisteo, N. Y., and Bel Air, Md., and Surveys for others.

C. H. MULLIGAN, Superintendent.

We Have Erected STAND-PIPES
at the following places:
Kankakee, Ill., 20 x 124; Cornwall, Ontario, 20 x 120; Salisbury, N. C., 20 x 100; Washington, Ind., 20 x 100; Beaver Dam, Wis., 20 x 84; Waterford, N. Y., 30 x 52; Berwick, Pa., 11 x 45; Homer, N. Y., 25 x 40; Cobourg, Ont., 18 x 116; Marshall, Mich., 20 x 100; Hamburg, N. Y., 16 x 125; Canton, N. Y., 20 x 72; Rhinelander, Wis., 20 x 65; Fort Smith, Ark., 30 x 100.

Sharon Boiler Works, Limited,
SHARON, PA.,

Stand-Pipes & Boilers
FOR WATER-WORKS.

E. B. JENNINGS, Engineer

R. F. HAWKINS IRON WORKS,
SPRINGFIELD, MASS.,
BUILDERS OF STAND PIPES,
Steam Boilers, Tanks, Iron Castings, Bridges, Bolts, Iron Roofs, etc.

R. F. HAWKINS, Proprietor.

TUERK HYDRAULIC POWER CO.,
SOLE MANUFACTURERS OF
Tuerk Water Motors
—AND—
ECLIPSE
Hydraulic Elevators,

237 BROADWAY, Corner Park Place, NEW YORK.

WM. H. MARSH, Prest.   T. C. DRAKE, Sec'y.
F. D. MALTBY, Mech. Engineer.

GEM WATER MOTOR.   DUPLEX GEM WATER MOTOR.

For supplying houses with pure water by pressure from water-works; also for supplying water to high buildings, where the ordinary pressure is not sufficient to carry it.

No Engineer. No Fire. No Wires. The Minimum of Expense.

Manufactured by GEORGE J. ROBERTS & CO.,
DAYTON, O.

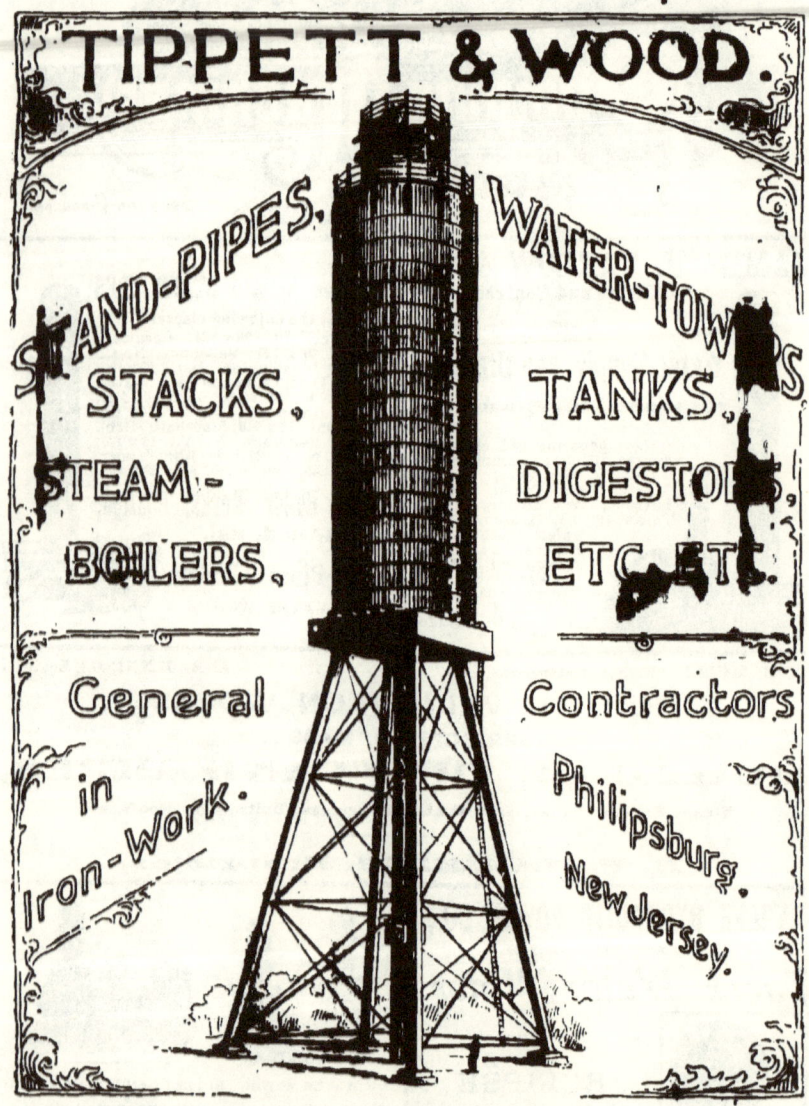

STAND-PIPE AT PRINCETON, N J.

Erected by Tippett & Wood in 1883.

METEOROLOGICAL OBSERVATORY ON TOP OF TOWER.

THE GASKILL-HOLLY
HIGH DUTY PUMPING ENGINE.

Under Contract to Supply the Buildings and Grounds of the
WORLD'S FAIR AT CHICAGO.

COMPARISON CHALLENGED WITH ANY OTHER IN THE WORLD.

146 of These Justly Celebrated Engines in Use and Under Contract, Average Daily Capacity of Which Is 5,048,000.

Consider the Points Which Demonstrate Its Superiority.

1. The first Gaskill crank and flywheel high duty pumping engine was built and put into service at Saratoga Springs, N. Y., in 1882, nine years ago.
2. The duty of the first engine of this type at Saratoga, on contract test, was 112,890,983 foot-pounds per 100 pounds of coal consumed. Other engines of the same type have shown as high as 131,120,226 foot-pounds.
3. The average duty of the above engine in current service from 1883 to 1890, inclusive—seven years—was 105,524,137 foot-pounds from coal actually consumed.
4. During the interval of seven years the cost of repairs was less than $100, as certified by officials in charge. Average cost of repairs on sixty-two engines, each per year, 27 cents. (Send for pamphlet giving cost of repairs.)
5. Another engine of the same type, of 8,000,000 gallons daily capacity, was constructed for Saratoga Springs, and tested and accepted in October, 1889. The duty obtained was 113,378,479 foot-pounds from coal consumed, and 117,936,698 on an evaporation of ten to one. This engine was operated six months of the year 1890, at about one-half its nominal capacity, and performed an average daily duty of 105,305,270 foot-pounds.
6. Other engines of the same type, in number about 40, have been tested by some of the most eminent engineers in the United States, all of them developing high duty beyond all former precedent; and at Dayton, O., eclipsing 10,000,000 engines in a duty of 124,782,157 foot-pounds on a ratio of ten to one evaporation.
7. The capacity of these engines manufactured thus far ranges from 1,000,000 to 20,000,000 gallons daily.
8. They are in use in Boston, Philadelphia, Chicago, Buffalo, Washington, Kansas City, Omaha, Denver, Cincinnati, Portland, Atlanta, Savannah, and other large cities in the United States, and we are building under contract two of 12,000,000 gallons each for the World's Columbian Exposition.
9. The aggregate capacity of these engines, constructed or contracted for to date, is 737,000,000 gallons daily, which, with other types, make a grand total of 405 engines and 1,256,120,000 gallons.
10. In symmetry of proportion, weight, strength, economy, durability, price, comparison is challenged with any other engine manufactured in this or any other country.
11. An illustrated book of 303 pages, giving full reports of duty tests in detail as made by about 30 mechanical experts eminent in their profession, will be mailed upon application to all who are interested in, or may have occasion to purchase, pumping engines, and who have not already been supplied. The book demonstrates beyond all dispute that the Gaskill compound crank and flywheel pumping engines, manufactured under patents by the Holly Mfg. Co., far surpass any other engine manufactured in the world.

The Gaskill Pumping Engine is made either of the horizontal or vertical type, as required, beside which the Holly Mfg. Co. make, of any capacity:
Vertical Rotative Beam Pumping Engines, Vertical Triple Expansion Pumping Engines, Duplex Direct-Acting Pumping Engines, Water Power Pumping Machinery, Straight Way Gate Valves and Check Valves, 4 in. to 36 in., Fire Hydrants all sizes, and cast iron flange pipe.

Address for Catalogue or Other Information,
The Holly Manufacturing Company,
Lockport, N. Y.

BRANCH OFFICES:
45 Broadway, New York City; 657 Washington St., Boston, Mass.; La Salle and Adams Sts., Chicago, Ill.; 94 Common St., New Orleans, La.; 315 Tenth St., Portland, Oregon.

www.ingramcontent.com/pod-product-compliance
Lightning Source LLC
Chambersburg PA
CBHW022103300426
44117CB00007B/564